A Faunal Review of Aleocharine Beetles in the Rapidly Changing Arctic and Subarctic Regions of North America (Coleoptera, Staphylinidae)

Jan Klimaszewski • Adam Brunke •
Derek S. Sikes • Mikko Pentinsaari •
Benoit Godin • Reginald P. Webster •
Anthony Davies • Caroline Bourdon •
Alfred F. Newton

A Faunal Review of Aleocharine Beetles in the Rapidly Changing Arctic and Subarctic Regions of North America (Coleoptera, Staphylinidae)

Springer

Jan Klimaszewski
Laurentian Forestry Centre
Natural Resources Canada
Canadian Forest Service
Quebec City, QC, Canada

Adam Brunke
Canadian National Collection of Insects
Agriculture and Agri-Food Canada
Ottawa, ON, Canada

Derek S. Sikes
University of Alaska Museum
Fairbanks, AK, USA

Mikko Pentinsaari
Centre for Biodiversity Genomics
University of Guelph
Guelph, ON, Canada

Benoit Godin
Whitehorse, YT, Canada

Reginald P. Webster
Charters Settlement, NB, Canada

Anthony Davies
Canadian National Collection of Insects
Agriculture and Agri-Food Canada
Ottawa, ON, Canada

Caroline Bourdon
Laurentian Forestry Centre
Natural Resources Canada
Canadian Forest Service
Quebec City, QC, Canada

Alfred F. Newton
Department of Zoology
Field Museum of Natural History
Chicago, IL, USA

ISBN 978-3-030-68190-6 ISBN 978-3-030-68191-3 (eBook)
https://doi.org/10.1007/978-3-030-68191-3

This Springer imprint is published by the registered company Springer Nature Switzerland AG.
The registered company address is: Gewerbestrasse 11, 6330 Cham, Switzerland

To all the little but megadiverse aleocharine beetles for helping us understand the natural world and the processes changing our planet.

Preface

Humans have long struggled to explore and understand the Arctic and Subarctic regions and found these irresistible. By the middle of this century, rising temperature could strip away the Arctic ice and thaw the tundra, affecting humans and our fauna and flora. The thawing tundra will accelerate warming of the climate, affecting people and animals, and putting many species at risk of extinction. An important task is to provide faunal inventory and tools for species identification to have baseline data for future comparisons/monitoring and to help protect and restore species-at-risk habitats. Aleocharine beetles are well represented in the North. However, they are among the most poorly known and difficult-to-identify groups of Coleoptera. Here, we present the first comprehensive synopsis of all known valid aleocharine rove beetle species (Coleoptera, Staphylinidae) from Arctic and Subarctic zones of Canada and Alaska. Two hundred thirty-eight species in 55 genera and 12 tribes are reported. Five new to science species are described. Forty-five new state distribution records for the USA (mainly Alaska) and 108 new provincial and territorial records for Canada (AB, BC, LB, MB, NT, NU, ON, QC, YT) are provided. Every treated species is presented with a diagnosis, including external and genital descriptions. For every species a plate with color habitus image and black and white images of genital structures are provided to aid in identification of species. Collection and habitat data, often new, are presented for each species including information on macrohabitat, microhabitat, collecting period, and collection methods. Keys to species, genera, and tribes are provided. A list of all valid recorded Arctic and Subarctic species with their currently known distribution in North America is presented at the end of the book (Table A.1, in Annex 2). A faunal analysis and discussion of Northern aleocharine beetles in a broader context of North America is provided. Aleocharines as indicator species of environmental changes and the impact of climate change on northern fauna are discussed.

TAXONOMIC SYNOPSIS. The following species are described as new to science: *Atheta* (*Datomicra*) *chadburnae* Klimaszewski and Godin, **sp. n.**; *Atheta* (*Dimetrota*) *takhini* Klimaszewski and Godin, **sp. n.**; *Hydrosmecta yukonensis* Klimaszewski and Godin, **sp. n.**, *Liparocephalus humblei* Klimaszewski, **sp. n.**; and *Philhygra donnellyana* Klimaszewski, **sp. n.** New synonyms are established as follows (first name being valid): *Atheta allocera* Eppelsheim, 1893 = *Dimetrota campbelli* Lohse, 1990 (=*Atheta*); *Atheta diversa* (Sharp, 1869, orig. *Homalota*) = *Dimetrota regissalmonis*

Lohse, 1990 (=*Atheta*); *Gyrophaena rugipennis* Mulsant et Rey, 1861 = *G. neonana* Seevers, 1951; *Oxypoda robusticornis* Bernhauer, 1907 = *O. madgei* Lohse, 1990; and *Philhygra ripicola* (Hanssen, 1932, orig. *Atheta*) = *P. ripicoloides* Lohse, 1990. New generic combinations: *Atheta* (*Dimetrota*) *onthophilides* (Newton, 2017), replacement name for preoccupied *Acrotona onthophila* Lohse 1990, a junior secondary homonym in *Acrotona* and *Atheta* of *Atheta onthophila* Cameron 1920; *Bellatheta microptera* (Lohse 1990), orig. *Pseudosipalia microptera* Lohse, in Lohse et al. 1990. A lectotype is designated for Palaearctic *Atheta aeneipennis* (Thomson 1856, orig. *Homalota*), which was determined to be a senior synonym of Palaearctic *A. parapicipennis* Brundin 1954. Identities of *Liogluta trapezicollis* Lohse and *L. vasta* (Mäklin) are revised. A neotype is designated for *Homalota planaris* Mäklin, 1852 (currently *Dinaraea*), and deposited in the CNC.

Corrections to Klimaszewski et al. 2020 (see Annex 3).

Quebec City, QC, Canada	Jan Klimaszewski
Ottawa, ON, Canada	Adam Brunke
Fairbanks, AK, USA	Derek S. Sikes
Guelph, ON, Canada	Mikko Pentinsaari
Whitehorse, YT, Canada	Benoit Godin
Charters Settlement, NB, Canada	Reginald P. Webster
Ottawa, ON, Canada	Anthony Davies
Quebec City, QC, Canada	Caroline Bourdon
Chicago, IL, USA	Alfred F. Newton

The original version of the book was revised: Annex 3 has been updated. A correction to the book can be found at https://doi.org/10.1007/978-3-030-68191-3_21

Acknowledgments

We thank the following institutions supporting this project: Laurentian Forestry Centre; Canadian National Collection of Insects, Arachnids and Nematodes; Centre for Biodiversity Genomics, Guelph, Ontario; and University of Alaska Fairbanks. The first author would like to thank his Director General Dominic St-Pierre and Research Director Elizabeth Gauthier for their support of this project under the Coleoptera Collection program. He would also like to thank his wife, Patricia Corvera, for continuous support at home, particularly during the COVID-19 pandemic, and to his technician, Caroline Bourdon, for her many years of outstanding work and support of his research projects including this book. We thank Greg Breed (University of Alaska Fairbanks) for helpful comments on Chap. 4. Credit for images goes mainly to Sylvain Roberge (LFC) and Caroline Bourdon (LFC). The following curators and individuals provided specimens including types and ordinary specimens, including those from Europe, and we gratefully acknowledge their assistance: Christopher M. Ernst (SFU), Christoffer Fägerström (BMLS), Christopher C. Grinter (CAS), Crystal Maier (FMNH), Jaakko Mattila (ZMH), Marko Mutanen (ZMUO), Floyd Shockley (USNM), and Volker Assing (Hannover, Germany). The fifth author would like to thank his wife Denise Horwood. For over a decade, she has been his enthusiastic partner in countless rove beetles "round-ups." Her insight into habitat has led to many discoveries. Our thanks to Craig Machtans who provided photographs of habitats in Northwest Territories and Maria Leung and Denise Horwood for habitats in the Yukon Territory. Funding for this project was provided by Natural Resources Canada, University of Alaska, and Alaskan funding agencies for collecting insects.

Contents

1 Introduction

The ecosystems of the Arctic and Subarctic of northern Canada and Alaska are rapidly changing due to faster than expected climate warming and thawing permafrost, reshaping the land and releasing carbon gases. In these regions, temperatures are warming at more than three times the global average. Northern ecosystems (including tundra, forests and shrublands) are being profoundly impacted, affecting the way of life for residents. Ecosystem integrity and function is dependent in large part on how constituent species adapt to environmental stressors (e.g., warming climate, melting permafrost and sea ice). Much attention has focused on species traditionally utilized by residents of the North (e.g., plants, mammals, birds, and fish). However, the vast majority of biodiversity is represented by less conspicuous and non-iconic species. Although relatively unseen, under-considered, and poorly studied, these species (e.g., invertebrate animals, fungi, lichens) play critical roles that contribute to the functioning of healthy ecosystems. Society must also be concerned about this more cryptic biodiversity as these organisms, to a large degree, drive these ecosystems from the bottom up. Understanding how climate change is affecting biodiversity in its broadest sense will provide a more comprehensive understanding of shifts in northern ecology and the implications for human well-being. The responses of vertebrates and plants to climate change are not a proxy for how other groups of organisms are impacted. Unfortunately, the biodiversity of invertebrates, fungi, lichens, etc. in the North is poorly documented, which provides an enormous challenge for documenting and predicting change in these assemblages. We propose a start to remedy this problem, focusing on aleocharine rove beetles. Aleocharine beetles, the largest group of rove beetles (Coleoptera), are being used in many Canadian Forest Service and other national and international projects as indicators of environmental change. Many species are specialists restricted to specific soil/litter and forest microhabitats, while some are generalists and others are open habitat specialists. They fill many ecological niches and as such are good indicators for some other insect groups sharing similar habitats. Aleocharine beetles are a proven ecological indicator group with species reacting to climate variables such as humidity and temperature (Klimaszewski et al. 2008, Pohl et al. 2007, 2008). In Canada, the majority of aleocharine beetle species (over 600 spp.), were treated by the first author and collaborators, setting a foundation for ecological work on this group (Klimaszewski et al. 2018, 2020). Most are species from southern, central, eastern and western Canada. However, the fauna of the Subarctic and Arctic regions of Canada remains poorly known and is poorly documented. Much material has been collected as a result of various studies and sampling in Arctic and Subarctic regions, and this represents an enormous effort and investment, yet

J. Klimaszewski et al., *A Faunal Review of Aleocharine Beetles in the Rapidly Changing Arctic and Subarctic Regions of North America (Coleoptera, Staphylinidae)*, https://doi.org/10.1007/978-3-030-68191-3_1

this material has not been identified nor ecological information synthesized, such that the full value of that investment has not been reaped. We provide new knowledge for this group of insects in Northern Canada and Alaska and synthesize existing information, in order to provide the baseline information and diagnostic tools to facilitate future ecological studies and assessment of the susceptibility of these organisms to climate change. We gathered and studied all available specimens and records on species of aleocharine occurring in the study area (see Chap. 2), creating a comprehensive baseline data set for these species. This information is largely extracted from specimens currently housed in collections in Canada and USA (Alaska), and from the literature. Comprehensive ecological work on vertebrates and plants in the North, including responses to climate change, is possible because identification tools are available for these species. However, this is often not the case for various groups of arthropods. Until diagnostic tools are available, many different arthropod groups will continue to be relegated to the realm of human ignorance. If we want to fully understand the ecological implications of climate change, consideration must be given to a broad cross-section of biodiversity, including the lesser-known groups that comprise the vast majority of species and play an important role in ecological functioning of the ecosystems. One such group is the aleocharine beetles. Our work will provide information and diagnostic tools to open up a new frontier of enquiry into Canada's and Alaskan northern biodiversity.

References

Klimaszewski J, Langor D, Work TT, Hammond JHE, Savard K (2008) Smaller and more numerous harvesting gaps emulate natural forest disturbances: a biological test case using rove beetles (Coleoptera, Staphylinidae). Divers Distrib 14:969–982

Klimaszewski J, Webster RP, Langor DW, Brunke A, Davies A, Bourdon C, Labrecque M, Newton AF, Dorval J-A, Frank JH (2018) Aleocharine rove beetles of eastern Canada (Coleoptera, Staphylinidae, Aleocharinae): a glimpse of megadiversity. Springer, Cham. xvi + 902 pp

Klimaszewski J, Hoebeke ER, Godin B, Davies A, Perry KI, Bourdon C, Winchester N (2020) Aleocharine rove beetles of British Columbia: a hotspot of Canadian biodiversity (Coleoptera, Staphylinidae). Springer, Cham. xvi + 631 pp

Pohl GR, Langor DW, Spence JR (2007) Rove beetles and ground beetles (Coleoptera: Staphylinidae, Carabidae) as indicators of harvest and regeneration practices in western Canadian foothills forests. Biol Conserv 137:294–307

Pohl GR, Langor DW, Klimaszewski J, Work TT, Paquin P (2008) Rove beetles (Coleoptera: Staphylinidae) in northern Nearctic forests. Can Entomol 140:415–436

2 A Historical Review of Research on Aleocharinae of the Arctic and Subarctic Ecoregions of North America and an Overview of the Study Region

Arctic and Subarctic North America

We define the arctic and subarctic regions in Canada and Alaska as comprising the following level II ecoregions as defined by CEC (2018) (see Fig. 2.1): Arctic Cordillera (1.1), all tundra regions (2.1–2.4), all taiga regions (3.1–3.4), the Hudson Plains (4.1) and the Boreal Cordillera (6.1). The study region includes Alaska (United States) and at least the north part of all Canadian provinces except the Maritimes. As the majority of samples from Labrador have been taken near the border and transitional zone between Taiga Shield (3.4) and the milder Softwood Shield ecoregion (5.1), we have included all Labrador records in the present study but excluded those from ecoregion 5.1 in the historical review (below) and faunal analysis of Chap. 6. Species known so far only from region 5.1 and southward are noted in the comments section of each species account. For a more detailed, regional description of each ecoregion, see Chap. 3.9.

Historical Review of Research on Aleocharinae of the Arctic and Subarctic

Taxonomic research on Nearctic Aleocharinae, beginning in 1830 (Say 1830), has left a nomenclaturally complex but undeniably valuable legacy that has been reviewed in detail by Klimaszewski et al. (2018) and Klimaszewski et al. (2020). During much of this time, the Arctic and Subarctic aleocharine (ASA) fauna of North America was virtually unknown. This was likely due to a paucity of material available to early taxonomists, as access to interior wilderness was impossible or extremely difficult. Just three localities, all coastal or up-river from the coast, appear in the historical literature: Nome and Unalaska, AK, and Glenora, BC. As could be expected, six of only seven ASA taxa published before 1979 were described in the massive treatments of Aleocharinae by T. L. Casey (1906, 1910, 1911): *Atheta nomadica* Casey, 1910 (AK: Nome) (syn.: *Atheta graminicola* (Gravenhorst, 1806)); *Atheta surgens* (Casey, 1910) (BC: Glenora); *Baryodma glenorana* Casey, 1906 (BC: Glenora) (syn.: *Aleochara castaneipennis* Mannerheim, 1843); *Baryodma mannerheimi* Casey, 1906) (BC: Glenora) (syn.: *Aleochara castaneipennis* Mannerheim, 1843); *Gnypeta brevicornis* Casey, 1906) (BC: Glenora); *Homalota granulata* Mannerheim, 1846 (AK: Unalaska) (syn.: *A. graminicola*; *Paradilacra glenorica* Casey, 1910 (BC: Glenora) (syn.: *P. densissima* (Bernhauer, 1909). Two species of *Oxypoda* described by Casey (1893) based on material from Glenora BC, *O. glenorae* and *O. impressa*, are of uncertain taxonomic status because their type specimens are missing genitalia (Klimaszewski et al. 2020). These species are not treated herein. As with much of his species-

J. Klimaszewski et al., *A Faunal Review of Aleocharine Beetles in the Rapidly Changing Arctic and Subarctic Regions of North America (Coleoptera, Staphylinidae)*,
https://doi.org/10.1007/978-3-030-68191-3_2

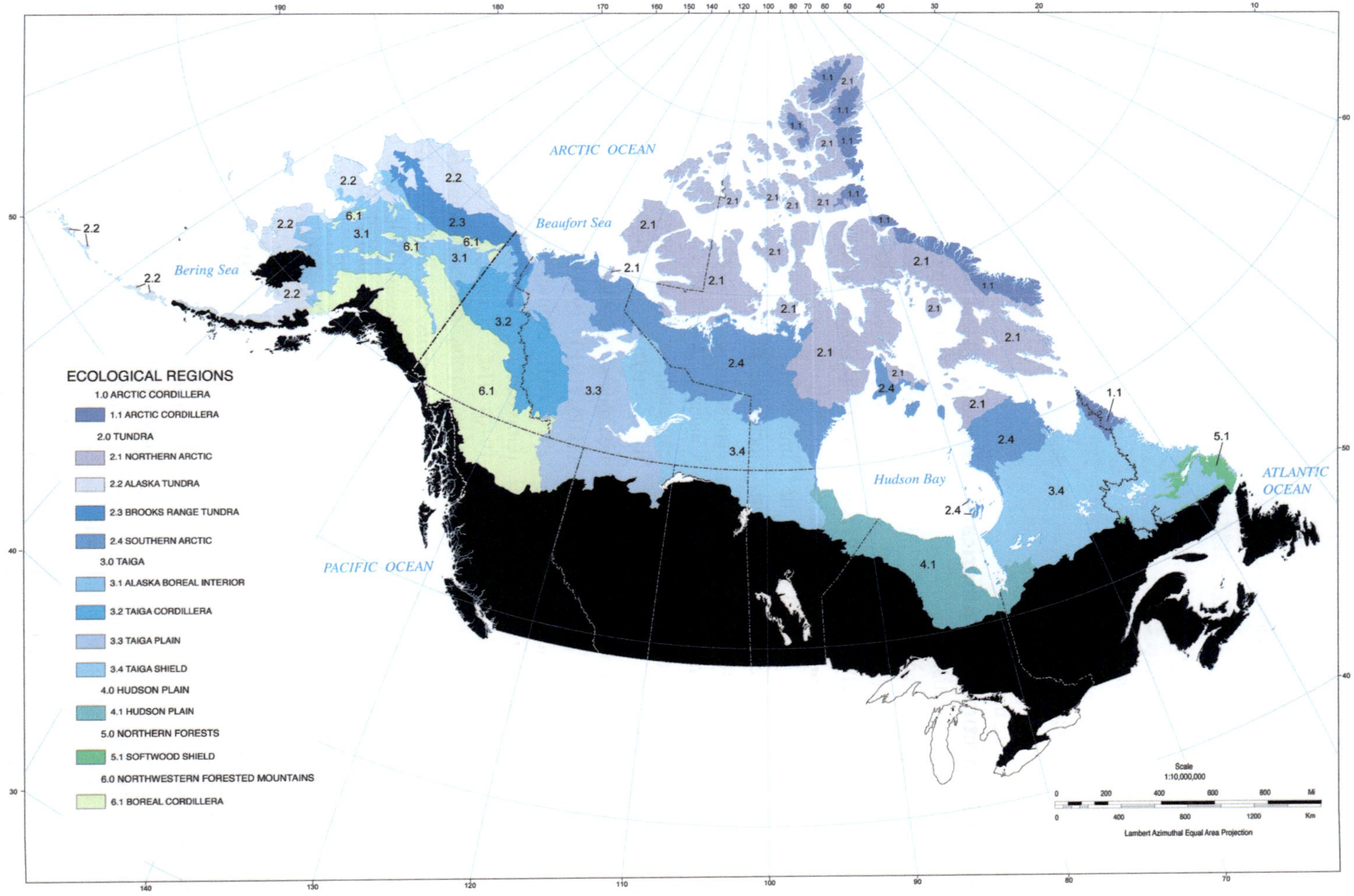

Fig. 2.1 Arctic and subarctic ecological regions (level II) of North America. Coloured area represents the study region. Map modified from Level II Ecoregions of North America (CEC 2018)

level taxonomic work on Coleoptera (reviewed in detail in Klimaszewski et al. 2020), Casey's underestimation of intraspecific morphological differences led to the later synonymy of all but two of his ASA taxa.

This long period of limited scientific progress was later countered with a large number of collecting expeditions to the study area. The Canadian Hudson Strait Expedition in 1927 (material from Port Burwell, NU; as QC in Lohse et al. 1990)) was the first to yield specimens of ASA taxa (F. Johansen). Material was also collected by W.J. Brown (CNC) from AK, MB and NT during the 1930s–1950s. During the 1940s–1980s, much of the material was collected by Canadian scientists, including those participating in the Northern Insect Survey (1947–1961) (Freeman 1959), which was conducted by staff at the Canadian National Collection of Insects, Arthropods and Nematodes (CNC) (Freeman 1959): A. Smetana, J.M. Campbell, J.R. Vockeroth, E. Becker, J.F. McAlpine, J.G. Chillcott, W.R.M. Mason, J.E.H. Martin, M. Wood and G.E. Shewell (all CNC; AK, LB, MB, NT, NU, QC, YT). Other significant Canadian collectors included R. E. Leech and B.S. Heming (University of Alberta, Edmonton, Alberta, Canada; AK, NT, YT), and A. and A. Morgan (University of Waterloo, Ontario, Canada; QC). Other material was collected by R. Madge in the 1960s (London, UK; AK) and C.H. Lindroth during the 1950s (Lund University, Sweden; AK).

The first taxonomic study to incorporate this wealth of aleocharine material was Klimaszewski (1979), later supplemented by Klimaszewski (1982a), which treated the world fauna of Gymnusini including ASA genera *Deinopsis* and *Gymnusa* (Fig. 2.2). Klimaszewski (1982b) revised the Nearctic *Myllaena*, which included

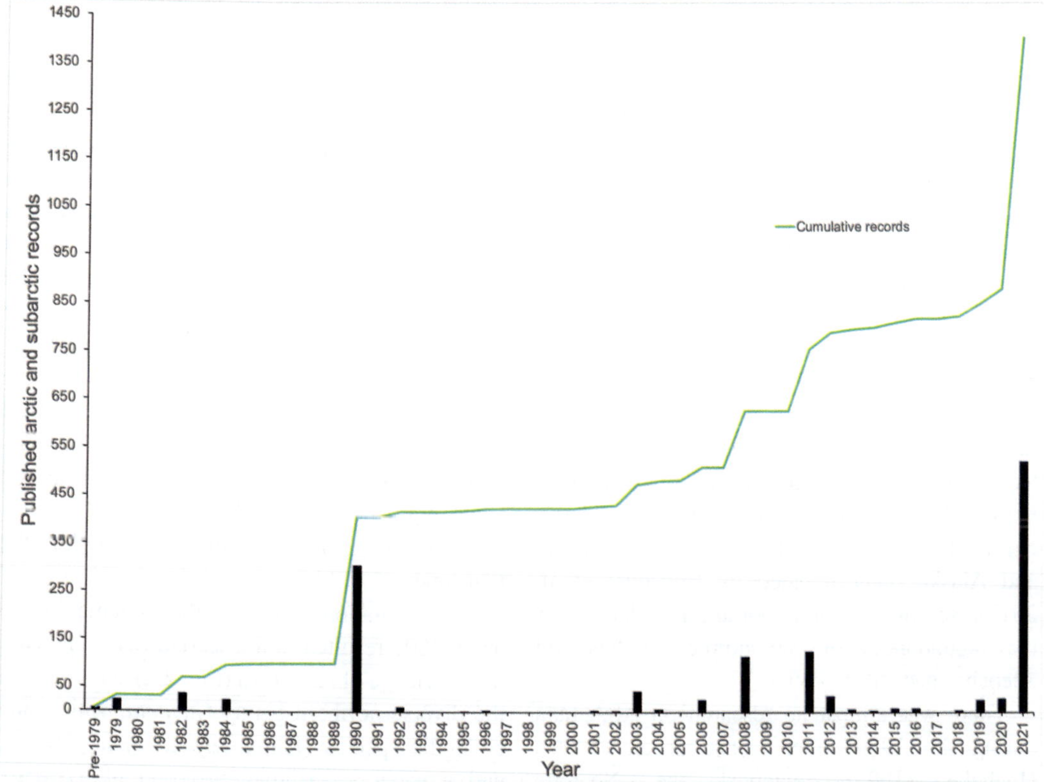

Fig. 2.2 Published records of Arctic and Subarctic Aleocharinae (Staphylinidae) over time. Included records are geographically distinct (to two decimal places, as given in decimal degrees) for a given species. Labrador records in ecoregion 5.1 (CEC 2018) are not considered and records from 2021 were entirely derived from the present contribution

three widespread species occurring in the study region. Shortly thereafter, a comprehensive revision of the Nearctic *Aleochara* by Klimaszewski (1984) brought another modest increase in the number of ASA records. Lohse and Smetana (1985) revised the type specimens of tribes Oxypodini and Athetini described from northwestern North America by Mannerheim (1843, 1846) and Mäklin (1852, 1853). In this paper the authors newly reported genera *Dinaraea* and *Liogluta* from the study region. While not a major contributor of ASA records as nearly all species were described from Sitka, AK (excluded here), this work formed an important taxonomic foundation for many species which were later discovered to occur in the study region.

Arguably, the most significant contribution to knowledge of ASA was the treatment of tribes Aleocharini, Oxypodini and Athetini by Lohse et al. (1990) (Fig. 2.2). Drawing upon the considerable material collected by the abovementioned expeditions and the previous study by Lohse and Smetana (1985), twenty genera and 73 species from the Arctic and Subarctic were treated, of which two genera and 50 species were described as new to science. Over 300 data points across the various species were published for the study region. An important achievement of this work was the recognition of thirteen species described from the Palaearctic region as naturally Holarctic species occurring also in North America. Their restricted taxonomic concept of the large and morphologically heterogeneous genus *Atheta* was useful and has been incorporated in most modern studies of Nearctic Aleocharinae thereafter as subgenus *Atheta* (including the present work). Campbell and Davies (1991) provided a checklist of the Aleocharinae recorded in Canada and Alaska (later updated by Bousquet et al. 2013), serving as an important foundation for the abundance of taxonomic studies on Aleocharinae still to come.

There was almost no research on ASA taxa during the 1990s after Lohse et al. (1990). Hoebeke (1992) reviewed the Nearctic *Lypoglossa* and reported this genus from the study region for the first time. Ahn and Ashe (1995), Ahn (1996a, b) and Ahn (1997) reviewed the seashore-dwelling genera *Amblopusa*, *Diaulota*, *Liparocephalus*, *Paramblopusa* and *Tarphiota*, and provided a few records from the study region. Eleven years later, a revision of *Placusa* (Placusini) was published by Klimaszewski et al. (2001), which included some widespread species occurring in the forested parts of the study zone. Gusarov (2002) revised the Nearctic species of *Earota* and provided the first record of the genus from the study region. Gusarov (2003) revised a large number of aleocharine type specimens, including those of Lohse et al. (1990) and synonymized five of these species. Here, the genera *Dochmonota* and *Schistoglossa* were reported for the first time from the study area. Gusarov (2003) also examined Arctic and Subarctic CNC specimens not included by Lohse et al. (1990), and contributed approximately 43 new data points across the various treated ASA taxa. In the same year, Klimaszewski et al. (2003) recorded *Silusa* (Homalotini: Silusina) from the study region for the first time. Shortly thereafter, Gusarov (2004) revised the species of *Lypoglossa*, and Klimaszewski and Pelletier (2004) revised the genera of the *Ocalea* group (Oxypodini), including the description of ASA genera *Neothetalia* and *Alfocalea*. In a review of the Canadian genera of the *Zyras* group (Lomechusini), Klimaszewski et al. (2005) reported the relatively thermophilic tribe from the study region for the first time (*Zyras obliquus*, far northern AB). In the following year, Klimaszewski et al. (2006) revised the species of diverse *Oxypoda* in Canada and Alaska, including a significant number of ASA species, and newly reported *Hylota* from the study region. Paśnik (2006) reported *Tachyusa* (Tachyusini) from the study region for the first time.

A combination of two 'northern-themed' studies in 2008 resulted in the second major increase in knowledge of ASA taxa (Fig. 2.2): a revision of the diverse northern genus *Gnypeta*, with many Holarctic species (Klimaszewski et al. 2008a), and a paper by Klimaszewski et al. (2008b), which reported new species and new records from YT and AK. The latter recorded the tribe Hypocyphtini (*Cypha*), subtribe Gyrophaenina of

Homalotini (*Gyrophaena*) and genus *Amischa* (Athetini) from the study region for the first time, with locality data. In the same year Majka and Klimaszewski (2008) reported two new jurisdictional records for the study region.

Several years later, a comprehensive treatment of the Aleocharinae of Newfoundland and Labrador was published (Klimaszewski et al. 2011), which included many records from the Taiga Shield (ecoregion 3.4) of Labrador (Fig. 2.1). This study represents the third major increase in data for ASA taxa (Fig. 2.2) as a result of the authors incorporating an abundance of newly collected samples from the region. The authors newly reported genera *Deinopsis*, *Paragoniusa* and *Parocyusa* from the study region. One year later, a second study on the Yukon aleocharine fauna was published (Klimaszewski et al. 2012), including first records of *Brachyusa*, *Calodera* and *Strigota* for the study region. Studies of biodiversity patterns in northern Canada by Ernst and Buddle (2013, 2015) brought new tundra and taiga locality records (NT and NU) for seven ASA taxa. Klimaszewski et al. (2013) revised the species of subcortical genus *Dinaraea* and provided a few relevant localities for the region, while the northern genera *Gnathusa*, *Mniusa* and *Ocyusa* were revised by Klimaszewski et al. (2014). Faunistic studies of Canadian Aleocharinae by Klimaszewski et al. (2015) and Webster et al. (2016) published new records of ASA taxa, including the first record for the study region of *Trichiusa* in the latter paper. The wet litter-associated species of *Liogluta*, with several ASA species, were revised by Klimaszewski et al. (2016) and later, the aleocharine fauna of eastern Canada was comprehensively reviewed by Klimaszewski et al. (2018), with a few new distributional records of ASA taxa.

Most recently, a revision of the northern genus *Boreophilia*, containing many Holarctic species was published (Klimaszewski et al. 2019). In addition to providing records from newly collected material, the authors also corroborated species concepts for many Holarctic taxa using DNA barcodes. A year later, a comprehensive review of Aleocharinae of BC was published, which contributed a number of new records from the study region based mostly on newly collected material from AK, NT, YT and northern BC. These records included the first evidence of *Meotica* in the study region. In the same year, Ahn et al. (2020) reported the intertidal species *P. eoa* from the Nearctic region (AK) and Yoo and Ahn (2020) provided several records of *Amblopusa breviceps* Casey from the study area.

Based mostly on newly collected material deposited in UAM, LFC and CBG and some previously unexamined material in CNC, we here contribute over 500 new data points on ASA species. These combine with previous data to form 1404 geographically distinct, species-level records that form the core of the present work as a whole. Recent collecting efforts associated with the above collections are described in detail in Chap. 3 and include the Canadian National Parks Malaise Program (http://biodiversitygenomics.net/projects/cnp/) and the Arctic BIOSCAN project (https://arcticbioscan.ca/). Although the present work contributes a considerable amount of new data, Arctic and Subarctic North America is vast and remains inadequately sampled (see Chap. 6). Many ASA species are known only from one or two localities and still little is known of the bionomics of most of these taxa. The present work as a whole provides a synthesis of existing knowledge and morphological identification tools. We hope that these will contribute to a better understanding of the changes happening to our fragile northern ecoregions, through the implementation of biodiversity and ecological impact studies that include this diverse subfamily.

References

Ahn KJ (1996a) A review of *Diaulota* Casey (Coleoptera: Staphylinidae: Aleocharinae), with description of a new species and known larvae. Coleopt Bull 50:270–290

Ahn K-J (1996b) Revision of the intertidal aleocharine genus *Tarphiota* (Coleoptera: Staphylinidae). Entomol News 107:177–185

Ahn K-J (1997) A review of *Liparocephalus* Mäklin (Coleoptera: Staphylinidae: Aleocharinae) with descriptions of larvae. Pan Pac Entomol 73(2):79–92

Ahn KJ, Ashe JS (1995) Revision of the intertidal aleocharine genus *Amblopusa* Casey and description of the new genus *Paramblopusa* (Coleoptera: Staphylinidae). J N Y Entomol Soc 103:138–154

Ahn K-J, Sikes DS, Klimaszewski J (2020) *Paramblopusa eoa* Ahn and Maruyama (Coleoptera: Staphylinidae: Aleocharinae), distribution range extension to Alaska. Coleopt Bull 74(2):324–326

Bernhauer M (1909) Neue Aleocharini aus Nordamerika. (Col.) (4. Stück.). Deutsche Entomologische Zeitschrift 1909(4):515–528

Bousquet Y, Bouchard P, Davies AE, Sikes DS (2013) Checklist of beetles (Coleoptera) of Canada and Alaska: second edition. Pensoft Series Faunistica No. 109. Pensoft, Sofia, Moscow, 402 p

Campbell JM, Davies A (1991) Family Staphylinidae rove beetles. In: Bousquet Y (ed) Checklist of beetles of Canada and Alaska. Publication 1861/E, Research Branch, Agriculture Canada, Ottawa, pp 86–124. vi + 430 pp

Casey TL (1893) Coleopterological notices. V. Ann N Y Acad Sci 7:281–606

Casey TL (1906) Observations on the staphylinid groups Aleocharinae and Xantholinini, chiefly of America. Trans Acad Sci St Louis 16(6):125–434

Casey TL (1910) New species of the staphylinid tribe Myrmedoniini. In: Memoirs on the Coleoptera. I. New Era Printing Co., Lancaster, PA, pp 1–183. 205 pp

Casey TL (1911) New American species of Aleocharinae and Myllaeninae. In: Memoirs on the Coleoptera. II. The New Era Printing Co., Lancaster, PA, pp 1–245. 259 pp

Commission for Environmental Cooperation (CEC) (2018) Terrestrial ecoregions – Level II ecoregions. http://www.cec.org/north-american-environmental-atlas/terrestrial-ecoregions-level-ii/. Accessed 8 September 2020

Ernst CM, Buddle CM (2013) Seasonal patterns in the structure of epigeic beetle (Coleoptera) assemblages in two subarctic habitats in Nunavut, Canada. Can Entomol 145:171–183

Ernst CM, Buddle CM (2015) Drivers and patterns of ground-dwelling beetle biodiversity across northern Canada. PLoS One 10:122163

Freeman T (1959) The Canadian northern insect survey, 1947–57. Polar Record 9(61):299–307. https://doi.org/10.1017/S0032247400066006

Gusarov VI (2002) A revision of Nearctic species of the genus *Earota* Mulsant & Rey, 1874 (Coleoptera: Staphylinidae: Aleocharinae). Zootaxa 92(1):16

Gusarov VI (2003) Revision of some types of north American aleocharines (Coleoptera: Staphylinidae: Aleocharinae), with synonymic notes. Zootaxa 353:1–134

Gusarov VI (2004) A revision of the genus *Lypoglossa* Fenyes, 1918 (Coleoptera: Staphylinidae: Aleocharinae). Zootaxa 747:1–36

Hoebeke ER (1992) Taxonomy and distribution of the athetine genus *Lypoglossa* Fenyes (Coleoptera: Staphylinidae: Aleocharinae) in North America, with description of a new species. J N Y Entomol Soc 100:381–398

Klimaszewski J (1979) A revision of the Gymnusini and Deinopsini of the world: Coleoptera, Staphylinidae, Aleocharinae, No. 25. Agriculture Canada, Research Branch. Monograph No. 25, 169 pp

Klimaszewski J (1982a) Studies of Myllaenini (Coleoptera: Staphylinidae, Aleocharinae): 1. Systematics, phylogeny, and zoogeography of Nearctic *Myllaena* Erichson. Can Entomol 114(3):181–242

Klimaszewski J (1982b) A revision of the Gymnusini and Deinopsini of the world (Coleoptera: Staphylinidae). Supplementum 2. Can Entomol 114 (4):317–335

Klimaszewski J (1984) A revision of the genus *Aleochara* Gravenhorst of America north of Mexico (Coleoptera: Staphylinidae, Aleocharinae). Mem Ent Soc Can 129. [2] + 211 pp

Klimaszewski J, Pelletier G (2004) Review of the *Ocalea* group of genera (Coleoptera, Staphylinidae, Aleocharinae) in Canada and Alaska: new taxa, bionomics, and distribution. Can Entomol 136:443–500

Klimaszewski J, Pelletier G, Germain C, Hebert C (2001) Diversity of *Placusa* (Coleoptera: Staphylinidae: Aleocharinae) in Canada, with descriptions of two new species. Can Entomol 133(1):1–47

Klimaszewski J, Pohl G, Pelletier G (2003) Revision of the Nearctic *Silusa* (Coleoptera, Staphylinidae, Aleocharinae). Can Entomol 135(2):159–186

Klimaszewski J, Pelletier G, Maruyama M, Hlaváč P (2005) Canadian species of the *Zyras* group of genera and review of types from America north of Mexico (Coleoptera, Staphylinidae, Aleocharinae). Rev Suisse Zool 112(3):703–733

Klimaszewski J, Pelletier G, Germain C, Work T, Hébert C (2006) Review of *Oxypoda* species in Canada and Alaska (Coleoptera, Staphylinidae, Aleocharinae): systematics, bionomics, and distribution. Can Entomol 138(6):737–852

Klimaszewski J, Savard K, Pelletier G, Webster R (2008a) Species review of the genus *Gnypeta* Thomson from Canada, Alaska and Greenland (Coleoptera, Staphylinidae, Aleocharinae): systematics, bionomics and distribution. ZooKeys 2:11–84

Klimaszewski J, Godin B, Pelletier G, Savard K (2008b) Six new species and records of aleocharine beetles from the Yukon and Alaska (Coleoptera: Staphylinidae: Aleocharinae). Can Entomol 140:265–291

Klimaszewski J, Langor D, Pelletier G, Bourdon C, Perdereau L (2011). Aleocharine beetles (Coleoptera, Staphylinidae) of the province of Newfoundland and Labrador, Canada. Pensoft Series Faunistica No. 98. Pensoft, Sophia-Moscow, 313 pp

Klimaszewski J, Godin B, Bourdon C (2012) Further contributions to the aleocharine fauna of the Yukon

Territory, Canada (Coleoptera, Staphylinidae). ZooKeys 186:207–237

Klimaszewski J, Webster RP, Langor DW, Bourdon C, Jacobs J (2013) Review of Canadian species of the genus *Dinaraea* Thomson, with descriptions of six new species (Coleoptera, Staphylinidae, Aleocharinae, Athetini). ZooKeys 327:65–101

Klimaszewski J, Webster RP, Langor DW, Bourdon C, Hammond HEJ, Pohl GR, Godin B (2014) Review of Canadian species of the genera *Gnathusa* Fenyes, *Mniusa* Mulsant & Rey and *Ocyusa* Kraatz (Coleoptera, Staphylinidae, Aleocharinae). ZooKeys 412:9–40

Klimaszewski J, Godin B, Langor D, Bourdon C, Lee S-I, Horwood D (2015) New distributional records for Canadian Aleocharinae (Coleoptera, Staphylinidae), and new synonymies for *Trichiusa*. ZooKeys 498:51–91

Klimaszewski J, Webster RP, Langor DW, Sikes D, Bourdon C, Godin B, Ernst C (2016) A review of Canadian and Alaskan species of the genus *Liogluta* Thomson, and descriptions of three new species (Coleoptera, Staphylinidae, Aleocharinae). ZooKeys 573:217–256

Klimaszewski J, Webster RP, Langor DW, Brunke A, Davies A, Bourdon C, Labrecque M, Newton AF, Dorval J-A, Frank JH (2018) Aleocharine rove beetles of eastern Canada (Coleoptera, Staphylinidae, Aleocharinae): a glimpse of megadiversity. Springer, Cham. xvi + 902 pp

Klimaszewski J, Sikes DS, Brunke A, Bourdon C (2019) Species review of the genus *Boreophilia* Benick from North America (Coleoptera, Staphylinidae, Aleocharinae, Athetini): systematics, habitat, and distribution. ZooKeys 848:57–102

Klimaszewski J, Hoebeke ER, Godin B, Davies A, Perry KI, Bourdon C, Winchester N (2020) Aleocharine rove beetles of British Columbia: a hotspot of Canadian biodiversity (Coleoptera, Staphylinidae). Springer, Cham. xvi + 631 pp

Lohse GA, Smetana A (1985) Revision of the types of species of Oxypodini and Athetini (sensu Seevers) described by Mannerheim and Mäklin from North America (Coleoptera: Staphylinidae). Coleop Bull 39 (3):281–300

Lohse GA, Klimaszewski J, Smetana A (1990) Revision of Arctic Aleocharinae of North America (Coleoptera: Staphylinidae). Coleopt Bull 44(2):121–202

Majka CG, Klimaszewski J (2008) New records of Canadian Aleocharinae (Coleoptera: Staphylinidae). ZooKeys 2:85–114

Mäklin FW (1852) [New species and notes]. In: Mannerheim, C.G.: Zweiter Nachtrag zur Kaefer-Fauna der Nord-Amerikanischen Laender des Russischen Reiches. Bulletin de la Société Impériale des Naturalistes de Moscou 25(2):283–372

Mäklin FW (1853) [New species and notes]. In: Mannerheim, C.G.: Dritter Nachtrag zur Kaefer-Fauna der Nord-Amerikanischen Laender des Russischen Reiches. Bulletin de la Société Impériale des Naturalistes de Moscou 26(3):95–273

Mannerheim CG (1843) Beitrag zur Kaefer-fauna der Aleutischen Inseln, der Insel Sitkha und Neu-Californiens. Bulletin de la Société Impériale des Naturalistes de Moscou 16(2):175–314

Mannerheim CG (1846) Nachtrag zur Kaefer-Fauna der Aleutischen Inseln und der Insel Sitkha. Bulletin de la Société Impériale des Naturalistes de Moscou 19:501–516

Paśnik G (2006) A revision of the World species of the genus *Tachyusa* Erichson, 1837 (Coleoptera, Staphylinidae: Aleocharinae). Zootaxa 1146:1–152

Say T (1830) Descriptions of new species of North American insects, and observations on some already described. New Harmony 1–41

Webster RP, Klimaszewski J, Bourdon C, Sweeney JD, Hughes CC, Labrecque M (2016) Further contribution to the Aleocharinae (Coleoptera, Staphylinidae) fauna of New Brunswick and Canada including description of 27 new species. In: Webster, R.P., Bouchard, P., Klimaszewski, J. (Eds) The Coleoptera of New Brunswick and Canada: providing baseline biodiversity and natural history data. ZooKeys [Special Issue] 573:85–216

Yoo I-S, Ahn K-J (2020) Discovery of a new intertidal species, *Amblopusa vancouverensis* from Vancouver Island, Canada with notes on *Amblopusa brevipes* Casey (Coleoptera, Staphylinidae, Aleocharinae). Zootaxa 4803:505–514

3 Material and Methods

Format

The classification of taxa used herein selectively follows concepts expressed by Seevers (1978), Klimaszewski (1979, 1982, 1984, 2018, 2020), Lohse et al. (1990), Ashe (2001), Gusarov (2003, 2011), Elven et al. (2011, 2012), Paśnik (2010), Bouchard et al. (2011), Schülke and Smetana (2015), and Yamamoto and Maruyama (2017). Distributional records and bionomic information (e.g., habitat associations, collection dates, collecting methods) are based on published records and on specimens in the collections of Laurentian Forestry Centre, the Canadian National Collection of Insects, Arachnids and Nematodes, the collection of University of Alaska, Centre for Biodiversity Genomics and the private collections of B. Godin (BGC) and R. Webster (RWC). This bionomic data often includes information from areas not treated by our book (see Chap. 2) for widespread taxa but should still be relevant to collecting these species. In the case of unpublished records, we provide specimen data under the respective species. Only records considered to be reliable are listed. Latitude/longitude data originally in degrees/minutes format on labels have been converted to decimal degrees.

ZooBank Registration

The present work has been registered in ZooBank, under the following identifier:

urn:lsid:zoobank.org:pub:08B3D0FD-6270-464C-9AE6-4427E5349A7D.

Abbreviations

Additional abbreviations are used in the main text and tables to indicate new records: **NPR** (New Provincial Record), **NTR** (New Territorial Record), **NSR** (New State Record), and **BOLD** (The Barcode of Life Data System).

Jurisdictions in boldface are included in Arctic/Subarctic North America, as defined in this work. We report separately on Newfoundland and Labrador even though together they constitute one province. Roughly, species occurring in LB, NT, NU, YT, and the northern parts of AB, BC, MB, ON, QC, and AK (excluding the Marine West Coast Forest ecozone) are included in this treatment (see Fig. 2.1).

AB	**Alberta**
AK	**Alaska**
BC	**British Columbia**

J. Klimaszewski et al., *A Faunal Review of Aleocharine Beetles in the Rapidly Changing Arctic and Subarctic Regions of North America (Coleoptera, Staphylinidae)*,
https://doi.org/10.1007/978-3-030-68191-3_3

LB **Labrador**
MB **Manitoba**
NB New Brunswick
NF **Newfoundland**
NS Nova Scotia
NT **Northwest Territories**
NU **Nunavut**
ON **Ontario**
PE Prince Edward Island
QC **Quebec**
SK **Saskatchewan**
YT **Yukon Territory**

State abbreviations for the United States of America follow those of the United States Postal Service.

Institution Codes

BGC Benoit Godin private collection, Whitehorse, Yukon, Canada
BMLS Biological Museum (Entomology), Lund, Sweden
CBG Centre for Biodiversity Genomics, University of Guelph, Guelph, Ontario, Canada
CNC Canadian National Collection of Insects, Arachnids, and Nematodes, Agriculture and Agri-Food Canada, Ottawa, Ontario, Canada.
FMNH Integrative Research Center, The Field Museum of Natural History, Chicago, Illinois, United States of America
LFC Natural Resources Canada, Canadian Forest Service, Laurentian Forestry Centre, R. Martineau Insectarium, Quebec City, Quebec, Canada.
NBP Northern Biodiversity Program, a colaborative research initiative.
NoFC Natural Resources Canada, Canadian Forest Service, Northern Forestry Centre, Edmonton, Alberta, Canada.
RWC Reginald Webster (private) collection, Charters Settlement, New Brunswick, Canada.
SFU Simon Fraser University, Burnaby, British Columbia, Canada.
UAM University of Alaska Museum Insect Collection, Fairbanks, Alaska, United States of America.
USNM United States National Museum, Washington, DC, United States of America.
ZMH Finnish Museum of Natural History, Zoological Museum, Helsinki, Finland.
ZMUO Zoological Museum, University of Oulu, Oulu, Finland.

Diagnostic Features

Most of the text below regarding diagnostic features is taken from Klimaszewski et al. (2018, 2020). It is important that readers of this book studying the aleocharine fauna of the North have all the information accessible in one source, thus increasing efficiency when working with this complex group. Aleocharine beetles are highly diverse taxonomically, morphologically (Fig. 3.3a–l), and ecologically. Significant challenges with aleocharine species identification are due to the poor state of knowledge of many genera still (e.g., *Acrotona*, *Amischa*, *Atheta*) and the lack of comprehensive diagnostic tools. These are the main obstacles to understanding species richness, assemblage composition, and ecological roles of aleocharines in ecosystems of the North. There are many species, often forming groups of cryptic species, which are externally similar, especially in the large tribes Aleocharini, Athetini, and Oxypodini.

External Body Structures Important for Identification

The terminology used herein follows that used by previous authors (Benick and Lohse 1974; Seevers 1978; Klimaszewski 1979, 1984; Klimaszewski et al. 2018, 2020; Lohse et al. 1990; Ashe 2001; Gusarov 2003). The body length of aleocharine beetles ranges from 1.0–13.0 mm (typically 3.0–5.0 mm) and habitus forms are diverse (Fig. 3.3a–l), reflecting different adaptations to a variety of microhabitats. They range from somewhat flat (e.g., *Dinaraea*, *Placusa*) to subcylindrical (e.g., some *Leptusa*), and from robust (e.g., *Aleochara*, *Gymnusa*, and *Oligota*) to slender (e.g., *Atheta*, *Myllaena*, and *Meotica*). The term forebody refers to head, pronotum, and elytra as a combined structure. The most important body structures used in identification of aleocharines are illustrated (Fig. 3.4). These relate to the **head** (Fig. 3.4): frontal suture, genae (postocular area, temples), infraorbital carina, neck; **antennae** (Fig. 3.4): proportion of length to width ratio of different antennomeres, length of antennae; **mouthparts** (Fig. 3.5a–f): labrum, labium, maxillae, labial palps, and ligula; **prothorax** (Figs. 3.4, 3.6): pubescence pattern, hypomeron (visible or not visible in lateral view of pronotum), pro-, meso- and metaventrites and their intercoxal processes, isthmus, and coxae; **elytra** (Fig. 3.4): size and shape, pubescence pattern, length of elytral suture, shape of basal margin, and presence or absence of lateral emargination; **abdomen** (Figs. 3.2, 3.7a–e) [10-segmented, position indicated by Roman numerals starting at the base; when describing features of the "first visible tergites" in the keys and diagnoses, we are referring to the basal segments following morphological segments I and II which are reduced and largely hidden by the elytra; segments IX and X are modified; transverse impressions of basal tergites (I-IV), structures inside impressions, and shape of the apical margin of male tergite VIII, and female tergite and sternite VIII; **legs** (Fig. 3.4): tarsal formula and length and proportions of the tarsomeres of the hind tarsus.

Genital Structures Important for Identification

The morphological diversification of the genitalia, particularly the shape of the median lobe of the aedeagus and its internal sac structures, including flagellum and internal sclerites (structures), morphology and macro-setal pattern on the apical lobe of parameres, and morphology, of the spermatheca, likely have played an important role in aleocharine evolution. These morphological features vary in different species and are generally the structures most important for species- and genus-level identification. The three-lobed aedeagus of aleocharines (median lobe, and two complex lateral lobes called parameres), is the most distinctive feature of the entire subfamily, and constitutes a synapomorphy for this group (Klimaszewski et al. 2018, 2020).

Aedeagus (Fig. 3.8a–d) The male copulatory organ consists of the median lobe (penis = bulbus+tubus) (Fig. 3.8a–b), and the two parameres (Fig. 3.8c–d).

Median lobe of aedeagus (Figs. 3.8a–b, 3.9a–i) This is a tubular, sclerotized, intromittent organ with an enlarged basal portion called the *bulbus* (Fig. 3.8a–b), and a narrow apical portion, subcylindrical or trough-like, called the *tubus* (Fig. 3.8a–b). The tubus is slender in most, and narrow towards the apex. The more or less narrow, basal or median part of lateral edges of the tubus are extended dorsally in Athetini/Lomechusini and form a complete or incomplete arched structure called the *athetine bridge* (Seevers 1978). The shape of the tubus, and particularly its apical portion in lateral view, varies from species to species and is a valuable diagnostic feature (Fig. 3.9a–i). The subspherical bulbus has a small, ventral round opening, the *foramen mediale*, which is an entrance for the ejaculatory duct, the *ductus ejaculatorius*, and one or more projections, the larger called *crista apicalis* or external carina of the bulbus. The *crista apicalis* provides attachment for the muscles connecting the median lobe and the two condylites of the parameres (Fig. 3.8c–d). The internal space of the bulbus contains a complex of mainly dorso-ventral muscles attached internally to the base of the bulbus and to the dorsal mobile compressor plate that is oval and weakly sclerotized in most. Contraction or expansion of the longitudinal and dorso-ventral muscle complexes are directly or indirectly responsible for creation of hydrostatic pressure that results in eversion or retraction of the *internal sac* and its associated internal structures (e.g., sclerites, membraneous structures, and flagellum). The *flagellum* (also known as the virga or copulatory piece)

(Fig. 3.8a–b) is a short or moderately-to-extremely elongate tubular structure (e.g., some *Aleochara*, *Oxypoda*) that can be narrow or broad, and it has a swollen basal part connected with the ejaculatory duct. The internal sac membranes frequently bear numerous spinules, denticulate plates, and other structures to aid in retention of the sac in the *vulva* (female copulatory tract). During eversion of the internal sac, the internal sclerites and flagellum are extended (everted) exteriorly filling the internal space of the female *vulva* and allowing safe transfer of sperm, or packages of sperm contained in spermatophores, with the aid of the flagellum. The internal sac is withdrawn by a set of longitudinal muscles that originate on the proximal surface of the base of the bulbus. The structures and function of the internal sac were previously explained and illustrated for *Aleochara curtula* by Peschke (1978).

The ventral side of the median lobe of the aedeagus is considered to be the side of the bulbus containing the foramen mediale, the entrance of the ductus ejaculatorius, and the adjacent ventral side of the tubus of the median lobe with the internal sac and its structures (this part is referred to as the parameral side in Gusarov 2003); the opposite side is referred to as the dorsal part.

Parameres (Fig. 3.8c–d) These are the tactile orientation organs, which enable the male to find the tip of the female abdomen just before copulation, using the setal pattern on the female abdomen as a guide (Peschke 1978). There are two symmetrical parameres, one on each side of the median lobe of the aedeagus, which are attached to the crista apicalis (represented by one, two or several carinae) of the bulbus of the median lobe by a complex system of muscles. Each paramere is subdivided into a *paramerite* and a narrowly elongate *condylite*. The paramerite is the main part of the paramere, and consists of a narrow apical lobe bearing a few macrosetae arranged in a particular pattern specific to different genera (Seevers 1978). Attached to the paramerite is a semimembraneous *velar sac* bearing microribbings, and the posterior, slat-shaped structure delimited from the apical portion by a weakly sclerotized strip (Seevers 1978; Klimaszewski 1984). The morphology of the paramere and the setal pattern of the apical lobe of the paramerite provide diagnostic characters at the generic and higher taxonomic levels.

Spermatheca (*receptaculum seminis*) (Figs. 3.8e, 3.10a–l) This is the female genital organ for storing and perhaps selecting sperm (Peschke 1978; Dybas and Dybas 1981). The spermatheca consists of a usually bulbous capsule and narrower and short neck connected to a tubular stem and a valve with thin and membranous, weakly sclerotized seminal canal (*ductus spermathecae*), which connects the spermatheca to the *vulva* and serves to transport sperm. The seminal canal is not considered to be a part of the spermatheca (Klimaszewski 1984). The capsule is a spherical or tubular structure that may be straight or bent, and bears apical invaginations of various shapes and sizes (Fig. 3.10a–l). The stem may be a simple narrow tube or more complex and subdivided into an elongate chamber and a narrower duct. The shape of the spermatheca varies from species to species and is valuable for species separation. In some genera (e.g., *Acrotona*, *Amischa*, some *Atheta*, *Gnypeta*, *Liogluta*, and *Mocyta*) the shape of the spermatheca may be a better diagnostic character than the shape of the median lobe of the aedeagus.

Terminalia (Fig. 3.7a–e) The terminalia herein refer to the male and female tergite VIII and sternite VIII (Fig. 3.7e) and the *pygidium*, the latter consisting of 2–3 modified terminal segments closing the end of the abdomen (Fig. 3.7a–e). Tergite VIII and sternite VIII provide useful additional morphological features for species level diagnostics. These structures can be elongate, subquadrate, or transverse and can have varying pubescence patterns. The *antecostal suture* is located subparallel to the base of tergite VIII and sternite VIII. The antecostal suture may be separated by a wide or narrow distance from the base of the disc and may be straight or sinuate in different species. The apical margin of tergite VIII of males is highly modified in many groups of aleocharines, and particularly in athetines, and often bears teeth, denticles, and other projections or emarginations, which provide important taxonomic identification features (Fig. 3.7e). The structure of the pygidium is slightly different between males (Fig. 3.7a–b) and females (Fig. 3.7c–d). In the male, the pygidium consists of tergite X, accompanied on both sides by tergite IX (completely subdivided or connected by a narrow dorsal strip), which is extended ventrally into two posterior narrow ventral struts (Fig. 3.7a–b). Ventrally located between the subdivided male tergite IX is a narrowly oval and flat structure which is regarded as the modified sternite IX (Fig. 3.7b). The pygidium of the female is generally similar to that of the male but is lacking ventral struts and an elongate sternite IX (Fig. 3.7c–d). The shape and pubescence pattern of tergite X is diagnostic for some species and genera. Specimens can be sexed by examination of the ventral part of the pygidium for presence (male) or absence (female) of the apical portion of sternite IX, which is present in males as a median rounded lobe (Fig. 3.7b).

Microdissections and Preparation of Genital Structures for Examination and Diagnostics

Identifying aleocharine species and genera can be challenging due to the small size of most specimens and minute size of many distinctive characters, coupled with a vast number of valid taxa (Ashe 2001; Hanley and Ashe 2003; Klimaszewski et al. 2018, 2020). There are many species, often forming groups of sibling species, which are externally similar, especially in the large tribes Athetini (e.g., *Atheta*), and Oxypodini (e.g., *Oxypoda*), and Homalotini (e.g., *Gyrophaena*). Thus, properly dissected genital structures (median lobe of aedeagus, parameres, spermatheca, and tergite VIII and sternite VIII of both sexes) carefully mounted in Canada balsam on plastic microslides for examination, are indispensable for identification of the majority of species and therefore essential for serious work in aleocharine systematics. The methodology of dissecting aleocharine species mouthparts is sufficiently described by Hanley and Ashe (2003) and is not discussed here because mouthparts are usually not essential for species identification.

Dissection and Dehydration (for Illustrations see Klimaszewski et al. 2020)
With the aid of a stereoscopic microscope, specimens are best dissected in water (room temperature), placed in a microscope depression slide (commercially available slides with impressions used for blood analysis). The abdomen is held in place by forceps while a hooked or straight minuten pin, mounted in a commercially available handle, is inserted between abdominal segment VII and VIII and pulled to dislodge the terminal segments from the rest of the abdomen. The genital structures (aedeagus or spermatheca) are usually inside the dislodged apical part of the abdomen but sometimes may remain in the basal part of the abdomen and should be removed from there. The dislodged apical part of the abdomen is dissected further by removing tergite VIII and sternite VIII from the pygidium using dissecting needles and then breaking the lateral internal muscles that fix the discs in place and pulling the sclerites apart. Subsequently, the aedeagus or spermatheca are also removed and detached from other structures so that tergite VIII, sternite VIII, aedeagus (or spermatheca), and pygidium are separated from each other. Remnants of muscles attached to structures may be removed mechanically using dissecting needles while structures are immersed in water or alternatively structures may be cleared in cold 10% potassium hydroxide for 1–4 min to digest the soft tissues and then washed in water. The aedeagus needs further dissection by removing the parameres from the median lobe using dissecting needles to break the muscles connecting the condylite of the paramere with the *crista apicalis* of the bulbus of the median lobe. All dissected structures need to be washed in clean water, transferred to 75% ethanol for about one minute and then to 100% ethanol for dehydration for about 1–2 min before mounting. Structures should then be mounted in Canada balsam on a celluloid microslide attached to the pin holding the card-mounted body of the dissected specimen. Celluloid microslides are

commercially available in different sizes; the size appropriate for Aleocharinae is 4.5 mm × 11 mm or 6.5 mm × 14.0 mm. One source of celluloid slides is: Entomowinkler—insects, entomological books and supplies, Dittesgasse 11, 1180 Wien, Austria (winkler@entowinkler.at). For beginners it is best to practice dissection on some unimportant specimens, starting with larger specimens.

The genital structures should be mounted on microslides in a consistent pattern of arrangement to facilitate easy comparison of homologous structures on different microslides. We follow the following arrangement of structures on the slide: median lobe of aedeagus in lateral view or spermatheca in right upper corner; separated parameres situated below the median lobe on the right side; tergite VIII in left upper corner, sternite VIII below tergite VIII, and pygidium below sternite VIII. For thinning Canada balsam, we use xylene or non-toxic terpineol, both commercially available. It is important to note that, at first, only a thin layer of Canada balsam is placed on a microslide to prevent floating of structures into undesirable positions. After the first layer of Canada balsam with the genital structures placed in their optimal position on a microslide is dried (after two days), a second layer of Canada balsam can be added to completely, but without excess, cover the structures. We regularly use the above procedure and we find it facilitates the microscopic examination of genital structures and their photography. The advantage of this system is that the dissected specimen and the genital structures are on the same pin in the collection instead of being stored in two separate collections which can be easily misplaced or lost. This approach is also superior to using genitalia vials which take up more room on the pins, may damage structures when taken out of the vial, and requires more handling time.

Image Production

Images of the entire body and the genital structures were taken using an image processing system (Nikon SMZ 1500 stereoscopic microscope; Nikon Digital Camera DXM 1200F, and Adobe Photoshop software). Images of the dorsal aspects of the head, pronotum, elytra, abdomen, legs and antennae were taken separately in the horizontal plane (except for the head image, which was taken while in a natural position) and fused together in Adobe Photoshop. Images of tergite VIII or sternite VIII that were damaged during dissection were 'repaired' by digitally duplicating the undamaged portion using Adobe Photoshop software. For details and illustrations see Klimaszewski et al. (2020).

DNA Barcoding

DNA barcoding (Fig. 3.11a–f) was proposed as a molecular approach to species identification in 2003 (Hebert et al. 2003). Species identification based on DNA sequences was not novel in itself, but the idea of standardization of genomic regions for this purpose across multiple taxa caught wind and has grown into a massive, global effort (https://ibol.org/). The 658 base pair fragment of the mitochondrial cytochrome c oxidase I gene (COI) initially proposed by Hebert et al. (2003) has been firmly established as the standard DNA barcode region for animals. Although the species-level resolution of this region is not perfect, DNA barcoding often allows the identification of e.g., immature stages, damaged specimens, or morphologically very similar sibling species, which are impossible to identify with certainty based on morphological characters. Males and females of species with pronounced sexual dimorphism can be difficult to associate with one another in the absence of direct observations of copulation, but these associations can be established through barcoding (Slowik and Blagoev 2012; Corley and Ferreira 2017). DNA barcode data are increasingly used in taxonomic revisions as an additional line of evidence and for initial sorting of the material into putative species (Butcher et al. 2012; Riedel et al. 2013; Brunke et al. 2020; Huemer 2020; Levesque-Beaudin and Mlynarek 2020). Development of new laboratory protocols (Prosser et al. 2016) has enabled the retrieval of barcode sequences even from old and poorly preserved type specimens, which has helped solve some taxonomic controversies and conundrums (Kaila 2015; Mutanen et al. 2015; Buchner and Stănescu 2019).

The Barcode of Life Data System (BOLD, http://boldsystems.org/, Fig. 3.12), developed and maintained at the Centre for Biodiversity Genomics at the University of Guelph (Ratnasingham and Hebert 2007), is the main public platform for DNA barcode research and data storage. It acts both as a data management and analysis platform for researchers working on various projects, including taxonomic ones, and as an identification database for unidentified sequences, which can be run against the BOLD ID engine to find the closest matches in the database. Another major public resource is GenBank, which has long served as the standard global repository for nucleotide sequence data. The barcode sequence content of BOLD and GenBank are largely overlapping because sequence records from BOLD are submitted to GenBank when they are published, and DNA barcode sequences submitted independently to GenBank are frequently mined and added into BOLD. As of April 2020, 11% of COI barcodes on BOLD originated from GenBank, while 75% of the COI barcodes on GenBank derived from BOLD (Pentinsaari et al. 2020a). One major advantage of BOLD over GenBank is that BOLD allows the storage of more extensive metadata than GenBank, including images of the sequenced specimens and their morphological details.

The Barcode Index Number (BIN) system was initially introduced as an interim taxonomic system to approximate species boundaries in DNA barcode datasets consisting of largely unidentified specimens (Ratnasingham and Hebert 2013). BIN clusters are formed through a two-phase process where COI DNA barcode sequences fulfilling quality criteria are initially clustered based on a fixed threshold of sequence divergence, and these clusters are then refined through Markov clustering into the final BINs (Ratnasingham and Hebert 2013). The process is repeated on BOLD several times per year as new sequence data are constantly added into the database. BINs represent a simplified approximation of species boundaries, but this approximation is often highly accurate. In North and Central European Coleoptera, ca. 90% of all studied species showed a perfect 1:1 correspondence with BINs (Pentinsaari et al. 2014; Hendrich et al. 2015). BINs have been widely used as a proxy for species in studies where morphological species identification of the material would be impractical or downright impossible due to the scale of the study, lack of taxonomic specialists, or largely undescribed target fauna (Young et al. 2012; D'Souza and Hebert 2018; deWaard et al. 2019). Comparing BINs to established species-level taxonomy has proved to be highly useful in revealing overlooked species diversity, new distribution records, and new synonymies among various taxa with a well-established backbone taxonomy (Mutanen et al. 2012; Landry et al. 2013; Blagoev et al. 2016).

The arthropod fauna of Canada and Alaska has been extensively sampled for DNA barcoding during the past decade both in the field and in natural history collections (Gwiazdowski et al. 2015; Hebert et al. 2016; Bouchard et al. 2017; Sikes et al. 2017; Steinke et al. 2017; Brunke et al. 2019). A handful of barcoding projects have focused on the subarctic and arctic arthropod fauna (Zhou et al. 2009; Woodcock et al. 2013; Wirta et al. 2016). The Canadian National Parks Malaise Program (http://biodiversitygenomics.net/projects/cnp/) targeted several arctic and subarctic national parks for DNA barcoding in 2014. The ongoing Arctic BIOSCAN project (https://arcticbioscan.ca/), led by CBG and funded by Polar Knowledge Canada, is aiming to build a DNA barcode reference library for the Kitikmeot region of Nunavut in order to monitor changes in the arctic arthropod fauna (Levesque-Beaudin et al. 2019; Pentinsaari et al. 2020b). No barcoding projects to date have specifically targeted North American Aleocharinae, but all the general ecological studies and biodiversity surveys coordinated from CBG, and the sampling efforts in natural history collections, have collectively added ca. 4000 DNA barcode records of Canadian Aleocharinae to BOLD.

Since 2017, representative specimens of Canadian Coleoptera BINs have been systematically examined in order to connect unnamed sequence clusters to known genera and species, and to validate specimen identifications in cases where named species are seemingly not distinguishable based on their barcode sequences. This work has revealed multiple new distribution records in Canada and new synonymies of species shared between the Nearctic and Palaearctic regions (Pentinsaari et al. 2019). DNA barcode data also played a role in detecting some of the new synonymies established in this book (e.g., *Atheta campbelli* = *A. allocera*, *A. regissalmonis* = *A. diversa*), and revealed the occurrence of *Gnathusa caribou*, previously only known from the Nearctic region, in northern Finland.

In gathering distribution data for this book, we have used specimens in the **BOLD** database identified based on BIN assignment as basis of new state, province, and territory records for many species. These records were only accepted if representative specimens of the BIN had been identified morphologically, the BIN was not shared by multiple species, and there was no obvious reason to suspect misidentification. These records based on sequence data alone (reported as **BOLD** records in the text) are distinguished in the species treatments from those based on morphologically examined specimens.

Collection Methods and Habitats

The Aleocharinae live in almost every conceivable terrestrial habitat from alpine tundra to the intertidal zone of oceans (Klimaszewski et al. 2018, 2020). The majority of species live in forest litter and wetlands and prefer moist habitats. However, many species are more specialized and occur in fungi, carrion, and subcortical habitats. Some species are associated with mammal and bird nests. To fully sample the species at a given site, one must use various collection methods and sample over time as many microhabitats as possible. Among the commonly used collection methods are traps such as pitfall traps, flight intercept traps, Lindgren funnel traps, Malaise traps, and light traps. Other productive sampling methods are sifting litter, treading wet or emergent vegetation in wetlands, hand collecting, and evening flight collections using an aerial net. Often the habitat will dictate the most effective collection technique(s) to use (Klimaszewski et al. 2018, 2020). Each collection technique will be discussed below.

Pitfall traps (Fig. 3.13a–b) Pitfall traps are cheap, simple to deploy, and collect a large number and diversity of Aleocharinae (and other epigeic arthropods) in many habitats, e.g., forests, fields, peatlands. They do not work well on highly rocky substrates (e.g., alvars) or in very wet substrates as the higher water table forces the trap out of the substrate. Pitfall traps may have different diameters, and there are some varieties that are commercially available. However, it is also easy to construct them from commonly available plastic containers (e.g., yogurt, ice-cream, 16 oz. beer cups). A variety of preservatives may be used (e.g., salt water, propylene glycol, 70% ethanol), and an elevated rain cover reduces dilution of the preservative and minimizes input of leaves, etc. There are many useful references about the use of pitfall traps (e.g., Martin 1977; Spence and Niemelä 1994; Skvarla et al. 2014).

Pitfall traps may be baited with carrion, dung, or decaying organic materials (e.g., fruit, vegetables) to attract diverse aleocharine assemblages associated with such substrates. Baits can be placed adjacent to the trap opening, in a small screened cage above the opening, or in a slurry of the killing agent which is later strained to remove the specimens (the latter method is less "pleasant" for sample processing).

Luminoc® pitfall-light traps use a small light built into the trap so that the trap has both a passive and active means of sampling insects. The cup is 12 cm in diameter, and lights are activated by a photocell with timer (Jobin and Coulombe 1992). The traps are commercially produced by Bio-Control Inc., Sainte-Foy, Quebec, Canada. These traps attract a greater diversity of species than unbaited pitfall traps and often catch rarely encountered species (Hébert et al. 2000), but they are more expensive to purchase and maintain.

Collecting at lights (Fig. 3.13c) Many Coleoptera are attracted to ultraviolet light at night. The most productive light-based method for capturing Aleocharinae is to use a white sheet as a background to a mercury vapour light, which is more attractive to most insects (including Aleocharinae) than ultraviolet light. In remote sites one can power the light by a portable generator. Most Staphylinidae and Aleocharinae come to the lights from just before sunset until about 1–2 h after sunset. Specimens can be collected off the sheet using an aspirator or collected directly into the collection vial. It is best to initially place the Staphylinidae into vials with sawdust (see below) without a killing agent (e.g. ethyl acetate); otherwise many individuals will die with their wings extended. After collecting is completed for the evening, a few drops of ethyl acetate and water should be added to the sawdust to kill the specimens. This method, however, is of limited use in the North as it does not get dark enough for the light to be effective in mid summer. However, it is effective in alpine areas in more southern areas.

Lindgren funnel traps and panel traps (Fig. 3.14a) Lindgren funnel traps of various lengths (typically with 8–12 funnels) present a vertical cylindrical profile that visually mimics tree trunks, a silhouette that is attractive to many insects associated with trees (Lindgren 1983). These traps are frequently deployed with semiochemicals (e.g., pheromones, monoterpenes, ethanol) for monitoring bark and ambrosia beetles (Curculionidae: Scolytinae), and other bark- and wood-boring beetles (e.g., Cerambycidae, Buprestidae), but they are often effective for sampling many other groups of Coleoptera that are associated with dead and dying trees (Lindgren 1983). A number of Aleocharinae species in eastern Canada were captured almost exclusively in Lindgren funnel traps (Webster et al. 2009, 2012, 2016b). As these traps sample only flying insects, they also provide useful data about flight capabilities and phenology of captured species. Traps can be suspended at any height above the ground but typically with the collecting cup at about 1.5 m (Webster et al. 2016a, c). In treeless areas one can string rope between two poles to hang traps. For details on Lindgren trap deployment, see Hughes et al. (2014). Black panel traps, a type of cross-vane trap, function in much the same way as Lindgren funnel traps and are less expensive. As well, the panels, made of corrugated plastic, are available in a variety of colours that may have differential attraction for different species. Both styles of trap are available at Synergy Semiochemicals Co., Burnaby, British Columbia, Canada.

Flight-intercept traps (Fig. 3.13d) Flight-intercept traps (or FITs) take advantage of typical beetle behaviour to drop upon hitting a barrier. FITs are typically composed of a clear plastic barrier or a tightly pulled vertical mesh barrier across a suspected flight corridor within a habitat (*e.g.*, Peck and Davies 1980). Wide collecting trays, e.g., aluminum catering trays, are placed underneath the barrier and filled with a high concentration killing agent (see options discussed below) to prevent degradation. FITs should be serviced about once a week (more frequently in sun-exposed hot environments) to prevent the decay of fragile specimens. Aleocharines that are flight-capable can be sampled in great numbers by FITs, but it is important to take a few precautions to prevent heavily damaging specimens. First, large beetles or other insects should be removed and placed in a separate sample bag or vial. Then a fine aquarium net can be used to sieve the remaining specimens from the preservative, and these are placed into a sample container or bag already filled with ethanol. Avoid pouring the contents of the trays into the net or using a squirt bottle to wash the net as these generally create a lot of damage to aleocharine specimens. There are varieties of FITs that can be attached to tree trunks for sampling saproxylic species (e.g., Hammond 1997), and these sometimes collect species that are not collected by funnel traps or panel traps (Langor, unpublished data).

Sifting (Figs. 3.13f, 3.14b) Leaf litter, moss, dung, or other kinds of debris can be placed in a large plastic box and hand-sifted by removing excess and larger debris, which concentrates specimens. Most Coleoptera, including Aleocharinae, move or fall (inactive species) to the bottom of the container and are easily collected using an aspirator as they become active (some species may take more than 10 min to become active, especially on cool days). This method is very effective for collecting species from forest litter, litter under bushes, mosses, grass and mosses in marshes and bogs, patches of debris in sandy habitats, flood debris, and drift material along the shores of rivers, ponds, and lakes, moist leaves along pond margins, seaweed along sea beaches, compost, dung, and from rotting logs. Aleocharinae are often abundant in various kinds of fungi. These are collected by removing fungi from the substrate (forest floor, logs, standing trees) and sifting them or placing them in the plastic box, breaking them into pieces, and then aspirating the specimens as they move. It is important to record data on the microhabitat that the specimens were collected from, including at least the major fungal group and stage of decay (fresh, decaying, rotten) when collected from fungi. Samples can also be concentrated using sifters with the finer material placed on a plastic sheet or in a plastic box for examination and specimen aspiration. For extremely small or inactive species, this sifted material can be placed in a Berlese funnel or Winkler extractor to extract specimens passively over time.

Winkler extractors are more flexible as they do not require electricity and can be hung from trees or other structures at collection sites (Brunke et al. 2012).

Treading water vegetation Many Aleocharinae live in floating or emergent vegetation along edges of lakes, ponds, rivers and streams and amongst vegetation in wet habitats such as marshes, bogs, and fens (e.g. *Gymnusa*, *Deinopsis*, *Myllaena*, *Philhygra*). Treading is the most effective method for collecting beetles in these habitats. Vegetation or moss is pressed into the water by walking on it. Beetles will float to the surface within 1–2 min and can be hand-picked from the surface of the water. It is best to repeat this once or twice in the same trodden spot to maximize the number and diversity of species collected as some species take more time and disturbance to reveal themselves.

Hand collecting (Figs. 3.13e, 3.14b, 3.21a) Many Aleocharinae species live on gravel bars and on gravel and cobblestone shorelines. These habitats are best sampled by turning over pebbles, cobblestones, or large rocks, and aspirating the specimens (Fig. 3.13e) as they move. It is important to sample shaded and sun-exposed sites and at different distances from the stream margins as different species can be found in each habitat. Collecting under rocks and pieces of dead wood can also be effective in drier habitats. Many Aleocharinae live in subcortical habitats. These are collected by first removing bark and checking underneath for revealed specimens. Bark and pieces of rotting wood should be placed into a plastic box, broken up somewhat, and then sifted or soaked as described above.

Splashing Some species of Aleocharinae live in habitats that are best sampled by 'splashing' the substrates they live in. A number of species such as *Gnypeta* and some *Philhygra* live in moss or in cracks on rocks in the middle of rivers or in moss in the splash zones of waterfalls (Webster et al. 2016b). Splashing the mosses with water will cause the beetles living there to move to the surface where they can be collected. It often takes repeated splashing for up to ten minutes before the beetles will leave the substrate. These same substrates may also be soaked as described above. Other species live in fine sand at stream and lake margins. Lightly splashing the area will force the beetles out of the sand. Often these are very small to minute species and close observation while on hands and knees is necessary to see the specimens as they leave the sand. Splashing gravel and cobblestone habitats along streams can also be effective.

Mammal and bird nests Some Aleocharinae are strongly associated with mammal and bird nests. Some species can be found in debris in the entrances to dens or tunnels of rodents, foxes, and other ground nesting mammals or in middens (Webster et al. 2016b; Brunke and Buffam 2018). These species are usually most common early in the season as they leave the dens for their dispersal flights. Some species are associated with microhabitats created by the activities of beavers, such as the nest material inside the lodge and debris in dams (Klimaszewski et al. 2016). Nests of birds such as ospreys and hawks, swallows, waterfowl, and especially nests in rookeries have not been properly sampled in North America and may have an interesting fauna (Brunke and Buffam 2018). Mammal and bird nests are a poorly sampled habitat in general, and new species likely remain undiscovered.

Compost One very effective substrate for attracting Aleocharinae and many other Coleoptera species is backyard composter or a pile of organic debris. Plant based kitchen wastes decay and become moldy over time, and start to attract a wide variety of species as the season progresses. Many new records including some new Canadian records were found in this substrate (Webster et al. 2009, 2012, 2016b). Establishment of a composter in northern areas could result in the discovery of many new species and records.

Arctic and Subarctic Habitats

The tundra is a sub-arctic zone characterized by long, cold winters and short, warm summers. It is sometimes referred to as a cold desert. Precipitation is low and in the form of rain and snow, but evapotranspiration is also minimal. The soil one meter down is permanently frozen and is called permafrost. Since water cannot easily drain through the soil because of this underlying permafrost, water tends to collect in bogs, fens, and shallow pools. This landscape, devoid of trees and tall shrubs, is dominated by low shrubs, mosses, lichens and small flowering herbaceous plants. **The taiga** generally lies to the south of the tundra, and is defined by very low winter temperatures, a longer growing season than the tundra, and more precipitation in the form of rain and snow. The soils are generally acidic. This biome is dominated by coniferous trees, especially balsam fir and black spruce, with white birch, trembling aspen, poplar, and mountain ash being the commonest deciduous trees.

a. LABRADOR (Figs. 2.1, 3.15–3.18)
Labrador, a north-western part of Province of Newfoundland (Fig. 2.1), is ecologically very diverse and is home to a wealth of native and non-native wildlife species. The aleocharine beetle fauna of the province was treated by Klimaszewski et al. (2011), and the fauna of Labrador is updated in this book as part of Arctic and Subarctic North America. Far northern Labrador is part of the Arctic Cordillera (Fig. 2.1, ecoregion 1.1) and southern Labrador is mostly taiga (Fig. 2.1, ecoregion 3.4). To a lesser extent, Labrador also features non-taiga forest along and near the southeastern coast (Fig. 2.1, ecoregion 5.1). There are four main habitats in Labrador: the forest, the coastline and ocean, ponds and lakes, and the Arctic Cordillera. The forest is the main ecosystem, covering most of Labrador and including various trees, such as spruce, birch, and fir. The coastline and ocean, includes very diversified coastal habitats, dunes, and ocean. It is very windy on the sea shore and because the wind often blows in off the water, the trees which grow there such as spruce or larch are often "slanted" and tilted. The Arctic Cordillera covers areas of Northern Labrador, and plant life consists only of very small bushes and shrubs. A few of the diverse habitat types are illustrated on the following photographs to show its enormous diversity: bog along Trans-Labrador Highway (Fig. 3.15a); patterned fen in central Labrador (Fig. 3.15b); boreal black spruce forest (Fig. 3.16a); slow stream with vegetated edges (Fig. 3.16b); dry forest floor with reindeer moss near Goose Bay (Fig. 3.17a); matt of *Sphagnum* moss in a bog (Fig. 3.17b); tundra with patches of snow (Fig. 3.18a); and coastal dunes (Fig. 3.18b).

b. NUNAVUT (Figs. 3.19–3.21)

Nunavut comprises a large portion of northern Canada, including most of Canada's Arctic Archipelago. The taiga zone (Fig. 2.1, ecoregion 3.4) reaches the southernmost parts of the territory, and parts of the northernmost islands (in ecoregion 1.1, Fig. 2.1) are under permanent ice caps, but the vast majority of the land area is covered by tundra (Fig. 2.1, ecoregions 2.1 and 2.4). The known beetle fauna of Nunavut is the least diverse of all Canadian provinces and territories: only 123 species were listed from Nunavut by Bousquet et al. (2013), compared to 1115 from Northwest Territories. Some older records of Coleoptera species reported from Northwest Territories may actually originate from Nunavut, since the territories have been officially separated only since 1999. Although significantly more sampling is needed to reveal its true diversity, Nunavut probably has a truly much lower beetle diversity as it is the most pronouncedly arctic territory of Canada. Most of Nunavut is extremely remote and difficult to access, and collecting efforts have mainly focused on the immediate surroundings of the larger settlements. The photographs in Figs. 3.19–3.21 show a selection of habitats around the hamlet of Kugluktuk: small freshwater stream between two lakes (Fig. 3.19a); willow thicket below a cliff just outside Kugluktuk (Fig. 3.19b); wet sedge fen between low hills (Fig. 3.20a); lush willow thicket with a thick layer of leaf litter in Kugluk (Bloody Falls) Territorial Park (Fig. 3.20b); and Coppermine River delta, with numerous ponds left over from the spring floods (Fig. 3.21b).

c. NORTHWEST TERRITORIES (Figs. 3.22–3.24)

The Northwest Territories is bordered by the territories of Nunavut to the east and Yukon to the west, and by the provinces of British Columbia, Alberta, and Saskatchewan to the south. Different combinations of climate, soils and terrain create a variety of landscapes across the Northwest Territories. These ecoregions include huge global biomes such as tundra and taiga as well as other kinds of habitats such as mixed wood forests, wetlands, and riparian habitats. There are five main ecoregions: Taiga Plains (ecoregion 3.3; Figs. 3.23a,b, 3.24a), Taiga Shield (ecoregion 3.4; Figs. 3.22a,b, 3.24b), Taiga Cordillera (ecoregion 3.2), Southern Arctic (ecoregion 2.4), and Northern Arctic (ecoregion 2.1) (Fig. 2.1). This territory also includes small portions of Brooks Range Tundra and Alaskan Tundra to the far northwest (Fig. 2.1). The Taiga Plain and Shield ecoregions are a northern extension of the flat Interior Plains which dominate the Prairie and Boreal Plains ecoregions to the south. The area is underlain by horizontal sedimentary rock, and is nearly level to gently rolling in topography. The broad lowlands and plateaus are incised by major rivers, including Mackenzie River, the longest river in Canada with the second largest drainage basin of all North American rivers (after Mississippi). The two largest lakes entirely within Canada, Great Bear Lake and Great Slave Lake, are also situated in the Northwest Territories.

The Taiga vegetation consists of open, very stunted stands of black spruce (*Picea mariana*), and the shrub component includes dwarf birch (*Betula* sp.), Labrador tea (*Ledum decumbens*), and willow (*Salix* spp.). Understory species include bearberry, mosses, and sedges. Upland and foothills areas, and southern regions are often better drained, are warmer, and support mixed wood forests characterized by white and black spruce (*Picea glauca*), lodgepole pine (*Pinus contorta*), tamarack (*Larix laricina*), white birch (*Betula* spp.), quaking aspen (*Populus tremuloides*), and balsam poplar (*Populus balsamifera*). Low-lying wetlands cover 25–50% of the ecoregion (Fig. 3.24a,b).

d. YUKON TERRITORY (Figs. 3.25–3.32)

The Yukon Territory is divided into seven ecoregions: Alaskan Tundra (ecoregion 2.2), Brooks Range Tundra (ecoregion 2.3), Alaska Boreal Interior (ecoregion 3.1), Taiga Cordillera (ecoregion 3.2), Taiga Plain (ecoregion 3.3), Boreal Cordillera (ecoregions 6.1, Fig. 2.1), and Pacific Maritime (Smith et al. 2004). The Arctic (as Alaskan Tundra) is only represented in the

Yukon by the Yukon Coastal Plain and constitutes only 1% of the Yukon's total surface area. Geographically this zone is characterised by undulating landscape with innumerable lakes, ponds, and wetlands, all part of the drainage directed towards the Beaufort Sea. The zone experiences long cold winters and short cool summers with cryosols of various types over continuous permafrost. The shallow active layers remain moist or wet throughout the summer due to poor drainage, and low evapotranspiration although the region receives very little precipitation. The vegetative cover is dominated by shrubs (dwarf birch, willow, and heath species), various herbs, lichens and mosses (80%). Wetlands and riparian habitats are common in the lowlying coastal areas consisting of a mixture of sedges and moss (20%).

The Taiga Plain ecoregion is mostly represented in the Northwest Territory but a small portion is represented in the Yukon in the extreme south east (Peel River Plateau), and northeast (Fort McPherson plain). The ecoregion represents 5% of the Yukon's total land cover. For the most part this ecoregion is a rather flat plateau dominated by subarctic coniferous and mixed forest (75%). The rest of the region is covered by forest tundra (9%), alpine tundra (9%), and lakes and wetlands (7%). Wetlands consisting primarily of peatlands are underlain by permafrost. The climate is marked by short cool summers and long cold winters, with low mean annual precipitation. Water in this ecoregion drains to the Mackenzie River basin.

The Taiga Cordillera ecoregion represents 37% of Yukon's total surface area and covers the subarctic regions of the western cordillera. Summers are warm to cool with long cold winters. Continuous permafrost is prevalent throughout the ecozone. Mountain ranges consisting of semi-parallel ridges with deep intervening valleys dominate the landscape of the ecozone. Its vegetation is dominated by subarctic coniferous forest (54%), and the rest of the area is covered by arctic and alpine tundra (27%), barren land and summits (16%), lakes and wetlands (2%), and mixed boreal forest (1%).

The Boreal Cordillera ecoregion represents 56% of Yukon's total surface area and is located in the midsection of the western cordillera system of Canada. Mountain ranges with numerous peaks, extensive plateaus separated by wide valleys characterise this ecozone. It encompasses major portions of the Yukon, Liard and Alsek river drainage systems. The region is characterised by discontinuous permafrost where more groundwater infiltration is possible. The climate ranges from sub-humid to semi-arid marked by long cold winters and short warm summers. The ecoregion has diverse vegetation comprising of boreal and subalpine coniferous forest (40%), arctic and alpine tundra (24%), boreal coniferous forest (20%), rock, glacier and snowfields (9%), lakes and wetlands (5%) and boreal mixed forest (2%).

The Marine West Coast Forest ecoregion is represented in the Yukon only by the Mount Logan subregion in the extreme southwest corner. The subregion is characterised by high elevation icefields and is the largest nonpolar icefield in the world. It is dominated by perennial snow and ice (85%) alpine tundra (5%) and barren ground and summits (10%). This ecoregion represents only 1% of the Yukon and is not included in the present study region (Fig. 2.1).

A few of the diverse habitat types are illustrated on the following photographs to show their diversity: alpine tundra (Fig. 3.26a); white spruce alluvial habitat (Fig. 3.30a,b); alluvial fan (Fig. 3.31a); boreal mixed forest (Fig. 3.27a); cotton-grass/moss habitat (Fig. 3.31b); upland heath habitat (Fig. 3.32); deciduous boreal forest (Fig. 3.27b); riparian habitat (Fig. 3.29); alkaline salt flats (Fig. 3.25a,b); mushrooms (Fig.3.28a,b); animal scats (Fig. 3.26b).

e. ALASKA (Figs. 3.33–3.46)

Alaska can be divided into various ecoregions (e.g. Gallant 1995) but four general ecozones are most relevant to its entomofauna: Boreal Cordillera, Alaska Boreal Interior, the arctic and maritime tundras of the Alaska Tundra ecoregion, and the Maritime West Coast Forest (Fig. 2.1). Each region has a relatively distinct entomofauna and, compared to the "lower 48" US states, relatively large and intact ecosystems with relatively few non-native arthropod species. Aleocharinae from all regions in Alaska are included in this book with the exception of the Maritime West Coast Forest ecozone that includes southcentral and southeast Alaska.

The Alaska Boreal Interior ecoregion, the boreal forested area east of the coastal tundra, south of the Brooks Range, and north of the Alaska Range (ecoregion 3.1, Fig. 2.1) was glacier-free (Kaufman et al. 2011) and acted as a glacial refugium connected with eastern Asia during glacial maxima when the land bridge was present. This is part of a larger region called "Beringia" which extended from the Lena River in northeastern Russia to the McKenzie River in northwestern Canada. Interior Alaska's dominant vegetation community is boreal forest with drier upland white spruce, aspen, balsam poplar, and birch (Fig. 3.33b), and wetter lowland black spruce (Figs. 3.33a, 3.34a,b) (Gallant 1995). This region includes a variety of less common habitats, including treeless alpine tundra on rolling hills or high elevation valleys (Figs. 3.35a,b, 3.36a,b, 3.38b), alpine snowfields, xeric south-facing bluffs, large river systems (Fig. 3.37a,b), and numerous ponds and lakes (Fig. 3.34a). Grasslands and meadows are rare (often resulting from human management or post fire recovery (Fig. 3.38a). Interior Alaska holds numerous Beringian species with biogeographic ties to Asia and species that have yet to be found outside of Alaska (Sikes and Allen 2016). The boreal forest of interior Alaska is considered a "fire forest" because fires set by lightning, sometimes burning over millions of acres, occur each summer and are an important part of its ecology. This region is likely the most biodiverse in Alaska because its summers are warmer, with less cloud cover, than coastal areas, and it was never glaciated.

Arctic tundra dominates the northernmost extent of the Alaskan Tundra ecoregion (ecoregion 2.2, Figs. 2.1, 3.39a,b) on Alaska's "north slope", also known as the Arctic Coastal Plain and occurring north of the Brooks Range and south of the Arctic Ocean. This treeless region has continuous permafrost, countless lakes, and the coldest climate and lowest rainfall of Alaska (Gallant 1995). Entomologically, however, it is not the region with the fewest species, despite it being the coldest. Alaska's arctic tundra has numerous taxa absent from the maritime tundra of the Aleutian Archipelago, such as dragonflies, butterflies, grasshoppers, and ants—highly vagile organisms capable of reaching the Aleutian Islands but likely excluded by the harsher climate of these islands (stronger and more frequent winds and less solar radiation due to more frequent cloud cover). The lack of Pleistocene glaciation in the Arctic Coastal Plain, its larger surface area, and its greater connectivity to Interior Alaska may also help explain its higher species richness relative to the Aleutians.

The maritime tundra and meadows of the Alaskan Tundra ecoregion (Fig. 2.1) dominate the western coasts (Figs. 3.40a,b, 3.41a,b, 3.42a, b), Bering Sea Islands (former high points on the Bering Land Bridge (Figs. 3.43a,b, 3.44a,b) and Aleutian Island Archipelago (Figs. 3.45a,b, 3.46a, b). The Aleutians were never part of the land bridge and even when sea levels were at their lowest, open water separated groups of islands from each other and the mainland. Most of the Aleutians were covered by glaciers during glacial maxima and they are highly volcanic. This leads to relatively frequent defaunation events, which combined with their harsh climate and geographic isolation, helps explain their low biodiversity.

South of the Alaska Range lies southcentral Alaska with the greatest density of people and agriculture. This region, including Kodiak Island, has biogeographic ties to southeastern Alaska, a region of islands and mountains covered with temperate rain forests. This Marine West Coast Forest ecoregion extends southwards through coastal British Columbia, terminating in California. This is the only ecoregion in Alaska excluded from this book (Fig. 2.1).

Fig. 3.3 (a–l). Examples of morphological diversity in Arctic and Subarctic regions of North America: (**a**) *Aleochara suffusa* (Casey); (**b**) *Aloconota cambrica* (Wollaston); (**c**) *Atheta subsinuata* (Erichson); (**d**) *Bellatheta microptera* (Lohse); (**e**) *Tarphiota geniculata* (Mäklin); (**f**) *Boreocypha websteri* Klimaszewski and Langor; (**g**) *Deinopsis canadensis* Klimaszewski; (**h**) *Gymnusa konopackii* Klimaszewski; (**i**) *Gyrophaena nana* (Paykull); (**j**) *Silusa californica* (Bernhauer); (**k**) *Cypha inexpectata* Klimaszewski and Godin; (**l**) *Paramblopusa eoa* Ahn and Maruyama. Scale bar = 1 mm

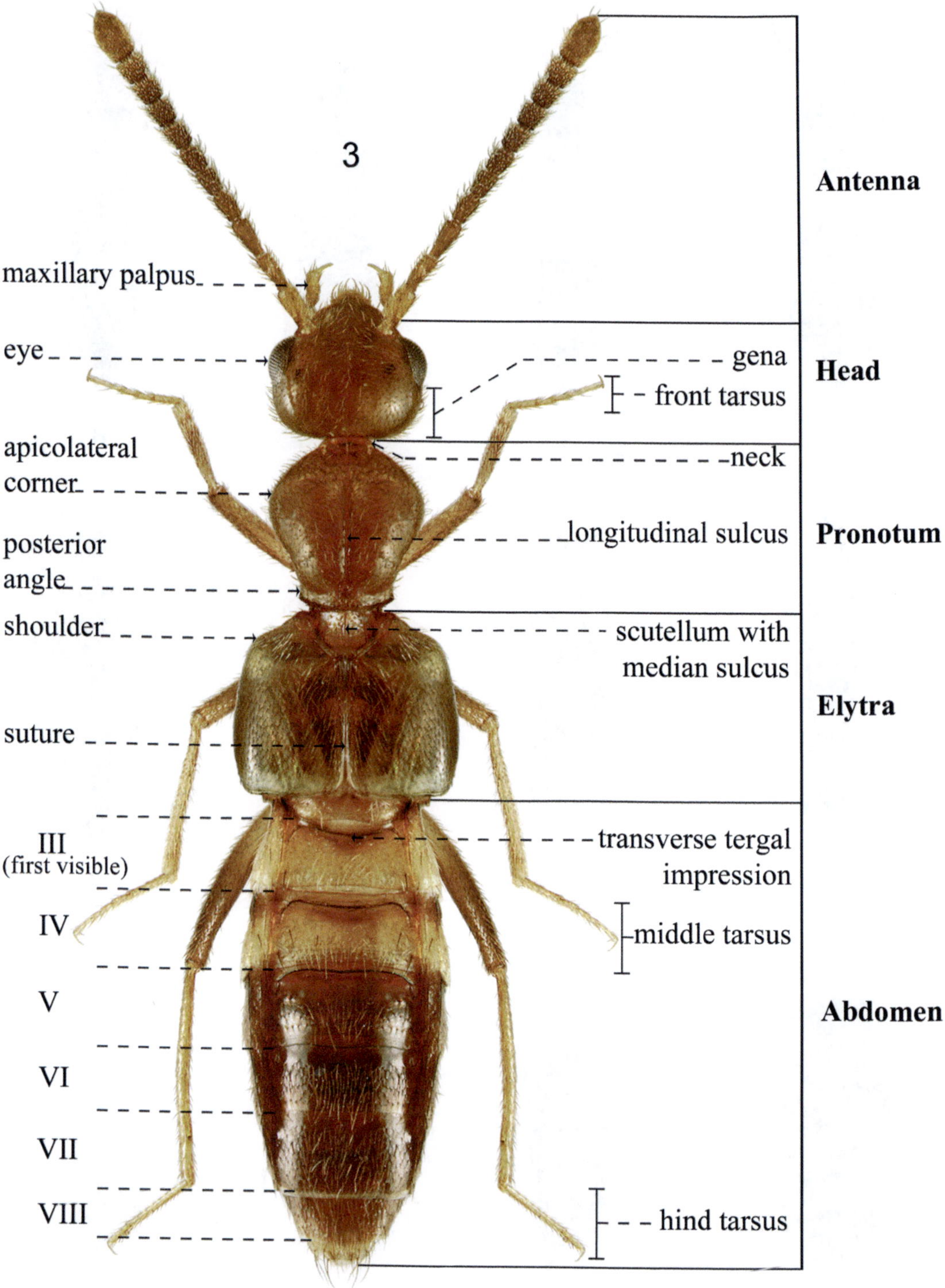

Fig. 3.4 Image of *Myrmecocephalus concinnus* (LeConte) in dorsal view (not occurring in present study area), with body parts important to identification. After Klimaszewski et al. (2018)

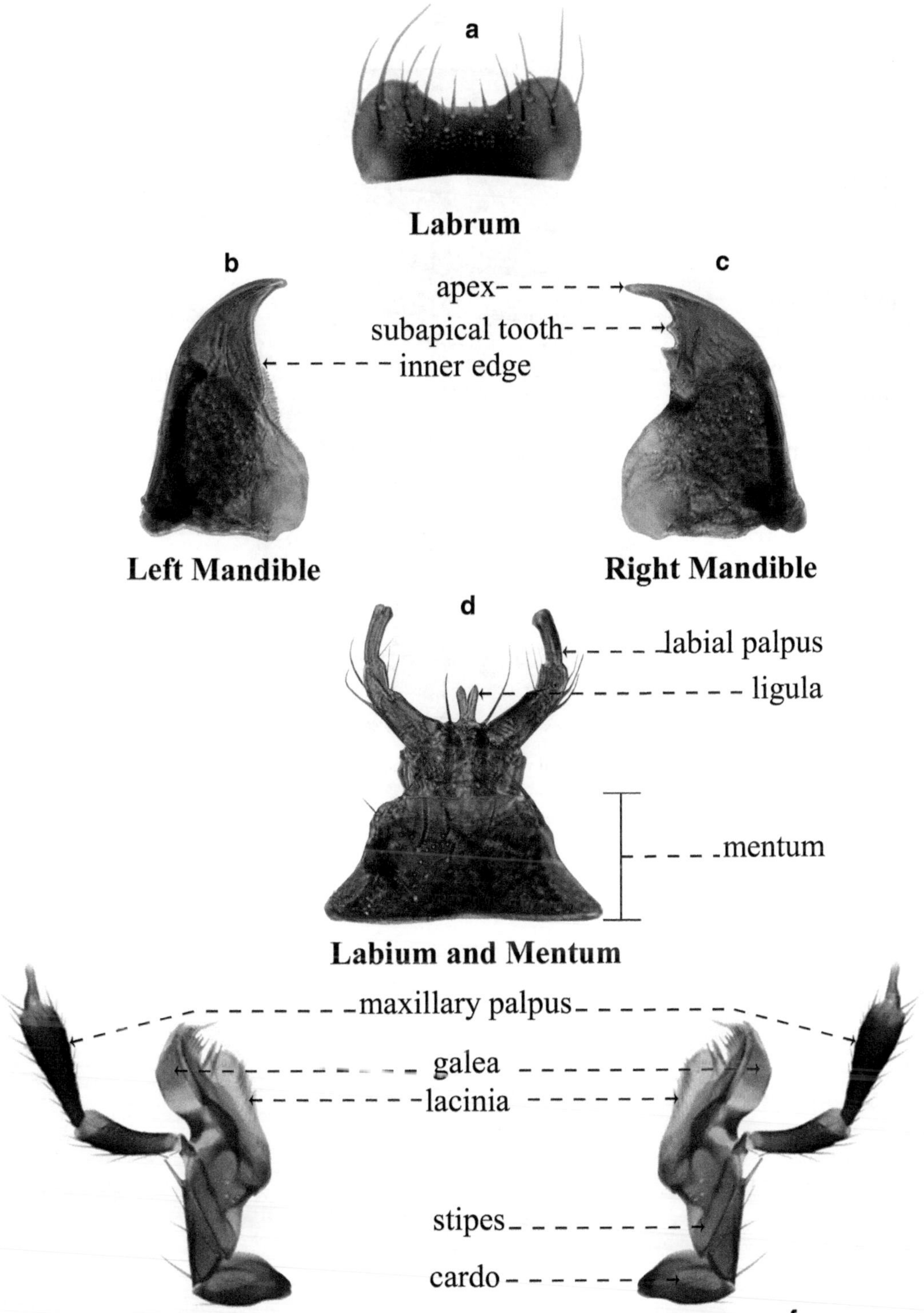

Fig. 3.5 (**a**–**f**). Mouthparts: (**a**) labrum; (**b**) left mandible; (**c**) right mandible; (**d**) labium and mentum; (**e**) left maxilla; (**f**) right maxilla. After Klimaszewski et al. (2018)

Fig. 3.6 Image of pronotum and thorax in ventral view, with important structures for identification. After Klimaszewski et al. (2018)

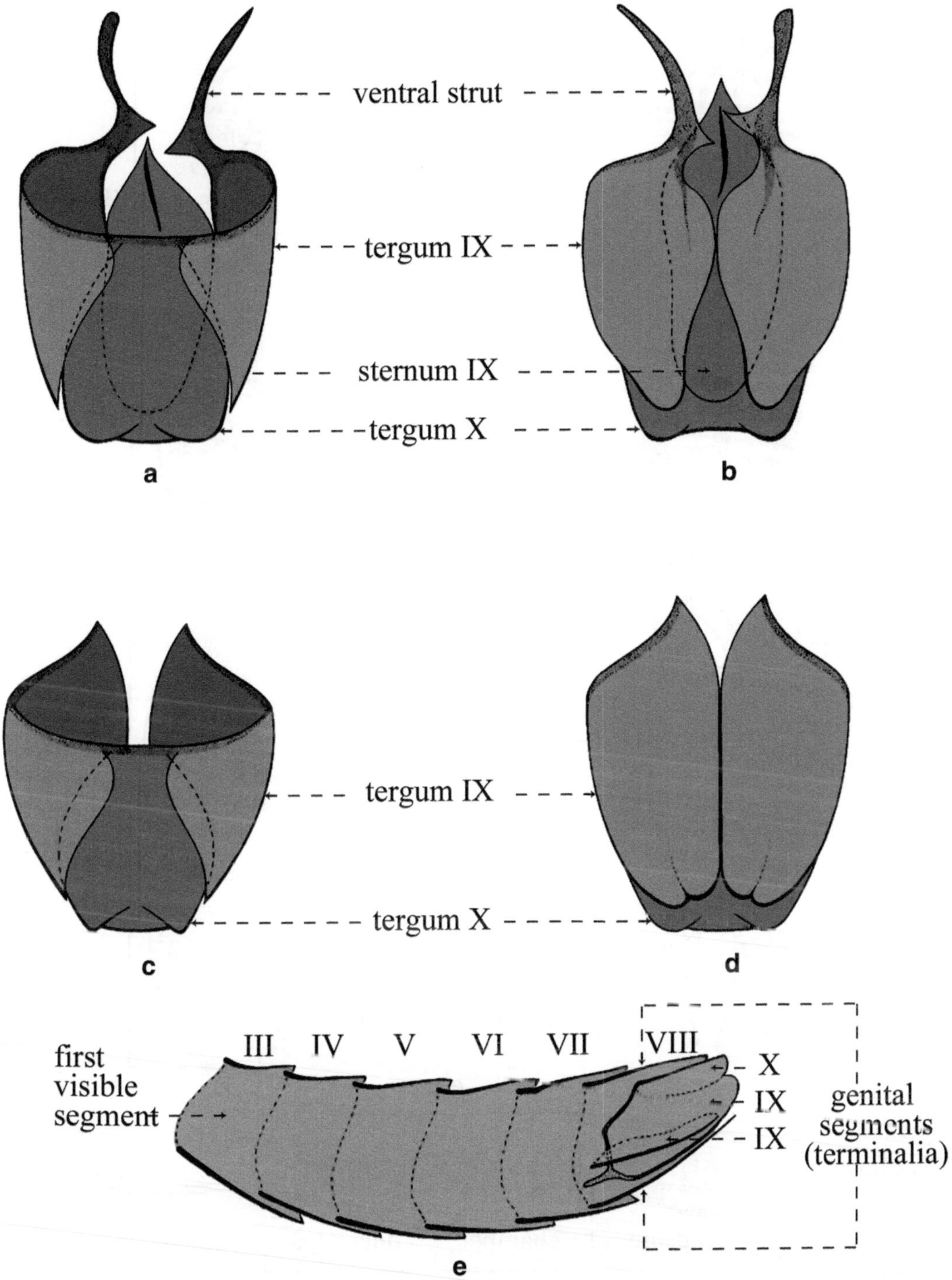

Fig. 3.7 (**a–e**). Morphological divisions of male and female terminal segments (**a,c,** dorsal view, **b,d,** ventral view), and (**e**) cross-section of abdomen in lateral view. After Klimaszewski et al. (2018)

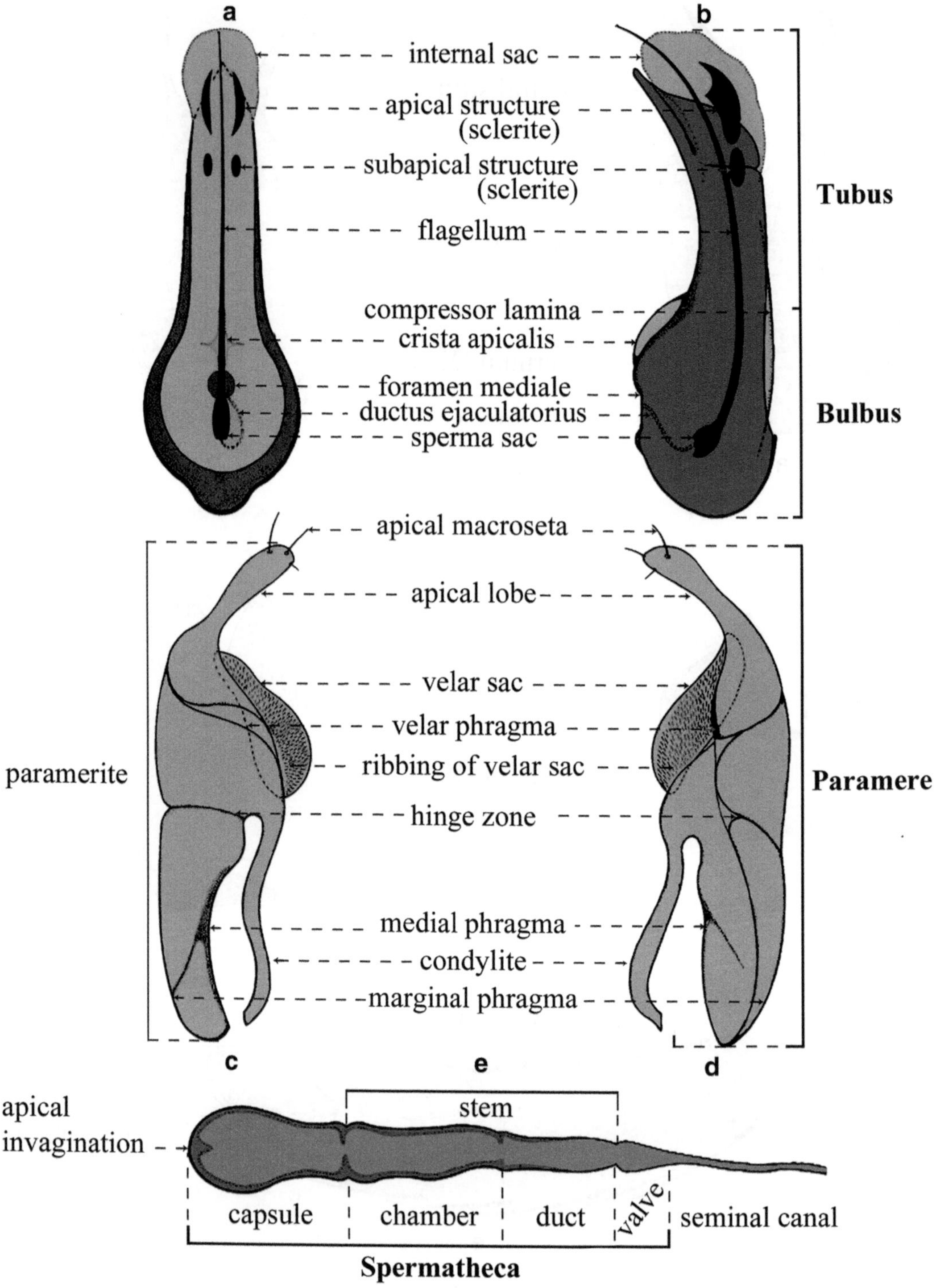

Fig. 3.8 (a–e). Morphological divisions of the median lobe of aedeagus, paramere, and spermatheca. After Klimaszewski et al. (2018)

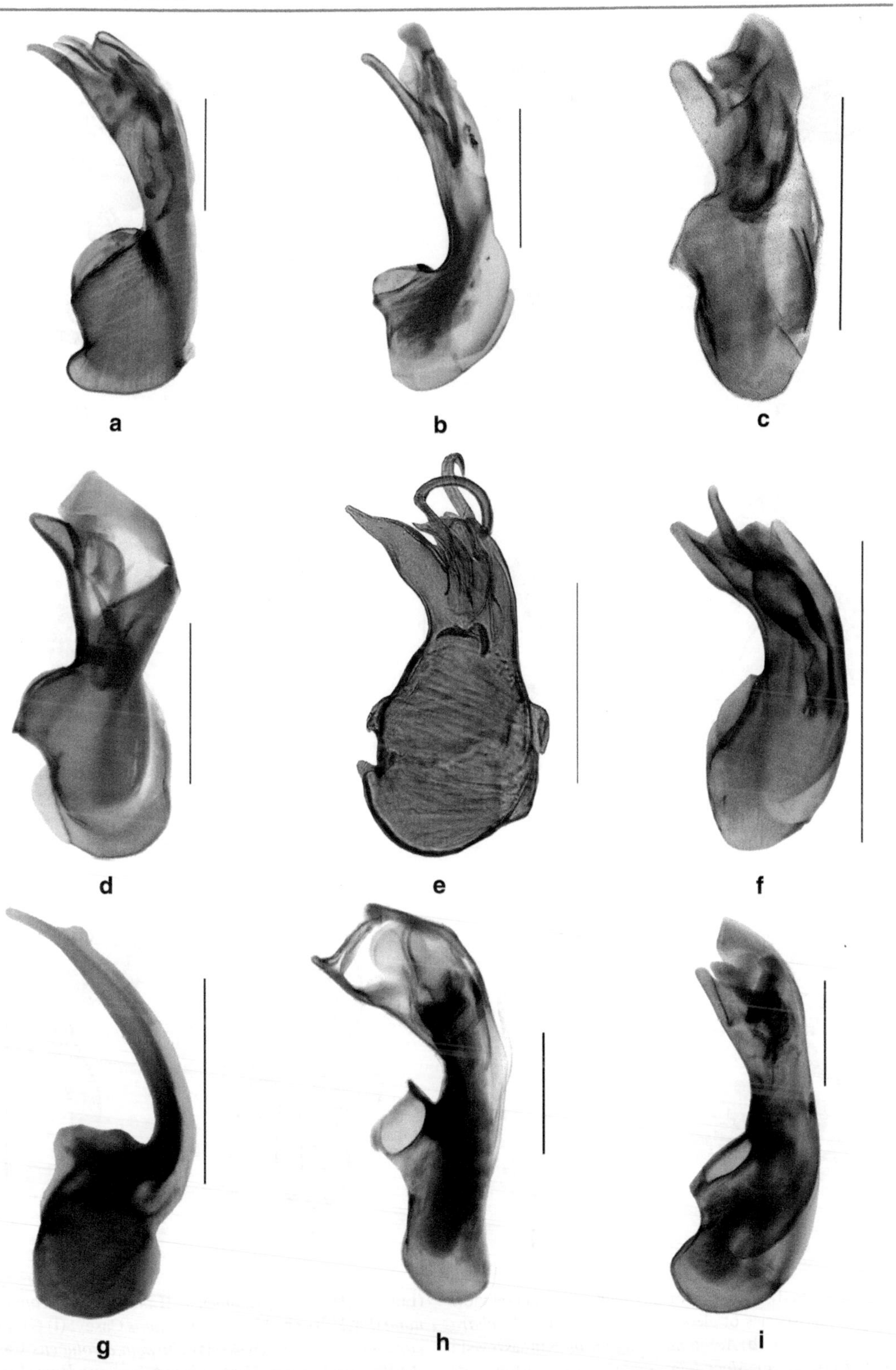

Fig. 3.9 (**a–i**). Examples of the median lobe of aedeagus in lateral view of different species of aleocharine beetles: (**a**) *Aleochara assiniboin* Klimaszewski; (**b**) *Atheta ventricosa* Bernhauer; (**c**) *Atheta munsteri* Bernhauer; (**d**) *Boreophilia beringi* Klimaszewski and Brunke; (**e**) *Gyrophaena keeni* Casey; (**f**) *Leptusa gatineauensis* Klimaszewski and Pelletier; (**g**) *Cypha inexpectata* Klimaszewski and Godin; (**h**) *Oxypoda lucidula* Casey; (**i**) *Mniusa yukonensis* (Klimaszewski and Godin). Scale bar = 0.2 mm

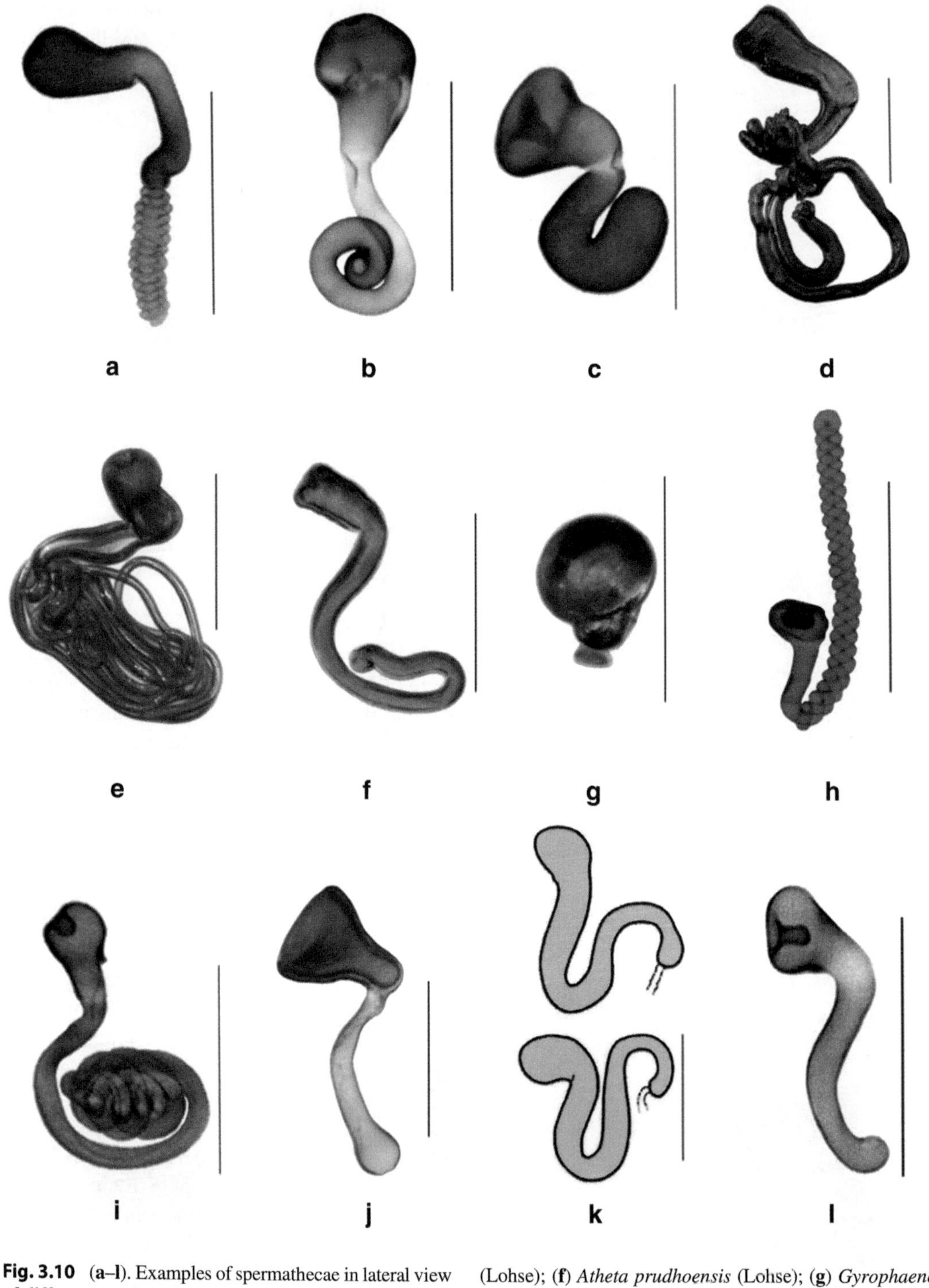

Fig. 3.10 (**a**–**l**). Examples of spermathecae in lateral view of different species of aleocharine beetles: (**a**) *Aleochara suffusa* (Casey); (**b**) *Acrotona sphagnorum* Klimaszewski and Webster; (**c**) *Aloconota cambrica* (Wollaston); (**d**) *Atheta borealides* (Newton); (**e**) *Atheta nearctica* (Lohse); (**f**) *Atheta prudhoensis* (Lohse); (**g**) *Gyrophaena nana* (Paykull); (**h**) *Myllaena insomnis* Casey; (**i**) *Oxypoda canadensis* Klimaszewski; (**j**) *Gnypeta atrolucens* Casey; (**k**) *Placusa tacomae* Casey; (**l**) *Ocyusa canadensis* Lohse. Scale bar = 0.2 mm

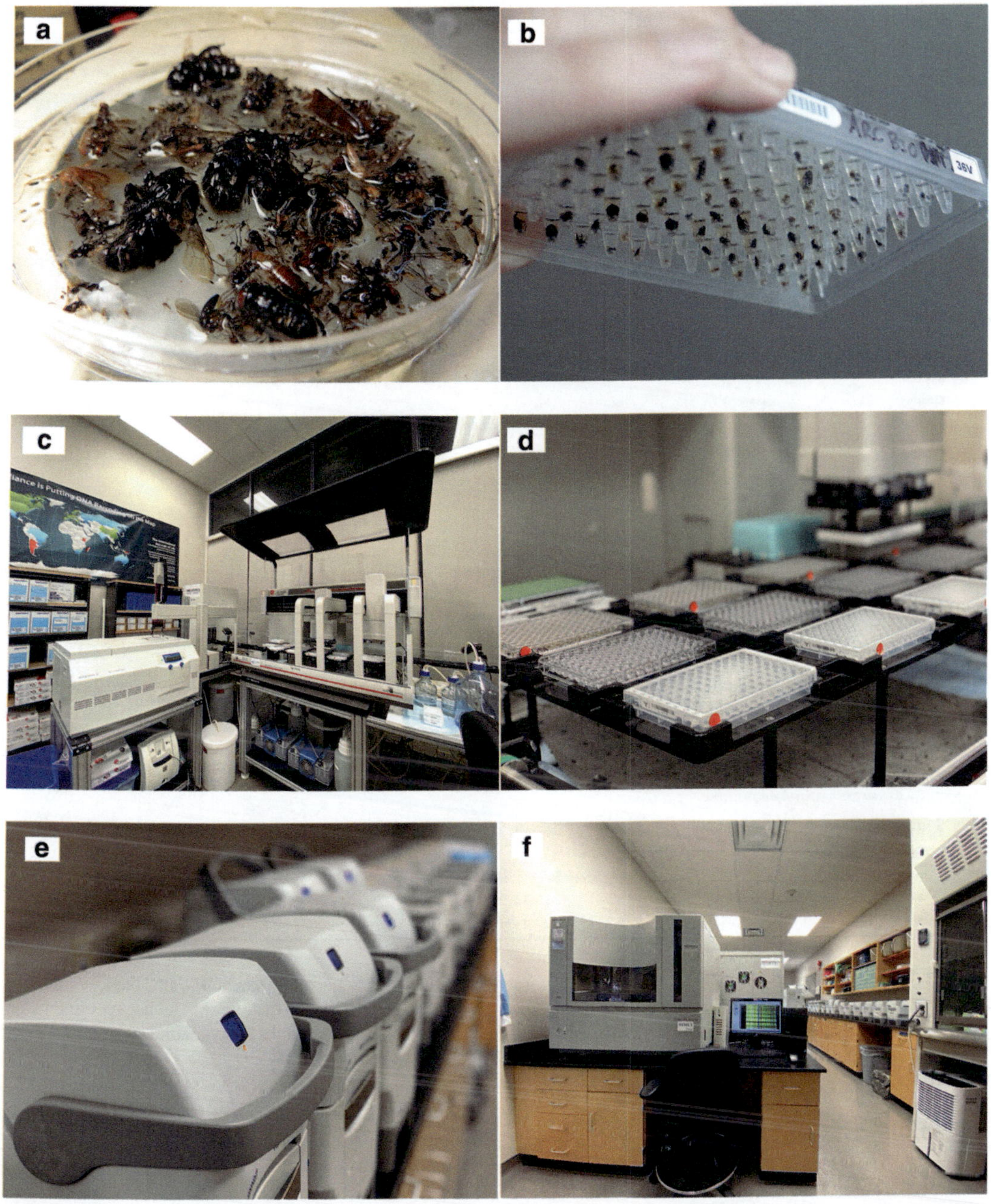

Fig. 3.11 (**a–f**). DNA barcoding procedures: (**a**) Malaise trap sample ready to be sorted for DNA barcoding; (**b**) small specimens are usually sorted individually in 96-well microplates, and whole specimens are used for non-destructive DNA extraction, larger specimens are pinned or preserved in ethanol, and a single leg is used for extraction; (**c–d**) DNA extraction is largely automatized at the Centre for Biodiversity Genomics; (**e**) thermocyclers are used for both PCR amplification and sequencing reactions in a Sanger sequencing workflow; (**f**) majority of the sequence data on BOLD has been generated though Sanger sequencing, using DNA analyzers such as this ABI 3730xl. These old sequencing workhorses are now gradually being replaced with newer technology. Photo credit: Centre for Biodiversity Genomics (**a**), M. Pentinsaari (**b**), Evgeny Zakharov (**c–f**)

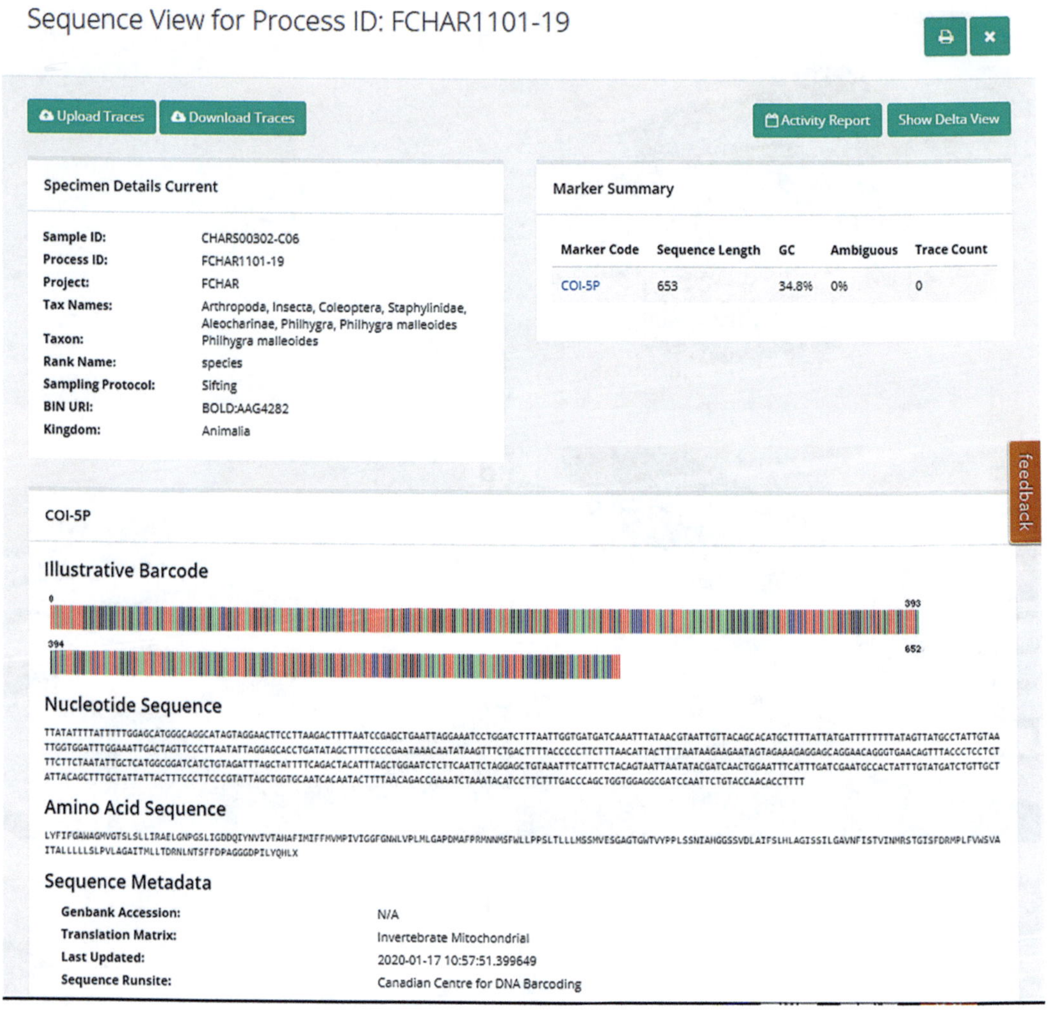

Fig. 3.12 Example of BOLD sequence page for *Philhygra malleoides* Lohse collected in Nunavut. This page shows the DNA barcode sequence associated with the specimen and some key metadata related to the sequence. A separate specimen data page displaying collecting data and information on voucher specimen storage, as well as photographs (if available), is also generated for every specimen on BOLD

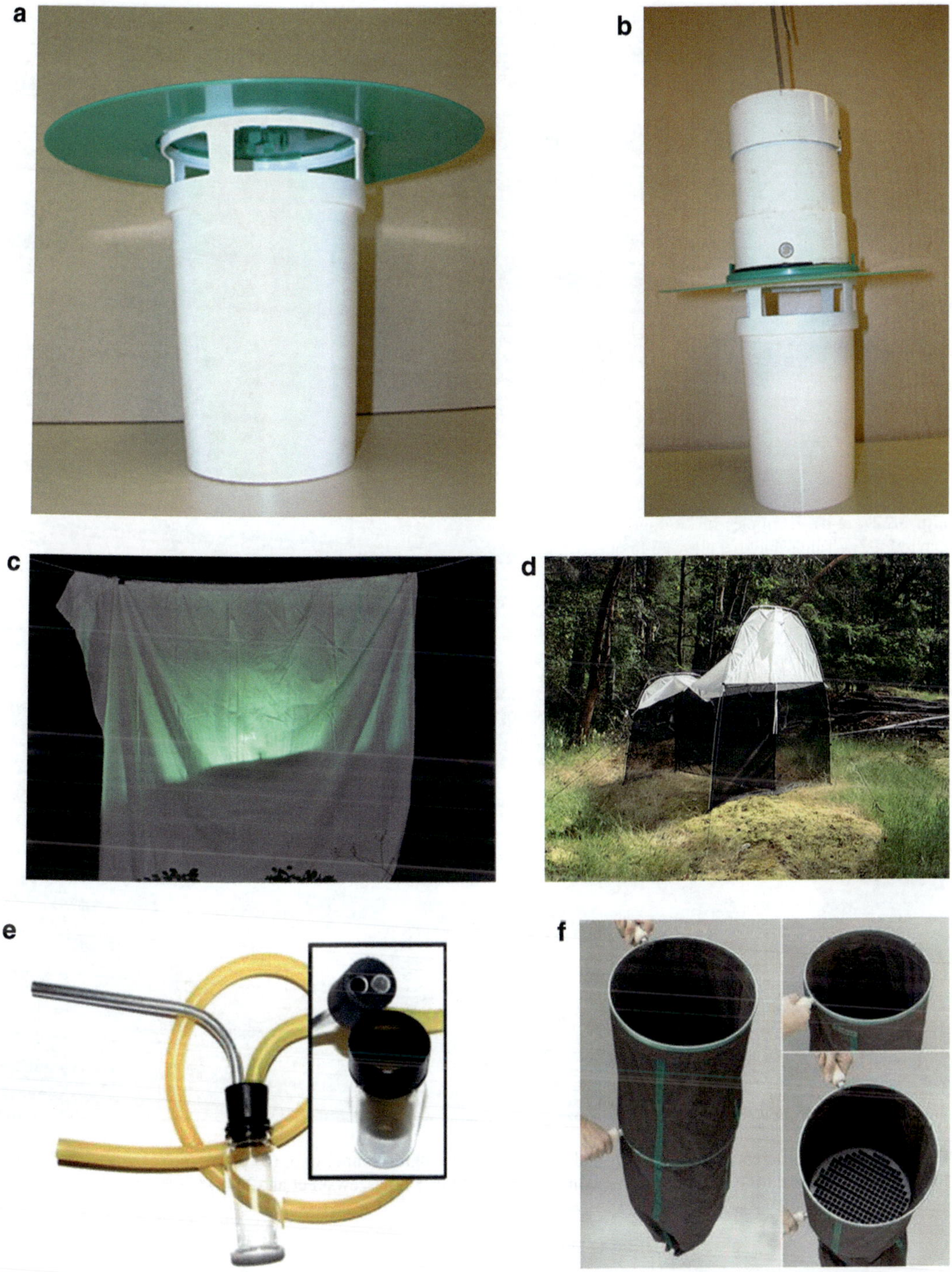

Fig. 3.13 (**a**–**f**). Different traps and tools for collecting alcocharine beetles; (**a**) pitfall trap; (**b**) Luminoc trap with UV light and timer; (**c**) light trap; (**d**) Interception trap (photo credit N. Winchester); (**e**) aspirator; (**f**) litter sifter. After Klimaszewski et al. (2020)

Fig. 3.14 (a-b). Different collecting tools and methods: (**a**) Lindgren funnel trap; (**b**) plastic boxes with sifted forest litter or plant material for collecting beetles using aspirator. After Klimaszewski et al. (2018)

Fig. 3.15 (**a–b**). Examples of Labrador habitats: (**a**) bog along Trans-Labrador Highway; (**b**) patterned fen in central Labrador. Photo credit G.R.Pohl, after Klimaszewski et al. (2011)

Fig. 3.16 (**a**–**b**). Examples of Labrador habitats: (**a**) Boreal black spruce forest; (**b**) shallow, rocky stream in open habitat. Photo credit G.R. Pohl, after Klimaszewski et al. (2011)

Fig. 3.17 (**a–b**). Examples of Labrador habitats: (**a**) dry forest floor with reindeer moss near Goose Bay; (**b**) matt of *Sphagnum* moss in bog. Photo credit G.R. Pohl, after Klimaszewski et al. (2011)

Fig. 3.18 (**a**–**b**). Examples of Labrador habitats: (**a**) tundra with patches of snow; (**b**) inland sand dunes near Goose Bay. Photo credit G.R. Pohl, after Klimaszewski et al. (2011)

Fig. 3.19 (**a–b**). Examples of Nunavut habitats: (**a**) small stream with exposed sandy banks connecting two lakes southwest of Kugluktuk; (**b**) willow thicket under a cliff just outside the hamlet of Kugluktuk. Photo credit M. Pentinsaari

Fig. 3.20 (**a-b**). Examples of Nunavut habitats: (**a**) wet sedge fen between hills in the tundra east of Kugluktuk; (**b**) lush willow thicket with a thick layer of leaf litter in Kugluk (Bloody Falls) Territorial Park. Photo credit M. Pentinsaari

Fig. 3.21 (**a–b**). Examples of Nunavut habitats: (**a**) one of the authors (M. Pentinsaari) sifting leaf litter at Kugluk Territorial Park; (**b**) Coppermine River delta, with ponds left behind by the spring floods. Photo credit M. Pentinsaari

Fig. 3.22 (**a–b**). Examples of Northwest Territories habitats: (**a**) boreal mist—Taiga shield NE of Yellowknife; (**b**) David Lake—Taiga shield, Yellowknife outskirt. Photo credit Craig Machtans

Fig. 3.23 (**a–b**). Examples of Northwest Territories habitats: (**a**) fern gully—Taiga plain, Fort Liard; (**b**) forest—Taiga plains, Fort Liard. Photo credit Craig Machtans

Fig. 3.24 (**a**–**b**). Examples of Northwest Territories habitats: (**a**) lakes—Taiga plain near Inuvik; (**b**) Taiga shield NE of Yellowknife. Photo credit Craig Machtans

Fig. 3.25 (**a–b**). Examples of Yukon habitats: (**a**) Takhini alkaline salt flats; (**b**) *Salicornia borealis* salt flats plant. Photo credit Denise Horwood

Fig. 3.26 (a–b). Examples of Yukon habitats: (**a**) alpine tundra; (**b**) moose droppings. Photo credit Denise Horwood

Fig. 3.27 (**a–b**). Examples of Yukon habitats: (**a**) Boreal mixed forest; (**b**) deciduous forest dominated with Aspen trees. Photo credit Denise Horwood

Fig. 3.28 (**a–b**). Examples of Yukon microhabitats: (**a**) fresh, and (**b**) dry mushrooms. Photo credit Denise Horwood

Fig. 3.29 Example of Yukon landscape: riparian habitat. Photo credit Denise Horwood

Fig. 3.30 (**a**–**b**). Example of Yukon habitats: (**a**) old white spruce trunk in alluvial habitat; (**b**) white spruce alluvial forest. Photo credit Denise Horwood

Fig. 3.31 (**a–b**). Example of Yukon habitats: (**a**) Alluvial fan with human settlement on Herschel Is.; (**b**) Cotton-grass/moss habitat on Herschel Is. Photo credit Maria Leung

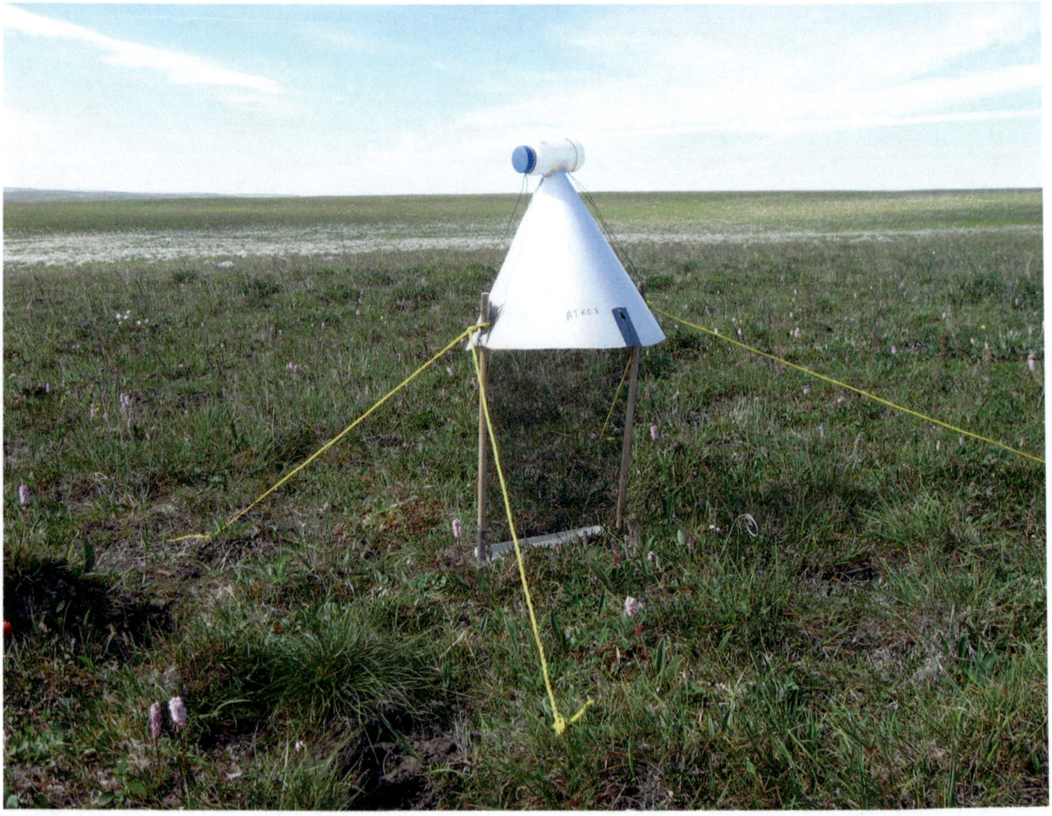

Fig. 3.32 Example of Yukon habitats: Insect trap in upland heath habitat, Herschel Is. Photo credit Maria Leung

Fig. 3.33 (**a**–**b**). Interior Alaska: (**a**) mixed forests; (**b**) primarily broadleaf trees, of uplands

Fig. 3.34 (**a–b**). Interior Alaska: (**a, b**) primarily black spruce forest of wet lowlands

Fig. 3.35 (**a**–**b**). Interior Alaska: (**a**) alpine, and (**b**) high elevation valley tundra

Fig. 3.36 (**a**–**b**). Interior Alaska: (**a**) alpine, and (**b**) high elevation valley tundra

Fig. 3.37 (**a–b**). Interior Alaska: (**a**) rivers and river floodplains, and (**b**) ponds

Fig. 3.38 (**a–b**). Interior Alaska: (**a**) post-fire meadow, and (**b**) Alaskan Arctic tundra

Fig. 3.39 (a–b). Alaskan Arctic tundra: (a) Muskox; (b) Toolik Field Station

Fig. 3.40 (**a–b**). Alaskan west coast maritime tundra

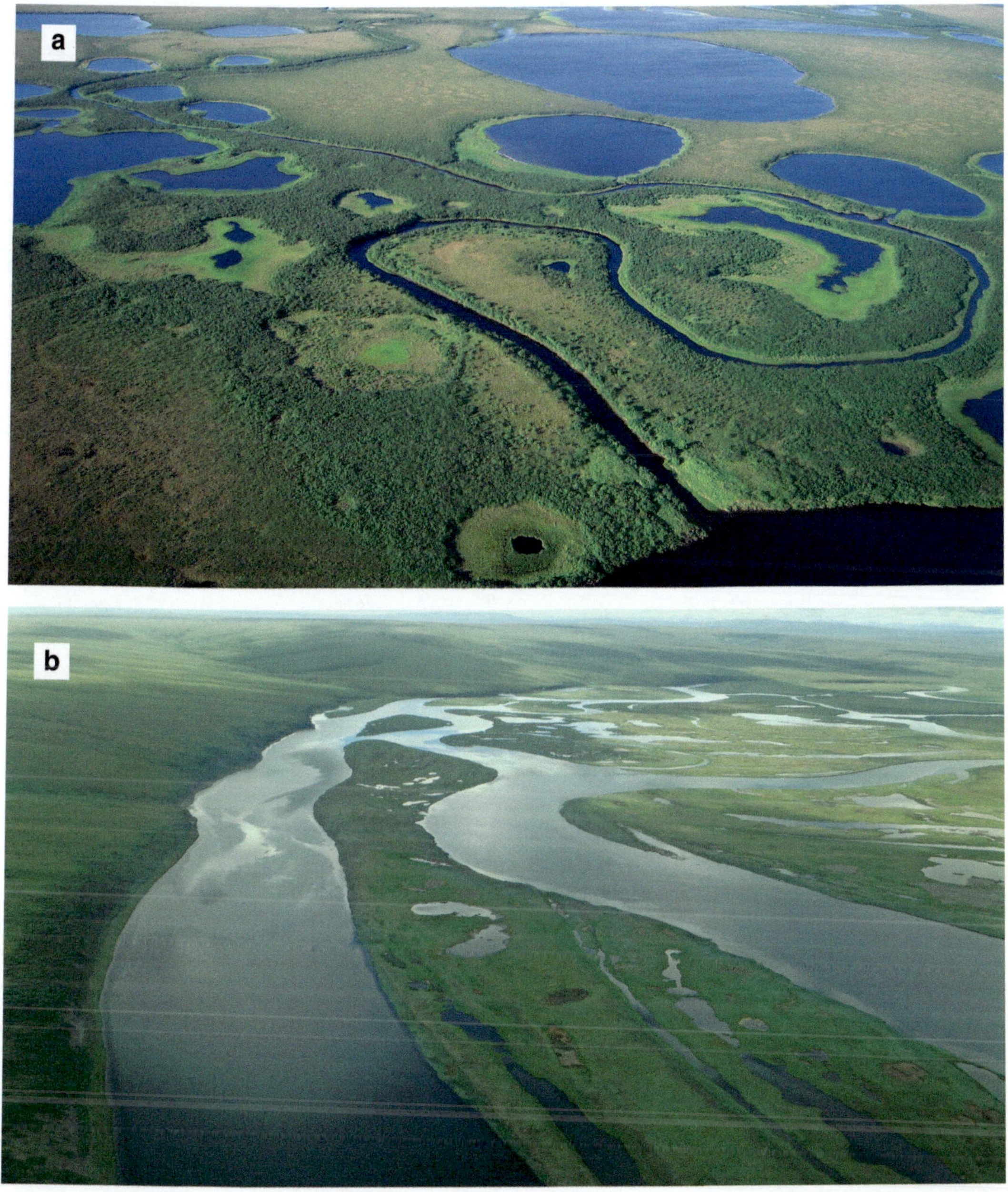

Fig. 3.41 (**a**–**b**). Alaskan west coast maritime tundra riverine systems

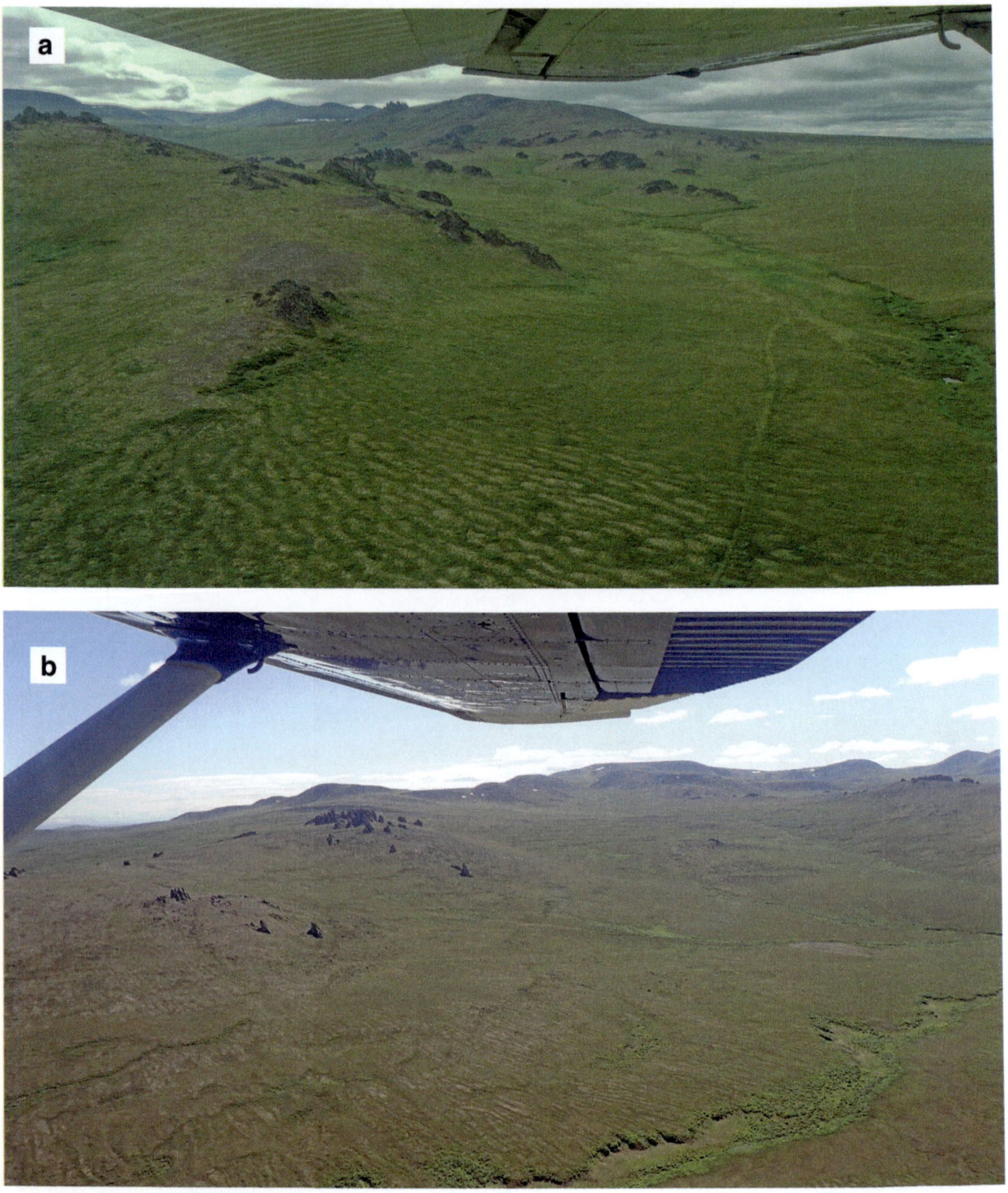

Fig. 3.42 (a–b). Alaskan west coast maritime tundra, Seward Peninsula

Fig. 3.43 (**a–b**). Bering Sea Island maritime tundra: Pribilof Islands: (**a**) St. Paul, and (**b**) St. George

Fig. 3.44 (**a**–**b**). Bering Sea Island maritime tundra. St. Matthew Island

Fig. 3.45 (**a–b**). Aleutian Island Archipelago maritime tundra with persistent snowfields

Fig. 3.46 (**a–b**). Aleutian Island Archipelago maritime tundra: (**a**) rocky beach with driftwood, and (**b**) inland and upland meadows

References

Ashe JS (2001) Key to the tribes and genera of Nearctic Aleocharinae. Viii. In: Arnett RH Jr, Thomas CM (eds) American beetles. Archostemata, Myxophaga, Adephaga, Polyphaga: Staphyliniformia. CRC Press, Boca Raton, London, New York, Washington, DC, pp 299–374. 443 pp [publ. 28 Dec 2000 according to CRC Press]

Benick G, Lohse GA (1974) Tribus 14 (Callicerini). In: Freude H, Harde KW, Lohse A (eds) Die Käfer Mitteleuropas. Band 5: Staphylinidae II (Hypocyphtinae und Aleocharinae) Pselaphidae. Goecke & Evers, Krefeld. 381 pp

Blagoev GA, deWaard JR, Ratnasingham S, deWaard SL, Lu L, Robertson J, Telfer AC, Hebert PDN (2016) Untangling taxonomy: a DNA barcode reference library for Canadian spiders. Mol Ecol Resour 16:325–341

Bouchard P, Bousquet Y, Davies AE, Alonso-Zarazaga MA, Lawrence JF, Lyal CHC, Newton AF, Reid CAM, Schmitt M, Ślipiński SA, Smith ABT (2011) Family-group names in Coleoptera (Insecta). ZooKeys 88:1–972

Bouchard P, Smith ABT, Douglas H, Gimmel ML, Brunke AJ, Kanda K (2017) Biodiversity of Coleoptera. In: Foottit RG, Adler PH (eds) Insect biodiversity: science and society. Wiley, Oxford, pp 337–417

Bousquet Y, Bouchard P, Davies AE, Sikes DS (2013) Checklist of beetles (Coleoptera) of Canada and Alaska. Second edition. Pensoft Publishers, Series Faunistica No. 109. Sofia-Moscow, 402 pp

Brunke JA, Buffam J (2018) A review of Nearctic rove beetles (Staphylinidae) specialized on the burrows and nests of vertebrates. In: Biology of rove beetles (Staphylinidae): life history, evolution, ecology and distribution. Springer, Cham. 351 pp

Brunke A, Klimaszewski J, Dorval J-A, Bourdon C, Paiero SM, Marshall SA (2012) New species and distributional records of Aleocharinae (Coleoptera, Staphylinidae) from Ontario, Canada, with a checklist of recorded species. In: Klimaszewski J., Anderson R. (Eds) Biosystematics and ecology of Canadian Staphylinidae (Coleoptera) II. ZooKeys [Special Issue] 186:119–206

Brunke AJ, Bouchard P, Douglas HB, Pentinsaari M (2019) Coleoptera of Canada. ZooKeys 819:361–376

Brunke AJ, Salnitska M, Hansen AK, Zmudzinska A, Smetana A, Buffam J, Solodovnikov A (2020) Are subcortical rove beetles truly Holarctic? An integrative taxonomic revision of north temperate *Quedionuchus* (Coleoptera: Staphylinidae: Staphylininae). Org Divers Evol 20:77–116

Buchner P, Stănescu M (2019) A revision of *Agonopterix lacteella* (Lepidoptera: Depressariidae) and establishing *A. pallidior* Stringer, 1930 as a junior synonym. Travaux du Muséum National d'Histoire Naturelle "Grigore Antipa" 62:87–98

Butcher BA, Smith MA, Sharkey MJ, Quicke DJ (2012) A turbo-taxonomic study of Thai *Aleiodes* (*Aleiodes*) and *Aleiodes* (*Arcaleiodes*) (Hymenoptera: Braconidae: Rogadinae) based largely on COI barcoded specimens, with rapid descriptions of 179 new species. Zootaxa 3457:1–232

Corley MFV, Ferreira S (2017) DNA barcoding reveals sexual dimorphism in *Isotrias penedana* Trematerra, 2013 (Lepidoptera: Tortricidae, Chlidanotinae). Zootaxa 4221:594–600

D'Souza ML, Hebert PDN (2018) Stable baselines of temporal turnover underlie high beta diversity in tropical arthropod communities. Mol Ecol 27:2447–2460

deWaard JR, Levesque-Beaudin V, DeWaard SL, Ivanova NV, McKeown JTA, Miskie R, Naik S, Perez KHJ, Ratnasingham S, Sobel CN, Sones JE, Steinke C, Telfer AC, Young AD, Young MR, Zakharov EV, Hebert PDN (2019) Expedited assessment of terrestrial arthropod diversity by coupling Malaise traps with DNA barcoding. Genome 62:85–95

Dybas LK, Dybas HS (1981) Coadaptation and taxonomic differentiation of sperm and spermathecae in featherwings beetles. Evolution 35(1):168–174

Elven H, Bachmann L, Gusarov VI (2011) Phylogeny of the tribe Athetini (Coleoptera: Staphylinidae) inferred from mitochondrial and nuclear sequence data. Mol Phylogenet Evol 57(1):84–100

Elven H, Bachmann L, Gusarov VI (2012) Molecular phylogeny of the Athetini-Lomechusini-Ecitocharini clade of aleocharine rove beetles (Insecta). Zool Scr 41(6):617–636

Gallant AL (1995) Ecoregions of Alaska (No. 1567). US Government Printing Office, Denver, CO

Gusarov VI (2003) Revision of some types of North American aleocharines (Coleoptera: Staphylinidae: Aleocharinae), with synonymic notes. Zootaxa 353:1–134

Gusarov VI (2011) Case 3537. Athetini Casey, 1910 and Geostibina Seevers, 1978 (Insecta, Coleoptera, Staphylinidae, Aleocharinae): proposed conservation. Bull Zool Nomencl 68(1):54–60

Gwiazdowski RA, Foottit RG, Maw HEL, Hebert PDN, Imondi R, LaRocca G (2015) The Hemiptera (Insecta) of Canada: constructing a reference library of DNA barcodes. PLoS One 10:e0125635

Hammond HEJ (1997) Arthropod biodiversity from *Populus* coarse woody material in north-central Alberta: a review of taxa and collection methods. Can Entomol 129:1009–1033

Hanley RS, Ashe JS (2003) Techniques for dissecting adult aleocharine beetles (Coleoptera: Staphylinidae). Bull Entomol Res 93(1):11–18

Hébert C, Jobin L, Fréchette M, Pelletier G (2000) An efficient pit-light trap to study diversity. J Insect Conserv 4:191–202

Hebert PDN, Cywinska A, Ball SL, deWaard JR (2003) Biological identifications through DNA barcodes. Proc R Socf Lond Series B: Biol Sci 270:313–321

Hebert PDN, Ratnasingham S, Zakharov EV, Telfer AC, Levesque-Beaudin V, Milton MA, Pedersen S, Jannetta P, DeWaard JR (2016) Counting animal species with DNA barcodes: Canadian insects. Philos Trans R Soc Lond Ser B Biol Sci 371:e1001127

Hendrich L, Morinière J, Haszprunar G, Hebert PDN, Hausmann A, Köhler F, Balke M (2015) A comprehensive DNA barcode database for central European beetles with a focus on Germany: adding more than 3,500 identified species to BOLD. Mol Ecol Resour 15:795–818

Huemer P (2020) Integrative revision of the Caryocolum schleichi species group – a striking example of a temporally changing species concept (Lepidoptera, Gelechiidae). Alpine Entomol 4:39–63

Hughes CC, Johns RC, Sweeney JD (2014) A technical guide to installing beetle traps in the upper crown of trees. J Acad Entomol Soc 10(4):12–18

Jobin LJ, Coulombe C (1992) The Luminoc® insect trap. Forestry Canada, Quebec Region, Sainte-Foy, Quebec. Information Leaflet LFC 26E. 12 pp

Kaila L (2015) The *Elachista dispunctella* (Duponchel) complex (Lepidoptera, Elachistidae) revisited, with exceptional level of synonymy. Zootaxa 3980:301–358

Kaufman DS, Young NE, Briner JP, Manley WF (2011) Alaska Palaeo-Glacier Atlas (Version 2). In: Developments in quaternary sciences. Elsevier, New York, pp 427–445

Klimaszewski J (1979) A revision of the Gymnusini and Deinopsini of the world (Coleoptera: Staphylinidae: Aleocharinae). Agriculture Canada Research Branch, Monograph No. 25:1–169

Klimaszewski J (1982) Studies of Myllaenini (Coleoptera: Staphylinidae, Aleocharinae). 1. Systematics, phylogeny, and zoogeography of Nearctic *Myllaena* Erichson. Can Entomol 114(3):181–242

Klimaszewski J (1984) A revision of the genus *Aleochara* Gravenhorst of America north of Mexico (Coleoptera: Staphylinidae, Aleocharinae). Memoirs of the Entomological Society of Canada No. 129, 211 pp

Klimaszewski J, Langor D, Pelletier G, Bourdon C Perdereau L (2011) Aleocharine beetles (Coleoptera, Staphylinidae) of the province of Newfoundland and Labrador, Canada. Pensoft Series Faunistica No. 98. Pensoft, Sophia-Moscow, 313 pp

Klimaszewski J, Webster RP, Langor D, Sikes D, Bourdon C, Godin B, Ernst C (2016) A review of Canadian and Alaskan species of the genus *Liogluta* Thomson, and descriptions of three new species (Coleoptera, Staphylinidae, Aleocharinae). In: Webster, R. P., Bouchard, P., Klimaszewski, J. (Eds) The Coleoptera of New Brunswick and Canada: providing baseline biodiversity and natural history data. ZooKeys [Special Issue] 573:217–256

Klimaszewski J, Webster RP, Langor DW, Brunke A, Davies A, Bourdon C, Labrecque M, Newton AF, Dorval J-A, Frank JH (2018) Aleocharine rove beetles of eastern Canada (Coleoptera, Staphylinidae, Aleocharinae): a glimpse of megadiversity. Springer, Cham. 902 pp

Klimaszewski J, Hoebeke ER, Godin B, Davies A, Perry KI, Bourdon C, Winchester N (2020) Aleocharine rove beetles of British Columbia: a hotspot of Canadian biodiversity (Coleoptera, Staphylinidae). Springer, Cham. 631 pp

Landry J-F, Nazari V, DeWaard JR, Mutanen M, Lopez-Vaamonde C, Huemer P, Hebert PDN (2013) Shared but overlooked: 30 species of Holarctic Microlepidoptera revealed by DNA barcodes and morphology. Zootaxa 3749:1

Levesque-Beaudin V, Mlynarek JJ (2020) Revision of Nearctic *Paramyia* Williston (Diptera: Milichiidae). Zootaxa 4732:1–56

Levesque-Beaudin V, James H, Pentinsaari M, Borisenko A (2019) Arctic Bioscan (ARCBIO): a new project surveying Arctic biodiversity in Nunavut employs DNA barcoding and community-based monitoring. Newslett Biol Survey Canada 38:25–26

Lindgren BS (1983) A multiple funnel trap for scolytid beetles (Coleoptera). Can Enomol 115(3):299–302

Lohse GA, Klimaszewski J, Smetana A (1990) Revision of Arctic Aleocharinae of North America (Coleoptera: Staphylinidae). Coleopt Bull 44(2):121–202

Martin JEH (1977) The insects and arachnids of Canada. Part 1. Collecting, preparing, and preserving insects, mites, and spiders. Research Branch. Canada Department of Agriculture. Publication 1643. Ministry of Supply and Services Canada 1978. Kromar Printing Ltd

Mutanen M, Hausmann A, Hebert PDN, Landry J-F, de Waard JR, Huemer P (2012) Allopatry as a Gordian knot for taxonomists: patterns of DNA barcode divergence in arctic-alpine Lepidoptera. PLoS One 7: e47214

Mutanen M, Kekkonen M, Prosser SWJ, Hebert PDN, Kaila L (2015) One species in eight: DNA barcodes from type specimens resolve a taxonomic quagmire. Mol Ecol Resour 15:967–984

Paśnik G (2010) Phylogeny and generic classification of Tachyusini (Coleoptera, Staphylinidae: Aleocharinae). Institute of Systematics and Evolution of Animals, Polish Academy of Sciences, Kraków. 129 pp

Peck SB, Davies A (1980) Collecting small beetles with large-area 'window' traps. Coleopt Bull 34:237–239

Pentinsaari M, Hebert PDN, Mutanen M (2014) Barcoding beetles: a regional survey of 1872 species reveals high identification success and unusually deep interspecific divergences. PLoS One 9:e108651

Pentinsaari M, Anderson R, Borowiec L, Bouchard P, Brunke A, Douglas H, Smith ABT, Hebert PDN (2019) DNA barcodes reveal 63 overlooked species of Canadian beetles (Insecta, coleoptera). ZooKeys 2019:53–150

Pentinsaari M, Ratnasingham S, Miller SE, Hebert PDN (2020a) BOLD and GenBank revisited – do identification errors arise in the lab or in the sequence libraries? Kuntner M (Ed). PLoS One 15:e0231814

Pentinsaari M, Blagoev GA, Hogg ID, Levesque-Beaudin-V, Perez K, Sobel CN, Vandenbrink B, Borisenko A (2020b) A DNA barcoding survey of an Arctic arthropod community: implications for future monitoring. Insects 11:46

Peschke K (1978) Funktionsmorphologische Untersuchungen zur Kopulation von *Aleochara curtula* Goeze (Coleoptera, Staphylinidae). Zoomorphologie 89(2):157–184

Prosser SWJ, DeWaard JR, Miller SE, Hebert PDN (2016) DNA barcodes from century-old type specimens using next-generation sequencing. Mol Ecol Resour 16:487–497

Ratnasingham S, Hebert PDN (2007) BOLD: the barcode of life data system (http://www.barcodinglife.org). Mol Ecol Notes 7:355–364

Ratnasingham S, Hebert PDN (2013) A DNA-based registry for all animal species: the barcode index number (BIN) system. PLoS One 8:e66213

Riedel A, Sagata K, Surbakti S, Tänzler R, Balke M (2013) One hundred and one new species of *Trigonopterus* weevils from New Guinea. ZooKeys 280:1–150

Schülke M, Smetana A (2015) Family Staphylinidae Latreille, 1802. In: Löbl I, Löbl D (eds) Catalogue of Palearctic Coleoptera, vol 2, Revised and updated edition. Hydrophiloidea-Staphylinoidea, Brill, Leiden & Boston, pp 304–1134

Seevers CH (1978) A generic and tribal revision of the north American Aleocharinae (Coleoptera: Staphylinidae) [with additions and annotations by Lee H. Herman]. Fieldiana: Zoology 71. vi + 289 pp

Sikes DS, Allen RT (2016) First Alaskan records and a significant northern range extension for two species of Diplura (Diplura, Campodeidae). ZooKeys 563:147–157

Sikes DS, Bowser M, Morton JM, Bickford C, Meierotto S, Hildebrandt K (2017) Building a DNA barcode library of Alaska's non-marine arthropods. Genome 60:248–259

Skvarla MJ, Larson JL, Dowling APG (2014) Pitfalls and preservatives: a review. J Entomol Soc Ontario 145:15–43

Slowik J, Blagoev GA (2012) First description of the male spider *Pacifiphantes magnificus* (Chamberlin & Ivie) (Araneae: Linyphiidae). Zootaxa 3481:73–81

Smith CAS, Meikle JC, Roots CF, (Eds) (2004). Ecoregions of the Yukon Territory: Biophysical properties of Yukon landscapes. Agriculture and Agri-Food Canada, PARC Technical Bulletin No. 04-01, Summerland, British Columbia, 313 pp

Spence JR, Niemelä JK (1994) Sampling carabid assemblages with pitfall traps: the madness and the method. Can Entomol 126:881–894

Steinke D, Breton V, Berzitis E, Hebert PDN (2017) The School Malaise trap program: coupling educational outreach with scientific discovery. PLoS Biol 15: e2001829

Webster RP, Klimaszewski J, Pelletier G, Savard K (2009) New Staphylinidae (Coleoptera) records with new collection data from New Brunswick, Canada. I. Aleocharinae. In: Majka, C.G., Klimaszewski, J. (Eds.) Biodiversity, biosystematics, and ecology of Canadian Coleoptera II. ZooKeys [Special Issue] 22:171–248

Webster RP, Klimaszewski J, Sweeney JD, DeMerchant I (2012) New Staphylinidae (Coleoptera) records with new collection data from New Brunswick, and an addition to the fauna of Quebec, Canada: Aleocharinae. In: Klimaszewski J., Anderson R (Eds) Biosystematics and ecology of Canadian Staphylinidae (Coleoptera) II. ZooKeys [Special Issue] 186:83–118

Webster RP, Anderson RS, Webster VL, Alderson CA, Hughes CC, Sweeney JD (2016a) New Curculionoidea records from New Brunswick, Canada with an addition to the fauna of Nova Scotia. In: Webster, R.P., Bouchard, P., Klimaszewski, J. (Eds) The Coleoptera of New Brunswick and Canada: providing baseline biodiversity and natural history data. ZooKeys [Special Issue] 573:367–386

Webster RP, Klimaszewski J, Bourdon C, Sweeney JD, Hughes CC, Labrecque M (2016b) Further contribution to the Aleocharinae (Coleoptera, Staphilinidae) fauna of New Brunswick and Canada including description of 27 new species. In: Webster, R.P., Bouchard, P., Klimaszewski, J. (Eds) The Coleoptera of New Brunswick and Canada: providing baseline biodiversity and natural history data. ZooKeys [Special Issue] 573:85–216

Webster RP, Webster VL, Alderson CA, Hughes CC, Sweeney JD (2016c) Further contributions to the Coleoptera fauna of New Brunswick with an addition to the fauna of Nova Scotia, Canada. In: Webster, R.P., Bouchard, P., and Klimaszewski, J. (Eds) The Coleoptera of New Brunswick and Canada: providing baseline biodiversity and natural history data. ZooKeys [Special Issue] 573:265–338

Wirta H, Várkonyi G, Rasmussen C, Kaartinen R, Schmidt NM, Hebert PDN, Barták M, Blagoev G, Disney H, Ertl S, Gjelstrup P, Gwiazdowicz DJ, Huldén L, Ilmonen J, Jakovlev J, Jaschhof M, Kahanpää J, Kankaanpää T, Krogh PH, Labbee R, Lettner C, Michelsen V, Nielsen SA, Nielsen TR, Paasivirta L, Pedersen S, Pohjoismäki J, Salmela J, Vilkamaa P, Väre H, von Tschirnhaus M, Roslin T (2016) Establishing a community-wide DNA barcode library as a new tool for arctic research. Mol Ecol Resour 16:809–822

Woodcock TS, Boyle EE, Roughley RE, Kevan PG, Labbee RN, Smith ABT, Goulet H, Steinke D, Adamowicz SJ (2013) The diversity and biogeography of the Coleoptera of Churchill: insights from DNA barcoding. BMC Ecol 13

Yamamoto S, Maruyama M (2017) Phylogeny of the rove beetle tribe Gymnusini sensu n. (Coleoptera: Staphylinidae: Aleocharinae): implications for the early branching events of the subfamily: Phylogeny of Gymnusini rove beetles. Syst Entomol 43 (1):183–189

Young MR, Behan-Pelletier VM, Hebert PDN (2012) Revealing the Hyperdiverse mite Fauna of subarctic Canada through DNA barcoding. Fenton B (Ed). PLoS One 7:e48755

Zhou X, Adamowicz SJ, Jacobus LM, Dewalt RE, Hebert PD (2009) Towards a comprehensive barcode library for arctic life – Ephemeroptera, Plecoptera, and Trichoptera of Churchill, Manitoba, Canada. Front Zool 6:30

Effects of Global Warming on the Distribution and Diversity of Arctic and Subarctic Insects

4

Anthropogenic climate change is causing Arctic and Subarctic regions to warm and dry faster than lower latitudes (Serreze et al. 2000, 2009; Post et al. 2009; IPCC 2014). Arthropods comprise the majority of animal species in the Arctic and Subarctic and are especially sensitive to abiotic environmental changes (Høye et al. 2007; Høye and Forchhammer 2008; Hodkinson 2013).

Warming is driving many alarming abiotic and biotic changes. Arctic and alpine tundra is shrinking as shrubs and treeline advance upslope and into higher latitudes (Lloyd and Fastie 2003; Dial et al. 2007; Myers-Smith et al. 2011). Much of the Subarctic is dominated by boreal forest, which covers about 17% of the planet (Bonan 1992), and has a fauna adapted to low temperatures, long winters, and intense seasonal temperature swings. Between 1984 and 2014 the growing season in boreal and Arctic regions lengthened by about 2.6 days per decade (Park et al. 2016). Because the development of most arthropod life history stages is affected more by accumulated thermal energy than photoperiod, lengthening of the growing season has resulted in phenology shifts with many Arctic and Subarctic species becoming active earlier in the season as the climate warms (Høye et al. 2014; Renner and Zohner 2018). However, not all arthropods are responding in the same way. For example, some exhibit delays in activity, while others have increased their number of generations per year (Forrest 2016).

There are particular concerns for overwintering insects in high latitudes which require cold temperatures for diapause. Exposure to warm weather in winter can result in loss of cold hardiness and/or needed snow cover. This may limit how far south some of these northern species can live. A warm spell in the winter can cause insects to break diapause. If a warm spell is followed by cold temperatures, the insects can freeze and die. Winter warm spells are becoming more frequent (Chapman et al. 2020). Winter warming is the mechanism identified by Harris et al. (2019) who found beetle capture rates declined significantly between the 1970s and the 2010s in a protected forest in New Hampshire, free from the influence of agriculture and deforestation. They found a strong relationship between warmth and significantly lower beetle abundance and diversity, specifically finding the lowest beetle counts following winters with the least snow.

Fire frequency and intensity in the boreal forest has reached levels unprecedented in the last 10,000 years (Kelly et al. 2013). Warming of the boreal forest appears to be causing more frequent large outbreaks of tree-feeding insects, further harming already moisture-stressed trees (Chapin et al. 2010). Thawing of permafrost is also accelerating boreal and tundra drying, as surface water drains, where it had been supported by ice, allowing the soil to dry, increasing water stress on vegetation and susceptibility to fire and herbivory

J. Klimaszewski et al., *A Faunal Review of Aleocharine Beetles in the Rapidly Changing Arctic and Subarctic Regions of North America (Coleoptera, Staphylinidae)*,
https://doi.org/10.1007/978-3-030-68191-3_4

(Barber et al. 2000; Helbig et al. 2016). These changes are causing shifts in post-fire successional trajectories in the boreal forest (Juday et al. 2015). Trends indicate a gradual replacement of conifers by shrubs, grasses, and more fire-resistant hardwoods (Hogg and Hurdle 1995; Chapin et al. 2010; Kelly et al. 2013). Conifer associated arthropods such as the aleocharine *Placusa tacomae* may, as a consequence, decline in abundance and diversity (Sikes 2019). The gradual loss and alteration (e.g. drying) of alpine, arctic, and maritime tundra communities from drying and shrubifications are similarly expected to cause declines in tundra associated species (Loboda et al. 2018). In this chapter, we review and discuss impacts of climate change on arthropods, including extinctions and extirpations, population declines, and range shifts. Despite a rapidly growing number of studies on these issues, for arthropods, the magnitude and pervasiveness of the problem is such that for key questions we have only incomplete answers and enormous data gaps remain (Gillespie et al. 2019; Montgomery et al. 2020).

Extinctions Looking three centuries into the future, under a "business as usual" model, with no reductions in the rate of human-produced greenhouse gas emissions, average Earth temperatures are predicted to reach levels not seen since the Eocene approximately 50 million years ago (Fig. 4.47) (Westerhold et al. 2020). During the Eocene the Earth was warmer than at any other period in the last 67 million years and the Arctic had a tropical climate and biota. It had taken many millions of years for the planet to reach that level of warmth (Westerhold et al. 2020), which gave organisms time to adapt and disperse and time for ecosystems to reach new equilibria. Reaching such levels of warmth in a mere 300 years would likely result in a much different outcome, and is, frankly, terrifying to contemplate. Many organisms unable to adapt physiologically, maintain needed ecological relationships, or disperse to suitable habitats to track their optimal temperatures, would likely suffer significant population declines, extirpations, and extinctions. Some studies suggest Earth's sixth mass extinction has already begun (Dunn et al. 2009; Barnosky et al. 2011; Dirzo et al. 2014; Ceballos et al. 2015, 2017; Raven and Miller 2020; Cardoso et al. 2020) caused by habitat destruction, overexploitation, invasive species, and pollution, in addition to climate change.

A number of studies, although few focussed on insects, have attempted to estimate the percent loss of species under different climate change scenarios. Thomas et al. (2004) used species-area relationships in their investigation of this question and concluded that by 2050 under a maximum warming scenario, 35% of the species in their sample would be committed to extinction. Urban (2015), who conducted a meta-analysis of over 130 such studies, predicts 16% of species will be lost to extinction with a 4.3 °C warming of the climate. Endemic species with small geographic ranges were found to be among those with the highest chance of extinction. However, none of the studies estimated extinction risks 300 years into the future with 10–12 °C of warming under a business-as-usual scenario. These estimates would almost certainly be far more dire.

Some studies suggest a minority of organisms will benefit from warming (e.g. Deutsch et al. 2008; Hunter et al. 2014). Deutsch et al. (2008) modeled ectotherm responses to warming and concluded that species closest to the equator were most likely to suffer fitness declines (i.e. losers), while those closest to the poles were most likely to see fitness increases (i.e. winners). However, this work was physiological and ignored ecological feedbacks. It is possible any such winners will eventually become losers under a business-as-usual warming scenario. A few degrees of warming might allow some species to increase their population sizes for a few decades but by the end of three centuries, 10–12 °C of global warming above the historic averages would have pushed many such former winners beyond their thermal maxima, eliminated their host species, or broken down other needed ecological relationships (Harvey et al. 2020).

For example, the spruce bark beetle, *Dendroctonus rufipennis*, in Alaska is thought to have benefitted from climate warming on the

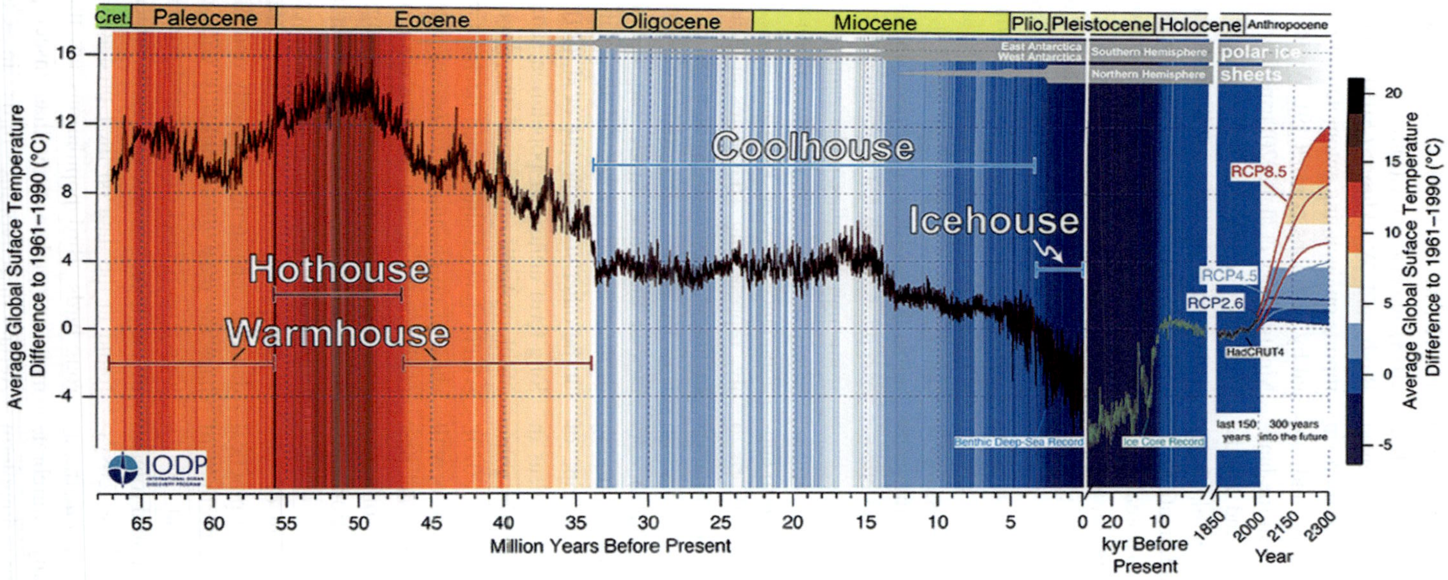

Fig. 4.47 Past and future trends in global temperature spanning the last 67 million years. Oxygen isotope values in deep-sea benthic foraminifera from sediment cores are a measure of global temperature and ice volume. Temperature is relative to the 1961–1990 global mean. Data are from ice core records of the last 25,000 years and illustrate the transition from the last glacial to the current warmer period, the Holocene. Historic data from 1850 to today show the distinct increase after 1950, marking the onset of the Anthropocene. Future projections for global temperature for three Representative Concentration Pathway (RCP) scenarios in relation to the benthic deep-sea record suggest that by 2100 the climate state will be comparable to the Miocene Climate Optimum (~16 million years ago), well beyond the threshold for nucleating continental ice sheets. If emissions are constant after 2100 and not stablized before 2250, global climate by 2300 might enter the hothouse world of the early Eocene (~50 million years ago) with its multiple global warming events and no large ice sheets at the poles. Republished from the press release for Westerhold et al. (2020) (CENOGRID) with permission of the authors

Kenai Peninsula, allowing an outbreak that spanned most of the 1990s (Berg 2015). This outbreak came to an end when most of the mature white spruce, covering more than three million acres on the Kenai had been killed. As reviewed in the introduction to this chapter, conifers in Alaska are declining in productivity and are predicted to eventually become rare on the landscape if warming continues (Hogg and Hurdle 1995; Chapin et al. 2010; Kelly et al. 2013; Juday et al. 2015). There are combinations of temperature and precipitation that no boreal forest white spruce has been found thriving in, and the climate of the boreal forest is shifting towards those combinations. In the short term, insects that feed on conifers may benefit from warming but continued warming is expected to greatly reduce or even eliminate their hosts. It is important to not be complacent if global warming leads to population increases in some ecosystems—such increases may be quite temporary.

Whether we avoid the worst-case scenario or not, it remains essential that we continue to support natural history museums and their efforts to document biodiversity. Science has given names to fewer than half and probably fewer than a fifth of our planet's species (Stork 2018). As populations and species disappear forever during Earth's 6th mass extinction, our best hope of learning about them is to preserve specimens in museums, and to support and maintain those museums indefinitely (Raven and Miller 2020). Sadly, many species have already been, or will be, lost without specimens having been preserved. Museum specimens will be the only evidence of many species driven to extinction.

Insect population declines Extinction of rarer species has been argued to be less of a practical concern for humanity than dramatic declines of abundant, common species (Dirzo et al. 2014). These common species often play central roles in ecosystems and perform the bulk of the ecosystem services. For insects, this includes acting as the primary prey base for many vertebrate populations. A number of worrisome long-term studies, reviews, and meta-analyses have recently been published concerning declines in insect abundance in Europe (Hallmann et al. 2017; Seibold et al. 2019), the United States (Wepprich et al. 2019; Harris et al. 2019), the tropics (Janzen and Hallwachs 2019), Greenland, (Loboda et al. 2018) and elsewhere (Dirzo et al. 2014). Sánchez-Bayo and Wyckhuys (2019) review 73 such reports on declining entomofaunas (but see rebuttal by Simmons et al. 2019). Wagner (2020) and Montgomery et al. (2020) carefully reviewed this literature on insect declines. The most well documented declines are in areas that have had high densities of humans and agriculture over long time spans. There is a dearth of long-term research on insect abundance from Earth's remaining wildlands. The causes of these declines have not been clearly identified in many studies, and certainly there are many potential drivers, in addition to climate change. Of the aforementioned studies, Loboda et al. (2018), based on research in eastern Greenland in the Arctic, is exceptional in that it took place in a region free from high densities of humans and agriculture. They found an 80% decrease in total muscid fly abundance between 1996 and 2014. This fly family, representing 16 species at their study site, includes important pollinators in the Arctic. Bowden et al. (2018), studying eight spider species at the same Arctic site, found significant declines in two species and no significant change in the other six. Gillespie et al. (2019) reviewed long-term arthropod population sampling in the North Atlantic region of the Arctic, some of which showed abundance declines while others showed no change or increases. However, their conclusion was that the data are too sparse and more extensive long-term sampling is required before arthropod biodiversity trends in the Arctic can be confidently understood.

Sampling many taxa over large areas for many decades consistently is costly and difficult. Perhaps the most geographically wide-ranging and long-term dataset of arthropod population trends in the boreal forest of North America can be found in the data on their predators—insectivorous birds. Bird populations are monitored far more comprehensively than insect populations. Aerial insectivorous birds in Canada have declined in abundance over 60% since 1970, more than any other bird group (Nebel et al. 2010; North American Bird Conservation

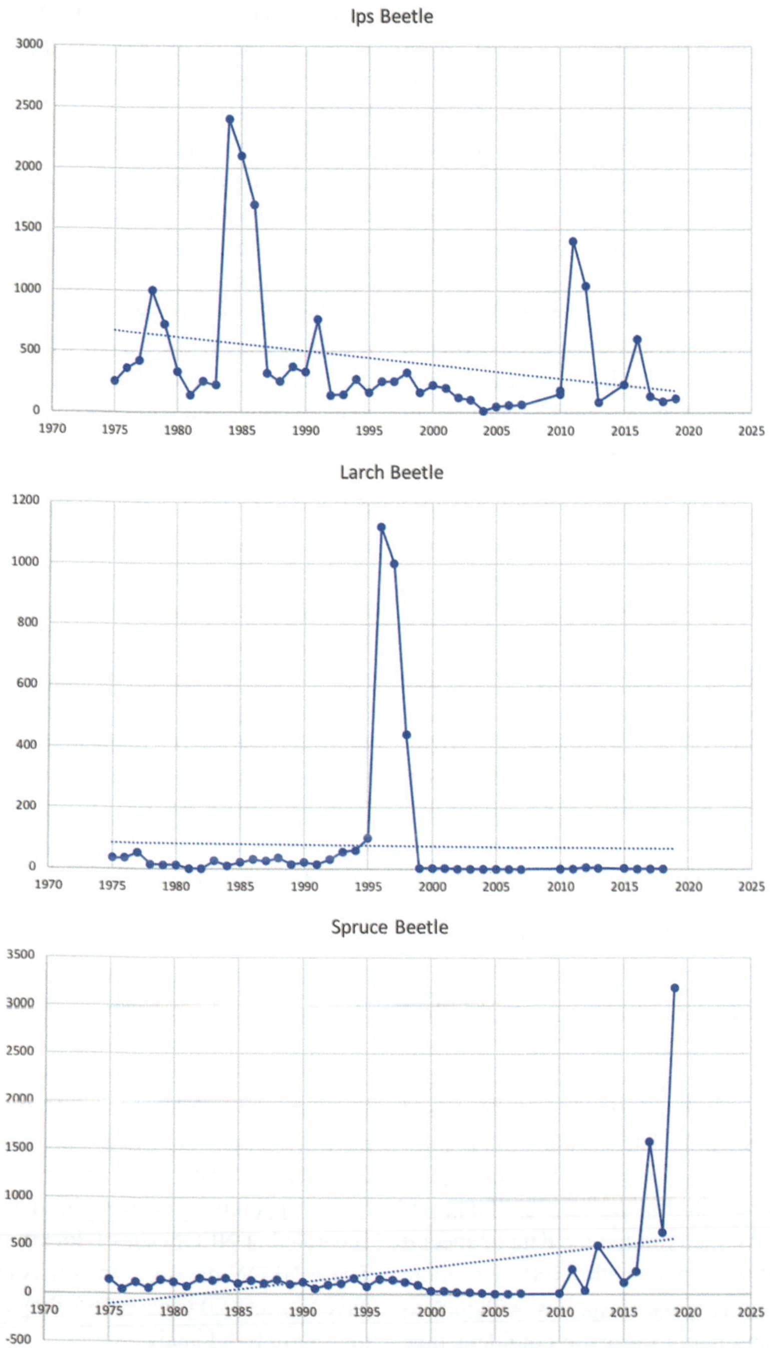

Fig. 4.48 Bonanza Creek Long Term Ecological Research site (LTER), interior Alaska, Bark Beetle data for 1975–2019 (including 7 years of recent data missing from Crossley et al. (2020) who categorized this dataset as showing decreases in abundance). Test for slope of linear regression differing from zero using R version 4.0.0 (2020-04-24). Ips sp. F-statistic: 3.263 on 1 and 41 DF, p-value: 0.07818, Larch beetle (*Dendroctonus simplex*) F-statistic: 0.02099 on 1 and 40 DF, p-value: 0.8855, Spruce beetle (*Dendroctonus rufipennis*) F-statistic: 7.232 on 1 and 41 DF, p-value: 0.01031

Initiative 2012). Declines of lesser magnitude in European aerial insectivores have been documented and numerous explanations explored, with a decline in insects being among the likely causes, in addition to agricultural intensification and habitat loss in the birds' overwintering sites, among others (Rioux Paquette et al. 2014; Bowler et al. 2019; Møller 2019).

Wagner (2020) described the difficulty of assessing the status of insect populations because most can fluctuate dramatically from year to year. Long-term datasets that can detect changes in such dramatically fluctuating populations are rare. Many datasets that show declines indicate the rate of decline is slow enough (1–2% per year) that short-term datasets would not have sufficient statistical power to detect them. As a case in point, a recent analysis of insect data from various long-term ecological research sites (LTERs) in the United States, with datasets spanning 4–36 years, reported no net decline in insect abundance or diversity (Crossley et al. 2020). However, the average time span for the 5376 datasets examined by Crossley et al. (2020) was 11.47 years. It is possible they detected no net decline because many of the datasets they examined covered too short a time period. We therefore decided to reanalyze a subset of the data used by Crossley et al. (2020).

Had Crossley et al. (2020) restricted their analyses to only the longest datasets, for example all those with 20 or more years of data, they might have come to a different conclusion. We used Crossley et al.'s online data to address this question by limiting the data to those with time spans 20 years or more, reducing the number of datasets from 22 to 6. We then limited the data to species (lowest identifications) with 100 or more specimens. This eliminated species with potentially insufficient data to provide confident time series trends and reduced the number of rows in the data from over 1000 to 286.

Analysis of the resulting data shows three of the remaining six datasets have increasing abundances and three have decreasing abundances (based on medians of Crossley et al. (2020) slopes of coefficients). Two of the three datasets with increasing abundance included only grasshoppers (Sevilleta LTER, New Mexico and Konza Prairie LTER Kansas), the other included aquatic insect taxa (North Temperate Lakes LTER, Wisconsin). The three decreasing datasets included one with bark beetles (Bonanza Creek LTER, Alaska), one with miscellaneous pitfall trap taxa (Central Arizona-Phoenix LTER), and one with Lepidoptera (Hubbard Brook LTER, New Hampshire).

The bark beetle data used by Crossley et al. (2020) spanned 1975–2012 which is the dataset available via the LTER website. That bark beetles would show population declines with increasing warming does not fit what we know about these beetles (Cudmore et al. 2010; Berg et al. 2006; Berg 2015). So, we obtained the complete Bonanza Creek LTER bark beetle dataset for 1975–2019 from the US Forest Health Protection. Of the three taxa of bark beetles (*Ips* sp., *Dendroctonus rufipennis*, and *Dendroctonus simplex*) two had slopes not significantly different from zero ($p > 0.07$) and one, *Dendroctonus rufipennis*, showed a dramatic increase in counts in recent years and a slope significantly different from zero ($p = 0.0103$) (Fig. 4.48). The Forest Health Protection personnel who ran this trapping program expressed concerns that trapping effort had not been consistent across the full-time span and some misidentifications may have happened. Despite these concerns, with the most recent data included, this LTER site would not be counted among those with insect population declines.

The only replication in the analysis is the two grasshopper datasets. It would be preferable to have the same taxa and sampling methods in each dataset from many different LTERs, such as is happening with the pitfall trap data being generated at NEON sites (Hoekman et al. 2017). These NEON datasets are still young so we will have to wait ~20 years before they can be used for long-term trend analyses.

This point was focused on by Welti et al. (2020), who also carefully reviewed the datasets from the 13 LTERs used by Crossley et al. (2020). They found problems with data from 7 of the 13 sites. Among the datasets Welti et al. (2020) found problems with were the five non-bark beetle datasets of the six we re-analyzed. They found Crossley et al. (2020)

had failed to account for inconsistent sampling effort and methods over time. Thus, all six datasets in Crossley et al. (2020) with the longest time spans, were found to be problematic. Welti et al. (2020) concluded that the results of Crossley et al. (2020) were not reliable.

But perhaps the most important take-away from our and Welti et al.'s reanalysis of Crossley et al. (2020) is not that these datasets, even if they were reliable, show "no net declines" but rather that there are winners and losers. This is not surprising (Wagner 2020) and thus makes for a less news worthy title. Unfortunately, many people will read the title of Crossley et al. (2020) and mistake the work as good news indicating the authors found no significant change to the insect fauna of the United States. A more careful reading of the work and exploration of the data leads to a different conclusion.

We add a personal comment on insect declines regarding the beetle fauna of Alaska. Intensive arthropod inventory work began in Alaska in 2006 to enlarge the University of Alaska Museum Insect Collection. Given the relatively small fauna of the state (1205 beetle species known in 1991 (Bousquet 1991) and 1448 species known in 2013 (Bousquet et al. 2013)) it should not have taken long before most species were represented in the collection. After 14 years of work the collection has 1091 species represented but these are only ~61% of the currently 1778 known beetle species of the state (Sikes et al. 2017a, b). It has taken a lot more time and effort than expected to get to this point, and given the nature of sampling, it will only get harder as returns on effort diminish and new records for the collection and state become harder to find. This would be true even if arthropod populations in Alaska are not declining (which is currently unknown).

One explanation that we are increasingly entertaining, in light of the Anthropocene defaunation (Dirzo et al. 2014), is that Alaskan beetle populations and diversity may have been significantly larger in the past. Notable declines in beetles, attributable to climate warming, have been detected in intact forests of northeastern North America between the 1970s and the 2010s (Harris et al. 2019). Perhaps the intuitive assumption that beetle populations in a region are more or less constant over time, which might have been true once, is now false?

Geographic range shifts Numerous studies have documented or modeled geographic range shifts as species with sufficient dispersal abilities track their optimal temperatures by moving to higher elevations and / or latitudes (Chen et al. 2011; Breed et al. 2013; Lenoir and Svenning 2015; Kerr et al. 2015; Warren et al. 2018). For many species with sufficient dispersal abilities, their success at tracking their preferred habitats will be influenced by the suitability of contiguous habitat through which they can disperse. Many regions with the most intact ecosystems are surrounded by highly altered (e.g. agricultural and urban environments) and / or mostly unsuitable habitats (e.g. oceans or land), which will limit the number of species, even among those with good dispersal abilities, that can successfully track warming while remaining in their preferred habitats as they disperse. Unfortunately, many of the species that are most adept at dispersing, and can tolerate the widest range of conditions, are predisposed towards becoming pestiferous invasive species.

The Arctic and Subarctic have relatively few non-native arthropod species (Simpson et al. 2019; Langor et al. 2008), compared to similarly sized regions closer to the equator. As the climate of northernmost latitudes warms, we expect to see an increase in the rate of establishment of non-native species in higher latitude regions. Of particular concern are non-native species of medical or economic importance, such as ticks (Durden et al. 2016). The tick, *Dermacentor variabilis,* a vector of the bacteria that cause Rocky Mountain spotted fever and tularemia, is non-native to Alaska and has apparently recently established in the state making it the first tick of medical concern in the region (Durden et al. 2016). The mosquito, *Culex tarsalis*, which is a known vector of West Nile Virus, Western Equine Encephalitis, and St. Louis Encephalitis, was recently recorded for the first time from Yukon Territory, Canada, although no evidence of these pathogens has been found in the Yukon (Peach 2019). This mosquito has also been recorded recently for the first time from Subarctic Alaska by the NEON project (GBIF 2020) making it the first mosquito in Alaska of potential medical concern. However, it is unknown if this mosquito has always been present in Alaska and escaped detection in prior survey efforts. Nonnative earthworms, although not arthropods, are also invading the boreal forest due to human introductions where they were previously absent (Reynolds 2015; Booysen et al. 2018). Some earthworms are considered ecosystem engineers and detrimental to previously worm-free forests. This is due to their tendency to accelerate the decomposition of boreal forest leaf litter which disrupts plant-fungal relationships thereby reducing the fitness of native plants and making it easier for non-native plants to take hold (Bohlen et al. 2004; Hale et al. 2006).

Not all species will be able to disperse such that they can track suitable habitat as climate change causes it to shift poleward. There are conservation concerns for species with limited dispersal abilities and small geographic ranges. For example, many of the more than three hundred potential endemic arthropods of Alaska are flightless, soil-and-litter-dwelling species with limited dispersal abilities (Sikes and Allen 2016). Species with such limited geographic ranges in the Arctic or Subarctic, potentially found nowhere else on Earth and likely unable to track their preferred temperatures, should be high priorities for study from a conservation perspective. However, little is known about the ecology or true distributions of these species. For example, *Chionotyphlus alaskensis* Smetana (1986) is a genus of blind and flightless leptotyphline staphylinid endemic to interior Alaska about which we know very little ecologically, aside from the habitats of the few sites from which it has been collected. Its closest phylogenetic and geographic relative is thought to be in California and it is thought to have survived the Pleistocene in the Beringian glacial refugium (Smetana 1986; Sikes and Allen 2016). Some of these potentially endemic flightless species lack scientific names (Sikes, unpub. data). If these species are to receive conservation attention they must first be formally named. This emphasizes the importance of taxonomy as the first step towards both understanding the role of high latitude arthropods and conserving this biodiversity (Høye and Sikes 2013). As George Barrowclough eloquently stated "A prerequisite to making any decisions concerning the preservation of populations, species, or higher taxa is knowledge of their existence" (Eldredge 1992: 124).

There are conservation concerns for species that have already dispersed to the highest latitudes or elevations they can reach. There are alpine snow-field and glacier associated insect species that may not survive long under a business-as-usual warming scenario (Kavanaugh and Schoville 2009; Muhlfeld et al. 2011) or may already be extinct (Mullen et al. 2018). Alpine snowfields are declining rapidly (Scalzitti et al. 2016) as are glaciers (Cauvy-Fraunié and Dangles 2019).

So again, there will be winners (at least temporarily), as some species expand their populations or ranges into new territories, and losers, as the ranges and populations of others shrink and disappear. There are many reasons to believe we will not stay on the business-as-usual climate change trajectory for the next three centuries, so hopefully the worst-case scenarios for ecosystem disruption and biodiversity loss will be avoided. However, we have already committed to a significantly different climate than humans have lived in for the past ten thousand years and thousands of species are already going extinct annually (Dirzo et al. 2014). Urgent action is needed to minimize the damage.

References

Barber VA, Juday GP, Finney BP (2000) Reduced growth of Alaskan white spruce in the twentieth century from temperature-induced drought stress. Nature 405:668–673

Barnosky AD, Matzke N, Tomiya S, Wogan GO, Swartz B, Quental TB, Marshall C, McGuire JL, Lindsey EL, Maguire KC, Mersey B (2011) Has the Earth's sixth mass extinction already arrived? Nature 471(7336):51–57

Berg E (2015) Warm summers prepare for spruce bark beetle return. USFWS Kenai National Wildlife Refuge Notebook 17(37):71–72

Berg EE, Henry JD, Fastie CL, De Volder AD, Matsuoka SM (2006) Spruce beetle outbreaks on the Kenai Peninsula, Alaska, and Kluane National Park and Reserve, Yukon Territory: relationship to summer temperatures and regional differences in disturbance regimes. For

Ecol Manag 227:219–232. https://doi.org/10.1016/j.foreco.2006.02.038

Bohlen PJ, Scheu S, Hale CM, McLean SM, Groffman PM, Parkinson D (2004) Non-native invasive earthworms as agents of change in northern temperate forests. Front Ecol Environ 2:427–435. https://doi.org/10.1890/1540-9295(2004)002[0427:NIEAAO]2.0.CO;2

Bonan G (1992) A simulation analysis of environmental factors and ecological processes in north American boreal forests. In: Shugart HH, Leemans R, Bonan GB (eds) A systems analysis of the global boreal forest. Cambridge University Press, New York. https://doi.org/10.1017/cbo9780511565489.018

Booysen M, Sikes D, Bowser M, Andrews R (2018) Earthworms (Oligochaeta: Lumbricidae) of interior Alaska. Biodivers Data J 6:e27427. https://doi.org/10.3897/BDJ.6.e27427

Bousquet Y (ed) (1991) Checklist of beetles of Canada and Alaska. Agriculture Canada, Ottawa. vi + 430 pp.

Bousquet Y, Bouchard P, Davies AE, Sikes DS (2013) Checklist of beetles (Coleoptera) of Canada and Alaska. Second edition. Pensoft Series Faunistica No 109, Sofia-Bulgaria, 402 pp. ISSN 1312-0174, ISBN 978-954-642-704-5

Bowden JJ, Hansen OL, Olsen K, Schmidt NM, Høye TT (2018) Drivers of inter-annual variation and long-term change in high-Arctic spider species abundances. Polar Biol 41(8):1635–1649. https://doi.org/10.1007/s00300-018-2351-0

Bowler DE, Heldbjerg H, Fox AD, de Jong M, Böhning-Gaese K (2019) Long-term declines of European insectivorous bird populations and potential causes. Conserv Biol 33(5):1120–1130

Breed GA, Stichter S, Crone EE (2013) Climate-driven changes in northeastern US butterfly communities. Nat Clim Chang 3(2):142–145

Cardoso P, Barton PS, Birkhofer K, Chichorro F, Deacon C, Fartmann T, Fukushima CS, Gaigher R, Habel JC, Hallmann CA, Hill MJ (2020) Scientists' warning to humanity on insect extinctions. Biol Conserv 242:108426. https://doi.org/10.1016/j.biocon.2020.108426

Cauvy-Fraunié S, Dangles O (2019) A global synthesis of biodiversity responses to glacier retreat. Nat Ecol Evol 3(12):1675–1685

Ceballos G, Ehrlich PR, Barnosky AD, García A, Pringle RM, Palmer TM (2015) Accelerated modern human–induced species losses: entering the sixth mass extinction. Sci Adv 1(5):E1400253

Ceballos G, Ehrlich PR, Dirzo R (2017) Biological annihilation via the ongoing sixth mass extinction signaled by vertebrate population losses and declines. Proc Natl Acad Sci 114(30):E6089–E6096

Chapin FS, McGuire AD, Ruess RW, Hollingsworth TN, Mack MC, Johnstone JF, Kasischke ES, Euskirchen ES, Jones JB, Jorgenson MT, Kielland K (2010) Resilience of Alaska's boreal forest to climatic change. Can J For Res 40(7):1360–1370

Chapman SC, Murphy EJ, Stainforth DA, Watkins NW (2020) Trends in winter warm spells in the central England temperature record. J Appl Meteorol Climatol 59(6):1069–1076

Chen I-C, Hill JK, Ohlemuller R, Roy DB, Thomas CD (2011) Rapid range shifts of species associated with high levels of climate warming. Science 333(6045):1024–1026. https://doi.org/10.1126/science.1206432

Crossley MS, Meier AR, Baldwin EM, Berry LL, Crenshaw LC, Hartman GL, Lagos-Kutz D, Nichols DH, Patel K, Varriano S, Snyder WE (2020) No net insect abundance and diversity declines across US long term ecological research sites. Nat Ecol Evol 4:1368–1376. https://doi.org/10.1038/s41559-020-1269-4

Cudmore TJ, Björklund N, Carroll AL, Staffan Lindgren B (2010) Climate change and range expansion of an aggressive bark beetle: evidence of higher beetle reproduction in naïve host tree populations. J Appl Ecol 47 (5):1036–1043

Deutsch CA, Tewksbury JJ, Huey RB, Sheldon KS, Ghalambor CK, Haak DC, Martin PR (2008) Impacts of climate warming on terrestrial ectotherms across latitude. Proc Natl Acad Sci 105(18):6668–6672

Dial RJ, Berg EE, Timm K, McMahon A, Geck J (2007) Changes in the alpine forest-tundra ecotone commensurate with recent warming in southcentral Alaska: evidence from orthophotos and field plots: tree line changes in southcentral Alaska. J Geophys Res Biogeo 112(G4). https://doi.org/10.1029/2007JG000453

Dirzo R, Young HS, Galetti M, Ceballos G, Isaac NJB, Collen B (2014) Defaunation in the Anthropocene. Science 345:401–406

Dunn RR, Harris NC, Colwell RK, Koh LP, Sodhi NS (2009) The sixth mass coextinction: are most endangered species parasites and mutualists? Proc R Soc B Biol Sci 276(1670):3037–3045

Durden LA, Beckmen KB, Gerlach RF (2016) New records of ticks (Acari: Ixodidae) from dogs, cats, humans, and some wild vertebrates in Alaska: invasion potential. J Med Entomol 53(6):1391–1395

Eldredge N (1992) Systematics, ecology, and the biodiversity crisis. Columbia University Press, New York

Forrest JR (2016) Complex responses of insect phenology to climate change. Curr Opin Insect Sci 17:49–54

GBIF.org (08 Oct 2020) GBIF occurrence download. https://doi.org/10.15468/dl.xbe734

Gillespie MAK, Alfredsson M, Barrio IC, Bowden JJ, Convey P, Culler LE, Coulson SJ, Krogh PH, Koltz AM, Koponen S, Loboda S, Marusik Y, Sandström JP, Sikes DS, Høye TT (2019) Status and trends of terrestrial arthropod abundance and diversity in the North Atlantic region of the Arctic. Ambio 49:718–731. https://doi.org/10.1007/s13280-019-01162-5

Hale CM, Frelich LE, Reich PB (2006) Changes in hardwood forest understory plant communities in response to European earthworm invasions. Ecology 87:1637–1649. https://doi.org/10.1890/0012-9658(2006)87[1637:CIHFUP]2.0.CO;2

Hallmann CA, Sorg M, Jongejans E, Siepel H, Hofland N et al (2017) More than 75 percent decline over 27 years in total flying insect biomass in protected areas. PLoS One 12(10):e0185809

Harris JE, Rodenhouse NL, Holmes RT (2019) Decline in beetle abundance and diversity in an intact temperate forest linked to climate warming. Biol Conserv 240:108219

Harvey JA, Heinen R, Gols R, Thakur MP (2020) Climate change-mediated temperature extremes and insects: from outbreaks to breakdowns. Glob Chang Biol 26 (12):6685–6701. https://doi.org/10.1111/gcb.15377

Helbig M, Pappas C, Sonnentag O (2016) Permafrost thaw and wildfire: equally important drivers of boreal tree cover changes in the Taiga Plains, Canada. Geophys Res Lett 43(4):1598–1606

Hodkinson ID (2013) Terrestrial and freshwater invertebrates. In: Meltofte H (ed) Arctic biodiversity assessment, status and trends in Arctic biodiversity. CAFF, Akureyri, pp 195–223

Hoekman D, LeVan KE, Gibson C, Ball GE, Browne RA, Davidson RL, Erwin TL, Knisley CB, LaBonte JR, Lundgren J, Maddison DR (2017) Design for ground beetle abundance and diversity sampling within the National Ecological Observatory Network. Ecosphere 8(4):e01744

Hogg EH, Hurdle PA (1995) The aspen parkland in western Canada: a dry-climate analogue for the future boreal forest? Water Air Soil Pollut 82(1–2):391–400

Høye TT, Forchhammer MC (2008) Phenology of high-arctic arthropods: effects of climate on spatial, seasonal and inter-annual variation. Adv Ecol Res 40:299–324

Høye TT, Sikes DS (2013) Arctic entomology in the 21st century. Can Entomol 145:125–130. https://doi.org/10.4039/tce.2013.14

Høye TT, Post E, Meltofte H, Schmidt NM, Forchhammer MC (2007) Rapid advancement of spring in the high Arctic. Curr Biol 17:R449–R451

Høye TT, Eskildsen A, Hansen RR, Bowden JJ, Schmidt NM, Kissling WD (2014) Phenology of high-arctic butterflies and their floral resources: species-specific responses to climate change. Curr Zool 60(2):243–251

Hunter MD, Kozlov MV, Itämies J, Pulliainen E, Bäck J, Kyrö EM, Niemelä P (2014) Current temporal trends in moth abundance are counter to predicted effects of climate change in an assemblage of subarctic forest moths. Glob Chang Biol 20(6):1723–1737

IPCC (2014) IPCC, 2014: climate change 2014: synthesis report. Contribution of working groups I, II and III to the fifth assessment report of the intergovernmental panel on climate change. IPCC, Geneva

Janzen DH, Hallwachs W (2019) Where might be many tropical insects? Biol Conserv 233:102–108

Juday GP, Alix C, Grant TA III (2015) Spatial coherence and change of opposite white spruce temperature sensitivities on floodplains in Alaska confirms early-stage boreal biome shift. For Ecol Manag 350 (2015):46–61. https://doi.org/10.1016/j.foreco.2015.04.016

Kavanaugh DH, Schoville SD (2009) A new and endemic species of *Nebria* Latreille (Insecta: Coleoptera: Carabidae: Nebriini), threatened by climate change in the Trinity Alps of northern California. Proc Calif Acad Sci 60(1):73–84

Kelly R, Chipman ML, Higuera PE, Stefanova I, Brubaker LB, Hu FS (2013) Recent burning of boreal forests exceeds fire regime limits of the past 10,000 years. Proc Natl Acad Sci 110(32):13055–13060

Kerr JT, Pindar A, Galpern P, Packer L, Potts SG, Roberts SM, Rasmont P, Schweiger O, Colla SR, Richardson LL, Wagner DL, Gall LF, Sikes DS, Pantoja A (2015) Climate change impacts on bumblebees converge across continents. Science 349(6244):177–180

Langor DW, DeHaas LJ, Foottit RG (2008) Diversity of non-nativeterrestrial arthropods on woody plants in Canada. In: Langor D, Sweeney J (eds) Ecological impacts of non-native invertebrates and Fungi on terrestrial ecosystems. Springer, Dordrecht, pp 5–19

Lenoir J, Svenning JC (2015) Climate-related range shifts–a global multidimensional synthesis and new research directions. Ecography 38(1):15–28

Lloyd AH, Fastie CL (2003) Recent changes in treeline forest distribution and structure in interior Alaska. Écoscience 10(2):176–185. https://doi.org/10.1080/11956860.2003.11682765

Loboda S, Savage J, Buddle CM, Schmidt NM, Høye TT (2018) Declining diversity and abundance of high Arctic fly assemblages over two decades of rapid climate warming. Ecography 41(2):265–277

Møller AP (2019) Parallel declines in abundance of insects and insectivorous birds in Denmark over 22 years. Ecol Evol 9(11):6581–6587

Montgomery GA, Dunn RR, Fox R, Jongejans E, Leather SR, Saunders ME, Shortall CR, Tingley MW, Wagner DL (2020) Is the insect apocalypse upon us? How to find out. Biol Conserv 241:108327

Muhlfeld CC, Giersch JJ, Hauer FR, Pederson GT, Luikart G, Peterson DP, Downs CC, Fagre DB (2011) Climate change links fate of glaciers and an endemic alpine invertebrate. Clim Chang 106 (2):337–345

Mullen LJ, Campbell JM, Sikes DS (2018) Taxonomic revision of the rove beetle genus *Phlaeopterus* Motschulsky, 1853 (Coleoptera: Staphylinidae: Omaliinae: Anthophagini). Coleopterists Bull 16:1–54. https://doi.org/10.1649/0010-065X-72.mo4.1. Patricia Vaurie Series Monograph

Myers-Smith IH, Forbes BC, Wilmking M, Hallinger M, Lantz T, Blok D, Tape KD, Macias-Fauria M, Sass-Klaassen U, Lévesque E, Boudreau S (2011) Shrub expansion in tundra ecosystems: dynamics, impacts and research priorities. Environ Res Lett 6(4):045509

Nebel S, Mills A, McCracken J, Taylor P (2010) Declines of aerial insectivores in North America follow a geographic gradient. Avian Conserv Ecol 5(2). https://doi.org/10.5751/ACE-00391-050201

North American Bird Conservation Initiative (2012) The state of Canada's birds, 2012. Environment Canada, Ottawa, ON

Park T, Ganguly S, Tømmervik H, Euskirchen ES, Høgda KA, Karlsen SR, Brovkin V, Nemani RR, Myneni RB (2016) Changes in growing season duration and productivity of northern vegetation inferred from long-term remote sensing data. Environ Res Lett 11 (8):084001

Peach DA (2019) First Record of *Culex tarsalis* (Diptera: Culicidae) in the Yukon. J Entomol Soc B C 115:123–125

Post E, Forchhammer MC, Bret-Harte MS, Callaghan TV, Christensen TR, Elberling B, Fox AD, Gilg O, Hik DS, Høye TT, Ims RA (2009) Ecological dynamics across the Arctic associated with recent climate change. Science 325(5946):1355–1358

Raven PH, Miller SE (2020) Here today, gone tomorrow. Science 370(6513):149. https://doi.org/10.1126/science.abf1185

Renner SS, Zohner CM (2018) Climate change and phenological mismatch in trophic interactions among plants, insects, and vertebrates. Annu Rev Ecol Evol Syst 49:165–182

Reynolds JW (2015) A checklist of earthworms (Oligochaeta: Lumbricidae and Megascolecidae) in western and northern Canada. Megadrilogica 17:141–156

Rioux Paquette S, Pelletier F, Garant D, Bélisle M (2014) Severe recent decrease of adult body mass in a declining insectivorous bird population. Proc R Soc B Biol Sci 281(1786):20140649

Sánchez-Bayo F, Wyckhuys KA (2019) Worldwide decline of the entomofauna: a review of its drivers. Biol Conserv 232:8–27

Scalzitti J, Strong C, Kochanski A (2016) Climate change impact on the roles of temperature and precipitation in western US snowpack variability. Geophys Res Lett 43 (10):5361–5369

Seibold S, Gossner MM, Simons NK, Blüthgen N, Müller J, Ambarlı D, Ammer C, Bauhus J, Fischer M, Habel JC, Linsenmair KE (2019) Arthropod decline in grasslands and forests is associated with landscape-level drivers. Nature 574(7780):671–674

Serreze MC, Walsh JE, Chapin FS, Osterkamp T, Dyugerov M, Romanovsky V, Oechel WC, Morison J, Zhang T, Barry RG (2000) Observational evidence of recent climate change in the northern high-latitude environment. Climate Change 46:159–207

Serreze MC, Barrett AP, Stroeve JC, Kindig DN, Holland MM (2009) The emergence of surface-based Arctic amplification. Cryosphere 3(1):11–19

Sikes DS (2019) Arthropods potentially associated with spruce (*Picea* spp.) in interior Alaska. Newslett Alaska Entomol Soc 12(1):2–5. https://doi.org/10.7299/X7RR1ZJT

Sikes DS, Allen RT (2016) First Alaskan records and a significant northern range extension for two species of Diplura (Diplura, Campodeidae). ZooKeys 563:147–157. https://doi.org/10.3897/zookeys.563.6404

Sikes DS, Bowser M, Morton JM, Bickford C, Meierotto S, Hildebrandt K (2017a) Building a DNA barcode library of Alaska's non-marine arthropods. Genome 60:248–259. https://doi.org/10.1139/gen-2015-0203

Sikes DS, Bowser M, Daly K, Høye TT, Meierotto S, Mullen L, Slowik J, Stockbridge J (2017b) The value of museums in the production, sharing, and use of entomological data to document hyperdiversity of the changing north. Arctic Sci 3:498–514. https://doi.org/10.1139/AS-2016-0038

Simmons BI, Balmford A, Bladon AJ, Christie AP, De Palma A, Dicks LV, Gallego-Zamorano J, Johnston A, Martin PA, Purvis A, Rocha R (2019) Worldwide insect declines: an important message, but interpret with caution. Ecol Evol 9(7):3678–3680

Simpson A, Eyler MC, Sikes D, Bowser M, Sellers E, Guala GF, Cannister M, Libby R, Kozlowski N (2019) A comprehensive list of non-native species established in three major regions of the United States: Version 2.0, (ver. 3.0, 2020). U.S. Geological Survey data release. https://doi.org/10.5066/P9E5K160

Smetana A (1986) *Chionotyphlus alaskensis* n. g., n. sp., a tertiary relict from unglaciated interior Alaska (Coleoptera, Staphylinidae). Nouvelle Revue d'Entomologie (N Sér) 3(2):171–187

Stork NE (2018) How many species of insects and other terrestrial arthropods are there on Earth? Annu Rev Entomol 63:31–45

Thomas C, Cameron A, Green R et al (2004) Extinction risk from climate change. Nature 427:145–148. https://doi.org/10.1038/nature02121

Urban MC (2015) Accelerating extinction risk from climate change. Science 348(6234):571–573

Wagner DL (2020) Insect declines in the Anthropocene. Annu Rev Entomol 65:457–480

Warren R, Price J, Graham E, Forstenhaeusler N, VanDerWal J (2018) The projected effect on insects, vertebrates, and plants of limiting global warming to 1.5 C rather than 2 C. Science 360(6390):791–795

Welti E, Joern A, Lightfoot DC, Record S, Rodenhouse N, Stanley EH, Kaspari M (2020) Meta-analyses of insect temporal trends must account for the complex sampling histories inherent to many long-term monitoring efforts. EcoEvoRxiv Preprint. https://ecoevorxiv.org/v3sr2/

Wepprich T, Adrion JR, Ries L, Wiedmann J, Haddad NM (2019) Butterfly abundance declines over 20 years of systematic monitoring in Ohio, USA. PLoS One 14(7): e0216270

Westerhold T, Marwan N, Drury AJ, Liebrand D, Agnini C, Anagnostou E, Barnet JS, Bohaty SM, De Vleeschouwer D, Florindo F, Frederichs T (2020) An astronomically dated record of Earth's climate and its predictability over the last 66 million years. Science 369(6509):1383–1387. https://doi.org/10.1126/science.aba6853

5 Aleocharine Beetles as Indicators of Environmental Change

The rove beetle family Staphylinidae is the most biologically diverse family of Coleoptera in North America north of Mexico (Poole and Gentili 1996), with over 4000 valid names (Newton et al. 2001). The world fauna consists of more than 63,495 described species of rove beetles, with over 1774 species recorded in Canada (Brunke et al. 2019). The subfamily Aleocharinae is the largest and comprises, depending on habitat and spatial distribution, about 30–33% of species in the family (Paquin and Dupérré 2001). There are over 600 valid aleocharine species in Canada but their true number including undescribed and unrecorded species may exceed 700 (Klimaszewski et al. 2018a, 2020). The species occur in all habitat types with the exception of extremely dry habitats (Newton et al. 2001).

The majority of aleocharine species are forest dwellers where they occur in leaf litter, under bark, in fungi, in moss, and in the nests of ants, termites, mammals, and birds. Elsewhere, they inhabit arctic and subarctic habitats, seashores, edges of water bodies, wetlands and prairies (Klimaszewski et al. 2018a). Many rove beetles (e.g. most Aleocharinae, Staphylininae, Paederinae, Scydmaeninae) are predators of other arthropods, but examples of other feeding relationships such as fungivory and detrivory are known and well documented. Aleocharine subtribe Gyrophaenina embraces species that are exclusively mycophagous, feeding on fungal spores and hyphae (Ashe 1984). However, information on feeding habits of many rove beetles is still expanding. For example, many species of the aleocharine tribe Athetini were historically considered predaceous, but this assumption was based on little empirical evidence. Recent molecular and microscopic analysis of the gut contents of abundant rove beetles (indicators of mature forests), including several athetine species in the boreal balsam fir forest of eastern Canada, revealed a lack of arthropod fragments or arthropod DNA in their posterior guts but presence of a diverse fungal community dominated by yeasts (Klimaszewski et al. 2013; Stefani et al. 2016). Predominance of yeasts and fungal spores in the posterior gut of rove beetles suggests that fungi may play an important role in rove beetle dietary requirements and as endosymbionts (Klimaszewski et al. 2013; Stefani et al. 2016). Aleocharine rove beetles represent key ecological indicators of changes in managed forests due to their highly specific microhabitat associations, their response to disturbances, their high local abundance, and their ease of collection (Pohl et al. 2008; Venier et al. 2017; Klimaszewski et al. 2018b). Often, rove beetles represent a large proportion of the total abundance and richness of epigeic organisms in biomonitoring studies. Unfortunately, rove beetle assemblages, and especially the Aleocharinae, are neglected in many studies because of the difficulties in identifying species. This may change with

J. Klimaszewski et al., *A Faunal Review of Aleocharine Beetles in the Rapidly Changing Arctic and Subarctic Regions of North America (Coleoptera, Staphylinidae)*,
https://doi.org/10.1007/978-3-030-68191-3_5

appropriate taxonomic tools including high resolution images of morphology and genomics (Klimaszewski et al. 2018b).

As indicators, the presence or absence of particular taxa of these beetles and their relative abundance may reveal details about the overall health of ecosystems and the impact of changing environment (e.g., caused by human factors or climate change) on them (Venier et al. 2017; Klimaszewski et al. 2018b, 2020). Thus far, in boreal ecosystems, the value of rove beetles has been assessed primarily in terms of their responses to stand-replacing wildfire, clear-cut, and forestry harvesting, and response to operational oil sands mine reclamation in northeastern Alberta (Hammond et al. 2018). However, given the apparent affinity of rove beetles for microhabitats, increasing attention has been focused on smaller-scale disturbances and silvicultural interventions that may leave residual standing trees, create small gaps and/or modify soils or deadwood (Klimaszewski et al. 2008, 2018b). At the landscape scale, species composition of rove beetles is dictated by vegetation type and other factors, and majority of species have strong affinities with forest microhabitats where moisture, temperature, substrate type, and debris accumulation are variables that promote high diversity (Pohl et al. 2008).

Small-scale disturbance events, such as gap dynamics, create heterogeneity in light and soil nutrient availability (Paré et al. 1993), and are important in the recruitment of structural features such as downed woody debris (Siitonen 2001). Ultimately, gaps influence future stand composition, typically by maintaining early-successional deciduous species that would otherwise be competitively replaced during succession (Lieffers et al. 1996; Kneeshaw and Bergeron 1998). Stand heterogeneity created by gaps is thought to have additional benefits for biodiversity and generally is more socially acceptable. Changes in the abundance, species richness and community composition of rove beetles were evaluated in response to ground scarification, three configurations of experimental gap cuts and clearcutting in an early-succession yellow birch-dominated boreal forest in Quebec, Canada (Klimaszewski et al. 2008). Rove beetle composition responded significantly to increasing gap size. Composition among intermediate-and small-sized gap treatments was more similar to that of uncut control forests than were those of larger gap treatments and clear-cuts. Differences in rove beetle assemblage (defined primarily by a loss of closed-canopy species and an increase in species with an affinity for open habitats) relative to control plots were more pronounced in the largest (two-gap treatments) and clear-cuts than in smaller ones (four- and eight-gap treatments). These assemblage differences were made greater by scarification. Assemblages of rove beetles were less affected by numerous small gaps with increased edge relative to fewer larger gaps with less edge. It may be that the smaller gaps, (8 gap treatment, each gap 28 m in diameter) were perceived by forest specialists as an extreme case of stand heterogeneity rather than perceptible gaps as rove beetles were much more abundant there than in larger gaps (Klimaszewski et al. 2008). Smaller gaps may also be considered more permeable for dispersion than larger gaps. Of the 116 species represented, 6 species dominated the total catch: *Tachinus fumipennis* (Say) (55.4% of the total catch), *Oxypoda convergens* Casey (10.6%), *Atheta pseudomodesta* Klimaszewski (6.6%), *Placusa tachyporoides* (Waltl) (3.7%), *Tachinus addendus* Horn (3.5%) and *Bisnius blandus* (Gravenhorst) (2.5%) (Klimaszewski et al. 2008). Together these species represented over 82% of the overall catch, but there were 27 additional significant indicator species associated with different treatments (Klimaszewski et al. 2008). When the total area of forest removed was held constant, smaller, more numerous gaps were more similar to uncut control stands than to larger gaps and fell more closely within the natural forest heterogeneity. The smaller and more numerous gaps were a better emulation of natural stand dynamics than the larger gaps and clear-cuts.

In lightly scarified plots, rove beetle richness was greater than in non-scarified plots. This may seem counter-intuitive, however, when applied to small gaps, scarification may actually increase overall habitat heterogeneity at the stand level by creating recently disturbed soils in close proximity to intact closed-canopy forest. Scarification also altered overall composition of rove beetles through an interaction with gap size. Scarification effectively nullified any compositional differences attributed to larger gaps, thus making large, scarified gaps indistinguishable from clearcuts (Klimaszewski et al. 2008).

Two studies using rove beetles were recently conducted to evaluate the impacts of different intensities of postharvest slash removal on rove beetle abundance, species richness and community composition: one in jack pine forest of the Island Lake Biomass Harvest Research and Demonstration Area, also known as Island Lake Site (ILS), in Ontario (Venier et al. 2017; Work et al. in preparation), and the other in a balsam fir-white birch forest at the Montmorency Teaching and Research Forest (MTRF), Quebec (Work et al. 2013; Klimaszewski et al. unpublished). The ILS project evaluated changes in species composition of litter-dwelling arthropod taxa, including rove beetles, in response to a gradient of forest disturbance including the following treatments, from the least to the most disturbed: 85-year-old postfire forest (old control), 45-year-old postharvest plantation forest (young control), 3-year postfire treatment (fire control), 3-year post stem-only harvesting (SOH) with branches and tops left at the stump, 3-year post whole-tree biomass harvesting treatment (WTH) with branches and tops moved to roadside, 3-year post WTH with stumps removed and 3-year post WTH with forest floor removed by blading (Venier et al. 2017). When rove beetle assemblages were compared between uncut stands and harvested stands with increasing removal of biomass, composition was markedly different between older uncut stands (85 years) and all other treatments, including the 45-year-old closed-canopy control. The vast majority of species and individuals were associated with the old-forest treatment, whereas very few individuals were captured in the other treatments. For example, of the 1231 individuals captured, 745 were captured in the five 85-year old control plots, whereas only 486 were captured in the other 30 plots of all other treatments combined. Species strongly associated with older, uncut forest included *Atheta capsularis*, *Atheta strigosula*, *Lordithon fungicola*, *Oxypoda sylvia*, and *Atheta remulsa*, and were found only in very small numbers or not at all in harvested, fire disturbed and young control treatments. Only two species were associated with the harvested sites or younger stands, i.e. *Aleochara fumata* and *Mycetoporus consors* (Venier et al. 2017). *Atheta capsularis*, *A. strigosula*, *L. fungicola* and *O. sylvia* are generally associated with forest macrofungi, while *M. consors* is an open habitat/edge-loving species and has been collected in agroecosystems (e.g., Brunke et al. 2014). *Aleochara fumata* is a more opportunistic and mobile species and its life history depends on the availability of habitats with dipteran larvae and pupae, which are more randomly distributed in the forest. Moist litter is a very important habitat for the majority of rove beetles because their small bodies are prone to desiccation. Removal of the litter layer and exposure of mineral soil likely reduce soil moisture and habitat suitability for rove beetles. Among the harvested treatments, rove beetle composition did not differ between tree-length and full-tree harvesting. This could result from the presence of only a few generalist or accidental species and lack of true microhabitat specialists. Lack of canopy and extensive biomass removal, including the removal of soil organic layers by blading, had strong effects on rove beetles. This treatment dramatically removed much of the habitat where a majority of rove beetles live. Removing the whole organic layer and deadwood from postharvest plots eliminates the buffer zone protecting subsequent layers of ground habitat from reduced humidity and increased temperature.

In a 2011 Quebec study, in a balsam fir-white birch forest at the Montmorency Teaching and Research Forest (MTRF), changes in rove beetle abundance and assemblages were compared between two treatments and a control: clearcuts where branches and non-merchantable portions of the stem were left on site (stem-only harvesting or SOH); whole-tree harvesting (WTH), in which stems, tops and branches are removed; and uncut control plots (C) in about 42-to 60-year-old mature balsam fir stands in Quebec (Work et al. 2013). Overall staphylinid catch rates were much greater in uncut plots (C) than in either SOH or WTH sites. The high overall abundance and differences in assemblage composition in uncut stands were attributable mainly to the presence of *Atheta capsularis*, *A. klagesi*, *A. strigosula* and *Tachinus fumipennis/frigidus* in controls. Among the clearcuts, catch rates in WTH were greater than in SOH sites, where three species (like *Pseudopsis subulata*, *Quedius labradorensis* and, to a lesser extent, *Gabrius brevipennis*) were more abundant when branches and finer woody debris had been removed. These differences between SOH and WTH in balsam fir-dominated stands contrast with the response of rove beetles in jack pine stands at the ILS experiment, where no differences in rove beetles were observed. In 2013 (3-year post-clearcut), rove beetles were resampled at the MRTF experiment, with an inclusion of the additional treatment (ADR), in which all woody debris were removed from new clearcut plots and samples were also taken from different chronosequences of the surrounding forests (11-, 22-, 42- and 73-year old stands) (Klimaszewski et al. in preparation). We collected 2069 specimens and 87 species of Staphylinidae in 10 subfamilies in harvesting treatments and 5291 specimens and 86 species in 12 subfamilies in chronosequence stands. The oldest chronosequence stand (73 years) had a distinct rove beetle community, and the communities in the younger chronosequence stands (11, 22, and 42 years) and in the control (C) stand were similar to each other and distinct from the harvesting treatments. Rove beetle abundance showed little relation to differences in fine or coarse woody material. However, staphylinid species richness was positively affected by higher fine woody debris volume.

In this 3-year post-clearcut resampling, *Atheta capsularis*, *A. klagesi*, *A. ventricosa*, *Liogluta terminalis*, *Lypoglossa franclemonti*, *Placusa tachyporoides*, *Acidota quadrata*, *Proteinus* sp., *Quedius rusticus*, *Q. densiventris* and *Tachinus* spp. (mainly *T. frigidus* and *T. fumipennis*) showed strong affinities for uncut forests, whereas several other species were more or equally abundant in treatments: *Gabrius picipennis* (SOH, ADR, WTH), *G. brevipennis* (SOH, WTH, ADR), *Quedius labradorensis* (SOH, ADR, WTH), *Q. peregrinus* (WTH, SOH, ADR), *Ischnosoma longicorne* (SOH, ADR, WTH), *I. splendidum* (SOH, ADR, WTH), *Parascydmus corpusculus* (SOH, WTH, ADR) and *Pseudopsis subulata* (WTH, SOH, ADR). Harvesting and biomass removal significantly reduced the mean abundance of several species of rove beetles in the subfamilies Aleocharinae, Omaliinae and Tachyporinae and increased or maintained the mean abundance of predatory species of subfamilies Staphylininae and Scydmaeninae, compared with uncut control plots. The species from the latter two subfamilies may find more diverse prey species in harvested forests. However, overall catch of rove beetles was not significantly affected by harvesting treatment, likely because the abundance of tolerant and open habitat species such as *Ischnosoma longicorne*, *Gabrius brevipennis*, *Pseudopsis subulata* and *Parascydmus corpusculus* increased and compensated for the decreased abundance of forest specialists such as *Atheta capsularis*, *A. klagesi* and *Tachinus frigidus* in harvested plots.

Fire is the dominant disturbance in the Nearctic boreal forest (Pohl et al. 2008). It has a profound effect on rove beetles (Pohl et al. 2008). The burn pattern results in a patchwork of different communities at various stages in the successional cycle. In contrast to fire, forest harvesting does not directly destroy the rove beetle community, but to a large extent, it destroys the forest habitat (Pohl et al. 2008). This results in a unique rove beetle community characterized by a mix of forest species and open-ground specialists and overall high diversity in this period of flux (Pohl et al. 2008). In the years following harvesting, the rove beetle community goes through successional changes and becomes more similar to the mature forest community, but it skips the early post fire stage and proceeds along the successional trajectory more rapidly than after fire (Pohl et al. 2008). Other less direct effects of harvesting on rove beetles are a decrease in the proportion of land-base suitable for communities associated with older successional stages, alteration of forests by postharvest site preparations and planting of exotic tree species, edge and fragmentation effects that are detrimental to the remaining forest surrounding harvested areas and an influx of exotic arthropod species with affinities for disturbed sites (Pohl et al. 2008). It was recommended that future work explore the effects of postharvest forestry activities, fragmentation, edges and changes in riparian zones and wetlands on rove beetles in forested habitats.

Several biodiversity studies conducted in partnership with John Spence and associates (e.g., Buddle et al. 2006; Pohl et al. 2007), have shown that Staphylinidae, are a useful indicator group, because they exhibit variation across multiple spatial scales, and they respond to successional gradients and disturbance (Pohl et al. 2008). With continued advances in identification tools such as taxonomic keys with high-resolution photos, aleocharine rove beetles have been increasingly incorporated into biodiversity research projects, thus increasing both the scope and inferences drawn from these studies (Klimaszewski et al. 2018b). Pohl et al. (2007, 2008) identified Aleocharinae from a large set of biodiversity samples, and assessed their usefulness in biodiversity research. In 2004, a large arthropod sampling program was set up in west-central Alberta, to determine how closely arthropod diversity patterns follow the Forest Ecosite Classification (FEC) system of Beckingham et al. (1996). That system, based on vegetation and soil characteristics, is useful for classifying and monitoring plant assemblages within western Canadian forests. It is widely used by the forest industry for management of their land base. In this study, they examined staphylinid and vegetation data from 40 forest stands that have been classified via FEC. The research examines the extent to which arthropod communities can be characterized by the FEC scheme and vegetation community. Aleocharinae exhibited higher differences than for all staphylinids combined, and also had the highest overall average distances. Staphylinids from the subxeric sites and rich mesic sites clustered somewhat along FEC lines, but for other forest types, the groupings evident in the vegetation data were fractured and scattered in the staphylinid data (Pohl et al. 2017). In the indicator value analysis, a total of 29 species had a significant ($p > 0.05$) indicator value of at least 50%. There were 10 Aleocharinae species with significant values. The Bracted Honeysuckle sites had the most indicator species.

The Arctic and Subarctic ecosystems in Canada and Alaska are rapidly changing due to environmental changes associated with warming climate. Northern ecosystems (including tundra, forests and shrublands) are being profoundly impacted, affecting ecosystem integrity and functioning. Much attention has focused on species traditionally utilized by residents of Canada's North (e.g., plants, mammals, birds, fish). However, the vast majority of biodiversity is represented by less conspicuous and non-iconic species. Although relatively unseen, underconsidered, and poorly studied, these species (e.g., invertebrate animals, fungi, lichens) play critical roles that contribute to the functioning of healthy ecosystems. This book is the first attempt to provide baseline comprehensive biodiversity data on aleocharine beetles recorded from the North, including identification tools to enable their use in environmental studies.

References

Ashe JS (1984) Generic revision of the subtribe Gyrophaenina (Coleoptera: Staphylinidae: Aleocharinae) with review of described subgenera and major features of evolution. Quest Entomol 20:129–349

Beckingham JD, Corns IGW, Archibald JH (1996) Field guide to ecosites of west-central Alberta. Canadian Forest Service, Northern Forestry Centre, Edmonton, AB, Special Report 9

Brunke AJ, Bahlai CA, Klimaszewsk J, Hallett RH (2014) Rove beetles (Coleoptera: Staphylinidae) in Ontario, Canada soybean agroecosystems: assemblage diversity, composition, seasonality, and habitat use. Can Entomol 146:652–670

Brunke AJ, Bouchard P, Douglas HB, Pentinsaari M (2019) Coleoptera of Canada. ZooKeys 819:361–376

Buddle CM, Spence JR, Langor DW, Pohl GR (2006) Arthropod responses to harvesting and wildfire: implications for emulation of natural disturbance in forest management. Biol Conserv 128:346–357

Hammond JHE, Hoffman PGK, Pinno BD, Pinzon J, Klimaszewski J, Dustin J, Hartley DJ (2018) Response of ground and rove beetles (Coleoptera: Carabidae, Staphylinidae) to operational oil sands mine reclamation in northeastern Alberta, a case study. J Insect Conserv 22:687–706

Klimaszewski J, Langor D, Work TT, Hammond JHE, Savard K (2008) Smaller and more numerous harvesting gaps emulate natural forest disturbances: a biological test case using rove beetles (Coleoptera, Staphylinidae). Divers Distrib 14:969–982

Klimaszewski J, Morency J-M, Labrie P, Séguin A, Langor D, Work T, Bourdon C, Thiffault E, Pare D, Newton AF, Thayer MK (2013) Molecular and microscopic analysis of the gut contents of abundant rove beetle species (Coleoptera, Staphylinidae) in the boreal balsam fir forest of Quebec, Canada. ZooKeys 353:1–24

Klimaszewski J, Webster RP, Langor DW, Brunke A, Davies A, Newton AF, Bourdon C, Labrecque M, Dorval J-A, Frank JH (2018a) Aleocharine rove beetles of eastern Canada (Coleoptera, Staphylinidae, Aleocharinae): a glimpse of megadiversity. Springer, Cham, xvi + 902 pp

Klimaszewski J, Brunke A, Work TT, Venier L (2018b) Rove beetles (Coleoptera, Staphylinidae) as bioindicators of change in boreal forests and their biological control Services in Agroecosystems: Canadian case studies. In: Betz OR, Irmler U, Klimaszewski J (eds) Biology of rove beetles (Staphylinidae) life history, evolution, ecology and distribution. Springer, Cham

Klimaszewski J, Hoebeke ER, Godin B, Davies A, Perry KI, Bourdon C, Winchester N (2020) Aleocharine rove beetles of British Columbia, a hotspot of Canadian biodiversity (Coleoptera, Staphylinidae). Springer, Cham, xvi + 631 pp

Kneeshaw DD, Bergeron Y (1998) Canopy gap characteristics and tree replacement in the southeastern boreal forest. Ecology 79:783–794

Lieffers VJ, Stewart JD, Macmillan RB, MacPherson D, Branter K (1996) Semi-natural and intensive silvicultural systems for the boreal mixedwood forest. For Chron 72:286–292

Newton AF, Thayer MK, Ashe JS, Chandler DS (2001) Staphylinidae. In: Arnett RH, Thomas MC (eds) American beetles, vol 1. CRC Press, Boca Raton, FL, pp 272–418

Paquin P, Dupérré N (2001) Beetles of the boreal forest: a faunistic survey carried out in western Québec. Proc Entomol Soc Ontario 132:57–98

Paré D, Bergeron Y, Camiré C (1993) Changes in the forest floor of Canadian southern boreal forest after disturbance. J Veg Sci 4:811–818

Pohl GR, Langor DW, Spence JR (2007) Rove beetles and ground beetles (Coleoptera: Staphylinidae, Carabidae) as indicators of harvest and regeneration practices in western Canadian foothills forests. Biol Conserv 137:294–307

Pohl GR, Langor DW, Klimaszewski J, Work TT, Paquin P (2008) Rove beetles (Coleoptera: Staphylinidae) in northern Nearctic forests. Can Entomol 140:415–436

Pohl G, Langor D, Hammond J, Hoffman P, Klimaszewski J (2017) Aleocharine rove beetles in forest biodiversity research. Poster presented at Entomological Society of Manitoba & Entomological Society of Canada Joint Annual Meeting (2017)

Poole RW, Gentili P (1996) Volume 1: Coleoptera and Strepsiptera. Nomina Insecta Nearctica: a check list of the insects of North America. Entomological Information Services, Rockville, ML. 827 pp

Siitonen J (2001) Forest management, coarse woody debris and saproxylic organisms: Fennoscandian boreal forests as an example. Ecol Bull 49:11–41

Stefani FOP, Klimaszewski J, Morency M-J, Bourdon C, Labrie P, Blais M, Venier L, Séguin A (2016) Fungal community composition in the gut of rove beetles (Coleoptera: Staphylinidae) from the Canadian boreal forest reveals possible endosymbiotic interactions for dietary needs. Fungal Ecol 23:164–171

Venier LA, Work TT, Klimaszewski J, Morris DM, Bowden J, Kwiaton MM, Webster K, Hazlett P (2017) Ground-dwelling arthropod response to fire and clearcutting in jack pine: implications for natural disturbance emulation. Can J For Res 47:1614–1631

Work TT, Klimaszewski J, Thiffault E, Bourdon C, Paré D, Bousquet Y, Venier L, Titus B (2013) Initial responses of rove and ground beetles (Coleoptera, Staphylinidae, Carabidae) to removal of logging residues following clearcut harvesting in the boreal forest of Quebec, Canada. ZooKeys 258:31–52

6 Faunal Analysis and Discussion

Aleocharinae is a hyperdiverse subfamily of Staphylinidae with 17 tribes together containing 623 species currently recorded from Canada and Alaska (updated from Klimaszewski et al. 2020 to include Pentinsaari et al. 2019; Ahn et al. 2020; Webster et al. 2020 and Yoo and Ahn 2020; but not Brunke et al. 2021). Aleocharines represent about 34% of the entire rove beetle fauna in this region (updated from Bousquet et al. 2013 and Brunke et al. 2019). We here record 231 species placed in 54 genera and 12 tribes of Aleocharinae from Subarctic and Arctic North America (ASA), and an additional seven species occurring in the transitional areas of Labrador. Although the aleocharine diversity of this geographically vast region is much lower than that of eastern Canada (411 spp., updated from Klimaszewski et al. 2018), it is still considerably rich and represents about 37% of all aleocharines known from Canada and Alaska. This fauna is approximately equal in species to the fauna of British Columbia (updated from Klimaszewski et al. 2020). No genera are globally endemic to the study region, though the single Nearctic species of *Bellatheta* (1 Nearctic sp.) has not been collected outside of this area. Forty-four species are thus far known only from the Subarctic and Arctic in North America, though many of these are Holarctic species. Of these, *Atheta nearctica*, *Boreophilia subplana*, *Boreostiba campbelliana* and *Parocalea pseudobaicalica* are also known from high elevation alpine refugia in Quebec, Canada and New Hampshire, United States.

Dominant Arctic and Subarctic Aleocharine (ASA) Tribes

Of the 17 aleocharine tribes found in Canada and Alaska, 12 are known from Subarctic and Arctic North America. The following tribes are absent from the study region (numbers represent Canadian species): Autaliini (3), Diglottini (2), Falagriini (13), Hoplandriini (5), Taxicerini (1). These lineages are either poor in species worldwide (Autaliini, Diglottini, Taxicerini) or species-poor in Canada but thermophilic and far more diverse in the tropics (Falagriini and Hoplandriini). The tribe level diversity of ASA taxa is extremely uneven, with Athetini comprising over half of the total species (Table 6.1). Oxypodini is second most dominant, with nearly 20% of all species (Table 6.1). Both Aleocharini and Tachyusini comprise approximately 6% of the species, while the remaining eight tribes comprise less than 5% of the total species. Both the faunas of eastern Canada (Klimaszewski et al. 2018) and British Columbia (Klimaszewski et al. 2020) are similar in that they are Athetini-, and

J. Klimaszewski et al., *A Faunal Review of Aleocharine Beetles in the Rapidly Changing Arctic and Subarctic Regions of North America (Coleoptera, Staphylinidae)*,
https://doi.org/10.1007/978-3-030-68191-3_6

secondarily, Oxypodini-dominant. Tribes that are poorly represented in the study area but well represented in BC and eastern Canada (Table 6.1) (Aleocharini, Placusini, Hypocyphtini, Lomechusini), are likely most affected by the latitudinal diversity gradient, with high numbers of species in the southern biodiversity hotspots of these areas. However, tribes that are poor in both the study region and BC but not eastern Canada (Homalotini, Myllaenini, Lomechusini), are likely constrained by biogeographic history where dispersal across the western Cordillera may have occurred only more recently. The Hypocyphtini follows the second pattern but this is probably driven by the high percentage of adventive species, most of which have been introduced east of the Rockies.

The Athetini are the largest aleocharine tribe (Elven et al. 2010; Newton 2019) and the dominant tribe in Canada and Alaska with 289 species in 41 genera (updated from Klimaszewski et al. 2020). The 127 ASA species of Athetini are placed in 21 genera with the majority in *Atheta* (50 spp.), *Boreophilia* (13 spp.), *Boreostiba* (6 spp.), *Liogluta* (11 spp.) and *Philhygra* (16 spp.). ASA genera *Bellatheta* and *Boreophilia* are pronouncedly Nordic, with at least half of their species known only from the Subarctic and Arctic in the Nearctic portion of their distribution.

The Oxypodini are the second most diverse tribe in Canada and Alaska, with 91 species in 26 genera (updated from Klimaszewski et al. 2020). The 44 ASA species of Oxypodini are placed in 13 genera with the majority in *Oxypoda* (23 spp.) and the remaining species spread rather evenly between 12 other, species-poor genera. ASA genera *Gnathusa* and *Parocalea* are pronouncedly Nordic with at least half of their species known only from the Subarctic and Arctic in the Nearctic portion of their distribution.

The Tachyusini are a relatively small component of the aleocharine fauna of Canada and Alaska, with only 29 species in five genera. This tribe is relatively well-represented in the study region with 14 species across all five genera. Most ASA Tachyusini are placed in the moderately diverse genus *Gnypeta* (10 spp.), which contains many Holarctic species. However, unlike *Boreophilia*, most Holarctic *Gnypeta* have very broad distributions that include localities far outside of the study region (Klimaszewski et al. 2008).

Ecoregional Differences in Aleocharine Diversity

It is technically impossible to perfectly delimit areas of northern Canada and Alaska into discrete regions, due to local variation (elevation, drainage etc.), and because ecoregions gradually transition into one another along a gradient. However, these subunits (CEC 2018) provide important structure to this work and allowed us to capture some very striking differences across major regions of Subarctic and Arctic North America. Here we treat ecoregions with forest (3.1–4.1, 6.1) as Subarctic, while the Arctic is here composed of ecoregions that generally do not have forest or only patches of sparse, low trees (1.1–2.4) (Fig. 2.1). Overall, the Arctic (103 species, 30 genera) is much less diverse than the Subarctic (167 species, 41 genera), especially considering that six genera are restricted to intertidal habitats that are unique to the Arctic in our study region but are all known from farther south, along the Pacific coast (Table 6.2) (Klimaszewski et al. 2020).

Subarctic Aleocharinae: A Transitional Faunal Between Boreal and Arctic

The most diverse ecoregion of the Subarctic is the Boreal Cordillera (6.1, CEC 2018), with 141 species in 39 genera (Table 6.2). Within the study region, the athetine genera *Amischa* and *Trichiusa,* the tachyusine genus *Paradilacra*, and oxypodine genera *Alfocalea*, *Meotica* and *Mniusa* are limited to the Boreal Cordillera (Fig. 2.1). For the purposes of this chapter, the Taiga biome was subdivided into a more southeastern area, grouping the Taiga Plain (ecoregion 3.3), Taiga Shield (ecoregion 3.4) and Hudson Plain (ecoregion 4.1), and a more northwestern area, grouping the Alaska Boreal Interior (ecoregion 3.1) and Taiga

Table 6.1 Proportion of species by tribe of total aleocharine richness reported from Subarctic and Arctic North America (SAN), and proportion of species by tribe of total tribe richness reported from entire Canada and Alaska (CANAK). Figures in parentheses indicate number of species for each tribe

		CANAK		
Tribe	SAN	SAN	BC	ECAN
Athetini	0.55 (127)	0.44 (127)	0.38 (111)	0.61 (176)
Oxypodini	0.19 (44)	0.48 (44)	0.49 (45)	0.63 (57)
Tachyusini	0.06 (14)	0.48 (14)	0.41 (12)	0.51 (15)
Aleocharini	0.06 (13)	0.30 (13)	0.53 (23)	0.81 (35)
Gymnusini	0.03 (8)	0.67 (8)	0.25 (3)	0.92 (11)
Homalotini	0.03 (7)	0.09 (7)	0.20 (15)	0.84 (62)
Liparocephalini	0.03 (7)	0.64 (7)	0.55 (6)	0.00 (0)
Myllaenini	0.02 (4)	0.33 (4)	0.33 (4)	0.83 (10)
Placusini	0.02 (4)	0.44 (4)	0.67 (6)	0.89 (8)
Boreocyphini	<0.01 (1)	1.00 (1)	0.00 (0)	1.00 (1)
Hypocyphtini	<0.01 (1)	0.11 (1)	0.33 (3)	0.67 (6)
Lomechusini	<0.01 (1)	0.06 (1)	0.19 (3)	0.81 (13)

Cordillera (ecoregion 3.2) (Fig. 2.1). Ecoregions were not treated separately in this chapter because of extreme differences in sampling effort, which prevented meaningful comparisons. The fauna of the Hudson Plain as far as known was found to be a subset of the Taiga fauna overall and was treated together with the geographically adjacent southeastern Taiga. The aleocharine fauna of the southeastern Taiga is slightly less diverse than that of the Boreal Cordillera, with 129 species in 35 genera (Table 6.2). The southeastern Taiga represents the northernmost known Nearctic limit of the generally myrmecophilous tribe Lomechusini (*Zyras*), the myrmecophilic athetine *Paragoniusa*, tachyusine genera *Brachyusa* and *Tachyusa*, and oxypodine genera *Hylota* and *Parocyusa*. The northwestern Taiga is slightly less diverse still, with 116 species in 33 genera (Table 6.2). The northwestern Taiga includes the northern limit of true Nearctic forests and the northernmost known Nearctic limits of tribes Homalotini (including subtribes Gyrophaenina,

Table 6.2 Ecoregional differences in aleocharine diversity across Subarctic and Arctic North America. Ecoregions (CEC 2018) with similar physical characteristics were grouped to avoid artifacts of uneven sampling effort in some regions. Numbers in parenthesis include intertidal species, restricted to ecoregion 2.2 in this study

Biome	Ecoregion (group)	Species	Genera	Tribes	Northern limits
Arctic	Tundra (1.1 + 2.1 + 2.2)	59(70)	14(20)	5(6)	–
	Transitional Tundra (2.3 + 2.4)	64	20	7	Boreocyphini, Myllaenini, some Athetini and Oxypodini: *Bellatheta, Calodera, Gnathusa, Hydrosmecta, Ocyusa, Parocalea, Phloeopora, Strigota*
	Total	103	30	8	
Subarctic					
	Northwestern Taiga (3.1 + 3.2)	116	33	11	Homalotini, Hypocyphtini, Placusini; some Athetini and Oxypodini: *Acrotona, Aloconota, Devia, Dinaraea, Earota*
	Southeastern Taiga (3.3 + 3.4 + 4.1)	129	35	12	Lomechusini; some Athetini, Oxypodini, Tachyusini: *Brachyusa, Hylota, Paragoniusa, Tachyusa, Parocyusa*
	Boreal Cordillera	141	39	12	*Alfocalea, Amischa, Meotica, Mniusa, Paradilacra, Trichiusa*
	Total	167	41	12	

Bolitocharina, Silusina), Hypocyphtini (*Cypha*) and Placusini (*Placusa*). Subcortical genus *Dinaraea* also reaches its limit here and, as could be expected, appears to be constrained by the distribution of forests. A further four genera are known only from the Subarctic: *Acrotona*, *Aloconota*, *Earota* and *Devia*.

Arctic Aleocharinae: A Fauna at the Extreme

The aleocharine fauna of the Arctic is impoverished, with only 8 of 12 ASA tribes present (Table 6.2). As its ecoregions have received uneven sampling effort, the Arctic was subdivided into transitional tundra, grouping Brooks Range Tundra (ecoregion 2.3) and Southern Arctic (ecoregion 3.4), and tundra, grouping Arctic Cordillera (ecoregion 1.1), Northern Arctic (ecoregion 2.1) and Alaskan Tundra (ecoregion 2.2) (Fig. 2.1). Transitional tundra includes the treeline in North America, and some limited patches of low, sparse trees, while 'tundra' groups together ecoregions that are clearly north of the treeline and experience more extreme climate. The transitional tundra ecoregions are inhabited by 64 aleocharine species in 20 genera (Table 6.2). Here the tribes Boreocyphini (*Boreocypha*) and Myllaenini (*Myllaena*) reach their northernmost known limit in the Nearctic, as do athetine genera *Bellatheta*, *Hydrosmecta* and *Strigota*, and oxypodine genera *Calodera*, *Gnathusa*, *Ocyusa*, *Parocalea* and *Phloeopora*. It is noteworthy that the generally subcortical genus *Phloeopora* has been collected in the Arctic (*P. arctica*) but this particular species' microhabitat preferences are not well-known, with a single record from 'sifting litter' (Klimaszewski et al. 2018). Considering only terrestrial species, the tundra ecoregions are the least diverse in Aleocharinae, with 59 species in 14 genera and classified in five tribes (Table 6.2). Intertidal microhabitats of the Alaskan Tundra (ecoregion 2.2) (Fig. 2.1), host an additional 11 species, some of which are placed in the intertidal specialist tribe Liparocephalini. Although comparisons are limited due to extremely low sampling effort, the aleocharine fauna of the most extreme Arctic ecoregions, Arctic Cordillera (1.1) and Northern Arctic (2.1) (Fig. 2.1), is thus far limited to 19 species in the following eight genera: *Atheta*, *Boreophilia*, *Boreostiba*, *Liogluta*, *Philhygra*, *Oxypoda* and *Gnypeta*.

Holarctic Species

Forty aleocharine species are here considered to be Holarctic. We expect further northern species to be identified as Holarctic, with increased sampling of remote areas of far eastern Russia and further detailed comparisons between Nearctic and Palaearctic populations with microdissections and molecular data. In the present study, *Gnathusa caribou* was detected in the Palaearctic region (Finland) for the first time via DNA barcoding (with subsequent morphological verification) and this species is now recognized as Holarctic. We are aware of other such cases in Aleocharinae that are still under study.

Holarctic Aleocharinae are overwhelmingly (35/40 species) either wetland species or those occurring in a wide variety of ground-based microhabitats in both open and forested settings. Aleocharinae specialized on forests, such as the subcortical genera *Dinaraea*, *Placusa* and *Phloeopora*, lack Holarctic species (see below regarding *Placusa*). This is consistent with the hypothesis proposed by Brunke et al. (2020), which states that forest-specialized beetles on either side of the Bering Strait have long since become isolated by a treeless tundra across the Bering Land Bridge and have not had dispersal opportunities for at least 5.4 Mya. Conversely, the poorly drained soils of the tundra have provided appropriate microhabitats for wetland and generalist species across the Bering Land Bridge as recently as 11,000 years ago (Jakobsson et al. 2017). Intertidal (halophile) insects, tolerant of saltwater are often proposed to have long distance dispersal capabilities via rafting on debris or simply by floating in seawater (e.g., Hansen et al. 2018; Ueno et al. 2020). Therefore, it is puzzling that only 3 of the 11 intertidal ASA taxa are Holarctic, despite multidirectional ocean currents to and from the Palaearctic region, along the Aleutian Island chain (Stabeno et al. 1999). The remaining two Holarctic Aleocharinae (*Gyrophaena*) are associated with macrofungi and have thus far only been collected in the Nearctic region as far north as the Taiga. We expect these species to occur on mushrooms growing on the tundra and further specialized collecting there is needed to confirm this.

Adventive Species in Subarctic and Arctic North America

Of the 51 adventive aleocharine species recorded from Canada and Alaska (updated from Klimaszewski and Brunke 2018; Klimaszewski et al. 2020), only eight are here recorded from Arctic and Subarctic North America, comprising just 3% of the fauna. Most of the adventive ASA species belong to the Athetini and Aleocharini, and all are thought to have become introduced from Europe via trade (Klimaszewski and Brunke 2018). Only two of these species have records from the Arctic and both need to be confirmed with further research. One of these, *Mocyta fungi*, needs comprehensive taxonomic revision using both morphological and molecular data. Based on preliminary barcode sequence data, it may be a complex of species, involving both native and adventive lineages. The second species, *Oxypoda brachyptera*, is only known from the study region based on a single female. Given the taxonomic difficulty of this species group (Assing 2012), males from this locality should be examined carefully as there could be a native, undescribed species near *O. brachyptera*. Of the six confirmed adventive ASA species, only *Aloconota cambrica* has been recorded from the Taiga but is also known from farther south in BC and California, USA (Klimaszewski et al. 2020).

The remaining five species are only known from as far north as the Boreal Cordillera (ecoregion 6.1) (Fig. 2.1): *Aleochara fumata*, *A. morion*, *A. subsinuata*, *Dinaraea angustula*, and *Meotica pallens*. The apparent resistance of more northern ecoregions to adventive species is likely to be a combination of remoteness from centres of commerce and human disturbance, and a lack of adaptation to the climatic extremes of the Subarctic and Arctic, which they do not experience in their native Central European distribution. Although we could not confirm here that adventive aleocharine species have become established in the North American Arctic, we do expect adventive species' ranges to extend northward with the rapid warming of Subarctic and Arctic North America.

Four ASA species were treated throughout this work as either Holarctic or adventive: *Aleochara villosa*, *Atheta atramentaria*, *Dochmonota rudiventris*, and *Placusa tachyporoides*. Current records for both *A. villosa* and *P. tachyporoides* show that these taxa occur as far north as the Taiga in AK but preliminary barcode sequence data place the Nearctic specimens in a separate BIN from the Palaearctic population and suggest that separate species pairs may be involved. Although there is no sequence data available for the Palaearctic population of *D. rudiventris*, its broad distribution, including some rather remote localities, suggest that it is a Holarctic wetland species. Barcode data for *A. atramentaria* place Nearctic and Palaearctic populations in the same BIN with no intraspecific divergence, strongly suggesting a single species. Although few Nearctic records exist for *A. atramentaria*, its southernmost record in North America is a classic alpine refugium on the Gaspé peninsula (QC) from where other Holarctic species are known to occur as relict populations.

Future Research and Sampling

Although approximately 1400 geographically distinct species occurrences have been published thus far, exploration of ASA biodiversity is still in its early stages. The least sampled ecoregions include the Southern Arctic (ecoregion 2.4), Hudson Plains (ecoregion 4.1) and the very difficult to access Arctic Cordillera and Northern Arctic (ecoregions 1.1, 2.1) (Fig. 2.1). Churchill, MB in the Hudson Plain is becoming better sampled with over 95% of the species occurrences in this ecoregion. No occurrences were available for

Subarctic SK and very few were available for Subarctic AB, ON and QC, and Arctic QC. Given their vast size, NT and NU are also inadequately sampled, though recent survey efforts (see Chapter 3) may improve this. Intertidal microhabitats on the western coast of AK (Bering Sea) have proven to be quite diverse in Aleocharinae but it is not known how far north and east these species occur. To our knowledge, halophilic aleocharines are unknown from the shores of the Beaufort Sea, Hudson Bay or the north Atlantic of Arctic QC and LB. Excluding the species described in the present work, the following 13 species are still known only from the type material collected in the study region (year described, distribution): *Acrotona horwoodae* (2012, YT), *Atheta cadeti* (2008, YT), *A. martini* (1990, YT), *A. onthophilides* (1990, YT, AK), *A. pecki* (2011, LB), *Boreostiba lagunae* (1990, YT), *Gnathusa alfacaribou* (2011, LB), *Gnypeta brevicornis* (1906, BC), *G. groenlandica* (2008, MB, NT, NU, YT, AK, Greenland), *Oxypoda leechi* (1990, YT), *O. volkeri* (2006, NT); *Liogluta trapezicollis* (1990, YT) and *Philhygra pseudoboreostiba* (1990, YT). Seven of these have not been collected since at least 1984. It is clear that much greater sampling effort is needed to fully uncover the true complexity of aleocharine diversity in the North American Arctic and Subarctic. It is our hope that the identification tools provided herein, including the reference DNA barcodes taxonomically improved through this study, will catalyze future efforts to document the hyperdiverse Aleocharine rove beetles of the North.

References

Ahn K-J, Sikes DS, Klimaszewski J (2020) *Paramblopusa eoa* Ahn and Maruyama (Coleoptera: Staphylinidae: Aleocharinae), distribution range extention to Alaska. The Coleopterist Bulletin 74(2):324–326

Assing V (2012) On the taxonomy and natural history of *Oxypoda brachyptera* and *O. tarda*. Beiträge zur Entomologie 62:207–224

Bousquet Y, Bouchard P, Davies AE, Sikes DS (2013) Checklist of beetles (Coleoptera) of Canada and Alaska: second edition. Pensoft Series Faunistica No. 109. Pensoft, Sofia, Moscow, 402 p

Brunke AJ, Bouchard P, Douglas HB, Pentinsaari M (2019) Coleoptera of Canada. ZooKeys 819:361–376

Brunke AJ, Salnitska M, Hansen AK, Zmudzinska A, Smetana A, Buffam J, Solodovnikov A (2020) Are subcortical rove beetles truly Holarctic? An integrative taxonomic revision of north temperate *Quedionuchus* (Coleoptera: Staphylinidae: Staphylininae). Organisms Diversity and Evolution 20:77–116

Brunke AJ, Pentinsaari M, Klimaszewski J (2021) Integrative taxonomy of Nearctic and Palaearctic Aleocharinae: new species, synonymies, and records (Coleoptera, Staphylinidae). Zookeys 1041:27–99

Commission for Environmental Cooperation (CEC) (2018) Terrestrial ecoregions – Level II ecoregions. http://www.cec.org/tools-and- resources/map-files/terrestrial-ecoregions-level-ii. Accessed 8 Sept 2020

Elven H, Bachmann L, Gusarov VI (2010) Phylogeny of the tribe Athetini (Coleoptera: Staphylinidae) inferred from mitochondrial and nuclear sequence data. Mol Phylogenet Evol 57(1):84–100

Hansen AK, Justesen MJ, Kepfer-Rojas S, Byriel DB, Pedersen J, Solodovnikov A (2018) Ecogeographic patterns in a mainland-island system in northern Europe as inferred from the rove beetles (Coleoptera: Staphylinidae) on Læsø island. European J Ent 115:256–263

Jakobsson M, Pearce C, Cronin TM, Backman J, Andersen LG, Barrientos N, Björk G, Coxall H, de Boer A, Mayer LA, Mörth CM, Nilsson J, Rattray JE, Stranne C, Semiletov I, O'Regan M (2017) Post-glacial flooding of the Bering land bridge dated to 11 cal ka BP based on new geophysical and sediment records. Clim Past 13:991

Klimaszewski J, Brunke AJ (2018) Chapter 5: Canada's adventive rove beetle (Coleoptera, Staphylinidae) Fauna: a long-term case study on the detection, origin, introduction pathways, and dynamic distribution of non-native beetles. In: Betz O, Irmler U, Klimaszewski J (eds) Biology of rove beetles (Staphylinidae). Springer, Cham, pp 65–79

Klimaszewski J, Savard K, Pelletier G, Webster R (2008) Species review of the genus *Gnypeta* Thomson from Canada, Alaska and Greenland (Coleoptera, Staphylinidae, Aleocharinae): systematics, bionomics and distribution. ZooKeys 2:11–84

Klimaszewski J, Webster RP, Langor DW, Brunke A, Davies A, Bourdon C, Labrecque M, Newton AF, Dorval J-A, Frank JH (2018) Aleocharine rove beetles of eastern Canada (Coleoptera, Staphylinidae, Aleocharinae): a glimpse of megadiversity. Springer, Cham. xvi + 902 pp

Klimaszewski J, Hoebeke ER, Godin B, Davies A, Perry KI, Bourdon C, Winchester N (2020) Aleocharine rove beetles of British Columbia: a hotspot of Canadian biodiversity (Coleoptera, Staphylinidae). Springer, Cham. xvi + 631 pp

Newton AF (2019) StaphBase: Staphyliniformia world catalog database (version Nov 2018). In: Roskov Y et al (eds) Species 2000 & ITIS catalogue of life, 2019 annual checklist. Species 2000, Leiden

Pentinsaari M, Anderson R, Borowiec L, Bouchard P, Brunke A, Douglas H, Smith ABT, Hebert PDN (2019) DNA barcodes reveal 63 overlooked species of Canadian beetles (Insecta, coleoptera). ZooKeys 894:53–150

Stabeno PJ, Schumacher JD, Ohtani K (1999) The physical oceanography of the Bering Sea. In: Loughlin TR, Ohtani K (eds) Dynamics of the Bering Sea: a summary of the physical, chemical and biological characteristics, and a synopsis of research on the Bering Sea. North Pacific Marine Science Organization, University of Alaska Sea Grant, Alaska, pp 1–28

Ueno H, Kitagawa K, Matsubayashi KW (2020) Unexpectedly long survivorship on seawater of multiple coastal beetles indicates the possibility of 'floating dispersal' for transoceanic migration. Entomological Science 23:294–296

Webster RP, de Tonnancour P, Sweeney J, Webster VL, Kostanowicz CA, Hughes C, Anderson RS, Klymko J, Chantal C, Vigneault R (2020) New Coleoptera records from eastern Canada, with additions to the fauna of Manitoba, British Columbia, and Yukon territory. ZooKeys 946:53–112

Yoo I-S, Ahn KJ (2020) Discovery of a new intertidal species, *Amblopusa vancouverensis* from Vancouver Island, Canada with notes on *Amblopusa brevipes* Casey (Coleoptera, Staphylinidae, Aleocharinae). Zootaxa 4803:505–514

7 List of Recorded Arctic and Subarctic Aleocharine Species of North America and Their Composition

GYMNUSINI [10]: *Deinopsis* (2), *Gymnusa* (8). **ALEOCHARINI [13]**: *Aleochara* (13). **OXYPODINI [44]**: *Alfocalea* (1), *Calodera* (2), *Devia* (1), *Gnathusa* (4), *Hylota* (1), *Meotica* (1), *Mniusa* (1), *Neothetalia* (3), *Ocyusa* (1), *Oxypoda* (23), *Parocalea* (2), *Phloeopora* (3), *Parocyusa* (1). **TACHYUSINI [14]**: *Brachyusa* (1), *Dasygnypeta* (1), *Gnypeta* (10), *Paradilacra* (1); *Tachyusa* (1). **BOREOCYPHINI [1]**: *Boreocypha* (1). **HYPOCYPHTINI [1]**: *Cypha* (1). **MYLLAENINI [4]**: *Myllaena* (4). **LIPAROCEPHALINI** [7]: *Amblopusa* (2), *Diaulota* (1), *Paramblopusa* (2), *Liparocephalus* (2). **HOMALOTINI [7]**: *Gyrophaena* (4), *Leptusa* (1), *Silusa* (2). **PLACUSINI [4]**: *Placusa* (4). **ATHETINI [132]**: *Acrotona* (2), *Adota* (1), *Aloconota* (2), *Amischa* (2), *Atheta* (54), *Bellatheta* (1), *Boreophilia* (13), *Boreostiba* (6), *Dinaraea* (4), *Dochmonota* (2); *Earota* (1), *Hydrosmecta* (2), *Liogluta* (11), *Lypoglossa* (2), *Mocyta* (4), *Paragoniusa* (1), *Philhygra* (16), *Schistoglossa* (3), *Strigota* (2), *Tarphiota* (2), *Trichiusa* (1). **LOMECHUSINI [1]**: *Zyras* (1).

[* Holarctic species, †adventive species, ‡ species of uncertain distribution status—adventive in Canada/Alaska, native or Holarctic, species without marks are native; numbers in brackets indicate number of species per taxon]

Family STAPHYLINIDAE Latreille, 1802 (Rove beetles)

Subfamily ALEOCHARINAE Fleming, 1821

Tribe Gymnusini Heer, 1839

Genus *DEINOPSIS* Matthews, 1838
1. *D. canadensis* Klimaszewski, 1979
2. *D. harringtoni* Casey, 1911
Genus *GYMNUSA* Gravenhorst, 1806
Brevicollis species group
3. *G. atra* Casey, 1911 *
4. *G. brevicollis* (Paykull, 1800) †
5. *G. lindrothi* Klimaszewski and Langor, in Klimaszewski et al. 2011
Grandiceps species group
6. *G. grandiceps* Casey, 1915
Variegata species group
7. *G. campbelli* Klimaszewski, 1979
8. *G. konopackii* Klimaszewski, 1979
9. *G. pseudovariegata* Klimaszewski, 1979
10. *G. smetanai* Klimaszewski, 1979 *

Tribe Aleocharini Fleming, 1821

Subtribe Aleocharina Fleming, 1821
Genus *ALEOCHARA* Gravenhorst, 1802
(Subgenus *Aleochara* Gravenhorst, 1802)
11. *A. assiniboin* Klimaszewski, 1985
12. *A. gracilicornis* Bernhauer, 1901b

J. Klimaszewski et al., *A Faunal Review of Aleocharine Beetles in the Rapidly Changing Arctic and Subarctic Regions of North America (Coleoptera, Staphylinidae)*,
https://doi.org/10.1007/978-3-030-68191-3_7

13. *A. sekanai* Klimaszewski, 1985
14. *A. tahoensis* Casey, 1906
(Subgenus *Calochara* Casey, 1906)
15. *A. villosa* Mannerheim, 1830 ‡
(Subgenus *Coprochara* Mulsant and Rey, 1874)
16. *A. bimaculata* Gravenhorst, 1802
17. *A. shelleyae* Klimaszewski and Langor, in Klimaszewski et al. 2011
18. *A. suffusa* (Casey, 1906)
19. *A. sulcicollis* Mannerheim, 1843
20. *A. verna* Say, 1836 *
(Subgenus *TINOTUS* Sharp, 1883
21. *A. morion* Gravenhorst, 1802 †
(Subgenus *Xenochara* Mulsant and Rey, 1874)
22. *A. castaneipennis* Mannerheim, 1843
23. *A. fumata* Gravenhorst, 1802 †

Tribe Oxypodini C.G. Thomson, 1859

Subtribe Meoticina Seevers, 1978
Genus *MEOTICA* Mulsant and Rey, 1873a
24. *M. pallens* (Redtenbacher, 1849) †
Subtribe Oxypodina C.G. Thomson, 1859
Genus *ALFOCALEA* Klimaszewski, in Klimaszewski and Pelletier, 2004
25. *A. montana* Klimaszewski, in Klimaszewski and Pelletier, 2004
Genus *CALODERA* Mannerheim, 1830
26. *C. caseyi* Assing, 2002
27. *C. parviceps* (Casey, 1893)
Genus *DEVIA* Blackwelder, 1952
28. *D. prospera* (Erichson, 1839) *
Genus *GNATHUSA* Fenyes, 1909
29. *G. alfacaribou* Klimaszewski and Langor, in Klimaszewski et al. 2011
30. *G. caribou* Lohse, in Lohse et al. 1990 *
31. *G. eva* Fenyes, 1909
32. *G. tenuicornis* Fenyes, 1921
Genus *HYLOTA* Casey, 1906
33. *H. ochracea* Casey, 1906
Genus *MNIUSA* Mulsant and Rey, 1875
34. *M. yukonensis* (Klimaszewski and Godin, in Klimaszewski et al. 2012)
Genus *NEOTHETALIA* Klimaszewski, in Klimaszewski and Pelletier, 2004
35. *N. canadiana* Klimaszewski, in Klimaszewski and Pelletier, 2004
36. *N. columbiana* (Klimaszewski and Winchester 2002)
37. *N. pallitarsis* (Kirby, in Kirby and Spence 1837)
Genus *OCYUSA* Kraatz, 1856
38. *O. canadensis* Lohse, in Lohse et al. 1990
Genus *OXYPODA* Mannerheim, 1830
Amica species group
39. *O. amica* Casey, 1906
Convergens species group
40. *O. canadensis* Klimaszewski, in Klimaszewski et al. 2006
41. *O. convergens* Casey, 1893
42. *O. manitobae* Casey, 1911
43. *O. nelsoni* Lohse, in Lohse et al. 1990
44. *O. pseudoconvergens* Klimaszewski and Godin, in Klimaszewski et al. 2008b
Gatosensis species group
45. *O. gatosensis* Bernhauer, 1905
Inimica species group
46. *O. inimica* Casey, 1911
47. *O. yukonensis* Klimaszewski and Godin, in Klimaszewski et al. 2008b
48. *O. volkeri* Klimaszewski, in Klimaszewski et al. 2006
Irrasa species group
49. *O. irrasa* Mäklin, 1853
Lacustris species group
50. *O. hiemalis* Casey, 1911
51. *O. lacustris* Casey, 1906
Lucidula species group
52. *O. demissa* Casey, 1911
53. *O. leechi* Lohse, in Lohse et al. 1990
54. *O. lucidula* Casey, 1906
55. *O. robusticornis* Bernhauer, 1907 [=*O. madgei* Lohse, in Lohse et al. 1990] **New Synonymy**
Nigriceps species group
56. *O. brachyptera* (Stephens, 1832) †
Operta species group
57. *O. operta* Sjöberg, 1950 *
Orbicollis species group
58. *O. frigida* Bernhauer, 1907
59. *O. orbicollis* Casey, 1911
Subpolaris species group
60. *O. subpolaris* Casey, 1911 [=*O. hemingi* Lohse, in Lohse et al. 1990]

Sylvia species group
61. *O. sylvia* Casey, 1906
Genus *PAROCALEA* Bernhauer, 1901a
62. *P. nearctica* Lohse, in Lohse et al. 1990
63. *P. pseudobaicalica* Lohse, in Lohse et al. 1990
Genus *PAROCYUSA* Bernhauer, 1902
[= *TETRALAUCOPORA* Bernhauer, 1928 (synonymized by Assing 2021)]
64. *P. fuliginosa* (Casey, 1906)
Subtribe Phloeoporina C.G. Thomson, 1859
Genus *PHLOEOPORA* C.G. Thomson, 1859
65. *P. arctica* Lohse, 1990, in Lohse et al. 1990
66. *P. canadensis* Klimaszewski and Langor, in Klimaszewski et al. 2011
67. *P. gilbertae* Klimaszewski and Webster, 2016, in Webster et al. 2016

Tribe Tachyusini C.G. Thomson, 1859

Genus *BRACHYUSA* Mulsant and Rey, 1873c
68. *B. helenae* (Casey, 1911)
Genus *DASYGNYPETA* Lohse, 1974
69. *D. velata* (Erichson, 1837) * [=*G. minuta* Klimaszewski and Webster, in Klimaszewski et al. 2008a (synonymized by Brunke et al. 2021)]
Genus *GNYPETA* C.G. Thomson, 1858
Caerulea species group
70. *G. caerulea* (C.R. Sahlberg, 1831) *
71. *G. lohsei* Klimaszewski, in Klimaszewski et al. 2008a
Crebrepunctata species group
72. *G. carbonaria* (Mannerheim, 1830) *
Sellmani species group
73. *G. ashei* Klimaszewski, in Klimaszewski et al. 2008a
74. *G. atrolucens* Casey, 1893
75. *G. brevicornis* Casey, 1906
76. *G. brincki* Palm, 1966 *
77. *G. dentata* Klimaszewski, in Klimaszewski et al. 2008a
78. *G. groenlandica* Lohse, 1989
79. *G. sellmani* Brundin, 1929 *
Genus *PARADILACRA* Bernhauer, 1909
80. *P. densissima* (Bernhauer, 1909) [=*G. saccharina* Klimaszewski and Webster, in Klimaszewski et al. 2008a (synonymized by Brunke et al. 2021)]

Genus *TACHYUSA* Erichson, 1837
(Subgenus *Tachyusa* Erichson, 1837)
81. *T. americanoides* Paśnik, 2006

Tribe Boreocyphini Klimaszewski and Langor, in Klimaszewski et al. 2011

Genus *BOREOCYPHA* Klimaszewski and Langor, in Klimaszewski et al. 2011
82. *B. websteri* Klimaszewski and Langor, in Klimaszewski et al. 2011

Tribe Hypocyphtini Laporte, 1835

Genus *CYPHA* Leach, 1819
83. *C. inexpectata* Klimaszewski and Godin, in Klimaszewski et al. 2008b

Tribe Myllaenini Ganglbauer, 1895

Genus *MYLLAENA* Erichson, 1837
Cuneata species group
84. *M. arcana* Casey, 1911
Insomnis species group
85. *M. audax* Casey, 1911
86. *M. insomnis* Casey, 1911
Potawatomi species group
87. *Myllaena ludificans* Casey, 1911

Tribe Liparocephalini Fenyes, 1918

Genus *AMBLOPUSA* Casey, 1893
88. *A. alaskana* Ahn and Ashe, 1995
89. *A. brevipes* Casey, 1893
Genus *DIAULOTA* Casey, 1893
90. *D. densissima* Casey, 1893
Genus *PARAMBLOPUSA* Ahn and Ashe, 1996
91. *P. borealis* (Casey, 1906) *
92. *P. eoa* Ahn and Maruyama, 2000 *
Genus *LIPAROCEPHALUS* Mäklin, 1853
93. *L. brevipennis* Mäklin, 1853
94. *L. humblei* Klimaszewski, **sp. n.**

Tribe Homalotini Heer, 1839

Subtribe Bolitocharina C.G. Thomson, 1859
Genus *LEPTUSA* Kraatz, 1856
(Subgenus *Gatineauleptusa* Klimaszewski, in Klimaszewski et al. 2018)

95. *L. gatineauensis* Klimaszewski and Pelletier, in Klimaszewski et al. 2004
Subtribe Gyrophaenina Kraatz, 1856
Genus *GYROPHAENA* Mannerheim, 1830
(Subgenus *Gyrophaena* Mannerheim, 1830)
Keeni species group
96. *G. keeni* Casey, 1911
Nana species group
97. *G. nana* (Paykull, 1800) *
98. *G. rugipennis* Mulsant and Rey, 1861 * [= *G. neonana* Seevers, 1951] **New Synonymy**
Pulchella species group
99. *G. insolens* Casey, 1906
Subtribe Silusina Fenyes, 1918
Genus *SILUSA* Erichson, 1837
Californica species group
100. *S. californica* (Bernhauer, 1905)
Vesperis species group
101. *S. densa* Fenyes, 1909

Tribe Placusini Mulsant and Rey, 1871

Genus *PLACUSA* Erichson, 1837
Canadensis species group
102. *P. tacomae* Casey, 1893
103. *P. vaga* Casey, 1911
Tachyporoides species group
104. *P. pseudosuecica* Klimaszewski, in Klimaszewski et al. 2001
105. *P. tachyporoides* (Waltl, 1838) ‡ [new distribution status]

Tribe Athetini Casey, 1910

Subtribe Athetina Casey, 1910
Genus *ACROTONA* C.G. Thomson, 1859
106. *A. sphagnorum* Klimaszewski and Webster, in Webster et al. 2016
107. *A. horwoodae* Klimaszewski and Godin, in Klimaszewski et al. 2012
Genus *ADOTA* Casey, 1910
108. *A. maritima* (Mannerheim, 1843) *
Genus *ALOCONOTA* C.G. Thomson, 1858
109. *A. cambrica* (Wollaston, 1855) †
110. *A. neocambrica* Klimaszewski and Langor, in Klimaszewski et al. 2011
Genus *AMISCHA* C.G. Thomson, 1858
111. *A. praelonga* (Casey, 1893)
112. *A. tersa* Casey, 1910 [tentative]
Genus *ATHETA* C.G. Thomson, 1858
(Subgenus *ALAOBIA* C.G. Thomson, 1858)
113. *A. ventricosa* Bernhauer, 1907
(Subgenus *ATHETA* C.G. Thomson, 1858)
114. *A. borealides* Newton, 2017 [= *A. borealis* Klimaszewski and Langor, in Klimaszewski et al. 2011]
115. *A. circulicollis* Lohse, in Lohse et al. 1990
116. *A. graminicola* (Gravenhorst, 1806) *
117. *A. martini* Lohse, in Lohse et al. 1990
118. *A. ringi* Klimaszewski, 2002, in Klimaszewski and Winchester 2002
(Subgenus *BESSOBIA* C.G. Thomson, 1858)
119. *A. cryptica* (Lohse, in Lohse et al. 1990)
(Subgenus *DATOMICRA* Mulsant and Rey, 1873c)
120. *A. chadburnae* Klimaszewski and Godin, **sp. n.**
121. *A. dadopora* C.G. Thomson, 1867 *
122. *A. whitehorsensis* Klimaszewski and Godin, in Klimaszewski et al. 2012
(Subgenus *DIMETROTA* Mulsant and Rey, 1873a)
Cadeti species group
123. *A. cadeti* Klimaszewski and Godin, in Klimaszewski et al. 2008b
Allocera (= *Campbelli*) species group
124. *A. allocera* Eppelsheim, 1893 * [=*Dimetrota campbelli* Lohse, in Lohse et al. 1990 (=*Atheta*)] **New Synonymy**
125. *A. diversa* (Sharp, 1869) * [= *Dimetrota regissalmonis* Lohse, in Lohse et al. 1990 (=*Atheta*)] **New Synonymy**
126. *A. remulsa* Casey, 1910
127. *A. smetanai* (Lohse, in Lohse et al. 1990)
Stercoris (=*Columbica*) species group
128. *A. stercoris* Fenyes, 1920
[=*A. columbica* Bernhauer, 1907]
Crassicornis (=*Modesta*) species group
129. *A. pseudometlakatlana* Klimaszewski and Godin, in Klimaszewski et al. 2008b
Crenuliventris species group
130. *A. crenuliventris* Bernhauer, 1907
131. *A. pecki* Klimaszewski and Langor, in Klimaszewski et al. 2011

132. *A. pseudocrenuliventris* Klimaszewski, in Klimaszewski et al. 2005
Districta species group
133. *A. districta* Casey, 1911
Fanatica species group
134. *A. fanatica* Casey, 1910
[=*A. irrita* Casey, 1911]
135. *A. munsteri* Bernhauer, 1902 *
Hampshirensis species group
136. *A. hampshirensis* Bernhauer, 1909
Picipennis species group
137. *A. altaica* Bernhauer, 1901a *
138. *A. atramentaria* (Gyllenhal, 1810) ‡
139. *A. nearctica* (Lohse, in Lohse et al. 1990)
140. *A. onthophilides* (Newton 2017) [replacement name for *Acrotona onthophila* Lohse, 1990]. **New Generic Combination.**
141. *A. picipennis* (Mannerheim, 1843)
[=*A. cursor* (Mäklin, 1852) = *A. carlottae* Casey, 1910]
Prudhoensis species group
142. *A. burwelli* (Lohse, in Lohse et al. 1990)
143. *A. caribou* Lohse, in Lohse et al. 1990
144. *A. capsularis* Klimaszewski, in Klimaszewski et al. 2005
145. *A. curvipennis* Klimaszewski and Langor, in Klimaszewski et al. 2011
146. *A. prudhoensis* (Lohse, in Lohse et al. 1990) *
147. *A. strigosula* Casey, 1910
148. *A. takhini* Klimaszewski and Godin, **sp. n.**
149. *A. terranovae* Klimaszewski and Langor, in Klimaszewski et al. 2011
Dimetrota incertae sedis
150. *A. avalon* Klimaszewski and Langor, in Klimaszewski et al. 2011
151. *A. gelida* J. Sahlberg, 1887*
(Subgenus *LAMIOTA* Casey, 1910)
152. *A. keeni* Casey, 1910
(Subgenus *MICRODOTA* Mulsant and Rey, 1873a)
153. *A. microelytrata* Klimaszewski and Godin, in Klimaszewski et al. 2012
154. *A. pennsylvanica* Bernhauer, 1907
155. *A. platonoffi* Brundin, 1948 *
156. *A. pratensis* (Mäklin, 1852)
157. *A. pseudosubtilis* Klimaszewski and Langor, in Klimaszewski et al. 2011
158. *A. ripariides* Newton, 2017 [replacement name for *Atheta riparia* Klimaszewski and Godin, in Klimaszewski et al. 2012]
159. *A. subtilis* (Scriba, 1866) †
160. *A. surgens* (Casey)
(Subgenus *PSEUDOTA* Casey, 1910)
161. *A. longiclava* Casey, 1910
162. *A. pseudoklagesi* Klimaszewski and Webster, in Webster et al. 2016
(Subgenus *RHAGOCNEME* Munster, 1923)
163. *A. subsinuata* (Erichson, 1839) †
(Subgenus *TETROPLA* Mulsant and Rey, 1873c)
164. *A. frosti* Bernhauer, 1909

Atheta Incertae Sedis

165. *A. brunswickensis* Klimaszewski, 2005
166. *A. sparreschneideri* Munster, 1923 *
Genus *BELLATHETA* Roubal, 1928 [=*Emmelostiba* Pace, 1982, = *Pseudosipalia* Lohse, 1990]
167. *B. microptera* (Lohse, in Lohse et al. 1990). **New Combination.**
Genus *BOREOPHILIA* Benick, 1973
Fusca species group
168. *B. eremita* (Rey, 1866) *
169. *B. fusca* (CR Sahlberg, 1831) *
170. *B. hyperborea* (Brundin, 1940) *
171. *B. islandica* (Kraatz, 1857) *
172. *B. nearctica* Lohse, in Lohse et al. 1990
173. *B. nomensis* (Casey, 1910)
Subplana species group
174. *B. beringi* Klimaszewski and Brunke, in Klimaszewski et al. 2019
175. *B. caseyi* Lohse, in Lohse et al. 1990
176. *B. davidgei* Klimaszewski and Godin, in Klimaszewski et al. 2012
177. *B. neoinsecuta* Klimaszewski, in Klimaszewski et al. 2019
178. *B. subplana* (J.R. Sahlberg, 1880) *
179. *B. vega* (Fenyes, 1920) *
180. *B. venti* Lohse, in Lohse et al. 1990
Genus *BOREOSTIBA* Lohse, in Lohse et al. 1990
Frigida species group

181. *B. frigida* (J.R. Sahlberg, 1880) *
182. *B. sibirica* (Mäklin, 1880) *
Parvipennis species group
183. *B. campbelliana* Lohse, in Lohse et al. 1990 *
184. *B. lagunae* Lohse, 1990
185. *B. parvipennis* (Bernhauer, 1907)
186. *B. websteri* Klimaszewski and Langor, in Klimaszewski et al. 2011
Genus *DINARAEA* C.G. Thomson, 1858
187. *D. angustula* (Gyllenhal, 1810) †
188. *D. borealis* Lohse, in Lohse et al. 1990
189. *D. pacei* Klimaszewski and Langor, in Klimaszewski et al. 2011
190. *D. planaris* (Mäklin, 1852). **Neotype designation**.
Genus *DOCHMONOTA* C.G. Thomson, 1859
191. *D. langori* Klimaszewski and Larson, in Klimaszewski et al. 2016
192. *D. rudiventris* (Eppelsheim, 1886) ‡
Genus *EAROTA* Mulsant and Rey, 1873c
193. *E. dentata* (Bernhauer, 1906)
Genus *HYDROSMECTA* Thomson, 1858
194. *H. pseudodiosica* Lohse, in Lohse et al. 1990
195. *H. yukonensis* Klimaszewski and Godin, **sp. n.**
Genus *LIOGLUTA* C.G. Thomson, 1858
Gigantea species group
196. *L. gigantea* Klimaszewski and Langor, in Klimaszewski et al. 2011
Granulosa species group
197. *L. granulosa* Lohse, in Lohse et al. 1990
Micrgranulosa species group
198. *L. atriventris* (Casey, 1906)
199. *L. intermedia* Klimaszewski and Langor, in Klimaszewski et al. 2011
Nigropolita species group
200. *L. nigropolita* (Bernhauer, 1907)
201. *L. nitens* (Mäklin, 1852)
Terminalis species group
202. *L. scudderi* Klimaszewski, in Klimaszewski et al. 2020
203. *L. terminalis* (Casey, 1906)
204. *L. trapezicollis* Lohse, in Lohse et al. 1990
205. *L. vasta* (Mäklin, 1853)
206. *L. wickhami* (Casey, 1893)
Genus *LYPOGLOSSA* Fenyes, 1918
207. *L. angularis* (Mäklin, 1853)
208. *L. franclemonti* Hoebeke, 1992
Genus *MOCYTA* Mulsant and Rey, 1873c
Amblystegii species group
209. *M. amblystegii* (Brundin, 1952) *
210. *M. breviuscula* (Mäklin, 1852)
211. *M. fungi* (Gravenhorst, 1806) †
Discreta species group
212. *M. sphagnorum* Klimaszewski and Webster, in Klimaszewski et al. 2015 ‡
Genus *PHILHYGRA* Mulsant and Rey, 1873a
213. *P. angusticauda* (Bernhauer, 1909) * [=*Atheta* (*Philhygra*) *pinegensis* Muona, 1983 (synonymized by Brunke et al. in 2021)]
214. *P. botanicarum* (Muona, 1983) *
215. *P. donnellyana* Klimaszewski, **sp. n.**
216. *P. falcifera* Lohse, in Lohse et al. 1990
217. *P. jarmilae* Klimaszewski and Langor, in Klimaszewski et al. 2011
218. *P. junii* Lohse, in Lohse et al. 1990
219. *P. leechi* Lohse, in Lohse et al. 1990
220. *P. malleoides* Lohse, in Lohse et al. 1990 *
221. *P. pseudoboreostiba* Lohse, in Lohse et al. 1990
222. *P. pseudolarsoni* Klimaszewski and Godin, in Klimaszewski et al. 2012
223. *P. pseudopolaris* Klimaszewski and Langor, in Klimaszewski et al. 2011
224. *P. ripicola* (Hanssen, 1932) * [=*P. ripicoloides* Lohse, in Lohse et al. 1990] **New Synonymy**
225. *P. rostrifera* Lohse, in Lohse et al. 1990
226. *P. sinuipennis* Klimaszewski and Langor, in Klimaszewski et al. 2011
227. *P. terrestris* Klimaszewski and Godin, in Klimaszewski et al. 2012
228. *P. varula* (Casey, 1911)
Genus *SCHISTOGLOSSA* Kraatz, 1856
(Subgenus *Boreomorpha* Klimaszewski and Webster, in Klimaszewski et al. 2009)
229. *S. blatchleyi* (Bernhauer and Scheerpeltz, 1926)
(Subgenus *Schistoglossa* Kraatz, 1856)
230. *S. campbelli* Klimaszewski, in Klimaszewski et al. 2009
231. *S. hampshirensis* Klimaszewski, in Klimaszewski et al. 2009

Genus *STRIGOTA* Casey, 1910
232. *S. ambigua* (Erichson, 1839)
233. *S. obscurata* Klimaszewski and Brunke, in Brunke et al. 2012
Genus *TARPHIOTA* Casey, 1893
234. *T. fucicola* (Mäklin, 1852)
235. *T. geniculata* (Mäklin, 1852)
Genus *TRICHIUSA* Casey, 1893
236. *T. pilosa* Casey, 1893
Athetini with Subtribe affiliation uncertain
Genus *PARAGONIUSA* Maruyama and Klimaszewski, 2004
237. *P. myrmicae* Maruyama and Klimaszewski, 2004

Tribe Lomechusini Fleming, 1821

Subtribe Myrmedonina C.G. Thomson, 1867
Genus *ZYRAS* Stephens, 1835
238. *Z. obliquus* (Casey, 1893)

References

Ahn K-J, Ashe S (1995) Revision of the intertidal aleocharine genus *Amblopusa* Casey and description of the new genus *Paramblopusa* (Coleoptera: Staphylinidae). J N Y Entomol Soc 103(2):138–154
Ahn K-J, Ashe S (1996) Phylogeny of the intertidal aleocharine tribe Liparocephalini (Coleoptera: Staphylinidae). Syst Entomol 21:99–114
Ahn K-J, Maruyama M (2000) *Paramblopusa eoa*, a new intertidal species (Coleoptera: Staphylinidae: Aleocharinae) from Kuril Islands. Coleopt Bull 54 (3):359–364
Assing V (2002) A taxonomic and phylogenetic revision of *Amarochara* Thomson. I. The species of the Holarctic region (Coleoptera: Staphylinidae, Aleocharinae, Oxypodini). Beiträge zur Entomologie 52(1):111–204
Assing V (2021) On the taxonomy of *Parocyusa*, *Tectusa*, and miscellaneous genera of Oxypodina (Insecta: Coleoptera: Staphylinidae: Aleocharinae: Oxypodini). Ann Naturhist Mus Wien, B 123:99–218
Benick G (1973) Die von Prof. Dr. Z. Kaszab in der Mongolei gesammelten Arten der Subfamilie Athetae (Col. Staphyl.). Nouvelle Revue d'Entomologie 3 (4):211–217
Bernhauer M (1901a) Die Staphyliniden der paläarktischen Fauna. Verhandlungen der Kaiserlich-Königlichen Zoologisch-Botanischen Gesellschaft in Wien 51(6):430–466. 51(7):467–506
Bernhauer M (1901b) Neue exotische Arten der Gattung *Aleochara* Gravenhorst. Steittiner Entomologishe Zeitschrift 62(3–4):366–373
Bernhauer M (1902) Die Staphyliniden der paläarktischen Fauna. 1. Tribus: Aleocharini (part II). Verhandlungen der kaiserlich-königlichen zoologisch-botanischen Gesellschaft in Wien 52:87–284
Bernhauer M (1905) Neue Aleocharinen aus Nordamerika. Deutsche Entomologische Zeitschrift 1905 (2):245–256
Bernhauer M (1906) Neue Aleocharinen aus Nordamerika. (II. Teil). Deutsche Entomologische Zeitschrift 1906 (2):337–348
Bernhauer M (1907) Neue Aleocharini aus Nordamerika. (Col.) (3. Stück.). Deutsche Entomologische Zeitschrift 1907(4):381–405
Bernhauer M (1909) Neue Aleocharini aus Nordamerika. (Col.) (4. Stück.). Deutsche Entomologische Zeitschrift 1909(4):515–528
Bernhauer M (1928) Neue Staphyliniden der palaearktischen Fauna. Kol Rundsch 14:8–23
Bernhauer M, Scheerpeltz O (1926) Staphylinidae VI. In: Schenkling S (ed) Coleopterorum Catalogus. W. Junk, Berlin, pp 499–988
Blackwelder RE (1952) The generic names of the beetle family Staphylinidae , with an essay on genotypy. US Natl Museum Bull 200(i–iv):1–483
Brundin L (1929) Eine neue *Gnypeta* aus dem Torneträskgebiete (Col. Staphylinidae). Entomologisk Tidskrift 50(1):14–15. pl.1
Brundin L (1940) Zwei neue nordische *Atheta*-Arten aus der *islandica*-Gruppe (Col. Staphylinidae). Vorläufige Mitteilung. Entomologisk Tidskrift 61:131–133
Brundin L (1948) *Microdota*-Studien. (Col. Staphylinidae). Entomologisk Tidskrift 69(1–2):8–66
Brundin L (1952) *Acrotona*-Studien (Gattung *Atheta*). Entomologisk Tidskrift 73:93–145
Brunke A, Klimaszewski J, Dorval J-A, Bourdon C, Paiero SM, Marshall SA (2012) New species and distributional records of Aleocharinae (Coleoptera, Staphylinidae) from Ontario, Canada, with a checklist of recorded species. In: Klimaszewski J., Anderson R. (Eds) Biosystematics and ecology of Canadian Staphylinidae (Coleoptera) II. ZooKeys [Special Issue] 186:119–206
Brunke AJ, Pentinsaari M, Klimaszewski J (2021) Integrative taxonomy of Nearctic and Palaearctic Aleocharinae: new species, synonymies, and records (Coleoptera, Staphylinidae). ZooKeys 1041:27–99
Casey TL (1893) Coleopterological notices. V. Annals of the New York Academy of Sciences 7[1894]:281–606. [these fascicles (Oct–Dec 1893) of the Annals were published in Mar 1894 (Trans. New York Acad. Sci. 13:152; Ent. News 5 (6):187); however, it was listed by Bertkau in his annual "Bericht über die wissenschaftlichen Leistungen im Gebiete der Entomologie während des Jahres 1893", which was published in Archiv für Naturgeschichte, 60. Jahrg., II. Band, Zweites Heft, which was issued in December 1893 (see Inhalt p. V and p. 198); it was also recorded on 24 Jan 1894, as a separate of 325 pp., 1893 (Ann. Soc. Ent. France 63: Bull. Ent.: xxxv), and in Jan 1894 (Naturae Novitates 16 1894 (2):72)]

Casey TL (1906) Observations on the staphylinid groups Aleocharinae and Xantholinini, chiefly of America. Trans Acad Sci St Louis 16(6):125–434

Casey TL (1910) New species of the staphylinid tribe Myrmedoniini. In: Memoirs on the coleoptera. I. New Era Printing Co., Lancaster, PA, pp 1–183. 205 pp

Casey TL (1911) New American species of Aleocharinae and Myllaeninae. In: Memoirs on the coleoptera. II. The New Era Printing Co., Lancaster, PA, pp 1–245. 259 pp

Casey TL (1915) Studies in some staphylinid genera of North America. In: Memoirs on the coleoptera. VI. The New Era Printing Co., Lancaster, PA, pp 395–450. 460 pp

Eppelsheim E (1886) Neue Staphylinen vom Amur. Deutsche Entomologische Zeitschrift 30(1):33–46

Eppelsheim E (1893) Beitrag zur Staphylinen-Fauna des sűdwestlichen Baikal-Gebietes. Deutsche Entomologische Zeitschrift 37:17–67

Erichson WF (1837) Die Käfer der Mark Brandenburg. Erster Band, Erste Abtheilung. F. H. Morin, Berlin, viii + 384 pp

Erichson WF (1839) Genera et species staphylinorum insectorum coleopterorum familiae. Erster Band. F. H. Morin, Berlin, viii, 400 pp

Fenyes A (1909) Two new species of Aleocharinae from California. Proc Entomol Soc Wash 11:197–199

Fenyes A (1918) Coleoptera. Family Staphylinidae. Subfamily Aleocharinae. In: Wytsman P (ed) Genera Insectorum, vol XXVII. Fascicule 173A. M. Nijhoff, Den Haag, pp 1–110. 453 pp. + 7 pls

Fenyes A (1920) Aleocharinae. Genera Insectorum, fasc 173B:111–414

Fenyes A (1921) New genera and species of Aleocharinae with a polytomic synopsis of the tribes. Bulletin of the Museum of Comparative. Zoology 65(2):17–36

Fleming J (1821) Insecta. In: Supplement to the fourth, fifth and sixth editions of the Encyclopaedia Britannica, with preliminary dissertations on the history of sciences, vol 5. A. Constable and Company, Edinburgh, pp 41–56. 584 pp

Ganglbauer L (1895) Die Käfer von Mitteleuropa. Die Käfer der österreichisch-ungarischen Monarchie, Deutschlands, der Schweiz, sowie des französischen und italienischen Alpengebietes, vol 2, Familienreihe Staphylinoidea, I. Theil: Staphylinidae, Pselaphidae. Carl Gerold's Sohn, Vienna. vi + 881 pp

Gravenhorst JLC (1802) Coleoptera Microptera Brunsvicensia nec non exoticorum quotquot exstant in collectionibus entomologorum Brunsvicensium in genera familias et species distribuit. Carolus Reichard, Brunsuigae, lxvi + 206 + [1, errata] pp

Gravenhorst JLC (1806) Monographia Coleopterorum Micropterorum. Henricus Dieterich, Göttingen: xvi + 236 + [12] pp

Gyllenhal L (1810) Insecta Svecica. Classis I: Coleoptera sive Eleuterata. Tomi I pars II. F.J. Leverentz, Scaris. xix + [1] + 660 pp

Hanssen HK (1932) Zwei neue Atheten (aus den nördlichen Norwegen). Norsk Entomologisk Tidsskrift 3[1932–1934]:72–78

Heer O (1839) (1838–1841) Fasciculus secundus. In: Fauna Coleopterorum Helvetica. Pars I. Turici [Zurich]. Orellii, Fuesslini et Sociorum, pp 145–360. xii + 652 pp

Hoebeke ER (1992) Taxonomy and distribution of the athetine genus *Lypoglossa* Fenyes (Coleoptera: Staphylinidae: Aleocharinae) in North America, with description of a new species. J N Y Entomol Soc 100 (2):381–398

Kirby W, Spence W (1837) Insects. Coleoptera. In Richardson, Fauna Boreali-Americana; or the zoology of the northern parts of British America, Norwich. 249 pp

Klimaszewski J (1979) A revision of the Gymnusini and Deinopsini of the world (Coleoptera: Staphylinidae: Aleocharinae). Research Branch Agriculture Canada Monograph 25:1–169

Klimaszewski J (1985) Nomenclatoral changes in the genus *Aleochara* Gravenhorst (Coleoptera: Staphylinidae: Aleocharinae). Coleopt Bull 39:376–376

Klimaszewski J, Pelletier G (2004) Review of the *Ocalea* group of genera (Coleoptera, Staphylinidae, Aleocharinae) in Canada and Alaska: new taxa, bionomics, and distribution. Can Entomol 136 (4):443–500

Klimaszewski J, Winchester NN (2002) Aleocharine rove beetles (Coleoptera Staphylinidae) of the ancient Sitka spruce forest on Vancouver Island, British Columbia, Canada. Mémoires de la Société royale belge d'Entomologie 40:3–126

Klimaszewski J, Pelletier G, Germain C, Hébert C, Humble LM, Winchester NN (2001) Diversity of *Placusa* (Coleoptera: Staphylinidae, Aleocharinae) in Canada, with descriptions of two new species. Can Entomol 133(1):1–47

Klimaszewski J, Pelletier G, Majka C (2004) A revision of Canadian *Leptusa* Kraatz (Col., Staphylinidae, Aleocharinae): new species, new distribution records, key and taxonomic considerations. Belgian. J Entomol 6:3–42

Klimaszewski J, Sweeney J, Price J, Pelletier G (2005) Rove beetles (Coleoptera: Staphylinidae) in red spruce stands, eastern Canada: diversity, abundance, and descriptions of new species. Can Entomol 137(1):1–48

Klimaszewski J, Pelletier G, Germain C, Work T, Hébert C (2006) Review of *Oxypoda* species in Canada and Alaska (Coleoptera, Staphylinidae, Aleocharinae): systematics, bionomics, and distribution. Can Entomol 138(6):737–852

Klimaszewski J, Savard K, Pelletier G, Webster R (2008a) Species review of the genus *Gnypeta* Thomson from Canada, Alaska and Greenland (Coleoptera, Staphylinidae, Aleocharinae): systematics, bionomics and distribution. In: Majka CG, Klimaszewski J (eds)

Biodiversity, biosystematics, and ecology of Canadian coleoptera. ZooKeys 2(Special Issue):11–84

Klimaszewski J, Godin B, Pelletier G, Savard K (2008b) Six new species and records of aleocharine beetles from the Yukon and Alaska (Coleoptera: Staphylinidae: Aleocharinae). Can Entomol 140 (3):265–291

Klimaszewski J, Webster RP, Savard K (2009) First record of the genus *Schistoglossa* Kraatz from Canada with descriptions of seven new species (Coleoptera, Staphylinidae, Aleocharinae). In: Majka, C.G. and Klimaszewski, J. (Eds.) Biodiversity, Biosystematics, and Ecology of Canadian Coleoptera II. ZooKeys [Special Issue] 22:45–79

Klimaszewski J, Langor D, Pelletier G, Bourdon C Perdereau L (2011) Aleocharine beetles (Coleoptera, Staphylinidae) of the province of Newfoundland and Labrador, Canada. Pensoft Series Faunistica No. 98. Pensoft, Sophia-Moscow, 313 pp

Klimaszewski J, Godin B, Bourdon C (2012) Further contributions to the aleocharine fauna of the Yukon Territory, Canada (Coleoptera, Staphylinidae). In: Klimaszewski J., Anderson, R. (Eds.): Biosystematics and Ecology of Canadian Staphylinidae (Coleoptera) II. ZooKeys [Special Issue] 186:207–237

Klimaszewski J, Webster RP, Bourdon C, Pelletier G, Godin B, Langor DW (2015) Review of Canadian species of the genus *Mocyta* Mulsant & Rey (Coleoptera, Staphylinidae, Aleocharinae), with the description of a new species and a new synonymy. Zookeys 487:111–139

Klimaszewski J, Larson DJ, Labrecque M, Bourdon C (2016) Twelve new species and fifty-three new provincial distribution records of Aleocharinae rove beetles of Saskatchewan, Canada (Coleoptera, Staphylinidae). ZooKeys 610:45–112

Klimaszewski J, Webster RP, Langor DW, Brunke A, Davies A, Bourdon C, Labrecque M, Newton AF, Dorval J-A, Frank JH (2018) Aleocharine rove beetles of eastern Canada (Coleoptera, Staphylinidae, Aleocharinae): a glimpse of megadiversity. Springer, Cham. 902 pp

Klimaszewski J, Sikes DS, Brunke AJ, Bourdon C (2019) Species review of the genus *Boreophilia* Benick from North America (Coleoptera, Staphylinidae, Aleocharinae, Athetini): systematics, habitat, and distribution. ZooKeys 848:57–102

Klimaszewski J, Hoebeke ER, Godin B, Davies A, Perry KI, Bourdon C, Winchester N (2020) Aleocharine rove beetles of British Columbia: a hotspot of Canadian biodiversity (Coleoptera, Staphylinidae). Springer, Cham, xvi + 631 pp

Kraatz G (1856) Staphylinii. Naturgeschichte der Insecten Deutschlands. Erste Abtheilung Coleoptera. Zweiter Band. Erste, zweite Lieferung. Bogen 1–24.Nicolai, Berlin, viii + 376 pp

Kraatz G (1857) Genera Aleocharinorum illustrata. Linnaea Entomologica 11:1–43. 2 pls

Laporte FL (1835) Études entomologiques, ou description d'insectes nouveaux, et observations sur leur synonymie. Méquignon-Marvis, Paris, 159 pp

Latreille PA (1802) Histoire naturelle, générale et particulière, des crustacés et des insectes. Familles naturelles des genres. Tome neuvième. F. Dufart, Paris, 416 pp. + pls. 74–80

Leach W E (1819) New genera. In: Samouelle (ed) The Entomologist's Useful Compendium; or an introduction to the knowledge of British insects, comprising the best means of obtaining and preserving them, and a description of the apparatus generally used; etc. Thomas Boys, London. 496 pp. + 12 pls

Lohse GA (1974) Staphylinidae II (Hypocyphtinae und Aleocharinae). In: Freude H, Harde KW, Lohse GA (eds) Die Käfer Mitteleuropas, Band 5. Goecke & Evers, Krefeld, 381 pp

Lohse GA (1989) In: Lohse, Lucht (ed) Die Käfer Mitteleuropas. 1. Supplementband, Krefeld, pp 121–240

Lohse GA, Klimaszewski J, Smetana A (1990) Revision of arctic Aleocharinae of North America (Coleoptera: Staphylinidae). Coleopt Bull 44(2):121–202

Mäklin FW (1852) [New species and notes]. In: Mannerheim, C.G.: Zweiter Nachtrag zur Kaefer-Fauna der Nord-Amerikanischen Laender des Russischen Reiches. Bulletin de la Société Impériale des Naturalistes de Moscou 25(2):283–372

Mäklin FW (1853) [New species and notes]. In: Mannerheim, C.G.: Dritter Nachtrag zur Kaefer-Fauna der Nord-Amerikanischen Laender des Russischen Reiches. Bulletin de la Société Impériale des Naturalistes de Moscou 26(3):95–273

Mäklin FW (1880) Ytterligare diagnoser öfver några nya sibirska Coleopter-arter. Öfversigt af Finska Vetenskaps-Societetens Förhandlingar 22:79–86

Mannerheim CG (1830) Précis d'un nouvel arrangement de la famille des brachélytres de l'ordre des insectes coléoptères. St. Petersburg, 87 pp. [separate, re-issued in Mémoires Présentées à l'Académie Impériale des Sciences de St.-Pétersbourg 1:415–501, in 1831]

Mannerheim CG (1843) Beitrag zur Kaefer-fauna der Aleutischen Inseln, der Insel Sitkha und Neu-Californiens. Bulletin de la Société Impériale des Naturalistes de Moscou 16(2):175–314

Matthews A (1838) Notice of some new genera and species of Brachelytra. Entomological Magazine 5:188–198

Mulsant E, Rey C (1861) Description de quelques insectes nouveaux ou peu connus. Opuscules Entomologiques 12:93–119

Mulsant E, Rey C (1871) Histoire naturelle des Coléoptères de France. Brévipennes. Aléochariens [Huitième Branche, Bolitocharaires]. Deyrolle Fils, Paris. [4] + 321 pp. + [5, explic.] + [4, table, erratum], pls. 1–5

Mulsant E, Rey C (1873a) Description de divers coléoptères brévipennes nouveaux ou peu connus. Opuscules Entomologiques 15:147–189

Mulsant E, Rey C (1873b) Histoire naturelle des Coléoptères de France: Brévipennes Aléochariens [Première branche dinardaires – cinquième branche oligotaires]. Deyrolle Fils, Paris. [4] + 155 pp., pls. 1–2

Mulsant E, Rey C (1873c) Histoire naturelle des Coléoptères de France. Brévipennes. Aléochariens (Suite) [Tableau... des Aléochariens + Septième Branche, Myrmédoniaires, part 1]. Deyrolle, Paris, [2] + 695 pp., pls. 1–5

Mulsant E, Rey C (1874) [Histoire naturelle des Coléoptères de France]. Tribu des brévipennes: Famille des aléochariens (suite): Sixième branche: Aléocharaires. Annales de la Société Linnéenne de Lyon (NS) 20(1873):285–447

Mulsant E, Rey C (1875) Histoire naturelle des Coléoptères de France: Brévipennes Aléochariens (suite). Myrmédoniaires (2eme partie). Deyrolle, Paris: [2] + 470 pp. + 4 pls

Munster T (1923) Bidrag til kjendskapen om slekten *Atheta* Thoms. (Col., Staph.) Norsk Entomologisk Tidsskrift 1[1920–1923](4):206–208

Muona J (1983) Two new Palearctic *Atheta* species (Coleoptera, Staphylinidae). Annales Entomologici Fennici 49(2):57–58

Newton AF (2017) Nomenclatural and taxonomic changes in Staphyliniformia (Coleoptera). Insecta Mundi 0595:1–52

Pace R (1982) *Oncosomechusa besucheti* n. gen., n. sp., dell India et *Emmelostiba besuchetti* n. gen., n. sp., del Libano (Coleoptera, Staphylinidae). Rev Suisse Zool 111(3):457–523

Palm T (1966) De svenska *Gnypeta*-arterna (Col. Staphylinidae, Aleocharinae). Norsk Entomologisk Tidskrift 87(3–4):136–141

Paśnik G (2006) A revision of the World species of the genus *Tachyusa* Erichson, 1837 (Coleoptera, Staphylinidae: Aleocharinae). Zootaxa 1146:1–152

Paykull G (1800) Fauna Suecica. Insecta. Tomus III. J.F. Edman, Upsala: 459 pp.

Redtenbacher L (1849) Fauna Austriaca. Die Käfer. Nach der analytischen Methode bearbeitet. C. Gerold, Wien: pp. I–xxvii + 641–883

Roubal J (1928) Zwei neue Coleopteren aus der Tschechoslovakei. Wiener Entomologische Zeitung 4(1):27–30

Rye EC (1866) Descriptions of new species of Brachelytra. Entomologist's Monthly Magazine 3 (11):121–125

Sahlberg CR (1831) [Pp. 361–408]. In: Insecta Fennica, dissertationibus academicis, A. 1817-1834 editis. Pars I: A. J.C. Frenckel, Helsingfors, VIII + 519 pp

Sahlberg JR (1880) Bidrag till Nordvestra Sibiriens Insektfauna. Coleoptera. Insamlade under Expeditionerna till obei och Jenessey 1876 och 1877. I. Cicindelidae, Carabidae, Dytiscidae, Hydrophilidae, Gyrynidae, Dryopidae, Georyssidae, Limnichidae, Heteroceridae, Staphylinidae och Micropeplidae. Kongliga Svenska Vetenskaps-Akademiens Handlingar 17(4):1–115

Say T (1836) Description of new North American insects and observations on some already described. Trans Am Philos Soc VI:155–190

Scriba W (1866) *Homalota subtilis* Scriba n. sp. Berliner. Entomologische Zeitschrift 10(1–3):128

Seevers CH (1951) A revision of the North American and European staphylinid beetles of the subtribe Gyrophaenae (Aleocharinae, Bolitocharini). Fieldiana Zoology 32(10):657–762

Seevers CH (1978) A generic and tribal revision of the North American Aleocharinae (Coleoptera: Staphylinidae) [with additions and annotations by Lee H. Herman]. Fieldiana Zoology 71(i–vi):1–289

Sharp D (1869) A revision of the British species of *Homalota*. Trans Entomol Soc London XIV:91–272

Sharp D (1883) Fam. Staphylinidae (in part). In: Biologia Centrali Americana. Insecta, Coleoptera, vol 1(2). Taylor and Francis, London, pp 145–312, pls. 5–7

Sjöberg O (1950) Drei neue Käfer aus der borealen Region Schwedens. Entomologisk Tidskrift 71(2):159–160. (3–4): 161–164

Stephens JF (1832) (1832–1835) Illustrations of British Entomology. Mandibulata, V. Baldwin & Cradock, London. pp. 1–240, pls. 24–26

Stephens JF (1835) Illustrations of British entomology; or, a synopsis of indigenous insects; containing their generic and specific distinctions; with an account of their metamorphoses, times of appearance, localities, food, and economy, as far as practicable, vol 5, pp 369–448

Thomson CG (1858) Försök till uppställning af Sveriges Staphyliner. Öfversigt af Kongliga Vetenskaps-Akademiens Förhandlingar 15(1):27–40

Thomson CG (1859) Skandinaviens Coleoptera, synoptiskt bearbetade. I. Tom. Berlingska Boktryckeriet, Lund, [3] + 290 pp

Thomson CG (1867) Skandinaviens Coleoptera, synoptiskt bearbetade. Supplementum. Tom IX. Lund: Lundbergska Boktryckeriet, 407 pp

Waltl J (1838) Verzeichniss der um Passau vorkommenden seltnern Käfer [cols. 263–267]; Um Passau vorkommende neue Arten von Käfern [cols. 267–273]. In: Beyträge zur nähern naturhistorischen Kenntniss des Unterdonaukreises in Bayern. Isis von Oken 1838(4):250–273

Webster RP, Klimaszewski J, Bourdon C, Sweeney JD, Hughes CC, Labrecque M (2016) Further contribution to the Aleocharinae (Coleoptera, Staphilinidae) fauna of New Brunswick and Canada including description of 27 new species. In: Webster, R.P., Bouchard, P., Klimaszewski, J. (Eds) The Coleoptera of New Brunswick and Canada: providing baseline biodiversity and natural history data. ZooKeys [Special Issue] 573:85–216

Wollaston TV (1855) Descriptions of two Coleopterous insects new to British fauna. Zoologists 13:205–207

8 Key to Aleocharinae Tribes of Arctic and Subarctic North America

Diagnosis Aleocharinae can be recognized by the following combination of characters (Fig. 8.49a–d): anterior tentorial pits located at inner corner of antennal fossae which may be connected by frontal suture (e.g. *Oxypoda*); antennal insertions exposed, usually located between eyes and posterior to a line drawn between anterior edges of eyes (except in species with reduced eyes); elytra transverse to slightly elongate; posterior coxae expanded laterally under the femora; male tergite IX usually subdivided (except in *Gymnusa*), and extends lateroposteriad forming two ventral struts; unique tergal defensive gland with associated structures present in higher Aleocharinae at base of segment VII in adults and at apex of segment VIII in larvae; aedeagus trilobate, with large and complex parameres (lateral lobes). The form of the parameres is unique to Aleocharinae (consisting of paramerite and condylite, with paramerite subdivided into several parts as illustrated in chapter 3, it is a shared derived feature that provides strong evidence for monophyly of this subfamily (Klimaszewski et al. 2018).

Key to Tribes

(Figs. 8.49–8.69) (Modified from Klimaszewski et al. 2011, 2018, 2020)

1. Body moderately convex; integument dull, with impressed and distinct punctation, microsculpture coarse and consisting of convex sculpticells (Fig. 8.50a, 8.62a); pronotum slightly broadened anteriad (Fig. 8.62a), abdomen broadest in apical one-third (Fig. 8.62a); glossae separated into two subparallel narrow lobes (Fig. 8.50b); labial palpi with basal palpomere swollen and terminal palpomere short and truncate apically (Fig. 8.50b); tarsal formula 5-5-5. ....**BOREOCYPHINI Klimaszewski and Langor**

– Body shaped differently; other characters different or in different combinations **2**

2. Body form distinctive (Fig. 8.62b–c), head with neck more than one-third as wide as head or without distinct neck (Fig. 8.62b–c); pronotum without median sulcus, median elongate depression, or basal foveae (Fig. 8.62b–c); for important characters see Fig. 8.51 .... **TACHYUSINI C.G. Thomson**

– Body form and other characters not as above **3**

3. Posterior margin of first two first visible abdominal tergites each with ctenidium of short comb-like cuticular teeth (projections) consisting of one (*Gymnusa*), or two types of teeth (*Deinopsis*) (Fig. 8.52b); tarsal formula 5-5-5 (*Gymnusa*) (Fig. 8.62d), or 3-3-3 (*Deinopsis*) (Fig. 8.62e); body as illustrated (Fig. 8.62d–e); for important characters see Fig. 8.52................. ....**GYMNUSINI Heer**

– Posterior margin of basal abdominal tergites without ctenidium; body not as above **4**

J. Klimaszewski et al., *A Faunal Review of Aleocharine Beetles in the Rapidly Changing Arctic and Subarctic Regions of North America (Coleoptera, Staphylinidae)*,
https://doi.org/10.1007/978-3-030-68191-3_8

4. Body robust (*Aleochara*) (Fig. 8.62f) to moderately slender (*Amarochara*, not recorded in the North); maxillary and labial palpi with pseudosegment on last palpomere, appearing as five and four palpomeres respectively (*Aleochara*, Fig. 8.53a), antennae moderately to distinctly robust or swollen; pronotal hypomera not or only narrowly visible in lateral view; tarsal formula 5-5-5 or 4-5-5 (subgenus *Tinotus*); for important characters see Fig. 8.53 .... **ALEOCHARINI Fleming**

– Body of a different shape; other characters in different combination **5**

5. Body wide and flattened (Fig. 8.63a); pronotum with anterolateral angles obtuse, pubescence short, dense and evenly distributed, directed straight posteriad on pronotum and elytra (Figs. 8.54a, 8.63a); tarsal formula 4-4-5; for important characters see Fig. 8.54 .... **PLACUSINI Mulsant and Rey**

– Body and other characters different; tarsal formula variable **6**

6. Labial palps very long and/or stylate (Figs. 8.55b, 8.56b), riparian or intertidal species **7**

– Characters not as above **8**

7. Body fusiform (Figs. 8.55a, 8.63b); pronotum widely rounded apically in dorsal view; hypomera not visible in lateral view; abdomen strongly tapering and bearing numerous robust macrosetae (Figs. 8.55a, 8.63b); tarsal claws slender without spatulate setae; riparian habitats; for important characters see Fig. 8.55 .... **MYLLAENINI Ganglbauer**

– Body form diverse, robust in *Liparocephalus* or slender in *Amblopusa*, *Paramblopusa* (Fig. 8.63c–d); abdomen dilated to subapex, without numerous robust macrosetae; tarsal claws robust with spatulate setae; intertidal species; for important characters see Fig. 8.56 .... **LIPAROCEPHALINI Fenyes**

8. Body minute and compact (Fig. 8.63e), length 0.8–1.6 mm; antennae with 10 antennomeres, antennomeres VIII–X enlarged and forming loose club (Fig. 8.57b); hind coxa with large ventral lamella covering base of femur; tarsal formula 4-4-4; for important characters see Fig. 8.57 .... **HYPOCYPHTINI Laporte**

– Body elongate, small to large-sized (1.0–4.5 mm); antennae with 11 antennomeres and without club; hind coxae without ventral lamella covering base of femur; tarsal formula variable **9**

9. Body narrowly elongate or ovoid, body compact or loosely articulated, moderately to distinctly flattened, usually compact, often well sclerotized and with coarse punctation (Figs. 8.63f, 8.64a–c); head with broad neck (not always well visible because head is contracted and covered by pronotum); mandibles with minute denticles in ventral molar region (Fig. 8.58a); prementum with bases of medial setae very close together, medial pseudopore field very narrow (Fig. 8.58b); ligula broadly rounded, completely split (divided from apex to base) or bifid (divided only apically) (Fig. 8.58b); mesocoxae broadly or narrowly separated; tarsal formula 4-4-5; most species associated with fungi or dead wood................... **HOMALOTINI Heer**

– Body shape different; other characters different or in different combinations **10**

10. Body shape diverse, spindle-shaped or tear-shaped in most (Figs. 8.64d–f, 8.65a–f, 8.66a–b); head with frontal suture present in many genera (appears as a fine suture/line between lower margins of eyes) (Fig. 8.59a); median lobe of aedeagus without athetine bridge (transverse sclerotized strip extending from subapical lateral edges of median lobe in front of compressor plate); tarsal formula 5-5-5 in Arctic and Subarctic taxa; for important characters see Fig. 8.59 .... **OXYPODINI C.G. Thomson**

– Body shape not as above, subparallel in most, a few species robust; head without frontal suture; median lobe of aedeagus with athetine bridge present (Figs. 8.60c, 8.61d); tarsal formula 4-5-5 .... **11**

11. Body rather robust (Fig. 8.69f); maxilla with galea and lacinia moderately to highly elongate (Fig. 8.61b); mesocoxae widely separated by mesoventral and metaventral processes, metaventral process longer than mesoventral

(Fig. 8.61a); compressor plate of median lobe of aedeagus covered by plate which extends from base of bulbus (Fig. 8.61d); most taxa associated with ants; for important characters see Fig. 8.61 .... **LOMECHUSINI Fleming**

– Body elongate, loosely articulated and very flexible, only some species robust (Figs. 8.66c–f, 8.67a–f, 8.68a–f, 8.69a–e); maxilla with galea and lacinia moderately elongate (Fig. 8.60b); mesocoxae narrowly separated by mesoventral process, which is narrow and moderate in length; compressor plate of median lobe of aedeagus exposed and not covered by a plate which extends from base of bulbus (Fig. 8.60c); some species associated with macro fungi but most found in decaying organic matter; for important characters see Fig. 8.60............................ **ATHETINI Casey**

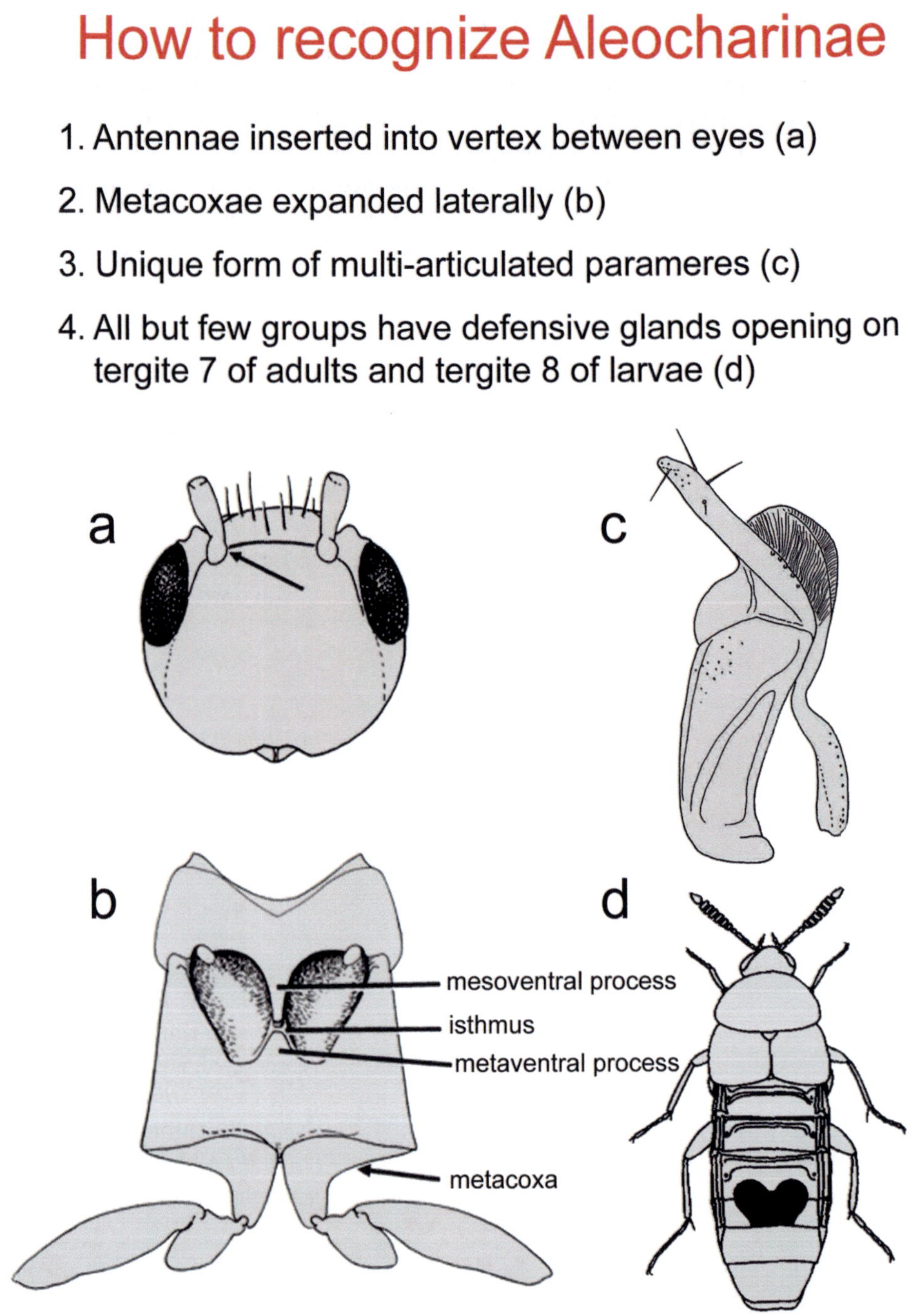

Fig. 8.49 (**a–d**). List and schematic illustrations of major characters for recognition of subfamily Aleocharinae of family Staphylinidae. (**a**) head, dorsal view; (**b**) meso- and metaventrite of ventral thorax (coxae removed), arrow indicating metacoxa; (**c**) paramere; (**d**) schematic aleocharine, solid black area indicating position of (internal) abdominal defensive glands. (**a–b**) modified from Newton et al. (2001)

Tribe BOREOCYPHINI Klimaszewski and Langor

1. Body with impressed punctation and coarse microsculpture of convex sculpticells (a)
2. Maxillary palpi with 4 palpomeres, apical palpomere much shorter and narrower than penultimate (a)
3. Labial palpi with 3 palpomeres, first swollen, second short and wide, third elongate with short and truncate microsegment; glossae separated as short subparallel lobes (b)

Fig. 8.50 (**a–b**). List and illustrations of major characters for recognition of the tribe Boreocyphini Klimaszewski and Langor. (**a**) dorsolateral habitus of *Boreocypha*, arrow indicating maxillary palpus; (**b**) labium, including labial palpi (L) and glossae (G). (**a–b**) after Klimaszewski et al. (2011)

Tribe TACHYUSINI C.G. Thomson

1. Metaventral process not longer than mesoventral process (a)
2. Medial pore field of epipharynx with less than 25 pseudopores (b)
3. Ligula Y-shaped, split at apex, with lobes diverging (c)
4. Apical third of lacinia with 6-8 contiguous spines and 2-5 clearly isolated (d)
5. Maxilla with median sclerite of stipes with 3 long setae
6. Tarsal formula 4-5-5
7. Median lobe of aedeagus with distinct, blade-like process (e)

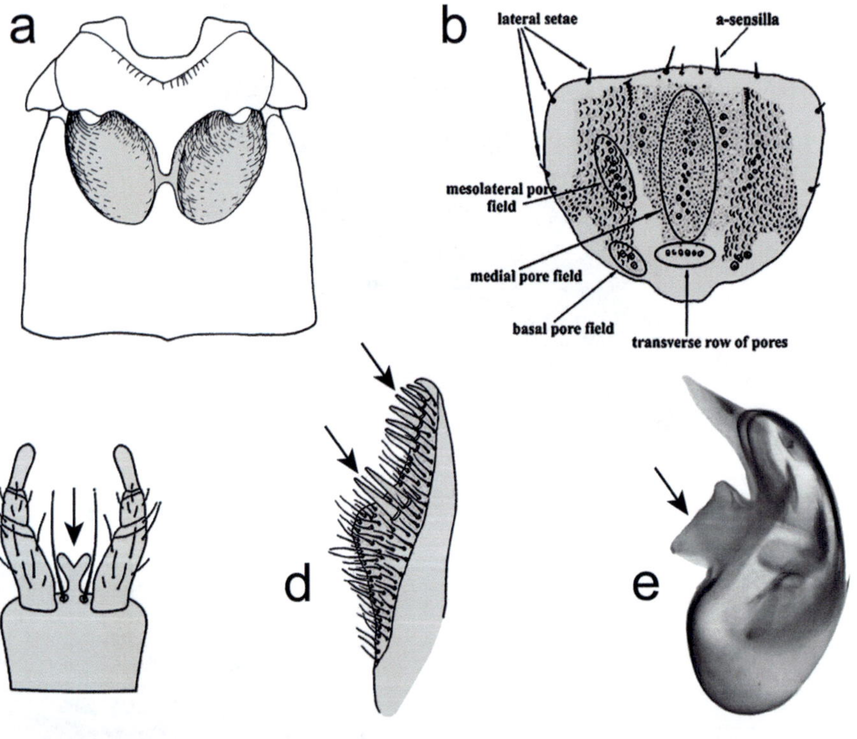

Fig. 8.51 (**a–e**). List and illustrations of major characters for recognition of the tribe Tachyusini C.G. Thomson. (**a**) meso- and metaventrite of ventral thorax (coxae removed); (**b**) epipharynx; (**c**) labium, including ligula (arrow); (**d**) lacinia of maxilla, showing spine groups (arrows); (**e**) aedeagus, lateral view, blade-like process (arrow). (**a–d**) modified from Newton et al. (2001)

Tribe GYMNUSINI Heer

1. Sublimuloid/tear-shaped body (a)
2. Antennae filiform (a)
3. Apex of abdominal segments III-VI (first four visible) with dense comb of short cuticular projections, consisting of one (*Gymnusa*) or two (*Deinopsis*) types of teeth (b)
4. Maxillary palps long, stylate (c)
5. Labial palpi elongate (d)
6. Tarsal formula: 3-3-3 (*Deinopsis*) or 5-5-5 (*Gymnusa*)

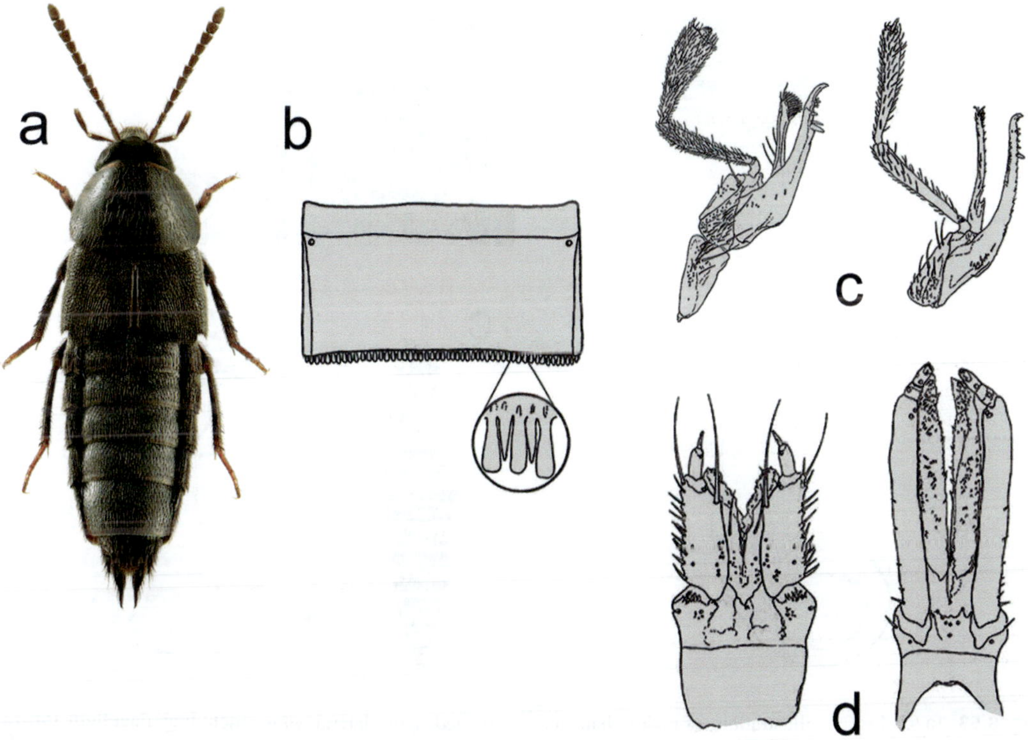

Fig. 8.52 (**a–d**). List and illustrations of major characters for recognition of the tribe Gymnusini Heer. (**a**) dorsal habitus; (**b**) apical margin of abdominal segment, showing cuticular projections (*Deinopsis*); (**c**) maxilla, with maxillary palpus (left—*Deinopsis*, right—*Gymnusa*); (**d**) labium, with labial palpi (left—*Deinopsis*, right—*Gymnusa*). (**b–d**) modified from Newton et al. (2001)

Tribe ALEOCHARINI Fleming

1. Body robust to moderately slender
2. Antennae moderately slender to dilated toward apex
3. Maxillary and labial palps each with pseudosegment on apical palpomere (appearing 5- and 4-segmented) (a)
4. Pronotal hypomeron not or narrowly visible in lateral view
5. Tarsal formula 5-5-5 or 4-5-5 in *Aleochara* (*Tinotus*)
6. Flagellum of median lobe of aedeagus long, longer than median lobe in many species (b)
7. All known larvae of *Aleochara* are ectoparasitoids of Dipteran pupae, some *Amarochara* associated with ants (c)

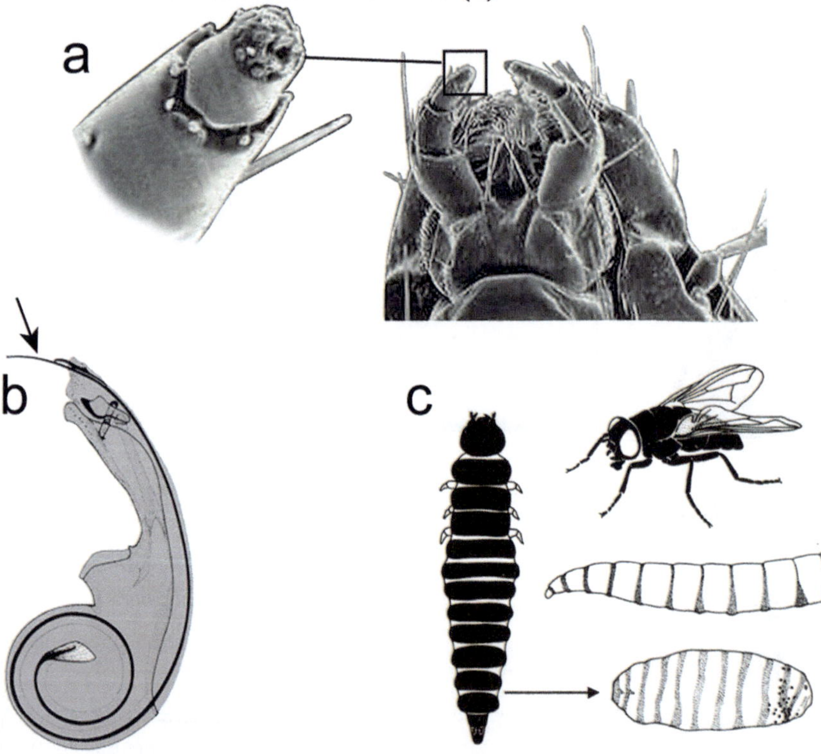

Fig. 8.53 (**a–c**). List and illustrations of major characters for recognition of the tribe Aleocharini Fleming. (**a**) labial palpus, including apical segment bearing pseudosegment; (**b**) aedeagus, lateral view, including flagellum (arrow); (**c**) parasitoidism of Diptera in *Aleochara*. (**a**) after Klimaszewski (1984)

Tribe PLACUSINI Mulsant and Rey

1. Body flattened, all pronotal pubescence directed posteriad (a)
2. Mandibles with dorso-basal velvety patch modified as transverse rows of larger teeth; without denticles in ventral molar region (b)
3. Ligula short, broad and rounded (c)
4. Prementum with bases of medial setae very close together, medial pseudopore field very narrow (c)
5. Tarsal formula 4-4-5

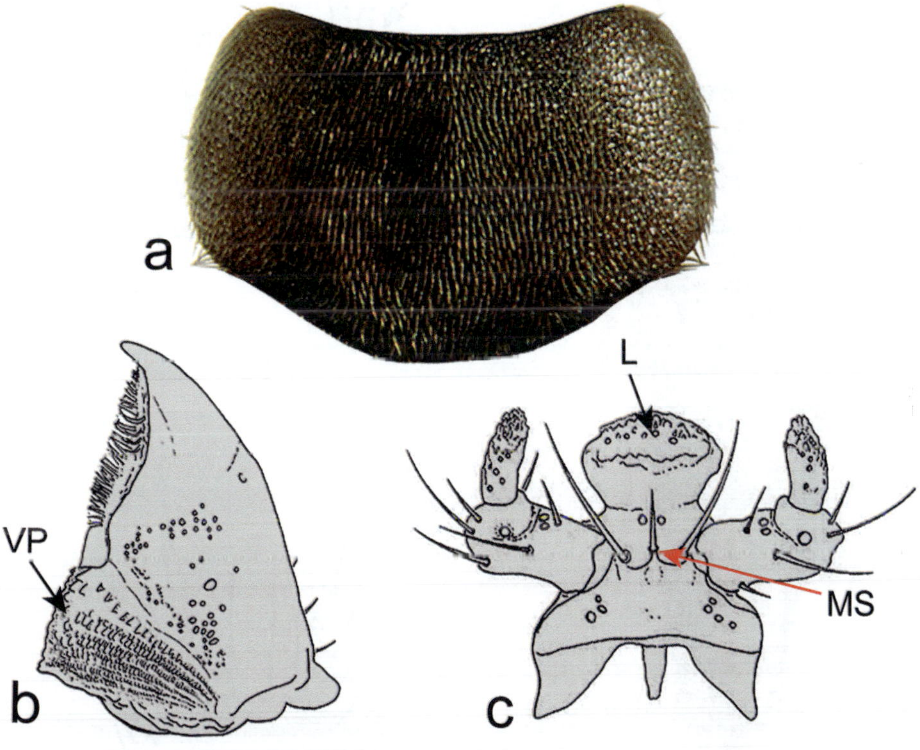

Fig. 8.54 (**a–c**). List and illustrations of major characters for recognition of the tribe Placusini Mulsant and Rey. (**a**) pronotum; (**b**) mandible, showing transverse rows of teeth of velvety patch (VP); (**c**) labial, including ligula (L) and medial setae (MS) of prementum. (**b–c**) modified from Newton et al. (2001)

Tribe MYLLAENINI Ganglbauer

1. Body fusiform, abdomen tapering apically (a)
2. Abdomen with coarse macrosetae (a)
3. Lacinia of maxilla long and stylate, with widely scattered teeth medially (b)
4. Maxillary palpi elongate and labial palpi stylate (b)
5. Mentum with anterolateral margins with spinose processes (c)
6. Tarsal formula 4-4-5

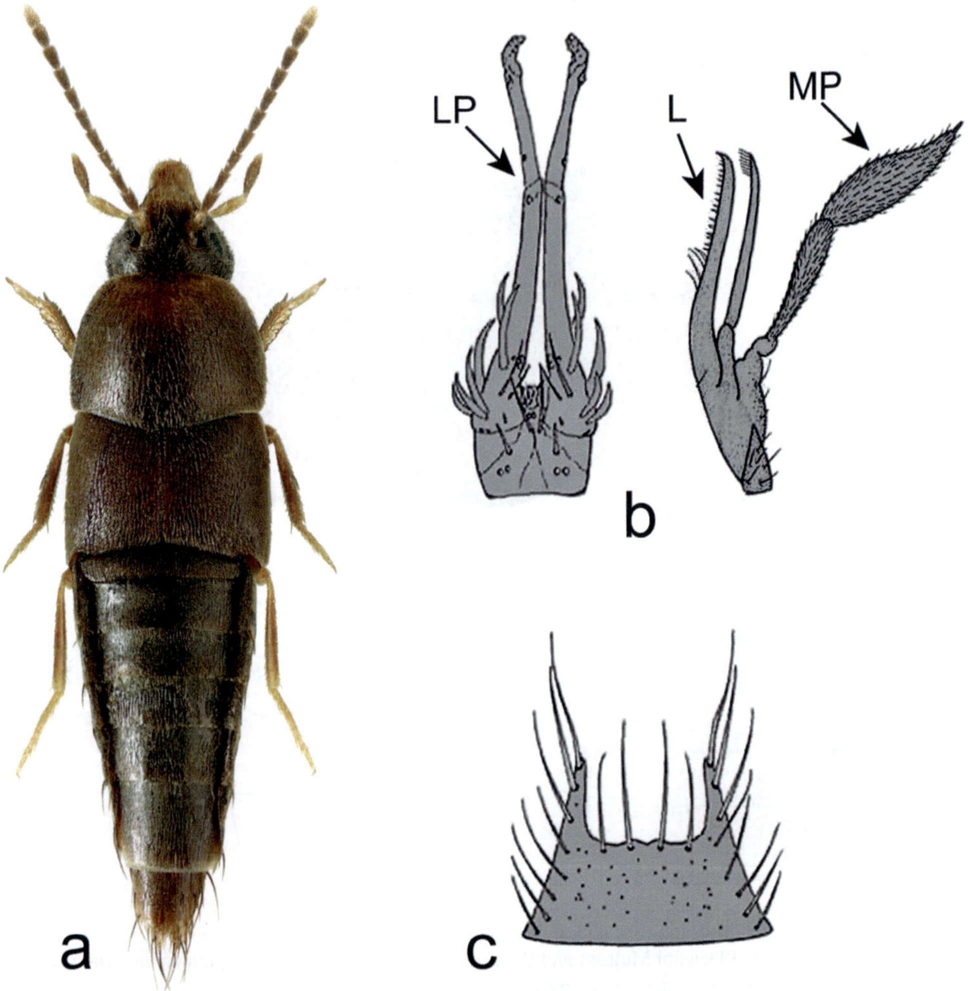

Fig. 8.55 (**a–c**). List and illustrations of major characters for recognition of the tribe Myllaenini Ganglbauer. (**a**) habitus; (**b**) labium (left), with labial palpus (LP) and maxilla (right) with lacinia (L) and maxillary palpus (MP); (**c**) mentum. (**b–c**) modified from Newton et al. (2001)

Tribe LIPAROCEPHALINI Fenyes

1. Body shape diverse, slender to robust; integument covered with short setae (a)
2. Elytra very short, at suture shorter than pronotum; wings absent (a)
3. Labial palpi elongate, with two palpomeres; ligula long, slender and entire at apex (b)
4. Prementum with one median seta and narrow pseudopore field (b)
5. Tarsal formula 4-4-5 or 4-4-4, tarsal claws large (a)

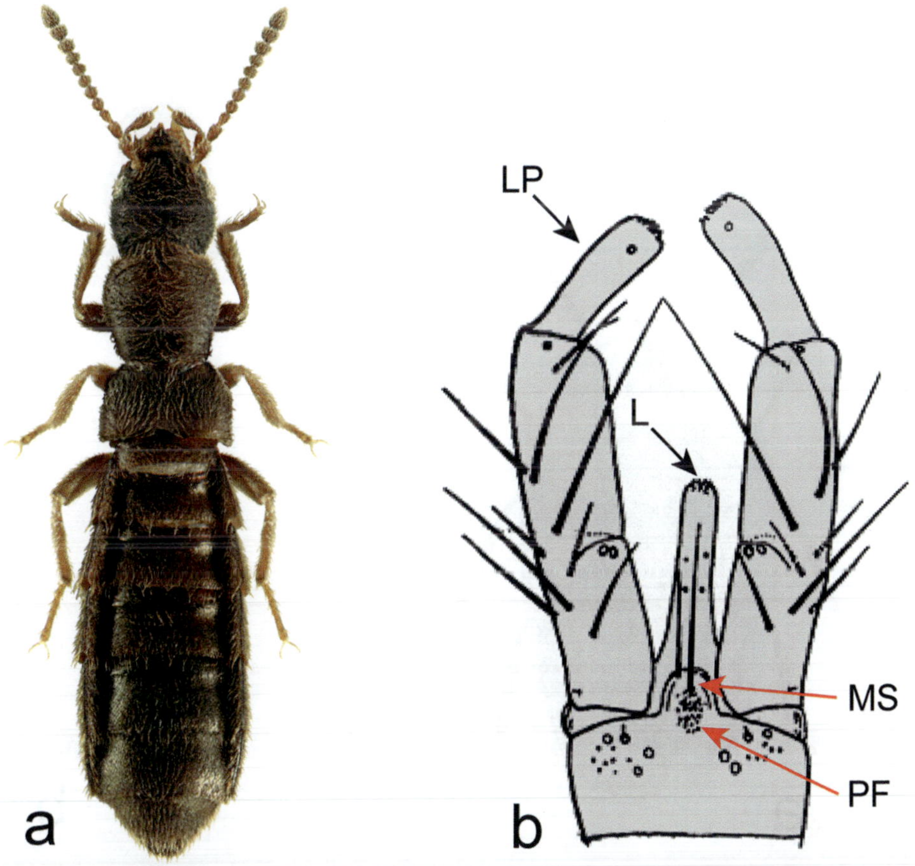

Fig. 8.56 (**a–b**). List and illustrations of major characters for recognition of the tribe Liparocephalini Fenyes. (**a**) habitus; (b) labium, with labial palpus (LP), ligula (L), median seta of prementum (MS) and pseudopore field (PF). (**b**) after Maruyama and Ahn (2000)

Tribe HYPOCYPHTINI Laporte

1. Body small, compact, rounded to subparallel, length 0.8-1.6 mm (a)
2. Antennae with 10 segments, VIII-X forming loose club (b)
3. Hind coxa with large ventral lamella covering base of femur
4. Pronotal hypomeron not (*Cypha*) or only partially visible (*Oligota*) in lateral view
5. Tarsal formula 4-4-4

Fig. 8.57 (**a–b**). List and illustrations of major characters for recognition of the tribe Hypocyphtini Laporte. (**a**) habitus; (**b**) antenna of *Oligota* (left) and *Cypha* (right)

Tribe HOMALOTINI Heer

1. Body shape diverse
2. Mandible with minute teeth in ventral molar region (a)
3. Prementum with bases of medial setae very close together, medial pseudopore field very narrow (b)
4. Tarsal formula 4-4-5

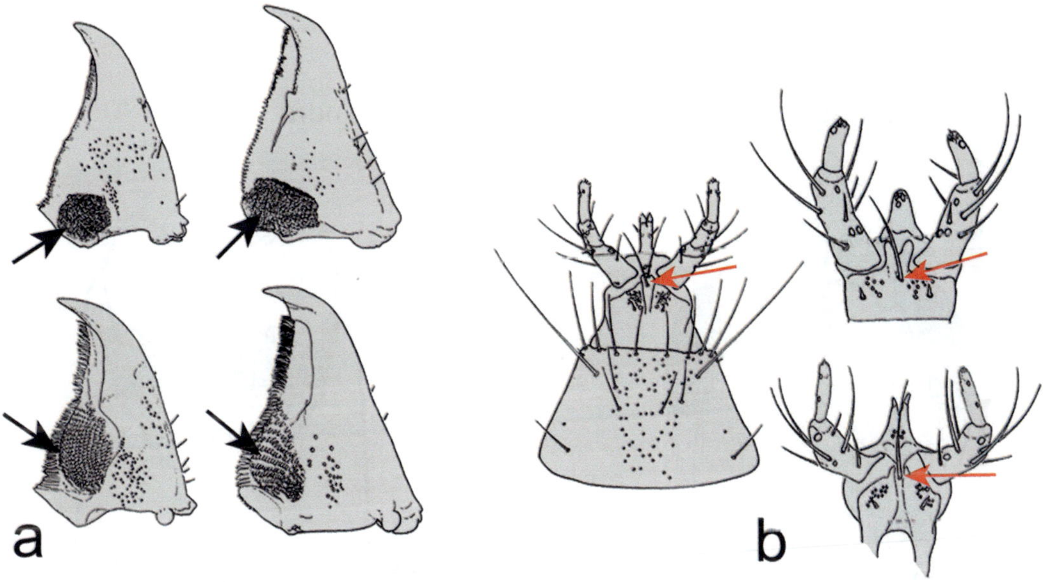

Fig. 8.58 (**a–b**). List and illustrations of major characters for recognition of the tribe Homalotini Heer. (**a**) examples of mandibles, with patch of minute teeth (arrows); (**b**) labium, with prementum bearing medial setae (arrows). (**a–b**) modified from Newton et al. (2001)

Tribe OXYPODINI C.G. Thomson

1. Body narrowly elongate or ovoid, segments loosely articulated
2. Head without distinct neck in most Arctic and Subarctic genera (a)
3. Frontal suture of head present in *Calodera*, *Devia*, *Oxypoda, Ocyusa, Parocyusa* (a)
4. Mesocoxae narrowly separated in most (b)
5. Pronotal hypomeron not visible in lateral view in *Devia*, *Hylota*, *Oxypoda* (c)
6. Tarsal formula 5-5-5 in Arctic and Subarctic genera

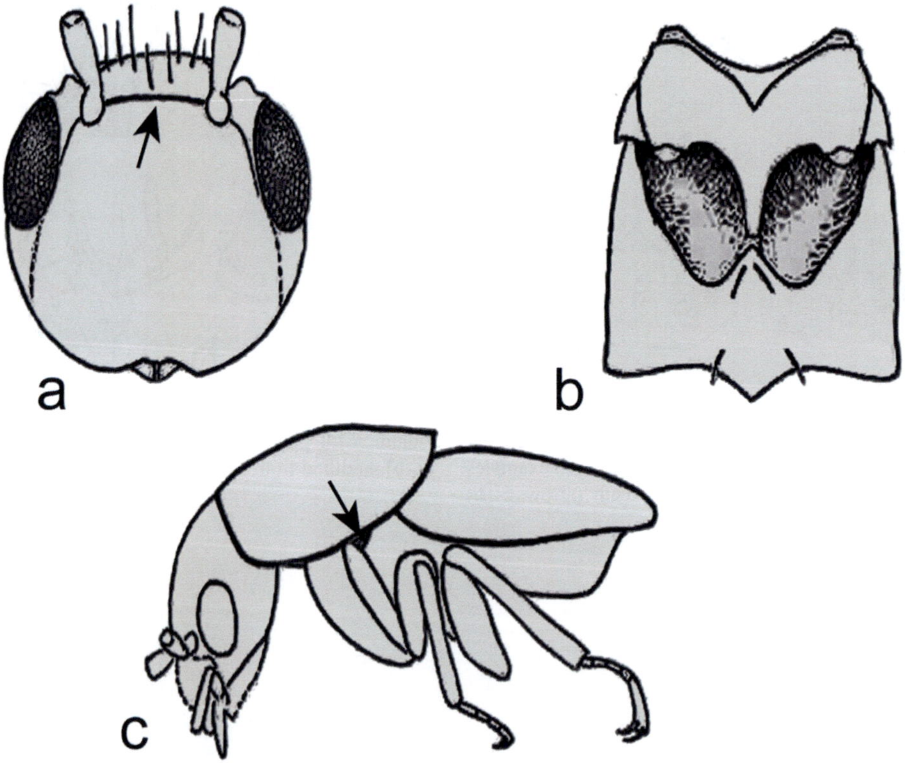

Fig. 8.59 (**a–c**). List and illustrations of major characters for recognition of the tribe Oxypodini C.G. Thomson. (**a**) head, showing frontal suture (arrow); (**b**) meso- and metaventrite of ventral thorax (coxae removed); (**c**) forebody in lateral view, hypomeron not visible in lateral view (arrow). (**a–c**) modified from Newton et al. (2001)

Tribe ATHETINI Casey

1. Body slender (except *Paragoniusa*), 0.9-6.0 mm
2. Head without distinct neck, postocular carina entire or incomplete
3. Pronotum with distinctive pubescence pattern (a)
4. Pronotal hypomeron visible in lateral view (except *Acrotona*, *Mocyta*, *Strigota*)
5. Mesocoxae narrowly to moderately separated; mesoventral process slender, extending to 1/3-1/2 length of mesocoxa
6. Maxillary palpi with four palpomeres, labial palpi with three palpomeres (b)
7. Tarsal formula 4-5-5

6. Aedeagus with 'athetine bridge' (c)

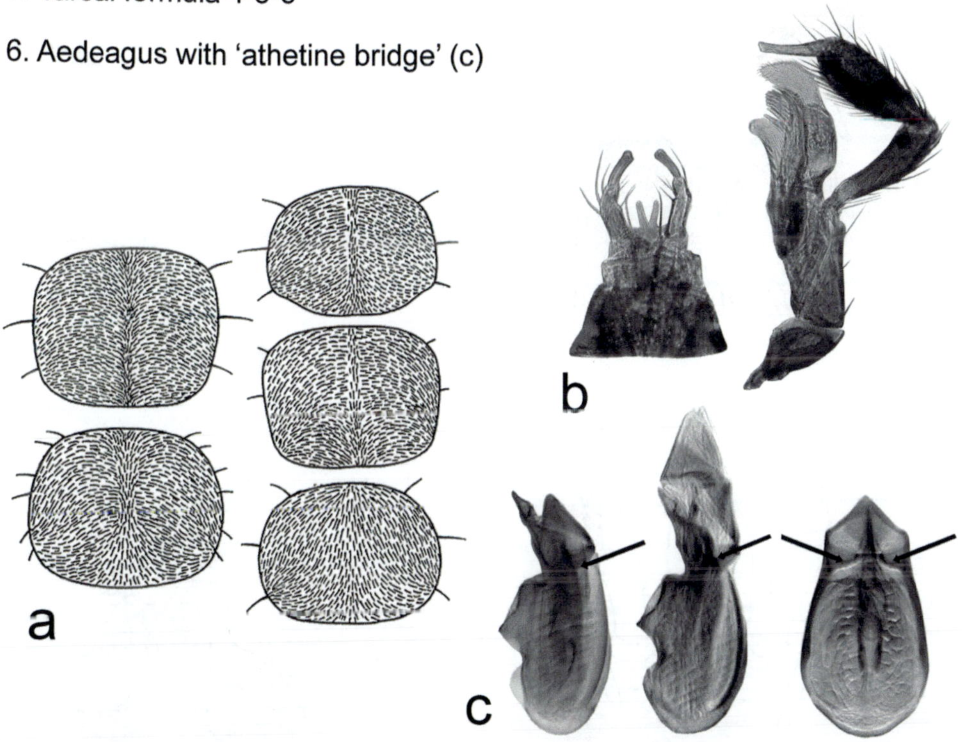

Fig. 8.60 (**a–c**). List and illustrations of major characters for recognition of the tribe Athetini Casey. (**a**) examples of pronotal pubescence pattern; (**b**) labium (left) and maxilla (right), with palpi; (**c**) aedeagi, with 'athetine bridge' (arrows). (**a**) after Lohse (1974)

Tribe LOMECHUSINI Casey

1. Body elongate but robust in single Arctic/Subarctic species, > 4 mm
2. Mesocoxae broadly separated by broad meso- and metaventral processes, metaventral process longer than mesoventral (a)
3. Galea and lacinia of maxilla distinctly elongate (b)
4. Ligula bifid (c)
5. Tarsal formula 4-5-5
6. Median lobe of aedeagus with 'athetine bridge', compressor plate covered by extension from base of bulbus (d)
7. Single Arctic/Subarctic species loosely associated with ants

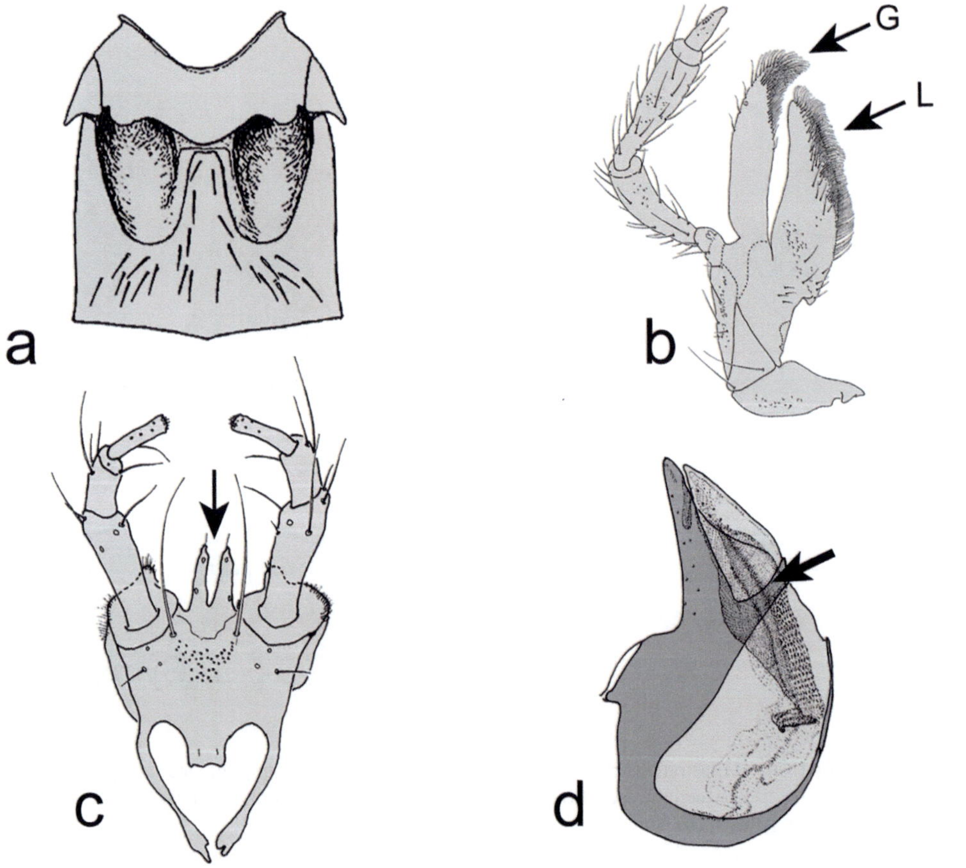

Fig. 8.61 (**a–d**). List and illustrations of major characters for recognition of the tribe Lomechusini Fleming. (**a**) meso- and metaventrite of ventral thorax (coxae removed); (**b**) maxilla, with galea (G) and lacinia (L); (**c**) labium with ligula (arrow); (**d**) aedeagus, with 'athetine bridge' of median lobe (arrow). (**a–d**) modified from Newton et al. (2001)

Fig. 8.62 (a–f). Habitus in dorsal view: (**a**) *Boreocypha websteri* Klimaszewski and Langor; (**b**) *Brachyusa helenae* (Casey); (**c**) *Gnypeta atrolucens* Casey; (**d**) *Gymnusa konopackii* Klimaszewski; (**e**) *Deinopsis canadensis* Klimaszewski; (**f**) *Aleochara gracilicornis* Bernhauer. Scale bar = 1 mm

Fig. 8.63 (**a–f**). Habitus in dorsal view: (**a**) *Placusa tacomae* Casey; (**b**) *Myllaena insomnis* Casey; (**c**) *Amblopusa breviceps* Casey; (**d**) *Diaulota densissima* Casey; (**e**) *Cypha inexpectata* Klimaszewski and Godin; (**f**) *Gyrophaena nana* (Paykull). Scale bar = 1 mm

Fig. 8.64 (**a–f**). Habitus in dorsal view: (**a**) *Leptusa gatineauensis* Klimaszewski and Pelletier; (**b**) *Silusa vesperis* Casey; (**c**) *Silusa californica* (Casey); (**d**) *Calodera parviceps* (Casey); (**e**) *Devia prospera* (Erichson); (**f**) *Gnathusa eva* Fenyes. Scale bar = 1 mm

Fig. 8.65 (**a–f**). Habitus in dorsal view: (**a**) *Hylota ochracea* Casey; (**b**) *Mniusa yukonensis* (Klimaszewski and Godin); (**c**) *Neothetalia canadiana* Klimaszewski; (**d**) *Ocyusa canadensis* Lohse; (**e**) *Oxypoda sylvia* Casey; (**f**) *Parocalea pseudobaicalica* Lohse. Scale bar = 1 mm

Fig. 8.66 (**a–f**). Habitus in dorsal view: (**a**) *Phloeopora gilbertae* Klimaszewski and Webster; (**b**) *Parocyusa fuliginosa* (Casey); (**c**) *Acrotona sphagnorum* Klimaszewski and Webster; (**d**) *Adota maritima* (Mannerheim); (**e**) *Aloconota cambrica* (Wollaston); (**f**) *Amischa praelonga* Casey. Scale bar = 1 mm

Fig. 8.67 (**a–f**). Habitus in dorsal view: (**a**) *Atheta ventricosa* Bernhauer; (**b**) *Atheta graminicola* (Gravenhorst); (**c**) *Atheta dadopora* C.G. Thomson; (**d**) *Atheta terranovae* Klimaszewski and Langor; (**e**) *Atheta atramentaria* (Gyllenhal); (**f**) *Atheta subsinuata* (Erichson). Scale bar = 1 mm

Fig. 8.68 (**a–f**). Habitus in dorsal view: (**a**) *Bellatheta micropter*a (Lohse); (**b**) *Boreophilia fusca* (CR Sahlberg); (**c**) *Dinaraea pacei* Klimaszewski and Langor; (**d**) *Earota dentata* (Bernhauer); (**e**) *Hydrosmecta pseudodiosica* Lohse; (**f**) *Liogluta nigropolita* (Bernhauer). Scale bar = 1 mm

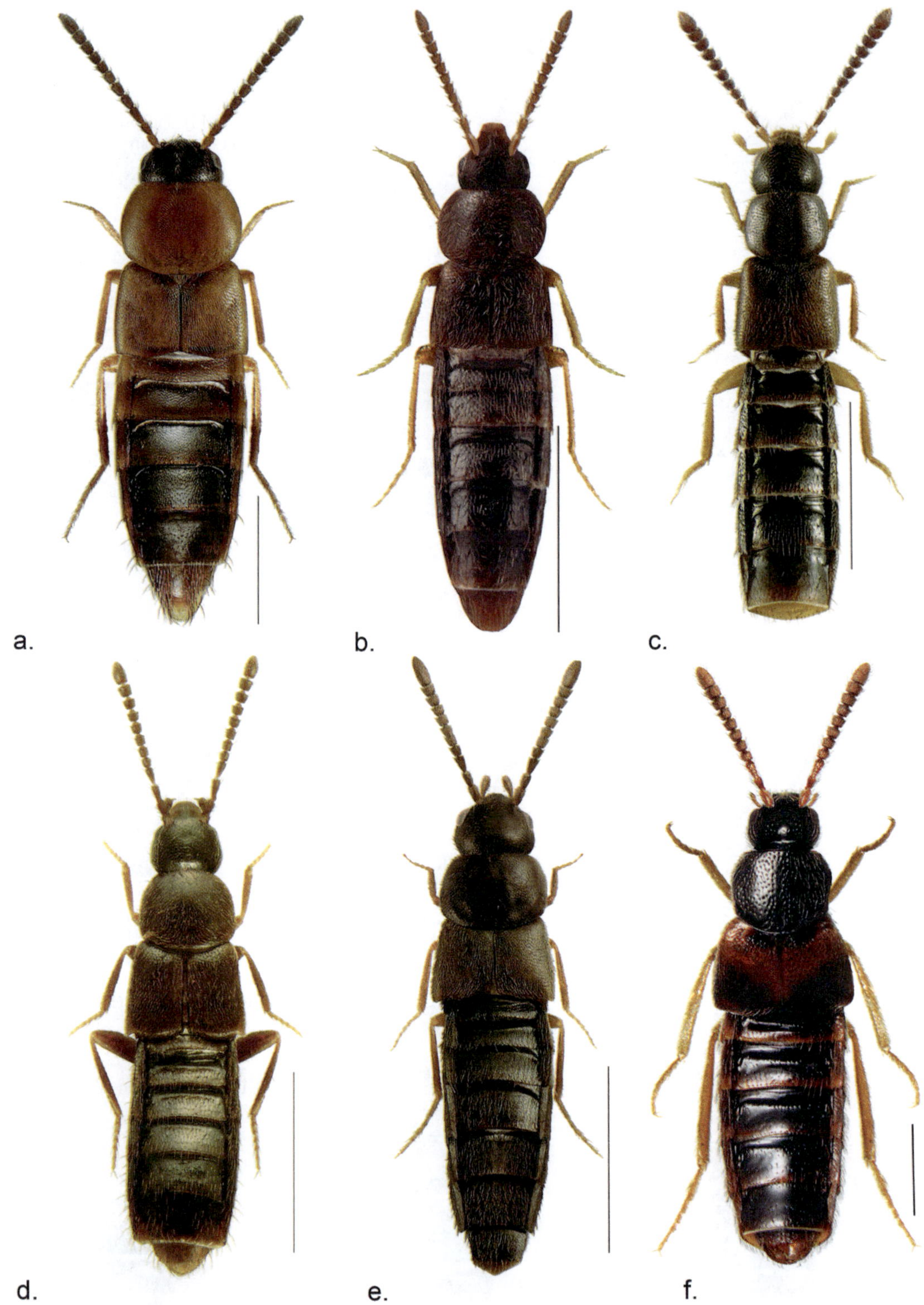

Fig. 8.69 (**a–f**). Habitus in dorsal view: (**a**) *Lypoglossa franclemonti* Hoebeke; (**b**) *Mocyta breviuscula* (Mäklin); (**c**) *Philhygra sinuipennis* Klimaszewski and Langor; (**d**) *Schistoglossa campbelli* Klimaszewski; (**e**) *Strigota obscurata* Klimaszewski and Brunke; (**f**) *Zyras obliquus* (Casey). Scale bar = 1 mm

References

Klimaszewski J (1984) A revision of the genus *Aleochara* Gravenhorst of America north of Mexico (Coleoptera: Staphylinidae, Aleocharinae). Memoirs of the Entomological Society of Canada No. 129, pp 1–211

Klimaszewski J, Langor D, Pelletier G, Bourdon C Perdereau L (2011) Aleocharine beetles (Coleoptera, Staphylinidae) of the province of Newfoundland and Labrador, Canada. Pensoft Series Faunistica No. 98. Pensoft, Sophia-Moscow, 313 pp

Klimaszewski J, Webster RP, Langor DW, Brunke A, Davies A, Bourdon C, Labrecque M, Newton AF, Dorval J-A, Frank JH (2018) Aleocharine rove beetles of eastern Canada (Coleoptera, Staphylinidae, Aleocharinae): a glimpse of megadiversity. Springer, Cham. 902 pp

Klimaszewski J, Hoebeke ER, Godin B, Davies A, Perry KI, Bourdon C, Winchester N (2020) Aleocharine rove beetles of British Columbia: a hotspot of Canadian biodiversity (Coleoptera, Staphylinidae). Springer, Cham. 631 pp

Lohse GA (1974) Staphylinidae II (Hypocyphtinae und Aleocharinae). In: Freude H, Harde KW, Lohse GA (eds) Die Käfer Mitteleuropas, Band 5. Goecke & Evers, Krefeld, 381 pp

Maruyama M, Ahn KJ (2000) Redescription of *Liparocephalus litoralis*, and key to the species of *Liparocephalus* (Coleoptera: Staphylinidae: Aleocharinae). Can Entomol 132:567–571

Newton A, Thayer MK, Ashe JS, Chandler DS (2001) Staphylinidae Latrielle, 1802. In: Arnett RH, Thomas MC (eds) American beetles: Archostemata, Myxophaga, Adephaga Polyphaga: Staphyliniformia. CRC Press, New York, pp 272–418

9 Tribe Gymnusini Heer, 1839

Tribe Gymnusini Heer, 1839

(Figs. 9.70–9.79)

(Klimaszewski, 1979; Ashe 2000, 2002, 2005; Ashe and Chatzimanolis 2003a, b; Wolf-Schwenninger 2004; Cai and Huang 2015; Yamamoto and Maruyama 2017 (revised definition of the tribe, Deinopsini merged with Gymnusini); Klimaszewski et al. 2018, 2020)

Diagnosis Body small to medium-sized (3.0–7.0 mm), sublimuloid to tear-drop shaped; densely pubescent throughout; uniformly blackish-brown to dark reddish-brown; head slightly hypognathous, moderately deflexed; vertex with (*Gymnusa*) or without (*Deinopsis*) six setigerous pores; antennae filiform, with 11 antennomeres, antennomere XI without papilla; mouthparts distinctive (see Klimaszewski 1979; Ashe 2000, 2002, 2005; Ashe and Chatzimanolis 2003a, b; Wolf-Schwenninger 2004; Cai and Huang 2015); mandibles with preapical tooth (*Gymnusa*) or two teeth (*Deinopsis*); maxillary palpi markedly long, slender, stylate to substylate; pronotum usually transverse, at least as long as wide, posterior margin truncate to bisinuate, narrowing apicad; elytra relatively short; posterolateral margins of elytra deeply emarginate; tarsal formula 3-3-3 (*Deinopsis*), or 5-5-5 (*Gymnusa*), terminal tarsomeres without two long setae surrounding the claws (*Gymnusa*); empodia with two setae; tibiae with irregularly scattered outer spines, front tibiae with preapical longitudinal ctenidium of setulae in shallow or deep depression (*Gymnusa*); metacoxae with medium to large ventral lamella; abdominal segments III–VI (i.e., first four visible segments) with tergites, sternites and paratergites each with a dense apical ctenidium of short comb-like cuticular projections, consisting of one (*Gymnusa*) or two (*Deinopsis*) types of teeth; abdominal segments III–VII each with two pairs of paratergites; tergite VII lacking tergal gland opening on anterior margin, at least for the extant genera; male tergite IX entire (*Gymnusa*) or divided (*Deinopsis*).

Key to Genera of Gymnusini (From Klimaszewski et al. 2018)

1. Body convex and robust (Figs. 9.72–9.79); length 4.2–7.0 mm, abdominal segments III–VI with apical ctenidium of comb-like cuticular projections along apical margin consisting of one type of teeth; disc of head with six large setigerous pores; tarsal formula 5-5-5 ....*Gymnusa* **Gravenhorst**

– Body moderately convex to flattened and moderately robust (Figs. 9.70–9.71); length 2.1–4.2 mm; abdominal segments III–VI with ctenidium of comb-like cuticular projections along apical margin consisting of two types of teeth; disc of head without large setigerous pores; tarsal formula 3-3-3............***Deinopsis*** **Matthews**

J. Klimaszewski et al., *A Faunal Review of Aleocharine Beetles in the Rapidly Changing Arctic and Subarctic Regions of North America (Coleoptera, Staphylinidae)*,
https://doi.org/10.1007/978-3-030-68191-3_9

Genus *Deinopsis* Matthews, 1838

(Figs. 9.70–9.71) (Klimaszewski 1979, Klimaszewski et al. 2018)

Diagnosis Body fusiform, length 2.1–4.2 mm, colour dark brown to black, pubescence uniformly piceous, integument finely punctate; head strongly deflexed, almost as wide as pronotum at anterior margin; frons obtusely and triangularly protruding; clypeus trapezoidal; labrum transverse; mandibles with two large subapical teeth; mentum trapezoidal in outline, widest behind middle, with acute lateral lobe on each side; maxillary palpi with three palpomeres, palpomere I inconspicuous, II and III elongate; labial palpi with three palpomeres; glossae subequal in length to labial palpi, wide and flat, weakly sclerotized, with apex narrow; antennae with 11 antennomeres, filiform, widest toward apex, antennomeres I and II subequal in length, each longer than antennomere III; tarsal formula 3-3-3; tarsomeres I and II subequal in length, tarsomere III longest, terminating with two claws and a few macrosetae; abdominal tergites III–VI each with dense apical comb consisting of both sharp-pointed and apically rounded teeth; proventrite reduced to narrow transverse plate in front of procoxae; procoxae conical and prominent; mesocoxae oblique; metacoxae flattened; male tergite IX divided, composed of densely pubescent valvula and ventral strut on each side; median lobe of aedeagus connected to parameres with ventral process; female tergite IX divided, similar to that of male but without ventral struts; spermatheca absent. Hydrophilous species, affilated with vegetation of swamps, ponds, small lakes, and sphagnum bogs.

Key to Species of *Deinopsis*

1. Antennomeres VII–IX subquadrate (Fig. 9.70a); elytra at suture at most as long as pronotum (Fig. 9.70a); female apical emargination of tergite VIII small and shallow (Fig. 9.70f); bulbus of median lobe of aedeagus large and round in dorsal view (Fig. 9.70c), tubus short, broadly triangular and with long apical projection at apex in dorsal view (Fig. 9.70c), in lateral view tubus short with ventral margin almost straight (Fig. 9.70b)... ....*Deinopsis canadensis* **Klimaszewski**

– Antennomeres VII–IX moderately elongate (Fig. 9.71a); elytra at suture at least as long as pronotum (Fig. 9.71a); female apical emargination of tergite VIII moderately sized and deep (Fig. 9.71f); bulbus of median lobe of aedeagus modarately sized and oval in dorsal view (Fig. 9.71c), tubus long, narrowly subparallel and with short apical projection in dorsal view (Fig. 9.71c), in lateral view tubus long with ventral margin arcuate (Fig. 9.71b) .... *Deinopsis harringtoni* **Casey**

1. *Deinopsis canadensis* Klimaszewski (Fig. 9.70a–h)

Deinopsis canadensis Klimaszewski, 1979. Klimaszewski and Frank 1992, Gouix and Klimaszewski 2007; Webster et al. 2009, Klimaszewski et al. 2011, 2018.

Diagnosis Body dark brown to almost black, length 3.2–3.6 mm; pronotum length: width = 0.6:1.0 (Fig. 9.70a); apical emargination of tergite VIII in both sexes small, v-shaped with rounded margins (Fig. 9.70d,f), shape of median lobe of aedeagus distinctive in lateral and dorsal views (Fig. 9.70b,c); and shape of female terminalia distinctive (Fig. 9.70h).

This species is externally similar to *D. harringtoni* Casey from which it differs in having antennomeres VII–IX subquadrate.

DISTRIBUTION Origin: Nearctic. **CANADA**: LB, NB, NF, ON. **USA**: not recorded.

COLLECTION AND HABITAT DATA

Habitat: black spruce bog margins in floating/saturated green sphagnum mats and floating vegetation. **Collecting period**: V–VII. **Collecting method**: treading sphagnum and vegetation into water.

2. *Deinopsis harringtoni* Casey (Fig. 9.71a–h)

Deinopsis harringtoni Casey, 1911. Klimaszewski 1979, 1982, Klimaszewski and Frank 1992, Gouix and Klimaszewski 2007, Klimaszewski et al. 2011, 2018.

Diagnosis Body dark brown to almost black, length 3.4–3.7 mm; pronotum length: width = 0.6:1.0 (Fig. 9.71a); emargination of tergite VIII acute v-shaped, in female very deep (Fig. 9.71f), in male moderately deep (Fig. 9.71d), shape of median lobe of aedeagus (Fig. 9.71b,c), and shape of female terminalia distinctive (Fig. 9.71h).

This species is externally similar to *D. canadensis* from which it differs in having antennomeres VII–IX more elongate (Fig. 9.71a).

DISTRIBUTION Origin: Nearctic. **CANADA**: LB, MB, NB, NF, NS, ON, QC. **USA**: AR, CA, MA, ME, MI, NH, WI.

COLLECTION AND HABITAT DATA
Habitat: alder (*Alnus*) swamps in leaf litter and moss, lake margins in emergent *Carex* and grasses, pond and brook margins in *Carex* and moist leaves, *Carex* marshes in *Carex* hummocks and areas with emergent *Carex*, old beaver flowages in emergent vegetation and grass litter. **Collecting period**: IV, V, VI, VIII. **Collecting method**: treading and sifting vegetation.

Genus *Gymnusa* Gravenhorst, 1806

(Fig. 9.72–9.79) (Klimaszewski 1979, Klimaszewski et al. 2018, 2020)

Diagnosis Body fusiform, length 4.2–7.0 mm; colour predominantly black, pubescence piceous or sometimes with patches of yellowish pubescence on elytra and abdomen; head strongly deflexed, almost as wide as pronotum at anterior margin; frons obtusely and triangularly protruding; vertex glabrous with three setigerous pores on each side of disc (one above antennal insertion, one at anteromedial margin of eye, and one behind eye); clypeus trapezoidal in shape; labrum transverse, broadly rounded; eye oviform; antennal insertion near anteromesal margin of eye, antennae almost filiform, widened toward apex; maxillary palpi with four palpomeres, palpomeres I and IV inconspicuous, II and III elongate and each twice as long as length of palpomere I; mandibles with one subapical tooth; labial palpi with three palpomeres; proventrite reduced to narrow transverse plate in front of procoxae; procoxae conical and prominent with cavities open behind; mesocoxae oblique; metacoxae flattened, longer than wide; abdominal tergites III–VII (basal five visible tergites) each with two laterosternites; tergites III–VII each with dense apical comb of teeth with rounded apices; tarsal formula 5-5-5, tarsomere I of all legs at least as long as tarsomere II and III combined; male tergite IX annular, with two lateral lobes and two ventral struts; median lobe of aedeagus with two ventral flaps serving as attachments to parameres; female tergite IX annular, without ventral struts; sclerotized spermatheca absent. Species associated with forested swamps, marshes, and sphagnum bogs.

Key to Species of *Gymnusa* (modified from Klimaszewski 1979, Klimaszewski et al. 2011)

1. Pubescence uniformly piceous; body length 5.0–6.5 mm **2**

– Pubescence bicoloured, mostly piceous with a few, small yellowish spots on elytra and often on abdomen; body length 4.2–6.0 mm **5**

2. Antennomeres I and II approximately subequal in length; body convex (Figs. 9.72a, 9.73a, 9.74a) [females can not be determined to species unless associated with males] **3**

– Basal antennomere longer than second; body flattened (Fig. 9.75a); genitalia and terminalia as illustrated (Fig. 9.75b–i).. ....***Gymnusa grandiceps* Casey**

3. Median lobe of aedeagus with apical part narrow and projecting ventrally in lateral view (Fig. 9.74b).. .***Gymnusa lindrothi* Klimaszewski and Langor**

– Median lobe of aedeagus with apex shaped differently (Figs. 9.72b, 9.73b) **4**

4. Median lobe of aedeagus with apex narrowly rounded in dorsal view and truncate apically in lateral view (Fig. 9.73b,c) .... *Gymnusa brevicollis* **(Paykull)**

– Median lobe of aedeagus with apex pointed in dorsal view and arcuate ventrally in lateral view (Fig. 9.72b,c) .... ***Gymnusa atra*** **Casey**

5. Male sternite IX extremally elongate apically; tubus of median lobe of aedeagus with small ventral median angular projection in lateral view (Fig. 9.77b); female tergite IX narrow medially (Fig. 9.77h) .... ***Gymnusa konopackii*** **Klimaszewski**

– Male sternite IX moderately elongate apically; tubus of median lobe of aedeagus with large and long ventral median projection in lateral view (Figs. 9.76b, 9.78b, 9.79b); female tergite IX broad medially (Figs. 9.76h, 9.78h, 9.79h) .. **6**

6. All antennomeres brown to black (Fig. 9.78a); body length 4.8–6.0 mm; median lobe of aedeagus widened subapically and with small ventral process in lateral view (Fig. 9.78b); female tergite VIII emarginate medially, emargination with, at most, a small membranous structure at middle (Fig. 9.78f); female terminalia as illustrated (Fig. 9.78f–h) .... ***Gymnusa pseudovariegata*** **Klimaszewski**

– Antennomere I reddish-brown, paler than other antennomeres (Figs. 9.76a, 9.79a); body length 4.2–5.5 mm; median lobe of aedeagus and female terminalia shaped differently **7**

7. Ventral process of median lobe of aedeagus widened toward apex in dorsal view (Fig. 9.76c); female tergite IX divided medially (Fig. 9.76h) .. .. ***Gymnusa campbelli*** **Klimaszewski**

– Ventral process of median lobe of aedeagus spatulate in dorsal view (Fig. 9.79c); female tergite IX connected medially by wide bridge that is membranous medially and often broken (Fig. 9.79h)... ***Gymnusa smetanai*** **Klimaszewski**

Brevicollis **species group** (Klimaszewski 1979)

Diagnosis Body fusiform, length 5.0–6.5 mm; pubescence uniformly piceous; punctation fine, coarser on elytra and abdomen.

3. *Gymnusa atra* Casey * (Fig. 9.72a–h)

Gymnusa atra Casey, 1911. Klimaszewski 1979, 1982, Klimaszewski and Génier 1985, Gouix and Klimaszewski 2007, Majka and Klimaszewski 2010, Klimaszewski et al. 2011, 2018, 2020.

Diagnosis Body convex, length 5.0–6.5 mm, colour black to dark brown, pubescence on elytra and abdomen uniformly piceous, punctation fine, coarser on elytra and abdomen, antennomeres I and II reddish-brown (Fig. 9.72a). MALE. Tergite and sternite VIII as illustrated (Fig. 9.72d,e); median lobe of aedeagus arcuate subapically in lateral view (Fig. 9.72b), with apex sharp in dorsal view (Fig. 9.72c). FEMALE. Tergite VIII with deep apical emargination (Fig. 9.72f); sternite VIII and pygidium as illustrated (Fig. 9.72g,h).

This species is not distinguishable externally from *G. brevicollis* but in *G. atra* the subapical part of the median lobe of the aedeagus is arcuate in lateral view (Fig. 9.72b) (truncate in *G. brevicollis*), and the apex of the median lobe is sharp in dorsal view (Fig. 9.72c). Females are not distinguishable and have similar terminalia.

DISTRIBUTION **Origin**: Holarctic. **CANADA**: AB, BC, LB, MB, NB, NF, NS, NT, ON, QC, YT. **USA**: AK, IL, IA, MA, ME, MI, MN, NH, NY. Eastern Russia, northeastern China.

COLLECTION AND HABITAT DATA

Habitat: tamarack (*Larix laricina* (Du Roi)) and black spruce (*Picea mariana* (Mill.) BSP)) bogs, marshy pond and lake margins, cattail/*Carex* marshes, and marshy margins of slow flowing rivers. Adults usually found in *Carex* hummocks, moist to wet sphagnum with *Carex*, and in saturated marsh vegetation. **Collecting period**: V–XI. **Collecting method**: sifting or treading vegetation (*Carex*), *Carex* hummocks, and wet sphagnum into water.

COMMENTS This species was listed from NU by Gouix and Klimaszewski (2007) and Klimaszewski et al. (2011) but no voucher data were published. We removed NU from this species' distribution until voucher specimens are available.

Additional records. See Annex 1.

4. *Gymnusa brevicollis* (Paykull) † (Figs. 9.73a–h)

Staphylinus brevicollis Paykull, 1800. Klimaszewski 1979, 1982, Klimaszewki and Génier, 1985, Gouix and Klimaszewski 2007, Klimaszewski et al. 2011, 2018.

Diagnosis Body length 5.0–6.5 mm, colour black to dark brown, pubescence uniformly piceous, punctation fine, coarser on elytra and abdomen; antennal articles I and II reddish-brown (Fig. 9.73a). MALE. Tergite and sternite VIII as illustrated (Fig. 9.73d,e); median lobe of aedeagus truncate subapically in lateral view (Fig. 9.73b), with apex rounded in dorsal view (Fig. 9.73c). FEMALE. Tergite VIII with deep apical emargination (Fig. 9.73f); sternite VIII and pygidium as illustrated (Fig. 9.73g,h).

This species is not distinguishable externally from *G. atra* but in *G. brevicollis* the subapical part of the median lobe of the aedeagus is truncate in lateral view (Fig. 9.73b) (arcuate in *G. atra* (Fig. 9.72b), and the apex of the median lobe is rounded in dorsal view (Fig. 9.73c). Females are not distinguishable and have similar terminalia.

DISTRIBUTION Origin: Palaearctic: northern Europe, western Siberia; adventive in North America; earliest North American record is from 1924 (Klimaszewski et al. 2013). **CANADA**: LB, NF, QC. **USA**: MA, ME., France (St. Pierre et Miquelon).

COLLECTION AND HABITAT DATA
Habitat: swamps, marshy edges of lakes and slow flowing rivers, and in sphagnum bogs. Occurs in moist to wet marsh vegetation and in wet sphagnum. **Collecting period**: VIII–IX. **Collecting method**: sifting or treading vegetation (e.g., wet sphagnum into water).

COMMENTS This species is known only from the softwood shield ecoregion (5.1) in Labrador and southward.

5. *Gymnusa lindrothi* Klimaszewski and Langor (Fig. 9.74a–h)

Gymnusa lindrothi Klimaszewski and Langor, in Klimaszewski et al. 2011. Klimaszewski et al. 2018.

Diagnosis Body length 4.9–6.6 mm, colour black to dark brown, antennal article I reddish-brown; pubescence uniformly piceous, punctation fine, coarser on elytra and abdomen (Fig. 9.74a). MALE. Tergite and sternite VIII as illustrated (Fig. 9.74d,e); median lobe of aedeagus in lateral view narrowly produced and sinuate subapically (Fig. 9.74b), apex rounded in dorsal view (Fig. 9.74c). FEMALE. Tergite VIII with deep and rounded emargination (Fig. 9.74f); sternite VIII with apical emargination very obtusely angular, rounded medially (Fig. 9.74g); tergite IX of pygidium with narrow median bridge (Fig. 9.74h).

DISTRIBUTION Origin: Nearctic. **CANADA**: LB, NF, NB. **USA**: not recorded.

COLLECTION AND HABITAT DATA
Habitat: in LB and NF in coastal barrens in vegetation, mosses and sedges at the edge of a small pond; the NB specimen captured in emergent *Carex* in a tamarack bog. **Collecting period**: V–VIII. **Collecting method**: treading mosses and sedges into water.

COMMENTS All known specimens of this species have been collected in more southerly ecozones: the Softwood Shield of Newfoundland and Labrador and the Acadian Highlands of New Brunswick. We include this species here as it may be eventually found in the transitional areas of the Taiga Shield of Labrador.

***Grandiceps* species group** (Klimaszewski 1979)

Diagnosis Body broadly oval, length 5.7–6.5 mm; integument with distinct meshed microsculpture; pubescence piceous; punctation shallow.

6. ***Gymnusa grandiceps*** **Casey** (Fig. 9.75a–i)

Gymnusa grandiceps Casey, 1915. Klimaszewski 1979, 1982, Klimaszewski and Génier 1985, Gouix and Klimaszewski 2007, Webster et al. 2009, Majka and Klimaszewski 2010, Klimaszewski et al. 2011, 2018, 2020.

Diagnosis Body widely oval, flattened, length 5.7–6.5 mm, colour dark brown to almost black, microsculpture distinct, pubescence on elytra and abdomen uniformly piceous, punctation shallow; antennal article I distinctly longer than II (Fig. 9.75a). MALE. Tergite and sternite VIII as illustrated (Fig. 9.75e,f); sternite IX deeply emarginate apically (an exception in *Gymnusa*) (Fig. 9.75d); median lobe of aedeagus in lateral view narrow, broadly arched (Fig. 9.75b); in dorsal view narrowly elongate and produced apically (Fig. 9.75c). FEMALE. Tergite VIII truncate apically (Fig. 9.75g); sternite VIII broadly produced medially at apex (Fig. 9.75h); valvulae of pygidium flat and emarginate apically (Fig. 9.75i).

DISTRIBUTION Origin: Nearctic. **CANADA**: BC, LB, MB, NB, NF, NS, ON, QC. **USA**: IL, MA, MD, ME, MI, NH, NY, RI.

COLLECTION AND HABITAT DATA
Habitat: wet forested biotypes such as black spruce bogs (usually in lag zone), alder swamps, spruce and eastern white cedar (*Thuja occidentalis* L.) forests, eastern white cedar swamps, vernal pond margins, and margins of small brooks under alders. Adults in moist leaf litter and moss in these habitats. **Collecting period**: III–IX. **Collecting method**: Sifting leaf litter and moss.

Additional records. See Annex 1.

Variegata **species group** (Klimaszewski 1979)

Diagnosis Body fusiform, length 4.2–6.0 mm; pubescence piceous with yellowish spots; interspaces of punctures on elytra and abdomen rugose.

7. ***Gymnusa campbelli*** **Klimaszewski** (Fig. 9.76a–h)

Gymnusa campbelli Klimaszewski, 1979. Klimaszewski 1982, Gouix and Klimaszewski 2007, Majka and Klimaszewski 2010, Klimaszewski et al. 2011, 2015, 2018.

Diagnosis Body length 4.5–5.5 mm, colour black to dark brown, pubescence piceous with yellowish spots, interspaces of punctures on elytra and abdomen rugose; antennal article I reddish-brown (Fig. 9.76a). MALE. Tergite and sternite VIII as illustrated (Fig. 9.76d,e); median lobe of aedeagus subdivided apically in lateral view (Fig. 9.76b), in dorsal view gradually narrowing apically, with apex pointed and with triangular ventral process (Fig. 9.76c). FEMALE. Tergite VIII deeply, emarginate apically (Fig. 9.76f); sternite VIII obtusely angular with apex rounded (Fig. 9.76g); pygidium (Fig. 9.76h).

This species is externally similar to *G. smetanai* from which it differs in the shape of the median lobe of the aedeagus (Fig. 9.76b,c), which bears an elongate ventral process that is widened apically in dorsal view (Fig. 9.76c).

DISTRIBUTION Origin: Nearctic. **CANADA**: MB, NB, NF, NT, ON, QC, SK, YT. **USA**: AK, MA.

COLLECTION AND HABITAT DATA
Habitat: black spruce bogs, tamarack bogs, and eastern white cedar swamps in saturated sphagnum with *Carex* and floating green sphagnum mats, in wet moss in an alder swamp, and wet moss on muddy shores of shallow lakes. **Collecting period**: V–VIII. **Collecting method**: treading saturated/wet moss and *Carex* into water.

Additional records. See Annex 1.

8. ***Gymnusa konopackii*** **Klimaszewski** (Fig. 9.77a–h)

Gymnusa konopackii Klimaszewski, 1979. Ernst and Buddle (2013).

Diagnosis Body length 4.5–5.5 mm, colour black, pubescence piceous with yellowish spots on elytra and abdomen, interspaces of punctures on elytra and abdomen rugose; antennal article I or I–II reddish-brown (Fig. 9.77a). MALE. Tergite VIII sternite VIII as illustrated (Fig. 9.77d,e); tubus of median lobe of aedeagus in lateral view with small angular projection on ventral margin (Fig. 9.77b), in dorsal view apex narrowly produced (Fig. 9.77c); sternite IX narowed and extremely elongate apically. FEMALE. Tergite VIII arcuate apically, without apical emargination (Fig. 9.77f); sternite VIII with apical margin obtusely angulate medially as illustrated (Fig. 9.77g); pygydium with tergite IX divided dorso-medially and dorsally consisting of two narrow lobes in contact medially (Fig. 9.77h).

Males of this species are extremely similar externally to *G. campbelli* and *G. smetanai* but may be easily distinguished by the unusually elongate apex of sternite IX and the shape of median lobe of aedeagus.

DISTRIBUTION Origin: Nearctic. **CANADA**: MB, NT, NU, YT. **USA**: AK.

COLLECTION AND HABITAT DATA
Habitat: wet moss on muddy shore of shallow lakes. **Collecting period**: V–VIII. **Collecting method**: treading wet moss into water.

9. *Gymnusa pseudovariegata* Klimaszewski (Fig. 9.78a–h)

Gymnusa pseudovariegata Klimaszewski, 1979. Klimaszewski 1982, Klimaszewski and Génier 1985, Caterino and Chatzimanolis 2007, Gouix and Klimaszewski 2007, Majka and Klimaszewski 2010, Klimaszewski et al. 2011, 2018, 2020.

Diagnosis Body length 4.8–6.0 mm, colour black to dark brown, pubescence on elytra and abdomen piceous with small yellow spots, basal antennomeres uniformly dark brown to black; elytra and abdomen with interspaces of punctures rugose (Fig. 9.78a). MALE. Tergite VIII sternite VIII as illustrated (Fig. 9.78d,e); tubus of median lobe of aedeagus in lateral view with small tooth on ventral margin (Fig. 9.78b), in dorsal view apex narrowly produced (Fig. 9.78c). FEMALE. Tergite VIII emarginate apically, without or with minute sclerotized lobe of variable shape and size in middle of apical emargination (Fig. 9.78f); sternite VIII as illustrated (Fig. 9.78g); pygydium with tergite IX having short and acute basal tooth (Fig. 9.78h).

DISTRIBUTION Origin: Nearctic. **CANADA**: AB, BC, LB, MB, NB, NF, NS, NT, NU [**BOLD**], ON, QC, YT. **USA**: AK, CA, MA, OR, WA.

COLLECTION AND HABITAT DATA
Habitat: sedge marshes, cattail/*Carex* marshes, boggy marshes and fens, tamarack bogs, pool margins in sphagnum bogs, vegetated edges of swamps, lakes, and slow flowing rivers, and small ponds on coastal barrens. Usually among emergent *Carex*, floating vegetation mats, and wet to saturated marsh vegetation, also under litter and stones on a sandy lakeshore. **Collecting period**: IV–XI. **Collecting method**: treading vegetation into water.

We report this species from NU (**NTR**) for the first time here, based on BOLD records: Canada, Nunavut, Kitikmeot, Kugluktuk, Kugluk (Bloody Falls) Territorial Park, 13 m cl. 67.745, -115.375, tundra pond, treading, 24.VII.2019, M. Pentinsaari (CBG), 1 specimen.

Additional records. See Annex 1.

10. *Gymnusa smetanai* Klimaszewski * (Fig. 9.79a–h)

Gymnusa smetanai Klimaszewski, 1979. Gouix and Klimaszewski 2007, Klimaszewski et al. 2011, 2018.

Diagnosis Body length 4.5–5.5 mm, colour black to dark brown, pubescence piceous with yellowish spots, antennal article I reddish-brown, interspaces of punctures on elytra and abdomen rugose (Fig. 9.79a). MALE. Tergite and sternite VIII as illustrated (Fig. 9.79d,e);

median lobe of aedeagus in lateral view gradually narrowing apically and with narrow median projection (Fig. 9.79b), in dorsal view with apex blunt and ventral process spatulate (Fig. 9.79c). FEMALE. Tergite and sternite VIII as illustrated (Fig. 9.79f,g); pygidium (Fig. 9.79h).

This species is externally very similar to *G. campbelli* from which it may be distingushed by the genital features.

DISTRIBUTION Origin: Holarctic. **CANADA**: MB, NF, NT, ON, YT. **USA**: AK.

COLLECTION AND HABITAT DATA
Habitat: margin of tundra lakes in saturated *Carex* and sphagnum. **Collecting period**: VI–VIII. **Collecting method**: treading *Carex* and sphagnum into water.

Additional records. See Annex 1.

Fig. 9.70 (**a–h**). Images of *Deinopsis canadensis* Klimaszewski: (**a**) habitus in dorsal view; (**b**) median lobe of aedeagus in lateral view; (**c**) aedeagus in dorsal view (median lobe and parameres); (**d**) male tergite VIII; (**e**) male sternite VIII; (**f**) female tergite VIII; (**g**) female sternite VIII; (**h**) female pygidium. Scale bar for habitus = 1 mm; remaining scale bars = 0.2 mm

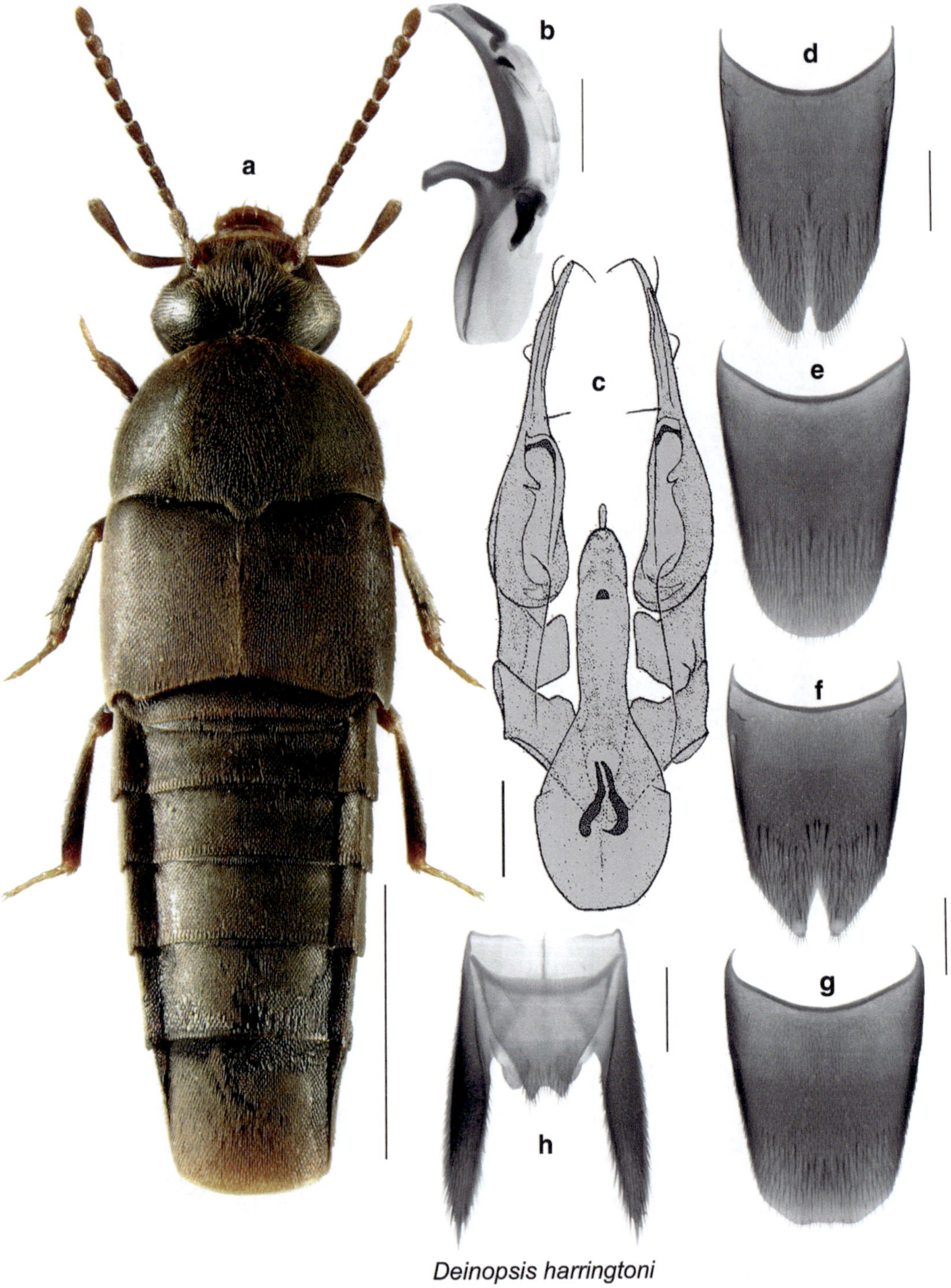

Fig. 9.71 (**a–h**). Images of *Deinopsis harringtoni* Casey: (**a**) habitus in dorsal view; (**b**) median lobe of aedeagus in lateral view; (**c**) aedeagus in dorsal view (median lobe and parameres); (**d**) male tergite VIII; (**e**) male sternite VIII; (**f**) female tergite VIII; (**g**) female sternite VIII; (**h**) female pygidium. Scale bar for habitus = 1 mm; remaining scale bars = 0.2 mm

Fig. 9.72 (**a–h**). Images of *Gymnusa atra* Casey: (**a**) habitus in dorsal view; (**b**) median lobe of aedeagus in lateral view; (**c**) median lobe of aedeagus in dorsal view; (**d**) male tergite VIII; (**e**) male sternite VIII; (**f**) female tergite VIII; (**g**) female sternite VIII; (**h**) female pygidium. Scale bar for habitus = 1 mm; remaining scale bars = 0.2 mm

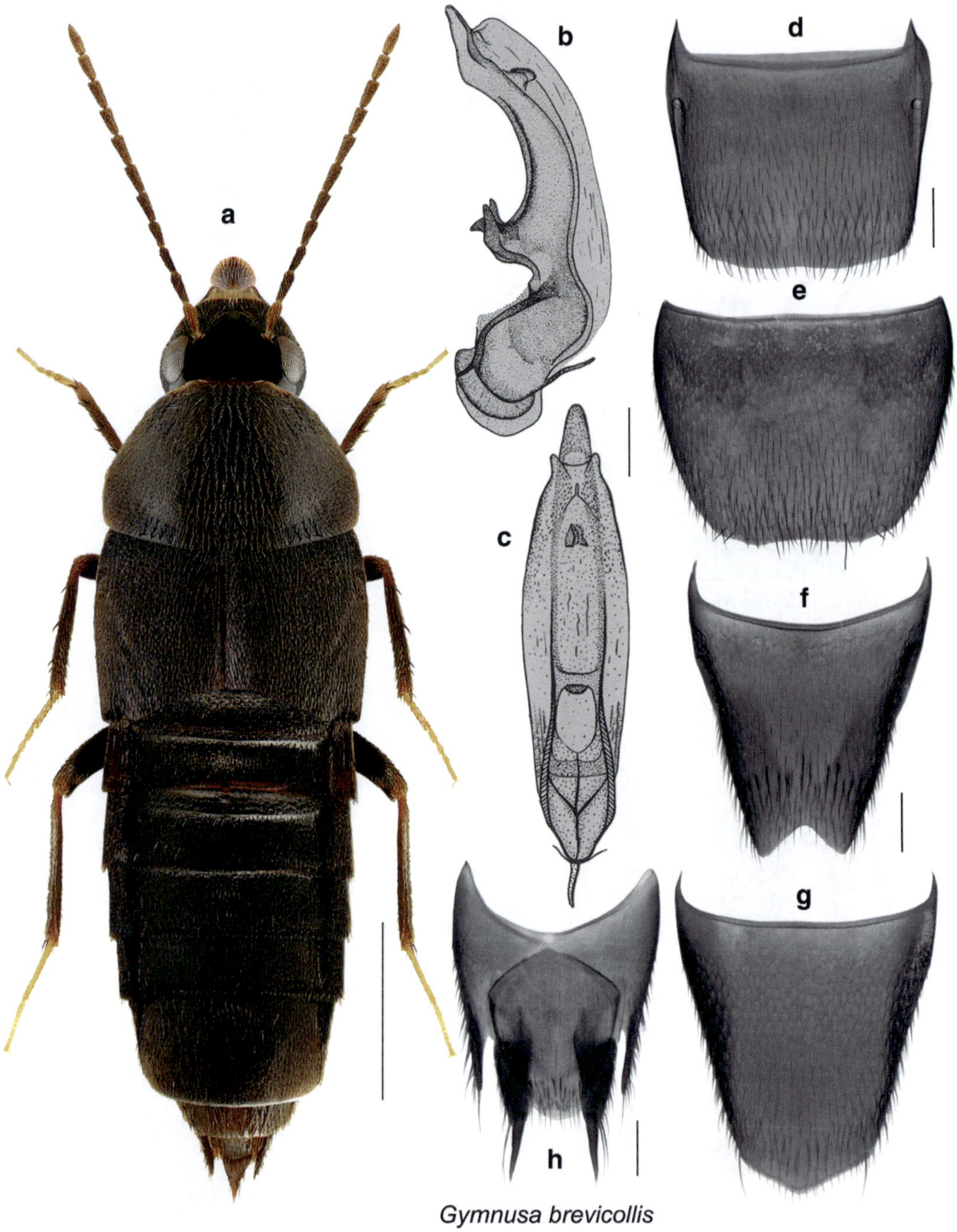

Fig. 9.73 (**a–h**). Images of *Gymnusa brevicollis* (Paykull): (**a**) habitus in dorsal view; (**b**) median lobe of aedeagus in lateral view; (**c**) median lobe of aedeagus in dorsal view; (**d**) male tergite VIII; (**e**) male sternite VIII; (**f**) female tergite VIII; (**g**) female sternite VIII; (**h**) female pygidium. Scale bar for habitus = 1 mm; remaining scale bars = 0.2 mm

Fig. 9.74 (**a–h**). Images of *Gymnusa lindrothi* Klimaszewski and Langor: (**a**) habitus in dorsal view; (**b**) median lobe of aedeagus in lateral view; (**c**) median lobe of aedeagus in dorsal view; (**d**) male tergite VIII; (**e**) male sternite VIII; (**f**) female tergite VIII; (**g**) female sternite VIII; (**h**) female pygidium. Scale bar for habitus = 1 mm; remaining scale bars = 0.2 mm

Fig. 9.75 (**a–i**). Images of *Gymnusa grandiceps* Casey: (**a**) habitus in dorsal view; (**b**) median lobe of aedeagus in lateral view; (**c**) median lobe of aedeagus in dorsal view; (**d**) male sternite IX; (**e**) male tergite VIII; (**f**) male sternite VIII; (**g**) female tergite VIII; (**h**) female sternite VIII; (**i**) female pygidium. Scale bar for habitus = 1 mm; remaining scale bars = 0.2 mm

Fig. 9.76 (**a–h**). Images of *Gymnusa campbelli* Klimaszewski: (**a**) habitus in dorsal view; (**b**) median lobe of aedeagus in lateral view; (**c**) median lobe of aedeagus in dorsal view; (**d**) male tergite VIII; (**e**) male sternite VIII; (**f**) female tergite VIII; (**g**) female sternite VIII; (**h**) female pygidium. Scale bar for habitus = 1 mm; remaining scale bars = 0.2 mm

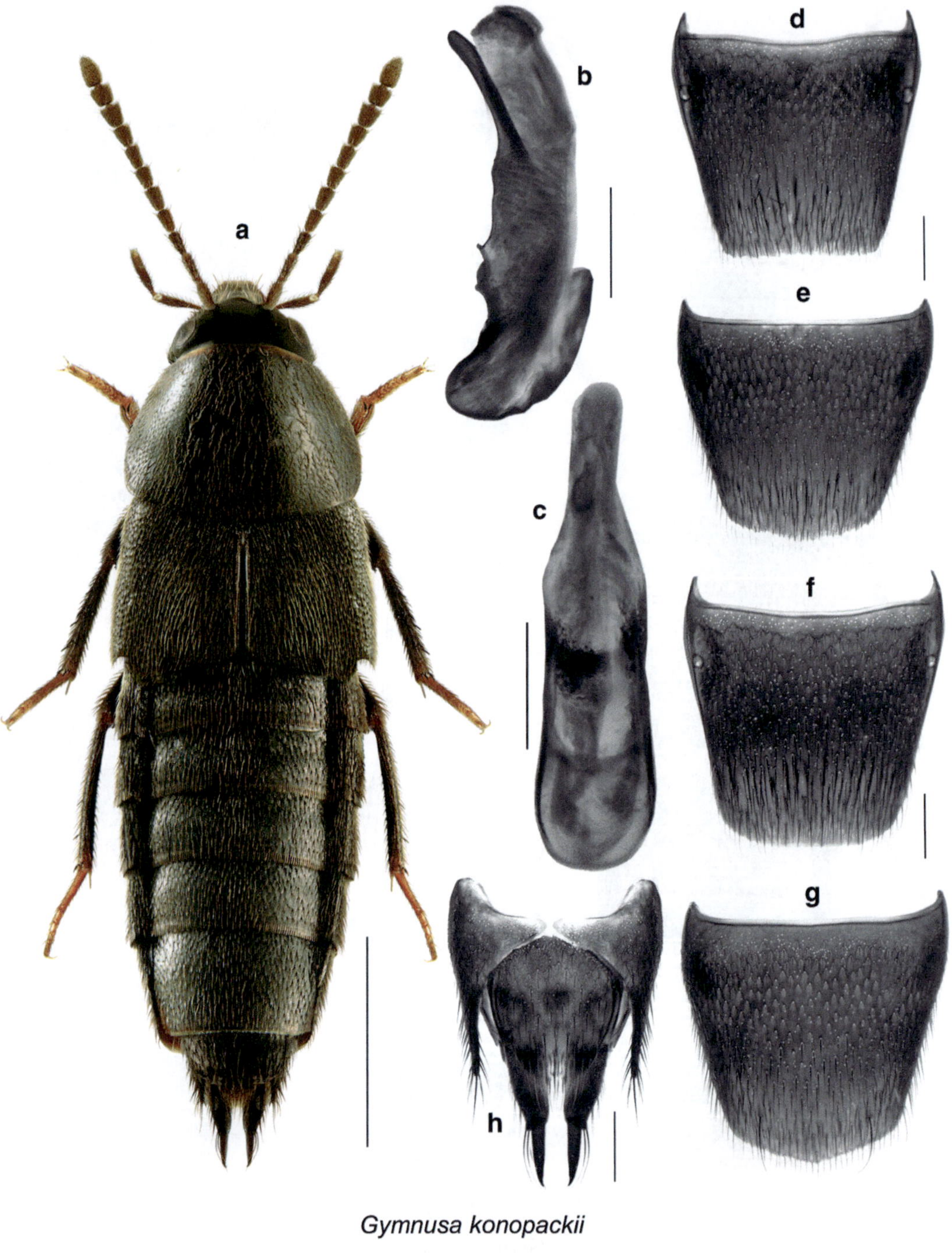

Fig. 9.77 (**a–h**). Images of *Gymnusa konopackii* Klimaszewski: (**a**) habitus in dorsal view; (**b**) median lobe of aedeagus in lateral view; (**c**) median lobe of aedeagus in dorsal view; (**d**) male tergite VIII; (**e**) male sternite VIII; (**f**) female tergite VIII; (**g**) female sternite VIII; (**h**) female pygidium. Scale bar for habitus = 1 mm; remaining scale bars = 0.2 mm

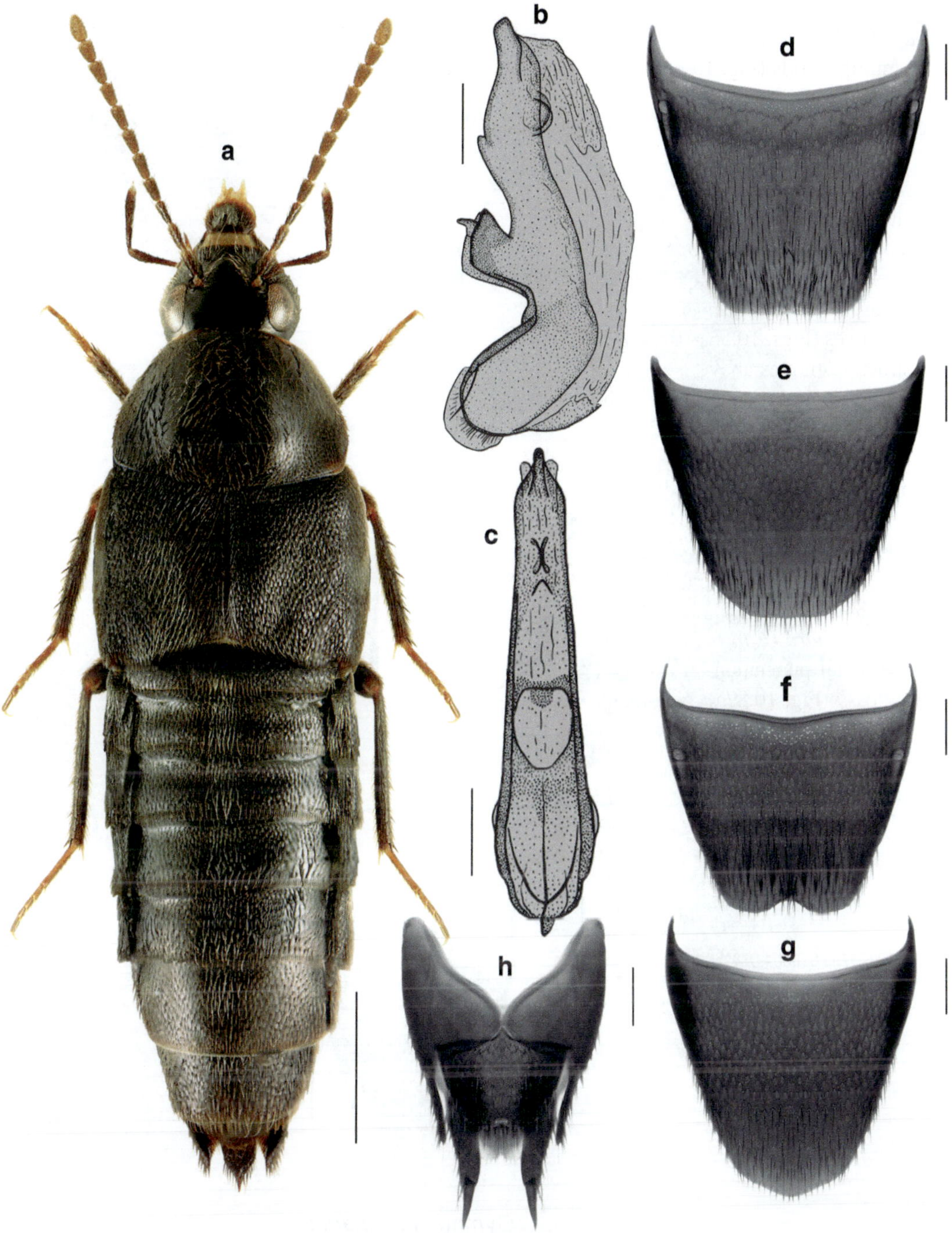

Fig. 9.78 (**a–h**). Images of *Gymnusa pseudovariegata* Klimaszewski: (**a**) habitus in dorsal view; (**b**) median lobe of aedeagus in lateral view; (**c**) median lobe of aedeagus in dorsal view; (**d**) male tergite VIII; (**e**) male sternite VIII; (**f**) female tergite VIII; (**g**) female sternite VIII; (**h**) female pygidium. Scale bar for habitus = 1 mm; remaining scale bars = 0.2 mm

Fig. 9.79 (**a–h**). Images of *Gymnusa smetanai* Klimaszewski: (**a**) habitus in dorsal view; (**b**) median lobe of aedeagus in lateral view; (**c**) median lobe of aedeagus in dorsal view; (**d**) male tergite VIII; (**e**) male sternite VIII; (**f**) female tergite VIII; (**g**) female sternite VIII; (**h**) female pygidium. Scale bar for habitus = 1 mm; remaining scale bars = 0.2 mm

References

Ashe JS (2000) Mouthpart structure of *Stylogymnusa subantarctica* Hammond, 1975 (Coleoptera: Staphylinidae: Aleocharinae) with a reanalysis of the phylogenetic position of the genus. Zool J Linnean Soc 130(4):471–498

Ashe JS (2002) *Allodinopsis*, new genus of deinopsine aleocharine from Central America, and a new species of *Metadeinopsis* Klimaszewski 1979 (Staphylinidae: Aleocharinae: Deinopsini). J Kansas Entomol Soc 75 (2):61–72

Ashe JS (2005) Phylogeny of the tachyporine group subfamilies and 'basal' lineages of the Aleocharinae (Coleoptera: Staphylinidae) based on larval and adult characteristics. Syst Entomol 30(1):3–37

Ashe JS, Chatzimanolis S (2003a) Gymnusini. Version 06 November 2003. Tree of life web project. http://tolweb.org/Gymnusini/9814/2003.11.06. Accessed 22 Sept 2016

Ashe JS, Chatzimanolis S (2003b) Deinopsini. Version 06 November 2003. http://tolweb.org/Deinopsini/9815/2003.11.06. Accessed 22 Sept 2016

Cai C, Huang D (2015) The oldest aleocharine rove beetle (Coleoptera, Staphylinidae) in Cretaceous Burmese amber and its implications for the early evolution of the basal group of hyper-diverse Aleocharinae. Gondwana Res 28(4):1579–1584

Casey TL (1911) New American species of Aleocharinae and Myllaeninae. In: Memoirs on the Coleoptera. II. The New Era Printing Co., Lancaster, PA, pp 1–245. 259 pp

Casey TL (1915) Studies in some staphylinid genera of North America. In: Memoirs on the Coleoptera. VI. The New Era Printing Co., Lancaster, PA, pp 395–450. 460 pp

Caterino MS, Chatzimanolis S (2007) Newly recorded and noteworthy California Staphylinidae. Coleopt Bull 61 (3):398–407

Ernst CM, Buddle CM (2013) Seasonal patterns in the structure of epigeic beetle (Coleoptera) assemblages in two subarctic habitats in Nunavut, Canada. Can Entomol 145(2):171–183

Gouix N, Klimaszewski J (2007) Catalogue of aleocharine rove beetles of Canada and Alaska (Coleoptera, Staphylinidae, Aleocharinae). Pensoft Publishers, Sofia-Moscow. 165 pp

Klimaszewski J (1979) A revision of the Gymnusini and Deinopsini of the world (Coleoptera: Staphylinidae: Aleocharinae). Research Branch Agriculture Canada Monograph 25:1–169

Klimaszewski J (1982) A revision of the Gymnusini and Deinopsini of the world (Coleoptera: Staphylinidae). Supplementum 2. Can Entomol 114(4):317–335

Klimaszewski J, Frank JH (1992) New distributional data for Nearctic *Myllaena* Erichson (Coleoptera: Staphylinidae). Supplement 2. Coleopt Bull 46 (4):397–402

Klimaszewski J, Génier F (1985) A revision of the Gymnusini and Deinopsini of the world (Coleoptera: Staphylinidae, Aleocharinae). Supplement 3. Coleopt Bull 39(1):60–66

Klimaszewski J, Langor D, Pelletier G, Bourdon C, and Perdereau L (2011) Aleocharine beetles (Coleoptera, Staphylinidae) of the province of Newfoundland and Labrador, Canada. Pensoft Series Faunistica No. 98. Pensoft, Sophia-Moscow, 313 pp

Klimaszewski J, Brunke A, Assing V, Langor DW, Newton AF, Bourdon C, Pelletier G, Webster RP, Herman L, Perdereau L, Davies A, Smetana A, Chandler DS, Majka C, Scudder GGE (2013) Synopsis of adventive species of Coleoptera (Insecta) recorded from Canada. Part 2: Staphylinidae. Pensoft, Sofia-Moscow, 360 pp

Klimaszewski J, Godin B, Langor D, Bourdon C, Lee S-I, Horwood D (2015) New distribution records for Canadian Aleocharinae (Coleoptera, Staphylinidae), and new synonymies for *Trichiusa*. ZooKeys 498:51–91

Klimaszewski J, Webster RP, Langor DW, Brunke A, Davies A, Bourdon C, Labrecque M, Newton AF, Dorval J-A, Frank JH (2018) Aleocharine rove beetles of eastern Canada (Coleoptera, Staphylinidae, Aleocharinae): a glimpse of megadiversity. Springer, Cham. 902 pp

Klimaszewski J, Hoebeke ER, Godin B, Davies A, Perry KI, Bourdon C, Winchester N (2020) Aleocharine rove beetles of British Columbia: a hotspot of Canadian biodiversity (Coleoptera, Staphylinidae). Springer, Cham. 631 pp

Majka CG, Klimaszewski J (2010) Contributions to the knowledge of the Aleocharinae (Coleoptera, Staphylinidae) in the Maritime Provinces of Canada. ZooKeys 46:15–39

Paykull G (1800) Fauna Suecica. Insecta. Tomus III. J.F. Edman, Upsala: 459 pp

Webster RP, Klimaszewski J, Pelletier G, Savard K (2009) New Staphylinidae (Coleoptera) records with new collection data from New Brunswick, Canada. I. Aleocharinae. In: Majka, C.G. and Klimaszewski, J. (Eds.) Biodiversity, Biosystematics, and Ecology of Canadian Coleoptera II. ZooKeys [Special Issue] 22:171–248

Wolf-Schwenninger K (2004) A new fossil staphylinid genus and species from Baltic amber (Coleoptera, Staphylinidae, Aleocharinae, Gymnusini), *Stuttgarter* Beiträge zur Naturkunde Serie B (Geologie und Paläontologie) 345:1-6

Yamamoto S, Maruyama M (2017) Phylogeny of the rove beetle tribe Gymnusini sensu n. (Coleoptera: Staphylinidae: Aleocharinae): implications for the early branching events of the subfamily: phylogeny of Gymnusini rove beetles. Syst Entomol 43 (1):183–199. https://doi.org/10.1111/syen.12267

10 Tribe Aleocharini Fleming, 1821

Tribe Aleocharini Fleming, 1821

(Figs. 10.80–10.93) (Klimaszewski 1984, Osswald et al. 2013, Klimaszewski et al. 2018, 2020)

Diagnosis Body robust to moderately slender (Figs. 10.80–10.93), length 2.0–13.0 mm; maxillary and labial palpi with a pseudosegment on last palpomere, appearing five- and four-segmented respectively (*Aleochara*), or without apparent pseudosegment (*Amarochara*, not recorded in Arctic and Subarctic); pronotal hypomera not visible or narrowly visible in lateral view; flagellum of median lobe of aedeagus long, longer than median lobe in most species; some species with velum of paramere reticulated; tarsal formula 5-5-5 in most species, 4-5-5 in the subgenus *Tinotus* (Klimaszewski et al. 2018).

Genus *Aleochara* Gravenhorst, 1802

(Figs. 10.80–10.93) (Klimaszewski 1984, Klimaszewski et al. 2018, 2020)

Diagnosis Body length very variable within same species, length within genus 1.8–13.0 mm; body shape broadly oval or subparallel, robust; tarsal formula 5-5-5 except 4-5-5 in *Tinotus*; last palpomere of maxillary and labial palpi with pseudosegment; laciniae with tuft of apical setae; ligula small, bilobed apically; mesoventrite varies from non-carinate to completely carinate, depending on subgenus; male tergite IX divided by tergite X, each half narrowly elongate anteriad and posteriad and bearing ventral strut; median lobe of aedeagus usually consisting of narrowly elongate tubus and moderate-sized bulbus (except in some species in subgenus *Aleochara*); internal sac with complex sclerites; spermatheca consisting of capsule (head and neck), chamber and duct (stem), variably modified in different subgenera. Adults are predacious on larvae, eggs, and pupae of cyclorrhaphous Diptera, and larvae are ectoparasitoids of dipteran pupae.

Key to Subgenera of *Aleochara* (Figs. 10.80a–h–10.93a–i)

1. Tarsal formula 4-5-5, body as illustrated (Fig. 10.91a)................................***Tinotus* Sharp**

– Tarsal formula 5-5-5, body not as above (Figs. 10.81–10.90, 10.92–10.93).....................2

2. Pronotum with two longitudinal, subparallel rows of more-or-less confluent setigerous punctures near midline of disc (Figs. 10.80c, 10.86a–10.90a); mesoventrite with complete longitudinal carina (Fig. 10.80d); spermatheca with multiple, small coils roughly regular in shape and size (Figs. 10.86h–10.90h)................................. ....***Coprochara* Mulsant and Rey**

– Pronotum evenly pubescent, without two longitudinal rows of setigerous punctures near

J. Klimaszewski et al., *A Faunal Review of Aleocharine Beetles in the Rapidly Changing Arctic and Subarctic Regions of North America (Coleoptera, Staphylinidae)*,
https://doi.org/10.1007/978-3-030-68191-3_10

midline of disc; mesoventrite either not carinate, or with partial or complete longitudinal carina; spermatheca without irregular-shaped posterior coils **2**

2. Body robust, broad and compact (Figs. 10.80e, 10.81a–10.84a); antennomeres V to X moderately to- distinctly transverse and forming an elongate club, each segment at least 2.2 times as broad as long (Figs. 10.80e, 10.81a–10.84a); mesoventrite not carinate (Fig. 10.80f); spermatheca relatively simple, narrowly elongate with small, spherical or sac-shaped capsule and thin stem (Figs. 10.81h–10.84h) .... ***Aleochara* Gravenhorst**

– Body spindle-shaped or subparallel in most (Fig. 10.80a,g); antennomeres V to X gradually broadening apically, slightly to moderately transverse, each antennomere 1.2–2.2 times as broad as long; mesoventrite either partially carinate or completely longitudinally carinate **3**

3. Mesoventrite completely longitudinally carinate (Fig. 10.80b); pronotum with pubescence on disc directed lateroposteriad (Fig. 10.80a); elytra at suture at most as long as pronotum................. .... ***Xenochara* Mulsant and Rey**

– Mesoventrite with short carina no longer than half mesoventrite length (Fig. 10.80h); pronotum with pubescence on disc directed laterad or obliquely or straight posteriad from midline (Figs. 10.80g, 10.85a); elytra at suture at least as long as pronotum .... ***Calochara* Casey**

Subgenus *Aleochara* Gravenhorst, 1802 (Figs. 10.80e,f, 10.81–10.84) (Klimaszewski 1984, Klimaszewski et al. 2018, 2020)

Diagnosis Body robust, length 3.0–9.0 mm; dorsal surface of body moderately to highly glossy, without distinct microsculpture (Figs. 10.81a–10.84a); antennomeres V–X moderately to distinctly transverse; pronotum evenly pubescent; mesoventrite wide and not carinate (Fig. 10.80f); first three visible abdominal tergites with deep transverse depressions; maxillary palpi with apical palpomere at least two-thirds as long as penultimate palpomere; species predominantly associated with carrion.

Key to Species of Subgenus *Aleochara*

1. Antennomere IV moderately transverse and V–X strongly transverse, more than twice as wide as long (Fig. 10.83a); elytra predominantly black with some red-brown sutural and posterior margin area (Fig. 10.83a); median lobe of aedeagus and spermatheca as in Fig. 10.83b,c,h ....***Aleochara* (*Aleochara*) *sekanai* Klimaszewski**

– Antennomere IV subquadrate or slightly elongate, and V–X subquadrate to moderately transverse (Figs. 10.81a, 10.82a, 10.84a); elytra predominantly red or red-brown with or without black lateral and basal margins (Figs. 10.81a, 10.82a, 10.84a); genitalic characters not as above **2**

2. Pronotum red-brown, sparsely pubescent and highly glossy (Fig. 10.81a); sclerites of internal sac of aedeagus as in Fig. 10.81b,c; shape of spermatheca as in Fig. 10.81h............***Aleochara* (*Aleochara*) *assiniboin* Klimaszewski**

– Pronotum brown or black, densely pubescent and moderately to highly glossy (Fig. 10.82a, 10.84a); genitalia not as above **3**

3. Body broad and moderately glossy (Fig. 10.84a); internal sac of aedeagus with reversed V-shaped median structure in dorsal view (Fig. 10.84c); spermatheca with elongate chamber (Fig. 10.84h).. .... ***Aleochara* (*Aleochara*) *tahoensis* Casey**

– Body narrow and highly glossy (Fig. 10.82a); internal sac of aedeagus with U-shaped median structure in dorsal view (Fig. 10.82c); spermatheca with short chamber (Fig. 10.82h).................. ***Aleochara* (*Aleochara*) *gracilicornis* Bernhauer**

11. *Aleochara* (*Aleochara*) *assiniboin* Klimaszewski (Fig. 10.81a–h)

Aleochara assiniboin Klimaszewski, 1985 [replacement name for preoccupied name *Aleochara rufonigra* Klimaszewski, 1984]. Gouix and Klimaszewski 2007, Ernst and Buddle 2015, Klimaszewski et al. 2018, 2020.

Diagnosis Body length 4.0–5.0 mm; colour dark brown, pronotum and elytra red-brown, highly glossy, or elytra dark with two lighter spots, appendages and two basal antennomeres reddish-brown; antennomere IV subquadrate or slightly transverse (Fig. 10.81a). MALE. Tergite VIII truncate apically (Fig. 10.81d); sternite VIII broadly arcuate apically (Fig. 10.81e); tubus of median lobe of aedeagus with ventral margin arcuate in lateral view (Fig. 10.81b); internal sac with three pairs of elongate sclerites (Fig. 10.81b, c). FEMALE. Tergite VIII truncate apically, apical margin not serrate (Fig. 10.81f); sternite VIII broadly arcuate apically (Fig. 10.81g); spermatheca with subovoid capsule, short neck, stem short and ovoid anteriorly and narrowly elongate posteriorly (Fig. 10.81h).

DISTRIBUTION **Origin**: Nearctic. **CANADA**: BC, MB, NT [**NTR**], ON, SK, YT. **USA**: not recorded.

COLLECTION AND HABITAT DATA
Habitat: some found among grasses, *Carex*, and mosses on the margin of a small lake margin; others found under a moose (*Alces alces* (L.) carcass, and in wet tundra. **Collecting period**: V–VIII. **Collecting method**: treading vegetation and hand-collecting from moose carcass, pitfall traps, yellow pan traps.

We report this species from NT for the first time here, based on the following: Canada: Northwest Territories, Norman Wells, 65.28901, –126.84077, Rep. 3 wet, pitfall trap. 7–11. VI.2011, NBP field party N3WP311 (LFC) 10 males, 12 females; Norman Wells, 65.25226, –126.66128, Rep. 2 wet, yellow pan, 11–14.VI.2011, NBP field party N2WY923 (LFC) 1 female; Norman Wells, 65.28901, –126.84077, Rep. 3 wet, pitfall trap, 11–14. VI.2011, NBP field party N3WP321 (LFC) 1 female; Norman Wells, 65.25226, –126.66128, Rep. 2 wet, pitfall trap, 11–14. VI.2011, NBP field party N2WP121 (LFC) 1 male; Norman Wells, 65.25226, –126.66128, Rep 2 wet, yellow pan, 14–17.VI.2011, NBP field party N2WP132, N2WP231, N2WP531, N2WP631, N2WP632, N2WP931, N2WY933 (LFC) 2 males. 5 females; Norman Wells, 65.28901, –126.84077, Rep 3 wet, yellow pan, 14–17.VI.2011, NBP field party N3WP632 (LFC) 1 female;

Yellowknife, 62.51923, –113.38494, Rep. 2 wet, pitfall trap, 11–15.VII.2011, NBP field party Y2WP521, Y2WP421, Y2WP122 (LFC) 1 male, 2 females; Yellowknife, 62.52110, –113.38174, Rep. 1 wet, yellow pan, 11–15. VII.2011, NBP field party Y1WY9210, Y1WY929, Y1WP721, Y1WP722, Y1WP423 (LFC) 1 male, 4 females; Yellowknife, 62.51923, –113.38494, Rep. 2 wet, pitfall trap, 11–15.VII.2011, NBP field party Y2WP121 (LFC) 1 male; Yellowknife, 62.51591 –113.39146, Rep. 3 wet, pitfall trap, 11–15. VII.2011, NBP field party Y3WP921 (LFC) 2 males; Yellowknife, 62.52110, –113.38174, Rep 1 wet, pitfall trap, 15–18.VII.2011, NBP field party Y1WP331, Y1WP531, Y1WP732, Y1WP733, Y1WP734, Y1WP831, (LFC) 1 male, 5 females; Yellowknife, 62.51591, –113.39146, Rep 3 wet, pitfall trap, 15–18. VII.2011, NBP field party Y3WP631, Y3WP631 (LFC) 2 females.

12. *Aleochara* (*Aleochara*) *gracilicornis* Bernhauer (Fig. 10.82a–h)

Aleochara gracilicornis Bernhauer, 1901. Klimaszewski 1984, Klimaszewski and Génier, 1987, Klimaszewski et al. 2005, Gouix and Klimaszewski 2007, Klimaszewski et al. 2018, 2020, Brattain et al. 2020.

Diagnosis Body narrowly oval, length 4.0–7.0 mm, integument highly glossy, colour dark brown to black, with elytra red brown except for darker margins and scutellar region, legs, palpi and 2–3 basal antennomeres yellow to reddish-brown; antennomere IV slightly elongate, and V–X moderately transverse; forebody and particularly abdomen glossy (Fig. 10.82a). MALE. Tergite VIII truncate apically (Fig. 10.82d); sternite VIII arcuate apically (Fig. 10.82e); median lobe of aedeagus in dorsal view with small oval bulbus and long subparallel tubus (Fig. 10.82c), in lateral view arcuate with apex narrow and slightly bent ventrad

(Fig. 10.82b); internal sac structures as illustrated (Fig. 10.82b,c). FEMALE. Tergite VIII truncate, with shallow median emargination (Fig. 10.82f); sternite VIII rounded apically (Fig. 10.82g); spermatheca sinuate, capsule small, ovoid, neck and slightly curved; stem narrow, long and sinuate (Fig. 10.82h).

DISTRIBUTION Origin: Nearctic. **CANADA**: BC, MB, NB, NF, NS, NT, ON, QC, SK. **USA**: AR, AZ, CO, FL, IL, IN, KS, LA, MA, MD, ME, MI, MN, MO, MT, NH, NJ, NM, NY, OH, PA, RI, SD, UT, VA, WI.

COLLECTION AND HABITAT DATA
Habitat: river margins in flood debris, drift material and decaying grass and litter, temporary creek margins in debris, lake margins in drift material, hardwood forests and swampy areas in leaves and debris, *Carex* marsh in leaf litter and moss, in an old American beaver (*Castor canadensis* Kuhl) lodge, in debris on beaver dam (common), and in carrion. Others from pitfall traps in mature red spruce forest. **Collecting period**: III–X. **Collecting method**: sifting organic debris.

13 *Aleochara* (*Aleochara*) *sekanai* Klimaszewski (Fig. 10.83a–h)

Aleochara (Aleochara) sekanai Klimaszewski, 1985 (replacement name for preoccupied name *A. unicolor* Klimaszewski, 1984). Klimaszewski 1984, Gouix and Klimaszewski 2007, Majka et al. 2009, Majka and Klimaszewski 2010, Klimaszewski et al. 2011, 2015, 2018, Schülke and Smetana 2015.

Diagnosis Body broadly oval, length 4.0–7.0 mm; colour uniformly black or black with red-brown diffuse spot in posterior sutural part of disc and apical margin of tergite V; antennomeres V-X distinctly transverse (Fig. 10.83a). MALE. Tergite VIII with apical margin truncate (Fig. 10.83d); sternite VIII with apical margin broadly rounded (Fig. 10.83e); median lobe of aedeagus in dorsal view with bulbus moderate-sized, oval, tubus long and subparallel (Fig. 10.83c); in lateral view tubus with ventral margin broadly arcuate with apex narrow and produced ventrad (Fig. 10.83b); internal sac structures complex, as illustrated (Fig. 10.83b,c). FEMALE. Tergite VIII arcuate apically (Fig. 10.83f); sternite VIII arcuate apically (Fig. 10.83g); spermatheca curved and S-shaped, capsule ovoid, neck short and slightly bent; stem sinuate, broadened anteriorly and narrowed posteriorly (Fig. 10.83h).

DISTRIBUTION Origin: Nearctic (see Comments). **CANADA**: AB, LB, MB, NB, NT, ON, SK, YT. **USA**: AK.

COLLECTION AND HABITAT DATA
Habitat: feather moss forest, *Carex* and moss near a lake, animal carcasses. **Collecting period**: V-VIII. **Collecting method**: carrion baited pitfall traps, flight intercept traps.

COMMENTS *Aleochara sekanai* was listed from Russia (Far East) by Schülke and Smetana (2015) but we hesitate to consider this species as Holarctic before voucher specimens can be carefully examined. Many cryptic species exist in *Aleochara*, differing only in the sclerites of the internal sac.

14. *Aleochara* (*Aleochara*) *tahoensis* Casey (Fig. 10.84a–h)

Aleochara tahoensis Casey, 1906. Hatch 1957, Moore and Legner 1975, Klimaszewski 1984, Klimaszewski and Génier 1987, Klimaszewski et al. 2005, Gouix and Klimaszewski 2007, Klimaszewski et al. 2016a, 2018, 2020.

Diagnosis Body broadly oval, length 4.5–7.0 mm; colour dark brown to black with elytra reddish-brown except antero-laterad margins dark brown, legs, maxillary palps, and often basal antennomere red-brown; antennomere IV slightly elongate and V-X distinctly transverse (Fig. 10.84a). MALE. Tergite VIII with apical margin slightly emarginate medially (Fig. 10.84d); sternite VIII with apical margin broadly rounded, obtusely angulate at apex (Fig. 10.84e); median lobe of aedeagus in dorsal view with bulbus moderately large, oval, tubus narrowly triangular (Fig. 10.84c); in lateral view tubus with ventral margin straight with apex narrow, slightly produced ventrad (Fig. 10.84b); internal sac structures complex, as illustrated (Fig. 10.84b,c). FEMALE. Tergite VIII truncate and slightly emarginate medially (Fig. 10.84f); sternite VIII arcuate apically (Fig. 10.84g); spermatheca thin, capsule narrowly ovoid, neck short and angular, stem long and sinuate (Fig. 10.84h).

This species is similar externally to *A. gracilicornis*, but in *A. tahoensis* the body is broader, antennomeres V-X are more transverse, the distance between the eyes is equal to 2.5 times the maximum diameter of the eye, and the genitalic features are distinctive.

DISTRIBUTION Origin: Nearctic. **CANADA**: AB, BC, MB, NB, NF, NS, NT, ON, SK, YT. **USA**: AR, CA, CO, ID, MT, NH, NM, NV, OR, SD, WA, WI.

COLLECTION AND HABITAT DATA
Habitat: forests and swampy areas in debris, under moose carcass, in leaf litter on a beaver dam, and along a lake shore in drift material. Some collected from pitfall traps in mature red spruce (*Picea rubens* Sarg.) forest. **Collecting period**: V-IX. **Collecting method**: sifting litter, pitfall traps.

Subgenus *Calochara* Casey, 1906 (Figs. 10.80g,h, 10.85a–h) (Klimaszewski 1984, 2018)

Diagnosis Pronotum evenly pubescent (Figs. 10.80g, 10.85a); mesoventrite narrow, wedge-shaped with carina no longer than half its length (Fig. 10.80h); male sternite VIII triangularly produced apically; median lobe of aedeagus narrow, strongly bent ventrally and with two pairs of elongate sclerites (Fig. 10.85b); spermatheca with capsule more-or-less spherical, stem usually thin and simple (Fig. 10.85h); associated with animal nests and burrows.

15. *Aleochara* (*Calochara*) *villosa* Mannerheim ‡ (Fig. 10.85a–h)

Aleochara villosa Mannerheim, 1830. Klimaszewski 1984, Klimaszewski and Frank 1992; Gouix and Klimaszewski 2007, Webster et al. 2009, Klimaszewski et al. 2018, 2020.

Diagnosis Body moderately broad, length 3.0–5.0 mm; colour dark brown to black, with tarsi, and last articles of maxillary palps often reddish-brown; pubescence sparse; forebody with meshed microsculpture; antennomeres V-X distinctly transverse; apical palpomere of maxillary palpus about as long as half of penultimate one (Fig. 10.85a). MALE. Tergite VIII sinuate apically, with distinct pattern of macrosetae (Fig. 10.85d); sternite VIII with apical margin broadly triangularly produced, sinuately emarginate on either side, with apex narrowly rounded (Fig. 10.85e); median lobe of aedeagus in dorsal view with bulbus large, oval, evenly coalescent with triangular tubus (Fig. 10.85c); in lateral view tubus with ventral margin straight, broadly curved ventrad at base, with apex narrow, slightly curved ventrad, dorsal part on internal sac strongly sclerotized (Fig. 10.85b); internal sac with dorsal part sclerotized in lateral view (Fig. 10.85b,c), with structures as illustrated (Figs. 10.85b,c). FEMALE. Tergite VIII truncate apically (Fig. 10.85f); sternite VIII rounded apically (Fig. 10.85g); spermatheca U-shaped, similar in shape to that of *A. speculicollis* (Fig. 10.85h).

DISTRIBUTION Origin: Palaearctic or Holarctic, if the former then adventive in North America. **CANADA**: AB, BC, NB, ON, QC, SK. **USA**: AK, CA, OK, OR, WA.

COLLECTION AND HABITAT DATA
Habitat: nest contents of barred owls (many sites in NB), a great horned owl, *Bubo virginianus* (Gmelin) and probably occurs in other tree hole nesting birds and mammals. Adults in moist smelly (urine smell) organic material consisting of small animal and bird remains, fur, feathers, and owl pellets, and many fly larvae. Nests were in tree holes in large trees in old mixed forests, old hardwood forests, an old silver maple forest (*Acer saccharinum* L.), and a black spruce forest. Other

individuals from old pile of hay and in carrion. **Collecting period**: II-VIII. **Collecting method**: sifting litter.

Additional records. See Annex 1.

COMMENTS The distribution status of this species is not clear. It was reported as an adventive species in North America but in fact it may be a Holarctic species. There are not enough distribution records from North America at this time to make a definitive statement.

Subgenus *Coprochara* Mulsant and Rey, 1874 (Figs. 10.80c,d, 10.86–10.90) (Klimaszewski 1984, Klimaszewski et al. 2018, 2020)

Diagnosis Pronotal disc with two longitudinal, subparallel rows of setigerous punctures laterad to median line (Fig. 10.80c), punctures fine or coarse, sometimes confluent, with interspace along midline glabrous or with fine punctation; mesoventrite completely carinate longitudinally (Fig. 10.80d); posterior stem of spermatheca with multiple coils (Fig. 10.86h).

Key to Species of Subgenus *Coprochara*

1. Pronotal subparallel rows of punctures situated in two prominent grooves (Fig. 10.89a); genital structures as illustrated (Fig. 10.89b,c,d,i) .... ***Aleochara* (*Coprochara*) *sulcicollis* Mannerheim**.

– Pronotal subparallel rows of punctures not in deep grooves (Figs. 10.86a–10.88a, 10.90a); genital structures not as above **2**

2. Elytra more or less uniformly reddish-brown (Figs. 10.87a, 10.88a) **3**

– Elytra black or dark brown with two distinct yellow or reddish-brown oval or round spots in posterior part of disc (Figs. 10.86a, 10.90a) **4**

3. Pronotal punctation weak, surface strongly glossy (Fig. 10.88a); genitalia as illustrated (Fig. 10.88b,c,h) .... ***Aleochara* (*Coprochara*) *suffusa* (Casey)**

– Pronotal punctation strong, surface moderately glossy (Fig. 10.87a); genitalia as illustrated (Fig. 10.87b,c,h) .... ***Aleochara shelleyae* Klimaszewski and Langor**

4. Body narrowly subparallel (Fig. 10.90a); elytral spots small, orange to rust-brown; pronotal punctation coarse (Fig. 10.90a); genitalia as illustrated (Fig. 10.90b,c,h) ....***Aleochara* (*Coprochara*) *verna* Say**

– Body narrowly oval (Fig. 10.86a); elytral spots large, yellowish-brown (Fig. 10.86a); genitalia as illustrated (Fig. 10.86b,c,h)..................... ....***Aleochara* (*Coprochara*) *bimaculata* Gravenhorst**

16. *Aleochara* (*Coprochara*) *bimaculata* Gravenhorst (Fig. 10.86a–h)

Aleochara bimaculata Gravenhorst, 1802. Klimaszewski 1984, Klimaszewski and Frank 1992, Gouix and Klimaszewski 2007, Majka and Klimaszewski 2010, Klimaszewski et al. 2011, 2018, 2020.

Diagnosis Body broadly oval, length 4.0–8.0 mm; colour piceous black, with tarsi, palps, and basal antennomere rusty-brown, each elytron with a large yellowish-brown spot in posterior median part of disc (Fig. 10.86a); pronotum glossy; setigerous punctures on head and pronotum slightly impressed; antennomeres V–X distinctly transverse (Fig. 10.86a). MALE. Tergite VIII with apical margin arcuate, with row of long setae (Fig. 10.86d); sternite VIII with apical margin arcuate (Fig. 10.86e); median lobe of aedeagus and internal sac structures as illustrated, including very long flagellum coiled in bulbus (Fig. 10.86b,c). FEMALE. Tergite VIII truncate apically (Fig. 10.86f); sternite VIII rounded apically (Fig. 10.86g); spermatheca with capsule spherical, with deep, broad apical invagination, neck short and angular, stem slightly sinuate with at least 8 tight posterior coils (Fig. 10.86h).

DISTRIBUTION Origin: Nearctic. **CANADA**: AB, BC, LB, MB, NB, NF, NS, NT, ON, QC, SK. **USA**: AL, AZ, CA, CO, CT, DC, FL, GA, IA, ID, IL, IN, KS, KY, LA, MA, MD, ME, MI, MN, MO, MT, NC, ND, NE, NH, NJ, NM, NV,

NY, OK, OR, PA, RI, SC, SD, TN, TX, UT, VA, VT, WA, WI, WV, WY.

COLLECTION AND HABITAT DATA
Habitat: diverse open and treed habitats; adults found in various kinds of dung (cow, horse, deer, dog, human), animal carcasses (fish and mammal), decaying fungi, moss and leaf litter along streams, and in plant debris in beaver lodges. **Collecting period**: I–XI (mainly summer months). **Collecting method**: sifting decaying organic matter, dung- and carrion-baited pitfall traps.

17. *Aleochara* (*Coprochara*) *shelleyae* Klimaszewski and Langor (Fig. 10.87a–h)

Aleochara (Coprochara) shelleyae Klimaszewski and Langor, 2011. Klimaszewski et al. 2018.

Diagnosis Body narrowly elongate, length 3.0–5.0 mm; colour dark brown, with elytra (except for dark brown scutellar region), tarsi, palps, antennomeres I-III and sometimes end of abdomen rusty-brown (Fig. 10.87a); setigerous punctures on head and pronotum slightly impressed; antennomeres V-X quadrate to slightly transverse; head with pubescence directed inwards, on disc of pronotum directed obliquely posteriad. MALE. Tergite VIII rounded apically (Fig. 10.87d); sternite VIII with apical margin slightly produced medially and rounded at apex (Fig. 10.87e); median lobe of aedeagus similar to that of *A. verna*, in lateral view tubus with ventral margin sinuate, apex narrowly triangular (Fig. 10.87b), sclerites of internal sac as illustrated (Fig. 10.87b,c). FEMALE. Tergite VIII truncate apically (Fig. 10.87f); sternite VIII rounded apically (Fig. 10.87g); spermatheca with capsule spherical, stem widely U-shaped with 4–5 tight posterior coils (Fig. 10.87h).

DISTRIBUTION Origin: Nearctic. **CANADA**: NF, ON. **USA**: not recorded.

COLLECTION AND HABITAT DATA
Habitat: coastal sand dunes in vegetation. **Collecting period**: VI-VIII. **Collecting method**: sifting dune vegetation and debris.

18. *Aleochara* (*Coprochara*) *suffusa* (Casey) (Fig. 10.88a–h)

Baryodma suffusa Casey, 1906. Klimaszewski 1984, Klimaszewski and Cervenka 1986, Klimaszewski and Frank 1992, Gouix and Klimaszewski 2007, Klimaszewski et al. 2016b, 2018, 2020, Maus 1998, Smetana 2004.

Diagnosis Body broadly oval, length 2.8–4.0 mm; colour dark brown, with elytra, tarsi, palps, basal antennomeres yellow-brown or reddish-brown (Fig. 10.88a); setigerous punctures on head and pronotum fine; pronotum sparsely pubescent and highly glossy, two subparellel rows of punctures slightly impressed; antennomeres V–X moderately transverse (Fig. 10.88a). MALE. Tergite VIII with apical margin truncate, with row of long setae (Fig. 10.88d); sternite VIII with apical margin slightly produced medially, rounded at apex (Fig. 10.88e); median lobe of aedeagus long, in lateral view with ventral margin of tubus broadly curved ventrad at middle, with apex narrow, slightly bent ventrad (Fig. 10.88b), sclerites of internal sac as illustrated (Fig. 10.88b,c), median lobe in dorsal view as illustrated (Fig. 10.88c). FEMALE. Tergite VIII truncate apically (Fig. 10.88f); sternite VIII slightly produced apically (Fig. 10.88g); spermatheca with capsule club-shaped, neck angular, stem straight medially and sinuate posteriad with up to 18 tight posterior coils (Fig. 10.88h).

DISTRIBUTION Origin: Nearctic. **CANADA**: AB, BC, MB, QC, SK. **USA**: AK, AZ, CO, ID, NM, WA, WY.

COLLECTION AND HABITAT DATA
Habitat: open habitats and woodlands. Adults found under rocks in a meadow. **Collecting period**: VII–VIII. **Collecting method**: hand picking from under rocks and rearing in laboratory from dipteran pupae collected in fields.

COMMENTS Maus (1998) examined the type material representing both Palaearctic and Nearctic populations of a species previously known as Holarctic *A. suffusa* (i.e., *sensu* Klimaszewski 1984). Based on this study, Maus (1998) reinstated *A. brundini* Bernhauer for the disjunct Palaearctic populations in Scandanavia and the Alps, while *A. suffusa* was restricted to the Nearctic. Recent treatments of *A. suffusa* (e.g., Klimaszewski et al. 2018) overlooked this. Finally, *A. brundini* may yet occur in the Nearctic, as some barcoded specimens from AK clustered with those from Finland identified as this species. More research is needed to determine the occurrence of *A. brundini* in North America as this is an extremely difficult group taxonomically.

19. *Aleochara (Coprochara) sulcicollis* Mannerheim (Fig. 10.89a–i)

Aleochara sulcicollis Mannerheim, 1843. Klimaszewski 1984, Navarrete-Heredia et al. 2002, Gouix and Klimaszewski 2007, Klimaszewski et al. 2020.

Diagnosis Body length 3.0–7.0 mm; colour piceous to black, with tarsi and antennae reddish brown; setigerous punctures on head and pronotum coarse and deeply impressed, those along midline of pronotum in grooves (Fig. 10.89a); antennae short, antennomeres V–X distinctly transverse (Fig. 10.89a). MALE. Tergite VIII with apical margin truncate, with row of very short setae (Fig. 10.89e); sternite VIII with apical margin slightly produced medially, rounded at apex (Fig. 10.89f); median lobe of aedeagus in lateral view with ventral margin of tubus broadly curved ventrad from base with apex narrowly triangular, slightly bent ventrad at tip in lateral view (Fig. 10.89b,c), sclerites of internal sac as illustrated (Fig. 10.89b,c), median lobe in dorsal view as illustrated (Fig. 10.89d). FEMALE. Tergite VIII truncate apically (Fig. 10.89g); sternite VIII slightly produced apically (Fig. 10.89h); spermatheca with capsule spherical, neck angular, stem sinuate with numerous (up to 80) tight posterior coils (Fig. 10.89i).

DISTRIBUTION Origin: Nearctic. **CANADA**: BC. **USA**: AK, CA, OR, WA. Mexico, Chile.

COLLECTION AND HABITAT DATA
Habitat: under drift wood on sandy beach, seaweed on sea beaches, in clear-cut. **Collecting period**: II–X. **Collecting method**: hand picking from seaweed, sifting seaweed and aspirating specimens, colored pan trap, and Lindgren trap.

Additional records. See Annex 1.

20. *Aleochara (Coprochara) verna* Say * (Fig. 10.90a–h)

Aleochara verna Say, 1833. Klimaszewski 1984, Hemachandra et al. 2005, Klimaszewski et al. 2005, Gouix and Klimaszewski 2007, Majka and Klimaszewski 2010, Maus 1998, Schülke and Smetana 2015, Klimaszewski et al. 2011, 2018, 2020, Brattain et al. 2020.

Diagnosis Body narrowly subparallel, length 2.0–4.0 mm; colour piceous to black, with tarsi, palps, and basal antennomere rusty-brown, each elytron with yellowish to rusty-brown spot in apico-median part of disc (Fig. 10.90a); setigerous punctures on head and pronotum coarse; two subparellel pronotal punctures not in grooves; antennomeres V–X distinctly transverse (Fig. 10.90a). MALE. Tergite VIII with apical margin truncate, with row of short setae (Fig. 10.90d); sternite VIII with apical margin slightly produced medially, rounded at apex (Fig. 10.90e); median lobe of aedeagus in lateral view with ventral margin of tubus broadly curved ventrad from base with apex narrowly triangular, slightly bent ventrad at tip (Fig. 10.90b), sclerites of internal sac as illustrated (Fig. 10.90b,c), median lobe in dorsal view as illustrated (Fig. 10.90c). FEMALE. Tergite VIII truncate apically (Fig. 10.90f); sternite VIII slightly produced apically (Fig. 10.90g); spermatheca with capsule club-shaped, neck angular, stem sinuate, with 9–11 tight posterior coils (Fig. 10.90h).

DISTRIBUTION Origin: Holarctic. **CANADA**: AB, BC, LB, MB, NB, NF, NS, ON, PE, QC, SK, YT. **USA**: AK, AZ, CA, CO, DC, FL, GA, IA,ID,

IL, IN, KS, MA, ME, MI, MN, MO, MT, NH, NC, NJ, NM, NY, NV, OK, OR, PA, SC, SD, TN, TX, UT, VA, VT, WA, WI, WY.

COLLECTION AND HABITAT DATA
Habitat: agricultural fields, pastures, forests, coastal limestone barrens, sea beaches, and on a sandy area near the ocean and forests, rocky alpine habitat, arctic tundra. Adults in various kinds of decaying organic matter such as mammal dung, carrion, beach drift material, rotting fungi, in moss and decaying vegetable matter; some specimens emerged from cabbage and radish maggots (Diptera, Anthomyiidae). Many specimens from pitfall traps in a mature red spruce forest in NB. **Collecting period**: VI–IX. **Collecting method**: pitfall traps, malt traps, light trap, rotary trap, sifting decaying organic matter.

Additional records. See Annex 1.

COMMENTS This species was recently considered as a Holarctic species by Maus (1998), based on careful examination of the internal sac of the aedeagus and the spermatheca, and the presence of *A. verna* in the Palaearctic region for more than 150 years (at the time of publication).

Subgenus *Tinotus* Sharp, 1883 (Fig. 10.91a–g) (Klimaszewski et al. 2002, 2018, 2020)

Diagnosis Species of this subgenus differ from all other *Aleochara* species in having a 4-5-5 tarsal formula (5-5-5 in all other *Aleochara*).

21. *Aleochara* (*Tinotus*) *morion* Gravenhorst † (Fig. 10.91a–g)

Aleochara morion Gravenhorst, 1802. Bernhauer 1906, Klimaszewski et al. 2002, Gouix and Klimaszewski 2007, Klimaszewski et al. 2005, Majka and Klimaszewski 2008, Bousquet et al. 2013 [as *Tinotus*], Klimaszewski et al. 2011, 2015, 2020.

Diagnosis Body narrowly oval, robust, length 1.8–3.0 mm; colour entirely black, rarely brown with appendages and sometimes elytra paler, reddish, surface densely pubescent, moderately glossy, integument with meshed microsculpture (Fig. 10.91a); head strongly deflexed; antennomeres IV–X slightly to highly transverse, about twice as wide as long; maximum width of elytra greater than that of pronotum, elytra at suture shorter than pronotum along midline; abdominal impressions with deep, round or longitudinally elongate pores. MALE. Tergite VIII with apical margin crenulate, slightly emarginate medially (Fig. 10.91c); sternite VIII rounded apically and vaguely angulate medially (Fig. 10.91d); median lobe of aedeagus in lateral view with tubus weakly produced apically, bearing large rounded basal projection (Fig. 10.91b). FEMALE. Tergite VIII with apical margin truncate, slightly emarginate (Fig. 10.91e); sternite VIII broadly rounded apically (Fig. 10.91f); spermatheca L-shaped, capsule sac-shaped and bent laterally, stem with 1–2 posterior coils (Fig. 10.91g).

DISTRIBUTION **Origin**: Palearctic, adventive in North America. **CANADA**: AB, BC, NB, NF, NS, ON, QC, SK. **USA**: CA, CT, NV.

COLLECTION AND HABITAT DATA
Habitat: open habitats and various forest types, in decaying organic matter, fungi, animal droppings, human feces, and carrion. **Collecting period**: II–IX. **Collecting method**: sifting organic debris, carrion baited pitfall traps.

COMMENTS This species is included based on a single record from BC: Liard River, 59.4288, −126.116, bison scat, 10.VI.2013, B. Godin (LFC, BGC) 1 male, 2 females.

Subgenus *Xenochara* Mulsant and Rey, 1874 (Figs. 10.92–10.93) (Klimaszewski 1984, Klimaszewski et al. 2017, 2018)

Diagnosis Body moderately to highly convex, spindle-shaped to broadly subparallel; antennomeres V–X gradually broadening apically, slightly to moderately transverse, each antennomere 1.2–2.2 times as broad as long; pronotum evenly pubescent, pubescence on disc directed lateroposteriad (Figs. 10.80a, 10.92a, 10.93a); mesoventrite with complete or almost complete longitudinal median carina (Fig. 10.80b); maxillary palpi with apical palpomere one-third to three-quarters as long as penultimate one; median lobe of aedeagus in dorsal view with bulbus moderate-sized, tubus narrowly elongate, subparallel basally and narrowly triangular apically (Figs. 10.92d, 10.93d), sclerites of internal sac as illustrtaed (Figs. 10.92c,d, 10.93c,d); spermatheca diverse in shape but capsule usually spherical with distinct invagination (Figs. 10.92i, 10.93i).

COMMENTS Habitats are diverse, including various decaying organic matter, rotting mushrooms, moss, animal droppings, and forest litter.

Key to Species of Subgenus *Xenochara*

1. Pronotum distinctly transverse, sides distinctly converging apically (Fig. 10.93a); pronotal microsculpture weak or absent (Fig. 10.93b); spaces betweeen setae smooth and highly glossy (Fig. 10.93b); median lobe of aedeagus long and strongly bent ventrad in lateral view, sclerites of internal sac distinctive (Fig. 10.93c); capsule of spermatheca spherical without apparent apical invagination (Fig. 10.93i) ....****Aleochara* (*Xenochara*) *fumata* Gravenhorst**

– Pronotum moderately transverse, sides moderately converging apically (Fig. 10.92a); pronotal microsculpture strong, spaces between setae covered with microsculpture and appearing less glossy (Fig. 10.92b); median lobe of aedeagus moderately long and moderately bent ventrad in lateral view, sclerites of internal sac distinctive (Fig. 10.92c); capsule of spermatheca ovoid with distinct apical invagination (Fig. 10.92i) ....***Aleochara* (*Xenochara*) *castaneipennis* Mannerheim**

22. *Aleochara* (*Xenochara*) *castaneipennis* Mannerheim (Fig. 10.92a–i)

Aleochara castaneipennis Mannerheim, 1843. Klimaszewski 1984, Klimaszewski and Peck 1986 (possible misidentification for *A. castaneimarmotae*), Gouix and Klimaszewski 2007, Klimaszewski et al. 2005, Majka and Klimaszewski 2010, Majka et al. 2011, Klimaszewski et al. 2011, 2017, 2018, 2020.

Diagnosis Body subparallel, length 2.8–6.0 mm; colour brown to dark brown, with tarsi and palps rusty-brown, elytra often paler than remainder of the body (Fig. 10.92a); forebody moderately glossy with coarse meshed microsculpture, sculpticells flat (Fig. 10.92b); antennomeres V–X distinctly transverse (Fig. 10.92a). MALE. Tergite VIII with apical margin truncate and slightly emarginate (Fig. 10.92e); sternite VIII triangularly produced apically (Fig. 10.92f); median lobe of aedeagus in lateral view with ventral margin of tubus arcuate (Fig. 10.92c), apical sclerites of internal sac directed anteriad (Fig. 10.92c), in dorsal view as illustrated (Fig. 10.92d). FEMALE. Tergite VIII with apical margin truncate, shallowly emarginate medially (Fig. 10.92g); sternite VIII with apical margin rounded, with short pubescence (Fig. 10.92h); spermatheca L-shaped, capsule ovoid with large apical invagination, neck angular, stem short and straight (Fig. 10.92i).

This species is often misidentified as the externally similar *A. fumata* but in *A. castaneipennis* the body is often larger, the pronotal punctation slightly coarser with the interstices between punctures less glossy, and the genital features are different.

DISTRIBUTION Origin: Nearctic. **CANADA**: AB, BC, LB, NB, NF, NS, NT, ON, QC, YT. **USA**: AK, AR, AZ, CO, WA, WY.

COLLECTION AND HABITAT DATA

Habitat: coniferous, deciduous and mixed wood forest types, coastal limestone barrens. Adults often found in rotting mushrooms, carrion, forest leaf litter, moss, and human feces. In NB, this species was often abundant in decaying mushrooms but was seldom found in other microhabitats. **Collecting period**: IV–XI. **Collecting method**: sifting decaying mushrooms, forest litter, using unbaited and carrion-baited pitfall traps, yellow pitfall traps, unbaited pitfall traps, flight intercept traps.

Additional records. See Annex 1.

23. *Aleochara* (*Xenochara*) *fumata* Gravenhorst † (Fig. 10.93a–i)

Aleochara fumata Gravenhorst, 1802. Klimaszewski 1984, Klimaszewski and Peck 1986, Klimaszewski and Génier 1987, Gouix and Klimaszewski 2007, Klimaszewski et al. 2005, Majka and Klimaszewski 2010, Klimaszewski et al. 2011, 2017, 2018, 2020.

Diagnosis Body subparallel, length 2.8–5.0 mm; colour brown to dark brown, with basal 1–3 antennomeres, legs/tarsi and palps rusty-brown (Fig. 10.93a); forebody strongly glossy with meshed microsculpture fine or absent (Fig. 10.93b); antennomeres V–X distinctly transverse (Fig. 10.93a). MALE. Tergite VIII with apical margin minutely crenulate, truncate, slightly broadly emarginate (Fig. 10.93e); sternite VIII slightly triangularly produced apically (Fig. 10.93f); median lobe of aedeagus in lateral view with ventral margin of tubus evenly arcuate (Fig. 10.93c), apical sclerites of internal sac directed ventrad (Fig. 10.93c), in dorsal view median lobe as illustrated (Fig. 10.93d). FEMALE. Tergite VIII arcuate apically (Fig. 10.93g); sternite VIII rounded apically (Fig. 10.93h); spermatheca L-shaped, capsule club-shaped without apparent apical invagination, neck angular, stem short and straight (Fig. 10.93i).

This species is externally very similar to *A. castaneipennis*, but in *A. fumata* the body is often smaller, the pronotal punctation finer, the microsculpture absent or fine, the elytra shorter, the pronotum highly glossy, and the genital features are different.

DISTRIBUTION Origin: Palearctic, adventive in North America. **CANADA**: AB, BC, LB, MB, NB, NF, NS, ON, PE, QC, YT. **USA**: AL, AR, AZ, CA, GA, ID, MA, MD, ME, MN, MO, NC, NH, NJ, NM, NY, OK, OR, PA, RI, TN, VA, VT, WA, WI, WV, WY. Brazil, Argentina.

COLLECTION AND HABITAT DATA

Habitat: hardwood, mixed, and coniferous forests. In rotting fungi (often very common), organic debris, carrion, and wet moss. **Collecting period**: IV–XI. **Collecting method**: sifting mushrooms and forest debris, pitfall traps, and flight intercept traps.

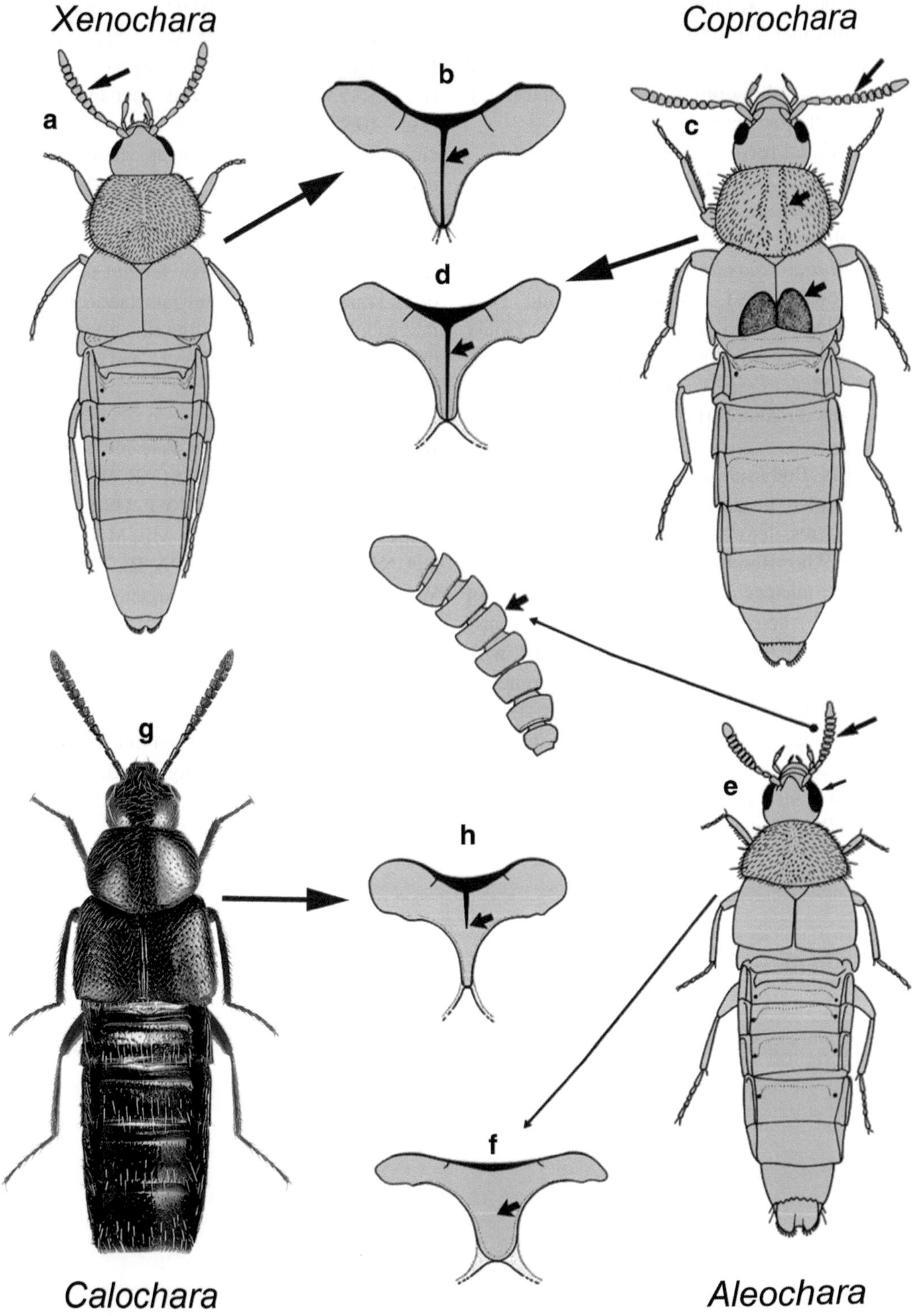

Fig. 10.80 (**a–h**). Representatives of subgenera of the genus *Aleochara* with important diagnostic features indicated by arrows (modified from Klimaszewski et al. (2020)): (**a**) habitus of subgenus *Xenochara*; (**b**) almost completely or completely carinate mesoventrite (mesosternum) of subgenus *Xenochara*; (**c**) habitus of subgenus *Coprochara*; (**d**) completely carinate mesoventrite of subgenus *Coprochara*; (**e**) habitus of subgenus *Aleochara*; (**f**) uncarinate mesoventrite of subgenus *Aleochara*; (**g**) habitus of subgenus *Calochara*; (**h**) partially carinate mesoventrite of subgenus *Calochara*

Fig. 10.81 (**a–h**). Images of *Aleochara (Aleochara) assiniboin* Klimaszewski: (**a**) habitus in dorsal view; (**b**) median lobe of aedeagus in lateral view; (**c**) median lobe of aedeagus in dorsal view; (**d**) male tergite VIII; (**e**) male sternite VIII; (**f**) female tergite VIII; (**g**) female sternite VIII; (**h**) spermatheca. Scale bar for habitus = 1 mm; remaining scale bars = 0.2 mm

Fig. 10.82 (**a–h**). Images of *Aleochara (Aleochara) gracilicornis* Bernhauer: (**a**) habitus in dorsal view; (**b**) median lobe of aedeagus in lateral view; (**c**) median lobe of aedeagus in dorsal view; (**d**) male tergite VIII; (**e**) male sternite VIII; (**f**) female tergite VIII; (**g**) female sternite VIII; (**h**) spermatheca. Scale bar for habitus = 1 mm; remaining scale bars = 0.2 mm

Fig. 10.83 (**a–h**). Images of *Aleochara (Aleochara) sekanai* Klimaszewski: (**a**) habitus in dorsal view; (**b**) median lobe of aedeagus in lateral view; (**c**) median lobe of aedeagus in dorsal view; (**d**) male tergite VIII; (**e**) male sternite VIII; (**f**) female tergite VIII; (**g**) female sternite VIII; (**h**) spermatheca. Scale bar for habitus = 1 mm; remaining scale bars = 0.2 mm

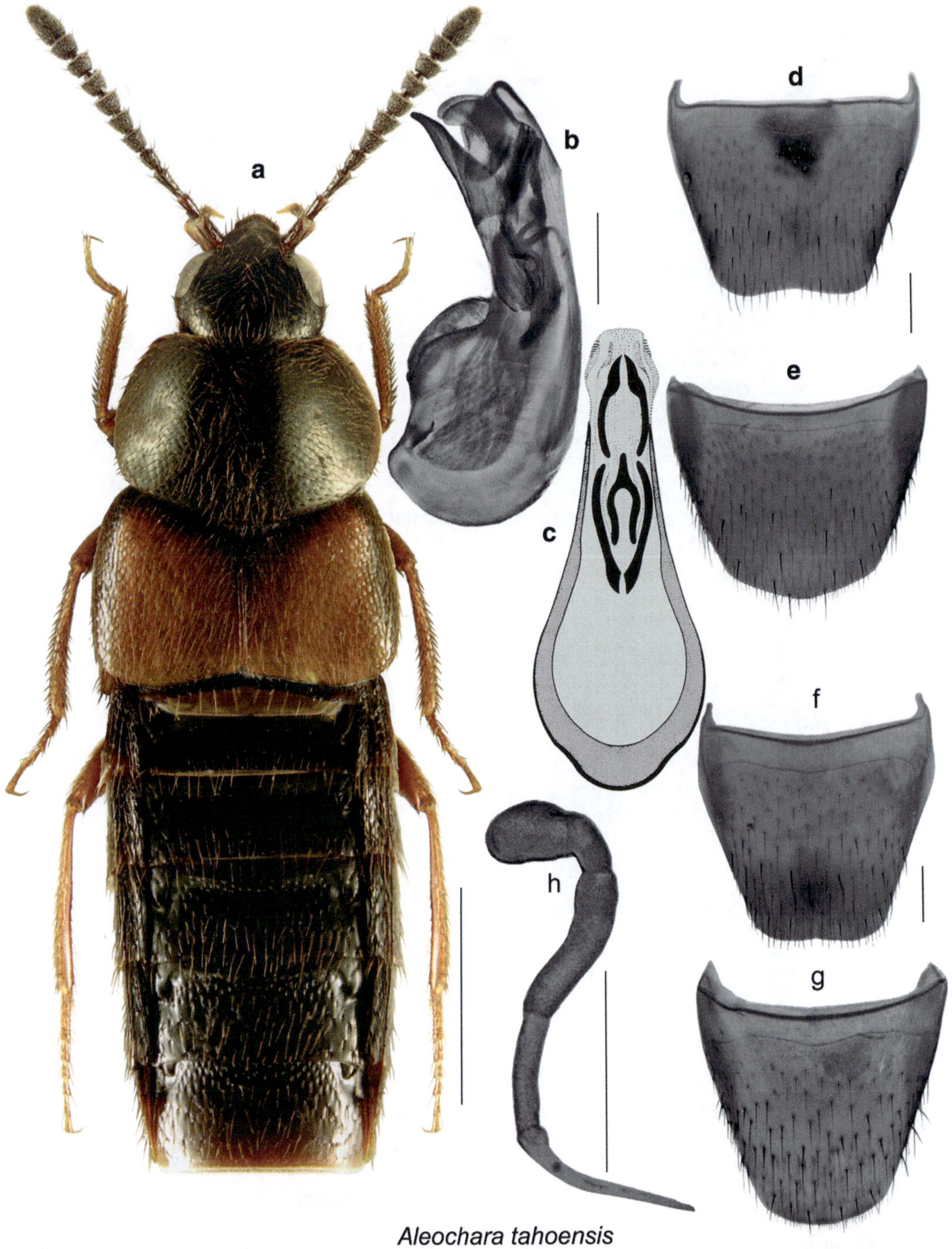

Fig. 10.84 (**a–h**). Images of *Aleochara (Aleochara) tahoensis* Casey: (**a**) habitus in dorsal view; (**b**) median lobe of aedeagus in lateral view; (**c**) median lobe of aedeagus in dorsal view; (**d**) male tergite VIII; (**e**) male sternite VIII; (**f**) female tergite VIII; (**g**) female sternite VIII; (**h**) spermatheca. Scale bar for habitus = 1 mm; remaining scale bars = 0.2 mm

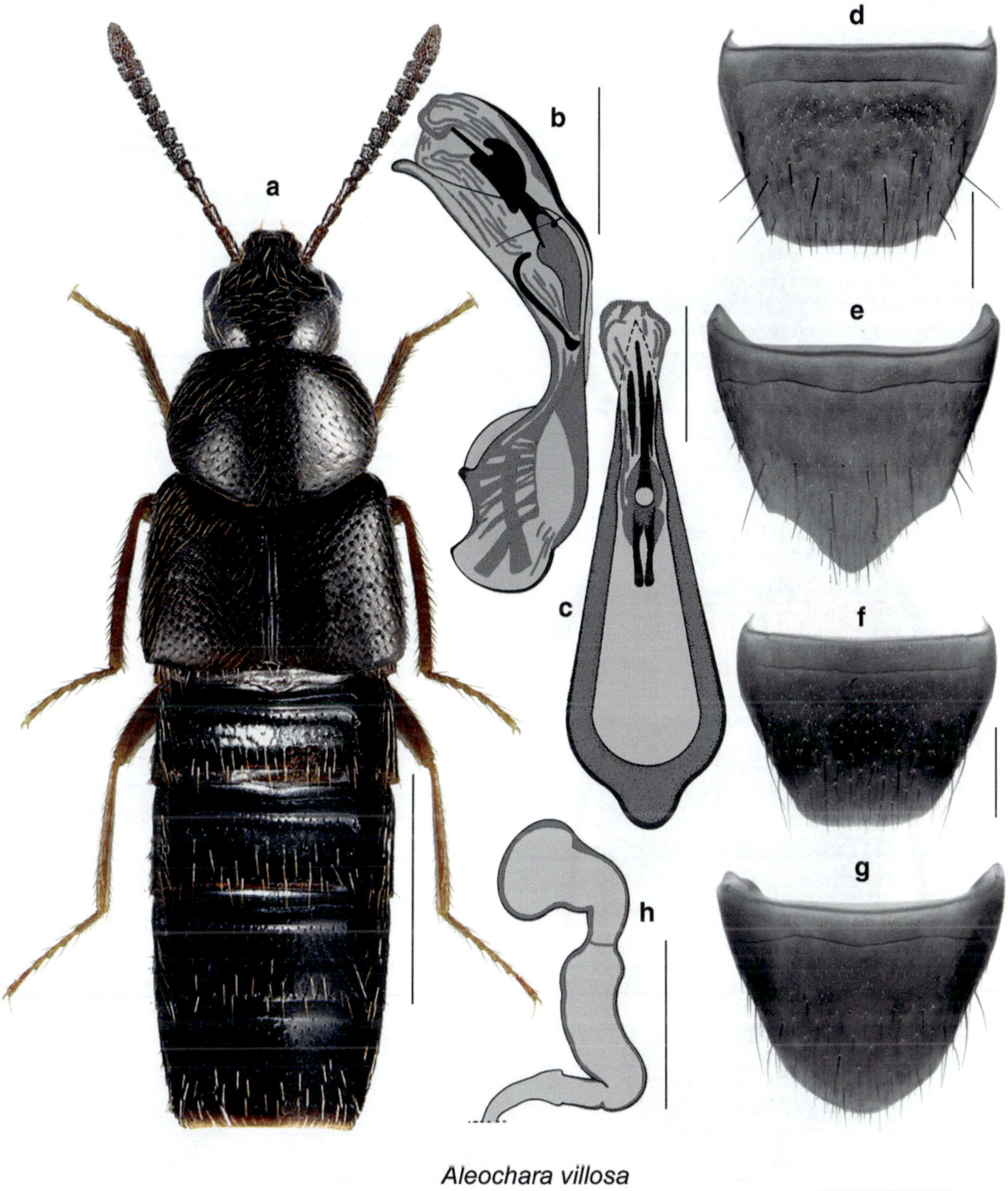

Fig. 10.85 (**a–h**). Images of *Aleochara (Calochara) villosa* Mannerheim: (**a**) habitus in dorsal view; (**b**) median lobe of aedeagus in lateral view; (**c**) median lobe of aedeagus in dorsal view; (**d**) male tergite VIII; (**e**) male sternite VIII; (**f**) female tergite VIII; (**g**) female sternite VIII; (**h**) spermatheca. Scale bar for habitus = 1 mm; remaining scale bars = 0.2 mm

Fig. 10.86 (**a–h**). Images of *Aleochara (Coprochara) bimaculata* Gravenhorst: (**a**) habitus in dorsal view; (**b**) median lobe of aedeagus in lateral view; (**c**) median lobe of aedeagus in dorsal view; (**d**) male tergite VIII; (**e**) male sternite VIII; (**f**) female tergite VIII; (**g**) female sternite VIII; (**h**) spermatheca. Scale bar for habitus = 1 mm; remaining scale bars = 0.2 mm

Fig. 10.87 (**a–h**). Images of *Aleochara (Coprochara) shelleyae* Klimaszewski and Langor: (**a**) habitus in dorsal view; (**b**) median lobe of aedeagus in lateral view; (**c**) median lobe of aedeagus in dorsal view; (**d**) male tergite VIII; (**e**) male sternite VIII; (**f**) female tergite VIII; (**g**) female sternite VIII; (**h**) spermatheca. Scale bar for habitus = 1 mm; remaining scale bars = 0.2 mm

Fig. 10.88 (**a–h**). Images of *Aleochara (Coprochara) suffusa* (Casey): (**a**) habitus in dorsal view; (**b**) median lobe of aedeagus in lateral view; (**c**) median lobe of aedeagus in dorsal view; (**d**) male tergite VIII; (**e**) male sternite VIII; (**f**) female tergite VIII; (**g**) female sternite VIII; (**h**) spermatheca. Scale bar for habitus = 1 mm; remaining scale bars = 0.2 mm

Fig. 10.89 (**a–i**). Images of *Aleochara (Coprochara) sulcicollis* Mannerheim: (**a**) habitus in dorsal view; (**b,c**) median lobe of aedeagus in lateral view; (**d**) median lobe of aedeagus in dorsal view; (**e**) male tergite VIII; (**f**) male sternite VIII; (**g**) female tergite VIII; (**h**) female sternite VIII; (**i**) spermatheca. Scale bar for habitus = 1 mm; remaining scale bars = 0.2 mm

Fig. 10.90 (**a–h**). Images of *Aleochara (Coprochara) verna* Say: (**a**) habitus in dorsal view; (**b**) median lobe of aedeagus in lateral view; (**c**) median lobe of aedeagus in dorsal view; (**d**) male tergite VIII; (**e**) male sternite VIII; (**f**) female tergite VIII; (**g**) female sternite VIII; (**h**) spermatheca. Scale bar for habitus = 1 mm; remaining scale bars = 0.2 mm

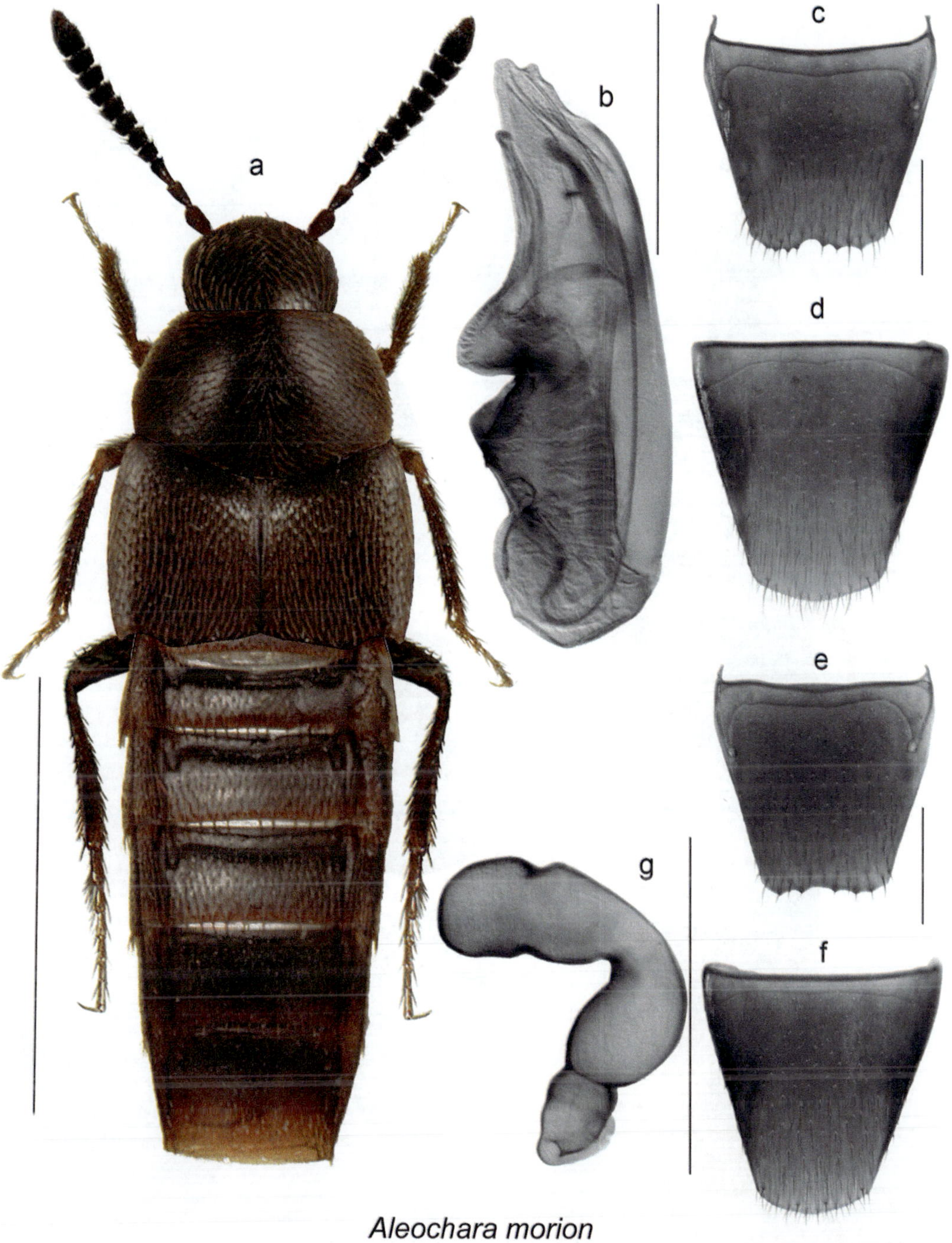

Fig. 10.91 (**a–g**). Images of *Aleochara (Tinotus) morion* Gravenhorst: (**a**) habitus in dorsal view; (**b**) median lobe of aedeagus in lateral view; (**c**) male tergite VIII; (**d**) male sternite VIII; (**e**) female tergite VIII; (**f**) female sternite VIII; (**g**) spermatheca. Scale bar for habitus = 1 mm; remaining scale bars = 0.2 mm

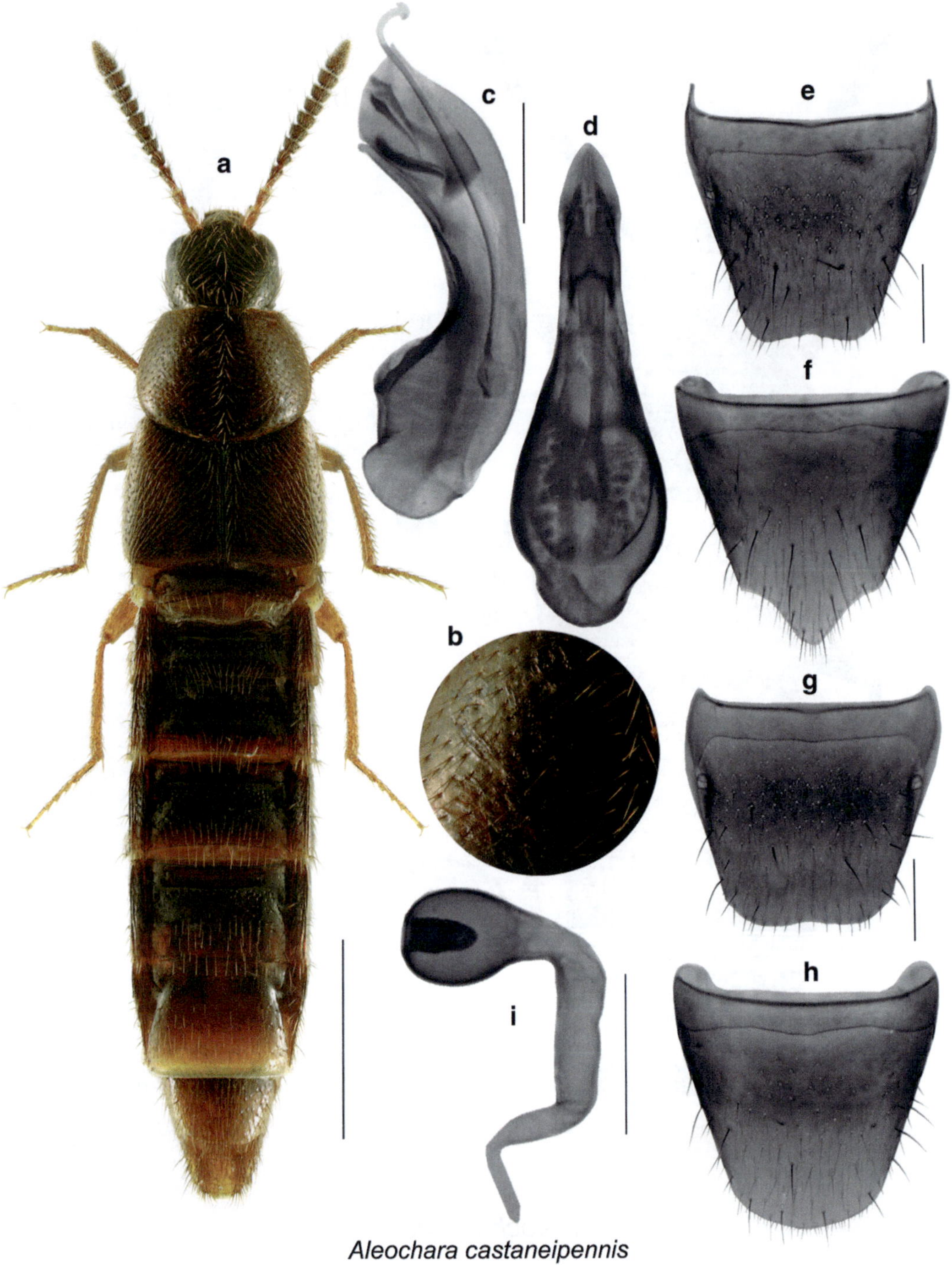

Fig. 10.92 (**a–i**). Images of *Aleochara (Xenochara) castaneipennis* Mannerheim: (**a**) habitus in dorsal view; (**b**) magnified part of pronotum showing pubescence and microsculpture; (**c**) median lobe of aedeagus in lateral view; (**d**) median lobe of aedeagus in dorsal view; (**e**) male tergite VIII; (**f**) male sternite VIII; (**g**) female tergite VIII; (**h**) female sternite VIII; (**i**) spermatheca. Scale bar for habitus = 1 mm; remaining scale bars = 0.2 mm

Fig. 10.93 (**a–i**). Images of *Aleochara (Xenochara) fumata* Gravenhorst: (**a**) habitus in dorsal view; (**b**) magnified part of pronotum showing pubescence and microsculpture; (**c**) median lobe of aedeagus in lateral view; (**d**) median lobe of aedeagus in dorsal view; (**e**) male tergite VIII; (**f**) male sternite VIII; (**g**) female tergite VIII; (**h**) female sternite VIII; (**i**) spermatheca. Scale bar for habitus = 1 mm; remaining scale bars = 0.2 mm

References

Bernhauer M (1901) Neue exotische Arten der Gattung *Aleochara* Gravenhorst. Steittiner Entomologishe Zeitschrift 62(3–4):366–373

Bernhauer M (1906) Neue Aleocharinen aus Nordamerika. (II. Teil). Deutsche Entomologische Zeitschrift 1906 (2):337–348

Bousquet Y, Bouchard P, Davies AE, Sikes DS (2013) Checklist of beetles (Coleoptera) of Canada and Alaska. Second edition. Pensoft Publishers, Series Faunistica No. 109. Sofia-Moscow, 402 pp

Brattain RM, Steury BW, Newton AF, Thayer MK, Holland JD (2020) The rove beetles (Coleoptera: Staphylinidae) of the George Washington Memorial Parkway, with a checklist of regional species. Banisteria 53:27–71

Casey TL (1906) Observations on the staphylinid groups Aleocharinae and Xantholinini, chiefly of America. Transactions of the Academy of Sciences of St. Louis 16(6):125–434

Ernst CM, Buddle CM (2015) Drivers and patterns of ground-dwelling beetle biodiversity across northern Canada. Plos One 10(4):e0122163

Gouix N, Klimaszewski J (2007) Catalogue of aleocharine rove beetles of Canada and Alaska (Coleoptera, Staphylinidae, Aleocharinae). Pensoft Publishers, Sofia-Moscow. 165 pp

Gravenhorst JLC (1802) Coleoptera Microptera Brunsvicensia nec non exoticorum quotquot exstant in collectionibus entomologorum Brunsvicensium in genera familias et species distribuit. Carolus Reichard, Brunsuigae, lxvi + 206 + [1, errata] pp

Hatch MS (1957) The Beetles of the Pacific Northwest. Part II: Staphyliniformia. University of Washington Publications in Biology 16. IX + 384 pp

Hemachandra KS, Holliday NJ, Klimaszewski J, Mason PG, Kuhlmann U (2005) Erroneous records of *Aleochara bipustulata* from North America: an assessment of the evidence. The Canadian Entomologist 137 (2):182–187

Klimaszewski J (1984) A revision of the genus *Aleochara* Gravenhorst of America north of Mexico (Coleoptera: Staphylinidae, Aleocharinae). Memoirs of the Entomological Society of Canada No.129, 211 pp

Klimaszewski J (1985) Nomenclatoral changes in the genus *Aleochara* Gravenhorst (Coleoptera: Staphylinidae: Aleocharinae). The Coleopterists Bulletin 39(4):376

Klimaszewski J, Cervenka VJ (1986) A revision of the genus *Aleochara* (Coleoptera: Staphylinidae) of America north of Mexico (Supplement 3). New distribution data. Entomological News 97(3):119–120

Klimaszewski J, Frank JH (1992) New distributional data for Nearctic *Myllaena* Erichson (Coleoptera: Staphylinidae). Supplement 2. The Coleopterists Bulletin 46(4):397–402

Klimaszewski J, Génier F (1987) A revision of the genus *Aleochara* Gravenhorst of America north of Mexico (Coleoptera: Staphylinidae, Aleocharinae). Supplement 4. New distribution data and description of two new species. Coleopterists Bulletin 41 (3):241–248

Klimaszewski J, Peck SB (1986) A review of the cavernicolous Staphylinidae (Coleoptera) of eastern North America: Part I. Aleocharinae. Quaestiones Entomologicae 22(2):51–113

Klimaszewski J, Pelletier G, Sweeney J (2002) Genus *Tinotus* (Coleoptera: Staphylinidae, Aleocharinae) from America north of Mexico: review of the types, distribution records, and key to species. The Canadian Entomologist 134:281–298

Klimaszewski J, Sweeney J, Price J, Pelletier G (2005) Rove beetles (Coleoptera: Staphylinidae) in red spruce stands, eastern Canada: diversity, abundance, and descriptions of new species. The Canadian Entomologist 137(1):1–48

Klimaszewski J, Langor D, Pelletier G, Bourdon C, Perdereau L (2011) Aleocharine beetles (Coleoptera, Staphylinidae) of the province of Newfoundland and Labrador, Canada. Pensoft Series Faunistica No. 98. Pensoft, Sophia-Moscow, 313 pp

Klimaszewski J, Godin B, Langor D, Bourdon C, Lee S-I, Horwood D (2015) New distribution records for Canadian Aleocharinae (Coleoptera, Staphylinidae), and new synonymies for *Trichiusa*. ZooKeys 498:51–91

Klimaszewski J, Langor DW, Bourdon C, Gilbert A, Labrecque M (2016a) Two new species and new provincial records of aleocharine rove beetles from Newfoundland and Labrador, Canada (Coleoptera, Staphylinidae, Aleocharinae). ZooKeys 593:49–89

Klimaszewski J, Larson DJ, Labrecque M, Bourdon C (2016b) Twelve new species and fifty-three new provincial distribution records of Aleocharinae rove beetles of Saskatchewan, Canada (Coleoptera, Staphylinidae). ZooKeys 610:45–112

Majka CG, Klimaszewski J (2008) Introduced Staphylinidae (Coleoptera) in the Maritime Provinces of Canada. The Canadian Entomologist 140(1):48–72

Maus C (1998) Taxonomical contributions to the subgenus *Coprochara* Mulsant & Rey, 1874 of the genus *Aleochara* Gravenhorst, 1802 (Coleoptera: Staphylinidae). Koleopterologische Rundschau 68:81–100

Klimaszewski J, Webster RP, Brunke A (2017) A new criptic species of *Aleochara* associated with *Marmota monax* (L.) burrows in North America (Coleoptera: Staphylinidae, Aleocharinae). Insecta Mundi 0600:1–11

Klimaszewski J, Webster RP, Langor DW, Brunke A, Davies A, Bourdon C, Labrecque M, Newton AF, Dorval J-A, Frank JH (2018) Aleocharine rove beetles of eastern Canada (Coleoptera, Staphylinidae, Aleocharinae): a glimpse of megadiversity. Springer, Cham. 902 pp

Klimaszewski J, Hoebeke ER, Godin B, Davies A, Perry KI, Bourdon C, Winchester N (2020) Aleocharine rove beetles of British Columbia: a hotspot of Canadian biodiversity (Coleoptera, Staphylinidae). Springer, Cham. 631 pp

Majka CG, Klimaszewski J (2010) Contributions to the knowledge of the Aleocharinae (Coleoptera, Staphylinidae) in the Maritime Provinces of Canada. ZooKeys 46:15–39

Majka CG, Klimaszewski J, Michaud J-P, Moreau G (2009) *Aleochara sekanai* Klimaszewski: A significant southern and eastern range extension (Coleoptera: Staphylinidae). The Coleopterists Bulletin 63 (4):454–455

Majka CG, Chandler DS, Donahue CP (2011) Checklist of the beetles of Maine, USA. Empty Mirrors Press, Halifax, NS. 328 pp

Mannerheim CG (1843) Beitrag zur Kaefer-fauna der Aleutischen Inseln, der Insel Sitkha und Neu-Californiens. Bulletin de la Société Impériale des Naturalistes de Moscou 16(2):175–314

Moore I, Legner EF (1975) A catalogue of the Staphylinidae of America north of Mexico (Coleoptera). University of California, Division of Agricultural Sciences. Special Publication 3015:1–514

Navarrete-Heredia JL, Newton AF, Thayer MK, Ashe JS, Chandler DS (2002) Guía ilustrada para los géneros de Staphylinidae (Coleoptera) de México. Illustrated guide to the genera of Staphylinidae (Coleoptera) of Mexico. Universidad de Guadalajara and CONABIO, México. xii + 401 pp

Osswald J, Bachmann L, Gusarov VI (2013) Molecular phylogeny of the beetle tribe Oxypodini (Coleoptera: Staphylinidae: Aleocharinae). Systematic Entomology 38(3):507–522. https://doi.org/10.1111/syen.12011

Schülke M, Smetana A (2015) Family Staphylinidae Latreille, 1802. In: Löbl I, Löbl D (eds) Catalogue of palearctic coleoptera: Hydrophiloidea – Staphylinoidea, vol 2. Revised and updated edition. Brill, Leiden, pp 304–1134

Say T (1833) Descriptions of new species of North American insects, and observations on some already described (cont'd.). T. Say, New Harmony, pp 58–65

Webster RP, Klimaszewski J, Pelletier G, Savard K (2009) New Staphylinidae (Coleoptera) records with new collection data from New Brunswick, Canada. I. Aleocharinae. In: Majka CG, Klimaszewski J (eds) Biodiversity, biosystematics, and ecology of canadian coleoptera II. ZooKeys [Special Issue] 22:171–248

11 Tribe Oxypodini C.G. Thomson, 1859

Tribe Oxypodini C.G. Thomson, 1859

(Figs. 11.94–11.137) (Seevers 1978, Ashe 2001, Klimaszewski et al. 2006, Osswald et al. 2013, Klimaszewski et al. 2018a, 2020)

Diagnosis Body narrowly elongate or ovoid, body very flexible with segments loosely articulated; head without distinct neck in most species; frontal suture of head absent or present; mouthparts generalized (similarly developed across the tribe and without distinctive features for the tribe); pronotal hypomera not visible in lateral view in *Devia*, *Hylota*, and *Oxypoda*; mesocoxae narrowly separated in most, moderately so in few; process of mesoventrite slender in most; abdominal tergite IX narrowly subdivided at base; tarsal formula in Arctic and Subarctic taxa 5-5-5, in some others outside this area 4-5-5 or 4-4-4. Highly heterogenous assemblage, difficult to characterize uniformly.

Key to Genera of Oxypodini (modified from Klimaszewski et al. 2018a, 2020)

1. Head with distinct broad neck [sometimes not visble in dry specimens when head concealed by pronotum] (Figs. 11.96a, 11.97a, 11.135a–11.137a) **2**

– Head without distinct neck.........................**3**

2. Head in dorsal view with lateral margins very angular posteriorly, body subparallel, all pronotal pubescence directed straight posteriad (Figs. 11.135a–11.137a); often found under bark of dead trees ....***Phloeopora* Erichson**

– Head in dorsal view with lateral margins rounded posteriorly, body and pronotal pubescence pattern not as above (Figs. 11.96a, 11.97a); found in habitats other than under bark ....***Calodera* Mannerheim**

3. Pronotal hypomera not visible in lateral view...**4**

– Pronotal hypomera visible in lateral view..**6**

4. Body broadly oval with large pronotum, abdomen subparallel and not strongly tapering apically (Fig. 11.98a); pronotum expanded laterally, subelliptical, without apparent hind angles (Fig. 11.98a); antennomere III distinctly longer than II; each mandible with a subapical tooth ....***Devia* Blackwelder**

– Body not as above, subparallel or ovate with abdomen tapering apically; pronotum of a different shape with posterior angles more or less distinctly defined; antennomere III as long as or shorter than II; mandibles without subapical tooth, right mandible may have basal median tooth **5**

5. Pronotum often horseshoe-shaped with posterior angles pointed or rounded, basal margin broadly arcuate (Figs. 11.109a–11.131a); male tergite VIII with apical margin truncate, rounded or shallowly emarginate, always smooth; spermathecal stem without multiple tight

J. Klimaszewski et al., *A Faunal Review of Aleocharine Beetles in the Rapidly Changing Arctic and Subarctic Regions of North America (Coleoptera, Staphylinidae)*,
https://doi.org/10.1007/978-3-030-68191-3_11

posterior coils (Fig. 11.109i).................................
....*Oxypoda* **Mannerheim**

– Pronotum not horseshoe-shaped, posterior angles never sharply pointed, base of disc distinctly sinuate laterally (Fig. 11.103a); male tergite VIII with apical margin broadly emarginate and dentate or crenulate (Fig. 11.103d); spermathecal stem with multiple tight posterior coils (Fig. 11.103h).......................*Hylota* **Casey**

6. Mandibles extremely long and sickle-shaped, apices very slender, each crossing the other in resting position, each mandible bearing a long spine or tooth (Fig. 11.101a); infraorbital carina strong and complete; ligula more-or-less deeply bilobed; terminal palpomere needle-shaped; frontal suture of head absent; body as illustrated (Figs. 11.99a–11.102a).......................
....*Gnathusa* **Fenyes**

– Mandibles short to moderately long and shaped differently, not crossing each other in resting position, other characters different.........**7**

7. Body compact and robust, subparallel, convex (Figs. 11.104a, 11.108a); integument of forebody with meshed microsculpture..............**8**

– Body not as above; integument of forebody with or without meshed microsculpture............**9**

8. All pronotal setae distributed along midline of disc directed posteriad (Fig. 11.108a); fronto-clypeal phragma present and visible externally as frontal suture between antennal pits (seen better in diffused light); spermatheca S-shaped with spherical capsule bearing deep and broad invagination, stem sinuate and moderately broad (Fig. 11.108g); median lobe of aedeagus with tubus markedly produced ventrally in lateral view (Fig. 11.108b); species known from wetland habitats.......................................*Ocyusa* **Kraatz**

– Pronotal setae along midline of disc directed anteriad over about apical one-third of midline and posteriad elsewhere (Fig. 11.104a); frontal suture absent; spermatheca capsule spherical or tubular and narrowly extended forming L-shaped neck, connected to thin and elongate stem (Fig. 11.104g); median lobe of aedeagus with tubus almost straight in lateral view (except basally) and slightly produced ventrally (Fig. 11.104b); Canadian species are known from forest habitats ...*Mniusa* **Mulsant and Rey**

9. Antennomeres VII–X each up to three times as wide as long (Fig. 11.94a); body length 1.0–2.0 mm; elytra at suture about as long as pronotum along midline; apical lobe of parameres strongly sclerotized and appearing black or dark brown; spermatheca U-shaped or sinuate, appearing as a simple tube, capsule without apical invagination (Fig. 11.94g,h)................................
.................................*Meotica* **Mulsant and Rey**

– Antennomeres VII–X each subquadrate, or at most two times as wide as long; body length over 3 mm; elytra at suture slightly longer than pronotum along midline; apical lobe of parameres not strongly sclerotized; spermatheca shaped differently...**10**

10. Pronotum distinctly transverse (Figs. 11.105a–11.107a) ....
..................................*Neothetalia* **Klimaszewski**

– Pronotum subquadrate to slightly elongate
....**11**

11. Antennomeres VII–X subquadrate (Fig. 11.134a), basal tarsomere of hind tarsi longer than two following tarsomeres combined (Fig. 11.134a).................*Parocyusa* **Bernhauer**

– Antennomeres VII–X distinctly transverse; basal tarsomere of hind tarsi about as long as two following tarsomeres combined.......................**12**

12. Maximum width of head distinctly narrower than maximum width of pronotum or elytra (Figs. 11.132a, 11.133a); median lobe of aedeagus and spermatheca as illustrated (Figs. 11.132b,g, 11.133b,g)...............................
....*Parocalea* **Bernhauer**

– Maximum width of head about equal to maximum width of pronotum or elytra (Fig. 11.95a); median lobe of aedeagus and spermatheca as illustrated (Fig. 11.95b,c,h).............
.....................................*Alfocalea* **Klimaszewski**

Subtribe Meoticina Seevers, 1978 (Seevers 1978, Ashe 2001, Klimaszewski et al. 2018a)

Diagnosis Body slender, dorsoventrally compressed, sides subparallel in most species; head with neck about one-third as wide as head, without frontal suture; terminal antennomere with pair of coeloconic sensillae; mesocoxae narrowly separated; process of mesoventrite slender and acutely pointed, not extended beyond middle of coxae; process of metaventrite very short, usually almost absent; mesocoxal acetabula not or weakly margined in most species; pronotal hypomera fully visible in lateral view; first two or three visible abdominal tergites with transverse impressions; tarsal claws scythe-shaped (falcate) [except in *Meotica*]; tarsal formula 4-4-4 or 5-5-5.

Genus *Meotica* Mulsant and Rey, 1873

(Fig. 11.94a–h) (Klimaszewski et al. 2007a, b, 2018a)

Diagnosis Body slender, parallel-sided, length 1.0–2.0 mm, colour light reddish-brown; antennomeres VI–X distinctly transverse, each up to three times as wide as long; eyes small, about one-third length of genae in dorsal view; pronotal hypomera visible in lateral view; mesocoxal cavities not margined or weakly so; process of mesoventrite short and acute; process of metaventrite short and poorly delimited; paramere with apical lobe small and heavily sclerotized; spermatheca U-shaped or L-shaped, forming simple tube.

24. *Meotica pallens* (Redtenbacher) † (Fig. 11.94a–h)

Homalota pallens Redtenbacher, 1849. Benick and Lohse 1974, Muona 1991, Gusarov 2002, Klimaszewski et al. 2007a, 2018a, 2020, Majka and Klimaszewski 2008c, Brunke et al. 2012, Webster et al. 2016, Assing and Vogel 2019.

Diagnosis Body narrow, subparallel, minute, length 1.3–1.6 mm, colour light rust-brown, integument moderately glossy, forebody with pubescence fine and sparse, head convex, eyes small; antennomeres VI–X distinctly transverse (Fig. 11.94a); pronotum subquadrate; elytra short and truncate posteriad; abdomen slightly broadening apicad. MALE. Tergite VIII broad, truncate apically, antecostal suture arcuate (Fig. 11.94c); sternite VIII transverse and truncate apically, antecostal suture almost straight (Fig. 11.94d); tubus of median lobe of aedeagus in lateral view with ventral margin slightly sinuate, apical part broadly triangular, internal sac structures of distinctive shape (Fig. 11.94b). FEMALE. Tergite VIII elongate with apical margin broadly truncate medially, rounded laterally, antecostal suture arcuate (Fig. 11.94e); sternite VIII elongate with apical margin narrowly truncate medially, broadly rounded laterally, antecostal suture almost straight (Fig. 11.94f); spermatheca U-shaped, capsule and stem without distinct separation (Fig. 11.94g,h).

DISTRIBUTION Origin: Palearctic, adventive in North America. **CANADA**: BC, NB, NS, ON, YT. **USA**: NJ, RI.

COLLECTION AND HABITAT DATA Habitat: old-growth northern hardwood forest, mixed forests (NB), and a *Sphagnum* bog (ON). Adults in muskrat nest, among sticks, debris and mud on beaver dam, in leaf litter on clay near small brook, one captured in Lindgren funnel trap in old-growth hardwood forest. **Collecting period**: V, VI. **Collecting method**: sweeping grasses near lake (new data); netted in a mixed forest opening during the evening, sifting from forest litter and among sticks, debris and clay on a beaver dam, and from litter on brook margin, Lindgren funnel trap.

COMMENTS This species was listed from QC and MA by Newton (2019) but we are not aware of any published voucher data.

Subtribe Oxypodina C.G. Thomson, 1859 (Ashe 2001)

Diagnosis Head with frontal suture present in many genera; pronotal hypomera visible or not in lateral view; tarsal formula 5-5-5. This is a heterogeneous assemblage that cannot be satisfactorily characterized (Ashe 2001).

Genus *Alfocalea* Klimaszewski, 2004

(Fig. 11.95a–h) (Klimaszewski and Pelletier 2004, Klimaszewski et al. 2020)

Diagnosis Adults of the genus *Alfocalea* may be distinguished from those of other aleocharine genera by the following combination of characters: body narrowly elongate, subparallel, dark brown with brown to rust brown pronotum, elytra, and appendages; size medium, length 4.3–4.7 mm, integument more or less densely punctate and pubescent, forebody with distinct isodiametric microsculpture; antennomere IV quadrate in female to slightly elongate in male, VII–X transverse; maxillary palpus with 4 palpomeres, apical palpomere narrowly elongate and one-third length of penultimate one; galea expanding slightly beyond apex of lacinia, its apical lobe bearing dense short pubescence; apex of lacinia with row of closely spaced spines; labial palpus with 3 palpomeres; ligula slender and shallowly emarginated apically; mentum trapezoidal, apical margin slightly concave, with three long apical setae on each side; labrum transverse and usually slightly projecting medially; right mandible with distinct medial tooth; head quadrate to slightly elongate and almost as wide and as large as pronotum; pronotum transverse, broadest in about middle, and with arcuate edges evenly converging apically and basally, pubescence directed anteriad medially except for very basal part, laterad on each side of disc and often forming broadly arcuate lines; mesoventrite about as long as maximum diameter of mesocoxa, with partial median carina, carina flattened and broad, about one-third of mesoventrite length, mesoventral process long, narrow, and acutely pointed, extending about two-thirds length of mesocoxal cavities, isthmus short and no more than one-fourth of maximum diameter of mesocoxa; elytra about as broad as pronotum, moderately transverse, at suture at most as long as pronotum; abdomen with three basal tergites shallowly impressed basally; basal tarsomere of mesotarsus slightly longer than following one, and basal tarsomere of metatarsus as long as two following segments combined. MALE. Median lobe of aedeagus with small and narrowly elongate crista apicalis at base of bulbus, tubus long, approximately arcuate medially, sinuate anteriorly, and strongly produced ventrally in lateral view, apex narrowly rounded. FEMALE. Spermatheca with small, spherical capsule and with short, narrow, apical invagination facing obliquely downwards, stem sinuate and slightly hooked posteriorly.

Adults were collected at higher altitudes ranging from 975–2133 m. The species apparently prefer cool, high mountain habitats (Klimaszewski and Pelletier 2004).

25. ***Alfocalea montana*** **Klimaszewski** (Fig. 11.95a–h)

Alfocalea montana Klimaszewski, in Klimaszewski and Pelletier 2004.

Diagnosis Body length 4.3–4.7 mm, dark brown with elytra, basal antennal segment, tarsi, and pronotum rust brown, forebody moderately glossy and abdomen strongly so (Fig. 11.95a); antennae stout, antennomere IV quadrate in female to slightly elongate in male, V–X slightly to strongly transverse, and terminal palpomere about as long as two preceding ones combined; pronotum strongly transverse (Pronotal Index = 1.3), lateral edges broadly arcuate, punctures sparse, distance between punctures about 2–3 times diameter of a puncture, space between punctures with strong, isodiametric microsculpture, pubescence directed anteriad along median line of disc and laterad, usually forming arcuate lines on both sides; elytra transverse (Elytral Index = 1.6), about as long as pronotum; abdomen arcuate laterally and broadest in apical third, three basal sternites with deep impressions. MALE. Tergite VIII transverse, truncate apically, and without median emargination (Fig. 11.95d); sternite VIII slightly transverse, apical part broadly triangularly produced (Fig. 11.95e); median lobe of aedeagus with small and narrowly elongate crista apicalis at base of bulbus, tubus long, arcuate medially, sinuate anteriorly and strongly produced ventrally in lateral view, apex narrowly rounded (Fig. 11.95b); median lobe in dorsal view as illustrated (Fig. 11.95c). FEMALE. Tergite VIII similar to that of male (Fig. 11.95f); sternite VIII transverse and broadly rounded apically (Fig. 11.95g); spermatheca (Fig. 11.95h) with small, approximately spherical capsule and with short, narrow apical invagination facing obliquely downwards, stem sinuate and slightly hooked posteriorly.

DISTRIBUTION Origin: Nearctic. **CANADA**: AB, BC. **USA**: not recorded.

COLLECTION AND HABITAT DATA Habitat: high mountain habitats at altitudes 975–2133 m. **Collecting period**: VI, VII, VIII. **Collecting method**: pitfall traps.

Genus *Calodera* Mannerheim, 1830

(Fig. 11.96–11.97) (Ashe 2001, Assing 2008a, b, Klimaszewski et al. 2018a)

Diagnosis Body robust, narrowly elongate, colour dark brown to black; head with neck about two-thirds as wide as maximum width of head; antennae thick, antennomeres V–X about 1.5–2.0 times as wide as long; ligula short, bifid at apex; eyes small, about half as long as genae in dorsal view; pronotum subquadrate, hypomera fully visible in lateral view; middle coxae narrowly separated; process of mesoventrite long and spine-like; process of metaventrite short and rounded apically; abdomen with first four visible tergites bearing deep, coarsely punctured basal impressions; median lobe of aedeagus simple with ventral margin of tubus arcuate; spermatheca with capsule small, spherical with large apical invagination, stem thin, long, and looped.

COMMENTS Adults of *Calodera* inhabit moist habitats such as swamps, moist leaf litter near water such as rivers, streams, lakes, and floodplain forests (Klimaszewski et al. 2018a).

Key to Species of *Calodera* (Klimaszewski et al. 2018a)

1. Body slender, subparallel, length < 2.3 mm; pronotum slightly narrower than elytra, elytra about as long as pronotum (Fig. 11.96a); male unknown; spermatheca with stem sinuate and broadly looped posteriorly (Fig. 11.96d) .... ***Calodera caseyi*** **Assing**

– Body moderately robust, length >2.7 mm; pronotum distinctly narrower than elytra, elytra slightly longer than pronotum (Fig. 11.97a); spermatheca with stem sinuate and narrowly looped posteriorly (Fig. 11.97g) .... ***Calodera parviceps*** **(Casey)**

26. *Calodera caseyi* Assing (Fig. 11.96a–d)

Calodera caseyi Assing, 2002 (replacement name for secondary homonym *C. humilis* (Casey, 1893)). Webster et al. 2016, Klimaszewski et al. 2018a.

Diagnosis Body slender, evenly narrowly elongate, subparallel (Fig. 11.96a), length 2.3 mm; head and abdomen dark brown except for paler apical part, remainder of body yellowish red-brown; integument moderately glossy, with punctation moderately coarse and dense, pubescence short; head slightly narrower than pronotum; antennomeres V–X distinctly transverse, about 1.5 times as broad as long; pronotum about as long as elytra at suture, only slightly narrower than maximum width of elytra; abdomen subparallel, almost as wide basally as elytra, four basal abdominal impressions with coarse punctation. MALE. Unknown. FEMALE. Tergite VIII broadly arcuate apically, with broad space between antecostal suture and base (Fig. 11.96b); sternite VIII evenly rounded apically, antecostal suture moderately sinuate (Fig. 11.96c); spermatheca with capsule spherical and long stem broadly looped posteriorly (Fig. 11.96d).

DISTRIBUTION Origin: Nearctic. **CANADA**: NB. **USA**: AK [**NSR**], PA.

COLLECTION AND HABITAT DATA
Habitat: Lindgren funnel trap in a mixed forest, and mixed birch, white spruce, grassy, disturbed

wood cutting area. **Collecting period**: V. **Collecting method**: Lindgren funnel trap, Malaise trap.

We report this species from Alaska for the first time here, based on the following:

USA: Alaska, Fairbanks, Bonanza Creek, 278m el., 64.72025, –148.38327, ±10m mixed birch, white spruce, grassy, disturbed wood cutting area, bottom catch, hanging Malaise 4.2m, 11-29.VII.2013, J. Hagelin, A. H. Scurr (UAM) 1 specimen.

COMMENTS This species was listed from MA and ME by Klimaszewski et al. (2018a, 2020) and Newton (2019) but we are not aware of any published voucher data. These records are likely in error.

27. *Calodera parviceps* (Casey) (Fig. 11.97a–g)

Nasirema parviceps Casey, 1893. Assing 2008a, b, Webster et al. 2009, Klimaszewski et al. 2012, 2018a, Webster et al. 2020.

Diagnosis Body slender, narrowly elongate, subparallel (Fig. 11.97a), length 2.8–3.5 mm; colour dark brown to almost black with basal antennomeres and legs/tarsi reddish-brown; integument moderately glossy, with punctation moderately coarse and dense, pubescence short; head distinctly narrower than pronotum; antennomeres V–X distinctly transverse, about 1.5 times as broad as long; pronotum as long as or shorter than elytra at suture, and about 1/4 narrower than maximum width of elytra; abdomen subparallel and basally slightly narrower than elytra, four basal abdominal impressions with coarse punctation. MALE. Tergite VIII broadly arcuate apically (Fig. 11.97c); sternite VIII with apical margin very obtusely angulate, broadly rounded at apex (Fig. 11.97d); median lobe of aedeagus in lateral view with ventral margin of tubus straight, moderately bent ventrad at base, apical part slightly arcuate and narrowly elongate, internal sac structures as illustrated (Fig. 11.97b). FEMALE. Tergite VIII broadly arcuate apically (Fig. 11.97e); sternite VIII rounded apically, antecostal suture sinuate (Fig. 11.97f); spermatheca with capsule compressed spherical, stem with single broad loop posteriorly (Fig. 11.97g).

DISTRIBUTION Origin: Nearctic. **CANADA**: BC, NB, NS, NT [**NTR**], ON, YT. **USA**: AK [**NSR**], RI.

COLLECTION AND HABITAT DATA

Habitat: adults in moist leaf litter and moss under alders on margins of vernal ponds, *Salix* leaf litter, spruce bog, boggy black spruce, open tussock and standing water, willow and bog birch, near brooks and small streams, and in *Carex* marshes. These habitats occurred in old-growth eastern white cedar swamps, silver maple swamps, alder swamps, and mixed forests. **Collecting period**: V–IX. **Collecting method**: sifting litter, Berlese funnels, Malaise trap, yellow pan traps.

We report this species from NT (**NTR**) and AK (**NSR**) for the first time here, based on the following: Canada, Northwest Territories, Norman Wells, 65.289, –126.84077, Rep. 3 wet yellow pan, 11-14.VI.2011, NBP field party N3WY122 (LFC) 1 male. USA: Alaska, Anaktuvuk Pass, 647m el., 68.14049, –151.74091, ±250m *Salix* leaf litter, Berlese, 19. V.2016, D.S. Sikes; K.M. Daly (UAM) 2 specimens; Eielson AFB, Moose Creek, 170m el., 64.71098, –147.09874, ±3m birch ~10m from creek, hanging Malaise, 4.95m, 15-24. VI.2016, J.C. Hagelin (UAM) 1 specimen; Fairbanks, Creamer's Field, Reindeer Lake, 146m el., 64.88118, –147.77643, ±10m b. spruce bog, hanging Malaise, 5.5m, 31.V-14. VI.2016, J.C. Hagelin (UAM) 1 specimen; Fairbanks, Creamer's Field, Reindeer Lake, 146m el., 64.88118, –147.77643, ±10m boggy black spruce, hanging Malaise, 5.5m, 27.V-10. VI.2015, J.C. Hagelin (UAM) 1 specimen; White Mountain Recreation Area, Nome Creek Site 5, 589m el., 65.34077, –146.9275, ±10m open tussock and standing water, willow and bog birch, hanging Malaise, 4.95m, 12.VI-3. VII.2015, J.C. Hagelin (UAM) 1 specimen; Mi

123 Taylor Hwy., 9.VII.1968, J.M. Campbell and A. Smetana (CNC) 1 male, 1 female.

Additional records. See Annex 1.

Genus *Devia* Blackwelder, 1952

(Fig. 11.98a–h) (Gusarov 2003, Klimaszewski et al. 2018a, 2020)

Diagnosis Body large and robust, narrowly oval, flattened; integument with pubescence short and moderately dense, appearing velvety; pronotum subelliptical, shield-like with hypomera not visible laterally; antennomere III longer than II, IV elongate, V–X moderately to disctinctly transverse; terminal palpomere thin and minute; mandibles bifid apically, right mandible with a small median tooth; abdomen with 3 first tergites shallowly impressed; genitalia distinctive (Fig. 11.98b,c,h).

28. *Devia prospera* (Erichson) * (Fig. 11.98a–h)

Oxypoda prospera Erichson, 1839. Lohse et al. 1990 [as D. congruens (Casey)] Gusarov 2003, Gouix and Klimaszewski 2007, Klimaszewski et al. 2007a, Webster et al. 2009, Klimaszewski et al. 2010, 2011, 2015, 2018a, 2020.

Diagnosis Body broadly oval, slightly flattened, length 2.8–3.3 mm, colour reddish-brown to brown, with appendages rusty-brown, head and posterior abdomen dark brown (Fig. 11.98a); pubescence short, evenly distributed; pronotum large, shield-shaped, slightly wider than maximum width of elytra, hypomera not visible in lateral view; antennae robust, antennomeres V–X elongate; abdomen with first three visible tergites deeply impressed; hind tarsus with basal article highly elongate. MALE. Tergite VIII arcuate apically (Fig. 11.98d); sternite VIII produced medially at apex (Fig. 11.98e); median lobe of aedeagus in dorsal view with moderately large bulbus, tubus elongate (Fig. 11.98c); in lateral view ventral margin of tubus almost straight, bent obliquely ventrad at base (Fig. 11.98b). FEMALE. Tergite VIII arcuate apically (Fig. 11.98f); sternite VIII broadly rounded apically (Fig. 11.98g); spermatheca with capsule small, spherical, stem highly sinuate, and thin (Fig. 11.98h).

DISTRIBUTION **Origin**: Holarctic. **CANADA**: AB, BC, LB, MB, NB, NT, ON, QC, SK, YT. **USA**: AK, CA, CO, MA, MI, MN, MT, NH, NM, OR, SD, UT, WA, WI, WY.

COLLECTION AND HABITAT DATA

Habitat. In NB, commonly found in moist leaf litter on the margins of vernal ponds in silver maple swamps and mixed forests. Elsewhere, in human settlements in stables, barns where it occurs in heaps of straw, haystacks. Also uncommon in a disturbed field with forbs and grasses as well as in rotting organic debris, mushrooms, willow humus and moss, and forest litter. In BC, specimens were found in *Salix* and *Alnus* litter with moss, 10 m from a lake. In AK, specimens collected from forest edge with *Populus tremuloides, Populus balsamea*; litter in sparse *P. tremuloides* forest, *Betula* litter, wet boreal forest, mesic *Betula* forest, black spruce with some white spruce forest, field edge of *Salix / Betula / Picea* forest and vernal ponds / stream, birch leaf litter among roots, and litter in in spruce, birch, alder, cottonwood. **Collecting period**: IV–X. **Collecting method**: sifting litter, pitfall traps, barrier trap, FIT traps, Berlese funnels, pollinator vane trap, yellow pan traps.

Additional records. See Annex 1.

Genus *Gnathusa* Fenyes, 1910

(Figs. 11.99–11.102) (Klimaszewski et al. 2014, 2018a, 2020)

Diagnosis Body compact, robust, subparallel, length 3.5–4.3 mm; colour dark brown to black, integument with distinct meshed microsculpture; head large with mandibles extremely long and sickle-shaped, apices very slender, crossing each other in resting position, each bearing long spine or tooth; postocular carina strong and complete;

ligula moderately deeply bilobed; apical palpomere needle-shaped; frontal suture of head absent; anterior margin of mesoventrite with short V-shaped basal carina, remaining mesoventrite not carinate; process of mesoventrite long, triangular basally, then narrowly produced and extending to about three-quarters length of mesocoxal cavities, process of metaventrite triangular in shape and short, isthmus short; median lobe of aedeagus with large crista apicalis, internal sac structures variable in shape, paramere with apical lobe narrowly elongate, bearing four macrosetae; spermatheca pipe-shaped with capsule spherical or tubular, stem elongate, narrow.

Key to Species of *Gnathusa* (Klimaszewski et al. 2014)

1. Apical margin of labrum with coarse spines (Fig. 11.101a); antennomeres VII–X strongly transverse (Fig. 11.101a); body length 3.5–4.0 mm; integument moderately glossy; genitalic structures as illustrated (Fig. 11.101b,g).. ***Gnathusa eva* Fenyes**

– Apical margin of labrum with fine setae (Figs. 11.99a, 11.100a, 11.102a); antennomeres VII–X subquadrate or slightly transverse (Figs. 11.99a, 11.100a, 11.102a); body length 2.5–3.7 mm; integument usually more glossy; genital structures differently shaped **2**

2. Antennomeres V–X subquadrate (Fig. 11.102a); genitalia as illustrated (Fig. 11.102b,g); body length 2.5–3.7 mm .... ***Gnathusa tenuicornis* Fenyes**

– Antennomeres V–X slightly transverse (Figs. 11.99a, 11.100a); genitalia differently shaped (Figs. 11.99b,g, 11.100b,g); body length 2.8–3.6 mm **3**

3. Head slightly narrower and smaller than pronotum in dorsal view (Fig. 11.99a); antennae brown or yellowish-brown; basal part of abdominal tergite III deeply impressed and moderately coarsely, densely punctate; genitalia as illustrated (Fig. 11.99b,g).................. ***Gnathusa alfacaribou* Klimaszewski and Langor**

– Head about as wide and large as pronotum in dorsal view (Fig. 11.100a); antennae light yellow; basal part of abdominal tergite III deeply impressed and coarsely and densely punctate; genitalia as illustrated (Fig. 11.100b,g)... ***Gnathusa caribou* Lohse**

29. *Gnathusa alfacaribou* Klimaszewski and Langor (Fig. 11.99a–g)

Gnathusa alfacaribou Klimaszewski and Langor, in Klimaszewski et al. 2011. Klimaszewski et al. 2014, 2018a.

Diagnosis Body length 2.8–3.6 mm, sides subparallel; colour dark brown, with tarsi lighter and antennae yellowish-brown, sometimes with reddish tinge (Fig. 11.99a); head round, equal in length to pronotum; antennomeres V–X quadrate to slightly transverse; pronotum transverse, about as wide as maximum width of elytra, anterior angles fairly evident; abdomen subparallel, middle segments as wide as elytra, slightly narrowing basad and apicad. MALE. Tergite VIII distinctly, very obtusely angulate apically (Fig. 11.99c); sternite VIII with apical margin broadly triangularly produced, apex slightly rounded (Fig. 11.99d); median lobe of aedeagus in lateral view with ventral margin of tubus arcuately bent ventrad basally, apex subtriangular (Fig. 11.99b). FEMALE. Tergite VIII broadly arcuate apically (Fig. 11.99e); sternite VIII parabolically rounded apically (Fig. 11.99f); spermatheca pipe-shaped, with capsule small, spherical, stem long, thin and slightly sinuate (Fig. 11.99g).

DISTRIBUTION Origin: Nearctic. **CANADA**: LB. **USA**: not recorded.

COLLECTION AND HABITAT DATA
Habitat: lichen and feather moss black spruce forests and old fir forests. Probably in forest litter and moss. **Collecting period**: VI–X. **Collecting method**: pitfall and flight intercept traps.

30. *Gnathusa caribou* Lohse * (Fig. 11.100a–g)

Gnathusa caribou Lohse, in Lohse et al. 1990. Klimaszewski et al. 2011, 2014.

Diagnosis Body length 2.8–3.6 mm, sides narrowly subparallel; body colour dark brown to almost black, with antennae bright yellow and tarsi rust-brown to yellowish; integumental microsculpture dense and surface strongly glossy; head round, about the same size as the pronotum, labrum lacking stout spines but with fine setae of unequal length; pronotum small, transverse, angular, slightly narrower than elytra; elytra at suture subequal in length to pronotum; abdomen subparallel; antennomeres VI–X subquadrate to slightly transverse, last one short and broadly oval (Fig. 11.100a). MALE. Tergite VIII widely truncate apically (Fig. 11.100c); sternite VIII obtusely angulate at apex (Fig. 11.100d); median lobe of aedeagus with tubus almost straight and apex directed ventrad in lateral view (Fig. 11.100b). FEMALE. Tergite VIII truncate apically (Fig. 11.100e); sternite VIII rounded apically (Fig. 11.100f); spermatheca pipe-shaped, with spherical capsule and long and straight stem, neck well sclerotized (Fig. 11.100g).

DISTRIBUTION Origin: Holarctic [**new distribution status**]. **CANADA**: NT, NU [**NTR**], YT. **USA**: AK.

COLLECTION AND HABITAT DATA
Habitat: tundra, in *Salix* litter, moss, and leaves stored by a rodent. **Collecting period**: VI–X. **Collecting method**: sifting organic litter under *Salix* and moss, or leaves in rodent nests.

We report this species from NU (**NTR**) and Finland (new country record, **NCR**) for the first time based on the following: Canada, Nunavut, Kitikmeot, Kugluktuk, S Lake, 9km SW of Kugluktuk, 22m el., 67.7812, −115.276, tundra, sifting, 20.VII.2019, G. Blagoev (CBG) 2 specimens. **Finland**: Lkor: Sodankylä, Viiankiaapa, 190m el., 67.576, 26.773 (YKJ grid 74988:34905), moist mixed forest, pitfall, 31.VII-27.VIII.2015, M. Pentinsaari, J. Raitanen (ZMUO) 1 male.

COMMENTS Until recently, this species was considered to be Nearctic, but one male specimen was discovered in Finland, suggesting that it is likely a Holarctic species.

31. *Gnathusa eva* Fenyes (Fig. 11.101a–g)

Gnathusa eva Fenyes, 1910. Klimaszewski et al. 2012, 2014, 2016, 2020.

Diagnosis Body length 2.9–3.6 mm, sides subparallel; colour dark brown to almost black, with tarsi and antennae reddish-brown (Fig. 11.101a); head round, equal in length to pronotum; anterior margin of labrum spinose; antennomeres V–X slightly to strongly transverse; pronotum transverse, slightly narrower than maximum width of elytra, anterior angles fairly evident; abdomen subparallel, middle segments as wide as elytra, slightly narrowing basad and apicad. MALE. Tergite VIII distinctly, very truncate apically (Fig. 11.101c); sternite VIII with apical margin broadly triangularly produced, apex slightly rounded (Fig. 11.101d); median lobe of aedeagus in lateral view with ventral margin of tubus arcuately bent ventrad basally, apex narrow and rounded (Fig. 11.101b). FEMALE. Tergite VIII broadly arcuate apically (Fig. 11.101e); sternite VIII parabolically rounded apically (Fig. 11.101f); spermatheca pipe-shaped, with capsule small, narrowly tubular, stem long, thin and slightly sinuate (Fig. 11.101g).

DISTRIBUTION Origin: Nearctic. **CANADA**: AB, BC, SK, YT. **USA**: CA.

COLLECTION AND HABITAT DATA
Habitat: in YT specimens were captured in coniferous woodchip pile. **Collecting period**: V, IX. **Collecting method**: pitfall traps.

32. *Gnathusa tenuicornis* Fenyes (Fig. 11.102a–g)

Gnathusa tenuicornis Fenyes 1921. Klimaszewski and Winchester 2002, Klimaszewski et al. 2012, 2014, 2018a.

Diagnosis Body length 2.5–3.7 mm, sides subparallel; colour light brown to almost black, with antennae, tarsi and often elytra and apical part of abdomen rust-brown (Fig. 11.102a); integument with microsculpture dense and surface highly glossy; head round to somewhat quadrate and almost as wide as pronotum, labrum with short fine setae but lacking coarse spines; pronotum transverse, subequal to slightly narrower than maximum width of elytra, slightly shorter than elytra at suture, angles somewhat evident; abdomen subparallel; antennomeres VI–X subquadrate, apical one short and broadly oval. MALE. Tergite VIII broadly arcuate apically (Fig. 11.102c); sternite VIII with apical margin obtusely produced, at apex broadly rounded (Fig. 11.102d); median lobe of aedeagus in lateral view with ventral margin of tubus bent obliquely ventrad at base, slightly sinuate, apex thin, pointed (Fig. 11.102b). FEMALE. Tergite VIII truncate apically (Fig. 11.102e); sternite VIII rounded apically (Fig. 11.102f); spermatheca pipe-shaped, with capsule spherical, stem long, thin, highly sinuate, neck weakly sclerotized, neck to capsule angle variable (Fig. 11.102g).

DISTRIBUTION **Origin**: Nearctic. **CANADA**: AB, BC, NB[?], MB [**BOLD**], SK [**BOLD**], YT. **USA**: AK, CA, OR.

COLLECTION AND HABITAT DATA
Habitat: in western Canada recorded from a Sitka spruce forest, Upper Cordilleran coniferous forests, including subxeric lodgepole pine forests, mesic white spruce and lodgepole pine stands, white spruce and poplar forest, spruce-dominated subhygric and hygric forests, and a subalpine meadow at 3000 feet in moss seepage on slope. At other sites in cold moss along a stream, from gravel at edge of stream, and from litter in a mixed aspen and spruce forest. In NB, in moss and leaves under alders near a brook in an eastern white-cedar swamp, from under cobblestones and gravel in sand on a partially shaded cobblestone bar near outflow of brook into a river, and in moss along a shaded spring-fed brook in a boreal forest area. In AK, found on arctic tundra and in clear-cut. **Collecting period**: V–VIII. **Collecting method**: sifting forest litter and cold moss, Berlese funnels, pitfall traps, aspirating from gravel and sand along stream margins.

We report this species from MB (**NPR**) and SK (**NPR**) for the first time based on **BOLD** records: Canada, Manitoba, Churchill, 16 km SW Churchill, Churchill River Pump House, 1m el., 58.626, –94.228, tundra (CBG) 1 specimen. Saskatchewan, Prince Albert National Park, Narrows Peninsula Trail, 530m el., 53.9872, –106.282, white spruce and poplar forest, pitfall, 9-14.VII.2012, BIOBus 2012 team (CBG) 2 specimens.

Additional records. See Annex 1.

COMMENTS The only eastern record of this species is based on tentatively identified females from NB (Klimaszewski et al. 2014).

Genus *Hylota* Casey, 1906

(Fig. 11.103a–h) (Klimaszewski et al. 2006)

Diagnosis Body narrowly oval, moderately convex, similar to that of *Oxypoda*; pronotum broad, strongly narrowed apically and sinuate basally, with posterior angles slightly obtusely angulate, as broad as elytra; ligula entire, narrowly elongate, with two apical sensillae; process of mesoventrite narrow with apex rounded, reaching two-thirds of mesocoxal length; tergite VIII in both sexes with antecostal suture widely separated from basal margin, with apical teeth or crenulation; median lobe of aedeagus with narrowly oval and long bulbus and moderately long simple tubus, structures of internal sac inconspicuous; spermatheca with capsule tubular and L-shaped, stem long with multiple tight coils.

33. *Hylota ochracea* Casey (Fig. 11.103a–h)
Hylota ochracea Casey 1906. Klimaszewski et al. 2006, Majka et al. 2006, Webster et al. 2009, Klimaszewski et al. 2016, 2018a.

Diagnosis Body narrowly oval, length 2.5–3.1 mm; colour light to rust brown, with antennae, tarsi and often elytra and apical part of abdomen rust-brown (Fig. 11.103a); integumental microsculpture dense and surface highly glossy; head round to somewhat quadrate and almost as wide as pronotum, labrum with short fine setae but lacking coarse spines; antennomeres VI–X transverse, apical one short and broadly oval; pronotum transverse, subequal to slightly narrower than maximum width of elytra, angles somewhat evident; elytra at suture subequal in length to pronotum; abdomen subparallel. MALE. Tergite VIII with apical margin oblique on either side of broad, shallow emargination with small teeth (Fig. 11.103d); sternite VIII with apical margin slightly produced, apex broadly rounded (Fig. 11.103e); median lobe of aedeagus in lateral view with ventral margin of tubus almost straight, bent obliquely ventrad at base, apex very narrow (Fig. 11.103b), in dorsal view with bulbus broadly elongate and tubus short, triangularly produced at apex (Fig. 11.103c); internal sac split apically in dorsal view (Fig. 11.103c). FEMALE. Tergite VIII truncate apically (Fig. 11.103f); sternite VIII evenly broadly rounded apically (Fig. 11.103g); spermatheca with capsule very small, spherical, stem L-shaped, with 15–17 tight coils posteriorly (Fig. 11.103h).

DISTRIBUTION **Origin**: Nearctic. **CANADA**: NB, NS, NT, ON, QC, SK. **USA**: NY, VT.

COLLECTION AND HABITAT DATA
Habitat: bird nests in forested habitats, agriculture farms (pigeon coop), barred owl nests in tree holes (usually in large trees) and in artificial nest boxes (nest contents, which usually consisted of rich decaying organic material with bones, fur, owl pellets, portions of dead prey items such as mice, squirrels and small birds, often with strong urine smell), nest contents of the great horned owl (*Bubo virginianus* (Gmelin)), boreal owl (*Aegolius funereus richardsoni* (Bonaparte)), and northern saw-whet owl (*Aegolius acadicus* (Gmelin)). Occurs also in decaying organic material such as carrion, and decaying vegetables inside a plastic compost bin, which in some respects mimics the conditions found within a tree hole occupied by tree hole nesting birds. **Collecting period**: V–IX. **Collecting method**: sifting bird nest contents and compost litter.

COMMENTS The genus *Hylota* likely reaches its northern limits in the southern taiga as a single record exists in our study region (Fort Simpson, NT). There is another species of this genus, *Hylota cryptica* Klimaszewski and Webster, occurring in Canada but in more southern localities (Klimaszewski et al. 2018a). It differs from *H. ochracea* in having antennomeres V–X slightly elongate, apical margin of male tergite VIII with minute crenulation, and spermatheca with 8–9 posterior coils (for illustrations see Klimaszewski et al. 2018a).

Genus *Mniusa* Mulsant and Rey, 1875

(Fig. 11.104a–g) (Klimaszewski et al. 2014, 2018a)

Diagnosis Body moderately robust with sides subparallel or body narrowly oval, length 2.0–3.2 mm; colour dark brown to black, integument with distinct meshed microsculpture and moderately dense punctation and pubescence; head large with mandibles broad and long, left mandible with a small tooth, right one with a slightly larger tooth at base of arcuate cutting edge, apices distinctly narrowly elongate [more than in *Ocyusa* and less than in *Gnathusa*]; postocular carina strong and complete; ligula shallowly split apically; penultimate maxillary palpomere minute, apical one needle-shaped; labrum narrow and transverse, apical edge entire; frontal suture of head absent; pronotal disc with pubescence along midline directed anteriad or obliquely anteriad in about apical one-third, and posteriad or obliquely posteriad medio-basally; process of mesoventrite without longitudinal carina, triangular basally and then produced and extending from one-third to almost two-thirds length of mesocoxal cavities; process of metaventrite triangular and short; isthmus short to long; median lobe of aedeagus strongly produced ventrally; spermatheca with capsule spherical or tubular, narrowed posteriorly into L-shaped neck, invagination small, stem narrowly elongate.

34. *Mniusa yukonensis* (Klimaszewski and Godin) (Fig. 11.104a–g)

Ocyusa yukonensis Klimaszewski and Godin, in Klimaszewski et al. 2012. Klimaszewski et al. 2014, 2018a.

Diagnosis Body narrowly elongate and broadest at elytra, length 2.8–3.0 mm, colour dark brown to almost black, elytra sometimes with reddish tinge, antennae and legs reddish-brown (Fig. 11.104a); forebody highly glossy, with dense microsculpture, dense punctation and pubescence; head round, narrower than pronotum; pronotum transverse, with sides distinctly arcuate, narrowed anteriad, slightly narrower than elytra; elytra at suture about as long as pronotum or slightly longer; abdomen broadly arcuate laterally, slightly narrower than elytra, first three visible tergites with basal impressions; antennomeres V–X slightly transverse. MALE. Tergite VIII truncate apically (Fig. 11.104c); sternite VIII with apical margin very broadly produced, very broadly rounded at apex (Fig. 11.104d); median lobe of aedeagus with ventral margin of tubus about straight and slightly bent ventrad near base, internal sac with complex structures (Fig. 11.104b). FEMALE. Tergite VIII truncate apically, lateral angles sharp (Fig. 11.104e); sternite VIII with apical margin very obtusely produced, broadly rounded at apex (Fig. 11.104f); spermatheca with capsule narrowly elongate, sac-shaped, neck L-shaped, stem narrow and long (Fig. 11.104g).

DISTRIBUTION **Origin**: Nearctic. **CANADA**: BC, NB, NS, ON [**NPR**], QC, YT. **USA**: AK [**NSR**].

COLLECTION AND HABITAT DATA **Habitat**: in AK, specimens were captured in rocky alpine tundra; alpine *Salix reticulata*; heath tundra with *Vaccinium caespitosum, Cassiope mertensiana, Harrimanella stelleriana*; tundra with *Luetkea pectinata, Harrimanella stelleriana, Vaccinium uliginosum*; rocky low tundra with *Vaccinium uliginosum, Luetkea pectinata, Phyllodoce glandulifera*; in *Nephrophyllidium, Carex nigricans*; meadow with *Nephrophyllidium crista-galli, Carex macrochaeta, Lupinus nootkatensis*; wet meadow with *Carex nigricans, Petasites frigidus, Ranunculus occidentalis*; creek side, corn lilly and *Rubus*; clear-cut; talus above / side of snowfield, crawling under or near rocks; subalpine, *Salix arctica, Vaccinium*, and subalpine *Vaccinium, Cornus*. In NB, all adults were captured in various kinds of traps in various forest types: rich Appalachian hardwood forest with some conifers, old-growth white spruce/balsam fir forest, balsam fir and sugar maple stands, mature white spruce with feather moss, and a red spruce forest. **Collecting period**: V–IX. **Collecting method**: hand picking/ ground search, pitfall traps, yellow pitfalls w/o rainroofs, flight intercept traps, and Lindgren funnel traps, Berlese funnels.

We report this species from ON (**NPR**) and AK (**NSR**) for the first time here, based on the following: Canada, Ontario, Lake Superior Prov. Pk., Algoma Co., 9.IX.1980, R. Baranowski (ZMLU) 1 male; Lake Superior Prov. Pk., Algoma Co., 11.IX.1980, R. Baranowski (ZMLU) 1 female; Algoma Co., Michipicoten River, south of Wawa, 5.IX.1980, R. Baranowski (ZMLU) 1 male. USA: Alaska, Chichagof Is. pl.05, 736m el., 57.9856, −135.20456, ±3m heath tundra, *Vaccinium caespitosum, Cassiope mertensiana, Harrimanella stelleriana*, pitfall, 15-16.VII.2013, S. Ridling (UAM) 1 specimen; Dall Isl., pl.00, 624m el., 54.99813, −133.0159, ±4m *Nephrophyllidium, Carex nigricans*, hand collected, 15.VII.2011, D.S. Sikes (UAM) 1 specimen; Denali NPP, nr Toklat, 1263m el., 63.50154, −150.10298, ±28m rocky alpine, 4 pitfalls w/ rainroofs, pooled, J.J. Rykken; 4-18.VIII.2016, D.S. Sikes; R. Stewart-Rogers (UAM) 1 specimen; Denali NPP, Primrose Ridge, 1276m el., 63.74095, −149.35963, ±28m rocky alpine, 10 minute ground search, 17.VIII.2016, D.S. Sikes (UAM) 1 specimen; Denali NPP., Mt. Healy, 1183m el., 63.75095, −148.96071, ±30m alpine tundra, 4 yellow pitfalls w/o rainroofs, pooled, 3-16.VIII.2016, J.J. Rykken; D.S. Sikes; R. Stewart-Rogers

(UAM) 1 specimen; Denali NPP., Mt. Healy, 1249m el., 63.75362, −148.96379, ±30m rocky alpine, 4 yellow pitfalls w/o rainroofs, pooled, 3-16.VIII.2016, J.J. Rykken; D.S. Sikes; R. Stewart-Rogers (UAM) 2 specimens; Haines, Flower Mtn. pl.07, 972m el., 59.38885, −136.31601, ±29m alpine *Salix reticulata*, pitfall, 18-19.VII.2010, D.S. Sikes (UAM) 1 specimen; Hawthorne Peak pl.05, 775m el., 58.24007, −134.24252, ±5m meadow, *Nephrophyllidium crista-galli, Carex macrochaeta, Lupinus nootkatensis*, pitfall, S. Ridling, 22-23.VII.2013, (UAM) 1 specimen; Hawthorne Peak pl.14, 811m el., 58.24704, −134.25217, ±6m wet meadow, *Carex nigricans, Petasites frigidus, Ranunculus occidentalis*, pitfall, 22-23.VII.2013, S. Ridling (UAM) 1 specimen; Mahoney Mt. pl.07, 881m el., 55.41548, −131.57814, ±3m rocky low tundra, *Vaccinium uliginosum, Luetkea pectinata, Phyllodoce glandulifera*, pitfall, 6-8.VII.2013, S. Ridling (UAM) 1 specimen; Mahoney Mt. pl.20, 784m el., 55.40266, −131.57668, ±5m tundra, *Luetkea pectinata, Harrimanella stelleriana, Vaccinium uliginosum*, pitfall, 6-7. VII.2013, S. Ridling (UAM) 1 specimen; Prince of Wales Island, Luck Point 2B, 264m el., 55.97939, −132.77216, ±25m clear-cut, Berlese, 6.VIII.2010, J. Stockbridge, C. Bickford (UAM) 1 specimen; Prince of Wales Island, Staney Ck. 1B, 77m el., 55.872, −133.06523, ±26m clear-cut, Lindgren 1, 16-31.V.2011, J. Stockbridge, B. Wong (UAM) 1 specimen; Prince of Wales, Hatchery Creek Falls, 47m el., 55.91675, −132.94337, ±18m creek side, corn lily + *Rubus*, hand, 17.V.2013, D.S. Sikes, Jozef A. Slowik, J. Stockbridge (UAM) 1 specimen; S. Baranof Is. E. of Snipe Bay, 720m el., 56.41095, −134.85593, ±14m talus above / side of snowfield, crawling under or near rocks, ground search, 30.VII.2013, S. Ridling (UAM) 1 specimen; S. Chilkat Range pl.10, 718m el., 58.42621, −135.1575, ±28m subalpine, *Salix arctica, Vacc. caesp.*, hand collected, 9. VII.2010, D.S. Sikes (UAM) 1 specimen; S. Chilkat Range pl.13, 687m el., 58.4252, −135.15205, ±29m subalpine *Vaccinium, Cornus*, pitfall, 8-9.VII.2010, D.S. Sikes (UAM) 1 specimen, 9-10.VII.2010, (UAM) 1 specimen.

Genus *Neothetalia* Klimaszewski, 2004

(Fig. 11.105–11.107) (Klimaszewski and Pelletier 2004, Klimaszewski et al. 2018a, 2020)

Diagnosis Adults of the genus *Neothetalia* may be distinguished from those of other aleocharine genera by the following combination of characters: body narrowly elongate, subparallel, dark brown to nearly black with brown to rust-brown elytra; size medium to large, length 4.0–6.0 mm, integument moderately glossy and moderately densely punctate and pubescent, forebody with distinct isodiametric microsculpture; antennomere IV quadrate to elongate, VII–X slightly transverse; maxillary palpus with 4 palpomeres, apical one narrowly elongate and one-third to nearly one-half length of penultimate palpomere; galea expanding slightly beyond apex of lacinia, its apical lobe bearing dense short pubescence; apex of lacinia with row of closely spaced spines; labial palpus with 3 palpomeres; ligula slender, deeply bifid (*N. nimia, N. pallitarsis*) or slightly emarginate apically, *N. smetanai, N. canadiana*, and *N. columbiana*; mentum trapezoidal, apical margin slightly concave, with three long apical setae on each side; labrum transverse and usually slightly projecting medially; right mandible with distinct medial tooth; head quadrate or slightly elongate and distinctly narrower than pronotum; pronotum broadest medially and rather abruptly narrowing anteriorly but slightly so posteriorly, moderately to strongly transverse; pubescence along median line of disc directed anteriad except for very basal part, laterad on each side of disc and often forming broadly arcuate lines; mesoventrite long, longer than maximum diameter of mesocoxa, with complete (*N. pallitarsis*) or partial median carina (remaining species), mesoventral process long, narrow, and acutely pointed, extending about two-thirds length of mesocoxal cavities, isthmus no more than one-fourth of maximum diameter of mesocoxa; elytra broader than pronotum, slightly transverse; abdomen with three basal tergites deeply impressed basally. MALE. Median lobe of aedeagus with tubus more or less narrowly elongate at apex in lateral view, bulbus approximately as long as tubus in lateral view, crista apicalis of bulbus broadly or narrowly oval, medium sized; internal sac with usually two pairs of inconspicuous subapical structures. FEMALE. Spermatheca with moderately sized, approximately spherical capsule bearing shallow or deep apical invagination, attachment of spermathecal gland located at capsule base near neck of spermatheca; basal, medial, and posterior portions of spermatheca consist of narrow tube-like stem, which is more or less broadly S-shaped or looped.

COMMENTS Adults of this genus are associated with cold wet moss found alongside streams, on vertical slopes, on logs, and on and under rocks. They are cold-loving creatures and have been captured at elevations of up to 2133 m (Klimaszewski and Pelletier, 2004).

Key to Species of the Genus *Neothetalia* (Modified from Klimaszewski et al. 2020)

1. Antennomere IV subquadrate (Fig. 11.106a); pronotum with edges strongly converging apically (Fig. 11.106a); median lobe of aedeagus and spermatheca as illustrated (Fig. 11.106b,c,h)... .... ***Neothetalia columbiana* (Klimaszewski)**

– Antennomere IV slightly elongate; pronotum with edges moderately converging apically; median lobe of aedeagus and spermatheca differently shaped **2**

2. Pronotum dark brown to virtually black (Fig. 11.107a); elytra 1.2 times as long as pronotum (Fig. 11.107a); tubus of median lobe of aedeagus sinuate in lateral view (Fig. 11.107b) and sharply pointed in dorsal view (Fig. 11.107c); spermatheca strongly sinuate, stem broadly looped (Figs. 11.107h)..................................... .... ***Neothetalia pallitarsis* (Kirby)**

– Pronotum brown with rust tinge (Fig. 11.105a); elytra 1.2–1.4 times as long as pronotum (Fig. 11.105a); tubus of median lobe of aedeagus and spermatheca not as above (Fig. 11.105b,c,h)... .... ***Neothetalia canadiana* Klimaszewski**

35. *Neothetalia canadiana* Klimaszewski (Fig. 11.105a–h)

Neothetalia canadiana Klimaszewski, in Klimaszewski and Pelletier 2004. Klimaszewski et al. 2011, 2018a, 2020, Webster et al. 2016.

Diagnosis Body length 4.8–5.5 mm, dark brown with elytra, basal antennomere, tarsi, and sometimes pronotum light brown with rust tinge, forebody moderately glossy and abdomen strongly so (Fig. 11.105a); antenna moderately stout, antennomere IV strongly elongate in both sexes, antennomeres V–X strongly elongate in male and slightly so in female, terminal antennomere about as long as two preceding ones combined; pronotum slightly transverse (Pronotal Index = 1.2), lateral edges slightly converging basally in basal half and moderately strong in apical half, length ratio of base to apex 1.5 times, pronotal punctures sparse, distance between punctures about 3 times diameter of a puncture, space between punctures with strong, isodiametric microsculpture, pubescence directed anteriad along median line of disc and laterad forming arcuate lines on both sides of disc; elytra transverse (Elytral Index = 1.5), 1.2 times as long as pronotum. MALE. Tergite VIII transverse, broadly arcuate apically (Fig. 11.105d); sternite VIII slightly transverse, apical part broadly triangularly produced (Fig. 11.105e); median lobe of aedeagus with small, oval crista apicalis at base of bulbus, tubus long, approximately straight medially, moderately produced ventrally at apex, apex narrowly rounded in lateral view (Fig. 11.105b), a subapical hemispherical carina occurs on each side of tubus and is visible in dorsal view (Fig. 11.105c), internal sac with two strong subapical structures (Fig. 11.105b,c). FEMALE. Tergite VIII similar to that of male (Fig. 11.105f); sternite VIII transverse and broadly rounded apically (Fig. 11.105g); spermatheca (Fig. 11.105h) with capsule approximately spherical and slightly compressed, bearing short apical invagination, stem forming characteristically broadly looped tube.

DISTRIBUTION Origin: Nearctic. **CANADA**: AB [**BOLD**], BC, LB, NB, QC, YT. **USA**: AK.

COLLECTION AND HABITAT DATA

Habitat: adults have been captured from gopher burrows, on carrion, and by sifting forest debris. The YT specimen was captured in a white spruce-feathermoss forest. The AK specimens were found in *Betula, Salix*, white spruce; alder, dwarf birch, grass, clubmoss; small stand with *Betula* and *Populus tremuloides*; and subalpine, *Salix arctica,* and *Vaccinium caespitosum.* **Collecting period**: IV–VII. **Collecting method**: flight intercept trap, unbaited pitfall traps, and yellow pitfalls w/o rainroofs.

We report this species from AB (**NPR**) for the first time here based on **BOLD** records: Canada, Alberta, Banff National Park, Corral Creek old road, 1539m el., 51.407, –116.154, wetland, flight-intercept trap, 16-20.VI.2012, BIOBus 2012 team (CBG) 1 specimen.

Additional records. See Annex 1.

36. ***Neothetalia columbiana*** **(Klimaszewski)** (Fig. 11.106a–h)

Ocalea columbiana Klimaszewski, in Klimaszewski and Winchester, 2002. Klimaszewski and Pelletier, 2004, Klimaszewski et al. 2020.

Diagnosis Body length 4.3–5.5 mm, dark brown with elytra, basal antennomere, tarsi, and sometimes pronotum light brown with rust tinge, forebody moderately glossy and abdomen strongly so (Fig. 11.106a); antenna stout, antennomere IV moderately elongate, IV slightly elongate, antennomeres VI–X quadrate to slightly transverse in male and distinctly transverse in female, and terminal antennomere about as long as two and a half preceding ones combined; pronotum strongly transverse (Pronotal Index = 1.3), lateral edges slightly converging basally in basal half and strongly so in apical half, length ratio of base to apex about 1.6 times, pronotal punctures sparse, distance between punctures about 2–3 times diameter of a puncture, space between punctures with strong, isodiametric microsculpture, pubescence directed anteriad along median line of disc and laterad on both sides; elytra transverse (Elytral Index = 1.5). MALE. Tergite VIII truncate apically and without median emargination (Fig. 11.106d); sternite VIII slightly transverse, apical margin broadly triangular, apex right-angled (Fig. 11.106e); median lobe of aedeagus with small, oval crista apicalis at base of bulbus, tubus long, approximately straight medially, moderately produced ventrally at apex, apex narrowly rounded in lateral view (Fig. 11.106b); median lobe in dorsal view as illustrated (Fig. 11.106c). FEMALE. Tergite VIII truncate apically (Fig. 11.106f); sternite VIII transverse and broadly rounded apically (Fig. 11.106g); spermatheca similar to that of *P. canadiana*, capsule approximately spherical and slightly compressed, with short apical invagination, spermathecal stem forming narrow, characteristically broadly looped tube (Fig. 11.106h).

DISTRIBUTION Origin: Nearctic. **CANADA**: AB [**BOLD**], BC. **USA**: AK.

COLLECTION AND HABITAT DATA

Habitat: in BC, specimens were captured in the forest interior and the transition zone of the Sitka spruce dominated forest. Some specimens were found on rotting piles of mushrooms and under deciduous shrubs around bases of cedar. One female was taken from a dung and carrion trap. One male was captured in the transition zone of Douglas-fir (*Pseudotsuga menziesii* (Mirb.)) dominated forest in in wetern part of Vancouver Island. In AK, adults were found in alpine zone under rocks; in old growth hemlock rotting logs, dung; old growth and 2nd growth forests; clear-cuts; and in subalpine habitats, *Salix arctica* and *Vaccinium caespitosum.* **Collecting period**: V–VIII. **Collecting method**: pitfall traps, primate dung baited pitfalls, Lindgren traps, Malaise and pan traps.

We report this species from AB (**NPR**) for the first time here based on **BOLD** records: Canada, Alberta, Waterton Lakes National Park, Akamina Parkway, Rowe Tamarack trail, 1640m el. 49.059, –114.014, subalpine conifer forest, pitfall, 22-27.VI.2012, BIOBus 2012 team (CBG) 3 specimens.

Additional records. See Annex 1.

37. *Neothetalia pallitarsis* (Kirby) (Fig. 11.107a–h)

Aleochara pallitarsis Kirby, in Kirby and Spence 1837. In *Atheta*: Fenyes 1920, Bernhauer and Scheerpeltz 1926. In undescribed genus near *Ocalea*: Klimaszewski 1984. As *Neothetalia*: Klimaszewski and Pelletier, 2004, Klimaszewski et al. 2020.

Diagnosis Body length 4.7–6.0 mm, dark brown to nearly black with elytra, basal antennal segment, and tarsi light brown with rust tinge, forebody moderately glossy and abdomen strongly so (Fig. 11.107a); antennae moderately stout, antennomere IV slightly elongate, V–VII quadrate in male and transverse in female, VIII–X quadrate to slightly elongate in male and transverse in female, terminal antennomere about as long as two preceding ones combined; pronotum slightly transverse (Pronotal Index = 1.2), lateral edges slightly converging basally in basal half and strongly so in apical half, length ratio of base to apex 1.6 times, pronotal punctures sparse, distance between punctures about 3 times diameter of a puncture, space between punctures with strong isodiametric microsculpture, pubescence directed anteriad along median line of disc and laterad forming slightly arcuate lines on both sides; elytra transverse (Elytral Index = 1.6). MALE. Tergite VIII truncate apically and bearing shallow median emargination (Fig. 11.107d); sternite VIII transverse, apical margin broadly triangularly produced, rounded medially (Fig. 11.107e); median lobe of aedeagus with short crista apicalis at base of bulbus, tubus slightly sinuate and strongly produced ventrally in lateral view (Fig. 11.107b), median lobe with internal sac bearing strongly sclerotized, reversed narrowly V-shaped structure in dorsal view (Fig. 11.107c). FEMALE. Tergite VIII similar to that of male (Fig. 11.107f); sternite VIII broadly rounded posteriorly (Fig. 11.107g); spermathecal capsule subspherical and with short apical invagination, stem forming narrow tube of a more or less semicircular shape, median portion of stem variably curved (Fig. 11.107h).

DISTRIBUTION Origin: Nearctic. **CANADA**: AB, BC. **USA**: AK.

COLLECTION AND HABITAT DATA

Habitat: in forest litter at altitudes of 1122–2,133 m. In AK, adults were captured in willow, dwarf birch, and shrub litter; alpine tundra; shrub, alder, dwarf birch, grass, clubmoss; under rocks above snow; on snow fields; subalpine rocks with *Alnus*; subalpine *Tsuga, Luetkea*, *Vaccinium*; and miscellaneous alpine habitats. **Collecting period**: V–IX. **Collecting method**: sifting forest litter, pitfall traps, Winkler Litter Extraction, Berlese funnels, yellow pitfalls w/o rainroofs.

Additional records. See Annex 1.

Genus *Ocyusa* Kraatz, 1856

(Fig. 11.108a–g) (Lohse et al. 1990, Klimaszewski et al. 2014, 2018a)

Diagnosis Body moderately robust, subparallel, narrowly oval, length 2.5–3.5 mm; colour dark brown to almost black, integument with distinct meshed microsculpture, punctation and pubescence moderate to dense; head large with mandibles broad and long, left mandible with a small tooth, right one with a slightly larger tooth at base of arcuate cutting edge, apices moderately narrowly elongate (less than in *Mniusa*); postocular carina strong and complete; ligula shallowly split apically; penultimate maxillary palpomere minute, apical one needle-shaped; labrum narrow and transverse, apical edge entire; frontal suture of head present; pronotal disc with pubescence directed posteriad or obliquely posteriad along midline; anterior margin of mesoventrite without longitudinal carina; process of mesoventrite triangular basally, then produced and extending to about two-thirds length of mesocoxal cavities, process of metaventrite narrowly triangular in shape and short, isthmus short; median lobe of aedeagus strongly bent ventrad at base; spermatheca S-shaped with capsule spherical, neck short, stem elongate and sinuate.

38. *Ocyusa canadensis* Lohse (Fig. 11.108a–g)
Ocyusa canadensis Lohse, in Lohse et al. 1990. Brunke et al. 2012, Klimaszewski et al. 2014, 2015, 2018a, 2020.

Diagnosis Body narrowly elongate, broadest at elytra, length 2.5–3.0 mm; colour dark brown to almost black, with tarsi, two basal antennal articles and tibiae rust-brown, rest of antennal segments brown (Fig. 11.108a); forebody highly glossy, with dense microsculpture, punctation and pubescence moderate; head round, narrower than pronotum; pronotum transverse, rectangular in shape with sides distinctly arcuate, narrower than elytra; elytra at suture about as long as pronotum; abdomen broadly arcuate laterally, slightly narrower than elytra, first three visible tergites with basal impressions; antennomeres V–X subquadrate to slightly elongate. MALE. Tergite VIII broadly arcuate apically (Fig. 11.108c); sternite VIII broadly parabolic apically (Fig. 11.108d); median lobe of aedeagus with ventral margin of tubus almost straight, strongly bent ventrad at base, internal sac with complex structures (Fig. 11.108b). FEMALE. Tergite VIII truncate apically (Fig. 11.108e); sternite VIII evenly rounded apically (Fig. 11.108f); spermatheca with capsule semi-spherical, apical invagination deep and evenly narrow, stem broadly S-shaped (Fig. 11.108g).

DISTRIBUTION **Origin**: Nearctic. **CANADA**: BC, NB, NF, NT [**NTR**], ON, SK, YT. **USA**: AK.

COLLECTION AND HABITAT DATA
Habitat: lake and pond margins in emergent *Carex* and grasses. Some also on moist soil/gravel among sedges. **Collecting period**: VI–VII. **Collecting method**: treading *Carex* and grasses into water, examining moist soil, and hand collecting (aspirating specimens), yellow pan traps.

We report this species from NT (**NTR**) for the first time here, based on the following: Canada: Northwest Territories, Yellowknife, 62.51591, –113.39146, Rep. 3 wet, yellow pan, 7-11.VII.2011, NBP field party Y3WY714 (LFC) 1 female; Yellowknife, 62.51591, –113.39146, Rep. 3 wet, yellow pan, 7-11.VII.2011, NBP field party Y3WY713 (LFC) 1 female.

Additional records. See Annex 1.

Genus *Oxypoda* Mannerheim, 1830

(Figs. 11.109–11.131) (Klimaszewski et al. 2006, 2018a, 2020)

Diagnosis Body subparallel, narrowly oval or fusiform, usually with abdomen strongly tapering; integument with pubescence fine, short and dense, in some species appearing silky, with numerous macrosetae especially on abdomen; head with frontal suture, postocular carina present, neck absent; mouthparts generalized (similarly developed across the genera); mandibles simple except right mandible with median tooth; pronotum usually highly convex with posterior angles very narrowly rounded, hypomera not visible in lateral view; process of mesoventrite slender and acute; tarsal formula 5-5-5; parameres with elongate apical lobe bearing four macrosetae: short (reduced) one at apex, moderately long one at middle, and two long ones at base; tergite VIII and sternite VIII usually similar in both sexes but in some taxa male sternite shaped differently, antecostal suture different in male and/or female in various species; median lobe of aedeagus with tubus well developed, in dorsal view usually subparallel for most of its length and gradually or strongly narrowed at apex, bulbus moderately large and bearing frontal and ventral carinae, the latter not always easily visible due to weak sclerotization, internal sac with complex structures, length of flagellum variable, usually shorter than tubus but occasionally longer than median lobe; spermatheca with capsule usually spherical but in some taxa elongate, stem long, sinuate and in some taxa looped posteriorly.

COMMENTS Some members of *Oxypoda* may appear superficially like some species of *Myllaena* Erichson. However, *Myllaena* differs in having extremely elongate glossae extending over the labrum, and a 4-4-5 tarsal formula.

Key to Species of *Oxypoda* (Modified from Klimaszewski et al. 2006, 2018a)

1. Body subparallel with dense and short pubescence (Figs. 11.122a–11.125a); antennomeres VII–X somewhat swollen in most (Figs. 11.122a, 11.124a); median lobe of aedeagus of characteristic shape and with apex bearing projecting membranous structure (Figs. 11.122b,c, 11.125b,c) **2**

– Body and other characters not as above..... **5**

2. Body length 2.5–2.8 mm **3**

– Body length 2.5–3.5 mm (greater than 2.8 mm in most) **4**

3. Body black, maximum width of elytra greater than maximum width of pronotum, elytra at suture longer than pronotum along midline (Fig. 11.124a); median lobe of aedeagus and spermatheca as illustrated (Fig. 11.124b,c,h) .... ***Oxypoda lucidula*** **Casey**

– Body dark brown, maximum width of elytra equal to maximum width of pronotum, elytra at suture as long as pronotum along midline (Fig. 11.122a); median lobe of aedeagus and spermatheca not as above (Fig. 11.122b,c,h,i) .... ***Oxypoda demissa*** **Casey**

4. Body narrow (Fig. 11.123a,b); pronotum strongly transverse (Fig. 11.123a,b); antennae not thick, antennomeres VII–X subquadrate (Fig. 11.123a,b); median lobe of aedeagus and spermatheca as illustrated (Fig. 11.123c,d,i)........ .... ***Oxypoda leechi*** **Lohse**

– Body broad (Fig. 11.125a); pronotum slightly transverse (Fig. 11.125a); antennae thick, antennomeres VII–X slightly transverse (Fig. 11.125a); median lobe of aedeagus as illustrated (Fig. 11.125b,c); spermatheca unknown...... ***Oxypoda robusticornis*** **Bernhauer**

5. Antennomere IV slightly to strongly elongate **6**

– Antennomere IV subquadrate to transverse.. **19**

6. Antennomere IV strongly elongate, at least 1 1/5 times as long as wide............................... **7**

– Antennomere IV slightly elongate, less than 1 1/5 times as long as wide............................. **15**

7. Body uniformly black (Fig. 11.118a); genitalia as illustrated (Fig. 11.118b,c,h).................... .... ***Oxypoda volkeri*** **Klimaszewski**

– Body light-brown to dark brown, genitalia not as above **8**

8. Pronotum strongly convex, broadest in basal half and abruptly converging apically (Fig. 11.116a), body length 2.5–2.7 mm, genitalia as illustrated (Fig. 11.116b,c,h) .... ***Oxypoda inimica*** **Casey**

– Pronotum moderately convex, broadest in basal third (except in *O. orbicollis*) and moderately strong converging apically, body length 2.8–4.6 mm, usually over 3.0 mm, genitalia differently shaped **9**

9. Body spindle-shaped, uniformly pale to dark brown, dull in lustre, length 3.0–3.8 mm, eyes small, distinctly shorter than postocular area (Fig. 11.129a); antennomeres V–X quadrate to slightly transverse (Fig. 11.129a), genitalia as illustrated (Fig. 11.129b,c,h,i)............................. .... ***Oxypoda orbicollis*** **Casey**

– Body with different combination of characters, length 3.0–4.6 mm **10**

10. Elytra length at shoulder longer than pronotum **11**

– Elytra length at shoulder about as long as or shorter than pronotum **12**

11. Maximum width of pronotum about the same as the maximum width of elytra, (Fig. 11.113a), genitalia as illustrated (Fig. 11.113b,c,h).......... ***Oxypoda nelsoni*** **Lohse**

– Maximum width of pronotum distinctly less than the maximum width of elytra (Fig. 11.131a), genitalia as illustrated (Fig. 11.131b,c,h)............ .... ***Oxypoda sylvia*** **Casey**

12. Pronotum large, shield-shaped, its maximum width slightly greater than maximum width of elytra (Fig. 11.127a); body broad; median lobe of aedeagus and spermatheca as illustrated (Fig. 11.127b,c,h)........ ***Oxypoda operta*** **Sjöberg**

– Pronotum moderately sized, its maximum width same as or less than maximum width of elytra; body moderately broad; median lobe of aedeagus and spermatheca not as above **13**

13. Eyes very small (Fig. 11.110a); body narrowly subparallel, length 2.8–3.5 mm, pronotum gradually converging apically (Fig. 11.110a), genitalia as illustrated (Fig. 11.110b,c,h) .... ***Oxypoda canadensis* Klimaszewski**

– Eyes not as above; body broadly fusiform, length 3.0–4.5mm, pronotum strongly converging apically, genitalia differently shaped **14**

14. Antennomere IV slightly elongate, VIII–X strongly transverse (Fig. 11.114a), body length 2.8–3.0 mm; elytra same color as the rest of the body; tubus of median lobe of aedeagus evenly narrowed dorsally in dorsal view (Fig. 11.114c), apex of tubus bent ventrad in lateral view (Fig. 11.114b); stem of spermatheca long, irregularly sinuate (Fig. 11.114h,i) .... ***Oxypoda pseudoconvergens* Klimaszewski and Godin**

– Antennomere IV elongate, VIII–X moderately transverse (Fig. 11.111a), body length 3.0–3.4 mm; elytra paler than rest of the body in most; tubus of median lobe of aedeagus swollen medially in dorsal view (Fig. 11.111c), and apex of tubus straight in lateral view (Fig. 11.111b); stem of spermatheca short, S-shaped (Fig. 11.111h)......... ***Oxypoda convergens* Casey**

15. Pronotum shield-shaped with sides semicircular, broadest at basal half and broader than elytra at base (Fig. 11.130a), abdomen swollen at midlength (Fig. 11.130a), median lobe of aedeagus as illustrated (Fig. 11.130b); female unknown.................. ***Oxypoda subpolaris* Casey**

– Pronotum differently shaped, broadest in basal third and about as broad as elytra at base, abdomen tapering apically, genitalia differently shaped **16**

16. Antennomeres VIII–X strongly transverse, about twice as wide as long (Fig. 11.121a); genitalia as illustrated (Fig. 11.121b,c,h,i)................. .... ***Oxypoda lacustris* Casey**

– Antennomeres VIII–X moderately transverse, less than twice as wide as long............. **17**

17. Elytra longer than pronotum (Fig. 11.119a), genitalia as illustrated (Fig. 11.119b,c,h).......... ***Oxypoda irrasa* Mäklin**

– Elytra as long as or shorter than pronotum, genitalia not as above...................................... **18**

18. Pronotum broad, at base as broad as maximum width of elytra (Fig. 11.112a); genitalia as illustrated (Fig. 11.112b,c,h).............................. .... ***Oxypoda manitobae* Casey**

– Pronotum narrow, at base narrower than maximum width of elytra (Fig. 11.115a); genitalia not as above (Fig. 11.115b,c,h) .... ***Oxypoda gatosensis* Bernhauer**

19. Antennomeres strongly transverse, VIII–X about 2 ½ as wide as long; body narrowly subparallel **20**

– Antennomeres moderately transverse, VIII–X at most 2 times as wide as long **22**

20. Body narrowly ovoid in shape (Fig. 11.120a); elytra along suture as long as or shorter than length of pronotum along midline of disc (Fig. 11.120a); pronotum 1/5 longer than wide (Fig. 11.120a); apex of median lobe of aedeagus broadened apically and truncate in lateral view (Fig. 11.120b); stem of spermatheca half looped (Fig. 11.120h) .... ***Oxypoda hiemalis* Casey**

– Body narrowly subparallel; elytra along suture shorter than length of pronotum along midline of disc; pronotum 1/5 wider than long; genitalia not as above...................................... **21**

21. Antennomeres VII–X strongly transverse (Fig. 11.128a); elytra brown to almost black; forebody with weak microsculpture; median lobe of aedeagus with crista apicalis of bulbus large, oval in shape in lateral view (Fig. 11.128b); stem of spermatheca coiled (Fig. 11.128h,i) ......***Oxypoda frigida* Bernhauer**

– Antennomeres VII–X less strongly transverse (Fig. 11.126a); elytra yellowish brown; forebody with strong microsculpture (Fig. 11.126a); median lobe of aedeagus with crista apicalis of bulbus narrowly elongate in lateral view (Fig. 11.126b); stem of spermatheca U-shaped (Fig. 11.126h) .... ***Oxypoda brachyptera* (Stephens)**

22. Body robust, broad, ovoid in dorsal view, dark brown to black (Fig. 11.109a); median lobe of aedeagus with long flagellum, longer than length of median lobe of aedeagus (Fig. 11.109b, c); stem of spermatheca fully looped (Fig. 11.109i) .... ***Oxypoda amica* Casey**

– Body not robust, narrowly elongate, light brown with dark brown head and abdomen (Fig. 11.117a); median lobe of aedeagus with short flagellum, shorter than length of median lobe (Fig. 11.117b); stem of spermatheca half looped (Fig. 11.117h)......... ***Oxypoda yukonensis* Klimaszewski and Godin**

***Amica* species group** (Klimaszewski et al. 2006, 2018a)

Diagnosis Body narrowly oval, medium-sized, length 2.5–2.9 mm, colour light brown to black; pronotum transverse; elytra moderately large and long; median lobe of aedeagus in dorsal view with tubus subparallel, narrowing apically and rounded at apex, in lateral view with ventral margin roughly straight or more-or-less sinuate, bulbus with crista apicalis narrowly elongate, flagellum extremely long, longer than median lobe, internal sac with small structures, apical lobe of paramere moderately elongate; spermatheca with capsule spherical, with large invagination, stem S-shaped, narrowly twisted posteriorly.

39. *Oxypoda amica* Casey (Fig. 11.109a–i)

Oxypoda amica Casey, 1906. Gusarov 2003, Klimaszewski et al. 2005, 2006, 2018a, Majka and Klimaszewski, 2010.

Diagnosis Body narrowly oval, broadest at apical margin of elytra, length 2.5–2.9 mm, colour variable, forebody light yellow brown to light brown with head and abdomen darker, or body entirely dark brown with head and abdomen darker, nearly black, basal antennomeres in some specimens rust-brown (Fig. 11.109a); bicoloured specimens usually have head and abdomen dark brown, elytra yellowish-brown or light brown mottled with dark brown spots, pronotum medium brown and mottled with darker irregular small spots; forebody with pubescence moderately dense, punctation slightly asperate, with isodiametric microsculpture; head broadest at base of eyes, two thirds as broad as pronotum; eyes as long as genae in dorsal view; antennae short and stout, antennomere IV subquadrate or slightly transverse, V–X transverse and slightly incrassate; pronotum transverse, slightly narrower than elytra, sides strongly converging apically, broadest in basal third, about one third broader than long, pubescence on disc directed obliquely posterolaterad from midline; elytra slightly longer at suture than pronotum, pubescence on disc directed approximately straight posteriad; abdomen gradually narrowed to apex. MALE. Tergite VIII arcuate apically, apical margin finely crenulate, antecostal suture broadly arcuate (Fig. 11.109e); sternite VIII with apical margin triangularly produced, rounded at apex, antecostal suture sinuate particularly medially (Fig. 11.109f); median lobe of aedeagus in dorsal view with tubus subparallel, gradually narrowing apically and rounded at apex (Fig. 11.109d), in lateral view tubus with ventral margin slightly sinuate, curved ventrad basally, apex narrowly rounded and not bent ventrad (Fig. 11.109b,c); flagellum extremely long, longer than median lobe, internal sac structures as illustrated (Fig. 11.109b,c,d). FEMALE. Tergite VIII truncate apically, antecostal suture straight (Fig. 11.109g); sternite VIII broadly arcuate apically, antecostal suture distinctly sinuate, particularly medially (Fig. 11.109h); spermatheca with capsule spherical, with deep apical invagination, stem narrow, sinuate, with broad posterior loop (Fig. 11.109i).

DISTRIBUTION Origin: Nearctic. **CANADA**: MB, NB, NS, NT, ON, QC. **USA**: IA, IN, KS, KY, LA, MO, NC, NJ, NY, TN.

COLLECTION AND HABITAT DATA

Habitat: in both forest and wetland habitats in leaf litter and moss. Found in moss and leaf litter in eastern white cedar swamps, in sphagnum in a black spruce forest, in moss and leaf litter in a calcareous cedar fen, in leaf litter and moss at bases of trees in a *Carex* marsh margin. Also collected from pitfall and other traps in red spruce dominated forest, dry zone of alvar habitat, sandy beaches at the edge of oak forests, and in maple forests and a Lindgren funnel trap in a red pine forest. One individual was collected from flowers of *Spiraea alba* in a *Carex* marsh. **Collecting period**: IV–VII. **Collecting method**: sifting forest litter, Lindgren funnel traps, Luminoc® pitfall-light traps, pitfall traps, flight intercept traps.

COMMENTS We could not locate any voucher specimens of this species from YT, listed by Brunke et al. (2012) and from MA listed by Klimaszewski et al. (2018b) and Newton (2019). YT and MA are here removed from the distribution of *O. amica*.

***Convergens* species group** (Klimaszewski et al. 2006, 2018a)

Diagnosis Body narrowly oval or subparallel, length 3.0–4.5 mm; antennomere IV elongate in both sexes; pronotum distinctly transverse; elytra at suture as long as pronotum or only slightly shorter; median lobe of aedeagus with tubus broadly subparallel in dorsal view, in lateral view bulbus moderately large, ventral margin of tubus arcuate or slightly sinuate with apex narrow and rounded, flagellum shorter than tubus, internal sac with complex structures, base of flagellum heart- or V-shaped in dorsal view, apical lobe of paramere moderately elongate; shape of spermatheca variable, with capsule roughly spherical or slightly elongate, with broad apical invagination, stem moderately long to long, sinuate or coiled posteriorly.

40. *Oxypoda canadensis* Klimaszewski (Fig. 11.110a–h)

Oxypoda canadensis Klimaszewski, in Klimaszewski et al. 2006. Klimaszewski et al. 2011, 2012, 2018a, 2020.

Diagnosis Body subparallel, broadest at middle of elytra, length 2.8–3.5 mm, colour uniformly dark brown with reddish tinge, or with legs, one to three basal antennomeres, and tip of abdomen slightly paler, forebody with pubescence and punctation dense, integument with isodiametric microsculpture (Fig. 11.110a); head broadest at basal third, slightly more than half as wide as pronotum; eyes very small, about one fourth/one fifth as long as genae in dorsal view; antennae moderately long and slender, antennomeres I–IV elongate, V–X quadrate to slightly transverse; pronotum moderately convex, moderately transverse and about one fourth broader than long, broadest at middle, pubescence on disc directed anteriad apically along midline and obliquely posteriad from midline elsewhere; elytra slightly broader than pronotum, about as long at suture as pronotum, pubescence on disc directed straight posteriad; abdomen gradually apically. MALE. Tergite VIII arcuate apically, antecostal suture slightly arcuate (Fig. 11.110d); sternite VIII with apical margin triangularly produced, apex narrowly rounded, antecostal suture almost straight (Fig. 11.110e); median lobe of aedeagus with tubus subparallel for most of its length in dorsal view (Fig. 11.110c), in lateral view ventral margin of tubus sinuate, at apex narrowly rounded and not bent (Fig. 11.110b), internal sac with complex structures (Fig. 11.110b,c). FEMALE. Tergite VIII arcuate apically, antecostal suture broadly arcuate (Fig. 11.110f); sternite VIII truncate apically (Fig. 11.110g); spermatheca with capsule spherical, with moderate, rounded apical invagination, neck straight; stem long, with several tightly clustered posterior coils (Fig. 11.110h).

DISTRIBUTION Origin: Nearctic. **CANADA**: AB, BC, LB, MB, NF, NT, ON, QC, SK [**NPR**], YT. **USA**: AK, NH.

COLLECTION AND HABITAT DATA
Habitat: mixed wood, deciduous birch-dominated or aspen-dominated forests, deciduous stands with developing white spruce understories, fir forests, mixed aspen and white spruce forest; also riparian habitats. Adults in forest litter, rotting fungi, mosses including *Sphagnum* sp., rotten logs, in lining of deserted beaver lodge, and in nest of *Larus marinus* L. (black-backed gull). **Collecting period**: V-IX, XII (most abundant in

VI). **Collecting method**: pitfall traps, sifting forest litter and processing through Berlese funnels, sifting mushrooms, Luminoc® pitfall-light traps, Malaise trap.

We report this species from SK (**NPR**) for the first time here, based on the following: Canada, Saskatchewan, Prince Albert National Park, Highway 263, 627m el., 53.85, −106.078, mixed aspen and white spruce forest, Malaise trap, 11-16.V.2012, Rod Thompson (CBG) 2 specimens; SK, Prince Albert, 14.IX.2014, 53.9386, −106.057, 556 m, mushroom sifting, coll. B. Godin and D. Horwood (LFC) 1 male.

41. *Oxypoda convergens* Casey (Fig. 11.111a–h)
Oxypoda convergens Casey, 1893. Klimaszewski et al. 2005, 2006, 2011, 2018a, b, 2020.

Diagnosis Body fusiform, broadest at middle of elytra, length 3.0–3.4 mm, colour entirely dark brown, or dark brown with legs, II–III basal antennomeres, and tip of abdomen rust-brown, in some specimens elytra and margins of pronotum also rust-brown (Fig. 11.111a); forebody with pubescence and punctation dense; integument with faint isodiametric microsculpture; head broadest at base, three-fifths as broad as pronotum; eyes about as long as genae in dorsal view; antennae moderately long, moderately stout, antennomere IV elongate, V–X slightly transverse, more so toward apex of antenna; pronotum moderately convex, moderately transverse, about one third broader than long, broadest in basal third, pubescence on disc directed anteriad apically and obliquely outward from midline elsewhere; elytra about as broad and as long as pronotum, pubescence on disc directed straight posteriad; abdomen tapering toward apex (Fig. 11.111a). MALE. Tergite VIII with apical margin truncate medially, broadly rounded laterally, antecostal suture almost straight except laterally (Fig. 11.111d); sternite VIII with apical margin triangularly produced, apex rounded, antecostal suture sinuate (Fig. 11.111e); median lobe of aedeagus in dorsal view with tubus swollen at middle, sides broadly arcuate (Fig. 11.111c), in lateral view ventral margin of tubus almost straight, bent obliquely ventrad at base, narrowly rounded apically (Fig. 11.111b), internal sac with complex structures as illustrated (Fig. 11.111b,c). FEMALE. Tergite VIII with apical margin truncate medially, broadly rounded laterally, antecostal suture slightly sinuate (Fig. 11.111f); sternite VIII similar to tergite VIII but antecostal suture distinctly sinuate (Fig. 11.111g); spermatheca with capsule large, roughly heart-shaped, apical invagination very deep, stem short, S-shaped (Fig. 11.111h).

DISTRIBUTION Origin: Nearctic. **CANADA**: AB, BC, LB, MB, NB, NF, NS, ON, QC, SK [**BOLD**]. **USA**: IA, MO, NY.

COLLECTION AND HABITAT DATA
Habitat: various forest types including white spruce, mature black spruce, red spruce, mature 50-year-old balsam fir, oak, maple, yellow birch/balsam fir white pine forests, red oak forests, mixed forests, boreal forest, and old eastern white cedar forest; often very common in decaying gilled mushrooms in the above forest habitats, especially those dominated with hardwoods; also in riparian habitats, and a meadow with *Salix* shrubs. Several collected from decaying corncobs and cornhusks. **Collecting period**: V–XI (most common in August and September). **Collecting method**: unbaited and carrion-baited pitfall traps, processing forest litter through Berlese funnels, Luminoc pitfall-light traps, Lindgren traps, intercept traps, and treading leaf litter at the edge of a pond, sifting decaying mushrooms.

We report this species from SK (**NPR**) for the first time here, based on the following: Canada: SK: Prince Albert NP, Red Deer Trail Red Loop, 53.907, −106.075, 578 m, grassland, pitfall trap (BIObus 2010), (CBG) 1 specimen.

42. *Oxypoda manitobae* Casey (Fig. 11.112a–h)

Oxypoda manitobae Casey, 1911. Klimaszewski et al. 2006, 2016, 2020; Lohse et al. 1990 (as *O. lata*).

Diagnosis Body broadly fusiform, length 3.0–3.8 mm, broadest at middle of elytra, uniformly dark brown or with legs, two to three basal antennal segments, posterior parts of tergites and tip of abdomen slightly paler, sometimes elytra and edges of pronotum also rust-brown, pubescence and punctation of forebody dense, integument with isodiametric microsculpture (Fig. 11.112a); head broadest at posterior level of eyes, about two thirds as broad as pronotum; eyes shorter than postocular region; antennae moderately long and moderately stout, antennomere IV quadrate, V quadrate to slightly transverse, VII–X moderately transverse and more so toward apex of antenna (Fig. 11.112a); pronotum, moderately convex, strongly transverse and about one third broader than long, broadest in basal part, pubescence directed anteriad apically along midline and elsewhere obliquely posteriad from midline of disc; elytra slightly broader than pronotum, about as long as pronotum at suture, pubescence directed straight posteriad; abdomen slightly tapering toward apex. MALE. Tergite VIII truncate apically, antecostal suture approximately straight (Fig. 11.112d); sternite VIII with apex triangularly produced, antecostal suture strongly sinuate (Fig. 11.112e); median lobe of aedeagus with tubus subparallel in dorsal view (Fig. 11.112c), broadly arcuate ventrally in lateral view (Fig. 11.112b); internal sac with complex structures (Fig. 11.112b,c). FEMALE. Tergite VIII truncate apically, antecostal suture approximately straight (Fig. 11.112f); sternite VIII truncate apically, antecostal suture approximately straight (Fig. 11.112g); spermatheca with moderately large spherical capsule bearing deep, broad invagination, stem sinuate and hooked posteriorly (Fig. 11.112h).

DISTRIBUTION Origin: Nearctic. **CANADA**: BC, MB, SK. **USA**: CO.

COLLECTION AND HABITAT DATA
Habitat: Hudson Plains of Churchill, MB [no further details], elevations 853–2896 m (Rocky Mts.). **Collecting period**: VI, VIII. **Collecting method**: sifting litter.

43. *Oxypoda nelsoni* Lohse (Fig. 11.113a–h)

Oxypoda (Oxypoda) nelsoni Lohse, in Lohse et al. 1990. Klimaszewski et al. 2006.

Diagnosis Body broadly subparallel, broadest at middle of elytra, length 3.8–4.0 mm, colour entirely dark brown, or dark brown with legs, antennae, and tip of abdomen rust-brown (Fig. 11.113a); forebody with pubescence and punctation dense; integument with faint isodiametric microsculpture; head broadest at eyes level, two-thirds as broad as pronotum; eyes about as long as genae in dorsal view; antennae moderately long, moderately stout, antennomere IV–V subquadrate, VI–X slightly transverse, more so toward apex of antenna; pronotum moderately convex, moderately transverse, about one third broader than long, broadest in basal third, pubescence on disc directed anteriad apically and obliquely outward from midline elsewhere; elytra about as broad and as long as pronotum, pubescence on disc directed straight posteriad; abdomen slightly tapering toward apex (Fig. 11.113a). MALE. Tergite VIII with apical margin truncate medially, broadly rounded laterally, antecostal suture arcuate (Fig. 11.113d); sternite VIII with apical margin triangularly produced, apex rounded, antecostal suture slightly sinuate (Fig. 11.113e); median lobe of aedeagus in dorsal view with tubus subparallel (Fig. 11.113c), in lateral view ventral margin of tubus arcuately bent ventrad, narrowly rounded apically (Fig. 11.113b), internal sac with complex structures as illustrated (Fig. 11.113b,c). FEMALE. Tergite VIII with apical margin truncate, slightly emarginate medially, antecostal suture straight medially (Fig. 11.113f); sternite VIII truncate apically and shallowly emarginate medially, antecostal straight medially (Fig. 11.113g); spermatheca with capsule roughly spherical, apical invagination broad and deep, stem long, broadly S-shaped (Fig. 11.113h).

DISTRIBUTION **Origin**: Nearctic. **CANADA**: YT [**NTR**]. **USA**: AK.

COLLECTION AND HABITAT DATA
Habitat: Salt flat, clay, very alkaline habitat dominated by *Salicornia europaea* . **Collecting period**: VII. **Collecting method**: pitfall traps.

We report this species from YT (**NTR**) for the first time here, based on the following: Canada, Yukon, Takhini Salt flats, 60.855, –135.717, 1-15.VIII.2014, alkaline clay, pitfall trap, *Salicornia* habitat, coll. B. Godin and B. Horwood (LFC) 1 male.

44. *Oxypoda pseudoconvergens* Klimaszewski and Godin (Fig. 11.114a–i)

Oxypoda pseudoconvergens Klimaszewski and Godin, in Klimaszewski et al. 2008.

Diagnosis Body narrowly elongate, length 2.8–3.0 mm, pronotum streamlined with elytra, dark brown and with elytra and legs paler (Fig. 11.114a); dorsal integument with dense, fine microsculpture; head approximately round, postocular region approximately equal to diameter of eye; antenna moderately robust, antennomere IV subquadrate, V–X distinctly transverse and increasingly more toward apex of antenna (Fig. 11.114a); pronotum slightly transverse, broadest in basal third, streamlined with elytra; elytra transverse, at suture about as long as pronotum; abdomen gradually tapering apically. MALE. Tergite VIII truncate apically (Fig. 11.114d); sternite VIII produced apically and rounded at apex, antecostal suture slightly sinuate (Fig. 11.114e); median lobe of aedeagus with large bulbus and subparallel tubus, apex tapering in dorsal view (Fig. 11.114c), tubus broadly arched with narrowly produced ventrad subapical part in lateral view (Fig. 11.114b); internal sac structures as illustrated (Fig. 11.114b,c). FEMALE. Tergite VIII truncate apically (Fig. 11.114f); sternite VIII truncate apically, antecostal suture sinuate (Fig. 11.114g); spermatheca with approximately spherical capsule bearing large apical invagination and long, sinuate stem (Fig. 11.114h,i).

Oxypoda pseudoconvergens can be separated from *O. convergens* by its smaller body, wider antennomere IV, and the differently shaped structures of the internal sac of median lobe of aedeagus and spermatheca.

DISTRIBUTION **Origin**: Nearctic. **CANADA**: AB [**BOLD**], YT. **USA**. Not recorded.

COLLECTION AND HABITAT DATA
Habitat: in YT, adults were found in white spruce (*Picea glauca* (Moench) Voss (Pinaceae)) (46%), Scouler's willow (*Salix scouleriana* Barratt *ex* Hook. (Salicaceae)) (29%), and lodgepole pine (*Pinus contorta* Douglas *ex* Louden (Pinaceae)) (25%), with no other tree species occurring within the plot; in AB, adults were collected in subalpine conifer forest and montane forest with douglas fir and lodgepole pine. **Collecting period**: VII–VIII. **Collecting method**: sifting leaf litter, pitfall traps.

We report this species from AB (**NPR**) for the first time based on **BOLD** records: Canada: Alberta, Waterton Lakes National Park, Akamina Parkway, Rowe Tamarack Trail, 1632m el., 49.059, –114.013, subalpine conifer forest, pitfall 6-11.VIII.2012, BIOBus 2012 team (CBG) 16 specimens; Waterton Lakes National Park, Highway 6 pullout, 1562m el., 49.065, –114.778, montane forest, pitfall, 6-11.VIII.2012, BIOBus 2012 team (CBG) 8 specimens; Waterton Lakes National Park, Highway 6, Foothills Parkland Region, 1338 m el., 49.083 –113.876, aspen forest, Malaise trap, 17-24.VII.2012, Cyndi Smith (CBG) 1 specimen; Banff National Park, Baker Creek picnic area, 1500m el., 51.35, –116.063, subalpine conifer forest, pitfall, 23-28.VII.2012, BIOBus 2012 team (CBG) 1 specimen; Jasper National Park, Pocahontas Campground, Site C21, 1131m el., 53.195, –117.914, subalpine conifer forest, pitfall, 16-21.VII.2012, BIOBus 2012 team (CBG)

6 specimens, Jasper National Park, Highway 16, Athabasca River, NE of Jasper Lake, 958m el., 53.193, −117.954, wetland, pitfall, 17-21. VII.2012, BIOBus 2012 team (CBG) 1 specimen, Jasper National Park, Miette Hotsprings, 1439m el., 53.124, −117.775, subalpine riverside forest, pitfall, 17-21.VII.2012, BIOBus 2012 team (CBG) 1 specimen.

***Gatosensis* species group** (Klimaszewski et al. 2006, 2020)

Diagnosis The *gatosensis* species group may be characterized by the following combination of characters: body medium-sized, length 2.8–3.8 mm, broadly oval; antennomere IV strongly elongate, V–X quadrate to slightly transverse; pronotum slightly transverse and about one fourth narrower than elytra; elytra large and at suture as long as or longer than pronotum; abdomen broadly curved laterally in dorsal view; median lobe of aedeagus consisting of narrowly subparallel tubus with swollen sides at the middle in dorsal view, ventral margin of tubus broadly arcuate to nearly straight; flagellum shorter than tubus; internal sac with complex structures; bulbus broadly oval, moderate in size, with more or less protruding crista apicalis; apical lobe of paramere strongly elongate; spermatheca with capsule spherical and bearing large invagination, stem S- or U-shaped and twisted posteriorly.

45. *Oxypoda gatosensis* Bernhauer (Fig. 11.115a–h)

Oxypoda gatosensis Bernhauer 1905, Klimaszewski et al. 2006, 2020.

Diagnosis Body length 2.8–3.8 mm, medium-sized, subparallel, approximately uniformly brown or rust-brown, with antennae, legs, most of elytra and posterior margins of tergites light brown (Fig. 11.115a); pubescence of forebody dense; integument moderately glossy, with isodiametric microsculpture; head broadest behind eyes, about two thirds as broad as pronotum; eyes slightly shorter than the postocular region; antennomere IV elongate, V–X quadrate to slightly transverse; pronotum (Pronotal Index = 1.2), strongly convex, slightly transverse, broadest below middle, about one-fifth broader than long, pubescence directed anteriad apically along midline and obliquely posteriad from midline of disc elsewhere; elytra (Elytral Index = 1.6), distinctly broader than pronotum, at suture longer than pronotum, pubescence directed straight posteriad; abdomen broadly arcuate laterally. MALE. Tergite VIII approximately as long as broad, rounded apically, antecostal suture approximately straight (Fig. 11.115d); sternite VIII with apical margin obtusely triangular, antecostal suture weakly sinuate and slightly inflected medially (Fig. 11.115e); median lobe of aedeagus swollen medially in dorsal view (Fig. 11.115c), and broadly arched with apex sharply pointed in lateral view (Fig. 11.115b). FEMALE. Tergite VIII elongate, apical margin rounded, truncate medially, antecostal suture approximately straight (Fig. 11.115f); sternite VIII broadly arcuate apically, antecostal suture slightly sinuate (Fig. 11.115g); spermatheca approximately S-shaped with moderately large spherical capsule, deep, broad invagination, posterior portion of stem twisted (Fig. 11.115h).

Oxypoda gatosensis is externally very similar to *O. longicarinata* from which it may be separated by the less strongly converging and less transverse pronotum, and the shape of genitalia.

DISTRIBUTION Origin: Nearctic. **CANADA**: BC. **USA**: CA.

COLLECTION AND HABITAT DATA

Habitat: in moss and alder litter at cold stream. **Collecting period**: VII. **Collecting method**: sifting litter and moss.

***Inimica* species group** (Klimaszewski et al. 2006, 2018a)

Diagnosis Body subparallel to narrowly oval, length 2.2–2.5 mm; antennae moderately slender, antennomeres I–III distinctly elongate, IV subquadrate to distinctly elongate, V–X quadrate or moderately transverse; pronotum moderately to distinctly transverse; elytra at suture about as long as pronotum or only slightly shorter; median lobe of aedeagus in dorsal view with tubus narrowly subparallel, oval, bulbus moderately large to large, in lateral view ventral margin of tubus arcuate, with apex more or less pointed, bulbus with large crista apicalis associated with small to large lateral wing-like projections, flagellum shorter than tubus, internal sac with complex structures, apical lobe of paramere moderately elongate; spermatheca with capsule spherical, with apical invagination medium-sized, stem S-shaped.

46. ***Oxypoda inimica*** **Casey** (Fig. 11.116a–h)

Oxypoda inimica Casey 1911. Klimaszewski et al. 2006, 2011, 2018a.

Diagnosis Body small, length 2.5–2.7 mm, roughly oval, broadest at middle of elytra, with abdomen subparallel; colour reddish-brown with head, elytra and in some specimens posterior section of tergites VI–VII of abdomen dark brown (Fig. 11.116a); forebody with pubescence and punctation dense; integument with isodiametric microsculpture; head broadest at basal third, distinctly more than half as wide as pronotum; eyes large, about as long as genae in dorsal view; antennae slender, antennomeres I–III distinctly elongate, IV elongate, V–X subquadrate to slightly elongate; pronotum moderately convex, distinctly transverse and about one fifth broader than long, broadest behind middle, pubescence on disc directed anteriad along midline apically and obliquely posteriad from midline of disc; elytra slightly broader at middle, at suture about as long as pronotum, pubescence on disc directed straight posteriad; abdomen subparallel and slightly tapering apically. MALE. Tergite VIII with apical margin slightly rounded on each side of very shallow median emargination, antecostal suture almost straight (Fig. 11.116d); sternite VIII parabolically rounded apically, antecostal suture straight (Fig. 11.116e); median lobe of aedeagus in dorsal view with tubus broadest medially, with two small angular lateral swellings (Fig. 11.116c), crista apicalis large with two narrowly elongate lateral projections (Fig. 11.116b); internal sac with distinct structures (Fig. 11.116b, c). FEMALE. Tergite VIII very similar to that of male (Fig. 11.116f); sternite VIII broadly, parabolically rounded apically, antecostal suture slightly sinuate, particularly medially (Fig. 11.116g); spermatheca with capsule spherical, apical invagination medium-sized, stem long, S-shaped (Fig. 11.116h).

DISTRIBUTION Origin: Nearctic. **CANADA**: NB, NF, NT, ON, QC. **USA**: MA.

COLLECTION AND HABITAT DATA Habitat: probably a wetland species. In NB, six individuals sifted from sphagnum hummocks and litter on the margin of tamarack bog. Others treaded from reeds in a marsh, two from drift material on a sandy lake margin near a seasonally flooded marsh, and one at a u.v. light in a mixed forest. **Collecting period**: IV, VI, VII, IX. **Collecting method**: sifting sphagnum and litter, treading marsh vegetation, u.v. light.

47. ***Oxypoda yukonensis*** **Klimaszewski and Godin** (Fig. 11.117a–h)

Oxypoda yukonensis Klimaszewski and Godin, in Klimaszewski et al. 2008.

Diagnosis Body narrowly elongate, length 2.8–3.0 mm, pronotum streamlined with elytra (Fig. 11.117a), dark or light brown and with pronotum, elytra, and legs paler; dorsal integument with dense and fine microsculpture; head approximately round, postocular region approximately equal to the diameter of eye; antenna slender, antennomere IV subquadrate, V–X slightly transverse and increasingly more toward the apex of antenna; pronotum slightly transverse, broadest in basal third, streamlined with elytra; elytra transverse, at suture shorter than pronotum; abdomen gradually tapering apically. MALE. Tergite VIII rounded apically (Fig. 11.117d); sternite VIII parabolic apically, antecostal suture slightly sinuate (Fig. 11.117e); median lobe of aedeagus with small bulbus, subparallel tubus and produced apex in dorsal view (Fig. 11.117c), in lateral view tubus broadly arched with slightly swollen subapical part (Fig. 11.117b); internal sac structures as illustrated (Fig. 11.117b,c). FEMALE. Tergite VIII arcuate apically (Fig. 11.117f); sternite VIII rounded apically, antecostal suture sinuate (Fig. 11.117g); spermatheca S-shaped, capsule approximately spherical with moderately broad invagination, stem S-shaped (Fig. 11.117h).

DISTRIBUTION **Origin**: Nearctic. **CANADA**: YT. **USA**: AK [**NSR**].

COLLECTION AND HABITAT DATA
Habitat: in YT, specimens were collected from different habitat types, in an old growth white spruce forest, with feathermoss and spruce needles as forest floor litter; and in a disturbed site composed mostly of gravel with sparse grasses along an access road in an urban area; and some on east-facing lodgepole pine forest slope adjacent to an urban area. In AK, captured in spruce, cottonwood, willow habitat. **Collecting period**: VII, VIII. **Collecting method**: pitfall traps.

We report this species from AK (**NSR**) for the first time here, based on the following: USA: Alaska, Cantwell, 660m el., 63.39058, −148.89932, ±533m spruce, cottonwood, willow, 28.VI.2011, S. Ridling, M. G. Wyatt (UAM) 1 specimen; Fairbanks, Creamer's Field, 154m el., 64.86607, −147.73817, ±8m pitfall, 19–21. VII.2008, D.S. Sikes, R. Spiers (UAM) 1 specimen.

48. ***Oxypoda volkeri*** **Klimaszewski** (Fig. 11.118a–h)

Oxypoda volkeri Klimaszewski, in Klimaszewski et al. 2006.

Diagnosis Body uniformly black, length 2.7–3.0 mm, compact, narrowly oval, broadest at middle part of elytra, abdomen arcuate laterally and slightly tapering (Fig. 11.118a); pubescence and punctation of forebody dense; integument with isodiametric microsculpture; head broadest at basal third, about two thirds as wide as pronotum; eyes large, at least as long as postocular area of head; antennae slender, antennomeres IV–X elongate; pronotum, moderately convex, strongly transverse and about one fourth broader than long, broadest in basal third, pubescence directed anteriad apically along midline and obliquely posteriad from midline of disc elsewhere; elytra, slightly broader medially than pronotum and at suture longer than pronotum, pubescence directed approximately straight posteriad (Fig. 11.118a); PEI = 0.9; abdomen arcuate laterally and slightly tapering apically. MALE. Tergite VIII broadly arcuate apically, antecostal suture arcuate (Fig. 11.118d); sternite VIII slightly produced apically, antecostal suture slightly sinuate (Fig. 11.118e); median lobe of aedeagus with tubus broadest medially and with two lateral tooth-like projections in dorsal view (Fig. 11.118c), subapical part of tubus swollen medially in lateral view (Fig. 11.118b); crista apicalis moderately large with two narrowly elongate lateral projections; internal sac with distinct structures (Fig. 11.118b,c). FEMALE. Tergite VIII arcuate apically, antecostal suture approximately sinuate (Fig. 11.118f); sternite VIII slightly produced apically, antecostal suture sinuate (Fig. 11.118g); spermatheca with capsule spherical and bearing medium-sized invagination, stem long and strongly sinate, S-shaped (Fig. 11.118h).

DISTRIBUTION **Origin**: Nearctic. **CANADA**: NT. **USA**: not recorded.

COLLECTION AND HABITAT DATA
Habitat: specimens were collected from moss and horsetail (*Equisetum* sp.). **Collecting period**: VII. **Collecting method**: treading moss and horsetail (*Equisetum* sp.).

***Irrasa* species group** (Klimaszewski et al. 2006)

Diagnosis Body narrowly oval, length 3.2–3.5 mm; antennae slender, antennomere IV elongate, V–VI quadrate or slightly transverse, VII–X moderately transverse; pronotum strongly transverse; elytra at suture slightly longer than pronotum; median lobe of aedeagus approximately narrowly oval, consisting of tubus slightly converging anteriad and streamlined with bulbus in dorsal view; ventral margin of tubus angularly connected with bulbus, straight or slightly sinuate medially and with apex narrow and produced ventrad in lateral view; flagellum shorter than tubus; internal sac with two subparallel, elongate subapical structures; apical lobe of paramere short and robust; spermatheca with capsule spherical and bearing small invagination, stem long and sinuate.

49. *Oxypoda irrasa* Mäklin (Fig. 11.119a–h)

Oxypoda irrasa Mäklin, 1853. Lohse and Smetana 1985 [types], Klimaszewski et al. 2006, 2008, 2016, 2020.

Diagnosis Body length 3.2–3.5 mm, narrowly oval, broadest at basal third of elytra, abdomen subparallel (Fig. 11.119a); uniformly dark brown or black, with basal two antennomeres, legs, elytra and apical part of abdomen reddish brown; pubescence and punctation of forebody dense; integument with isodiametric microsculpture; head broadest at base, distinctly broader than half of pronotal width; eyes moderately large, about as long as postocular area; antennae slender, antennomere IV moderately to strongly elongate, V–VI quadrate, and VII–X moderately transverse; pronotum moderately convex, strongly transverse and about one third broader than long, broadest in basal third, pubescence directed anteriad apically along midline and obliquely posteriad from midline of disc elsewhere; elytra slightly broader than pronotum and at suture slightly longer than pronotum, pubescence directed approximately straight posteriad; abdomen subparallel and slightly tapering apically. MALE. Tergite VIII arcuate apically, antecostal suture approximately straight (Fig. 11.119d); sternite VIII produced apically, antecostal suture broadly arcuate (Fig. 11.119e); median lobe of aedeagus approximately narrowly oval, consisting of anteriorly converging tubus, which is streamlined with bulbus in dorsal view (Fig. 11.119c); ventral margin of tubus angularly connected with bulbus, slightly sinuate and with apex narrow and produced ventrad in lateral view (Fig. 11.119b); flagellum shorter than tubus; internal sac with two elongate subapical structures (Fig. 11.119b,c); bulbus with large crista apicalis (Fig. 11.119b). FEMALE. Tergite VIII similar to that of male (Fig. 11.119f); sternite VIII with apical margin truncated, antecostal suture sinuate (Fig. 11.119g); spermatheca with capsule spherical and bearing small invagination, stem long and sinuate (Fig. 11.119h).

DISTRIBUTION Origin: Nearctic. **CANADA**: AB, BC, NT [**BOLD**], ON [**BOLD**], SK, YT. **USA**: AK, OR.

COLLECTION AND HABITAT DATA

Habitat: in AB, adults occurred in all forest cover types and all retention treatments, but were most abundant in unharvested stands. In BC, one specimen was captured from mushrooms, one was found in alder soil, and a few from mature and cut forests. In ON, specimens were found in temperate mixed forest, and in YT from mixed forest litter (white spruce, willow and logepole pine), white spruce /feather moss litter, and in salt flat (*Salicornia* habitat). **Collecting period**: V–VIII, most abundant in VIII. **Collecting method**: hand collectiong, pitfall traps, sifting forest litter and processing it through Berlese funnels, sifting mushrooms, flight intercept traps, Malaise trap.

We report this species from NT (**NTR**) and ON (**NPR**) for the first time here, based on **BOLD** records: Canada, Northwest Territories, Nahanni National Park Reserve, Nailicho, Virginia Falls, 578m el., 61.606, –125.758, 01-08. VIII.2014, Malaise trap, Parks Canada staff (CBG) 1 specimen; ON, Wellington County, Puslinch Township, Concession 11/Hume Rd, 320m el., 43.537, - 80.134, 02-08.VIII.2010, temperate mixed forest, Malaise trap, Paul Hebert (CBG) 1 specimen.

Additional records. See Annex 1.

***Lacustris* species group** (Klimaszewski et al. 2006, 2018a)

Diagnosis Body narrowly oval, length 1.6–2.5 mm; antennomere IV subquadrate to moderately elongate; pronotum transverse; elytra at suture as long as pronotum or distinctly shorter; median lobe of aedeagus with tubus narrowly subparallel in dorsal view, in lateral view ventral margin of tubus arcuate at base and nearly straight anteriorly, apex bluntly truncate (except for *O. nimbata*), bulbus moderately large, flagellum shorter than tubus, internal sac with semicircular apical structure consisting of several interconnected elements in lateral view and falciform or evenly narrow subapical sclerite, apical lobe of paramere slender and extremely elongate; spermatheca capsule roughly spherical, with apical invagination moderate, constricted at apex, stem moderately long, sinuate and spirally looped posteriorly.

50. *Oxypoda hiemalis* Casey (Fig. 11.120a–h)

Oxypoda hiemalis Casey, 1911. Klimaszewski et al. 2006, 2008, 2011, 2015, 2018a, b.

Diagnosis Body narrowly fusiform, broadest at middle of elytra with abdomen slightly tapering apically (Fig. 11.120a), body length 1.8–2.5 mm; colour variable, usually head dark brown and remainder of body light brown to reddish light brown or yellowish light brown, in some specimens head and abdomen dark brown and pronotum and elytra light brown or entire body uniformly light brown; forebody with pubescence and punctation dense; integument with isodiametric microsculpture; head broadest at basal third, two thirds as broad as pronotum; eyes small, about half as long as genae in dorsal view; antennomere IV subquadrate to slightly elongate, V slightly transverse, and VI–X distinctly so; pronotum moderately convex, slightly transverse, about one fifth broader than long, broadest in basal third, pubescence on disc directed anteriad along midline apically, and obliquely posteriad from midline elsewhere; elytra as broad as pronotum, at suture distinctly shorter than pronotum, pubescence on disc directed straight posteriad; abdomen slightly tapering posteriad. MALE. Tergite VIII elongate, broadly rounded apically, antecostal suture almost straight (Fig. 11.120d); sternite VIII elongate and broadly, parabolically rounded apically, antecostal suture sinuate (Fig. 11.120e); median lobe of aedeagus with tubus narrowly subparallel in dorsal view (Fig. 11.120c), in lateral view ventral margin curved ventrad basally, straight to near apex, apex bluntly truncate, slightly bent ventrad (Fig. 11.120b); internal sac with subapical sclerite elongate and evenly narrow (Fig. 11.120b,c), apical structure semicircular (Fig. 11.120b); flagellum highly sinuate (Fig. 11.120b,c). FEMALE. Tergite and sternite VIII similar to those of male (Fig. 11.120f,g); spermatheca very similar to that of *O. lacustris*, with capsule spherical, apical invagination moderate, constricted at apex, stem long, sinuate and broadly looped posteriorly (Fig. 11.120h).

This species is externally similar to *O. pseudolacustris*, but in *O. hiemalis* the size is smaller (length 1.8–2.5 mm), antennomeres VII–X are more transverse, the pronotum is more elongate, narrower and even with the elytra in dorsal view, the eyes are small, the tubus of the median lobe of the aedeagus is narrower and more elongate, the subapical sclerite of the internal sac is very thin and long (Fig. 11.120b), and the spermathecal stem is more broadly looped (Fig. 11.120h).

DISTRIBUTION Origin: Nearctic. **CANADA**: AB, LB, MB, NB, NF, NS, NT, ON, QC, YT. **USA**: AK, NH.

COLLECTION AND HABITAT DATA

Habitat: hardwood (silver maple, sugar maple, red oak, birch, alder) and mixed forests with some in fir, pine forests, mature spruce and eastern white cedar forest, and a tamarack bog margin. Adults in leaf litter, moss, litter and *Sphagnum* near small body of water, in moist leaves on vernal pond margins, in litter at base of tree in *Carex* marsh, in lining of deserted beaver lodge, in moist grass litter on abandoned beaver pond margin, *Microtus* nests, marmot (*Marmota* sp.) burrows, *Sphagnum* moss, litter on barren, litter around raspberry, in frass in tree holes of birch and maple, and in leaf litter on river margin. One individual in lichen on tree trunk. **Collecting period**: III–X. **Collecting method**: sifting various kinds of litter, pitfall traps, Luminoc® pitfall-light traps, Malaise traps.

Additional records. See Annex 1.

51. *Oxypoda lacustris* Casey (Fig. 11.121a–i)

Oxypoda lacustris Casey 1906. Webster et al. 2009, Klimaszewski et al. 2005, 2006, 2011, 2015, 2018a, 2020, Work et al. 2013.

Diagnosis Body broad, subparallel, broadest at middle of elytra (Fig. 11.121a), length 2.4–3.0 mm; colour dark brown with antennae, tip of abdomen, legs and rarely elytra lighter, in some specimens outline of body with lighter reddish tinge or black (most northern specimens); forebody with pubescence and punctation dense; integument with isodiametric microsculpture; head broadest at basal third, slightly more than half as wide as pronotum; eyes moderately large; antennae moderately slender, antennomere IV slightly elongate, V–X transverse; pronotum slightly flattened medially, distinctly transverse and about one third broader than long, broadest at about middle, pubescence on disc directed anteriad along midline apically and obliquely posteriad from midline elsewhere; elytra slightly broader than pronotum, at suture at least as long as pronotum, pubescence on disc directed straight posteriad; abdomen gradually tapering apically. MALE. Tergite VIII elongate, broadly rounded apically, antecostal suture virtually straight (Fig. 11.121d); sternite VIII elongate, parabolically rounded apically, antecostal suture sinuate (Fig. 11.121e); median lobe of aedeagus with tubus in dorsal view gradually narrowing from bulbus, subparallel toward apex (Fig. 11.121c), in lateral view ventral margin curved obliquely ventrad basally, straight to bluntly truncate apex (Fig. 11.121b); internal sac with falciform subapical sclerite angular ventrally at base, with elongate posterior projection (Fig. 11.121b); flagellum moderately long (Fig. 11.121b); apical structure semicircular, consisting of several articulately connected elements (Fig. 11.121b,c). FEMALE. Tergite VIII similar to that of male but truncate apically (Fig. 11.121f); sternite VIII similar to that of male but antecostal suture almost straight (Fig. 11.121g); spermatheca with capsule spherical, apical invagination moderate, constricted at apex, stem long, sinuate and spirally looped posteriorly (Fig. 11.121h,i).

Oxypoda lacustris differs from *O. hiemalis* and *O. pseudolacustris* in having the elytra at least as long as the pronotum (shorter than the pronotum in the two latter species).

DISTRIBUTION Origin: Nearctic. **CANADA**: AB, BC, LB, MB, NB, NF, NS [**BOLD**], NT, ON, QC, SK, YT. **USA**: AK, WA.

COLLECTION AND HABITAT DATA

Habitat: in predominantly wet habitats; in moist leaves on vernal pond margins in mixed forests and seasonally flooded silver maple forests and swamps, in leaf and grass litter at base of red maple in a *Carex* marsh, in moist leaf and grass litter in *Carex*, cattail marshes,in drift material on a lake margin and salt flat. Also in deciduous forests (oak, birch, aspen), fir and conifer dominated forests, old silver maple swamp, coastal dunes and coastal barrens. Specimens were also found in gopher burrows and muskrat nests. In AK, specimens were captured in forest edge with *Populus tremuloides and P. balsamea*; black spruce, grass, aspen; *Betula papyrifera and P. tremuloides*; mesic *Betula*; wet boreal forest; mixed black and white spruce forest with open/boggy areas and dry zones; open black spruce; open tussock and standing water; willow and bog birch rich meadow with *Geranium erianthum, Artemisia arctica, Sanguisorba canadensis*; spruce bog; creekside/ocean beach confluence, under boards and drift wood; under rocks, dung, wood in meadows.**Collecting period**: III–X. **Collecting method**: Malaise traps, sifting forest litter, pitfall traps, yellow pan traps, Lindgren funnel traps, and sifting debris from burrows and nests of rodents.

Majka and Klimszewski (2010) recorded the species from NS based on specimens from Queens County, but they did not provide collection data, we confirm its presence in NS based on BOLD records: Canada: Nova Scotia, Cape Breton Highlands National Park, Benjies Lake Trail, 425m el., 46.746, –60.817, black spruce and balsam fir forest, forest litter, Berlese funnel extraction, 29.VII.2013, BIObus 2013 team (CBG) 1 specimen.

Additional records. See Annex 1.

***Lucidula* species group** (Klimaszewski et al. 2006, 2018a)

Diagnosis Body subparallel, length 2.1–2.7 mm; antennae moderately to very robust, antennomere IV subquadrate to elongate or transverse, VI–X usually transverse, but rarely slightly (*O. lucidula*) to distinctly (*O. sunpokeana*) elongate; pronotum moderately transverse; elytra at suture at least as long as pronotum; median lobe of aedeagus with tubus usually evenly coalescent with elongate bulbus in dorsal view, ventral margin of tubus arcuate basally, straight medially and slightly sinuate apically with apex bearing two narrow, symmetrical, butterfly-like, lateral projections, flagellum shorter than tubus, internal sac with complex structures, apical lobe of paramere short and robust; spermatheca (known only in three species) with capsule slightly elongate, with small invagination, stem moderately long, sinuate and narrowly looped posteriorly.

52. *Oxypoda demissa* Casey (Fig. 11.122a–i)

Oxypoda demissa Casey, 1911. Klimaszewski et al. 2006, 2008, 2011, 2016, 2018a, Webster et al. 2009.

Diagnosis Body narrowly subparallel, broadest at middle of elytra, with abdomen subparallel for most of its length then gradually narrowed apically (Fig. 11.122a), body length 2.8–3.2 mm; slightly bicoloured with head and most of abdomen (except apical part) dark brown, pronotum and elytra reddish, rest of body light brown except at least antennomeres I–III reddish brown; forebody with pubescence and punctation dense; integument with isodiametric microsculpture; head broadest at basal third, as broad as two thirds width of pronotum; eyes large, about as long as genae in dorsal view; antennae thick, antennomeres I–III elongate, IV subquadrate or slightly transverse, V–X slightly transverse; pronotum moderately convex, moderately transverse, about one fourth broader than long, broadest in basal third, pubescence on disc directed anteriad along midline apically, obliquely posteriad from midline elsewhere; elytra slightly broader than pronotum, at suture about as long as pronotum, pubescence on disc directed approximately straight posteriad (Fig. 11.122a). MALE. Tergite VIII broadly rounded apically, antecostal suture virtually straight (Fig. 11.122d); sternite VIII elongate, parabolically rounded apically, antecostal suture virtually straight (Fig. 11.122e); median lobe of aedeagus in dorsal view with bulbus narrowly oval, tubus broadly arcuate laterally lacking angular lateral projections at middle (Fig. 11.122c), in lateral view ventral margin of tubus bent obliquely ventrad basally, practically straight medially and slightly sinuate apically (Fig. 11.122b); internal sac with complex paired structures: apical structure narrow and inconspicuous, subapical sclerite narrowly elongate, slightly diffused medially and arcuate basally and dorsally in lateral view (Fig. 11.122b); flagellum short. FEMALE. Tergite VIII with apex broadly arcuate, antecostal suture almost straight (Fig. 11.122f); sternite VIII elongate and broadly parabolically rounded apically, antecostal suture almost straight (Fig. 11.122g); spermatheca with capsule roughly spherical, slightly elongate, with small invagination; stem moderately long, sinuate, narrowly spirally looped posteriorly (Fig. 11.122h,i).

DISTRIBUTION Origin: Nearctic. **CANADA**: AB [**BOLD**], LB, MB [**BOLD**], NB, NF, NS, ON, QC, SK, YT. **USA**: AK [**NSR**].

COLLECTION AND HABITAT DATA

Habitat: Associated with various wet habitats. Adults in moist leaf litter on vernal pond margins in mixed forests, among leaves and sedges on pond margin, in moist grass litter and sphagnum in *Carex* marsh, among sedges along margin of small spring-fed brook in a mature hardwood forest, among leaf and grass on hummocks in a wet alder (*Alnus* sp.) swamp, among shrubs and forbs on a coastal limestone barrens, and from litter of *Alnus* clumps. In AB, found in emergent white birch/willow/trembling aspen forest. In AK, collected in rocky alpine habitat, and deciduous forest, birch, rose and willow. **Collecting period**: IV–X. **Collecting method**: Malaise trap, sifting forest litter, ground search, pitfall traps, Luminoc® pitfall-light traps, Malaise trap, and flying adults collected with butterfly net during evening.

We report this species from AB and MB (**NPR**) for the first time here, based on **BOLD** records, and from AK based on CNC, and UAM specimens: Canada, Alberta, Elk Island National Park, Astotin Lake, 719m el., 53.685, –112.86, emergent white birch, willow and trembling aspen, Malaise trap, 20-27.VII.2012, Stephanie Church (CBG) 1 specimen; Manitoba, 16 km E Churchill, Krummholz forest at Launch Rd., 58.754, –93.913, pitfall trap, 16.VII.2009, P. Kevan/T. Woodcock (CBG), 1 specimen. USA, Alaska, Tangle Lakes, VI-VII, 18, 1969, J. Matthews (CNC) 1 female; Denali NPP., Igloo Canyon, 1123m el., 63.59198, –149.6042, ±30m rocky alpine, 10 minute ground search, 17. VIII.2016, D.S. Sikes (UAM) 1 specimen; Fairbanks, Kiak Radio 2, 156m el., 64.87154, –147.68457, ±10m deciduous forest, birch, rose and willow, hanging Malaise, 6.15m, 10-24. VI.2015, J.C. Hagelin (UAM) 1 specimen.

Additional records. See Annex 1.

53. *Oxypoda leechi* Lohse (Fig. 11.123a–i)

Oxypoda (Demosoma) leechi Lohse, in Lohse et al. 1990. Klimaszewski et al. 2006.

Diagnosis Body length 2.5–2.7 mm, narrowly subparallel, abdomen subparallel for most of its length and then gradually narrowed apically (Fig. 11.123a,b); uniformly dark brown with appendages red-brown (holotype male), or dark brown with pronotum, elytra, apical part of abdomen and appendages red-brown (female non-paratype); pubescence and punctation of forebody dense; integument with isodiametric microsculpture; head broadest at basal third, as broad as two thirds of pronotal width; eyes slightly shorter than postocular area; antennae with antennomeres VI–X subquadrate to slightly transverse; pronotum moderately convex, moderately transverse and 1.4 times broader than long, broadest in basal third, pubescence directed anteriad apically along midline and obliquely posteriad from midline of disc elsewhere; elytra about as long and as broad as pronotum, pubescence directed approximately straight posteriad; abdomen subparallel except for the apical part. MALE. Tergite VIII transverse and truncate apically, antecostal suture arcuate (Fig. 11.123e); sternite VIII slightly truncate apically, antecostal suture gently sinuate (Fig. 11.123f); median lobe of aedeagus with tubus narrowly oval with complex structures in dorsal view (Fig. 11.123d), ventral margin of tubus strongly bent ventrad in lateral view (Fig. 11.123c); internal sac with complex structures as illustrated (Fig. 11.123d,c). FEMALE. Tergite VIII transverse with apex truncate medially, antecostal suture slightly sinuate (Fig. 11.123g); sternite VIII broadly rounded apically, antecostal suture arcuate (Fig. 11.123h); spermatheca with approximately spherical capsule bearing small invagination, neck elbowed; stem sinuate (damaged posteriorly) (Fig. 11.123i).

DISTRIBUTION Origin: Nearctic. **CANADA**: YT. **USA**: not recorded.

COLLECTION AND HABITAT DATA

Habitat: in moss, leaf litter along stream, *Alnus* and *Sphagnum* litter near pond, and in sedges around edge of pond. **Collecting period**: IV–VI. **Collecting method**: sifting moss, leaf litter along stream, *Alnus* and *Sphagnum* litter near pond, and treading sedges around edge of pond.

COMMENTS The holotype male was captured in Yukon, Firth River, 3.VII.1956, R.L. Leech

(CNC). A female paratype was captured in Yukon, Komakuk Bch., 4.VII.1984, V. Behan (CNC). The male holotype and female paratype are dark brown with reddish appendages but the non-paratype female, is differently colored, dark brown with pronotum, elytra and apex of abdomen yellow-red brown. This female may or may not belong to this species.

54. *Oxypoda lucidula* Casey (Fig. 11.124a–h)
Oxypoda lucidula Casey 1906. Klimaszewski et al. 2006, 2011, 2018a, Webster et al. 2009.

Diagnosis Body narrowly subparallel, broadest at middle of elytra (Fig. 11.124a), length 2.5–3.5 mm, abdomen subparallel for most of its length, then gradually narrowed apically; colour uniformly dark brown to black, with legs or only tarsi, basal 2–3 antennomeres, and occasionally elytra, reddish brown; forebody with pubescence and punctation dense; integument with isodiametric microsculpture; head broadest at basal third, two thirds as broad as pronotum; eyes large, about as long as genae in dorsal view; all antennomeres slightly elongate; pronotum moderately convex, moderately transverse and about one third broader than long, broadest in basal third, pubescence on disc directed anteriad along midline apically, and obliquely posteriad from midline elsewhere; elytra broader and longer than pronotum, pubescence on disc directed approximately straight posteriad; abdomen subparallel except in apical part. MALE. Tergite VIII arcuate apically, antecostal suture straight (Fig. 11.124d); sternite VIII broadly parabolically rounded apically, antecostal suture sinuate (Fig. 11.124e); median lobe of aedeagus with bulbus broadly oval, tubus with two distinct angular projections laterally at middle in dorsal view (Fig. 11.124c), in lateral view ventral margin of tubus bent obliquely laterad basally, virtually straight medially and slightly sinuate apically (Fig. 11.124b); internal sac with complex paired structures: apical structure broad and strongly sclerotized in lateral view (Fig. 11.124b); subapical sclerite narrowly elongate, diffused medially and angular basally and dorsally in lateral view (Fig. 11.124b); flagellum short. FEMALE. Tergite VIII with apical margin broadly truncate medially, broadly arcuate laterally, antecostal suture almost straight (Fig. 11.124f); sternite VIII slightly elongate, with apical margin unevenly rounded, antecostal suture distinctly sinuate, obtusely angulate medially (Fig. 11.124g); spermatheca with capsule roughly spherical, slightly elongate, with small invagination, stem long, sinuate and narrowly looped posteriorly, with section proximal to capsule variable in length and degree of angular bend (Fig. 11.124h).

DISTRIBUTION Origin: Nearctic. **CANADA**: AB, LB, MB, NB, NF, NT, NU [**BOLD**], ON, QC, YT. **USA**: AK, IA, MO, NH, NY.

COLLECTION AND HABITAT DATA
Habitat: most adults were captured in litter near various wetlands; sifted from moss and leaf litter along a stream, from *Alnus* and *Sphagnum* litter near a pond, treading sedges around edge of pond, in gilled mushrooms in spruce and fir forest, in drift material on upper river margins, and in vegetation on coastal sand and dunes (NF). Some specimens captured near sphagnum bogs, among shrubs and forbs on coastal limestone barrens, and a few were captured flying at dusk. In AK, found in *Salix* leaf litter. **Collecting period**: IV–X. **Collecting method**: sifting forest litter, Berlese funnels, pitfall traps, and flight intercept traps.

We report this species from NU [**NTR**] for the first time here, based on BOLD records: Canada, Nunavut, Kitikmeot, Kugluktuk, Kugluk (Bloody Falls) Territorial Park, 20m el., 67.7431, -115.372, *Salix* leaf litter, sifting, 20.VII.2019, G. Blagoev, M. Pentinsaari (CBG) 1 specimen.

Additional records. See Annex 1.

COMMENTS This species was listed from ME by Newton (2019) but we are not aware of any published voucher data.

55. *Oxypoda robusticornis* Bernhauer (Fig. 11.125a–g)

Oxypoda robusticornis Bernhauer, 1907. Lohse et al. 1990 [lectotype], Klimaszewski et al. 2006.

Oxypoda madgei Lohse, in Lohse et al. 1990. **New synonymy**.

Diagnosis Body length 2.7–2.8 mm, narrowly subparallel, abdomen subparallel for most of its length and then gradually narrowed apically (Fig. 11.125a); dark brown with paler and slightly reddish pronotum, patches of elytra, posterior abdomen and appendages; pubescence and punctation of forebody dense; integument with isodiametric microsculpture; head broadest at basal third, as broad as three fourths of pronotal width; eyes about as long as postocular area; antennae moderately thick, antennomere IV subquadrate, and V–X slightly transverse; pronotum moderately convex, moderately transverse and about 1.2 times as broad as long, broadest in basal third, pubescence directed anteriad apically along midline and obliquely posteriad from midline of disc elsewhere; elytra about as broad as pronotum and at suture about as long as pronotum, pubescence directed approximately straight posteriad; abdomen subparallel except for the apical part. MALE. Tergite VIII slightly transverse and arcuate apically, antecostal suture arcuate (Fig. 11.125d); sternite VIII slightly parabolic apically, antecostal suture slightly sinuate (Fig. 11.125e); median lobe of aedeagus with bulbus narrowly oval and tubus broad and subparallel with complex structures of internal sac in dorsal view (Fig. 11.125c), in lateral view ventral margin of tubus arcuate baso-medially and slightly sinuate apically and strongly projecting ventrad (Fig. 11.125b); internal sac with complex structures as illustrated (Fig. 11.125b,c); flagellum short. FEMALE. Tergite and sternite VIII as illustrated (Fig. 11.125f, g); spermatheca unknown.

DISTRIBUTION Origin: Nearctic. **CANADA**: not recorded. **USA**: AK, NH.

COLLECTION AND HABITAT DATA Habitat: in dwarf *Salix*, *Vaccinium*, leatherleaf litter. **Collecting period**: IV–IX. **Collecting method**: sifting litter.

Additional records. See Annex 1

COMMENTS This species was not recorded from Canada but it is most likely transcontinental in the north. An isolated population of this species is known from Mt. Washington in NH. We could not find any significant morphological differences between types of *O. robusticornis* and *O. madgei* and therefore consider them conspecific. *Oxypoda robusticornis* is similar to *O. leechi* but differs from it by its broader body, less transverse pronotum, thicker antennae with more transverse apical antennomeres. The median lobe of aedeagus is very similar in the two species.

***Nigriceps* species group** (Klimaszewski et al. 2006, 2018a)

Diagnosis Body narrowly subparallel, highly elongate length 2.0–2.6 mm; integument of forebody with punctation slightly asperate; antennomere IV subquadrate to slightly elongate, V–X moderately to distinctly transverse and progressively more so toward the apex; pronotum slightly transverse, with sides gently converging in apical one-third; elytra at suture as long as or slightly shorter than pronotum; abdomen long and subparallel; median lobe of aedeagus in dorsal view with tubus narrowly subparallel, bulbus oval, in lateral view ventral margin of tubus arcuate or slightly sinuate, apex more-or-less pointed, bulbus with large crista apicalis, flagellum shorter than tubus, internal sac with complex structures, apical lobe of paramere moderately elongate; spermatheca with capsule spherical, with invagination moderate to large, stem U- or S-shaped.

56. *Oxypoda brachyptera* (Stephens) † (Fig. 11.126a–h)

Aleochara brachyptera Stephens, 1832. Klimaszewski et al. 2006, 2007b, Majka and Klimaszewski 2008b, 2010, Klimaszewski et al. 2011, 2018a.

Diagnosis Body narrowly subparallel, small, length 2.0–2.2 mm, colour uniformly dark brown or brown with pronotum, elytra, base of abdomen, antennomeres and legs light yellowish-brown (Fig. 11.126a); pubescence on forebody dense, elytra with strong isodiametric microsculpture, punctation coarsely asperate; head broadest behind eyes, two thirds as broad as pronotum; eyes slightly shorter than genae in dorsal view; antennomere IV subquadrate to slightly transverse, V–X moderately transverse and increasingly more so toward apex of antenna; pronotum moderately convex, slightly transverse, broadest in basal third, about one third broader than long, pubescence on disc directed straight or slightly obliquely posteriad; elytra about as broad as pronotum, pubescence on disc directed straight posteriad; abdomen subparallel and tapering apically. MALE. Tergite VIII broadly arcuate apically, antecostal suture almost straight (Fig. 11.126d); sternite VIII with apical margin triangularly produced, apex rounded, antecostal suture sinuate (Fig. 11.126e); median lobe of aedeagus in dorsal view with tubus subparallel, gradually tapering apically (Fig. 11.126c), in lateral view tubus broadly bent ventrad basally, straight to narrowly rounded apex (Fig. 11.126b), internal sac structures complex (Fig. 11.126b,c), bulbus with crista apicalis almost flat (Fig. 11.126b). FEMALE. Tergite VIII broadly arcuate apically, antecostal suture virtually straight (Fig. 11.126f); sternite VIII elongate, apical margin almost truncate medially, arcuate laterally, antecostal suture distinctly sinuate (Fig. 11.126g); spermatheca with capsule small, elongate spherical, invagination small, stem acutely bent behind capsule, U-shaped (Fig. 11.126h).

DISTRIBUTION Origin: Palaearctic, Europe, Asia; adventive in Canada. **CANADA**: NB, NF, NS, NU [**NTR**], ON, QC. **USA**: not recorded.

COLLECTION AND HABITAT DATA
Habitat: in clearcut of yellow birch dominated forest and agricultural fields where collected from forest leaf litter, at base of crucifer plant, and litter in lightly scarified clearcut. **Collecting period**: VI–VIII. **Collecting method**: sifting forest litter, pitfall traps.

We report this species from NU [**NTR**] for the first time here based on the following: Canada, Nunavut, Kugluktuk, 67.78182, –115.27699, Rep. 1 mesic, pitfall trap, 22-26.VI.2011, NBP field party K1MP111 (LFC) 1 female.

COMMENTS The discovery of this adventive species in NU is unexpected and suggests a secondary introduction to the North. There is a possibility that this species could be Holarctic but this is unlikely, as the first North American record is from 1977 (NB) (Klimaszewski et al. 2013) and all other records until now have been restricted to eastern Canada. Additionally, only the West Palaearctic distribution of this species has been confirmed (Assing 2012) due to the presence of a complex of closely related species in the Palaearctic. This species was among the top eight most abundant species in southern Ontario soybean fields (Brunke et al. 2014).

Operta **species group** (Klimaszewski et al. 2006)

Diagnosis Body narrowly oval, length 4.0–4.5 mm; antennae slender, antennomeres I–VI elongate, VII–X slightly transverse; pronotum moderately transverse; elytra at suture distinctly shorter than pronotum; median lobe of aedeagus narrowly oval, in dorsal view tubus with lateral margins distinctly arcuate, bulbus streamlined, in lateral view tubus with ventral margin connected with bulbus at an angle, straight or slightly sinuate, apex narrow, flagellum about as long as tubus, internal sac with complex structures, apical lobe of paramere short and robust; spermatheca with capsule spherical, with invagination moderately large, stem long, broadly looped, coiled posteriorly.

57. *Oxypoda operta* Sjöberg * (Fig. 11.127a–h)

Oxypoda (Oxypoda) operta Sjöberg, 1950. Klimaszewski et al. 2006, 2011, 2012, Brunke et al. 2012, Klimaszewski et al. 2011, 2018a, Work et al. 2013.

Diagnosis Body narrowly oval, broadest at basal thirds of pronotum, with lateral margins of abdomen broadly arcuate, body large, length 4.0–4.5 mm, colour uniformly light reddish-brown (Fig. 11.127a); forebody with pubescence and punctation dense; integument with isodiametric microsculpture; head broadest at basal third, slightly more than half as wide as pronotum; eyes small, about half as long as postocular region in dorsal view; antennae slender, antennomeres I–III highly elongate, IV–VI moderately elongate, VII–X slightly transverse; pronotum moderately convex, moderately transverse, about one fifth broader than long, broadest in basal third, pubescence on disc directed anteriad apically along midline and obliquely posteriad from midline elsewhere; elytra slightly narrower than pronotum in most, at suture shorter than pronotum, pubescence on disc directed approximately straight posteriad; abdomen with lateral margins broadly arcuate. MALE. Tergite VIII arcuate apically, antecostal suture virtually straight (Fig. 11.127d); sternite VIII with apical margin moderately produced, apex broadly rounded, antecostal suture slightly sinuate (Fig. 11.127e); bulbus of median lobe of aedeagus in dorsal view narrowly oval, tubus broadly swollen laterally along middle (Fig. 11.127c), in lateral view tubus with ventral margin sinuate, with apex very narrow, slightly bent ventrad (Fig. 11.127b); flagellum about as long as tubus; internal sac with complex structures (Fig. 11.127b,c). FEMALE. Tergite VIII rounded apically, antecostal suture virtually straight (Fig. 11.127f); sternite VIII with apical margin broadly truncate medially, arcuate laterally, antecostal suture virtually straight (Fig. 11.127g); spermatheca with capsule compressed spherical, invagination moderately large and deep, neck curved, stem long, broadly irregularly looped posteriorly (Fig. 11.127h).

DISTRIBUTION Origin: Holarctic. **CANADA**: AB, LB, NS, ON, QC, YT. **USA**: NH.

COLLECTION AND HABITAT DATA
Habitat: many forest types and associated with low shrubs on the Forteau Barrens of southern LB. Adults in *Sphagnum* under a spruce tree near spring and in litter of dwarf *Betula* sp. **Collecting period**: V–IX. **Collecting method**: pitfall traps, sifting forest litter and deciduous debris and soil.

COMMENTS In the Palaearctic region, *Oxypoda operta* is known only from Finland, Sweden and recently, the Far East of Russia (Schülke and Smetana 2015). We are almost certain that this is a Holarctic but rarely collected species with an incompletely known distribution.

***Orbicollis* species group** (Klimaszewski et al. 2006)

Diagnosis Median lobe of aedeagus in dorsal view with tubus narrow, subparallel, strongly narrowing in apical part, apex rounded, bulbus small, oval, in lateral view tubus with ventral margin broadly arcuate, narrowly elongate and rounded apically, bulbus with crista apicalis enlarged, flagellum about as long as tubus, internal sac with complex structures, apical lobe of paramere moderately elongate; spermatheca with capsule spherical, with medium to large invagination, stem sinuate with large posterior loop.

58. *Oxypoda frigida* Bernhauer (Fig. 11.128a–i)
Oxypoda frigida Bernhauer, 1907. Klimaszewski et al. 2006, 2011, 2018a, 2020, Work et al. 2013.

Diagnosis Body subparallel, small, length 2.0–2.6 mm, colour brown or rust-brown with head dark brown and elytra and abdomen darker than pronotum and appendages (Fig. 11.128a); forebody with pubescence dense; integument moderately glossy with fine isodiametric microsculpture; head broadest behind eyes, less than two thirds as wide as pronotum; eyes small, shorter than genae in dorsal view; antennomere IV subquadrate to slightly transverse, V–X moderately to distinctly transverse; pronotum highly convex, slightly transverse, broadest in basal third, about one fourth broader than long, pubescence on disc directed anteriad along midline apically, and obliquely posteriad from midline elsewhere; elytra about as broad as pronotum, at suture distinctly shorter than pronotum, pubescence on disc directed straight posteriad; abdomen subparallel, slightly tapering apically. MALE. Tergite VIII slightly produced apically, antecostal suture sinuate (Fig. 11.128d); sternite VIII with apical margin obtusely angulate, apex broadly rounded, antecostal suture sinuate (Fig. 11.128e); median lobe of aedeagus in dorsal view subparallel, tubus tapering apically (Fig. 11.128c), in lateral view tubus bent obliquely ventrad at base, almost straight to narrowly rounded apex, bulbus with crista apicalis very large (Fig. 11.128b); internal sac structures complex (Fig. 11.128b,c), flagellum shorter than tubus (Fig. 11.128b). FEMALE. Tergite VIII rounded apically, antecostal suture arcuate (Fig. 11.128f); sternite VIII broadly rounded apically, antecostal suture virtually straight (Fig. 11.128g); spermatheca with capsule roughly spherical, with invagination moderately large, stem with broad, slightly spiral posterior loop (Fig. 11.128h,i).

DISTRIBUTION Origin: Nearctic. **CANADA**: AB, BC, LB, NB, NF, NS, NT, ON, QC, SK [**BOLD**], YT. **USA**: AK, NH.

COLLECTION AND HABITAT DATA

Habitat: in hardwood-dominated stands with white spruce understories, balsam fir, old fir, black spruce, white spruce, maple forests, *Salix* and *Alnus* stands, white spruce and poplar forests, alpine forests, and shrubland habitats, among shrubs in coastal limestone barrens. Adults in moss and litter near a pond, in leaf litter, in *Sphagnum*, and in mushrooms. In SK, found in white spruce and poplar forest and in shrubland. In AK, captured in gravel bars, mature forest, low bluff with sage; forest edge with *Populus tremuloides, P. balsamea*; meadow, *Nephrophyllidium crista-galli, Geranium erianthum, Artemisia arctica*; meadow heath, *Phyllodoce glandulifera, Senecio triangularis, Luetkea pectinata*; forest, mostly spruce, some aspen; shrub, alder, dwarf birch, grass, clubmoss litter; spruce litter; *Populus, Betula, Alnus* leaf litter; mixed *Betula, Picea, Alnus* forest; spruce, birch, alder, cottonwood; alpine *Salix, Valariana* shrubs; alpine meadow litter; old growth and unthinned 2nd growth forests; clear-cuts; subalpine *Tsuga, Luetkea*; and under rocks, dung, wood, meadows w/lupine. **Collecting period**: V–X. **Collecting method**: Luminoc pitfall-light traps, sifting forest litter and processing it through Berlese funnels, Winkler Litter Extraction, yellow pitfalls w/o rainroofs, Malaise trap, sweeping vegetation, and Lindgren funnel traps.

We report this species from SK (**NPR**) for the first time here, based on **BOLD** records: Canada, Saskatchewan, Prince Albert National Park, Narrows Peninsula Trail, 530m el., 53.987, –106.282, white spruce and poplar forest, soil and leaf litter, Berlese funnel extraction, 09.VII.2012, BIOBus 2012 team (CBG) 1 specimen; Grasslands National Park, East Block, 889m el., 49.001, –106.557, shrubland, Malaise trap, 10-20.VI.2014, L. Gardiner (CBG) 1 specimen.

Additional records. See Annex 1.

59. *Oxypoda orbicollis* Casey (Fig. 11.129a–i)

Oxypoda orbicollis Casey, 1911. Klimaszewski et al. 2006, Brunke et al. 2012, Webster et al. 2012, Klimaszewski et al. 2011, 2015, 2018a, 2020.

Diagnosis Body narrowly spindle-shaped, medium-sized, length 3.0–3.8 mm, colour uniformly brown or rust-brown (Fig. 11.129a); forebody with pubescence dense; integument dull in luster with fine isodiametric microsculpture; head broadest behind eyes, about two thirds as broad as pronotum; eyes small, shorter than genae in dorsal view; antennomere IV distinctly elongate, V–X quadrate to slightly transverse; pronotum large, highly convex, slightly transverse, broadest at middle, about one fifth broader than long, pubescence on disc directed anteriad along midline apically, obliquely posteriad from midline elsewhere; elytra slightly broader and at suture shorter than pronotum, pubescence on disc directed straight posteriad; abdomen tapering apically. MALE. Tergite VIII elongate, rounded apically, with minute medial emargination, antecostal suture arcuate (Fig. 11.129d); sternite VIII broadly parabolically rounded apically, antecostal suture slightly sinuate (Fig. 11.129e); median lobe of aedeagus in dorsal view subparallel, strongly narrowing apically (Fig. 11.129c), in lateral view tubus broadly arched ventrad in basal half, straight to narrowly rounded apex (Fig. 11.129b); flagellum with base heart-shaped (Fig. 11.129b). FEMALE: tergite VIII rounded apically, antecostal suture straight (Fig. 11.129f); sternite VIII broadly rounded apically, antecostal suture slightly sinuate (Fig. 11.129g); spermatheca with capsule compressed spherical, with invagination deep and very wide, stem with single broad loop (Fig. 11.129h,i).

DISTRIBUTION Origin: Nearctic. **CANADA**: AB, BC, LB, NB, NS, ON, QC, SK, YT. **USA**: WI.

COLLECTION AND HABITAT DATA

Habitat: hardwood-dominated stands, old-growth balsam fir forests, 50-year-old balsam fir forest, in maple forests, and in clearcuts and uncut stands. Adults in moss and litter near a pond, leaf litter, in *Sphagnum*, in saturated moss on margin of small shaded brook, and in mushrooms. **Collecting period**: VI–IX. **Collecting method**: pitfall traps, Luminoc pitfall-light traps, Lindgren traps, sifting forest litter and *Sphagnum* and processing it through Berlese funnels, sifting saturated moss.

Subpolaris species group

The *subpolaris* species group may be characterized by the following combination of characters: body small, length 2.3–2.5 mm, narrowly oval with large, shield-shaped pronotum; antennae with antennomere IV slightly elongate, V–X slightly to moderately transverse; pronotum large, strongly transverse and at least as broad as elytra; elytra short, at suture shorter than pronotum; abdomen arcuate laterally and broadest at middle of its length; median lobe of aedeagus consisting of subparallel tubus with swollen subapical part in dorsal view, and oval, moderate in size bulbus with small crista apicalis; ventral margin of tubus sinuate, apex produced ventrally; flagellum shorter than tubus; internal sac with complex structures; apical lobe of paramere moderately elongate.

60. *Oxypoda subpolaris* Casey (Fig. 11.130a–d)

Oxypoda subpolaris Casey 1911. Lohse et al. 1990, Klimaszewski et al. 2006.

Oxypoda hemingi Lohse, in Lohse et al. 1990. Synonymized by Klimaszewski et al. 2006.

Diagnosis Body small, length 2.3–2.5 mm, subparallel, uniformly dark brown or reddish brown with slightly paler pronotum and apex of abdomen (Fig. 11.130a); pubescence of forebody dense; integument moderately glossy with dense isodiametric microsculpture; head broadest behind eyes, about two thirds of pronotum; eyes small, shorter than postocular region; antennae with antennomere IV slightly elongate, V–X slightly to moderately transverse; pronotum large, shield-shaped, with rounded margins, strongly transverse, broadest in middle, about one third broader than long, pubescence directed anteriad apically along midline and obliquely posteriad from midline of disc elsewhere; elytra much shorter than pronotum, pubescence directed straight posteriad; abdomen arcuate laterally and slightly swollen medially. MALE. Tergite VIII elongate and truncate apically, antecostal suture sinuate (Fig. 11.130c); sternite VIII elongate, broadly triangularly produced apically, antecostal suture arcuate (Fig. 11.130d); median lobe of aedeagus with ventral margin of tubus sinuate and apex produced ventrad (Fig. 11.130b); flagellum shorter than tubus; internal sac with complex structures (Fig. 11.130b). FEMALE. Undescribed.

DISTRIBUTION Origin: Nearctic. **CANADA**: not recorded. **USA**: AK.

COLLECTION AND HABITAT DATA
Habitat: *Betula, Alnus* litter. **Collecting period**: VI. **Collecting method**: Winkler Litter Extraction.

Additional records. See Annex 1.

***Sylvia* species group** (Klimaszewski et al. 2006 [as *grandipennis* group], 2018, 2020)

Diagnosis Body length 3.5–4.6 mm, narrowly oval, pronotum small, elytra large and long; median lobe of aedeagus in dorsal view with tubus broad, gradually narrowing apically, base of bulbus bearing elongate projection, which is split along median line and forms two subparallel arms in ventral view, bulbus oval, evenly coalescent with tubus dorsally, with crista apicalis narrowly elongate, flagellum about as long as tubus, internal sac with complex structures; apical lobe of paramere moderately elongate; spermatheca with capsule spherical, with large invagination, stem with large posterior loop, twisted posteriorly.

61. *Oxypoda sylvia* Casey (Fig. 11.131a–h)

Oxypoda sylvia Casey 1906. Casey 1911 [as *Moluciba grandipennis*], as *Oxypoda grandipennis*: Gusarov 2003, Klimaszewski et al. 2005, 2006, 2011, 2015, *Oxypoda sylvia*: Klimasewski et al. 2018a [synonymy], 2018b, 2020.

Diagnosis Body large, length 3.5–4.6 mm, narrowly oval, broadest at apex of elytra, colour brown with yellowish tinge, head and posterior abdomen usually dark brown, with apical antennomeres, marginal parts of pronotum, basal and posterior portions of elytra and base of abdomen yellow-brown (Fig. 11.131a); forebody with pubescence moderately dense and punctation fine; integument with isodiametric microsculpture; head broadest at level of eyes, two thirds as broad as pronotum; eyes about as long as genae in dorsal view; antennae long and slender, antennomere IV moderately to distinctly elongate, V–X subquadrate to slightly transverse; pronotum flattened, transverse, broadest in basal third, about one fifth broader than long, pubescence on disc directed obliquely apicad along apical part of median line and posterolaterad from midline elsewhere; elytra distinctly elongate, longer than pronotum at suture, pubescence on disc directed slightly obliquely outward from suture; abdomen gradually narrowed to apex. MALE. Tergite VIII slightly elongate and rounded apically, antecostal suture almost straight (Fig. 11.131d); sternite VIII elongate, with apex produced, antecostal suture sinuate (Fig. 11.131e); median lobe of aedeagus broad dorsally (Fig. 11.131c), with long narrow projection diverging from base of tubus in lateral view (Fig. 11.131b). FEMALE. Tergite VIII slightly elongate, rounded apically, antecostal suture arcuate (Fig. 11.131f); sternite VIII rounded apically, antecostal suture distinctly sinuate (Fig. 11.131g); last two abdominal segments with internal sclerotized structure; spermatheca with capsule moderately large, spherical, apical invagination broad, stem broadly looped posteriorly (Fig. 11.131h).

DISTRIBUTION Origin: Nearctic. **CANADA**: AB, BC, LB, MB, NB, NF, NS, ON, QC, SK, YT. **USA**: AK, ME, NH.

COLLECTION AND HABITAT DATA
Habitat: many forest types, including hardwood, red oak, mixed, mature red spruce, red spruce and birch, old red pine, spruce and balsam fir forests, and old-growth eastern white cedar forests. Adults common in decaying and rotten mushrooms in many of the above forest types. Some in group of *Pholiota* mushrooms of trunk of dead *Populus*; another under coyote dung on forest road. In AK, specimens were captured in alpine tundra, mostly spruce, some aspen forest; shrub, alder, dwarf birch, grass, clubmoss; shrub, dwarf birch; creek edge; *Betula* sp. and *Populus tremuloides* forests; disturbed *P. glauca* forest;

white spruce, *Betula* forest; old growth, and thinned and unthinned 2nd growth forests. **Collecting period**: V–X (in NB, most common in September). **Collecting method**: flight intercept trap, sifting mushrooms, forest litter, dung, and moss; some of above material also processed through Berlese funnels, ground search, Luminoc® pitfall-light traps, unbaited pitfall traps, pitfalls w/ rainroofs, yellow pitfalls w/o rainroofs, UV trap.

Additional records. See Annex 1.

Genus *Parocalea* Bernhauer, 1902

(Fig. 11.132–11.133) (Lohse et al. 1990, Klimaszewski et al. 2018a, 2020)

Diagnosis Body narrowly elongate, subparallel, head rounded, slightly narrower than pronotum; integument of forebody moderately glossy and with strong isodiametric microsculpture consisting of convex sculpticells, punctation fine; head and pronotum only slightly narrower than elytra; eyes small, infraorbital carina complete; antennae slightly incrassate, antennomere IV quadrate or slightly transverse, V–X distinctly transverse; pubescence on midline of disc of pronotum directed anteriad along anterior half to two-thirds, laterad on each side, and posteriad in basal one-third to half; mesocoxae contiguous or narrowly separated; elytra at suture slightly longer than pronotum along midline; abdomen subparallel, bases of first four visible tergites with deep transverse impressions; legs moderately long, basal tarsomere at least as long as two following ones combined, tarsal formula 5-5-5.

Key to Species of *Parocalea*

1. Punctation in the impressions of the basal tergites dense; elytra uniformly dark brown, colored as pronotum (Fig. 11.133a); tubus of the median lobe of the aedeagus evenly arcuate in lateral view (Fig. 11.133b); apical sclerites of internal sac of aedeagus well defined (Fig. 11.133b) ....***Parocalea pseudobaicalica* Lohse**

– Punctation in the impressions of the basal tergites sparser; elytra reddish-brown, paler than pronotum (Fig. 11.132a); tubus of the median lobe of the aedeagus feebly sinuate medially in lateral view (Fig. 11.132b); apical sclerites of internal sac of aedeagus weakly defined (Fig. 11.132b)...........***Parocalea nearctica* Lohse**

62. *Parocalea nearctica* Lohse (Fig. 11.132a–g)

Parocalea nearctica Lohse, in Lohse et al. 1990. Klimaszewski and Pelletier 2004.

Diagnosis Body medium-sized, narrowly linear, length 3.5–4.1 mm, colour brown or brownish-black, elytra and apex of abdomen paler (reddish-brown), appendages light to dark brown (Fig. 11.132a); head one tenth narrower than length of pronotum, broadest in apical half; eyes small, each about half length of genae in dorsal view; antennae slightly longer than head and pronotum combined, antennomeres I–III elongate, IV subquadrate, V–X transverse and incrassate; pronotum subquadrate, broadest in apical half, hind angles obtuse; elytra at suture as long as pronotum along midline and 1 1/3 as broad as pronotum; abdomen subparallel. MALE. Tergite VIII truncate apically (Fig. 11.132c); sternite VIII triangularly produced apically (Fig. 11.132d); median lobe of aedeagus in dorsal view broad, narrowly oval, with two pairs of elongate sclerites; in lateral view bulbus moderately large, with large and oval crista apicalis, tubus with ventral margin strongly arched ventrad and sinuate medially (Fig. 11.132b). FEMALE. Tergite VIII arcuate apically (Fig. 11.132e); sternite VIII with apical margin broadly triangular, vaguely angulate medially (Fig. 11.132f); spermatheca with capsule subspherical, stem irregularly coiled and twisted posteriorly (Fig. 11.132g).

DISTRIBUTION Origin: Nearctic. **CANADA**: NT, YT. **USA**: AK.

COLLECTION AND HABITAT DATA

Habitat: one specimen in AK taken from under rocks. **Collecting period**: VI, VII. **Collecting method**: sifting litter and soil, collecting with aspirator.

Additional records. See Annex 1.

63. ***Parocalea pseudobaicalica*** **Lohse** (Fig. 11.133a–g)

Parocalea pseudobaicalica Lohse, in Lohse et al. 1990. Bernhauer 1907 [as *P. baicalica*], Klimaszewski and Pelletier 2004, Klimaszewski et al. 2018a, 2020.

Diagnosis Body medium-sized, narrowly linear, length 3.5–4.2 mm, colour brown or brownish-black, apex of abdomen paler, appendages light to dark brown (Fig. 11.133a); head one tenth narrower than length of pronotum, broadest in apical half; eyes small, each about half length of genae in dorsal view; antennae about as long as head and pronotum combined, antennomeres I–III elongate, IV quadrate, V–X transverse and incrassate; pronotum subquadrate, broadest in apical half, hind angles obtuse; elytra at suture as long as pronotum along midline and 1 1/3 as broad as pronotum; abdomen subparallel; antennae with basal antennomere as long as antennomeres II–IV combined. MALE. Tergite VIII truncate apically (Fig. 11.133c); sternite VIII triangularly produced apically (Fig. 11.133d); median lobe of aedeagus in dorsal view broad, narrowly oval, with two pairs of elongate sclerites; in lateral view bulbus moderately large, with large and oval crista apicalis, tubus with ventral margin strongly arched ventrad (Fig. 11.133b). FEMALE. Tergite VIII arcuate apically (Fig. 11.133e); sternite VIII slightly produced at apex (Fig. 11.133f); spermatheca with capsule subspherical, stem irregularly coiled and twisted posteriorly (Fig. 11.133g).

DISTRIBUTION **Origin**: Nearctic. **CANADA**: BC, NT, NU [**BOLD**], QC, YT. **USA**: AK, NH.

COLLECTION AND HABITAT DATA
Habitat: near *Salix*, *Larix*, and *Picea* trees, edges of standing or running water, under stones on silt and gravel, among grasses, moss, and *Ranunculus*. The BC specimens were found in soil under alders. **Collecting period**: VI–VIII. **Collecting method**: sifting forest litter, soil and moss.

We report this species from NU (**NTR**) for the first time here, based on BOLD records: Canada, Nunavut, Kitikmeot, S Lake stream, 9km SW of Kugluktuk,17 m el., 67.7806, –115.272, mosses on stream banks, sifting, 16.VII.2019, G. Blagoev (CBG) 2 specimens; Kitikmeot, S Lake, 9km SW of Kugluktuk, 17 m el., 67.7726, –115.269, plant debris on sandy lake shore, sifting, 22.VII.2019, M. Pentinsaari (CBG) 1 specimen.

Additional records. See Annex 1.

Genus *Parocyusa* Bernhauer, 1902

(= *Tetralaucopora* Bernhauer, 1928, synonymyzed by Assing 2021)

Tetralaucopora Bernhauer, 1928. Seevers 1978, Klimaszewski et al. 2018a. As *Parocyusa*: Ashe 2001, Brunke et al. 2012, Webster et al. 2016, Assing 2021.

Diagnosis Body narrowly elongate, length 3.5–4.0 mm, colour reddish-brown, with head, scutellar part of elytra, and posterior part of abdomen dark brown, integument moderately glossy; forebody with punctation dense, fine and slightly asperate, pubescence fine and dense; head elongate, strongly narrowed basally, neck scarcely discernable, frontal suture present, postocular carina fine; eyes moderately large, about as long as genae in dorsal view; antennae long, antennomeres I–III distinctly elongate, IV subquadrate to slightly elongate, V–X slightly elongate to transverse, XI as long as two preceding antennomeres combined; ligula elongated, slightly incised at apex; right mandible with prominent median tooth; pronotum subquadrate to slightly transverse, broadest in apical one-third, base truncate, pubescence on disc directed posteriad on midline and obliquely laterad on each side; elytra broader than pronotum, with pubescence on disc directed posteriad, forming wavy pattern in middle of disc; process of mesoventrite three-fourths as long as mesocoxae; first three visible abdominal tergites with deep impressions; hind tarsi very long, only slightly shorter than tibiae; median lobe of aedeagus in dorsal view with broad tubus, bulbus of moderate size with long crista apicalis, in lateral view tubus arcuate; spermatheca with capsule spherical, apical invagination large, stem thin, long sinuate, and broadly looped posteriorly.

COMMENTS Assing and Schülke (2007) synonymized *Parocyusa* Bernhauer, 1902 with *Tectusa* Bernhauer, 1899 based on a re-evaluation of the type species. This was accepted in the most recent Palaearctic catalog (Schülke and Smetana 2015), though the Nearctic species were not directly mentioned in either work. However, Assing and Schülke (2007) placed most other world *Parocyusa* in the former subgenus *Tetralaucopora*, within which Seevers (1978) placed the Nearctic species. Assing (2021) revised *Tetralaucopora* Bernhauer, 1928, *Tectusa* Bernhauer, 1899, and various other genera of Oxypodina, and revalidated *Parocyusa* Bernhauer, 1902, previously a synonym of *Tectusa*. He synonymized *Tetralaucopora* with *Parocyusa* and proposed new combinations for several species.

64. *Parocyusa fuliginosa* (Casey)
(Fig. 11.134a–i)

Chilopora fuliginosa Casey, 1906. In *Tetralaucopora*: Moore and Legner 1975, Seevers, 1978, in *Parocyusa*: Klimaszewski et al. 2011, Brunke et al. 2012, Webster et al. 2016 Klimaszewski et al., in *Tetralaucopora*: Klimaszewski et al. 2018a.

Diagnosis Body subparallel, slightly flattened, elongate, length 3.5–4.0 mm, colour dark brown, pubescence short and evenly distributed (Fig. 11.134a); head and pronotum with meshed reticulation and coarse punctation; pronotum slightly elongate, narrower than elytra; antennae slender and long, antennomeres VIII–X quadrate to slightly transverse; abdomen with first three visible tergites deeply impressed (Fig. 11.134a); hind tarsus highly elongate, 2/3 to 3/4 as long as hind tibia. MALE. Tergite VIII broadly rounded apically (Fig. 11.134d); sternite VIII rounded apically from near lateral margins (Fig. 11.134e); median lobe of aedeagus in dorsal view with tubus broad, only slightly narrower than bulbus (Fig. 11.134c); in lateral view tubus bent obliquely ventrad at base, ventral margin arcuate in basal half, with apex broad, blunt (Fig. 11.134b). FEMALE. Tergite VIII rounded apically, with antecostal suture widely separated from basal margin (Fig. 11.134f); sternite VIII with apical margin broadly truncate medially, arcuate laterally (Fig. 11.134g); spermatheca with capsule spherical, apical invagination broad, stem long, thin, sinuate, and partially looped posteriorly (Fig. 11.134h,i).

DISTRIBUTION Origin: Nearctic. **CANADA**: LB, NB, ON, SK. **USA**: MA, NC, PA.

COLLECTION AND HABITAT DATA
Habitat: in LB, adults collected from under rocks and gravel along a stream margin. In NB, one specimen collected by splashing gravel margin of a clear rocky stream. **Collecting period**: VII–VIII. **Collecting method**: splashing gravel margin of rocky stream and hand collecting from gravel and rocks.

Subtribe Phloeoporina C.G. Thomson, 1859
(Ashe 2001, Osswald et al. 2013)

Diagnosis Body slender, subparallel, convex, colour brown to black, forebody with pubescence short and dense; head nearly as broad as pronotum or broader, strongly constricted basally, with distinct neck, eyes as long as or slightly shorter than genae in dorsal view, postocular carina absent; antennomeres V–X moderately to distinctly transverse; pronotum angular, subquadrate to slightly transverse, pubescence on disc directed obliquely inward basally, straight posteriad elsewhere; elytra broader than pronotum, posterior margin distinctly sinuate; first three to four visible abdominal tergites with basal impressions deep, transverse, with very coarse punctation.

COMMENTS Osswald et al. (2013), based on molecular studies, considered *Phloeopora* to be a separate lineage with unresolved relationships to the other clades of Oxypodini, and based on inferred phylogeny they treated *Phloeopora* as the only member of the subtribe Phlocoporina.

Genus *Phloeopora* Erichson, 1837

(Figs. 11.135–11.137) (Lohse et al. 1990, Ashe 2001, Klimaszewski et al. 2018a, 2020)

Diagnosis Body slender, subparallel, convex, length 2.2–3.5 mm; colour brown to black, forebody with pubescence short and dense; head nearly as broad as pronotum or broader, strongly constricted basally, with distinct neck, eyes as long as or slightly shorter than genae in dorsal view, postocular carina absent; antennomeres V–X moderately to distinctly transverse; pronotum angular, subquadrate to slightly transverse, pubescence on disc directed obliquely inward basally, straight posteriad elsewhere; elytra broader than pronotum, posterior margin distinctly sinuate; first three to four visible abdominal tergites with basal impressions deep, transverse, with very coarse punctation; male sternite VIII triangularly produced apically; median lobe of aedeagus narrow, tubus subparallel, bulbus oval and small; spermatheca with capsule spherical, with invagination deep, moderately broad, stem sinuate with posterior coils and loops.

COMMENTS All species of this genus are externally similar to each other and differ in small details such as body proportions (e.g., width of pronotum, length of elytra), density of punctation and pubescence on forebody, body colour, shape of the apical part of the median lobe of the aedeagus in lateral view, and shape of the spermatheca.

Key to Species of *Phloeopora*

1. Body uniformly black, except for reddish tarsi, length 2.5–3.0 mm (Fig. 11.137a); tubus of median lobe of aedeagus weakly arcuate with apex slightly produced ventrally in lateral view (Fig. 11.137b); spermatheca with thick stem with one posterior loop (Fig. 11.137g) ..... ***Phloeopora gilbertae* Klimaszewski and Webster**

– Body brown to reddish-brown with paler appendages, length 2.3–3.5 mm (Figs. 11.135a, 11.136a); tubus of median lobe of aedeagus distinctly arcuate with apex strongly produced ventrally in lateral view (Figs. 11.135b, 11.136b); spermatheca with thin stem with several posterior loops or twists (Figs. 11.135h, 11.136h) **2**

2. Body fusiform, head slightly and elytra distinctly broader than pronotum (Fig. 11.135a); forebody with pubescence long and sparse; integument glossy; median lobe of aedeagus with tubus thin apically and moderately strong produced ventrally in lateral view (Fig. 11.135b); spermatheca stem with several broad coils and loops (Fig. 11.135h).. ***Phloeopora arctica* Lohse**

– Body narrowly subparallel, head at most as broad as pronotum and elytra slightly broader than pronotum (Fig. 11.136a); forebody with pubescence short and dense; integument less glossy; median lobe of aedeagus with tubus moderately thin apically and strongly produced ventrally in lateral view (Fig. 11.136b); spermatheca stem with few tighter coils and loops (Fig. 11.136h)................. ***Phloeopora canadensis* Klimaszewski and Langor**

65. *Phloeopora arctica* Lohse (Fig. 11.135a–h)

Phloeopora arctica Lohse, in Lohse et al. 1990. Majka and Klimaszewski 2008a, Klimaszewski et al. 2018a, 2020.

Diagnosis Body moderately broad, elongate, length 3.0–3.5 mm, colour dark brown with tarsi, maxillary palpi, labrum and antennae reddish-brown (Fig. 11.135a); integument moderately glossy, moderately densely punctate and pubescent, pubescence long, adhering to body; head slightly wider and longer than pronotum, pubescence on disc directed inward from sides to midline; eyes large, slightly shorter than genae in dorsal view; antennae slightly incrassate, antennomeres V–X distinctly transverse; pronotum trapezoidal in shape, broadest subapically, ¼ narrower than width of elytra, pubescence on disc directed almost straight posteriad; elytra elongate, sides arcuate, with shoulders distinct, narrowly rounded, pubescence on disc directed obliquely and straight posteriad; abdomen subparallel with first three visible tergites deeply impressed basally MALE. Tergite VIII with apical margin very broadly arcuate medially, sinuate laterally (Fig. 11.135d); sternite VIII with apical margin strongly triangularly produced, rounded at apex (Fig. 11.135e); median lobe of aedeagus in lateral view with ventral margin of tubus curved obliquely ventrad near base, thin and slightly curved ventrad apically (Fig. 11.135b), in dorsal view bulbus round and tubus narrowly subparallel (Fig. 11.135c). FEMALE. Tergite VIII broadly arcuate apically (Fig. 11.135f); sternite VIII with apical margin obtusely angularly produced, apex arcuate (Fig. 11.135g); spermatheca with capsule spherical, acutely bent laterad at base, stem with several broad, irregular loops (Fig. 11.135h).

This species differs from its Nearctic congeners in having the body fusiform, the colour uniformly dark brown except for the appendages, and the shapes of the median lobe of the aedeagus in lateral view and the spermatheca are distinctive.

DISTRIBUTION Origin: Nearctic. **CANADA**: AB, BC, NT, ON, YT. **USA**: not recorded.

COLLECTION AND HABITAT DATA
Habitat: one specimen sifted from litter, otherwise nothing is known about the habitat association. **Collecting period**: VIII. **Collecting method**: sifting litter.

66. *Phloeopora canadensis* Klimaszewski and Langor (Fig. 11.136a–h)

Phloeopora canadensis Klimaszewski and Langor, in Klimaszewski et al. 2011. Webster et al. 2016, 2020, Klimaszewski et al. 2018a, 2020.

Diagnosis Body narrowly elongate, length 2.4–3.0 mm, subparallel, colour brown with head dark brown and antennae, legs, palps and apical part of abdomen light brown with rusty-brown tinge (Fig. 11.136a); integument moderately glossy with meshed microsculpture; head as wide and as long as pronotum; pronotum 1/5 narrower than elytra; antennomeres V–X moderately transverse; elytra narrow, arcuate laterally, slightly longer than pronotum; abdomen subparallel, slightly narrower than elytra. MALE. Tergite VIII arcuate apically (Fig. 11.136d); sternite VIII with apical margin strongly triangularly produced, slightly emarginate on each side of narrowly rounded apex (Fig. 11.136e); tubus of median lobe of aedeagus in lateral view broadly bent obliquely ventrad near base, slightly elongate and slightly curved ventrad subapically, apex narrowly rounded (Fig. 11.136b), in dorsal view bulbus oval and tubus subparallel (Fig. 11.136c). FEMALE. Tergite VIII broadly arcuate apically (Fig. 11.136f); sternite VIII with apical margin very obtusely produced, apex broadly rounded (Fig. 11.136g); spermatheca with capsule spherical, stem irregularly coiled posteriorly (Fig. 11.136h).

DISTRIBUTION Origin: Nearctic. **CANADA**: BC, NB, NF, NS, YT [**BOLD**]. **USA**: AK [**NSR**].

COLLECTION AND HABITAT DATA
Habitat: in NB, specimens were captured in Lindgren funnel traps in conifer and mixed forests. These were an old black spruce forest, an old-growth white spruce and balsam fir forest, a mature red and white spruce forest, an old white pine stand, an old mixed forest, and a red oak forest adjacent to a black spruce stand. The only specimens with microhabitat data were collected from under bark of red spruce infested with *Dendroctonus* in a red spruce stand. In NF, this species was found under the bark of tamarack, *Larix laricina* (Du Roi) K. Koch, recently killed by the eastern larch beetle, *Dendroctonus simplex* LeConte. In AK, specimens were collected in boggy and flat black spruce habitat, and black spruce with a few young birch and young spruce in understory coming in in response to blowdown disturbance. **Collecting period**: V–VII, X. **Collecting method**: Lindgren funnel traps, aspirating from under bark.

We report this species from YT (**NTR**) for the first time here based on **BOLD** records and from AK (**NSR**) based on UAM specimens: Canada, Yukon, Kluane National Park and Reserve, Auriol Trail, 801 m el., 60.714, –137.432, spruce forest, Malaise trap, 30.VII-12.VIII.2014, M. Trotter, Monica J, Craig M (CBG) 1 specimen; Nilsson-Lammers Research Forest, 682m el.,

60.841, −135.208, Malaise trap , 27.VI-06. VII.2014, Mary Whitley (CBG) 1 specimen. USA, Alaska, Anchorage, JBER, Fish Lake, 91m el., 61.27446, −149.80235, ±3m boggy and flat black spruce, hanging Malaise, 5.0m, 28.V-12.VI.2015, J.C. Hagelin (UAM) 1 specimen; Tetlin NWR, Fish Camp Lake, 511m el., 63.14107, −142.26184, ±10m black spruce w/ few young birch and young spruce in understory coming in in response to blowdown disturbance, hanging Malaise, 6m (estimated), 27.VI-19. VII.2016, J.C. Hagelin (UAM) 1 specimen.

COMMENTS This species was listed from QC by Newton (2019) but we are not aware of any published voucher data.

67. *Phloeopora gilbertae* Klimaszewski and Webster (Fig. 11.137a–g)

Phloeopora gilbertae Klimaszewski and Webster, in Webster et al. 2016, Klimaszewski et al. 2018a.

Diagnosis Body narrowly elongate, subparallel, length 2.5–2.8 mm, colour black except tarsi and bases of antennae reddish-brown (Fig. 11.137a); integument highly glossy, rather sparsely punctate and pubescent, pubescence short, adhering to body; head slightly larger than pronotum, pubescence on disc directed inward from sides to midline; eyes large, slightly shorter than genae in dorsal view; antennomeres V–X moderately transverse; pronotum trapezoidal in shape, broadest subapically, 1/5 narrower than elytra, pubescence on disc directed almost straight posteriad; elytra elongate, sides subparallel, with shoulders narrowly rounded, pubescence on disc directed obliquely and straight posteriad; abdomen subparallel with first three visible tergites deeply impressed basally. MALE. Tergite VIII truncate apically (Fig. 11.137c); sternite VIII distinctly transverse, with apical margin broadly arcuate, triangularly produced at middle 1/5 (Fig. 11.137d); median lobe of aedeagus in lateral view with tubus arcuately bent obliquely ventrad in basal half, slightly elongate and slightly arcuate apically, very narrowly rounded at apex (Fig. 11.137b). FEMALE. Tergite VIII with apical margin broadly arcuate except slightly sinuate at lateral margins (Fig. 11.137e); sternite VIII with apical margin obtusely triangularly produced, broadly rounded medially (Fig. 11.137f); spermatheca with capsule spherical, bent at right angles to stem, stem forming loose coil in apical half, sinuate and narrower basally (Fig. 11.137g).

This species differs from its Nearctic congeners in having the body colour uniformly black except for the appendages, different body proportions, and the shapes of the median lobe of the aedeagus in lateral view and the spermatheca are distinctive.

DISTRIBUTION Origin: Nearctic. **CANADA**: NB, NT, QC [**NPR**]. **USA**: AK [**NSR**].

COLLECTION AND HABITAT DATA

Habitat: All specimens from NB were collected from Lindgren funnel traps, most in hardwood forests. These included traps in canopy of a *Populus tremuloides* Michx. (trembling aspen) stand, in the canopy of *Populus tremuloides, Juglans cinerea* L. (butternut) and traps under *Tilia americana* L. (American basswood) in a hardwood forest on an island in a river, in Lindgren traps under *Populus balsamifera* L. in a *P. balsamifera* stand near a river, in a red oak stand with *Populus*, a hardwood stand, and in a jack pine (*Pinus banksiana* Lamb.) forest. One specimen from NT was collected in a Lindgren funnel trap in a mixed *Picea–Betula* forest. The QC specimens were collected in a Malaise trap in a mixed *Picea-Populus* forest. Nothing is known about the micro-habitat association of the species but it may occur in subcortical habitats as its congeners. **Collecting period**: V–VIII. **Collecting method**: Lindgren funnel traps.

We report this species from QC (**NPR**) and AK (**NSR**) for the first time based on the following: Canada, Quebec, Forillon National Park, operation centre, 135m el., 48.857, –64.376, spruce and poplar forest, Malaise trap, 10-17. VI.2013, F.Tremblay (CBG) 2 specimens. United States: Alaska, Fairbanks, 930 Fitz Ct., 94.901, –147.529, flight intercept trap, D.S. Sikes (UAM) 1 specimen.

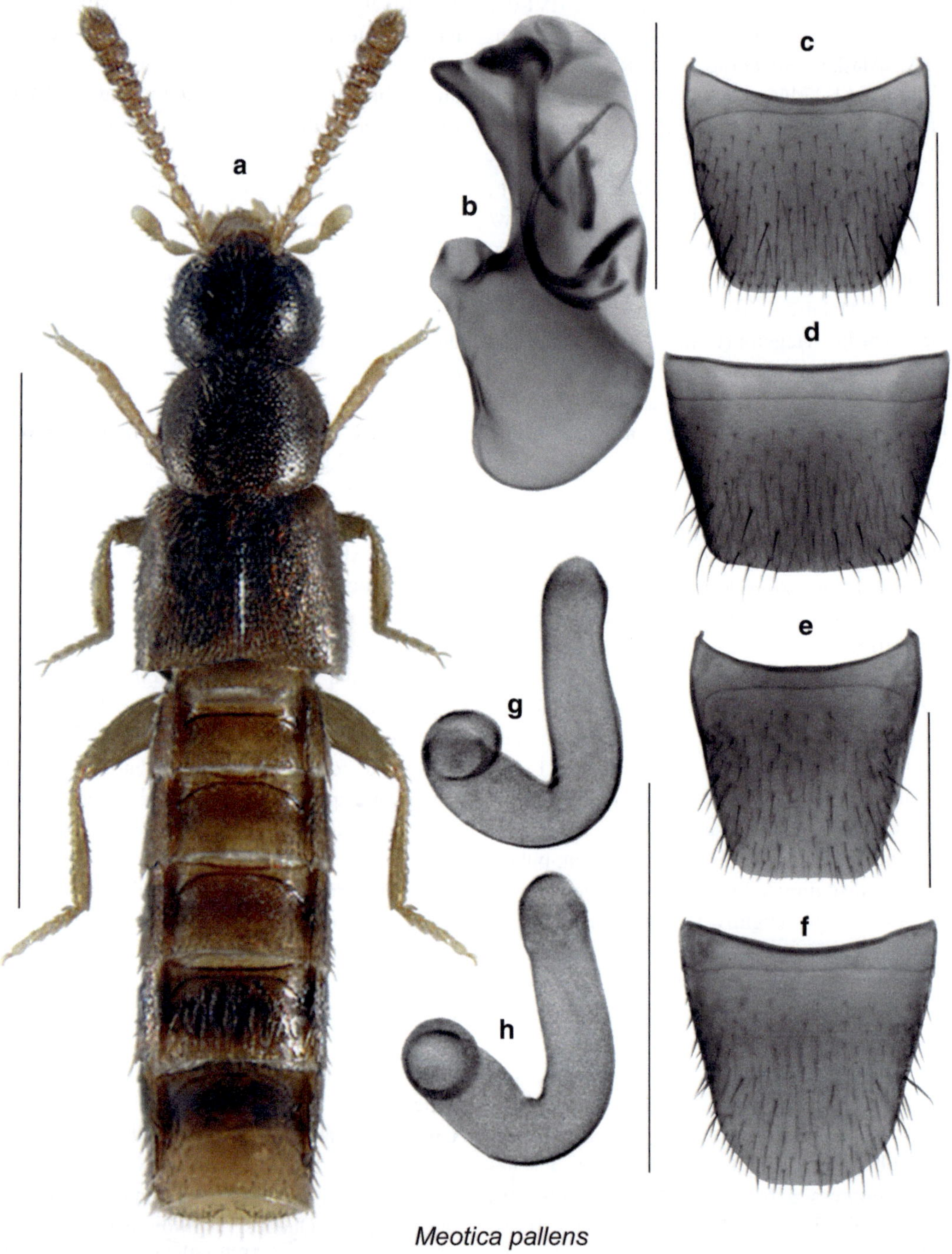

Fig. 11.94 (**a–h**). Images of *Meotica pallens* (Redtenbacher): (**a**) habitus in dorsal view; (**b**) median lobe of aedeagus in lateral view; (**c**) male tergite VIII; (**d**) male sternite VIII; (**e**) female tergite VIII; (**f**) female sternite VIII; (**g–h**) spermatheca. (**a–h**) after Klimaszewski et al. (2018a). Scale bar for habitus = 1 mm; remaining scale bars = 0.2 mm

Fig. 11.95 (**a–h**). Images of *Alfocalea montana* Klimaszewski: (**a**) habitus in dorsal view; (**b**) median lobe of aedeagus in lateral view; (**c**) median lobe of aedeagus in dorsal view; (**d**) male tergite VIII; (**e**) male sternite VIII; (**f**) female tergite VIII; (**g**) female sternite VIII; (**h**) spermatheca. (**a–h**) after Klimaszewski et al. (2020). Scale bar for habitus = 1 mm; remaining scale bars = 0.2 mm

Fig. 11.96 (**a–d**). Images of *Calodera caseyi* Assing: (**a**) habitus in dorsal view; (**b**) female tergite VIII; (**c**) female sternite VIII; (**d**) spermatheca. (**a–d**) after Klimaszewski et al. (2018a). Scale bar for habitus = 1 mm; remaining scale bars = 0.2 mm

Fig. 11.97 (**a–g**). Images of *Calodera parviceps* (Casey): (**a**) habitus in dorsal view; (**b**) median lobe of aedeagus in lateral view; (**c**) male tergite VIII; (**d**) male sternite VIII; (**e**) female tergite VIII; (**f**) female sternite VIII; (**g**) spermatheca. (**a–g**) after Klimaszewski et al. (2018a). Scale bar for habitus = 1 mm; remaining scale bars = 0.2 mm

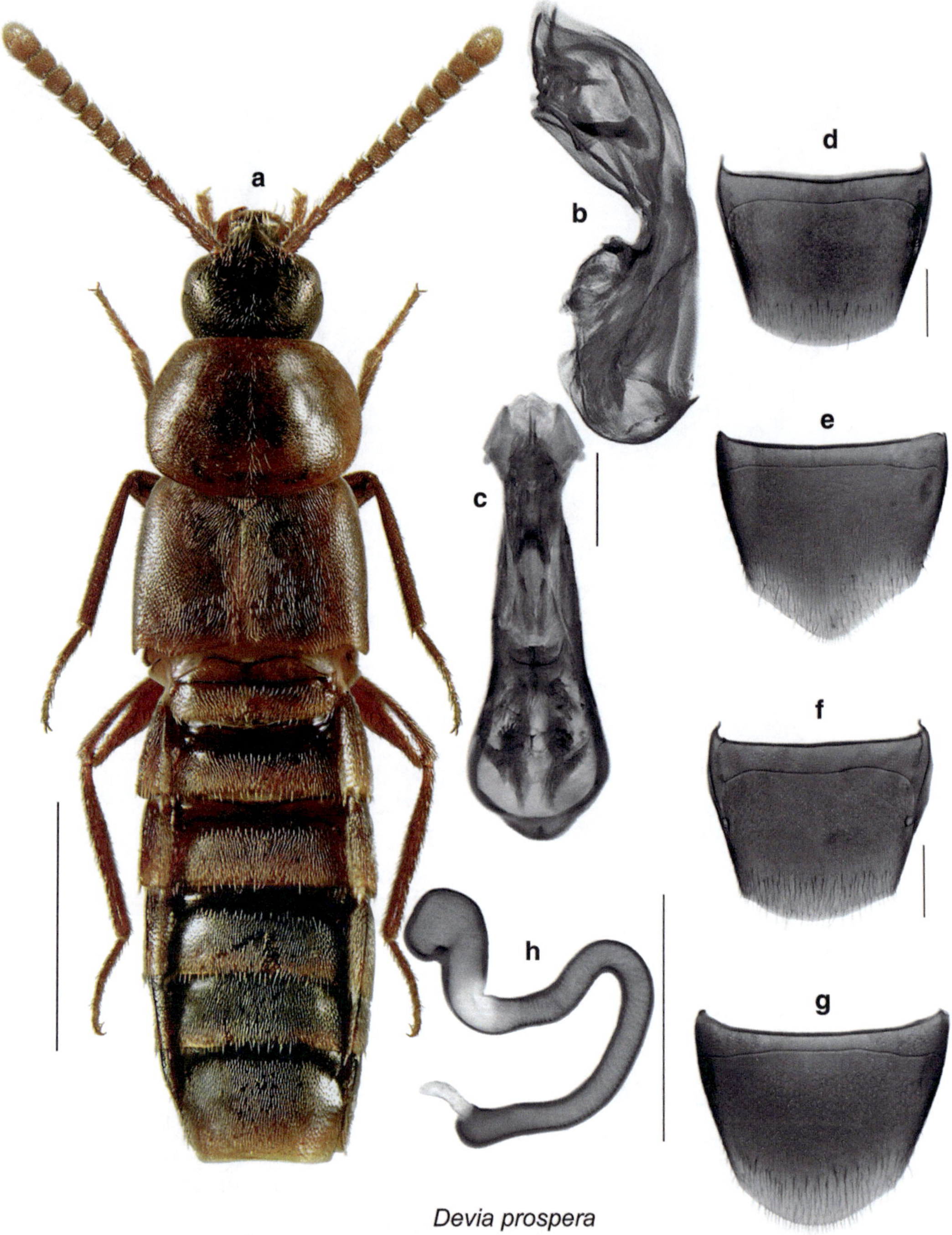

Fig. 11.98 (**a–h**). Images of *Devia prospera* (Erichson): (**a**) habitus in dorsal view; (**b**) median lobe of aedeagus in lateral view; (**c**) median lobe of aedeagus in dorsal view; (**d**) male tergite VIII; (**e**) male sternite VIII; (**f**) female tergite VIII; (**g**) female sternite VIII; (**h**) spermatheca. (**a–h**) after Klimaszewski et al. (2018a). Scale bar for habitus = 1 mm; remaining scale bars = 0.2 mm

Fig. 11.99 (**a–g**). Images of *Gnathusa alfacaribou* Klimaszewski and Langor: (**a**) habitus in dorsal view; (**b**) median lobe of aedeagus in lateral view; (**c**) male tergite VIII; (**d**) male sternite VIII; (**e**) female tergite VIII; (**f**) female sternite VIII; (**g**) spermatheca. (**a–g**) after Klimaszewski et al. (2011, 2018a). Scale bar for habitus = 1 mm; remaining scale bars = 0.2 mm

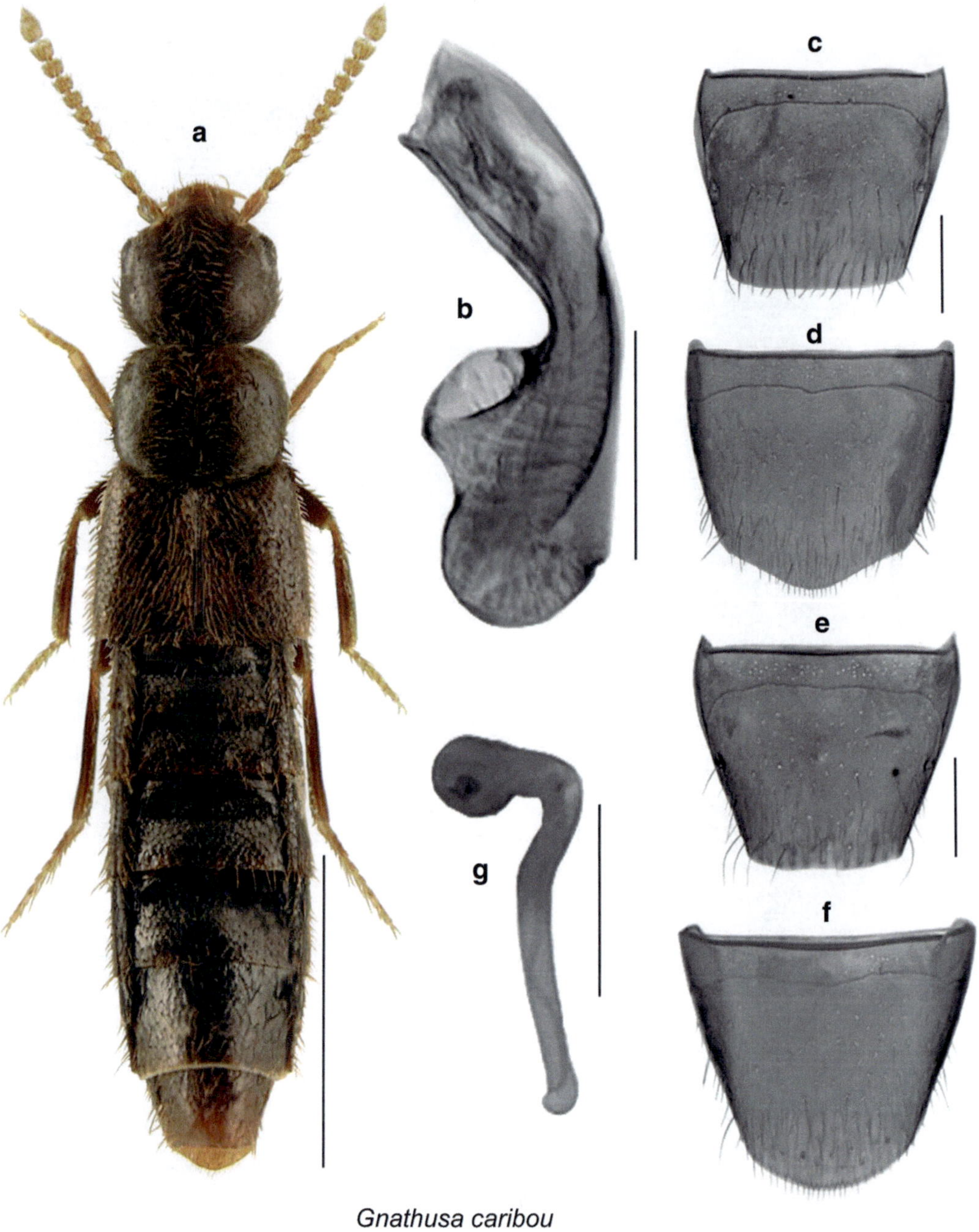

Fig. 11.100 (**a–g**). Images of *Gnathusa caribou* Lohse: (**a**) habitus in dorsal view; (**b**) median lobe of aedeagus in lateral view; (**c**) male tergite VIII; (**d**) male sternite VIII; (**e**) female tergite VIII; (**f**) female sternite VIII; (**g**) spermatheca. (**a–g**) after Klimaszewski et al. (2011, 2018a). Scale bar for habitus = 1 mm; remaining scale bars = 0.2 mm

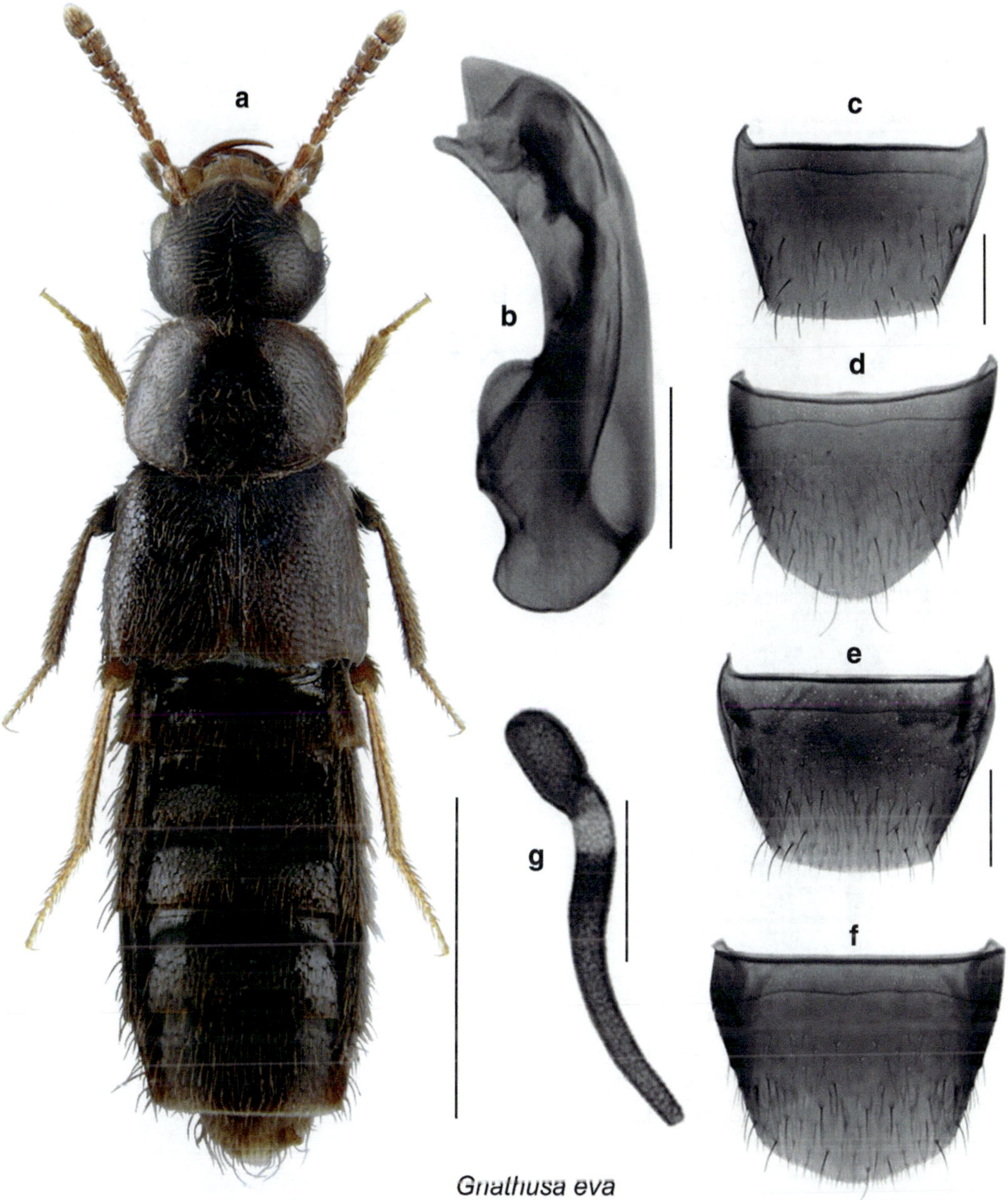

Fig. 11.101 (**a–g**). Images of *Gnathusa eva* Fenyes: (**a**) habitus in dorsal view; (**b**) median lobe of aedeagus in lateral view; (**c**) male tergite VIII; (**d**) male sternite VIII; (**e**) female tergite VIII; (**f**) female sternite VIII; (**g**) spermatheca. (**a–g**) after Klimaszewski et al. (2018a). Scale bar for habitus = 1 mm; remaining scale bars = 0.2 mm

Fig. 11.102 (**a–g**). Images of *Gnathusa tenuicornis* Fenyes: (**a**) habitus in dorsal view; (**b**) median lobe of aedeagus in lateral view; (**c**) male tergite VIII; (**d**) male sternite VIII; (**e**) female tergite VIII; (**f**) female sternite VIII; (**g**) spermatheca. (**a–g**) after Klimaszewski et al. (2018a). Scale bar for habitus = 1 mm; remaining scale bars = 0.2 mm

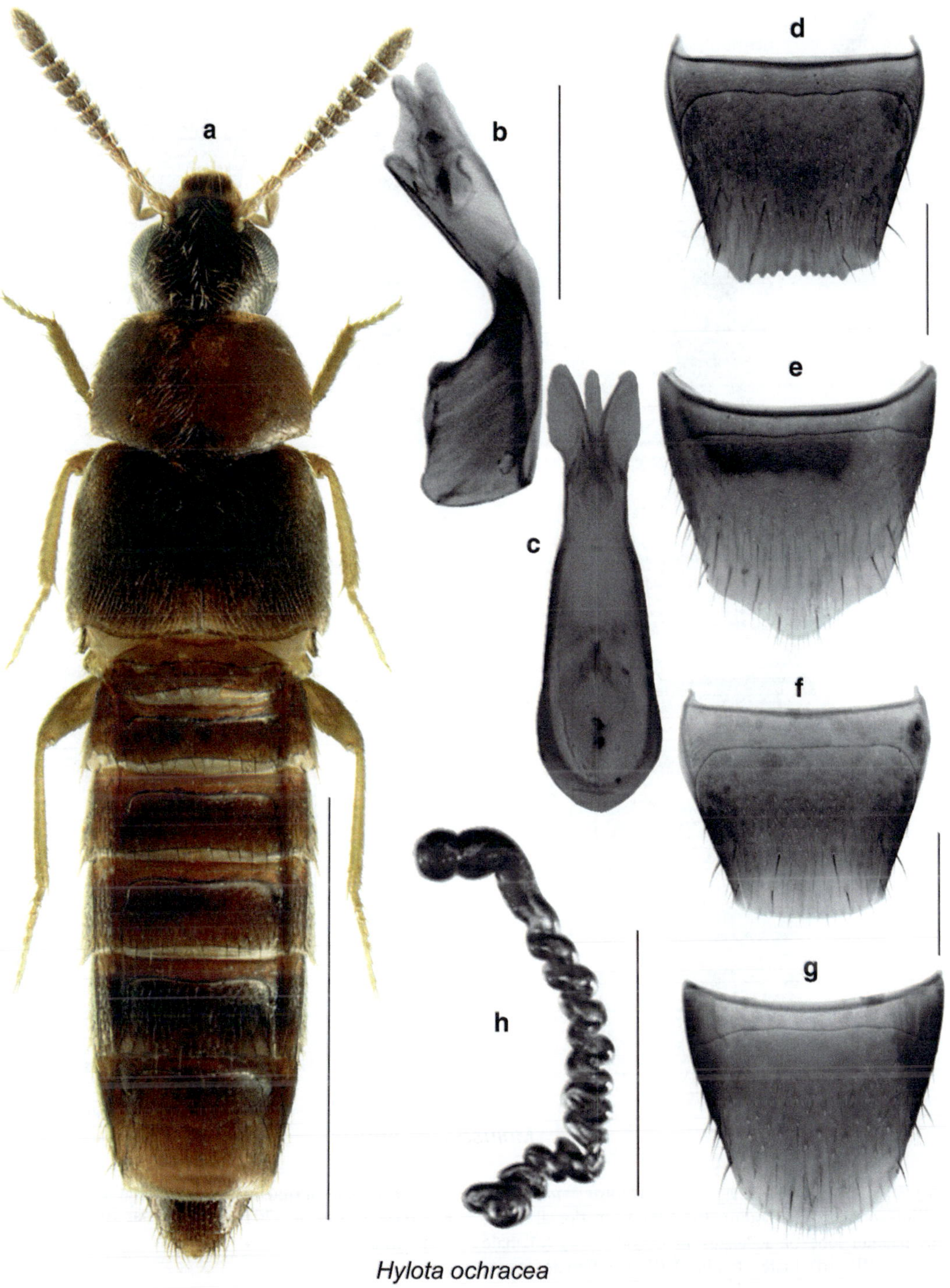

Fig. 11.103 (**a–h**). Images of *Hylota ochracea* Casey: (**a**) habitus in dorsal view; (**b**) median lobe of aedeagus in lateral view; (**c**) median lobe of aedeagus in dorsal view; (**d**) male tergite VIII; (**e**) male sternite VIII; (**f**) female tergite VIII; (g) female sternite VIII; (**h**) spermatheca. (**a–h**) after Klimaszewski et al. (2018a). Scale bar for habitus = 1 mm; remaining scale bars = 0.2 mm

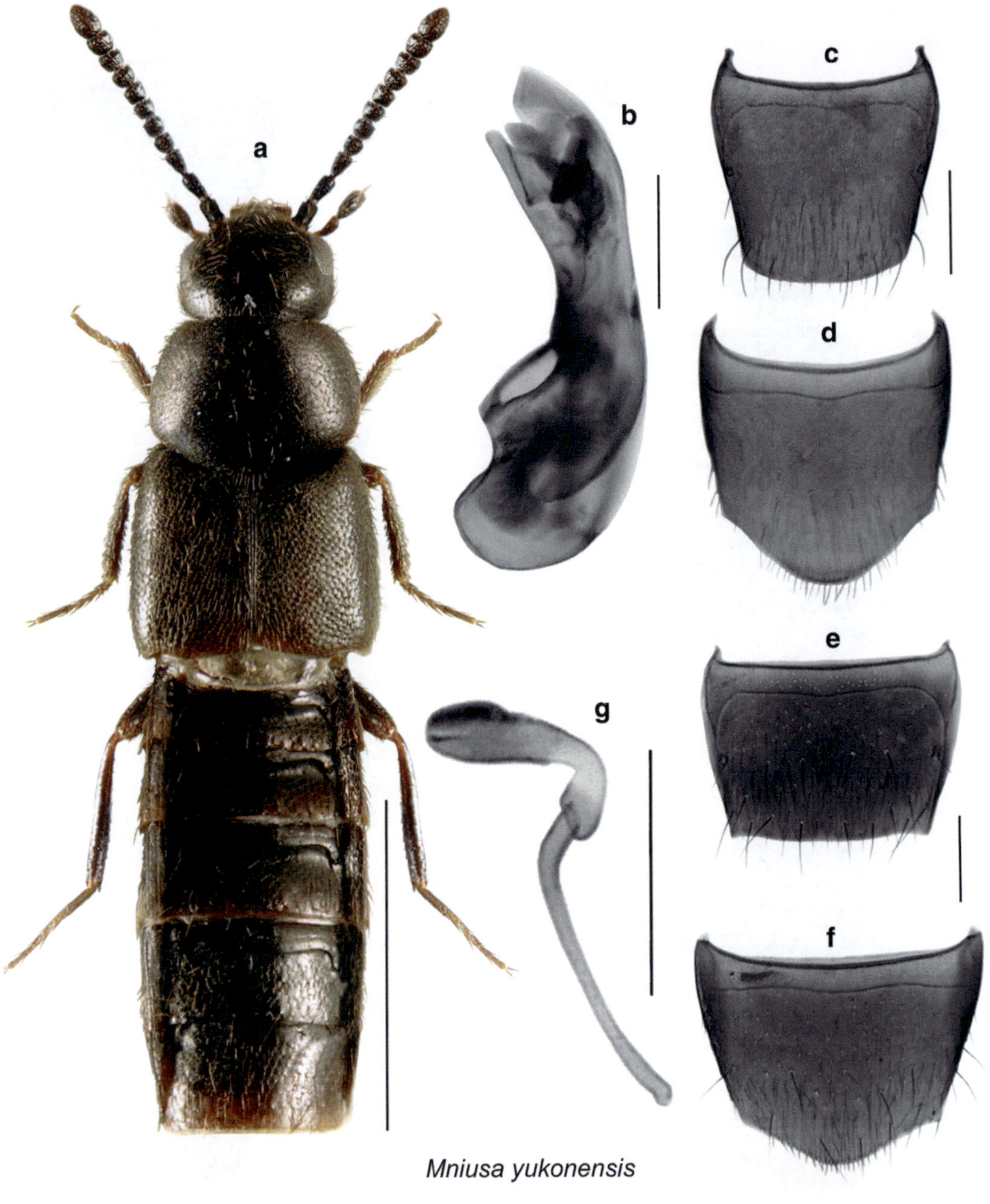

Fig. 11.104 (**a–g**). Images of *Mniusa yukonensis* (Klimaszewski and Godin): (**a**) habitus in dorsal view; (**b**) median lobe of aedeagus in lateral view; (**c**) male tergite VIII; (**d**) male sternite VIII; (**e**) female tergite VIII; (**f**) female sternite VIII; (**g**) spermatheca. (**a–g**) after Klimaszewski et al. (2018a). Scale bar for habitus = 1 mm; remaining scale bars = 0.2 mm

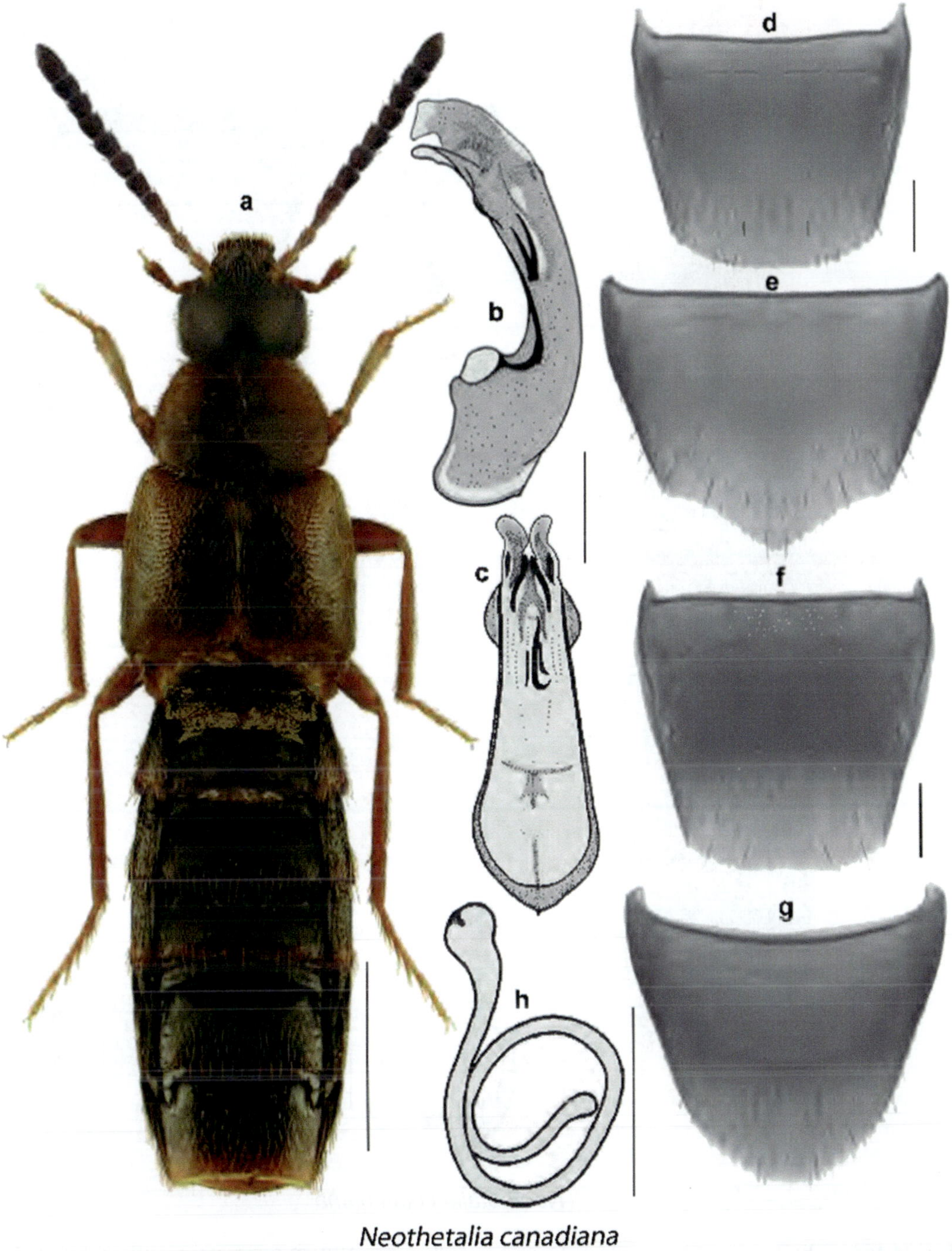

Fig. 11.105 (**a–h**). Images of *Neothetalia canadiana* Klimaszewski: (**a**) habitus in dorsal view; (**b**) median lobe of aedeagus in lateral view; (**c**) median lobe of aedeagus in dorsal view; (**d**) male tergite VIII; (**e**) male sternite VIII; (**f**) female tergite VIII; (**g**) female sternite VIII; (**h**) spermatheca. (**a–h**) after Klimaszewski et al. (2018a). Scale bar for habitus = 1 mm; remaining scale bars = 0.2 mm

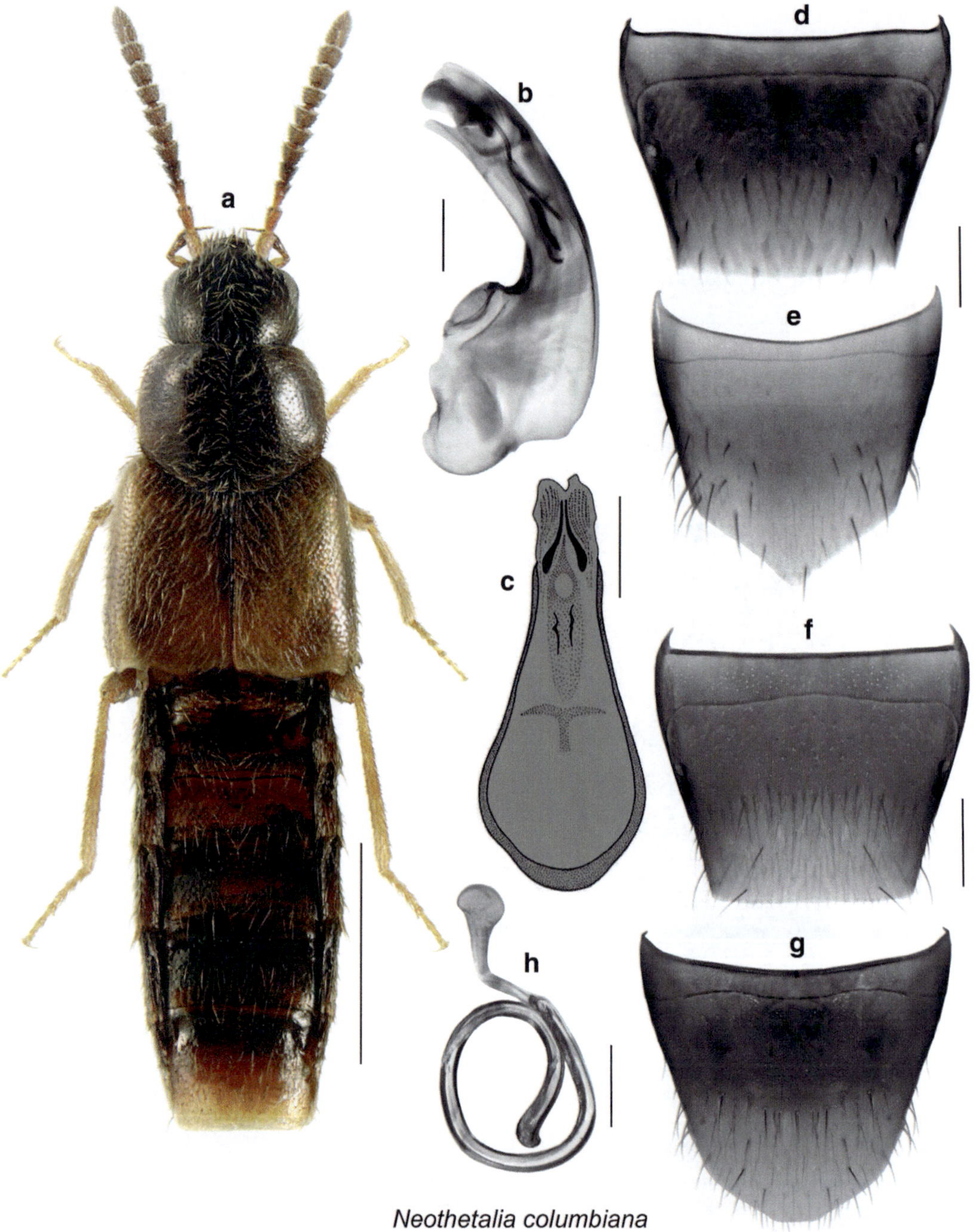

Fig. 11.106 (**a–h**). Images of *Neothetalia columbiana* (Klimaszewski): (**a**) habitus in dorsal view; (**b**) median lobe of aedeagus in lateral view; (**c**) median lobe of aedeagus in dorsal view; (**d**) male tergite VIII; (**e**) male sternite VIII; (**f**) female tergite VIII; (**g**) female sternite VIII; (**h**) spermatheca. (**a–h**) after Klimaszewski et al. (2020). Scale bar for habitus = 1 mm; remaining scale bars = 0.2 mm

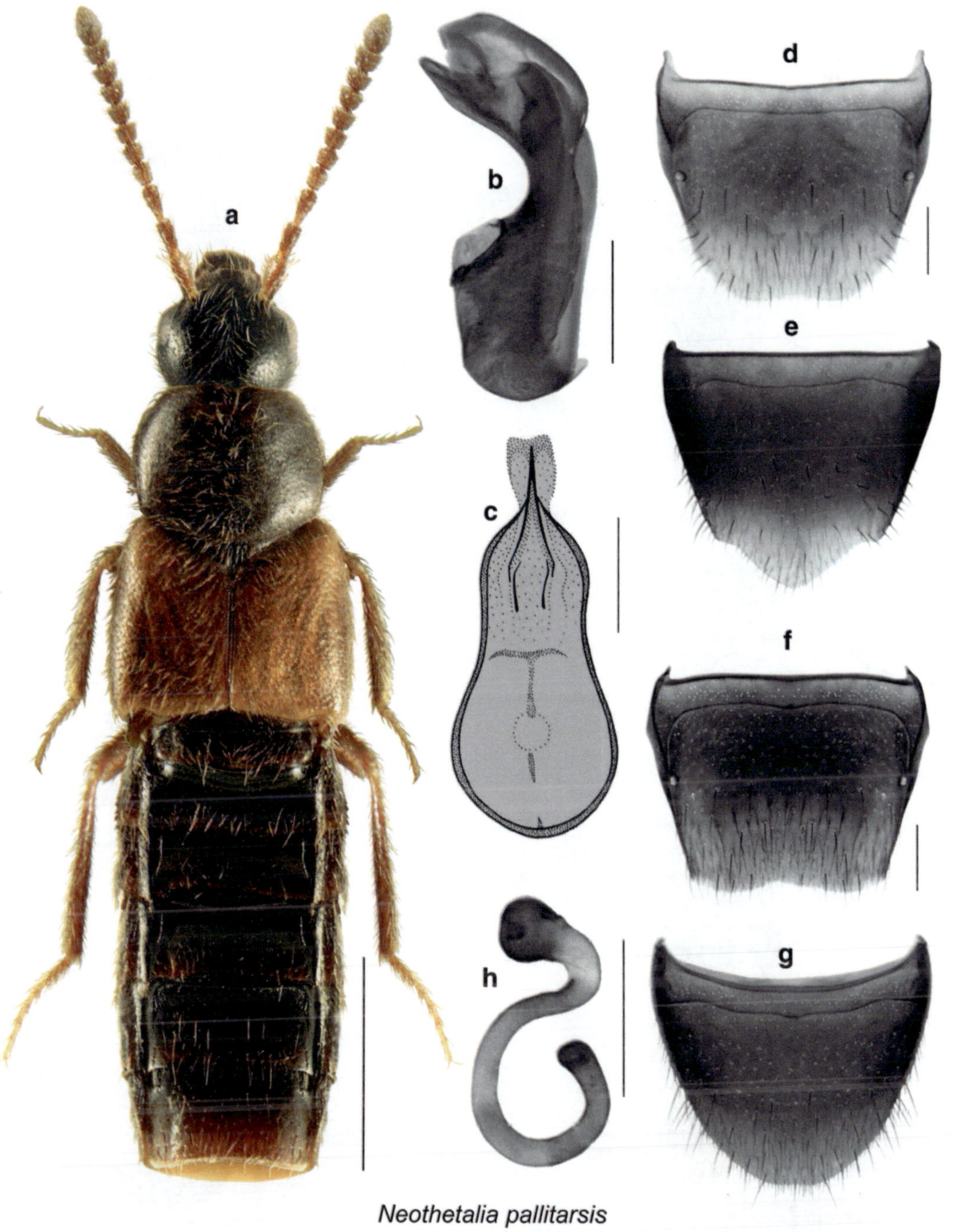

Fig. 11.107 (**a–h**). Images of *Neothetalia pallitarsis* (Kirby): (**a**) habitus in dorsal view; (**b**) median lobe of aedeagus in lateral view; (**c**) median lobe of aedeagus in dorsal view; (**d**) male tergite VIII; (**e**) male sternite VIII; (**f**) female tergite VIII; (**g**) female sternite VIII; (**h**) spermatheca. (**a–h**) after Klimaszewski et al. (2020). Scale bar for habitus = 1 mm; remaining scale bars = 0.2 mm

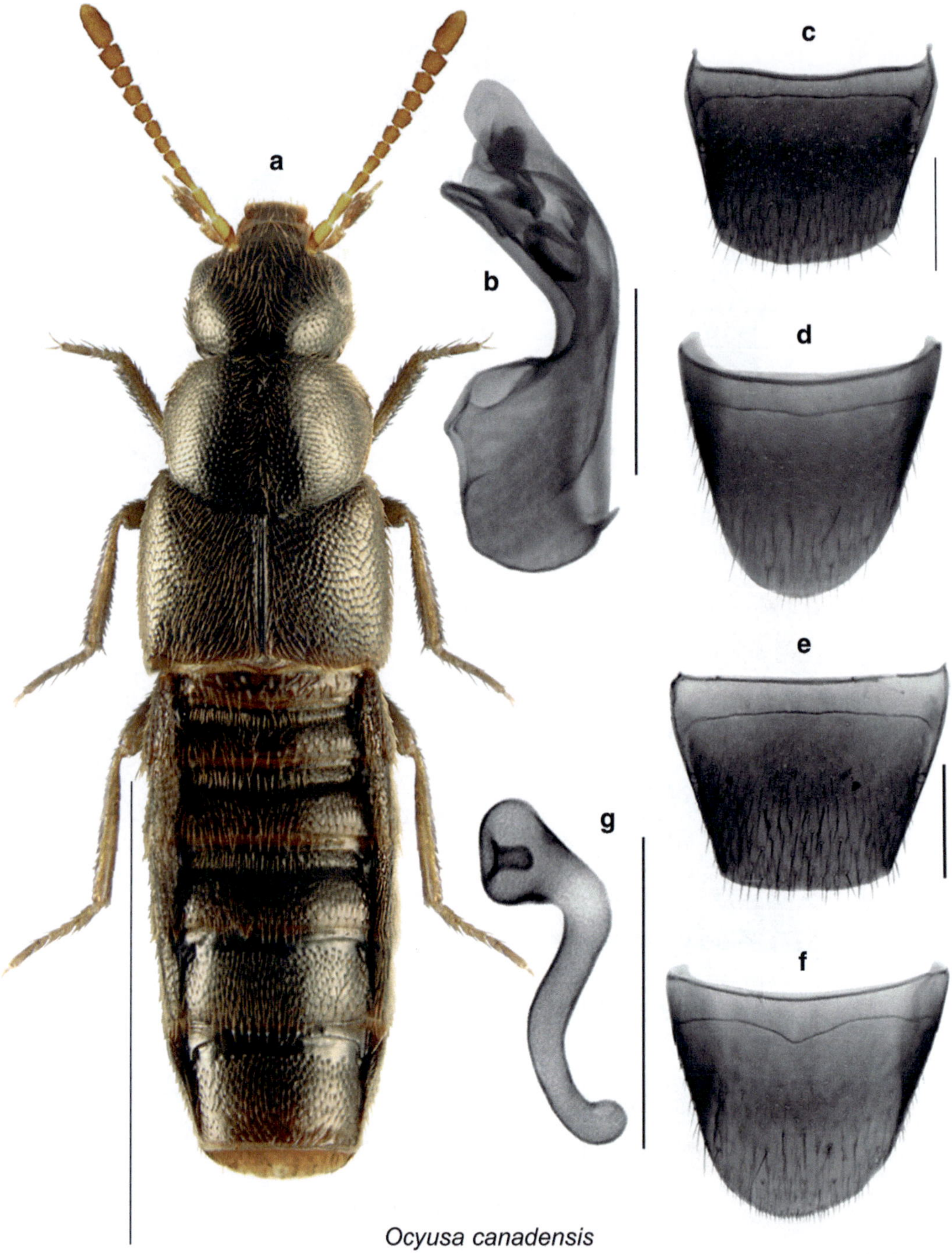

Fig. 11.108 (**a–g**). Images of *Ocyusa canadensis* Lohse: (**a**) habitus in dorsal view; (**b**) median lobe of aedeagus in lateral view; (**c**) male tergite VIII; (**d**) male sternite VIII; (**e**) female tergite VIII; (**f**) female sternite VIII; (**g**) spermatheca. (**a–g**) after Klimaszewski et al. (2018a). Scale bar for habitus = 1 mm; remaining scale bars = 0.2 mm

Fig. 11.109 (**a–i**). Images of *Oxypoda amica* Casey: (**a**) habitus in dorsal view; (**b–c**) median lobe of aedeagus in lateral view; (**d**) median lobe of aedeagus in dorsal view; (**e**) male tergite VIII; (**f**) male sternite VIII; (**g**) female tergite VIII; (**h**) female sternite VIII; (**i**) spermatheca. (**a–i**) after Klimaszewski et al. (2018a). Scale bar for habitus = 1 mm; remaining scale bars = 0.2 mm

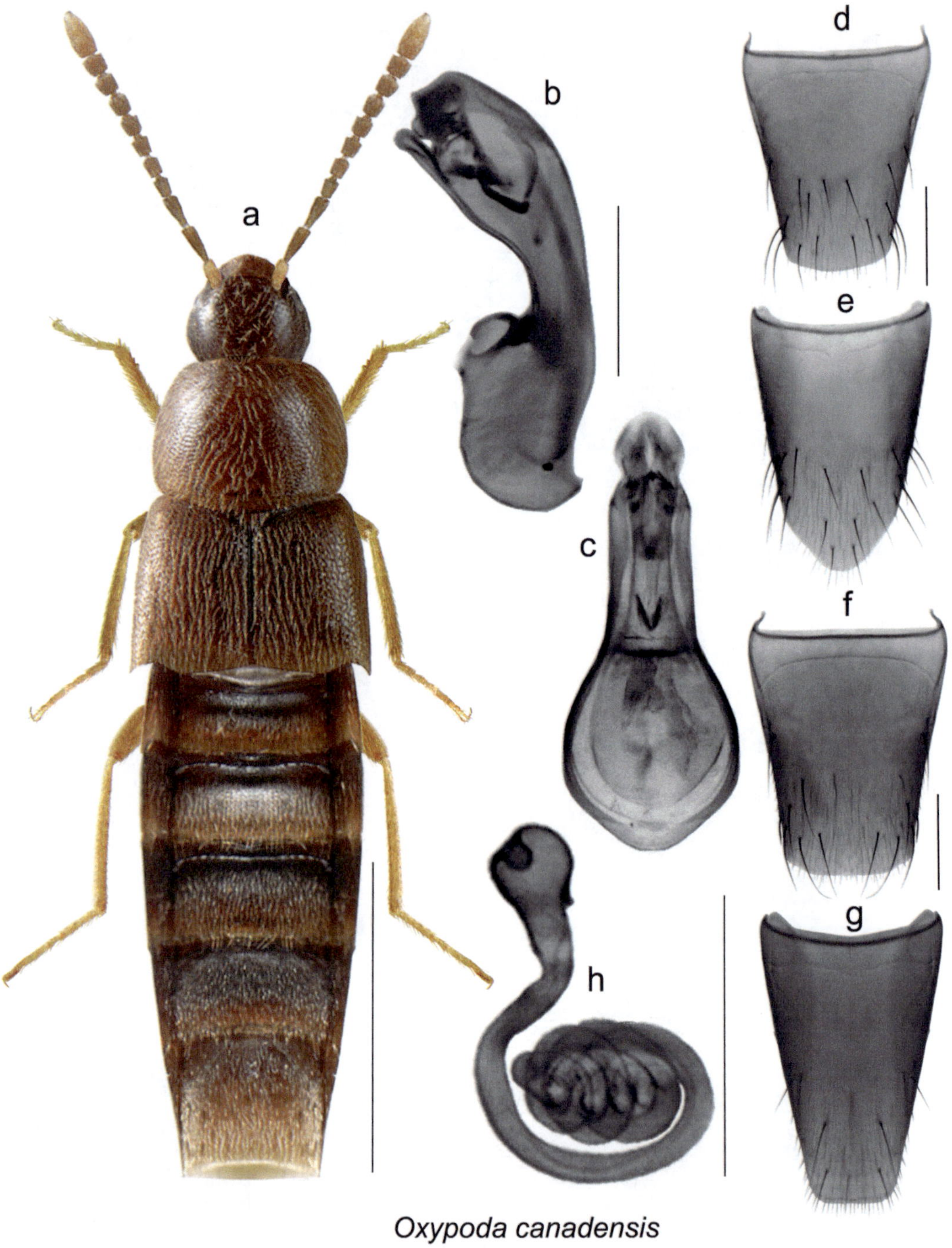

Fig. 11.110 (**a–h**). Images of *Oxypoda canadensis* Klimaszewski: (a) habitus in dorsal view; (**b**) median lobe of aedeagus in lateral view; (**c**) median lobe of aedeagus in dorsal view; (**d**) male tergite VIII; (**e**) male sternite VIII; (**f**) female tergite VIII; (**g**) female sternite VIII; (**h**) spermatheca. (**a–h**) after Klimaszewski et al. (2018a). Scale bar for habitus = 1 mm; remaining scale bars = 0.2 mm

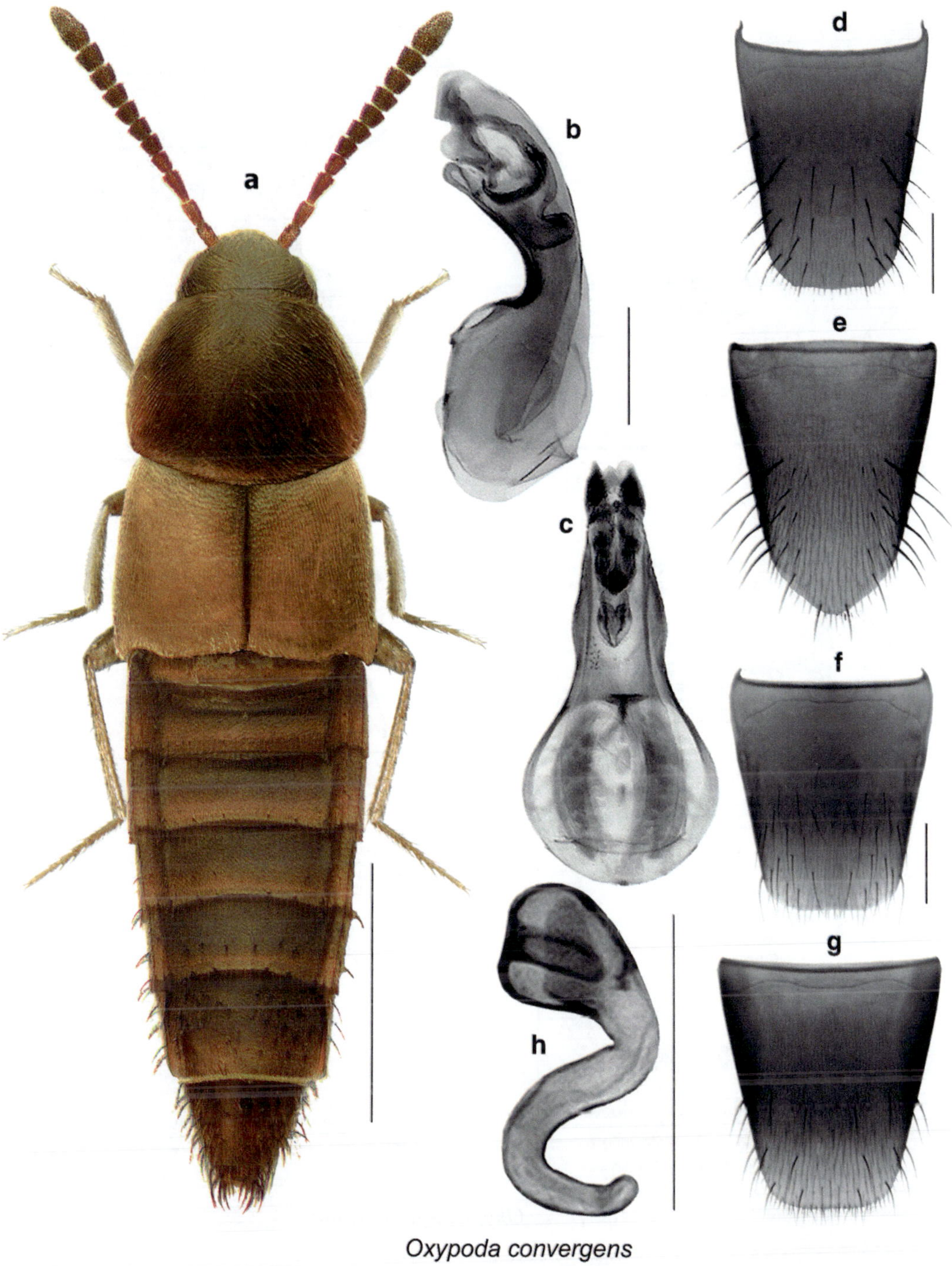

Fig. 11.111 (**a–h**). Images of *Oxypoda convergens* Casey: (**a**) habitus in dorsal view; (**b**) median lobe of aedeagus in lateral view; (**c**) median lobe of aedeagus in dorsal view; (**d**) male tergite VIII; (**e**) male sternite VIII; (**f**) female tergite VIII; (**g**) female sternite VIII; (**h**) spermatheca. (**a–h**) after Klimaszewski et al. (2018a). Scale bar for habitus = 1 mm; remaining scale bars = 0.2 mm

Fig. 11.112 (**a–h**). Images of *Oxypoda manitobae* Casey: (**a**) habitus in dorsal view; (**b**) median lobe of aedeagus in lateral view; (**c**) median lobe of aedeagus in dorsal view; (**d**) male tergite VIII; (**e**) male sternite VIII; (**f**) female tergite VIII; (**g**) female sternite VIII; (**h**) spermatheca. (**a–h**) after Klimaszewski et al. (2020). Scale bar for habitus = 1 mm; remaining scale bars = 0.2 mm

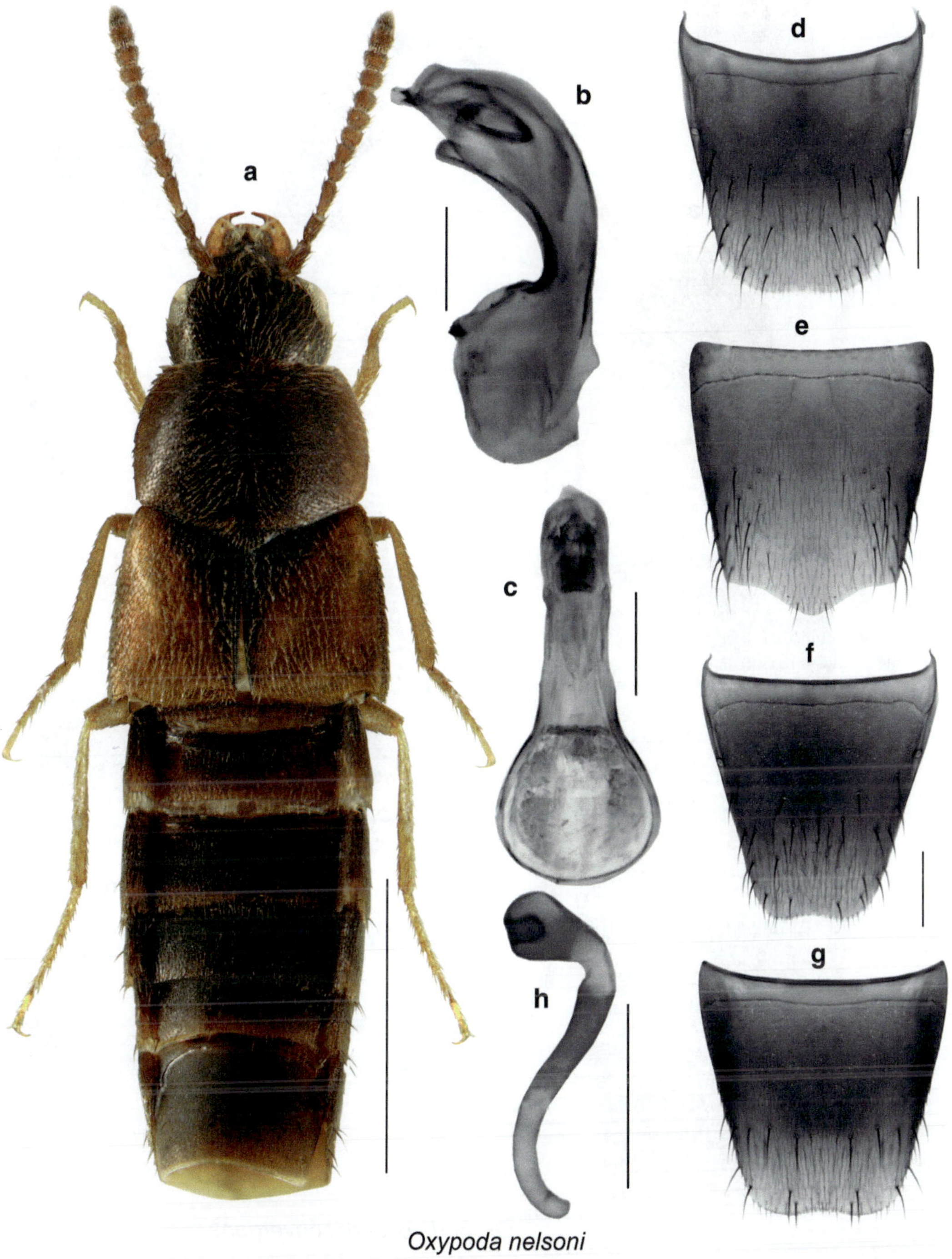

Fig. 11.113 (**a–h**). Images of *Oxypoda nelsoni* Lohse: (**a**) habitus in dorsal view; (**b**) median lobe of aedeagus in lateral view; (**c**) median lobe of aedeagus in dorsal view; (**d**) male tergite VIII; (**e**) male sternite VIII; (**f**) female tergite VIII; (**g**) female sternite VIII; (**h**) spermatheca. Scale bar for habitus = 1 mm; remaining scale bars = 0.2 mm

Fig. 11.114 (**a–i**). Images of *Oxypoda pseudoconvergens* Klimaszewski and Godin: (**a**) habitus in dorsal view; (**b**) median lobe of aedeagus in lateral view; (**c**) median lobe of aedeagus in dorsal view; (**d**) male tergite VIII; (**e**) male sternite VIII; (**f**) female tergite VIII; (**g**) female sternite VIII; (**h–i**) spermatheca. (**a–i**) after Klimaszewski et al. (2008). Scale bar for habitus = 1 mm; remaining scale bars = 0.2 mm

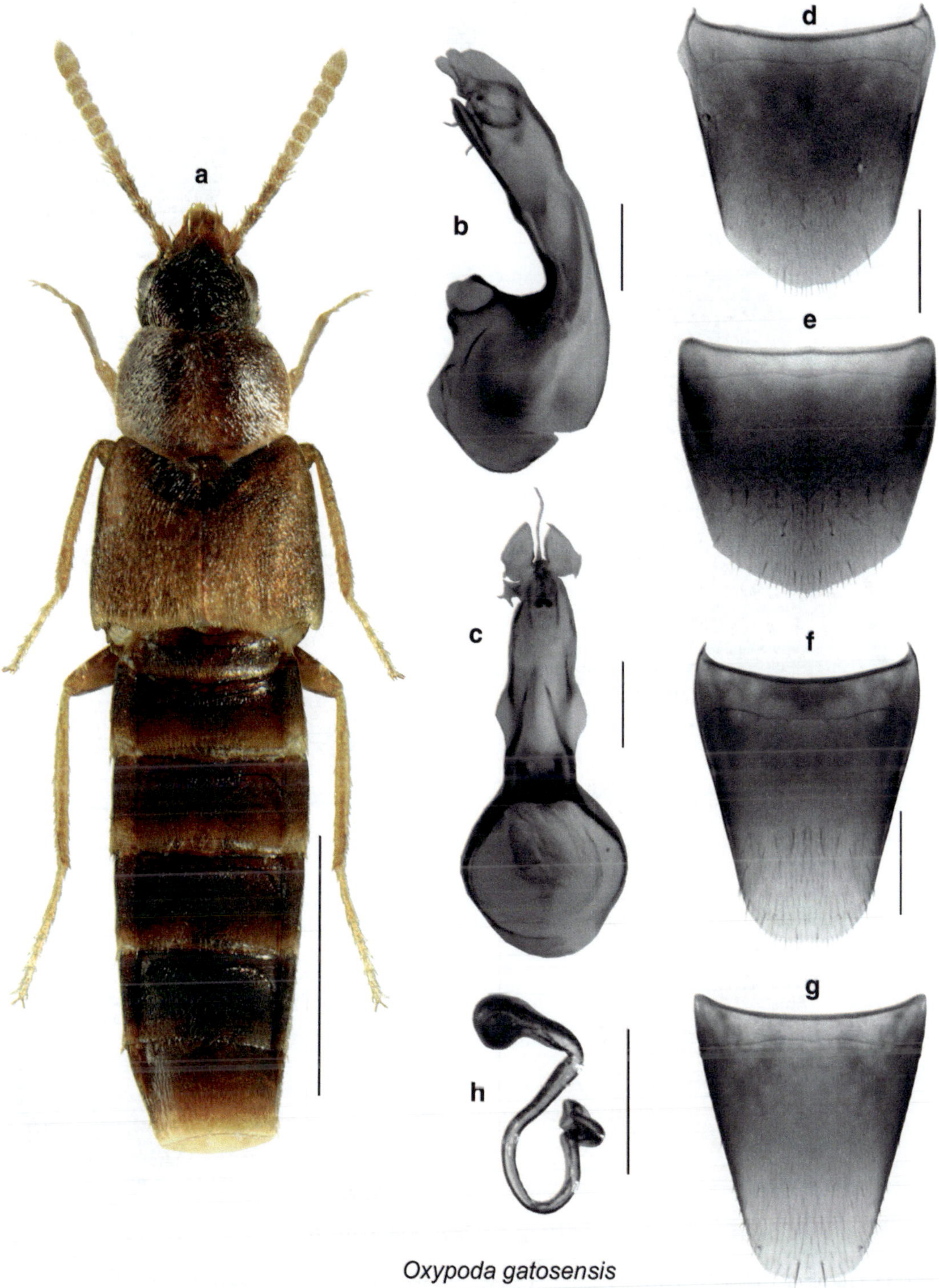

Fig. 11.115 (**a–h**). Images of *Oxypoda gatosensis* Bernhauer: (**a**) habitus in dorsal view; (**b**) median lobe of aedeagus in lateral view; (**c**) median lobe of aedeagus in dorsal view; (**d**) male tergite VIII; (**e**) male sternite VIII; (**f**) female tergite VIII; (**g**) female sternite VIII; (**h**) spermatheca. After Klimaszewski et al. (2020). Scale bar for habitus = 1 mm; remaining scale bars = 0.2 mm

Fig. 11.116 (**a–h**). Images of *Oxypoda inimica* Casey: (**a**) habitus in dorsal view; (**b**) median lobe aedeagus of in lateral view; (**c**) median lobe of aedeagus in dorsal view; (**d**) male tergite VIII; (**e**) male sternite VIII; (**f**) female tergite VIII; (**g**) female sternite VIII; (**h**) spermatheca. (**a–h**) after Klimaszewski et al. (2018a). Scale bar for habitus = 1 mm; remaining scale bars = 0.2 mm

Fig. 11.117 (**a–h**). Images of *Oxypoda yukonensis* Klimaszewski and Godin: (**a**) habitus in dorsal view; (**b**) median lobe of aedeagus in lateral view; (**c**) median lobe of aedeagus in dorsal view; (**d**) male tergite VIII; (**e**) male sternite VIII; (**f**) female tergite VIII; (**g**) female sternite VIII; (**h**) spermatheca. (**a–h**) after Klimaszewski et al. (2008). Scale bar for habitus = 1 mm; remaining scale bars = 0.2 mm

Fig. 11.118 (**a–h**). Images of *Oxypoda volkeri* Klimaszewski: (**a**) habitus in dorsal view; (**b**) median lobe of aedeagus in lateral view; (**c**) median lobe of aedeagus in dorsal view; (**d**) male tergite VIII; (**e**) male sternite VIII; (**f**) female tergite VIII; (**g**) female sternite VIII; (**h**) spermatheca. (**a–h**) after Klimaszewski et al. (2006). Scale bar for habitus = 1 mm; remaining scale bars = 0.2 mm

Fig. 11.119 (**a–h**). Images of *Oxypoda irrasa* Mäklin: (**a**) habitus in dorsal view; (**b**) median lobe of aedeagus in lateral view; (**c**) median lobe of aedeagus in dorsal view; (**d**) male tergite VIII; (**e**) male sternite VIII; (**f**) female tergite VIII; (**g**) female sternite VIII; (**h**) spermatheca. (**a–h**) after Klimaszewski et al. (2020). Scale bar for habitus = 1 mm; remaining scale bars = 0.2 mm

Fig. 11.120 (**a–h**). Images of *Oxypoda hiemalis* Casey: (**a**) habitus in dorsal view; (**b**) median lobe of aedeagus in lateral view; (**c**) median lobe of aedeagus in dorsal view; (**d**) male tergite VIII; (**e**) male sternite VIII; (**f**) female tergite VIII; (**g**) female sternite VIII; (**h**) spermatheca. (**a–h**) after Klimaszewski et al. (2018a). Scale bar for habitus = 1 mm; remaining scale bars = 0.2 mm

Fig. 11.121 (**a–i**). Images of *Oxypoda lacustris* Casey: (**a**) habitus in dorsal view; (**b**) median lobe of aedeagus in lateral view; (**c**) median lobe of aedeagus in dorsal view; (**d**) male tergite VIII; (**e**) male sternite VIII; (**f**) female tergite VIII; (**g**) female sternite VIII; (**h–i**) spermatheca. (**a–i**) after Klimaszewski et al. (2018a). Scale bar for habitus = 1 mm; remaining scale bars = 0.2 mm

Fig. 11.122 (**a–i**). Images of *Oxypoda demissa* Casey: (a) habitus in dorsal view; (**b**) median lobe of aedeagus in lateral view; (**c**) median lobe of aedeagus in dorsal view; (**d**) male tergite VIII; (**e**) male sternite VIII; (**f**) female tergite VIII; (**g**) female sternite VIII; (**h–i**) spermatheca. (**a–i**) after Klimaszewski et al. (2018a). Scale bar for habitus = 1 mm; remaining scale bars = 0.2 mm

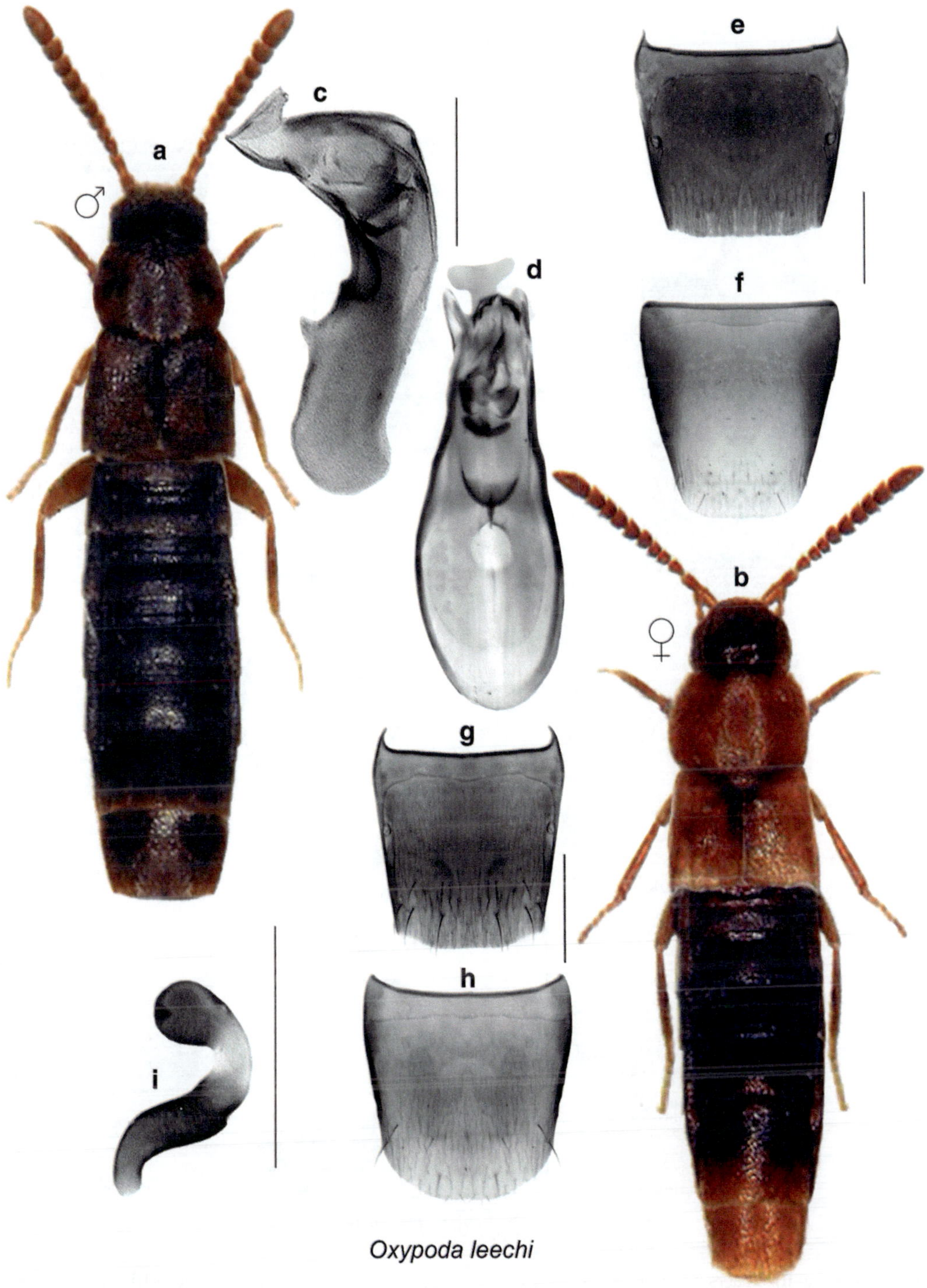

Fig. 11.123 (**a–i**). Images of *Oxypoda leechi* Lohse: (**a**) male habitus in dorsal view (holotype); (**b**) female habitus in dorsal view (non-paratype); (**c**) median lobe of aedeagus in lateral view; (**d**) median lobe of aedeagus in dorsal view; (**e**) male tergite VIII; (**f**) male sternite VIII; (**g**) female tergite VIII; (**h**) female sternite VIII; (**i**) spermatheca. Scale bar for habitus = 1 mm; remaining scale bars = 0.2 mm

Fig. 11.124 (**a–h**). Images of *Oxypoda lucidula* Casey: (**a**) habitus in dorsal view; (**b**) median lobe of aedeagus in lateral view; (**c**) median lobe of aedeagus in dorsal view; (**d**) male tergite VIII; (**e**) male sternite VIII; (**f**) female tergite VIII; (**g**) female sternite VIII; (**h**) spermatheca. (**a–h**) after Klimaszewski et al. (2018a). Scale bar for habitus = 1 mm; remaining scale bars = 0.2 mm

Fig. 11.125 (**a–g**). Images of *Oxypoda robusticornis* Bernhauer: (**a**) habitus in dorsal view; (**b**) median lobe of aedeagus in lateral view; (**c**) median lobe of aedeagus in dorsal view; (**d**) male tergite VIII; (**e**) male sternite VIII; (**f**) female tergite VIII; (**g**) female sternite VIII. Scale bar for habitus = 1 mm; remaining scale bars = 0.2 mm

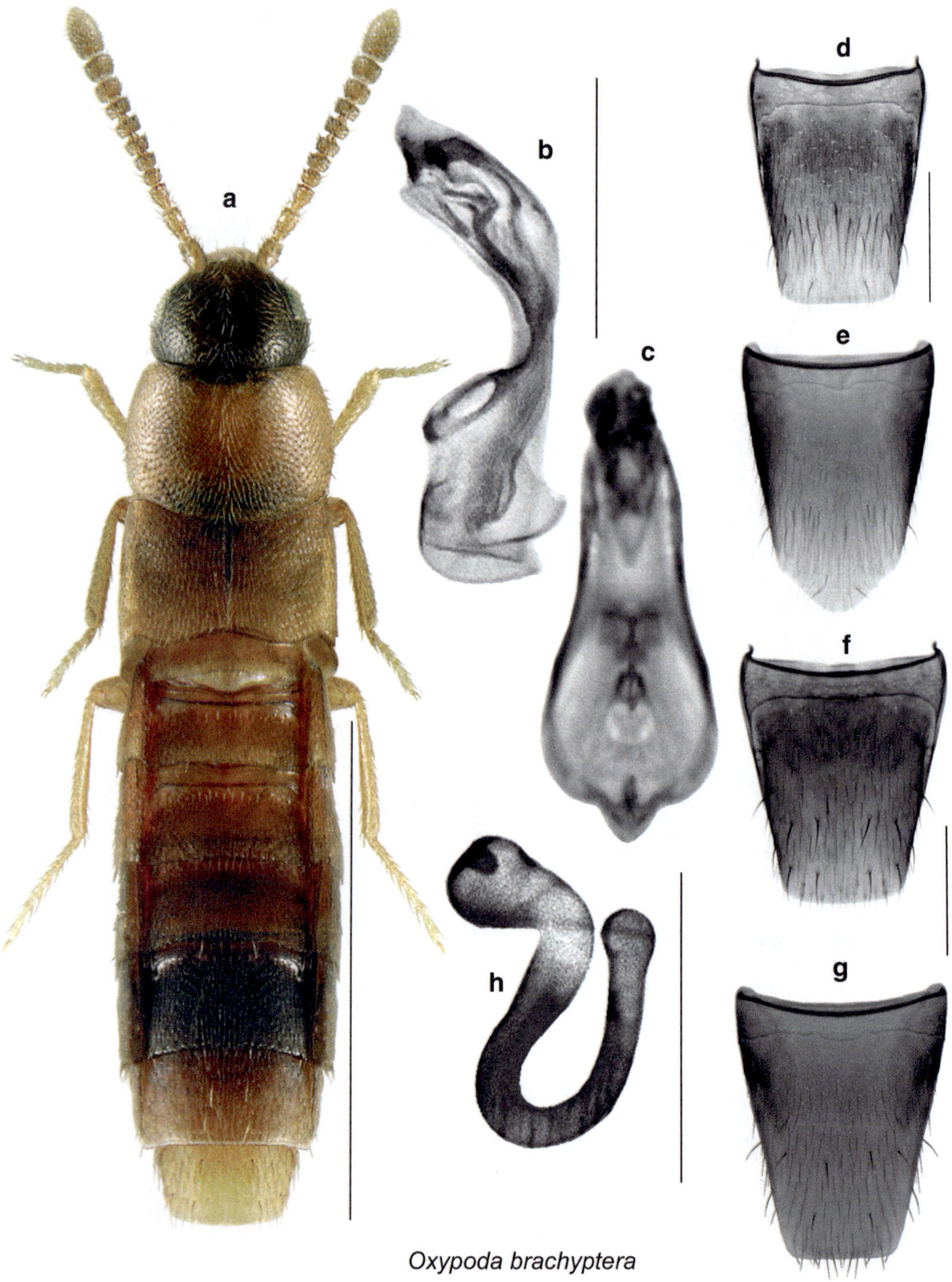

Fig. 11.126 (**a–h**). Images of *Oxypoda brachyptera* (Stephens): (**a**) habitus in dorsal view; (**b**) median lobe of aedeagus in lateral view; (**c**) median lobe of aedeagus in dorsal view; (**d**) male tergite VIII; (**e**) male sternite VIII; (**f**) female tergite VIII; (**g**) female sternite VIII; (**h**) spermatheca. (**a–h**) after Klimaszewski et al. (2018a). Scale bar for habitus = 1 mm; remaining scale bars = 0.2 mm

Fig. 11.127 (**a–h**). Images of *Oxypoda operta* Sjöberg: (**a**) habitus in dorsal view; (**b**) median lobe of aedeagus in lateral view; (**c**) median lobe of aedeagus in dorsal view; (**d**) male tergite VIII; (**e**) male sternite VIII; (**f**) female tergite VIII; (**g**) female sternite VIII; (**h**) spermatheca. (**a–h**) after Klimaszewski et al. (2018a). Scale bar for habitus = 1 mm; remaining scale bars = 0.2 mm

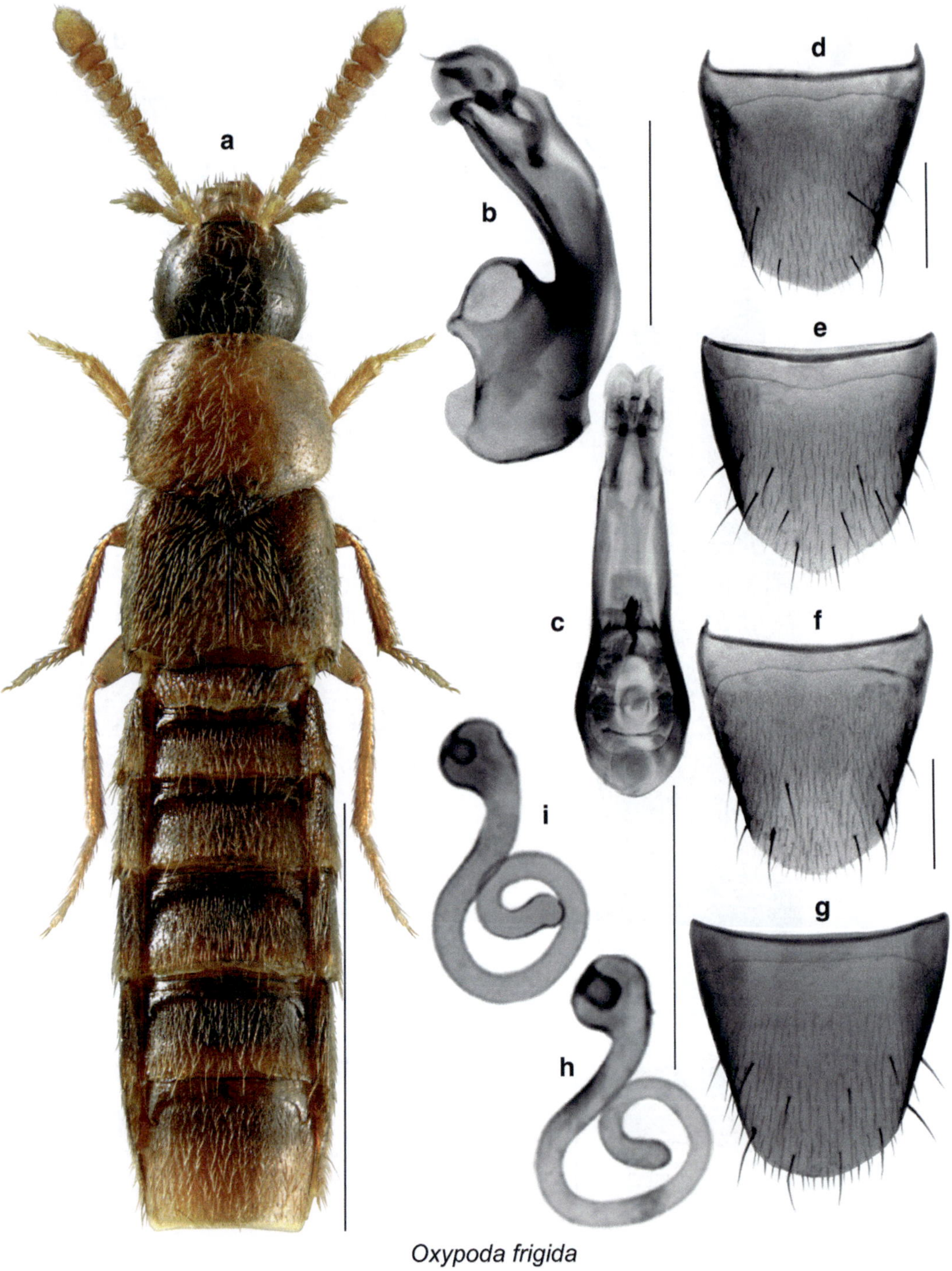

Fig. 11.128 (**a–i**). Images of *Oxypoda frigida* Bernhauer: (**a**) habitus in dorsal view; (**b**) median lobe of aedeagus in lateral view; (**c**) median lobe of aedeagus in dorsal view; (**d**) male tergite VIII; (**e**) male sternite VIII; (**f**) female tergite VIII; (**g**) female sternite VIII; (**h–i**) spermatheca. (**a–i**) after Klimaszewski et al. (2018a). Scale bar for habitus = 1 mm; remaining scale bars = 0.2 mm

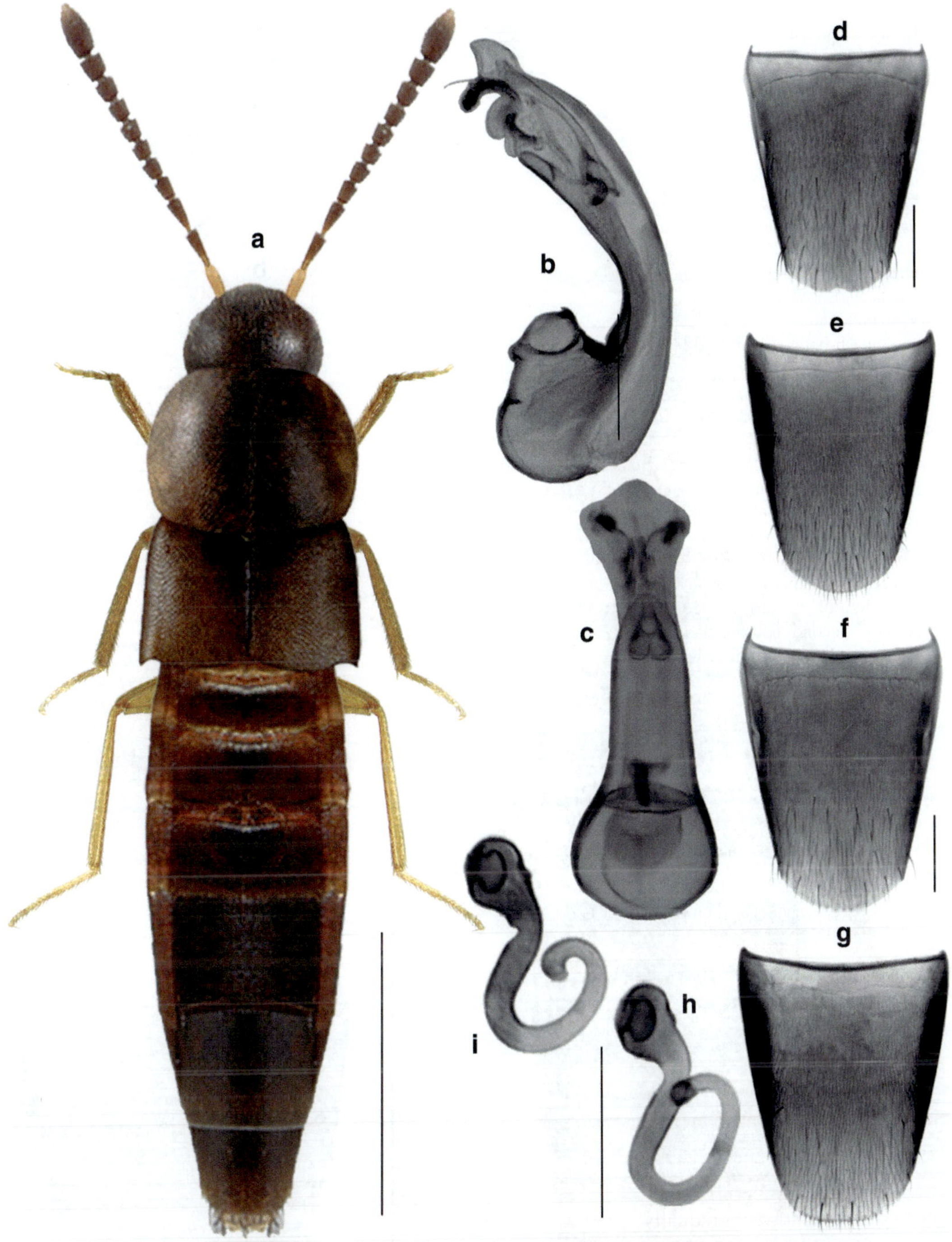

Fig. 11.129 (**a–i**). Images of *Oxypoda orbicollis* Casey: (**a**) habitus in dorsal view; (**b**) median lobe of aedeagus in lateral view; (**c**) median lobe of aedeagus in dorsal view; (**d**) male tergite VIII; (**e**) male sternite VIII; (**f**) female tergite VIII; (**g**) female sternite VIII; (**h–i**) spermatheca. (**a–i**) after Klimaszewski et al. (2018a). Scale bar for habitus = 1 mm; remaining scale bars = 0.2 mm

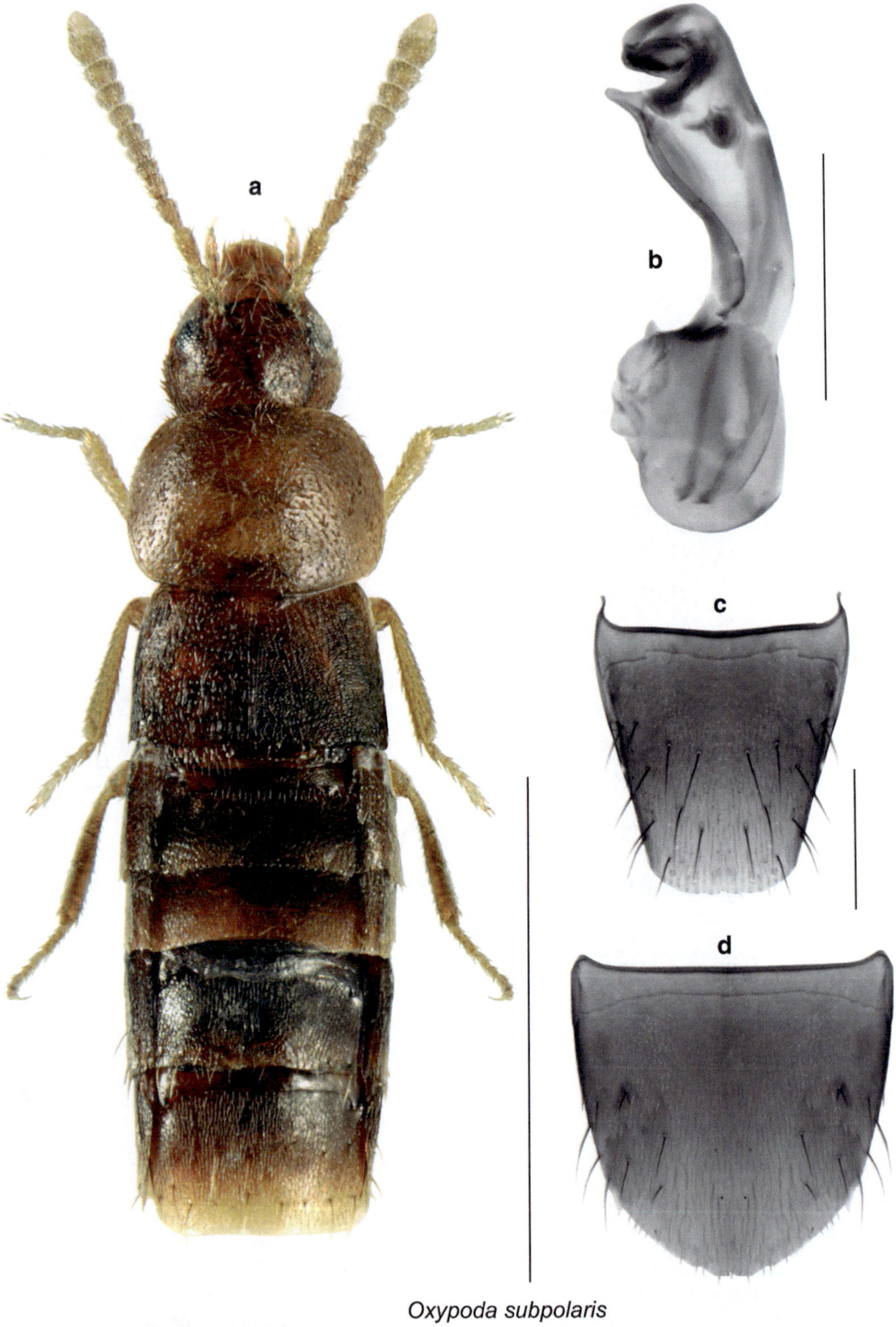

Fig. 11.130 (**a–d**). Images of *Oxypoda subpolaris* Casey: (**a**) habitus in dorsal view; (**b**) median lobe of aedeagus in lateral view; (**c**) male tergite VIII; (**d**) male sternite VIII. Scale bar for habitus = 1 mm; remaining scale bars = 0.2 mm

Fig. 11.131 (**a–h**). Images of *Oxypoda sylvia* Casey: (**a**) habitus in dorsal view; (**b**) median lobe of aedeagus in lateral view; (**c**) median lobe of aedeagus in dorsal view; (**d**) male tergite VIII; (**e**) male sternite VIII; (**f**) female tergite VIII; (**g**) female sternite VIII; (**h**) spermatheca. (**a–h**) after Klimaszewski et al. (2018a). Scale bar for habitus = 1 mm; remaining scale bars = 0.2 mm

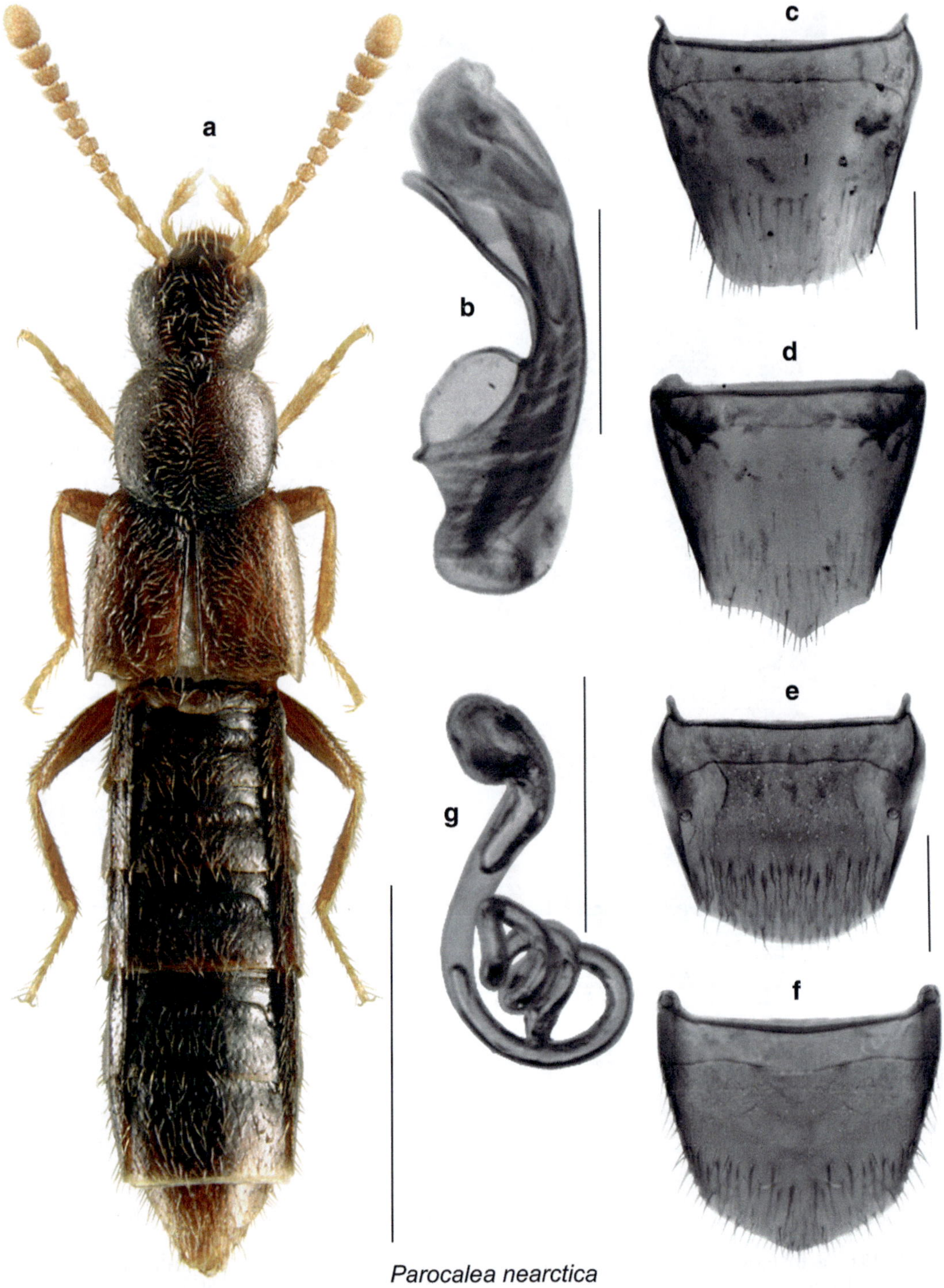

Fig. 11.132 (**a–g**). Images of *Parocalea nearctica* Lohse: (**a**) habitus in dorsal view; (**b**) median lobe of aedeagus in lateral view; (**c**) male tergite VIII; (**d**) male sternite VIII; (**e**) female tergite VIII; (**f**) female sternite VIII; (**g**) spermatheca. Scale bar for habitus = 1 mm; remaining scale bars = 0.2 mm

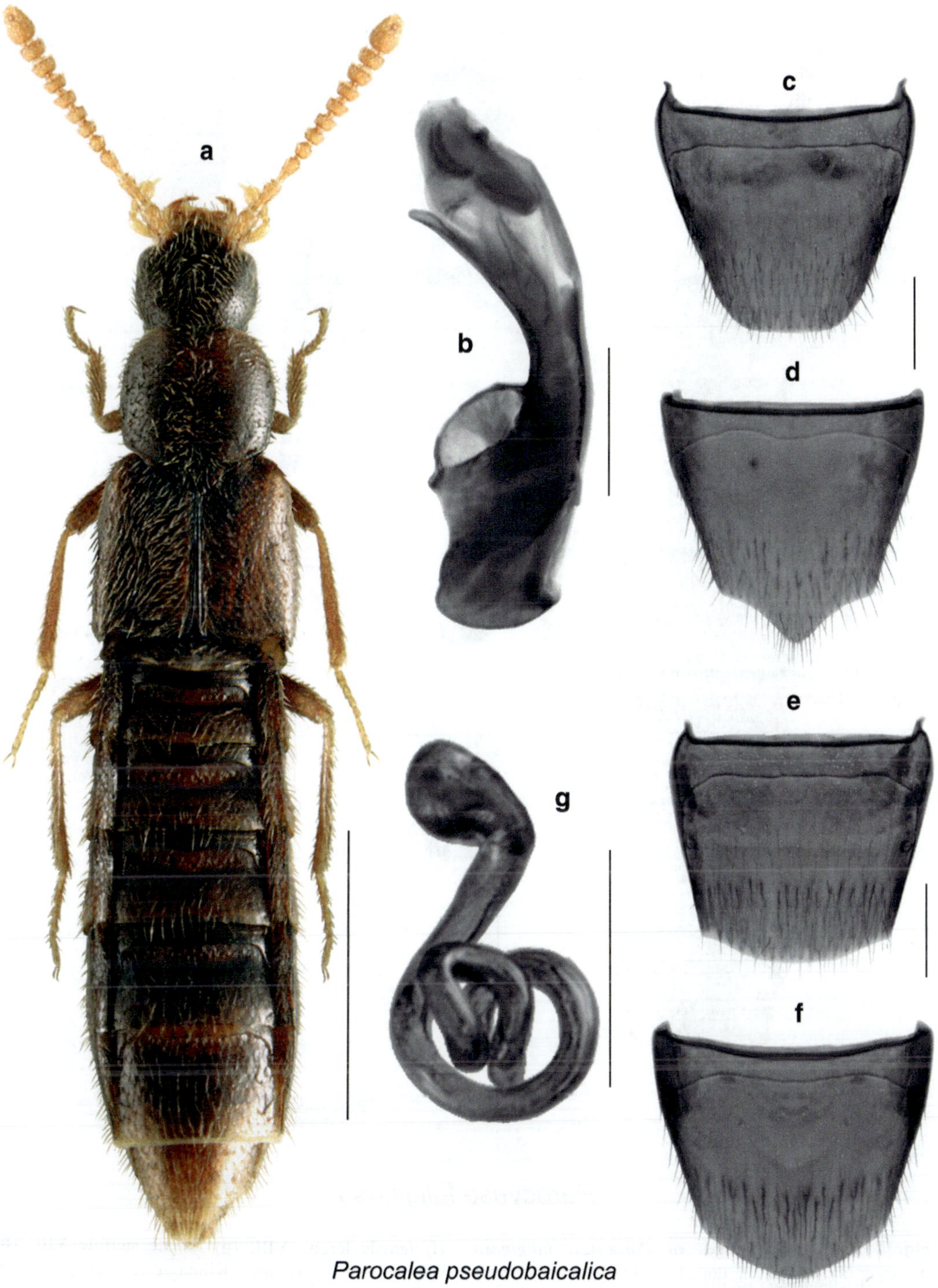

Fig. 11.133 (**a–g**). Images of *Parocalea pseudobaicalica* Lohse: (**a**) habitus in dorsal view; (**b**) median lobe of aedeagus in lateral view; (**c**) male tergite VIII; (**d**) male sternite VIII; (**e**) female tergite VIII; (**f**) female sternite VIII; (**g**) spermatheca. Scale bar for habitus = 1 mm; remaining scale bars = 0.2 mm

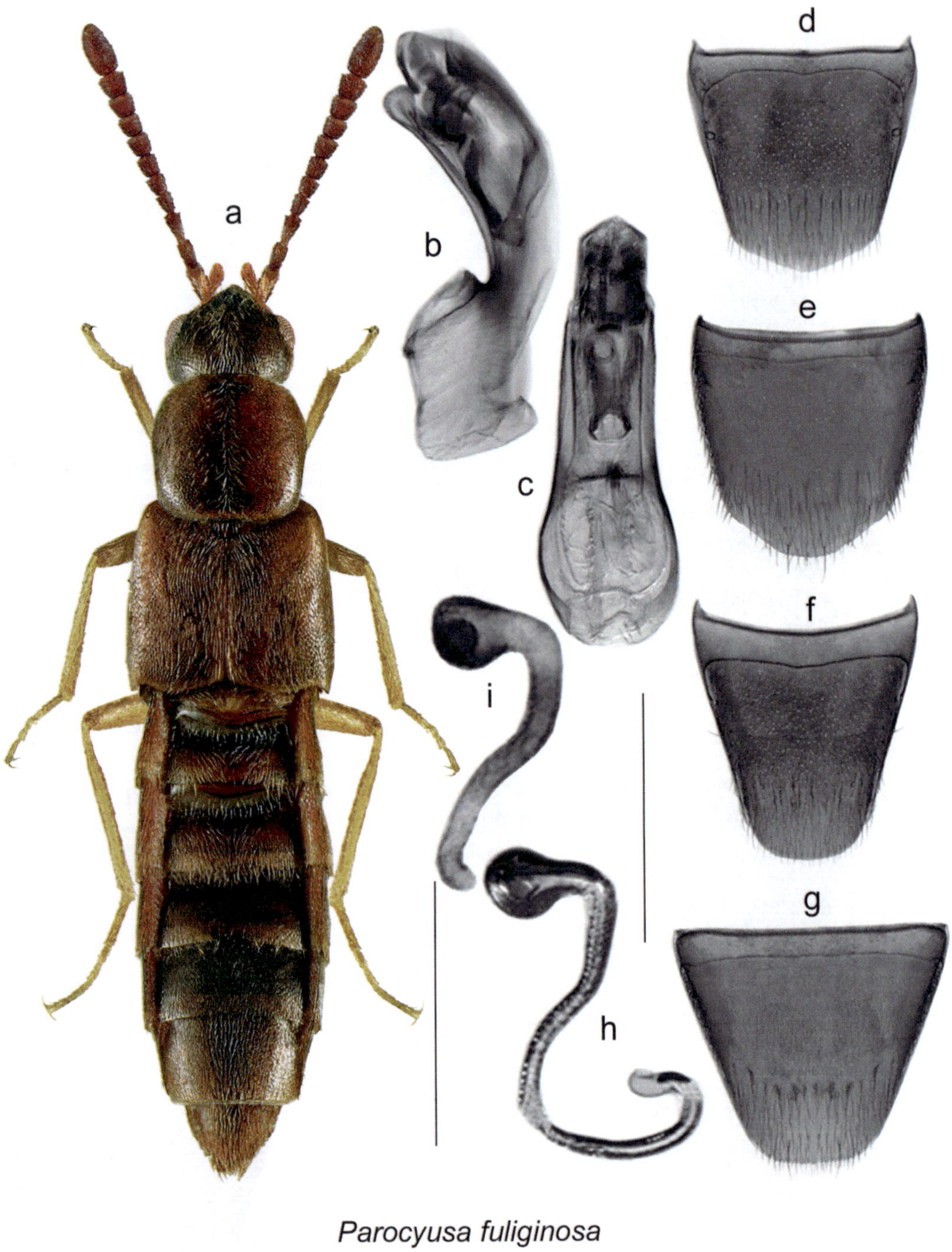

Fig. 11.134 (**a–i**). Images of *Parocyusa fuliginosa* (Casey): (**a**) habitus in dorsal view; (**b**) median lobe of aedeagus in lateral view; (**c**) median lobe of aedeagus in dorsal view; (**d**) male tergite VIII; (**e**) male sternite VIII; (**f**) female tergite VIII; (**g**) female sternite VIII; (**h–i**) spermatheca. (**a–i**) after Klimaszewski et al. (2018a). Scale bar for habitus = 1 mm; remaining scale bars = 0.2 mm

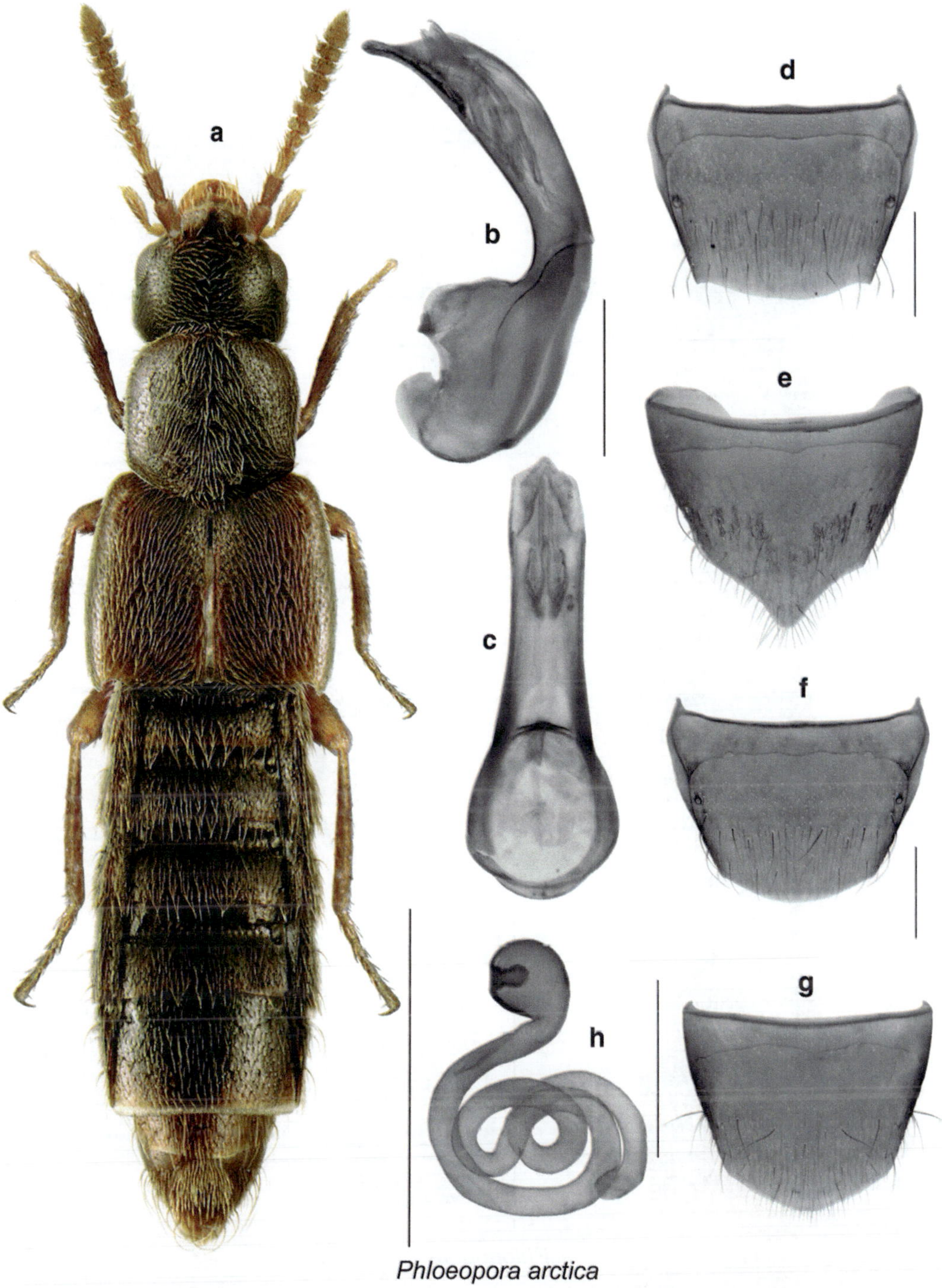

Fig. 11.135 (**a–h**). Images of *Phloeopora arctica* Lohse: (**a**) habitus in dorsal view; (**b**) median lobe of aedeagus in lateral view; (**c**) median lobe of aedeagus in dorsal view; (**d**) male tergite VIII; (**e**) male sternite VIII; (**f**) female tergite VIII; (**g**) female sternite VIII; (**h**) spermatheca. (**a–h**) after Klimaszewski et al. (2018a). Scale bar for habitus = 1 mm; remaining scale bars = 0.2 mm

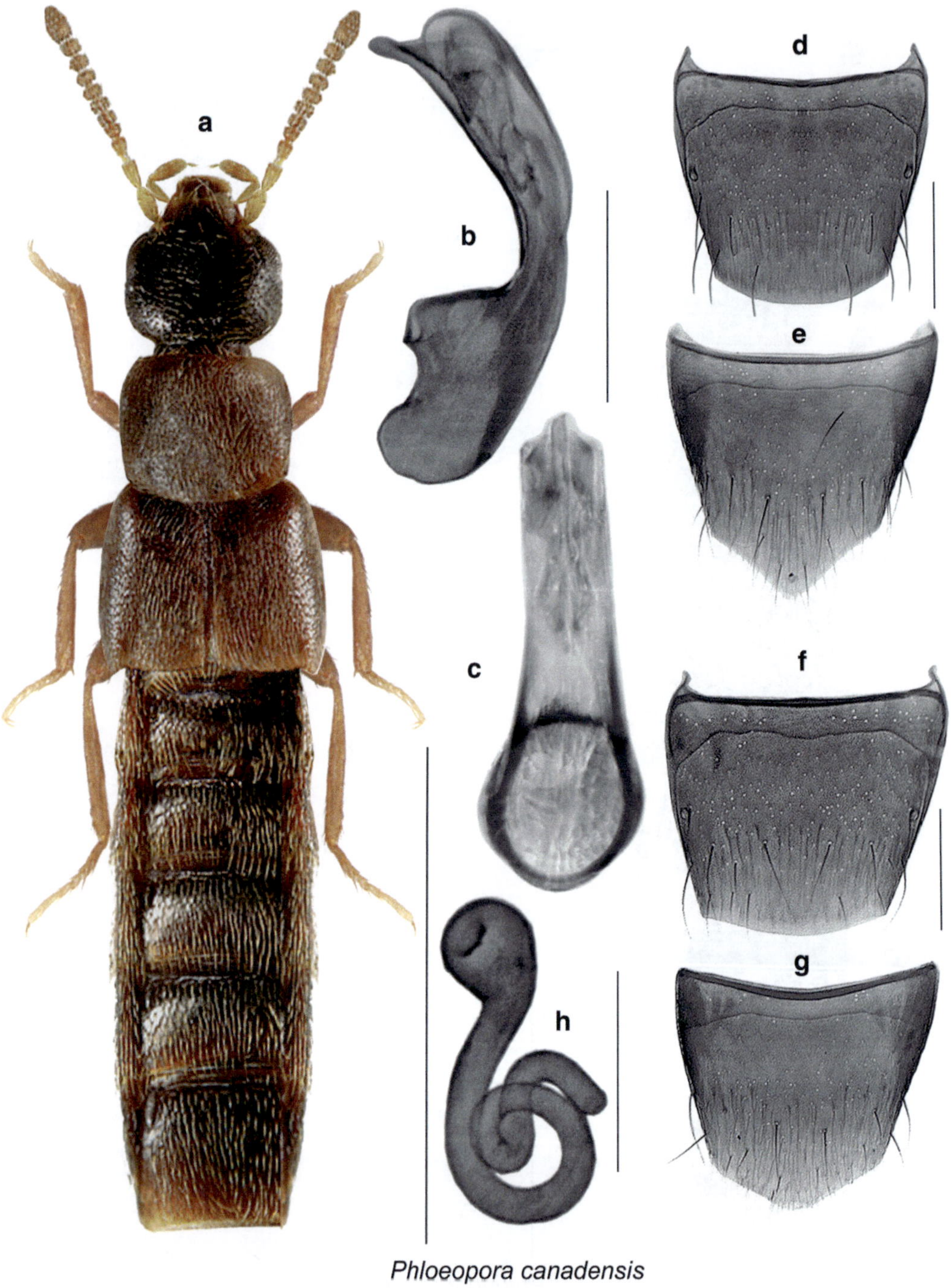

Fig. 11.136 (**a–h**). Images of *Phloeopora canadensis* Klimaszewski and Langor: (**a**) habitus in dorsal view; (**b**) median lobe of aedeagus in lateral view; (**c**) median lobe of aedeagus in dorsal view; (**d**) male tergite VIII; (**e**) male sternite VIII; (**f**) female tergite VIII; (**g**) female sternite VIII; (**h**) spermatheca. (**a–h**) after Klimaszewski et al. (2011, 2018a). Scale bar for habitus = 1 mm; remaining scale bars = 0.2 mm

Fig. 11.137 (**a–g**). Images of *Phloeopora gilbertae* Klimaszewski and Webster: (**a**) habitus in dorsal view; (**b**) median lobe of aedeagus in lateral view; (**c**) male tergite VIII; (**d**) male sternite VIII; (**e**) female tergite VIII; (**f**) female sternite VIII; (**g**) spermatheca. (**a–g**) after Webster et al. (2016). Scale bar for habitus = 1 mm; remaining scale bars = 0.2 mm

References

Ashe JS (2001) Key to the tribes and genera of Nearctic Aleocharinae. 8. In: Arnett RH Jr, Thomas CM (eds) American Beetles. Archostemata, Myxophaga, Adephaga, Polyphaga: Staphyliniformia. CRC Press, Washington, DC, pp 299–374. 443 pp. [publ. 28 Dec 2000 acccording to CRC Press]

Assing V (2002) A taxonomic and phylogenetic revision of *Amarochara* Thomson. I. The species of the Holarctic region (Coleoptera: Staphylinidae, Aleocharinae, Oxypodini). Beiträge zur Entomologie 52(1):111–204

Assing V (2008a) The genus *Calodera* Mannerheim in Canada (Insecta, Coleoptera, Staphylinidae, Aleocharinae). In: Majka CG and Klimaszewski J (Eds.) Biodiversity, Biosystematics, and Ecology of Canadian Coleoptera. Zookeys [Special Issue] 2:203–208

Assing V (2008b) The genus *Calodera* Mannerheim in Canada (Insecta, Coleoptera, Staphylinidae, Aleocharinae). Zookeys (Special Issue) 2:203–208

Assing V (2012) On the taxonomy and natural history of *Oxypoda brachyptera* and *O. tarda*. Beiträge zur Entomologie 62:207–224

Assing V (2021) On the taxonomy of *Parocyusa, Tectusa*, and miscellaneous genera of oxypodina (Insecta: Coleoptera: Staphylinidae: Aleocharinae: Oxypodini). Ann Naturhist Mus Wien, B 123:99–218

Assing V, Schülke M (2007) Supplemente zur mitteleuropäischen Staphylinidenfauna (Coleoptera, Staphylinidae). III. Entomologische Blätter für Biologie und Systematik der Käfer 102(1–3):1–78

Assing V, Vogel J (2019) The mother of synonyms: on the *Meotica* species of the Palaearctic Region (Coleoptera, Staphylinidae, Aleocharinae, Oxypodini). Linzer Biol Beitr 51(2):731–772

Benick G, Lohse GA (1974) Tribus 14 (Callicerini). In: Freude H, Harde KW, Lohse A (eds) Die Käfer Mitteleuropas. Band 5: Staphylinidae II (Hypocyphtinae und Aleocharinae) Pselaphidae. Goecke & Evers, Krefeld, pp 72–221. 381 pp

Bernhauer M (1902) Die Staphyliniden der paläarktischen Fauna. I. Tribus: Aleocharini (II. Theil). Verhandlungen der kaiserlich-königlichen zoologisch-botanischen Gesellschaft in Wien 52 (Beiheft):87–284

Bernhauer M (1907) Neue Aleocharini aus Nordamerika. (Col.) (3. Stück.). Deutsche Entomologische Zeitschrift 1907(4):381–405

Bernhauer M (1928) Neue Staphyliniden der palaearktischen Fauna. Kol Rundsch 14:8–23

Bernhauer M, Scheerpeltz O (1926) Staphylinidae (Pars 82) (pp. 499-988). In: Junk W, Schenkling S (eds) Coleopterorum catalogus. V. W. Junk, Berlin

Bousquet Y, Bouchard P, Davies AE, Sikes DS (2013) Checklist of beetles (Coleoptera) of Canada and Alaska. Second edition. Pensoft Publishers, Series Faunistica No. 109. Sofia-Moscow, 402 pp

Brunke A, Klimaszewski J, Dorval J-A, Bourdon C, Paiero SM, Marshall SA (2012) New species and distributional records of Aleocharinae (Coleoptera, Staphylinidae) from Ontario, Canada, with a checklist of recorded species. *In*: Klimaszewski J, Anderson R (Eds) Biosystematics and Ecology of Canadian Staphylinidae (Coleoptera) II. Zookeys [Special Issue] 186:119–206

Brunke AJ, Bahlai CA, Klimaszewski J, Hallett RH (2014) Rove beetles (Coleoptera: Staphylinidae) in Ontario, Canada soybean agroecosystems: assemblage diversity, composition, seasonality, and habitat use. Can Entomol 146:652–670

Brunke A, Pentinsaari M, Klimaszewski J (2021) Integrative taxonomy of Nearctic and Palaearctic Aleocharinae: new species, synonymies, and records (Coleoptera, Staphylinidae). Zookeys 1041:27–99

Casey TL (1893) Coleopterological notices. V. Ann N Y Acad Sci 7(1894):281–606. [see reference in Chapter 5 for explanation of date]

Casey TL (1906) Observations on the staphylinid groups Aleocharinae and Xantholinini, chiefly of America. Trans Acad Sci St Louis 16(6):125–434

Casey TL (1911) New American species of Aleocharinae and Myllaeninae [pp. 1-245] In: Memoirs on the Coleoptera. II. The New Era Printing Co., Lancaster, PA, p 259

Erichson WF (1839) Genera et species staphylinorum insectorum coleopterorum familiae. Erster Band. F. H. Morin, Berlin. viii, 400 pp

Fenyes A (1910) Two new species of Aleocharinae from California. Proc Entomol Soc Wash 11:197–199

Fenyes A (1920) Coleoptera: Fam Staphylinidae, subfam. Aleocharinae. In: Wytsman P (Ed) Genera Insectorum, Fasc 173 A. L. Desmet-Verteneuil, Bruxelles, 1–110

Fenyes A (1921) New genera and species of Aleocharinae with a polytomic synopsis of the tribes. Bulletin of the Museum of Comparative. Zoology 65:17–36

Gouix N, Klimaszewski J (2007) Catalogue of aleocharine rove beetles of Canada and Alaska (Coleoptera, Staphylinidae, Aleocharinae). Pensoft Publishers, Sofia-Moscow. 165 pp

Gusarov VI (2002) A revision of Nearctic species of the genus *Geostiba* Thomson, 1858 (Coleoptera: Staphylinidae: Aleocharinae). Zootaxa 81:1–88

Gusarov VI (2003) Revision of some types of North American aleocharines (Coleoptera: Staphylinidae: Aleocharinae), with synonymic notes. Zootaxa 353:1–134

Kirby W, Spence W (1837) Insects. Coleoptera. In: Richardson, Fauna Boreali-Americana; or the zoology of the northern parts of British America. Norwich. 249 pp

Klimaszewski J, Pelletier G (2004) Review of the *Ocalea* group of genera (Coleoptera, Staphylinidae, Aleocharinae) in Canada and Alaska: new taxa, bionomics, and distribution. Can Entomol 136 (4):443–500

Klimaszewski J, Winchester NN (2002) Aleocharine rove beetles (Coleoptera Staphylinidae) of the ancient Sitka spruce forest on Vancouver Island, British Columbia, Canada. Mémoires de la Société royale belge d'Entomologie 40:3–126

Klimaszewski J, Sweeney J, Price J, Pelletier G (2005) Rove beetles (Coleoptera: Staphylinidae) in red spruce stands, eastern Canada: diversity, abundance, and descriptions of new species. Can Entomol 137(1):1–48

Klimaszewski J, Pelletier G, Germain C, Work T, Hébert C (2006) Review of *Oxypoda* species in Canada and Alaska (Coleoptera, Staphylinidae, Aleocharinae): systematics, bionomics, and distribution. Can Entomol 138(6):737–852

Klimaszewski J, Assing V, Majka CG, Pelletier G, Webster RP, Langor D (2007a) Records of adventive aleocharine beetles (Coleoptera: Staphylinidae: Aleocharinae) found in Canada. Can Entomol 139 (1):54–79

Klimaszewski J, Langor D, Savard K, Pelletier G, Chandler DS, Sweeney J (2007b) Rove beetles (Coleoptera: Staphylinidae) in yellow birch-dominated stands of southeastern Quebec, Canada: diversity, abundance, and description of a new species. Can Entomol 139 (6):793–833

Klimaszewski J, Godin B, Pelletier G, Savard K (2008) Six new species and records of aleocharine beetles from the Yukon and Alaska (Coleoptera: Staphylinidae: Aleocharinae). Can Entomol 140 (3):265–291

Klimaszewski J, Langor D, Majka CG, Bouchard P, Bousquet Y, LeSage L, Smetana A, Sylvestre P, Pelletier G, Davies A, DesRochers P, Goulet H, Webster R, Sweeney J (2010) Review of adventive species of Coleoptera (Insecta) recorded from eastern Canada. Pensoft Series Faunistica No. 94. Pensoft, Sofia-Moscow, 272 pp

Klimaszewski J, Langor D, Pelletier G, Bourdon C, Perdereau L (2011) Aleocharine beetles (Coleoptera, Staphylinidae) of the province of Newfoundland and Labrador, Canada. Pensoft Series Faunistica No. 98. Pensoft, Sophia-Moscow, 313 pp

Klimaszewski J, Godin B, Bourdon C (2012) Further contributions to the aleocharine fauna of the Yukon Territory, Canada (Coleoptera, Staphylinidae). In: Klimaszewski J, Anderson R (Eds.): Biosystematics and Ecology of Canadian Staphylinidae (Coleoptera) II. Zookeys [Special Issue] 186:207–237

Klimaszewski J, Brunke A, Assing V, Langor D, Newton AF, Bourdon C, Pelletier G, Webster R, Herman L, Perdereau L, Davies A, Smetana A, Chandler D, Majka C, Scudder G (2013) Synopsis of adventive species of Coleoptera (Insecta) recorded from Canada. Part 2: Staphylinidae. Pensoft Series *Faunistica*. Pensoft Publishers, Sofia

Klimaszewski J, Webster RP, Langor DW, Bourdon C, Hammond HEJ, Pohl GR, Godin B (2014) Review of Canadian species of the genera *Gnathusa* Fenyes, *Mniusa* Mulsant & Rey and *Ocyusa* Kraatz (Coleoptera, Staphylinidae, Aleaocharinae). ZooKeys 412:9–40

Klimaszewski J, Godin B, Langor D, Bourdon C, Lee S-I, Horwood D (2015) New distribution records for Canadian Aleocharinae (Coleoptera, Staphylinidae), and new synonymies for *Trichiusa*. ZooKeys 498:51–91

Klimaszewski J, Larson DJ, Labrecque M, Bourdon C (2016) Twelve new species and fifty-three new provincial distribution records of Aleocharinae rove beetles of Saskatchewan, Canada (Coleoptera, Staphylinidae). ZooKeys 610:45–112

Klimaszewski J, Webster RP, Langor DW, Brunke A, Davies A, Bourdon C, Labrecque M, Newton AF, Dorval J-A, Frank JH (2018a) Aleocharine rove beetles of eastern Canada (Coleoptera, Staphylinidae, Aleocharinae): a glimpse of megadiversity. Springer, Cham, xvi + 902 pp

Klimaszewski J, Godin B, Davies A, Bourdon C, Horwood D (2018b) Forty new records of aleocharine beetles, and two new species in the genera *Acrotona* Thomson and *Atheta* Thomson, for the province of Manitoba, Canada (Coleoptera: Staphylinidae: Aleocharinae). Insecta Mundi 641:1–33

Klimaszewski J, Hoebeke ER, Godin B, Davies A, Perry KI, Bourdon C, Winchester N (2020) Aleocharine rove beetles of British Columbia: a hotspot of Canadian biodiversity (Coleoptera, Staphylinidae). Springer, Cham. 631 pp

Lohse GA, Smetana A (1985) Revision of the types of species of Oxypodini and Athetini (*sensu* Seevers) described by Mannerheim and Mäklin from North America (Coleoptera: Staphylinidae). Coleopt Bull 39:281–300

Lohse GA, Klimaszewski J, Smetana A (1990) Revision of arctic Aleocharinae of North America (Coleoptera: Staphylinidae). Coleopt Bull 44(2):121–202

Majka CG, Klimaszewski J (2008a) New records of Canadian Aleocharinae (Coleoptera: Staphylinidae). *In*: Majka CG, Klimaszewski J (Eds.) Biodiversity, Biosystematics, and Ecology of Canadian Coleoptera. ZooKeys [Special Issue] 2:85–114

Majka CG, Klimaszewski J (2008b) Introduced Staphylinidae (Coleoptera) in the Maritime Provinces of Canada. Can Entomol 140(1):48–72

Majka CG, Klimaszewski J (2008c) Adventive Staphylinidae (Coleoptera) of the maritime provinces of Canada: further contributions. ZooKeys 2:151–174

Majka CG, Klimaszewski J (2010) Contributions to the knowledge of the Aleocharinae (Coleoptera, Staphylinidae) in the Maritime Provinces of Canada. ZooKeys 46:15–39

Majka CG, Klimaszewski J, Lauff RF (2006) New coleoptera records from owl nests in Nova Scotia, Canada. Zootaxa 1194:33–47

Mannerheim CG (1830) Précis d'un nouvel arrangement de la famille des brachélytres de l'ordre des insectes coléoptères. St. Petersburg, 87 pp. [separate, re-issued in Mémoires Présentées à l'Académie Impériale des Sciences de St.-Pétersbourg 1: 415–501, in 1831]

Moore I, Legner EF (1975) A catalogue of the Staphylinidae of America north of Mexico (Coleoptera). University of California, Division of Agricultural Sciences, Special Publication No. 3015:1–514

Muona J (1991) The north European and British species of the genus *Meotica* Mulsant & Rey (Coleoptera, Staphylinidae). Deutsche entomologische Zeitschrift 38(1–3):225–246

Newton AF (2019) StaphBase: staphyliniformia world catalog database (version Nov 2018). In: Roskov Y, Ower G, Orrell T, Nicolson D, Bailly N, Kirk PM, Bourgoin T, DeWalt RE, Decock W, Nieukerken E. van, Zarucchi J, Penev L (Eds.) Species 2000 & ITIS Catalogue of Life, 2019 Annual Checklist. Digital resource at http://www.catalogueoflife.org/annual-checklist/2019. Species 2000: Naturalis, Leiden, the Netherlands. ISSN 2405-884X

Osswald J, Bachmann L, Gusarov VI (2013) Molecular phylogeny of the beetle tribe Oxypodini (Coleoptera: Staphylinidae: Aleocharinae). Syst Entomol 38 (3):507–522

Redtenbacher L (1849) Fauna Austriaca. Die Käfer. Nach der analytischen Methode bearbeitet. C. Gerold, Wien. i-xxvii + 641–883

Schülke M, Smetana A (2015) Family Staphylinidae Latreille, 1802. In: Löbl I, Löbl D (eds) Catalogue of Palaearctic Coleoptera: Hydrophiloidea – Staphylinoidea, vol 2. Brill, Leiden & Boston, pp 304–1134. Revised and updated edition

Seevers CH (1978) A generic and tribal revision of the North American Aleocharinae (Coleoptera: Staphylinidae) [with additions and annotations by Lee H. Herman]. Fieldiana Zoology 71. vi + 289 pp

Sjöberg O (1950) Drei neue Käfer aus der borealen Region Schwedens. Entomologisk Tidskrift 71(2):159–60, (3/4):161–164

Stephens JF (1832) (1832–1835) Illustrations of British entomology... Mandibulata, vol V. Baldwin & Cradock, London, pp 1–240. pls. 24-26

Thomson CG (1859) Skandinaviens Coleoptera, synoptiskt bearbetade. I. Tom. Berlingska Boktryckeriet, Lund, [3] + 290 pp

Webster RP, Klimaszewski J, Pelletier G, Savard K (2009) New Staphylinidae (Coleoptera) records with new collection data from New Brunswick, Canada. I. Aleocharinae. *In*: Majka CG, Klimaszewski J (Eds.) Biodiversity, Biosystematics, and Ecology of Canadian Coleoptera II. ZooKeys [Special Issue] 22:171–248

Webster RP, Klimaszewski J, Sweeney JD, DeMerchant I (2012) New staphylinidae (Coleoptera) records with new collection data from new brunswick, and an addition to the fauna of Quebec, Canada: Aleocharinae. In: Klimaszewski J, Anderson R (eds) Biosystematics and ecology of canadian staphylinidae (Coleoptera) II. ZooKeys [Special Issue] 186:83–118

Webster RP, Klimaszewski J, Bourdon C, Sweeney JD, Hughes CC, Labrecque M (2016) Further contribution to the Aleocharinae (Coleoptera, Staphilinidae) fauna of New Brunswick and Canada including description of 27 new species. *In*: Webster RP, Bouchard P, and Klimaszewski J (Eds) The Coleoptera of New Brunswick and Canada: providing baseline biodiversity and natural history data. ZooKeys [Special Issue] 573:85–216

Webster RP, deTonnancour P, Sweeney JD, Webster VL, Kostanowicz CA, Hughes C, Anderson RS, Klymko J, Chatal C, Vigneault R (2020) New Coleoptera records from eastern Canada, with additions to the fauna of Manitoba, British Columbia, and Yukon Territory. ZooKeys 946:53–112

Work TT, Klimaszewski J, Thiffault E, Bourdon C, Paré D, Bousquet Y, Venier L, Titus B (2013) Initial responses of rove and ground beetles (Coleoptera, Staphylinidae, Carabidae) to removal of logging residues following clearcut harvesting in the boreal forest of Quebec, Canada. ZooKeys 258:31–52

Tribe Tachyusini C.G. Thomson, 1859 12

Tribe Tachyusini C.G. Thomson, 1859

(Figs. 12.138–12.151) (Lohse 1989, Ashe 2001, Paśnik 2006, Osswald et al. 2013, Klimaszewski et al. 2018, 2020, Brunke et al. 2021)

Diagnosis Body form diverse, from robust, to slender; head with neck more than one-third as wide as head or without distinct neck; pronotum without median sulcus, median elongate depression, or basal foveae; abdomen slightly narrower than elytra; or constricted at base; maxillae with galea and lacinia moderately elongate, length of galea usually shorter than, or subequal to, distance between base of galea to cardo; ligula Y-shaped, split at apex, with lobes diverging; mesocoxal cavities in most species moderately to narrowly separated or contiguous in some; process of metaventrite no longer than that of mesoventrite; tarsal formula 4-5-5.

Key to Genera of Tachyusini (from Brunke et al. 2021)

1. Elytra at humerus, only slightly broader than pronotum at base (Figs. 12.138a, 12.150a); impressions of abdominal tergites shallow, with punctation similar to that of disc (Figs. 12.138a, 12.150a) **2**

– Elytra at humerus, distinctly broader than pronotum at base (Figs. 12.139a–12.149a, 12.151a); impressions of abdominal tergites with at least a few coarse punctures and glossy areas, punctation distinctly different from that of disc (Figs. 12.139a–12.149a, 12.151a)......................**3**

2. Pronotum with pubescence directed straight posteriad; hind tarsus subequal in length to hind tibia or longer (Fig. 12.138a) ....***Brachyusa* Mulsant and Rey**

– Pronotum with pubescence directly posteriolaterad from midline; hind tarsus shorter, slightly longer than half the length of hind tibia or shorter (Fig. 12.150a)....***Paradilacra* Bernhauer**

3. Abdomen clavate, at base distinctly narrower than head (Fig. 12.151a); tergite III about as long as wide or longer; tergal impressions with median carina (Fig. 12.151a)... ....***Tachyusa* Erichson**

– Abdomen at most, slightly constricted at base, subequal to or wider than head (Figs. 12.139a–12.149a); tergite III about twice as wide as long or wider; tergal impressions never with median carina (Figs. 12.139a–12.149a).....**4**

4. Abdomen, at base, elongate and moderately constricted, about as wide as head (Fig 12.139a); tergite III, at most, twice as wide as long; tergal impressions deep and sharply delineated from strongly convex disc (Fig. 12.139a).................... .....***Dasygnypeta* Lohse** *sensu* Brunke et al. 2021

– Abdomen, at base at most slightly constricted, wider than head (Figs. 12.140a–12.149a); tergite III strongly transverse, about 2.5 times wider than long or wider; tergal

J. Klimaszewski et al., *A Faunal Review of Aleocharine Beetles in the Rapidly Changing Arctic and Subarctic Regions of North America (Coleoptera, Staphylinidae)*,
https://doi.org/10.1007/978-3-030-68191-3_12

impressions shallower, gradually sloping to disc at base (Figs. 12.140a–12.149a) .... *Gnypeta* **Thomson**

Genus *Brachyusa* Mulsant and Rey, 1873

(Fig. 12.138a–g) (Casey 1911 [as *Tetralina*], Seevers, 1978, Ashe 2001, Bousquet et al. 2013, Klimaszewski et al. 2011, 2016, 2018, 2020)

Diagnosis Body robust, elongate, arcuate laterally; integument covered with dense, fine, velvet-like pubescence, surface slightly glossy; head with genae short in dorsal view (only narrow strip visible behind eyes), eyes large, frons abruptly deflexed; clypeus strongly deflexed; antennomeres V–X elongate, subquadrate or transverse; terminal palpomere of maxillary palpus short and needle-shaped; pronotum trapezoidal in shape, slightly transverse; elytra slightly transverse, about as long as pronotum, posterior margin truncate and not emarginate laterally; abdomen moderately narrower than elytra, with first three visible tergites with basal impressions shallow; hind tarsus long, almost as long as hind tibia, basal tarsomere at least as long as next two combined; male sternite VIII extremely elongate; median lobe of aedeagus with tubus triangularly elongate, bulbus swollen, large, with large crista apicalis; spermatheca L- or S-shaped, capsule club-shaped, stem short and thick. Tarsal formula 4-5-5.

68. *Brachyusa helenae* (Casey) (Fig. 12.138a–g)

Tetralina helenae Casey, 1911. Klimaszewski et al. 2011, 2012, 2016, 2018, 2020, Brunke et al. 2012, Webster et al. 2012.

Diagnosis Body oval, slightly flattened, length 2.2–2.5 mm; colour predominantly black but sometimes brown, with appendages light brown; integument moderately glossy with pubescence short, silky (Fig. 12.138a); head narrower than elytra, with eyes large; antennomeres I–VII elongate, VIII–X quadrate to transverse; hind tarsus with basal article about as long as next two articles combined; pronotum wider than head but much narrower than elytra; elytra at suture longer than pronotum; abdomen strongly narrowed posteriorly. MALE. Tergite VIII truncate apically (Fig 12.138c); sternite VIII distinctly elongate, with apical margin broadly shallowly emarginate, lateral angles rounded (Fig. 12.138d); tubus of median lobe of aedeagus in lateral view short, bent acutely ventrad basally, acutely tapering to apex, apical triangular part varying in width (Fig. 12.138b); in dorsal view, basal half of median lobe broadly rounded / swollen. FEMALE. Tergite VIII with apical margin very obtusely angulate (Fig. 12.138e); sternite VIII with apical margin broadly rounded on either side of shallow medial emargination (Fig. 12.138f); spermatheca loosely S-shaped, with capsule elongate, bent at acute angle to sinuate stem (Fig. 12.138g).

DISTRIBUTION **Origin**: Nearctic. **CANADA**: BC, LB, NB, NF, NT, ON, SK, YT. **USA**: AK, MT.

COLLECTION AND HABITAT DATA
Habitat: found along shores of rivers and streams near flowing water; adults in pine/spruce litter near stream, on sand and gravel on the banks of a river, near lake and river shorelines; on clay, sand, and gravel beaches, on sandy and silty river margins, in moss near the splash zone of a waterfall, in moss on rocks in the middle of a river, in gravel on a gravel bar along a shaded brook, and in gravel on a gravel/clay margin of a partially shaded brook, among cobblestones along a large shaded brook, in flood debris resting on a sand/clay mix along an upper river margin, treading emergent *Carex* and grasses on the margin of a lake near the outflow of a stream and in debris on the margin of a vernal pond in a silver maple (*Acer saccharinum* L.) swamp near a river, and debris on a American beaver (*Castor canadensis* Kuhl) dam near the outflow area (flowing water) from the dam; one specimen from debris within wall of beaver lodge. In BC, specimens were collected in sedges along creek, from river debris and coarse woody debris. **Collecting period**: IV–

IX. **Collecting method**: pitfall traps, sifting litter near bodies of water, splashing water on moss, gravel, and sand or turning gravel and aspirating beetles. It took 5–10 min. before adults appeared after repeated splashing of moss near splash zone of a waterfall.

Additional records. See Annex 1.

Genus *Dasygnypeta* Lohse, 1974

(redefined by Brunke et al. 2021)

(Fig. 12.139a–h) (Lohse 1974, Klimaszewski et al. 2011, 2016, 2018, Brunke et al. 2021)

Diagnosis (redefined by Brunke et al. 2021). Members of *Dasygnypeta* may be distinguished by their characteristic abdomen (Fig. 12.139a): base of abdomen about as wide as head; basal half of abdomen elongate, tergite III (first visible) only moderately transverse, about twice as wide as long (at least 2.5 times in *Gnypeta*); tergites III–V with very deep basal impressions, each creating strongly convex areas on the disc. Members of *Dasygnypeta* could be confused with *Tachyusa*, which also bears an elongate basal abdomen, but in the latter genus tergites III–V are far more elongate and the abdominal base is narrower than the head (Fig. 12.151a).

69. *Dasygnypeta velata* (Erichson) (Fig. 12.139a–h)

Homalota velata Erichson, 1837. *Gnypeta minuta* Klimaszewski and Webster, in Klimaszewski et al. 2008. Klimaszewski et al. 2011, 2016, 2018. As *Dasygnypeta velata*: Brunke et al. 2021.

Diagnosis Body moderately narrow, subparallel, length 2.3 mm, colour dark brown; integument moderately glossy (Fig. 12.139a); pubescence yellowish grey, moderately short and dense; antennomeres V–VII moderately elongate, VIII–X subquadrate; head slightly narrower than pronotum; elytra wider than either head or pronotum; head rounded posteriorly; pronotum broadest in apical third, pubescence on disc directed obliquely posterolaterad from midline; elytra at suture longer than pronotum, about 1/6 wider than maximum width of pronotum, pubescence on disc directed obliquely posterolaterad, forming weak wavy pattern medially on each side; abdomen broadening posteriorly, distinctly narrower at base than elytra; hind tarsus with basal article slightly longer than following article. MALE. Tergite VIII sinuately arcuate apically (Fig. 12.139d); sternite VIII, broadly rounded apically (Fig. 12.139e); median lobe of aedeagus in lateral view short, with entire ventral margin expanded ventrad from tip of prominent, angular crista apicalis to broadly triangular apex, arcuately curved obliquely ventrad in apical half (Fig. 12.139b); in dorsal view bulbus moderately large and tubus broad (Fig. 12.139c); internal sac with complex structures (Fig. 12.139b,c). FEMALE. Tergite VIII arcuate apically (Fig. 12.139f); sternite VIII with apical margin broadly rounded, with shallow emargination medially (Fig. 12.139g); spermatheca pipe-shaped (Fig. 12.139h).

DISTRIBUTION Origin: Holarctic. **CANADA**: LB, MB, NB, NT, SK. **USA**: AK.

COLLECTION AND HABITAT DATA

Habitat: Silver maple forest under debris on muddy soil near a small pool, river margin, under debris on clay and sand mix. In NL, specimens from sandy shore of river. In AK, found in mud bank and shoreline. **Collecting period**: VI–VII. **Collecting method**: sifting litter, shore washing.

We here provide the following additional data for this species: Canada, Northwest Territories, Reindeer St., Caribou Hills, 2.VII.1972, A. Smetana (CNC) 1 male. USA, Alaska, Selawik NWR, Kugarak Riv., 2m el., 66.5611, -158.99625, ±14m mud bank, shore washing, 23.VI.2010, D.S. Sikes (UAM) 4 specimens; Tanana River, 197m el., 64.54204, -147.02621, ±24m shoreline, 18.VI.2008, D.S. Sikes, D. R. Maddison (UAM) 1 specimen; Mi. 1252 Alaska Hwy., 7.VII.1968, J.M. Campbell and A. Smetana (CNC) 1 female.

Genus *Gnypeta* C.G. Thomson, 1859

(Figs. 12.140–12.149) (Klimaszewski et al. 2008, 2018, 2020)

Diagnosis Body robust, length 2.3–3.6 mm, integument moderately to highly glossy (often with bluish metallic reflections), finely punctate and pubescent, pubescence moderately long and silky in appearance in some; head and pronotum distinctly narrower than elytra; infraorbital carina partial or absent; ligula divided to base, or to near base, lobes parallel-sided; galea subequal in length to lacinia; pronotum broadest in apical one-third or at middle of disc, with sides converging apically and basally, pubescence on disc directed latero-posteriad from midline; elytra much broader than either head or pronotum, pubescence on disc often forming wavy pattern on each side; process of mesoventrite moderately broad, reaching middle of mesocoxae, its apex truncate; process of metaventrite broadly rounded; legs long and slender; tarsal formula 4-5-5; hind tarsus with basal article moderately elongate, usually shorter than two following ones combined; abdomen as broad as elytra at base or moderately narrower, subparallel, or broadening apically, first three visible tergites with deep basal impressions with coarse punctures but without pronounced longitudinal ridges; median lobe of aedeagus with tubus short, triangularly produced, bulbus enlarged, swollen, with crista apicalis large, subtriangular in shape laterally; paramere broad with short apical lobe bearing three long subapical and one short apical macrosetae; spermatheca of four types: S-shaped, C-shaped, hatchet-shaped, or club-shaped; capsule tubular with apical part roughly spherical, tubular, or funnel-shaped; stem tubular, elongate, more-or-less sinuate and moderately swollen basally; sternite VIII of male and female with basal margin broadly separate from antecostal suture.

COMMENTS Adults are associated with riparian habitats and debris along the margins of marshes, ponds, lakes, and streams. Some Canadian species were found in cold wet moss alongside streams. They may also occur in vegetation and litter along edges of streams, river, and lakes, in grass tussocks on mud flats, in gravel, wooded bogs, beaver lodges, and decaying fungi. This genus consists of mainly northern, cold-loving species, and some have Holarctic distributions.

Key to Species of *Gnypeta*

1. Body broad and robust (Fig. 12.148a), pronotum at base slightly narrower than elytra at base (Fig. 12.148a), abdomen broadest medially and at base as broad as elytra (Fig. 12.148a); median lobe of aedeagus and spermatheca as illustrated (Fig. 12.148b,c,h) .... ***Gnypeta groenlandica*** **Lohse**

– Body narrow to moderately broad and less robust, pronotum at base distinctly narrower than elytra, abdomen subparallel and at base narrower than elytra; genitalia differently shaped **2**

2. Elytra at suture approximately as long as pronotum **3**

– Elytra at suture longer than pronotum **7**

3. Antennomeres VII–X strongly transverse (Fig. 12.145a); median lobe of aedeagus and spermatheca as illustrated (Figs. 12.145b,c,h) .... ***Gnypeta brevicornis*** **Casey**

– Antennomeres VII–X elongate, subquadrate or slightly transverse; genitalia differently shaped **4**

4. Body length 3.0–3.4 mm; median lobe of the aedeagus with apical part narrowly produced in lateral view (Fig. 12.149b); spermatheca with capsule funnel-shaped and stem strongly sinuate anteriorly (Fig. 12.149h) .... ***Gnypeta sellmani*** **Brundin**

– Body length 2.5–3.0 mm; median lobe of aedeagus and spermatheca differently shaped **5**

5. Body strongly glossy with metallic blue reflection (Fig. 12.147a); apical margin of male tergite VIII with four sharp teeth (Fig. 12.147d); median lobe of aedeagus and spermatheca as illustrated (Fig. 12.147b,c,h) .... *Gnypeta dentata* **Klimaszewski**

– Body glossy but without strong metallic blue reflection; apical margin of male tergite VIII without teeth; median lobe of aedeagus and spermatheca not as above **6**

6. Body length 2.5–2.8 mm; spermatheca with capsule funnel-shaped (Fig. 12.143h); median lobe of aedeagus with apical part broadly triangular with straight margin in lateral view (Fig. 12.143b)....... *Gnypeta ashei* **Klimaszewski**

- Body length 2.7–3.0 mm; spermatheca with capsule approximately spherical (Fig. 12.142h); median lobe of aedeagus with apical part moderately narrowly triangular, with sinuate margin in lateral view (Fig. 12.142b).................................. .... *Gnypeta carbonaria* **(Mannerheim)**

7. Antennomeres VII–X distinctly elongate ...**8**

– Antennomeres VII–X subquadrate to slightly transverse... **9**

8. Pronotum evenly rounded laterally (Fig. 12.140a); apex of median lobe of aedeagus broadly triangular in lateral view (Fig. 12.140b); capsule of spermatheca tubular with narrow apical capsule (Fig. 12.140h) .... *Gnypeta caerulea* **(Sahlberg)**

– Pronotum broadest at base with sides strongly converging apically (Fig. 12.144a); apex of median lobe of aedeagus narrowly triangular in lateral view (Fig. 12.144b); capsule of spermatheca pitcher-shaped (Fig. 12.144h) .... *Gnypeta atrolucens* **Casey**

9. Abdomen broad, at base approximately as broad as elytra (Fig. 12.141a); genital structures as illustrated (Fig. 12.141b,c,h) .... *Gnypeta lohsei* **Klimaszewski**

– Abdomen moderately broad, at base distinctly narrower than elytra (Fig. 12.146a); genital structures differently shaped (Fig. 12.146b,c,h) .. .. *Gnypeta brincki* **Palm**

***Caerulea* species group** (Klimaszewski et al. 2008)

Diagnosis Male tergite VIII with two small lateral teeth at apical margin; spermatheca S-shaped, apical part of capsule mushroom-shaped, stem narrow and sinuate, slightly swollen basally; median lobe of aedeagus with short tubus broadly triangular apically in lateral view.

70. *Gnypeta caerulea* (C.R. Sahlberg) *
(Fig. 12.140a–h)

Aleochara caerulea C.R. Sahlberg 1831. Klimaszewski et al. 2008, Majka and Klimaszewski 2008, Klimaszewski et al. 2011, 2015, 2018, 2020.

Diagnosis Body length 3.0–3.5 mm; colour uniformly dark brown to black, sometimes middle of elytra, tarsi and sometimes entire legs rust brown; integument highly glossy with metallic reflection; pubescence yellowish-grey, moderately long and dense (Fig. 12.140a); antennomere IV distinctly elongate, V–X moderately to distinctly elongate, increasingly less so toward apex of antenna; head and pronotum of about same width; elytra and abdomen wider than either head or pronotum; head rounded posteriorly; pronotum about two-thirds (0.8) as wide as elytra, broadest in apical third, pubescence on disc directed anteriad along midline and laterad elsewhere; elytra at suture slightly longer than pronotum, pubescence on disc directed obliquely posterolaterad, forming wavy pattern medially on each side; abdomen subparallel, narrower than elytra at base, first three visible tergites with basal impressions deep, with large punctures; hind tarsus with two basal tarsomeres of about same length, third one slightly shorter. MALE. Tergite VIII transverse and truncate apically with two minute apical teeth (Fig. 12.140d); sternite VIII about as long as wide, rounded apically, a little more narrowly so at apex (Fig. 12.140e); tubus of median lobe of aedeagus in lateral view very short, broadly triangular apically, with ventral margin curved obliquely ventrad in basal half (Fig. 12.140b); in dorsal view bulbus moderately large with two small lateral projections (Fig. 12.140c); internal sac with structures as illustrated (Fig. 12.140b,c). FEMALE. Tergite VIII truncate apically (Fig. 12.140f); sternite VIII broadly rounded apically (Fig. 12.140g); spermatheca with capsule elongate, mushroom-shaped, stem sinuately club-shaped, gradually widening posteriad (Fig. 12.140h).

DISTRIBUTION **Origin**: Holarctic. **CANADA**: AB, BC, LB, MB, NF, NS, NB, NT, ON, PE, QC, SK, YT. **USA**: AK.

COLLECTION AND HABITAT DATA
Habitat: riparian and wetland habitats. In NB, adults collected from drift material along river margins, from gravel and under cobblestones on margins of cool, clear shaded brooks, treading emergent *Carex* and grasses on lake margin, sifted from leaf litter on beaver pond margin and from sticks and debris from outflow area of beaver dam, sifted from leaves and sedges from pond margin, and sifted from sphagnum from swamp edges. Elsewhere, collected from margins of streams, ponds, and lakes in moss, among rocks, gravel, sand, and driftwood, sweeping marsh vegetation, and sifted from vegetation at beaver pond. **Collecting period**: V–IX. **Collecting method**: splashing, treading, sifting litter, and turning gravel and cobblestones, and aspirating specimens, sweeping vegetation.

COMMENTS The illustrations for *G. caerulea* (Klimaszewski et al. 2008) and *G. carbonaria* (Klimaszewski et al. 2008) were cited as "*brincki*" in the captions.

71. *Gnypeta lohsei* Klimaszewski (Fig. 12.141a–h)
Gnypeta lohsei Klimaszewski, 2008.

Diagnosis Body length 2.7–3.0 mm; uniformly dark brown to black, sometimes central part of elytra and tarsi rust brown; integument moderately glossy, without bluish metallic reflection; pubescence yellowish grey and moderately long and dense (Fig. 12.141a); antennal article IV moderately elongate, V–VII subquadrate, VIII–X slightly transverse (Fig. 12.141a); head and pronotum of about the same width; elytra and abdomen wider than either head or pronotum; head rounded posteriorly; pronotum broadest in apical third, about two-thirds as wide as elytra, pubescence directed anteriad along midline and laterad elsewhere; elytra at suture slightly longer than pronotum, pubescence directed obliquely posterolaterad, in wavy pattern on medial parts of disc; abdomen subparallel, almost as broad as elytra at base; metatarsus with two basal tarsomeres of about the same length and the third one slightly shorter. MALE. Tergite VIII transverse, apical margin truncate with two minute sublateral teeth (Fig. 12.141d); sternite VIII as broad as long, apical margin parabolically rounded, vaguely angulate medially (Fig. 12.141e); median lobe of aedeagus with triangularly produced apical part of tubus and edges sinuate in lateral view (Fig. 12.141b); bulbus moderately enlarged with two anterior projections in dorsal view (Fig. 12.141c); internal sac with structures as illustrated (Fig. 12.141b,c). FEMALE. Tergite VIII truncate apically (Fig. 12.141f); sternite VIII broadly rounded posteriorly (Fig. 12.140g); spermatheca with capsule mushroom-shaped, elongate, stem sinuate and slightly swollen basally (Fig. 12.141h).

DISTRIBUTION **Origin**: Nearctic. **CANADA**: AB, BC. **USA**: AK, WA.

COLLECTION AND HABITAT DATA
Habitat: not recorded, other than collections

from high altitudes sites, up to 2300 m. **Collecting period**: VI–VIII. **Collecting method**: not recorded.

***Crebrepunctata* species group** (Klimaszewski et al. 2008, 2018)

Diagnosis Male tergite VIII truncate; spermatheca C-shaped, capsule and stem short, broad, and tubular or club-shaped; median lobe of aedeagus with bulbus enlarged.

72. *Gnypeta carbonaria* (Mannerheim) * (Fig. 12.142a–h)

Bolitochara carbonaria Mannerheim 1830. Palm 1966, Klimaszewski et al. 2008, 2011, 2018.

Diagnosis Body length 2.7–3.0 mm; colour dark brown to black, elytra often rust brown medially, legs and/or tarsi rust brown; integument highly glossy (Fig. 12.142a); pubescence yellowish-grey, moderately long and dense; antennomere IV distinctly elongate, V–VII moderately elongate, VIII–X subquadrate to slightly transverse; head and pronotum of about same width; elytra wider than either head or pronotum; head rounded posteriorly; pronotum broadest in apical third, pubescence on disc directed anteriad along midline and laterad elsewhere; elytra at suture slightly longer than pronotum and about 1.25 times as wide as maximum width of pronotum, pubescence on disc directed obliquely posterolaterad, forming wavy pattern medially on each side; abdomen subparallel, distinctly narrower than elytra; hind tarsus with basal article slightly longer than following one. MALE. Tergite VIII transverse, with apical margin arcuately truncate on each side of very shallow medial emargination (Fig. 12.142d); sternite VIII elongate, broadly parabolically rounded apically (Fig. 12.142e); median lobe of aedeagus in lateral view with tubus very short, gradually sinuately tapering to narrowly rounded apex, ventral margin broadly curved ventrad in basal half (Fig. 12.142b); in dorsal view bulbus moderately large and tubus broad (Fig. 12.142c); internal sac with complex structures (Fig. 12.142b,c). FEMALE. Tergite VIII sinuately truncate apically (Fig. 12.142f); sternite VIII broadly rounded posteriorly with very shallow emargination medially (Fig. 12.142g); spermatheca with capsule spherical, and stem short, tubular and broadening posteriorly (Fig. 12.142h).

DISTRIBUTION Origin: Holarctic. **CANADA**: AB, MB, NB, NF, NT, ON, QC, SK. **USA**: AK.

COLLECTION AND HABITAT DATA

Habitat: mostly in riparian habitats. In NB, river margins in drift material (mostly maple seeds) and flood debris, river margin on clay sand mix on steep bank (splashing), collected by splashing moss near splash zone of waterfalls (common), splashing gravel along small shaded brook, treading reeds in a marsh next to river. Elsewhere in Canada, collected from wet ground in a bog, from rotten fungus, and a UV light trap. **Collecting period**: VI–IX. **Collecting method**: splashing moss and gravel, and steep clay river bank, treading marsh vegetation, sifting litter, and UV light trap.

***Sellmani* species group** (Klimaszewski et al. 2008)

Diagnosis Median lobe of aedeagus in lateral view with tubus short, narrowly or broadly triangular apically, crista apicalis prominent, angular; spermatheca with capsule widely funnel-shaped to elongate bell-shaped, with broad invagination, stem club-shaped.

73. *Gnypeta ashei* Klimaszewski (Fig. 12.143a–h)

Gnypeta ashei Klimaszewski, in Klimaszewski et al. 2008.

Description Body length 2.5–2.8 mm; dark brown to black, sometimes central part of elytra and tarsi reddish brown (Fig. 12.143a); integument strongly glossy; pubescence yellowish grey and moderately long and dense; antennomere IV slightly shorter than preceding article, IV–VI subquadrate to slightly elongate in male, and VII–IX transverse; head and pronotum of about the same width; elytra and abdomen wider than either head or pronotum; elytra small and short; head rounded basally; pronotum broadest in apical third, pubescence directed anterad along midline and laterad elsewhere; elytra (at suture) about as long as pronotum and 1/4 broader than maximum width of pronotum, pubescence directed obliquely postero-laterad, with wavy pattern on medial part of disc; abdomen subparallel, slightly narrower than elytra at base; metatarsus with two basal articles of about the same length and the third one slightly shorter. MALE. Tergite VIII transverse and truncate apically (Fig. 12.143d); sternite VIII about as long as wide and slightly produced posteriorly (Fig. 12.143e); median lobe of aedeagus with broadly triangular apical part of tubus in lateral view (Fig. 12.143b); bulbus moderately sized (Fig. 12.143b,c); internal sac with structures as illustrated (Fig. 12.143b,c). FEMALE. Tergite VIII similar to that of male (Fig. 12.143f); sternite VIII broadly rounded posteriorly (Fig. 12.143g); spermatheca with capsule wider than long and broadly funnel-shaped, stem sinuate and swollen basally (Fig. 12.143h).

DISTRIBUTION Origin: Nearctic. **CANADA**: MB, NT, NU, QC, YT. **USA**: AK.

COLLECTION AND HABITAT DATA

Habitat: marshes, sedges around edge of pond, one specimen was captured on snow on Devon Island (NU). **Collecting period**: VI–VIII. **Collecting method**: treading sedges around edges of pond, sifting litter at shore of a lagoon.

Additional records. See Annex 1.

COMMENTS Quebec was accidentally omitted from the distribution of this species in Klimaszewski et al. (2008), despite a paratype listed from Payne Bay, QC in the Appendix of that paper. This omission has been repeated in the literature since.

74. *Gnypeta atrolucens* Casey (Fig. 12.144a–h)

Gnypeta atrolucens Casey, 1893. Fenyes 1918, Klimaszewski et al. 2008, 2011, 2018, Webster et al. 2012.

Diagnosis Body length 3.2–3.4 mm, colour dark brown to black with metallic bluish reflection, central part of elytra and tarsi or entire legs reddish-brown (Fig. 12.144a); integument highly glossy; pubescence yellowish-grey, moderately long and dense; antennomere IV slightly shorter than preceding one, V–VII elongate, VIII–IX subquadrate or slightly elongate; head, pronotum and abdomen of about same width, elytra large and distinctly broader than remainder of body; head rounded posteriorly; pronotum broadest in apical third, pubescence on disc directed anteriad along midline and laterad elsewhere; elytra at suture about 1/5 longer and 1/3 broader than maximum width of pronotum, pubescence on disc directed obliquely posterolaterad; abdomen subparallel; hind tarsus with two basal articles of about same length. third one slightly shorter. MALE. Tergite VIII transverse, truncate apically (Fig. 12.144d); sternite VIII elongate, with apical margin parabolically rounded, narrowly rounded medially (Fig. 12.144e); tubus of median lobe of aedeagus very short in lateral view, with ventral margin expanded ventrad, about evenly curved from tip of large angular crista apicalis to triangular apex (Fig. 12.144b); in dorsal view bulbus enlarged with rounded angular projection on each side apically, tubus gradually narrowing apicad (Fig. 12.144c); internal sac with structures as illustrated (Fig. 12.144b,c). FEMALE. Tergite VIII similar to that of male but broadly shallowly emarginate apically (Fig. 12.144f), sternite VIII broadly rounded apically (Fig. 12.144g); spermatheca with capsule elongate bell-shaped, stem curved, slightly swollen posteriorly (Fig. 12.144h).

DISTRIBUTION Origin: Nearctic. **CANADA**: LB, NB, NF, QC. **USA**: NH, NY, VT.

COLLECTION AND HABITAT DATA

Habitat: riparian habitats often in moss near water falls or on rocks in fast flowing streams.

Adults in moss near splash zone of waterfall (common), on rocks in middle of rivers and brooks (adults in moss and in cracks in rocks). One was collected from gravel on a gravel bar on river margin. Others collected by splashing exposed bedrock with moss patches in the middle of a fast flowing river above a waterfall. Some were collected from moss and gravel along edges of forest streams and shallow pools at 1100 m elevation in shrubby barrens. **Collecting period**: VI–IX. **Collecting method**: splashing water on moss near the splash zone of a waterfall, splashing water on moss and rocks in the middle of a rivers and brooks. It took up to 10 min and repeated splashing before adults appeared on rocks and moss surface in the above habitats. Sifted from moss and gravel on stream margin.

75. *Gnypeta brevicornis* Casey (Fig. 12.145a–h)

Gnypeta brevicornis Casey, 1906. Moore and Legner 1975, Klimaszewski et al. 2008, 2020.

Diagnosis Body length 2.7–2.9 mm; rust brown with darker head and tip of abdomen (Fig. 12.145a); integument moderately glossy; pubescence yellowish grey and sparse; antennal article IV elongate but distinctly shorter than antennomere III, antennomeres V–VI slightly and VII–IX strongly transverse; head and pronotum of about the same width; elytra and abdomen slightly wider than either head or pronotum; elytra small and moderately elongate; head rounded posteriorly; pronotum broadest in apical third, pubescence directed postero-laterad from midline of disc; elytra at suture approximately as long as pronotum, 1/5 broader than maximum width of pronotum, pubescence directed obliquely posterolaterad from midline of disc; abdomen swollen posteriorly, as broad as elytra at base; metatarsus with basal article of about the same length as the two following articles combined. MALE. Tergite VIII transverse and truncate apically, slightly concave medially (Fig. 12.145d); sternite VIII elongate and parabolic posteriorly (Fig. 12.145e); median lobe of aedeagus with narrowly triangular apical part of tubus in lateral view (Fig. 12.145b), in dorsal view base of tubus with two lateral projections (Fig. 12.145c); internal sac with structures as illustrated (Fig. 12.145b,c). FEMALE. Tergite VIII truncate apically (Fig. 12.145f); sternite VIII broadly rounded posteriorly (Fig. 12.145g); spermatheca with capsule transverse, funnel-shaped with sides arcuate, stem slightly sinuate and swollen posteriorly (Fig. 12.145h).

DISTRIBUTION **Origin**: Nearctic. **CANADA**: BC (Glenora and Kamloops). **USA**: not recorded.

COLLECTION AND HABITAT DATA
Habitat: banks of rivers and lakes and from a eutrophic pond edge; some from silt around *Typha* (cattails) on a river bank. **Collecting period**: VI–VIII. **Collecting method**: sifting litter.

76. *Gnypeta brincki* Palm * (Fig. 12.146a–h)

Gnypeta brincki Palm, 1966. Klimaszewski et al. 2008, 2018.

Diagnosis Body length 3.2–3.6 mm, colour black, sometimes central part of elytra and tarsi reddish brown (Fig. 12.146a); integument highly glossy; pubescence yellowish grey and moderately long and dense; antennomere IV slightly shorter than preceding one, IV–VI elongate in males and subquadrate in females, VII–IX subquadrate or slightly transverse; head and pronotum of about the same width, abdomen wider than either head or pronotum; elytra large and distinctly broader than remainder of body; head rounded posteriorly; pronotum broadest in apical third, pubescence on disc directed anteriad along midline and laterad elsewhere; elytra (at suture) slightly longer than pronotum and 1/3 broader than maximum width of pronotum, pubescence on disc directed obliquely laterad; abdomen subparallel; hind tarsus with two basal articles of about same length, third one slightly shorter. MALE. Tergite VIII transverse, truncate apically (Fig. 12.146d); sternite VIII slightly transverse, broadly rounded apically

(Fig. 12.146e); median lobe of aedeagus in lateral view with large fin-shaped crista apicalis, tubus short, slightly bent ventrad at base, straight to broadly triangular apex (Fig. 12.146b), in dorsal view bulbus enlarged, with elongate arcuate angular projection on each side apically, tubus moderately narrow (Fig. 12.146c), internal sac with structures as illustrated (Fig. 12.146b,c). FEMALE. Tergite VIII transverse, with apical margin truncate, broadly shallowly emarginate medially (Fig. 12.146f); sternite VIII broadly rounded posteriorly (Fig. 12.146g); spermatheca with capsule very broadly funnel-shaped, stem sinuate, distinctly swollen posteriorly (Fig. 12.146h).

DISTRIBUTION Origin: Holarctic. **CANADA**: NT, QC, YT. **USA**: AK.

COLLECTION AND HABITAT DATA

Habitat: the only data recorded on micro-habitat association is one specimen from QC in organic mat on sand and gravel in an estuary of a river, 1–2 m above high tide line. In AK, one specimen was captured in area with *Spiraea*, dwarf birch, and *Equisetum*. **Collecting period**: VI–VIII. **Collecting method**: unspecified.

Additional records. See Annex 1.

77. *Gnypeta dentata* Klimaszewski (Fig. 12.147a–h)

Gnypeta dentata Klimaszewski, in Klimaszewski et al. 2008. Klimaszewski et al. 2018.

Description Body length 2.5–2.7 mm; dark brown to black, sometimes central part of elytra and legs or tarsi only reddish brown (Fig. 12.147a); integument moderately strongly glossy; pubescence yellowish grey and moderately long and dense; antennomere IV slightly shorter than preceding one, IV–VI subquadrate to moderately elongate, and VII–IX slightly transverse; head and pronotum of about the same width; elytra and abdomen wider than either head or pronotum; elytra small and short; head rounded basally; pronotum broadest in apical third, pubescence directed anterad along midline and laterad elsewhere; elytra (at suture) about as long as pronotum and 1/4 broader than maximum width of pronotum, pubescence directed obliquely postero-laterad, wavy pattern occurs on each side of disc; abdomen subparallel, slightly narrower than elytra at base; metatarsus with two basal articles of about the same length and the third one slightly shorter. MALE. Tergite VIII transverse and truncate apically, with four apical teeth (Fig. 12.147d); sternite VIII transverse and slightly produced posteriorly (Fig. 12.147e); median lobe of aedeagus with narrowly produced apical part of tubus in lateral view (Fig. 12.147b); bulbus enlarged without pronounced two anterior projections in dorsal view (Fig. 12.147c); internal sac with structures as illustrated (Fig. 12.147b,c). FEMALE. Tergite VIII truncate apically (Fig. 12.147f); sternite VIII broadly rounded posteriorly and bearing median emargination (Fig. 12.147g); spermatheca with capsule funnel-shaped and cylindrical, stem curved and swollen basally (Fig. 12.147h).

DISTRIBUTION Origin: Nearctic. **CANADA**: AB, MB [**NPR**], NT, SK. **USA**: not recorded.

COLLECTION AND HABITAT DATA

Habitat: low tundra vegetation, deformed vegetation encountered in the subarctic and subalpine tree line landscapes. **Collecting period**: VI–VIII. **Collecting method**: not recorded.

We report this species from MB (**NPR**) for the first time here, based on the following: Canada, Manitoba, Churchill, 4 km SE Churchill, Dene Village, 5m el., 58.734, -94.112, sweep net, 14. VII.2008, P. Kevan & T. Woodcock (CBG) 1 specimen; Churchill, 16 km E Churchill, krummholz formation, 5m el., 58.754, -93.913, pitfall trap, 16.VII.2009, P. Kevan & T. Woodcock (CBG) 1 specimen.

Additional records. See Annex 1.

78. *Gnypeta groenlandica* Lohse (Fig. 12.148a–h)

Gnypeta groenlandica Lohse, 1989. Klimaszewski et al. 2008.

Diagnosis Body length 3.8–4.2 mm, colour dark brown to black, tarsi reddish brown (Fig. 12.148a); integument moderately glossy; pubescence yellow-grey and moderately long and dense; antennomere IV–V slightly elongate, VI subquadrate, and VII–IX slightly transverse; head and pronotum of about the same width; elytra and abdomen wider than either head or pronotum; elytra as long as pronotum; head rounded posteriorly; pronotum broadest in apical third, pubescence on disc directed anteriad apically and posteriad basally along midline, laterad and posteriad elsewhere; elytra at suture about as long as pronotum and 1/5 broader than maximum width of pronotum, pubescence on disc directed obliquely posterolaterad; abdomen subparallel, about as wide as elytra at base; hind tarsus with two basal articles of about same length, third one slightly shorter. MALE. Tergite VIII transverse, truncate apically and slightly emarginate medially (Fig. 12.148d); sternite VIII broadly parabolic posteriorly (Fig. 12.148e); median lobe of aedeagus in lateral view with tubus short, sinuate, curved obliquely ventrad in basal half, abruptly narrowed apically to very narrowly rounded apex, crista apicalis very large, fin-shaped (Fig. 12.148b); in dorsal view bulbus rounded, moderately enlarged, with minute anterior projection on each side, tubus moderately wide (Fig. 12.148c); internal sac with structures as illustrated (Fig. 12.148b,c). FEMALE. Tergite VIII similar to that of male (Fig. 12.148f); sternite VIII truncate apically (Fig. 12.148g); spermatheca with capsule tubular, longer than wide, invagination broad and deep, stem sinuate, distinctly swollen posteriorly (Fig. 12.148h).

DISTRIBUTION Origin. Nearctic. **CANADA**: MB, NU, YT. **USA**: AK. **GREENLAND**.

COLLECTION AND HABITAT DATA
Habitat: not recorded. **Collecting period**: not recorded. **Collecting method**: not recorded.

79. ***Gnypeta sellmani*** **Brundin *** (Fig. 12.149a–h)

Gnypeta sellmani Brundin, 1929. Palm 1966, Klimaszewski et al. 2008, 2011, 2018.

Diagnosis Body length 3.0–3.4 mm, colour dark brown to black, sometimes central part of elytra and tarsi reddish brown (Fig. 12.149a); integument moderately glossy; pubescence yellowish grey and moderately long and dense; antennomere IV slightly shorter than preceding one, V–VI elongate, and VII–IX about as long as wide to slightly transverse; head and pronotum of about the same width; elytra and abdomen wider than either head or pronotum; elytra small and short; head rounded posteriorly; pronotum broadest in apical third, pubescence on disc directed anteriad along midline and laterad elsewhere; elytra at suture about as long as pronotum and 1/4 broader than maximum width of pronotum, pubescence on disc directed obliquely posterolaterad; abdomen subparallel, about as wide as elytra at base; hind tarsus with two basal articles of about same length, third one slightly shorter. MALE. Tergite VIII transverse, truncate apically (Fig. 12.149d); sternite VIII broadly rounded posteriorly (Fig. 12.149e); median lobe of aedeagus in lateral view with tubus very short, curved obliquely ventrad in basal half, abruptly narrowed apically to very narrowly rounded apex, crista apicalis very large, fin-shaped (Fig. 12.149b); in dorsal view bulbus moderately enlarged, with collar-shaped anterior projection on each side, tubus moderately wide (Fig. 12.149c); internal sac with structures as illustrated (Fig. 12.149b,c). FEMALE. Tergite VIII similar to that of male (Fig. 12.149f); sternite VIII broadly rounded apically (Fig. 12.149g); spermatheca with capsule about as wide as long, bell-shaped, stem sinuate, distinctly swollen posteriorly (Fig. 12.149h).

DISTRIBUTION Origin: Holarctic. **CANADA**: LB, MB, NF, NT, QC, SK, YT. **USA**: AK.

COLLECTION AND HABITAT DATA
Habitat: in YT, found in moss in water in polygon channels and treaded from sedges on temporary pond margin,. In NF, adults in moss along a

seep. **Collecting period**: VI–VIII. **Collecting method**: treading sedges and sifting moss.

Additional records. See Annex 1.

COMMENTS Disjunct populations were found on Mont Albert (1233 m) and Mont Jacques Cartier (1333 m) in Quebec (LFC), where other insect species with boreal and arctic distribution patterns are known to occur.

Genus *Paradilacra* Bernhauer, 1909

(Fig. 12.150a–h) (Bernhauer 1909, Casey 1910, Fenyes 1918, Seevers 1978, Gusarov 2003, Brunke et al. 2021. As *Gnypeta*: Klimaszewski et al. 2008, 2011, 2016, 2018)

Diagnosis Body subparallel, without macrosetae or bristles, integument with dense, minute asperulate punctules and fine microreticulation; pubescence pale, dense and short giving body a semidull appearance; midline pubescence of pronotum directed posteriad in apical 1/2 to 1/3 and anteriorly in posterior 1/2 to 1/3, on sides of disc pubescence directed lateroposteriad and laterad at base of disc; elytral pubescence longitudinally subparallel except directed laterad basally and apically; head about as wide as pronotum; antennomeres I–II strongly elongate, IV–VI slightly elongate, and VII–X subquadrate; pronotum is two-fifths broader than long; elytra slightly wider than pronotum, at suture about as long as pronotum along midline. MALE. Tergite and sternite VIII densely pubescent; median lobe of aedeagus of *Gnypeta* type, with triangular crista apicalis, and large bulbus and narrow triangular tubus in lateral view. FEMALE. Tergite and sternite VIII densely pubescent; spermatheca with globular capsule bearing wide but shallow apical invagination; stem narrowly elongate, slightly sinuate.

80. ***Paradilacra densissima*** **(Bernhauer)** (Fig. 12.150a–h)

Atheta densissima Bernhauer, 1909. Casey 1910, Fenyes 1912 [records], 1920 [synonymies], Seevers 1978, Ashe 2001, Gusarov 2003 [synonymies], Klimaszewski et al. 2020.

Gnypeta saccharina Klimaszewski and Webster, in Klimaszewski et al. (2008). Synonymized by Brunke et al. (2021).

Diagnosis Body subparallel, length 2.6–3.0 mm, dark brown with slightly reddish central part of elytra, tarsi and antennae slightly paler; integument with dense, minute asperulate punctules and fine microreticulation; pubescence pale, dense and short (Fig. 12.150a); head slightly narrower than elytra, eyes moderately large; antennomeres VII–X subquadrate; pronotum rounded laterally, pubescence pattern described under genus diagnosis; elytra distinctly wider than pronotum, at suture about as long as pronotum; abdomen subparallel. MALE. Tergite VIII truncate apically (Fig. 12.150d); sternite VIII parabolic apically (Fig. 12.150e); bulbus of median lobe of aedeagus large and swollen, tubus short, about arcuately curved ventrad, narrowly triangular at apex in lateral view (Fig. 12.150b), in dorsal view bulbus broad, and tubus short, narrowly triangular apically (Fig. 12.150c). FEMALE. Tergite VIII arcuate apically and truncate medially (Fig. 12.150f); sternite VIII with small median apical emargination (Fig. 12.150g); spermatheca with globular capsule bearing wide but shallow apical invagination, stem narrowly elongate, slightly sinuate (Fig. 12.150h).

DISTRIBUTION Origin: Nearctic. **CANADA**: AB, BC (Glenora), NB, ON, SK. **USA**: CA, MT, NV, ND, OR, UT.

COLLECTION AND HABITAT DATA
Habitat: banks of lakes, rivers, and a reservoir, beaver pond margin. **Collecting period**: VI, VII, VIII. **Collecting method**: treading vegetation in wet habitats, malaise and intercept traps.

Genus *Tachyusa* Erichson, 1837

(Fig. 12.151a–k) (Paśnik 2006, Klimaszewski et al. 2008, 2018, 2020)

Diagnosis Body elongate with slender and clavate abdomen, convex and glossy, length

2.3–3.9 mm; head without infraorbital carina; pronotal hypomera fully visible in lateral view; elytra laterally distinctly arcuate, widest at middle, posterolateral angles distinctly emarginate or broadly rounded; abdomen clavate, distinctly constricted at base, base narrower than width of elytra, strongly swollen posteriorly, tergite V (third visible) moderately transverse, 1.2–1.8 times as wide as long; tergites III–V (first to third visible) or III–VI (first to fourth visible) distinctly transversely impressed at base, impressions coarsely punctate, interstices between punctures forming longitudinal ridges, some species with distinct median carina; pubescence on tergites III–V (first to third visible) directed inward.

COMMENTS Species affiliated with wetland and riparian habitats along margins of ponds, lakes, streams, and marshes.

81. *Tachyusa americanoides* Paśnik (Fig. 12.151a–k)

Tachyusa americanoides Paśnik, 2006. Webster et al. 2012, Klimaszewski et al. 2011, 2018, 2020.

Diagnosis Body convex, parallel-sided, length 2.5–3.2 mm, integument glossy, colour dark brown to black, abdomen black with first three visible tergites reddish brown, legs reddish brown, antennae brown with antennomeres I–III red (Fig. 12.151a); head circular in outline, convex, glossy, narrower than pronotum; eyes relatively small, moderately protruding from lateral contours of head, slightly shorter than genae in dorsal view; surface of head without microsculpture, punctation fine and moderately dense, pubescence on disc short and moderately dense, directed inward, antennae short, slightly incrassate apically, antennomeres II and III subequal in length, antennomeres IV–VII subquadrate to slightly elongate, antennomeres VIII–X sexually dimorphic, quadrate in male and slightly transverse in female, antennomere XI nearly conical; pronotum quadrate, convex, glossy, widest in apical third, lateral sides rounded in front, gradually narrowed in straight line to obtuse hind angles, with small and shallow transverse basal impression, surface lacking microsculpture, punctation fine, dense and asperate, pubescence at midline directed posteriad; elytra longer than wide, slightly wider than pronotum, lateral sides arcuate, at suture as long as pronotal length at midline; surface lacking microsculpture; punctation fine, dense and asperate; abdomen parallel-sided, first three visible tergites with deep transverse basal impression, impressions with 7–8 longitudinal ridges and distinct median carina, tergal punctation fine and moderately dense, punctures moderately small and well visible, surface without microsculpture. MALE. Tergite broadly arcuate apically (Fig. 12.151f); sternite VIII rounded apically (Fig. 12.151g); tubus of median lobe of aedeagus arcuately curved slightly ventrad in basal half, then straight to narrowly triangular and narrowly rounded apex in lateral view (Fig. 12.151b,c), in dorsal view bulbus ovoid in shale, tubus short, triangular (Fig. 12.151d,e). FEMALE. Tergite arcuate apically (Fig. 12.151h); sternite VIII truncate apically with minute median emargination (Fig. 12.151i); spermatheca with stem thick, slightly sinuate (Fig. 12.151j,k).

DISTRIBUTION Origin: Nearctic. **CANADA**: AB, BC, MB, NB, NF, NS, NT, ON, QC. **USA**: IL, MA, MT, NH, NY, PA, VT.

COLLECTION AND HABITAT DATA

Habitat: under/in debris in riparian habitats and along pond margins; under debris on muddy soil near a small pool, under debris on clay and sand at a river margin, in grassy debris near the overflow area of a beaver dam, among sticks, debris, and clay on a beaver dam, under debris on muddy soil along a slow-flowing river in a silver maple swamp, in debris along the margin of a rocky river, under debris on muddy soil near a small pool in a silver maple forest, and among gravel on gravel bar along a small shaded brook in a northern coniferous forest. **Collecting period**: V–VII. **Collecting method**: sifting debris and, aspirating adults from debris on muddy soils or from soil surface.

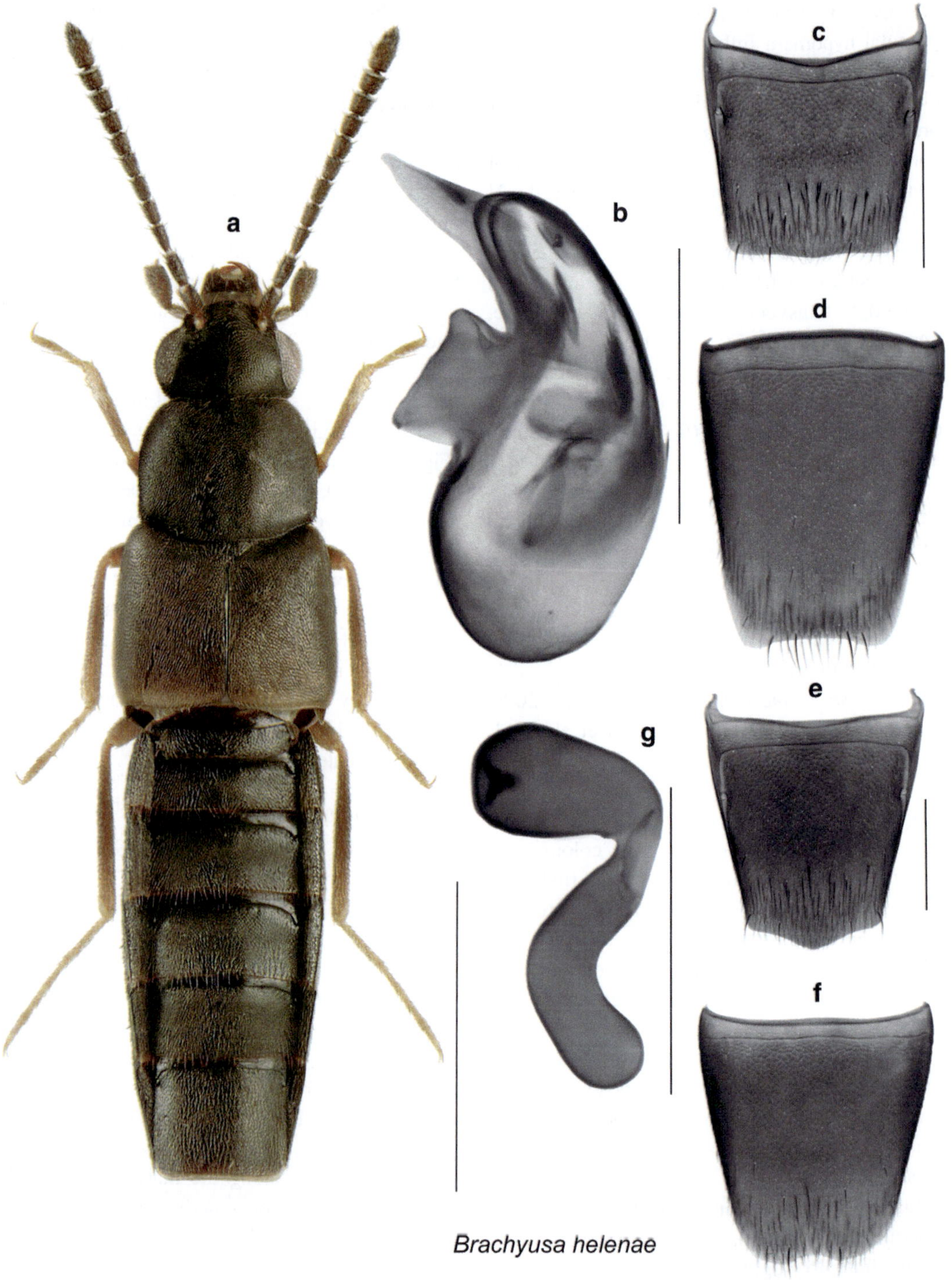

Fig. 12.138 (**a–g**). Images of *Brachyusa helenae* (Casey): (**a**) habitus in dorsal view; (**b**) median lobe of aedeagus in lateral view; (**c**) male tergite VIII; (**d**) male sternite VIII; (**e**) female tergite VIII; (**f**) female sternite VIII; (**g**) spermatheca. (**a–g**) after Klimaszewski et al. (2018). Scale bar for habitus = 1 mm; remaining scale bars = 0.2 mm

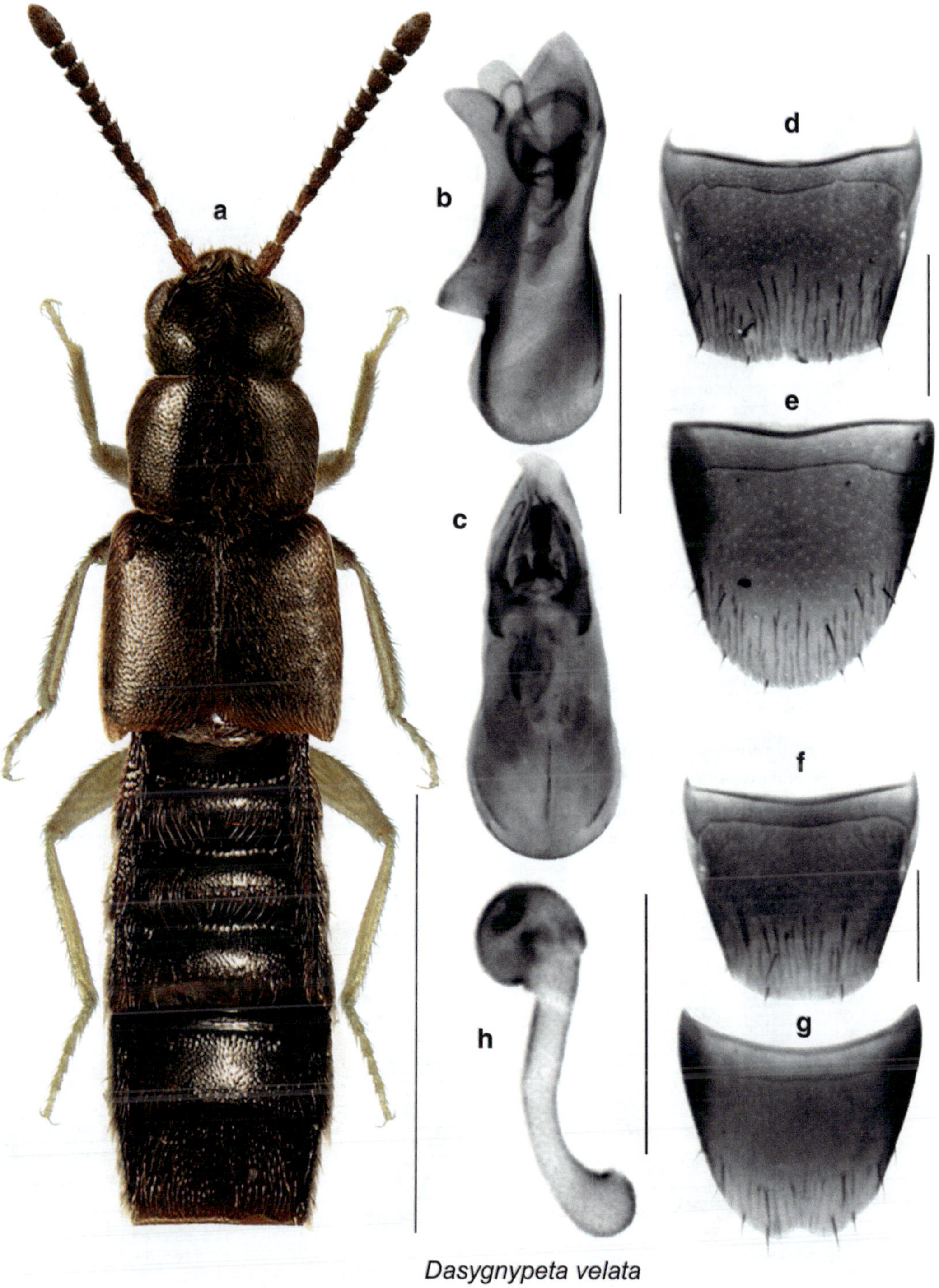

Fig. 12.139 (**a–h**). Images of *Dasygnypeta velata* (Erichson): (**a**) habitus in dorsal view; (**b**) median lobe of aedeagus in lateral view; (**c**) median lobe of aedeagus in dorsal view; (**d**) male tergite VIII; (**e**) male sternite VIII; (**f**) female tergite VIII; (**g**) female sternite VIII; (**h**) spermatheca. (**a–h**) after Klimaszewski et al. (2008). Scale bar for habitus = 1 mm; remaining scale bars = 0.2 mm

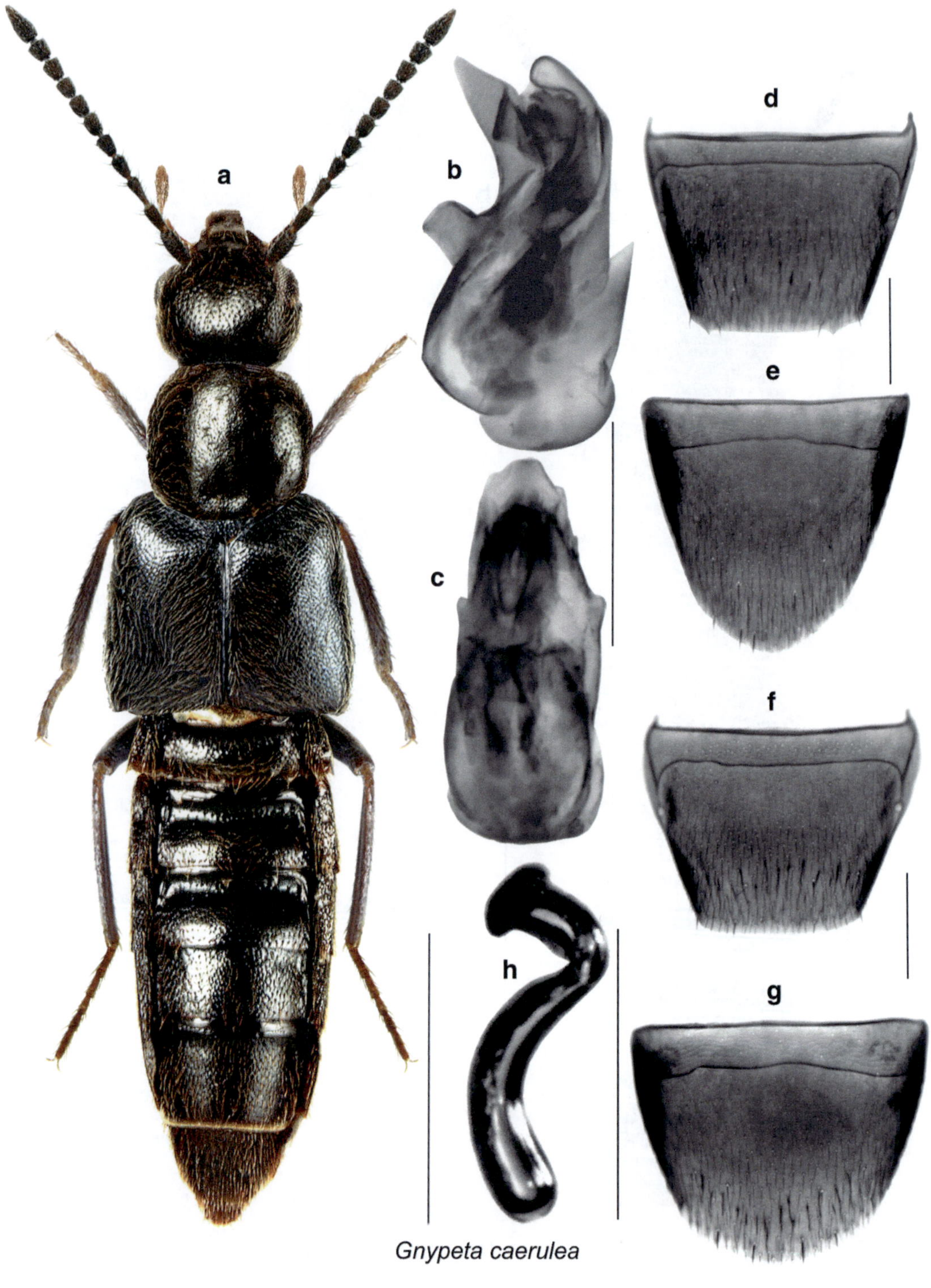

Fig. 12.140 (**a–h**). Images of *Gnypeta caerulea* (C.R. Sahlberg): (**a**) habitus in dorsal view; (**b**) median lobe of aedeagus in lateral view; (**c**) median lobe of aedeagus in dorsal view; (**d**) male tergite VIII; (**e**) male sternite VIII; (**f**) female tergite VIII; (**g**) female sternite VIII; (**h**) spermatheca. (**a–h**) after Klimaszewski et al. (2008). Scale bar for habitus = 1 mm; remaining scale bars = 0.2 mm

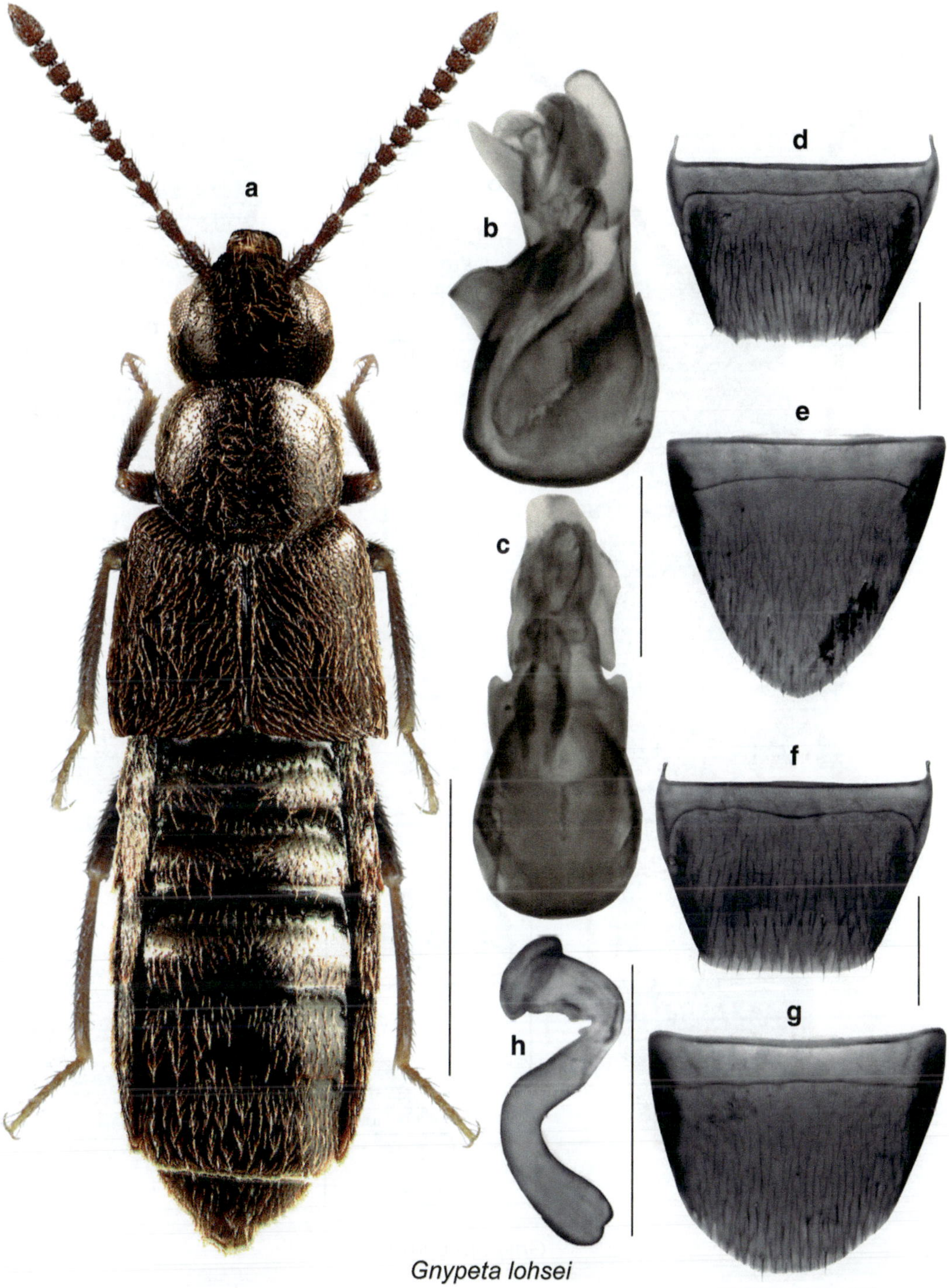

Fig. 12.141 (**a–h**). Images of *Gnypeta lohsei* Klimaszewski: (**a**) habitus in dorsal view; (**b**) median lobe of aedeagus in lateral view; (**c**) median lobe of aedeagus in dorsal view; (**d**) male tergite VIII; (**e**) male sternite VIII; (**f**) female tergite VIII; (**g**) female sternite VIII; (**h**) spermatheca. (**a–h**) after Klimaszewski et al. (2008). Scale bar for habitus = 1 mm; remaining scale bars = 0.2 mm

Fig. 12.142 (**a–h**). Images of *Gnypeta carbonaria* (Mannerheim): (**a**) habitus in dorsal view; (**b**) median lobe of aedeagus in lateral view; (**c**) median lobe of aedeagus in dorsal view; (**d**) male tergite VIII; (**e**) male sternite VIII; (**f**) female tergite VIII; (**g**) female sternite VIII; (**h**) spermatheca. (**a–h**) after Klimaszewski et al. (2008). Scale bar for habitus = 1 mm; remaining scale bars = 0.2 mm

Fig. 12.143 (**a–h**). Images of *Gnypeta ashei* Klimaszewski: (**a**) habitus in dorsal view; (**b**) median lobe of aedeagus in lateral view; (**c**) median lobe of aedeagus in dorsal view; (**d**) male tergite VIII; (**e**) male sternite VIII; (**f**) female tergite VIII; (**g**) female sternite VIII; (**h**) spermatheca. (**a–h**) after Klimaszewski et al. (2008). Scale bar for habitus = 1 mm; remaining scale bars = 0.2 mm

Fig. 12.144 (**a–h**). Images of *Gnypeta atrolucens* Casey: (**a**) habitus in dorsal view; (**b**) median lobe of aedeagus in lateral view; (**c**) median lobe of aedeagus in dorsal view; (**d**) male tergite VIII; (**e**) male sternite VIII; (**f**) female tergite VIII; (**g**) female sternite VIII; (**h**) spermatheca. (**a–h**) after Klimaszewski et al. (2008). Scale bar for habitus = 1 mm; remaining scale bars = 0.2 mm

Fig. 12.145 (**a–h**). Images of *Gnypeta brevicornis* Casey: (**a**) habitus in dorsal view; (**b**) median lobe of aedeagus in lateral view; (**c**) median lobe of aedeagus in dorsal view; (**d**) male tergite VIII; (**e**) male sternite VIII; (**f**) female tergite VIII; (**g**) female sternite VIII; (**h**) spermatheca. (**a–h**) after Klimaszewski et al. (2008). Scale bar for habitus = 1 mm; remaining scale bars = 0.2 mm

Fig. 12.146 (**a–h**). Images of *Gnypeta brincki* Palm: (**a**) habitus in dorsal view; (**b**) median lobe of aedeagus in lateral view; (**c**) median lobe of aedeagus in dorsal view; (**d**) male tergite VIII; (**e**) male sternite VIII; (**f**) female tergite VIII; (**g**) female sternite VIII; (**h**) spermatheca. (**a–h**) after Klimaszewski et al. (2008). Scale bar for habitus = 1 mm; remaining scale bars = 0.2 mm

Fig. 12.147 (**a–h**). Images of *Gnypeta dentata* Klimaszewski: (**a**) habitus in dorsal view; (**b**) median lobe of aedeagus in lateral view; (**c**) median lobe of aedeagus in dorsal view; (**d**) male tergite VIII; (**e**) male sternite VIII; (**f**) female tergite VIII; (**g**) female sternite VIII; (**h**) spermatheca. (**a–h**) after Klimaszewski et al. (2008). Scale bar for habitus = 1 mm; remaining scale bars = 0.2 mm

Fig. 12.148 (**a–h**). Images of *Gnypeta groenlandica* Lohse: (**a**) habitus in dorsal view; (**b**) median lobe of aedeagus in lateral view; (**c**) median lobe of aedeagus in dorsal view; (**d**) male tergite VIII; (**e**) male sternite VIII; (**f**) female tergite VIII; (**g**) female sternite VIII; (**h**) spermatheca. (**a–h**) after Klimaszewski et al. (2008). Scale bar for habitus = 1 mm; remaining scale bars = 0.2 mm

Fig. 12.149 (**a–h**). Images of *Gnypeta sellmani* Brundin: (**a**) habitus in dorsal view; (**b**) median lobe of aedeagus in lateral view; (**c**) median lobe of aedeagus in dorsal view; (**d**) male tergite VIII; (**e**) male sternite VIII; (**f**) female tergite VIII; (**g**) female sternite VIII; (**h**) spermatheca. (**a–h**) after Klimaszewski et al. (2008). Scale bar for habitus = 1 mm; remaining scale bars = 0.2 mm

Fig. 12.150 (**a–h**). *Paradilacra densissima* (Bernhauer): (**a**) habitus in dorsal view; (**b**) median lobe of aedeagus in lateral view; (**c**) median lobe of aedeagus in dorsal view; (**d**) male tergite VIII; (**e**) male sternite VIII; (**f**) female tergite VIII; (**g**) female sternite VIII; (**h**) spermatheca. Scale bar for habitus = 1 mm; remaining scale bars = 0.2 mm

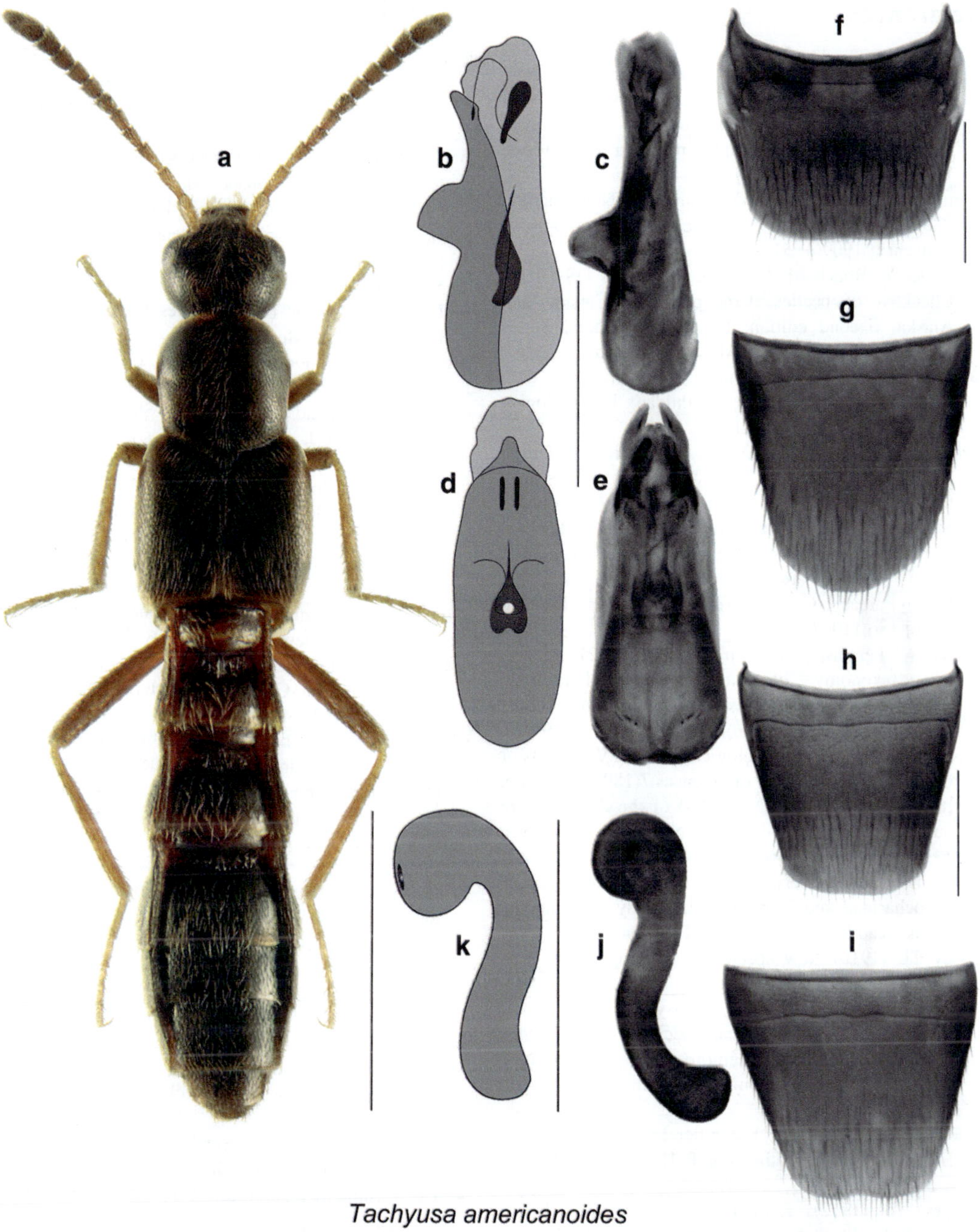

Fig. 12.151 (**a–k**). Images of *Tachyusa americanoides* Paśnik: (**a**) habitus in dorsal view; (**b–c**) median lobe of aedeagus in lateral view; (**d–e**) median lobe of aedeagus in dorsal/ventral view; (**f**) male tergite VIII; (**g**) male sternite VIII; (**h**) female tergite VIII; (**i**) female sternite VIII; (**j–k**) spermatheca. (**b, d, k**) after Paśnik (2006); remaining figures after Klimaszewski et al. (2018). Scale bar for habitus = 1 mm; remaining scale bars = 0.2 mm

References

Ashe JS (2001) Key to the tribes and genera of Nearctic Aleocharinae. 8. In: Arnett RH Jr, Thomas CM (eds) American Beetles. Archostemata, Myxophaga, Adephaga, Polyphaga: Staphyliniformia. CRC Press, Washington, DC, pp 299–374. 443 pp

Bernhauer M (1909) Neue Aleocharini aus Nordamerika. (Col.) (4. Stück.). Deutsche Entomologische Zeitschrift 1909(4):515–528

Bousquet Y, Bouchard P, Davies AE, Sikes DS (2013) Checklist of beetles (Coleoptera) of Canada and Alaska. Second edition. Pensoft Publishers, Series Faunistica No. 109. Sofia-Moscow, 402 pp

Brundin L (1929) Eine neue *Gnypeta* aus dem Torneträskgebiete (Col. Staphylinidae). Entomologisk Tidskrift 50(1):14–15. pl.1

Brunke A, Klimaszewski J, Dorval J-A, Bourdon C, Paiero SM, Marshall SA (2012) New species and distributional records of Aleocharinae (Coleoptera, Staphylinidae) from Ontario, Canada, with a checklist of recorded species. In: Klimaszewski J, Anderson R (Eds) Biosystematics and Ecology of Canadian Staphylinidae (Coleoptera) II. ZooKeys [Special Issue] 186:119–206

Brunke AJ, Pentinsaari M, Klimaszewski J (2021) Integrative taxonomy of Nearctic and Palaearctic Aleocharinae: new species, synonymies, and records (Coleoptera, Staphylinidae). ZooKeys 1041:27–99

Casey TL (1893) Coleopterological notices. V. Annals of the New York Academy of Sciences 7(1894):281–606. [these fascicles (Oct–Dec 1893) of the Annals were published in Mar 1894 (Trans. New York Acad. Sci. 13:152; Ent. News 5 (6):187)

Casey TL (1906) Observations on the staphylinid groups Aleocharinae and Xantholinini, chiefly of America. Trans Acad Sci St Louis 16(6):125–434

Casey TL (1910) New species of the staphylinid tribe Myrmedoniini. Memoirs on the Coleoptera 1. New Era Printing Co., Lancaster, Pennsylvania, 184 pp.

Casey TL (1911) New American species of Aleocharinae and Myllaeninae. In: Memoirs on the Coleoptera. II. The New Era Printing Co., Lancaster, PA, pp 1–245. 259 pp

Erichson WF (1837) Die Käfer der Mark Brandenburg. Erster Band, Erste Abtheilung. F. H. Morin, Berlin, viii + 384 pp

Fenyes A (1912) Remarks on *Gnypeta* Thoms. (Staphylinidae Col.). Can Entomol 44(4):105–112

Fenyes A (1918) Coleoptera. Family Staphylinidae. Subfamily Aleocharinae. In: Wytsman P (ed) Genera Insectorum, vol XXVII. Fascicule 173A. M. Nijhoff, Den Haag, pp 1–110. 453 pp. + 7 pls

Fenyes A (1920) Coleoptera. Family Staphylinidae. Subfamily Aleocharinae. In: Wytsman P (ed) Genera Insectorum, vol XXVII. Fascicule 173B. L. Desmet-Verteneuil, Bruxelles, pp 111–414. 453 pp. + 7 pls

Gusarov VI (2003) Revision of some types of North American aleocharines (Coleoptera: Staphylinidae: Aleocharinae), with synonymic notes. Zootaxa 353:1–134

Klimaszewski J, Savard K, Pelletier G, Webster R (2008) Species review of the genus *Gnypeta* Thomson from Canada, Alaska and Greenland (Coleoptera, Staphylinidae, Aleocharinae): systematics, bionomics and distribution. In: Majka CG, Klimaszewski J (Eds.) Biodiversity, biosystematics, and ecology of Canadian Coleoptera. ZooKeys 2([Special Issue]):11–84

Klimaszewski J, Langor D, Pelletier G, Bourdon C, Perdereau L (2011) Aleocharine beetles (Coleoptera, Staphylinidae) of the province of Newfoundland and Labrador, Canada. Pensoft Series Faunistica No. 98. Pensoft, Sophia-Moscow, 313 pp

Klimaszewski J, Godin B, Bourdon C (2012) Further contributions to the aleocharine fauna of the Yukon Territory, Canada (Coleoptera, Staphylinidae). In: Klimaszewski J, Anderson R (Eds.): Biosystematics and Ecology of Canadian Staphylinidae (Coleoptera) II. Zookeys [Special Issue] 186:207–237

Klimaszewski J, Godin B, Langor D, Bourdon C, Lee S-I, Horwood D (2015) New distribution records for Canadian Aleocharinae (Coleoptera, Staphylinidae), and new synonymies for *Trichiusa*. ZooKeys 498:51–91

Klimaszewski J, Larson DJ, Labrecque M, Bourdon C (2016) Twelve new species and fifty-three new provincial distribution records of Aleocharinae rove beetles of Saskatchewan, Canada (Coleoptera, Staphylinidae). ZooKeys 610:45–112

Klimaszewski J, Webster RP, Langor DW, Brunke A, Davies A, Bourdon C, Labrecque M, Newton AF, Dorval J-A, Frank JH (2018) Aleocharine rove beetles of eastern Canada (Coleoptera, Staphylinidae, Aleocharinae): a glimpse of megadiversity. Springer, Cham. 902 pp

Klimaszewski J, Hoebeke ER, Godin B, Davies A, Perry KI, Bourdon C, Winchester N (2020) Aleocharine rove beetles of British Columbia: a hotspot of Canadian biodiversity (Coleoptera, Staphylinidae). Springer, Cham. 631 pp

Lohse GA (1974) 18. Tribus: Oxypodini. In: Freude H, Harde KW, Lohse GA (eds) Band 5 Staphylinidae II (Hypocyphtinae und Aleocharinae) Pselaphidae. Die Kafer Mitteleuropas. Goecke and Evers, Krefeld, 381 pp

Lohse GA (1989) Familie: Staphylinidae. In: Lohse GA, Lucht WH (eds) Die Käfer Mitteleuropas, Band 12, 1 Suplementband mit Katalogteil. Goecke and Evers, Krefeld, pp 185–240

Majka CG, Klimaszewski J (2008) Introduced Staphylinidae (Coleoptera) in the Maritime Provinces of Canada. Can Entomol 140(1):48–72

Mannerheim CG (1830) Précis d'un nouvel arrangement de la famille des brachélytres de l'ordre des insectes coléoptères. St. Petersburg, 87 pp. [separate, re-issued in Mémoires Présentées à l'Académie Impériale des Sciences de St.-Pétersbourg 1:415–501, in 1831]

Mulsant E, Rey C (1873) Histoire naturelle des Coléoptères de France. Brévipennes. Aléochariens (Suite) [Tableau... des Aléochariens + Septième

Branche, Myrmédoniaires, part 1]. Deyrolle, Paris, [2] + 695 pp., pls. 1–5

Osswald J, Bachmann L, Gusarov VI (2013) Molecular phylogeny of the beetle tribe Oxypodini (Coleoptera: Staphylinidae: Aleocharinae). Systematic Entomology 38(3):507–522. https://doi.org/10.1111/syen.12011

Palm T (1966) De svenska *Gnypeta*-arterna (Col. Staphylinidae, Aleocharinae). Norsk Entomologisk Tidskrift 87(3–4):136–141

Paśnik G (2006) A revision of the World species of the genus *Tachyusa* Erichson, 1837 (Coleoptera, Staphylinidae: Aleocharinae). Zootaxa 1146:1–152

Sahlberg CR (1831) [Pp. 361–408]. In: Insecta Fennica, dissertationibus academicis, A. 1817–1834 editis. Pars I: A. J.C. Frenckel, Helsingfors, VIII + 519 pp

Seevers CH (1978) A generic and tribal revision of the North American Aleocharinae (Coleoptera: Staphylinidae) [with additions and annotations by Lee H. Herman]. Fieldiana: Zoology 71. vi + 289 pp

Thomson CG (1859) Skandinaviens Coleoptera, synoptiskt bearbetade. I. Tom. Berlingska Boktryckeriet, Lund, [3] + 290 pp

Webster RP, Klimaszewski J, Sweeney JD, DeMerchant I (2012) New Staphylinidae (Coleoptera) records with new collection data from New Brunswick, and an addition to the fauna of Quebec, Canada: Aleocharinae. In: Klimaszewski J, Anderson R (Eds) Biosystematics and Ecology of Canadian Staphylinidae (Coleoptera) II. ZooKeys [Special Issue] 186:83–118

Tribe Boreocyphini Klimaszewski and Langor, 2011

13

Tribe Boreocyphini Klimaszewski and Langor, 2011

(Figs. 13.152–13.153) (Klimaszewski et al. 2011, 2018)

Diagnosis Body with impressed punctation and coarse microsculpture consisting of convex sculpticells; labial palpi with three palpomeres, basal palpomere swollen, second short and wide, and terminal one elongate with short and bluntly ended microsegment; glossae separated as short subparallel lobes; maxillary palpi with four palpomeres, apical palpomere short and with blunt apex.

Genus *Boreocypha* Klimaszewski and Langor, 2011

(Figs. 13.152–13.153) (Klimaszewski et al. 2011, 2018)

Diagnosis Body broadly elongate and matte; antennomeres I–II elongate and equal in length, III–VII slightly elongate, VIII–X subquadrate, and XI widely oval; pronotal hypomera visible laterally, pubescence on disc short, dense and directed almost straight laterad from midline; abdomen wide, widest in apical one-third, with four deep tergal impressions; mesocoxae contiguous, process of mesoventrite widely triangular and short, isthmus long; process of metaventrite short; median lobe of aedeagus in dorsal view with wide and long tubus and small bulbus; paramere subparallel-sided and long, apical lobe moderately elongate with three macrosetae.

82. *Boreocypha websteri* Klimaszewski and Langor (Figs. 13.152, 13.153a–h)

Boreocypha websteri Klimaszewski and Langor, in Klimaszewski et al. 2011.

Diagnosis Body broadly elongate, moderately convex, length 2.5–2.8 mm; integument matte, colour dark brown to almost black (Figs. 13.152, 13.153a); head circular, slightly narrower than pronotum, widest across eyes, eyes small, as long as postocular area; pronotum rectangular, widest in apical third, widely arcuate apically and with distinct angular apical angles, gradually widened posteriorly and with slightly angular posterior angles, punctation fine, dense and asperate; elytra transverse, at suture slightly shorter than pronotum; abdomen at base narrower than elytra but becoming as wide or slightly wider than elytra at apical third, first four visible tergites with deep transverse impressions bearing fine punctation; antennae slender, antennomeres VIII–X roughly bead-shaped. MALE. Tergite VIII broadly arcuately emarginate apically (Fig. 13.153d); sternite VIII with lateral margins

J. Klimaszewski et al., *A Faunal Review of Aleocharine Beetles in the Rapidly Changing Arctic and Subarctic Regions of North America (Coleoptera, Staphylinidae)*,
https://doi.org/10.1007/978-3-030-68191-3_13

Fig. 13.152 Image of *Boreocypha websteri* Klimaszewski and Langor: habitus in dorso-lateral view. After Klimaszewski et al. (2011). Body length 3.0 mm

obliquely converging apicad, apical margin obtusely angulately produced, rounded apically, antecostal suture very widely separated from basal margin (Fig. 13.153e); median lobe of aedeagus elongate and subparallel in dorsal view (Fig. 13.153c), in lateral view narrow and arcuate with apex slightly bent ventrad (Fig. 13.153b). FEMALE. Spermatheca not found, probably very small or lacking, tergite and sternite VIII and pygidium as illustrated (Fig. 13.153f–h).

DISTRIBUTION Origin: Nearctic. **CANADA**: LB, NB, PE. **USA**: AK [**NSR**].

COLLECTION AND HABITAT DATA

Habitat: shaded riparian habitats in forests. Treading moss along wooded stream, under small rocks at margin of cold shaded spring-fed brook in rich Appalachian hardwood forest, splashing gravel on gravel bar along small cold shaded brook in boreal forest area, and along a stream. In AK, one specimen was captured under rocks. **Collecting period**: VI–VII. **Collecting method**: splashing water on rocks/gravel, hand collecting from under small rocks, and treading moss.

We report this species from AK (**NSR**) for the first time here based on the following: USA, Alaska, Toolik Field Station, 724m el., 68.6286, −149.59772, ±36m, under rocks, 1-3. VI.2008, D.S. Sikes (UAM) 4 specimens.

Fig. 13.153 (**a–h**). Images of *Boreocypha websteri* Klimaszewski and Langor: (**a**) habitus in dorsal view; (**b**) median lobe of aedeagus in lateral view; (**c**) median lobe of aedeagus in dorsal view; (**d**) male tergite VIII; (**e**) male sternite VIII; (**f**) female tergite VIII; (**g**) female sternite VIII; (**h**) female pygidium. (**a–h**) after Klimaszewski et al. (2011). Scale bar for habitus = 1 mm; remaining scale bars = 0.2 mm

References

Klimaszewski J, Langor D, Pelletier G, Bourdon C, Perdereau L (2011) Aleocharine beetles (Coleoptera, Staphylinidae) of the province of Newfoundland and Labrador, Canada. Pensoft Series Faunistica No. 98. Pensoft, Sophia-Moscow, 313 pp

Klimaszewski J, Webster RP, Langor DW, Brunke A, Davies A, Bourdon C, Labrecque M, Newton AF, Dorval J-A, Frank JH (2018) Aleocharine rove beetles of eastern Canada (Coleoptera, Staphylinidae, Aleocharinae): a glimpse of megadiversity. Springer, Cham. 902 pp

14 Tribe Hypocyphtini Laporte, 1835

Tribe Hypocyphtini Laporte, 1835

(Fig. 14.154a–g) (Ashe 2001)

Diagnosis Body minute and compact, length 0.8–1.6 mm; antennae with ten antennomeres, antennomeres VIII–X enlarged and forming loose club; hind coxae with large ventral lamella covering base of femur; tarsal formula 4-4-4.

Genus *Cypha* Leach, 1819

(Fig. 14.154a–g) (Ashe 2001, Klimaszewski et al. 2008, 2018)

Diagnosis Body subovate, length 0.8–1.25 mm, colour brownish to black, integument moderately glossy, with coarse sculpture and inconspicuous pubescence; head hypognathous (strongly deflexed), narrower than elytra; eyes large; antennae slender, antennomeres VIII–X forming club, terminal antennomere as long as two preceding ones combined; pronotum narrow and distinctly transverse, twice as wide as long, apex narrower than base and broadly emarginate adjacent to head, sides rounded, hind angles obtusely rounded, posterior margin slightly bisinuate, maximum width of pronotum at least as maximum width of elytra; pronotal hypomera not visible in lateral view; elytra at suture distinctly shorter then pronotum along midline; mesocoxae narrowly separated; process of mesoventrite long and obtusely pointed, extending to three-quarters length of mesocoxae and overlapping apex of short and broader process of metaventrite; abdomen acuminate, tergites not impressed; hind tarsi with tarsomere I subequal in length to II and III combined; median lobe of aedeagus very long, narrow, gradually tapering apicad, paramere evenly narrowly elongate with small apical lobe bearing short apical setae; spermatheca not found.

83. *Cypha inexpectata* Klimaszewski and Godin (Fig. 14.154a–g)

Cypha inexpectata Klimaszewski and Godin, in Klimaszewski et al. 2008. Majka and Klimaszewski 2008, Klimaszewski et al. 2016, 2018.

Diagnosis Body length 0.8–1.0 mm, small, subovate, colour light brown to black (Fig. 14.154a); pubescence long, sparse; head broad, retracted under pronotum; pronotum distinctly transverse, markedly narrowed apically; elytra distinctly transverse, shorter at suture than pronotum; abdomen shortly triangular, usually about as long as pronotum (individual specimens

J. Klimaszewski et al., *A Faunal Review of Aleocharine Beetles in the Rapidly Changing Arctic and Subarctic Regions of North America (Coleoptera, Staphylinidae)*,
https://doi.org/10.1007/978-3-030-68191-3_14

may exhibit different degree of abdominal contraction). MALE. Tergite VIII distinctly transverse, with apical margin broadly truncate, arcuate laterally, microsetae long and elevated on microtubercules (Fig. 14.154d); sternite VIII transverse with lateral margins obliquely converging posteriad to narrowly truncate apex, antecostal suture broadly separated from basal margin, pubescence as on tergite VIII (Fig. 14.154e); median lobe of aedeagus in lateral view with bulbus arched abruptly dorsad apically so that base of tubus situated on dorsoapical margin of bulbus, tubus bent obliquely ventrad at base, gradually arcuately tapering to very fine apex (Fig. 14.154b), in dorsal view tubus narrow basally, very long, evenly tapering to narrowly rounded apex (Fig. 14.154c), internal sac lacking sclerites (Fig. 14.154b,c). FEMALE. Tergite and sternite VIII as illustrated (Fig. 14.154f,g). Females can only be positively identified by association with males.

Cypha inexpectata is extremely similar externally to the western *C. crotchii* (Horn), the only other recorded species of this genus from Canada. It differs from *C. crotchii* in having the median lobe of the aedeagus with the tubus broadly arcuate, with the apex very fine in lateral view and the internal sac lacking sclerites.

DISTRIBUTION **Origin**: Nearctic. **CANADA**: AB [**BOLD**], MB [**BOLD**], NT [**NPR**], ON, QC [**NPR**], SK, YT. **USA**: CA [**BOLD**], one female of *Cypha* from AK is housed in UAMC but male is essential for positive specific identification.

COLLECTION AND HABITAT DATA

Habitat: mossy hummocks at the border between a marsh and spruce forest, and mossy hummocks near creek, otherwise nothing is known about habitat associations of this species. In AB, found in subalpine conifer forest in mountain valley, lodgepole pine and fir, birch in understory; and in MB collected in meadow dominated by sedges. **Collecting period**: VI, VIII-IX. **Collecting method**: unbaited and uncovered pitfall traps

We report this species from AB (**NPR, BOLD**), MB (**NPR, BOLD**), NT (**NTR**), QC (**NPR**), and CA (**NSR, BOLD**) for the first time here, based on the following: Canada, Alberta, Waterton Lakes National Park, Akamina Parkway, Rowe Tamarack Trail, 1632m el., 49.059, –114.013, subalpine forest of lodgepole pine and fir, birch in understory, pan trap, 6-11. VIII.2012, BIOBus 2012 team (CBG) 1 specimen; Calgary, Our Lady of Peace school, 1088m el., 50.9182, –114.089, Malaise trap, 19-30.IX.2016, Brian Briggs (CBG) 1 specimen; Calgary, Monsignor J.S. Smith school, 1036m el., 50.941, –113.991, Malaise trap, 21.IX-02.X.2015, Wanda Paslawski (CBG) 1 specimen; Drayton Valley, Evergreen Elementary School, 876m el., 53.218, –114.97, Malaise trap, 20.IV-08. V.2015, Barbara Gregory (CBG) 1 specimen. Manitoba, Wapusk National Park, 2m el., 58.723, –93.460 , meadow dominated by sedges, Malaise trap, 15-22.VI.2014, D. Iles (CBG) 1 specimen. Northwest Territories, Yellowknife, 62.50293, –113.40479, Rep. 2 mesic, pitfall trap, 11-15.VII.2011, NBP Field Party Y2MP211, Y2MP221 (LFC) 2 males. Quebec, Schefferville, 54.90310, –67.14765, Rep. 3, mesic, pitfall trap, 6-9.VII.2010, NBP Field Party S3MP231 (LFC) 1 male. USA, California, San Diego, 9001 Spectrum Center Blvd. 150m el., 32.826, –117.134, urban area, Malaise trap, 02-09.V.2014, Peter Farrell (CBG) 1 specimen; same data but 09-16. V.2014 (CBG) 1 specimen; 16-23.V.2014 (CBG) 1 specimen; 23-30.V.2014 (CBG) 1 specimen; 27.VI-03.VII.2014 (CBG) 1 specimen; 01-08. VIII.2014 (CBG) 2 specimens; 19-26.VI.2015 (CBG) 2 specimens; 03-10.VI.2016 (CBG) 2 specimens; 18-25.VII.2016 (CBG) 2 specimens; 01-15.VIII.2016 (CBG), 2 specimens; 20-28. XII.2016 (CBG) 1 specimen.

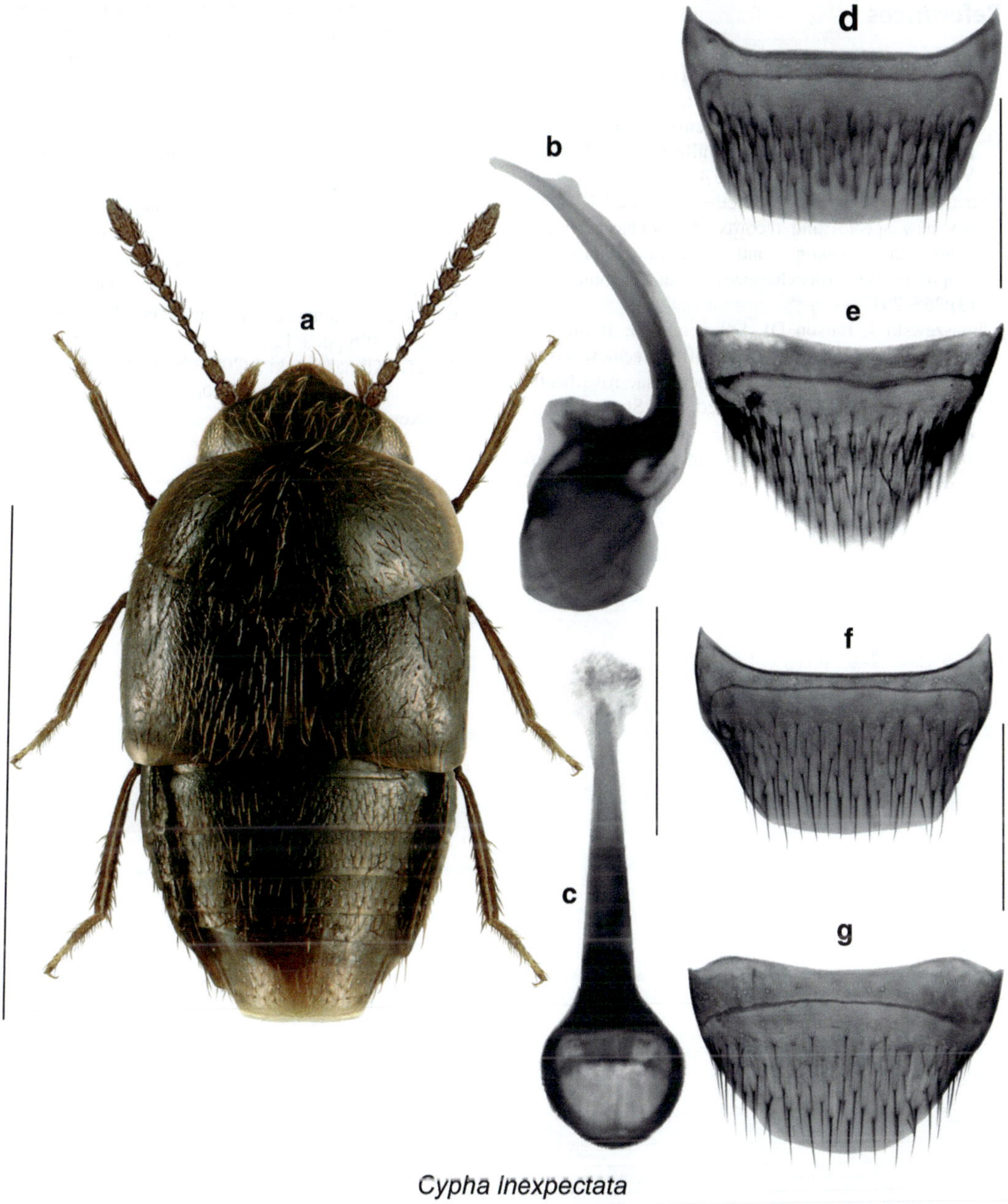

Fig. 14.154 (a–g). Images of *Cypha inexpectata* Klimaszewski and Godin: (**a**) habitus in dorsal view; (**b**) median lobe of aedeagus in lateral view; (**c**) median lobe of aedeagus in dorsal view; (**d**) male tergite VIII; (**e**) male sternite VIII; (**f**) female tergite VIII; (**g**) female sternite VIII. (**a–g**) after Klimaszewski et al. (2018). Scale bar for habitus = 1 mm; remaining scale bars = 0.2 mm

References

Ashe JS (2001) Key to the tribes and genera of Nearctic Aleocharinae. 8. In: Arnett RH Jr, Thomas CM (eds) American beetles. Archostemata, myxophaga, adephaga, polyphaga: staphyliniformia. CRC Press, Washington, DC, pp 299–374. 443 pp.

Klimaszewski J, Godin B, Pelletier G, Savard K (2008) Six new species and records of aleocharine beetles from the Yukon and Alaska (Coleoptera: Staphylinidae: Aleocharinae). Can Entomol 140 (3):265–291

Klimaszewski J, Larson DJ, Labrecque M, Bourdon C (2016) Twelve new species and fifty-three new provincial distribution records of Aleocharinae rove beetles of Saskatchewan, Canada (Coleoptera, Staphylinidae). ZooKeys 610:45–112

Klimaszewski J, Webster RP, Langor DW, Brunke A, Davies A, Bourdon C, Labrecque M, Newton AF, Dorval J-A, Frank JH (2018) Aleocharine rove beetles of eastern Canada (Coleoptera, Staphylinidae, Aleocharinae): a glimpse of megadiversity. Springer, Cham. 902 pp

Laporte FL (1835) Études entomologiques, ou description d'insectes nouveaux: et observations sur leur synonymie. Méquignon-Marvis Père et Fils, Paris. 159 pp. + 4 pls

Leach WE (1819) New genera. In: Samouelle (ed) The entomologist's useful compendium; or an introduction to the knowledge of British insects, comprising the best means of obtaining and preserving them, and a description of the apparatus generally used; etc. Thomas Boys, London. 496 pp. + 12 pls

Majka CG, Klimaszewski J (2008) New records of Canadian Aleocharinae (Coleoptera: Staphylinidae). ZooKeys 2:85–114

15 Tribe Myllaenini Ganglbauer, 1895

Tribe Myllaenini Ganglbauer, 1895

(Figs. 15.155–15.158) (Klimaszewski 1982, Ashe 2001, Klimaszewski et al. 2018a, 2020)

Diagnosis Body fusiform, pronotal hypomera not visible in lateral view, abdomen tapering apically, with coarse macrosetae; labial and maxillary palpi very long and stylate; galeae very slender with setae at apex; laciniae long and stylate with widely scattered teeth internally; mentum with anterolateral margins produced into spinose processes; tarsal formula 4-4-5, tarsal claws slender.

COMMENTS Species affiliated with riparian and wetland habitats.

Genus *Myllaena* Erichson, 1837

(Figs. 15.155–15.158) (Klimaszewski 1982, Klimaszewski et al. 2018a, 2020)

Diagnosis Body narrowly ovoid or fusiform, length 0.6–3.4 mm; integument covered with short and dense pubescence; head hypognathous; maxillary palpi with four palpomeres, palpomeres I and IV vestigial; glossae styliform; labial palpi and paraglossae reduced; mandibles with inner margins smooth; pronotum broadly rounded anteriad, basal margin arcuate or sinuately truncate, posterior lateral angles well defined, hypomera not visible in lateral view; prosternum reduced to narrow, transverse plate in front of procoxae; procoxae conical; tarsal formula 4-4-5; elytra transverse, laterally subparallel or rounded, sinuate at base; scutellum small, covered by pronotum; abdomen tapering from base, with coarse macrosetae; median lobe of aedeagus narrow, bulbus moderately large, internal sac with simple or complex sclerites, paramere with apical lobe narrow and very long, with two long basal and two short apical setae; spermatheca variable in shape, capsule spherical or sac-shaped with apical invagination small to large, stem short and not coiled or long and tightly multi-coiled.

COMMENTS Adults are found along margins of fresh water streams, rivers, ponds, lakes, and bogs.

Key to Species of *Myllaena*

1. Antennomeres VII–X distinctly transverse (Fig. 15.158a); median lobe of aedeagus and spermatheca as illustrated (Fig. 15.158b,c,h) ....***Myllaena ludificans*** **Casey**

– Antennomeres VII–X distinctly elongate; median lobe of aedeagus and spermatheca not as above **2**

2. Antennomeres VII–X about 1.5 times as long as wide (Fig. 15.155a); median lobe of

J. Klimaszewski et al., *A Faunal Review of Aleocharine Beetles in the Rapidly Changing Arctic and Subarctic Regions of North America (Coleoptera, Staphylinidae)*,
https://doi.org/10.1007/978-3-030-68191-3_15

aedeagus and spermatheca as illustrated (Fig. 15.155b,c,h) .... ***Myllaena arcana* Casey**

– Antennomeres VII–X about two times as long as wide; median lobe of aedeagus and spermatheca not as above **3**

3. Spermathecal capsule tubular, without apical invagination, instead with a small apical projection (Fig. 15.156h); median lobe of aedeagus with characteristic arrangement and shape of internal sac sclerites, as illustrated (Fig. 15.156b,c) .... ***Myllaena audax* Casey**

– Spermathecal capsule spherical, with large apical invagination (Fig. 15.157h); median lobe of aedeagus with arrangement and shape of internal sac sclerites as illustrated (Fig. 15.157b,c) .... ***Myllaena insomnis* Casey**

***Cuneata* species group** (Klimaszewski 1982)

Diagnosis Body length 1.7–2.5 mm; antennomeres VII–X no more than 1.5 times as long as maximum width; tergite X of both sexes almost completely divided; sternite VIII of both sexes with four pairs of setae; median lobe of aedeagus in lateral view narrow apically, apex sharp.

84. *Myllaena arcana* Casey (Fig. 15.155a–h)

Myllaena arcana Casey, 1911. Klimaszewski 1982, Klimaszewski and Génier 1986, Klimaszewski and Frank 1992, Bishop et al. 2009, Majka and Klimaszewski 2010, Klimaszewski et al. 2011, 2015, 2018a, b.

Diagnosis Body narrowly fusiform, length 1.9–2.5 mm, colour dark brown with palpi, legs and tip of abdomen light brown (Fig. 15.155a); head small and deflexed, two-thirds as wide as maximum width of pronotum; all antennomeres elongate, at least two times as long as wide; pronotum broad, at maximum width slightly wider than elytra; elytra very short, at suture about two-thirds maximum width of pronotum; abdomen gradually tapering apically. MALE. Tergite VIII with apical margin very obtusely triangularly produced, arcuately emarginate on either side of slightly acute apex (Fig. 15.155d); sternite VIII evenly rounded apically (Fig. 15.155e); median lobe of aedeagus in dorsal view with tubus elongate, slightly sinuately tapering from base to acute apex (Fig. 15.155c), in lateral view tubus with ventral margin arcuate basally, almost straight to narrow, slightly curved apex (Fig. 15.155b). FEMALE. Tergite VIII with apical margin obtusely triangularly produced, apex rounded (Fig. 15.155f); sternite VIII with apical margin broadly arcuate medially, rounded laterally (Fig. 15.155g); spermatheca with capsule large, spherical, lacking apical invagination, stem thin and narrowly twisted posteriorly (Fig. 15.155h).

DISTRIBUTION Origin: Nearctic. **CANADA**: AB, LB, MB, NB, NF, NS, ON, QC, SK. **USA**: AL, FL, IA, IL, MA, ME, NH, NJ, TN, WV. **MEXICO**.

COLLECTION AND HABITAT DATA

Habitat: lives in debris in various riparian and wetland habitats; in leaf litter in areas with *Carex* on brook margin, river margin among cobblestones near outflow of brook, margin of large brook under alders among gravel on gravel bar, in leaf litter near seepage and brook in hardwood forest, in moist leaves on vernal pond margin in mixed forest, lake margins in wet leaf litter, inland margin of salt marsh in leaf litter, in *Carex* hummocks in marsh near small stream, in sphagnum hummocks on margin of tamarack bog, one from sifted from dead grass on pond margin, one from beaver lodge, some collected near streams, and in beaver lodge debris. **Collecting period**: II–XI (mainly V–VIII). **Collecting method**: sifting leaf litter, turning gravel and cobblestones and aspirating specimens.

***Insomnis* species group** (Klimaszewski 1982)

Diagnosis Body length 1.9–2.6 mm; antennomeres VII–X at least twice as long as maximum width; tergite X of both sexes nearly completely divided; sternite VIII of both sexes with 5–7 pairs of setae; median lobe of aedeagus in lateral view with apex narrowly rounded; stem of spermatheca with many intertwined coils.

85. *Myllaena audax* Casey (Fig. 15.156a–i)

Myllaena audax Casey, 1911. Klimaszewski 1982, Klimaszewski and Génier 1986, Klimaszewski and Frank, 1992, Klimaszewski et al. 2011, 2018a, 2020.

Diagnosis Body moderately broadly fusiform, length 1.9–2.6 mm, colour dark brown with appendages and tip of abdomen light brown (Fig. 15.156a); head small and deflexed, about half as wide as maximum width of pronotum; antennomeres elongate, VII–X at least twice as long as wide; pronotum broad, maximum width about equal to that of elytra; elytra at suture slightly shorter than pronotum along midline of disc; abdomen tapering apically. MALE. Tergite VIII slightly obtusely triangularly produced apically (Fig. 15.156d); sternite VIII broadly parabolically rounded apically (Fig. 15.156e); median lobe of aedeagus in dorsal view with tubus elongate, subparallel, truncate apically (Fig. 15.156c), in lateral view tubus with ventral margin arcuate basally, straight to narrow and rounded apex (Fig. 15.156b). FEMALE. Tergite and sternite VIII similar to those of male (Fig. 15.156f,g); spermatheca with capsule pipe-shaped, with small apical projection, stem long, with about 24 tight intertwined coils (Fig. 15.156h,i).

DISTRIBUTION **Origin**: Nearctic. **CANADA**: BC, LB, NB, NF, NT, ON, QC. **USA**: AK, CA, GA, IL, LA, MA, NH, NJ, NY, OR, RI, UT, WA.

COLLECTION AND HABITAT DATA
Habitat: various wetland habitats; in leaf litter and grass on hummocks in a wet alder swamp, sedge marshes in leaf litter and moss at base of trees, tamarack bog in sphagnum hummocks with dense leatherleaf, in debris near creeks, rivers, and lakes, in beaver lodge debris, in leaf litter on beach, on a pond margin, and in black spruce forest (no specific micro-habitat data). **Collecting period**: III–XII (mainly V–VII). **Collecting method**: sifting leaf litter, moss, sphagnum, and other organic debris.

COMMENTS This species was listed from CO by Newton (2019) but we are not aware of any published voucher data.

86. *Myllaena insomnis* Casey (Fig. 15.157a–i)

Myllaena insomnis Casey, 1911. Klimaszewski 1982, Klimaszewski and Frank 1992, Majka and Klimaszewski 2010, Klimaszewski et al. 2011, 2015, 2018a.

Diagnosis Body moderately broadly fusiform, length 1.9–2.6 mm, colour dark brown with appendages and tip of abdomen light brown; head small and deflexed, about one-third as wide as maximum width of pronotum (Fig. 15.157a); all antennomeres elongate, VII–X at least twice as long as wide; pronotum broad, maximum width about equal to that of elytra; elytra at suture slightly shorter than pronotum along midline of disc; abdomen tapering apically. MALE. Tergite VIII with apical margin slightly acutely triangularly produced, with apex narrowly rounded (Fig. 15.157e); sternite VIII broadly parabolically rounded apically (Fig. 15.157f); median lobe of aedeagus in dorsal view with tubus elongate, subparallel, broadly rounded apically (Fig. 15.157d), in lateral view tubus slightly arcuate ventrally with apex narrow, rounded, slightly curved dorsad (Fig. 15.157b,c). FEMALE. Tergite and sternite VIII similar to those of male (Fig. 15.157g,h); spermatheca with capsule pipe-shaped, apical invagination large, stem long with about 16–24 tight intertwined coils (Fig. 15.157i).

DISTRIBUTION **Origin**: Nearctic. **CANADA**: AB, BC, LB, MB, NB, NF, NS, NT, ON, QC, SK, YT. **USA**: AK, ID, MA, MI, MN, MT, NY, WI.

COLLECTION AND HABITAT DATA
Habitat: various wetland habitats; in NB, along lake margin/dune in flood debris, treading emergent *Carex* and grasses along a lake margin, on

bare mud along margin of slow flowing stream, in moist leaves along vernal pond margin in a mixed forest, calcareous fen in sphagnum moss and litter. In AK, one specimen was found under beach rocks on sandy beach. Elsewhere, in debris near margins of ponds, streams, rivers, and lakes, in marsh refuse, and treading in a in bog. **Collecting period**: III–XI. **Collecting method**: treading marsh and bog vegetation, sifting various kinds of plant debris, Malaise trap.

Additional records. See Annex 1.

***Potawatomi* species group** (Klimaszewski 1982)

Diagnosis Body length 1.6–3.0 mm; antennomeres VII–X transverse or less than twice as long as maximum width in most; tergite X of both sexes emarginate apically; sternite VIII of both sexes with four pairs of setae; median lobe of aedeagus in lateral view with apex narrowly rounded; stem of spermatheca with variable numbers of intertwined coils.

87. *Myllaena ludificans* Casey (Fig. 15.158a–h)

Myllaena ludificans Casey, 1911. Klimaszewski 1982, Klimaszewski and Génier 1986, Webster et al. 2009, Klimaszewski et al. 2018a.

Diagnosis Body narrowly fusiform, length 1.6–1.7 mm, colour dark brown, some slightly grayish, with posterior part of elytra, appendages and tip of abdomen light brown (Fig. 15.158a); head small and deflexed, about one-third as wide as maximum width of pronotum; antennomeres VII–X slightly transverse; pronotum narrow, maximum width about equal to that of elytra; elytra at suture slightly shorter than pronotum along midline of disc; abdomen gradually tapering apically. MALE. Tergite VIII broadly rounded apically (Fig. 15.158d); sternite VIII with apical margin parabolically rounded, more narrowly so apically (Fig. 15.158e); median lobe of aedeagus in dorsal view with tubus subparallel, slightly swollen near middle, truncate apically (Fig. 15.158c), in lateral view tubus sinuate ventrally with apex narrow, rounded (Fig. 15.158b). FEMALE. Tergite VIII very obtusely angulate apically (Fig. 15.158f); sternite VIII parabolically rounded apically (Fig. 15.158g); spermatheca with capsule spherical, apical invagination small, stem thin, sinuate, with about three loosely intertwined coils (Fig. 15.158h).

DISTRIBUTION Origin: Nearctic. **CANADA**: NB, ON, QC. **USA**: AR, MA, OK, RI, WI.

COLLECTION AND HABITAT DATA

Habatat: in NB, in gravel on the margin of a large brook on a gravel bar partially shaded by alders, in grass litter on muddy soil along a sun-exposed stream margin through a former beaver pond, and in boggy marsh with cattails near a salt marsh; elsewhere, this species was collected from pond margins. **Collecting period**: IV, V–X. **Collecting method**: hand collecting (turning over cobblestones and pebbles, sifting plant litter, treading marsh vegetation).

COMMENTS This species was listed from LA by Newton (2019) and MI by Klimaszewski (2018a) but we are not aware of any published voucher data.

Fig. 15.155 (**a–h**). Images of *Myllaena arcana* Casey: (**a**) habitus in dorsal view; (**b**) median lobe of aedeagus in lateral view; (**c**) median lobe of aedeagus in dorsal view; (**d**) male tergite VIII; (**e**) male sternite VIII; (**f**) female tergite VIII; (**g**) female sternite VIII; (**h**) spermatheca. (**a–h**) after Klimaszewski et al. (2018). Scale bar for habitus = 1 mm; remaining scale bars = 0.2 mm

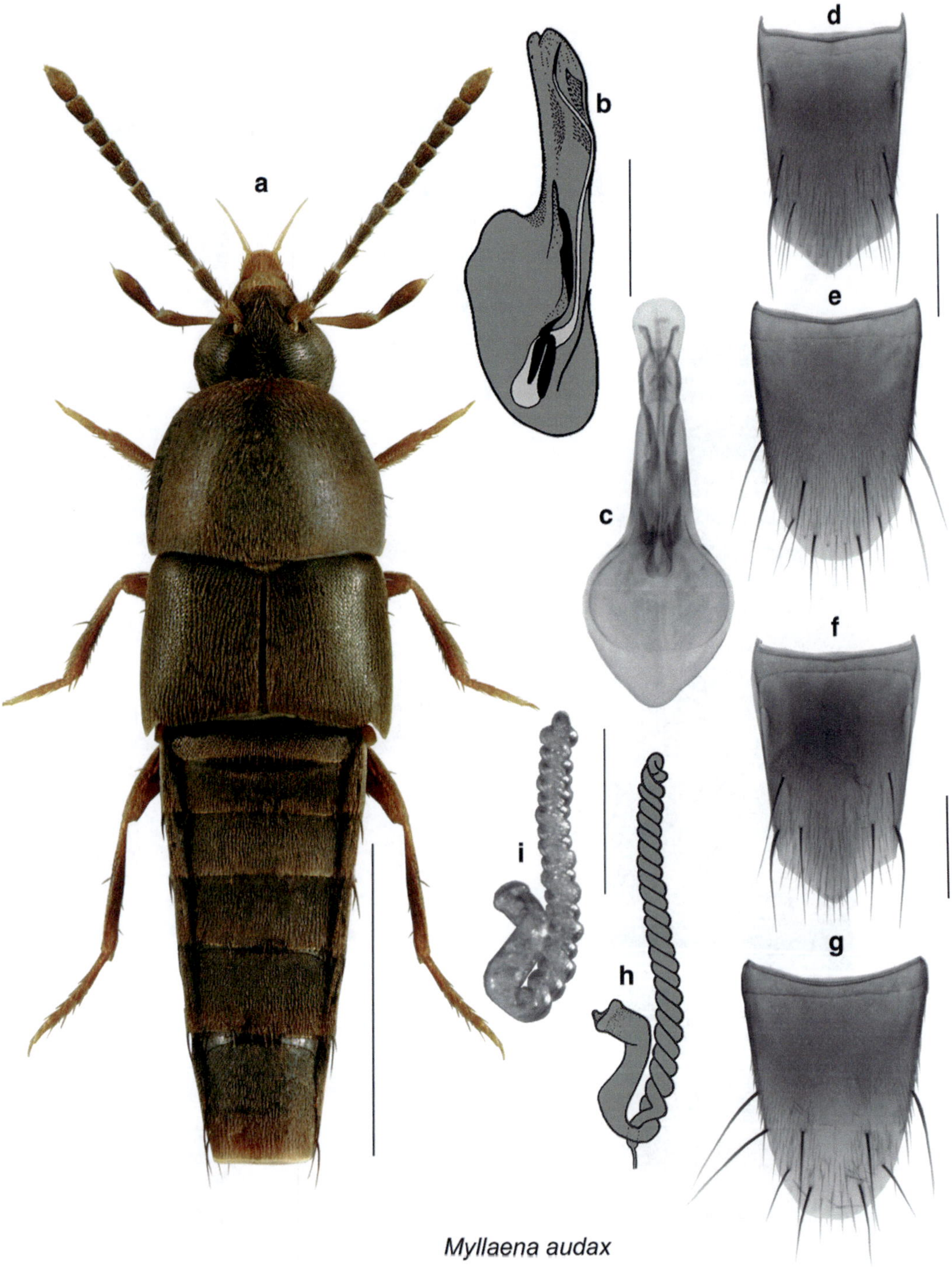

Fig. 15.156 (**a–i**) Images of *Myllaena audax* Casey: (**a**) habitus in dorsal view; (**b**) median lobe of aedeagus in lateral view; (**c**) median lobe of aedeagus in dorsal view; (**d**) male tergite VIII; (**e**) male sternite VIII; (**f**) female tergite VIII; (**g**) female sternite VIII; (**h–i**) spermatheca. (**b, h**) after Klimaszewski, 1982; remaining figures after Klimaszewski et al. (2018). Scale bar for habitus = 1 mm; remaining scale bars = 0.2 mm

Fig. 15.157 (**a–i**). Images of *Myllaena insomnis* Casey: (**a**) habitus in dorsal view; (**b–c**) median lobe of aedeagus in lateral view; (**d**) median lobe of aedeagus in dorsal view; (**e**) male tergite VIII; (**f**) male sternite VIII; (**g**) female tergite VIII; (**h**) female sternite VIII; (**i**) spermatheca. (**b**) after Klimaszewski (1982); remaining images after Klimaszewski et al. (2018). Scale bar for habitus = 1 mm; remaining scale bars = 0.2 mm

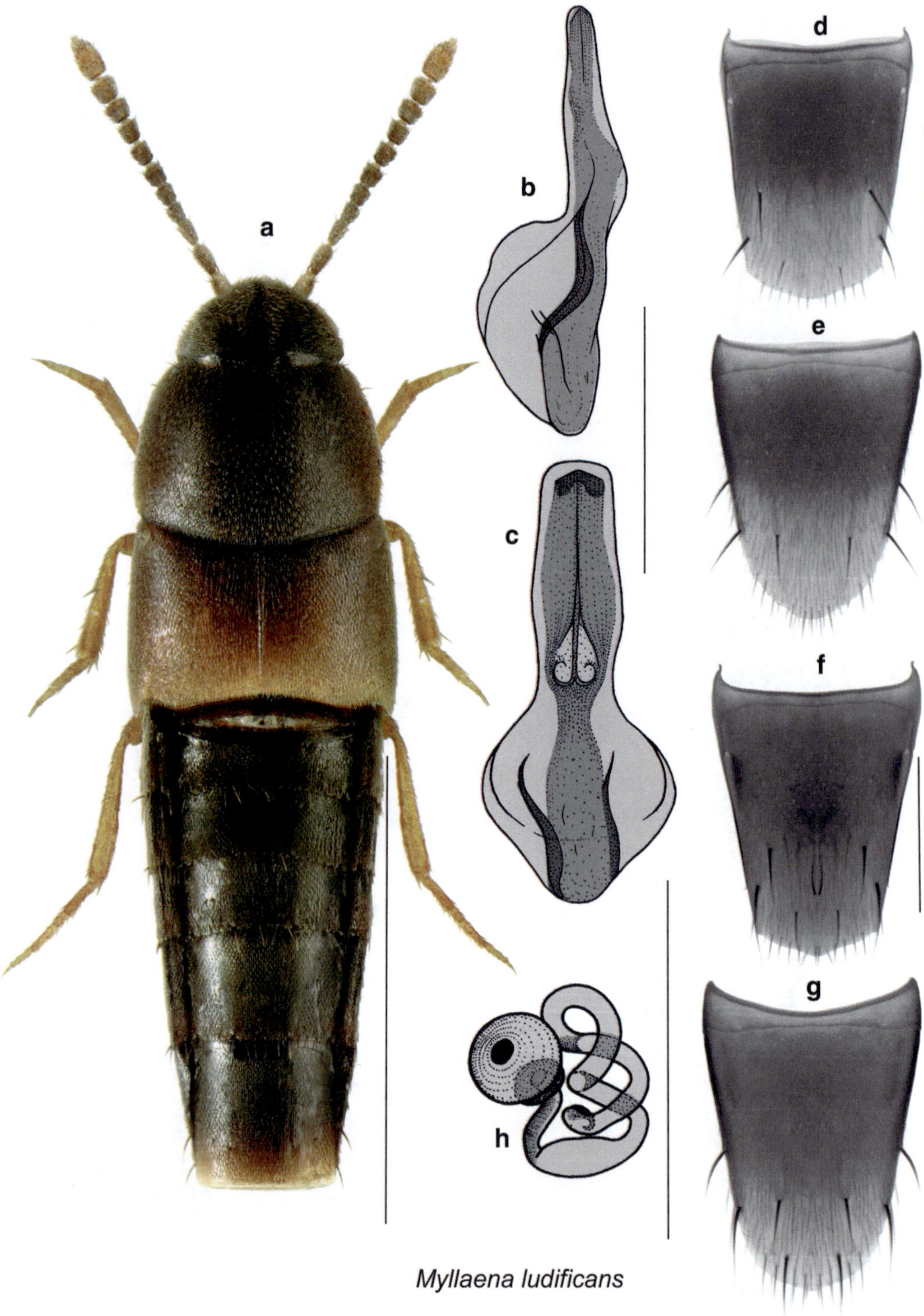

Fig. 15.158 (**a–h**). Images of *Myllaena ludificans* Casey: (**a**) habitus in dorsal view; (**b**) median lobe of aedeagus in lateral view; (**c**) median lobe of aedeagus in dorsal view; (**d**) male tergite VIII; (**e**) male sternite VIII; (**f**) female tergite VIII; (**g**) female sternite VIII; (**h**) spermatheca. (**b, c,h**) after Klimaszewski (1982); remaining images after Klimaszewski et al. (2018). Scale bar for habitus = 1 mm; remaining scale bars = 0.2 mm

References

Ashe JS (2001) Key to the tribes and genera of Nearctic Aleocharinae. 8. In: Arnett RH Jr, Thomas CM (eds) American beetles. Archostemata, myxophaga, adephaga, polyphaga: staphyliniformia. CRC Press, Washington, DC, pp 299–374. 443 pp

Bishop DJ, Majka CG, Bondrup-Nielsen S, Peck SB (2009) Deadwood and saproxylic beetle diversity in naturally disturbed and managed spruce forests in Nova Scotia. In: Majka, C.G. and Klimaszewski, J. (Eds) Biodiversity, Biosystematics, and Ecology of Canadian Coleoptera II. ZooKeys [Special Issue] 22:309–340

Casey TL (1911) New American species of Aleocharinae and Myllaeninae. In: Memoirs on the Coleoptera. II. The New Era Printing Co., Lancaster, PA, pp 1–245. 259 pp

Erichson WF (1837) Die Käfer der Mark Brandenburg. Erster Band, Erste Abtheilung. F. H. Morin, Berlin. viii + 384 pp

Ganglbauer L (1895) Die Käfer von Mitteleuropa. Die Käfer der österreichisch-ungarischen Monarchie, Deutschlands, der Schweiz, sowie des französischen und italienischen Alpengebietes. Vol. 2, Familienreihe Staphylinoidea, I. Theil: Staphylinidae, Pselaphidae. Carl Gerold's Sohn, Vienna. vi + 881 pp

Klimaszewski J (1982) Studies of Myllaenini (Coleoptera: Staphylinidae, Aleocharinae). 1. Systematics, phylogeny, and zoogeography of Nearctic *Myllaena* Erichson. Can Entomol 114(3):181–242

Klimaszewski J, Frank JH (1992) New distributional data for Nearctic *Myllaena* Erichson (Coleoptera: Staphylinidae). Supplement 2. Coleopt Bull 46 (4):397–402

Klimaszewski J, Génier F (1986) New distribution and collection data for Nearctic *Myllaena* Erichson: studies of Myllaenini (Coleoptera: Staphylinidae: Aleocharinae). Supplement 1. Coleopt Bull 40 (1):33–36

Klimaszewski J, Langor D, Pelletier G, Bourdon C, Perdereau, L (2011) Aleocharine beetles (Coleoptera, Staphylinidae) of the province of Newfoundland and Labrador, Canada. Pensoft Series Faunistica No. 98. Pensoft, Sophia-Moscow, 313 pp

Klimaszewski J, Godin B, Langor D, Bourdon C, Lee S-I, Horwood D (2015) New distribution records for Canadian Aleocharinae (Coleoptera, Staphylinidae), and new synonymies for *Trichiusa*. ZooKeys 498:51–91

Klimaszewski J, Webster RP, Langor DW, Brunke A, Davies A, Bourdon C, Labrecque M, Newton AF, Dorval J-A, Frank JH (2018a) Aleocharine rove beetles of eastern Canada (Coleoptera, Staphylinidae, Aleocharinae): a glimpse of megadiversity. Springer, Cham. 902 pp

Klimaszewski J, Godin B, Davies A, Bourdon C, Horwood D (2018b) Forty new records of aleocharine beetles, and two new species in the genera Acrotona Thomson and Atheta Thomson, for the province of Manitoba, Canada (Coleoptera: Staphylinidae: Aleocharinae). Insecta Mundi 0641:1–33

Klimaszewski J, Hoebeke ER, Godin B, Davies A, Perry KI, Bourdon C, Winchester N (2020) Aleocharine rove beetles of British Columbia: a hotspot of Canadian biodiversity (Coleoptera, Staphylinidae). Springer, Cham. 631 pp

Majka CG, Klimaszewski J (2010) Contributions to the knowledge of the Aleocharinae (Coleoptera, Staphylinidae) in the Maritime Provinces of Canada. ZooKeys 46:15–39

Newton AF (2019) StaphBase: staphyliniformia world catalog database (version Nov 2018). In: Roskov Y, Ower G, Orrell T, Nicolson D, Bailly N, Kirk PM, Bourgoin T, DeWalt RE, Decock W, Nieukerken E. van, Zarucchi J, Penev L (Eds.) Species 2000 & ITIS Catalogue of Life, 2019 Annual. Digital resource at http://www.catalogueoflife.org/annual-checklist/2019. Species 2000: Naturalis, Leiden, the Netherlands. ISSN 2405-884X

Webster RP, Klimaszewski J, Pelletier G, Savard K (2009) New Staphylinidae (Coleoptera) records with new collection data from New Brunswick, Canada. I. Aleocharinae. In: Majka CG, Klimaszewski J (eds) Biodiversity, biosystematics, and ecology of canadian coleoptera II. ZooKeys [Special Issue] 22:171–248

16 Tribe Liparocephalini Fenyes, 1918

Tribe Liparocephalini Fenyes, 1918

(Figs. 16.159–16.165) (Ahn and Ashe 1996a, b, Ahn and Maruyama 2000a, b, Frank and Ahn 2011, Klimaszewski et al. 2020, Ahn et al. 2020)

Diagnosis Body densely covered with fine and short setae; elytra very short, at suture distinctly shorter than pronotum, flight wings absent; labial palpus elongate with two palpomeres; ligula long, slender and entire at apex; prementum with one median seta, and with narrow medial pseudopore field; mandibles with denticles in ventral molar area; metaventrite greatly shortened; middle coxae contiguous; tarsal formula 4-4-5 or 4-4-4 in some *Diaulota*, tarsal claws robust. Includes genera with species found exclusively in the intertidal zones of Pacific shores.

COMMENTS There are several genera of this tribe, with Nearctic species restricted in distribution to Pacific seashores of North America. Many are recorded from Alaska but only from coastal areas of the Marine West Coast Forest ecoregion, which is not included in this book. For the review of genera and species see Klimaszewski et al. 2020.

Key to Genera of Liparocephalini (from Klimaszewski et al. 2020)

1. Body robust, heavily sclerotized, length 2.5–5.2 mm (Figs. 16.161a, 16.164a, 16.165a) ... **2**

– Body slender, weakly to moderately sclerotized, length 1.9–3.4 (Figs. 16.159a, 16.160a, 16.162a, 16.163a) ... **3**

2. Head subquadrate, broader than maximum width of pronotum (Figs. 16.164a, 16.165a); infraorbital region of head about as long as 2–3 times width of an eye; antennomers V–X subquadrate to moderately transverse (Figs. 16.164a, 16.165a); basal abdominal tergites with shallow basal impressions ... ***Liparocephalus* Mäklin**

– Head elongate, about as wide as maximum width of pronotum (Fig. 16.161a); infraorbital region of head about as long as 3–4 times with of an eye; antennomers V–X slightly to moderately transverse (Fig. 16.161a); basal abdominal tergites with deep basal impressions ... ***Diaulota* Casey**

3. Eyes vestigial, scarcely visible, each consisting of 4–5 ommatidia; pronotum narrow, maximum width of pronotum slightly less than maximum width of head, apical corners slightly angulate (Figs. 16.159a, 16.160a) .... ***Amblopusa* Casey**

– Eyes small but clearly visible, each consisting of multiple ommatidia (Figs. 16.162a, 16.163a); pronotum wide, maximum width of

J. Klimaszewski et al., *A Faunal Review of Aleocharine Beetles in the Rapidly Changing Arctic and Subarctic Regions of North America (Coleoptera, Staphylinidae)*,
https://doi.org/10.1007/978-3-030-68191-3_16

pronotum equal to or greater than maximum width of head, apical corners rounded (Figs. 16.162a, 16.163a) ....*Paramblopusa* **Ahn and Ashe**

Genus *Amblopusa* Casey, 1893

(Figs. 16.159–16.160)

Amblopusa Casey 1893. Casey 1906, 1911, Fenyes 1918, Ahn and Ashe 1996a, Frank and Ahn 2011, Yoo and Ahn 2020, Klimaszewski et al. 2020.

Diagnosis Body small, narrow, subparallel, flattened, length 1.9–2.2 mm; integument with evenly distributed short pubescence, and scattered macrosetae; head elongate, slightly deflexed; eyes vestigial, scarcely visible, each consisting of 4–5 ommatidia; labrum transverse; mandibles with apices acute and slightly turned downward, prostheca membranous, well developed; lacinia and galea elongate; lacinia with a large subapical spine, a row of short subapical spines; maxillary palps with four palpomeres, subapical one swollen apically; labial palps styliform and elongate; ligula elongate, thin, entire at apex; mentum trapezoidal with broadly arcuate apical emargination, apical corners produced, spine-like; pronotum narrow, maximum width of pronotum slightly less than maximum width of head, apical corners slightly angulate; elytra with posterior margin almost straight and slightly sinuate laterally, hind wings absent, abdomen broad at base and sides uniformly converging to rounded apex, three basal tergites strongly impressed at base; tarsal formula 4-4-5. Species confined to the rocky seashores of Pacific coast of North America

Key to Species of *Amblopusa* (modified from Ahn and Ashe 1996a)

1. Forebody brown-yellow (Fig. 16.160a), body length 1.9–2.2 mm; ligula of labium short, about 0.3 times length of first labial palpomere; pronotum not longitudinally impressed; abdominal sternites IV-VI constricted at base; median lobe of aedeagus as illustrated (Fig. 16.160b).. ... . ***Amblopusa brevipes*** **Casey**

– Forebody dark brown to almost balck (Fig. 16.159a), body length 2.4–3.0 mm; ligula of labium long, about half length of first labial palpomere; pronotum slightly longitudinally impressed; abdominal sternites IV-VI not constricted at base; median lobe of aedeagus not as above (Fig. 16.159b) ....***Amblopusa alaskana*** **Ahn and Ashe**

88. *Amblopusa alaskana* Ahn and Ashe (Fig. 16.159a–e)

Amblopusa alaskana Ahn and Ashe, 1996a.

Diagnosis Body length 2.4–3.0 mm, forebody dark brown with legs red-brown (Fig. 16.159a); antennomeres I and II strongly elongate and of about the same length, IV–V subquadrate, VI slightly elongate, VI–X subquadrate to moderately transverse, antennae slightly incrassate apically; head elongate, slightly wider than maximum width of pronotum; pronotum subquadrate, narrowest at base and widest in apical third; elytra about one third wider than long along suture, and 0.8 times as long as pronotum, at suture shorter than pronotum; abdomen slightly broadening apically, at base narrower than elytra. MALE. Tergite VIII truncate apically, antecostal suture broadly separated from base of disc (Fig. 16.159c); sternite VIII triangularly produced apically, apex rounded (Fig. 16.159d); bulbus of median lobe of aedeagus ovoid, tubus strongly sinuate and bent ventrad apically in lateral view, subapical part triangular with basal rounded projection (Fig. 16.159b). FEMALE. Tergite VIII truncate apically, similar to that of male; sternite VIII slightly produced apically; spermatheca with multiple and tight coils (Fig. 16.159e).

DISTRIBUTION Origin: Nearctic. **CANADA**: not recorded. **USA**: AK.

COLLECTION AND HABITAT DATA

Habitat: confined to the rocky seashores of Pacific coast of Alaska; one specimen from AK found under driftwood. **Collecting period**: VIII. **Collecting method**: hand collected with forceps.

Additional records. See Annex 1.

89. ***Amblopusa brevipes*** **Casey** (Fig. 16.160a–g)

Amblopusa brevipes Casey, 1893. Ahn and Ashe 1996a, Klimaszewski et al. 2020, Yoo and Ahn 2020.

Diagnosis Body length 1.9–2.2 mm, forebody pale, yellow brown and abdomen dark brown (Fig. 16.160a); antennomeres I and II strongly elongate and of about the same length, IV–V subquadrate, VI–X moderately to strongly transverse, antennae incrassate apically; head elongate, slightly wider than maximum width of pronotum; pronotum subquadrate, narrowest at base and widest in apical third; elytra about one third wider than long along suture, and 0.8 times as long as pronotum, at suture shorter than pronotum; abdomen slightly broadening apically, at base almost as wide as elytra. MALE. Tergite VIII truncate apically, antecostal suture broadly separated from base of disc (Fig. 16.160c); sternite VIII triangularly produced apically, apex rounded (Fig. 16.160d); bulbus of median lobe of aedeagus ovoid, tubus strongly sinuate and bent ventrad apically, subapical part with rounded subapical projection, apical part narrowly triangular in lateral view (Fig. 16.160b). FEMALE. Tergite VIII truncate apically, similar to that of male (Fig. 16.160e); sternite VIII slightly produced apically with rounded apex (Fig. 16.160f); spermatheca small, capsule club-shaped, apical invagination not apparent, stem narrow, short, and slightly hooked posteriorly (Fig. 16.160g).

DISTRIBUTION **Origin**: Nearctic. **CANADA**: BC. **USA**: AK, CA.

COLLECTION AND HABITAT DATA
Habitat: confined to the rocky seashores of Pacific coast of North America. **Collecting period**: III. **Collecting method**: not recorded.

COMMENTS Yoo and Ahn (2020), provided a new record of the species from Aleutian Islands: Agattu Is.

Genus *Diaulota* Casey, 1893

(Fig. 16.161a–h)

Diaulota Casey, 1893. Ahn 1996, Ahn and Ashe 1996b, Frank and Ahn 2011, Klimaszewski et al. 2020.

Diagnosis Body robust, narrowly elongate, length 2.5–2.9 mm; colour variable, black, dark brown, reddish brown, or brown; integument with relatively long and evenly distributed microsetae; head slightly deflexed, 0.8–1.2 times as long as wide; eyes small, 0.2 times length of head, tempora very long, neck absent; antennomeres V–X slightly to moderately transverse; labrum semicircular; mandibles asymmetrical or nearly symmetrical, apex acute and turned downward, right mandible with median tooth, internal teeth absent, prostheca well developed; maxillary palps with 4 palpomeres; labial palpus with 2–3 palpomeres; ligula simple, elongate; mentum without V seta; elytra shorter than pronotum; hind wings absent; abdomen enlarged, sides arcuate, broader than maximum width of head, tergites III-VI impressed at base; sternite not constricted at base; tarsal formula 4-4-5, 4-4-4, or 3-3-4. Species confined to the intertidal zone of rocky shores of the Pacific.

90. ***Diaulota densissima*** **Casey** (Fig. 16.161a–h)

Diaulota densissima Casey, 1893. Ahn and Ashe 1996b, Klimaszewski et al. 2020.

Diagnosis Body robust, heavily sclerotized, length 2.5–2.9 mm, colour black or dark brown with appendages and tip of abdomen often reddish brown (Fig. 16.161a); pubescence uniformly brownish gray; head large, elongate, as long as pronotum, at base about as broad as maximum width of pronotum; antennomeres VII–X strongly transverse; pronotum elongate, trapezoidal in shape, broadest apically, about as long as head; elytra 1.1 times as long as wide, 0.7 times as long as pronotum; abdomen large, broader than forebody, tergite impression shallow, paratergites evenly narrow. MALE. Tergite VIII with apical margin truncate, slightly emarginate medially (Fig. 16.161d); sternite VIII with apical margin strongly sinuate laterally and broadly triangularly produced medially, apex rounded (Fig. 16.161e); median lobe of aedeagus with bulbus narrowly ovoid, tubus moderately produced ventrad in lateral view (Fig. 16.161b), in dorsal view median lobe as illustrated (Fig. 16.161c). FEMALE. Tergite VIII truncate apically and shallowly emarginated medially (Fig. 16.161f); sternite VIII rounded apically (Fig. 16.161g); spermatheca small, capsule spherical, apical invagination not apparent, stem narrowly elongate and sinuate (Fig. 16.161h).

DISTRIBUTION Origin: Nearctic. **CANADA**: BC. **USA**: AK, CA, OR, WA.

COLLECTION AND HABITAT DATA
Habitat: confined to the rocky seashores of the Pacific coast of North America, found in rock crevices of intertidal zones; ex *Fucus* zone; in wrack, in rocky shore with mud and sand, between green algae in oligohaline water; in rocky shore, euhaline water, between *Fucus*, barnacles and lichens; in vertical crevices of rocks in splash zone (Ahn and Ashe 1996b). **Collecting period**: I–IX. **Collecting method**: hand collecting from rock crevices in intertidal zones.

Genus *Paramblopusa* Ahn and Ashe, 1995

(Fig. 16.162–16.163) (Ahn and Ashe 1996a, b, Ahn and Maruyama 2000b, Frank and Ahn 2011, Ahn 1996, Klimaszewski et al. 2020)

Diagnosis Body narrowly parallel-sided, flattened; length 2.5–3.4 mm; forebody black, reddish or dark brown and abdomen dark brown in most; integument with short microsetae about uniformly distributed and macrosetae scattered; head slightly deflexed, broader than remaining parts of body; eyes small, consisting of multiple ommatidia; mandibles with small teeth between apex and median tooth; lacinian setae only on mesal surface; maxillary palpus with four palpomeres; labial palpus styliform; ligula elongate, entire at apex; pronotum wide, maximum width of pronotum equals to maximum width of head, apical corners rounded; elytra with posterior margin arcuate and slightly sinuate laterally; hind wings absent; abdomen broadening apically, basal three abdominal tergites strongly impressed at base, anterior margins of abdominal tergites III–VI straight, not emarginate; tarsal formula 4-4-5. Species confined to intertidal zones of Pacific.

Key to Species of *Paramblopusa*

1. Body uniformly black (Fig. 16.163a); tubus of median lobe of aedeagus moderately long with narrowly rounded apex, ventral margin sinuate in subapical part, curved in basal part (Fig. 16.163b); spermatheca with moderately thick stem, coils tight and close or touching, capsule with neck bent or curved (Fig. 16.163h,i) .... ***Paramblopusa eoa* Ahn and Maruyama**

– Body bicolored, head and abdomen dark brown, and pronotum and elytra yellow-brown (Fig. 16.162a); tubus of median lobe of aedeagus long with sharply pointed apex, ventral margin straight in subapical part and in basal part (Fig. 16.162b); spermatheca with thin stem, coils loose, not touching, capsule with straight neck (Fig. 16.162g) .... ***Paramblopusa borealis* (Casey)**

91. *Paramblopusa borealis* (Casey) * (Fig. 16.162a–g)

Amblopusa borealis (Casey, 1906). Moore and Legner 1975, Ahn and Ashe 1996a, b, Ahn and Maruyama 2000b, Ahn et al. 2020, Klimaszewski et al. 2020.

Diagnosis Body narrowly linear, length 2.5–3.4 mm, colour variable, reddish brown with head and most of abdomen dark brown in most; pubescence uniformly yellow brown; head large, slightly deflexed, at least as broad as maximum width of pronotum (Fig. 16.162a); antennomeres I–II strongly elongate, IV subquadrate, V–X subquadrate to slightly transverse; head large, transverse; eyes small but distinct with multiple ommatidia; pronotum trapezoidal in shape, slightly elongate, as wide as maximum width of head, basal angles distinct, pronotal punctures fine, dense; elytra with posterior margin arcuate and slightly sinuate laterally; abdomen broadening apically. MALE. Tergite VIII with apical margin very broadly arcuate (Fig. 16.162c); sternite VIII with apical margin broadly triangularly produced mediad, apex angulate (Fig. 16.162d); bulbus narrowly ovoid, tubus of median lobe of aedeagus strongly bent ventrad at base in lateral view, subapex broad (lancet-shaped), apex pointed (Fig. 16.162b). FEMALE. Tergite VIII truncate apically (Fig. 16.162e); sternite VIII with apical margin obtusely triangularly produced, broadly rounded medially (Fig. 16.162f); spermatheca small and slender, capsule spherical, neck straight, stem thin, with two loose coils, twisted posteriorly (Fig. 16.162g).

DISTRIBUTION **Origin**: Holarctic.
CANADA: BC. **USA**: AK, CA, OR, WA.
JAPAN: Hokkaido

COLLECTION AND HABITAT DATA
Habitat: confined to beaches of the Pacific coast; found in organic debris on beaches and on mud flats under rocks. **Collecting period**: V, VIII, X. **Collecting method**: hand collecting from organic debris on beaches.

92. *Paramblopusa eoa* Ahn and Maruyama * (Fig. 16.163a–i)

Paramblopusa eoa Ahn and Maruyama, 2000b. Ahn et al. 2020.

Diagnosis Body narrowly linear with slightly swollen abdomen, length 2.8–3.1 mm, colour black with tibiae and tarsi reddish; pubescence uniformly yellow brown; head slightly elongate at least as broad as maximum width of pronotum (Fig. 16.163a); antennomeres I–III strongly elongate, IV subquadrate to slightly elongate, V–II monoliform, VII-X moderately transverse; eyes small but distinct with multiple ommatidia; pronotum trapezoidal in shape, slightly elongate, slightly wider than maximum width of head, basal angles vague, pronotal punctures moderate, well separated; elytra shorter than pronotum, with posterior margins oblique; abdomen broadening apicad, slightly swollen medially. MALE. Tergite VIII with apical margin very broadly arcuate (Fig. 16.163d); sternite VIII with apical margin broadly triangularly produced, apex obtusely angulate (Fig. 16.163e); median lobe of aedeagus with bulbus narrowly ovoid, tubus in lateral view with ventral margin strongly bent ventrad in basal half, apex narrowly rounded (Fig. 16.163b,c). FEMALE. Tergite VIII with apical margin truncate (Fig. 16.163f); that of sternite VIII obtusely triangularly produced (Fig. 16.163g); spermatheca with capsule spherical, neck arcuate or bent, stem moderately thick with coils tight or touching, twisted posteriorly (Fig. 16.163h,i).

DISTRIBUTION **Origin**: Holarctic.
CANADA: not recorded. **USA**: AK. RUSSIA: Kuril Island, Urup Is., Paramushir Is.

COLLECTION AND HABITAT DATA
Habitat: under rocks below cliff, under rocks and pebbles in the mid- to high tide zone along sandy beaches. In AK, specimens were captured in splash zone rocks with *Verrucaria* lichens, in crevices; splash zone cliff cracks; mid beach/ splash zone boulder with cracks; under drift wood; under rocks; and under mid-intertidal rocks. **Collecting period**: III, VI, VIII. **Collecting method**: hand collecting from under rocks and pebbles.

Additional records. See Annex 1.

Genus *Liparocephalus* Mäklin, 1853

(Figs. 16.164–16.165) (Ahn and Ashe 1996b, Ahn and Maruyama 2000a, Ahn 1997, Ahn et al. 2010, Frank and Ahn 2011)

Diagnosis Body length 3.8–5.2 mm, robust, heavily sclerotized, integument with moderately long setae densely and uniformly distributed; head slightly deflexed, large and robust; antennomeres V–X subquadrate to slightly transverse; eyes small, 0.2 times length of head, and without setae between facets; labrum transverse; mandibles almost symmetrical with 4–5 teeth between apex and median tooth; maxillary palpi with four palpomeres; labial palpi with two palpomeres; ligula simple, elongate; pronotum trapezoidal in shape, broadest apically, about size of head; elytra reduced to small rounded lobes; membranous wings absent; abdomen broad and robust, sides arcuate, abdominal tergites without deep transverse impressions; tarsi with large spatulate setae; tarsi with spatulate setae, claws narrow, long and sickle-shaped, tarsal formula 4-4-5. Male sternite VIII prolonged apically as broad triangular projection. Includes intertidal species confined to the rocky intertidal zones of Pacific shores.

Key to Species of *Liparocephalus*

1. Pronotum subquadrate to slightly transverse (Fig. 16.164a); apical margin of male sternite VIII broadly, deeply emarginate on each side of rounded middle projection (Fig. 16.164d); apical margin of female sternite VIII not emarginate, obtusely triangularly produced mediad (Fig. 16.164f); apex of tubus of median lobe of aedeagus sharply pointed (Fig. 16.164b)............ ....***Liparocephalus brevipennis*** **Mäklin**

– Pronotum strongly transverse (Fig. 16.165a); apical margin of male sternite VIII shallowly emarginate on each side of narrow, triangular middle portion (Fig. 16.165d); apical margin of female tergite VIII deeply emarginate on each side of small median projection (Fig. 16.165f); apex of tubus of median lobe of aedeagus truncate (Fig. 16.165b)... ... ***Liparocephalus humblei*** **Klimaszewski, sp.n.**

93. *Liparocephalus brevipennis* Mäklin (Fig. 16.164a–g)

Liparocephalus brevipennis Mäklin, 1853. Ahn 1997.

Diagnosis Body robust, heavily sclerotized, length 4.2–4.4 mm, colour black or dark brown with appendages and tip of abdomen often reddish brown (Fig. 16.164a); pubescence uniformly brownish gray; head large, slightly deflexed, about as wide as maximum width of pronotum; antennomeres VII–X slightly transverse; pronotum trapezoidal in shape, broadest apically, about as long as wide; elytra reduced to two small lobes, at suture much shorter than pronotum; abdomen large, tergal impressions scarcely visible, paratergites narrow anteriorly and broadening posteriorly creating distinct broadly serrate margin. MALE. Tergite VIII with apical margin arcuate laterally, truncate medially (Fig. 16.164c); sternite VIII with apical margin broadly, deeply emarginate on each side of broadly rounded median projection of same shape and size as emarginations (Fig. 16.164d); median lobe of aedeagus with bulbus large and ovoid, tubus narrow and strongly bent ventrad in lateral view (Fig. 16.164b). FEMALE. Tergite VIII similar to that of male (Fig. 16.164e); sternite VIII with apical margin very broadly triangularly produced, obtusely angulate medially (Fig. 16.164f); spermatheca with capsule spherical, apical invagination not apparent, stem narrowly elongate and slightly sinuate (Fig. 16.164g).

DISTRIBUTION Origin: Nearctic. **CANADA**: BC. **USA**: AK.

COLLECTION AND HABITAT DATA

Habitat: confined to the rocky intertidal zones of Pacific shores, found under beach wrack, under rocks, in rock crevices on mud flat at low tide; in rocky shore, among *Enteromorpha*, *Fucus* algae and barnacles. **Collecting period**: V–IX.

Collecting method: hand collecting from rock crevices in intertidal zones.

Additional records. See Annex 1.

94. ***Liparocephalus humblei*** **Klimaszewski, sp. n.** (Fig. 16.165a–g)

urn:lsid:zoobank.org:act:14A3268A-DEEA-4DA1-BC95-4C7A06588405

Holotype (male): USA, Alaska, Amchitka Is., Banjo Pt., 51.288, −179.083, 14.VIII.1971, C.E. O'Clair (CNC). Paratypes: Alaska, Amchitka Is., Banjo Pt., 51.288, −179.083, 14. VIII.1971, C.E. O'Clair (CNC) three males, three females, seven sex undetermined.

Etymology This species is named after our colleague Dr. Lee Humble. Lee's research in the Canadian Arctic on overwintering strategies of willow gall-forming and catkin-feeding nematine sawflies and their parasitoids attests to his impact on our understanding of Arctic biodiversity. Lee's expertise and passion for insects and the many contributions that he made to science established his reputation as a well-respected expert in entomology in Canada and around the world.

Diagnosis Body robust, heavily sclerotized, length 3.8–4.0 mm, colour dark brown to almost black or red-brown with appendages and tip of abdomen often paler (Fig. 16.165a); pubescence uniformly brownish gray; head large, slightly deflexed, about as wide as maximum width of pronotum; antennomeres VII–X subquadrate; pronotum trapezoidal in shape, broadest apically, about as long as wide; elytra reduced to two small lobes, at suture much shorter than length of pronotum; abdomen large, tergite impression scarcely visible, paratergites narrow anteriorly and broadening posteriorly creating distinct broadly serrate margin. MALE. Tergite VIII with apical margin sinuately truncate (Fig. 16.165c); sternite VIII with entire apical margin deeply emarginate and with small obtuse triangular median projection (Fig. 16.165d); median lobe of aedeagus with bulbus large and ovoid, tubus narrow, sinuate, and strongly bent ventrad in lateral view, apex truncate (Fig. 16.165b). FEMALE. Tergite VIII truncate apically with dense apical macrosetae (Fig. 16.165e); sternite VIII with entire apical margin deeply emarginate, with short, broadly-rounded projection medially (Fig. 16.165f); spermatheca with capsule spherical, neck short, stem narrowly elongate and slightly sinuate (Fig. 16.165g).

DISTRIBUTION **Origin**: Nearctic. **USA**: AK.

COLLECTION AND HABITAT DATA

Habitat: not specified, probably confined to the rocky intertidal zones of Pacific shores of Alaska. **Collecting period**: VIII. **Collecting method**: not specified.

COMMENTS This new species, is similar to *Liparocephalus litoralis* Kirshenblatt from the Kuril Islands and Bering Island (Russia) and northern Japan, but in *L. humblei* the tubus of median lobe of aedagus in lateral view is gradually narrowed and more sinuate apicad, with the apex broadly truncate, and the median projection of male sternite VIII is smaller compared to that of *L. litoralis*. Spermatheca of the two species are similar in shape. For illustrations of *L. litoralis* see Maruyama and Ahn (2000). *Liparocephalus humblei* is also similar in general habitus to *L. cordicollis* LeConte, in which the apex of the tubus is very short and acute, and the median triangular projection of male sternite VIII extends well beyond the lateral angles. Four recently collected specimens from Little Sitkin Island (UAM), not far from the type locality of *L. humblei*, represent *L. brevipennis* (see Annex 1).

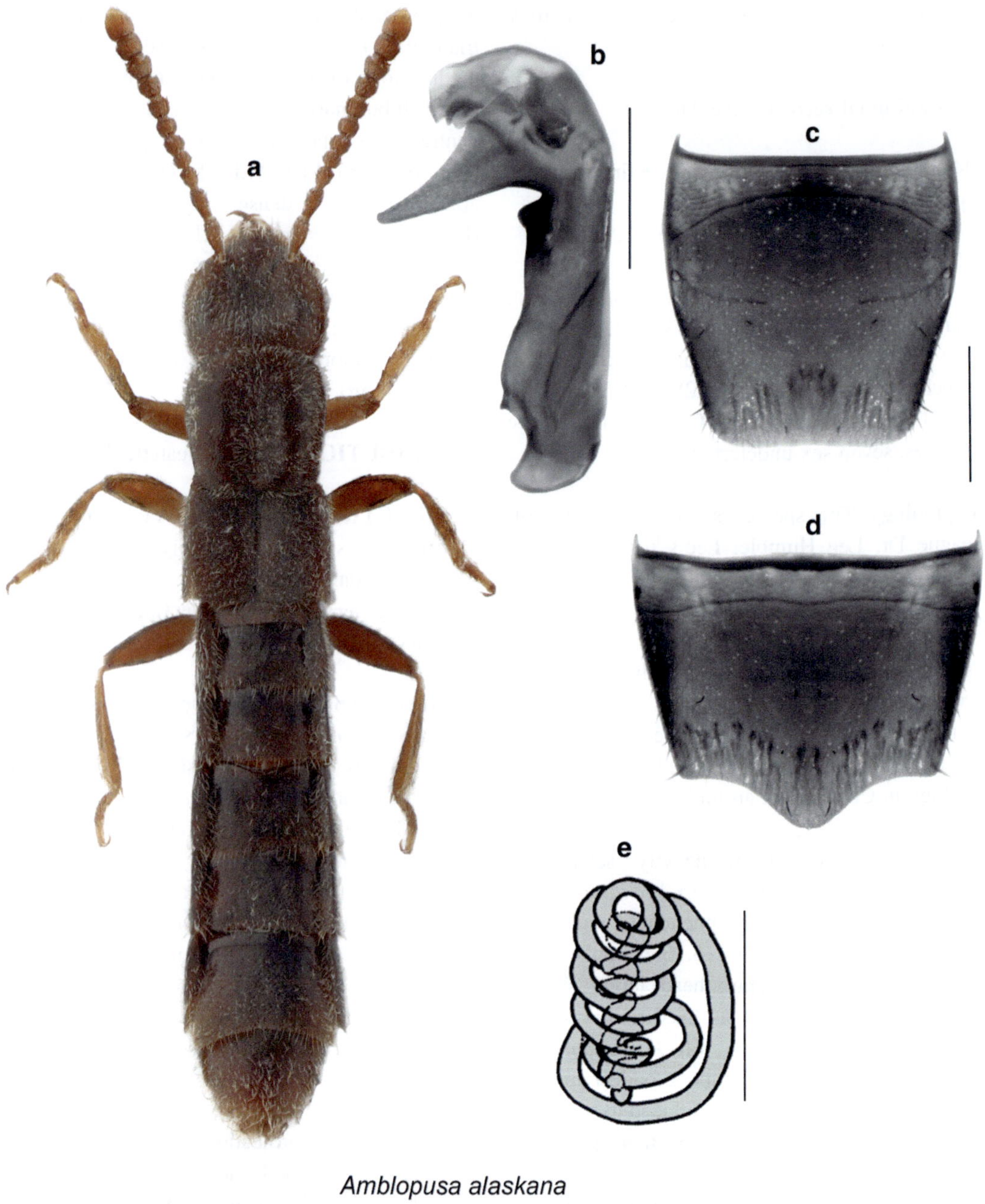

Fig. 16.159 (**a–e**). Images of *Amblopusa alaskana* Ahn and Ashe: (**a**) habitus in dorsal view; (**b**) median lobe of aedeagus in lateral view; (**c**) male tergite VIII; (**d**) male sternite VIII; (**e**) spermatheca. (**e**) after Ahn and Ashe (1995). Scale bar for habitus = 1 mm; remaining scale bars = 0.2 mm

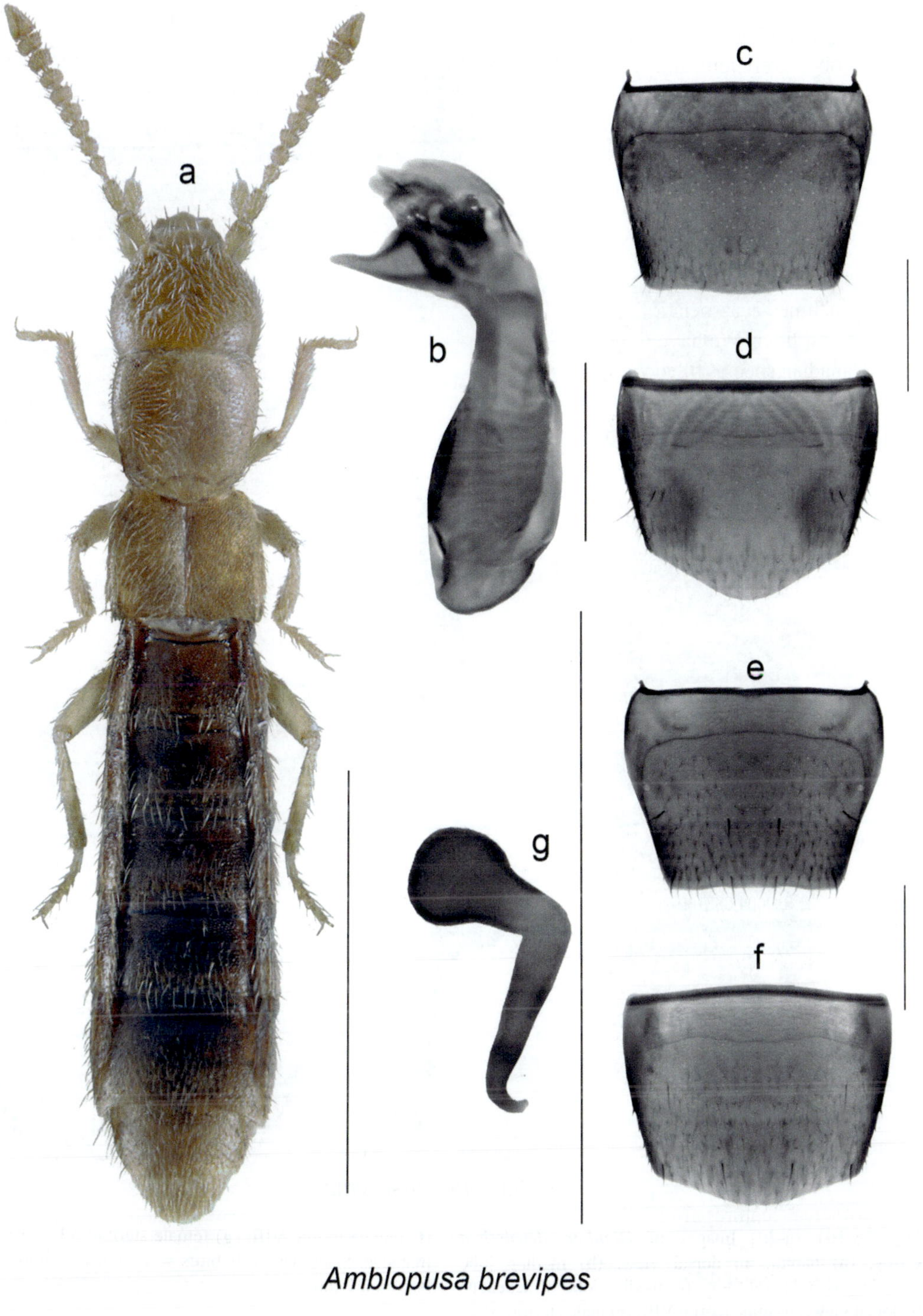

Fig. 16.160 (**a–g**). Images of *Amblopusa brevipes* Casey: (**a**) habitus in dorsal view; (**b**) median lobe of aedeagus in lateral view; (**c**) male tergite VIII; (**d**) male sternite VIII; (**e**) female tergite VIII; (**f**) female sternite VIII; (**g**) spermatheca. Scale bar for habitus = 1 mm; remaining scale bars = 0.2 mm

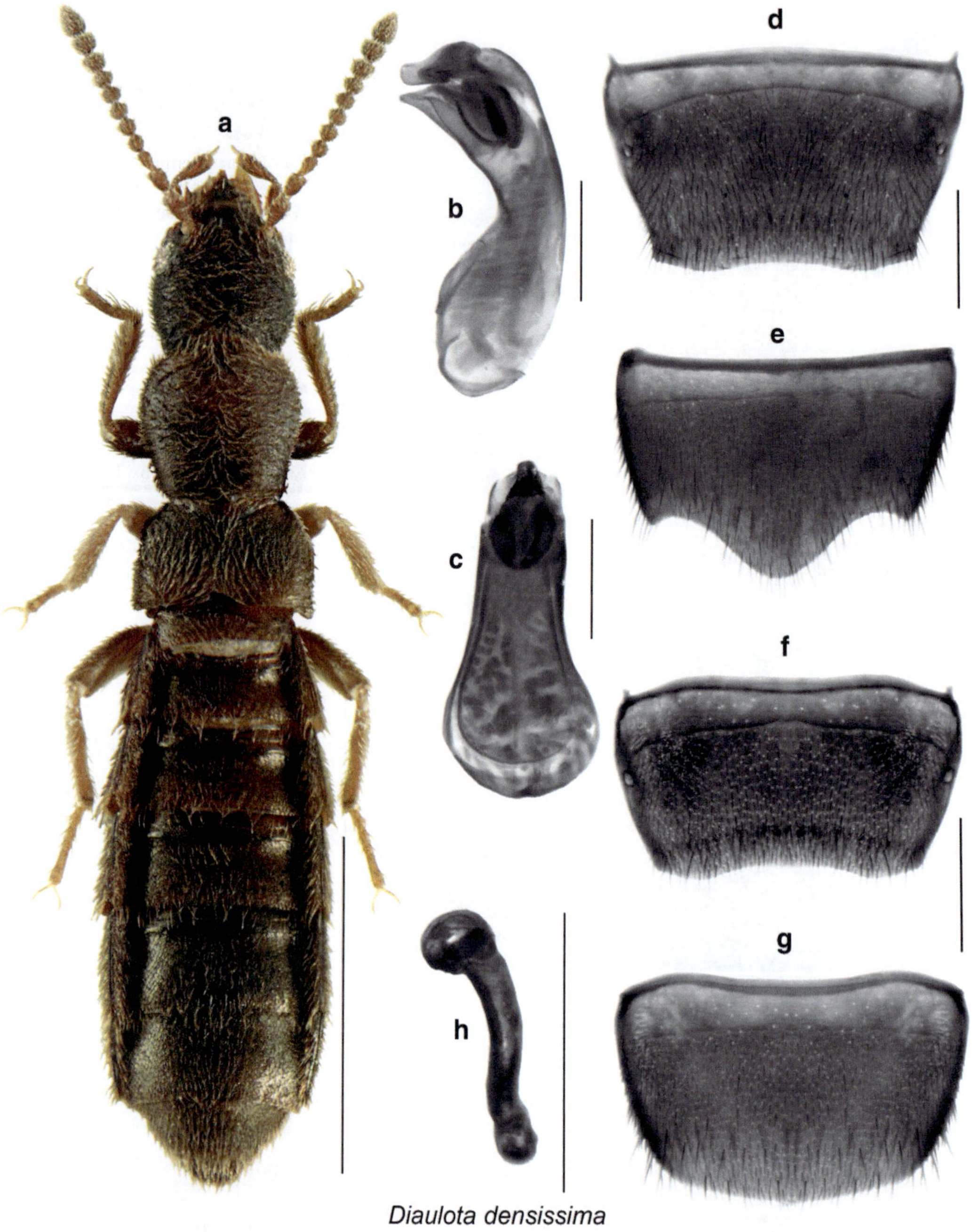

Fig. 16.161 (**a–h**). Images of *Diaulota densissima* Casey: (**a**) habitus in dorsal view; (**b**) median lobe of aedeagus in lateral view; (**c**) median lobe of aedeagus in dorsal view; (**d**) male tergite VIII; (**e**) male sternite VIII; (**f**) female tergite VIII; (**g**) female sternite VIII; (**h**) spermatheca. Scale bar for habitus = 1 mm; remaining scale bars = 0.2 mm

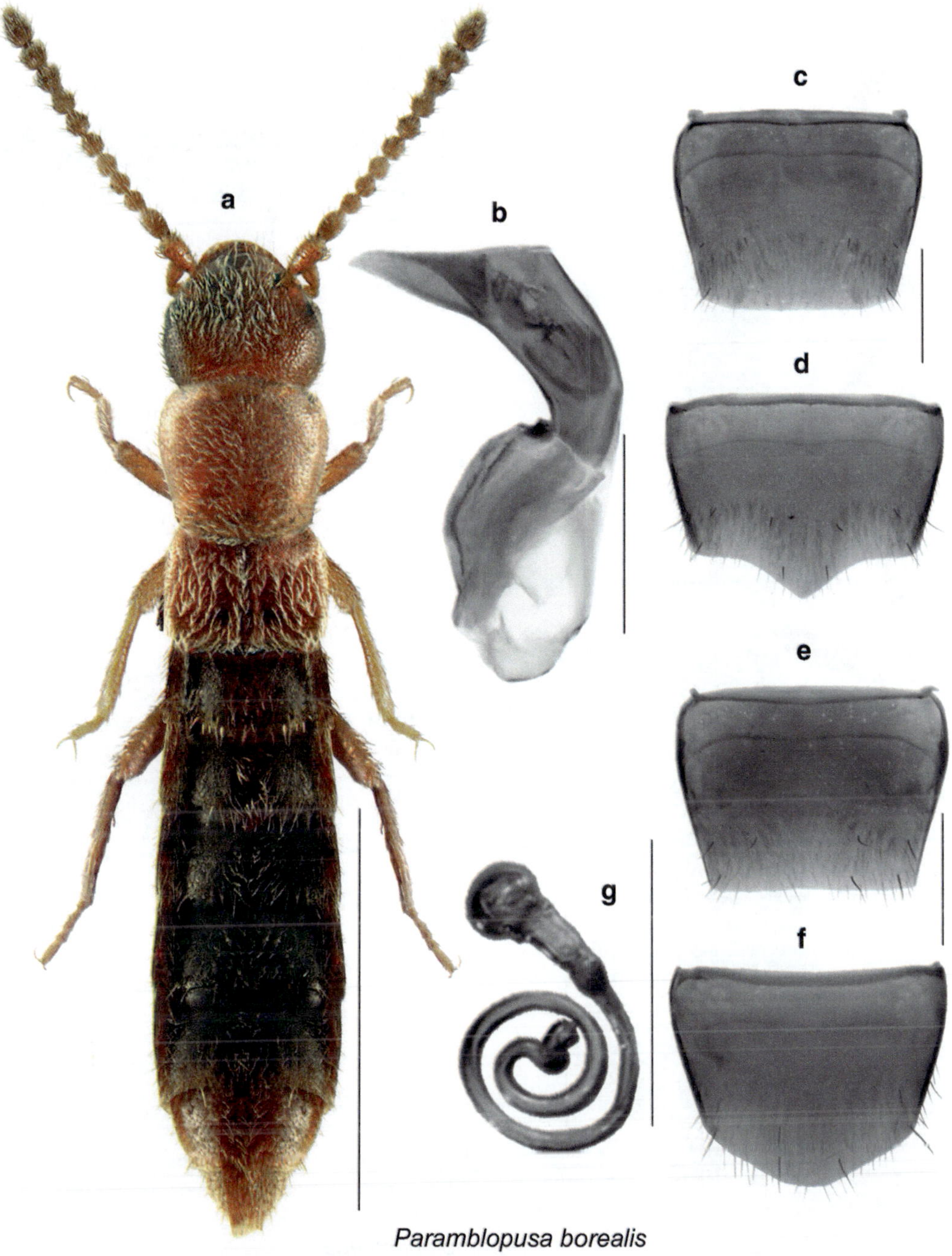

Fig. 16.162 (**a–g**). Images of *Paramblopusa borealis* (Casey): (**a**) habitus in dorsal view; (**b**) median lobe of aedeagus in lateral view; (**c**) male tergite VIII; (**d**) male sternite VIII; (**e**) female tergite VIII; (**f**) female sternite VIII; (**g**) spermatheca. Scale bar for habitus = 1 mm; remaining scale bars = 0.2 mm

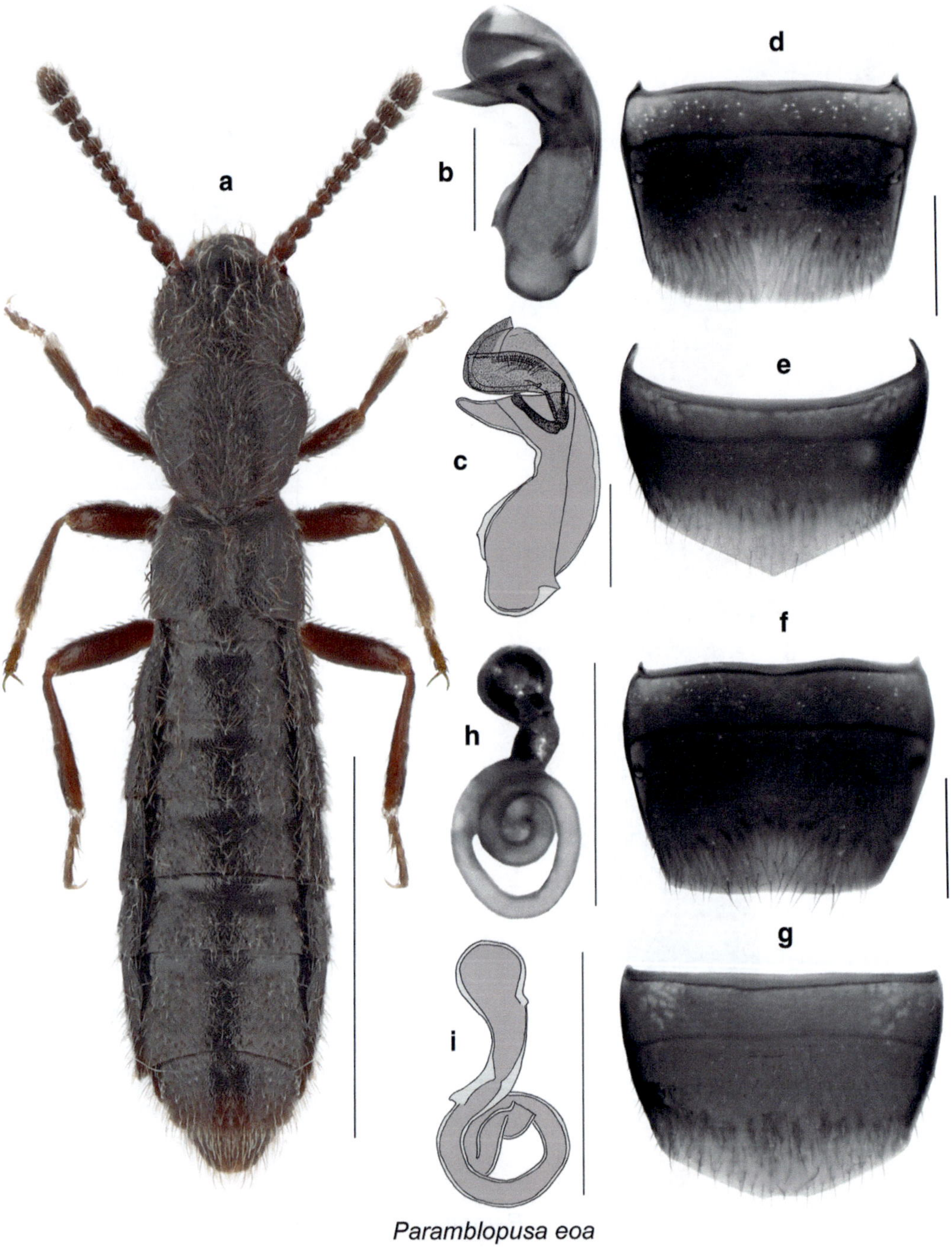

Fig. 16.163 (**a–i**). Images of *Paramblopusa eoa* Ahn and Maruyama: (**a**) habitus in dorsal view; (**b,c**) median lobe of aedeagus in lateral view; (**d**) male tergite VIII; (**e**) male sternite VIII; (**f**) female tergite VIII; (**g**) female sternite VIII; (**h, i**) spermatheca. (**c,i**) after Ahn and Maruyama (2000b). Scale bar for habitus = 1 mm; remaining scale bars = 0.1 mm

Fig. 16.164 (**a–g**). Images of *Liparocephalus brevipennis* Mäklin: (**a**) habitus in dorsal view; (**b**) median lobe of aedeagus in lateral view; (**c**) male tergite VIII; (**d**) male sternite VIII; (**e**) female tergite VIII; (**f**) female sternite VIII; (**g**) spermatheca. Scale bar for habitus = 1 mm; remaining scale bars = 0.2 mm

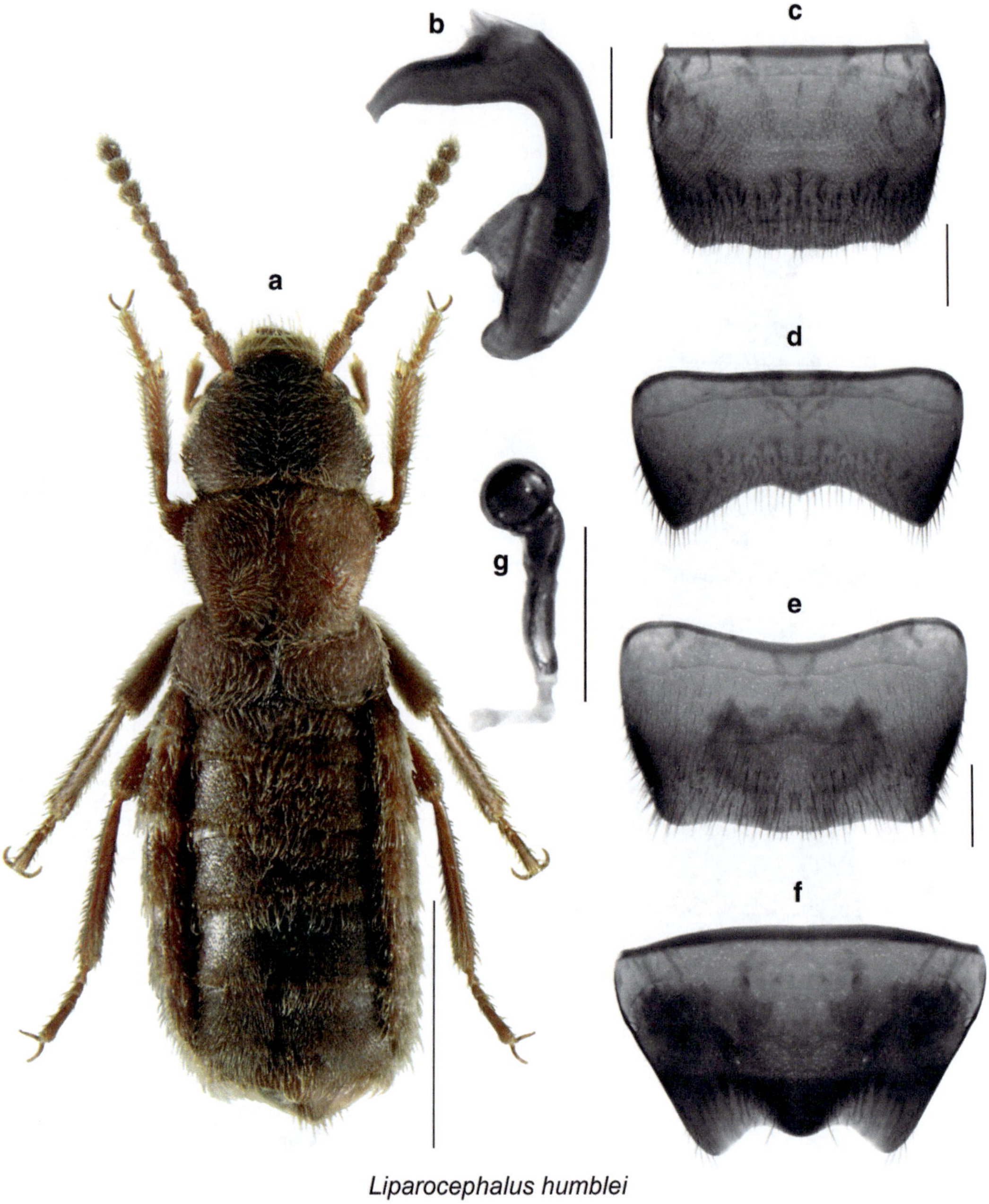

Fig. 16.165 (**a–g**). Images of *Liparocephalus humblei* Klimaszewski, **sp. n.**: (**a**) habitus in dorsal view; (**b**) median lobe of aedeagus in lateral view; (**c**) male tergite VIII; (**d**) male sternite VIII; (**e**) female tergite VIII; (**f**) female sternite VIII; (**g**) spermatheca. Scale bar for habitus = 1 mm; remaining scale bars = 0.2 mm

References

Ahn K-J (1996) A review of *Diaulota* Casey (Coleoptera: Staphylinidae: Aleocharinae), with description of a new species and known larvae. Coleopt Bull 50 (3):270–290

Ahn K-J (1997) A review of *Liparocephalus* Mäklin (Coleoptera: Staphylinidae: Aleocharinae) with descriptions of larvae. Pan Pac Entomol 73(2):79–92

Ahn K-J, Ashe JS (1996a) Revision of the intertidal aleocharine genus *Amblopusa* Casey and description of the new genus *Paramblopusa* (Coleoptera: Staphylinidae). J N Y Entomol Soc 103[1995] (2):138–154

Ahn K-J, Ashe JS (1996b) Phylogeny of the intertidal aleocharine tribe Liparocephalini (Coleoptera: Staphylinidae). Syst Entomol 21:99–114

Ahn K-J, Maruyama M (2000a) Redescription of *Liparocephalus litoralis*, and a key to species of *Liparocephalus* (Coleoptera: Staphylinidae: Aleocharinae). Can Entomol 132(5):567–571

Ahn K-J, Maruyama M (2000b) *Paramblopusa eoa*, a new intertidal species (Coleoptera: Staphylinidae: Aleocharinae) from Kuril Islands. Coleopt Bull 54 (3):359–364

Ahn K-J, Jeon M-J, Branham MA (2010) Phylogeny, biogeography and the stepwise evolutionary colonization of intertidal habitat in the Liparocephalini based on morphological and molecular characters (Coleoptera: Staphylinidae: Aleocharinae). Cladistics 26 (4):344–358

Ahn K-J, Sikes DS, Klimaszewski J (2020) *Paramblopusa eoa* Ahn and Maruyama (Coleoptera: Staphylinidae: Aleocharinae), distributional range extention to Alaska. Coleopt Bull 74(2):324–326

Casey TL (1893) Coleopterological notices. V. Ann N Y Acad Sci 7(1894):281–606

Casey TL (1906) Observations on the staphylinid groups Aleocharinae and Xantholinini, chiefly of America. Trans Acad Sci St Louis 16(6):125–434

Fenyes A (1918) Coleoptera. Family Staphylinidae. Subfamily Aleocharinae. In: Wytsman P (ed) Genera Insectorum, vol XXVII. Fascicule 173A. M. Nijhoff, Den Haag, pp 1–110. 453 pp. + 7 pls

Frank JH, Ahn K-J (2011) Coastal Staphylinidae (Coleoptera): A worldwide checklist, biogeography and natural history. ZooKeys 107:1–98

Klimaszewski J, Hoebeke ER, Godin B, Davies A, Perry KI, Bourdon C, Winchester N (2020) Aleocharine rove beetles of British Columbia: a hotspot of Canadian biodiversity (Coleoptera, Staphylinidae). Springer, Cham, xvi + 631 pp

Mäklin FW (1853) [New species and notes]. In: Mannerheim, C.G.: Dritter Nachtrag zur Kaefer-Fauna der Nord-Amerikanischen Laender des Russischen Reiches. Bulletin de la Société Impériale des Naturalistes de Moscou 26(3):95–273

Yoo I-S, Ahn K-J (2020) Discovery of a new intertidal species, *Amblopusa vancouverensis* from Vancouver Island, Canada with notes on *Amblopusa brevipes* Casey (Coleoptera, Staphylinidae, Aleocharinae). Zootaxa 4803(3):505–514

17 Tribe Homalotini Heer, 1839

Tribe Homalotini Heer, 1839

(Figs. 17.166–17.172) (Ashe 2001, Klimaszewski et al. 2018a, 2020)

Diagnosis Body of diverse forms, pronotum and elytra without foveae and grooves; tarsal formula 4-4-5; mandibles with minute teeth in ventral molar region; prementum with bases of medial setae very close together, medial pseudopore field very narrow; ligula broadly rounded, completely split (divided from apex to base) or bifid (divided only apically); mesocoxae narrowly to broadly separated.

COMMENTS Morphologically diverse group with many species associated with fungi or dead wood.

Key to Genera of Homalotini

1. Labial palpi extremely elongate, extending beyond labrum, forming a V-shape structure.................................... ***Silusa*** **Erichson**

– Labial palpi not as above...........................**2**

2. Body subparallel, moderately convex (Fig. 17.166a); pronotum moderately glossy, with punctation and pubescence moderately dense (Fig. 17.166a); mesocoxae narrowly separated.................................... ***Leptusa*** **Kraatz**

– Body short and compact (Figs. 17.167a–17.170a); pronotum moderately to highly glossy, with punctation and pubescence sparse (Figs. 17.167a–17.170a); mesocoxae broadly separated.................. ***Gyrophaena*** **Mannerheim**

Subtribe Bolitocharina C.G. Thomson, 1859

Diagnosis Integument often coarsely asperate; body usually convex, sometimes subcylindrical; labial palpi with three palpomeres, not styliform, basal two palpomeres indistinctly separated in *Leptusa*; ligula slender and elongate; mandibles with large, dense patch of denticles in ventral molar region; prementum with medial setae arranged one behind the other, medial "pseudopore field" of prementum narrow, without pseudopores; male tergites VII and VIII often with secondary sexual structures consisting of carinae or knobs, tergite VIII often emarginate apically and bearing teeth or denticles (Seevers 1978, Ashe 2001).

COMMENTS Adults and larvae occur in association with macrofungi, especially the Polyporaceae and lignicolous Agaricales (Ashe 2001).

J. Klimaszewski et al., *A Faunal Review of Aleocharine Beetles in the Rapidly Changing Arctic and Subarctic Regions of North America (Coleoptera, Staphylinidae)*,
https://doi.org/10.1007/978-3-030-68191-3_17

Genus *Leptusa* Kraatz, 1856

(Fig. 17.166a–h) (Pace 1989, Klimaszewski et al. 2004, 2018a, 2020)

Diagnosis Body convex, subparallel, integument usually with asperate punctation; pronotum often granulose, broadest at apical one-third; elytra variable in length, sometimes shorter than pronotum; abdomen subparallel or broadening apically; eyes small not protruding from outline of head; antenna with three basal antennomeres elongate, remaining ones quadrate to transverse; ligula slender and entire; maxillary palpi with four palpomeres, apical palpomere narrow, about one-third length of penultimate one; labial palpi with three palpomeres but basal two indistinctly separated, last one narrow, glossae slender and entire; mandibles robust with small molar areas; labrum transverse; mesocoxae narrowly separated by narrow and pointed process of mesoventrite which extends about two-thirds length of mesocoxae; tarsal formula 4-4-5, hind tarsus with basal article about as long as two following ones together or slightly longer; median lobe of aedeagus with flagellum broadly tubular, predominantly shorter than length of median lobe, rarely long and coiled within bulbus (subgenus *Adoxopisalia*), internal sac with two well-defined pairs of basal and apical sclerites, tubus subparallel and moderately long; spermatheca with capsule spherical or ovoid, stem usually narrow and short.

COMMENTS Adults occur on fungi or under bark (sometimes in association with bark beetles), and in forest litter. We have observed the guts of some Canadian species filled with fungal flesh. At least some species overwinter in the adult stage in bark beetle galleries or under bark. Species are known from coniferous and hardwood forests.

Subgenus *Gatineauleptusa* Klimaszewski (Fig. 17.166a–h) (Klimaszewski et al. 2018a, 2020)

Diagnosis Body narrowly subparallel; forebody sparsely pubescent, with integument deeply coarsely punctate; antennomeres IV–X transverse and incrassate, VII–X distinctly transverse, each antennomere at least three times as wide as long; male tergite VIII truncate and crenulate apically; median lobe of aedeagus with apical sclerites of internal sac not apparent, basal sclerites in form of two elongate structures, apical part of membrane of internal sac with minute denticles, flagellum short and thick, forming long ovoid base; spermatheca of a unique mushroom shape: capsule kidney-shaped, without apparent invagination, stem long, thin, with one posterior loop.

Evenly narrow body (head, pronotum, elytra and abdomen of about the same width), pronounced body microsculpture, elytra with setae on short, transverse line-like elevated structures, and the unique shape of spermatheca define this subgenus.

95. *Leptusa* (*Gatineauleptusa*) *gatineauensis* Klimaszewski and Pelletier (Fig. 17.166a–h)

Leptusa gatineauensis Klimaszewski and Pelletier, in Klimaszewski et al. 2004. Bishop et al. 2009, McLean et al. 2009, Majka and Klimaszewski 2008, Webster et al. 2012, Klimaszewski et al. 2011, 2016, 2018a, 2020.

Diagnosis Body length 1.5–1.9 mm, colour brown, head and abdomen darker than remainder of body, elytra and pronotum with dark reddish tinge, antennae dark yellowish-brown (Fig. 17.166a); integument with punctation moderate in size, with distinct isodiametric microsculpture, pronotum and elytra slightly granulose; antenna with antennomeres I–III distinctly elongate, III a little shorter and slimmer than II, IV subquadrate, V–X transverse and VII–X distinctly so, each about three times as broad as long; pronotum transverse, about 1.4 times as wide as long, with sides gradually converging basad and broadest about middle of its length, shorter than elytra at suture, punctation moderate in size, distance between punctures greater than diameter of a puncture; elytra slightly transverse, at suture about 1.2 times as long as pronotum;

abdomen subparallel, with first three visible tergites deeply impressed basally and punctate. MALE. Tergite VIII with apical margin broadly truncate medially, minutely crenulate (Fig. 17.166d); sternite VIII with apical margin very obtusely produced, broadly arcuate medially (Fig. 17.166e); median lobe of aedeagus in lateral view with crista apicalis of bulbus narrowly oval, tubus moderately elongate, ventral margin curved ventrad in basal half, distinctly sinuate to narrow and narrowly rounded apex (Fig. 17.166b,c); internal sac with flagellum shorter than median lobe, lamina basalis in form of two elongate structures, lamina apicalis inconspicuous, with minute denticles (Fig. 17.166b,c). FEMALE. Tergite VIII truncate apically, antecostal suture sinuate, moderately narrowly separated from basal margin (Fig. 17.166f); sternite VIII broadly rounded apically, antecostal suture narrowly separated from basal margin (Fig. 17.166g); spermatheca mushroom-shaped, capsule kidney-shaped, without apparent invagination, stem long, thin, with one posterior loop (Fig. 17.166h).

DISTRIBUTION **Origin**: Nearctic. **CANADA**: AB, BC, NB, NF, NS, ON, QC, SK. **USA**: AK [**NSR**].

COLLECTION AND HABITAT DATA
Habitat: adults in NB, collected from Lindgren funnel traps in an old red pine forest, a mature (110-year-old) red spruce forest, an old red oak forest, a rich Appalachian hardwood forest, a trembling aspen stand with a few conifers, and a jack pine forest. Also reported from deciduous and mature red spruce-hemlock and mixed forests in other areas. Some collected from forest litter, in/on *Polyporus betulinus* (Bull.) Fr. In AK, one specimen found in a *Betula papyrifera* and *Populus tremuloides* forest. **Collecting period**: V–VI. **Collecting method**: Lindgren funnel traps, flight intercept traps, pitfall traps, sifting litter, and fungi.

We report this species from AK (**NSR**) for the first time here, based on the following: USA, Alaska, Fairbanks, 930 Fitz Ct, 280m el., 64.9013, −147.52861, ±30m, *Betula pap., Pop. trem.*, flight intercept trap, 20.V-3.VI.2007, D.S. Sikes (UAM) 1 specimen.

Subtribe Gyrophaenina Kraatz, 1856

(Ashe 2001, 1984, Klimaszewski et al. 2018a, 2020)

Diagnosis Mesocoxae broadly separated by width of mesocoxal cavity or slightly less by broad and broadly contiguous processes of meso- and metaventrite; laciniae distinctive, apex obliquely truncate and covered with numerous spinules, with setae and spines of inner surface reduced or absent (for illustrations see Ashe, 2001).

COMMENTS Associated with fresh fruiting bodies of basidiomycete fungi.

Genus *Gyrophaena* Mannerheim, 1830

(Figs. 17.167–17.170) (Seevers 1951, Klimaszewski et al. 2018a, 2020)

Diagnosis Body broadly oval and dorsoventrally flattened, length 1.2–3.5 mm, moderately or highly glossy; color flavate, brown, piceous, or black, usually bicoloured but sometimes uniformly coloured; integument of forebody with large insertion pores, often elevated (asperities), forming distinct patterns on pronotum and head; isodiametric meshed microsculpture usually present; head with genae well developed; infraorbital carina present; eyes large and often prominent, finely faceted and broadly separated; labrum broadly oval or trapezoidal with sensory setae forming patterns; mandibles robust, the right mandible bearing one, usually well-developed internal tooth, and a well-developed internal membranous lobe (the prostheca), which bears rows of small teeth; maxillae with laciniae obliquely truncate and with numerous closely spaced teeth (spore brush), and rows of setae on the outer lobe, galeae provide a cup-like cap over the apex of the lacinial comb; ligula short, entire, produced and broadly rounded at apex; antennomere IV small, V slightly broader than VI and IV–X usually incrassate apically; pronotum usually transverse with hypomera visible from the side; mesocoxae broadly separated by about width of mesocoxal cavity by broad and broadly contiguous processes of meso- and metaventrite; elytra usually broad-shouldered; abdomen tapering apically; process of mesoventrite truncate apically, at least as long as that of metaventrite; spermatheca sclerotized, with lateral fin-like flange on the short stem, capsule spherical; median lobe of aedeagus variably shaped, often tubular or trough-shaped, often with ventral projections of various shapes and sizes forming a complex structure; apical portion of internal sac rigid and forming elongate projection from where flagellum is normally exerted, the internal sac lacks internal and external spines and sclerites (unlike most other aleocharine genera); paramere with broad apical lobe bearing four setae, usually two longer and two shorter ones; apical margin of male tergite VIII usually with two relatively large sublateral teeth bordering emargination with smaller teeth, rarely larger sublateral teeth reduced and smaller teeth lacking, or sublateral teeth replaced with large lobes; species strictly associated with non-woody macrofungi (mushrooms).

The species groups in *Gyrophaena* are mainly defined by the similarities of the median lobe of the aedeagus.

Key to Species of *Gyrophaena*

1. Pronotum broad, one-fifth narrower than elytra, moderately glossy (Figs. 17.168a, 17.169a); elytral punctures coarse, deep and dense (Figs. 17.168a, 17.169a); median lobe of aedeagus with tubus simple, triangularly elongate, sharply pointed apically, with or without short vestigial narrow projection in lateral view (Figs. 17.168b, 17.169b) **2**

– Pronotum narrow, one-third narrower than elytra, strongly glossy; elytral punctures and tubus of median lobe of aedeagus not as above **3**

2. Elytra more or less uniformly dark red-brown, surface strongly rugose (Fig. 17.169a); apical margin of male tergite VIII with large, broad submedian tooth on each side of two small median teeth (Fig. 17.169c); tubus of median lobe of aedeagus in lateral view with prominent basal swelling (Fig. 17.169b) .... ***Gyrophaena rugipennis* Mulsant et Rey**

– Elytra bicolored, yellow with irregular dark brown spots in scutellar and postero-laterad regions, surface not rugose (Fig. 17.168a); apical margin of male tergite VIII with two narrow, short sublateral teeth and two narrower, shorter median teeth (Fig. 17.168c); tubus of median lobe of aedeagus in lateral view without basal swelling (Fig. 17.168b) ....***Gyrophaena nana* (Paykull)**

3. Antennomeres VII–X subquadrate (Fig. 17.170a); male tergite VIII with two sublateral lobate teeth, margin between teeth emarginated and with two minute median teeth (Fig. 17.170c); median lobe with two pairs of narrow projections (Fig. 17.170b); apical margin of female sternite VIII evenly rounded medially (Fig. 17.170f) ..... ***Gyrophaena insolens* Casey**

– Antennomeres VII–X clearly transverse (Fig. 17.167a); male tergite VIII with two narrow lateral and two narrow and slightly shorter median teeth (Fig. 17.167c); median lobe of aedeagus with tubus sinuate ventrally in lateral view (Fig. 17.167b); apical margin of female sternite VIII obtusely angulate medially (Fig. 17.167f) .... ***Gyrophaena keeni* Casey**

(Subgenus *Gyrophaena* Mannerheim, 1830)

***Keeni* species group** (Seevers 1951)

Diagnosis Antennomeres V–X moderately to distinctly transverse; pronotum sparsely pubescent and highly glossy; male tergite VIII with two large lateral teeth and two small approximate median teeth; tubus of median lobe of aedeagus narrowly triangular and sinuate ventrally in lateral view, flagellum everted, coiled and twisted.

96. *Gyrophaena (Gyrophaena) keeni* Casey (Fig. 17.167a–g)

Gyrophaena keeni Casey, 1911. Seevers 1951, Klimaszewski and Winchester 2002, Klimaszewski et al. 2005, 2009, 2011, 2018a, 2020.

Diagnosis Body roughly narrowly ovoid, length 1.6–1.9 mm; head about as broad as pronotum, rufo-piceous to piceous; pronotum reddish-brown or medium dark brown, elytra light reddish-brown or light brown with posterior angles dark brown, abdomen reddish-brown or light brown with posterior portion piceous (Fig. 17.167a). Vertex of head with at least six small punctures on each side, disc of pronotum usually with median rows of three punctures and a small cluster medially near base, elytra sparsely and irregularly punctate; microsculpture distinct, finely meshed on head and pronotum; antennomeres VI–X moderately transverse; pronotum 1.4 times as wide as long. MALE. Tergite VIII with apical margin truncate, with two long, narrow sublateral teeth and two shorter, finer medial teeth (Fig. 17.167c); sternite VIII with margins almost evenly arcuate from base to apex (Fig. 17.167d); median lobe of aedeagus in lateral view with tubus moderately long and broad, sinuately tapering to acute apex, curved obliquely ventrad basally, apical projection of internal sac tubular and short, flagellum everted and curled, compressor plate elevated (Fig. 17.167b). FEMALE. Tergite VIII truncate apically (Fig. 17.167e); sternite VIII with apical margin obtusely angulate (Fig. 17.167f); spermatheca as illustrated (Fig. 17.167g).

DISTRIBUTION **Origin**: Nearctic. **CANADA**: AB, BC, LB, MB, NB, NF, NS, ON, QC, SK [**NPR**], YT. **USA**: AK [**NSR**], FL, MA, ME, MT, NH, NY, OR, TN, WA, WY.

COLLECTION AND HABITAT DATA

Habitat: mature mixed forest, 8.5-year-old regenerating mixed forest, red oak forest, eastern white cedar (*Thuja occidentalis* L.) swamps, red spruce (*Picea rubens* Sarg.) and red maple (*Acer rubrum* L.) forest (80–120 years old), black spruce and *Populus* forests, recently burned coniferous forest, and an eastern hemlock forest [*Tsuga canadensis* (L.)] 120+ years old. Adults in in gilled mushrooms, in gilled fungi on rotten log, in gilled mushrooms on stump, in moss near brook, on polypore fungi on dead standing *Populus* sp. In AK, specimens were found in clear-cut, thinned 2nd growth forest, and *Picea glauca* forest. **Collecting period**: VI–IX. **Collecting method:** sifting mushrooms, sifting moss, flight intercept trap, and pitfall trap.

We report this species from SK (**NPR**) and AK (**NSR**) for the first time here, based on the following: Canada, Saskatchewan, Prince Albert, 53.9726, −106.2902, 533m, 15.IX.2014, mushrooms, sifting, coll. B. Godin and D. Horwood (BGC, LFC) 3 males. USA, Alaska, Prudhoe Bay Rd., Fish Cr., 66.5333, −150.8333, 900 feet, 2.VII.1978, J.M. Campbell and A. Smetana (CNC) 2 males, 11 sex undetermined; Prudhoe Bay Rd., Bonanza Cr., 66.6666, −150.6666, 900 feet, 2.VII.1978, A. Smetana and J.M. Campbell (CNC) 1 sex undetermined; Fairbanks, Creamer's Field, 154m el., 64.86607, −147.73817, ±108m, 15.VII.2008, D.S. Sikes (UAM) 19 specimens; Prince of Wales Island, Hatchery Ck.2, 134m el., 55.89356, −132.9437, ±13m thinned 2nd growth, Lindgren 1, 25.VI-9.VII2012, J. Stockbridge, S. Meierotto, I. MacDougall (UAM) 1 specimen; Prince of Wales Island, Luck Point 2°, 245m el., 55.97953, −132.77156, ±24m clear-cut, Lindgren 1, 14-25.VII.2010, J. Stockbridge, C. Bickford (UAM) 1 specimen; Sterling, Corey St, 90m el., 60.53953, −150.83232, ±150m *P. glauca* forest, uv light, 13.VII.2004, D.M. Collet (UAM) 1 specimen.

Additional records. See Annex 1.

COMMENTS (Klimaszewski et al. 2018a). Strand (1938) described *Gyrophaena orientalis* from the Palearctic region, and it is known to occur in central and northern Europe, Caucasus and Siberia (Enushchenko and Semenov 2016). *Gyrophaena orientalis* is extremely similar to *Gyrophaena keeni* but differs from it by antennomeres VII–X subquadrate (transverse in *G. keeni*), pronotum broader (narrower in *G. keeni*), and tubus wider with strongly ventral sinuate margin in lateral view (slightly in *G. keeni*). For images of *G. orientalis* see Enushchenko and Semenov (2016). Additional studies are needed to understand variation of morphological features from a broad geographic area. For the time being we retain the status of the two species. Although this species was reported from YT by Campbell and Davies (1991), we here provide voucher data for the first time (see Annex 1). This species was listed from WI by Newton (2019) but we are not aware of published voucher data for this record.

***Nana* species group** (Seevers 1951)

Diagnosis Antennomeres V–X moderately to distinctly transverse; head with punctures strongly impressed; apical margin of male tergite VIII with two long, apically rounded sublateral teeth and two shorter, finer, apically rounded teeth medially; tubus of median lobe of aedeagus simple in lateral view, roughly triangular apically, dorsal margin usually with minute, subapical denticle; internal sac and flagellum everted and straight or sinuate.

97. *Gyrophaena (Gyrophaena) nana* (Paykull) * (Fig. 17.168a–g)

Staphylinus nana Paykull, 1800. Campbell and Davies 1991, Klimaszewski et al. 2009, 2011, 2012, 2018a, 2020, Webster et al. 2012, Enushchenko and Semenov 2016.

Diagnosis Body widely oval, length 1.7–2.0 mm; head piceous, pronotum reddish-brown with central part of disc darker, elytra piceous, paler anteriorly and along lateral margins, abdomen dark brown or lighter with some tergites dark brown (Fig. 17.168a); antennomeres VI–X transverse and incrassate apically; head slightly narrower than pronotum, pronotum distinctly transverse, widest at middle, distinctly narrower than elytra; elytra at suture longer than pronotum; abdomen wide, basally almost as wide as elytra, punctures coarse, deep and dense. MALE. Apical margin of tergite VIII with two sublateral, moderately long teeth and two finer, slightly shorter, approximate teeth medially (Fig. 17.168c); sternite VIII semicircularly rounded from base (Fig. 17.168d); median lobe of aedeagus in lateral view with tubus simple, curved very obliquely ventrad before middle of ventral margin, gradually tapering to acute apex, apical projection of internal sac curved dorsad (Fig. 17.168b). FEMALE. Tergite VIII virtually truncate apically (Fig. 17.168e); sternite VIII similar to that of male but slightly obtuse apically (Fig. 17.168f); spermatheca as illustrated (Fig. 17.168g).

DISTRIBUTION Origin: Holarctic. Europe, Caucasus, Russian Far East, and North America. **CANADA**: AB, BC, MB, NB, NF, NT [**NTR**], ON, QC [**NPR**], SK [**NPR**], YT. **USA**: AK, CA, MA, ME, MI, MT, WI, WY.

COLLECTION AND HABITAT DATA

Habitat: one specimen from NB from mushroom on log in mixed forest; specimens from NL and YK from rotting mushrooms in unspecified forest type. In AK, specimens were found in *Betula* and *Populus tremuloides* forest. **Collecting period**: IV–IX. **Collecting method:** sifting mushrooms, yellow pans, Malaise traps.

We report this species from NT (**LFC, BOLD**), QC, and SK for the first time here, based on the following: Canada, Northwest Territories, Yellowknife, 62.52110, –113.38174, Rep. 1 wet, yellow pan, 7-11.VII.2011, NBP field party Y1WY515, Y1WY516, Y1WY517 (LFC) 2 males, 1 female; 62.52110, –113.38174, Rep. 1 wet, yellow pan, 11-15.VII.2011, NBP field party Y1WY2210, Y1WY2211, Y1WY9211 (LFC) 2 males, 1female; Nahanni National Park Reserve, Nailicho (Virginia Falls), 578m el., 61.606, –125.758, Malaise trap, 27.VI-04. VII.2014, Parks Canada staff (CBG) 2 specimens. Quebec, Abitibi, lac Duparquet, 22.IX.1997, Microhabitat: Berlese Champ. Lamellés sol, Echant. S-124, 1994, Peupleraie, P. Paquin (LFC) 2 males; Quebec, Terr. Nouveau-Québec, Chemin Selbaie, 15.X.1997, Echantillon sol, Echantillon P-3440 F4, Pessière noir, P. Paquin (LFC) 1 male. Saskatchewan, Prince Albert, 53.9789, –106.2859, 534m, 13.IX.2014, mushrooms, sifting, coll. B. Godin and D. Horwood (LFC) 2 males, 1 female; same data except 53.9726, –106.2902, 533m (LFC) 1 female.

Additional records. See Annex 1.

COMMENTS Most specimens from NT and YT were slightly smaller and more glossy than other studied specimens of *G. nana*, had slightly shorter elytra and tubus of median lobe was slightly more sinuate ventrally. We consider these differences as morphological variation within the same species. Although this species was previously reported from YT (Campbell and Davies 1991), voucher specimen data is here provided for the first time (see Annex 1). This species was listed from AZ by Newton (2019) but we are not aware of published voucher data for this state.

98. ***Gyrophaena (Gyrophaena) rugipennis* Mulsant et Rey *** (Fig. 17.169a–g)

Gyrophaena rugipennis Mulsant et Rey, 1861.

Gyrophaena neonana Seevers, 1951. Klimaszewski et al. 2008, 2011, 2018a, Brunke et al. 2012, Webster et al. 2012. **New Synonymy**.

Diagnosis Body wide, elongate-ovoid, length 1.7–2.0 mm; head piceous, disc of pronotum dark brown with margins rufo-flavate, elytra light brown to reddish-brown, abdomen dark brown or lighter with some tergites dark brown (Fig. 17.169a); head almost as wide as pronotum; antennomeres VI–X transverse and incrassate apically; pronotum distinctly transverse, widest at middle, distinctly narrower than elytra; elytra at suture about as long as pronotum along midline of disc; abdomen wide and tapering apicad from near base, basally almost as wide as elytra. MALE. Tergite VIII with broad, rounded tooth on each side of and close to two very small medial teeth (Fig. 17.169c); sternite VIII with margins arcuate from base to broadly rounded apex (Fig. 17.169d); tubus of median lobe of aedeagus in lateral view curved obliquely ventrad at base, with ventral margin highly swollen in basal half, then straight to slightly bent apex, dorsal margin evenly arcuately tapering, projection of internal sac sinuate, whip-like (Fig. 17.169b). FEMALE. Tergite VIII truncate apically (Fig. 17.169e); sternite VIII roughly semicircular from base to apex (Fig. 17.169f); spermatheca as illustrated (Fig. 17.169g).

DISTRIBUTION Origin: Holarctic. Europe, Caucasus, Siberia, Russian Far East, and North America. **CANADA**: NB, NF, ON, SK [**NPR**], YT [tentative]. **USA**: IN, NC, PA, WI.

COLLECTION AND HABITAT DATA
Habitat: mixed forests on/in gilled mushrooms, in fungus on log, and in a fleshy polypore mushroom on an American beech log. **Collecting period**: VIII–IX. **Collecting method**: intercept traps, sifting mushrooms.

We report this species from SK (**NPR**) for the first time here, based on the following: Canada, Saskatchewan, Meadow Lake, 54.4826, –109.6584, 513m, 10.IX.2014, mushrooms, sifting, coll. B. Godin and D. Horwood (LFC) 2 males.

COMMENTS We have compared external morphology, median lobe of aedeagus, and male tergite VIII of *Gyrophaena neonana* with those of *G. rugipennis* and did not find significant differences between the two species. We therefore formally synonymize the former species with *G. rugipennis*.

***Pulchella* species group** (Seevers 1951)

Diagnosis Antennomeres V–X subquadrate to slightly elongate, rarely moderately transverse; dominant body colour is yellow; pronotum slightly transverse, yellow or reddish-yellow; male tergite VIII with semicircular median emargination and usually lobe-like lateral teeth/ projections; tubus of median lobe of aedeagus narrowly tubular with one or more digital ventral processes. Species in this group are externally very similar to each other.

99. ***Gyrophaena (Gyrophaena) insolens* Casey** (Fig. 17.170a–g)

Gyrophaena insolens Casey, 1906. Seevers 1951, Klimaszewski et al. 2009, 2011, 2015, 2018a.

Diagnosis Body subparallel, length 1.9–2.5 mm; head rufo-piceous to piceous, pronotum rufo-testaceous, elytra piceous posteriorly with testaceous humeri, or uniformly testaceous, abdomen rufo-testaceous, apical part of abdomen sometimes darker (Fig. 17.170a); vertex of head with about ten large umbilicate punctures on each side, pronotum with two irregular median rows of large punctures, scattered punctures elsewhere, elytra with fine, sparse punctures; forebody with microsculpture: reticulate throughout except on pronotum; antennomere IV small and subquadrate, V–VII elongate, VIII–X subquadrate or slightly transverse; pronotum 1.2 times as wide as long. MALE. Tergite VIII with apical margin obliquely produced laterally, forming a triangular, inward-facing lobe-liketooth on each side of deep, rounded medial emargination with two, or occasionally one, small denticles (Fig. 17.170c); sternite VIII with apical margin broadly rounded laterally, broadly truncate medially (Fig. 17.170d); tubus of median lobe of aedeagus in lateral view elongate, tubular, bearing three projections, a long straight basal one and two shorter, narrower, pointed, approximate ante-basal ones, which often overlap, the middle one with an angular ante-basal elevation, apical part of internal sac tubular, often extending beyond tubus, with flagellum projecting externally from apex (Fig. 17.170b). FEMALE. Tergite VIII truncate apically, finely crenulate (Fig. 17.170e); sternite VIII with margins semicircularly rounded from base to apex (Fig. 17.170f); spermatheca as illustrated (Fig. 17.170g).

DISTRIBUTION **Origin**: Nearctic. **CANADA**: AB [**BOLD**], LB, MB, NB, NF, ON, SK. **USA**: AK [**NSR**], MI.

COLLECTION AND HABITAT DATA
Habitat: mixed forest, white and red spruce forest, old-growth spruce and balsam fir forest, spruce-poplar forests, eastern white cedar swamps, yellow birch and spruce forest, on ridge with oak in silver maple forest, red oak forest, and red oak and red maple forest. Adults on/in fresh gilled mushrooms on forest floor, in fresh *Tricholoma* sp., in *Russula virescens* on forest floor. In AK, one specimen was collected from *Tricholoma sapanaceum* (Fr.: Fr.) P. Kumm. **Collecting period**: VIII–X. **Collecting method**: sifting mushrooms, flight intercept traps, hand collecting.

We report this species for the first time from AB (**BOLD** records) and AK (**UAM** specimens) as follow: Canada, Alberta, Jasper National Park, Pocahontas Campground, 1131m el., 53.195, –117.914, birch and spruce forest, flight-intercept trap, 16-21.VII.2012, BIOBus 2012 team (CBG) 1 specimen; Banff National Park, Baker Creek picnic area, 1500m el., 51.35, –116.063, mature lodgepole pine forest, flight-intercept trap, 23-28. VII.2012, BIOBus 2012 team (CBG) 1 specimen. USA, Alaska, Fairbanks 148m el., 64.85117, –147.84203, ±50m ex. *Tricholoma sapanaceum* det. G. Laursen, hand collected, 19.VIII.2006, D.S. Sikes (UAM) 7 specimens.

COMMENTS The former BC record of this species was based on Moore and Legner's (1975) misinterpretation of "British America" [=-Canada] in Seevers (1951: 705); this error was copied in Klimaszewski et al. 2009, 2011, 2015, 2016.

Subtribe Silusina Fenyes, 1918 (Ashe 2001, Klimaszewski et al. 2003, 2018a).

Diagnosis Labial palpi extremely elongate, filiform or stylate, appearing 2-segmented but borders between segments are often diffuse; ligula elongate, entire apically; mesocoxae narrowly separated by narrow processes of meso- and metaventrite (fig. 217, in Ashe 2001).

Genus *Silusa* Erichson, 1837

(Figs. 17.171–17.172) (Klimaszewski et al. 2003, 2018a, 2020)

Diagnosis Body elongate and subparallel, robust, length 2.5–4.5 mm; integument usually coarsely punctate, pubescence fine and sparse to dense; head narrowly protruding in front of eyes, postocular carina strong and complete; antennomeres VII–X 1.3-2.0 times as wide as long; labial palpi extremely elongate, filiform or stylate, forming an acute V; maxillary palpi with four palpomeres, terminal one narrow and pointed; mentum with anterolateral margins not produced as spiniform processes; pronotum transverse and broadest about middle or in apical one-third, pubescence on disc directed anteriad along midline apically and posteriad elsewhere, on each side directed obliquely lateroposteriad; pronotal hypomera visible in lateral view; elytra more-or-less elongate with posterolateral angles distinctly, sinuately emarginate, pubescence on disc directed slightly obliquely posteriad; first four visible abdominal tergites deeply impressed basally; hind tarsus with basal article slightly longer than following one; apical margin of male tergite VIII denticulate; median lobe of aedeagus strongly curved or bent ventrad at some point, internal sac with distinct subapical sclerites and with flagellum strongly sclerotized, with basal part swollen; spermatheca with capsule large, usually globose, lacking apical invagination, stem thin and short.

Key to Species of *Silusa*

1. Pronotum convex, body surface highly glossy (Fig. 17.171a); colour dark reddish brown with margins of pronotum, elytra and abdominal tergites paler (Fig. 17.171a); in lateral view median lobe of aedeagus with tubus abruptly bent almost perpendicularly ventrad at middle of ventral margin (Fig. 17.171b); spermatheca with capsule spherical, stem long, thin, sinuate, with accessory sclerites at base of seminal channel (Fig. 17.171h)....***Silusa californica* (Bernhauer)**

– Pronotum flattened, body surface moderately glossy (Fig. 17.172a); colour dark brown with sutural area of elytra slightly reddish (Fig. 17.172a); in lateral view median lobe of aedeagus with tubus sinuate, apex not so strongly produced ventrad (Fig. 17.172b); spermatheca with capsule ovoid, stem narrow and short, without accessory sclerites at base of seminal channel (Fig. 17.172h).................... ***Silusa densa* Fenyes**

***Californica* species group** (Klimaszewski et al. 2003, 2018a)

Diagnosis Median lobe of aedeagus in lateral view with tubus moderately narrow and elongate, about as long as bulbus, curved or abruptly bent ventrad at about middle of ventral margin; flagellum broad and moderately long, about as long as tubus, with basal part highly enlarged, funnel-shaped.

100. *Silusa californica* (Bernhauer) (Fig. 17.171a–h)

Stenusa (Silusa) californica Bernhauer, 1905. Klimaszewski and Winchester 2002, Klimaszewski et al. 2003, 2005, 2008, 2011, 2016, 2018b, 2020, Majka and Klimaszewski 2008, 2010, Brunke et al. 2012.

Diagnosis Body length 3.0–4.0 mm; colour dark reddish brown with margins of pronotum, elytra, abdominal tergites and basal antennomeres paler (Fig. 17.171a); body surface highly glossy, moderately densely punctate and pubescent; antennomere IV elongate, VIII–X distinctly transverse; pronotum distinctly transverse, pronotal punctures dense and coarse, usually separated by diameter of a puncture at least basally, the intervals with fine meshed microsculpture; elytra moderately transverse; first four visible abdominal tergites with deep transverse impressions bearing moderately coarse punctation. MALE. Tergite VIII broadly shallowly emarginate apically bearing distinct, rounded, slightly uneven teeth (Fig. 17.171d); sternite VIII with apical margin obtusely triangularly produced with apex rounded (Fig. 17.171e); median lobe of aedeagus in lateral view with tubus narrowly elongate, bent obtusely ventrad at base, abruptly bent again at middle of ventral margin so that apex directed slightly basad (Fig. 17.171b), in dorsal view median lobe as illustrated (Fig. 17.171c). FEMALE. Tergite VIII transverse, with apical margin broadly truncate, arcuately produced laterally (Fig. 17.171f); sternite VIII with apical margin obliquely produced laterally, apex slightly emarginate (Fig. 17.171g); spermatheca with capsule large, globose, stem sinuate, thin, duct with auxiliary posterior sclerites (Fig. 17.171h).

DISTRIBUTION Origin: Nearctic. **CANADA**: AB, BC, LB, MB, NB, NF, NS, ON, PE, QC, SK, YT. **USA**: AK, CA, MN.

COLLECTION AND HABITAT DATA
Habitat: in many hardwood and conifer forest types; unspecified hardwood forest types, rich Appalachian hardwood forest, red oak forest, old mixed forest with red oak, mixed forests, northern hardwood forest, coastal red spruce and birch forest, mature red spruce forest with eastern white cedar, old red spruce forest, spruce forest, white spruce forest, yellow birch/balsam fir forest, trembling aspen forest with a small amount of eastern balsam poplar, white birch forest, white spruce, and willow forests. In NB, adults common in decaying mushrooms in late summer; in fleshy fungi, in decaying gilled mushrooms, baited with pile of decaying mushrooms, fleshy fungi in various stages of decay, in gilled mushrooms, in rotten fungi. Elsewhere, in wet moss on forest floor, American marten (*Martes americana*) dung on moss, and in mushrooms. In AK, specimens were captured in shrub, alder, dwarf birch, grass, clubmos; *Betula* and *Populus tremuloides* forest; aspen litter; old growth, unthinned and thinned 2nd growth forest; and clear-cuts. **Collecting period**: VII–X. **Collecting method**: sifting mushrooms, baiting with mushrooms, pitfall traps, Luminoc pit-light traps, Malaise traps, flight intercept traps, carrion baited traps, Berlese funnels, and sifting forest litter.

Additional records. See Annex 1.

***Vesperis* species group** (Klimaszewski et al. 2003, 2018b)

Diagnosis Median lobe of aedeagus in lateral view with tubus moderately narrow and elongate, about as long as bulbus, ventral margin obliquely to perpendicularly curved ventrad near middle, apex acutely triangular; flagellum broad and moderately long, about as long as tubus, with basal part highly enlarged, funnel-shaped.

101. *Silusa densa* Fenyes (Fig. 17.172a–h)

Silusa densa Fenyes, 1909. Klimaszewski et al. 2003, 2011, 2018a, Webster et al. 2009.

Diagnosis Body length 2.2–2.5 mm; colour dark brown with posterior part of elytra, abdomen, and legs reddish brown, or entire body dark brown (Fig. 17.172a); forebody slightly glossy; antennomere IV slightly transverse, V–X distinctly so; pronotum moderately transverse, pronotal punctures dense and fine, usually separated by the diameter of a puncture or less, intervals with fine meshed microsculpture; elytra slightly transverse, at suture distinctly longer than pronotum; first four visible abdominal tergites with deep transverse impressions bearing moderately coarse punctation. MALE. Tergite eight truncate apically bearing several (8–9) short denticles (Fig. 17.172d); sternite VIII with apical margin obtusely triangularly produced with apex rounded (Fig. 17.172e); tubus of median lobe of aedeagus curved obliquely ventrad before middle of ventral margin, then sinuate to narrowly triangular apex (Fig. 17.172b), in dorsal view median lobe as illustrated (Fig. 17.172c). FEMALE. Tergite VIII transverse, broadly arcuate apically (Fig. 17.172f); sternite VIII with apical margin obtusely triangularly with apex rounded (Fig. 17.172g); spermatheca with capsule small, ovoid, stem short, L-shaped (Fig. 17.172h).

DISTRIBUTION Origin: Nearctic. **CANADA**: AB, LB, NB, NF, ON [**NPR**], QC [**NPR**]. **USA**: CA.

COLLECTION AND HABITAT DATA
Habitat: coniferous forests, mixed forests, clearcuts. Adults under bark of spruce logs infested with Scolytinae, under tight bark of a dead standing balsam fir, under bark of tamarack infested by the eastern bark beetle, and in lodgepole pine (*Pinus contorta* Douglas ex Louden logs) (AB). **Collecting**: III–VII. **Collecting method**: hand searching from subcortical habitats.

We report this species from ON (**NPR**) and QC (**NPR**) for the first time here, based on the following: Canada, Ontario, Wellington County, Guelph, Eramosa River Trail, 310m el., 43.539, –80.236, mixed forest, under bark of fallen trees, 14.IV.2017, M. Pentinsaari (CBG) 2 specimens; same data but 07.V.2017 (CBG) 2 specimens. Quebec, Dosquet Co., 23.IV.1984, C. Chantal (LFC) 1 specimen.

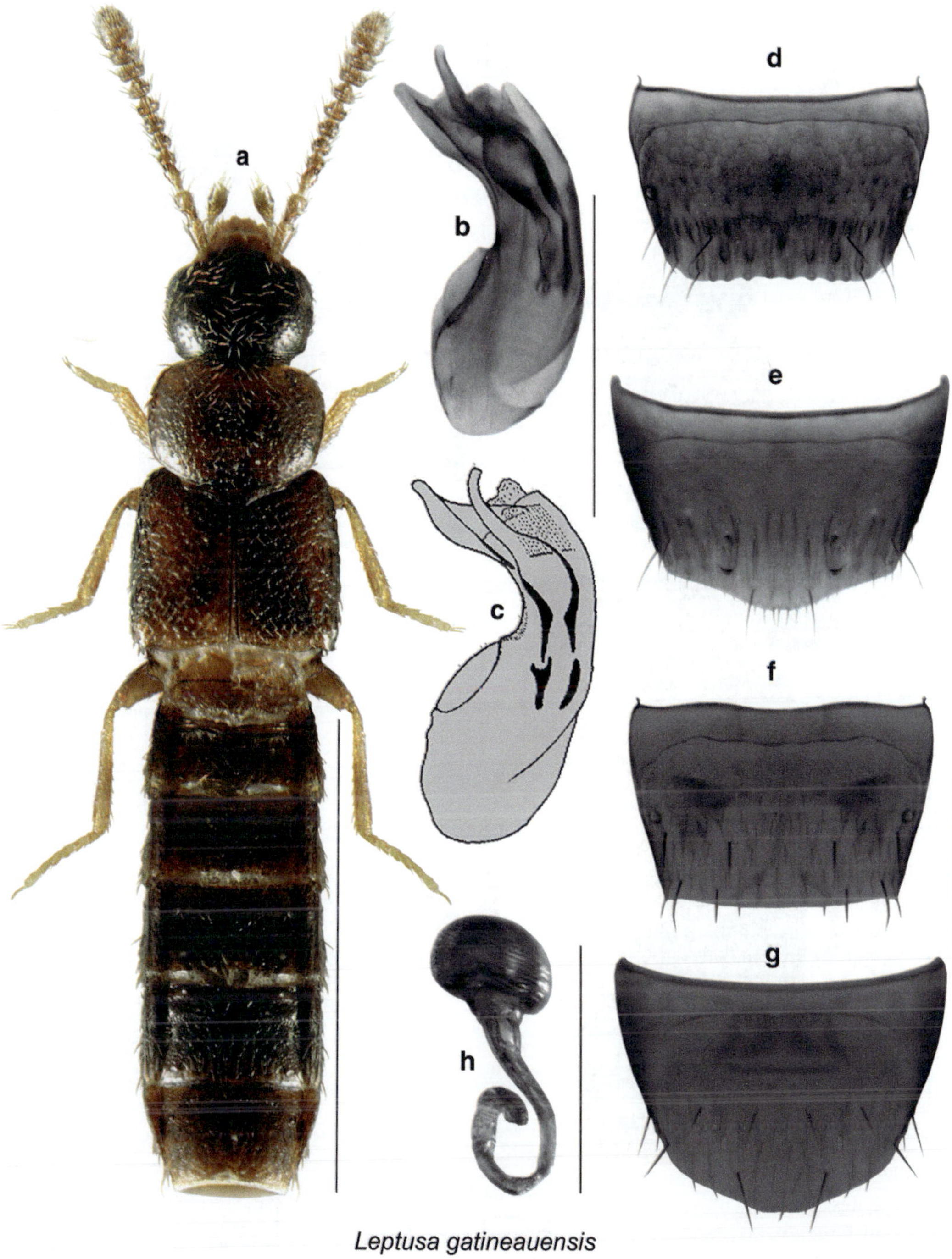

Fig. 17.166 (**a–h**). Images of *Leptusa gatineauensis* Klimaszewski and Pelletier: (**a**) habitus in dorsal view; (**b–c**) median lobe of aedeagus in lateral view; (**d**) male tergite VIII; (**e**) male sternite VIII; (**f**) female tergite VIII; (**g**) female sternite VIII; (**h**) spermatheca. (**c**) after Klimaszewski et al. (2004b); remaining images after Klimaszewski et al. (2018a). Scale bar for habitus = 1 mm; remaining scale bars = 0.2 mm

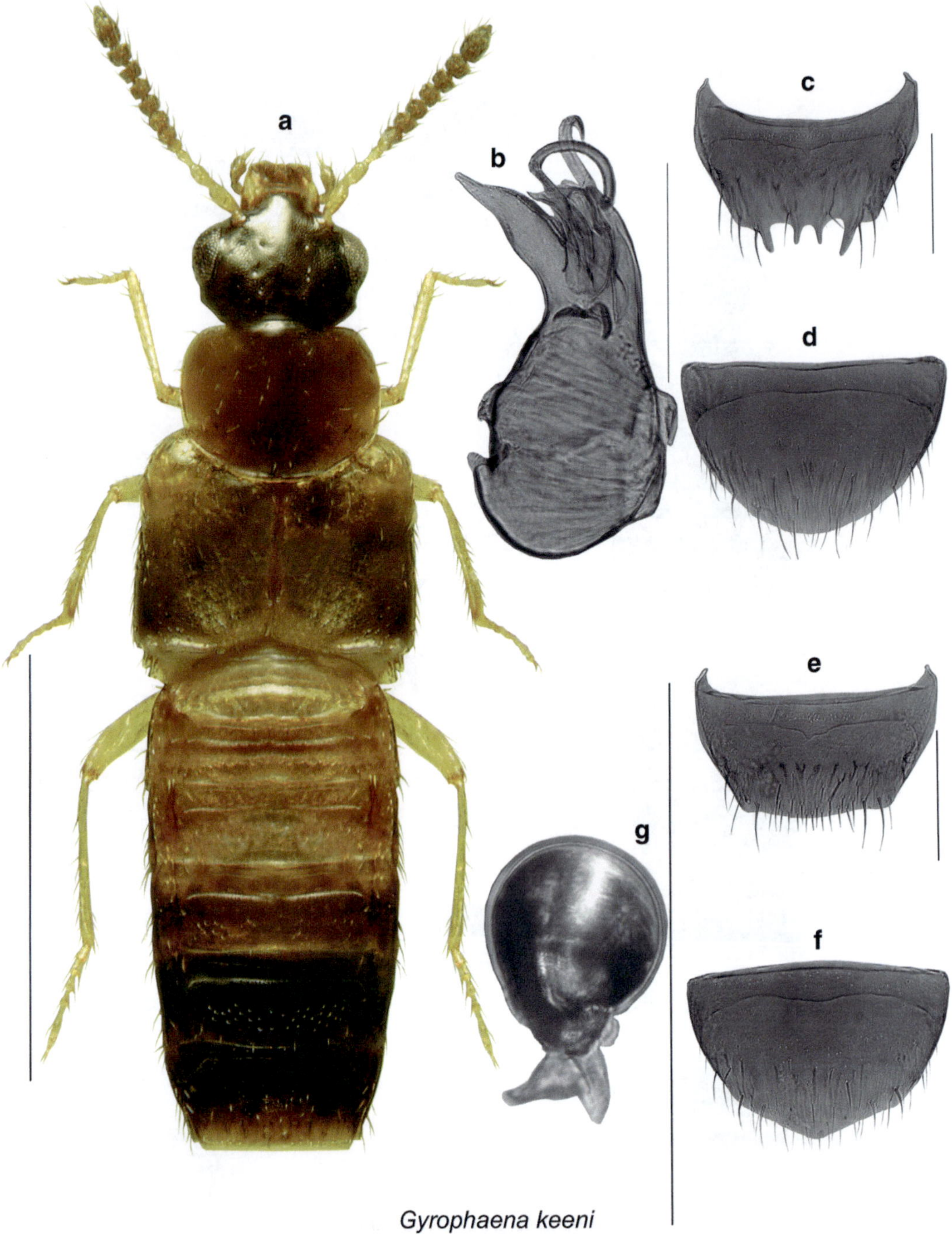

Fig. 17.167 (**a–g**). Images of *Gyrophaena keeni* Casey: (**a**) habitus in dorsal view; (**b**) median lobe of aedeagus in lateral view; (**c**) male tergite VIII; (**d**) male sternite VIII; (**e**) female tergite VIII; (**f**) female sternite VIII; (**g**) spermatheca. (**a–g**) after Klimaszewski et al. (2009). Scale bar for habitus = 1 mm; remaining scale bars = 0.2 mm

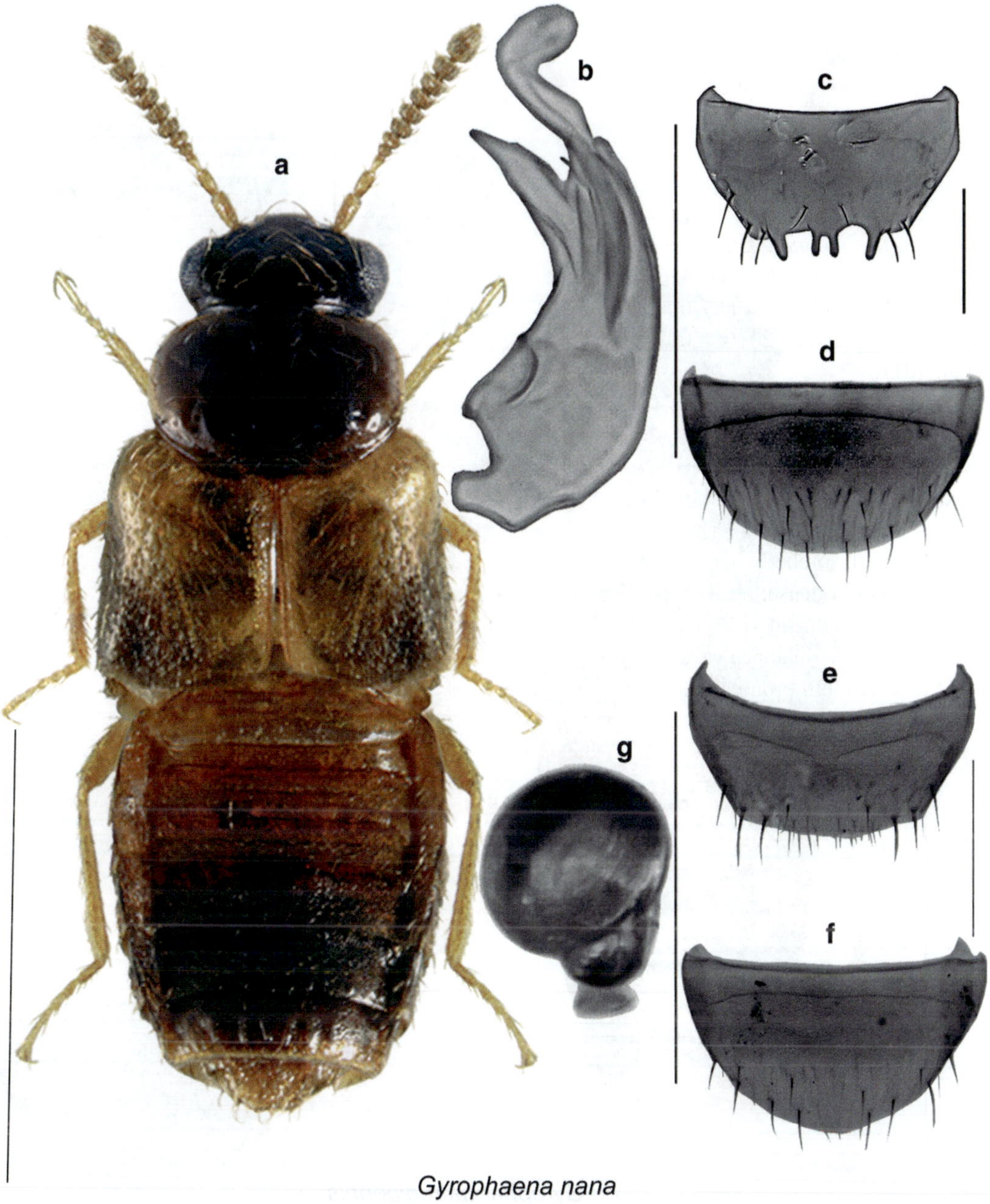

Fig. 17.168 (**a–g**). Images of *Gyrophaena nana* (Paykull): (**a**) habitus in dorsal view; (**b**) median lobe of aedeagus in lateral view; (**c**) male tergite VIII; (**d**) male sternite VIII; (**e**) female tergite VIII; (**f**) female sternite VIII; (**g**) spermatheca. (**a–g**) after Klimaszewski et al. (2009). Scale bar for habitus = 1 mm; remaining scale bars = 0.2 mm

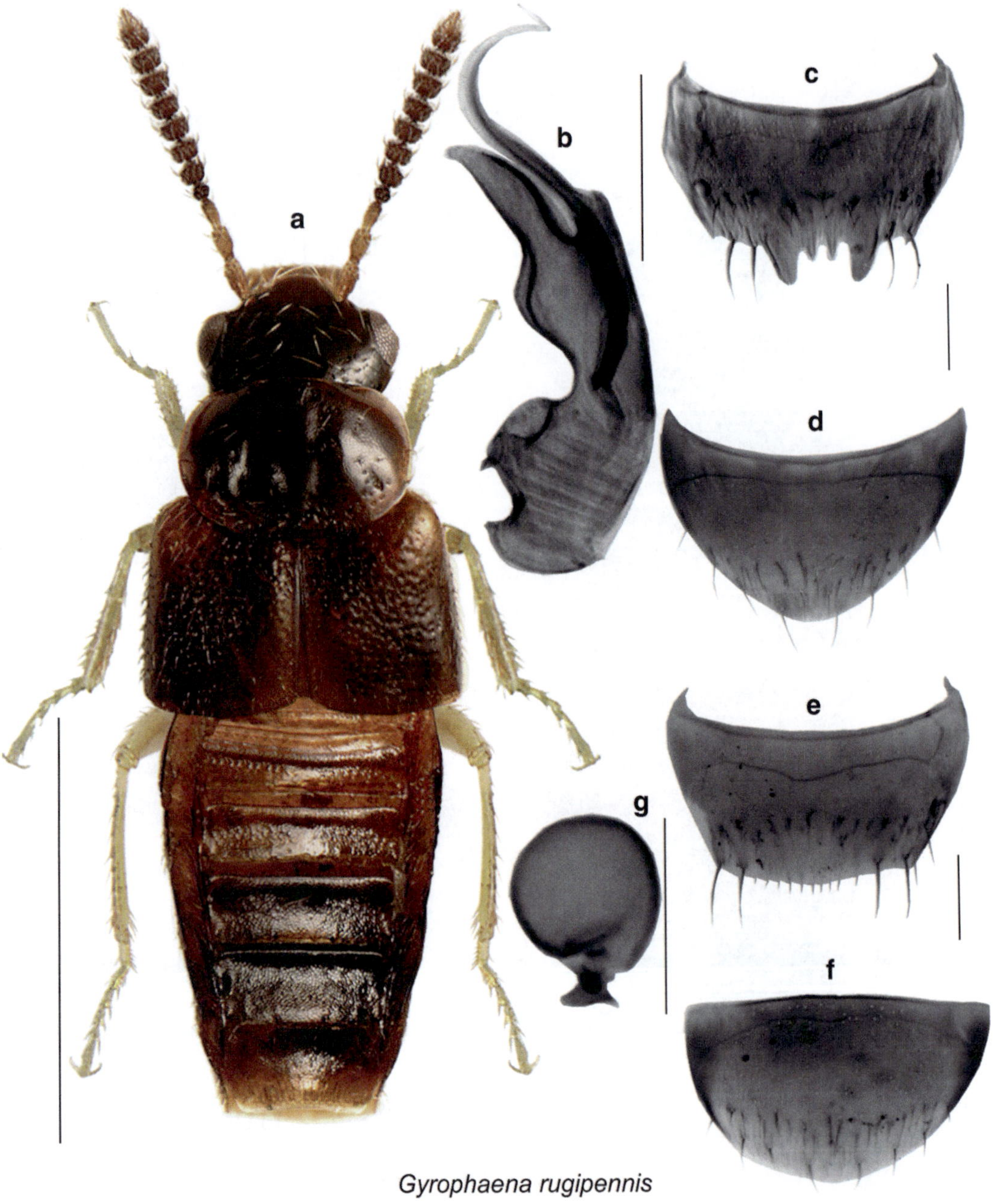

Fig. 17.169 (**a–g**). Images of *Gyrophaena rugipennis* Mulsant and Rey: (**a**) habitus in dorsal view; (**b**) median lobe of aedeagus in lateral view; (**c**) male tergite VIII; (**d**) male sternite VIII; (**e**) female tergite VIII; (**f**) female sternite VIII; (**g**) spermatheca. (**a–g**) after Klimaszewski et al. (2009). Scale bar for habitus = 1 mm; remaining scale bars = 0.2 mm

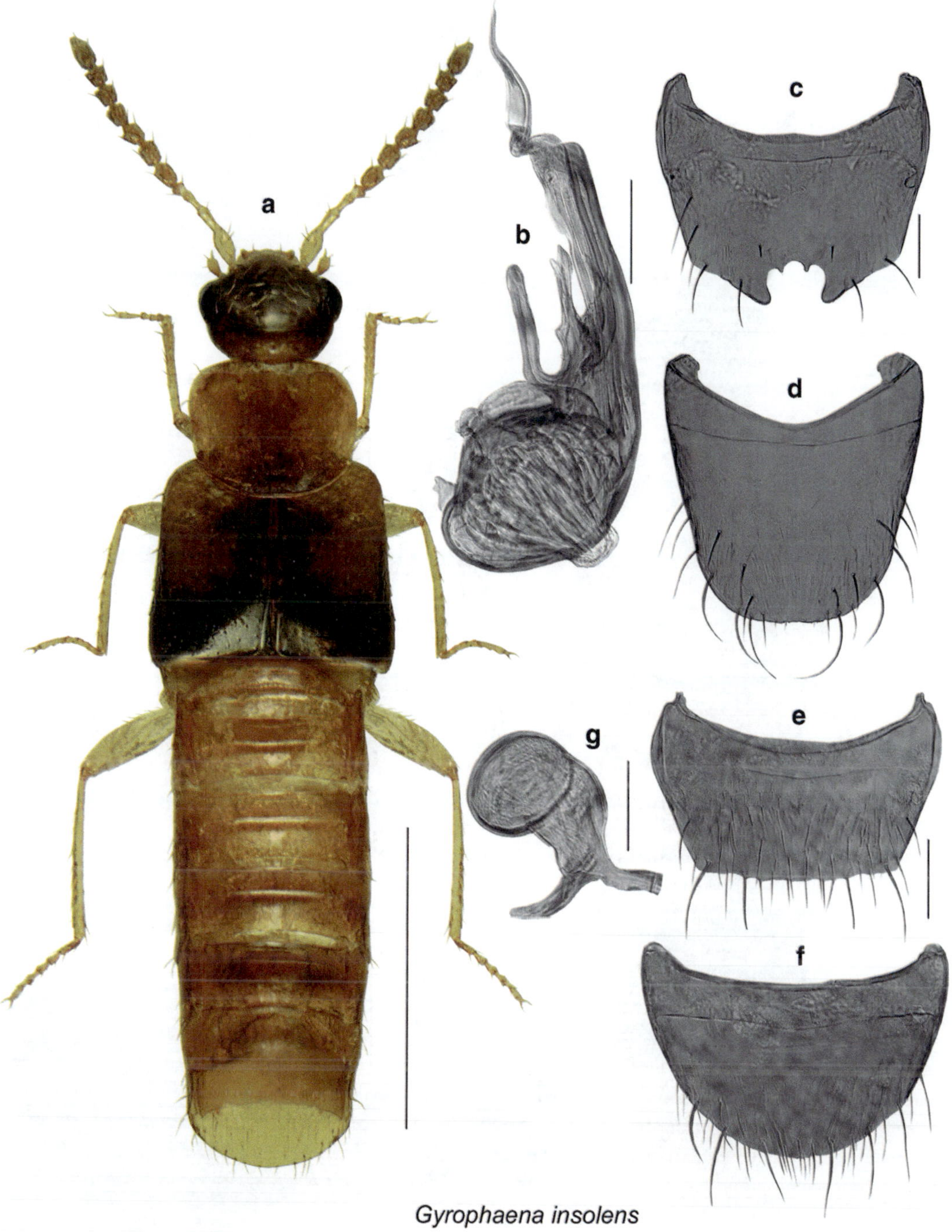

Fig. 17.170 (**a–g**). Images of *Gyrophaena insolens* Casey: (**a**) habitus in dorsal view; (**b**) median lobe of aedeagus in lateral view; (**c**) male tergite VIII; (**d**) male sternite VIII; (**e**) female tergite VIII; (**f**) female sternite VIII; (**g**) spermatheca. (**a–g**) after Klimaszewski et al. (2009). Scale bar for habitus = 1 mm; remaining scale bars = 0.2 mm

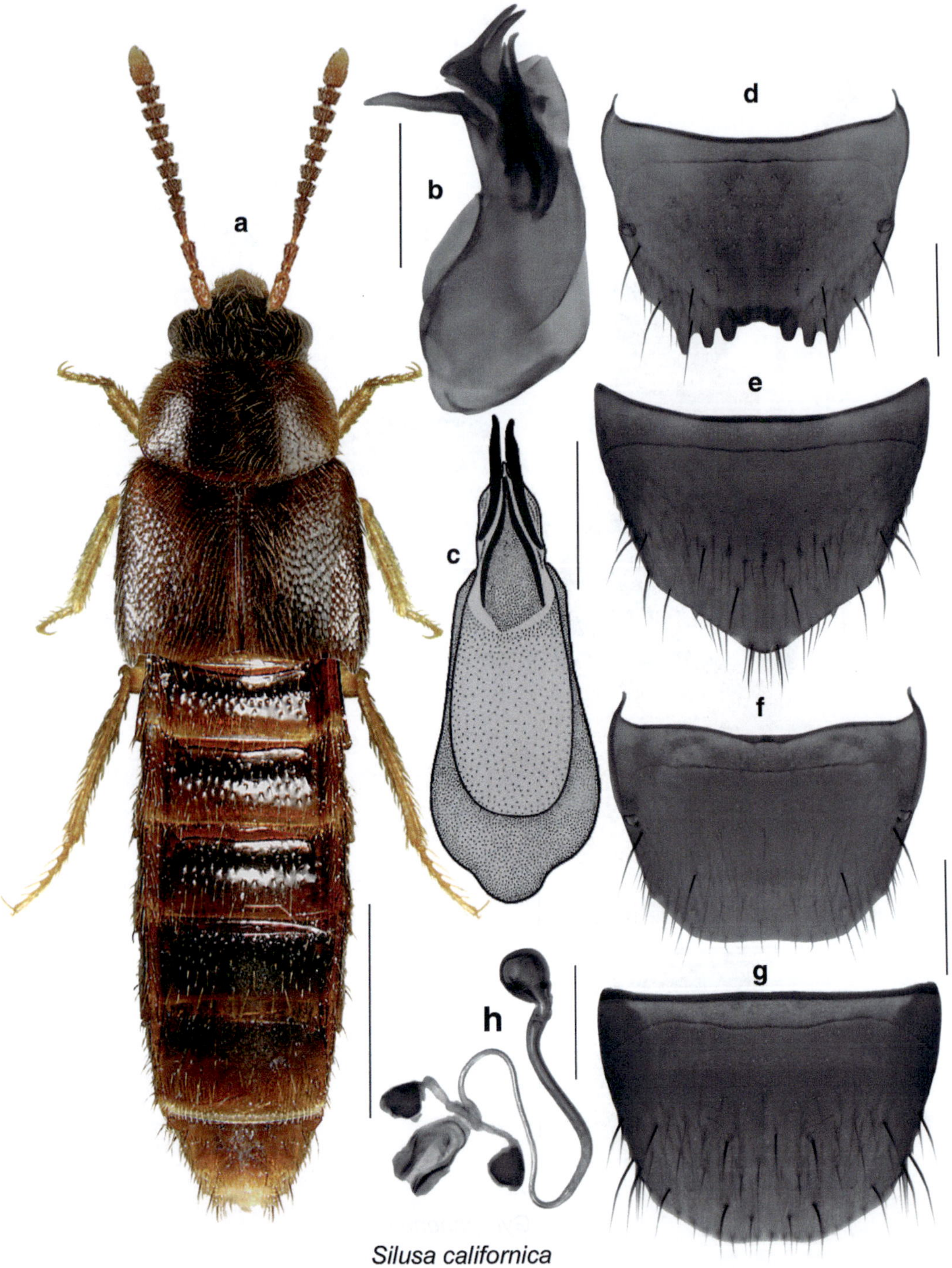

Fig. 17.171 (**a–h**). Images of *Silusa californica* (Bernhauer): (**a**) habitus in dorsal view; (**b**) median lobe of aedeagus in lateral view; (**c**) median lobe of aedeagus in dorsal view; (**d**) male tergite VIII; (**e**) male sternite VIII; (**f**) female tergite VIII; (**g**) female sternite VIII; (**h**) spermatheca. (**a–h**) after Klimaszewski et al. (2018a). Scale bar for habitus = 1 mm; remaining scale bars = 0.2 mm

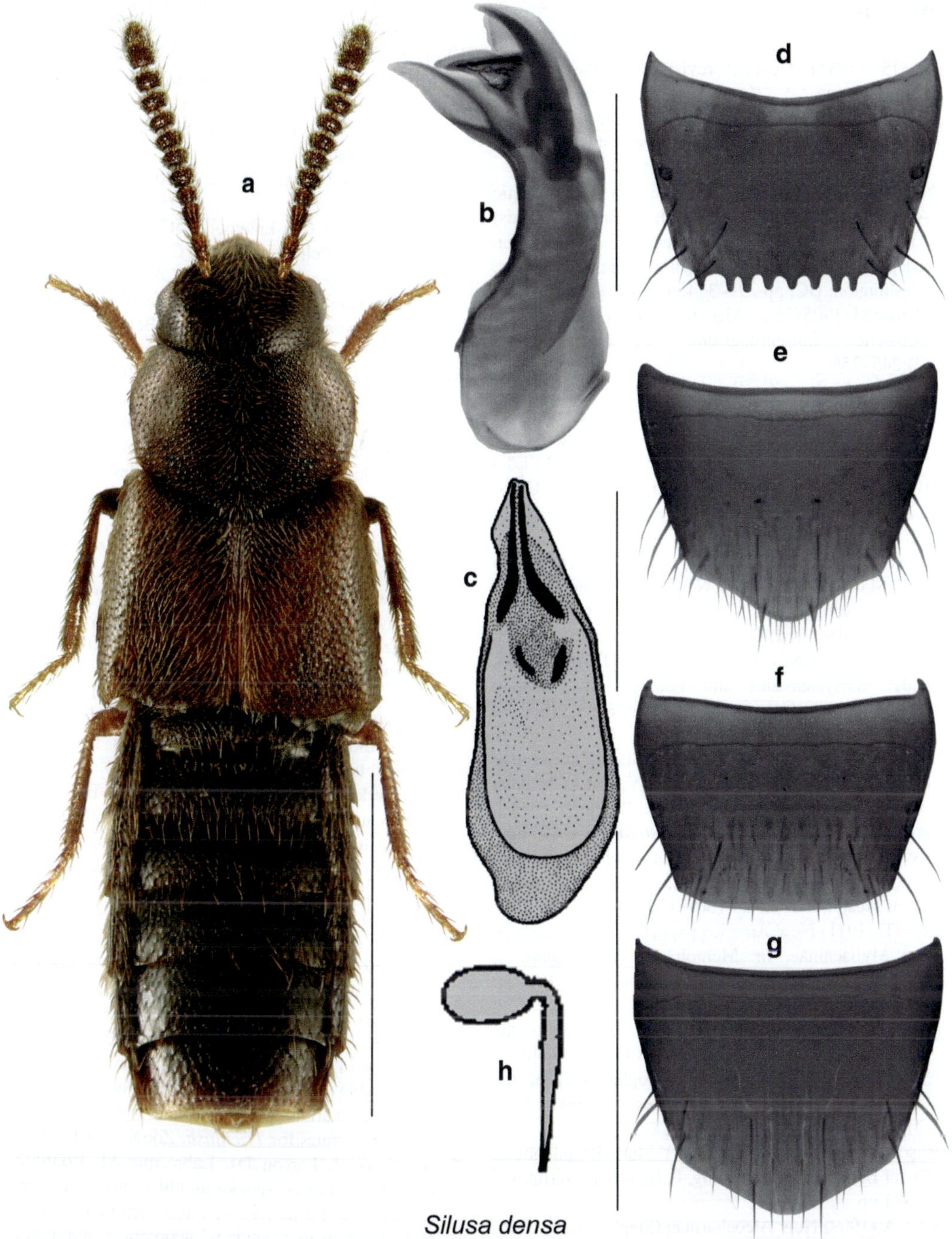

Fig. 17.172 (**a–h**). Images of *Silusa densa* Fenyes: (**a**) habitus in dorsal view; (**b**) median lobe of aedeagus in lateral view; (**c**) median lobe of aedeagus in dorsal view; (**d**) male tergite VIII; (**e**) male sternite VIII; (**f**) female tergite VIII; (**g**) female sternite VIII; (**h**) spermatheca. (**a–h**) after Klimaszewski et al. (2018a). Scale bar for habitus = 1 mm; remaining scale bars = 0.2 mm

References

Ashe JS (1984) Generic revision of the subtribe Gyrophaenina (Coleoptera: Staphylinidae: Aleocharinae) with a review of the described subgenera and major features of evolution. Quaestiones Entomologicae 20(3):129–349

Ashe JS (2001) Key to the tribes and genera of Nearctic Aleocharinae. 8. In: Arnett RH Jr, Thomas CM (eds) American Beetles. Archostemata, Myxophaga, Adephaga, Polyphaga: Staphyliniformia. CRC Press, Washington, DC, pp 299–374. 443 pp

Bernhauer M (1905) Neue Aleocharinen aus Nordamerika. Deutsche Entomologische Zeitschrift 1905 (2):245–256

Bishop DJ, Majka CG, Bondrup-Nielsen S, Peck SB (2009) Deadwood and saproxylic beetle diversity in naturally disturbed and managed spruce forests in Nova Scotia. *In*: Majka CG, Klimaszewski J (Eds) Biodiversity, Biosystematics, and Ecology of Canadian Coleoptera II. ZooKeys [Special Issue] 22:309–340

Brunke A, Klimaszewski J, Dorval J-A, Bourdon C, Paiero SM, Marshall SA (2012) New species and distributional records of Aleocharinae (Coleoptera, Staphylinidae) from Ontario, Canada, with a checklist of recorded species. In: Klimaszewski J, Anderson R (Eds) Biosystematics and Ecology of Canadian Staphylinidae (Coleoptera) II. ZooKeys [Special Issue] 186:119–206

Campbell JM, Davies A (1991) Family Staphylinidae rove beetles. Pp. 86–124. In: Bousquet, Y. (Ed.) Checklist of beetles of Canada and Alaska. Research Branch Agriculture Canada Publication 1861/E, vi + 430 pp

Casey TL (1906) Observations on the staphylinid groups Aleocharinae and Xantholinini, chiefly of America. Trans Acad Sci St Louis 16(6):125–434

Casey TL (1911) New American species of Aleocharinae and Myllaeninae. In: Memoirs on the Coleoptera. II. The New Era Printing Co., Lancaster, PA, pp 1–245. 259 pp

Enushchenko IV, Semenov VB (2016) A review of the genus *Gyrophaena* Mannerheim 1830 (Coleoptera: Staphylinidae: Aleocharinae: Gyrophaenina) of the Caucasus and adjacent territories. Zootaxa 4126 (3):301–337

Erichson WF (1837) Die Käfer der Mark Brandenburg. Erster Band, Erste Abtheilung. F. H. Morin, Berlin. viii + 384 pp

Fenyes A (1909) New Aleocharinae (Staphylinidae, Col.) of the U. S. Entomological News and Proceedings of the Entomological Section of the Academy of Natural Sciences of Philadelphia 20(10):418-425

Fenyes A (1918) Coleoptera. Family Staphylinidae. Subfamily Aleocharinae. In: Wytsman P (ed) Genera Insectorum, vol XXVII. Fascicule 173A. M. Nijhoff, Den Haag, pp 1–110. 453 pp. + 7 pls

Heer O (1839) (1838–1841) Fasciculus secundus. In: Fauna coleopterorum Helvetica. Pars I. Turici [Zurich]. Fuesslini et Sociorum, Orellii, pp 145–360. xii + 652 pp

Klimaszewski J, Winchester NN (2002) Aleocharine rove beetles (Coleoptera Staphylinidae) of the ancient Sitka spruce forest on Vancouver Island, British Columbia, Canada. Mémoires de la Société royale belge d'Entomologie 40:3–126

Klimaszewski J, Pohl G, Pelletier G (2003) Revision of the Nearctic *Silusa* (Coleoptera, Staphylinidae, Aleocharinae). Can Entomol 135(2):159–186

Klimaszewski J, Pelletier G, Majka C (2004) A revision of Canadian *Leptusa* Kraatz (Col., Staphylinidae, Aleocharinae): new species, new distribution records, key and taxonomic considerations. Belgian. J Entomol 6:3–42

Klimaszewski J, Sweeney J, Price J, Pelletier G (2005) Rove beetles (Coleoptera: Staphylinidae) in red spruce stands, eastern Canada: diversity, abundance, and descriptions of new species. Can Entomol 137(1):1–48

Klimaszewski J, Godin B, Pelletier G, Savard K (2008) Six new species and records of aleocharine beetles from the Yukon and Alaska (Coleoptera: Staphylinidae: Aleocharinae). Can Entomol 140 (3):265–291

Klimaszewski J, Webster RP, Savard K (2009) Review of the rove beetle species of the subtribe Gyrophaenina Kraatz (Coleoptera, Staphylinidae) from New Brunswick, Canada: new species, provincial records and bionomic information. *In*: Majka CG, Klimaszewski J (Eds.) Biodiversity, Biosystematics, and Ecology of Canadian Coleoptera II. ZooKeys [Special Issue] 22:81–170

Klimaszewski J, Langor D, Pelletier G, Bourdon C, Perdereau L (2011) Aleocharine beetles (Coleoptera, Staphylinidae) of the province of Newfoundland and Labrador, Canada. Pensoft Series Faunistica No. 98. Pensoft, Sophia-Moscow, 313 pp

Klimaszewski J, Godin B, Bourdon C (2012) Further contributions to the aleocharine fauna of the Yukon Territory, Canada (Coleoptera, Staphylinidae). In: Klimaszewski J, Anderson R (Eds.): Biosystematics and Ecology of Canadian Staphylinidae (Coleoptera) II. Zookeys [Special Issue] 186:207–237

Klimaszewski J, Godin B, Langor D, Bourdon C, Lee S-I, Horwood D (2015) New distribution records for Canadian Aleocharinae (Coleoptera, Staphylinidae), and new synonymies for *Trichiusa*. ZooKeys 498:51–91

Klimaszewski J, Larson DJ, Labrecque M, Bourdon C (2016) Twelve new species and fifty-three new provincial distribution records of Aleocharinae rove beetles of Saskatchewan, Canada (Coleoptera, Staphylinidae). ZooKeys 610:45–112

Klimaszewski J, Godin B, Davies A, Bourdon C, Horwood D (2018a) Forty new records of aleocharine beetles, and two new species in the genera *Acrotona* Thomson and *Atheta* Thomson, for the province of Manitoba, Canada (Coleoptera: Staphylinidae: Aleocharinae). Insecta Mundi 0641:1–33

Klimaszewski J, Webster RP, Langor DW, Brunke A, Davies A, Bourdon C, Labrecque M, Newton AF, Dorval J-A, Frank JH (2018b) Aleocharine rove beetles of eastern Canada (Coleoptera, Staphylinidae, Aleocharinae): a glimpse of megadiversity. Springer, Cham, xvi + 902 pp

Klimaszewski J, Hoebeke ER, Godin B, Davies A, Perry KI, Bourdon C, Winchester N (2020) Aleocharine rove beetles of British Columbia: a hotspot of Canadian biodiversity (Coleoptera, Staphylinidae). Springer, Cham, xvi + 631 pp

Kraatz G (1856) Staphylinii. Naturgeschichte der Insecten Deutschlands. Erste Abtheilung Coleoptera. Zweiter Band. Erste, zweite Lieferung. Bogen 1-24.Nicolai, Berlin, viii + 376 pp

Majka CG, Klimaszewski J (2008) New records of Canadian Aleocharinae (Coleoptera: Staphylinidae). *In*: Majka CG, Klimaszewski J (Eds.) Biodiversity, Biosystematics, and Ecology of Canadian Coleoptera. ZooKeys [Special Issue] 2:85–114

Majka CG, Klimaszewski J (2010) Contributions to the knowledge of the Aleocharinae (Coleoptera, Staphylinidae) in the Maritime Provinces of Canada. ZooKeys 46:15–39

Mannerheim CG (1830) Précis d'un nouvel arrangement de la famille des brachélytres de l'ordre des insectes coléoptères. St. Petersburg, 87 pp. [separate, re-issued in Mémoires Présentées à l'Académie Impériale des Sciences de St.-Pétersbourg 1: 415–501, in 1831]

McLean JA, Klimaszewski J, Li A, Savard K (2009) Survey of rove beetles (Coleoptera, Staphylinidae) from Stanley Park, Vancouver, British Columbia, Canada, with new records and description of a new species. Part 1. In: Majka CG, Klimaszewski J (eds.) Biodiversity, Biosystematics, and Ecology of Canadian Coleoptera II. ZooKeys [Special Issue] 22:5–17

Moore I, Legner EF (1975) A catalogue of the Staphylinidae of America north of Mexico (Coleoptera). University of California, Division of Agricultural Sciences, Special Publication No. 3015:1–514

Mulsant E, Rey C (1861) Description de quelques insectes nouveaux ou peu connus. Opuscules Entomologiques 12:93–119

Paykull G (1800) Fauna Suecica. Insecta. Tomus III. J.F. Edman, Upsala. 459 pp

Pace R (1989) Monografia del genere *Leptusa* Kraatz (Coleoptera, Staphylinidae). Memorie del Museo Civico di Storia Naturale di Verona (2) (A: Biologica) 8:1–307

Seevers CH (1951) A revision of the North American and European staphylinid beetles of the subtribe Gyrophaenae (Aleocharinae, Bolitocharini). Fieldiana Zoology 32(10):657–762

Seevers CH (1978) A generic and tribal revision of the North American Aleocharinae (Coleoptera: Staphylinidae) [with additions and annotations by Lee H. Herman]. Fieldiana Zoology 71. vi + 289 pp

Strand A (1938) *Gyrophaena orientalis* n. sp. (Col., Staph.). Notulae Entomologicae 18(2):39–40

Thomson CG (1859) Skandinaviens Coleoptera, synoptiskt bearbetade. I. Tom. Berlingska Boktryckeriet, Lund, [3] + 290 pp

Webster RP, Klimaszewski J, Pelletier G, Savard K (2009) New Staphylinidae (Coleoptera) records with new collection data from New Brunswick, Canada. I. Aleocharinae. In: Majka CG, Klimaszewski J (Eds.) Biodiversity, biosystematics, and ecology of Canadian Coleoptera II. ZooKeys [Special Issue] 22:171–248

Webster RP, Klimaszewski J, Sweeney JD, DeMerchant I (2012) New Staphylinidae (Coleoptera) records with new collection data from New Brunswick, and an addition to the fauna of Quebec, Canada: Aleocharinae. In: Klimaszewski J, Anderson R (Eds) Biosystematics and ecology of Canadian Staphylinidae (Coleoptera) II. ZooKeys [Special Issue] 186:83–118

18 Tribe Placusini Mulsant and Rey, 1871

Tribe Placusini Mulsant and Rey, 1871

(Figs. 18.173–18.176) (Ashe 2001, Klimaszewski et al. 2001, 2018, 2020)

Diagnosis Body broad to moderately broad, moderately convex or flattened; pronotum with anterolateral angles obtuse, pubescence short, dense, on disc evenly distributed, directed approximately straight posteriad; labrum rounded medially with small a-sensillum; mandible with dorsobasal "velvety patch" modified to form a transverse row of large teeth, and without rows of denticles in ventral molar region; ligula short, broad and apically rounded; labial palpi short, each with two palpomeres; mesocoxae narrowly separated or contiguous; tarsal formula 4-4-5.

Adults occur in subcortical habitats.

Genus *Placusa* Erichson, 1837

(Figs. 18.173–18.176) (Klimaszewski et al. 2001, 2018, 2020)

Diagnosis Body broad and strongly flattened; pronotum transverse, 1.5–1.7 times as broad as long, base arcuate, pubescence on disc usually short, dense, directed approximately straight posteriad, sometimes obliquely laterad, particularly along posterior part of median line and basally; terminal antennomere flattened, broad, moderately elongate; procoxal cavities open behind; process of mesoventrite short, projecting to a maximum of half the mesocoxal length, narrowly triangular and pointed; mesocoxae contiguous medially; hind tarsi with basal segment elongate, at least as long as two following tarsomeres combined; male tergite VIII with a variable number of small to large spines or teeth, always with one incurved spine near each lateral margin; male sternite VIII obtusely angulate apically; median lobe of aedeagus in lateral view with a narrowly elongate, curved process of crista apicalis; flagellum long, with multiple coils basally (*Tachyporoides* group), or short and inconspicuous (*Canadensis* group); spermatheca simple, capsule small and roughly elongate-spherical, stem sinuate, sometimes with irregular posterior coils.

Key to Species of *Placusa*

1. Body black; pronotum distinctly narrower than elytra at base (Fig. 18.174a); male tergite VIII with two long and narrow lateral teeth and three shorter median teeth (Fig. 18.174c); spermatheca as illustrated (Fig. 18.174g–i)... ***Placusa vaga*** **Casey**

– Body light to dark brown; pronotum about as wide as elytra at base or only slightly narrower; male tergite VIII and spermatheca not as above **2**

2. Pronotum broadest closer to middle, flat and dull, granulation fine and very close, elytra with

J. Klimaszewski et al., *A Faunal Review of Aleocharine Beetles in the Rapidly Changing Arctic and Subarctic Regions of North America (Coleoptera, Staphylinidae)*,
https://doi.org/10.1007/978-3-030-68191-3_18

sides subparallel (Fig. 18.173a); male and female tergite VIII with highly elongate lateral teeth and lacking median teeth (Fig. 18.173c); median lobe of aedeagus and spermatheca as illustrated (Fig. 18.174b,g,h)..........***Placusa tacomae*** **Casey**

– Pronotum broadest behind middle, convex and glossy, granulation fine and distant, elytra distinctly widening posteriad; male tergite VIII with short to moderately elongate sublateral teeth, and with smaller dents between; female tergite VIII entire and without projections; spermatheca as illustrated **3**

3. Elytra slightly wider at apex than base of abdomen (Fig. 18.176a); male tergite VIII with numerous, elongate teeth medially (Fig. 18.176c); spermatheca as illustrated (Fig. 18.176g–i) ....***Placusa tachyporoides*** **(Waltl)**

– Elytra distinctly wider at apex than base of abdomen (Fig. 18.175a); male tergite VIII with three very short median teeth (Fig. 18.175c); spermatheca not as above (Fig. 18.175g–i)............... ....***Placusa pseudosuecica*** **Klimaszewski**

Canadensis **species group** (Klimaszewski et al. 2001)

Diagnosis Apical margin of male tergite VIII with sublateral spines long, curved (Figs. 18.173–18.174); median lobe of aedeagus small, flagellum no longer than median lobe, bulbus with basal process wide, apico-ventral portion of bulbus broadly rounded.

102. *Placusa tacomae* Casey (Fig. 18.173a–h)

Placusa tacomae Casey, 1893. Criddle 1922, Hatch 1957, Webster et al. 2009, Klimaszewski et al. 2001, 2011, 2016, 2018, 2020, Majka et al. 2011.

Diagnosis Body flattened, length 2.1–3.0 mm; colour dark brown or dark brown with legs, bases of antennae, elytra, and often pronotum lighter, rust-brown (Fig. 18.173a); pubescence fine and short; forebody finely and closely granulate, punctation dense; head with genae about as long as eyes in dorsal view, arcuate and strongly narrowed basally in dorsal view, head 0.8 times as wide as maximum pronotal width; antennomeres VII–X usually about twice as wide as long; pronotum transverse, broadest near middle, maximum width 0.5–0.6 mm, length 0.3–0.4 mm, pronotal index = 1.5; pubescence on disc directed approximately straight posteriad on both sides of median line; elytra transverse, maximum width 0.5–0.7 mm, length 0.3–0.4 mm, elytral index = 1.7, pubescence on disc directed obliquely posteriad apically, approximately straight posteriad elsewhere. MALE. Tergite VIII with apical margin sinuately truncate between two long, curved, sharp sublateral teeth (Fig. 18.173c); sternite VIII with apical margin triangularly produced, apex rounded (Fig. 18.173d); median lobe of aedeagus in lateral view with bulbus moderately large, tubus arcuate, apex truncate, flagellum short, coils in bulbus not apparent, process of crista apicalis wide, broadly arcuate dorsally, slightly sinuate ventrally, and sharply pointed apically (Fig. 18.173b). FEMALE. Tergite VIII unique in having apical margin with two long blunt lateral spine-like teeth and a broad subrectangular projection medially (Fig. 18.173e); sternite VIII very obtusely produced apically (Fig. 18.173f); spermatheca with capsule small and roughly spherical, stem with middle and posterior portion sinuate and S-shaped (Fig. 18.173g,h).

DISTRIBUTION Origin: Nearctic. **CANADA**: AB, BC, NB, NF, NS, NT, ON, QC, SK, YT. **USA**: AK [**NSR**], AZ, CO, MA, ME, MN, NY, VA, WA, WI.

COLLECTION AND HABITAT DATA

Habitat: in various conifer and hardwood forests with conifers; white pine, red pine, jack pine, white spruce, sugar maple stands, rich Appalachian hardwood stand with some conifers, and mature red spruce and red maple forests. Adults under bark of conifers; under bark of fire-killed red pine (NL), under bark of red spruce infested with *D. rufipennis* and *P. rufipennis* (NB), and in bark of log pile of *Pinus contorta*; and in galleries of *Dendroctonus ponderosae* (Alberta) and *Ips pini* (BC), presumably from conifers. In BC, one specimen was collected in log pile, in *Pinus contorta* Dougl. ex Loud. bark (CNC). In AK, adults were captured in clear-cut; *Betula, Salix,* white spruce forest; cut spruce forest; boggy black spruce forest; birch and white spruce with wild rose; deciduous forest, birch, rose and willow; and *Picea glauca* disturbed forest. **Collecting period**: V–IX. **Collecting method**: alpha-pinene baited Lindgren funnel traps, hand searching and collecting from under bark.

We report this species from AK (**NSR**) for the first time here, based on the following: USA, Alaska, 35.5 Mi Dalton Hwy, 271m el., 65.74425, −149.34564, ±4m *Betula, Salix,* white spruce, flight intercept trap, 31.V-5. VI.2008, D.S. Sikes (UAM) 1 specimen; Fairbanks, 930 Fitz CT, 280-293m el., 64.90099, −147.52808, ±15m yard; cut spruce, hand, 1-2.VI.2013, D.S. Sikes (UAM) 6 specimens; Fairbanks, Creamer's Field, Reindeer Lake, 146m el., 64.88118, −147.77643, ±10m boggy black spruce, hanging Malaise, 5.5m, 10-24.VI.2015, J.C. Hagelin (UAM) 1 specimen; Fairbanks, Goldstream Creek, 184m el., 64.93695, −147.75876, ±10m birch and white spruce, wild rose, ~20m from creek, hanging Malaise, 5.4m, 10-24.VI.2015, J.C. Hagelin (UAM) 1 specimen; Fairbanks, Kiak Radio 2, 156m el., 64.87154, −147.68457, ±10m deciduous forest, birch, rose and willow, hanging Malaise, 6.15m, 10-24.VI.2015, J.C. Hagelin (UAM) 1 specimen; Prince of Wales Island, Staney Ck. 1A, 89m el., 55.87134, −133.06755, ±24m clear-cut, Lindgren 2, 28.V.-11.VI.2012, J. Stockbridge, S. Meierotto, I. MacDougall (UAM) 1 specimen; Prince of Wales Island, Staney Ck. 1B, 77m el., 55.872, −133.06523, ±26m clear-cut, 28.V-11.VI.2012, Lindgren 1, J. Stockbridge, S. Meierotto, I. MacDougall (UAM) 1 specimen; Sterling, Corey St, 90m el., 60.53953, −150.83232, ±150m *P. glauca,* disturbed, Lindgren, 12.VI.2005, D.M. Collet (UAM) 1 specimen; Sterling, Corey St, 90m el., 60.53953, −150.83232, ±150m *Picea glauca,* sweep, 26.IV.2005, D.M. Collet (UAM) 1 specimen.

COMMENTS This species was listed from CA by Newton (2019) but we are not aware of published voucher data for this record.

103. *Placusa vaga* Casey (Fig. 18.174a–i)

Placusa vaga Casey, 1911. Gusarov 2003, Majka and Klimaszewski 2008, Webster et al. 2009, Klimaszewski et al. 2001, 2008, 2011, 2018, 2020, Brunke et al. 2012.

Diagnosis Body moderately flattened, length 1.8–2.6 mm; colour uniformly dark brown to black, tarsi and palps reddish-brown (Fig. 18.174a); pubescence fine and short; forebody finely and closely granulate, punctation dense; head with genae about as long as eyes in dorsal view, subparallel behind eyes, then arcuately narrowed basad, 0.7–0.8 times as wide as maximum pronotal width; antennomeres VII–X usually less than twice as wide as long; pronotum moderately flattened, distinctly narrower than elytra at base, transverse, maximum width 0.4 mm, length 0.2–0.3 mm, pronotal index = 1.6; pubescence on disc directed slightly obliquely posteriad on both sides of median line and approximately laterad near base; elytra transverse, maximum width 0.5–0.6 mm, length 0.4 mm, elytral index = 1.3, pubescence on disc directed obliquely posteriad apically, approximately straight posteriad elsewhere. MALE. Tergite VIII with two long, sharp, incurved lateral teeth, and three shorter median teeth (rarely two), blunt and subequal in length (Fig. 18.174c); sternite VIII obtusely triangularly produced apically, apex rounded (Fig. 18.174d); median lobe of aedeagus in lateral view with bulbus moderately large, tubus curved obtusely ventrad basally, ventral margin nearly straight to blunt apex, flagellum short, coils in bulbus not apparent, process of crista apicalis moderately wide, broadly arcuate dorsally, sharply pointed apically (Fig. 18.174b). FEMALE. Tergite VIII broadly arcuate apically (Fig. 18.174e); sternite VIII slightly produced apically, apex rounded (Fig. 18.174f); spermatheca with capsule small, roughly elongate-spherical, stem with median portion narrowly

U-shaped, posterior portion curled (Fig. 18.174g–i).

DISTRIBUTION Origin: Nearctic. **CANADA**: AB, BC, NB, NS, NT, ON, QC, SK, YT. **USA**: AK [**NSR**], CA.

COLLECTION AND HABITAT DATA
Habitat: mixed forests, old growth *Abies balsamea*, *Picea glauca*, *Populus tremuloides* with *Picea glauca* stand, oak savannah, and *Populus* forest. In scolytid galleries, under bark of *Populus* species (common), at a sap flow on a recently cut *Populus* sp., on sappy *Populus* wood, in decaying (moldy) corncobs and cornhusks, and in drift material on a river margin. In AK, specimens were found in forest edge with *Populus tremuloides, Salix* spp.; cut spruce; mixed birch, white spruce, grassy, disturbed wood cutting area; and ±61m summit with *Populus tremuloides*. **Collecting period**: V–IX. **Collecting method**: Multi-Pher 7 traps, pitfall traps; Lindgren funnel traps baited with alpha-pinene and 95% ethanol, and with 70% ethanol as preservative; Malaise traps, hand searching subcortical habitats, sifting litter and decaying corncobs and cornhusks.

We report this species from AK (**NSR**) for the first time here, based on the following: USA, Alaska, Chena Hot Springs Rd, 200m el., 64.89227, –147.50857, ±30m forest edge *Populus tremuloides, Salix* spp., pitfall17-24. VII.2007, B. Mortensen (UAM) 1 specimen; Fairbanks, 930 Fitz CT, 280-293m el., 64.90099, –147.52808, ±15m yard; cut spruce, hand1-2.VI.2013, D.S. Sikes (UAM) 4 specimens; Fairbanks, Bonanza Creek, 278m el., 64.72025, –148.38327, ±10m mixed birch, white spruce, grassy, disturbed wood cutting area, hanging Malaise, 4.2m height, bottom catch, 17. VI-2.VII.2013, S. Busby, M. Roed (UAM) 1 specimen; Fairbanks, Ester Dome, 719m el., 64.8768, –148.06427, ±61m summit, *Populus tremuloides*, pitfall, 17-24.VII.2007, B. Mortensen (UAM) 41 specimens.

***Tachyporoides* species group** (Klimaszewski et al. 2001)

Diagnosis Apical margin of male tergite VIII with sublateral teeth usually very short; median lobe of aedeagus large, flagellum long, coiled in bulbus, at least twice as long as median lobe when everted, aedeagal process thin (except in *P. despecta*), and apico-ventral portion of bulbus keel-shaped.

104. *Placusa pseudosuecica* Klimaszewski (Fig. 18.175a–h)

Placusa pseudosuecica Klimaszewski, in Klimaszewski et al. 2001. Klimaszewski et al. 2015, 2016, 2018, 2020.

Diagnosis Body moderately flatened, length 2.0–2.7 mm; colour uniformly light to dark brown or dark brown with legs, basal antennal segments, elytra, base and tip of abdomen light brown to rust-brown (Fig. 18.175a); pubescence moderately long; forebody finely and closely granulate, punctation moderately dense; head with genae about as long as eyes in dorsal view, broadly arcuate and gradually narrowed basally (not dilated), head 0.6 times as wide as maximum pronotal width; antennomeres VII–X usually less than twice as wide as long (average 1.8 times as wide as long); pronotum broadest near middle, moderately flattened, transverse, maximum width 0.5–0.6 mm, length 0.3–0.4 mm, pronotal index = 1.5; pubescence on disc directed slightly obliquely posteriad on both sides of median line; elytra transverse, wider apically than at base, maximum width 0.7–0.8 mm, length 0.3–0.4 mm, elytral index = 2.1; pubescence on disc directed obliquely apically, approximately straight posteriad elsewhere. MALE. Apical margin of tergite VIII truncate, with two short, acutely triangular sublateral teeth, and three minute median teeth (Fig. 18.175c); sternite VIII obtusely angulately produced apically, apex rounded (Fig. 18.175d); median lobe of aedeagus in lateral view with enlarged bulbus, tubus arcuate, apex blunt, flagellum long, coiled 1.5–5 times in bulbus, process of crista apicalis narrow, broadly arcuately tapering to sharply pointed apex (Fig. 18.175b). FEMALE. Tergite VIII arcuate apically (Fig. 18.175e); sternite VIII obtusely triangularly produced apically, apex rounded (Fig. 18.175f); spermatheca with capsule large, spherical, stem (duct) narrowly U-shaped, slightly sinuate posteriorly (Fig. 18.175g,h).

DISTRIBUTION Origin: Nearctic. **CANADA**: AB, BC, ON, QC, SK. **USA**: AK [**NSR**].

COLLECTION AND HABITAT DATA

Habitat: balsam fir stands, black spruce stands, white spruce stands, a spruce/fir stand, trembling aspen dominated stands, and a sugar maple stand; in BC—western hemlock, Sitka spruce (*Picea sitchensis* (Bongard)), and Pacific silver fir (*Abies amabilis* (Douglas ex Louden)) forest. In general, the species seems to prefer mature conifer forests in eastern as well as western Canada. In AK, adults were found in cut spruce and on dried carcass. **Collecting period**: VI–X. **Collecting method**: Malaise traps, Lindgren funnel traps, and ethanol-baited Lindgren funnel traps.

We report this species from AK (**NSR**) for the first time here, based on the following: USA, Alaska, Fairbanks, 930 Fitz CT, 280-293m el., 64.90099, −147.52808, ±15m yard; cut spruce, hand, 1-2.VI.2013, D.S. Sikes (UAM) 11 specimens; Fairbanks, 930 Fitz Ct, 280m el., 64.9013, −147.52861, ±30m dried carcass, 5. V.2009, D.S. Sikes (UAM) 1 specimen.

105. ***Placusa tachyporoides*** **(Waltl)** ‡
(Fig. 18.176a–i)

Aleochara tachyporoides Waltl, 1838. Horion 1967, Moore and Legner 1975, Klimaszewski et al. 2001, 2007, 2015, 2016, 2018, Majka and Klimaszewski 2008.

Diagnosis Body moderately convex, length 1.8–3.0 mm; colour uniformly dark brown or dark brown with legs, basal antennal segments and elytra light brown to yellowish/reddish brown (Fig. 18.176a); pubescence fine and short; forebody finely and closely granulate, punctation moderately dense; head with genae shorter than eyes in dorsal view, broadly arcuate and gradually narrowed basally (not dilated); head 0.6 times as wide as maximum pronotal width; antennomeres VII–X usually less than twice as wide as long; pronotum moderately flattened, transverse, maximum width 0.5 mm, length 0.3 mm, pronotal index = 1.6, pubescence on disc directed slightly obliquely posteriad on both sides of median line; elytra transverse, maximum width 0.5–0.7 mm, length 0.3–0.4 mm, elytral index = 1.7, pubescence on disc directed obliquely posteriad apically, approximately straight posteriad elsewhere. MALE. Apical margin of tergite VIII truncate, with two larger lateral incurved teeth and several (3–8, usually 5, in combinations: 1+1+2, 2+1+2, 1+1+1, 2+1+3, 2 +2+3, 3+1+3, 3+2+3) medial teeth varying in length from short and sharply pointed to long, rounded apically and approaching the length of the lateral teeth (Fig. 18.176c); sternite VIII obliquely triangularly produced apically, apex rounded (Fig. 18.176d); median lobe of aedeagus with enlarged bulbus, tubus arcuate, ventral margin slightly sinuate, apex blunt in lateral view, flagellum long, coiled 1.5–3 times in bulbus, process of crista apicalis narrow, slightly sinuate and sharply pointed apically (Fig. 18.176b). FEMALE. Tergite VIII broadly arcuate apically (Fig. 18.176e); sternite VIII broadly rounded apically (Fig. 18.176f); spermatheca with capsule asymmetrical, stem U-shaped, doubled back to irregularly coiled base (Fig. 18.176g–i).

DISTRIBUTION Origin: Palearctic, adventive in North America or Holarctic. **CANADA**: AB, BC, LB [**NPR**], NB, NS, ON, QC, SK. **USA**: AK [**NSR**], CA, MA, MN.

COLLECTION AND HABITAT DATA

Habitat: sugar maple stands, balsam fir, and black spruce stands, mixed trembling aspen, white spruce, and white pine stands, and a yellow birch forest in QC. In NB, collected from Lindgren funnel traps in old-growth northern hardwood forest, a red oak forest, an old jack pine forest, and in a rich Appalachian hardwood forest. In Europe, in *Acer*, *Alnus*, *Betula*, *Carpinus*, *Fagus*, *Populus*, *Quercus*, *Ulmus*, *Sorbus*, and coniferous forests. Two individuals from NB sifted from pile of fairly fresh corncobs and cornhusks. In AK, specimens were found in *Picea glauca* forest and on dried carcass. **Collecting period**: V–IX. **Collecting method**: Lindgren funnel traps baited with alpha-pinene and 95% ethanol or with 95% ethanol only, Luminoc pit-light traps, some in unbaited Multi-Pher pitfall traps, and flight interception traps, yellow pan traps, and pitfall traps. There are some indications that this species, like *P. incompleta*, is attracted to ethanol-baited traps and to light.

We report this species from LB (**NPR**) and AK (**NSR**) for the first time here, based on the following: Canada, Labrador, Goose Bay, 53.32036, –60.29691, Rep. 2, mesic, yellow pan, 21-24.VI.2010, NBP Field Party G2MY431 (LFC) 1 female. USA, Alaska, Fairbanks, 930 Fitz Ct, 280m el., 64.9013, –147.52861, ±30m dried carcass, 5.V.2009, D.S. Sikes (UAM) 1 specimen; Sterling, Corey St, 90m el., 60.53953, –150.83232, ±150m *P. glauca* forest, Lindgren, 12.VI.2005, D.M. Collet (UAM) 1 specimen.

COMMENTS DNA barcoded Palaearctic and Nearctic specimens of *P. tachyporoides* form separate BIN clusters that are ~4% divergent, casting serious doubt on the status of this species as introduced in North America. This species was confirmed from MN by Klimaszewski et al. (2001) [specimen record in their Appendix 1] but this was overlooked in recent publications (e.g., Klimaszewski et al. 2018). We were unable to verify voucher data for the records (CO, ME, NC, NM) in Newton (2019).

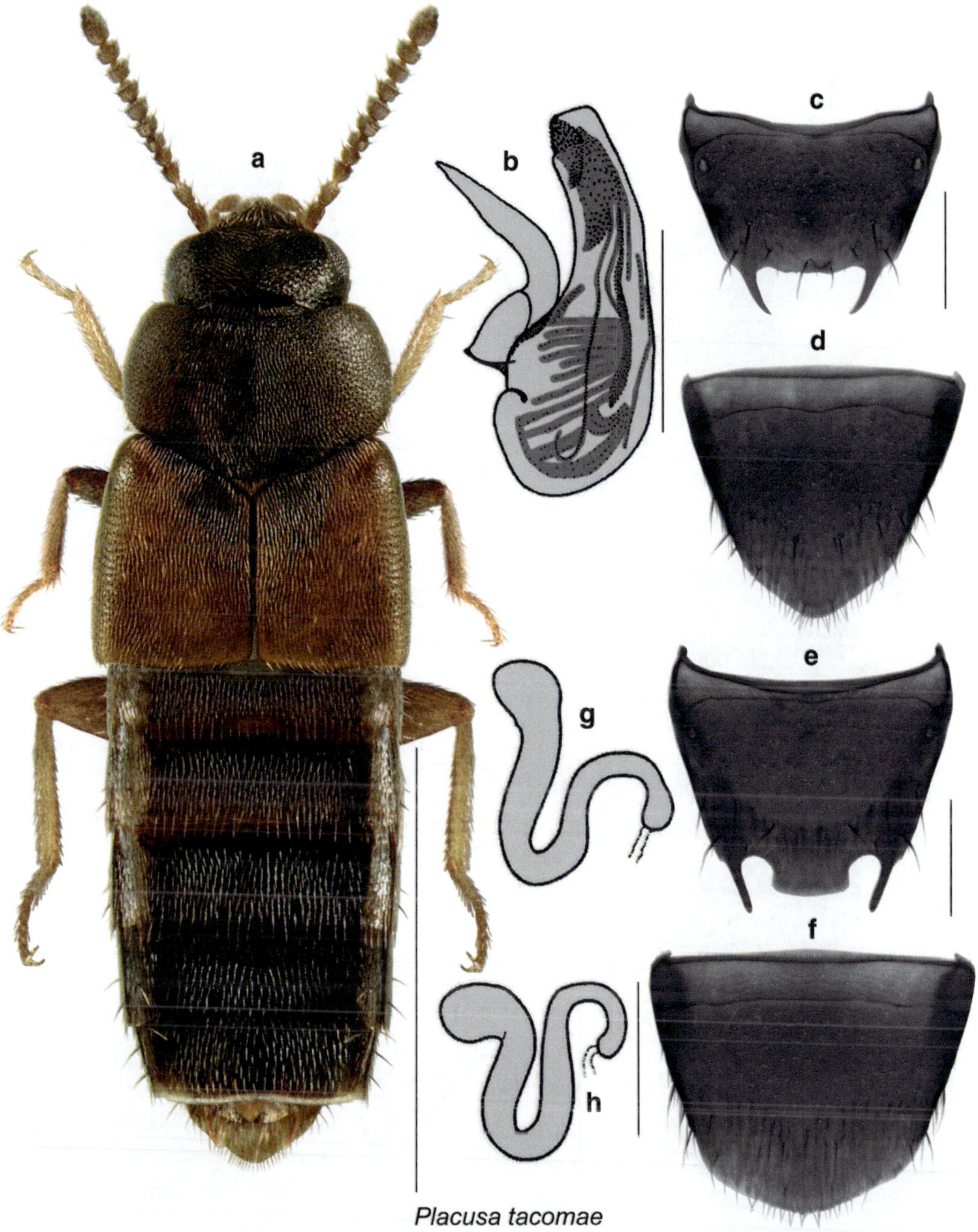

Fig. 18.173 (**a–h**). Images of *Placusa tacomae* Casey: (**a**) habitus in dorsal view; (**b**) median lobe of aedeagus in lateral view; (**c**) male tergite VIII; (**d**) male sternite VIII; (**e**) female tergite VIII; (**f**) female sternite VIII; (**g– h**) spermatheca. (**b, g, h**) after Klimaszewski et al. (2001); remaining images after Klimaszewski et al. (2020). Scale bar for habitus = 1 mm; remaining scale bars = 0.2 mm

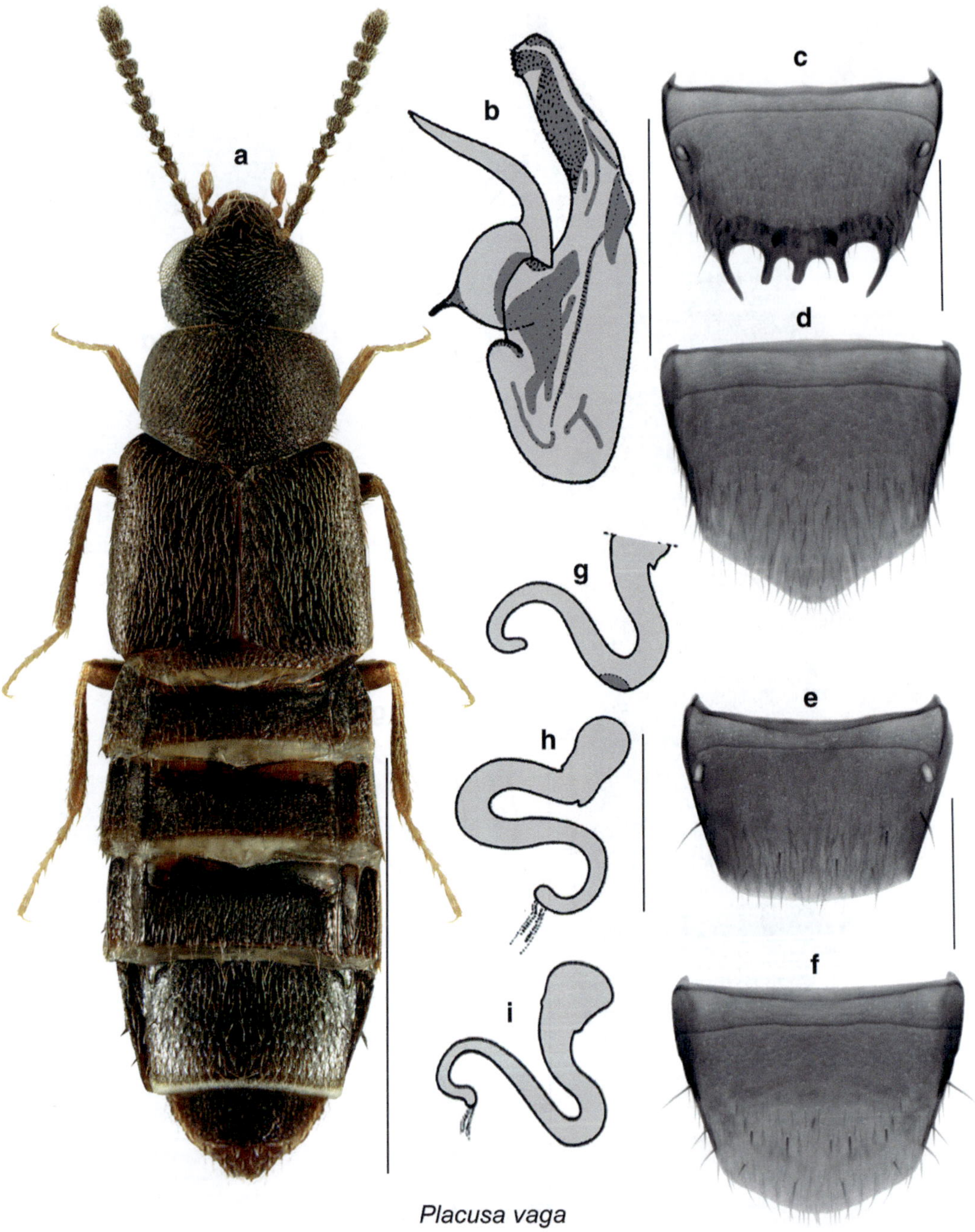

Fig. 18.174 (**a–i**). Images of *Placusa vaga* Casey: (**a**) habitus in dorsal view; (**b**) median lobe of aedeagus in lateral view; (**c**) male tergite VIII; (**d**) male sternite VIII; (**e**) female tergite VIII; (**f**) female sternite VIII; (**g–i**) spermatheca. (**b, g, h, i**) after Klimaszewski et al. (2001); remaining images after Klimaszewski et al. (2020). Scale bar for habitus = 1 mm; remaining scale bars = 0.2 mm

Fig. 18.175 (**a–h**). Images of *Placusa pseudosuecica* Klimaszewski: (**a**) habitus in dorsal view; (**b**) median lobe of aedeagus in lateral view; (**c**) male tergite VIII; (**d**) male sternite VIII; (**e**) female tergite VIII; (**f**) female sternite VIII; (**g–h**) spermatheca. (**b**, **g–h**) after Klimaszewski et al. (2001); remaining images after Klimaszewski et al. (2020). Scale bar for habitus = 1 mm; remaining scale bars = 0.2 mm

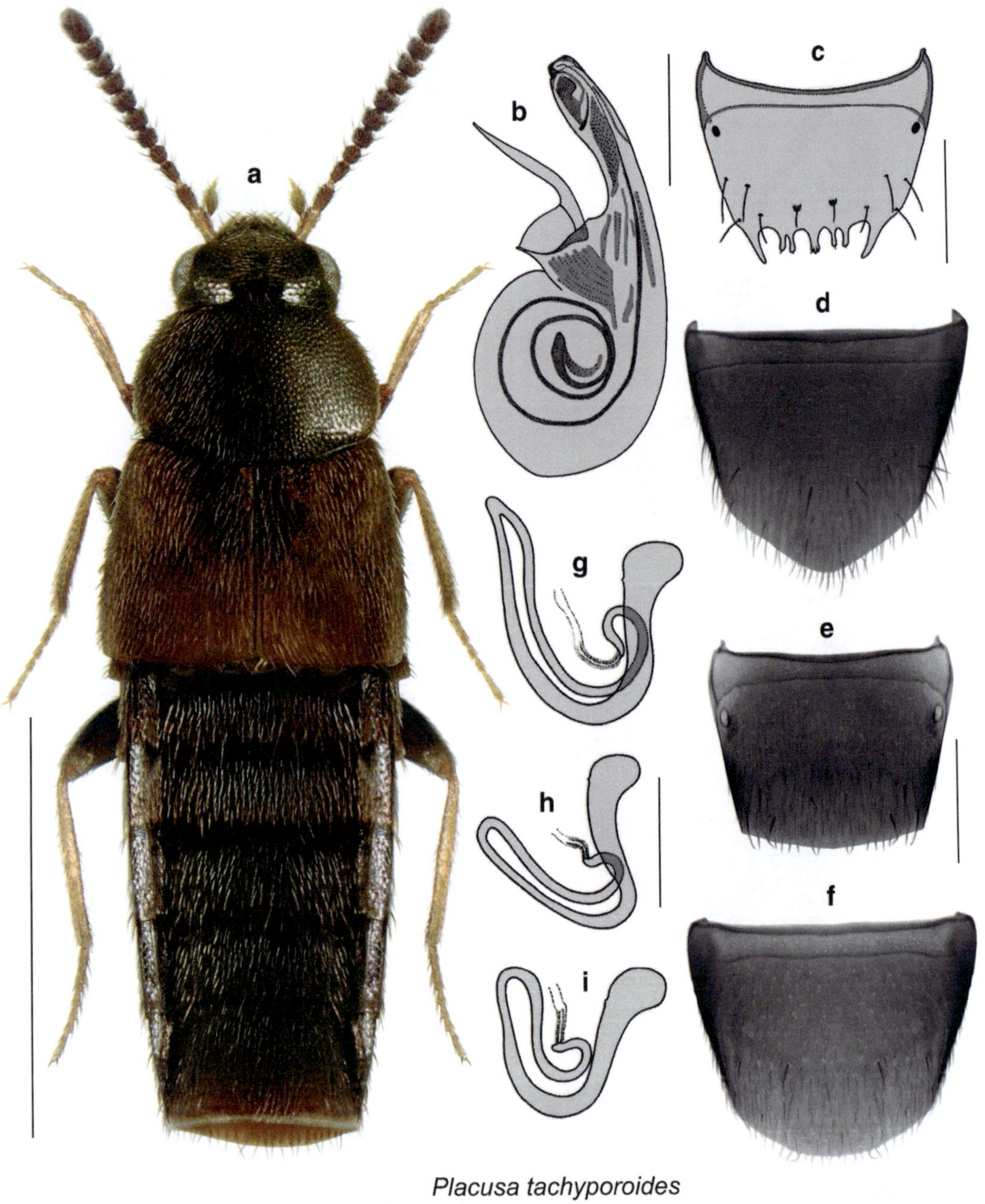

Fig. 18.176 (**a–i**). Images of *Placusa tachyporoides* (Waltl): (**a**) habitus in dorsal view; (**b**) median lobe of aedeagus in lateral view; (**c**) male tergite VIII; (**d**) male sternite VIII; (**e**) female tergite VIII; (**f**) female sternite VIII; (**g–i**) spermatheca. (**b, c, g–i**) after Klimaszewski et al. (2001); remaining images after Klimaszewski et al. (2018). Scale bar for habitus = 1 mm; remaining scale bars = 0.2 mm

References

Ashe JS (2001) Key to the tribes and genera of Nearctic Aleocharinae. 8. In: Arnett RH Jr, Thomas CM (eds) American Beetles. Archostemata, Myxophaga, Adephaga, Polyphaga: Staphyliniformia. CRC Press, Washington, DC, pp 299–374. 443 pp

Brunke A, Klimaszewski J, Dorval J-A, Bourdon C, Paiero SM, Marshall SA (2012) New species and distributional records of Aleocharinae (Coleoptera, Staphylinidae) from Ontario, Canada, with a checklist of recorded species. In: Klimaszewski J, Anderson R (Eds) Biosystematics and ecology of Canadian Staphylinidae (Coleoptera) II. Zookeys [Special Issue] 186:119–206

Casey TL (1893) Coleopterological notices. V. Annals of the New York Academy of Sciences 7 [1894]:281–606 [see reference in Chapter 7 for explanation of date]

Casey TL (1911) New American species of Aleocharinae and Myllaeninae. In: Memoirs on the Coleoptera. II. The New Era Printing Co., Lancaster, PA, pp 1–245. 259 pp

Criddle N (1922) The Entomological Record, 1921. Ann Rep Entomol Soc Ont 52(1921):57–70

Erichson WF (1837) Die Käfer der Mark Brandenburg. Erster Band, Erste Abtheilung. F. H. Morin, Berlin. viii + 384 pp

Gusarov VI (2003) Revision of some types of North American aleocharines (Coleoptera: Staphylinidae: Aleocharinae), with synonymic notes. Zootaxa 353:1–134

Hatch MH (1957) The Beetles of the Pacific Northwest. Part II: Staphyliniformia. Univ Washing Pub Biol 16. IX + 384 pp

Horion A (1967) Faunistik der Mitteleurop�ischen K�fer. Bd. XI: Staphylinidae. 3. Teil: Habrocerinae bis Aleocharinae (ohne Subtribus Athetae). Űberlingen -Bodensee, P.C.V. Schmidt, xxiv + 419 pp

Klimaszewski J, Pelletier G, Germain C, Hébert C, Humble LM, Winchester NN (2001) Diversity of *Placusa* (Coleoptera: Staphylinidae, Aleocharinae) in Canada, with descriptions of two new species. Can Entomol 133(1):1–47

Klimaszewski J, Langor D, Savard K, Pelletier G, Chandler DS, Sweeney J (2007) Rove beetles (Coleoptera: Staphylinidae) in yellow birch-dominated stands of southeastern Quebec, Canada: diversity, abundance, and description of a new species. Can Entomol 139 (6):793–833

Klimaszewski J, Langor DW, Work TT, Hammond JHE, Savard K (2008) Smaller and more numerous harvesting gaps emulate natural forest disturbances: a biodiversity test case using rove beetles (Coleoptera, Staphylinidae). Divers Distrib 14(6):969–982

Klimaszewski J, Langor D, Pelletier G, Bourdon C, Perdereau L (2011) Aleocharine beetles (Coleoptera, Staphylinidae) of the province of Newfoundland and Labrador, Canada. Pensoft Series Faunistica No. 98. Pensoft, Sophia-Moscow, 313 pp

Klimaszewski J, Godin B, Langor D, Bourdon C, Lee S-I, Horwood D (2015) New distribution records for Canadian Aleocharinae (Coleoptera, Staphylinidae), and new synonymies for *Trichiusa*. ZooKeys 498:51–91

Klimaszewski J, Larson DJ, Labrecque M, Bourdon C (2016) Twelve new species and fifty-three new provincial distribution records of Aleocharinae rove beetles of Saskatchewan, Canada (Coleoptera, Staphylinidae). ZooKeys 610:45–112

Klimaszewski J, Webster RP, Langor DW, Brunke A, Davies A, Bourdon C, Labrecque M, Newton AF, Dorval J-A, Frank JH (2018) Aleocharine rove beetles of eastern Canada (Coleoptera, Staphylinidae, Aleocharinae): a glimpse of megadiversity. Springer, Cham, xvi + 902 pp

Klimaszewski J, Hoebeke ER, Godin B, Davies A, Perry KI, Bourdon C, Winchester N (2020) Aleocharine rove beetles of British Columbia: a hotspot of Canadian biodiversity (Coleoptera, Staphylinidae). Springer, Cham, xvi + 631 pp

Majka CG, Klimaszewski J (2008) New records of Canadian Aleocharinae (Coleoptera: Staphylinidae). In: Majka CG, Klimaszewski J (Eds.) Biodiversity, biosystematics, and ecology of Canadian Coleoptera. ZooKeys [Special Issue] 2:85–114

Majka CG, Chandler DS, Donahue CP (2011) Checklist of the beetles of Maine, USA. Empty Mirrors Press, Halifax, NS. 328 pp

Moore I, Legner EF (1975) A catalogue of the Staphylinidae of America north of Mexico (Coleoptera). University of California, Division of Agricultural Sciences, Special Publication No. 3015:1–514

Mulsant E, Rey C (1871) Histoire naturelle des Coléoptères de France. Brévipennes. Aléochariens [Huitième Branche, Bolitocharaires]. Deyrolle Fils, Paris. [4] + 321 pp. + [5, explic.] + [4, table, erratum], pls. 1–5

Newton AF (2019) StaphBase: staphyliniformia world catalog database (version Nov 2018). In: Roskov Y, Ower G, Orrell T, Nicolson D, Bailly N, Kirk PM, Bourgoin T, DeWalt RE, Decock W, Nieukerken E. van, Zarucchi J, Penev L (eds) Species 2000 & ITIS Catalogue of Life, 2019 Annual. Digital resource at http://www.catalogueoflife.org/annual-checklist/2019. Species 2000: Naturalis, Leiden, the Netherlands. ISSN 2405-884X

Waltl J (1838) Verzeichniss der um Passau vorkommenden seltnern Käfer [cols. 263–267]; Um Passau vorkommende neue Arten von Käfern [cols. 267–273]. In: Beiträge zur nähern naturhistorischen Kenntniss des Unterdonaukreises in Bayern. Isis von Oken 1838(4):250–273

Webster RP, Klimaszewski J, Pelletier G, Savard K (2009) New Staphylinidae (Coleoptera) records with new collection data from New Brunswick, Canada. I. Aleocharinae. In: Majka CG, Klimaszewski J (Eds.) Biodiversity, biosystematics, and ecology of Canadian Coleoptera II. ZooKeys [Special Issue] 22:171–248

19 Tribe Athetini Casey, 1910

Tribe Athetini Casey, 1910

(Figs. 19.177–19.313) (Seevers 1978, Ashe 2001, Klimaszewski and Winchester 2002, Elven et al. 2010, 2012, Klimaszewski et al. 2018a, 2020)

Diagnosis Body small to medium sized (0.9–6.0 mm), slender and very flexible, subparallel or ovoid; pronotum with pubescence pattern variable but microsetae often directed obliquely laterad from midline of disc; integument brown, light brown, yellowish-brown, or black with appendages usually paler; forebody often with meshed microsculpture and fine to asperate punctation; head without distinct neck in most (except e.g., *Amischa*), postocular carina present, entire or incomplete; maxillary palpi with four palpomeres; labial palpi with three palpomeres in most species; antennae long, usually extending at least to middle of elytra; pronotal hypomera visible in lateral view in most taxa; mesocoxae narrowly or moderately narrow separated; process of mesoventrite slender, extending to one-third or half of length of mesocoxae; male tergite VIII often modified with teeth, crenulations or emarginations; median lobe of aedeagus with compressor plate oval, and with transverse sclerotized strip called the "athetine bridge", which may be entire or subdivided medially; tarsal formula 4-5-5.

COMMENTS The tribe Athetini is by far the largest evolutionary lineage of aleocharines. It is one of the most difficult and poorly known tribes of beetles in North America and elsewhere. This situation is caused by various factors: enormous species diversity, overwhelming number of species descriptions without illustrations, the small size of the majority of species, difficult-to-observe diagnostic features, convergent character states, extensive literature, and the lack of studies of genitalic structures by most previous authors due to a lack of awareness of the value of genitalic characters. Most Nearctic species were described by Casey (1885, 1893, 1906, 1910, 1911), and are grouped in a large number of genera (subgenera), which are poorly defined with overlapping characteristics. Ranking at the generic or subgeneric level is a serious problem, particularly in the large genus *Atheta* (*sensu lato*). At the present stage of knowledge of this genus, it is neither practical nor possible to elevate subgenera of *Atheta* to a generic status and place all the species in generic taxa because they are little understood and inadequately defined (Klimaszewski et al. 2018a). In this contribution we will treat the genus *Atheta* in a broad sense (*sensu lato*), and will continue, like most European authors (e.g., Palm 1970, Benick and Lohse 1974), to use subgeneric classification with the exceptions of those groups that are clearly delimited e.g., *Acrotona*, *Amischa*, *Boreophilia*, *Earota*, *Philhygra*. Elven et al. (2012) conducted

J. Klimaszewski et al., *A Faunal Review of Aleocharine Beetles in the Rapidly Changing Arctic and Subarctic Regions of North America (Coleoptera, Staphylinidae)*,
https://doi.org/10.1007/978-3-030-68191-3_19

a molecular phylogenetic analysis that included many Nearctic genera (*Amischa*, *Atheta*, *Boreostiba*, *Boreophilia*, *Dalotia*, *Hydrosmecta*, *Liogluta*, *Lypoglossa*, *Meronera*, *Mocyta*, *Myrmedonota*, *Stethusa*, *Strigota*, and *Tarphiota*) and confirmed their position within Athetini. The current classification of Nearctic athetines is inadequate and many genus and subgenus-level taxa currently recognized cannot be clearly delimited and diagnosed.

Subtribe Athetina Casey, 1910

(Ashe 2001)

This subtribe is very poorly defined so a meaningful diagnosis is not possible. Currently, the subtribe includes many different groups of athetines and likely does not represent a monophyletic group. We have adopted a broad definition of this subtribe to include diverse groups of the genus *Atheta* and those genera/subgenera that do not belong to tribe Lomechusini. The genera and species within species groups are listed in alphabetical order. In recent molecular studies (Elven et al. 2010), a diverse group of athetine genera (*Alevenota*, *Aloconota*, *Callicerus* and *Earota*), were included in a newly proposed tribe Geostibini, which was considered a sister group of Lomechusini. Morphologically, the tribe Geostibini was defined by a simple autapomorphy of a reduced sensillum "a" of the epipharynx. We believe that more morphological evidence should be presented before this new classification can be accepted and therefore in this book these genera are listed alphabetically under a broadly defined tribe Athetini.

Key to Genera of Athetini

1. Body extremely broad (Fig. 19.313a); pronotum with all pubescence on disc directed posteriad or posterolaterad (Fig. 19.313a), disc about 1.6 times as wide as long with basal angles slightly angulate and basal margin distinctly sinuate (Fig. 19.313a) ....***Paragoniusa* Maruyama and Klimaszewski**

– Body moderately broad; pronotum with pubescence and other characters not as above...**2**

2. Body form distinctive, head and pronotum much narrower than elytra, which are at least 1.2 times as wide as the pronotum (Fig. 19.312a); abdomen oval in outline (Fig. 19.312a); forebody with pubescence long and shaggy, erect on pronotum (Fig. 19.312a) ....***Trichiusa* Casey**

– Body form and pubescence not as above...**3**

3. Integument of forebody strongly granulose due to fine-meshed, raised reticulation (Figs. 19.310a, 19.311a); head without infraorbital carinae; eyes large; pronotum 1.2–1.3 times as wide as long (Figs. 19.310a, 19.311a); mandibles each with blunt apex; anterior and middle tibiae each with a row of spines; species inhabit mid to upper littoral zone of fine-grained sandy beaches.............. ***Tarphiota* Casey**

– Integument of forebody not strongly granulose, other characters not as above.....................**4**

4. Elytra at suture distinctly shorter than pronotum along midline of disc (Fig. 19.241a); median lobe of aedeagus and spermatheca as illustrated (Fig. 19.241b,g).....***Bellatheta* Roubal**

– Elytra at suture as long as or distinctly longer than pronotum along midline in most, except for some species of *Acrotona*, *Atheta* and *Liogluta*, but then genitalia not as above..........................**5**

5. Pronotum with hypomera not visible in lateral view...**6**

– Pronotal hypomera fully or partially visible in lateral view...**8**

6. Antennae thick (Figs. 19.308a, 19.309a); body narrowly elongate, densely punctate, particularly on abdomen, dorsal surface with fine white pilose pubescence (Figs. 19.308a, 19.309a); pronotum about as broad as maximum width of elytra (Figs. 19.308a, 19.309a); tergite VIII in both sexes with antecostal suture joining basal margin, not lateral margins as in other aleocharines (Figs. 19.308d,f, 19.309d,f); apical margin of female tergite VIII obtusely triangular or with triangular projection medially (Figs. 19.308g, 19.309g).......***Strigota* Casey**

– Antennae thinner in most (e.g., *Mocyta*); body broadly to narrowly elongate, sparsely to moderately punctate, pubescence different; pronotum often broader than maximum width of elytra; tergite VIII in both sexes with antecostal suture joining lateral margins (Fig. 19.177d,f);

apical margin of female tergite VIII rounded, truncate or emarginate..**7**

7. Antennae very thin and pale and in most specimens contrasting in colour with head, antennomeres V–X subquadrate, transverse or slightly elongate (Figs. 19.284a–19.287a); pronotum broad and shield-shaped, often broader than maximum width of elytra, pubescence on disc moderately dense and directed straight posteriad or obliquely posterolaterad from midline (Figs. 19.284a–19.287a); spermatheca with capsule hemispherical or elongate and ovoid in shape, usually with small apical invagination and short neck, stem thin and regularly or irregularly coiled posteriorly (Figs. 19.284i,j, 19.285h, 19.286f–j, 19.287h)... ***Mocyta* Mulsant and Rey**

– Antennae normally developed and appearing not very thin, usually not strongly contrasting in colour with head (Figs. 19.177a, 19.178a); pronotum subquadrate to transverse, approximately as wide as elytra in most, pubescence on disc usually very dense and directed lateroposteriad from midline (Figs. 19.177a, 19.178a); spermatheca shaped differently, capsule more or less spherical and extended to elongate neck, stem broader than those in *Mocyta*, regularly coiled and often swollen posteriorly (Figs. 19.177h, 19.178g).. ***Acrotona* C.G. Thomson**

8. Body subdepressed, slender (Figs. 19.269a, 19.270a); antennomeres VI–X elongate in most; mandibles slender; size small, 2.0–2.7 mm .... ***Hydrosmecta* C.G. Thomson**

– Body moderately to highly convex (except in some *Dinaraea* but then aedeagus without projecting crista apicalis on aedeagus), more or less robust; antennomeres VI–X transverse in most; mandibles moderately robust............**9**

9. Head strongly constricted basally and distinctly expanded posteriad of eyes to create triangular shape (Figs. 19.182a, 19.183a).....***Amischa* Thomson**

– Head gradually narrowed basally.............**10**

10. Pronotum exceptionally broad and shield-like, at least 1.6 times as wide as long (Figs. 19.282a, 19.283a).......***Lypoglossa* Fenyes**

– Pronotum not as above, at most 1.5 times as wide as long...**11**

11. Mandibles appearing bifid apically (left mandible with tip bifid and right mandible bears one apical and one subapical tooth creating illusion of bifid mandible) (Fig. 19.304a–d); body subparallel; pronotum slightly transverse (at most 1.3 times as wide as long) to nearly subquadrate, integument glossy, microsculpture weak (Figs. 19.305a–19.307a)............................ ....***Schistoglossa* Kraatz**

– Mandibles acutely pointed and not bifid; other characters not as above..........................**12**

12. Body narrowly subparallel in most (Figs. 19.180a, 19.181a, 19.288a); head about as large as pronotum or slightly smaller, genae at least as long as eyes in dorsal view **13**

– Characters or combination of characters not as above..**15**

13. Empodial bristles much longer than tarsal claws; pronotum often with distinctive pattern of pubescence (wavy, V-shaped) in males; male tergite VIII with apical margin bearing two lateral teeth and two median smaller teeth or protuberances (Figs. 19.180c, 19.181d); spermatheca S-shaped (Figs. 19.180g, 19.181h); flagellum of median lobe of aedeagus at least as long as length of median lobe ....***Aloconota* C.G. Thomson**

– Characters or combination of characters not as above..**14**

14. Pronotum with pubescence on disc directed posteriad along midline, either posteriad or posterolaterad elsewhere (Fig. 19.281a); apical margin of male tergite VIII often with broad rectangular projection, often crenulate, with two lateral teeth, punctation often asperate near apex of disc (Figs. 19.271c, 19.272d, 19.273c, 19.274c); spermatheca large (Figs. 19.271g, 19.272h........................***Liogluta* C.G. Thomson**

– Pronotum with pubescence on disc directed anteriad along midline, laterad elsewhere (Figs. 19.292a, 19.293a); male tergite VIII truncate apically, without additional structures, punctation not asperate (Figs. 19.292d, 19.293e); spermatheca vestigial (very small) and difficult to find...................***Philhygra* Mulsant and Rey**

15. Postocular carina of head absent; body flattened (Figs. 19.262a, 19.263a, 19.264a, 19.265a); forebody with coarse microsculpture; mainly subcortical species........... ***Dinaraea*** **C.G. Thomson**

– Postocular carina of head partial or complete but always present; body convex to moderately flattened; forebody with microsculpture fine to coarse; mainly litter-dwelling species, not subcortical **16**

16. Pubescence on pronotum and elytra dense, appearing silky (Figs. 19.179a, 19.268a) (except *Dochmonota*, Figs. 19.266a, 19.267a); apical antennomere almost as long as previous three combined (except *Dochmonota*, Figs. 19.266a, 19.267a); ligula with wide base, split into two lobes in apical third; mesocoxae broadly separated by distance about equal to length of isthmus **17**

– Combination of characters not as above .. **19**

17. Body small, length 2.2–2.5 mm, narrowly ovoid, with dense punctation (Figs. 19.266a, 19.267a); antennomeres VI–X subquadrate to slightly transverse (Figs. 19.266a, 19.267a); male sternite VIII highly elongate, apical part triangularly produced, apex obtuse (Figs. 19.266e, 19.267e); median lobe of aedeagus with large sac-shaped crista apicalis on apical margin of prominent triangular projection of bulbus (Figs. 19.266b, 19.267b); spermathecal capsule hemispherical, with narrow, very deep apical invagination (Figs. 19.266h, 19.267h) ....***Dochmonota*** **C.G. Thomson**

– Body large, over 2.5 mm long, other combination of characters not as above..................... **18**

18. Infraorbital carinae normally developed; antennomeres VII–X strongly transverse (Fig. 19.268a); pubescence on pronotum and elytra short, dense, appearing silky, directed approximately straight posteriad (Fig. 19.268a); male tergite VIII produced apically and not serrate (Fig. 19.268d); median lobe of aedeagus and spermatheca as illustrated (Fig. 19.268b,c,h); occurs in terrestrial habitats ....***Earota*** **Mulsant and Rey**

– Infraorbital carinae vestigial, very short; antennomeres VII–X subquadrate to slightly transverse (Fig. 19.179a); pubescence on pronotum and elytra moderately long and dense, slightly silky, directed mostly laterad and forming wavy pattern on elytra (Fig. 19.179a); male tergite VIII truncate apically and with four rounded teeth (Fig. 19.179d); spermatheca capsule tubular and flattened (Fig. 19.179h); occurs in debris on sea beaches.....................................***Adota*** **Casey**

19. Procoxal cavities separated from lateral margin of pronotum by long and wide hypomeron; apical margin of male tergite VIII entire, without teeth or crenulation; median lobe of aedeagus in most with enlarged, about rectangular in shape bulbus and short tubus in dorsal view (Figs. 19.244c, 19.245d,e, 19.246c); body broad in most (Figs. 19.244a, 19.245a) **20**

– Procoxal cavities reaching lateral margin of pronotum or separated from lateral margin of pronotum by a narrow part of hypomeron; apical margin of male tergite VIII in some subgenera of *Atheta* (e.g. *Datomicra*, *Dimetrota*) with teeth or some degree of crenulation in most (Figs. 19.195d, 19.196d, 19.197d); median lobe of aedeagus with moderately large to small ovoid in shape bulbus and moderately long tubus in dorsal view (Figs. 19.184c, 19.186c, 19.207c, 19.208c); body moderately broad or narrow (Figs. 19.184a–19.240a) .... ***Atheta*** **C.G. Thomson**

20. Ligula divided to base into two divergent lobes (Fig. 19.242b); antennomeres VII–X quadrate, elongate, or moderately transverse ....***Boreophilia*** **Benick**

– Ligula divided into two lobes only in apical half; antennomeres VII–X moderately to distinctly transverse ....***Boreostiba*** **Lohse**

Genus *Acrotona* Thomson, 1859

(Figs. 19.177–19.178) (Brundin 1952, Ashe 2001, Klimaszewski et al. 2011, 2018a)

Diagnosis Body subparallel or ovoid, length 2.1–2.8 mm; head subquadrate, broadest at eye level, postocular carina complete; ligula bifid; antennomeres I–III elongate, IV subquadrate, VI–X subquadrate or slightly to distinctly transverse; pronotum subquadrate to transverse, approximately as wide as elytra, pubescence on disc usually very dense and directed lateroposteriad from midline, hypomera not visible in lateral view; elytra transverse, often at least as long as pronotum but exceptionally may be shorter; abdomen tapering apicad, often acutely so near apex, with strong lateral macrosetae; process of mesoventrite reaching beyond middle of mesocoxae, its apex acute; process of metaventrite short; spermatheca with capsule more-or-less spherical and extended to elongate neck, stem broader than that in *Mocyta*, regularly coiled posteriorly and often with swollen apex (Figs. 19.177h, 19.178g).

Key to Species of *Acrotona*

1. Elytra reduced, at suture distinctly shorter than pronotum along midline (Fig. 19.177a); male tergite VIII emarginate medially (Fig. 19.177d); median lobe of aedeagus and spermatheca as illustrated (Fig. 19.177b,c,h) ....***Acrotona sphagnorum* Klimaszewski and Webster**

– Elytra normally developed, at suture as long as or longer than pronotum along midline (Fig. 19.178a); male tergite VIII, median lobe of aedeagus and spermatheca not as above (Fig. 19.178b,c,g) .... ***Acrotona horwoodae* Klimaszewski and Godin**

106. *Acrotona sphagnorum* Klimaszewski and Webster (Fig. 19.177a–h)

Acrotona sphagnorum Klimaszewski and Webster, in Webster et al. 2016. Klimaszewski et al. 2018a.

Diagnosis Body moderately broad, subparallel, length 2.7–3.0 mm; colour uniformly dark brown except for light-brown elytra, apex of abdomen, legs and basal antennomeres (Fig. 19.177a); integument highly glossy, moderately densely punctate and pubescent, pubescence short and adhering to body; head about one-third narrower than pronotum, rounded laterally, eyes small, about 1/2 as long as genae in dorsal view; antennomeres V–X transverse, slightly incrassate; pronotum broad, transverse, slightly broader than elytra, truncate posteriorly; elytra at suture shorter than pronotum along midline; abdomen subparallel. MALE. Tergite VIII with apical margin with broad, obtusely angulate emargination (Fig. 19.177d); sternite VIII elongate, apical margin arcuately rounded (Fig. 19.177e); median lobe of aedeagus in dorsal view with bulbus narrowly oval, tubus narrowly elongate to narrowly triangular at apex (Fig. 19.177c); in lateral view bulbus narrowly oval, tubus arcuate basally, straight medially and narrowly rounded apically (Fig. 19.177b); internal sac structures as illustrated (Fig. 19.177b,c). FEMALE. Tergite VIII with apical margin broadly shallowly emarginate (Fig. 19.177f); sternite VIII with apical margin broadly arcuate, feebly angulate medially (Fig. 19.177g); spermatheca with capsule funnel-shaped basally and irregularly spherical apically, stem sinuate, tightly coiled posteriorly (Fig. 19.177h).

Acrotona sphagnorum differs from *A. brachyptera* in having a broader and more robust body, differently shaped spermatheca and median lobe of aedeagus, the apical margin of male tergite VIII broadly, obtusely emarginate and the apical margin of female sternite VIII broadly arcuate, and feebly angulate medially (see images in Klimaszewski et al. 2018a).

DISTRIBUTION Origin: Nearctic. **CANADA**: LB [**NTR**], NB, ON [**NTR**], QC [**NTR**]. **USA**: not recorded.

COLLECTION AND HABITAT DATA

Habitat: In NB, found in moist sphagnum in forest black spruce bogs, black spruce and tamarack bogs, eastern white cedar swamps and forests. **Collecting period**: IV, V, VI, VII. **Collecting method**: Sifting moist sphagnum, yellow pan traps, pitfall traps.

We report this species from LB, ON, QC (**NPR**) for the first time based on the following: Canada: Labrador, Goose Bay, 53.21199, –60.45062, Rep. 1 wet, yellow pan, 15-18. VI.2010, NBP field party G1WY111, G1WY112, G1WP521 G2WY612 (LFC, CNC) 1 male, 3 females; Goose Bay, 53.21199, –60.45062, Rep. 1 wet, yellow pan, 21-24. VI.2010, NBP field party G1WY431 (LFC) 1 male; Goose Bay, 53.21408, –60.45044, Rep.

1 wet, yellow pan, 15-18.VI.2010, NBP field party G2WP211, G2WY613 (LFC) 1 female, 1 sex undetermined; Goose Bay, 51.27717, –80.64778, Rep. 3 wet, yellow pan, 16-19. VI.2010, NBP field party M3WY211 (LFC) 1 female; Goose Bay, 53.21408, –60.45044, Rep. 2 wet, pitfall trap, 15-18.VI.2020, NBP field party G2WP212 (CNC) 1 female. Quebec, Schefferville, 54.75970, –66.71120, Rep. 1, wet, yellow pan, 6-9.VII.2010, NBP field party S1WY932 (LFC) 1 female; Schefferville, 54.85350, –66.67043, Rep. 3, wet, pitfall trap, 30.VI-3.VII.2010, NBP field party S3WP311 (LFC) 1 female. Ontario, Moosonee, 51.277, –80.647, Rep. 3 wet, yellow pan, 16-19.VI.2010, NBP field party M3WY211 (CNC) 1 female.

107. ***Acrotona horwoodae*** **Klimaszewski and Godin** (Fig. 19.178a–g)

Acrotona horwoodae Klimaszewski and Godin, in Klimaszewski et al. 2012.

Diagnosis Body narrowly oval, moderately convex, length 2.4 mm (Fig. 19.178a); uniformly black, punctation on forebody fine, dense and not asperate, microsculpture fine but not pronounced; head narrower than pronotum, ratio of maximum width of head to maximum width of pronotum 0.7; antennal articles 7–10 slightly transverse; pronotum moderately transverse, ratio of maximum width to length 1.4, about as wide as elytra; elytra at suture about as long as pronotum; abdomen slightly narrowed posteriad. MALE. Tergite VIII moderately elongate and truncate apically (Fig. 19.178c); sternite VIII widely arcuate apically (Fig. 19.178d); median lobe of aedeagus as illustrated (Fig. 19.178b). FEMALE. Tergite VIII moderately elongate and truncate apically, base not sinuate (Fig. 19.178e); sternite VIII widely arcuate apically, base not sinuate (Fig. 19.178f); spermatheca with capsule tulip-shaped and stem coiled posteriorly (Fig. 19.178g).

DISTRIBUTION Origin: Nearctic. **CANADA**: YT. **USA**: not recorded.

COLLECTION AND HABITAT DATA
Habitat: mixed aspen and white spruce forest. **Collecting period**: V. **Collecting method**: sifting forest litter.

Genus *Adota* Casey, 1910

(Fig. 19.179a–h)

(Casey 1910, Fenyes 1920, Seevers 1978, Lohse and Smetana 1985, Ashe 2001, Gusarov 2003b, Lee et al. 2013, Klimaszewski et al. 2020)

Diagnosis Body length 2.2–3.2 mm, body parallel-sided and flat, colour brown to black, elytra dark brown with some reddish tinge, antennae black to brown, legs brown to yellowish brown, entire body with strong and dense isodiametric microsculpture; head as long as wide; eyes large, temple length to eye length ratio 0.8–1.2; infraorbital carina very short, ends in posterior portion of temples; antennomere IV–VI subquadrate, VII–X subquadrate to slightly transverse, terminal one without coeloconic sensilla, about as long as articles IX–X combined; labrum transverse, with concave anterior margin; mandibles broad, right mandible with a small medial tooth, dorsal molar area with velvety patch consisting of very small denticles; maxilla with galea projecting slightly beyond apex of lacinia; apical lobe of galea covered with numerous fine and short setae; internal margin of galea with long subapical setae; apical 1/3 of lacinia with row of closely spaced spines, middle portion produced medially and covered with numerous setae; ligula short, with broad base and split apex; medial area of prementum with 2 pores and with 8–23 pseudopores, lateral areas each with two asetose pores, single setose pore and 7–12 pseudopores; labial palpus with setae a, b, g and d present (Fig. 10, in Gusarov 2003a, b); mentum with concave anterior margin; pronotum slightly transverse, with microsetae directed anteriorly in midline, in lateral portions of disc microsetae directed laterally, macrosetae short; hypomera fully visible in lateral view; mesosternal process narrow, extending about 3/5 length of mesocoxal cavities, metasternal process short, mesoventrite and mesosternal process not carinate medially; relative lengths of mesoventral process: isthmus: metaventral process in ratio of about 3:1:1; mesocoxal cavities margined posteriorly; mesocoxae narrowly separated; tarsal formula 4-5-5, metatarsomere I as long as II; all tarsomeres with long setae; one empodial seta, as long as claws; claws strong of equal length; elytral posterior margin straight; wings fully developed; abdominal tergites III–VI with moderate basal impressions, tergite VII as long as tergite VI, punctation on tergites VI–VII only slightly sparser than on III–V, tergite VII with wide white palisade fringe.

Adota can be distinguished from *Atheta* by the concave anterior margin of the labrum; by basal impression on the tergite VI, and by lacking distinct medial lamellae (sclerites) of internal sac.

Adota differs from *Psammostiba* in having apical margin of male tergite VIII with 2–4 teeth (in *Psammostiba* serrate), and tarsi with single empodial seta and equally long claws.

108. *Adota maritima* (Mannerheim) * (Fig. 19.179a–h)

Homalota maritima Mannerheim, 1843. Casey 1910 [as three synonyms], 1911 [as three synonyms], Fenyes 1920, Lohse and Smetana 1985 (type material), Gusarov 2003a, b [list of synonyms], Sazhnev 2018, Klimaszewski et al. 2020.

Diagnosis (adapted from Gusarov 2003b). Body length 2.5–3.4 mm, colour dark brown or black with light brown tarsi (Fig. 19.179a); head surface dull, with strong and dense isodiametric microsculpture, with fine punctation, distance between punctures equals 1–2 times their diameter; frons with sparser punctation, in both sexes with longitudinal impression; temple length to eye length ratio 1.0–1.2; antennomere II about 1.1 times as long as antennomere III, IV–V slightly elongate (length to width ratio 1.2–1.3), VI–VII as long as wide, VIII–X slightly transverse (length to width ratio 0.8); pronotum subquadrate or slightly transverse, 1.1 times as wide as head, width 0.47–0.61 mm, length 0.41–0.51 mm, width to length ratio 1.2, with strong and dense isodiametric microsculpture; punctation finer than on head, poorly visible; elytra much wider (0.63–0.89 mm) and longer (0.57–0.79 mm; measured from humeral angle) than pronotum (elytral length to pronotal length ratio 1.5), 1.1 times wider than long, surface matte, with strong and dense isodiametric microsculpture; punctation poorly visible; abdominal tergites dull with strong and dense isodiametric microsculpture and fine punctation, distance between punctures equals 2–3 times their diameter on tergites III–V and 3–5 times on tergites VI–VII, apical margin of tergite VII with white palisade fringe. MALE. Apical margin of VIII with broad, short projection bounded laterally by two rounded triangular teeth, and with a rounded tooth on each side of middle (Fig. 19.179d); sternite VIII parabolic posteriorly (Fig. 19.179e); tubus of median lobe of aedeagus slightly sinuate ventrally, curved obliquely ventrad basally, with narrowly rounded apex in lateral view (Fig. 19.179b); in dorsal view tubus only slightly narrower than bulbus (Fig. 19.179c). FEMALE. Tergite VIII unevenly arcuate apically (Fig. 19.179f); sternite VIII broadly, evenly rounded from base to apex (Fig. 19.179g); spermatheca with tubular capsule bearing broad and short apical invagination, stem narrow, sinuate and narrowly looped posteriorly (Fig. 19.179h).

DISTRIBUTION Origin: Holarctic (Japan and Russian Far East, adventive in Great Britain and Ireland). **CANADA**: BC. **USA**: AK, CA.

COLLECTION AND HABITAT DATA Habitat: mainly litoral habitats, found in decomposing seaweed on sandy seashores, under seaweed on rocks with Diptera larvae (Gusarov 2003a, b); beach drift wood, under back beach rocks and drift, under rock at river edge, treading *Carex* marsh and sifting *Alnus* litter, and under rock at river edge. **Collecting period**: V–VIII, X. **Collecting method**: sifting decomposing seaweed and algae; treading *Carex* marsh and sifting *Alnus* litter.

Additional records. See Annex 1.

Genus *Aloconota* C.G. Thomson, 1858

(Figs. 19.180–19.181)

(Klimaszewski and Peck 1986, Gusarov 2003a, Klimaszewski et al. 2011, 2018a, 2020)

Diagnosis Body narrowly subparallel; pronotum and elytra often with complex pubescence pattern in males (wavy, V-shaped); ligula divided into two lobes at base; prementum with two medial, widely separated setae, lateral pseudopores absent; twin pores of labial palpomere I small; mesocoxal cavities narrowly separated; process of mesoventrite pointed at apex; hind tarsus with tarsomere I longer than II to about as long as II and III combined, empodial seta distinctly longer than tarsal claw; male tergite VIII with apical margin obliquely produced laterally, with two small, rounded medial teeth or tubercles as well as two sublateral teeth or projections; median lobe of aedeagus narrowly elongate and without distinct sclerites, athetine bridge wide but incomplete medially, flagellum thin and longer than median lobe; spermatheca very short, swollen toward apex.

Key to Species of *Aloconota*

1. Antennomeres VII–X slightly transverse (Fig. 19.180a); spermatheca (Fig. 19.180g); apical margin of male tergite VIII with two elongate, sharp lateral teeth, and two large, triangular projections medially (Fig. 19.180c); median lobe of aedeagus as illustrated (Fig. 19.180b) ....***Aloconota cambrica* (Wollaston)**

– Antennomeres VII–X slightly elongate (Fig. 19.181a); spermatheca (Fig. 19.181h); apical margin of male tergite VIII with two small lateral teeth, and two short, rounded teeth medially (Fig. 19.181d); median lobe of aedeagus not as above (Fig. 19.181b) ....***Aloconota neocambrica* Klimaszewski and Langor**

109. *Aloconota cambrica* (Wollaston) † (Fig. 19.180a–g)

Homalota cambrica Wollaston, 1855. In *Aloconota*: Klimaszewski and Winchester 2002, Smetana 2004, Klimaszewski et al. 2020.

Diagnosis Body length 2.6–2.8 mm, flattened, narrowly subparallel, colour brown to dark brown, legs, and posterior part of elytra slightly paler; forebody densely punctate and pubescent, punctation slightly asperate, microsculpture meshed, particularly well visible on pronotum, sculpticells slightly convex (Fig. 19.180a); head elongate and strongly depressed medially, nearly as broad as pronotum, postocular region elongate, twice as long as maximum diameter of eye, tempora with vestigial carinae, not visible laterally when head slightly retracted into thorax or visible only at very base of tempora; antennae slender, as long as pronotum and elytra combined, basal three antennomeres strongly elongate, IV–X slightly elongate, and terminal one strongly elongate and about as long as two preceding antennomeres combined; pronotum narrowly trapezoidal in shape, flattened, slightly elongate, hypomeron visible across basal two-thirds of pronotal length in lateral view, pubescence radiating from midline of disk anteriad, laterad along broadly arched lines and obliquely posteriad; elytra flat, slightly elongate, broader than pronotum, posterior margins broadly arcuate and without angular emarginations, pubescence directed obliquely posteriad; abdomen subparallel, four first visible tergites impressed at base; basal metatarsomere only slightly longer than the following one. MALE. Apical margin of tergite VIII with two elongate, sharp lateral teeth, and two large triangular projections medially (Fig. 19.180c); sternite VIII rounded apically (Fig. 19.180d); tubus of median lobe of aedeagus long, nearly straight in lateral view (Fig. 19.180b). FEMALE. Tergite VIII broadly arcuate apically (Fig. 19.180e); sternite VIII broadly arcuate apically (Fig. 19.180f); spermatheca S-shaped, capsule compressed facing laterad (Fig. 19.180g).

DISTRIBUTION Origin: Palearctic, adventive in Canada. **CANADA**: BC. **USA**: AK [**NSR**], CA.

COLLECTION AND HABITAT DATA (Klimaszewski and Winchester 2002). **Habitat:** in BC adults were captured in Sitka spruce forest, in forest interior, tranzition zone, and clearcut. Majority of specimens were collected in clearcut zone. In AK, specimens were captured under rocks on gravel creek shore, and in clear-cut. **Collecting period**: V, VI. **Collecting method**: Malaise traps. Lindgren trap.

We report this species from AK (**NSR**) for the first time here, based on the following: USA, Alaska, Dalton Hwy Mi 135, Prospect Crk, 265m el., 66.79376, −150.73164, ±24m gravel creek shore, under rocks, 31.V.2008, D. S. Sikes (UAM) 2 specimens; Prince of Wales Island, Staney Ck. 1B, 77m el., 55.872, −133.06523, ±26m, clear-cut, Lindgren 1, 14-22.V.2012, J. Stockbridge, S. Meierotto, I. MacDougall (UAM) 1 specimen; Prince of Wales Island, Staney Ck. 1B, 77m el., 55.872, −133.06523, ±26m clear-cut, Lindgren 1, 16-31.V.2011, J. Stockbridge, B. Wong (UAM) 1 specimen.

110. *Aloconota neocambrica* Klimaszewski and Langor (Fig. 19.181a–h)

Aloconota neocambrica Klimaszewski and Langor, in Klimaszewski et al. 2011, 2018a.

Diagnosis Body narrowly elongate, slightly flattened, length 2.9–3.0 mm; colour dark brown with palps, legs, and sometimes pronotum and elytra light brown (Fig. 19.180a); forebody moderately glossy with meshed microsculpture; pubescence on disc of pronotum uniform, directed anterio-laterad from midline, forming arcuate lines; head almost as wide as pronotum; antennae elongate, reaching apex of elytra, antennomeres I–VI elongate, VII–X subquadrate to slightly elongate; pronotum slightly transverse, widest at middle, nearly as wide as elytra; elytra at suture longer than pronotum; abdomen wide at base and arcuate laterally. MALE. Apical margin of tergite VIII obliquely produced laterally, broadly truncate medially with two small, sharp lateral denticles and one short, rounded tooth each side of middle (Fig. 19.180d); sternite VIII broadly parabolically rounded from base to slightly more narrowly rounded apex (Fig. 19.180e); median lobe of aedeagus with large bulbus and narrowly triangular tubus in dorsal view (Fig. 19.180c); ventral margin of tubus straight with apex slightly swollen in lateral view (Fig. 19.180b). FEMALE. Tergite VIII truncate apically (Fig. 19.180f); sternite VIII similar to that of male but broader (Fig. 19.180g); capsule of spermatheca with apex flattened, perpendicular to basal part, stem short, broad, folded at middle, swollen basally (Fig. 19.180h).

DISTRIBUTION Origin: Nearctic. **CANADA**: LB, NB, NF. **USA**: not recorded.

COLLECTION AND HABITAT DATA
Habitat: riparian along shaded brooks, streams and rivers in forests; old-growth mixed forest in gravel on gravel bar along shaded brook, mature mixed forest in gravel and under cobblestones along cool shaded medium sized stream, among gravel on firm sand/clay/gravel mix on margin of shaded spring-fed brook in a rich Appalachian hardwood forest, under rocks and gravel at river margin, and among cobblestones near outflow of brook into a river. **Collecting period**: VI–VIII. **Collecting method**: aspirating specimens from gravel and cobblestones along shaded streams and rivers.

COMMENT This species is known only from the softwood shield ecoregion (5.1) in Labrador and southward.

Genus *Amischa* C.G. Thomson, 1858

(Figs. 19.182–19.183) (Muona 1990, Ashe 2001, Klimaszewski et al. 2018a, b, 2020)

Diagnosis Body ovoid or subparallel, rather loosely articulated, length 2.5–4.8 mm in most; integument dark brown to black, often with paler elytra, usually with meshed microsculpture;

pronotum with pubescence on disc forming distinct pattern, usually directed lateroposteriad from midline, but with specific patterns in different subgenera; hypomera completely visible in lateral view; mesocoxae narrowly separated; process of mesoventrite slender; tarsal formula 4-5-5; athetine bridge located in dorso-anterior part of median lobe of aedeagus; spermatheca with capsule club-shaped, perpendicularly bent at middle, stem long, tightly coiled posteriorly.

Comments Species of this genus are difficult to distinguish, and thus a taxonomic revision is necessary to fully understand interspecific variability and differences among species. We have observed that the median lobe of the aedeagus offers few diagnostic characters for species in this genus and that the shape of the spermatheca and tergite VIII of both sexes offer much better characters for species identification.

Key to Species of *Amischa*

1. Antennomere IV subquadrate (Fig. 19.183a); body light brown with darker head and apical part of abdomen (Fig. 19.183a).................................. .... ***Amischa tersa* Casey**

– Antennomere IV distinctly transverse (Fig. 19.182a); body uniformly dark brown with some reddish tinge on elytra (Fig. 19.182a)........ .... ***Amischa praelonga* (Casey)**

111. *Amischa praelonga* (Casey) (Fig. 19.182a–d)

Colposura praelonga Casey, 1893. Klimaszewski et al. 2012.

Diagnosis Body narrowly subparallel, moderately convex, length 2.0 mm; colour dark brown with antennae, palpi, legs and apical part of abdomen paler (Fig. 19.182a); forebody moderately glossy, pubescence fine and sparse; head abruptly constricted at base, almost as wide as pronotum, genae subparallel, head broadest basally; antennomere IV distinctly transverse, V–X slightly to distinctly transverse; pronotum slightly transverse and as wide as elytra; elytra at suture about as long as pronotum; abdomen subparallel for most of its length. MALE. Undescribed. FEMALE. Tergite VIII slightly transverse, arcuate apically (Fig. 19.182b); sternite VIII moderately elongate, slightly arcuate apically (Fig. 19.182c); spermatheca with capsule spherical, short neck slightly sinuate, apical invagination broad and deep; stem long, and tightly coiled posteriorly (Fig. 19.182d).

DISTRIBUTION Origin: Nearctic. **CANADA**: YT. **USA**: WY.

COLLECTION AND HABITAT DATA
Habitat: willow stand adjacent to creek bank. **Collecting period**: VII. **Collecting method**: litter sifting, flight intercept trap.

112. *Amischa tersa* Casey (Fig. 19.183a–e)

Amischa (Colposura) tersa Casey, 1910. Klimaszewski et al. 2008a, 2012.

Diagnosis Body narrowly subparallel, moderately convex, length 1.9 mm; colour dark brown with antennae, palpi, legs and apical part of abdomen paler (Fig. 19.183a); forebody slightly glossy, pubescence fine and sparse; head abruptly constricted at base, almost as wide as pronotum, genae subparallel; antennomere IV subquadrate, V–X subquadrate to lightly transverse; pronotum slightly transverse and as wide as elytra; elytra at suture about as long as pronotum; abdomen subparallel for most of its length. MALE. Tergite VIII transverse, apical margin truncate, slightly rounded laterally (Fig. 19.183d); sternite VIII distinctly elongate, truncate apically, with long apical macrosetae (Fig. 19.183e). Median lobe of aedeagus as illustrated (Fig. 19.183b,c). FEMALE. Undescribed.

DISTRIBUTION Origin: Nearctic. **CANADA**: YT. **USA**: CO.

COLLECTION AND HABITAT DATA
Habitat: not recorded. **Collecting period**: VII. **Collecting method**: flight intercept trap.

COMMENTS The only known specimen from YT is tentatively associated with this species. It has a uniformly brown body, whereas in the holotype specimen of *A. tersa* the pronotum is light brown and the rest of the body dark brown. For now, we attribute the colour differences to intraspecific variation, but more specimens are needed from a broader geographical range to understand colour variability.

Genus *Atheta* C.G. Thomson, 1858 [sensu lato]

(Figs. 19.184–19.240)

(Strand and Vik 1964, Benick and Lohse 1974, Seevers 1978, Lohse et al. 1990, Ashe 2001, Klimaszewski and Winchester 2002, Gusarov 2003a, Elven et al. 2010, Klimaszewski et al. 2018a, 2020)

Diagnosis Body ovoid or subparallel, rather loosely articulated with very flexible abdomen, length 2.5–4.8 mm in most species; colour dark brown to black, integument usually with meshed microsculpture; suborbital carina partial or complete; ligula split; pronotum with pubescence on disc forming distinct pattern, usually directed lateroposteriad from midline, but with specific patterns in different subgenera; hypomera completely visible in lateral view; mesocoxae narrowly separated; process of mesoventrite slender; tarsal formula 4-5-5; athetine bridge of median lobe of aedeagus partial (divided dorsally), or complete and located in dorso-basal part of median lobe of aedeagus; spermatheca with club-shaped capsule and long stem that is narrowly looped posteriorly in most species (there are many spermathecal forms present in *Atheta*).

COMMENTS This genus is treated here as *Atheta sensu lato*, similar as in Klimaszewski et al. (2018a, 2020), and embraces a wide range of species with diverse morphology forming the *Atheta* complex. It includes several subgenera that are not sufficiently delimited to justify elevating them to generic status. The subgenera of *Atheta* are listed below in alphabetical order because the relationships between them are not clear. For species identification, dissection and examination of genitalic structures is necessary.

Key to Subgenera of *Atheta*

(Species of *Atheta* currently considered to be **incertae sedis** are not included in the key to subgenera of *Atheta*. Should the key not work well for the specimens at hand, the user should compare genital structures of their specimens to those species listed as incertae sedis).

1. Body usually subparallel, highly glossy (Figs. 19.185a–19.189a); length 3.5–4.4 mm; forebody with microsculpture fine or not evident; elytra often with granulose texture; antennomeres VI–X slightly transverse, quadrate or elongate (Figs. 19.185a–19.189a); spermatheca with stem irregularly twisted and bearing small tuberculate projections (Fig. 19.185d)............... ***Atheta s. str.***

– Combination of characters not as above **2**

2. Pronotum rounded, highly convex and glossy, all pubescence on disc directed straight or obliquely posteriad (Fig. 19.184a); pronotal hypomera narrowly visible in lateral view; male tergite VIII truncate apically, with or without teeth on apical margin (Fig. 19.184a); median lobe of aedeagus in dorsal view with small bulbus and long, narrow tubus (Fig. 19.184c), ventral margin of tubus long, arcuate in lateral view, internal sac without distinct sclerites (Fig. 19.184b)......................... ***Alaobia* Thomson**

– Combination of characters not as above **3**

3. Body length 2.5–2.7 mm, pronotum rectangular in shape (Fig. 19.190a); abdomen broadening apically; postocular carinae incomplete; antennomeres VII–X distinctly transverse and incrassate; male tergite VIII with entire apical margin subrectangularly produced, rounded laterally, shallowly sinuate along margin, without teeth (Fig. 19.190e); tubus of median lobe of aedeagus in lateral view with ventral margin irregularly, shallowly sinuate, apex narrowly triangular (Fig. 19.190b,c)....... ***Bessobia* Thomson**

– Combination of characters not as above **4**

4. Punctation of forebody or at least that of pronotum dense and asperate in most (Figs. 19.191a–19.193a, 19.194a–19.224a, 19.237a).... **5**

– Punctation not as above **7**

5. Body small, length 0.9–2.5 mm; spermatheca as illustrated (Figs. 19.191g, 19.192h, 19.193h), median lobe of aedeagus as illustrated (Figs. 19.191b, 19.192b,c, 19.193b,c) .... ***Datomicra* Mulsant and Rey**

– Body large, length 2.2–4.0 mm (2.8–3.0 mm in most); genital characters not as above **6**

6. Lateral margins of pronotum, elytra with stout bristles; elytra of the same colour as pronotum or paler but not strongly contrasting with rest of the body; antennomeres VII–X subquadrate, slightly transverse, or rarely slightly elongate; apical margin of male tergite VIII obliquely produced laterally, to broadly truncate apex which is usually bounded by teeth or angulate projections (Figs. 19.195d, 19.197d, 19.199d, 19.204d); apical part of tubus of median lobe of aedeagus in dorsal view narrow, triangular in shape (Figs. 19.195c, 19.197c, 19.199c, 19.204c) .... ***Dimetrota* Mulsant and Rey**

– Lateral margins of pronotum and elytra without stout bristles (Figs. 19.237a); body uniformly dark brown to black; antennomeres VII–X transverse; apical margin of male tergite VIII truncate and without teeth; apical part of tubus of median lobe of aedeagus in dorsal view broad, rounded apically (Fig. 19.237c) .... ***Rhagocneme* Munster**

7. Body length 3.0–4.0 mm; body broad, subparallel, broadest at elytra; antennomere IV subquadrate, II–X subquadrate to moderately transverse; abdomen broad, arcuate laterally (Fig. 19.225a); tubus of the median lobe of aedeagus with two lateral narrow and diverging projections in dorsal view (Fig. 19.225c); spermatheca with small spherical capsule and stem thin, very long, sinuate, and narrowly looped posteriorly (Fig. 19.225h) .... ***Lamiota* Casey**

– Body size variable, smaller in most, length 2.8–3.7 mm; other character not as above or in different combination **8**

8. Elytra yellowish to yellowish-red, strongly contrasting with dark brown pronotum, frequently darker baso-medially (Figs. 19.235a, 19.236a); antennomeres V–X distinctly transverse (Figs. 19.235a, 19.236a); apical margin of male tergite VIII truncate, strongly sinuate or crenulate, with sublateral teeth (Figs. 19.235c, 19.236d); genitalia as illustrated (Figs. 19.235b, g, 19.236b,c,h) .... ***Pseudota* Casey**

– Combination of characters not as above **9**

9. Head, pronotum and abdomen dark brown, elytra yellowish-brown (Fig. 19.238a); antennomeres V–X moderately transverse (Fig. 19.238a); apical margin of male tergite VIII abruptly produced in middle 2/5, apex sinuately truncate, angulate laterally (Fig. 19.238e); median lobe of aedeagus as illustrated (Fig. 19.238b–d); spermatheca with basal part of capsule almost as broad as small spherical apex, stem elbowed at base of capsule and near middle, with single tight curl at base (Fig. 19.238i,j) .... ***Tetropla* Mulsant and Rey**

– Head, pronotum and abdomen uniformly dark brown to black (Figs. 19.226a, 19.228a), except for orange pronotum and yellowish elytra in *A. pennsylvanica* (Fig. 19.227a); apical margin of male tergite VIII truncate (Figs. 19.226d, 19.227f, 19.228d); median lobe of aedeagus as illustrated (Figs. 19.226b,c, 19.227b–d, 19.228b, c); spermatheca with capsule tubular, broadest at rounded apex, stem angulately bent at base of capsule (Figs. 19.226h,i, 19.227j,k, 19.228h)...... .... ***Microdota* Mulsant and Rey**

(Subgenus *Alaobia* C.G. Thomson, 1858)

(Fig. 19.184a–h) (Gusarov 2003a)

Diagnosis Body narrowly ovoid, length 2.5–3.0 mm; integument highly glossy, unevenly coloured; antennomere IV subquadrate, VII–X distinctly transverse; cervical carina not diverged; ligula fusiform; pronotum rounded, with all pubescence on disc directed straight or obliquely posteriad; pronotal hypomera narrowly visible in lateral view; male tergite VIII truncate apically; median lobe of aedeagus in dorsal view with small bulbus and very elongate narrow tubus; in lateral view ventral margin of tubus long, arcuate, internal sac without distinct sclerites; spermatheca with moderately large club-shaped capsule with narrow and moderately long apical invagination, stem angulately bent at base of capsule, narrowly looped posteriorly and irregularly twisted at base.

113. ***Atheta (Alaobia) ventricosa*** **Bernhauer** (Fig. 19.184a–h)

Atheta (Atheta) ventricosa Bernhauer, 1907. Gusarov 2003a, Klimaszewski et al. 2005, 2007b, 2008a, b, Majka and Klimaszewski 2008b, 2010, McLean et al. 2009, Klimaszewski et al. 2011, 2015a, 2018a, 2020, Stefani et al. 2016.

Diagnosis Body narrowly ovoid, moderately convex, length 2.5–3.0 mm; colour brown to dark brown, with antennomeres I–III, palps, legs, often pronotum and part of elytra, and basal and apical part of abdomen paler, yellowish-brown (Fig. 19.184a); integument highly glossy; pubescence fine and sparse, denser on elytra; head slightly narrower than pronotum; antennae moderately thick, antennomere IV subquadrate, VII–X distinctly transverse, each about 1.2–1.4 times as wide as long; pronotum round, highly convex, moderately transverse, widest at middle, distinctly narrower than elytra, all pubescence on disc directed latero-posteriad; elytra at suture about as long as pronotum; abdomen mostly subparallel. MALE. Tergite VIII with apical margin obliquely produced laterally, broadly shallowly emarginate medially (Fig. 19.184d); sternite VIII parabolically rounded from base to apex (Fig. 19.184e); median lobe of aedeagus in dorsal view with bulbus small, roughly rhomboid, tubus long, subparallel and gradually tapering apicad (Fig. 19.184c), in lateral view tubus long, ventral margin curved obliquely ventrad basally, almost straight to middle, then thin and slightly arcuate to narrowly rounded apex (Fig. 19.184b). FEMALE. Tergite VIII with apical margin obliquely produced laterally, truncate in about apical 1/3 (Fig. 19.184f); sternite VIII broadly parabolically rounded from base to apex (Fig. 19.184g); spermatheca with capsule large, club-shaped with apex broadly cone-shaped, stem arcuate, looped and irregularly twisted posteriorly (Fig. 19.184h).

COMMENTS Although Bernhauer (1906, 1907) provided records of this species from CA, MA and WA, these have not been verified based on examination of genitalia. We treat these records as tentative (marked as [?]) until vouchers can be examined.

DISTRIBUTION **Origin**: Nearctic. **CANADA**: AB, BC, LB, NB, NF, NS, ON, QC, SK, YT. **USA**: AK, CA[?], DC, IN, MA[?], NC, NH, NJ, NY, PA, VT, WA[?].

COLLECTION AND HABITAT DATA

Habitat: in NB, found in various forest types usually in decaying fungi (often common) and leaf litter; mature mixed forest in polypore fungi, mixed forest with cedar in litter at margin of vernal pond, old spruce forest in *Tricholoma* sp. in various stages of decay, spruce and balsam fir forests in decaying lobster mushrooms, mature red spruce and eastern white cedar forests in decaying mushrooms and leaf litter, old-growth eastern white cedar swamp in moss and litter at base of tree (6 April, probably overwintering site), balsam fir stand in moldy conifer duff, and in *Carex* marsh in litter at base of tree. In NF, in unbaited and carrion-baited pitfall traps and flight intercept traps in various coniferous, mixed wood, and riparian forests. Elsewhere, in conifer forests and organic litter in mixed forests. In AK, specimens were captured in *Betula, Salix*, white spruce forest; *Betula* and *Populus tremuloides* forest; decaying mushrooms in *Betula* forest; forest edge, *Populus* litter; old growth conifer and thinned and unthinned 2nd growth forests; *Picea glauca* forest; clear-cuts; coastal rain forest. **Collecting period**: IV–X. **Collecting method**: sifting fungi and various kinds of forest litter, flight intercept traps, Winkler Litter Extractor, Berlese funnels, unbaited and carrion-baited pitfall traps, flight intercept traps, sweeping vegetation, Lindgren traps.

Additional records. See Annex 1.

COMMENTS This species was recorded from CA, MA, and WA by Bernhauer (1907) but these records were not confirmed by Gusarov (2003a) and are therefore considered uncertain.

(Subgenus *Atheta* C.G. Thomson, 1858)

(Figs. 19.185–19.189)

(Benick and Lohse 1974, Muona 1979, Klimaszewski and Winchester 2002, Klimaszewski et al. 2018a, 2020)

Diagnosis Body length 3.5–4.4 mm, narrowly subparallel, elongate, usually highly glossy; pronotum with pubescence on disc directed laterad from midline; pronotal hypomera fully visible in lateral view; antennomere IV slightly elongate to subquadrate, IV–X slightly transverse, quadrate or slightly elongate; spermatheca with stem thin, irregularly twisted and bearing small tuberculate projections.

COMMENTS The shape of spermatheca constitutes an apomorphy of this group. This subgenus contains mostly northern species.

Key to Species of Subgenus *Atheta*

1. Antennomeres VII–X distinctly transverse (Fig. 19.189a); median lobe of aedeagus and spermatheca as illustrated (Fig. 19.189b,c,d,i) ....***Atheta (Atheta) ringi* Klimaszewski**

– Antennomeres VII–X elongate, subquadrate or slightly transverse; median lobe of aedeagus and spermatheca not as above...........................2

2. Antennomeres VII–X distinctly elongate (Figs. 19.185a, 19.187a) **3**

– Antennomeres VII–X subquadrate or transverse (Figs. 19.188a, 19.186a)..........................**4**

3. Head almost as wide as pronotum, elytra with coarse asperate punctation (Fig. 19.187a); body compact and moderately robust (Fig. 19.187a); male tergite VIII with apical margin truncate, crenulate (Fig. 19.187d); male sternite VIII notched apically (Fig. 19.187e); median lobe of aedeagus with tubus sinuate along ventral margin in lateral view (Fig. 19.187b); spermatheca as illustrated (Fig. 19.187h)....................... ....***Atheta graminicola* (Gravenhorst)**

– Head distinctly narrower than pronotum, elytra with fine punctation (Fig. 19.185a); body narrowly elongate (Fig. 19.185a); male unknown; female spermatheca as illustrated (Fig. 19.185d). ....***Atheta borealides* Newton**

4. Antennomeres VII–X subquadrate or slightly transverse (Fig. 19.188a); body black with elytra yellowish-brown; male tergite VIII with apical margin crenulate between triangular lateral teeth (Fig. 19.188e); tubus of median lobe of aedeagus as illustrated (Fig. 19.188b,c,d) ....***Atheta martini* Lohse**

– Antennomeres VII–X subquadrate (Fig. 19.186a); body dark brown with appendages and elytra reddish-brown (Fig. 19.186a); male tergite VIII with apical margin entire, lacking teeth and crenulation (Fig. 19.186d); tubus of aedeagus as illustrated (Fig. 19.186b,c,h)........... ....***Atheta circulicollis* Lohse**

114. *Atheta* (*Atheta*) *borealides* Newton

(Fig. 19.185a–d)

Atheta (*Atheta*) *borealides* Newton, 2017 (replacement name). *Atheta* (*Atheta*) *borealis* Klimaszewski and Langor, in Klimaszewski et al. 2011 (preoccupied, junior secondary homonym of *Atheta borealis* (C.R. Sahlberg, 1831)). Brunke et al. 2012, Ernst and Buddle 2013, Klimaszewski et al. 2016b, 2018a [as *A. borealides*].

Diagnosis Body narrowly ovoid, moderately convex, length 4.8–5.0 mm; colour piceous to black, with tibiae and tarsi rusty-brown, elytra posteriorly with some diffuse, irregularly shaped, small reddish spots (Fig. 19.185a); entire body highly glossy; pubescence fine and sparse; head elongate, narrower than pronotum; antennae slender, antennomeres IV–VI elongate, each about twice as long as wide, VII–X subquadrate to slightly elongate; pronotum slightly transverse, widest before middle, slightly narrower than elytra; elytra at suture slightly longer than pronotum; abdomen with sides arcuate, gradually narrowed posteriorly. MALE. Undescribed. FEMALE. Tergite VIII unevenly truncate/arcuate apically (Fig. 19.185b); sternite VIII broadly rounded apically (Fig. 19.185c); spermatheca with broad, pitcher-shaped capsule, stem long, thin, irregularly twisted and coiled, with basal part highly granulate (Fig. 19.185d).

DISTRIBUTION Origin: Nearctic. **CANADA**: AB, NF, NU, ON. **USA**: not recorded.

COLLECTION AND HABITAT DATA
Habitat: in NL, from pitfall traps in riparian forest and a coastal limestone barren; in ON, sweeping in wet sedge meadow. Nothing else has been reported about the specific microhabitat associations. **Collecting period**: VII–VIII, X. **Collecting method**: pitfall traps and sweeping.

115. *Atheta* (*Atheta*) *circulicollis* Lohse (Fig. 19.186a–h)

Atheta circulicollis Lohse, in Lohse et al. 1990. Klimaszewski et al. 2011, 2018a, Brunke et al. 2012, Webster et al. 2012.

Diagnosis Body narrowly ovoid, moderately convex, length 3.8–4.0 mm; colour dark brown to piceous with elytra, bases of antennae, palps, legs, and often apical part of abdomen light brown to rusty brown; integument glossy, with head, pronotum and abdomen highly so (Fig. 19.186a); pubescence fine and sparse, denser on elytra; head narrower than pronotum; antennae thick, antennomeres IV–X subquadrate, antennomere XI about 3 times as long as wide; pronotum slightly transverse, widest at apical third, narrower than elytra; elytra at suture as long as pronotum; abdomen gradually narrowed posteriorly. MALE. Tergite VIII with apical margin entire, very obtusely produced laterally, broadly truncate medially (Fig. 19.186d); sternite VIII parabolically rounded from base to apex (Fig. 19.186e); median lobe of aedeagus in dorsal view with bulbus oval, slightly compressed laterally, tubus short, triangularly shaped (Fig. 19.186c), in lateral view tubus curved slightly ventrad basally, ventral margin broadly arcuate to narrowly rounded apex (Fig. 19.186b). FEMALE. Tergite VIII with apical margin arcuately produced laterally, broadly shallowly emarginate medially (Fig. 19.186f); sternite VIII broadly rounded apically (Fig. 19.186g); spermatheca with capsule roughly S-shaped, tubular, with broader pitcher-shaped apical part, stem irregularly twisted and coiled with small granulate projections (Fig. 19.186h).

DISTRIBUTION Origin: Nearctic. **CANADA**: LB, NB, NF, NT [**NTR**], ON, QC. **USA**: AK [**NSR**].

COLLECTION AND HABITAT DATA
Habitat: a forest species probably associated with leaf litter. In NB, collected in old-growth eastern white cedar swamps in moss and leaves under alders on brook margin and among moist leaves at base of tree, and in mixed forest in a *Pleurotus* sp. on the side of a log. In NF, in fir forests, riparian forests, and a recently burned coniferous forest. Specimens from ON from flooded mixed forest and other unspecified forests. Lohse et al. (1990) reported it from northern boreal forest localities in QC as far north as tree line. In AK, found in shrub meadow, *Hylocomius splendens*; open tussock and standing water, willow and bog birch; burned black spruce forest; old growth hemlock forest; old growth and thinned and unthinned 2nd growth forests; and clear-cuts. **Collecting period**: V–X. **Collecting method**: Berlese funnels, Malaise traps, pitfall traps, sifting litter.

We report this species from NT (**NTR**) and AK (**NSR**) for the first time here, based on the following: Canada, Northwest Territories, Yellowknife, 62.51923, −113.38494, Rep. 2, wet, pitfall trap, 7–11.VII.2011, NBP Field Party Y2WP812 (LFC) 1 male; Yellowknife, 62.51923, −113.38494, Rep. 2, wet, pitfall trap, 11-15.VII.2011, NBP Field Party Y2WP623 (LFC) 1 female; Yellowknife, 62.51923, −113.38494, Rep. 2, wet, yellow pan, pitfall traps, 15-18.VII.2011, NBP Field Party Y2WY931, Y2WP831, Y2WP832 (LFC) 2 males, 1 female. USA, Alaska, Dall Isl., pl.16, 557m el., 54.99485, −133.01144, ±4m shrub meadow, *Hylocomius splendens,* pitfall, 13-14. VII.2011 (UAM) 1 specimen, (UAM) 1 specimen; Elliot Highway Mile 87, 334m el., 65.36793, −148.95032, ±10m burned b. spruce, hanging Malaise, 4.54m, 30.VI-14.VII.2016 (UAM)

1 specimen; Peninsula NWR 1484 Ugashik River, 29m el., 57.702, –156.52213, ±8m pitfall trap, 29.VIII.2007 (UAM) 1 specimen; Prince of Wales Island, Coffman Cv, 67m el., 55.9795, –132.86256, ±101m thinned 2nd growth, pitfall 4, 25.VI-9.VII.2012 (UAM) 1 specimen; 28.VI-12.VII.2011 (UAM) 1 specimen; Prince of Wales Island, Coffman Cv, 78m el., 55.98053, –132.8607, ±15m unthinned 2nd growth, pitfall 1, 10-30.VI.2010 (UAM) 2 specimens; same data except, pitfall 4, 11-25.VI.2012 (UAM) 1 specimen; pitfall 1, 11-25.VI.2012 (UAM) 1 specimen; pitfall 3, 24.VII-11.VIII.2010 (UAM) 1 specimen; Prince of Wales Island, Hatchery Ck.1, 49m el., 55.92654, –132.95645, ±10m unthinned 2nd growth, pitfall 4, 11-25.VI.2012 (UAM) 2 specimens, same data except, 25.VI-9.VII.2012 (UAM) 1 specimen; pitfalls 3, 4, 27.VI-10.VII.2010 (UAM) 5 specimens; pitfall 2, 6-19.VII.2011 (UAM) 1 specimen; Prince of Wales Island, Hatchery Ck.1, 50m el., 55.92444, –132.93938, ±4m old growth, pitfall 1, 24.V-5.VI.2011 (UAM) 1 specimen; Prince of Wales Island, Hatchery Ck.1, 57m el., 55.92524, –132.95054, ±10m thinned 2nd growth, pitfall 1, 4-27.VI.2010 (UAM) 2 specimens; Prince of Wales Island, Hatchery Ck.2, 134m el., 55.89356, –132.9437, ±13m thinned 2nd growth, pitfalls 3, 4, 11-25.VI.2012 (UAM) 2 specimens; same data except, pitfall 3, 2-29.VI.2010 (UAM) 5 specimens; Berlese, 25.VI.2012 (UAM) 1 specimen; pitfall 4, 9-24.VII.2012 (UAM) 1 specimen; Prince of Wales Island, Hatchery Ck.4, 78m el., 55.88602, –132.8607, ±11m old growth, pitfall 3, 25.VI-9.VII.2012 (UAM) 1 specimen; pitfall 2, 28.VI-12.VII.2011 (UAM) 1 specimen; Prince of Wales Island, Hatchery Ck.4, 82m el., 55.88285, –132.89795, ±27m unthinned 2nd growth, pitfall 4, 11-25.VI.2012 (UAM) 1 specimen; same data except, pitfalls 2, 3, 12-27.VII.2011 (UAM) 4 specimens; pitfalls 1, 4, 15-27.VI.2010 (UAM) 3 specimens; pitfalls 1, 2, 3, 25.VI-9.VII.2012 (UAM) 7 specimens; pitfall 4, 28.VI-12.VII.2011 (UAM) 1 specimen; pitfall 2, 31.V-14.VI.2011 (UAM) 1 specimen; pitfalls 1, 2, 9-24.VII.2012 (UAM) 7 specimens; 25.VI.-9.VII.2012 (UAM) 2 specimens; Prince of Wales Island, Luck Lk. 1 Rd., 101m el., 55.97805, –132.75456, ±10m old growth, pitfall 3, 13-27.VI.2012 (UAM) 2 specimens; same data except, pitfalls 1, 2, 3, 4, 27.VI-11.VII.2012 (UAM) 15 specimens; pitfalls 1, 2, 29.VI-8.VII.2010 (UAM) 2 specimens; pitfall 4, 4-29.VI.2010 (UAM) 1 specimen; pitfalls 1, 3, 4, 8-30.VII.2010 (UAM) 3 specimens; Prince of Wales Island, Luck Lk. 2 Rd., 105m el., 55.96855, –132.75615, ±10m old growth, pitfall 2, 13-27.VI.2012 (UAM) 2 specimens; same data except, pitfalls 1, 2, 27.VI-11.VII.2012 (UAM) 9 specimens; pitfall 1, 4-29.VI.2010, (UAM) 1 specimen; pitfall 4, 4-29.VI.2010, (UAM) 1 specimen; pitfall 2, 5-21.VI.2011 (UAM) 1 specimen; Prince of Wales Island, Luck Lk. 3 Rd., 117m el., 55.95347, –132.7708, ±5m old growth, Berlese, 18.VI.2012 (UAM) 1 specimen; same data except, 4.VI.2012 (UAM) 1 specimen; pitfalls 3, 4, 13-27.VI.2012 (UAM) 10 specimens; pitfall 3, 31.V-13.VI.2012 (UAM) 1 specimen; pitfall 4, 4-29.VI.2010 (UAM) 1 specimen; Prince of Wales Island, Luck Point 1A, 174m el., 55.98452, –132.78786, ±26m clear-cut, pitfall 3, 27.VI-11.VII.2012 (UAM) 1 specimen; Prince of Wales Island, Luck Point 1B, 169m el., 55.98497, –132.787, ±25m clear-cut, Berlese, 5.VI.2011 (UAM) 1 specimen; same data except, pitfall 2, 14-27.VI.2012 (UAM) 2 specimens; pitfalls 1, 2, 27.VI-11.VII.2012 (UAM) 2 specimens; 31.V-14.VI.2012 (UAM) 1 specimen; Prince of Wales Island, Luck Point, 205m el., 55.98261, –132.77986, ±6m thinned 2nd growth, pitfall 4, 14-27.VI.2012 (UAM) 1 specimen; same data except, pitfalls 1, 3, 4, 27.VI-11.VII.2012 (UAM) 16 specimens; Prince of Wales Island, Luck Point 2A, 245m el., 55.97953, –132.77156, ±24m clear-cut, pitfall 3, 2-27.VI.2010 (UAM) 4 specimens; same data except, pitfall 3, 20.VI-5.VII.2011 (UAM) 1 specimen; pitfalls 3, 4, 27.VI-11.VII.2012 (UAM) 3 specimens; Prince of Wales Island, Luck Point 2B, 264m el., 55.97939, –132.77216, ±25m clear-cut, Berlese, 5.VI.2011 (UAM) 2 specimens;same data except, pitfall 3, 27.VI-9.VII.2010 (UAM) 1 specimen; pitfall 2, 31.V-14.VII.2012 (UAM) 1 specimen; pitfall 2, 9.VII-1.VIII.2010, (UAM) 1 specimen;

Prince of Wales Island, Luck Pt., 129m el., 55.98256, −132.77943, ±12m unthinned 2nd growth, pitfalls 1, 3, 4, 14-27.VI.2012 (UAM) 10 specimens; same data except, pitfall 3, 18.VII-2.VIII.2011 (UAM) 1 specimen; pitfall 2, 2-27.VI.2010 (UAM) 14 specimens; pitfalls 1, 2, 3, 4, 27.VI-11.VII.2012 (UAM) 4 specimens; pitfalls 1, 2, 27.VI-9.VII.2010 (UAM) 4 specimens; pitfall 3, 31.V-14.VI.2012 (UAM) 1 specimen; pitfall 2, 5-18.VII.2011 (UAM) 1 specimen; Berlese, 5.VI.2011 (UAM) 2 specimens; pitfalls 1, 4, 9.VII-1.VIII.2010 (UAM) 1 specimen; pitfall 4, 9.VII-1.VIII.2010 (UAM) 1 specimen; Prince of Wales Island, Staney Ck, 63m el., 55.79723, −133.13467, ±51m unthinned 2nd growth, pitfall 4, 11-25.VI.2012 (UAM) 2 specimens; same data except, pitfall 4, 14-24.VII.2010 (UAM) 1 specimen; pitfall 3, 14-30.V.2010 (UAM) 1 specimen; pitfalls 1, 3, 4, 25.VI-9.VII.2012 (UAM) 9 specimens; pitfalls 1, 3, 27.VI-14.VII.2010 (UAM) 3 specimens; pitfalls 2, 4, 9-24.VII.2012 (UAM) 9 specimens; Prince of Wales Island, Staney Ck., 89m el., 55.87134, −133.06755, ±24m clear-cut, pitfalls 1, 2, 3, 11-25.VI.2012 (UAM) 4 specimens; same data except, pitfall 3, 12-27.VI.2010 (UAM) 1 specimen; pitfall 4, 14-28.V.2012 (UAM) 1 specimen; pitfalls 1, 4, 25.VI-9.VII.2012 (UAM) 3 specimens; pitfall 2, 27.VI-12.VII.2011 (UAM) 1 specimen; pitfalls 2, 4, 28.V-11.VI.2012 (UAM) 1 specimen; pitfall 3, 30.V-12.VI.2010, (UAM) 2 specimens; pitfall 4, 31.V-13.VI.2011 (UAM) 1 specimen; Prince of Wales Island, Staney Ck. 1B, 77m el., 55.872, −133.06523, ±26m clear-cut, pitfall 4, 11-25.VI.2012 (UAM) 1 specimen; pitfalls 3, 4, 12-27.VII. 2011 (UAM) 3 specimens; pitfalls 1, 3, 4, 13-27.VI.2011 (UAM) 4 specimens; pitfall 3, 15-25.VII.2010 (UAM) 1 specimen; pitfall 2, 16-31.V.2011 (UAM) 1 specimen; pitfalls 2, 3, 27.VI-12.VII.2011 (UAM) 2 specimens; pitfall 1, 27.VI-15.VII.2010 (UAM) 1 specimen; pitfalls 1, 3, 4, 28.V-11.VI.2012 (UAM) 3 specimens; Prince of Wales Island, Staney Ck., 41-45m el., 55.79901, −133.11782, ±20m old growth, Berlese, 2.VI.2011 (UAM) 1 specimen; same data except, pitfall 1, 10-27.VI.2010 (UAM) 1 specimen; Prince of Wales Island, Staney Ck., 50m el., 55.79726, −133.1363, ±20m thinned 2nd growth, pitfall 4, 24.VII-11.VIII.2010 (UAM) 2 specimens; same data except, pitfall 1, 27.VI-12.VII.2011 (UAM) 1 specimen; White Mountain Recreation Area, Nome Creek Site 5, 589m el., 65.34077, −146.9275, ±10m open tussock and standing water, willow and bog birch, hanging Malaise, 4.95m, 12.VI-3.VII.2015 (UAM) 1 specimen; Wrangell, near rifle range, 56.47307, −132.34599, ±50m old growth hemlock, pitfall, 26.VI-8.VII.2009, (UAM) 1 specimen.

COMMENTS The following collector names were cut from label data to save space: C. Bickford, J.C. Hagelin, S.E, Hawks, A. Hutton, I. MacDougall, S. Meierotto, D.S. Sikes, J. Stockbridge, S. Wise-Eagle and B. Wong.

116. *Atheta* (*Atheta*) *graminicola* (Gravenhorst) * (Fig. 19.187a–h)

Aleochara graminicola Gravenhorst, 1806. As *Atheta granulata* (Mannerheim): Lohse and Smetana 1985 [lectotype design.], Lohse et al. 1990, as *A. graminicola*: Bernhauer 1907, Casey 1910 [as two synonyms], Gusarov 2003a [synonymy], Webster et al. 2009, Sazhnev 2018, Klimaszewski et al. 2011, 2015a, 2018a.

Diagnosis Body narrowly ovoid, moderately convex, length 3.5–4.4 mm; colour dark brown, often with antennae, palps, legs, pronotum/elytra (or elytra only) and apical part of abdomen light reddish-brown or entire body black, elytra with some red-brown tinge and appendages brown (Fig. 19.187a); integument strongly glossy; pubescence fine and sparse, denser on elytra; head slightly narrower than pronotum; antennae robust, with all antennomeres elongate, each about 1.5 times as long as wide; pronotum rounded, moderately transverse, widest at middle, narrower than elytra; elytra at suture at least as long as pronotum; abdomen arcuately narrowed to apex posteriorly. MALE. Tergite VIII with apical margin obliquely produced laterally, broadly truncate and coarsely crenulate apically between two rounded angulate teeth

(Fig. 19.187d); sternite VIII parabolically rounded from base to apex except narrowly emarginate at apex (Fig. 19.187e); median lobe of aedeagus in dorsal view with bulbus broad and short, abruptly constricted at base of triangular tubus (Fig. 19.187c); in lateral view tubus deeply constricted at base, ventral margin highly sinuate to curved, moderately narrowly rounded apex (Fig. 19.187b). FEMALE. Tergite VIII with apical margin broadly truncate, sinuate laterally (Fig. 19.187f); sternite VIII broadly rounded apically, less so mediad (Fig. 19.187g); spermatheca with capsule U-shaped tubular, with apical part broader pitcher-shaped, stem irregularly twisted and coiled with small granulate projections (Fig. 19.187h).

DISTRIBUTION Origin: Holarctic.

CANADA: AB, BC, LB, MB, NB, NF, NT, ON, QC, SK, YT. **USA**: AK, NH [**NSR**], OR.

COLLECTION AND HABITAT DATA

Habitat: in NB, mostly associated with various wetland habitats; *Carex* marsh in emergent vegetation (treading, very common), splashing exposed bedrock next to waterfalls, in leaf litter and moss under alders next to small brook in *Carex* marsh, among cobblestones, gravel, and sand on gravel bar near outflow of brook into a river, in hardwood forest in moist leaves on vernal pond margin, upper margin of sea beach in drift material (mostly grass litter), in leaf litter at base of tree in red oak and red maple forest, in leaf litter in crotch of silver maple in silver maple swamp in early April (probably overwintering site); two individuals collected with aerial net in late afternoon (15:00–18:00 h) when snow still present in shaded areas of woods. In QC, found in leaf litter next to stream and in *Salix* litter at two localities in YT. In NL, collected using a flight intercept trap in a mixed forest. In AK, found in maritime tundra; beach, with *Heracleum lanatum*; streamside; in CTS-*Salix*, and fern and cow parsnip; under CTS-*Salix* near the edge with crowberry; in edges of crowberry, ~3–5 m of closed tall shrub; mesic *Betula and Picea* stand; small stand with *Betula* and *Populus tremuloides*; black and white spruce, wild rose, ~20m from creek; *Salix* leaf litter; *Salix*, tundra hillside; lake shore and pond; 20m lakeshore sand, detritus; cutting area with spruce, birch, grass; forest edge, meadow, grass, forbs; spruce bog; *Eriophorum* tussock shrubby wetland; brackish shoreline, under rocks, wrack; back beach, *Leymus* grass; sandy lakeshore, alder detritus; muddy river bank; and shoreline vegetation, under log. **Collecting period**: IV–IX. **Collecting method**: sifting litter, treading marsh vegetation, aerial net, flight intercept traps, pitfall traps, sweeping vegetation, and yellow pan trap.

We report this species from NH (**NSR**) for the first time here, based on the following: USA, New Hampshire, Coos Co., Mt. Washington, Alpine Garden, 5300 feet, 12.IX.1987, dwarf *Salix*, *Vaccinium*, leatherleaf litter, J.M. Campbell and A. Davies (CNC) 1 male, 9 sex undetermined.

Additional records. See Annex 1.

COMMENTS We followed Gusarov (2003a) who considered *A. granulata* Lohse to be synonymous with *A. graminicola*. However, more research is needed to determine the status of the Nearctic population as there are a few Nearctic specimens from NT with barcode sequences that cluster with those of Palearctic ones, but most (ON, MB etc.) form a separate cluster.

117. *Atheta* (*Atheta*) *martini* Lohse (Fig. 19.188a–f)

Atheta martini Lohse, in Lohse et al. 1990.

Diagnosis Body narrowly ovoid, moderately convex, length 4.0 mm; colour black, elytra yellow except brown scutellar and lateral regions, bases of antennae light brown, legs yellowish (Fig. 19.188a); integument forebody moderately glossy, abdomen strongly so; pubescence fine and sparse, denser on elytra; head transverse, as wide as length of pronotum at middle; eyes subequal in length to postocular region of head in dorsal view; antennae robust, antennomere IV elongate, antennomeres VIII–X, each about 1.5 times as long as wide; pronotum transverse, widest in apical third, 11/4 as wide as long; elytra 1 1/3 as broad as pronotum, at suture 1 1/5 as long as pronotum; abdomen subparallel. MALE. Tergite

VIII with apical margin coarsely crenulate between two large sublateral teeth (Fig. 19.188e); sternite VIII broadly parabolically rounded from base to apex (Fig. 19.188f); median lobe of aedeagus in dorsal view with bulbus narrowly elongate, gradually constricted at base of triangular tubus (Fig. 19.188d); in lateral view tubus slightly constricted at base, ventral margin straight basally and gently curved ventrad before narrowly rounded apex (Fig. 19.188b,c); internal sac structures as illustrated. FEMALE. Undescribed.

DISTRIBUTION **Origin**: Nearctic. **CANADA**: YT. **USA**: not recorded.

COLLECTION AND HABITAT DATA
Habitat: not recorded. **Collecting period**: VIII. **Collecting method**: not recorded.

COMMENTS The only known specimen of this species, the male holotype, was captured at Rampart House, Yukon (CNC).

118. *Atheta* (*Atheta*) *ringi* Klimaszewski (Fig. 19.189a–i)

Atheta (Atheta) ringi Klimaszewski, in Klimaszewski and Winchester, 2002. McLean et al. 2009, Klimaszewski et al. 2020.

Diagnosis Body length 3.2–4.0 mm, moderately convex, narrowly subparallel, dark brown, sometimes head and basal part of abdomen nearly black, elytra, legs and basal antennal segment rust-brown (Fig. 19.189a); forebody moderately sparsely punctate and pubescent, pubescence long, punctation slightly asperate, microsculpture meshed and particularly well visible on pronotum, sculpticells slightly convex; head slightly elongate, round, slightly depressed medially, distinctly narrower than pronotum, postocular region as long as or slightly shorter than the maximum diameter of eye, tempora carinate, carinae strong basally and diffused apically; antenna slender and long, slightly widening apically, approximately as long as pronotum and elytra combined, antennomeres I–III strongly elongate, IV slightly elongate, V quadrate, VI–X slightly transverse (up to about one third broader than long), terminal one elongate and nearly as long as two preceding antennomeres combined, basal antennomere IV glossy and remaining matte; pronotum impressed mesobasally, slightly transverse, approximately rectangular in shape, hypomeron visible for almost entire pronotal length in lateral view, pubescence directed anteriad at one third of apical midline, obliquely laterad across arcuate lines medially and obliquely posteriad elsewhere; elytra moderately convex, slightly transverse, distinctly broader than pronotum, posterior margins broadly arcuate and with small but distinct shallow lateral emarginations, pubescence directed obliquely posteriad, forming wavy pattern basally; abdomen subparallel, three first visible tergites strongly impressed at base; basal metatarsomere about one third longer than the following one. MALE. Tergite VIII with apical margin obliquely produced laterally, broadly truncate medially (Fig. 19.189e); sternite VIII distinctly elongate, apical margin parabolic except vaguely angulate medially (Fig. 19.189f); median lobe of aedeagus with broadly oval large bulbus and short triangular tubus in dorsal view (Fig. 19.189d), in lateral view tubus straight ventrally, apical part very narrow, slightly curved ventrad (Fig. 19.189b,c). FEMALE. Tergite VIII transverse, truncate apically (Fig. 19.189g); sternite VIII transverse, with apical margin broadly rounded except briefly straight medially (Fig. 19.189h); spermatheca with pitcher-shaped and sinuate capsule, stem irregularly twisted and finely granulate (Fig. 19.189i).

DISTRIBUTION **Origin**: Nearctic. **CANADA**: BC. **USA**: AK.

COLLECTION AND HABITAT DATA
Habitat: adults were captured in forest litter in Sitka spruce forest, in forest interior, transition zone, and clearcut, with most captured in transition zone. **Collecting period**: VI–VIII. **Collecting method**: pitfall traps.

COMMENTS A list of localities is published in Klimaszewski and Winchester (2002).

(Subgenus *Bessobia* C.G. Thomson, 1858)

(Fig. 19.190a–i) (Smetana 2004)

Diagnosis Body narrowly subparallel, length 2.5–2.7 mm; integument slightly glossy, unevenly coloured; antennomere IV slightly transverse, VII–X distinctly transverse, about twice as wide as long; pronotum rectangular with all pubescence on disc directed obliquely laterad from midline; pronotal hypomera fully visible in lateral view; male tergite VIII with entire apical margin subrectangularly produced, rounded laterally, broadly sinuate along margin, without teeth; median lobe of aedeagus in dorsal view with large oval bulbus and very short, right-angled triangular tubus, in lateral view ventral margin of tubus irregularly shallowly sinuate; spermathecal capsule small, club-shaped, with small apical invagination, sinuate stem narrowly looped posteriorly.

119. *Atheta* (*Bessobia*) *cryptica* (Lohse) (Fig. 19.190a–i)

Bessobia cryptica Lohse, in Lohse et al. 1990. Klimaszewski et al. 2008b, 2011, 2015a, 2018a, 2020.

Diagnosis Body elongate, narrowly subparallel, moderately convex, length 2.5–2.7 mm; colour brown, often with reddish tinge, basal three antennomeres, palps, legs, pronotum, and elytra light brown to reddish-brown, moderately glossy (Fig. 19.190a); pubescence fine and dense, denser on elytra; head slightly narrower than pronotum, impressed between eyes; antennae moderately thick and incrassate, antennomere IV slightly transverse, V–X distinctly transverse, each at least 1.5 times as wide as long; pronotum rectangular in shape, transverse, widest at middle, slightly narrower than elytra; elytra at suture about as long as pronotum along midline; abdomen mostly subparallel and widening apicad to about 3/4 of its length, then moderately narrowing apically. MALE. Tergite VIII with entire apical margin subrectangularly produced, lateral angles rounded and posterior margin shallowly sinuate (Fig. 19.190e); sternite VIII rounded apically (Fig. 19.190f); median lobe of aedeagus with large oval bulbus and very short triangular tubus in dorsal view (Fig. 19.190d), in lateral view ventral margin of tubus bent obtusely ventrad at base, irregularly, shallowly sinuate to narrowly triangular apex (Fig. 19.190b,c). FEMALE. Tergite VIII arcuate apically (Fig. 19.190g); sternite VIII broadly rounded apically (Fig. 19.190h); spermatheca with small club-shaped capsule, with small apical invagination, and sinuate stem narrowly looped posteriorly (Fig. 19.190i).

DISTRIBUTION Origin: Nearctic. **CANADA**: AB [**NPR**], BC, NF, QC, YT. **USA**: AK [**NSR**].

COLLECTION AND HABITAT DATA
Habitat: collected using pitfall traps in fir forests (NL); sifted from willow (*Salix* sp.) leaf litter (YT); in AB, found in wetland habitats and in birch and spruce forests. In AK, collected in wetland, on slope of birch and spruce forest, and forest edge *Populus tremuloides* and *Salix* spp. **Collecting period**: V–VII. **Collecting method**: sifting forest litter, pitfall traps.

We report this species from AB [**NPR** and **BOLD**] and from AK [**NSR**] for the first time here, based on the following: Canada, Alberta, Elk Island National Park, Wood bison trail, 722m el., 53.567, −112.851, wetland alongside aspen forest, pitfall, 29.VI-03.VII.2012, BIOBus 2012 team (CBG) 1 specimen; Jasper National Park, Pocahontas campground, 1131m el., 53.195, −117.914, birch and spruce forest, flight-intercept trap, 16-21.VII.2012, BIOBus 2012 team (CBG) 1 specimen; Waterton Lakes National Park, Highway 6 pullout, 1569m el., 49.065, −113.779, montane forest, douglas fir and lodgepole pine stand with aspen and birch understory, flight-intercept trap, 21-27.VI.2012, BIOBus 2012 team (CBG) 1 specimen. USA, Alaska, 25 mi. S Delta Junction, 552m el., 63.67343, −145.88283, ±4m old glaciated; *Populus tremuloides, P. balsamea, Picea,* Berlese, 7.IX.2009, G. Henry (UAM) 1 specimen; Chena Hot Springs Rd, 200m el., 64.89227,

−147.50857, ±30m forest edge *Pop. trem. Salix* spp., pitfall, 17-24.VII.2007, B. Mortensen (UAM) 1 specimen.

COMMENTS This species was listed from AK and OR by Newton (2019) but we are not aware of published voucher data for OR, or voucher data for AK published before the present contribution.

(Subgenus *Datomicra* Mulsant and Rey, 1873b)

(Figs. 19.191–19.193) (Smetana 2004, Klimaszewski et al. 2011, 2018a, 2020, Lee and Ahn, 2017)

Diagnosis Body length 0.9–2.5 mm; pronotum distinctly transverse, usually densely and coarsely punctate, matte with finely reticulate microsculpture, setae along midline directed anteriad; antennomeres VII–X distinctly transverse. This subgenus generally is similar to *Dimetrota* but the body size is smaller.

COMMENTS Species of this subgenus were recently reviewed for South Korea, including some species occurring in Canada (Lee and Ahn 2017).

Key to Species of Subgenus *Datomicra*

1. Pronotal punctures very coarse and deep (Fig. 19.193a); median lobe of aedeagus and spermatheca as illustrated (Fig. 19.190b,c,h) ....***Atheta (Datomicra) whitehorsensis* Klimaszewski and Godin**

– Pronotal punctures finer and more shallow (Figs. 19.191a, 19.192a); median lobe of aedeagus and spermatheca not as above **2**

2. Elytra slightly broader than pronotum at base, shoulders rounded (Fig. 19.192a); male tergite VIII with four small apical teeth (Fig. 19.192d); apex of median lobe of aedeagus broad in lateral view (Fig. 19.192b); spermathecal stem not looped posteriorly (Fig. 19.192h).......... ....***Atheta (Datomicra) dadopora* Thomson**

– Elytra distinctly broader than pronotum at base, shoulders slightly angular (Fig. 19.191a); male tergite VIII without apical teeth (Fig. 19.191c); apex of median lobe of aedeagus narrow in lateral view (Fig. 19.191b); spermathecal stem looped and coiled posteriorly (Fig. 19.191g).....................***Atheta (Datomicra) chadburnae* Klimaszewski and Godin, sp. n.**

120. *Atheta (Datomicra) chadburnae* Klimaszewski and Godin, sp. n. (Fig. 19.191a–g)

urn:lsid:zoobank.org:act:14B035D2-E1C1-4F3D-81C6-4A35CBF192C1

Holotype (male): Canada, Yukon, Whitehorse, Chadburn Lake, 60.6543, −134.9505, 667 m, 7.VII.2013, riparian area, soil sifting, coll. B. Godin and D. Horwood (CNC) 1 male. Paratype: same label data as the holotype (LFC) 1 female.

Etymology The name of this species is derived from Chadburn Lake in Yukon, the type locality for this species.

Diagnosis Body small, subparallel, length 2.3–2.5 mm; colour uniformly dark brown, with tarsi and posterior section of elytra slightly paler, rusty-brown, integument moderately glossy, forebody with coarse and dense asperate punctation and meshed microsculpture (Fig. 19.191a); head slightly narrower than pronotum, slightly transverse; antennae slender, antennomere slightly IV transverse, V–X distinctly transverse; pronotum distinctly transverse, narrower than elytra; elytra at suture about as long as pronotum; abdomen subparallel and strongly glossy posteriad. MALE. Tergite VIII sharply truncate apically, slightly concave medially, without teeth (Fig. 19.191c); sternite VIII parabolic apically (Fig. 19.191d); median lobe of aedeagus with moderately-sized bulbus and tubus sinuate ventrally in lateral view (Fig. 19.191b). FEMALE. Tergite VIII with apical margin very shallowly, broadly emarginate, slightly obliquely produced laterally (Fig. 19.191e); sternite VIII rounded apically (Fig. 19.191f); spermatheca with tubular sinuate capsule, stem strongly sinuate and half-looped with tight spiral posteriorly (Fig. 19.191g).

This species may be differentiated from the other congeners in the shape of median lobe of aedeagus, shape of spermatheca, and in having the apical margin of male tergite VIII sharply truncate.

DISTRIBUTION Origin: Nearctic. **CANADA**: YT. **USA**: not recorded.

COLLECTION AND HABITAT DATA
Habitat: riparian area. **Collecting period**: VI, VII. **Collecting method**: sifting litter, soil, and old caribou dung.

COMMENTS The following paratype of *Atheta smetanai* (Lohse) represents misidentification for *Atheta chadburnae*: Yukon, British Mts., Firth R., 250 m, 69°13', –140°04', 18.VI.1984, 84-9B, sifting under old caribou dung, Paratype - *Dimetrota smetanai* Lohse, G.A. Lohse 1985, CNC No. 20331 (CNC) 1 female.

121. *Atheta (Datomicra) dadopora* Thomson *
(Fig. 19.192a–h)

Atheta dadopora Thomson, 1867. Casey 1910 [as *Datomicra decolorata*, *D. inopia*, *D. schematica* and *D. stilla*], Gusarov 2003a, Klimaszewski et al. 2005, Majka and Klimaszewski 2008a, 2008b, 2008c, Klimaszewski et al. 2007b, 2008a, 2011, 2015a, 2018a, b, 2020, Majka and Sikes 2009, Lee and Ahn, 2017.

Diagnosis Body small and narrow, subparallel, length 1.8–2.2 mm; colour uniformly dark brown to black, with tibiae and tarsi rusty-brown, integument slightly glossy, forebody with coarse and dense asperate punctation and meshed microsculpture (Fig. 19.192a); head slightly narrower than pronotum, slightly transverse; antennae slender, antennomere IV subquadrate, and V–X slightly transverse; pronotum distinctly transverse, about as wide as elytra; elytra at suture about as long as pronotum; abdomen subparallel. MALE. Tergite VIII truncate apically, with two sublateral, sharply pointed teeth and two small submedian rounded teeth (Fig. 19.192d); sternite VIII slightly elongate, apical margin broadly triangular, apex rounded (Fig. 19.192e); median lobe of aedeagus with large oval bulbus, tubus short and broad, rounded apically in dorsal view (Fig. 19.192c), tubus straight and short with broad triangular apex in lateral view (Fig. 19.192b). FEMALE. Tergite VIII with apical margin obliquely produced laterally, broadly, subangulately emarginate medially (Fig. 19.192f); sternite VIII with apical margin broadly, slightly emarginate medially, rounded laterally (Fig. 19.192g); spermatheca with capsule slightly tapering to rounded apex, stem short and sinuate (Fig. 19.192h).

DISTRIBUTION Origin: Holarctic, recorded from Europe, China, India, Japan and South Korea. **CANADA**: AB, BC, LB, MB, NB, NF, NS, NT [**BOLD**], ON, PE, QC, SK, YT. **USA**: AK, NY, PA, RI.

COLLECTION AND HABITAT DATA
Habitat: in various mixed, coniferous, and hardwood forest types, usually in decaying materials such as decaying fungi and carrion; from Lindgren funnel trap in red spruce forest with red maple and balsam fir, unbaited pitfall traps in red spruce forest, spruce forest in decaying fleshy fungi, mixed forests on *Pleurotus* sp. on *Populus* log, red oak and red maple forest in gilled mushrooms, alder swamp on remains of dead deer, hardwood forests in gilled fungi, and rich Appalachian hardwood forest in fleshy fungi in various stages of decay; and also in compost, poplar/spruce litter, carrion, bison and caribou dung, and under the bark of decaying spruce logs. In AK, found in *Betula, Salix*, white spruce; *Betula* and *Populus tremuloides*; white spruce, *Betula* forest; OTS-*Salix*, crowberry hummock; and *Salix barclayi* meadow. **Collecting period**: V–IX. **Collecting method**: sifting various kinds of organic materials, flight intercept traps, Lindgren funnel traps, unbaited and carrion-baited pitfall traps, sweeping vegetation.

We report this species from NT (**NTR**) for the first time here, based on **BOLD** records: Canada, Northwest Territories, Norman Wells, 77m el. 65.2791, –126.83, urban area, Malaise trap 06, 13.VIII.2014, S. Behrens, R. Popko (CBG) 1 specimen.

Additional records. See Annex 1.

COMMENTS This species was listed from MA by Newton (2019) but we are not aware of any published voucher data for this record.

122. *Atheta* (*Datomicra*) *whitehorsensis* Klimaszewski and Godin (Fig. 19.193a–h)

Atheta (Datomicra) whitehorsensis Klimaszewski and Godin, in Klimaszewski et al. 2012. Webster et al. 2016, Klimaszewski et al. 2018a.

Diagnosis Body narrowly oval, length 1.9–2.0 mm; colour dark brown to black, with bases of antennae and legs rust-brown, surface matte, forebody with dense asperate punctation and coarse meshed microsculpture (Fig. 19.193a); head narrower than pronotum and elytra; eyes large, slightly protruding, slightly longer than genae in dorsal view; antennae slender, slightly incrassate apically, antennomere IV subquadrate, V slightly elongate and VI–X slightly to distinctly transverse; pronotum distinctly transverse and broadest in the middle; elytra transverse, at suture longer than pronotum along midline; abdomen broadly arcuate laterally. MALE. Tergite VIII transverse and truncate apically (Fig. 19.193d); sternite VIII widely rounded apically (Fig. 19.193e); median lobe of aedeagus with venter of tubus straight and short, apex narrow and rounded in lateral view (Fig. 19.193b), in dorsal view bulbus large, oval and tubus short and triangular (Fig. 19.193c). FEMALE. Tergite VIII with apical margin arcuate on each side of broad, shallow medial emargination, sternite VIII truncate apically (Fig. 19.193f,g); spermatheca with club-shaped capsule and long stem curled posteriorly (Fig. 19.193h).

DISTRIBUTION Origin: Nearctic. **CANADA**: NB, YT. **USA**: not recorded.

COLLECTION AND HABITAT DATA

Habitat: in NB, near a mixed forest in a pile of decaying corncobs; in YT sifted from soil in a black spruce stand, otherwise little else is known about the habitat associations of this species. **Collecting period**: VIII–IX. **Collecting method**: sifting litter and soil.

(Subgenus *Dimetrota* Mulsant and Rey, 1873a)

(Figs. 19.194–19.224)

(Benick and Lohse 1974, Seevers 1978, Lohse et al. 1990, Klimaszewski et al. 2011, 2018a, 2020)

Diagnosis Forebody with distinct meshed or irregularly hexagonal microsculpture, punctation asperate; head slightly narrower than pronotum; eyes of moderate size, subequal in length to genae in dorsal view, postocular carina present; glossae divided into two lobes in apical half; pronotum transverse with hypomeron partially or fully visible in lateral view, pronotum with pubescence on disc directed mostly anteriad at midline, laterad or obliquely posteriad elsewhere, lateral bristles thick and long; elytra at suture subequal in length to length of pronotum along midline in most, hind margins without sinuation, pubescence on disc directed obliquely posteriad, sometimes forming wavy patterns; legs moderately long, tibiae with long bristles; basal tarsomere of hind tarsus about subequal in length to second one; abdomen sparsely punctate and glossy.

COMMENTS Canadian species are arranged into species groups, and species within groups are arranged in alphabetical order. Only a key to species groups is provided for this subgenus because it contains clusters of morphologically similar species within species groups and positive identification requires examination of genitalic characters. The reader should first identify the taxon to species group and then examine and compare genitalic characters to those for all known species in the group for specific identification.

Key to Species Groups of the Subgenus *Dimetrota* of *Atheta*

(Modified from Klimaszewski et al. 2018a, 2020)

1. Apical margin of male tergite VIII deeply (Fig. 19.194d) and of female shallowly emarginate medially (Fig. 19.194f); base of tubus of median lobe of aedeagus as broad as bulbus in dorsal view (Fig. 19.194c) .... ***Cadeti*** **species group**

– Apical margin of male and female tergite VIII not as above; base of tubus of median lobe of aedeagus narrower than bulbus in dorsal view **2**

2. Median lobe of aedeagus in dorsal view with tubus long, narrowly triangular, constricted subapically and spatulate at apex (Fig. 19.201c), in lateral view tubus straight or slightly produced ventrally in apical one-third to one-half (Figs. 19.201b, 19.202b, 19.203b); internal sac with two highly elongate sclerites in bulbus (Figs. 19.201b,c, 19.202b,c, 19.203b,c); male tergite VIII with apical margin shallowly emarginate, with very small lateral teeth (Fig. 19.202d), or crenulate, with moderately large lateral teeth (Figs. 19.201d, 19.203d) .... ***Crenuliventris*** **species group**

– Median lobe of aedeagus and male tergite VIII not as above **3**

3. Antennomeres VIII–X subquadrate to transverse (species groups: *Allocera*, *Stercoris*, *Crassicornis*, *Fanatica*, *Hampshirensis*, *Picipennis*, *Prudhoensis*) **4**

– Antennomeres VIII–X elongate (Fig. 19.204a); median lobe of aedeagus in dorsal view with bulbus moderate-sized, tubus long, subparallel, triangular apically (Fig. 19.204c), in lateral view tubus straight and produced ventrally at apex (Fig. 19.204b); male tergite VIII with apical margin approximately straight or slightly angularly emarginate with small lateral teeth (Fig. 19.204d); spermatheca with capsule long, tubular, with tubular appendix and small apical invagination, stem sinuate (Fig. 19.204h) .... ***Districta*** **species group**

4. Body small, length 2.0–2.6 mm, punctation of forebody dense and asperate, punctures separated by less than two diameters of a puncture (Fig. 19.207a); microsculpture of forebody weak; male tergite VIII with two long and thin spine-like lateral teeth (Fig. 19.207d); internal sac of median lobe of aedeagus with two narrowly elongate sclerites in bulbus (Fig. 19.207c); spermatheca with capsule tubular, strongly curved medially and narrowed apically, apex directed ventrad, apical invagination absent, stem sinuate and looped posteriorly (Fig. 19.207h)
Hampshirensis **species group**

– Body medium-sized to larger, length over 2.6 mm in most; punctation of forebody less dense; other characters not as above **5**

5. Body strongly glossy, punctation of forebody dense and asperate (Fig. 19.205a); median lobe of aedeagus in dorsal view with bulbus oval in outline and streamlined with base of tubus, tubus short and triangular (Fig. 19.205c); in lateral view tubus broad, slightly to strongly sinuate ventrally and apex broadly rounded not produced ventrad (Fig. 19.205b); male tergite VIII with apical margin truncate, with small lateral teeth, apical margin slightly crenulate (Fig. 19.205d); male sternite VIII with long macrosetae (Fig. 19.205e); spermatheca with capsule spherical or broadly tubular and rounded apically, stem sinuate and looped posteriorly (Fig. 19.205h) .... ***Fanatica*** **species group**

– Body moderately glossy; genital characters not as above **6**

6. Median lobe of aedeagus with large oval bulbus and apically acute tubus in dorsal view (Figs. 19.195c–19.200c), in lateral view apical part of tubus moderately to strongly produced ventrad (Figs. 19.195b–19.200b) **7**

– Median lobe of aedeagus not as above **9**

7. Median lobe of aedeagus in lateral view with apical part of tubus moderately long and apical part strongly, angularly produced ventrally (Figs. 19.195b–19.198b); male tergite VIII with apical margin straight or shallowly emarginate, teeth variable in size and shape (Figs. 19.195d–19.198d).... ***Allocera*** **(=*****Campbelli*****) species group**

– Median lobe of aedeagus in lateral view with tubus almost straight in basal half and moderately produced ventrally at apex, apical part of tubus not as above; other characters not as above **8**

8. Apical margin of male tergite VIII with two lateral teeth and strongly sinuate margin between teeth (Fig. 19.199d); bulbus of median lobe with very elongate and thin sclerites (Fig. 19.199c);

spermatheca with very long stem and narrow and long posterior loop (Fig. 19.199h) ....
..............***Stercoris* (= *Columbica*) species group**

– Apical margin of male tergite VIII with two lateral teeth and margin between teeth truncate/crenulate (Fig. 19.200d); bulbus of median lobe with thicker and shorter sclerites as illustrated (Fig. 19.200c); spermatheca with moderately long stem and narrow and short posterior loop (Fig. 19.200h)...
.............***Crassicornis* (=*Modesta*) species group**

9. Tubus of median lobe of aedeagus in dorsal view gradually or triangularly narrowing from apex of bulbus (Figs. 19.213c–19.217c, 19.220c, 19.222c), in lateral view ventral margin of tubus distinctly curved ventrad only in apical half (Figs. 19.213b–19.217b, 19.220b–19.222b); male tergite VIII with apical margin straight to obtusely emarginate, lateral teeth not extending beyond margin (Figs. 19.213d–19.217d, 19.220d, 19.221c, 19.222d); spermatheca with stem with only a single long, sinuate loop posteriorly (Figs. 19.213h, 19.215h, 19.216h).............
.... ***Prudhoensis* species group**

– Tubus of median lobe of aedeagus in dorsal view with sides parallel to slightly constricted from apex of bulbus (Figs. 19.208c–19.212c), in lateral view tubus with entire ventral margin arcuate, straight, or abruptly bent ventrad near apex (Figs. 19.208b–19.212b); internal sac with two medium-sized, elongate sclerites in bulbus and additional structures in tubus (Figs. 19.208b,c–19.212b,c); spermatheca with stem long, thin, with multiple dense, irregular coils posteriorly (spaghetti-like) (Figs. 19.208h–19.212h).............
....................................***Picipennis* species group**

***Cadeti* species group**

Diagnosis Antennae robust with antennomeres V–X distinctly transverse; integument of forebody moderately glossy with dense microsculpture of irregularly hexagonal sculpticells; elytra with punctation asperate; tubus of median lobe of aedeagus as broad as base of bulbus basally in dorsal view, in lateral view tubus angularly bent ventrad; male and female tergite VIII emarginate medially and with no teeth.

123. *Atheta* (*Dimetrota*) *cadeti* Klimaszewski and Godin (Fig. 19.194a–h)

Atheta (Dimetrota) cadeti Klimaszewski and Godin, in Klimaszewski et al. 2008a.

Diagnosis Body length 2.7–3.0 mm, dark brown to black, with legs and elytra dark brownish yellow, darkened at scutellum and on anterior angles (Fig. 19.194a); punctation fine, microsculpture present and consisting of irregularly hexagonal sculpticells; head round, nearly as large as pronotum, postocular area shorter than diameter of eye; antennae moderately robust, shorter than pronotum and elytra combined, antennomeres IV–X distinctly transverse; pronotum small, strongly transverse, 1.5 times as broad as long, with pubescence directed laterad from the mid-line of the disc; elytra at suture longer than pronotum with pubescence directed obliquely laterad; abdomen subparallel and tapering apically. MALE. Tergite VIII with apical margin deeply emarginate medially (Fig. 19.194d); sternite VIII rounded apically and with wide distance between basal margin and antecostal suture (Fig. 19.194e); median lobe of the aedeagus broad dorsally, tubus broad basally, short and triangular in dorsal view (Fig. 19.194c), venter of tubus sinuate, angularly bent ventrad medially in lateral view (Fig. 19.194b). FEMALE. Tergite VIII slightly emarginate at apex (Fig. 19.194f); sternite VIII with wide distance between base and antecostal suture (Fig. 19.194g), spermatheca generalized in shape as illustrated (Fig. 19.194h).

DISTRIBUTION Origin: Nearctic. **CANADA**: YT. **USA**: not recorded.

COLLECTION AND HABITAT DATA

Habitat: adults were captured in white spruce forest and a mixed white spruce – lodgepole pine forest. **Collecting period**: VI–VIII. **Collecting method**: sifting litter.

***Allocera* species group (=*Campbelli* Klimaszewski et al. 2018a, 2020)** (new name, present designation)

Diagnosis Antennae slender, antennomeres V–X moderately to distinctly transverse; integument of forebody moderately glossy, with microsculpture moderate to coarse, consisting of irregularly hexagonal sculpticells; median lobe of aedeagus in dorsal view with bulbus enlarged, tubus short and triangular, in lateral view tubus straight and strongly produced ventrally at apex; internal sac with two coarse, falciform sclerites in bulbus; male tergite VIII with apical margin shallowly concave, with very small lateral teeth, or truncate with large apical teeth and apical margin crenulate or emarginate; spermatheca with capsule club-shaped, tubular with apical invagination narrow and moderately deep, stem sinuate.

124. *Atheta* (*Dimetrota*) *allocera* Eppelsheim * (Figs. 19.195a–h)

Atheta (Dimetrota) allocera Eppelsheim, 1893. Palm 1970, Smetana 2004.

Dimetrota campbelli Lohse, in Lohse et al. 1990. Klimaszewski et al. 2011, 2018a, Brunke et al. 2012, Webster et al. 2016. **New Synonymy**.

Diagnosis Body subparallel, slightly flattened, length 2.8–3.0 mm; colour dark brown to black, with tarsi, basal parts of antennae and elytra light yellowish-brown or reddish-brown; integument highly glossy (Fig. 19.195a); forebody with punctation sparse and finely asperate, with fine meshed microsculpture; pubescence on disc of pronotum directed laterad and obliquely posteriad from midline, forming arcuate pattern, that on elytra directed obliquely posteriad; head slightly narrower than pronotum; antennae slender, antennomeres V–X slightly to distinctly transverse; pronotum transverse, about 1/7 narrower than width of base of elytra; elytra transverse, at suture as or slightly longer than pronotum; abdomen subparallel. MALE. Tergite VIII with apical margin truncate, evenly crenulate between two small sublateral obtuse teeth (Fig. 19.195d), sternite VIII broadly rounded apically (Fig. 19.195e); tubus of median lobe of aedeagus in lateral view with ventral margin arcuate, curved ventrad before moderately narrow apical part (Fig. 19.195b); in dorsal view tubus shortly triangular, evenly coalescent with large and oval bulbus (Fig. 19.195c). FEMALE. Tergite VIII with apical margin arcuate on each side of broad, shallow medial emargination (Fig. 19.195f); sternite VIII with apical margin rounded except very shallowly emarginate medially (Fig. 19.195g); spermatheca with capsule club-shaped, stem long, thin, with a few broad, irregular coils posteriorly (Fig. 19.195h).

DISTRIBUTION Origin: Holarctic [**new status**]. **CANADA**: AB [**BOLD**], LB, NB, NF, ON, QC [**NPR**]. YT. **USA**: AK.

COLLECTION AND HABITAT DATA

Habitat: in NB, adults in dung and litter in and near entrances of woodchuck (*Marmota monax*) burrows in a meadow, jack pine forest adjacent to a field, and a hardwood forest next to a meadow. Elsewhere, adults captured in unbaited and carrion-baited pitfall traps in balsam fir forests and in rotting mushrooms in a mixed forest (NF), a hedgerow (ON), and from bear and caribou dung. In AB, found in subalpine lodgepole pine and fir forest with birch in understory. In AK, collected in *Betula, Salix*, white spruce forest; Alder, white spruce, *Betula* forest; and maritime tundra, moist valleys. **Collecting period**: V–VIII. **Collecting method**: sifting litter and dung, unbaited and carrion baited pitfall traps, flight intercept trap.

We report this species from AB (**BOLD**) and QC (**LFC**) for the first time here, based on the following: Canada, Alberta, Waterton Lakes National Park, Akamina Parkway, Rowe Tamarack Trail, 1640m el. 49.059, –114.014, subalpine lodgepole pine and fir forest with birch in understory, pitfall, 22-27.VI.2012, BIOBus 2012 team (CBG) 2 specimens. Quebec, Gaspé, Mt. Albert, Sommet nord, 1000m, F. Génier and J. Klimaszewski (LFC) 1 female.

Additional records. See Annex 1.

COMMENTS We have compared specimens of *Atheta allocera* from Europe (Finland: Sodankylä, Kuusamo, Utsjoki) with specimens of *A. campbelli* from Canada and did not find any significant morphological differences in external morphology and genitalia. We consider the two populations as belonging to the same species, and the name which has priority is *A. allocera*. DNA barcoded material is available from Finland, Alaska, British Columbia, Alberta, Yukon, and Newfoundland. All specimens have identical barcode sequences. *Atheta allocera* is extremely similar externally and genitally to *A. alabama* Klimaszewski and Peck, which is known only from caves in Morgan Co., Alabama (Klimaszewski and Peck 1986). *Atheta alabama* is either a cryptic species closly affiliated with *A. allocera* or conspecific with it and representing relict population occurring only in caves in southern distribution areas. DNA study is needed to compare populations of *A. alabama* with those of *A. allocera* from northern North America to unravel the true identity of *A. alabama*.

125. *Atheta* (*Dimetrota*) *diversa* Sharp * (Fig. 19.196a–h)

Homalota diversa Sharp, 1869. Palm 1970.

Dimetrota regissalmonis Lohse, in Lohse et al. 1990. Work et al. 2013, Klimaszewski et al. 2011, 2018a. **New Synonymy**.

Diagnosis Body narrowly oval, slightly flattened, length 3.0–3.5 mm; colour dark piceous brown, with legs/tarsi, basal parts of antennae and elytra brownish or reddish-brown, scutellar region of elytra darker; integument moderately glossy (Fig. 19.196a), forebody with punctation moderately dense and finely asperate, with coarse meshed microsculpture; pubescence on disc of pronotum directed laterad and obliquely posteriad from midline, that on elytra directed obliquely posteriad, forming wavy pattern posteriorly; head slightly narrower than pronotum; antennae moderately robust, antennomeres V–X subquadrate to slightly transverse; pronotum transverse, about 1/6 narrower than width of base of elytra; elytra transverse, at suture as long as pronotum; abdomen arcuate laterally. MALE. Apical margin of tergite VIII truncate and slightly crenulate between two small sublateral teeth (Fig. 19.196d); sternite VIII rounded apically (Fig. 19.196e); tubus of median lobe of aedeagus narrowly elongate and strongly bent ventrad before middle, apex pointed in lateral view (Fig. 19.196b); in dorsal view tubus broadly oval, and triangular apically (Fig. 19.196c). FEMALE. Tergite VIII with apical margin broadly truncate, rounded laterally (Fig. 19.196f); sternite VIII broadly rounded apically (Fig. 19.196g); spermatheca with small spherical capsule, stem long, thin, sinuate, narrowly looped posteriorly (Fig. 19.196h).

The median lobe of the aedeagus of this species is similar to that of *A. smetanai*, but in *A. diversa* the body proportions are different, antennomeres V–X are subquadrate (transverse in *A. smetanai*), and male tergite VIII and the spermatheca are shaped differently.

DISTRIBUTION Origin: Holarctic [**new status**]. **CANADA**: LB, NF, QC. **USA**: AK.

COLLECTION AND HABITAT DATA

Habitat: specimens were captured in unbaited and carrion-baited pitfall traps in a 60-year old fir forest and in spruce/moss and spruce/lichen forests in North America. In northern Europe, this species has been collected from various types of decaying organic material, especially carrion and dung, both above and below the tree line (Palm 1970, M. Pentinsaari pers.obs.). **Collecting period**: VI–VIII. **Collecting method**: pitfall traps.

COMMENTS Two specimens of *A. diversa* from Alaska have been DNA barcoded. Their genetic divergence from two specimens barcoded from Finland is 0.005 (uncorrected p-distance). We have compared external morphology and genitalia of specimens from Finland with those from North America and found no significant difference between these populations. We therefore offically synonymize the Nearctic name of this species with the older name from Europe.

126. *Atheta* (*Dimetrota*) *remulsa* Casey (Fig. 19.197a–h)

Atheta (*Atheta*) *remulsa* Casey, 1910. Casey 1910 [as the synonyms *A. asperiola* and *A. frugalis*], Klimaszewski et al. 2005, 2007b, Dollin et al. 2008, Majka and Klimaszewski 2008b, Klimaszewski et al. 2011, 2016d, in *Atheta* (*Dimetrota*): Klimaszewski et al. 2018a, b, 2020.

Diagnosis Body narrowly subparallel, slightly flattened, length 2.4–2.8 mm; colour dark brown to black, with legs/tarsi, basal parts of antennae and elytra yellowish-brown or reddish-brown; integument moderately glossy (Fig. 19.197a); forebody with punctation dense, coarse and finely asperate, with coarse meshed microsculpture; pubescence on disc of pronotum directed laterad and obliquely posteriad, forming arcuate lines from midline, that on elytra directed obliquely posteriad forming wavy pattern posteriorly; head slightly narrower than pronotum; antennae moderately robust, antennomeres V–X distinctly transverse; pronotum transverse, about 1/5 narrower than width of base of elytra; elytra transverse, at suture as long as pronotum; abdomen subparallel. MALE. Tergite VIII truncate apically, with two blunt sublateral teeth and medially emarginated broad projection (Fig. 19.197d); sternite VIII rounded apically (Fig. 19.197e); tubus of median lobe of aedeagus straight basally and strongly curved ventrad in apical half in lateral view (Fig. 19.197b); in dorsal view bulbus broadly oval and tubus broadly triangular apically (Fig. 19.197c). FEMALE. Tergite VIII with apical margin broadly, shallowly emarginate medially, rounded laterally (Fig. 19.197f); sternite VIII broadly rounded apically (Fig. 19.197g); spermatheca with tubular, clubbed-shaped capsule with asymmetrically located apical invagination, and with long sinuate stem narrowly looped posteriorly (Fig. 19.197h).

DISTRIBUTION Origin: Nearctic. **CANADA**: AB, BC, LB, MB, NB, NF, NS, ON, QC, SK, YT. **USA**: AK, NY.

COLLECTION AND HABITAT DATA
Habitat: in various forest types, usually in mushrooms; yellow birch/balsam fir forest, red spruce forest, old-growth hemlock forest, mixed forests, deciduous forests, and a red oak forest; most individuals in gilled mushrooms and fleshy fungi in various stages of decay, in leaf litter and moss and in *Carex* marsh. In AK, found in *Betula* and *Populus tremuloides* forest, and in mushrooms. **Collecting period**: V–X. **Collecting method**: sifting litter and fungi, pitfall traps, pitfall Luminoc traps, flight intercept traps.

Additional records. See Annex 1.

127. *Atheta* (*Dimetrota*) *smetanai* (Lohse) (Fig. 19.198a–h)

Dimetrota smetanai Lohse, in Lohse et al. 1990. Klimaszewski et al. 2018a.

Diagnosis Body narrowly subparallel, slightly flattened, length 2.8–3.0 mm; colour dark brown to almost black, with legs/tarsi, and often elytra reddish brown; forebody with punctation moderately dense and asperate, with moderate meshed microsculpture (Fig. 19.198a); integument moderately glossy, pubescence on disc of pronotum directed laterad and obliquely posteriad from midline, that on elytra directed obliquely posteriad forming wavy patterns posteriorly; head slightly narrower than pronotum; antennae slender, antennomeres V–X transverse; pronotum transverse, about 1/6 narrower than width of base of elytra; elytra transverse, at suture as long as pronotum; abdomen subparallel. MALE. Tergite VIII with apical margin broadly emarginate, angular laterally, with more or less distinct crenulation (Fig. 19.198d); sternite VIII with apical margin broadly parabolically rounded (Fig. 19.198e); tubus of median lobe of aedeagus in lateral view straight basally, strongly bent ventrad at middle, tapering to narrow apex (Fig. 19.198b); in dorsal view tubus short, broadly triangular, bulbus large and oval with complex elongate structures of internal sac (Fig. 19.198c). FEMALE. Tergite VIII with apical margin broadly truncate medially, oblique laterally (Fig. 19.198f); sternite VIII unevenly rounded apically (Fig. 19.198g); spermatheca with tubular, club-shaped capsule with narrow invagination, stem sinuate and narrowly looped posteriorly (Fig. 19.198h).

DISTRIBUTION Origin: Nearctic. **CANADA**: LB, MB, NT, QC, YT. **USA**: AK.

COLLECTION AND HABITAT DATA
Habitat: specimens from under bear and caribou dung and in litter on tundra. **Collecting period**: VI–VIII. **Collecting method**: sifting litter and dung.

COMMENTS The following paratypes of *A. smetanai* represent misidentification: *Atheta chadburnae* Klimaszewski and Godin: Yukon, British Mts., Firth R., 250 m, 69°13', –140°04', 18.VI.1984, 84-9B, sifting under old caribou dung, Paratype- *Dimetrota smetanai* Lohse, G.A. Lohse 1985, CNC No. 20331 (CNC) 1 female. *Atheta campbelli* Lohse: YT, Dempster Hwy mi 139, 5.VII.1978, 1900 feet, J.M. Campbell and A. Smetana, CNC No. 20331 (CNC) 1 male; Alaska, Unalaklet, 21.VI.1961, R. Madge, CNC No. 20331 (CNC) 1 male.

***Stercoris* species group** (Klimaszewski et al. 2020 [as Columbica])

Diagnosis Antennae slender, antennomeres V–X moderately transverse; integument of forebody moderately glossy, with microsculpture moderate to coarse, consisting of irregularly hexagonal sculpticells; median lobe of aedeagus in dorsal view with bulbus large, oval in shape, tubus short and triangular, in lateral view tubus straight basally and apically strongly produced ventrad, apical part triangular; internal sac with two long, falciform sclerites in bulbus; male tergite VIII with apical margin variable in shape; spermatheca with capsule club-shaped, stem thin, very long, sinuate and narrowly looped posteriad.

128. *Atheta* (*Dimetrota*) *stercoris* Fenyes [replacement name for *A. columbica* Bernhauer, 1907] (Fig. 19.199a–h)

Atheta (Dimetrota) stercoris Fenyes, 1920. Klimaszewski et al. 2020.

Diagnosis Body moderately convex, broad, subparallel, length 3.7 mm, dark brown to black, elytra with some reddish tinge, apical part of abdomen, and legs paler, antennae uniformly dark brown (Fig. 19.199a); forebody moderately densely punctate and pubescent, pubescence long, punctation slightly asperate, microsculpture meshed and particularly well visible on pronotum, sculpticells slightly convex; head subquadrate, round posteriorly, distinctly narrower than pronotum, postocular region shorter than the maximum diameter of eye, tempora carinate, carinae complete, strong throughout; antennae slender, shorter than pronotum and elytra combined, antennomere IV subquadrate to slightly transverse, VII–X slightly transverse, basal 3 antennomeres glossy and remaining matte; pronotum impressed mesobasally, transverse, approximately rectangular in shape, basal margin sinuate, hypomeron visible for two thirds of pronotal length in lateral view, pubescence directed from midline of disc anteriad at apical third, on sides nearly horizontally laterad medially, and obliquely laterad posteriorly; elytra slightly convex, distinctly transverse, broader than pronotum, at suture slightly longer than pronotum along midline, posterior margins feebly sinuate and with shallow, scarcely visible, lateral emarginations, pubescence directed straight or slightly obliquely posteriad; abdomen subparallel, strongly glossy, four first visible tergites strongly impressed at base; basal metatarsomere as long as the following one. MALE. Tergite VIII with apical margin highly sinuate between two distinct arcuate lateral teeth (Fig. 19.199d); sternite VIII parabolic apically (Fig. 19.199e); median lobe of aedeagus broadly oval, apex narrowly triangular in dorsal view (Fig. 19.199c); in lateral view tubus with ventral margin about straight basally and arcuately curved ventrad before narrowly triangular apex (Fig. 19.199b). FEMALE. Tergite VIII transverse, apical margin broadly rounded on each side of shallow medial emargination

(Fig. 19.199f): sternite VIII transverse, apical margin broadly rounded (Fig. 19.199g); spermatheca with capsule small, hemispherical apically and with elbowed neck posteriorly, stem thin, long, broadly U-shaped medially, and narrowly looped posteriorly (Fig. 19.199h).

DISTRIBUTION **Origin**: Nearctic. **CANADA**: BC. **USA**: AK [**NSR**].

COLLECTION AND HABITAT DATA

Habitat: found in transition zone (between clear-cut and forest interior) of Sitka spruce forest, in gravel along edge of stream; miscellaneous alpine and subalpine habitats, maritime tundra, subalpine wet tundra on fresh deer, carnivore/bear and fresh marmot scats, tundra with *Cassiope mertensiana*, *Empetrum nigrum*, *Salix arctica*, on ground in rich meadow with *Geranium erianthum*, *Artemisia arctica*, *Sanguisorba canadensis*, wet meadow with *Senecio triangularis*, *Ranunculus occidentalis*, *Valeriana sitchensis*, flood meadow, *Athyrium*, *Caltha leptosepala*, subalpine meadow to alpine zone, meadow with *Leptarrhena pyrolifolia*, *Veratrum viride*, *Heracleum maximum*, in wetland with *Eriophorum angustifolium*, *Nephrophyllidium crista-galli*, *Carex nigricans*; shrub, willow, dwarf birch; heath, stream, shallow pond, snow fields; subalpine heath with *Empetrum*, *Vaccinium*; and ridgetops, tundra, on snowfields. **Collecting period**: VI–IX. **Collecting method**: yellow bowl traps, pitfall traps, hand picking from ground between vegetation, sifting organic litter, sifting bear dung, ex rotting *Boletus*, hand picking from ground, gravel along edge of stream.

We report this species from AK (**NSR**) for the first time here, based on the following: USA, Alaska, Baranof Is. pl.08, 758m el., 56.41191, –134.85774, ±4m wetland, *Eriophorum angustifolium*, *Nephrophyllidium crista-galli*, *Carex nigricans*, pitfall, 28-29.VII.2013, S. Ridling (UAM) 1 specimen; Baranof Is. wpt.171, 667m el., 56.41022, –134.86116, ±5m subalpine wet tundra, fresh deer scat, forceps, 27.VII.2013, S. Ridling (UAM) 1 specimen; Baranof Is. wpt.172, 752m el., 56.41158, –134.85837, ±5m subalpine wet tundra, fresh deer scat, forceps, 27.VII.2013, S. Ridling (UAM) 2 specimens; Baranof Is. wpt.174, 505m el., 56.41296, –134.84076, ±9m subalpine wet tundra, fresh carnivore scat, forceps, 27.VII.2013, S. Ridling (UAM) 1 specimen; Chichagof Is., 642-1015m el., 57.98567, –135.20573, ±1.5km subalpine meadow to alpine zone, hand collected, 14.VII.2013, S. Ridling (UAM) 1 specimen; Chichagof Is., 671m el., 57.9878, –135.20276, ±5m subalpine meadow, on/in bear scat w/ lots of berries, hand collected, 14.VII.2013, S. Ridling (UAM) 6 specimens; Chichagof Is., 692m el., 57.9872, –135.20332, ±5m subalpine meadow, on/in deer scat, hand collected, 14.VII.2013, S. Ridling (UAM) 1 specimen; Chichagof Is., 741m el., 57.98567, –135.20573, ±34m alpine meadow, 20 yellow pan traps, 14-15.VII.2013, S. Ridling (UAM) 1 specimen; Chichagof Is., 767m el., 57.9841, –135.20654, ±6m subalpine meadow, bear scat w/ lots of grass, hand collected, 14.VII.2013, S. Ridling (UAM) 2 specimens; Chichagof Is. pl.04, 728m el., 57.98643, –135.20468, ±4m rich meadow, *Geranium erianthum*, *Artemisia arctica*, *Sanguisorba canadensis*, ground search, 15.VII.2013, S. Ridling (UAM) 1 specimen; Chichagof Is. pl.12, 851m el., 57.98392, –135.21161, ±4m, meadow, *Leptarrhena pyrolifolia*, *Veratrum viride*, *Heracleum maximum*, ground search, 15.VII.2013, S. Ridling, (UAM) 1 specimen; Chichagof Is. pl.13, 874m el., 57.9832, –135.21288, ±5m low tundra, *Cassiope mertensiana*, *Empetrum nigrum*, *Salix arctica*, ground search, 16.VII.2013, S. Ridling (UAM) 1 specimen; Chichagof Is. pl.16, 963m el., 57.98159, –135.21707, ±4m rocky pavement, *Petasites frigidus*, *Parnassia fimbriata*, *Pedicularis oederi*, pitfall, 15-16.VII.2013, S. Ridling (UAM) 1 specimen; Chichagof Is. pl.17, 979m el., 57.9809, –135.21817, ±3m wet meadow, *Senecio triangularis*, *Ranunculus occidentalis*, *Valeriana sitchensis*, pitfall, 14-15.VII.2013, 17-18.VII.2013, S. Ridling (UAM) 3 specimens; Dall Isl., pl.17, 548m el., 54.99555, –133.01039, ±5m flood meadow, *Athyrium*, *Caltha leptosepala*, hand collected, 14.VII.2011, D. S. Sikes (UAM) 1 specimen;

Denali NPP., nr Toklat, 1047m el., 63.50232, –150.06764, ±29m shrub, willow, dwarf birch, 4 pitfalls w/ rainroofs, pooled, 2-21.VII.2016, J.J. Rykken, D. S. Sikes, A. Haberski, Z. Snelson (UAM) 1 specimen; Etolin Is., 754m el., 56.14261, –132.33534, ±1km heath, stream, shallow pond, snow fields, sweep, hand, 17-22. VII.2011, D. S. Sikes (UAM) 2 specimens; Etolin Isl., 760m el., 56.14311, –132.33711, ±5m alpine heath, sparse Tsuga, primate dung , pitfall, 19-22.VII.2011, D. S. Sikes(UAM) 17 specimens; Haines, Flower Mtn., 1010m el., 59.38636, –136.31877, ±1km misc. alpine habitats, hand collected, 15-19.VII.2010, D. S. Sikes (UAM) 1 specimen; Haines, Flower Mtn., 1117m el., 59.37817, –136.32945, ±505m under alpine rocks, hand collected, 15-19.VII.2010, D. S. Sikes (UAM) 1 specimen; Haines, Flower Mtn., 981m el., 59.38751, –136.31448, ±54m alpine heath, forbs, 20 yellow bowls, 17-19.VII.2010, D. S. Sikes (UAM) 1 specimen; Hawthorne Peak, 788m el., 58.24706, –134.2513, ±5m cirque valley, fresh marmot scat, hand collected, 20.VII.2013, S. Ridling (UAM) 8 specimens; Heintzleman Ridge, 810m el., 58.42327, –134.44302, ±504m alpine meadow, hand collected, 29.VI-3.VII.2010, D. S. Sikes (UAM) 5 specimens; Juneau, Heintzleman Ridge, 695m el., 58.42668, –134.43564, ±6m Marmota dung, 2J.VII.2010, D. S. Sikes (UAM) 1 specimen; Mahoney Mt., 55.40391, –131.57674, ±1m black bear scat, hand collected, 7.VII.2013, Karen Blejwas (UAM) 9 specimens; Mahoney Mt., 855-907m el., 55.41093, –131.58089, ±1.21km rocky ridgetops, tundra, on snowfields, hand collected, 7.VII.2013, S. Ridling (UAM) 1 specimen; Mahoney Mt. pl.02, 783m el., 55.41679, –131.5656, ±5m krummholz, *Chamaecyparis nootkatensis*, *Tsuga mertensiana*, *Nephrophyllidium* crista-galli, ground search, 9.VII.2013, S. Ridling (UAM) 2 specimens; Mahoney Mt. pl.07, 881m el., 55.41548, –131.57814, ±3m rocky low tundra, *Vaccinium uliginosum*, *Luetkea pectinata*, *Phyllodoce glandulifera*, pitfall, 6-8.VII.2013, S. Ridling (UAM) 1 specimen; Mahoney Mt. pl.20, 784m el., 55.40266, –131.57668, ±5m tundra, *Luetkea pectinata*, *Harrimanella stelleriana*, *Vaccinium uliginosum*, pitfall, 7-8. VII.2013, S. Ridling, (UAM) 1 specimen; Prince of Wales Isl. nr Black Lk, 858m el., 55.58892, –132.89342, ±17m alpine meadow vegetation, sweep, 9.VII.2011, Casey Bickford (UAM) 6 specimens; Prince of Wales Isl. nr Black Lk pl.02, 780m el., 55.58988, –132.89548, ±2m wet meadow, Calt. lept., Athy. filix-fem., hand, 8. VII.2011, Casey Bickford (UAM) 6 specimens; Prince of Wales Island, Coffman Cv, 78m el., 55.98053, –132.8607, ±15m unthinned 2nd growth, pitfall 3, 24.VII-7.VIII.2012, J. Stockbridge, S. Meierotto, I. MacDougall (UAM) 1 specimen; Prince of Wales Island, Coffman Cv, 78m el., 55.98053, –132.8607, ±15m unthinned 2nd growth, pitfall 4, 9-24. VII.2012, J. Stockbridge, S. Meierotto, I. MacDougall (UAM) 1 specimen; Prince of Wales Island, Hatchery Ck.2, 134m el., 55.89356, –132.9437, ±34m unthinned 2nd growth, pitfall 4, 12-27.V.2013, J. A. Slowik, A. Hutton (UAM) 1 specimen; Prince of Wales Island, Luck Lk. 1 Rd., 101m el., 55.97805, –132.75456, ±10m old growth, Lindgren 1, 11-26. VII.2012, J. Stockbridge, S. Meierotto, I. MacDougall (UAM) 1 specimen; Prince of Wales Island, Luck Point, 205m el., 55.98261, –132.77986, ±6m thinned 2nd growth, pitfall 1, 26.VII-9.VIII.2012, J. Stockbridge, S. Meierotto, I. MacDougall (UAM) 1 specimen; S. Baranof Is. E. of Snipe Bay, 688m el., 56.40948, –134.86179, ±17m subalpine tundra, near edge of krummholz patch, primate dung, baited pitfall, 29-30.VII.2013, S. Ridling (UAM) 1 specimen; S. Chilkat Range, 638m el., 58.42384, –135.14883, ±6m subalpine, primate dung, pitfall, 9-10.VII.2010, D. S. Sikes (UAM) 41 specimens; S. Chilkat Range, 658m el., 58.42489, –135.14287, ±6m brown bear scat, alpine, hand collected, 9.VII.2010, D. S. Sikes (UAM) 5 specimens; S. Chilkat Range, 780m el,. 58.43079, –135.16963, ±24m alpine heath, 24 yellow bowls, 9-10.VII.2010, D. S. Sikes (UAM) 1 specimen; S. Chilkat Range pl.02, 773m el., 58.43019, –135.16829, ±29m alpine *Carex*, *Cassiope*, pitfall, 8-9.VII.2010, D. S. Sikes (UAM) 3 specimens; S. Chilkat Range pl.09, 743m el., 58.427, –135.15852, ±30m subalpine heath, *Empetrum*, *Vaccinium*, pitfall, 11-12.VII.2010, D. S. Sikes (UAM) 1 specimen; Sanak Isl, 1-22m el., 54.37127, −162.61283, ±180m maritime tundra, pitfall trap, 24.VI-4.VII.2006, Joseph Cornell (UAM) 1 specimen.

***Crassicornis* (=*Modesta*) species group** (Gusarov 2003a, Klimaszewski et al. 2008a, 2018a, 2020)

Diagnosis Antennae slender, antennomeres V–X moderately to distinctly transverse; integument of forebody highly glossy, with coarse microsculpture consisting of irregularly shaped hexagonal sculpticells; pubescence on disc of elytra directed approximately straight posteriad; median lobe of aedeagus in dorsal view with bulbus large, elongate, oval, tubus moderately long and triangular apically, in lateral view tubus arcuate or straight in basal half and more or less produced ventrally at apex, bulbus of internal sac with two highly elongate sclerites and additional complex structures; male tergite VIII with apical margin straight or convex, with lateral teeth small or large and rounded; spermatheca with capsule moderately large, club-shaped, with small apical invagination, stem long, thin, sinuate, and coiled and twisted posteriorly.

129. *Atheta* (*Dimetrota*) *pseudometlakatlana* Klimaszewski and Godin (Fig. 19.200a–h)

Atheta (*Hypatheta*) *pseudometlakatlana* Klimaszewski and Godin, in Klimaszewski et al. 2008a. In *Atheta* (*Dimetrota*): Klimaszewski et al. 2012, 2016d, 2020.

Diagnosis Body moderately convex, narrowly oval, length 3.5–4.0 mm, dark brown, elytra light brown to yellowish or rust brown, legs yellowish brown, antenna uniformly dark brown with basal segment light brown to yellowish brown (Fig. 19.200a); forebody densely punctate and pubescent, pubescence long, punctation strongly asperate, microsculpture meshed and particularly well visible on pronotum, sculpticells slightly convex; head as broad as long or slightly elongate, round, distinctly narrower than pronotum, postocular region shorter than the maximum diameter of eye, tempora carinate, carinae complete, strong throughout; antennae slender and long, slightly incrassate, as long as the pronotum and elytra combined, antennomeres VII–X subquadrate to slightly transverse; pronotum strongly transverse, approximately rectangular in shape, pubescence directed anteriad at one third of apical midline, and obliquely laterad across arcuate lines elsewhere; elytra moderately convex, slightly elongate or about as long as wide, broader than pronotum, posterior margins broadly arcuate and with shallow lateral emarginations, pubescence directed obliquely posteriad; abdomen subparallel, strongly glossy; basal mesotarsomere as long as the following one. MALE. Tergite VIII transverse, apical margin truncate or shallowly emarginate, crenulate between two short, broad, and rounded lateral teeth (Fig. 19.200d); sternite VIII about as long as wide, and parabolic apically (Fig. 19.200e); median lobe of aedeagus in dorsal view with enlarged, broadly suboval bulbus and tubus short, triangular (Fig. 19.200c), in lateral view tubus short, straight mesomedially, bent obliquely ventrad at base, slightly more so before very narrow apex (Fig. 19.200b). FEMALE. Tergite VIII transverse, apical margin with broad, short, truncate process medially (Fig. 19.200f); sternite VIII transverse, apical margin broadly, shallowly emarginate, with fringe of long and dense setae (Fig. 19.200g); spermatheca with capsule hemispherical and round apically, stem narrowly elongate and broadly looped posteriorly with two small coils at apex (Fig. 19.200h).

Externally, this species is hardly distinguishable from its sister species *A. metlakatlana* (see description in Klimaszewski et al. 2020). There are minute external differences between the two species, and A. *pseudometlakatlana* may be distinguished from *A. metlakatlana* by the following: slightly more transverse antennomeres VII–X (Fig. 19.200a), and slightly shorter and more transverse elytra (Fig. 19.200a). The two species may be easily differentiated by the shape of tubus of median lobe of aedeagus and the internal sac structures. In *A. metlakatlana* tubus of median lobe is longer, almost straight and about horizontal in lateral view (for illustrations see Klimaszewski et al. 2020), and in *A. pseudometlakatlana* tubus is shorter, and strongly produced ventrad in lateral view (Fig. 19.200b). The terminal segments of male and female and spermatheca are very similar in the two species. Females may be identified with certainty only by associations with males and slight differences in the shape of the capsule of the spermatheca (Fig. 19.200h).

DISTRIBUTION Origin: Nearctic. **CANADA**: AB [**BOLD**], BC, SK, YT. **USA**: AK [**NSR**].

COLLECTION AND HABITAT DATA
Habitat: sub-boreal spruce forest, forest litter, fungi, Douglas fir montane forest. In AK, found in *Betula* and *Populus tremuloides* forest, and in maritime tundra with *Rumex, Calamagrostis,* and *Cerastium.* **Collecting period**: VI–IX. **Collecting method**: sifting forest litter and fungi, pitfall traps, flight intercept trap.

We report this species for the first time here from AB (**NPR**) based on **BOLD** records, and from AK based on UAM specimens: Canada, Alberta, Jasper National Park, Pyramid Lake, 1181m el., 52.916, −118.104, Douglas fir forest, pitfall, 25-28.VI.2010, BIOBus 2010 team (CBG) 2 specimens; Banff National Park, Baker Creek picnic area, 1500m el., 51.35, −116.063, mature lodgepole pine forest, pitfall, 15-20.VI.2012, BIOBus 2012 team (CBG) 1 specimen; Banff National Park, Storm Mountain, 1462m el., 51.282, −115.944, low alpine forest, 23-28. VII.2012, BIOBus 2012 team (CBG) 1 specimen; Jasper National Park, Pocahontas Campground, 1131m el., 53.195, −117.914, birch and spruce forest, pitfall, 16-21.VII.2012, BIOBus 2012 team (CBG) 6 specimens; Waterton Lakes National park, Highway 6 pullout, 1569m el., 49.065, −113.779, montane Douglas fir forest, pitfall, 21-27.VI.2012, BIOBus 2012 team (CBG) 10 specimens; Waterton Lakes National Park, Akamina Parkway, Rowe Tamarack Trail, 1640m el., 49.059, −114.014, subalpine lodgepole pine and fir forest with birch in understory, pitfall, 22-27.VI.2012, BIOBus 2012 team (CBG) 35 specimens: Waterton Lakes National Park, Highway 6 pulloff East of 2 Flags Lookout, 1562m el., 49.065, −113.778, douglas fir and lodgepole pine forest with aspen and birch understory, pitfall, 06-11.VIII.2012, BIOBus 2012 team (CBG) 24 specimens. USA, Alaska, Fairbanks, 930 Fitz Ct, 280m el., 64.9013, −147.52861, ±30m *Betula* & *P. tremuloides*, flight intercept trap, 16.VIII-22. IX.2007, D.S. Sikes (UAM) 4 specimens; same data except, 20.V-3,VI.2007 (UAM) 1 specimen; 29.V-4.VII.2008 (UAM) 2 specimens; 29.IV-29. V.2008 (UAM) 1 specimen.

***Crenuliventris* species group** (Klimaszewski et al. 2018a)

Diagnosis Antennae thick, antennomeres V–X subquadrate to moderately transverse; integument of forebody moderately glossy, with fine microsculpture consisting of irregularly shaped hexagonal sculpticells; median lobe of aedeagus in dorsal view with enlarged bulbus and long and triangular tubus that is constricted subapically, in lateral view tubus straight or slightly produced ventrally in apical one-third to one-half; internal sac with two highly elongate sclerites in bulbus in most taxa; male tergite VIII with shallowly concave apical margin and very small lateral teeth, or with large apical teeth and crenulated apical margin; spermatheca with club-shaped tubular capsule bearing shallow apical invagination, and with stem sinuate and narrowly looped posteriorly.

130. *Atheta* (*Dimetrota*) *crenuliventris* Bernhauer (Fig. 19.201a–h)

Atheta crenuliventris Bernhauer, 1907. Lohse et al. 1990 [as *Dimetrota bradorensis*], Gusarov 2003a [synonymy], Klimaszewski et al. 2005, Majka and Klimaszewski 2008b, Klimaszewski et al. 2011, 2016d, 2018a, b, Webster et al. 2016.

Diagnosis Body narrowly oval, slightly flattened, length 2.8–3.2 mm; colour dark brown to black, with legs/tarsi, apical part of abdomen and elytra yellowish-brown or reddish-brown (Fig. 19.201a), first two segments of antennae yellowish-brown to reddish-brown, remaining segments black; integument moderately glossy; forebody with punctation moderately dense and finely asperate, with coarse microsculpture; pubescence on disc of pronotum directed laterad and obliquely posteriad from midline, that on elytra directed obliquely posteriad forming wavy patterns near suture; head slightly narrower than pronotum; antennae slender, antennomeres V-X subquadrate; pronotum transverse, about 1/9 narrower than width of base of elytra; elytra transverse, at suture slightly longer than pronotum; abdomen arcuate laterally. MALE. Tergite VIII with apical margin truncate, irregularly crenulate between two rounded sublateral teeth (Fig. 19.201d); sternite VIII rounded apically (Fig. 19.201e); tubus of median lobe of aedeagus in lateral view bent obliquely ventrad basally, then very long, broadly arcuate, gradually tapering to briefly expanded and truncate apex, crista apicalis of bulbus small and narrow (Fig. 19.201b); in dorsal view tubus narrow and spatulate apically, bulbus oval, moderately large (Fig. 19.201c). FEMALE. Tergite VIII with apical margin broadly truncate, rounded laterally (Fig. 19.201f); sternite VIII with apical margin very obtusely triangularly produced, broadly rounded medially (Fig. 19.201g); spermatheca with small spherical capsule bearing shallow apical invagination, stem long, sinuate, and narrowly looped posteriorly (Fig. 19.201h).

Atheta crenuliventris is very similar externally to *A. pseudocrenuliventris* but differs in having the apical margin of male tergite VIII crenulate, the apical part of the median lobe of the aedeagus narrow in lateral view, the crista apicalis of the bulbus smaller, and the spermatheca less sinuate. *Atheta crenuliventris* also appears to be an eastern species, while *A. pseudocrenuliventris* is more northern transcontinental in distribution.

COLLECTION AND HABITAT DATA
Origin: Nearctic. **CANADA**: LB, MB, NB, NF, NS [**BOLD**], ON, QC, SK. **USA**: ME.

NATURAL HISTORY Habitat: in NB, found in various forested and open habitats, often in decaying organic materials; in compost near mixed forest, in dung in entrance to fox den in a pasture, in decaying sea wrack on sea beach, under dung on margin of lake, in litter of roof of an abandoned beaver lodge occupied by muskrats, in decaying moist litter in *Carex* marsh, and from unbaited pitfall traps in a red spruce forest. In NS, found in forested area and in wetland. In SK, in compost, wrack on lakeshore; in NF, captured in carrion-baited pitfall traps and flight intercept traps in conifer-dominated and mixed forests, and on coastal barrens in LB. **Collecting period**: IV–IX. **Collecting method**: sifting litter, pitfall traps, flight intercept traps.

We report this species from NS (**NPR**) for the first time here, based on **BOLD** records: Canada, Nova Scotia, Cape Breton Highlands National Park, Clyburn Valley Road, near golf course, 21m el., 46.655, –60.428, forested area, Malaise trap, 19–26.VII.2013, CBHNP staff, (CBG) 1 specimen; Kejimkujik National Park, Eel Weir Road, 96m el., 44.348, –65.189, bog, pitfall, 17-22.VI.2013, BIOBus 2013 team (CBG) 1 specimen.

131. ***Atheta* (*Dimetrota*) *pecki* Klimaszewski and Langor** (Fig. 19.202a–e)

Atheta (Dimetrota) pecki Klimaszewski and Langor, in Klimaszewski et al. 2011. Klimaszewski et al. 2018a.

Diagnosis Body narrow, subparallel, slightly flattened, length 3.3 mm; colour dark brown, with legs/tarsi, bases of antennae and elytra slightly paler, yellowish brown, area around scutellum darker (Fig. 19.202a); integument moderately glossy; forebody with punctation sparse and finely asperate, with fine microsculpture; pubescence on disc of pronotum directed laterad and obliquely posteriad from midline, forming arcuate lines; that on elytra directed obliquely posteriad; head slightly narrower than pronotum; antennae slender, antennomeres V–X subquadrate to slightly transverse; pronotum transverse, about

1/7 narrower than width of base of elytra; elytra transverse, at suture as long as pronotum; abdomen arcuate laterally. MALE. Apical margin of tergite VIII sinuately truncate, smooth between two very small, rounded lateral teeth (Fig. 19.202d); sternite VIII rounded apically (Fig. 19.202e); tubus of median lobe of aedeagus in lateral view with ventral margin straight basally, gradually curved ventrad to narrowly elongate subapical part, rounded at apex (Fig. 19.202b); in dorsal view bulbus broadly subparallel with broader base, and tubus subparallel basally and triangular apically (Fig. 19.202c); internal sac without distinct sclerites (Fig. 19.202b,c). FEMALE. Undescribed.

Atheta pecki differs from *A. crenuliventris* and *A. pseudocrenuliventris* in having the apical margin of male tergite VIII sinuate and not crenulate, and in shape of median lobe of the aedeagus.

DISTRIBUTION Origin: Nearctic. **CANADA**: LB. **USA**: not recorded.

COLLECTION AND HABITAT DATA

Habitat: only known specimen captured in carrion-baited pitfall trap in a spruce-moss forest. **Collecting period**: VIII. **Collecting method**: carrion-baited pitfall trap.

132. *Atheta* (*Dimetrota*) *pseudocrenuliventris* Klimaszewski (Fig. 19.203a–h)

Atheta (Dimetrota)? pseudocrenuliventris Klimaszewski, in Klimaszewski et al. 2005. In *Atheta*: Klimaszewski et al. 2008a, in *Atheta (Dimetrota)*: Majka and Klimaszewski 2010, Klimaszewski et al. 2011, 2018a, Brunke et al. 2012.

Diagnosis Body subparallel, length 2.8–3.2 mm; colour dark brown to piceous, with elytra and legs light yellowish-brown or reddish-brown, first segment of antennae reddish brown to dark brown, the remaining segments black (Fig. 19.203a); forebody with asperate punctation and meshed microsculpture, abdomen moderately to highly glossy; pubescence fine and sparse, denser on elytra; head slightly narrower than pronotum, slightly transverse; antennae robust, with antennomeres IV–X subquadrate to slightly elongate; pronotum moderately transverse, widest at middle, distinctly narrower than elytra; elytra at suture about as long as pronotum; abdomen widely arcuate laterally. MALE. Tergite VIII with apical margin truncate, irregularly crenulate between two obtuse lateral teeth (Fig. 19.203d); sternite VIII rounded apically, feebly angulate medially (Fig. 19.203e); median lobe of aedeagus in lateral view with large bulbus bearing large crista apicalis, subapical part of tubus wide (Fig. 19.203b), in dorsal view median lobe as illustrated (Fig. 19.203c). FEMALE. Tergite VIII with apical margin sinuately truncate, shallowly emarginate medially (Fig. 19.203f); sternite VIII evenly rounded apically (Fig. 19.203g); spermatheca with club-shaped capsule, angularly bent at base of neck, stem sinuate and narrowly looped posteriorly (Fig. 19.203h).

This species is hardly distinguishable externally from *A. crenuliventris*, but it differs in having the median lobe of the aedeagus in lateral view with the tubus wide subapically and the crista apicalis of the bulbus large (tubus narrow subapically and crista apicalis of bulbus small in *A. crenuliventris*).

DISTRIBUTION Origin: Nearctic. **CANADA**: AB [**NPR**], BC [**NPR**], LB, NB, NF, NS, NT [**NTR**], ON, YT. **USA**: AK [**NSR**].

COLLECTION AND HABITAT DATA

Habitat: captured in pitfall traps in a red spruce forest (NB) and in NL from carrion-baited pitfall and flight intercept traps in mixed and coniferous forest types, on coastal barrens in LB, and in owl nests in NS. In NB, specimens sifted from litter, grasses and moss on hummocks near water in a wet old-growth eastern white cedar swamp, in leaf litter on vernal pond margin in a mature red spruce and eastern white cedar forest, leaf litter at bases of trees in a red maple swamp, grass and litter at bases of alders in a marsh, in drift material in seepage area on river margin, and in grass litter in seepage area with *Carex* and *Spartina patens* on upper margin of salt marsh. It is possible that this is a wetland species associated with decaying litter. In AK, adults were found in *Betula, Salix,* white spruce forest; *Salix* spp., pruned branch tips; low shrub in crowberry, wet, mossy; OTS-*Salix*, crowberry hummock; *Betula* and *Populus tremuloides*; boggy black spruce and birch; sand, *Honkenya peploides*; and mixed black and white spruce with willow. **Collecting period**: IV–VIII. **Collecting method**: sifting litter, carrion-baited pitfall traps, flight intercept traps, pollinator vane trap, yellow pan trap.

We report this species from AB, BC (**NPR**), NT (**NTR**), and AK (**NSR**) for the first time here, based on the following: Canada, Northwest Territories, 65.25226, –126.661128, Rep. 2, wet, yellow pan, 14-17.VI.2011, NBP Field Party N2WY231 (LFC) 1 female; British Columbia, Burnt Cabin Bog Ecological Reserve, 54.761, –126.931, wetland, intercept trap, 12.VII.2014, BIObus 2014 (CBG), 2 specimens unknown sex; Alberta, Banff National Park, Corral Creek, old road, wetland, intercept trap, 20.VI.2012, BIObus 2012 (CBG), 1 specimen unknown sex. USA, Alaska, 35.5 Mi Dalton Hwy, 271m el., 65.74425, –149.34564, ±4m *Betula, Salix,* white spruce, flight intercept trap, 31.V-5.VI.2008, D.S. Sikes (UAM) 4 specimens; Alaska Peninsula / Becharof NWR, plot 1362 (Dog Salmon), 4m el., 57.40319, –157.21692, ±8m open low shrub in crowberry, wet, mossy, pitfall trap, 28.VI-22.VIII.2007, S. Savage (UAM) 5 specimens; Alaska Peninsula / Becharof NWR, plot 1428 (Ugashik Lake East), 8m el., 57.59805, –156.63251, ±9m OTS-*Salix*, crowberry hummock, pitfall trap, 19.VI-8.VIII.2007, S. Savage (UAM) 2 specimens; Anchorage, Campbell Creek, 61.15957, –149.79493, ±10m *Salix* spp., pruned branch tips, 4.VIII.2014, R. Progar; S. Bresney (UAM) 1 specimen; Anchorage, JBER, Otter Lake, 31m el., 61.28137, –149.75098, ±3m boggy black spruce and birch, pollinator vane trap, 0.5m, 13-27.VI.2016, J.C. Hagelin (UAM) 1 specimen; same data except, 27.VI-11.VII.2016 (UAM) 6 specimens; Fairbanks, 930 Fitz Ct, 280m el., 64.9013, –147.52861, ±30m *Betula* & *P. tremuloides*, flight intercept trap, 16.VIII-22.IX.2007, D.S. Sikes (UAM) 1 specimen; same data except, 29.IV-29.V.2008 (UAM) 1 specimen; Kiska Is., Sirius Pt., 15m el., 52.13293, –177.59668, ±15m sand, *Honkenya peploides*, 5 pitfalls, 15-29.VII.2008, A. Bond (UAM) 1 specimen; same data except, 2-15.VII.2008 (UAM) 3 specimens; White Mountains Recreation Area, Nome Creek Site 4, 589m el., 65.3335, –146.83023, ±10m mixed black and white spruce with willow, pollinator vane trap, 0.5m, 12.VI-3.VII.2015, J.C. Hagelin (UAM) 1 specimen; White Mountains Recreation Area, Nome Creek Site 7, 589m el., 65.34084, –146.93216, ±10m boggy black spruce, pollinator vane trap, 0.5m, 12.VI-2.VII.2015, J.C. Hagelin (UAM) 1 specimen; Tangle Lakes, 23.VI-2.VII.1969, J. Matthews (CNC) 2 males, 12 females.

***Districta* species group** (Klimaszewski et al. 2018a, 2020)

Diagnosis Antennae thick and robust, antennomeres V–X elongate or subquadrate; integument of forebody moderately glossy, with fine microsculpture consisting of irregularly shaped hexagonal sculpticells; median lobe of aedeagus in dorsal view with medium-sized bulbus and long and apically triangular tubus, in lateral view tubus straight or slightly produced ventrally at apex; internal sac with two moderately elongate sclerites in bulbus; male tergite VIII with approximately straight apical margin and small lateral teeth; spermatheca with long tubular capsule bearing tubular extention and small apical invagination, and with sinuate stem.

133. *Atheta* (*Dimetrota*) *districta* Casey (Fig. 19.204a–h)

Atheta (Atheta) districta Casey, 1911. Klimaszewski et al. 2005, 2007b, in *Atheta (Dimetrota)*: Majka and Klimaszewski 2008b, Klimaszewski et al. 2011, 2016d, 2018a, b, 2020.

Diagnosis Body narrowly elongate, slightly flattened, length 2.8–3.3 mm; colour dark brown to black, with legs/tarsi, basal parts of antennae and elytra yellowish-brown (Fig. 19.204a); integument moderately glossy; forebody with punctation dense, finely asperate on elytra; with fine isodiametric microsculpture; pubescence on disc of pronotum directed laterad and obliquely posteriad from midline, that on elytra directed obliquely or straight posteriad; head distinctly narrower than pronotum; antennae thick, antennomeres V–X slightly elongate; pronotum transverse, about 1/7 narrower than width of base of elytra; elytra transverse, at suture as long as pronotum; abdomen arcuate laterally. MALE. Tergite VIII with almost entire apical margin moderately produced, truncate to slightly emarginate apically, angular laterally (Fig. 19.204d); sternite VIII parabolic apically except briefly straight mediad (Fig. 19.204e); in lateral view tubus of median lobe of aedeagus long, arcuate basally, almost straight to moderately narrow and rounded apical part (Fig. 19.204b); in dorsal view bulbus moderately large and suboval in shape, tubus long, gradually tapering, parabolically rounded apicad (Fig. 19.204c). FEMALE. Tergite VIII unevenly arcuate apically (Fig. 19.204f); sternite VIII rounded with dense fringe of short setae apically (Fig. 19.204g); spermatheca with capsule in shape of an arched tube with small apical invagination, neck long, slightly curved, bent at base, stem thin, recurved with globular swelling posteriorly (Fig. 19.204h).

The median lobe of the aedeagus and particularly the spermatheca of this species are similar to its sibling species *A. pseudodistricta*, from which it differs in having antennomeres V–X elongate and the median lobe of the aedeagus shaped differently.

DISTRIBUTION Origin: Nearctic. **CANADA**: AB, BC, LB, MB, NB, NF, NS, ON, QC, SK. **USA**: not recorded.

COLLECTION AND HABITAT DATA
Habitat: in NB, in various forest types usually in dung or decaying mushrooms; eastern white cedar swamp in moose dung, mature red spruce and eastern white cedar forest in deer dung, mixed forest in bear dung, in decaying (moldy) corncobs and cornhusks near mixed forest, regenerating mixed forest in decaying mushrooms, near lakeshore in decaying fungi, old jack pine forest in rotten boletus mushrooms, red maple swamp in leaf litter near stream, in unbaited pitfall traps in red spruce forest, and a few collected at u.v. light. Elsewhere, captured in flight intercept traps and carrion-baited pitfall traps in coniferous and mixed forests (NL) and on coastal barrens (LB); and from spruce litter, and dry or decaying mushrooms. **Collecting period**: VI–IX. **Collecting method**: sifting litter, dung, and fungi, pitfall traps, flight intercept traps.

COMMENTS *Atheta districta*, as far as known, reaches its northern limit in the softwood shield (ecoregion 5.1) of Labrador and is not yet known from the arctic/subarctic.

***Fanatica* species group** (Klimaszewski et al. 2020)

Diagnosis Body length 2.5–3.3 mm, subparallel, antennae slender, antennomeres V–X subquadrate to moderately transverse; integument of forebody glossy, with microsculpture distinct; punctation of forebody dense and asperate; median lobe of aedeagus in dorsal view with bulbus oval in outline and streamlined with base of tubus, tubus short and triangular; in lateral view tubus broad, slightly to strongly sinuate ventrally and apex broadly rounded not produced ventrad; internal sac with two complex sclerites in bulbus; male tergite VIII with apical margin truncate, with small lateral teeth, apical margin slightly crenulate; male sternite VIII with long macrosetae; spermatheca with capsule spherical or broadly tubular and rounded apically, apical invagination present, stem sinuate and looped posteriorly.

134. *Atheta* (*Dimetrota*) *fanatica* Casey (Fig. 19.205a–h)

Atheta (Atheta) fanatica Casey, 1910. As synonym *A. omissa*: Casey 1910, as synonym *A. irrita*: Casey 1911, Webster et al. 2009, as

A. fanatica: Klimaszewski et al. 2011, 2012, 2015a, 2018a, 2020.

Diagnosis Body narrowly subparallel, slightly flattened, length 2.0–2.3 mm; colour dark brown to almost black, with legs/tarsi, and maxillary palpi yellowish-brown or reddish-brown (Fig. 19.205a); integument moderately glossy; forebody with punctation dense and asperate, with fine microsculpture; pubescence on disc of pronotum directed laterad and obliquely posteriad from midline, forming transversely arcuate pattern, that on elytra directed obliquely posteriad forming wavy pattern posteriorly; head slightly narrower than pronotum; antennae slender, antennomeres V–X slightly to distinctly transverse; pronotum transverse, about 1/7 narrower than width of base of elytra; elytra transverse, at suture slightly longer than pronotum; abdomen arcuate laterally. MALE. Apical margin of tergite VIII sharply truncate, minutely crenulate, (Fig. 19.205d); sternite VIII broadly parabolic apically (Fig. 19.205e); tubus of median lobe of aedeagus highly sinuate medio-ventrally and with broadly rounded apex in lateral view (Fig. 19.205b); in dorsal view bulbus large and oval, tubus short, broadly triangular (Fig 19.205c). FEMALE. Tergite VIII with apical margin broadly truncate, obtusely produced laterally (Fig. 19.205f); sternite VIII evenly rounded apically (Fig. 19.205g); spermatheca with sac-shaped, tubular capsule, and sinuate stem narrowly curled posteriorly (Fig. 19.205h).

DISTRIBUTION Origin: Nearctic. **CANADA**: AB, BC, LB, NB, NS, NT [**BOLD**], ON [**BOLD**], QC, SK, YT. **USA**: AK [**NSR**], NV.

COLLECTION AND HABITAT DATA

Habitat: *Atheta fanatica* was found in owl nests and other decaying organic material on standing trees. In NB, this species was a common inhabitant of nests of barred owls that nest in tree holes in large trees and artificial nest boxes in hardwood, silver maple, and mixed forests. The material in nests usually consisted of rich decaying organic material with bones, fur, owl pellets and portions of dead prey items and the contents often had a strong urine smell. This species was also found in nest contents of the great horned owl (NB), boreal owl, and northern saw-whet owl (NS). It also has been found in *Pleurotus* sp. (oyster mushrooms) in the early stages of decay on dead standing trembling aspen trees in oak forests, *Populus* stands, hardwood forests, and mixed forests. One specimen from Lindgren funnel trap in red pine forest (NB) and one from NF captured in a carrion-baited pitfall trap in a spruce/moss forest. In NT, it was found in mature mixed forest with aspen undergrowth. **Collecting period**: V–VIII. **Collecting method**: sifting owl nest contents, mushrooms, and litter, carrion-baited pitfall traps, Lindgren funnel trap, and Malaise traps.

We report this species from NT (**NTR**), and ON (**NPR**) for the first time here, based on **BOLD** records, and from AK (previous AK records without voucher data) based on UAM specimens: Canada, Northwest Territories, Fort Smith, 204m el., 60.002, −111.906, mature mixed forest with 10 ft aspen undergrowth, Malaise trap, 05-23.IX.2014, K. Cox (CBG) 1 specimen, Yellowknife, Pontoon Lake, 220m el., 62.534, −113.976, forested area, Malaise trap, 27.VII-03. VIII.2014, S. Carriere (CBG) 1 specimen; Ontario, Wellington Co., Guelph, Hanlon Preservation Park, 43.51, −80.22, mixed forest, beating and sweeping trees and bushes, 2.VII.2017 (CBG), 1 specimen unknown sex. USA, Alaska, Tetlin NWR, 822m el., 63.36124, −143.47868, ±10m black spruce, hanging Malaise, 2.9m, 24. VII-10.VIII.2015, J.C. Hagelin (UAM) 1 specimen.

135. *Atheta (Dimetrota) munsteri* Bernhauer *
(Fig. 19.206a–h)

Atheta (*Dimetrota*) *munsteri* Bernhauer, 1902. Lohse et al. 1990 [in *Dimetrota*]; Smetana 2004 [in *Boreophilia*]; Schülke and Smetana 2015 [in *Boreophilia*]; in *Atheta* (*Dimetrota*): Klimaszewski et al. 2012, 2019, 2020.

Diagnosis Body narrowly subparallel, slightly flattened, length 2.6–3.0 mm; colour uniformly black (Fig. 19.206a); integument moderately glossy; forebody with punctation sparse and asperate, with fine microsculpture; pubescence on disc of pronotum directed laterad and obliquely posteriad from midline, and that on elytra directed obliquely posteriad forming slightly wavy pattern posteriorly; head slightly narrower than pronotum; antennae slender, antennomeres V–X subquadrate to slightly transverse; pronotum transverse, about 1/3 narrower than width of base of elytra; elytra transverse, at suture slightly longer than pronotum; abdomen arcuate laterally. MALE. Apical margin of tergite VIII truncate, minutely crenulate, slightly obtusely angulate laterally (Fig. 19.206d); sternite VIII evenly rounded apically with long, coarse macrosetae (Fig. 19.206e); tubus of median lobe of aedeagus in lateral view with ventral margin bent obtusely ventrad basally, then straight to broadly rounded apex (Fig. 19.206b); in dorsal view bulbus large and oval, tubus short, broadly triangular (Fig 19.206c). FEMALE. Tergite VIII with apical margin obtuse laterally, with broad, arcuate emargination medially (Fig. 19.206f); sternite VIII broadly rounded apically (Fig. 19.206g); spermatheca S-shaped, with spherical capsule, and strongly sinuate stem moderately widely looped posteriorly (Fig. 19.206h).

DISTRIBUTION **Origin**: Holarctic.

CANADA: BC [?], MB, NT, NU [**BOLD**], YT. **USA**: AK.

COLLECTION AND HABITAT DATA
Habitat: alpine tundra, fungi/litter in unspecified forest. **Collecting period**: VI. **Collecting method**: sweeping tundra vegetation, sifting forest litter and fungi, yellow pitfalls w/o rainroofs.

We report this species from NU (**NTR**) for the first time here, based on **BOLD** records: Canada, Nunavut, Kitikmeot, Kugluktuk, S Lake, 9 km SW of Kugluktuk, 22m el., 67.781, –115.276, tundra, sweep net, 24.VII.2019, G. Blagoev, M. Pentinsaari (CBG) 1 specimen.

Additional records. See Annex 1.

COMMENTS The BC record should be considered as a tentative identification because it is based on a single female (Klimaszewski et al. 2020). A male from BC is essential to validate this record.

***Hampshirensis* species group** (Klimaszewski et al. 2020)

Diagnosis Body small, length 2.0–2.6 mm, antennae slender, antennomeres V–X moderately transverse; integument of forebody slightly glossy, with microsculpture weak; punctation of forebody dense and asperate; median lobe of aedeagus in dorsal view with bulbus oval in outline, tubus short and triangular; in lateral view tubus nearly straight ventrally and apex not produced ventrad; internal sac with two narrowly elongate sclerites in bulbus; male tergite VIII with apical margin truncate, slightly crenulate, with thin and long lateral teeth; spermatheca with capsule tubular, strongly recurved near middle and tapering apicad, apical invagination absent, stem sinuate and looped posteriorly. The taxonomic position of this group is unsettled and it is tentatively assigned to the subgenus *Dimetrota*.

136. *Atheta* (*Dimetrota*) *hampshirensis* Bernhauer (Fig. 19.207a–h)

Atheta (Dimetrota) hampshirensis Bernhauer, 1909. Mäklin 1852 [as *Homalota moesta*], Bernhauer 1907 [as *A. moesta*], Casey 1910 [as six synonyms], 1911 [as one synonym], Hatch 1957 [as *Ischnopoda* (*Coproceramius*) *maeklini*], Lohse and Smetana 1985 [as *Dimetrota moesta* (Mäklin 1852)], Klimaszewski and Winchester 2002, Gusarov 2003a [list of synonyms including *Dimetrota retrusa* Casey from Metlakatla, BC], Klimaszewski et al. 2005, 2007b, Majka and Klimaszewski 2008b, Majka and Sikes 2009, McLean et al. 2009, Klimaszewski et al. 2011, 2016b, 2018a, b, 2020.

Diagnosis Body narrowly oval, slightly flattened, length 2.2–2.6 mm; colour dark brown to black, with legs/tarsi, basal parts of antennae yellowish brown and elytra light brown with rusty tinge (Fig. 19.207a); integument moderately glossy; forebody with punctation dense and asperate, with fine microsculpture; pubescence on disc of pronotum directed laterad from midline, that on elytra directed obliquely posteriad; head slightly narrower than pronotum; antennae slender, antennomeres V–X slightly transverse; pronotum transverse, about 1/6 narrower than width of base of elytra; elytra transverse, at suture as long as pronotum; abdomen arcuate laterally and strongly tapering apically. MALE. Tergite VIII irregularly truncate, slightly crenulate apically with two long, thin, sharp lateral spine-like teeth (Fig. 19.207d); sternite VIII rounded apically (Fig. 19.207e); tubus of median lobe of aedeagus straight ventrally and with thin, rounded apex in lateral view (Fig. 19.207b); in dorsal view bulbus large and oval, tubus narrowly triangular (Fig. 19.207c). FEMALE. Tergite VIII truncate apically (Fig. 19.207f); sternite VIII rounded apically (Fig. 19.207g); spermatheca with capsule tubular, recurved and tapering apically, stem sinuate and looped posteriorly (Fig. 19.207h).

Externally, this species is superficially similar to *Atheta (Datomicra) celata*, *A. (D.) dadopora*, *A. (D.) whitehorsensis* and *Strophogastra penicillata*, but it differs in having the body more compact and robust, the teeth of male tergite VIII long, and median lobe of the aedeagus and spermatheca are differently shaped. The placement of *A. hampshirensis* in *Dimetrota* is not settled and its affiliation with *Datomicra* is unclear (Klimaszewski et al. 2020). COI sequence data of this species form a group with those of *A.* (*Microdota*) *formicaensis* and an unindentified species, suggesting that these form a natural group.

DISTRIBUTION Origin: Nearctic. **CANADA**: AB, BC, LB, MB, NB, NF, NS, ON, QC. **USA**: AK, CA, NC, NH, NY, PA, RI, WA.

COLLECTION AND HABITAT DATA

Habitat: in eastern and central Canada, it occurs in various coniferous, mixed, and hardwood forests, usually in decaying organic materials such as rotting fungi, dung, and carrion; in NB, specimens from black spruce forest on gilled mushrooms, Jack pine forest in decaying mushrooms, from pitfall traps in red spruce forest, old-growth hemlock forest in decaying bracket fungi (polypore), various mixed forests in gilled mushrooms and decaying fleshy and gilled mushrooms in various stages of decay, in decaying corncobs and cornhusks near mixed forest, alder swamp on remains of dead deer, rich Appalachian hardwood forest in fleshy fungi of various stages of decay, some from leaf litter and scat under great horned owl nest in hardwood forest; in NS, in compost, carrion, and mushrooms in coniferous and deciduous forests and open habitats. In NL, captured in carrion traps and flight intercept traps in mixed and coniferous forests. In BC, captured in rotting *Boletus*, in *Sphagnum*, and in gravel along edge of stream. In AK, adults were found in decaying mushrooms in *Betula* forest; old growth and thinned and unthinned 2nd growth forests; clearcuts; old growth Hemlock, coastal rain forest; and forest edge with *Populus tremuloides*. **Collecting period**: V–IX. **Collecting method**: sifting fungi and litter, unbaited and carrion-baited pitfall traps, flight intercept traps, Lindgren traps.

Additional records. See Annex 1.

***Picipennis* species group** (Klimaszewski et al. 2008a)

Diagnosis Antennae slender, antennomeres V–X subquadrate to slightly transverse; integument of forebody highly glossy, with coarse microsculpture consisting of irregularly shaped hexagonal sculpticells; pubescence on disc of elytra directed approximately straight or obliquely posteriad; median lobe of aedeagus in dorsal view with bulbus moderate-sized, narrowly elongate, tubus long and apically triangular, in lateral view tubus arcuate, slightly sinuate or straight, with apical part slightly curved ventrad; internal sac with two medium-sized, elongate sclerites in bulbus; male tergite VIII with straight or convex and sinuate or crenulate apical margin, and with small or large lateral teeth; spermatheca very distinct, with capsule narrow, tubular, club-shaped, with small apical invagination, stem long, thin, twisted, with multiple coils.

137. *Atheta* (*Dimetrota*) *altaica* Bernhauer *
(Fig. 19.208a–h)

Atheta (Dimetrota) altaica Bernhauer, 1901. Lohse et al. 1990 [in *Dimetrota*], Gusarov 2003a, Smetana 2004 [Palaearctic distrib.], Klimaszewski et al. 2008a, 2011, 2018a.

Diagnosis Body narrowly oval, slightly flattened, length 3.0–3.5 mm; colour dark brown to black, with legs/tarsi, parts of antennae and medio-posterior part of elytra reddish-brown (Fig. 19.208a); integument moderately glossy; forebody with punctation dense and finely asperate; microsculpture coarse; pubescence on disc of pronotum directed obliquely anteriad, laterad and obliquely posteriad from midline, that on elytra directed obliquely posteriad; head distinctly narrower than pronotum; antennae slender, antennomeres V–X subquadrate to slightly transverse; pronotum transverse, about 1/5 narrower than width of base of elytra; elytra transverse, at suture slightly longer than pronotum; abdomen subparallel. MALE. Tergite VIII with apical margin truncate, crenulate between two rounded lateral teeth (Fig. 19.208d); sternite VIII broadly parabolically rounded apically (Fig. 19.208e); tubus of median lobe of aedeagus slightly sinuate ventrally and with broad, rounded, apical part curved slightly ventrad in lateral view (Fig. 19.208b); in dorsal view tubus tapering apicad, triangular apically (Fig. 19.208c). FEMALE. Tergite VIII unevenly truncate apically (Fig. 19.208f); sternite VIII rounded apically, briefly straight mediad (Fig. 19.208g); spermatheca with capsule tubular, broadly L-shaped, apically clubbed, stem long, forming a mass of twisted loops posteriorly (Fig. 19.208h).

Females of this species may be confused with those of *A. atramentaria* and *A. picipennis* (occurring only in AK, BC) but *A. altaica* differs in having coarser microsculpture on the forebody, antennomeres V–X subquadrate or slightly transverse, and the apical margin of male tergite VIII truncate (not emarginate).

DISTRIBUTION Origin: Holarctic: Central Asia, northern Scandinavia, northern Russia, and northern North America. **CANADA**: AB [**NPR**], BC [**BOLD**], LB, NF, NT, QC [**NPR**], YT. **USA**: AK.

COLLECTION AND HABITAT DATA

Habitat: from carrion-baited pitfall traps and in spruce- and fir-dominated boreal forests (NF) and from caribou dung, gopher burrows, elsewhere. In AB, found in forested area. In AK, collected in wetland, *Eriophorum angustifolium, Nephrophyllidium crista-galli, Carex nigricans*; on fresh deer scat in subalpine wet tundra; forest, mostly spruce with some aspen; shrub, alder, dwarf birch, grass, clubmoss; *Betula* sp., pruned branch tips; old growth forest; subalpine tundra, near edge of krummholz patch, primate dung; and in open tussuck and standing water, some willow and bog birch. **Collecting period**: VIII. **Collecting method**: sifting litter and dung, pitfall traps, yellow pitfalls w/o rainroofs, flight intercept traps, hanging Malaise traps.

We report this species for the first time here, from AB (**NPR**) based on DNA barcoded specimens identified morphologically, from BC (**NPR**) based on **BOLD** records matching the AB specimens, and from QC (**NPR**) based on LFC specimens: Canada, Alberta, Waterton Lakes National Park, Akamina Parkway, Rowe Tamarack Trail, 1640m el. 49.059, –114.014, subalpine lodgepole pine and fir forest with birch understory, sweep net, 25.VI.2012, BIOBus 2012 team (CBG) 1 specimen; Waterton Lakes National Park, East of 2 Flags Lookout, Highway 6 pulloff, 1569m el. 49.065, –113.779, douglas fir and lodgepole pine forest with aspen and birch understory, flight-intercept trap, 21-27.VI.2012, BIOBus 2012 team (CBG) 1 specimen; same data but sweep net, 21.VI.2012 (CBG) 1 specimen. British Columbia, Glacier National Park, Hemlock Grove Boardwalk, 1094m el. 51.243, –117.649, old-growth hemlock and cedar forest, pitfalls, pan traps and flight-intercept trap, 04-10.VIII.2014, BIObus 2014 team (CBG) 20 specimens; Kinaskan Lake Provincial Park, Kinaskan Lake Trail, 833m el. 57.532, –130.202, spruce and aspen forest, flight-intercept trap, 26.VII-01.VIII.2014, BIObus 2014 team (CBG) 1 specimen. Quebec, Gaspé, Mt. Albert, Sommet nord, 1000m, 21.VII.1985, human feces, F. Génier, J. Klimaszewski (LFC) 3 males, 4 females; Gaspè, Mt. Jacques-Cartier, 24. VII.1985, caribou dung, F. Génier, J. Klimaszewski (LFC) 3 males, 6 females.

Additional records. See Annex 1.

COMMENTS The Quebec localities of this species at Mt. Albert and Mt. Jacques-Cartier represent the most southern isolated population of this species.

138. *Atheta* (*Dimetrota*) *atramentaria* (Gyllenhal) ‡ (Fig. 19.209a–h)

Aleochara atramentaria Gyllenhal, 1810. Klimaszewski et al. 2011, 2018a.

Diagnosis Body narrowly oval, slightly flattened, length 2.8–3.3 mm; colour dark brown to black, with tarsi, parts of antennae and latero-posterior part of elytra yellowish-brown or reddish-brown (Fig. 19.209a); integument moderately glossy; forebody with punctation moderately dense and finely asperate; microsculpture fine; pubescence on disc of pronotum directed laterad and obliquely posteriad, that on elytra directed obliquely posteriad; head slightly narrower than pronotum; antennae slender, antennomeres V–X subquadrate to slightly transverse; pronotum transverse, about 1/7 narrower than width of base of elytra; elytra transverse, at suture as long as pronotum; abdomen arcuate laterally. MALE. Tergite VIII with apical margin sinuately truncate between two short, sharp lateral teeth (Fig. 19.209d); sternite VIII broadly parabolically rounded apically (Fig. 19.209e); tubus of median lobe of aedeagus in lateral view slightly sinuate ventrally, with thin apical part curved ventrad (Fig. 19.209b); in dorsal view tubus broad, triangular apically, margin angulate at juncture with bulbus (Fig. 19.209c). FEMALE. Apical margin of tergite VIII with rounded projection on each side of same size and shape as broad medial emargination (Fig. 19.209f); sternite VIII arcuate apically (Fig. 19.209g); spermatheca with capsule tubular, apically clubbed, stem long, twisted, with multiple irregular loops posteriorly (Fig. 19.209h).

Females of this species may be confused with those of *A. altaica* and *A. picipennis* (AK, BC), but *A. atramentaria* differs in having feeble microsculpture on the forebody, antennomeres V–X subquadrate, and female tergite VIII emarginate.

DISTRIBUTION Origin: Palaearctic, adventive in North America or Holarctic.

CANADA: LB, NF, QC [**NPR**]. **USA**: not recorded.

COLLECTION AND HABITAT DATA

Habitat: one specimen from LB collected from litter in a spruce forest with moss substrate. Other specimens (QC) were collected from dung. **Collecting period**: VI–IX. **Collecting method**: sifting litter.

We report this species from QC (**NPR**) for the first time here, based on the following: Canada, Quebec, Gaspé, Mt. Albert, Sommet nord, 1000m, 20.VII.1985, human feces, F. Génier and J. Klimaszewski (LFC) 1 female; Gaspé, Mt. Jacques-Cartier, 24.VIII.1985, caribou dung, F. Génier and J. Klimaszewski (LFC) 2 males, 2 females.

COMMENTS The localities in QC are from a classic alpine refugium for arctic species and represent the most southern known records of this species in North America. This is not typical of an adventive species and we here treat *A. atramentaria* as adventive/Holarctic uncertain status until more records are available.

139. *Atheta* (*Dimetrota*) *nearctica* (Lohse) (Figs. 19.210a–h)

Dimetrota nearctica Lohse, in Lohse et al. 1990. Klimaszewski et al. 2011, 2018a.

Diagnosis Body narrowly subparallel, slightly flattened, length 3.0–3.5 mm; colour dark brown to almost black, with legs/tarsi, and maxillary palps reddish-brown (Fig. 19.210a); integument moderately glossy; forebody with punctation sparse and finely asperate; microsculpture fine; pubescence on disc of pronotum directed laterad and obliquely posteriad from midline, that on elytra directed obliquely posteriad; head distinctly narrower than pronotum; antennae moderately robust, antennomeres V–X distinctly transverse; pronotum transverse, about 1/10 narrower than width of base of elytra; elytra transverse, at suture as long as pronotum; abdomen subparallel. MALE. Tergite VIII with apical margin obliquely produced laterally, broadly truncate medially (Fig. 19.210d); sternite VIII rounded apically (Fig. 19.210e); tubus of median lobe of aedeagus in lateral view straight ventrally and with apical part elongate, curved slightly ventrad, rounded at apex (Fig. 19.210b); in dorsal view bulbus round basally, almost evenly tapering to moderately broad, triangular apex (Fig. 19.210c). FEMALE. Tergite VIII truncate and slightly emarginate apically (Fig. 19.210f); sternite VIII with apical margin slightly arcuate medially, rounded laterally (Fig. 19.210g); spermatheca with capsule L-shaped, broadly tubular, clubbed apically, stem long, forming a mass of twisted, irregular coils (Fig. 19.210h).

The spermatheca of *A. nearctica*, has the stem narrow, twisted, and with multiple coils as in *A. picipennis*, *A. altaica*, and *A. atramentaria*, but it differs by having the capsule more robust, and the margin of male tergite VIII truncate and straight.

DISTRIBUTION **Origin**: Nearctic. **CANADA**: LB, MB [**NPR**], NT, QC [**NPR**], YT. **USA**: AK.

COLLECTION AND HABITAT DATA
Habitat: NF specimens captured in pitfall traps in recently burned coniferous forest; one specimen from YT sifted from moss and litter, MB specimens captured in Tamarak open forest, and in spruce near stream (AK). **Collecting period**: VI, VIII. **Collecting method**: sifting litter, pitfall traps and Malaise traps.

We report this species from MB (**NPR**) and QC (**NPR**) for the first time here, based on the following: Canada, Manitoba, Churchill, 26 km SE Churchill, Twin Lakes burn site, 32m el., 58.618, −93.829, open tamarak forest, 20. VIII.2006, C. Rosarius (CBG) 1 specimen; Churchill, Vicinity of Churchill, welcome sign, 5m el., 58.758, −94.152, Malaise trap, 24-VII-2010, P. Kevan and T. Woodcock (CBG) 1 specimen. Quebec, Gaspé, Mt. Albert, Sommet nort, 1000m, 21.VII.1985, F. Génier, J. Klimaszewski (LFC) 1 male.

Additional records. See Annex 1.

COMMENTS The Mt. Albert locality represents the most southern distribution of this species.

140. ***Atheta*** **(*Dimetrota*)** ***onthophilides*** **(Newton), New Combination.** (Fig. 19.211a–h)

Acrotona onthophila Lohse, in Lohse et al. 1990 (junior secondary homonym in *Acrotona* of *Atheta onthophila* Cameron, 1920, replacement name *A. onthophilides* Newton, 2017) [see below].

Diagnosis Body subparallel, moderately convex, length 2.8–3.2 mm; colour black, with reddish-brown legs/tarsi, and some reddish tinge on black elytra (Fig. 19.211a); integument moderately glossy; forebody with punctation dense and asperate; microsculpture fine; pubescence on disc of pronotum directed laterad and obliquely posteriad from midline, and on elytra directed more or less straight posteriad; head distinctly narrower than pronotum; antennae moderately robust, antennomere IV subquadrate, V–X slightly to distinctly transverse; pronotum transverse, about 1/10 narrower than width of base of elytra; elytra transverse, at suture slightly longer than pronotum; abdomen subparallel. MALE. Tergite VIII truncate apically with multiple, irregular small teeth (Fig. 19.211d); sternite VIII broadly parabolic apically (Fig. 19.211e); tubus of median lobe of aedeagus in lateral view broadly arcuate ventrally from base to elongate apical part, narrowly rounded at apex (Fig. 19.211b); in dorsal view bulbus narrowly oval, and tubus triangular apically (Fig. 19.211c).

FEMALE. Tergite VIII transverse, with apical margin almost entirely truncate, obliquly emarginate at lateral angles (Fig. 19.211f); sternite VIII broadly rounded apically (Fig. 19.211g); spermatheca with capsule L-shaped, broadly tubular, stem long, twisted, forming a dense mass of irregular coils (Fig. 19.211h).

DISTRIBUTION **Origin**: Nearctic. **CANADA**: YT. **USA**: AK.

COLLECTION AND HABITAT DATA
Habitat: in caribou and bear dung, in *Salix* litter. **Collecting period**: VI–VII. **Collecting method**: sifting litter.

COMMENTS Lohse (in Lohse et al. 1990), described the new species *Acrotona onthophila*, and provided genitalic illustrations. However, this species does not belong to the genus *Acrotona,* which is defined by different body characteristics and different type of genitalia, particularly the morphology of the spermatheca. Spermathecae of *Acrotona* have a broadly pitcher-shaped capsule with short neck and moderately long stem with one or few regular posterior coils. The spermatheca of *A. onthophilides* has a narrow tubular and sinuate capsule and irregularly twisted stem (spaghetti-type), typical of the Picipennis group of *Atheta* (*Dimetrota*). The male apical margin of tergite VIII in *Acrotona* is entire and not dentate as in *A. onthophilides*. Therefore, we have transferred *Acrotona onthophilides* to the subgenus *Dimetrota* of the genus *Atheta*. The replacement name *Atheta onthophilides* must still be used since, by the above combination, *Acrotona onthophila* Lohse is now a junior secondary homonym of *Atheta onthophila* Cameron in *Atheta*, and is not to be resurrected under ICZN Article 59.4.

141. *Atheta* (*Dimetrota*) *picipennis* (Mannerheim) (Fig. 19.212a–h)

Homalota picipennis Mannerheim, 1843. Casey 1910 (as two synonyms), 1911 (as four synonyms), as *Atheta*: Bernhauer 1907, Moore and Legner 1975, Lohse and Smetana 1985 [lectotype, in *Dimetrota*], in *Atheta* (*Dimetrota*): Klimaszewski and Winchester, 2002, Gusarov 2003a, Majka and Klimaszewski 2008b, Klimaszewski et al. 2020, Voinkov 2020 (misidentification of *A. aeneipennis*, *A. picipennis sensu* Brundin 1954)

Diagnosis Body moderately convex, narrowly subparallel, length 2.9–3.5 mm, dark brown to nearly black, elytra usually paler, rust-brown with darker base near scutellum, legs and I–II basal antennomeres rust-brown or yellowish brown (Fig. 19.212a); forebody moderately sparsely punctate and pubescent, pubescence long, punctation strongly asperate, microsculpture meshed and particularly well visible on pronotum, sculpticells slightly convex; head slightly elongate, round, narrower than pronotum, postocular region shorter than the maximum diameter of eye, tempora carinate, carinae strong throughout; antennae slender and moderately long, filiform, shorter than the pronotum and elytra combined, antennomeres VII–X subquadrate to slightly transverse, basal 4 antennomeres elongate, glossy and remaining matte; pronotum impressed mesobasally, strongly transverse, approximately rectangular in shape, hypomeron visible for two thirds of pronotal length in lateral view; pubescence directed anteriad at one third of apical midline, and obliquely laterad across arcuate lines elsewhere; elytra moderately convex, slightly elongate, and broader than pronotum, posterior margins broadly arcuate and with small and scarcely visible lateral emarginations, pubescence directed obliquely posteriad; abdomen arcuate laterally, four first visible tergites strongly impressed at base; basal metatarsomere as long as the following one. MALE. Apical margin of tergite VIII obliquely produced laterally, with broad, obtusely angular, slightly crenulate emargination between two rounded, roughly triangular processes (Fig. 19.212d); sternite VIII evenly parabolically rounded apically (Fig. 19.212e); median lobe of aedeagus in dorsal view with bulbus narrowly oval, tapering slightly sinuately to elongate, subparallel tubus, apex rounded triangular (Fig. 19.212c); in lateral view tubus of median lobe obliquely bent ventrad basally, slightly sinuate to moderately broad, rounded apex (Fig. 19.212b); internal sac with two pairs of large, elongate structures in bulbus (Fig. 19.212b,c). FEMALE. Tergite VIII transverse, with apical margin obliquely produced laterally, truncate and slightly emarginate medially (Fig. 19.212f); sternite VIII transverse and broadly rounded apically (Fig. 19.212g); spermatheca with narrowly elongate, tubular capsule, subspherical apically, stem extremely fine, long, forming irregularly twisted mass posteriorly (Fig. 19.212h).

DISTRIBUTION **Origin**: Nearctic. **CANADA**: BC. **USA**: AK, CA [see below], WA.

COLLECTION AND HABITAT DATA

Habitat: in BC, found in transition zone and interior of Sitka spruce forest, under dung in spruce forest, in bear dung, moss along seepage and waterfall, rotting pile of mushrooms, under deciduous shrubs around base of cedar, and under debris and kelp on beach. In AK, collected in *Betula, Salix*, white spruce; alpine heath, sparse *Tsuga* sp.; *Betula papilifera, Populus tremuloides*; wet meadow; old growth and thinned 2nd growth forests; brown bear scat in alpine habitat; primate dung in subalpine habitat; subalpine meadow with *Gymnocarpium* sp.; subalpine *Tsuga, Vaccinium*; and old growth Hemlock, coastal rain forest. **Collecting period**: V–IX. **Collecting method**: sifting forest litter, kelp, moss, fungi, bear and martin dung, primate dung baited pitfalls, flight intercept traps, barrier traps, Lindgren traps, yellow bowls.

Additional records. See Annex 1.

COMMENTS Moore and Legner (1975) recorded this species from AK, BC, NH, and southern CA. The CA record was probably based on the type localities of the synonyms *A. aperta* Casey 1910, *A. intacta* Casey 1911, *A. morbosa* Casey 1911, all from CA; their NH record was probably based on Fauvel (1889), who listed "White Mountains" in the distribution, which Moore and Legner probably inferred to be in NH. This record is doubtful, and should be regarded as such unless proven otherwise. The male median lobe of aedeagus presented here is based on non-type specimens. We have found that there are two forms of median lobe of aedeagus, one elongate (type series) (see Fig. 20.178h in Klimaszewski et al. (2020)), and another slightly shorter and more rounded apically, most common in specimens from other parts of AK and BC (Klimaszewski and Winchester 2002, and Fig. 19.212b in present book). We believe that the two forms may represent intraspecific morphological variation. However, the DNA comperative study of the types of *A. picipennis* from Sitka Is., and the population with shorter and rounded apex of median lobe of aedeagus should be analysed to confirm that they belong to the same species. *Atheta picipennis* sensu Brundin 1954, is actually the species currently called *A. aeneipennis* in Europe, not the true *A. picipennis* (Lohse and Smetana 1985). A recent record of *A. picipennis* from Russia (Voinkov 2020) represents this confusion. See the comments under *A. prudhoensis* Lohse.

***Prudhoensis* species group** (Klimaszewski et al. 2008a)

Diagnosis Antennae slender, antennomeres V–X subquadrate to slightly transverse; forebody with integument moderately glossy, with moderately impressed microsculpture consisting of irregularly shaped hexagonal sculpticells, with punctation moderately dense and asperate; pubescence on disc of elytra directed obliquely posteriad; median lobe of aedeagus in dorsal view with large, oval bulbus and moderately long and apically triangular tubus, in lateral view tubus arcuate or straight basally, with ventrally produced broad apical part; internal sac with two large, falciform sclerites in bulbus; male tergite VIII with apical margin straight to very obtusely emarginate, and with small or large lateral teeth; spermatheca S-shaped, with narrow, club-shaped capsule bearing small apical invagination, and with long, sinuate stem, narrowly looped posteriorly.

142. ***Atheta*** **(*Dimetrota*)** ***burwelli*** **(Lohse)** (Fig. 19.213a–h)

Dimetrota burwelli Lohse, in Lohse et al. 1990. Majka and Klimaszewski 2008b, 2010, Brunke et al. 2012, Klimaszewski et al. 2008a, 2011, 2018a.

Diagnosis Body narrowly oval, slightly flattened, length 2.8–3.5 mm; colour dark brown to black, with legs/tarsi, basal parts of antennae and elytra yellowish-brown (Fig. 19.213a); integument highly glossy; forebody with punctation dense and finely asperate, microsculpture fine; pubescence on disc of pronotum directed obliquely anteriad, laterad and obliquely posteriad from midline, that on elytra directed obliquely posteriad; head distinctly narrower than pronotum; antennae slender, antennomeres V–X subquadrate to slightly transverse; pronotum slightly transverse, about 1/5 narrower than width of base of elytra; elytra transverse, at suture as long as pronotum; abdomen subparallel. MALE. Tergite VIII with broadly V-shaped apical emargination between two sublateral broad triangular processes (Fig. 19.213d), sternite VIII with apical margin broadly parabolically rounded (Fig. 19.213e); tubus of median lobe of aedeagus in lateral view arcuately curved ventrally, with broadly oval apical part abruptly tapering near apex (Fig. 19.213b); in dorsal view tubus narrow and short, and triangular apically, bulbus large, broadly oval (Fig. 19.213c). FEMALE. Tergite VIII with apical margin broadly truncate, oblique laterally (Fig. 19.213f); sternite VIII with apical margin rounded, truncate medially (Fig. 19.213g); spermatheca S-shaped, with tubular, clubbed apical capsule, and long sinuate stem narrowly looped posteriorly (Fig. 19.213h).

Females of this species may be confused with those of *A. curvipennis*, *A. prudhoensis* and *A. terranovae*, and a positive identification may be achieved by association with male specimens which are more easily identifiable.

DISTRIBUTION Origin: Nearctic. **CANADA**: AB [**BOLD**], BC [**BOLD**], LB, NB, NF, NT [**BOLD**], NU, ON, YT. **USA**: AK [**NSR**].

COLLECTION AND HABITAT DATA
Habitat: from unbaited and carrion-baited pitfall traps, flight intercept traps, and rotting mushrooms in conifer dominated forests in NL; some from *Russula virescens* (Schaeff.) Fr. mushrooms in a white spruce forest in NS; and in spruce and aspen forest with shrubs in BC. In AK, found in *Betula* and *Populus tremuloides* forest. **Collecting period**: VII–IX. **Collecting method**: sifting litter and mushrooms, pitfall traps, flight intercept traps.

We report this species for the first time here, from AB, BC [**NPR**] and NT (**NTR**) based on **BOLD** records, and from AK based on a UAM specimen: Canada, British Columbia, Kinaskan Lake Provincial Park, Kinaskan Lake Trail, 833m el., 57.532, –130.202, spruce and aspen forest with shrubs, pitfall, 26.VII-01.VIII.2014, BIObus 2014 team (CBG) 1 specimen; Alberta, Jasper National Park, Miette Hotsprings, 53.124, –117.775, 1439 m, pan trap in riparian zone, 21.VII.2012 (BIOBus 2012) (CBG) 1 specimen; Northwest Territories, Norman Wells, ENR Regional Office, 65.279, –126.83, malaise trap in urban area, S. Behrens, R. Popko, 13.VIII.2014 (CBG), 1 specimen. USA, Alaska, Fairbanks, 930 Fitz Ct, 280m el., 64.9013, –147.52861, ±30m *Betula* and *P. tremuloides*, flight intercept trap, 16.VIII-22.IX.2007, D.S. Sikes (UAM) 1 specimen; 29.V-4.VII.2008 (UAM) 1 specimen.

COMMENTS Lohse in Lohse et al. 1990, wrongly associated Port Burwell with province of Quebec. It is located in Nunavut and therefore the QC record was removed from distribution of this species as no QC specimen data have been published.

143. ***Atheta*** **(*Dimetrota*)** ***caribou*** **(Lohse)** (Fig. 19.214a–e)

Dimetrota caribou Lohse, in Lohse et al. 1990. Klimaszewski et al. 2011, 2012, 2018a.

Diagnosis Body narrowly oval, slightly flattened, length 2.4–2.6 mm; colour dark brown to black, with elytra lightly reddish brown and legs/

tarsi, antennae yellowish-brown or reddish-brown (Fig. 19.214a); integument moderately glossy; forebody with punctation moderately dense and finely asperate, with fine meshed microsculpture; pubescence on disc of pronotum directed laterad and obliquely posteriad from midline, that on elytra directed obliquely posteriad; head slightly narrower than pronotum; antennae slender, antennomeres V-X subquadrate; pronotum transverse, about 1/7 narrower than width of base of elytra; elytra transverse, at suture as or slightly longer than pronotum; abdomen arcuate laterally. MALE. Tergite VIII with shallow broadly V-shaped apical emargination between two short triangular lateral teeth (Fig. 19.214d); sternite VIII parabolically rounded apically (Fig. 19.214e); tubus of median lobe of aedeagus in lateral view almost straight ventrally, slightly arched ventrad before narrowly triangular apical part (Fig. 19.214b); in dorsal view tubus triangular apically and constricted basally, bulbus broadly oval (Fig. 19.214c). FEMALE: unknown.

This species may be positively identified by the shape of the median lobe of the aedeagus and the shape of the apical margin of male tergite VIII.

DISTRIBUTION **Origin**: Nearctic. **CANADA**: LB, YT. **USA**: not recorded.

COLLECTION AND HABITAT DATA
Habitat: specimens from LB from carrion-baited pitfall traps in boreal spruce forests with moss substrate; in YT, sifted from litter under *Salix* and *Alnus* bushes and from old caribou dung. **Collecting period**: VI, VIII. **Collecting method**: sifting litter and dung, carrion-baited pitfall traps.

144. ***Atheta*** **(*Dimetrota*)** ***capsularis*** **Klimaszewski** (Fig. 19.215a–h)

Atheta capsularis Klimaszewski, in Klimaszewski et al. 2005. Work et al. 2013, Klimaszewski et al. 2007b, 2008a, 2011, in *Atheta* (*Dimetrota*): Brunke et al. 2012, Klimaszewski et al 2018a, b, 2020.

Diagnosis Body narrowly oval, moderately convex, length 2.8–3.2 mm; colour dark brown to black, with legs/tarsi, basal parts of antennae, elytra, and basal and apical parts of abdomen yellowish-brown, some specimens entirely yellowish brown (Fig. 19.215a); integument moderately glossy; forebody with punctation dense and asperate, with fine meshed microsculpture; pubescence on disc of pronotum directed laterad and obliquely posteriad from midline, forming arcuate pattern, that on elytra directed obliquely posteriad, forming wavy pattern laterally; head slightly narrower than pronotum; antennae slender, antennomeres V–X subquadrate; pronotum transverse, about 1/5 narrower than width of base of elytra; elytra transverse, at suture as long as pronotum; abdomen subparallel. MALE. Apical margin of tergite VIII angulately produced laterally, apex with broad, shallow, slightly sinuate medial emargination (Fig. 19.215d); sternite VIII rounded apically (Fig. 19.215e); tubus of median lobe of aedeagus in lateral view straight ventrally, slightly arched ventrad before broad, rounded apical part (Fig. 19.215b); in dorsal view tubus triangular, bulbus large, and narrowly oval (Fig. 19.215c). FEMALE. Tergite VIII with apical margin broadly truncate medially, sinuately obtuse laterally (Fig. 19.215f); sternite VIII broadly rounded apically (Fig. 19.215g); spermatheca with club-shaped capsule with narrow apical invagination, stem sinuate and with gradually expanding posterior loop (Fig. 19.215h).

This species may be confused externally with *Atheta terranovae*, but it differs in being generally larger, in having the elytra yellowish and the body paler, in the shape and structures of the median lobe of the aedeagus, and in the unique shapes of the spermatheca and male tergite VIII.

DISTRIBUTION **Origin**: Nearctic. **CANADA**: AB [**BOLD**], BC, LB, MB, NB, NF, NT [**BOLD**], ON, QC, SK [**NPR**], YT. **USA**: AK [**NSR**].

COLLECTION AND HABITAT DATA
Habitat: red spruce, balsam fir, jack pine, mixed and riparian type forests. Most specimens were from pitfall traps and flight intercept traps; some from forest litter. In AB, found in montane and

subalpine conifer forest, Douglas fir and lodgepole pine forest, mature forest, birch and spruce forest, thinned out lodgepole pine stand, riparian area, rocky and mossy habitat. In NT, found in urban area. In AK, found in forest edge, *Populus tremuloides* and grass; *Betula* and *P. tremuloides*; Alder, white spruce, *Betula* forest; decaying mushrooms in *Betula* forest; mostly spruce with some aspen forest; and shrub, alder, dwarf birch, grass, and clubmoss. The posterior gut of this species revealed several species of yeasts and fungal spores. **Collecting period**: VI–VIII. **Collecting method**: sifting litter, pitfall traps, yellow pitfalls w/o rainroofs, flight intercept traps (Malaise), ground search, barrier traps.

We report this species for the first time here, from AB (**NPR**), and NT (**NTR**) based on **BOLD** records, from SK (**NPR**) based on LFC specimens, and from AK (**NSR**) based on UAM specimens: Canada, Alberta, Waterton Lakes National Park, Highway 6 pulloff, 1562m el., 49.065, –113.778, douglas fir and lodgepole pine forest with aspen and birch understory, pitfall, 06-11.VIII.2012, BIOBus 2012 team (CBG) 2 specimens; Jasper National Park, Highway 16 / 93A Junction, 1060m el., 52.867, –118.077, thinned out lodgepole pine stand, Malaise trap, 03-09.VIII.2012, B. Sharp (CBG) 1 specimen; Jasper National Park, Miette Hotsprings, 1439m el., 53.124, –117.775, along forest creek on a valley bed, pitfall, 17-21.VII.2012, BIOBus 2012 team (CBG) 1 specimen; Jasper National Park, Pocahontas Campground, 1131m el., 53.195, –117.914, birch and spruce forest, pitfall, 17-21.VII.2012, BIOBus 2012 team (CBG) 12 specimens. Northwest Territories, Norman Wells, 77m el., 65.2791, –126.83, urban area, Malaise trap, 06-13.VIII.2014, S. Behrens, R. Popko (CBG) 1 specimen. Saskatchewan, Meadow Lake, 53.9386, –108.8900, 487m, 11. IX.2014, mushrooms, sifting, coll. B. Godin and D. Horwood (LFC) 1 male; Prince Albert, 53.8995, –106.155, 533m, 15.IX.2014, mushrooms, sifting, coll. B. Godin and D. Horwood (LFC) 1 male; Prince Albert, 53.8995, –106.057, 556m, 14.IX.2014, mushrooms, sifting, coll. B. Godin and D. Horwood (LFC) 1 male. USA, Alaska, Bonanza Crk LTER, 225m el., 64.70963, –148.32647, ±61m forest edge, *Pop. trem.* grass, pitfall, 17-24.VII.2007, B. Mortensen (UAM) 1 specimen; Denali National Park and Preserve, Stampede Creek, Kantishna Hills, 63.75, –150.35, ±3km, 1.VI-30.IX.1984 (UAM) 1 specimen; Denali NPP., Mt. Healy, 595m el., 63.73478, –148.92911, ±30m forest, mostly spruce, some aspen, 10 minute ground search, 16.VIII.2016, D.S. Sikes (UAM) 2 specimens; same data except, 4 pitfalls w/ rainroofs, pooled, 2-19.VII.2016, J.J. Rykken, D.S. Sikes, A. Haberski, Z. Snelson (UAM) 1 specimen; 4 yellow pitfalls w/o rainroofs, pooled, 3-16. VIII.2016, J.J. Rykken, D.S. Sikes, R. Stewart-Rogers (UAM) 4 specimens; Denali NPP., Mt. Healy, 927m el., 63.74016, –148.95346, ±30m shrub, alder, dwarf birch, grass, clubmoss, 4 yellow pitfalls w/o rainroofs, pooled, 3-16. VIII.2016, J.J. Rykken, D.S. Sikes, R. Stewart-Rogers (UAM) 1 specimen; Denali NPP., Toklat, 960m el., 63.52373, –150.04829, ±30m spruce, 4 yellow pitfalls w/o rainroofs, pooled, 2-21. VII.2016, J.J. Rykken, D.S. Sikes, A. Haberski, Z. Snelson (UAM) 1 specimen; Fairbanks, 930 Fitz Ct, 280m el., 64.9013, –147.52861, ±30m *Betula* & *P. tremuloides*, flight intercept trap, 16.VIII.–22.IX.2007, D.S. Sikes (UAM) 16 specimens; 26.VIII-10.IX.2006 (UAM) 2 specimens; Fairbanks, Creamer's Field, 152m el., 64.86864, –147.7381, ±6m barrier trap pair, 15-18.VII.2008, D.S. Sikes (UAM) 1 specimen; Fairbanks, Creamer's Field, 154m el., 64.86607, –147.73817, ±8m pitfall, 19-21.VII.2008, D.S. Sikes, R. Spiers (UAM) 1 specimen; Fairbanks, Creamer's Field, 158m el., 64.86823, –147.73758, ±12m mesic *Betula*, pitfall, 26.VI-18.VII.2008, D.S. Sikes, R. Davidson (UAM) 1 specimen; Fairbanks; Yak Estates, 230-237m el., 64.84651, –147.8962, ±155m pitfall, 18-20.VIII.1977, M.W. Oswood, H.C. Oswood (UAM) 4 specimens; Fairbanks: Creamer's Field, 140m el., 64.86883, –147.73065, ±104m *Betula* forest, decaying mushrooms, 31.VII.2010, D.S. Sikes (UAM) 1 specimen; Mi 179 Dalton Hwy, 365m el.,

67.29803, −150.15699, ±7m 3/4 Alder, white spruce, *Betula* forest, 5-31.VII.2008, D.S. Sikes (UAM) 3 specimens; near Kasilof, 60.3375, −151.2744, ±50m in mushrooms, 21.VII.2005, D. Langor (UAM) 1 specimen.

145. *Atheta* (*Dimetrota*) *curvipennis* Klimaszewski and Langor (Fig. 19.216a–h)

Atheta (Dimetrota) curvipennis Klimaszewski and Langor, in Klimaszewski et al. 2011. Klimaszewski et al. 2018a.

Diagnosis Body narrowly oval, slightly flattened, length 2.8–3.5 mm; head, abdomen except for apical part and antennomeres III–XI dark brown to black, remainder of body yellowish- or reddish-brown (Fig. 19.216a); integument moderately glossy; forebody with punctation moderately dense and finely asperate, microsculpture fine; pubescence on disc of pronotum directed laterad and obliquely posteriad from midline, forming arcuate pattern, that on elytra directed obliquely posteriad; head slightly narrower than pronotum; antennae slender, antennomeres V–X subquadrate; pronotum transverse, about 1/9 narrower than width of base of elytra; elytra transverse, at suture as long as pronotum; abdomen arcuate laterally. MALE. Tergite VIII with apical margin sharply truncate laterally, feebly, broadly emarginate (Fig. 19.216d); sternite VIII elongate, with apical margin parabolically rounded except narrowly truncate medially (Fig. 19.216e); tubus of median lobe of aedeagus in lateral view straight ventrally, angularly bent ventrad before roughly triangular apical part (Fig. 19.216b); in dorsal view tubus short and broadly triangular, bulbus large and oval in outline (Fig. 19.216c). FEMALE tergite VIII truncate apically and slightly concave (Fig. 19.216f); sternite VIII with apical margin truncate, broadly rounded laterally (Fig. 19.216g); spermatheca with clubbed-shaped capsule elbowed at base, stem sinuate, narrowly looped posteriorly (Fig. 19.216h).

Females of this species may be confused with those of other species of the group, and may be best identified by association with males which are more easily identifiable.

DISTRIBUTION Origin: Neractic. **CANADA**: LB, NF. **USA**: AK [**NSR**].

COLLECTION AND HABITAT DATA

Habitat: specimens collected in carrion-baited pitfall traps, and mushrooms in spruce-poplar forest, spruce lichen forest, spruce-moss forest, mixed forest, and spruce-birch forest (LB, NF). In AK, found in krummholz, *Tsuga mertensiana, Elliottia pyroliflora, Vaccinium ovalifolium*, and inky cap mushroom. **Collecting period**: VI–IX. **Collecting method**: sifting mushrooms, flight intercept traps, carrion-baited pitfall traps.

We report this species from AK (**NSR**) for the first time here, based on the following: USA, Alaska, Chichagof Is. pl.06, 725m el., 57.98495, −135.20358, ±4m krummholz, *Tsuga mertensiana, Elliottia pyroliflora, Vaccinium ovalifolium*, pitfall, 15-16.VI.2013, S. Ridling (UAM) 2 specimens; Fairbanks, 930 Fitz Ct, 280m el., 64.9013, −147.52861, ±30m inky cap mushroom, 27.VIII.2009, D.S. Sikes (UAM) 2 specimens.

146. *Atheta* (*Dimetrota*) *prudhoensis* (Lohse) * (Figs. 19.217a–h)

Dimetrota prudhoensis Lohse, in Lohse et al. 1990. In *Atheta* (*Dimetrota*): Klimaszewski and Winchester 2002, Gusarov 2003a [synonymy of *D. dempsterensis* Lohse], Klimaszewski et al. 2007b, 2008a, Majka and Klimaszewski 2008b, Webster et al. 2009, Klimaszewski et al. 2011, 2015a, 2018a, 2020.

Diagnosis Body narrow, subparallel, slightly flattened, length 3.3–3.5 mm; colour dark brown to black, with legs, antennae, elytra, and posterior part of abdomen yellowish brown or reddish brown (Fig. 19.217a); integument moderately glossy; forebody with punctation dense and asperate, with fine microsculpture; pubescence on disc of pronotum directed laterad and obliquely posteriad from midline, that on elytra directed obliquely and straight posteriad; head slightly narrower than pronotum; antennae slender, antennomeres V–X subquadrate to slightly transverse; pronotum transverse, about 1/5 narrower than width of base of elytra; elytra transverse, at suture as long as pronotum; abdomen arcuate laterally. MALE. Tergite VIII with a sharp triangular sublateral tooth on each side of broadly V-shaped apical emargination (Fig. 19.217d); sternite VIII rounded apically (Fig. 19.217e); tubus of median lobe of aedeagus in lateral view arcuately arched ventrad throughout to moderately broad apical part, apex narrowly rounded (Fig. 19.217b); in dorsal view bulbus large and oval, and tubus parallel-sided to roughly triangular apex (Fig. 19.217c). FEMALE. Tergite VIII truncate apically, rounded laterally (Fig. 19.217f); sternite VIII semicircularly rounded apically (Fig. 19.217g); spermatheca S-shaped, with tubular, club-shaped capsule, and sinuate stem narrowly looped posteriorly (Fig. 19.217h).

Atheta prudhoensis is closely related to *A. burwelli*, *A. curvipennis*, and possibly *A. caribou*, from which it differs in the shape of the male genitalia and in some external features. Females of this species may be confused with those of *A. burwelli* and *A. curvipennis*, but may be recognized by association with males which are more easily identifiable.

DISTRIBUTION Origin: Holarctic [**New Status**]. **CANADA**: AB [**BOLD**], BC, LB, NB, NF, NS, ON, QC, SK, YT. **USA**: AK, NY, VT.

COLLECTION AND HABITAT DATA

Habitat: in many kinds of forests such as conifer dominated forests, spruce, white spruce, spruce and fir, red pine, jack pine, birch, maple, oak, northern hardwood, hemlock, and mixed forests. Adults often common in decaying mushrooms and *Russula* sp. (gilled mushroom) but also were found in deer dung, carrion, compost, and leaf litter. In AB, found in riparian area, rocky and mossy. In AK, specimens were captured in field/ forest edge; mesic forest; forest, mostly black spruce, some aspen; *Betula, Salix*, white spruce forest; *Salix* leaf litter; *Picea, Populus, Betula, Salix*; *Populus, Betula, Alnus* leaf litter; *Betula* and *Populus tremuloides*; black spruce bog; old growth Hemlock, coastal rain forest; shrubland, *Nephrophyllidium crista-galli, Elliottia pyrolifora, Luetkea pectinata*; willow, dwarf birch, shrub; tundra; maritime tundra with *Leymus, Heracleum*; bottom and at the edge of closed tall shrub-alder, crowberry; alpine tundra; fresh deer scat in subalpine wet tundra; subalpine meadow; creek edge; beaver pond shore; treading vegetation; black spruce bog; meadow, *Nephrophyllidium crista-galli, Geranium erianthum, Artemisia arctica*; rich meadow, *Geranium erianthum, Artemisia arctica, Sanguisorba canadensis*; *Chamaedaphne*, sedge, non-sphagnum moss innundated meadow; herbaceous heath, *Luetkea pectinata, Cassiope mertensiana, Lupinus nootkatensis*; mossy area near petrel burrows; krummholz, *Tsuga mertensiana, Elliottia pyroliflora, Vaccinium ovalifolium*; shrub, alder, dwarf birch, grass, clubmoss; krummholz, *Tsuga mertensiana, Vaccinium ovalifolium, Chamaecyparis nootkatensis*; inky cap and other mushrooms. **Collecting period**: VI–X. **Collecting method**: sifting litter and mushrooms, Berlese funnels, pitfall traps, yellow pitfalls w/o rainroofs, flight intercept traps (Malaise), sweeping vegetation, colored bowls transect, yellow bowls, barrier traps.

We report this species from AB (**NPR**) for the first time here, based on **BOLD** records: Canada, Alberta, Jasper National Park, Miette Hotsprings, 1439m el., 53.124, –117.775, along forest creek on a valley bed, pitfall, 17-21.VII.2012, BIOBus 2012 team (CBG) 3 specimens; Waterton Lakes National Park, Highway 6 pulloff East of 2 Flags Lookout, 1562m el., 49.065, –113.778, douglas fir and lodgepole pine forest with aspen and birch understory, pitfall, 06-11.VIII.2012, BIOBus 2012 team (CBG) 7 specimens; Waterton Lakes National Park, Akamina Parkway, Rowe Tamarack Trail, 1640m el., 49.059, –114.014, subalpine lodgepole pine and fir forest with birch in understory, pitfall, 07-11.VIII.2012, BIOBus 2012 team (CBG) 4 specimens.

Additional records. See Annex 1.

COMMENTS *Atheta prudhoensis* (Lohse, in Lohse et al. 1990, orig. *Dimetrota*) was described from Alaska and Yukon and until now was considered as a Nearctic species (Lohse et al. 1990, Klimaszewski et al. 2018a). However, this species is morphologically not distinguishable from the Palaearctic species currently known as *Atheta aeneipennis* (Thomson 1856, orig. *Homalota*) in the Palaearctic region (Schülke and Smetana 2015) (*A. picipennis* sensu Brundin 1954, nec Mannerheim, 1843), described from northern Europe (Sweden). The illustrations of median lobe of aedeagus and spermatheca provided by Brundin (1954, as *A. picipennis*) match those of the Nearctic species *A. prudhoensis*. We have compared morphologically (external morphology and genitalia) specimens from North America and Finland (Oulanka, ZMUO, http://id.zmuo.oulu.fi/ZMUO.001.252) and found no significant diffrences between these populations. Lohse and Smetana (1985) stated that *A. aeneipennis* should be used as the valid name for *A. picipennis* sensu Brundin 1954, but apparently did not study the types of *A. aeneipennis*. We have examined 9 syntypes of *A. aeneipennis* (BMLS), and they represent two different, already described European species: *A. parapicipennis* Brundin, 1954 (Fig. 19.219a–h) (Brundin 1954, Fig. 34, 35), and *A. picipennoides* H.K. Hanssen, 1932 (Fig. 19.218a–h) (Brundin 1954, Fig. 32, 33). We designate the following specimen from Thomson's original collection as lectotype of *A. aeneipennis* (male, dissected, genitalia on plastic microslide in Canada balsam, green rectangular label, 939, 2019, MZLU), it is morphologically the same as *A. parapicipennis*. Therefore *A. aeneipennis* becomes a new (senior) synonym of *A. parapicipennis*. The remaining syntypes of *A. aeneipennis* represent mixed series of *A. parapicipennis* [=*A. aeneipennis*] (green rectangular labels: 940, 2019, 1 male MZLU; 941, 2019, 1 female MZLU; 942, 2019, 2 slightly damaged undissected specimens MZLU), and *A. picipennoides* (green rectangular labels: 937, 2019, male MZLU; 938, 2019, female MZLU; 941, 2019, male MZLU; 943, 2019, male MZLU). The above two European species (*A. aeneipennis* and *A. picipennoides*) have different genitalia (Figs. 19.218a–h, 19.219a–h) from those of *A. prudhoensis* from North America, therefore *A. aeneipennis* cannot be used as the correct name for *A. prudhoensis*. We know that *A. prudhoensis* is a Holarctic species occurring widely in North America and Eurasia (Europe to East Siberia and Japan), and that one of the names listed as synonyms of *A. aeneipennis* by Schülke and Smetana (2015) is probably the valid name for this species. We cannot at present determine which one of the Palaearctic species names is the valid name for *A. prudhoensis* because we have not studied any of the type material of the listed synonyms of *A. aeneipennis* (*A. immunda* (Brisout de Barneville, 1860); *A. cursor* (Motschulsky, 1858) [homonym replaced by *A. cursoria* Gemminger and Harold, 1868]; *A. laetipennis* G. Benick, 1982; and *A. tenuiducta* (Sawada, 1970)). Therefore, we tentatively use the name *A. prudhoensis* in this contribution. The DNA barcoded specimens of *A. prudhoensis* (ca. 70 specimens from multiple localities across Canada) share the same BIN with Finnish material identified as *A.* "*aeneipennis*" (6 specimens) and *A. lapponica* J. Sahlberg, 1876 (5 specimens).

147. *Atheta* (*Dimetrota*) *strigosula* Casey (Fig. 19.220a–i)

Atheta (Atheta) strigosula Casey, 1910. Klimaszewski et al. 2005, 2007b, 2008a, Majka and Klimaszewski 2008b; Klimaszewski et al. 2011; Brunke et al. 2012, Stefani et al. 2016, Klimaszewski et al. 2016d, 2018a, b, 2020.

Diagnosis Body widely subparallel, length 3.6–4.0 mm, head and pronotum flattened medially; colour dark brown, with elytra, legs and bases of antennae light yellowish-brown or reddish-brown (Fig. 19.220a); integument moderately glossy, forebody with asperate punctation and meshed microsculpture, pubescence fine and sparse and denser on elytra; head slightly narrower than pronotum, subquadrate, disc impressed medially; antennae slender, with articles IV–V slightly elongate, VI–X subquadrate to slightly transverse; pronotum distinctly transverse, widest at middle, slightly narrower than elytra (wider in male than in female); elytra at suture slightly longer than pronotum; abdomen widely arcuate laterally. MALE. Tergite VIII with apical margin truncate, finely crenulate between two rounded sublateral teeth (Fig. 19.220d); sternite VIII broadly parabolically rounded apically (Fig. 19.220e); tubus of median lobe of aedeagus with ventral margin straight to broad and triangular apical part in lateral view (Fig. 19.220b), in dorsal view bulbus large, oval, and tubus moderately long and narrowly triangularly shaped (Fig. 19.220c). FEMALE. Tergite VIII with apical margin truncate medially, rounded laterally (Fig. 19.220f); sternite VIII evenly rounded apically (Fig. 19.220g); spermatheca small, capsule club-shaped, stem thin and sinuate with single narrow loop posteriad (Fig. 19.220h,i).

DISTRIBUTION Origin: Nearctic. **CANADA**: AB [**BOLD**], BC, LB, MB, NB, NF, ON, QC, SK, YT. **USA**: AK, NY.

COLLECTION AND HABITAT DATA

Habitat: occurs in various forest types such as red spruce, mature red spruce and eastern white cedar, yellow birch and balsam fir, mixed, red oak, rich Appalachian hardwood, northern hardwood, and other deciduous forests (NB). Adults usually found in decaying mushrooms (often common) on forest floor but also in leaf litter and possibly carrion. In AB, collected in montane Douglas fir and lodgepole pine forest, low alpine dry slope temperate forest, white birch/willow/trembling aspen forest, and wetland. In AK, captured in low alpine dry slope, temperate forest; white birch/willow/trembling aspen, *Salix*, *Betula* and *Populus tremuloides* forest; mesic *Betula, Picea* forest; disturbed *P. glauca* forest; thinned and unthinned 2nd growth; pruned tree branch; rocky alpine; and in mushrooms. **Collecting period**: IV–X (most common in IX in NB). **Collecting method**: sifting mushrooms and litter, carrion-baited and unbaited pitfall traps, flight intercept traps, Malaise trap, sweeping vegetation, UV light trap.

We report this species from AB (**NPR**) for the first time here, based on **BOLD** records: Canada, Alberta, Elk Island National Park, Astotin Lake, 719m el., 53.685, −112.860, emergent white birch/willow/trembling aspen, Malaise trap, 06-13.VII.2012, Stephanie Church (CBG) 1 specimen; same data but 03-10.VIII.2012 (CBG) 1 specimen; Waterton Lakes National Park, Akamina Parkway, Rowe Tamarack Trail, 1640m el,. 49.059, −114.014, subalpine lodgepole pine and fir forest with birch in understory, pitfall, 22-27.VII.2012, BIOBus 2012 team (CBG) 1 specimen; Waterton Lakes National Park, Highway 6 pullout, 1562m el., 49.065, −114.778, douglas fir and lodgepole pine forest with aspen and birch understory, pitfall, 06-11.VIII.2012, BIOBus 2012 team (CBG) 2 specimens.

Additional records. See Annex 1.

148. ***Atheta (Dimetrota) takhini*** **Klimaszewski and Godin, sp. n.** (Fig. 19.221a–d)

urn:lsid:zoobank.org:act:F89225AF-73EC-46A6-AB87-DC6126C3A79D

Holotype (male): Canada, Yukon, Takhini Salt flats, 15-29.VIII.2014, Alkaline clay, 60.855, −135.717, 663 m, pitfall trap, Salicornia habitat, coll. B. Godin and D. Horwood (CNC) 1 male.

Etymology The name of this species derives from the Takhini Salt flats near Whitehorse, Yukon, where the holotype was captured.

Diagnosis Body narrowly subparallel, moderately convex, length 3.0 mm; dark brown with appendages and elytra yellowish brown (Fig. 19.221a); integument of forebody moderately glossy and strongly so on abdomen; forebody with punctation dense and slightly asperate, with weak microsculpture; pubescence of pronotum directed laterad from midline of disc, and on elytra directed obliquely posteriad; head slightly narrower than pronotum; antennae slender, antennomere IV subquadrate, V–X slightly to distinctly transverse; pronotum subquadrate; elytra transverse, at suture as long as pronotum; abdomen subparallel. MALE. Tergite VIII with apical margin truncate, finely crenulate between two small lateral teeth (Fig. 19.221c); sternite

VIII broadly parabolic apically (Fig. 19.221d); tubus of median lobe of aedeagus in lateral view with ventral margin straight, apical part slightly curved ventrad, beak-shaped, narrowly-rounded at apex (Fig. 19.221b). FEMALE. Unknown.

This species may be distinguished from its congeners by the unique shape and structures of the median lobe of aedeagus combined with crenulate truncate margin of male tergite VIII.

DISTRIBUTION Origin: Nearctic. **CANADA**: YT. **USA**: not recorded.

COLLECTION AND HABITAT DATA
Habitat: Alkaline clay, habitat dominated by *Salicornia borealis* Wolff and Jeffries, a halophyte plant **Collecting period**: VIII. **Collecting method**: pitfall trap.

149. *Atheta* (*Dimetrota*) *terranovae* Klimaszewski and Langor (Fig. 19.222a–h)

Atheta (Dimetrota) terranovae Klimaszewski and Langor, in Klimaszewski et al. 2011. Brunke et al. 2012, Webster et al. 2012, Klimaszewski et al. 2012, Work et al. 2013, Klimaszewski et al. 2016d, 2018a, b, 2020.

Diagnosis Body narrowly subparallel, moderately convex, length 2.5–3.0 mm; head, pronotum, and entire/or posterior of abdomen dark brown to nearly black, with appendages and elytra paler, yellowish brown mottled with dark brown (Fig. 19.222a); integument moderately glossy; forebody with punctation dense and asperate, with microsculpture moderately coarse; pubescence on disc of pronotum directed laterad from midline of disc, that on elytra directed obliquely posteriad; head distinctly narrower than pronotum; antennae slender, antennomeres V–X slightly transverse; pronotum transverse, about 1/8 narrower than width of base of elytra; elytra transverse, at suture at least as long as pronotum; abdomen subparallel. MALE. Tergite VIII with lateral margins distinctly converging apicad, entire apical margin shallowly, obtusely angulately emarginate, lateral angles rectangular (Fig. 19.222d); sternite VIII parabolically rounded from base to apex (Fig. 19.222e); tubus of median lobe of aedeagus in lateral view slightly arched ventrad basally, straight to slightly curved apex, with apical part narrowly rounded (Fig. 19.222b); in dorsal view tubus broad, triangular apically with narrowly oval bulbus (Fig. 19.222c). FEMALE. Tergite VIII with apical margin broadly truncate medially, obliquely shallowly emarginate laterally (Fig. 19.222f); sternite VIII broadly rounded apically (Fig. 19.222g); spermatheca S-shaped, with club-shaped tubular capsule and sinuate and stem narrowly looped posteriorly (Fig. 19.222h).

This species bears some superficial external resemblance to *Atheta capsularis*, but in *A. terranovae* the body is slightly smaller and narrower, the pronotum is less transverse, the punctation is denser, antennomeres VII–X are more transverse, the elytra are darker, and the shapes of the median lobe of the aedeagus and spermatheca are different.

DISTRIBUTION Origin: Nearctic. **CANADA**: AB [**BOLD**], BC, LB, MB, NB, NF, ON, QC, SK, YT. **USA**: AK.

COLLECTION AND HABITAT DATA
Habitat: occurs in many coniferous, mixed and hardwood/deciduous forest types in Canada. Adults commonly found in decaying mushrooms (polypore and gilled) and sometimes in fresh mushrooms. In NB, in an oak forest, rich Appalachian hardwood forest, northern hardwood forest, various mixed forests, red spruce forest, mature red spruce forest with red maple or yellow birch, black spruce forest, and eastern white cedar forests and swamps; specimens in fresh and decaying gilled mushrooms of various species, rotting lobster mushrooms, and a coral mushroom on spruce log. Also captured in unbaited and carrion-baited pitfall traps and flight intercept traps in various coniferous (spruce-feathermoss, spruce-lichen forests), mixed (spruce/aspen, riparian), and deciduous forests in NF and LB including some from decaying mushrooms; also from red maple, pine/willow, lodgepole pine, balsam fir/white birch forests, elsewhere. In AK,

adults were found in gravel bars, mature forest, low bluff with sage; Alder, white spruce, *Betula* forest; *Salix, Alnus, Picea balsamea* forest; disturbed *Picea glauca* forest; forest, mostly spruce, some aspen; mixed *Betula, Picea, Alnus* forest; *Betula, Salix*, white spruce; *Populus tremuloides, P. balsamea,* and *Picea* forest; bottom and at the edge of closed tall shrub-alder, near eds-crowberry; low alpine dry slope; rocky alpine habitat; willow, dwarf birch, shrub; alpine tundra; shrub, alder, dwarf birch, grass, clubmoss; and decaying mushrooms in *Betula* forest. **Collecting period**: VI–X (most common in VIII and IX). **Collecting method**: sifting litter and mushrooms, carrion-baited and unbaited pitfall traps, pitfalls w/ rainroofs, yellow pitfalls w/o rainroofs, flight intercept traps, Lindgren funnel traps, Berlese funnels; sweeping vegetation.

We report this species from AB (**NPR**) for the first time here, based on **BOLD** records: Canada, Alberta, Banff National Park, Baker Creek picnic area, 1500m el., 51.35, −116.063, mature lodgepole pine forest, pitfall, 23–28.VII.2012, BIOBus 2012 team (CBG) 1 specimen; Banff National Park, Storm Mountain, 1462m el., 51.282, −115.944, low alpine forest, pitfall, 23-28.VII.2012, BIOBus 2012 team (CBG) 13 specimens; Jasper National Park, Miette Hotsprings, 1439m el., 53.124, −117.775, along forest creek on a valley bed, pitfall, 17-21. VII.2012, BIOBus 2012 team (CBG) 1 specimen; Waterton Lakes National Park, Highway 6 pulloff, 1562m el., 49.065, −113.778, douglas fir and lodgepole pine forest with aspen and birch understory, pitfall, 06-11.VIII.2012, BIOBus 2012 team (CBG) 5 specimens; Waterton Lakes National Park, Akamina Parkway, Rowe Tamarack Trail, 1632m el., 49.059, −114.013, subalpine lodgepole pine and fir forest with birch in understory, pitfall, 06-11.VIII.2012, BIOBus 2012 team (CBG) 10 specimens.

Additional records. See Annex 1.

(Subgenus *Dimetrota*) *Incertae sedis*

150. ***Atheta* (*Dimetrota*) *avalon* Klimaszewski and Langor** (Fig. 19.223a–e)

Atheta avalon Klimaszewski and Langor, in Klimaszewski et al. 2011. Klimaszewski et al. 2018a.

Diagnosis Body narrowly subparallel, moderately convex, length 2.2–2.4 mm; colour reddish brown, with head, antennae and middle parts of most tergites dark brown (Fig. 19.223a); integument moderately glossy; forebody with punctation fine and dense, with fine meshed microsculpture; pubescence on disc of pronotum directed laterad and obliquely posteriad from midline of disc, that on elytra directed obliquely posteriad; head distinctly narrower than pronotum; antennae moderately slender, antennomeres V–X transverse; pronotum distinctly transverse, broadest at middle, about 1/9 narrower than width of base of elytra; elytra transverse, at suture slightly longer than pronotum; abdomen subparallel. MALE. Apical margin of tergite VIII with broadly V-shaped medial emargination between two small rounded lateral teeth (Fig. 19.223d); sternite VIII truncate apically (Fig. 19.223e); tubus of median lobe of aedeagus straight subapically with rounded apex and sinuate basally in lateral view (Fig. 19.223b); in dorsal view tubus triangular, evenly coalescent with large oval bulbus (Fig. 19.223c). FEMALE. Undescribed.

DISTRIBUTION Origin: Nearctic. **CANADA**: LB, NF. **USA**: not recorded.

COLLECTION AND HABITAT DATA
Habitat: one specimen from carrion-baited trap in spruce-moss boreal forest; nothing else has been recorded for habitat association of this species. **Collecting period**: V–VI. **Collecting method**: carrion-baited pitfall trap.

COMMENTS *Atheta avalon*, as far as known, occurs in the softwood shield (5.1) of Labrador and Newfoundland and is not yet known from the arctic/subarctic.

151. ***Atheta* (*Dimetrota*) *gelida* J Sahlberg *** (Fig. 19.224a–h)

Atheta gelida J. Sahlberg, 1887. In *Boreophilia*: Lohse et al. 1990, Smetana 2004, Schülke and Smetana 2015, Klimaszewski et al. 2018a. In *Atheta (Dimetrota)*: Klimaszewski et al. 2019.

Diagnosis Body length 3.0–3.9 mm, head and basal part of abdomen dark brown to black, pronotum chestnut brown, elytra yellowish-brown, appendages paler, yellowish-brown (Fig. 19.224a); integument moderately glossy; head round, large, about as large as pronotum, slightly transverse; eyes moderately large, slightly shorter than genae in dorsal view; antennomeres IV–V subquadrate, VI–X slightly transverse; pronotum 1.3 times as broad as long, narrowest at base and widest at apical third, hind angles obtusely rounded; elytra about 11/3 times as broad as pronotum, at suture as long as pronotum along midline of disc, sides slightly divergent apicad; abdomen as broad as elytra at base, subparallel. MALE. Tergite VIII with apical margin broadly truncate medially, obliquely shallowly emarginate laterally (Fig. 19.224d); sternite VIII elongate, parabolically rounded from base to apex, basal margin broadly emarginate (Fig. 19.224e); median lobe of aedeagus with narrowly oval bulbus and moderately broad tubus in dorsal view (Fig. 19.224c); tubus in lateral view broadly curved ventrad basally, slightly sinuate before narrow, rounded apex (Fig. 19.224b); internal sac structures as illustrated (Fig. 19.224b,c). FEMALE. Tergite VIII transverse, apical margin broadly truncate medially, oblique laterally (Fig. 19.224f); sternite VIII rounded apically, antecostal suture sinuate (Fig. 19.224g); spermatheca club-shaped, stem sinuate and narrowly looped posteriorly (Fig. 19.224h).

DISTRIBUTION Origin: Holarctic, recorded from North America and Russia: Chukotka Peninsula. **CANADA**: MB, NT, NU, QC, YT. **USA**: AK.

COLLECTION AND HABITAT DATA
Habitat: open habitats; many specimens from gopher (*Thomomys* sp.) burrows in AK. **Collecting period**: VI-VIII. **Collecting method**: sifting litter in entrances of gopher burrows, sifting *Alnus* litter.

Additional records. See Annex 1.

(Subgenus *Lamiota* Casey, 1910)

(Fig. 19.225a–h) (Casey 1910, Seevers 1978, Gusarov 2003a, Klimaszewski and Winchester 2002, Klimaszewski et al. 2020)

Diagnosis Large species, length 3.0–4.0 mm; body subparallel, flattened, broad, broadest at elytra; pronotum moderately densely punctate, punctation slightly asperate, pubescence on midline of disc directed posteriad basally and anteriad in apical half, on sides pubescence directed laterad and obliquely posteriad, several macrosetae distributed on lateral margins; antennomeres IV–V elongate or slightly transverse, VI subquadrate, slightly elongate, or transverse, VII–X subquadrate to slightly transverse; abdomen broad, arcuate laterally; male tergite VIII with or without teeth; apical part of internal sac of median lobe of aedeagus with two narrow, diverging apically, lateral projections; subapical part of median lobe rod-shaped in lateral view; spermatheca with small spherical capsule and stem thin, very long, sinuate, and narrowly looped posteriorly.

152. *Atheta* (*Lamiota*) *keeni* Casey (Fig. 19.225a–h)

Atheta (*Lamiota*) *keeni* Casey, 1910. Gusarov 2003a [synonymy], Klimaszewski and Winchester 2002 [as *Atheta* (*Liogluta*) *vasta*], Klimaszewski et al. 2020.

Diagnosis Body moderately convex, broad, subparallel, length 3.7–4.0 mm, dark brown to nearly black, elytra, apical part of abdomen, and legs red-brown, antennae uniformly dark brown with antennomeres I–III paler (Fig. 19.225a); forebody moderately densely punctate and pubescent, pubescence long, punctation slightly asperate, microsculpture meshed and particularly well visible on pronotum, sculpticells slightly convex; head elongate, round posteriorly, distinctly narrower than pronotum, postocular region shorter than the maximum diameter of eye, tempora carinate, carinae complete, strong throughout; antennae slender, as long as pronotum and elytra combined, antennomere IV slightly elongate, VII–X subquadrate to slightly transverse, basal 4 antennomeres glossy and remaining matte; pronotum impressed mesobasally, transverse, approximately rectangular in shape, hypomeron visible for two thirds of pronotal length in lateral view, pubescence directed from midline of disc anteriad at apical third, on sides nearly horizontally laterad medially, and obliquely laterad posteriorly; elytra flattened, distinctly transverse, broader than pronotum, at suture as long as pronotum along midline, posterior margins feebly sinuate and with shallow, scarcely visible, lateral emarginations, pubescence directed straight or slightly obliquely posteriad; abdomen subparallel, strongly glossy, four first visible tergites strongly impressed at base; basal metatarsomere as long as the following one. MALE. Apical margin of tergite VIII slightly produced, crenulate between two distinct arcuate sublateral teeth (Fig. 19.225d); sternite VIII parabolic apically (Fig. 19.225e); bulbus of median lobe of aedeagus narrowly oval, tubus with two narrow lateral projections in dorsal view (Fig. 19.225c); tubus in lateral view with ventral margin about straight basally, curved ventrad at middle, straight and narrowly elongate to rounded apex (Fig. 19.225b). FEMALE. Tergite VIII broadly truncate and slightly emarginate medially (Fig. 19.225f): sternite VIII broadly rounded apically, with row of distinct setae on apical margin (Fig. 19.225g); spermatheca with capsule small, hemispherical apically and with elbowed neck posteriorly, stem thin, long, broadly U-shaped medially, and narrowly looped posteriorly with a single small coil at apex (Fig. 19.225h).

DISTRIBUTION **Origin**: Nearctic. **CANADA**: AB [**NPR**], BC. **USA**: AK, OR.

COLLECTION AND HABITAT DATA

Habitat: in AK, found in forest litter, *Alnus*, *Verathrum*, ferns. In BC, found in transition zone of Sitka spruce, in dung and carrion, in *Alnus* litter, in moss, fern, and *Amelanchier* litter at edge of upland pond, and in in moss and leaf litter near creek. Specimens from AB in Douglas fir and lodgepole pine stand with aspen and birch understory. **Collecting period**: VII. **Collecting method**: sifting *Alnus* litter, moss and leaf litter, moss, fern, and *Amelanchier* litter; dung and carrion baited pitfall traps.

We report this species from AB (**NPR**) for the first time here, based on the following: Canada, Alberta, Waterton Lakes National park, Highway 6 pullout, 1562m el., 49.065, –113.778, montane Douglas fir and lodgepole pine forest with aspen and birch understory, pitfall, 21-27.VI.2012, BIOBus 2012 team (CBG) 11 specimens.

(Subgenus *Microdota* Mulsant and Rey, 1873a)

(Figs. 19.226–19.234) (Klimaszewski et al. 2015c, 2018a, 2020, Lee and Ahn 2015)

Diagnosis Body subparallel, length 1.5–2.5 mm; antennomere I enlarged but not extremely swollen, longer than II, II longer than III, V–X slightly to distinctly transverse; median region of prementum very narrow, without pseudopores; labial palpi with three palpomeres; glossae split apically forming Y-shaped structure; maxillary palpi with three palpomeres, last one narrowly elongate; pronotum transverse, more than 1.2 times as wide as long, pubescence on disc directed anteriad along midline in most species, and laterad elsewhere; pronotal hypomera fully visible laterally; elytra in some species with pubescence forming wavy pattern in posterosutural area of disc; abdomen subparallel; male tergite VIII in most species simple, truncate apically, without teeth, sometimes with minute crenulation and pairs of small teeth; median lobe of aedeagus in dorsal view with large bulbus, tubus triangularly shaped apically, internal sac with well-developed complex structures; spermatheca of variable shape, L- or S-shaped, capsule usually narrowly sac-shaped, or club-shaped, stem often sinuate.

Some species of *Microdota*, due to small body size and some superficial external similarity, may be confused in collections with members of the subgenus *Datomicra* Mulsant and Rey, from which they differ in having the pronotal hypomera fully exposed in lateral view, while they are only partially visible in *Datomicra*. Many species of *Datomicra* also have the forebody more densely and coarsely punctate than that of *Microdota*.

Microdota may be distinguished from *Dimetrota* by the following combination of characters: body usually parallel-sided, on average 2 mm long (1.5–2.8 mm in *Microdota*; 1.8–3.8 mm in *Dimetrota*), glossae Y-shaped (deeply split in *Dimetrota*); pronotum with sparse to moderately dense and slightly asperate punctation (dense and coarsely asperate in *Dimetrota*); lateral margins of pronotum and elytra, and middle and hind tibiae with moderately coarse macrosetae (strong bristles in *Dimetrota*); hypomera fully visible in lateral view (partially to less often fully visible in *Dimetrota*); and male tergite VIII truncate apically, rarely crenulate, and usually without large lateral teeth (with two large lateral teeth and margin between them often distinctive or with pattern of smaller teeth in *Dimetrota*). Details on diagnostics of *Microdota* are provided by Brundin (1948) and Lee and Ahn (2015). Species of Nearctic *Microdota* need revision.

Key to Species of Subgenus *Microdota* of *Atheta*

1. Body contrasting bicoloured, head and abdomen dark brown, and pronotum and elytra, or pronotum only, yellowish-brown or orange brown (Fig. 19.227a); genitalia as illustrated (Fig. 19.227b–e, j,k)..
...***Atheta* (*Microdota*) *pennsylvanica* Bernhauer**

– Body approximately uniformly brown to black, elytra may be brown with reddish tinge, but not contrasting with rest of the body; genitalia shaped differently...**2**

2. Elytra at suture shorter than pronotum along midline of disc (Fig. 19.226a); median lobe of aedeagus and spermatheca distinctive in shape (Fig. 19.226b,c,h,i)..
***Atheta* (*Microdota*) *microelytrata* Klimaszewski and Godin**

– Elytra at suture as long as or longer than pronotum along midline of disc; median lobe of aedeagus and spermatheca not as above............**3**

3. Shoulders of elytra strongly pronounced, base of elytra distinctly wider than base of pronotum (about 1/4 times wider) (Figs. 19.228a,19.232a, 19.233a).......................**4**

– Shoulders of elytra moderately pronounced, base of elytra as wide as or slightly wider than base of pronotum (much less then 1/4 times wider) (Figs. 19.229a, 19.230a, 19.231a, 19.234a) **6**

4. Pronotal punctures fine and sparse (Fig. 19.233a); integument strongly glossy (Fig. 19.233a); antennomeres VII–X moderately transverse (Fig. 19.233a); median lobe of aedeagus and spermatheca as illustrated (Figs. 19.233b,c,h) ....
..................***Atheta* (*Microdota*) *subtilis* (Scriba)**

– Pronotal punctures course and dense (Figs. 19.228a, 19.232a), integument moderately glossy (Figs. 19.228a, 19.232a); antennomeres VII–X strongly transverse (Figs. 19.228a, 19.232a); median lobe of aedeagus and spermatheca not as above...**5**

5. Maximum width of elytra 1/6 times greater than maximum width of pronotum (Fig. 19.232a); eyes not protruding from outline of head in dorsal view (Fig. 19.232a); pubescence of sutural region of elytra directed obliquely posteriad (Fig. 19.232a); median lobe of aedeagus and spermatheca as illustrated (Fig. 19.232b,c,h) ....
...............***Atheta* (*Microdota*) *ripariides* Newton**

– Maximum width of elytra 1/4 times greater than maximum width of pronotum (Fig. 19.228a); eyes protruding from outline of head in dorsal view (Fig. 19.228a); pubescence of sutural region of elytra forming wavy pattern (Fig. 19.228a); median lobe of aedeagus and spermatheca differently shaped (Fig. 19.228b,c,h)..........................
.............***Atheta* (*Microdota*) *platonoffi* Brundin**

6. Antennomeres VII–X subquadrate or slightly transverse (Fig. 19.234a); elytra distinctly broader than base of pronotum (Fig. 19.234a); ventral margin of tubus of median lobe of aedeagus almost straight in lateral view (Fig. 19.234b)..
..................***Atheta* (*Microdota*) *surgens* (Casey)**

– Antennomeres VII–X distinctly transverse transverse; elytra about as wide as base of pronotum; ventral margin of tubus of median lobe of aedeagus differently shaped..................**7**

7. Pronotum and base of elytra of the same width (Fig. 19.231a); median lobe of aedeagus with apex strongly produced ventrad (Fig. 19.231b); spermatheca with broadly tubular capsule and club-shaped stem (Fig. 19.231h)......
.....................***Atheta* (*Microdota*) *pseudosubtilis* Klimaszewski and Langor**

– Pronotum narrower than width of base of elytra (Fig. 19.229a,b); median lobe of aedeagus and spermatheca differently shaped (Fig. 19.229c,d,i)..
.... ***Atheta* (*Microdota*) *pratensis* (Mäklin)**

153. *Atheta* (*Microdota*) *microelytrata* Klimaszewski and Godin (Fig. 19.226a–i)

Atheta (*Microdota*) *microelytrata* Klimaszewski and Godin, in Klimaszewski et al. 2012.

Diagnosis Body narrowly subparallel, length 1.9–2.0 mm; dark brown, with bases of antennae and legs rust-brown; strongly glossy, with fine and moderately dense punctation on forebody and strong, meshed microsculpture (Fig. 19.226a); head as wide as pronotum and elytra, with long postocular area, eyes moderately small and slightly protruding; antennae slender, slightly incrassate apicad, antennomeres IV–V subquadrate, VI–X slightly to strongly transverse; pronotum narrower at base and broadening apicad; elytra transverse, shorter than pronotum; abdomen widest subapically. MALE. Tergite VIII broadly truncate apically and with scarcely visible crenulation (Fig. 19.226d); sternite VIII widely rounded apically (Fig. 19.226e); tubus of median lobe of aedeagus in lateral view curved obtusely ventrad basally, then straight to slightly curved, narrowly triangular apex (Fig. 19.226b); in dorsal view bulbus broadly oval and tubus short, triangular in shape (Fig. 19.226c). FEMALE. Tergite VIII truncate apically (Fig. 19.226f); sternite VIII with apical margin broadly, shallowly emarginate medially, oblique laterally (Fig. 19.226g); spermatheca with pipe-shaped capsule and long stem, strongly sinuate, slightly swollen at middle and narrowly curled posteriorly (Fig. 19.226h,i).

DISTRIBUTION Origin: Nearctic. **CANADA**: NT [**NTR**], YT. **USA**: AK [**NSR**].

COLLECTION AND HABITAT DATA

Habitat: adults were found in aspen, alder and willow litter. In AK, found in forest edge *Populus tremuloides, Salix* spp.; *Populus, Betula, Alnus* leaf litter; mixed *Betula, Picea, Alnus* forest; *Salix, Alnus, P. balsam*; and spruce, aspen, leaf litter. **Collecting period**: III, V, X. **Collecting method**: sifting litter, pitfall traps.

We report this species from NT (**NTR**), and AK (**NSR**) for the first time here, based on the following: Canada, Northwest Territories, Norman Wells, 65.30494, –126.70873, 11-14. VI.2011, rep 3, mesic, pitfall trap, NBP field Party N3MP421 (LFC) 1 female. USA: Alaska, Chena Hot Springs Rd, 200m el., 64.89227, –147.50857, ±30m forest edge *Populus tremuloides, Salix* spp., pitfall, 17-24.VII.2007, B. Mortensen (UAM) 1 specimen; Delta, 336m el., 63.98769, –145.31691, ±100m pitfall, 11-18. VI.2010, B. Fleshman (UAM) 1 specimen; Fairbanks, 930 Fitz Ct, 280-293m el., 64.90127, –147.52797, ±30m *Populus, Betula, Alnus* leaf litter, Berlese, 10.X.2013, D.S. Sikes (UAM) 9 specimens; Fairbanks, 930 Fitz Ct, 280m el., 64.9013, –147.52847, ±25m spruce, aspen, leaf litter, 15-17.X.2013, M. C. McHugh (UAM) 2 specimens; Fairbanks, Murphy Dome Rd., 635m el., 64.9428, –148.29016, ±5m unglaciated; mixed *Betula, Picea, Alnus* forest, Berlese, 12.IX.2009, G. Henry (UAM) 1 specimen; Fairbanks, Murphy Dome Rd., 638m el., 64.94277, –148.29033, ±5m unglaciated; *Betula,*

Picea, *Alnus* forest, Berlese, 12.IX.2009, G. Henry (UAM) 10 specimens; mile 217.2 Richardson Highway, Castner Glacier, 781m el., 63.40134, −145.71895, ±5m low *Salix, Alnus, P. balsam*, Berlese, 7.IX.2009, G. Henry (UAM) 1 specimen.

Additional records. See Annex 1.

154. ***Atheta* (*Microdota*) *pennsylvanica* Bernhauer** (Fig. 19.227a–k)

Atheta (Microdota) pennsylvanica Bernhauer, 1907. Blatchley 1910, Casey 1910 [in *Atheta* (*Hilara*)], in *Atheta* (*Microdota*): Gusarov 2003a, Klimaszewski et al. 2005, 2007b, Dollin et al. 2008, Bishop et al. 2009, Majka and Klimaszewski 2008b, Klimaszewski et al. 2011, 2015c, 2018a, b.

Diagnosis Body narrow, subparallel, length 1.5–1.8 mm; head, elytra and posterior part of abdomen dark brown to black, elytra sometimes dark reddish brown, basal 3 antennomeres orange, base and apex of abdomen and legs yellowish-brown or reddish (Fig. 19.227a); integument highly glossy, forebody with fine, sparse punctation and meshed microsculpture; head elongate, head, pronotum, elytra and abdomen subparallel; antennomeres VI–X strongly transverse. MALE. Tergite VIII with apical margin obtuse laterally, broadly truncate mediad (Fig. 19.227f); sternite VIII with apical margin arcuately rounded from near base to apex (Fig. 19.227g); median lobe of aedeagus with tubus short, almost straight ventrally and narrowly triangular apically in lateral view (Fig. 19.227b,c); in dorsal/ventral view bulbus oval, tubus short, and triangular (Figs. 19.227d, e); structures of internal sac complex as illustrated (Fig. 19.227b–e). FEMALE. Tergite VIII arcuate apically (Fig. 19.227h); sternite VIII broadly rounded apically (Fig. 19.227i); spermatheca S-shaped, acutely bent at base of capsule, capsule club-shaped with small apical invagination, stem with single loop posteriorly (Fig. 19.227j,k).

DISTRIBUTION **Origin**: Nearctic. **CANADA**: LB, MB, NB, NF, NS, ON, QC. **USA**: IN, MA, MN, NY, PA, RI, VA, VT.

COLLECTION AND HABITAT DATA
Habitat: this species is found in various conifer, mixed and hardwood/deciduous forests such as red spruce dominated forests, black spruce forests, balsam fir-spruce forests, jack pine forests, red pine forests, old-growth hemlock forests, riparian forests, red oak and maple forests, red oak forests, various mixed forests, northern hardwood forests, rich Appalachian hardwood forest, *Populus* stands, and yellow birch forests. Adults are often abundant in various kinds of decaying and fresh mushrooms (gilled, polypore; *Polyporus varius* on side of log, stalked polypores) on forest floor and also occur in forest litter, especially in the spring; often captured in abundance in pitfall traps. **Collecting period**: IV–X. **Collecting method**: sifting forest litter, pitfall traps, yellow pan.

Additional records. See Annex 1.

155. *Atheta* (*Microdota*) *platonoffi* Brundin * (Fig. 19.228a–h)

Atheta (Microdota) platonoffi Brundin, 1948. Smetana 2004, Klimaszewski et al. 2005, Majka and Klimaszewski 2008b, Klimaszewski et al. 2012, 2015a, 2015c, 2018a, 2020.

Diagnosis Body small and narrow, subparallel, length 1.5–1.9 mm; colour dark brown to almost black, with legs and elytra except for scutellar region, often paler and yellowish- or reddish-brown (Fig. 19.228a); integument moderately glossy, forebody with fine, dense, slightly asperate punctation and meshed microsculpture; head almost as wide as pronotum, pronotum slightly narrower than elytra; eyes large, much longer than genae in dorsal view; antennae slender, slightly incrassate apically, antennomere IV subquadrate, V–X slightly to distinctly transverse; elytra slightly broader than pronotum, at suture longer than pronotum; abdomen subparallel, highly glossy. MALE. Tergite VIII with apical margin broadly truncate, rounded laterally (Fig. 19.228d); sternite VIII with apical margin obliquely produced, arcuately rounded apically (Fig. 19.228e); median lobe of aedeagus in lateral view with tubus acutely bent ventrad basally, then sinuate to narrowly elongate apical part which is slightly arched dorsad (Fig. 19.228b); in dorsal view bulbus large, obtusely angulate basolaterally, tubus short, parallel-sided to roughly triangular apex (Fig. 19.228c), internal sac split apically (Fig. 19.228b,c). FEMALE. Tergite VIII with apical margin broadly truncate mediad, slightly arcuate laterally (Fig. 19.228f); sternite VIII evenly rounded apically (Fig. 19.228g); spermatheca highly sinuate, acutely bent at base of capsule, capsule tubular with long and narrow apical invagination, stem narrowly looped posteriorly and slightly twisted at base (Fig. 19.228h).

DISTRIBUTION **Origin**: Holarctic.

CANADA: AB, BC, LB, MB [**NPR**], NB, NF, NS, ON, SK, YT. **USA**: AK.

COLLECTION AND HABITAT DATA

Habitat: this species occurs in decaying organic material (compost, rotting mushrooms, dung) and litter in forests and open habitats; in NF, specimens from pitfall carrion-baited and flight intercept traps in various mixed and coniferous forests; in NB, from unbaited pitfall traps in a red spruce forest, well rotted mushrooms in jack pine forest, in decaying corncobs (moldy) and cornhusks near a mixed forest (common), in grass debris on river margin, under dung on lake margin; a few flying adults collected in late afternoon in mixed forest opening. In BC, specimens were found in moss along seepage, in gravel along edge of stream, in moss ex under deciduous shrubs around base of cedar, in moss under deciduous shrubs around base of cedar, and in mushrooms, moss, and compost. In AK, found in forest edge, *Populus tremuloides* and grass; *Betula* and *P. tremuloides*; mesic *Betula*; and 61m summit with *Populus tremuloides*. **Collecting period**: V–IX. **Collecting method**: sifting forest litter, sweeping, compost, rotting mushrooms, and dung, unbaited and carrion-baited pitfall traps, flight intercept traps, sweep netting.

We report this species from MB (**NPR**) for the first time here, based on the following: Canada, Manitoba, Riding Mountain National Park, Brule Trail, 660m el. 50.873, –100.05, coniferous forest and adjacent meadow, sweep net, 14.VIII.2008, BIOBus 2008 team (CBG) 1 specimen.

Additional records. See Annex 1.

156. *Atheta* (*Microdota*) *pratensis* (Mäklin) (Fig. 19.229a–i)

Homalota pratensis Mäklin, 1852. In *Atheta* (*Microdota*): Bernhauer 1907, Lohse and Smetana, 1985, Klimaszewski et al. 2011 [in part], 2012, 2015c.

Diagnosis Body small, length 2.0–2.2 mm; colour dark brown, with elytra and appendages slightly paler (Fig. 19.229a,b); integument highly glossy, forebody with fine, sparse punctation and weak meshed microsculpture; head round and as wide as pronotum, depressed medially; eyes as long as postocular area in dorsal view; pronotum small, round, equal in size to head, and as wide as two thirds of maximum width of elytra; elytra broad, at suture distinctly longer than pronotum along median line of disc, posterior margin truncate; abdomen distinctly narrower than elytra, subparallel; antennae slender, slightly incrassate apically, scape swollen, antennomeres IV subquadrate and VII–X slightly transverse. MALE. Tergite VIII with apical margin rounded apically except slightly emarginate medially, lacking teeth (Fig. 19.229e); sternite VIII slightly elongate, with apical margin broadly arcuate medially, rounded laterally (Fig. 19.229f); median lobe of aedeagus in lateral view with large oval bulbus and tubus narrow and curved obliquely ventrad, apex narrowly rounded (Fig. 19.229c), in dorsal view median lobe with large ovoid bulbus, apical part of tubus short, roughly triangular (Fig. 19.229d). FEMALE. Tergite VIII with apical margin truncate (Fig. 19.229g); sternite VIII evenly rounded apically, antecostal suture strongly sinuate medially (Fig. 19.229h); spermatheca S-shaped, capsule club-shaped with neck acutely bent, stem strongly sinuate and looped posteriad (Fig. 19.229i).

DISTRIBUTION **Origin**: Nearctic. **CANADA**: YT. **USA**: AK, WA.

COLLECTION AND HABITAT DATA
Habitat: in NF, specimens were captured from spruce-birch forest and mossy spruce forest. Some other specimens were found in mushrooms. **Collecting period**: VII–X. **Collecting method**: sifting litter, pitfall traps.

COMMENTS We have studied the type series of this species but the median lobe of aedeagus was dried and slightly deformed (Fig. 19.230c,d). The record of *A. pratensis* from NF is based on a male specimen that should be regarded as a tentative identification. It probably represents misidentification for *A. subtilis* (Fig. 19.230a–c). We removed NF from distribution of *A. pratensis* until more specimens, including females, will become available for study.

157. *Atheta* (*Microdota*) *pseudosubtilis* Klimaszewski and Langor (Fig. 19.231a–h)

Atheta (*Microdota*) *pseudosubtilis* Klimaszewski and Langor, in Klimaszewski et al. 2011. Webster et al. 2012, Klimaszewski et al. 2015a,c, 2018a.

Diagnosis Body small and narrow, subparallel, length 1.8–2.0 mm; colour dark brown, with bases of antennae, legs and elytra yellowish-brown (Fig. 19.231a); integument highly glossy, forebody with fine, dense punctation and meshed microsculpture; head almost as wide as pronotum, depressed medially; pronotum arcuate laterally; pronotum, elytra and base of abdomen of about same width; abdomen distinctly broadening to segment VI; antennae slender, slightly incrassate apically, scape swollen, antennomeres IV–V subquadrate and VI–X slightly to distinctly transverse. MALE: tergite VIII arcuately truncate apically and lacking teeth (Fig. 19.231d); sternite VIII broadly parabolically rounded from base to apex (Fig. 19.231e); median lobe of aedeagus in lateral view with ventral margin of tubus sinuate, apical part curved obliquely ventrad, apex narrowly rounded (Fig. 19.231b), in dorsal view median lobe with bulbus ovoid, gradually narrowing to short, triangular tubus (Fig. 19.231c). FEMALE. Tergite VIII with apical margin broadly rounded except shallowly emarginate medially (Fig. 19.231f); sternite VIII similar to tergite VIII but moderately deeply emarginate medially (Fig. 19.231g); capsule of spermatheca broadly tubular, L-shaped, neck acutely bent at base of capsule, stem short, club-shaped, distinctly swollen toward base (Fig. 19.231h).

DISTRIBUTION Origin: Nearctic. **CANADA**: AB, LB, NB, NF. **USA**: not recorded.

COLLECTION AND HABITAT DATA
Habitat: in NF, captured in flight intercept traps, unbaited- and carrion-baited pitfall traps in mixed and coniferous forests; one specimen was taken from *Larix laricina* recently killed by the eastern larch beetle. NB specimens sifted from moldy conifer duff at the bases of white pine in a balsam fir stand (common) and from leaf litter and moss near a shaded brook in a mixed forest. **Collecting period**: IV–VIII. **Collecting method**: sifting forest litter, flight intercept traps, unbaited and carrion-baited pitfall traps.

COMMENTS *Atheta pseudosubtilis*, as far as known, reaches its northern limit in the softwood shield (5.1) of Labrador and is not yet known from the arctic/subarctic. This species was listed in error from QC and YT by Klimaszewski et al. (2015a, c) as there were no voucher specimens.

158. *Atheta* (*Microdota*) *ripariides* Newton (Fig. 19.232a–h)

Atheta (*Microdota*) *riparia* Klimaszewski and Godin, in Klimaszewski et al. 2012. Klimaszewski et al. 2016c, 2018b.

Atheta (*Microdota*) *ripariides* Newton, 2017, replacement name for *Atheta riparia* Klimaszewski and Godin 2012, a junior primary homonym of *Atheta riparia* Cameron 1939. Klimaszewski et al. 2020.

Diagnosis Body length 1.9–2.0 mm, narrow, subparallel; black, with tarsi reddish-brown; moderately glossy, with fine, dense punctation and meshed microsculpture on forebody (Fig. 19.232a); head approximately as wide as pronotum, depressed medially, eyes slightly protruding; antennae slender, slightly incrassate apicad, antennomeres IV–X slightly to strongly transverse; pronotum emarginate laterally; elytra broader and longer at suture than pronotum; head, pronotum and base of abdomen of the same width; sides of abdomen subparallel. MALE. Tergite VIII with apical margin broadly truncate, rounded laterally, without teeth (Fig. 19.232d); sternite VIII broadly parabolic apically (Fig. 19.232e); median lobe of aedeagus in lateral view with small bulbus, ventral margin of tubus curved obtusely ventrad basally, then long, sinuate to narrow subapical part, apex slightly curved ventrad (Fig. 19.232b), in dorsal view bulbus oval, narrowed apically, and tubus narrowly triangular (Fig. 19.232c). FEMALE. Tergite VIII very broadly arcuate apically (Fig. 19.232f); sternite VIII broadly rounded apically (Fig. 19.232g); spermatheca with club-shaped capsule and posteriorly-twisted stem (Fig. 19.232h).

This species differs from other Nearctic *Microdota* by the combination of body shape, strongly punctate surface and the shape of median lobe of aedeagus and spermatheca.

DISTRIBUTION Origin: Nearctic. **CANADA**: BC, MB, SK, YT. **USA**: not recorded.

COLLECTION AND HABITAT DATA

Habitat: litter in mixed aspen and white spruce forest, wet organic forest litter and mushrooms. In BC, one specimen was found in shoreline habitat and one inland. **Collecting period**: IX. **Collecting method**: sifting forest litter and mushrooms, pitfall traps.

Additional records. See Annex 1.

159. ***Atheta* (*Microdota*) *subtilis* (Scriba)** † (Fig. 19.233a–h)

Homalota subtilis Scriba, 1866. Palm 1970, Benick and Lohse 1974, Smetana 2004, Klimaszewski et al. 2011 [misidentified as *Atheta pratensis*, in part], 2015c, 2018a.

Diagnosis Body subparallel, flattened, length 1.5–2.0 mm; colour dark brown, head and abdomen darker than pronotum and elytra in some specimens, legs yellowish brown (Fig. 19.233a); integument moderately glossy, densely punctate and densely pubescent on forebody, less so on abdomen, microsculpture fine; head slightly narrower than pronotum, strongly narrowed posteriad, eyes large, slightly longer than genae in dorsal view; pronotum transverse, narrower than elytra; elytra wider and at suture longer than pronotum; abdomen subparallel, gradually widening to segment VIII. MALE. Tergite VIII with apical margin arcuate laterally, shallowly emarginate medially (Fig. 19.233d); sternite VIII broadly rounded apically (Fig. 19.233e); tubus of median lobe of aedeagus in lateral view with ventral margin arcuately curved ventrad, apical half narrow to narrowly rounded apex (Fig. 19.233b), in dorsal view bulbus large, oval, and tubus short and triangular (Fig. 19.233c); internal sac structures complex (Fig. 19.233b,c). FEMALE. Tergite VIII with apical margin broadly, shallowly emarginate medially, obtuse laterally (Fig. 19.233f); sternite VIII rounded and slightly emarginate apically (Fig. 19.233g); spermatheca L-shaped, acutely bent at base of capsule, capsule long, broad, tubular with wide, shallow apical invagination, stem short, club-shaped, distinctly swollen toward base (Fig. 19.233h).

DISTRIBUTION Origin: Palaearctic, adventive in Canada. **CANADA**: LB, NB, NF, QC. **USA**: not recorded.

COLLECTION AND HABITAT DATA

Habitat: in LB, captured in spruce-poplar forest (flight intercept trap) and a spruce-birch forest; in NB, adults in decaying gilled mushrooms in a red spruce and yellow birch forest, in *Pleurotus* sp. on dead standing *Populus tremuloides*, an old spruce and fir forest in gilled mushrooms in various stages of decay, and on flowers of *S. alba* in a mixed forest. **Collecting period**: VII–IX. **Collecting method**: sifting mushrooms and litter, flight intercept trap, pitfall trap.

COMMENTS *Atheta subtilis*, as far as known, reaches its northern limit in the softwood shield

(5.1) of Labrador and is not yet recorded from the arctic/subarctic. Adults of *A. subtilis* from LB were captured in association with *A. pseudosubtilis* Klimaszewski and Langor. Some poorly preserved females of the former species, because of their similarly-shaped spermatheca, were misidentified as the latter species (Klimaszewski et al. 2015c, 2018a). Records of *A. pratensis* from NF and LB (Klimaszewski et al. 2011) were later discovered to represent *A. subtilis*.

160. *Atheta* (*Microdota*) *surgens* (Casey) (Fig. 19.234a–e)

Datomicra surgens Casey 1910. Klimaszewski et al. 2020 (new subgeneric placement).

Diagnosis Body length 1.7–2.0 mm, stout, small and narrow, subparallel, broadest at elytra; colour uniformly dark brown with appendages paler or head and pronotum slightly darker, integument strongly glossy, forebody with coarse and dense asperate punctation and meshed microsculpture (Fig. 19.234a); head slightly narrower than pronotum, slightly transverse; antennae slender, antennomere IV subquadrate, and V–X subquadrate to slightly transverse; pronotum transverse, distinctly narrower than elytra at base; elytra distinctly broader than pronotum, at suture slightly longer than pronotum, shoulders angular; abdomen distinctly narrower than elytra, subparallel. MALE. Tergite VIII sinuately truncate apically, without teeth (Fig. 19.234d); sternite VIII semicircularly rounded from base to apex (Fig. 19.234e); tubus of median lobe of aedeagus in lateral view obliquely bent ventrad at base, almost straight ventrally to short, acutely triangular apical part (Fig. 19.234b); in dorsal view tubus as illustrated (Fig. 19.234c). FEMALE. Unknown.

DISTRIBUTION Origin: Nearctic. **CANADA**: BC. **USA**: AK.

COLLECTION AND HABITAT DATA
Habitat: not recorded. **Collecting period**: not recorded. **Collecting method**: not recorded.

COMMENTS This species is distinguishable from *A. platonoffi* by broader elytra and antennomeres VII–X less transverse (subquadrate to slightly transverse). The two species differ in the shape of the tubus of the median lobe of the aedeagus in lateral view. In *A. platanoffi*, the ventral margin of the tubus in lateral view is consistently strongly sinuate and strongly arcuate medially (slightly so, almost straight in *A. surgens*), and the apical triangular part is longer and more projecting upward in *A. platonoffi* (Fig. 19.228b), while it is almost straight and less projecting upward in *A. surgens* (Fig. 19.234b). These are clearly different species.

(Subgenus *Pseudota* Casey, 1910)

(Figs. 19.235–19.236) (Seevers 1978, Gusarov 2003a, Klimaszewski et al. 2018a, 2020)

Diagnosis Pronotum and elytra asperately punctate; pronotum 1.3–1.4 times as wide as long; pronotum with pubescence directed straight posteriad or obliquely posteriad along midline and obliquely posteriad elsewhere on disc; elytra 1.25–1.40 times as long as pronotum; eyes large; infraorbital carinae complete but feeble; antennomere V subquadrate to slightly transverse, VI–X transverse, and moderately to highly incrassate; isthmus between intercoxal processes long; apical margin of male tergite VIII with four or more teeth.

Key to Species of Subgenus *Pseudota* of *Atheta*

1. Antennomeres VII–X strongly transverse (Fig. 19.235a); apical margin of male tergite VIII with two lateral large teeth and two lobe-shaped projections inward from lateral teeth, median margin sinuate (Fig. 19.235c); median lobe of aedeagus as illustrated (Fig. 19.235b); stem of spermatheca broadly looped posteriorly (Fig. 19.235g).. .... ***Atheta (Pseudota) longiclava*** **(Casey)**

– Antennomeres VII–X moderately strongly transverse (Fig. 19.236a); apical margin of male tergite VIII with two large lateral teeth and multiple smaller teeth, median margin truncate

(Fig. 19.236d); median lobe of aedeagus not as above (Fig. 19.236b,c); stem of spermatheca tightly coiled posteriorly (Fig. 19.236h) .... ***Atheta (Pseudota) pseudoklagesi* Klimaszewski and Webster**

161. *Atheta (Pseudota) longiclava* (Casey) (Fig. 19.235a–g)

Sableta (Canastota) longiclava Casey, 1910. Transferred to *Atheta*: Ashe and Gusarov 2004. In *Atheta* (*Pseudota*): Brunke et al. 2012 [misidentified as *Atheta nescia* (Casey)], Klimaszewski et al. 2018a.

Diagnosis Body subparallel, length 2.4 mm; head and abdomen dark brown and pronotum, elytra and appendages light yellow-brown (Fig. 19.235a); punctation moderately dense, coarser on elytra; microsulpture of forebody moderately strong; head round, distinctly narrower than pronotum, postocular area shorter than eyes in dorsal view; pronotum transverse, arcuate laterally, broadest in apical half, pubescence directed laterad from midline of disc; elytra distinctly broader than maximum width of pronotum, as long as pronotum along midline, pubescence directed posteriad; abdomen subparallel. MALE. Tergite VIII with apical margin oblique, sinuate laterally, broadly truncate medially, with a small, obtusely rounded tooth adjacent to each moderate, triangular lateral tooth (Fig. 19.235c); sternite VIII broadly parabolically rounded from base to apex (Fig. 19.235d); median lobe of aedeagus in lateral view with large ovoid bulbus, tubus slightly curved ventrad before middle of ventral margin, then almost straight to acutely triangular apex, dorsal margin obliquely truncate (Fig. 19.235b). FEMALE. Tergite VIII transverse, arcuate apically and truncate medially (Fig. 19.235e); sternite VIII transverse, broadly rounded apically (Fig. 19.235f); spermatheca S-shaped, capsule club-shaped and acutely bent at base, invagination deep and narrow; stem sinuate, broadly looped posteriorly and swollen (Fig. 19.235g).

DISTRIBUTION Origin: Nearctic. **CANADA**: LB [**NPR**], MB [**NPR**], ON, QC[**NPR**]. **USA**: RI.

COLLECTION AND HABITAT DATA
Habitat: sparsly treed open habitats including oak savanah/alvar, a savanah, and an agricultural headgrow. **Collecting period**: V–VI, VIII. **Collecting method**: pitfall traps, yellow pan traps, flight intercept traps.

We report this species from MB (**NPR**), LB (**NPR**), and QC (**NPR**) for the first time here, based on the following: Canada, Manitoba, Churchill, 58.73235, –93.79028, rep. 2 mesic, pitfall trap, 1-4.VII.2010, NBP field party H2MP411 (LFC) 1 female. Newfoundland, Labrador, Goose Bay, 53.32035, –60.29268, rep. 3, mesic, yellow pan, 21-24.VI.2010, NBP field party G3MY831 (LFC) 1 female. Quebec, Schefferville, 54.75970, –66.71120, Rep 1, wet, yellow pan, 6-9.VII.2010, NBP Field Party S1WY231 (LFC) 1 female; Schefferville, 54.90310, –67.14765, Rep 3, mesic, pitfall pan, 6-9.VII.2010, NBP Field Party S3MP231 (LFC) 1 female.

COMMENTS *Atheta longiclava* is externally and genitally similar to *A. nescia* Casey, which does not occur west of Rocky Mts. (Klimaszewski et al. 2020).

162. *Atheta* (*Pseudota*) *pseudoklagesi* Klimaszewski and Webster (Fig. 19.236a–h)

Atheta (Pseudota) pseudoklagesi Klimaszewski and Webster, in Webster et al. 2016. Klimaszewski et al. 2016b, c, 2018a, b, 2020.

Diagnosis Body narrowly oval, length 2.9–3.5 mm; head, pronotum, and posterior part of abdomen dark brown to nearly black, elytra dark brown with oblique yellowish-brown band extending from shoulder to posterior part of suture, legs, bases of antennae and maxillary palpi yellow, basal part of abdomen dark brown to nearly black, light brown in some individuals (Fig. 19.236a); integument highly glossy, with meshed microsculpture; forebody with punctation and pubescence minute and dense, less so on head; head rounded posterolaterally, eyes moderately large, about as long as genae in dorsal view; antennomere IV subquadrate, V–X moderately to distinctly transverse; pronotum broadly rounded laterally, broadest just anterior of middle of its length, slightly transverse, distinctly wider than head, distinctly narrower than elytra, pubescence on disc directed laterad from midline; elytra distinctly transverse, with pubescence on disc directed posterolaterad; abdomen subparallel, narrower than elytra. MALE. Tergite VIII truncate apically, with two larger sublateral rounded teeth and multiple smaller narrowly rounded ones between (Fig. 19.236d); sternite VIII rounded apically (Fig. 19.236e); median lobe of aedeagus in lateral view with tubus long, arcuately curved ventrad, with apical part elongate, dorsal margin sinuate to rounded apex (Fig. 19.236b); in dorsal view bulbus moderately broad, ovoid, tubus elongate, tapering to rounded triangular apex (Fig. 19.236c), internal sac with weak structures (Fig. 19.236b,c). FEMALE. Tergite VIII with apical margin arcuate laterally, truncate medially (Fig. 19.236f); sternite VIII broadly rounded apically (Fig. 19.236g); spermatheca with capsule large, spherical with deep apical invagination, stem long, narrow, tightly coiled posteriorly (Fig. 19.236h).

This is the sibling species of *A. klagesi* and was confused with the latter in collections. It may be distinguished from *A. klagesi* by the following combination of characters: size slightly larger, body less glossy, legs, bases of antennae, maxillary palps and oblique bands on elytra less intensely yellowish in coloration, body colour less contrasting overall; median lobe of aedeagus with tubus longer, with apex shaped slightly differently in lateral view; the shape of the spermatheca is very similar in the two species, and females may be difficult to identify without accompanying males.

DISTRIBUTION Origin: Nearctic. **CANADA**: AB, BC, MB, NB, NF, QC, SK, YT. **USA**: AK [**NSR**].

COLLECTION AND HABITAT DATA

Habitat: in NB, in mature mixed forest, old-growth and old white spruce and balsam fir forests, a mature red spruce forest, and in a wet alder swamp; adults in coral fungi on a *Populus* log, fleshy polypore fungi at base of a dead standing *Populus*, in decaying gilled mushrooms, in gilled mushrooms, and under bark of red spruce and dead aspen. In NF, from pitfall traps in boreal forests. In QC, in yellow birch/balsam fir forest, using pitfall traps, Luminoc pit/light traps, and intercept traps. In SK, in aspen woodland, in bracket/gilled fungi, under bark of dead aspen, and on fungus on dead lodgepole pine stump. In YT, in mushrooms. In AK, found in *Betula, Salix*, white spruce forest; *Betula* and *Populus tremuloides*; and in *Betula* forest, decaying mushrooms.

Collecting period: V–X. **Collecting method**: sifting mushrooms.

We report this species from AK for the first time here based on UAM specimens: USA, Alaska, 35.5 Mi Dalton Hwy, 271m el., 65.74425, −149.34564, ±4m *Betula, Salix*, white spruce, flight intercept trap, 31.V-5.VI.2008, D.S. Sikes (UAM) 2 males; Fairbanks, 930 Fitz Ct, 280m el., 64.9013, −147.52861, ±30m *Betula & P. tremuloides*, flight intercept trap, 20.V.–3.VI.2007, D.S. Sikes (UAM) 5 males; 29.V-4.VII.2008 (UAM) 2 males; 29.IV-29.V.2008 (UAM) 6 males; Fairbanks: Creamer's Field, 140m el., 64.86883, −147.73065, ±104m *Betula* forest, decaying mushrooms, 31.VII.2010, D.S. Sikes (UAM) 1 male.

Additional records. See Annex 1.

(Subgenus *Rhagocneme* Munster, 1922)

(Figs. 19.237a–h) (Munster 1922)

Diagnosis Body narrowly subparallel; forebody asperately punctate; head rounded basally, eyes large, longer than genae in dorsal view; antennomere V subquadrate, VIII–X distinctly transverse; pronotum at base as wide as elytra, pubescence on disc directed laterad and obliquely posteriad from midline; elytra at suture slightly shorter than pronotum along midline, posterior margin arcuate, with lateral emargination; apical margin of male tergite VIII truncate and without teeth; tubus broad in dorsal view and broadly rounded apically; spermatheca with capsule pipe-shaped, sinuate, with small apical invagination, stem strongly sinuate, looped, twisted and slightly swollen posteriorly.

163. *Atheta* (*Rhagocneme*) *subsinuata* (Erichson, 1839) † (Fig. 19.237a–h)

Homalota subsinuata Erichson, 1839. As *Atheta* (*Rhagocneme*): Smetana 2004, Klimaszewski et al. 2008a, 2013a [list of synonyms], 2016d.

Diagnosis Body length 2.3–2.6 mm; body uniformly black, integument with punctation dense, microsculpture of forebody distinct and consisting of isodiametric sculpticells; head narrower than pronotum; pronotum narrower than elytra, abdomen broadly arcuate laterally and glossier than remainder of the body (Fig. 19.237a). MALE. Tergite VIII truncate apically (Fig. 19.237d); sternite VIII parabolic apically (Fig. 19.237e); median lobe of aedeagus with narrow bulbus, nearly streamlined with base of tubus, tubus broad and rounded apically in dorsal view (Fig. 19.237c), tubus straight ventrally with broad and rounded apex in lateral view (Fig. 19.237b), internal sac structures simple and inconspicuous (Fig. 19.237b,c). FEMALE. Tergite VIII with apical margin broadly, shallowly emarginate (Fig. 19.237f); sternite VIII rounded apically except slightly emarginate medially (Fig. 19.237g); spermatheca with capsule pipe-shaped, sinuate, with small apical invagination, stem strongly sinuate, looped, twisted and slightly swollen posteriorly (Fig. 19.237h).

DISTRIBUTION Origin: Palaearctic, adventive in Canada. **CANADA**: SK, YT. **USA**: not recorded.

COLLECTION AND HABITAT DATA
Habitat: compost pile. **Collecting period**: IX. **Collecting method**: pitfall traps.

COMMENTS We consider this species adventive in Canada, but additional distribution records are needed from a broader range to confirm its status. Ádám questioned the synonymy of *Neorhagocneme* under *Rhagocneme* and implied the removal of *A. subsinuata* to the genus *Neorhagocneme*, but he did not make any formal changes.

(Subgenus *Tetropla* Mulsant and Rey, 1873b)

(Figs. 19.238a–j) (Gusarov 2003a, Klimaszewski et al. 2018a, b)

Diagnosis Body narrowly subparallel; pronotum and elytra asperately punctate; head rounded basally, eyes large, longer than genae in dorsal view; antennomere V subquadrate to slightly transverse, VIII–X 1.3 times as wide as long; pronotum about four-fifths as wide as elytra, pubescence on disc directed obliquely posteriad from midline; elytra at suture slightly longer than pronotum along midline, posterior margin truncate and straight, without lateral emargination; apical margin of male tergite VIII angulately produced in middle 2/3, sinuate medially; median lobe of aedeagus with two elongate structures in bulbus; spermatheca with capsule club-shaped, straight, with narrow and deep apical invagination, stem angularly looped, twisted and slightly swollen posteriorly.

164. *Atheta* (*Tetropla*) *frosti* Bernhauer (Fig. 19.238a–j)

Atheta (Atheta) frosti Bernhauer, 1909. Casey 1910, 1911, in *Atheta* (*Tetropla*): Gusarov 2003a, Klimaszewski et al. 2005, Majka and Klimaszewski 2008b, Klimaszewski et al. 2011, 2016d, 2018a, b, 2020.

Diagnosis Body narrowly subparallel, moderately convex, length 2.0–2.8 mm; colour brown to dark brown, with antennomeres I–III, palps, legs, elytra and apical part of abdomen light brown to yellowish-brown, integument moderately glossy (Fig. 19.238a); forebody with pubescence fine and sparse, denser on elytra; head narrower than pronotum, genae shorter than eyes in dorsal view; antennae moderately thick, antennomere IV subquadrate to slightly transverse, V–X moderately transverse, each about 1.5 times as wide as long; pronotum distinctly transverse, widest at middle, slightly narrower than elytra; elytra at suture slightly longer than pronotum, posterior margin straight, without lateral emarginations; abdomen mostly subparallel. MALE. Apical margin of tergite VIII angulately produced in middle 2/3, sinuate medially (Fig. 19.238e); sternite VIII with sides almost evenly arcuate to slightly more rounded apex (Fig. 19.238f); median lobe of aedeagus in lateral view with tubus moderately long, arcuate ventrally, with apex narrow and rounded (Fig. 19.238b,c); in dorsal view bulbus oval, tubus broad and triangular apically (Fig. 19.238d); internal structures elongate (Fig. 19.238b–d). FEMALE. Tergite VIII with apical margin obtuse laterally, broadly truncate medially (Fig. 19.238g); sternite VIII semicircularly rounded from antecostal suture to apex (Fig. 19.238h); spermatheca with capsule club-shaped, straight, with narrow and deep apical invagination, stem angularly looped, twisted and slightly swollen basally (Fig. 19.238i,j).

DISTRIBUTION **Origin**: Nearctic. **CANADA**: BC, LB, MB, NB, NS, ON, QC, SK. **USA**: AK [**NSR**], MA, NC, NH, NY, PA, RI, VT, WV.

COLLECTION AND HABITAT DATA
Habitat: in NB, in various coniferous, mixed, and hardwood forests in decaying fungi where it is often common; white pine and spruce forest in decaying mushrooms, old-growth eastern hemlock forest in decaying mushrooms, red spruce and eastern white cedar forest in decaying mushrooms, red pine forest in decaying mushrooms, Jack pine forest in decaying mushrooms, mixed forests in decaying mushrooms, red oak forest in gilled mushrooms, rich Appalachian hardwood forest in decaying/rotten fleshy fungi; one from pitfall trap in red spruce forest and some in decaying (moldy) corncobs and cornhusks near a mixed forest. Elsewhere, from pitfall traps in open spruce forest with sandy soil and *Cladina* lichen cover (LB), birch-dominated forests (QC), in polypore fungus in coniferous forest, and on forest floor in red oak and deciduous forests (NS). **Collecting period**: VII–X. **Collecting method**: sifting fungi and forest litter, pitfall traps.

We report this species from AK (**NSR**) for the first time here, based on the following UAM specimen: United States, Alaska, Fairbanks, 930 Fitz Ct, 94.901, –147.529, *Betula* and *P. tremulo.*, FIT, 16.VIII.2007, D.S. Sikes (UAM), 1 specimen unknown sex.

COMMENTS This species was listed from NJ by Newton (2019) but we are not aware of published voucher data for this record.

Genus *Atheta* Incertae Sedis

165. *Atheta brunswickensis* Klimaszewski (Fig. 19.239a–h)

Atheta brunswickensis Klimaszewski, in Klimaszewski et al. 2005. Majka and Klimaszewski 2008b, Klimaszewski et al. 2008a, 2018a, b, 2020.

Diagnosis Body narrowly subparallel, length 1.9–2.4 mm, colour yellowish brown or rust-brown, with head, antennae (except for basal two or three segments), and apical part of abdomen dark brown, or entire body dark brown with legs lighter, yellowish-brown (Fig. 19.239a); integument moderately glossy, forebody with isodiametric microsculpture and fine punctation; pubescence moderately dense, sparser on pronotum than on elytra; head rounded posteriad, slightly transverse, broadest at level of eyes, strongly flattened medially; antennomere IV subquadrate, V–X slightly to moderately transverse, slightly incrassate apically, and terminal one elongate; pronotum convex, moderately transverse (wider in females than in males), broadest at middle, pubescence on disc directed laterad from midline; elytra transverse, at suture about as long as pronotum along midline, pubescence on disc directed slightly obliquely posteriad; abdomen subparallel. MALE. Tergite VIII transverse, apical margin truncate between two broad, rounded, sublateral teeth (Fig. 19.239d); sternite VIII broadly rounded apically, with antecostal suture sinuate, widely separated from basal margin (Fig. 19.239e); median lobe of aedeagus in lateral view with crista apicalis of bulbus large, tubus broadly curved ventrad in basal half, apical part moderately narrow, tapering to rounded apex (Fig. 19.239b); in dorsal view bulbus oval in outline and evenly coalescent with narrowly elongate tubus, apex triangular (Fig. 19.239c); internal sac with two inconspicuous round structures (Fig. 19.239b,c). FEMALE: Tergite VIII transverse, apical margin truncate medially, oblique laterally (Fig. 19.239f); sternite VIII broadly rounded posteriorly, briefly truncate medially, with apical fringe of microsetae, antecostal suture widely separate from basal margin (Fig. 19.239g); spermatheca with capsule club-shaped, neck elbowed, stem thick, sinuate, narrowly looped and twisted posteriorly (Fig. 19.239h).

DISTRIBUTION **Origin**: Nearctic. **CANADA**: BC, MB, NB, NS, NT [**NTR**], ON, QC, SK [**NPR**], YT. **USA**: AK [**NSR**].

COLLECTION AND HABITAT DATA

Habitat: in NB, from various forest types; upland black spruce forest, red spruce and yellow birch forest, mature red spruce and eastern white cedar forest, balsam fir stand, red pine forest, Jack pine forest, various mixed forests, red oak forest, and various hardwood forests. Adults in various kinds of mushrooms (mostly gilled mushroom species) in various stages of decay (fresh to well rotted) on forest floor; also found in forest litter (moldy conifer duff, leaf litter). Some from pitfall traps in regenerating red spruce forest and yellow birch forest. Others from decaying (moldy) corncobs and cornhusks near mixed forest. In AK, found in *Betula*, *Salix*, white spruce, and *Betula* and *Populus tremuloides* forests. **Collecting period**: V–IX. **Collecting method**: flight intercept trap, sifting fungi and forest litter, pitfall traps, yellow pan traps.

We report this species for the first time from NT (**NTR**), SK (**NPR**), and AK (**NSR**) for the first time here, based on the following: Canada, Northwest Territories, Yellowknife, 62.51070, –113.39479, Rep. 1, mesic, pitfall trap, 7-11. VII.2011, NBP Field Party Y1MP611 (LFC) 1 male; Yellowknife, 62.51070, –113.39479, Rep.1, mesic, pitfall trap, yellow pan, 11-15. VII.2011, NBP Field Party Y1MP121, Y1MY621 (LFC) 1 male, 1 female; Norman Wells, 65.30494, –126.70873, Rep.3, mesic, yellow pans, pitfall traps, 11-14.VI.2011, NBP Field Party N1MY422, N3MY221, N3MY522, N3MP121, N3MP222 (LFC) 3 males, 3 females; Norman Wells, 65.30494, –126.70873, Rep. 3, mesic, yellow pan, 14-17.VI.2011, NBP Field Party N3MY232, N3MY233, N3MY234 (LFC) 1 male, 2 females. Saskatchewan, Paradise Hill, 53.836, –109.558, 620m, 16.IX.2014, mushrooms, sifting, coll. B. Godin and D. Horwood (LFC) 1 male, 1 female; Meadow Lake, 54.4012, –108.8014, 483m, mushrooms, lakeshore, sifting, coll. B. Godin and D. Horwood (BGC, LFC) 2 males, 2 females; Prince Albert, 53.9789, –106.2859, 534m, 13. IX.2014, mushrooms, sifting, coll. B. Godin and D. Horwood (BGC, LFC) 2 males, 2 females. USA, Alaska, 35.5 Mi Dalton Hwy, 271m el., 65.74425, –149.34564, ±4m *Betula, Salix*, white spruce, flight intercept trap, 31.V-5. VI.2008, D.S. Sikes (UAM) 1 specimen; Fairbanks, 930 Fitz Ct, 280m el., 64.9013, –147.52861, ±30m *Betula* & *P. tremuloides*, flight intercept trap, D.S. Sikes, 16.VIII-22.IX.2007 (UAM) 7 specimens; same data except, 26.VIII-10.IX.2006 (UAM) 1 specimen; 20.V-3.VI.2007 (UAM) 5 specimens; 29.V-4.VII.2008 (UAM) 1 specimen; 29.IV-29.V.2008, (UAM) 4 specimens.

COMMENTS This species bears a certain resemblance to *A. irrupta* (occurring south of the study area) in the similar shape of male tergite VIII and the female spermatheca. This similarity is corroborated by CO1 sequence data, where the two species form sister clusters.

166. ***Atheta sparreschneideri*** **Munster** * (Fig. 19.240a–g)

Atheta (*Dimetrota*) *sparreschneideri* Munster, 1923. Lohse et al. 1990 [in *Alaobia*], in *Atheta* (*Oreostiba*): Smetana 2004, Klimaszewski et al. 2012.

Diagnosis Body black with tarsi paler, surface moderately glossy (Fig. 19.240a), length 2.2-3.0 mm; punctation fine, dense on pronotum and elytra, asperate on elytra; head transverse, at level of eyes as broad as postocular region; antennae slender, antennomere III slightly shorter than II, VII–X transverse; pronotum transverse, sides arcuate, broadest at middle, hind angles obtuse, 1 ¼ as broad as head; elytra at suture as long as pronotum along median line, at apex 1 ¼ as broad as pronotum; abdomen subparallel. MALE. Tergite VIII with apical margin truncate and distinctly crenulate (Fig. 19.240e); sternite VIII arcuately rounded apically (Fig. 19.240f); bulbus of median lobe of aedeagus narrowly oval, tubus subparallel basally and broadly triangular apically in dorsal view (Fig. 19.240c,d); in lateral view tubus almost straight ventrally, apex narrowly rounded (Fig. 19.240b). FEMALE. Tergite and sternite VIII nearly truncate apically; spermatheca with capsule club-shaped, apical invagination small, neck elbowed, stem with simple loop, slightly swollen posteriorly (Fig. 19.240g).

DISTRIBUTION Origin: Holarctic: Europe, Russia, North Korea, North America. **CANADA**: YT. **USA**: AK.

COLLECTION AND HABITAT DATA
Habitat: under bear dung and dead bird, and in gopher burrows. **Collecting period**: VI, VII. **Collecting method**: sifting litter and dung.

Genus *Bellatheta* Roubal, 1928

(Fig. 19.241a–g)
(Roubal, 1928, as *Emmelostiba* Pace, 1982, as *Pseudosipalia* Lohse, in Lohse et al. 1990)

Diagnosis Body narrow, subparallel, brown to black, strongly glossy, head about as big as pronotum; eyes small, distinctly shorter than postocular area of head, temples without carina; antennomeres VII–X strongly transverse; glossa divided into two parallel lobes; pronotum small, trapezoidal in shape; elytra transverse, very short, along suture shorter than pronotum along midline, pubescence sparse, wingless; abdomen slightly swollen, broadest at middle; male tergite VIII entire, without teeth; median lobe of aedeagus with tubus long, arcuate ventrally and narrowly elongate apically; spermatheca S-shaped.

167. ***Bellatheta microptera*** **(Lohse), New Combination** (Fig. 19.241a–g)

Pseudousipalia microptera Lohse, in Lohse et al. 1990. In *Emmelostiba*: Gusarov 2003a, Klimaszewski et al. 2012.

Diagnosis Body narrow, subparallel, brown, appendages slightly paler (Fig. 19.241a); length 2.2–2.8 mm; integument strongly glossy; forebody sparsely pubescent and punctate; head circular, as big as pronotum, postocular area of head almost twice as long as eyes in dorsal view; eyes small; antennomere I long and swollen, IV subquadrate to slightly transverse, VII–X strongly transverse; pronotum small, trapezoidal in shape; elytra short, at suture shorter than pronotum along midline; abdomen swollen and broadest at middle. MALE. Tergite VIII with apical margin entire, without teeth, arcuate (Fig. 19.241c); sternite VIII elongate, rounded apically, antecostal suture widely separate from basal margin (Fig. 19.241d); median lobe of aedeagus with tubus long, arcuate ventrally and narrowly elongate apically in lateral view (Fig. 19.241b). FEMALE. Tergite VIII rounded apically (Fig. 19.241e); sternite VIII rounded apically, antecostal suture sinuate, widely separate from basal margin (Fig. 19.241f); spermatheca S-shaped, capsule club-shaped, apical invagination broad, stem strongly sinuate, broadly looped posteriorly (Fig. 19.241g).

DISTRIBUTION Origin: Nearctic. **CANADA**: YT. **USA**: AK.

COLLECTION AND HABITAT DATA
Habitat: tundra, in moss, grass and leaf litter. In AK, found in litter from shrub, alder, dwarf birch, grass, clubmoss; *Betula nana* and *Salix* leaf litter, and *Alnus crispa* and *Picea mariana* litter. **Collecting period**: VI–VIII. **Collecting method**: sifting litter and moss, Winkler litter processing.

Additional records. See Annex 1.

Genus *Boreophilia* Benick, 1973

(Figs. 19.242–19.255) (Lohse et al. 1990; Klimaszewski et al. 2011, 2018a, 2019, 2020)

Diagnosis Body moderately narrow to broad, subparallel; average body length 3.0 mm; head narrower or nearly as broad as pronotum; eyes moderate in size, as long as or shorter than postocular region of head, tempora partially feebly carinate at base of head; antennae often reaching posterior margin of elytra, antennomeres V–X subquadrate, slightly transvers, or elongate; ligula divided into two separate and diverging lobes (Fig. 19.242b); mandibles simple (not bifid at apex as in *Schistoglossa*) (Fig. 19.242c); maxillary palps with 4 palpomeres, terminal one needle-shaped (Fig. 19.242a); pronotum transverse, broadest in the middle or in apical fourth, hind angles rounded, hypomeron clearly visible in lateral view; integument smooth and moderately glossy, weak microsculpture present, punctation fine, pubescence on midline of disc directed anteriad at least in apical ¾, and elsewhere directed laterally; mesocoxae contiguous; median lobe of aedeagus broad, flattened latero-ventrally, bulbus enlarged and tubus narrow, short or moderately elongate, approximately triangularly shaped in dorsal view, crista apicalis narrowly elongate, internal sac with two large, elongate sclerites in bulbus, and additional smaller sclerites in apical part of internal sac; apical margin of male tergite VIII entire, and not modified; spermatheca diverse in shape, with elongate tubular capsule and sinuate stem, capsule's invagination small or absent; cold loving species occurring in arctic habitats, in temperate regions usually confined to fens and bogs or alpine zones.

Key to Species of *Boreophilia*
(modified from Klimaszewski et al. 2019)

1. Body moderately broad, elytra at base about as broad as maximum width of pronotum (Figs. 19.243a, 19.248a, 19.249a, 19.252a) **2**

– Body distinctly broad, elytra at base distinctly broader than maximum width of pronotum **5**

2. Elytra at suture shorter than pronotum at midline (Figs. 19.248a, 19.249a) **3**

– Elytra at suture as long as or longer than pronotum at midline **4**

3. Legs moderately long, hind legs much shorter than abdomen (Fig. 19.248a); median lobe of aedeagus as illustrated (Fig. 19.248b–d), female unknown .... ***Boreophilia nomensis*** **(Casey)**

– Legs extremely long, hind legs almost as long as abdomen (Fig. 19.249a); median lobe of aedeagus and spermatheca as illustrated (Fig. 19.249b,c,h) .... ***Boreophilia beringi*** **Klimaszewski and Brunke**

4. Tubus of median lobe of aedeagus without basal projection on each side in dorsal view (Fig. 19.252d); capsule of spermatheca club-shaped and with small apical invagination, stem broadly looped posteriorly (Fig. 19.252i,j) .... ***Boreophilia neoinsecuta*** **Klimaszewski**

– Tubus of median lobe of aedeagus angulately explanate basally in dorsal view (Fig. 19.243e,d); capsule of spermatheca elongate, tubular, and tapering apically, without apical invagination, stem narrowly looped posteriorly (Fig. 19.243j) .... ***Boreophilia eremita*** **(Rye)**

5. Maximum width of elytra one-fourth wider than pronotum (Figs. 19.250a, 19.255a) **6**

– Maximum width of elytra one-fifth wider than pronotum **7**

6. Median lobe of aedeagus with ventral margin of tubus arcuate basally and apex broad and rounded in lateral view (Fig. 19.255b); sclerites of internal sac broad, curved and with small median projection in lateral view (Fig. 19.255b); capsule of spermatheca narrowly club-shaped, spherical apically, stem coiled posteriorly (Fig. 19.255h) .. ..***Boreophilia venti* (Lohse)**

– Median lobe of aedeagus with ventral margin of tubus straight basally and apex broad and angular in lateral view (Fig. 19.250b); sclerites of internal sac long, narrow and straight, without small median projection in lateral view (Fig. 19.250b); capsule of spermatheca broadly club-shaped, hemispherical apically, stem hooked posteriorly (Fig. 19.250i) ....***Boreophilia caseyi* Lohse**

7. Antennomeres VIII–X elongate, less so in females (Figs. 19.247a, 19.253a) **8**

– Antennomeres VIII–X subquadrate to slightly transverse...**9**

8. Median lobe of aedeagus with apical part narrow, finely tapering in lateral view (Fig. 19.247b, c), bulbus broad, slightly narrowing apicad in dorsal view (Fig. 19.247d); female sternite VIII truncate apically and slightly emarginate medially (Fig. 19.247h); capsule of spermatheca elongate, curved, stem broadly sinuate without posterior loop (Fig. 19.247i,j) ***Boreophilia nearctica* Lohse**

– Median lobe of aedeagus broadly rounded apically in lateral view (Fig. 19.253b), bulbus broad, parallel-sided to near apex in dorsal view (Fig. 19.253c); sclerites of internal sac narrow (Fig. 19.253b,c); capsule of spermatheca large, subspherical apically, stem narrow, abruptly bent before capsule, narrowly looped posteriorly (Fig. 19.253h) ....***Boreophilia subplana* (J. Sahlberg)**

9. Pronotum width to length ratio 1.3 (Figs. 19.245a, 19.251a, 19.254a) **10**

– Pronotum width to length ratio 1.4–1.5 (Figs. 19.244a, 19.246a) **12**

10. Median lobe of aedeagus narrowly rounded apically in lateral view (Fig. 19.245b, c), bulbus oval and narrowed apically in dorsal view (Fig. 19.245d,e); sclerites of internal sac narrow (Fig. 19.245b–e); capsule of spermatheca evenly broad apically, apex rounded, stem broad, sinuate, and looped posteriorly (Fig. 19.245j,k) .. ..***Boreophilia hyperborea* (Brundin)**

– Genitalic structures not as above **11**

11. Capsule of spermatheca with apical part ovoid, apical invagination not apparent, stem narrow and hooked posteriorly (Fig. 19.251d); male undescribed .... ***Boreophilia davidgei* Klimaszewski and Godin**

– Capsule of spermatheca with apical part spherical, apical invagination present, stem moderately wide (Fig. 19.254h); median lobe of aedeagus as illustrated (Fig. 19.254b,c).............. .... ***Boreophilia vega* (Fenyes)**

12. Tubus of median lobe of aedeagus with rounded baso-lateral projection in lateral view (Fig. 19.246b); capsule of spermatheca with narrowly rounded apical part (Fig. 19.246h) ....***Boreophilia islandica* (Kraatz)**

– Tubus of median lobe of aedeagus without basal projections in lateral view (Fig. 19.244a); capsule of spermatheca with apical part abruptly tapering and curved laterad (Fig. 19.244h).......... ....***Boreophilia fusca* (C.R. Sahlberg)**

Fusca species group

[Modified from Lohse et al. 1990, Klimaszewski et al. 2019]

This group contains mostly broad and large species (except *B. nomensis*), and is defined based on similarity of median lobe of aedeagus and spermatheca. Bulbus of aedeagus is moderately broad to broad, oval in shaped in dorsal view, with two prominent and elongate structures bearing ventrally a short, angular projection; tubus of median lobe is triangular in dorsal view, and in lateral view arcuate or straight and in most species narrow apically. Spermatheca: capsule pitcher shaped basally and tubular apically, apical invagination lacking; stem long, strongly sinuate and looped or arcuate posteriorly.

168. *Boreophilia eremita* (Rye) * (Fig. 19.243a–j)

Homalota eremita Rye 1866. In *Boreophilia*: Brundin 1952 [as *B. hercynica*], Palm 1970 [as *B. hercynica*], Lohse et al. 1990, Smetana 2004, Webster et al. 2012, Sazhnev 2018, Klimaszewski et al. 2011, 2018a, 2019.

Diagnosis Body moderately broad, forebody slightly and abdomen strongly glossy (Fig. 19.243a); length 3.0–3.6 mm; dark brown to almost black, appendages light brown or reddish brown; antennomeres VIII–X subquadrate to slightly elongate; pronotum as long as elytra at suture, maximum width of pronotum slightly less than maximum width of elytra, pronotum in males is longer and less transverse than in females. MALE. Tergite VIII arcuate apically (Fig. 19.243f); sternite VIII elongate, parabolic apically (Fig. 19.243g); tubus of median lobe of aedeagus in dorsal view angulately explanate basally (Fig. 19.243d,e), in lateral view ventral part of tubus angulately, obliquely produced at base, then arcuately elongate, narrow to narrowly rounded apex (Fig. 19.243b,c), bulbus broadly oval in dorsal view, with two elongate narrow sclerites of internal sac (Fig. 19.243d,e). FEMALE. Tergite VIII arcuate apically (Fig. 19.243h); sternite VIII rounded apically, antecostal suture moderately sinuate (Fig. 19.243i); spermatheca: neck pitcher-shaped, capsule elongate, curved basally, tapering to rounded apex, without apical invagination, stem sinuate, narrowly curled posteriorly (Fig. 19.243j).

The median lobe of aedeagus of *B. eremita* is similar to that of *B. islandica*, but tubus is more elongate, narrower, less arcuate, and with basal projection angular in lateral view (Fig. 19.243b, c), in dorsal view tubus with two pointed latero-basal projections (Fig. 19.243d,e). In *B. islandica*, the tubus is broader, shorter, more arcuate ventrally, and with basal arcuate projection rounded in lateral view (Fig. 19.246b), in dorsal view tubus with two smaller and less angular basal projections (Fig. 19.246c). Spermathecae of the two species is variable in shape and very similar, female tergite and sternite VIII are similar in shape but sternite VIII in *B. eremita* has medial part of antecostal suture feebly sinuate (Fig. 19.243i), while it is strongly sinuate in *B. islandica* in the majority of examined specimens (Fig. 19.246g). On average, the body of *B. eremita* is narrower and elytra shorter than that of *B. islandica*.

DISTRIBUTION Origin: Holarctic; recorded from north and central Europe, Ireland, Ukraine, Russia (West and East Siberia), the Russian Far East and North America. **CANADA**: LB, MB, NB, QC [**NPR**]. **USA**: AK.

COLLECTION AND HABITAT DATA
Habitat: in NB, old silver maple forest with green ash and seasonally flooded marsh; silver maple swamp, margin of vernal pond, found in moist leaves; in AK, creekside/ocean beach confluence, under boards and drift wood; black and white spruce, willow; subalpine habitat with *Veratrum*, and *Calamagrositis*. **Collecting period**: V–VIII. **Collecting method**: sifting litter, Lindgren funnel trap, Malaise trap, hand collecting.

We report this species from QC (**NPR**) for the first time here, based on the following: Canada, Quebec, Blanc Sablon, 9.VIII.1972, J.M. Campbell (CNC) 2 females.

169. *Boreophilia fusca* (C.R. Sahlberg) *
(Fig. 19.244a–h)

Aleochara fusca C.R. Sahlberg 1831. In *Boreophilia*: Lohse et al. 1990, Smetana 2004, Sazhnev 2018, Klimaszewski et al. 2019.

Diagnosis Body broad, forebody moderately and abdomen slightly more glossy (Fig. 19.244a); length 3.4–3.8 mm; head, pronotum and abdomen except for its apex dark brown, elytra dark-reddish brown, appendages light brown, or body entirely dark brown to almost black with tarsi reddish-brown; antennomeres VIII–X subquadrate; pronotum shorter than elytra at suture, maximum width of pronotum distinctly less than maximum width of elytra. MALE. Tergite VIII arcuate apically (Fig. 19.244d); sternite VIII elongate, parabolic apically (Fig. 19.244e); tubus of median lobe of aedeagus evenly arcuate laterally, apex narrow and slightly pointed in lateral view (Fig. 19.244b), bulbus oval, broad basally and narrowed apically in dorsal view, and with two elongate sclerites of internal sac (Fig. 19.244c). FEMALE. Tergite VIII arcuate apically (Fig. 19.244f); sternite VIII rounded apically, antecostal suture slightly sinuate (Fig. 19.244g); spermatheca: capsule pitcher-shaped basally with broadly tubular apical projection, moderately long, abruptly tapered and curved laterad at apex, stem with single irregular loop posteriorly (Fig. 19.244h).

DISTRIBUTION Origin: Holarctic; recorded from Fennoscandia, Russia (West and East Siberia, the Far Easr) and North America. **CANADA**: MB, NT. **USA**: AK.

COLLECTION AND HABITAT DATA
Habitat: tundra. **Collecting period**: VI, VII. **Collecting method**: sweeping tundra vegetation.
Additional records. See Annex 1.

170. ***Boreophilia hyperborea*** **(Brundin)** * (Fig. 19.245a–k)
Atheta hyperborea Brundin, 1940. In *Boreophilia*: Lohse et al. 1990, Smetana 2004, Ernst and Buddle 2013, Klimaszewski et al. 2019.

Diagnosis Body broad, forebody glossy; length 2.8–3.5 mm; black with tarsi reddish-brown (Fig. 19.245a); antennomeres VIII–X subquadrate; pronotum as long as or slightly shorter than elytra at suture, maximum width of pronotum distinctly less than maximum width of elytra. MALE. Tergite VIII arcuate apically (Fig. 19.245f); sternite VIII triangular, with apical margin rounded (Fig. 19.245g); tubus of median lobe of aedeagus narrow, broadly arcuate in lateral view, apex narrow and rounded (Fig. 19.245b,c), bulbus large, oval, broad basally and narrowed apically in dorsal view, and with two elongate sclerites (Fig. 19.245d,e). FEMALE. Tergite VIII broadly rounded apically (Fig. 19.245h); sternite VIII rounded apically except shallowly emarginate mediad, antecostal suture straight medially and slightly sinuate laterally (Fig. 19.245i); spermatheca: neck pitcher-shaped, capsule moderately elongate, curved basally, slightly tapering to broadly rounded apex, stem sinuate, with single curl posteriorly (Fig. 19.245j,k).

DISTRIBUTION Origin: Holarctic; recorded from Fennoscandia, Greenland, Russia (North European Territory) and North America. **CANADA**: MB, NT, NU. **USA**: AK.

COLLECTION AND HABITAT DATA
Habitat: tundra, under rocks. In AK, found in *Salix*, tundra hillside, cobble; *Salix, Betula, Alnus*, grasses; vegetation lake shore near pond, and under rocks. **Collecting period**: V–VII. **Collecting method**: hand collected from under rocks, netting, pollinator cup/pitfall, sweeping vegetation near body of water.
Additional records. See Annex 1.

171. ***Boreophilia islandica*** **(Kraatz)** * (Fig. 19.246a–h)
Homalota islandica Kraatz 1857. In *Boreophilia*: Lohse et al. 1990, Smetana 2004, Sazhnev 2018, Klimaszewski et al. 2011, 2015a, 2018a, 2019.

Diagnosis Body broad, forebody moderately and abdomen highly glossy (Fig. 19.246a); length 2.8–3.5 mm; uniformly black with paler, reddish-brown appendages, or head, pronotum and VI–VII basal segments of abdomen dark brown, remainder of the body reddish brown, appendages light brown, sometimes elytra with some reddish tinge; antennomeres VIII–X subquadrate, bead-like; pronotum as long as elytra at suture, maximum width of pronotum distinctly less than maximum width of elytra; elytral length variable, as long as pronotum or slightly longer. MALE. Tergite VIII arcuate apically (Fig. 19.246d); sternite VIII elongate, parabolic apically (Fig. 19.246e); tubus of median lobe of aedeagus obtusely angulately explanate basolaterally in dorsal view (Fig. 19.246c), appearing as a rounded projection in lateral view (Fig. 19.246b); bulbus moderately broadly oval with two elongate sclerites of internal sac in dorsal view (Fig. 19.246c). FEMALE. Tergite VIII arcuate apically (Fig. 19.246f); sternite VIII rounded apically, antecostal suture strongly sinuate in most examined specimens (Fig. 19.246g); spermatheca: shape of capsule highly variable, usually elongate, pitcher-shaped basally, gradually narrowing to rounded apex, stem with single coil posteriorly (Fig. 19.246h).

DISTRIBUTION Origin: Holarctic; recorded from Fennoscandia, Estonia, Faeroe Islands, Great Britain, Iceland, Russia (North European Territory) and North America. **CANADA**: AB, NF, NT, NU, YT. **USA**: AK.

COLLECTION AND HABITAT DATA
Habitat: Alder, white spruce, *Betula* forest; *Betula* and *Salix* litter, *Salix* tundra hillside; *Salix/Betula/Alnus*/grasses, black/white spruce, willow, vegetation at lakeshore pond; maritime tundra, *Leymus, Heracleum*; *Leymus, Heracleum, Angelica* and subalpine habitat with *Veratrum, Calamagrostis*, and *Leymus, Heracleum, Angelica*. **Collecting period**: V–VIII. **Collecting method**: hanging Malaise trap, pitfall traps, sweeping vegetation with net.

Additional records. See Annex 1.

COMMENTS Females of this species may be confused with other species of *Boreophilia* and particularly those closely related to *B. eremita*. Associating females with males is considered here to be the most reliable way of identifying females of this and the previous species. At present, *B. islandica* is considered a somewhat variable species. Specimens vary from moderately robust and narrower to more robust and broader, with elytra as long as pronotum or slightly longer, all with the same genitalic morphology.

172. *Boreophilia nearctica* Lohse (Fig. 19.247a–j)

Boreophilia nearctica Lohse, in Lohse et al. 1990. Klimaszewski et al. 2011, 2018a, 2019.

Diagnosis Body moderately broad, forebody moderately glossy, abdomen slightly more so (Fig. 19.247a); length 3.0–3.5 mm; head, pronotum and abdomen except for its apex dark brown, elytra reddish brown, appendages light brown, or body dark brown to almost black with elytra partially dark reddish brown and tarsi reddish brown; antennomeres VIII–X subquadrate to slightly elongate; pronotum about as long as elytra at suture, maximum width of pronotum distinctly less than maximum width of elytra. MALE. Tergite VIII with apical margin arcuate, scarcely angulate medially (Fig. 19.247e); sternite VIII elongate, parabolic apically (Fig. 19.247f); tubus of median lobe of aedeagus in lateral view almost straight ventrally, apical part narrow and finely tapering, slightly curved ventrad (Fig. 19.247b,c), bulbus in dorsal view broadly oval, slightly tapering apicad, internal sac with two elongate sclerites (Fig. 19.247d). FEMALE. Tergite VIII arcuate apically (Fig. 19.247g); sternite VIII truncate apically and slightly emarginate medially, antecostal suture almost straight medially and slightly sinuate laterally (Fig. 19.247h); spermatheca: capsule moderately elongate, slightly curved, pitcher-shaped basally, rounded apically, stem broadly sinuate and without posterior loop (Fig. 19.247i,j).

DISTRIBUTION Origin: Nearctic. **CANADA**: AB, LB, MB, NF, QC, YT. **USA**: AK.

COLLECTION AND HABITAT DATA
Habitat: black spruce forest, alpine meadow. In AK, in wood cutting area, white spruce, birch, grass; creekside/ocean beach confluence, under boards and drift wood; and *Spirea*, dwarf birch. **Collecting period**: VI–IX. **Collecting method**: hanging Malaise trap, pitfall traps, hand collecting under rocks and litter.

Additional records. See Annex 1.

COMMENTS The southernmost record of this species in the Rocky Mountains of southern Alberta suggests that *B. nearctica* probably occurs continuously along the western cordilleras, at successively higher elevation sites with decreasing latitude (Klimaszewski et al. 2019).

173. *Boreophilia nomensis* (Casey) (Fig. 19.248a–f)

Dinaraea nomensis Casey, 1910. In *Boreophilia*: Lohse and Smetana 1985; Lohse et al. 1990; Gusarov 2003a; Klimaszewski et al. 2019, 2020. As *Boreophilia caseyiana* Lohse 1990, *in* Lohse et al. 1990, synonymized with *B. nomensis* by Gusarov 2003a.

Diagnosis Body narrow, subparallel, moderately glossy, abdomen slightly more so (Fig. 19.248a); length 3.0–3.8 mm; head, pronotum and abdomen dark brown, elytra reddish-brown, legs paler, yellowish-brown; antennomeres VIII–X subquadrate; pronotum longer than elytra at suture, maximum width of pronotum about the same as maximum width of elytra. MALE. Tergite VIII arcuate apically (Fig. 19.248e); sternite VIII elongate, apical margin parabolic (Fig. 19.248f); tubus of median lobe of aedeagus with ventral margin slightly arcuate in lateral view, apical part triangular, apex narrowly rounded (Fig. 19.248b), bulbus broad in dorsal view and with two elongate narrow sclerites as illustrated (Fig. 19.248c,d). FEMALE. Undescribed.

DISTRIBUTION **Origin**: Nearctic. **CANADA**: BC, YT. **USA**: AK.

COLLECTION AND HABITAT DATA
Habitat: spruce and aspen forest with horsetail/shrub/grass undergrowth, and under rocks near snowfield. In AK, found in alpine tundra, and under rocks near snow field. **Collecting period**: VI–VIII. **Collecting method**: pitfall trap, pitfalls w/ rainroofs, hand collecting under rocks.

Additional records. See Annex 1.

Subplana **species group** (Klimaszewski et al. 2019)

This newly defined group contains species defined by the similarity of the median lobe of aedeagus and capsule of spermatheca with a small apical invagination. Bulbus of median lobe of aedeagus broad apically, oval in shape in dorsal view, with two prominent elongate structures of slightly different shape in different species; tubus of median lobe of aedeagus arcuate or straight and broad or narrow in lateral view; internal sac broadly open apically and supported by two narrow arcuate structures. Spermatheca: capsule pitcher-shaped basally and globular or ovoid apically with small apical invagination in most species; stem long strongly sinuate and looped or coiled posteriorly.

174. ***Boreophilia beringi*** **Klimaszewski and Brunke** (Fig. 19.249a–h)

Boreophilia beringi Klimaszewski and Brunke, in Klimaszewski et al. 2019.

Diagnosis Body narrow, subparallel, glossy, abdomen slightly more so; microsculpture of forebody strong (Fig. 19.249a); length 2.2–2.4 mm; head, head, pronotum and abdomen dark brown, elytra with reddish-brown tinge, legs paler, yellowish-brown; antennomeres VIII–X subquadrate; pronotum slightly longer than elytra at suture, maximum width of pronotum about the same as maximum width of elytra. MALE. Tergite VIII broadly arcuate apically (Fig. 19.249d); sternite VIII elongate, with apical margin parabolically rounded (Fig. 19.249e); tubus of median lobe of aedeagus in lateral view straight basally and curved ventrad apically, apex narrowly triangular in shape (Fig. 19.249b); bulbus in dorsal view broadly oval, distinctly narrowing apicad, with two elongate sclerites narrowed apicad (Fig. 19.249c). FEMALE. Tergite VIII arcuate apically (Fig. 19.249f); sternite VIII broadly rounded apically, antecostal suture strongly sinuate medially (Fig. 19.249g); spermatheca: capsule club-shaped, moderately long and rounded apically, with small apical invagination; stem sinuate with single coil posteriorly (Fig. 19.249h).

DISTRIBUTION **Origin**: Nearctic. **CANADA**: YT [**NTR**]. **USA**: AK.

COLLECTION AND HABITAT DATA
Habitat: snowfield, tundra, under rocks, on moss. **Collecting period**: VII. **Collecting method**: aspirating specimens from moss.

We report this species from YT (**NTR**) and Canada for the first time here, based on the following: Canada, Yukon Territory, 8 mi NW Keno

Mts., 6000 feet, 19.VII.1968, A. Smetana (CNC) 18 males, 12 females.

Additional records. See Annex 1.

175. *Boreophilia caseyi* Lohse (Fig. 19.250a–i)

Boreophilia caseyi Lohse, in Lohse et al. 1990. Klimaszewski et al. 2019. As *Boreophilia manitobensis*: Lohse 1990, *in* Lohse et al. 1990, synonymized with *B. caseyi* by Klimaszewski et al. 2019.

Diagnosis Body narrow, subparallel, moderately glossy, abdomen slightly more so (Fig. 19.250a); length 3.5–3.8 mm; head, pronotum and abdomen, except for its apex, dark brown, elytra reddish-brown, legs yellowish red-brown, or body uniformly piceous with tarsi and tibiae reddish-brown; antennomeres VIII–X subquadrate (females) to slightly elongate (males); pronotum as long as elytra at suture or slightly shorter, maximum width of pronotum slightly less than maximum width of elytra. MALE. Tergite VIII arcuate apically (Fig. 19.250e); sternite VIII elongate, parabolic apically (Fig. 19.250f); tubus of median lobe of aedeagus straight basally and strongly projecting ventrad at apex, apex broad and angular in lateral view (Fig. 19.250b), in dorsal view bulbus broad and angular apico-laterally, with two elongate narrow sclerites of internal sac (Fig. 19.250c,d). FEMALE. Tergite VIII arcuate apically (Fig. 19.250g); sternite VIII rounded apically, antecostal suture straight medially and sinuate laterally (Fig. 19.250h); spermatheca: capsule pitcher-shaped basally with subspherical apical part bearing small invagination, stem sinuate, narrow, looped posteriorly (Fig. 19.250i).

The spermatheca of *B. caseyi* was illustrated in Lohse et al. (1990). It is slightly deformed and based on a female captured in a different locality than that of the male holotype. It may belong to *B. subplana*, a species with a very similar shaped spermatheca. The spermathecal stem of *B. caseyi* has a broad posterior loop (Fig. 19.250i), and the apical margin of female sternite VIII is rounded (Fig. 19.250h), and not triangularly produced and angulate medially as in *B. subplana* (Fig. 19.253g).

DISTRIBUTION Origin: Nearctic. **CANADA**: MB, NU, YT. **USA**: AK.

COLLECTION AND HABITAT DATA
Habitat: arctic tundra. **Collecting period**: VI, VII. **Collecting method**: pitfall traps.

Additional records. See Annex 1.

176. *Boreophilia davidgei* Klimaszewski and Godin (Fig. 19.251a–d)

Boreophilia davidgei Klimaszewski and Godin 2012, in Klimaszewski et al. 2012, 2015a, 2019.

Diagnosis Body moderately broad, subparallel, moderately glossy, abdomen slightly more so (Fig. 19.251a); length 2.8–2.9 mm; uniformly dark brown to almost black, basal sternite slightly reddish-brown, legs paler, yellowish-brown; antennomeres VIII–X subquadrate to slightly transverse; pronotum slightly shorter than elytra at suture, maximum width of pronotum distinctly less than maximum width of elytra. MALE. Undescribed. FEMALE. Tergite VIII truncate apically (Fig. 19.251b); sternite VIII rounded apically, antecostal suture arcuate and slightly sinuate laterally (Fig. 19.251c); spermatheca: capsule ovoid apically and pitcher-shaped basally, without apical invagination; stem short, narrow, sinuate and looped posteriad (Fig. 19.251d). This species may be easily distinguished by the unique shape of its spermatheca.

DISTRIBUTION Origin: Nearctic. **CANADA**: AB, YT. **USA**: not recorded.

COLLECTION AND HABITAT DATA
Habitat: white spruce and feathermoss forest, mixed pine and willow forest, black spruce stand, mixed aspen and white spruce forest. **Collecting period**: V–IX. **Collecting method**: pitfall traps and litter sifting.

COMMENTS This species is tentatively assigned to this group, because the male is unknown and morphology of median lobe of aedeagus could not be analysed.

177. ***Boreophilia neoinsecuta*** **Klimaszewski** (Fig. 19.252a–j)

Boreophilia neoinsecuta Klimaszewski, in Klimaszewski et al. 2019, 2020. As *Boreophilia insecuta*: Lohse, in Lohse et al. 1990 (misidentification).

Diagnosis Body moderately broad, subparallel, forebody moderately glossy, abdomen slightly more so (Fig. 19.252a); length 3.2–3.5 mm; head, pronotum and abdomen except for its apex dark brown, elytra reddish brown, mottled with some darker spots, appendages reddish brown; antennomeres VIII–X subquadrate; pronotum about as long as elytra at suture, maximum width of pronotum about equals to maximum width of elytra. MALE. Tergite VIII arcuate apically (Fig. 19.252e); sternite VIII elongate, parabolic apically (Fig. 19.252f); tubus of median lobe of aedeagus in lateral view almost straight, apex short, narrow and rounded, bulbus broad and with two elongate and posteriorly bent sclerites (Fig. 19.252b,c), in dorsal view median lobe elongate oval (Fig. 19.252d). FEMALE. Tergite VIII arcuate apically (Fig. 19.252g); sternite VIII with apical margin broadly triangular, obtusely angulate medially, antecostal suture slightly sinuate laterally (Fig. 19.252h); spermatheca: capsule club-shaped, moderately long, rounded apically, with small apical invagination; stem slightly sinuate, coiled one and a half times posteriorly (Fig. 19.252i,j).

DISTRIBUTION **Origin**: Nearctic. **CANADA**: MB, NT [**NTR**], YT. **USA**: AK.

COLLECTION AND HABITAT DATA
Habitat: tundra, under rocks. **Collecting period**: V, VI. **Collecting method**: collected from under rocks using forceps/aspirator.

We report this species from NT (**NTR**) for the first time here based on the following: Canada, Northwest Territories, Reindeer Sta., Caribou Hills, 30.V.1972, A. Smetana (CNC) 1 female.

Additional records. See Annex 1.

COMMENTS Lohse, in Lohse et al. (1990) reported *Boreophilia insecuta* (Eppelsheim) in North America as a Holarctic species from AK, MB, and YT. However, these records represented a different, new Nearctic species, *B. neoinsecuta* Klimaszewski. The two species, *B. insecuta* and *B. neoinsecuta*, have similarly shaped genitalia, which explains why they were confused. The true *B. insecuta* (Eppelsheim) was illustrated by Brundin (1954) and has the apex of the median lobe of the aedeagus distinctly dilated in lateral view, while it is narrower in *B. neoinsecuta* (Figs. 19.252b,c), and the two sclerites of internal sac of median lobe of aedeagus are broader and differently shaped than those of *B. neoinsecuta*. Spermathecae of both species are very similar in shape, but the apical margin of female sternite VIII in *B. insecuta* is rounded and truncate medially, while in *B. neoinsecuta* it is triangularly produced and obtusely angulate medially (Fig. 19.252h).

178. ***Boreophilia subplana*** **(J. Sahlberg)** * (Fig. 19.253a–h)

Atheta subplana J. Sahlberg, 1880. As *Boreophilia*: Lohse et al. 1990; Gusarov 2003a [as *B. angusticornis* (Bernhauer)]; Smetana 2004; Klimaszewski et al. 2019.

Diagnosis Body moderately broad, strongly glossy, abdomen slightly more so (Fig. 19.253a); length 2.8–3.5 mm; black, elytra with some reddish tinge, tarsi yellowish-brown; antennomeres VIII–X slightly elongate; pronotum shorter than elytra at suture, maximum width of pronotum distinctly less than maximum width of elytra. MALE. Tergite VIII truncate or slightly concave apically (Fig. 19.253d); sternite VIII elongate, broadly parabolic apically (Fig. 19.253e); tubus of median lobe of aedeagus in lateral view with ventral margin almost straight and slightly curved ventrad, apex broad and rounded (Fig. 19.253b), in dorsal view bulbus broad, parallel-sided to near apex, with two elongate sclerites of distinctive shape (Fig. 19.253c). FEMALE. Tergite VIII slightly arcuate apically (Fig. 19.253f); sternite VIII with apical margin triangularly produced, slightly obtusely angulate medially, antecostal suture straight medially, obliquely deflected laterally (Fig. 19.253g); spermatheca: capsule broadly club-shaped, large, subspherical apically, with small apical invagination; stem abruptly bent before capsule, straight to small posterior loop (Fig. 19.253h).

DISTRIBUTION Origin: Holarctic; recorded from Spitsbergen (Norway), Fennoscandia, Russia (West and East Siberia) and North America. **CANADA**: NT, NU. **USA**: AK, NH.

COLLECTION AND HABITAT DATA
Habitat: *Salix* leaf litter, tundra hillside, black spruce forest, brackish shoreline, under rocks and wrack. **Collecting period**: VI–VIII. **Collecting method**: hanging Malaise traps, aspirating from under rocks/cobbles, sweeping low vegetation, pitfall traps, pollinator cup/pitfall, Berlese funnels.

Additional records. See Annex 1.

COMMENTS Bernhauer (1907) described *Atheta (Metaxya) angusticornis* and Casey (1910) described *Metaxya plutonica* from Mount Washington, NH, USA. Gusarov (2003a) transferred *A. angusticornis* to the genus *Boreophilia*, with *Metaxya plutonica* as a synonym, and considered it closely related to *B. subplana*, from which he differentiated it by the "shape of aedeagus, particularly a narrower apex of the median lobe in parameral view". We have studied the median lobe of AK and NH specimens and found no differences warranting different species recognition. We therefore consider the two populations as belonging to the same species. The Mount Washington, NH, population represents an isolated, southernmost population of this species known only from higher elevations.

179. *Boreophilia vega* Fenyes * (Fig. 19.254a–h)

Boreophilia vega Fenyes, 1920. Brundin 1954, Smetana 2004, Klimaszewski et al. 2019. As *Boreophilia herschelensis* Klimaszewski and Godin 2012, in Klimaszewski et al. 2012. Synonymized by Klimaszewski et al. 2019.

Diagnosis Body broad, narrowly oval, moderately glossy, abdomen slightly more so (Fig. 19.254a); length 2.8–30 mm; uniformly dark brown to almost black, tarsi paler, yellowish-brown; antennomeres VIII–X subquadrate to slightly transverse; pronotum as long as elytra at suture, maximum width of pronotum distinctly shorter than maximum width of elytra. MALE. Tergite VIII truncate apically (Fig. 19.254d); sternite VIII elongate, parabolic apically (Fig. 19.254e); tubus of median lobe of aedeagus slightly arcuate in lateral view, apex narrowly rounded, bulbus broad and with two elongate narrow sclerites (Fig. 19.254b), in dorsal view median lobe of aedeagus elongate oval (Fig. 19.254c). FEMALE. Tergite VIII sinuately truncate apically (Fig. 19.254f); sternite VIII rounded apically except truncate medially, antecostal suture arcuate and slightly sinuate laterally (Fig. 19.254g); spermatheca: capsule broadly club-shaped, moderately long and rounded apically, with small apical invagination; stem short, sinuate with single curl posteriad (Fig. 19.254h).

DISTRIBUTION Origin: Holarctic; known from West and East Siberia, Russian Far East, North Korea and North America. **CANADA**: Herschel Island, YT, NT [**NPR**], NU [**BOLD**]. **USA**: not recorded.

COLLECTION AND HABITAT DATA
Habitat: on alluvial fan, tundra along freshwater creek, riverside flood debris. **Collecting period**: VI, VII. **Collecting method**: sifting litter and moss.

We report this species from NT (**NTR)** for the first time here, based on a specimen in LFC, and from NU (**NTR)** based on BOLD records:

Canada, Northwest Territories, Banks Island, 73.23284, −119.54719, Rep 2, wet, pitfall trap, 7-11.VII.2011, NBP Field Party B2WP113 (LFC) 1 female. Nunavut, Kitikmeot, Cambridge Bay, 1m el., 69.1294, −104.996, tundra, in flood debris along river bank, sifting, 13.VII.2018, M. Pentinsaari, A. Borisenko, C. Sobel (CBG) 3 specimens; Kitikmeot, 2km W of Cambridge Bay, 1m el., 69.1208, −105.105, tundra, along small stream, 13.VII.2018, M. Pentinsaari, A. Borisenko, C. Sobel (CBG) 1 specimen.

COMMENTS *Boreophilia vega* has a median lobe of aedeagus similar to that of *B. neoinsecuta* (Fig. 19.252b–d), but the spermatheca of *B. vega* compared to the latter species is differently shaped (Figs. 19.252i,j, 19.254h). Externally, *B. vega* differs from *B. neoinsecuta* in elytra distinctly broader than the maximum width of pronotum. Female sternite VIII is differently shaped in the two species (Figs. 19.252h, 19.254g). *B. vega* was an unnecessary replacement name for *Atheta atricornis* J. Sahlberg 1887 since it is no longer congeneric with its senior homonym, but *B. vega* is in use and *A. atricornis* was replaced before 1961 and so is permanently invalid (ICZN Article 59.3).

180. *Boreophilia venti* (Lohse) (Fig. 19.255a–h)
Dimetrota venti (Lohse), in Lohse et al. 1990. Klimaszewski et al. 2019.

Diagnosis Body narrowly subparallel, forebody moderately glossy, abdomen slightly more so (Fig. 19.255a); length 3.0–3.4 mm; uniformly dark brown to black, appendages yellowish brown or reddish brown; antennomeres VIII–X subquadrate; pronotum about as long as elytra at suture, maximum width of pronotum slightly less than maximum width of elytra. MALE. Tergite VIII arcuate apically (Fig. 19.255d); sternite VIII elongate, parabolic apically (Fig. 19.255e); tubus of median lobe of aedeagus in lateral view slightly arcuate basally, apical part straight, bent obliquely ventrad, apex broad and rounded (Fig. 19.255b), in dorsal view bulbus broad, slightly more so apicad, and with two large sclerites, each apically curved, sharply pointed, and with small projection at middle (Fig. 19.255c). FEMALE. Tergite VIII with apical margin arcuate except truncate medially (Fig. 19.255f); sternite VIII with apical margin broadly parabolically rounded except truncate medially, antecostal suture distinctly sinuate (Fig. 19.255g); spermatheca: capsule club-shaped, tubular basally and rounded apically, with small apical invagination; stem angularly bent at base of capsule, with one an a half coils posteriorly (Fig. 19.255h). The spermatheca of this species is very similar to that of *B. neoinsecuta* (Fig. 19.252i,j), but the shape of the apex of female sternite VIII is different in the two species (Figs. 19.252h, 19.255g).

DISTRIBUTION Origin: Nearctic. **CANADA**: YT. **USA**: AK.

COLLECTION AND HABITAT DATA
Habitat: *Salix* litter, *Salix* tundra hillside, lakeshore debris. **Collecting period**: V, VII. **Collecting method**: pitfall traps, pollinator cup/pitfall, hand/aspirator collecting from under rocks, sifting lakeshore debris and *Salix* litter.

Additional records. See Annex 1.

COMMENTS Lohse, in Lohse et al. 1990, described *Dimetrota venti* from Yukon. This species is extremely similar externally and genitally to *Boreophilia insecuta* illustrated by Brundin (1954). In *B. venti*, the tubus of the median lobe is distinctly arcuate in lateral view (Fig. 19.255b), but in *B. insecuta* it is always straight basally and the two main sclerites are slightly different in shape. However, the remaining genital characters are similar and females of the two species can only be separated by association with conspecific males.

Genus *Boreostiba* Lohse, 1990

(Figs. 19.256–19.261) (Lohse et al. 1990, Klimaszewski et al. 2011, 2018a)

Diagnosis Body elongate, subparallel; integument brown to black, elytra dark brown to black or in some species reddish-brown; ligula divided into two lobes in apical half; head with postocular carinae incomplete, never reaching hind border of eye; elytra at suture at most as long as pronotum along midline; mesocoxae narrowly separated; procoxal cavities separated from lateral margin of pronotum by long hypomera; median lobe of aedeagus in lateral view with tubus strongly produced ventrally, thin subapically; condylite of paramere greatly enlarged in most species; male tergite VIII with apical margin truncate, entire and without teeth; spermatheca with capsule short, club-shaped, stem thin, sinuate, narrowly looped posteriorly.

COMMENTS Gusarov (2003a) recommended transfer of *Boreostiba parvipennis* (Bernhauer), *B. campbelliana* Lohse, and *B. lagunae* Lohse to the genus *Atheta*. We believe that it is better to keep these species in a separate group within *Boreostiba* because they have more external body features in common with *Boreostiba* than with *Atheta sensu lato*.

Key to Species of *Boreostiba*

1. Elytra at suture shorter than pronotum along midline of disc (Figs. 19.257a–19.260a) **2**

– Elytra at suture longer than pronotum along midline of disc..**5**

2. Body black, elytra dark reddish-brown or black (Fig. 19.257a); median lobe of aedeagus with narrow and strongly sinuate subapical part of tubus in lateral view (Fig. 19.257b); spermatheca with spherical capsule and short and straight neck, stem sinuate and narrowly looped posteriorly (Fig. 19.257i).. ....***Boreostiba sibirica* (Mäklin)**

– Body dark brown, elytra light reddish-brown; median lobe and spermatheca not as above..**3**

3. Body length 3.5 mm, eyes small, shorter than half length of postocular region of head in dorsal view (Fig. 19.259a); male undescribed; spermatheca very thin, long, capsule almost evenly tubular (Fig. 19.259d) ....***Boreostiba lagunae* Lohse**

– Body length 2.2-2.5 mm; eyes moderately large, longer than half of postocular region of head in dorsal view; spermatheca not as above...**4**

4. Tubus of median lobe of aedeagus moderately produced ventrad in apical half, subapical part moderately broad in lateral view (Fig. 19.260b), in dorsal view tubus evenly triangular without basal constriction (Fig. 19.260c)... ....***Boreostiba parvipennis* (Bernhauer)**

– Tubus of median lobe of aedeagus strongly produced ventrad in apical half, subapical part very thin in lateral view (Fig. 19.258b), in dorsal view tubus triangular and constricted basally (Fig. 19.258c)...***Boreostiba campbelliana* Lohse**

5. Size of pronotum smaller than size of head (Fig. 19.256a); body moderately glossy; maximum width of head greater than maximum width of pronotum (Fig. 19.256a); median lobe of aedeagus with long athetine bridge forming continuous tube in basal half of tubus in lateral view (Fig. 19.256b); spermatheca with hemispherical capsule (Fig. 19.256i).......................... ....***Boreostiba frigida* (J. Sahlberg)**

– Size of pronotum greater than size of head (Fig. 19.261a); body strongly glossy; maximum width of head smaller than maximum width of pronotum (Fig. 19.261a); median lobe of aedeagus with short athetine bridge forming narrow strip in lateral view (Fig. 19.261b); spermatheca with narrow pitcher-shaped capsule (Fig. 19.261h)............ ...***Boreostiba websteri* Klimaszewski and Langor**

***Frigida* species group** (Klimaszewski et al. 2018a)

Diagnosis Median lobe of aedeagus with tubus having long, tubular, and completely fused athetine bridge.

181. *Boreostiba frigida* (J. Sahlberg) *
(Fig. 19.256a–i)

Atheta frigida J. Sahlberg, 1880. Brundin 1940, Lohse et al. 1990 [as *B. lamellifera*], in *Boreostiba*: Gusarov 2003a, Klimaszewski et al. 2011, 2018a.

Diagnosis Body elongate, subparallel, length 2.5–3.0 mm; colour dark brown to black, mouthparts and legs/tarsi reddish-brown (Fig. 19.256a); integument moderately glossy; forebody with hexagonal microsculpture and fine punctation, on elytra slightly asperate; head at least as broad as pronotum, eyes about as long as genae in dorsal view, both prominent; antennae slightly incrassate, antennomere IV subquadrate, V–X slightly to moderately transverse; pronotum small, rectangular in outline and narrower than base of elytra, posterior angles obtusely arcuate; elytra at suture slightly longer than pronotum; abdomen subparallel. MALE. Tergite VIII with apical margin arcuate except broadly, shallowly emarginate medially (Fig. 19.256e); sternite VIII rounded apically (Fig. 19.256f); tubus of median lobe of aedeagus in lateral view strongly curved ventrad basally, apical part narrowly triangular (Fig. 19.256b); in dorsal view tubus moderately broad without elongate median structures in internal sac (Fig. 19.256c); paramere with broad and large condylite, and triangular process on inner side of paramerite (Fig. 19.256d). FEMALE. Tergite VIII with apical margin arcuate except broadly, shallowly emarginate medially (Fig. 19.256g); sternite VIII rounded apically (Fig. 19.256h); spermatheca with capsule club-shaped, stem highly sinuate, narrowly looped posteriorly (Fig. 19.256i).

Boreostiba frigida differs from *B. sibirica* in having the subapical part of the tubus of the median lobe of the aedeagus not sinuate in lateral view, and in having a triangular process on the inner side of the paramerite (Fig. 19.256d).

DISTRIBUTION **Origin**: Holarctic. **CANADA**: LB, QC, YT. **USA**: AK [**NSR**].

COLLECTION AND HABITAT DATA
Habitat: tundra; snowfield, tundra; under rocks, on moss; splash zone rocks, driftwood, creekside rocks. **Collecting period**: VI–VIII. **Collecting method**: pitfall traps.

Additional records. See Annex 1.

COMMENTS When Gusarov (2003a) resurrected *B. frigida* from synonmy with *B. sibirica*, verified records of the former species were known only from QC and YT. Klimaszewski et al. (2011) newly reported this species from LB. We here provide the first verified records of this species from AK (UAM) (see Annex 1).

182. *Boreostiba sibirica* (Mäklin) * (Fig. 19.257a–i)

Homalota (Alianta) sibirica Mäklin, 1880. In *Atheta* (*Oreostiba*): Bernhauer 1907, Brundin 1940, in *Boreostiba*: Lohse et al. 1990, Gusarov 2003a, Smetana 2004, Klimaszewski et al. 2018a.

Diagnosis Externally similar to *A. frigida* except eyes and genae not prominent, pronotum longer, elytra narrower and shorter (Fig. 19.257a). MALE. Tergite VIII with apical margin arcuate on each side of very obtuse medial emargination (Fig. 19.257e); sternite VIII parabolically rounded apically (Fig. 19.257f); median lobe of aedeagus in lateral view with tubus arcuate, strongly produced ventrally, narrow and highly sinuate subapically (Fig. 19.257b); in dorsal view tubus moderately broad and evenly narrowed apicad (Fig. 19.257c), internal sac with two narrowly elongate median structures (Fig. 19.257b,c); paramere with large, broad, condylite, and dark, rounded protuberance on inner side of paramerite (Fig. 19.257d). FEMALE. Tergite VIII with apical margin arcuate on each side of very obtuse medial emargination (Fig. 19.257g); sternite VIII rounded apically (Fig. 19.257h); spermatheca with capsule club-shaped, stem highly sinuate, narrowly looped posteriorly (Fig. 19.257i).

Boreostiba sibirica differs from *B. frigida* in having the subapical part of the tubus of the median lobe of the aedeagus sinuate in lateral view, and in having a dark, rounded protuberance on the inner side of the paramerite (Fig. 19.257d).

DISTRIBUTION **Origin**: Holarctic. **CANADA**: LB, NT, NU, YT. **USA**: AK, NH [?].

COLLECTION AND HABITAT DATA
Habitat: in moss and litter on tundra and under rocks at margin of snowfield; in *Ptasites*, *Salix* litter, moss, *Saxifraga* by stream; in rich *Alnus*, *Ribes*, *Salix* litter on bluffs above river; in *Geum*

glaciale, *Primula tschuktschorum* by stream from snow melt. **Collecting period**: VI–XI (most VI–VII). **Collecting method**: sifting moss and litter, pitfall traps.

Additional records. See Annex 1.

COMMENTS The record from NH listed by Bernhauer (1907) needs verification. Lohse in Lohse et al. 1990, wrongly associated Port Burwell with the province of Quebec. It is located in Nunavut and therefore the QC record was removed from the distribution of this species.

***Parvipennis* species group** (Klimaszewski et al. 2018a)

Diagnosis Median lobe of aedeagus with tubus having short, and narrow bridge.

183. *Boreostiba campbelliana* Lohse * (Fig. 19.258a–h)

Boreostiba campbelliana Lohse, in Lohse et al. 1990. Gusarov 2003a [to *Atheta*], in *Boreostiba*: Sazhnev 2018.

Diagnosis Body subparallel, length 2.4–2.5 mm (Fig. 19.258a); punctation of elytra asperate; color black, elytra red brown, legs paler, tarsi brown; head slightly transverse; eyes small, two-thirds length of postocular region of head in dorsal view; antennomeres V–X increasingly more transverse apically; pronotum slightly transverse 1 1/5 as broad as long, sides weakly arcuate, hind angles rounded; elytra at suture shorter than pronotum along midline of disc, about 1 1/3 as broad as length at sides; abdomen subparallel. MALE. Tergite VIII with apical margin broadly, shallowly emarginate medially, rounded laterally (Fig. 19.258d); sternite VIII parabolic apically (Fig. 19.258e); median lobe of aedeagus strongly bent ventrad in apical half, apical part narrow in lateral view (Fig. 19.258b); in dorsal view bulbus broadly suboval, tubus short, constricted basally and narrowly triangular and rounded apically (Fig. 19.258c). FEMALE. Tergite VIII truncate apically and slightly emarginate medially (Fig. 19.258f); sternite VIII rounded apically, with dense fringe of short setae, antecostal suture strongly sinuate (Fig. 19.258g); spermathecal capsule club-shaped, neck short and arcuate, stem strongly sinuate and narrowly looped posteriad (Fig. 19.258h).

DISTRIBUTION Origin: Holarctic (reported from Russian Far East by Sazhnev 2018). **CANADA**: NU [**NTR**], YT. **USA**: AK, NH [**NSR**].

COLLECTION AND HABITAT DATA
Habitat: in litter under dwarf *Populus* sp. and *Salix* bushes **Collecting period**: VII. **Collecting method**: sifting litter.

We report this species for the first time from NU (**NTR,**) and NH (**NSR**) based on the following: Nunavut, Kitikmeot, Kugluktuk, Coppermine River delta, 5m el., 67.8157, –115.106, wet leaf litter at the water's edge, sifting, 16-.VII.2019, M.Pentinsaari (CBG) 2 specimens; Kugluktuk, Kugluk (Bloody Falls) Territorial Park, 20m el. 67.7431, –115.372, leaf litter under *Salix* bushes, sifting, 20.VII.2019, G. Blagoev, M.Pentinsaari (CBG) 3 specimens; Kitikmeot, Kugluktuk, S Lake, 9km SW of Kugluktuk, 17m el. 67.7726, –115.269, sandy lake shore, leaf litter and plant debris, sifting, 22.VII.2019, M.Pentinsaari (CBG) 1 specimen; Kitikmeot, Kugluktuk, 2m el. 67.8252, –115.081, river bank, flood debris and leaf litter under *Salix* bushes, 27.VII.2019, M.Pentinsaari (CBG) 3 specimens. USA, New Hampshire, Coos Co., Mt. Washington, Alpine Garden, 5300 feet, 12.IX.1987, dwarf Salix, Vaccinium, leatherleaf litter, J.M. Campbell and A. Davies (CNC) 4 males, 5 females, 7 sex undetermined.

Additional records. See Annex 1.

COMMENTS This species is not distinguishable externally from *B. parvipennis*, from which it may separated by the shape of tubus of median lobe of aedeagus in dorsal and lateral view. In lateral view, tubus of *B. campbelliana* is thinner and slightly longer than that of *B. parvipennis* and in dorsal view it is constricted basally while it is evenly triangular in *B. parvipennis*. These two species can be easily misidentified. The DNA barcode divergence between the two species is 0.048 (uncorrected p-distance).

The NH locality at Mt. Washington represents alpine habitat and the southern most disjunct distribution of *Boreostiba campbelliana*. This locality is known to host other boreal species, e.g. *Boreophilia subplana*.

184. *Boreostiba lagunae* Lohse (Fig. 19.259a–d)

Boreostiba lagunae Lohse, in Lohse et al. 1990. Gusarov 2003a [to *Atheta*], in *Boreostiba*: Klimaszewski et al. 2012.

Diagnosis Body subparallel, length 3.5 mm; punctation of elytra slightly asperate; color black, legs brown with femora and middle of tibiae darkened; head slightly transverse; eyes small, each as long as half of postocular region in dorsal view (Fig. 19.259a); antennomere IV–V subquadrate, VI–X transverse and broadening apically; head circular, about as broad as pronotum; eyes shorter than postocular region; pronotum trapezoidal in shape, broadest apically, maximum width as elytra at base; elytra at suture shorter than pronotum; abdomen subparallel, broadest at middle of its length. MALE. Undescribed. FEMALE. Tergite VIII rounded apically with very shallow median emargination (Fig. 19.259b); sternite VIII with apical margin truncate medially, rounded laterally, antecostal suture strongly sinuate (Fig. 19.259c); spermatheca very thin and long, capsule narrowly tubular slightly broader apically with inconspicuous small invagination, neck short and arcuate, stem long and strongly sinuate, half looped posteriad (Fig. 19.259d).

DISTRIBUTION Origin: Nearctic. **CANADA**: YT. **USA**: not recorded.

COLLECTION AND HABITAT DATA
Habitat: in litter on shore in moss. **Collecting period**: VII. **Collecting method**: sifting litter.

185. *Boreostiba parvipennis* (Bernhauer) (Fig. 19.260a–h)

Atheta (Ousipalia) parvipennis Bernhauer, 1907. Lohse et al. 1990 [as *B. hudsonica*], in *Atheta*: Gusarov 2003a, Klimaszewski et al. 2008a, b, in *Boreostiba*: Klimaszewski et al. 2011, 2015a, 2018a, 2020.

Diagnosis Body narrowly elongate, subparallel, length 2.0–2.2 mm; colour dark brown to black, with antennae, elytra except for scutellar region, apex of abdomen and legs reddish-brown (Fig. 19.260a); integument moderately glossy, forebody with fine, dense punctation and meshed microsculpture; head subquadrate, almost as wide as pronotum and elytra, with eyes shorter than genae in dorsal view; antennae moderately robust, antennomere IV subquadrate to slightly elongate, V–X slightly to moderately transverse; pronotum narrowest at base, slightly widening apicad, arcuate laterally, all angles broadly rounded; elytra at suture slightly shorter than pronotum; abdomen slighty widening apicad. MALE. Tergite VIII with apical margin truncate medially, very obtuse laterally, lacking teeth or denticles (Fig. 19.260d); sternite VIII rounded apically (Fig. 19.260e); median lobe of aedeagus in lateral view with tubus straight basally and strongly arched ventrad at middle, apical half acutely triangular, pointed (Fig. 19.260b), in dorsal view tubus elongate, narrowly triangular (Fig. 19.260c). FEMALE. Tergite VIII with apical margin truncate medially, very obtuse laterally (Fig. 19.260f); sternite VIII broadly rounded apically (Fig. 19.260g); spermatheca with capsule club-shaped and stem highly sinuate and narrowly looped posteriorly (Fig. 19.260h).

DISTRIBUTION Origin: Nearctic. **CANADA**: AB, BC, MB [**BOLD**], LB, NF, NT, QC, YT. **USA**: AK, NH.

COLLECTION AND HABITAT DATA
Habitat: found in tamarak open forest, fir and spruce forests; one from detritus on a sandy beach and on rotting mushroom. In AK, found in *Salix, Alnus, Populus balsamea*; litter in forest, mostly spruce, and some aspen; shrub, alder, dwarf birch, grass, clubmoss, litter. **Collecting period**: VI–VIII. **Collecting method**: sifting

organic debris, Berlese funnels, Winkler Litter Processing, pitfall traps, pitfalls w/ rainroofs.

We report this species for the first time from MB (**NPR**) based on **BOLD** records: Canada, Manitoba, Churchill, 26 km SE Churchill, Twin Lakes, 35m el., 58.622, –93.827, open tamarak forest, 19.VIII.2006, C. Rosarius (CBG) 1 specimen; Churchill, 10 km SE Churchill, Farnworth Lake, 15m el., 58.704, –94.054, 16-VIII-2006, Arctic and Boreal Entomology course (CBG) 1 specimen.

Additional records. See Annex 1.

COMMENTS See under *B. campbelliana*.

186. *Boreostiba websteri* Klimaszewski and Langor (Fig. 19.261a–h)

Boreostiba websteri Klimaszewski and Langor, in Klimaszewski et al. 2011, 2018a.

Diagnosis Body narrowly elongate, subparallel, length 2.4–2.5 mm; colour black, with tibiae and tarsi reddish-brown; integument highly glossy particularly on abdomen, forebody with fine, moderately dense punctation and meshed microsculpture (Fig. 19.261a); head round, slightly narrower than pronotum, eyes as long as genae in dorsal view; pronotum subquadrate, widest at middle, slightly narrower than elytra at base; elytra at suture as long as pronotum or slightly longer; abdomen subparallel and broadening apicad to apical third, arcuately tapering to apex; antennae slender, articles 4-7 subquadrate, 5–10 slightly transverse. MALE. Tergite VIII with apical margin broadly, shallowly emarginate medially, oblique laterally (Fig. 19.261d); sternite VIII broadly rounded apically (Fig. 19.261e); median lobe of aedeagus similar to that of *B. parvipennis*, narrow subapically and strongly produced ventrally in lateral view (Fig. 19.261b), in dorsal view bulbus enlarged, oval, and tubus small, triangular (Fig. 19.261c). FEMALE. Tergite VIII with apical margin broadly, shallowly emarginate medially, obtuse laterally (Fig. 19.261f); sternite VIII rounded apically, antecostal suture sinuate (Fig. 19.261g); spermatheca with capsule narrowly tubular, neck elbowed, stem very long, thin, and narrowly looped posteriorly (Fig. 19.261h).

This species differs from *B. parvipennis* in having the body broader and more convex, the elytra as long as or slightly longer than the pronotum, the shape of male tergite VIII is distinctive, and the stem of the spermatheca extremely thin and elongate. The shape of the spermatheca of *B. websteri* is similar to that of *B. lagunae* Lohse, but the loop is much longer and thinner, and capsule broader apically in the former species.

DISTRIBUTION Origin: Nearctic. **CANADA**: LB, NB. **USA**: AK [**NSR**].

COLLECTION AND HABITAT DATA

Habitat: in NB, in northern conifer forest (boreal forest; spruce, balsam fir, some yellow birch), old-growth northern hardwood forest, old mixed forest; adults in wet moss on rocks in or on margins of small shaded spring-fed brooks with mossy margins, where they are sometimes common. In LB, in moss along brook. In AK, found in *Populus tremuloides, P. balsamea,* and *Picea* forest; on snow fields; in alpine *Caltha, Salix* shrubs; alpine *Salix, Valariana* shrubs; and under rocks near snow field. **Collecting period**: VI–IX. **Collecting method**: sifting wet moss, Berlese funnels, hand collecting on snow and under rocks, pitfall traps.

We report this species from AK (**NSR**) for the first time here, based on the following: USA: Alaska, 25 mi. S Delta Junction, 552m el., 63.67342, –145.88287, ±4m old glaciated; *Populus tremuloides, P. balsamea, Picea*, Berlese, 7.IX.2009, G. Henry (UAM) 1 specimen; Haines, Flower Mtn., 1116m el., 59.37788, –136.32814, ±1km on snow fields, 15-19. VII.2010, D.S. Sikes (UAM) 1 specimen; Haines, Flower Mtn. pl.03, 913m el., 59.39175, –136.3121, ±28m alpine *Caltha, Salix* shrubs, hand collected, 16.VII.2010, D.S. Sikes (UAM) 1 specimen; Haines, Flower Mtn. pl.04, 917m el., 59.39156, –136.31377, ±29m alpine *Salix, Valariana* shrubs, pitfall, 15-16.VII.2010, D.S. Sikes (UAM) 2 specimens; same data except, 16-17.VII.2010 (UAM) 1 specimen, 17-18.VII.2010 (UAM) 3 specimens, 18-19. VII.2010 (UAM) 1 specimen; Haines, Flower Mtn. pl.05, 939m el., 59.39038, –136.31345, ±29m alpine *Salix, Valariana*, shrubs, hand collected, 16.VII.2010, D.S. Sikes (UAM) 1 specimen; same data except, pitfall, 17-18.VII.2010, D.S. Sikes (UAM) 1 specimen; S. Chilkat Range, 730m el., 58.42705, −135.15805, ±17m karst talus adjacent to snowfield, 9.VII.2010, D.S. Sikes (UAM) 1 specimen; Thompson Pass, 796m el., 61.13731, −145.74487, ±105m under rocks nr. snow field, hand, 28.VII.2010, D.S. Sikes, A.B. Sikes (UAM) 1 specimen.

Genus *Dinaraea* C.G. Thomson, 1858

(Figs. 19.262–19.265) (Klimaszewski et al. 2013b, 2018a, 2020)

Diagnosis Body subparallel, flattened, on average 3.0 mm long; integument with coarse meshed microsculpture, punctation distinct; head large, robust, subquadrate to slightly elongate, eyes usually shorter than genae in dorsal view, infraorbital carina absent; clypeus long, nearly horizontal, not depressed as in other genera of Athetini; mandibles strong, slightly curved apically, left one with margins smooth, inner margin of right one with small subapical tooth in apical one-third; maxillary palpi with four palpomeres, basal one small and apical one short and needle-shaped; laciniae and galeae moderately elongate, laciniae with fringe of long hairs subapically; labial palpi with three palpomeres; ligula V-shaped with each arm narrow; pronotum usually trapezoidal, widest in apical one-third in most species, disc with pubescence at midline directed posteriad in apical two-thirds, anteriad in basal one-third, and laterad elsewhere; tarsal formula 4-5-5, hypomera visible in lateral view; elytra short, at suture as long as pronotum or only slightly longer; first three visible abdominal tergites with distinct basal impression; apical margin of male tergite VIII without teeth or with 2–4 small teeth; median lobe of aedeagus of a simple form, with elongate structures in internal sac; spermatheca with capsule with subspherical apical part and very short, narrower basal part, stem long, narrow, coiled or looped posteriorly.

A number of *Dinaraea* species are known from the subcortical galleries of other insects but their diet is unknown. The adults and larvae may be either predatory on larvae of insects inhabiting galleries or feed on fungus growing inside galleries.

Key to Species of *Dinaraea*

1. Body shape as illustrated (Fig. 19.265a), length 2.0–2.1 mm; median lobe of aedeagus with two short arcuate sclerites (Fig. 19.265b) .... ***Dinaraea planaris* (Mäklin)**

– Body shape not as above, length 2.2–3.7 mm; median lobe of aedeagus and sclerites of different shape **2**

2. Body broad, moderately convex, length 3.3–3.7 mm, dark brown with elytra paler in most (Fig. 19.262a); male tergite VIII with four large apical teeth (Fig. 19.262c); median lobe of aedeagus as illustrated (Fig. 19.262b); spermatheca with capsule spherical, neck short and angular, stem long and tightly coiled posteriorly (Fig. 19.262g) ...***Dinaraea angustula* (Gyllenhal)**

– Body narrow, flattened, length 2.2–3.0 mm, uniformly dark brown to black (Fig. 19.264a); male tergite VIII without or with four very small apical teeth (Fig. 19.264c); median lobe of aedeagus and spermatheca not as above **4**

4. Antennomeres VII–X subquadrate to slightly transverse (Fig. 19.264a), head broader than pronotum, elytral shoulders strongly angular (Fig. 19.264a); male tergite VIII with 4 vestigial apical teeth (Fig. 19.264c); median lobe of aedeagus with enlarged bulbus and strongly sinuate ventral margin of tubus in lateral view (Fig. 19.264b); spermatheca S-shaped, capsule club-shaped, stem broadly looped (Fig. 19.264g) ....***Dinaraea pacei* Klimaszewski and Langor**

– Antennomeres VII–X strongly transverse (Fig. 19.263a); head narrower than pronotum, elytral shoulders rounded, less angular (Fig. 19.263a); male tergite VIII without apical teeth (Fig. 19.263c); median lobe of aedeagus with bulbus moderately sized, and ventral margin of tubus straight in lateral view (Fig. 19.263b); spermatheca with capsule club-shaped, stem tightly coiled (Fig. 19.263g) ....*Dinaraea borealis* Lohse

187. *Dinaraea angustula* (Gyllenhal) † (Fig. 19.262a–g)

Aleochara angustula Gyllenhal, 1810. Fauvel 1889 [in *Thectura*]; in *Dinaraea*: Muona 1984, Smetana 2004, Klimaszewski et al. 2007a, Webster et al. 2009, Majka and Klimaszewski 2008b, Klimaszewski et al. 2011, 2012, 2013a,b, 2016d, 2018a, 2020.

Diagnosis Body length 3.3–3.7 mm; colour dark brown with legs, antennae at least basally, labial palpi and most of elytra yellowish- or reddish-brown, pronotum and abdomen lighter than head, sometimes entire body appears dark brown (Fig. 19.262a); head, pronotum and elytra slightly glossy, with dense microsculpture; abdominal microsculpture less dense and integument more glossy than that of pronotum and elytra; head almost as broad as pronotum, eyes slightly shorter than prominent genae in dorsal view; antennomeres VII–X moderately transverse; pronotum broadest at middle, slightly transverse, slightly longer than elytra at suture; elytra transverse, truncate posteriorly; abdomen subparallel. MALE. Tergite VIII with four small ridge-like teeth protruding slightly beyond apical margin and additional smaller subapical ridges on disc (Fig. 19.262c); sternite VIII parabolically rounded from base to more narrowly rounded apex (Fig. 19.262d); tubus of median lobe of aedeagus with ventral margin straight, apico-dorsal part curved, slightly bent ventrad at apex in lateral view (Fig. 19.262b). FEMALE. Tergite VIII arcuate apically (Fig. 19.262e); sternite VIII broadly rounded apically, antecostal suture arcuate (Fig. 19.262f); spermatheca with short pear-shaped asymmetrical capsule with small apical invagination, stem long, rectangularly bent at base of capsule, straight to tight posterior coil, base slightly swollen (Fig. 19.262g).

DISTRIBUTION Origin: Palearctic, adventive in Canada and the USA. **CANADA**: AB, BC, LB, NB, NF, NS, ON, PE, QC, SK, YT. **USA**: CA, NY, PA.

COLLECTION AND HABITAT DATA

Habitat: in variety of habitats such as open dry area with *Purshia* association (BC), agricultural fields and disturbed urban meadows, cattail and sedge marshes, mixed forests, coastal dunes, and lakeshores. Adults in compost (decaying corncobs and cornhusks and vegetables), soil and organic debris, amongst vegetation on a coastal sand dune, in litter in a cattail marsh, in leaf litter along a vernal pond, and in drift material along a lakeshore. **Collecting period**: IV–IX. **Collecting method**: sifting organic litter, pitfall traps.

188. *Dinaraea borealis* Lohse (Fig. 19.263a–g)

Dinaraea borealis Lohse, in Lohse et al. 1990. Klimaszewski et al. 2013b, 2018a.

Diagnosis Body length 2.8–3.0 mm; colour variable, either entirely black with appendages and part of elytra brown or reddish, or dark brown with legs, antennae at least basally, and labial palpi appearing reddish-brown (Fig. 19.263a); head, pronotum and elytra slightly glossy, the latter more so, with dense microsculpture; abdominal microsculpture less dense and integument more glossy than that of pronotum and elytra; head moderately large, almost as broad as pronotum, eyes slightly shorter than genae in dorsal view; antennomeres VII–X moderately transverse; pronotum broadest in apical third, slightly transverse, usually as long as elytra at suture; elytra transverse, truncate posteriorly; abdomen arcuate laterally. MALE. Apical margin of tergite VIII obliquely produced laterally, very slightly, broadly emarginate medially, without teeth (Fig. 19.263c); sternite VIII parabolically rounded from base to apex (Fig. 19.263d); median lobe of aedeagus in lateral view with tubus curved slightly ventrad at base, long and straight to thin, sharp, slightly curved apex (Fig. 19.263b). FEMALE. Tergite VIII arcuate apically (Fig. 19.263e); sternite VIII rounded apically, antecostal suture distinctly sinuate (Fig. 19.263f); spermatheca elongate with pear-shaped capsule, apical invagination deep and narrow, stem moderately long, with double tight coil posteriorly, base highly swollen (Fig. 19.263g).

DISTRIBUTION **Origin**: Nearctic. **CANADA**: NB, NT [**BOLD**], ON, QC, SK [**BOLD**], YT [**BOLD**]. **USA**: AK [**NSR**].

COLLECTION AND HABITAT DATA
Habitat: forested areas in NT and YT, in old-growth eastern white cedar forests, old growth white spruce forest, mixed forest with alder, and a black spruce forest; adults from under bark of large fallen spruce logs, in leaf litter under alders near a small stream, and in moss (several sites). In AK, found in *Salix, Alnus, Abies balsamea* forest. **Collecting period**: IV–VIII. **Collecting method**: sifting moss and leaf litter, Lindgren funnel traps, Malaise traps.

We report this species for the first time from NT (**NTR**), YT (**NTR**), SK (**NPR**) based on **BOLD** records, and from AK (**NSR**) based on UAM specimens: Canada, Northwest Territories, Nahanni National Park Reserve, Nailicho, Virginia Falls, 578m el., 61.606, –125.758, Malaise trap, 27.VI-04.VII.2014, Parks Canada staff (CBG) 1 specimen. Yukon, Kluane National Park and Reserve, Auriol Trail, 801m el., 60.714, –137.432, spruce forest, Malaise trap, 15-30. VII.2014, S. Davidson (CBG) 1 specimen. Saskatchewan, Whitewood, Whitewood School, 596m el., 50.328, –102.258, Malaise trap, 20.IV-08.V.2015, John Pollock (CBG) 1 specimen. USA, Alaska, mile 217.2 Richardson Hwy, Castner Glacier, 786m el., 63.40134, –145.71895, ±5m Low *Salix, Alnus, A. balsamea*, Berlese, 7.IX.2009, G. Henry (UAM) 1 specimen.

189. *Dinaraea pacei* Klimaszewski and Langor (Fig. 19.264a–g)

Dinaraea pacei Klimaszewski and Langor, in Klimaszewski et al. 2011. Klimaszewski et al. 2013b, 2015a, 2016d, 2018a, 2020.

Diagnosis Body length 2.2–2.5 mm; colour dark brown with legs, antennae (at least basally), palpi and elytra slightly paler and appearing reddish-brown, pronotum and abdomen lighter than head, sometimes entire body appears brown (Fig. 19.264a); head, pronotum and elytra slightly glossy, with dense microsculpture; abdominal microsculpture less dense and integument more glossy than pronotum and elytra; head about as large and broad as pronotum, eyes slightly shorter than genae in dorsal view; antennomeres VII–X moderately transverse; pronotum broadest in apical third, slightly transverse, slightly shorter than elytra at suture; elytra transverse, truncate posteriorly; abdomen subparallel. MALE. Apical margin of tergite VIII oblique laterally, obtusely angulate and slightly projecting on each side of broadly truncate apex, with two low ridge-like elevations medially, barely protruding at margin (Fig. 19.264c); apical margin of sternite VIII arcuate, slightly truncate medially (Fig. 19.264d); median lobe of aedeagus in lateral view with tubus short, sinuate basally, broadly arched ventrad at middle, apical half very narrow, slightly curled dorsad at apex (Fig. 19.264b); internal sac with elongate, sinuate structures (Fig. 19.264b). FEMALE. Apical margin of tergite VIII arcuate, slightly emarginate medially (Fig. 19.264e), sternite VIII broadly rounded apically, antecostal suture slightly sinuate (Fig. 19.264f); spermatheca with short pear-shaped capsule with small apical invagination, stem short, curved, with single broad loop posteriorly, slightly swollen at base (Fig. 19.264g).

DISTRIBUTION **Origin**: Nearctic. **CANADA**: BC, LB, NB, NF, QC, SK, YT. **USA**: AK.

COLLECTION AND HABITAT DATA
Habitat: in NF and LB, from pitfall traps and flight intercept traps in various coniferous forest types; one specimen from under bark of red pine; in NB, adults under bark of large fallen spruce in old-growth eastern white cedar swamp, under tight bark of American elm in a silver maple forest, in fleshy polypore fungi at base of dead standing *Populus* sp. in a wet alder swamp, and in

a group of *Pholiota* sp. at base dead standing *Populus* sp. in a mixed forest. Adults also from Lindgren funnel traps in old-growth white spruce and balsam fir forest, an old mixed forest, and a rich Appalachian hardwood forest with some conifers. In BC, from emergence traps attached to the trunks of lodgepole pine (*Pinus contorta* Dougl. ex Loud. *latifolia* Engelm.) infested by mountain pine beetle (*Dendroctonus ponderosae* Hopkins. **Collecting period**: III–IX. **Collecting method**: hand searching under bark, sifting fungi, Lindgren funnel traps, pitfall traps, flight intercept traps (Malaise trap), emergence traps attached to tree trunks.

COMMENTS Klimaszewski et al. (2015a, 2016d) listed this species from AB, ON, PE, but we could not find voucher specimens representing these records and therefore we have removed these provinces from the official distribution list for this species.

190. *Dinaraea planaris* (Mäklin) (Fig. 19.265a–d)

Homalota planaris Mäklin, 1852. In *Dinaraea*: Lohse and Smetana 1985.

Neotype (male): Canada, Yukon, Dempster Hwy. mi 42, N. Klondike R., 18.VII.1978, 3300 feet, A. Smetana and J.M. Campbell (CNC). Present designation, labelled by Klimaszewski 2020.

Diagnosis Body length 2.0–2.1 mm; colour dark brown almost black, elytra brownish medially and dark brown laterally; antennae and legs dark brown (Fig. 19.265a); forebody moderately glossy with distinct microsculpture; abdominal microsculpture less dense and integument more glossy than on forebody; head round, about as large and broad as pronotum, eyes about as long as genae in dorsal view; antennomeres VII–X moderately transverse; pronotum broadest in apical third, slightly transverse, shorter than elytra at suture; elytra transverse, truncate posteriorly; abdomen subparallel. MALE. Apical margin of tergite VIII truncate with two sharp, minute sublateral teeth (Fig. 19.265c); sternite VIII rounded apically, antecostal suture slightly sinuate, widely separated from basal margin (Fig. 19.265d); median lobe of aedeagus in lateral view with tubus straight, apical part narrow, slightly curved ventrad at apex (Fig. 19.265b); in dorsal view tubus subparallel; internal sac with two apical sclerites. FEMALE. Not described.

DISTRIBUTION Origin: Nearctic. **CANADA**: YT. **USA**: AK.

COLLECTION AND HABITAT DATA
Habitat: not recorded. **Collecting period**: VII. **Collecting method**: not recorded.

COMMENTS Lohse and Smetana (1985) treated *Homalota planaris* in their revision of type species of the tribes Athetini and Oxypodini described by Mannerheim and Mäklin from North America. The authors transferred it to the genus *Dinaraea*, redescribed it, and illustrated the median lobe of the aedeagus in dorsal view (Fig. 16, *ibid.*). They also designated a lectotype specimen from Sitka Is., type No. 2247 housed in ZMH. We tried to borrow the lectotype from the collection of ZMH, but the specimen and two other cotypes could not be located in the collection nor in the loan system. The type specimens are presumably lost and therefore we have designated a male specimen as the neotype, which was compared with the lectotype by Lohse in 1983 (CNC).

Genus *Dochmonota* C.G. Thomson, 1859

(Figs. 19.266–19.267) (Gusarov 2003a, Klimaszewski et al. 2011, 2018a, 2020)

Diagnosis Body elongate, moderately broad; head subquadrate, eyes large, much longer than genae in dorsal view; antennomere II as long as III, IV subquadrate, V–X moderately transverse, last antennomere as long as three preceding ones combined; mandibles broad, medial tooth of right mandible inconspicuous; ligula with wide base and split into two lobes in apical one-third; pronotum transverse and widest at middle, with microsetae directed posteriad along midline of disc and slightly obliquely posteriad elsewhere, pronotal macrosetae moderately long, pronotal hypomera fully visible in lateral view; mesocoxae broadly separated; mesolateral macroseta of middle tibia inconspicuous, shorter than tibial width; hind tarsi with article I slightly longer than II, each tarsus with one empodial seta; first three visible abdominal tergites each with moderately transverse basal impression, tergite VII as long as tergite VI; punctation on tergites VI–VII sparser than on tergites III–V, tergite VII with white palisade fringe; median lobe of aedeagus with apical process straight or slightly bent ventrally in lateral view; spermatheca with capsule mushroom-shaped, stem thick and tightly coiled posteriorly.

Key to Nearctic Species of *Dochmonota*

1. Head about as broad as pronotum (Fig. 19.266a); body narrow with elytra at base only slightly broader than pronotum (Fig. 19.266a); male sternite VIII notched medially at apex (Fig. 19.266e); ventral margin of tubus of median lobe of aedeagus straight with base slightly sinuate in lateral view (Fig. 19.266b); spermathecal capsule with broad apical invagination, stem narrowing to posterior coil (Fig. 19.266h) .... ***Dochmonota langori*** **Klimaszewski and Larson**

– Head distinctly narrower than pronotum (Fig. 19.267a); body broad with elytra at base distinctly broader than pronotum (Fig. 19.267a); male sternite VIII with apex entire (Fig. 19.267e); ventral margin of tubus of median lobe of aedagus shaped differently (Fig. 19.267b); spermathecal capsule with long, narrow apical invagination, stem thick through posterior coil (Fig. 19.267h) . ... ***Dochmonota rudiventris*** **(Eppelsheim)**

191. *Dochmonota langori* Klimaszewski and Larson (Fig. 19.266a–h)

Dochmonota langori Klimaszewski and Larson, in Klimaszewski et al. 2016d.

Diagnosis Body narrowly subparallel (Fig. 19.266a), length 2.2–2.5 mm, uniformly black; head, pronotum and elytra finely and densely punctate, punctures small; pubescence dense; integument moderately glossy, more so on abdomen, with meshed microsculpture; antenna with articles V–X subquadrate to slightly transverse; head about as wide as pronotum; pronotum transverse, slightly narrower than elytra at base, with pubescence directed obliquely laterad from median line of disc and in basal part of median line directed anteriad and laterad, base of disc with small oval impression; elytra at suture about as long as pronotum and slightly wider at base than pronotum; abdomen subparallel. MALE. Tergite VIII truncate apically (Fig. 19.266d); sternite VIII elongate, apical margin parabolic except with a small notch medially (Fig. 19.266e); median lobe of aedeagus in lateral view with tubus sinuate basally and straight to angulate apex (Fig. 19.266b); in dorsal view bulbus large and broad, tubus narrowly triangular (Fig. 19.266c), internal sac structures as illustrated (Fig. 19.266b,c). FEMALE. Tergite VIII very broadly arcuate apically (Fig. 19.266f); sternite VIII with apical margin broadly rounded, arcuate medially (Fig. 19.266g); spermatheca with pitcher-shaped capsule bearing broad and deep apical invagination, stem broad, gradually narrowing to tight coil posteriorly (Fig. 19.266h).

DISTRIBUTION Origin: Nearctic. **CANADA**: MB [**NPR**], SK. **USA**: not recorded.

COLLECTION AND HABITAT DATA

Habitat: adults were found in wind-drift on a lake. **Collecting period**: V–VIII. **Collecting method**: sifting wrack on lakeshore beach.

We report this species from MB (**NPR**) for the first time here, based on the following: Canada, Manitoba, 11 km S Churchill, Churchill river weir, 58.682 –94.152, 20.VIII.2006, C. Rosarius (CBG), 1 specimen; 16 km SW Churchill, River Pump House, 58.626 –94.228, Artic and Boreal Entomology course, 11.VIII.2006 (CBG), 1 specimen; 2 km S Churchill, Beech Bay, 58.747 –

94.134, Artic and Boreal Entomology course, 10. VIII.2006 (CBG), 1 specimen.

192. *Dochmonota rudiventris* (Eppelsheim) ‡ (Fig. 19.267a–h)

Homalota rudiventris Eppelsheim, 1886. Casey, 1910 [as *Dimetrota revoluta* and *Datomicra vaciva*], in *Dochmonota*: Muona 1984, Gusarov 2003a [synonyms], Webster et al. 2009, Majka and Klimaszewski 2008b, Klimaszewski et al. 2011, 2013a, 2016d, 2018a, 2020.

Diagnosis Body narrowly elongate, subparallel, length 2.2–2.5 mm; colour dark brown to black, often with antennae, legs and posterior elytra reddish-brown (Fig. 19.267a), moderately glossy, punctation fine, pubescence brownish, microsculpture present on forebody; head round, distinctly narrower than pronotum, eyes moderately large; antennomeres IV–X subquadrate to slightly transverse; pronotum slightly transverse, widest at middle; elytra distinctly wider than pronotum, at suture as long as pronotum; abdomen broadly arcuate laterally. MALE. Tergite VIII truncate apically (Fig. 19.267d); sternite VIII with apical margin subtriangularly produced, obtusely angulate medially (Fig. 19.267e); in lateral view median lobe of aedeagus with large bulbus and short tubus, tubus sinuate basally and very narrow apically, crista apicalis enlarged, bulbous (Fig. 19.267b), in dorsal view bulbus elongate, narrowly oval, tubus short and wide (Fig. 19.267c). FEMALE. Tergite VIII truncate apically, with shallow medial emargination (Fig. 19.267f); apical margin of sternite VIII oblique laterally, with shallow emargination medially (Fig. 19.267g); spermatheca with small mushroom-shaped capsule, apical invagination deep and narrow, stem thick and tightly coiled posteriorly (Fig. 19.267h).

DISTRIBUTION Origin: Possibly Holarctic. **CANADA**: BC, NB, NF, NT, ON [**BOLD**], QC, YT. **USA**: ID, MA.

COLLECTION AND HABITAT DATA
Habitat: mixed forests and silver maple swamps (flood plain forests) in moist leaf litter near the margins of vernal ponds and in flood debris along rivers, in moss in swamp; similar to habitats reported from Europe. **Collecting period**: IV–VIII. **Collecting method**: sifting leaf litter.

We report this species from ON (**NPR**) for the first time here, based on BOLD records: Canada, Ontario, Guelph, Hanlon Preservation Park, 335m el., 43.51, –80.221, mixed forest, UV light sheet, 30.VI.2018, M. Pentinsaari (CBG) 1 specimen; Guelph, University of Guelph Arboretum, 345m el., 43.537, –80.213, wet leaf litter at pond margin, sifting, 12.VI.2019, M. Pentinsaari (CBG) 3 specimens.

COMMENTS In view of its wide distribution in the Palearctic region (Smetana 2004), its wide distribution in Canada (BC, NF, NB, ON, QC, west to NT and YT), and its occurrence in several natural wetland habitats, which are not typical of adventive species, it is likely that this is a Holarctic species. This species was listed from Manitoba by Klimaszewski et al. (2018b) in error as no voucher data were presented.

Genus *Earota* Mulsant and Rey, 1873b

(Fig. 19.268a–h) (Gusarov 2002, Klimaszewski et al. 2018a, 2020)

Diagnosis Body elongate, moderately broad, length 3.0–4.5 mm (average 3.0 mm); head subquadrate, eyes large, longer than genae in dorsal view; antennomere II as long as III, IV subquadrate, V–X moderately transverse, last antennomere as long as three preceding ones combined; mandibles broad, medial tooth of right mandible inconspicuous; ligula with wide base and split into two lobes in apical one-third; infraorbital carina present; pronotum transverse and widest at middle, with pubescence short and dense, appearing silky, directed posteriad along midline of disc and slightly obliquely posteriad elsewhere, pubescence, pronotal hypomera fully visible in lateral view; mesocoxae broadly separated; mesolateral macroseta of middle tibia inconspicuous, shorter than tibial width; hind tarsus with article I slightly longer than II, each tarsus with one empodial seta; first three visible abdominal tergites with moderately transverse basal impression, tergite VII as long as tergite VI; punctation on tergites VI–VII sparser than on tergites III–V, tergite VII with white palisade fringe; median lobe of aedeagus with apical process straight or slightly bent ventrally in lateral view; spermatheca with dorso-ventrally compressed capsule, stem sinuate and highly swollen posteriad. For illustrations of mouthparts see Gusarov (2002).

193. ***Earota dentata*** **(Bernhauer)** (Fig. 19.268a–h)

Atheta dentata Bernhauer, 1906. Casey 1906 [as *Macroterma alutacea* and *M. borealis*], Bernhauer 1907 [in *Atheta* (*Earota*)], Casey 1910 [as *Atheta* (*Macroterma*) *iowensis*], in *Earota*: Gusarov 2002, Klimaszewski and Winchester 2002, Klimaszewski et al. 2005, 2007b, 2008a,b, Majka and Klimaszewski 2008b, Webster et al. 2009, Klimaszewski et al. 2011, 2016d, 2018a, 2020.

Diagnosis Body narrowly elongate, subparallel, slightly flattened, length 3.0–4.5 mm; colour dark brown, elytra except for scutellar region, legs, and antennomeres I–II rusty-brown (Fig. 19.268a); integument of forebody moderately glossy, with fine, sparse punctation, short pubescence and meshed microsculpture; head subquadrate, distinctly narrower than pronotum; antennae slender and long, antennomere IV subquadrate, V–X moderately transverse, last antennomere as long as three preceding ones combined; pronotum impressed medially, moderately transverse, widest at middle; elytra wider than pronotum, at suture slightly longer than pronotum, with no apparent emargination near angular posterior angles; abdomen subparallel, tergites glossy and with fine punctation and microsculpture. MALE. Tergite VII with short medial carina before posterior margin; apical margin of tergite VIII obtusely angulate, with shallow notch on each side halfway to apex and a shallow angular notch at apex (Fig. 19.268d); sternite VIII parabolically rounded from base to apex (Fig. 19.268e); median lobe of aedeagus in lateral view with tubus evenly arcuately curved ventrad, narrow apically, apex narrowly rounded (Fig. 19.268b), in dorsal view bulbus oval and tubus triangular apically (Fig. 19.268c). FEMALE. Tergite VIII arcuate apically (Fig. 19.268f); apical margin of sternite VIII semicircularly rounded (Fig. 19.268g); spermatheca with capsule broad, flattened, stem short, looped and distinctly swollen posteriad (Fig. 19.268h).

DISTRIBUTION Origin: Nearctic. **CANADA**: AB, BC, MB, NB, NF, NS, ON, QC, SK, YT. **USA**: AK, AL, AZ, CA, CO, DC, IA, ID, IL, IN, KS, MA, ME, MD, MT, NC, NH, NJ, NM, NV, NY, OH, OR, PA, VA, WA.

COLLECTION AND HABITAT DATA
Habitat: occurs in various habitats; in NB, adults from leaf litter near brook margin in red maple swamp, moss and leaf litter along small shaded brooks in mixed forests, under sea wrack on sea beach, among decaying corncobs and cornhusks (common); flying adults collected with butterfly net late in afternoon (16:00–18:00 h) early in spring (April), and some from pitfall traps in red spruce forest. Elsewhere, it has been reported from leaf litter, river debris, and pocket gopher (*Geomys* sp.) burrows. In AK, found in *Betula* and *Populus* forest. **Collecting period**: IV–IX. **Collecting method**: sifting organic debris, pitfall traps, flight intercept traps.

Additional records. See Annex 1.

COMMENTS This species was reported from AZ by Gusarov (2002) but this record was overlooked by recent publications (e.g., Klimaszewski et al. 2018a). The record from MI in Newton (2019) is not supported by published voucher data as far as we are aware.

Genus *Hydrosmecta* C.G. Thomson, 1858

(Figs. 19.269–19.270) (Ashe 2001, Lohse et al. 1990, Klimaszewski et al. 2017, 2018a)

Diagnosis Body slender, narrow, parallel-sided, depressed, length 1.4–3.0 mm; integument light to dark brown; antennae long, reaching pronotal base or slightly shorter, antennomeres VI–X elongate or may be subquadrate to slightly transverse in some; head subquadrate, eyes moderate to large, usualy prominent, genae long; pronotum almost wide as elytra, subquadrate to transverse, pubescence on disc directed latero-cephalad and latero-posteriad from median line; elytra flattened, elongate, at suture as long as or distinctly longer than pronotum, shoulders angular, pubescence on disc directed straight or obliquely latero-posteriad; abdomen subparallel; mesocoxae contiguous; processes of meso- and metaventrite short; tergite VII distinctly longer than VI; median lobe of aedeagus in lateral view with tubus simple or with narrow basal process, crista apicalis usually prominent, fin-shaped, athetine bridge narrow, bulbus large and crista apicallis situated on triangularly shaped projection of bulbus; spermatheca either S-shaped or irregularly twisted, capsule spherical, stem elongate and coiled or twisted posteriorly.

COMMENTS Riparian species associated with gravel and sand along stream margins.

Key to Species of *Hydrosmecta*

1. Antennomeres V–X distinctly elongate (Fig. 19.269a); elytra at suture distinctly longer than pronotum along midline (Fig. 19.269a); median lobe of aedeagus and spermatheca as illustrated (Fig. 19.269b,c,h,i)............................. ....***Hydrosmecta pseudodiosica* Lohse**

– Antennomeres V–X distinctly transverse (Fig. 19.270a); elytra at suture as long as pronotum along midline (Fig. 19.270a); median lobe of aedeagus and spermatheca as illustrated (Fig. 19.270b,c,h)... .....***Hydrosmecta yukonensis* Klimaszewski and Godin, sp. n.**

194. *Hydrosmecta pseudodiosica* Lohse (Fig. 19.269a–i)

Hydrosmecta pseudodiosica Lohse, in Lohse et al. 1990. Majka and Klimaszewski 2008b, Webster et al. 2012, Klimaszewski et al. 2017, 2018a.

Diagnosis Body length 2.5–2.7 mm; colour brown to piceous, elytra and legs dark brown (Fig. 19.269a); forebody dull, with minute microsculpture of rounded meshes; head subquadrate, at eye level about as wide as at level of genae; eyes about as long as genae in dorsal view; antennae slender, antennomeres II and III subequal in length, following ones elongate but decreasing in length toward apex; glossa thin and long, Y-shaped, deeply split apically; each mandible with subapical tooth; pronotum transverse, 1 1/4 times as broad as long and 1/6 broader than head, broadest near apex, sides converging basally to obtuse posterior angles; elytra at suture 1 1/3 as long as pronotum along midline of disc, about 1 1/5 as broad as pronotum; abdomen subparallel, tergites becoming longer toward apex. MALE. Tergite VIII with apical margin arcuate, slightly truncate medially (Fig. 19.269d); sternite VIII arcuate apically (Fig. 19.269e); median lobe of aedeagus in lateral view with tubus narrow and apical part projecting slightly ventrad, crista apicalis prominent, fin-shaped (Fig. 19.269b); in dorsal view bulbus narrowly oval, evenly coalescent with short triangular tubus (Fig. 19.269c); internal sac with two elongate structures in bulbus (Fig. 19.269b,c). FEMALE. Tergite VIII truncate apically (Fig. 19.269f); apical margin of sternite VIII obtuse laterally, with broadly rounded median emargination (Fig. 19.269g); spermatheca with small spherical capsule bearing short and moderately wide apical invagination, stem long, irregularly twisted, swollen and recurved posteriorly (Fig. 19.269h,i).

DISTRIBUTION Origin: Nearctic. **CANADA**: NB, ON, YT. **USA**: AK [**NSR**].

COLLECTION AND HABITAT DATA

Habitat: riparian habitats such as sandy and cobblestone shores of brooks and rivers; in NB, adults under small cobblestones and gravel set in sand on a partially shaded cobblestone bar near the outflow of a brook into a clear rocky river; edges of running water in YT. In AK, found in gravel creek shore, under rocks. **Collecting period**: V, VI. **Collecting method**: hand collecting or aspirating specimens from near and under cobblestones at edges of streams.

We report this species from AK (**NSR**) for the first time here, based on the following: USA, Alaska, Dalton Hwy Mi 135, Prospect Crk

265m el., 66.79376, −150.73164, ±24m gravel creek shore, under rocks, 31.V.2008, D.S. Sikes (UAM) 3 specimens; Kenai Pen., Anchor River at Hwy. 1, 450 feet, 4.VI.1978, A. Smetana and E. Becker (CNC) 1 female.

195. *Hydrosmecta yukonensis* Klimaszewski and Godin, sp. n. (Fig. 19.270a–h)

urn:lsid:zoobank.org:act:B412C66A-53ED-421F-9567-284465F138AB

Holotype (male): Canada, Yukon, Takhini Salt flats, 17.VI-1.VII.2014, alkaline clay, 60.855, −135.717, 663m, pifall trap, *Salicornia* habitat, coll. B. Godin and D. Horwood (CNC). **Paratypes**: Canada, Yukon, Takhini Salt flats, 17.VI-1.VII.2014, alkaline clay, 60.855, −135.717, 663m, pifall trap, *Salicornia* habitat, coll. B. Godin and D. Horwood (BGC, LFC) 4 males, 2 females; Takhini Salt flats, 1-15. VIII.2014, alkaline clay, 60.855, −135.717, 663m, pifall trap, *Salicornia* habitat, coll. B. Godin and D. Horwood (BGC, LFC) 1 male, 2 females; Takhini Salt flats, 2-16. VI.2014, alkaline clay, 60.855, −135.717, 663m, pifall trap, *Salicornia* habitat, coll. B. Godin and D. Horwood (CNC, BGC, LFC) 5 males, 4 females; Takhini Salt flats, 14-28.V.2014, alkaline clay, 60.855, −135.717, 663m, pifall trap, *Salicornia* habitat, coll. B. Godin and D. Horwood (CNC, LFC) 2 males, 1 female.

Etymology The specific name is derived from Yukon Territory, where the type series was captured.

Diagnosis Body length 2.0–2.3 mm; colour dark brown to piceous, elytra slightly paler with reddish tinge, appendages yellow-brown (Fig. 19.270a); head, pronotum and abdomen strongly glossy and elytra less so; forebody with minute microsculpture of rounded meshes; head subquadrate, at eye level and slightly less wide than at level of genae; eyes about as long as genae in dorsal view; antennae slender, antennomeres II slightly longer than III, following ones transverse and increasing so toward apex; glossa thin and long, Y-shaped, deeply split apically; each mandible with subapical tooth; pronotum subquadrate and as long and broad as head, broadest in apical third, sides converging basally to obtuse posterior angles; elytra at suture as long as pronotum along midline of disc, slightly broader than pronotum; abdomen subparallel, tergites becoming longer toward apex. MALE. Tergite VIII with apical margin broadly truncate, obtusely angulate sublaterally (Fig. 19.270d); sternite VIII with apical margin obliquely produced, broadly rounded medially (Fig. 19.270e); median lobe of aedeagus in lateral view with ventral margin of tubus slightly sinuate to triangular apical part, directed slightly ventrad at apex (Fig. 19.270b); in dorsal view bulbus narrowly oval, evenly coalescent with short triangular tubus (Fig. 19.270c); internal sac without distinct structures in bulbus (Fig. 19.270b,c). FEMALE. Tergite VIII arcuate apically (Fig. 19.270f); sternite VIII with apical margin obliquely produced on each side of broad, shallow medial emargination (Fig. 19.270g); spermatheca S-shaped, with tubular sinuate capsule bearing short apical invagination, stem with single tight coil, tightly twisted posteriorly (Fig. 19.270h).

This species is distinguishable from all its congeners by having a very narrow, minute body, distinctive shape of median lobe of aedeagus and spermatheca, and emarginate apex of female sternite VIII.

DISTRIBUTION Origin: Nearctic. **CANADA**: **YT**. **USA**: not recorded.

COLLECTION AND HABITAT DATA
Habitat: Salt flat, clay, very alkaline habitat dominated by *Salicornia borealis* Wolff and Jeffries, a halophyte plant. **Collecting period**: V–VIII. **Collecting method**: pitfall traps.

Genus *Liogluta* C.G. Thomson, 1858

(Figs. 19.271–19.281)

(Klimaszewski et al. 2016a, 2018a, 2020)

Diagnosis Body narrowly subparallel, moderately flattened, length 2.8–5.4 mm; colour uniformly dark brown or reddish-brown with head and posterior part of abdomen dark brown to almost black; forebody with coarse meshed microsculpture, elytra often granulose; head with incomplete infraorbital carinae, genae moderately long to long; ligula apically split into two lobes; antennae long with antennomeres VI–X subquadrate, slightly transverse or rarely slightly elongate; pronotum with hypomera fully visible in lateral view; pubescence on disc of pronotum directed posteriad on midline and lateroposteriad to laterad elsewhere; elytra and abdomen wider than head and pronotum; mesocoxae narrowly separated, process of metaventrite short and acute; legs long, three basal tarsomeres of hind tarsi highly elongate in most species, tarsal formula 5-5-5; apical margin of male tergite VIII often with broad, variably-shaped rectangular projection, with edge entire or bearing crenulation or variably shaped structures, with two lateral teeth and sometimes with additional median tooth; integument of disc often with broadly distributed asperate punctation near apex; female sternite VIII with apical margin rounded or medially emarginate, antecostal suture arcuate, or slightly to distinctly sinuate; median lobe of aedeagus simple with apical part variably shaped in lateral view; spermatheca with capsule narrowly club-shaped or tubular, with apical invagination moderate to deep, stem long, sinuate, and looped or twisted posteriorly.

Key to Species of *Liogluta*

1. Eyes large, each about subequal in length to postocular region of head in dorsal view (Figs. 19.271a, 19.272a, 19.275a) **4**

– Eyes small, each shorter than postocular region of head in dorsal view (Figs. 19.274a, 19.278a, 19.281a).. **2**

2. Antennae dark brown to black, thick, antennomeres VIII–X slightly elongate, pronotum transverse (Fig. 19.281a); median lobe of aedeagus with large crista apicalis of bulbus in lateral view (Fig. 19.281b); spermathecal stem with broad posterior loop and apical coils not tight (Fig. 19.281g) .... ***Liogluta wickhami* (Casey)**

– Antennae yellow-brown, not thick, antennomeres VIII–X subquadrate to slightly transverse, pronotum subquadrate(Figs. 19.274a, 19.278a); median lobe of aedeagus with small crista apicalis of bulbus in lateral view (Figs. 19.274b, 19.278b); spermathecal stem with narrow posterior loop with apical coils not tight (Figs. 19.274g, 19.278g) **3**

3. Body not uniform in colour, head, pronotum and apical part of abdomen dark brown and remainder of body yellow-brown (Fig. 19.274a); apical margin of male tergite VIII truncate, crenulate and often somewhat arcuate medially (Fig. 19.274c); apical margin of female sternite VIII slightly emarginate (Fig. 19.274f); spermatheca with simple sinuate loop posteriorly (Fig. 19.274g)..
***Liogluta intermedia* Klimaszewski and Langor**

– Body uniformly red-brown (Fig. 19.278a); apical margin of male tergite VIII truncate apically and not crenulate (Fig. 19.278c); apical margin of female sternite VIII truncate medially (Fig. 19.278f); spermatheca with twisted coil posteriorly (Fig. 19.278g).......................................
...............................***Liogluta terminalis* (Casey)**

4. Body strongly glossy; apical margin of male tergite VIII arcuate or slightly broadly rectangularly produced and sinuate medially but not crenulate (Figs. 19.275d, 19.276c, 19.277c) **5**

– Body moderately glossy; apical margin of male tergite VIII truncate/crenulate or triangularly produced and angulate medially **7**

5. Antennomeres VII–X subquadrate to slightly transverse, body usually uniformly dark brown to black (Fig. 19.275a); apical margin of male tergite VIII arcuate (Fig. 19.275d); remaining terminalia, median lobe of aedeagus and spermatheca as illustrated (Fig. 19.275b,c,e–i) .... ***Liogluta nigropolita* (Bernhauer)**

– Antennomeres VII–X strongly transverse, body dark brown to black with brownish elytra; other characters not as above **6**

6. Body broad, elytra light yellow-brown, contrasting with darker color of pronotum and abdomen (Fig. 19.276a); male tergite VIII angularly produced apically and slightly sinauate (Fig. 19.276c); tubus of median lobe of aedeagus with apex not produced ventrad in lateral view (Fig. 19.276b); spermatheca not found ....
.... ***Liogluta nitens* (Mäklin)**

– Body narrow, elytra slightly paler than pronotum and abdomen and not contrasting with the remaining body (Fig. 19.277a); apical margin of male tergite VIII arcuate apically (Fig. 19.277c); tubus of median lobe of aedeagus with apex produced ventrad in lateral view (Fig. 19.277b); spermatheca as illustrated (Fig. 19.277g) ...***Liogluta scudderi*** **Klimaszewski**

7. Punctures on pronotum and elytra sparse, distance between punctures greater than diameter of a puncture, surface of disc strongly granulose (Fig. 19.272a,b); male tergite VIII broadly triangularly produced apically and finely angulate medially (Fig. 19.272d); genitalia as illustrated (Fig. 19.272c,h)......... ***Liogluta granulosa*** **Lohse**

– Punctures on pronotum and elytra dense, distance between punctures about equal to diameter of a puncture, surface of disc slightly granulose; other characters not as above **8**

8. Body large and broad, length 4.2–5.0 mm (Fig. 19.271a); genitalia and terminalia as illustrated (Fig. 19.271b–g)................................ ...***Liogluta gigantea*** **Klimaszewski and Langor**

– Body smaller and narrower, length 2.8–4.5 mm; genitalia and terminalia not as above **9**

9. Body length 2.8 mm; pronotum distinctly transverse, with weakly defined microsculpture (Fig. 19.273a); median lobe of aedeagus as illustrated (Fig. 19.273b)...... ***Liogluta atriventris*** **(Casey)**

– Body length 3.8–4.5 mm; median lobe not as above **10**

10. Male tergite VIII with broad, arcuate medioapical projection, apical margin oblique laterally (Fig. 19.279e); ventral margin of tubus of median lobe of aedeagus sinuate in lateral view (Fig. 19.279b)......... ***Liogluta trapezicollis*** **Lohse**

– Male tergite VIII with apical margin truncate and crenulate between two small, triangular sublateral teeth (Fig. 19.280d); ventral margin of tubus of median lobe of aedeagus arcuate in lateral view (Fig. 19.280b)...................................... .... ***Liogluta vasta*** **(Mäklin)**

Gigantea **species group** (Klimaszewski et al. 2016a, 2018a)

Diagnosis Body broad, eyes large and bulging, about as long as genae in dorsal view; integument of forebody glossy; elytra not granulose, about 1.2 times as wide as pronotum, at suture about as long as pronotum; apical margin of male tergite VIII rounded with broad crenulations; female with apical margin of sternite VIII broadly truncate, with a row of microsetae; median lobe of aedeagus with tubus arched slightly ventrad near apex, apical part narrow in lateral view; spermatheca vaguely S-shaped.

196. ***Liogluta gigantea*** **Klimaszewski and Langor** (Fig. 19.271a–g)

Liogluta gigantea Klimaszewski and Langor, in Klimaszewski et al. 2011. Klimaszewski et al. 2016a, 2018a.

Diagnosis Body robust, broad, length 4.2–5.0 mm; colour dark brown, with pronotum, elytra (except for scutellar region), and legs reddish-brown (Fig. 19.271a); forebody moderately glossy, with fine and dense punctation, short pubescence and meshed microsculpture; head subquadrate, slightly narrower than pronotum, eyes large, as long as genae in dorsal view; antennae thin, all antennomeres elongate to subquadrate; pronotum transverse, widest at apical third; elytra wider than pronotum, at suture as long as or slightly longer than pronotum, with posterior margin almost rectangular; abdomen broad and flattened. MALE. Tergite VIII with apical margin unevenly arcuate, with faint crenulations, more prominent one at middle (Fig. 19.271c); sternite VIII with apical margin parabolic (Fig. 19.271d); median lobe of aedeagus short and stout, with tubus arched slightly ventrad near apex, apical part narrow in lateral view (Fig. 19.271b). FEMALE. Tergite VIII with apical margin rounded, and vaguely angulate medially (Fig. 19.271e); sternite VIII with apical margin truncate, with dense fringe of microsetae (Fig. 19.271f); spermatheca vaguely S-shaped, capsule club-shaped, stem sinuate with single spiral posteriorly (Fig. 19.271g).

DISTRIBUTION Origin: Nearctic. **CANADA**: LB, ON, QC. **USA**: not recorded.

COLLECTION AND HABITAT DATA

Habitat: from interior debris of a deserted beaver lodge (QC) and carrion-baited pitfall traps in spruce-lichen forest (LB). **Collecting period**: VI–X. **Collecting method**: carrion baited pitfall traps, Berlese funnel extraction of beaver lodge debris.

***Granulosa* species group** (Klimaszewski et al. 2016a)

This group of species is characterized by medium-sized and subparallel body, large and bulging eyes, diameter of eye about as long as postocular area of head in dorsal view; integument of forebody strongly glossy; elytra at suture about as long as pronotum, and elytra about one-fifth broader than pronotum; elytra sparsely and strongly granulose; male tergite VIII with apical margin obtusely triangularly produced, slightly angulate medially; tubus of median lobe of aedeagus arched in lateral view, moderately narrow apically.

197. *Liogluta granulosa* Lohse (Fig. 19.272a–h)

Liogluta (Liogluta) granulosa Lohse, in Lohse et al. 1990. Klimaszewski et al. 2008a, 2016a.

Diagnosis Body broadly subparallel, dark brown, with elytra, tarsi and tibiae often reddish brown (Fig. 19.272a,b) (one specimen from northern Yukon was entirely black); length 2.8–3.3 mm; integument of forebody with moderately pronounced meshed microsculpture; head about one-eighth narrower than maximum width of pronotum; antenna slender, antennomeres I–III elongate, IV subquadrate, VI–X slightly transverse; pronotum transverse, about evenly wide in basal one-third of its length, then strongly narrowed apically; elytra at suture about as long as pronotum, its surface strongly and coarsely granulose; basal two articles of metatarsus about the same length, each shorter than fifth article. MALE. Tergite VIII with apical margin oblique laterally, obtusely triangularly produced and slightly angulate medially, finely sinuate or with micro-denticulation (Fig. 19.272d); sternite VIII rounded apically (Fig. 19.272e); median lobe of aedeagus with tubus strongly arched ventrally and with narrow and rounded apex (Fig. 19.272c). FEMALE. Tergite VIII unevenly truncate apically (Fig. 19.272f); sternite VIII arcuate apically with strongly sinuate antecostal suture (Fig. 19.272g); spermatheca vaguely S-shaped with single spiral posteriorly as illustrated (Fig. 19.272h).

DISTRIBUTION Origin: Nearctic. **CANADA**: YT. **USA**: AK.

COLLECTION AND HABITAT DATA

Habitat: tundra between *Rubus* sp. hummocks and at creekside/ocean beach confluence, under boards and driftwood; in forest, mostly spruce, some aspen; shrub, alder, dwarf birch, grass, clubmoss; salt water marsh, old growth forest edge, grasses, *Heracleum* flowers. **Collecting period**: VI–VIII, X. **Collecting method**: Berlese funnels, yellow pan trap, pitfall traps, pitfalls w/ rainroofs, yellow pitfalls w/o rainroofs, hand collecting, sweeping vegetation.

Additional records. See Annex 1.

***Microgranulosa* species group** (Klimaszewski et al. 2016a)

Diagnosis Body subparallel, length 2.8–5.4 mm; eyes large and bulging, about as long as genae in dorsal view; integument of forebody moderately glossy; elytra about 1.20–1.25 times as wide as pronotum, at suture at least as long as pronotum; elytra densely and finely granulose in most species; apical margin of male tergite VIII usually with well-developed broad projection, with margin straight, entire, serrate or obtusely angulate; median lobe of aedeagus with tubus straight to distinctly arched ventrad and narrow to moderately wide apically in lateral view; spermatheca S-shaped, capsule with invagination short or deep.

198. *Liogluta atriventris* (Casey) (Fig. 19.273a–d)

Athetota atriventris Casey 1906. In *Liogluta*: Bernhauer and Scheerpeltz 1926, Klimaszewski et al. 2016a, 2020.

Diagnosis (based on male lectotype). Body small, length 2.8 mm; head, and abdomen dark brown, and pronotum, elytra and legs reddish yellow (Fig. 19.273a); integument strongly glossy with weak meshed microsculpture; moderately dense punctation and pubescence on pronotum and elytra, elytra with very fine micro-granulation; head subquadrate, slightly narrower than pronotum; antenna slender, antennomeres I–III elongate, VI subquadrate, VII–X slightly transverse; large eyes, each about as long as postocular region in dorsal view; antennae of the lectotype are partially damaged and cannot be described; pronotum slightly transverse, widest near the apical third; elytra wider and slightly longer than pronotum; abdomen subparallel, about as wide as elytra. MALE. Tergite VIII with apical margin truncate, crenulate between two small, obtuse sublateral processes (Fig. 19.273c); sternite VIII broadly rounded apically (Fig. 19.273d); median lobe of aedeagus in lateral view with tubus arched slightly ventrad at middle of ventral margin, apical part narrowly elongate, apex rounded (Fig. 19.273b). FEMALE. Unknown.

DISTRIBUTION **Origin**: Nearctic. **CANADA**: BC. **USA**: AK [**NSR**].

COLLECTION AND HABITAT DATA

Habitat: in EDS-crowberry, ~3–5 m of closed tall shrub to East and West. **Collecting period**: VII, VIII. **Collecting method**: pitfall traps.

We report this species from AK (**NSR**) for the first time here, based on the following: USA: Alaska, Alaska Peninsula / Becharof NWR, plot 1692 (Becharof Outlet), 20m el., 58.04911, -156.84003, ±4m In EDS-crowberry, ~3–5 m of closed tall shrub to E & W, pitfall trap, 5.VII–15.VIII.2007, S. Savage (UAM) 1 specimen.

199. *Liogluta intermedia* Klimaszewski and Langor (Fig. 19.274a–g)

Liogluta intermedia Klimaszewski and Langor, in Klimaszewski et al. 2011. Klimaszewski et al. 2016a.

Diagnosis Body length 4.2–4.5 mm; colour dark reddish-brown, with head dark brown, and legs and at least basal three antennomeres reddish-yellow (Fig. 19.274a); integument glossy; pronotum with dense punctation and pubescence; elytra with dense punctation and pubescence with very fine micro-granulation; head subquadrate, slightly narrower than pronotum, eyes large, about as long as genae in dorsal view; antenna slender, antennomeres I–VI elongate, VII–X slightly transverse; pronotum subquadrate, widest at apical third; elytra subparallel, as wide as pronotum and at suture about as long as pronotum; abdomen subparallel, about as wide as elytra. MALE. Tergite VIII with apical margin with broad, moderately long projection, with apical margin crenulate, antecostal suture sinuate, curved laterally (Fig. 19.274c); sternite VIII with apical margin broadly parabolic (Fig. 19.274d); median lobe of aedeagus with tubus short and straight, apical part narrowly rounded in lateral view (Fig. 19.274b). FEMALE. Tergite VIII rounded apically and vaguely angulate medially (Fig. 19.274e); sternite VIII with apical margin broadly, shallowly emarginate medially (Fig. 19.274f); spermatheca S-shaped, capsule short, club-shaped, stem broad, sinuate, slightly twisted posteriorly (Fig. 19.274g).

DISTRIBUTION **Origin**: Nearctic. **CANADA**: LB, NF, NS, ON, QC. **USA**: NH.

COLLECTION AND HABITAT DATA

Habitat: adults from litter and moss, treaded from flooded *Carex* and grasses, sifted from *Alnus* litter and sphagnum near pond; others from pitfall traps in a conifer forest, carrion trap in spruce-moss forest, and highbush blueberry field. **Collecting period**: V–X. **Collecting method**: treading *Carex* and grasses, sifting *Alnus* litter and *Sphagnum* moss, unbaited and carrion-baited pitfall traps.

Nigropolita **species group** (Klimaszewski et al. 2016a, 2018a)

Diagnosis Body shape non-typical for *Liogluta*, resembling that of some *Atheta* (*Dimetrota*) species with elytra distinctly wider than head and pronotum; eyes large and bulging, about as long as genae in dorsal view; integument of forebody highly glossy; elytra at suture at least as long as pronotum, and elytra about 1.3 times as wide as pronotum; male tergite VIII with apical margin truncate, entire or slightly produced with rounded lateral teeth; median lobe of the aedeagus in lateral view with tubus arched or almost straight, moderately narrow apically; spermatheca with spherical capsule bearing moderately long apical invagination, stem sinuate.

200. *Liogluta nigropolita* (Bernhauer) (Fig. 19.275a–i)

Atheta (*Pseudomegista*) *nigropolita* Bernhauer, 1907. In *Liogluta*: Lohse and Smetana 1985, Lohse et al. 1990, Klimaszewski et al. 2011, 2016a, 2018a, Ernst and Buddle 2013.

Diagnosis Body elongate, length 3.5–4.0 mm, with elytra and abdomen broad, moderately robust; colour dark brown to black with tarsi and tibiae often reddish-brown, elytra sometimes with reddish tinge (Fig. 19.275a); integument of forebody highly glossy, with moderately coarse meshed microsculpture; head about one-eighth narrower than maximum width of pronotum; antenna slender, antennomeres I–IV elongate, V–VI subquadrate, VII–X slightly transverse; pronotum transverse, about evenly wide in posterior half, then distinctly narrowing apicad, forming round apical angles; elytra at suture slightly longer than pronotum; hind tarsus with four basal tarsomeres of about same length, each shorter than fifth article. MALE. Tergite VIII with apical margin broadly arcuate, without teeth or crenulations (Fig. 19.275d); sternite VIII broadly parabolic apically (Fig. 19.275e); median lobe of aedeagus in lateral view with tubus broadly arcuate ventrally, apical part narrow, tapering to narrowly rounded apex (Fig. 19.275b); in dorsal view tubus broad and triangular apically (Fig. 19.275c). FEMALE. Tergite VIII broadly, unevenly arcuate apically (Fig. 19.275f); sternite VIII evenly rounded apically, with antecostal suture distinctly sinuate (Fig. 19.275g); spermatheca with capsule spherical, apical invagination deep and narrow, stem S-shaped, gradually becoming very narrow with small curl posteriad (Fig. 19.275h,i).

DISTRIBUTION Origin: Nearctic. **CANADA**: LB, NF, NT, NU, QC, YT. **USA**: AK [**NSR**], NH.

COLLECTION AND HABITAT DATA

Habitat: in LB, reported from pond and stream margins in moss and clay and under rocks, and from pitfall traps in shrubby coastal barrens in NL. In AK, found in *Salix*, lake shore and pond, and boggy black spruce. **Collecting period**: V. VII–IX. **Collecting method**: sifting moss and litter, yellow pan traps, pitfall traps, sweeping vegetation, hanging Malaise.

We report this species from AK (**NSR**) for the first time here, based on the following: USA, Alaska, Anaktuvuk Pass, 647m el., 68.14049, −151.74091, ±250m *Salix*, sweeping, 19.V.2016, D.S. Sikes, K.M. Daly (UAM) 1 specimen; Anaktuvuk Pass, Eleanor lake, 649m el., 68.14742, −151.71854, ±100m lake shore and pond, sweep and dipnet, 20.V.2016, D.S. Sikes, K.M. Daly (UAM) 2 specimens; White Mountains Recreation Area, Nome Creek Site 6, 589m el., 65.33446, −146.84001, ±10m boggy black spruce, hanging Malaise, 4.95m, 12.VI-3.VII.2015, J.C. Hagelin (UAM) 1 specimen.

201. *Liogluta nitens* (Mäklin) (Fig. 19.276a–f)

Homalota nitens Mäklin, 1852. Bernhauer 1907 [in *Atheta* (*Philhygra*)], Hatch 1957 [in *Ischnopoda* (*Philhygra*)], in *Liogluta*: Lohse and Smetana 1985 [types], Klimaszewski and Winchester 2002 [in *Atheta*], Gusarov 2003a [synonymy], Klimaszewski et al. 2016a, 2020.

Diagnosis Body length 2.8–3.3 mm, narrowly elongate, *Atheta*-like, moderately robust, dark brown to black with elytra, tarsi and tibiae often yellow- or red-brown (Fig. 19.276a); integument of forebody with moderately pronounced meshed microsculpture, surface strongly glossy; head about one-eighth narrower than maximum width of pronotum; antenna slender, antennomeres I–III elongate, IV subquadrate or slightly elongate, V–X slightly to distinctly transverse; pronotum transverse, about evenly wide until basal third of its length then becoming strongly narrowed apically (Fig. 19.276a); elytra at suture about as long as pronotum; basal four metatarsomeres about the same length and each shorter than fifth one. MALE. Tergite VIII with apical margin obtusely produced laterally, with broad, truncate projection medially with rounded lateral angles, apical margin smooth or with micro-crenulation (Fig. 19.276c); sternite VIII broadly parabolic apically (Fig. 19.276d); median lobe of aedeagus with tubus slightly sinuate ventrally and with moderately narrow and rounded apex, crista apicalis prominent, subquadrate (Fig. 19.276b). FEMALE. Tergite VIII arcuate apically except briefly straight medially (Fig. 19.276e); sternite VIII evenly rounded apically and with moderately sinuate antecostal suture (Fig. 19.276f). Spermatheca not found.

DISTRIBUTION **Origin**: Nearctic. **CANADA**: AB, BC, YT (see comments below). **USA**: AK, OR, WA.

COLLECTION AND HABITAT DATA
Habitat: in BC, adults were found in the interior and transition zones of a Sitka spruce forest and in moss on forest floor and edge of an old road, in mushrooms, and in *Carex* marsh and *Alnus*; several adults were collected from moss at the edge of an old road. Adults in Alaska were collected in a wide variety of habitats spanning lowland forests to alpine zones: alpine meadow litter, lowland forest clearcuts, floodplain meadows with *Athyrium, Caltha,* and *Rubus*, under rocks, in krummholz alpine habitats of *Tsuga mertensiana*, near bear dung in alpine habitats, old growth temperate rain coniferous forests, alpine heath with *Empetrum,* and *Vaccinium,* subalpine habitats with *Salix,* and *Veratrum*. **Collecting period**: VI–IX. **Collecting method**: pitfall traps, sifting moss and mushrooms, and treading *Carex* marsh and sifting *Alnus*.

COMMENTS There is considerable variation in length and width of the elytra in specimens from Vancouver Island, BC and OR (having broader and longer elytra), and those with narrow and shorter elytra from the Queen Charlotte Islands, BC, AB, and AK. The genitalic features were the same in those of the typical form with the longer and broader elytra, and those with narrower and shorter elytra. Therefore, we consider this as intraspecific variation. Additional studies, including DNA comparison, are needed to reveal the relationship between these two morphotypes. Two UAM Alaskan specimens (UAM:Ento:152502, UAM:Ento:232546) were DNA barcoded (UAMIC2665-15, UAMIC2701-15) and they cluster closely with two specimens of this species. The YT record (from Lohse and Smetana 1985) citing specimens (without label data) deposited in the CNC needs verification as there were no YT specimens located in the CNC or otherwise.

***Terminalis* species group** (Klimaszewski et al. 2016a, 2018a, 2020)

Diagnosis Eyes small, distinctly shorter than genae in dorsal view; integument of forebody moderately to highly glossy; elytra short, at suture about as long as pronotum; male tergite VIII truncate apically, with or without minute crenulations; median lobe of aedeagus with tubus usually arched and moderately narrow apically in lateral view; capsule of spermatheca with apical part compressed, with apical invagination small and short.

202. *Liogluta scudderi* Klimaszewski (Fig. 19.277a–g)

Liogluta scudderi Klimaszewski, in Klimaszewski et al. 2020.

Diagnosis Body length 3.7–4.0 mm, narrowly subparallel, colour dark brown to almost black, elytra slightly paler and legs reddish-brown (Fig. 19.277a); integument of entire body strongly glossy, with meshed microsculpture; head elongate, about one-quarter narrower than maximum width of pronotum; antennomere IV subquadrate, VII–X distinctly transverse, last antennomere as long as two preceding ones combined; pronotum strongly margined, trapezoidal in shape, slightly transverse, narrower at base and widest in apical third; elytra at suture about as long as pronotum; abdomen broad and arcuate laterally; hind tarsus with four basal tarsomeres about the same length. MALE. Tergite VIII with apical margin arcuate and smooth (Fig. 19.277c); sternite VIII with apical margin rounded, briefly straight medially (Fig. 19.277d); tubus of median lobe of aedeagus in lateral view straight basally, arcuate before narrow and rounded apex (Fig. 19.277b). FEMALE. Tergite VIII arcuate apically (Fig. 19.277e); sternite VIII with apical margin arcuate except broadly straight medially (Fig. 19.277f); spermatheca with club-shaped narrow capsule and arcuate neck, stem narrow, strongly sinuate, narrowly looped and twisted posteriorly (Fig. 19.277g).

This species may be distinguished from its congeners by the narrow and strongly glossy body, short elytra, by the arcuate apical margin of male tergite VIII, by the shape of the median lobe and the strongly sinuate stem of the spermatheca.

DISTRIBUTION Origin: Nearctic. **CANADA**: BC. **USA**: not recorded.

COLLECTION AND HABITAT DATA
Habitat: mature unspecified forest and peatland. **Collecting period**: VI. **Collecting method**: pitfall traps.

203. *Liogluta terminalis* (Casey) (Fig. 19.278a–g)

Anepsiota terminalis Casey, 1906. As *L. aloconotoides* Lohse, in Lohse et al. 1990: Majka and Klimaszewski 2008b, Klimaszewski et al. 2008b, 2011, Webster et al. 2012, Work et al. 2013, Klimaszewski et al. 2015a, Stefani et al. 2016. As *L. terminalis*: Klimaszewski et al. 2016a, 2018a, 2020.

Diagnosis Body subparallel, length 3.9–4.5 mm; colour entirely reddish-brown or with head and posterior abdomen chestnut brown (Fig. 19.278a); integument of forebody moderately glossy, with meshed microsculpture; head about one-quarter narrower than maximum width of pronotum; antenna slender, antennomeres I–VI elongate, VII subquadrate, VIII–X slightly transverse;; pronotum transverse, narrower at base and widest in apical third; elytra at suture about as long as pronotum; hind tarsus with three basal tarsomeres elongate, first longest, second about as long as third, fourth shorter than either of preceding articles. MALE. Tergite VIII with apical margin with broad, short, truncate projection and rounded lateral angles, apical margin smooth or finely crenulate, antecostal suture straight, abruptly deflected posterad at lateral margin (Fig. 19.278c); sternite VIII broadly parabolic apically (Fig. 19.278d); median lobe of aedeagus in lateral view arcuate ventrally, subapical part narrow, apex narrowly rounded (Fig. 19.278b). FEMALE. Tergite VIII arcuate apically (Fig. 19.278e); sternite VIII shallowly emarginate at apex (Fig. 19.278f); spermatheca with broad, club-shaped capsule, stem moderately looped, twisted posteriorly (Fig. 19.278g).

DISTRIBUTION Origin: Nearctic. **CANADA**: AB, BC, LB, NB, NF, NS, ON, QC, SK [**BOLD**], YT. **USA**: MT, NH.

COLLECTION AND HABITAT DATA
Habitat: in NF, collected from pitfall traps in various forest types such as birch-dominated forests, spruce-fir forests, black spruce/lichen forests, fir forests, a recently burned forest, and from coastal limestone barrens; in NB, from dung in a coastal red spruce forest, Lindgren funnel trap in a rich Appalachian hardwood forest with some conifers, and treaded from sedges on lake margin. In AB, some reared from well-decayed white spruce logs; in SK collected from mixed wood forest, poplar/white spruce, with thick understory,

and mixed wood, aspen/white spruce forest. **Collecting period**: VII–IX. **Collecting method**: pitfall traps, treading sedges along a small lake margin, Lindgren funnel trap.

We report this species from SK (**NPR**) for the first time here, based on **BOLD** and LFC records: Canada, Saskatchewan, Prince Albert National Park, Spruce River Highlands Trail, 541m el. 53.712, –106.048, poplar and white spruce forest, thick understory, flight-intercept trap, 10-14.VII.2012, BIOBus 2012 team (CBG) 1 specimen; Prince Albert National Park, Hwy 263, 627m el. 53.85, –106.078, aspen and white spruce forest, Malaise trap, 17-26.VII.2012, F. Moreland (CBG) 1 specimen; Prince Albert, 53.9726, –106.2902, 533m, 15.IX.2014, mushrooms, sifting, coll. B. Godin and D. Horwood (LFC) 1 female.

204. *Liogluta trapezicollis* Lohse, New Concept (Fig. 19.279a–f)

Liogluta (Anepsiota) trapezicollis Lohse, in Lohse et al. 1990. [see Comments below].

Diagnosis Body length 4.5 mm, subparallel, slender, black with pronotum dark brown, elyta yellowish-brown, and legs yellowish (Fig. 19.279a); integument of forebody with meshed microsculpture moderately pronounced, surface moderately strongly glossy; head about as wide as maximum width of pronotum; antenna slender, antennomeres I–IV elongate, V–VI subquadrate, VII–X slightly transverse; pronotum trapezoidal in shape, transverse, narrow at base and widest apically; elytra at suture as long as pronotum along midline; basal three metatarsomeres elongate, equal in length and each slightly longer than fourth one. MALE. Tergite VIII with broad, moderately long medioapical projection, apical margin arcuate and smooth, oblique laterally (Fig. 19.279e); sternite VIII broadly parabolic apically (Fig. 19.279f); tubus of median lobe of aedeagus sinuate ventrally, apical part narrow, apex rounded in lateral view (Fig. 19.279b,c), in dorsal view as illustrated (Fig. 19.279d). FEMALE. Unknown.

DISTRIBUTION Origin: Nearctic. **CANADA**: YT. **USA**: not recorded.

COLLECTION AND HABITAT DATA

Habitat: unspecified, holotype collected at 3500ft. **Collecting period**: VII. **Collecting method**: not recorded but probably sifting organic litter.

COMMENTS During publication of Klimaszewski et al. (2020), we were not able to study the holotype of *L. trapezicollis* Lohse (CNC) and compare it with the specimens we examined and determined as *L. trapezicollis,* as it had been borrowed several years earlier. Since then, we were able to examine the holotype and we conclude that *L. trapezicollis* (sensu Klimaszewski et al. 2016a, 2020) is conspecific with *L. vasta* (Mäklin) described from Sitka Is., AK.

205. *Liogluta vasta* (Mäklin) (Fig. 19.280a–h)

Homalota vasta Mäklin, 1853. In *Liogluta*: Lohse and Smetana 1985, Lohse et al. 1990. Klimaszewski et al. 2016a, 2020 (as *L. trapezicollis*).

Diagnosis Body length 3.8–4.4 mm, subparallel, slender, dark brown to black with pronotum brown and paler than head, elytra with yellow-brown irregular spots (Fig. 19.280a); integument of forebody with meshed microsculpture moderately pronounced, surface moderately strongly glossy; head about one-quarter narrower than maximum width of pronotum; antenna slender, antennomeres I–IV elongate, V–VI subquadrate, VII–X distinctly transverse; pronotum transverse, narrow at base and widest at middle (width of pronotum is variable, some specimens have pronotum markedly narrower than base of elytra and some have pronotum nearly as wide as base of elytra); elytra at suture slightly shorter than pronotum; basal three metatarsomeres elongate, equal in length and each slightly longer than fourth one. MALE (**revised redescription**). Tergite VIII with apical margin truncate between two short triangular sublateral teeth, apical margin variably sculptured, smooth, crenulate, or denticulate (Fig. 19.280d); sternite VIII broadly parabolic apically (Fig. 19.280e); tubus of median lobe of aedeagus in lateral view with ventral margin arcuate, apical part narrow, abruptly tapering to narrowly rounded apex (Fig. 19.280b,c). FEMALE. Tergite VIII arcuate apically (Fig. 19.280f); sternite VIII arcuately narrowed to shallowly emarginate apical margin (Fig. 19.280g); spermatheca with club-shaped capsule bearing shallow invagination, stem narrow and sinuate, narrowly looped and twisted posteriorly (Fig. 19.280h).

DISTRIBUTION Origin: Nearctic. **CANADA**: BC, YT. **USA**: AK.

COLLECTION AND HABITAT DATA
Habitat: the BC specimens were collected from moss, and some from boreal spruce forest; the AK specimens were collected from alpine zones between 453–1071 m elevation; none were collected in lowland forests. Habitats include alpine flood meadows, under rocks, herbaceous heath with *Luetkea, Cassiopes,* and *Lupinus*, low rocky tundra with *Dryas*, meadow heath with *Phyllodoce, Senecio*, and *Luetkea*, shrubby krummholz with *Elliottia* and *Tsuga*, wet meadows with *Carex, Petasites*, *Senecio* and *Ranunculus*. **Collecting period**: V, VII–IX. **Collecting method**: pitfall traps, treading sedges along a small lake margin, Lindgren funnel trap, processing moss and forest debris through Berlese funnels.

206. *Liogluta wickhami* (Casey) (Fig. 19.281a–g)

Anepsiota wickhami Casey 1893. In *Liogluta*: Klimaszewski et al. 2016a, 2020.

Diagnosis Body length length 4.0–4.2 mm mm, broadly subparallel; pronotum, elytra, legs and basal antennomeres red-brown, head and most of abdomen chestnut brown (Fig. 19.281a); integument of forebody with meshed microsculpture, moderately glossy; head about one-third narrower than maximum width of pronotum; antennae robust, antennomere IV slightly elongate, VII–X elongate or subquadrate; pronotum more or less evenly arcuate laterally; elytra at suture about as long as pronotum; abdomen broad, at base as broad as elytra, sides arcuate, basal metatarsomere the longest, the following two slightly shorter, and fourth shorter than penultimate one. MALE. Apical margin of tergite VIII with broad, truncate projection medially with apical margin faintly crenulate (Fig. 19.281c); sternite VIII parabolically rounded apically (Fig. 19.281d); tubus of median lobe of aedeagus in lateral view short, arcuate ventrally, with apex narrow and rounded (Fig. 19.281b). FEMALE. Tergite VIII with apical margin obtusely produced, truncate medially (Fig. 19.281e); sternite VIII with apical margin rounded except broadly, shallowly emarginate medially (Fig. 19.281f); spermatheca with tubular capsule and deep apical invagination, stem thin, long, and strongly sinuate, looped posteriorly and twisted posteriorly (Fig. 19.281g).

This species is somewhat similar to *L. terminalis* but has thick dark brown antennae, head and pronotum (antennae, head, and pronotum are uniformly reddish brown or only slightly darker than remaining parts of the body in *L. terminalis*). The spermatheca of *L. wickhami* is smaller and with a differently shaped capsule, with a deep apical invagination and has the stem shorter and differently looped posteriorly. *Liogluta wickhami* is also very similar to *L. vasta* but can be distinguished by the shape of the pronotum which has evenly arcuate sides and is broadest at the middle, while it is trapezoidal in shape and is broadest in the apical third in the latter species.

DISTRIBUTION Origin: Nearctic. **CANADA**: **BC**. **USA**: not recorded.

COLLECTION AND HABITAT DATA
Habitat: spruce forest. **Collecting period**: VII, IX. **Collecting method**: pitfall traps.

COMMENTS The female holotype was captured in the Stickeen [Stikine] River Valley of British Columbia (Casey 1893), and subsequently one male was found in Three Guardsmen Pass, British Columbia, with additional records in Klimaszewski et al. (2020).

Genus *Lypoglossa* Fenyes, 1918

(Figs. 19.282–19.283) (Hoebeke 1992, Gusarov 2004, Klimaszewski et al. 2018a, b, 2020)

Diagnosis Body broad, ovoid, length 3.2–4.6 mm; colour uniformly brown to dark brown with appendages paler, pronotum and elytra reddish-brown in some specimens; head transverse, infraorbital carinae complete; *a* sensillum of epipharynx long; ligula with broad base, split apically into two lobes; labial palpi with setae α β, γ and δ present; antennomere II as long III, IV elongate, V elongate to subquadrate, VI–X elongate to transverse, apical antennomere without coeloconic sensilla and as long as IX and X combined; pronotum exceptionally broad, 1.5 times as wide as long, disc with microsetae directed anteriad along midline, laterad and posteriad elsewhere, macrosetae short and thin, hypomera incompletely visible or totally obscured in lateral view; posterior margin of elytra near posterolateral angle straight or slightly concave; mesocoxae narrowly separated; process of mesoventrite long and slender, acute at apex; hind tarsi with basal article much longer than second one, each tarsus with single empodial seta shorter than claws, mesolateral macroseta of middle tibia thin, much shorter than tibial width; first three visible abdominal tergites with weak transverse basal impression, densely and finely punctate with short, silky, dense apressed pubescence, tergites VI–VIII sparsely punctate and pubescent; median lobe of aedeagus broad in dorsal view with base of tubus about as wide as bulbus; spermatheca S-shaped, capsule with large apical invagination, stem long and sinuate, narrowly looped posteriad. For illustrations of mouthparts and terminology see Hoebeke (1992) and Gusarov (2004).

Species of *Lypoglossa* share an overall resemblance to *Devia prospera* (Oxypodini) but differ in having a 4-5-5 tarsal formula (5-5-5 in *Devia*).

Key to Species of *Lypoglossa*
(From Klimaszewski et al. 2020)

1. Pronotal width to length ratio 1.2–1.3 (Fig. 19.283a); median lobe of aedeagus in ventral/dorsal view with apical part of tubus broadest basad, not constricted and without projections (Fig. 19.283c,d); spermatheca larger, posterior loop of stem wide, almost circular (Fig. 19.283i) .... ***Lypoglossa franclemonti*** **Hoebeke**

– Pronotal width to length ratio 1.3–1.4 (Fig. 19.282a); median lobe of aedeagus in ventral/dorsal view with apical part of tubus with dentiform projection on each side of constriction basad (Fig. 19.282c,d); spermatheca smaller, with posterior loop of stem narrow (Fig. 19.282i) ***Lypoglossa angularis*** (Mäklin)

207. *Lypoglossa angularis* (Mäklin) (Fig. 19.282a–i)

Myrmedonia angularis Mäklin, 1853. In *Lypoglossa*: Lohse and Smetana 1985 (types, new synonymy), Hoebeke 1992 [as *L. angularis*, *ex parte*], Gusarov 2004 [new synonymy, with three subspecies], as *L. angularis obtusa*: Majka and Klimaszewski 2010, Klimaszewski et al. 2011, 2018a, Webster et al. 2012. As *L. angularis*: Klimaszewski et al. 2020.

Diagnosis Body robust, wide, oval, length 3.8–4.2 mm; head and abdomen brown to dark brown and pronotum, elytra, legs and antennae light rusty-brown; integument moderately glossy, forebody with fine and dense punctation, short pubescence and meshed microsculpture (Fig. 19.282a); head transverse, distinctly narrower than pronotum; antennae thin, antennomeres IV–VIII elongate, IX–X subquadrate; pronotum widely oval, transverse, widest at basal third; elytra as wide as pronotum, at suture about as long as pronotum, posterior margin arcuate; abdomen subparallel, as wide as elytra and pronotum. MALE. Tergite VIII arcuate apically (Fig. 19.282e); sternite VIII parabolic apically (Fig. 19.282f); median lobe of aedeagus in lateral view with wide bridge, ventral margin of tubus arcuate (Fig. 19.282b), in lateral view apical part with elongate projection on each side of constriction basad (Fig. 19.282c,d). FEMALE. Tergite VIII arcuate apically (Fig. 19.282g);

sternite VIII rounded apically (Fig. 19.282h); spermatheca with posterior loop moderately narrow (Fig. 19.282i).

DISTRIBUTION **Origin**: Nearctic. **CANADA**: AB, BC, LB, NB, NF, NS, QC, SK [**NPR**], YT. **USA**: AK, AZ, CA, CO, ME, MT, NH, NM, OR, WA, WY [distribution including all subspecies of *angularis* in Gusarov 2004; MA listed by Klimaszewski et al. 2020 was in error].

COLLECTION AND HABITAT DATA
Habitat: gopher burrows, moss at edge of forest, forest litter in *Betula*, *Picea* and *Alnus*, *Verathrum*, ferns forest litter and, treeline subalpine meadow. In AK, found in boreal forest; subalpine *Tsuga, Vaccinium*; black and white spruce with willow; mesic *Betula, Picea* forest; *Betula, Salix, Picea, Populus, Alnus* leaf litter; forest, mostly spruce with some aspen; under CTS-*Salix* near the edge with crowberry, bottom edge of closed tall shrub-alder, near eds-crowberry; in EDS-crowberry, ~3–5 m of closed tall shrub; forest edge *Populus tremuloides, P. balsamea;* herbaceous tundra, *Vaccinium caespitosum, Phyllodoce glanduliflora, Geranium erianthum*, in soil; forest edge, meadow, grass, forbs; shrubby krummholz, *Elliottia pyroliflora, Tsuga mertensiana, Nephrophyllidium crista-galli*; heath, *Cassiope mertensiana, Luetkea pectinata, Harrimanella stelleriana*; meadow, *Nephrophyllidium crista-galli, Geranium erianthum, Artemisia arctica*; rich meadow, *Geranium erianthum, Artemisia arctica, Sanguisorba canadensis*; herbaceous heath, *Vaccinium caespitosum, Vaccinium uliginosum, Empetrum nigrum*; shrub, alder, dwarf birch, grass, clubmoss; krummholtz, *Tsuga mertensiana*; tundra, *Vaccinium uliginosum, Luetkea pectinata, Phyllodoce glandulifera*; herbaceous tundra, *Vaccinium caespitosum, Phyllodoce glandulifera, Geranium erianthum*; subalpine meadow, *Athyrium, Carex macrocephala*; shrub, alder, dwarf birch, grass, clubmoss; flood meadow, *Athyrium, Caltha leptosepala*; wet meadow, *Carex nigricans, Petasites frigidus, Ranunculus occidentalis*; creekside/ocean beach confluence, under boards and drift wood; near bear dung. **Collecting period**: IV–X. **Collecting method**: sifting forest litter and moss, Berlese funnels, sifting litter in gopher borrows, pitfall traps, yellow pitfalls w/o rainroofs, and hanging Malaise.

We report this species from SK (**NPR**) for the first time here, based on the following: Canada, Saskatchewan, La Ronge, 55.1180, –105.2457, 366m, 6.VI.2013, wet spruce litter sifting, coll. B. Godin and D. Horwood (LFC) 1 female.

Additional records. See Annex 1.

208. ***Lypoglossa franclemonti*** **Hoebeke** (Fig. 19.283a–i)

Lypoglossa franclemonti Hoebeke, 1992. Gusarov 2004, Webster et al. 2009, Majka and Klimaszewski 2010, Klimaszewski et al. 2011, 2012, 2015a, 2020, Work et al. 2013, Stefani et al. 2016.

Diagnosis Body robust, wide, oval, length 3.8–4.6 mm; head and abdomen dark brown, pronotum, elytra, legs and antennae light rusty-brown (Fig. 19.283a); integument moderately glossy, forebody with fine and dense punctation, short pubescence and meshed microsculpture; head transverse, distinctly narrower than pronotum; antennae thin, antennomeres IV–VIII elongate (more elongate than in *L. angularis*), IX–X subquadrate; pronotum widely oval, transverse, widest at basal third (less transverse than in *L. angularis*); elytra as wide as pronotum, at suture about as long as pronotum, posterior margin arcuate; abdomen subparallel, as wide as elytra and pronotum. MALE. Tergite VIII arcuate apically (Fig. 19.283e); sternite VIII parabolic (Fig. 19.283f); tubus of median lobe of aedeagus narrowly oval, widest at middle, not constricted, and without projections in dorsal/ventral view (Fig. 19.283c,d); in lateral view athetine bridge narrow, ventral margin of tubus almost straight (Fig. 19.283b). FEMALE. Tergite VIII arcuate apically (Fig. 19.283g); sternite VIII rounded apically (Fig. 19.283h); spermatheca with stem forming broader posterior loop (Fig. 19.283i).

DISTRIBUTION Origin: Nearctic. **CANADA**: AB, BC, MB, NB, NF, NS, NT, ON, QC, SK, YT. **USA**: ME, NH, NY, VT.

COLLECTION AND HABITAT DATA

Habitat: coniferous and deciduous forests, balsam fir stand, spruce bogs, birch bogs; adults in forest litter, in moss and lichens; in NB, this species was found under sea wrack under alders on sea beach in May (4 individuals) and in gilled mushrooms in an adjacent red spruce and yellow birch forest in August and September (4 individuals). **Collecting period**: V–IX. **Collecting method**: sifting sea wrack and mushrooms, pitfall traps.

Genus *Mocyta* Mulsant and Rey, 1873b

(Figs. 19.284–19.287) (Klimaszewski et al. 2015b, 2018a, 2020)

Diagnosis Antennae very thin and pale, in most specimens contrasting with body colour, antennomeres V–X subquadrate or slightly transverse; pronotum glossy, moderately convex, broad and shield-shaped, widest at or near middle, at least as broad as the base of the elytra but in most specimens slightly broader, pubescence on disc directed posteriad along midline (or throughout central section) and posterolaterad elsewhere, hypomera not visible in lateral view; male tergite VIII arcuate apically, without teeth and other secondary sexual characters; male sternite VIII with longer macrosetae than those of females; median lobe of aedeagus of a simple form, tubus about half length of median lobe, narrowly tapering and rounded apically in dorsal view, straight and narrow apicad in lateral view, internal sac structures inconspicuous, usually elongate and not pronounced; spermatheca with capsule hemispherical, or elongate and sac-shaped, usually with small apical invagination and short neck, stem thin and regularly or irregularly coiled posteriorly.

Key to Species of *Mocyta*
(Modified from Klimaszewski et al. 2015b)

1. Elytra longer than pronotum in most specimens (Figs. 19.284a, 19.286a) (some specimens have elytra of variable length and may appear as long as pronotum along midline of disc); maximum width of pronotum slightly greater than width of elytra at base (Figs. 19.284a, 19.286a) **2**

– Elytra as long as or slightly shorter than pronotum; maximum width of pronotum at most as great as width of elytra at base **3**

2. Pronotum approximately rectangular in shape, gradually narrowed in apical third of its length (Fig. 19.286a); spermathecal capsule pear-shaped, rounded apically and gradually narrowed posteriorly, apical invagination small and shallow, stem thin and irregularly twisted posteriorly (Fig. 19.286f–j); only females are present in North America.......***Mocyta fungi* (Gravenhorst)**

– Pronotum approximately trapezoidal in shape, strongly narrowed apically from basal third of its length (Fig. 19.284a); spermathecal capsule narrowly elongate and sac-shaped, apical invagination small and shallow; stem sinuate and irregularly twisted or coiled posteriorly (Fig. 19.284i,j)..... ***Mocyta amblystegii* (Brundin)**

3. Pronotum as broad as elytra (Fig. 19.285a); spermatheca with capsule pitcher-shaped and flat apically with elongate apical invagination, stem broadly coiled posteriorly (Fig. 19.285h) ....***Mocyta breviuscula* (Mäklin)**

– Pronotum broader than elytra in most (Fig. 19.287a); spermatheca with capsule pear-shaped with short apical invagination, stem broadly irregularly coiled posteriorly (Fig. 19.287h)....................***Mocyta sphagnorum* Klimaszewski and Webster**

***Amblystegii* species group** (Klimaszewski et al. 2015b, 2018a)

Diagnosis Body entirely dark brown to black except for light-coloured appendages; pronotum approximately as broad as elytra or slightly broader, sides arcuate, pubescence on disc directed posteriad along midline and obliquely posteriad elsewhere; elytra in most specimens longer than pronotum, except for *M. breviuscula*.

209. *Mocyta amblystegii* (Brundin) * (Fig. 19.284a–j)

Atheta amblystegii Brundin 1952: 135; in *Mocyta*: Lohse et al. 1990, Smetana 2004, Klimaszewski et al. 2015b.

Diagnosis Body narrowly oval, length 2.5–3.0 mm; uniformly brown to black, appendages light brown (Fig. 19.284a); antennomeres I–IV elongate and V–X subquadrate or slightly transverse; pronotum broad, strongly transverse, rounded laterally and arcuate basally; elytra transverse and at least as long as pronotum; broadly arcuate laterally. MALE. Tergite VIII truncate apically (Fig. 19.284e); sternite VIII produced apically, with numerous macrosetae and with a broad space between base of disc and antecostal suture, the suture nearly straight or slightly sinuate (Fig. 19.284f); median lobe of aedeagus as illustrated (Fig. 19.284b–d). FEMALE. Tergite and sternite VIII truncate apically (Fig. 19.284g, h); spermatheca with capsule pear-shaped, as illustrated (Fig. 19.284i,j).

Adults are externally similar to those of *M. fungi* and may be identified with certainty by the pear-shaped capsule of spermatheca. The presence of males in Canadian populations of *M. amblystegii* and lack of males in Canadian populations of *M. fungi* may also aid in identification of this species.

DISTRIBUTION Origin: Holarctic; Europe (Scandinavia), North America. **CANADA**: MB, NT, NU, YT. **USA**: AK.

COLLECTION AND HABITAT DATA

Habitat: under leaf litter and in moss; elevation 200–3100 feet. In AK, found in black and white spruce; *Salix* leaf litter; under rocks; in birch bark; spruce bog; and open tussock and standing water, with willow and bog birch. **Collecting period**: IV–VIII. **Collecting method**: sifting leaf litter and moss, ground search forceps, aspirator, hanging Malaise.

Additional records. See Annex 1.

210. *Mocyta breviuscula* (Mäklin) (Fig. 19.285a–h)

Homalota breviuscula Mäklin, 1852. Bernhauer 1907 [in *Atheta* (*Acrotona*)], Casey 1910 [as seven synonyms in *Acrotona*], Lohse and Smetana 1985 [as *Atheta* (*Mocyta*) *breviuscula*]. In *Mocyta*: Gusarov 2003a, Klimaszewski et al. 2007b, 2008a,b, Majka and Klimaszewski 2008b, Webster et al. 2009, Brunke et al. 2012, Work et al. 2013, Klimaszewski et al. 2011, 2015b, 2016d, 2018a, b, 2020.

Diagnosis Body narrowly oval, length 2.4–3.0 mm; colour uniformly dark brown to almost black and often with reddish tinge, appendages yellowish to reddish-brown (Fig. 19.285a); antennomeres I–IV elongate and V–X subquadrate; pronotum transverse, arcuate laterally and arcuate basally; elytra transverse and nearly as long as pronotum; abdomen broadly arcuate laterally. MALE. Tergite VIII broadly arcuate apically (Fig. 19.285d); sternite VIII parabolic apically, antecostal suture sinuate, widely separated from basal margin (Fig. 19.285e); median lobe of aedeagus in lateral view with ventral margin of tubus straight, apex narrowly rounded (Fig. 19.285b); in dorsal view bulbus narrowly oval and tubus narrowly triangular apically, apex rounded (Fig. 19.285c). FEMALE. Tergite and sternite VIII broadly arcuate apically (Fig. 19.285f,g); spermatheca with capsule pitcher-shaped and flat apically, apical invagination deep, stem broadly coiled posteriorly (Fig. 19.285h).

The combination of uniform body colour, elytra no longer than pronotum, distinct shape of spermatheca with deep capsular invagination, and shape of male sternite VIII with broad space between base of disc and antecostal suture, can distinguish *M. breviuscula* from the remaining Nearctic congeners.

DISTRIBUTION Origin: Nearctic. **CANADA**: AB, BC, LB, MB, NB, NF, NS, ON, QC, SK, YT. **USA**: AK, CA, NV, OR.

COLLECTION AND HABITAT DATA
Habitat: in NF, frequently captured in pitfall traps in various forest types such as birch, spruce-lichen, spruce-poplar, fir; also in vegetation on coastal dunes, in disturbed fields among grass and weeds, and on shrubby limestone barrens (Klimaszewski et al. 2018a). In QC, from pitfall traps in yellow birch/balsam fir forests, and sifted from litter in an oak savanna/alvar in ON (Klimaszewski et al. 2018a). One individual from sphagnum and litter in an eastern white cedar swamp and another captured in Lindgren funnel trap in an old red pine forest in NB (Klimaszewski et al. 2018a). In BC, specimens were found in aspen litter beside high pond, in moss and alder litter at cold stream, and in sedges at edge of creek. In AK, found in wet boreal forests; spruce, birch, alder, cottonwood forest; forest edge *Populus tremuloides* and *Salix* spp.; *Populus, Betula, Alnus* leaf litter; birch, black spruce on northern slope; mesic *Betula* forest; birch bark; alder and birch litter near dry vernal pool; *Betula* forest, decaying mushrooms; ant nest under rock, alpine tundra; alpine *Salix, Valariana*, shrubs; brackish shoreline; under rocks, wrack; creekside/ocean beach confluence, under boards and drift wood; *Salix barclayi* meadow; under rocks, dung, wood, meadows w/lupine; and in fireweed, grass habitat. **Collecting period**: IV–IX. **Collecting method**: pitfall traps, Lindgren funnel trap, Berlese funnels; sifting forest litter, sweeping vegetation, flight intercept trap, hanging Malaise.

Additional records. See Annex 1.

COMMENTS This species was listed for NT by Newton (2019) but we are not aware of published voucher data.

211. *Mocyta fungi* (Gravenhorst) † (Fig. 19.286a–j)

Aleochara fungi Gravenhorst, 1806. Casey 1893, 1910 [as four synonyms], Bernhauer 1907 [in *Atheta* (*Acrotona*)], Clark 1949 [in *Acrotona*], Muona 1984 [in *Atheta* (*Xenota*)], in *Mocyta*: Gusarov 2003a [synonymy], Smetana 2004 [synonymy], Klimaszewski et al. 2007a,b, 2008a,b, Majka and Klimaszewski 2008a,b, McLean et al. 2009, Klimaszewski et al. 2011, 2015b, 2018a, b, 2020, Ernst and Buddle 2013.

Diagnosis Body oval, length 2.4–3.0 mm; colour uniformly dark brown to black, in some specimens body black and posterior or central part of elytra with reddish tinge, appendages light brown (Fig. 19.286a); antennomeres I–IV elongate and V–X subquadrate or slightly transverse; pronotum broad, transverse, rounded laterally and arcuate basally; elytra transverse, variable in length, and about as long as pronotum or longer; abdomen broadly arcuate laterally. MALE. Median lobe of aedeagus as illustrated (Fig. 19.286b,c) [males absent in North America]. FEMALE. Tergite VIII broadly arcuate apically (Fig. 19.286d); sternite VIII truncate apically with fringe of microsetae, distance between antecostal suture and base of disc narrow, antecostal suture sinuate (Fig. 19.286e); spermatheca variable in shape, with capsule pear-shaped, and stem variably twisted posteriorly (Fig. 19.286f–j).

This species is externally very similar to *M. amblystegii* and may only be identified with certainty by the shape of the spermatheca. The presence of males in Canadian populations of *M. amblystegii* and lack of males in Canadian populations of *M. fungi* may also aid in the identification of these species.

DISTRIBUTION Origin: Palearctic, adventive in North America. **CANADA**: AB, BC, LB, MB, NB, NF, NS, NU, ON, PE, QC, SK, YT. **USA**: AK, CA, ME, MA, MN, NY, OR, RI, WA.

COLLECTION AND HABITAT DATA
Habitat: found in various forest types and wetlands, usually in litter. In NB, in silver maple swamps in leaf litter, in leaf litter in eastern white cedar swamps and forests, under alders in leaf litter and moss in a marsh, treading vegetation in boggy marsh and a cattail marsh, river margin on moist clay soil among tall grasses, a sea beach in

decaying sea wrack, in pitfall traps in a red spruce forest. In NF, from pitfall traps in cut and burned balsam fir, birch, spruce-poplar and riparian forests, in agricultural fields and among vegetation on coastal sand dunes. In QC, in mixed forests and yellow birch forests. In AK, in forest edge *Populus tremuloides, Rubus idaeus*; *Spirea*, dwarf birch; *Salix* leaf litter; clear-cuts; creekside/ocean beach confluence, under boards and drift wood; and under rocks. **Collecting period**: V–IX. **Collecting method**: pitfall traps, sifting forest litter, treading marsh vegetation.

Additional records. See Annex 1.

COMMENTS This species was listed from CO, NH, NV, SC by Newton (2019) but we are not aware of published voucher data for these records nor have examined specimens from these localities.

***Discreta* species group** (Klimaszewski et al. 2015b)

Diagnosis Pronotum large, transversely orbicular with rounded lateral edges and hind angles, usually much broader than elytra, pubescence on disc directed posteriad along midline and obliquely posteriad elsewhere; elytra short, as long as pronotum or shorter.

212. *Mocyta sphagnorum* Klimaszewski and Webster ‡ (Fig. 19.287a–h)

Mocyta sphagnorum Klimaszewski and Webster, in Klimaszewski et al. 2015b. Klimaszewski et al. 2016c,d, 2018a.

Diagnosis Body ovoid, length 2.4–2.7 mm; colour uniformly brown to almost black, legs and palps yellowish to reddish-brown and antennae uniformly light brown to brown (Fig. 19.287a,b); antennomeres I–IV elongate and V–X variable in length from subquadrate to slightly transverse; pronotum transverse, variable in width, from slightly to distinctly broader than elytra [pronotum usually broader in females than in males], highly rounded laterally, and arcuate basally; elytra transverse and slightly shorter than pronotum; abdomen broadly arcuate laterally and with strong macrosetae apically. MALE. Tergite VIII arcuate apically (Fig. 19.287d); sternite VIII parabolic apically and with numerous coarse macrosetae in apical part of disc, space between base of disc and antecostal suture narrow, antecostal suture arcuate (Fig. 19.287e); median lobe of aedeagus in lateral view with narrowly elongate bulbus and tubus straight ventrally, subapical part narrow, apex rounded, internal sac structures as illustrated (Fig. 19.287c). FEMALE. Tergite and sternite VIII truncate apically (Fig. 19.287f,g); spermatheca pear-shaped with small and shallow apical invagination, stem thin and irregularly coiled posteriorly (Fig. 19.287h).

This species differs from other *Mocyta* species in having the pronotum larger and dark brown to black, and in the shape of the spermatheca.

DISTRIBUTION Origin: Nearctic or Holarctic (status not certain, see Comments). **CANADA**: NB, NF, ON, QC, SK. **USA**: AK [**NSR**].

COLLECTION AND HABITAT DATA

Habitat: black spruce forests, old-growth eastern white cedar swamps, and calcareous fens in moist sphagnum moss and sphagnum moss and litter; also in moldy conifer duff at the base of a large white pine in spruce forest, in moist moss in a mixed forest; on summit of Mt-Albert (~1000 m) specimens collected by sweeping grasses/*Carex* of tundra. In AK, found in birch bark. **Collecting period**: IV–VII. **Collecting method**: sifting forest litter, pitfall traps, sweeping field vegetation.

We report this species from AK (**NSR)** for the first time here based on the following: USA, Alaska, Fairbanks, 166m el., 64.88281, −147.78452, ±104m pitfall slotcup, 2–9.VI.2009, Jozef A. Slowik (UAM) 2 specimens; Fairbanks, 930 Fitz Ct, 280m el., 64.9013, −147.52861, ±30m birch bark, 25.IV.2011, D.S. Sikes (UAM) 5 specimens.

COMMENTS The status of this species in North America is not clear. It may be Holarctic because there are specimens from Finland that share a CO1 BIN cluster with a Paratype from NB, and specimens from Connecticut, USA.

Genus *Philhygra* Mulsant and Rey, 1873a

(Figs. 19.288–19.303) (Lohse et al. 1990. Klimaszewski et al. 2011, 2018a, 2020)

Diagnosis Body elongate, subparallel; colour dark brown to black; head orbicular or subquadrate, more-or-less narrower than pronotum; antennae long and slender, antennomeres V–X elongate to slightly transverse; infraorbital carina partial or complete; pronotum at most one quarter wider than long, broadest at middle of its length, sides arcuate, hind angles rounded, pubescence on disc directed anteriad on apical two-thirds of midline and posteriad on rest of midline, laterad elsewhere; elytra at suture about as long as pronotum or longer; abdomen subparallel; apex of process of mesoventrite rounded; coxae nearly contiguous; male tergite VIII truncate apically and without secondary sexual characters; male sternite VIII very elongate, often projecting posteriorly under pygidium; median lobe of aedeagus with tubus enlarged and of complex forms, bulbus large, internal sac often with dentate structures; spermatheca vestigial in most species and difficult to find; female pygidium with often complex membranous structures.

Comments This genus is rich in species in Canada and many species remain undescribed. Species are affiliated with riparian and wetland habitats, and one species is affiliated with fungi.

Key to Species of *Philhygra*

1. Eyes small, streamlined with outline of head, length of each seen dorsally shorter than length of postocular region of head in dorsal view (Fig. 19.293a); tubus of median lobe of aedeagus evenly narrow and strongly projecting ventrad apically in lateral view, ventral margin entire without teeth (Fig. 19.293b), in dorsal view apex of tubus split apically (Fig. 19.293c,d) ....***Philhygra junii*** **Lohse**

– Eyes moderately large, slightly protruding from outline of head in most, length of each greater than postocular region of head in dorsal view; genital structures not as above **2**

2. Antennomeres VII–X distinctly elongate (Fig. 19.289a); median lobe of aedeagus with tubus highly sinuate subapically in lateral view (Fig. 19.289b); female pygidium as illustrated (Fig. 19.289h)... ....***Philhygra botanicarum*** **(Muona)**

– Antennomeres VII–X subquadrate or transverse; other characters not as above..................**3**

3. Antennae moderately robust, antennomeres VII–X distinctly transverse (Figs. 19.294a, 19.297a, 19.299a, 19.301a **4**

– Antennae slender, antennomeres VII–X subquadrate to slightly transverse (Figs. 19.291a, 19.292, 19.298a) **7**

4. Elytra at suture slightly longer than pronotum along midline (Fig. 19.299a); median lobe of aedeagus with tubus moderately narrow and straight ventrally in lateral view (Fig. 19.299b,c)....***Philhygra ripicola*** **(Hanssen)**

– Elytra at suture about as long as pronotum along midline; median lobe of aedeagus with tubus differently shaped **5**

5. Forebody with head dark brown and elytra and pronotum pale brown (Fig.19.294a); male sternite VIII with basal margin emarginate at middle (Fig. 19.294e); tubus of median lobe of aedeagus narrow and strongly produced ventrad (Fig. 19.294b).................***Phlihygra leechi*** **Lohse**

– Forebody more or less evenly dark brown (Figs. 19.297a, 19.301a); male sternite VIII and tubus of median lobe of aedeagus not as above...**6**

6. Body robust, length 2.7–2.8 mm (Fig. 19.297a); apical margin of male tergite VIII crenulate (Fig. 19.297c); median lobe of aedeagus with tubus broad and strongly narrowed to acute apex in lateral view (Fig. 19.297b); female terminalia as illustrated (Fig. 19.297e-g). ......***Philhygra pseudolarsoni*** **Klimaszewski and Godin**

– Body slender, length over 2.8 mm (Fig. 19.301a); apical margin of male tergite VIII entire and without crenulation (Fig. 19.301d); median lobe of aedeagus with tubus shaped differently (Figs. 19.301b); female terminalia different (Fig. 19.301f-h) ....

..........*Philhygra sinuipennis* **Klimaszewski and Langor**

7. Body length 4.0–4.5 mm; median lobe of aedeagus with tubus broad and bearing subapical tooth in lateral view (Fig. 19.298b); female terminalia as illustrated (Fig. 19.298f-h).............
......*Philhygra pseudopolaris* **Klimaszewski and Langor**

– Body length 1.8–3.8 mm; other characters not as above **8**

8. Tubus of median lobe of aedeagus extremely elongate, rod-like, sinuate and gradually produced ventrad in lateral view (Fig. 19.291b); female terminalia as illustrated (Fig. 19.291f–h) .... *Philhygra falcifera* **Lohse**

– Tubus of median lobe of aedeagus and female terminalia not as above **9**

9. Body slender, length 1.8–2.7 mm (Fig. 19.292a)...**10**

– Body robust, length 3.2–3.8 mm **11**

10. Tubus of median lobe of aedeagus abruptly and angularly bent ventrad in apical fifth, apical part narrow, subacute in lateral view (Fig. 19.296a); male sternite VIII truncate apically (Fig. 19.296f).........................*Philhygra pseudoboreostiba* **Lohse**

– Tubus of median lobe of aedeagus broad, slightly swollen apically, apex rounded, with one small subapical tooth in lateral view (Fig. 19.292b); male sternite VIII broadly rounded apically (Fig. 19.292e)..........................
. *Philhygra jarmilae* **Klimaszewski and Langor**

11. Pronotum slightly transverse (Figs. 19.288a, 19.290a, 19.300a) **12**

– Pronotum distinctly transverse, at least 1.2 times as wide as long **14**

12. Body uniformly dark brown (Fig. 19.290a); tubus of median lobe of aedeagus straight and narrow, slightly widening apicad with apex hooked and pointed ventrad in lateral view (Fig. 19.290b); female terminalia as illustrated (Fig. 19.290e-g)..................................
......*Philhygra donnellyana* **Klimaszewski, sp.n.**

– Body bicolored, head, pronotum and abdomen dark brown to black and elytra reddish brown or pale brown..**13**

13. Head, pronotum and abdomen dark brown to black and elytra pale reddish brown (Fig. 19.288a); tubus of median lobe of aedeagus hook-like, with dorsal margin curved ventrad in apical half in lateral view (Fig. 19.288b); female sternite VIII broadly rounded apically (Fig. 19.288f); terminalia as illustrated (Fig. 19.288e-g) .... *Philhygra angusticauda* **(Bernhauer)**

– Head, pronotum and abdomen dark brown and elytra pale brown (Fig. 19.300a); tubus of median lobe of aedeagus claw-like, with dorsal margin almost straight, perpendicular to bulbus in lateral view (Fig. 19.300b), and female sternite VIII obtusely angulate medioapically (Fig. 19.300g); terminalia not as above (Fig. 19.300f-h)........*Philhygra rostrifera* **Lohse**

14. Body highly glossy, more-or-less uniformly dark brown to black (Fig. 19.295a); median lobe of aedeagus and terminalia of both sexes as illustrated (Fig. 19.295b–h)...................
...............................*Philhygra malleoides* **Lohse**

– Body moderately glossy, dark brown with elytra yellowish-brown, reddish-brown, or dark brown mottled with reddish-brown patches (Figs. 19.302a, 19.303a); genital structures shaped differently...**15**

15. Median lobe of aedeagus with tubus broad in lateral view (Fig. 19.302b); male tergite VIII subquadrate (Fig. 19.302c); female terminalia as illustrated (Fig. 19.302e–g)..................................
.. *Philhygra terrestris* **Klimaszewski and Godin**

– Median lobe of aedeagus with tubus narrow in lateral view (Fig. 19.303b); male tergite VIII elongate (Fig. 19.303d); female terminalia shaped differently (Fig. 19.303f-h) ....
..................................*Philhygra varula* **(Casey)**

213. *Philhygra angusticauda* (Bernhauer) * (Fig. 19.288a–g)

Atheta (Metaxya) angusticauda Bernhauer, 1909. Hatch 1957 [cited as *Ischnopoda (Brundinia)*], in *Philhygra*: Seevers 1978, Klimaszewski and Winchester 2002, Webster et al. 2009, Klimaszewski et al. 2018a, 2020, Brunke et al. 2021 [Holarctic; *Atheta (Philhygra) pinegensis* Muona, 1983 as synonym].

Diagnosis Body moderately flattened, narrowly subparallel, length 3.0–3.2 mm; colour uniformly dark brown to nearly black, elytra with rust tinge, legs usually slightly paler and reddish brown, antenna uniformly dark brown or reddish brown (Fig. 19.288a); integument moderately glossy, forebody moderately densely punctate and pubescent, pubescence long, punctation fine and slightly asperate, entire forebody with distinct microsculpture, sculpticells consist of round and slightly convex meshes; head slightly elongate, round, approximately as wide as pronotum, eyes about as long as genae in dorsal view, strong basally, diffused posteriorly; antennae slender and long, about as long as the pronotum and elytra combined, basal 3 antennomeres distinctly elongate, IV–V moderately elongate, VI–VII quadrate, VIII–X slightly transverse, terminal one elongate and as long as two preceding antennomeres combined; pronotum impressed mesally, slightly transverse, emarginate laterally and posteriorly, approximately rectangular in shape, broadest at apical third, hypomeron visible for nearly entire length of pronotum in lateral view, pubescence on disc directed anteriad and anterolaterad along midline, laterad basally, and obliquely laterad along arcuate lines elsewhere; elytra moderately convex, slightly elongate, and insignificantly broader than pronotum, posterior margins broadly arcuate and with shallow lateral emarginations, pubescence on disc directed slightly obliquely posterolaterad to nearly straight laterad at base; abdomen subparallel, slightly glossy, three first visible tergites deeply impressed at base; hind tarsus with basal article as long as following one. MALE. Tergite VIII truncate apically (Fig. 19.288c); sternite VIII highly elongate, broadly parabolic apically, antecostal suture widely separated from basal margin (Fig. 19.288d); median lobe of aedeagus in lateral view with tubus narrowly elongate, apical half hook-like, strongly curved ventrad, apex sharply pointed (Fig. 19.288b), internal sac sclerites as illustrated (Fig. 19.288b), in dorsal view median lobe elongate, with broadly oval bulbus and narrow tubus. FEMALE. Tergite VIII unevenly truncate apically (Fig. 19.288e); sternite VIII with apical margin arcuately produced except at lateral angles, antecostal suture sinuate, widely separated from basal margin (Fig. 19.288f); pygidium as illustrated (Fig. 19.288g); spermatheca (not illustrated) minute with short sac-shaped capsule without apical invagination and with short and narrow stem.

DISTRIBUTION Origin: Holarctic. Europe, Russia (European, Far East), North America. **CANADA**: AB, BC, NB. **USA**: AK, NH.

COLLECTION AND HABITAT DATA
Habitat: in NB, specimens collected from gravel on gravel bars along shaded brooks in a northern hardwood forest (common), a boreal forest (white spruce, black spruce, balsam fir), a mature mixed forest, and in an alder swamp in debris on muddy soil near a brook flowing into a river; in BC, from pitfall traps in clearcut and transition zone of Sitka spruce forest; in AB in Malaise trap in valley with creek. **Collecting period**: V–VIII. **Collecting method**: splashing gravel and turning gravel on gravel bars along brooks, sifting litter on brook margin, pitfall traps, Lindgren traps, Malaise trap.

Additional records. See Annex 1.

COMMENTS This species was recently discovered to be Holarctic, with Palaearctic populations previously known under the synonym *Atheta* (*Philhygra*) *pinegensis* Muona (Brunke et al. 2021).

214. *Philhygra botanicarum* (Muona) * (Fig. 19.289a–h)

Atheta (*Philhygra*) *botanicarum* Muona, 1983. Muona 1984 [in *Atheta* (*Philhygra*)], in *Philhygra*: Klimaszewski et al. 2008a, Majka and Klimaszewski 2008a, 2008b, Webster et al. 2009, Klimaszewski et al. 2011, 2015a, 2018a, 2020.

Diagnosis Body narrowly oval, broadest at elytra, length 3.8–4.5 mm; colour dark brown to piceous, with bases of antennae, legs, elytra except for scutellar region, and tip of abdomen rusty-brown or brownish yellow (Fig. 19.289a); integument moderately glossy, forebody with fine, dense punctation and meshed microsculpture; head round, slightly narrower than pronotum; eyes longer than genae in dorsal view; antennae slender with all antennomeres elongate (more elongate in females); pronotum slightly transverse and almost as wide as elytra; elytra at suture slightly longer than pronotum; abdomen subparallel. MALE. Tergite VIII shallowly, widely emarginate medioapically (Fig. 19.289d); sternite VIII highly elongate, apical margin truncate medially (Fig. 19.289e); median lobe of aedeagus with tubus broadening apically and with finely hooked apex in lateral view (Fig. 19.289b); in dorsal view bulbus broadly oval, tubus broad, triangular apically (Fig. 19.289c); internal sac structures complex as illustrated (Fig. 19.289b,c). FEMALE. Tergite VIII with apical margin arcuate on each side of obtusely angulate medial emargination (Fig. 19.289f); sternite VIII with apical margin triangularly produced, apex broadly rounded medially (Fig. 19.289g); pygidium as illustrated (Fig. 19.289h); spermatheca (not illustrated) minute with short sac-shaped capsule without apical invagination and with short and narrow stem.

DISTRIBUTION Origin: Holarctic. Europe (Finland), North America. **CANADA**: AB [**BOLD**], BC, LB, NB, NF, NS, ON, SK, YT. **USA**: AK [**NSR**].

COLLECTION AND HABITAT DATA
Habitat: typically found in moist leaves (often on muddy soils) near margins of vernal ponds or small brooks in alder swamps, mixed forests, northern hardwood forests, a coniferous forest, a spruce forest, and a wet willow stand; also in drift material on a lakeshore, in moist leaves under a sap flow of yellow birch, and collected by treading vegetation in a seasonally flooded marsh on pond margin. In AB, found in montane forest, douglas fir and lodgepole pine stand with aspen and birch understory. In AK, found in mesic *Betula, Picea* forest and near a creek. **Collecting period**: V–VIII. **Collecting method**: sifting moist leaf litter and drift material, treading marsh vegetation, flight intercept traps, sweeping vegetation.

We report this species for the first time from AB (**NPR**) based on **BOLD** records, and from AK (**NSR**) based on UAM specimens: Canada, Alberta, Waterton Lakes National park, Highway 6 pullout, 1569m el., 49.065, –113.779, montane douglas fir and lodgepole pine forest with aspen and birch understory, sweep net, 21.VI.2012, BIOBus 2012 team (CBG) 1 specimen. USA, Alaska, Fairbanks, Creamer's Field, 132m el., 64.86805, –147.73764, ±20m mesic *Betula, Picea*, pitfall, 15-18.VII.2008, D.S. Sikes, C. Eshleman (UAM) 1 specimen; Horner Hot Springs Crk, by jct w/ Yukon R., 61m el., 64.91582, –154.83471, ±6m falls by creek, 15-16.VII.2010, J. A. Slowik (UAM) 1 specimen.

215. *Philhygra donnellyana* Klimaszewski sp. n. (Fig. 19.290a–g)

urn:lsid:zoobank.org:act:9EF6F9E9-36BA-49D5-B00E-15917C8933B4

Holotype (male): Alaska, Donnelly Campground [Donnelly Creek State Recreation Site and Campground], Rich. [Richardson] Highway, 10.VIII.1974, J. Matthews and A. Sigovin (CNC). Paratype: same label data as the holotype (CNC) 1 female.

Etymology The species name derives from the name Donnelly Campground, where the holotype was captured.

Diagnosis Body robust, uniformly dark brown to almost black, appendages of same color, length 3.6–3.8 mm (Fig. 19.290a); head circular; antennae slender, antennomeres VII–X subquadrate; pronotum slightly transverse; elytra at suture as long as pronotum along midline; abdomen subparallel. MALE. Tergite VIII unevenly truncate apically (Fig. 19.290c); sternite VIII elongate with apical margin parabolic except broadly arcuate medially (Fig. 19.290d); median lobe of aedeagus in lateral view with bulbus narrowly oval, tubus of median lobe of aedeagus narrow, straight ventrally, slightly widening apicad, apex hooked and pointed ventrad (Fig. 19.290b), internal sac with coarse membraneous structures (Fig. 19.290b). FEMALE. Tergite VIII truncate apically (Fig. 19.290e); sternite VIII with apical margin broadly rounded (Fig. 19.290f); pygidium as illustrated (Fig. 19.290g).

This species may be distinguished from all its congeners by the shape of the median lobe and structures of the internal sac of the aedeagus.

DISTRIBUTION Origin: Nearctic. **CANADA**: not recorded. **USA**: AK.

COLLECTION AND HABITAT DATA
Habitat: not recorded. **Collecting period**: VIII. **Collecting method**: not recorded.

216. *Philhygra falcifera* Lohse (Fig. 19.291a–h)

Philhygra falcifera Lohse, in Lohse et al. 1990, Klimaszewski et al. 2016d.

Diagnosis Body robust, brownish-black, appendages yellowish, length 3.2–3.7 mm (Fig. 19.291a); head circular; antennae slender; pronotum slightly transverse; elytra at suture as long as pronotum along midline; abdomen subparallel. MALE. Tergite VIII truncate apically (Fig. 19.291d); sternite VIII elongate and arcuate apically (Fig. 19.291e); median lobe of aedeagus with bulbus narrowly oval, narrowed apically, tubus extremely long, narrowly triangular in dorsal view (Fig. 19.291b); in lateral view bulbus with large crista apicalis, tubus curved obliquely ventrad basally, then extremely long, arcuate, narrowly tapering to acutely pointed apex (Fig. 19.291c), internal sac with coarse membraneous structures (Fig. 19.291b,c). FEMALE. Tergite VIII broadly arcuate apically (Fig. 19.291f); sternite VIII with apical margin obtusely triangular, obtusely angulate medially (Fig. 19.291g); pygidium as illustrated (Fig. 19.291h). Externally, similar to *P. polaris*, but has a smaller body and differently shaped median lobe of aedeagus.

DISTRIBUTION Origin: Nearctic. **CANADA**: MB, SK. **USA**: not recorded.

COLLECTION AND HABITAT DATA
Habitat: not recorded. **Collecting period**: VI, VIII. **Collecting method**: sifting litter.

217. *Philhygra jarmilae* Klimaszewski and Langor (Fig. 19.292a–h)

Philhygra jarmilae Klimaszewski and Langor, in Klimaszewski et al. 2011. Webster et al. 2012, Brunke et al. 2012, Klimaszewski et al. 2012, 2015a, 2018a, b.

Diagnosis Body narrowly subparallel, length 2.5–2.7 mm; head and abdomen dark brown, pronotum, elytra, bases of antennae light reddish-brown (Fig. 19.292a); integument moderately glossy, forebody with fine, dense punctation and meshed microsculpture; head round, slightly narrower than pronotum, eyes as long as genae in dorsal view; pronotum slightly transverse, widest at middle, margined laterally, and almost as wide as elytra; elytra at suture longer than pronotum; antennae slender, antennomeres IV–VI subquadrate, and VII–X slightly transverse. MALE. Tergite VIII broadly arcuate apically (Fig. 19.292d); sternite VIII very broadly parabolic apically (Fig. 19.292e); median lobe of aedeagus in lateral view with tubus evenly broad, elongate, subapical part slightly swollen with a basal tooth, apex rounded (Fig. 19.292b), in dorsal view bulbus roughly oval, tubus narrow and rounded at apex (Fig. 19.292c). FEMALE. Tergite VIII arcuate apically (Fig. 19.292f); sternite VIII broadly rounded apically (Fig. 19.292g); pygidium as illustrated (Fig. 19.292h); spermatheca (not illustrated) minute with short sac-shaped capsule without apical invagination and with short narrow stem.

DISTRIBUTION Origin: Nearctic. **CANADA**: AB [**BOLD**], MB, NB, NF, NT [**BOLD**], ON, SK, YT. **USA**: AK [**NSR**].

COLLECTION AND HABITAT DATA
Habitat: occurs in various riparian and wetland habitats; adults sifted from moist leaves on vernal pond margins in mixed forests, hardwood forests, and a silver maple swamp, sifted from sedges and leaves on pond margin, treaded from *Carex* and grasses on lake margin, treaded from vegetation in seasonally flooded marsh, sifted from moss and leaves in seepage area in hardwood forest and

debris on muddy soil near brook, from drift material on river margin, and sifted from leaf litter and moss under alders near brook. In AB, found in montane forest, Douglas fir and lodgepole pine stand with aspen and birch understory and in wetland. In AK, found on sandy beach + under beach rocks and in *Spiraea*, dwarf birch. **Collecting period**: V–VII, IX–X. **Collecting method**: sifting litter, treading vegetation, splashing moss, hand collecting from cobblestones on partially shaded cobblestone bar along stream, from moss near splash zone of waterfalls by splashing, flight intercept traps, pitfall traps.

We report this species for the first time from AB (**NPR**), NT (**NTR**) based on **BOLD** records, and from AK (**NSR**) based on UAM specimens: Canada, Alberta, Waterton Lakes National park, Highway 6 pullout, 1569m el., 49.065, −113.779, montane Douglas fir and lodgepole pine forest with aspen and birch understory, pitfall, 21-27. VI.2012, BIOBus 2012 team (CBG) 3 specimens; Elk Island National Park, Wood bison trail, 722m el., 53.567, −112.851, wetland alongside aspen forest, flight-intercept trap, 29.VI-03.VII.2012, BIOBus 2012 team (CBG) 1 specimen. Northwest Territories, Fort Providence, Deh Gah Elementary & Secondary School, 152m el., 61.358, −117.66, Malaise trap, 22.IX-03.X.2014, James Hatch (CBG) 1 specimen. USA, Alaska, Norutak Lake, 229m el., 66.79609, −154.27638, ±19m on sandy beach + under beach rocks, 23.VI.2010, D.S. Sikes (UAM) 5 specimens; Selawik NWR, 8m el., 66.56137, −158.99834, ±304m *Spirea*, dwarf birch, 23-24.VI.2010, D.S. Sikes (UAM) 1 specimen.

218. *Philhygra junii* Lohse (Fig. 19.293a–i)
Philhygra junii Lohse, in Lohse et al. 1990.

Diagnosis Body subparallel, length 2.3–2.8 mm; black, antennae and elytra dark brown, mouthparts and legs yellow-brown (Fig. 19.293a); integument weakly glossy, forebody with punctation invisible on head, and distinct and slightly asperate on elytra, fine and sparse on abdomen; rounded meshed microsculpture present; head round, in male with large and deep impression, eyes small, as long as 2/3 of genae in dorsal view; pronotum transverse, 1 1/6 as broad as head, widest in apical third, margined laterally; elytra at suture slightly shorter than pronotum along midline, sides slightly divergent toward apex; abdomen subparrallel; antennae slender, antennomeres IV–VI subquadrate, and VII–X slightly transverse, more so in the female. MALE. Tergite VIII with apical margin broadly arcuate, less so mediad (Fig. 19.293e); sternite VIII elongate, with apical margin truncate medially, broadly rounded laterally (Fig. 19.293f); median lobe of aedeagus in lateral view with tubus evenly narrow, elongate, apical part obtusely triangular, produced ventrad, apex rounded (Fig. 19.293b), in dorsal view bulbus roughly oval, tubus narrow, broadly expanding apicad, apex emarginate, appearing vaguely heart-shaped (Fig. 19.293c,d). FEMALE. Tergite VIII very broadly arcuate apically (Fig. 19.293g); sternite VIII arcuate apically (Fig. 19.293h); pygidium as illustrated (Fig. 19.293i); spermatheca not found.

DISTRIBUTION Origin: Nearctic. **CANADA**: NU [**NTR**], YT. **USA**: not recorded.

COLLECTION AND HABITAT DATA
Habitat: in leaf litter among low vegetation, wet leaf litter and flood debris along a freshwater creek, moist meadow, tundra near river. **Collecting period**: VI–VII. **Collecting method**: sifting organic litter.

We report this species from NU (**NTR**) for the first time here, based on the following: Canada, Nunavut, Kitikmeot, Kugluktuk, 7.5km SW of Kugluktuk, 30m el., 67.786, −115.244, tundra, leaf litter among low vegetation, sifting, 18. VII.2019, G. Blagoev (CBG) 1 specimen; Kitikmeot, Kugluktuk, Four Mile Bay, 8.6km West of Kugluktuk, 1m el., 67.853, −115.277, wet leaf litter and flood debris along a freshwater creek, sifting, 21.VII.2019, M. Pentinsaari (CBG) 1 specimen; Kitikmeot, Kugluktuk, 40m el., 67.817, −115.109, leaf litter at the base of a cliff, sifting, 25.VII.2019, M. Pentinsaari (CBG) 2 specimens.

219. *Philhygra leechi* Lohse (Fig. 19.294a–e)

Philhygra leechi Lohse, in Lohse et al. 1990. Klimaszewski et al. 2012.

Diagnosis Body subparallel, dark brown with pronotum, elytra, hind borders of abdominal tergites and apex of abdomen and appendages light brown (Fig. 19.294a); length 2.2–2.4 mm; surface slightly glossy; punctation on sides of head and pronotum fine and dense; weak microsculpture present; head slightly transverse, sides rounded, broadest at middle of postocular region; eyes moderately sized, each about ¼ shorter length of postocular region in dorsal view; suborbical carinae incomplete; antennomeres V–X transverse, becoming broader to apex; pronotum about 1 ¼ as broad as head, slightly transverse, broadest in apical third, arcuately narrowed posteriad to rounded hind angles; elytra subparallel, 1 ¼ as broad as pronotum at suture as long as pronotum; abdomen subparallel. MALE. Tergite VIII with apical margin truncate medially, rounded laterally (Fig. 19.294d); sternite VIII with apical margin broadly parabolically rounded, basal margin broadly emarginate medially (Fig. 19.294e); median lobe of aedeagus with oval bulbus, and narrow, subparallel tubus in dorsal view (Fig. 19.294c); in lateral view bulbus broadly elongate, tubus narrow sinuate and strongly bent ventrad in apical third (Fig. 19.294b). FEMALE (not examined). Tergite and sternite VIII moderately arcuate apically (description from Lohse et al. 1990).

DISTRIBUTION Origin: Nearctic. **CANADA**: MB, NT, YT. **USA**: not recorded.

COLLECTION AND HABITAT DATA
Habitat: not recorded. **Collecting period**: VI. **Collecting method**: not recorded.

Additional records. See Annex 1.

220. *Philhygra malleoides* Lohse* (Fig. 19.295a–h)

Philhygra malleoides Lohse, in Lohse et al. 1990. Klimaszewski et al. 2011, 2018a, Sazhnev 2018.

Diagnosis Body narrowly subparallel, length 2.9–3.3 mm; colour dark brown to black, with elytra except for scutellar region, and antennae often dark reddish-brown, legs yellowish-brown (Fig. 19.295a); integument moderately glossy; forebody with meshed microsculpture, fine, dense punctation, asperate on pronotum and elytra; head round, about as wide as pronotum; eyes large, subequal in length to genae in dorsal view; antennae slender, antennomeres IV–X subquadrate to slightly transverse; pronotum moderately transverse, widest at apical third, narrower than elytra; elytra at suture distinctly longer than pronotum; abdomen subparallel. MALE. Tergite VIII arcuate apically (Fig. 19.295d); sternite VIII elongate, broadly arcuate apically (Fig. 19.295e); median lobe of aedeagus with tubus narrowly elongate and strongly bent ventrally in fourth of its length in lateral view (Fig. 19.295b), in dorsal view bulbus moderate-sized, oval, tubus narrowly elongate, apex spatulate (Fig. 19.295c). FEMALE. Tergite VIII unevenly arcuate apically (Fig. 19.295f); sternite VIII with apical margin obtusely triangularly produced, arcuate medially (Fig. 19.295g); pygidium as illustrated (Fig. 19.295h); spermatheca not found.

DISTRIBUTION Origin: Holarctic; reported from Russia (Far East (Bering Is.)) by Sazhnev (2018). **CANADA**: MB, NF, NT, NU [**BOLD**], QC, YT. **USA**: AK.

COLLECTION AND HABITAT DATA
Habitat: tundra, *Salix* leaf litter, under rocks, black spruce forest, and black spruce bog. **Collecting period**: VI–VIII. **Collecting method**: litter processing through Berlese, sweeping vegetation in tundra, hanging Malaise, aspirating from under rocks.

We report this species from NU (**NTR**) for the first time here, based on **BOLD** records: Canada, Nunavut, Kitikmeot, Kugluktuk, S Lake, 9 km

SW of Kugluktuk, 23m el., 67.781, −115.276, tundra, soil and leaf litter, Berlese funnel extraction, 18.VII.2019, G. Blagoev, M. Pentinsaari (CBG) 1 specimen; Kitikmeot, Kugluktuk, S Lake, 9 km SW of Kugluktuk, 17m el., 67.7726, −115.269, plant debris on sandy lake shore, sifting, 22.VII.2019, M. Pentinsaari (CBG) 1 specimen; Kitikmeot, Kugluktuk, Kugluk (Bloody Falls) Territorial Park, 20m el., 67.7431, −115.372, leaf litter under Salix bushes, sifting, 20.VII.2019, G. Blagoev, M. Pentinsaari (CBG) 1 specimen; Kitikmeot, 4.5km West of Kugluktuk, 1m el., 67.8346, −115.2000, sand dunes at the mouth of a freshwater creek, leaf litter and flood debris, sifting, 21.VII.2019, M. Pentinsaari (CBG) 2 specimens.

Additional records. See Annex 1.

221. ***Philhygra pseudoboreostiba*** **Lohse** (Fig. 19.296a–f)

Philhygra pseudoboreostiba Lohse, in Lohse et al. 1990.

Diagnosis Body slender, black, slightly glossy, length 3.2–3.7 mm; head circular, broadest at middle of postocular region of disc, punctation weak (Fig. 19.296a); eyes slightly convex; antennal slender, antennomeres VIII–X slightly transverse; pronotum distinctly broader than head (1 1/5 as broad as head), broadest at middle, sides arcuate, narrowed posteriorly to rounded hind angles, punctation fine and moderately dense; elytra 1 ¼ as broad as pronotum, at suture about as long as pronotum along midline of disc, punctation dense, scattered near apex; abdomen subparallel. MALE. Tergite VIII truncate apically (Fig. 19.296e); sternite VIII moderately elongate, with apical margin shallowly emarginate medially, rounded laterally (Fig. 19.296f); in dorsal view median lobe of aedeagus pear-shaped, tubus moderately broad and rounded apically (Fig. 19.296d); in lateral view tubus curved obliquely ventrad basally, then slightly sinuate, abruptly and angularly bent ventrad in apical fifth, apical part narrow, subacute (Fig. 19.296b,c). FEMALE. Tergite as in male, and sternite VIII slightly arcuate apically (original description from Lohse et al. 1990).

DISTRIBUTION Origin: Nearctic. **CANADA**: YT. **USA**: not recorded.

COLLECTION AND HABITAT DATA

Habitat: moss on open flat meadow or at edge of body of water. **Collecting period**: VI–VII. **Collecting method**: sifting moss and litter.

222. ***Philhygra pseudolarsoni*** **Klimaszewski and Godin** (Fig. 19.297a–g)

Philhygra pseudolarsoni Klimaszewski and Godin, in Klimaszewski et al. 2012. Webster et al. 2016, Klimaszewski et al. 2018a.

Diagnosis Body narrowly subparallel, length 2.7–2.8 mm; colour uniformly black or dark brown with legs and sutural part of elytra reddish-brown (Fig. 19.297a); integument moderately glossy, forebody with fine, dense punctation and meshed microsculpture; head round, distinctly narrower than pronotum; eyes as long as genae in dorsal view; antennae slender with antennomeres IV–V elongate, VI–X subquadrate to slightly transverse; pronotum slightly transverse and almost as wide as elytra; elytra at suture as long as or slightly longer than pronotum; abdomen subparallel. MALE. Tergite VIII truncate apically, finely crenulate (Fig. 19.297c); sternite VIII parabolic apically (Fig. 19.297d); in lateral view tubus of median lobe of aedeagus with ventral margin long, straight, with apical part broadly triangular, apex curved ventrad to a sharp point (Fig. 19.297b). FEMALE. Tergite VIII arcuate apically (Fig. 19.297e); sternite VIII with apical margin broadly triangular, apex obtusely angulate (Fig. 19.297f); pygidium as illustrated (Fig. 19.297g); spermatheca (not illustrated) minute with short sac-shaped capsule without apical invagination, stem short and narrow.

This species is similar in both external morphology and genitalia to *P. larsoni* Klimaszewski and Langor, but in *P. pseudolarsoni* the body is smaller and darker, antennomeres VI–X are quadrate or transverse, and the tubus of the median lobe of the aedeagus is more elongate apically in lateral view.

DISTRIBUTION **Origin**: Nearctic. **CANADA**: AB [**NPR**], MB [**BOLD**], NB, YT. **USA**: not recorded.

COLLECTION AND HABITAT DATA
Habitat: in wetland habitats in NB; specimens from litter, grasses, and moss on hummocks near water in old-growth eastern white cedar swamp, among emergent vegetation, grasses, and *Carex* hummocks on lake margins, in sphagnum and litter near brook in eastern white cedar swamp; one adult was collected with a butterfly net between 16:30 and 18:00 h in a mixed forest opening during a warm evening. Specimens from YT sifted from soil litter from deciduous and mixed forests. **Collecting period**: V–VI. **Collecting method**: treading emergent vegetation and *Carex* hummocks, sifting soil litter, sphagnum, and leaf litter, collecting with butterfly net during warm evening.

We report this species from AB (**NPR)** for the first time here, based on DNA barcoded specimens identified morphologically, and from MB (**NPR)** based on BOLD records: Canada, Alberta, Elk Island National Park, Astotin Lake, 719m el., 53.685, –112.86, emergent white birch/willow/trembling aspen forest, Malaise trap, 08-16.X.2012, Brett Sarchuk (CBG) 2 specimens. Manitoba, Churchill, 16 km E Churchill, krummholz formation, 5m el., 58.754, –93.913, pitfall, 16.VII.2009, P. Kevan, T. Woodcock (CBG) 1 specimen.

223. *Philhygra pseudopolaris* Klimaszewski and Langor (Fig. 19.298a–h)

Philhygra pseudopolaris Klimaszewski and Langor, in Klimaszewski et al. 2011. As *P. polaris* [misidentification]: Bernhauer 1906, 1907 [both in *Atheta*], Lohse et al. 1990. Klimaszewski et al. 2018a, Webster et al. 2020, Schülke and Smetana (2015)

Diagnosis Body narrowly subparallel, length 4.0–4.5 mm; colour uniformly dark brown, or head black, with pronotum, elytra except for scutellar region, legs and bases of antennae reddish-brown (Fig. 19.298a); integument moderately glossy, forebody with fine, dense punctation, and meshed microsculpture; head round, slightly narrower than pronotum; eyes slightly longer than genae in dorsal view; antennae slender, antennomeres IV–VI elongate and VII–X subquadrate to slightly transverse; pronotum moderately transverse, widest at apical third, slightly narrower than elytra and abdomen; elytra at suture slightly longer than pronotum; abdomen subparallel. MALE. Tergite VIII with posterior margin broadly arcuate (Fig. 19.298d); sternite VIII elongate, apical margin parabolic except truncate medially (Fig. 19.298e); median lobe of aedeagus in lateral view with tubus swollen apically, irregularly shaped, apex somewhat angular and rounded with subapical spine directed ventroposteriad (Fig. 19.298b); in dorsal view bulbus broadly, roughly oval, tubus narrow and sinuate apicad (Fig. 19.298c) [the median lobe of similar Palaearctic species, *P. polaris* (Bernhauer), has apical part of tubus in lateral view semicircular in shape and sharply pointed, and subapically there is only short dent pointed ventrally, see Fig. 19.338 in Klimaszewski et al. 2011]. FEMALE. Tergite VIII arcuate apically except slightly emarginate medially (Fig. 19.298f); sternite VIII moderately elongate, evenly rounded apically (Fig. 19.298g); pygidium as illustrated (Fig. 19.298h); spermatheca (not illustrated) minute with short sac-shaped capsule without apical invagination, stem short and narrow.

Externally this species is similar to the Palaearctic *P. polaris*, from which it differs in having the apical part of the tubus of the median lobe of the aedeagus shaped differently. For illustrations of *P. polaris*, see Palm (1970), Strand and Vik (1964), and Klimaszewski et al. (2011).

DISTRIBUTION **Origin**: Nearctic. **CANADA**: AB [**BOLD**], BC, MB, NF, NT, QC, YT. **USA**: AK, CA (not verified).

COLLECTION AND HABITAT DATA
Habitat: treaded from emergent grasses and sedges from small pond margin, open marsh with willows; *Salix* leaf litter, under rocks, and

lakeshore grasses. **Collecting period**: V–VIII. **Collecting method**: Berlese funnels, treading emergent vegetation, pitfall traps, aspirating from under rocks.

We report this species from AB [**NPR**] for the first time here, based on the following: Canada, Alberta, 13 km N Burmis, Chain Pond, marsh, 11. VII.1987, G.R. Pohl (NoFC) 1 specimen; Maybelle River Provincial Park, jack pine forest, 12. VI.2000, G.R. Pohl (NoFC) 1 specimen.

Additional records. See Annex 1.

COMMENTS This species was listed from Russia (Far East) by Schülke and Smetana (2015) but voucher specimens need to be carefully examined to ensure that the Russian specimens do not represent similar *P. polaris*.

224. *Philhygra ripicola* (Hanssen) * (Fig. 19.299a–h)

Atheta (Metaxya?) ripicola Hanssen, 1932. Brundin 1944, Smetana 2004.

Philhygra ripicoloides Lohse, in Lohse et al. 1990. Klimaszewski et al. 2011, 2015a, 2018a. **New Synonymy**.

Diagnosis Body narrowly subparallel, length 2.4–2.6 mm; head and median part of abdomen dark brown, pronotum dark reddish-brown, and elytra, legs and antennae yellowish-brown (Fig. 19.299a); integument moderately glossy, forebody with fine, dense punctation, and meshed microsculpture; head, pronotum, elytra and abdomen subequal in width; head round, eyes as long as genae in dorsal view; antennae slender, with article IV elongate, V–VI subquadrate, and VII–X subquadrate in male and slightly transverse in female, pronotum slightly transverse, subquadrate, widest at apical third; elytra at suture distinctly longer than pronotum; abdomen subparallel. MALE. Tergite VIII with apical margin broadly, arcuately produced except near lateral angles (Fig. 19.299d); sternite VIII elongate, with apical margin broadly parabolic except arcuate medially (Fig. 19.299e); median lobe of aedeagus in lateral view with narrowly oval bulbus, tubus narrowly elongate, straight ventrally, broader in apical half, apex narrowly rounded (Fig. 19.299b,c). FEMALE. Tergite VIII rounded apically and truncated medially (Fig. 19.299f); sternite VIII with apical margin obtusely produced, arcuate medially (Fig. 19.299g); pygidium as illustrated (Fig. 19.299h).

DISTRIBUTION Origin: Holarctic; Europe, Russia (East and West Siberia), Mongolia, North America. **CANADA**: NF, NT, SK, YT. **USA**: AK [**NSR**].

COLLECTION AND HABITAT DATA

Habitat: small stand with *Betula* and *Populus tremuloides*. **Collecting period**: IV–VIII. **Collecting method**: flight intercept trap.

We report this species from AK (**NSR**) for the first time here based on the following: USA: Alaska, Fairbanks, 930 Fitz Ct, 280m el., 64.9013, −147.52861, ±30m small stand *Betula* & *Pop. trem.*, flight intercept trap, 29.IV.29. V.2008, D.S. Sikes (UAM) 1 specimen.

Additional records. See Annex 1.

COMMENTS We have compared external morphology and the morphology of median lobe of aedeagus of *Philhygra ripicoloides* Lohse from NT (Fig. 19.299b), and the Palearctic species *Philhygra ripicola* (Hanssen) from Europe (Fig. 19.299c), and found no significant differences between them. On the original drawing of median lobe of *P. ripicoloides* (Lohse et al. 1990) the tubus of median lobe is shorter than that of the European species but we compared additional males from NT and these have length and shape of tubus of median lobe exactly like those of the European population. We therefore consider *Philhygra ripicoloides* Lohse as a synonym of *Philhygra ripicola* (Hanssen).

225. *Philhygra rostrifera* Lohse (Fig. 19.300a–h)

Philhygra rostrifera Lohse, in Lohse et al. 1990. Majka and Klimaszewski, 2008b, Klimaszewski et al. 2011, 2015a, 2018a.

Diagnosis Body very narrow, subparallel, length 3.5 mm; head, pronotum, median part of abdomen and antennomeres III–XI dark brown, and elytra with exception of scutellar region, legs and bases of antennae and apical part of abdomen rusty-brown (Fig. 19.300a); integument moderately glossy, forebody with fine, dense punctation, and meshed microsculpture; head slightly narrower than pronotum; pronotum, elytra and abdomen subequal in width; head round, eyes as long as genae in dorsal view; antennae slender, antennomeres IV–X subquadrate; pronotum slightly transverse, almost subquadrate, widest at apical third; elytra at suture as long as pronotum; abdomen subparallel. MALE. Tergite VIII transverse, with apical margin arcuate (Fig. 19.300d); sternite VIII elongate, with apical margin parabolic except broadly arcuate medially (Fig. 19.300e); median lobe of aedeagus in lateral view with tubus claw-shaped, long, angularly bent ventrad at middle, gradually narrowing in an arc to a fine point (Fig. 19.300b); in dorsal view bulbus large, broadest near base, tubus narrowly elongate (Fig. 19.300c). FEMALE. Tergite VIII truncate apically (Fig. 19.300f); sternite VIII broadly triangularly produced apically, apex obtusely subangulate (Fig. 19.300g); pygidium as illustrated (Fig. 19.300h); spermatheca not found.

DISTRIBUTION **Origin**: Nearctic. **CANADA**: LB, MB [**BOLD**], NT, NU [**BOLD**], SK, YT [**BOLD**]. **USA**: AK.

COLLECTION AND HABITAT DATA
Habitat: one specimen from NF collected by treading vegetation and sphagnum moss in a boggy area. In SK, specimens were found in wet willow stand; spruce/alder litter; fern and scat; and deciduous leaf litter in a ditch. In NT, Birch/spruce litter. In AK, found in mesic *Betula, Picea* forest. **Collecting period**: VI–VIII. **Collecting method**: treading vegetation and sifting moss and leaf litter.

We report this species for the first time from NU (**NTR**), and MB (**NPR**) based on DNA barcoded specimens identified morphologically, and confirm its presence in YT based on **BOLD** records matching the NU specimens: Canada, Nunavut, Kitikmeot, Kugluktuk, 20m el., 67.782, −115.276, lake shore, leaf litter, sifting, 18.VII.2019, M. Pentinsaari (CBG) 1 specimen; Kitikmeot, Kugluktuk, Four Mile Bay, 8.6km West of Kugluktuk, 1m el., 67.8531, −115.277, along freshwater creek in wet leaf litter and flood debris, sifting, 21.VII.2019, M. Pentinsaari (CBG) 1 specimen; Kitikmeot, Kugluktuk, S Lake, 9km SW of Kugluktuk, 23m el., 67.781, −115.276, tundra, soil and leaf litter, Berlese funnel extraction, 06.VIII.2019, K.Perez, V. Levesque-Beaudin (CBG) 1 specimen; Manitoba, 12 km S Churchill, Goose Creek Marina, open substrate, 9.VIII.2010, J. Witt (CBG), 1 specimen; same except: vegetation (CBG) 1 specimen; 16 km E Churchill, tundra ponds, dip net, 9. VIII.2010, J. Witt (CBG), 1 specimen; same except: ponds near krummholz, malaise trap, 31. VII.2005, R.E. Roughly (CBG), 1 specimen; Churchill, Twin Lakes Rd., N of Malcolm Ramsay, 58.7306, −93.7807, 9.VII.2015, M. Pyle & A.M. Solecki (CBG), 1 specimen; Wapusk National Park, 58.723, −93.46, small lawn dominated by sedges, malaise trap, 6. VII.2014, D. Iles (CBG), 1 specimen; Yukon, Kluane National Park and Reserve, Dezadeash River Trail, 582m el., 60.748, −137.513, wetland, pitfall, 15-24.VII.2014, BIObus 2014 team (CBG) 1 specimen.

Additional records. See Annex 1.

226. *Philhygra sinuipennis* Klimaszewski and Langor (Fig. 19.301a–h)

Philhygra sinuipennis Klimaszewski and Langor, in Klimaszewski et al. 2011. Klimaszewski et al. 2012, 2015a, 2018a, Webster et al. 2012.

Diagnosis Body narrowly oval, length 3.0 mm; head and abdomen dark brown, almost piceous, pronotum chestnut brown, and elytra except for scutellar region, legs and bases of antennae light-brown (Fig. 19.301a); integument moderately glossy, forebody with fine, dense punctation and meshed microsculpture; head transverse, angular, narrower than pronotum; eyes longer than genae in dorsal view; antennae moderately thick, incrassate, antennomeres IV–X slightly to distinctly transverse; pronotum distinctly transverse, widest at middle, margined laterally, narrower than elytra; elytra at suture longer than pronotum; abdomen subparallel, arcuate laterally. MALE. Tergite VIII broadly, unevenly arcuate apically (Fig. 19.301d); sternite VIII elongate, parabolic apically, apex narrowly truncate (Fig. 19.301e); median lobe of aedeagus with tubus arcuate ventrally, subapical part broadly subparallel to apex, with minute tooth ventrad in lateral view (Fig. 19.301b); in dorsal view tubus elongate, narrow, gradually tapering apically (Fig. 19.301c). FEMALE. Tergite VIII unevenly arcuate apically (Fig. 19.301f); sternite VIII with apical margin broadly triangular, subangulate medially, antecostal suture distinctly sinuate (Fig. 19.301g); pygydium as illustrated (Fig. 19.301h); spermatheca not found.

DISTRIBUTION Origin: Nearctic. **CANADA**: AB [**NPR**], NB, NF, SK, YT. **USA**: not recorded.

COLLECTION AND HABITAT DATA
Habitat: wetland habitats; in NF, specimen from litter and stones on the sandy shore of a lake; NB specimens from moist litter and grass on hummocks surrounded by water in an alder swamp (common) and in litter from *Carex* hummocks in *Carex* marsh (common); one from emergent vegetation on lake shore; in AB found in riparian area, rocky and mossy. **Collecting period**: V–VI. **Collecting method**: sifting litter from *Carex* hummocks, sifting litter among stones and deciduous litter, treading vegetation on lake margin, interception trap.

We report this species from AB (**NPR**) for the first time here based on the following: Canada, Alberta, Jasper National Park, Miette Hotsprings, 1439m el., 53.124, –117.775, along forest creek on a valley bed, flight-intercept trap, 17-21. VII.2012, BIOBus 2012 team (CBG) 1 specimen.

Additional records. See Annex 1.

COMMENTS This species was listed from LB in error by Klimaszewski et al. (2015a), as no voucher data were provided and only a holotype (from NF) was mentioned.

227. *Philhygra terrestris* Klimaszewski and Godin (Fig. 19.302a–g)

Philhygra terrestris Klimaszewski and Godin, in Klimaszewski et al. 2012. Klimaszewski et al. 2015a, 2018a, 2020, Webster et al. 2016.

Diagnosis Body narrowly subparallel, length 2.9–3.0 mm; head and abdomen dark brown to black, pronotum and particularly elytra slightly paler, basal antennomeres and legs yellowish (Fig. 19.302a); integument highly glossy, forebody with fine, dense punctation and meshed microsculpture; head round, slightly narrower than pronotum; eyes as long as genae in dorsal view; antennae slender, antennomeres IV–VI elongate, VII–X subquadrate to slightly transverse; pronotum slightly transverse and almost as wide as elytra; elytra at suture longer than pronotum; abdomen long, subparallel. MALE. Tergite VIII arcuate apically (Fig. 19.302c); sternite VIII moderately elongate, broadly parabolic apically (Fig. 19.302d); median lobe of aedeagus with tubus broadly elongate, constricted in basal half, subapical part broadly compressed oval in lateral view (Fig. 19.302b). FEMALE. Tergite VIII rounded apically (Fig. 19.302e); sternite VIII with apical margin obtusely triangularly produced, arcuate medially (Fig. 19.302f); pygidium as illustrated (Fig. 19.302g); spermatheca (not illustrated) minute with short sac-shaped capsule without apical invagination, stem short, narrow.

DISTRIBUTION Origin: Nearctic. **CANADA**: AB [**BOLD**], BC, NB, SK, YT. **USA**: AK.

COLLECTION AND HABITAT DATA
Habitat: riparian; in NB, in gravel/clay on margin of partially shaded brooks in old-growth mixed forests and near lake, in drift material (mostly small sticks and conifer buds) on river margin, and in clay/sand under alders on river margin; SK specimens collected in wet willow stand, deciduous ditch litter; YT specimen found in litter in a mixed forest, BC specimen found in

gravel along edge of stream, in sedges along a creek, and in moss at creek and in swamp. **Collecting period**: V–VII. **Collecting method**: splashing clay and fine sand on sand bars along shaded brooks, and splashing drift material along a river margin, sifting litter, pitfall traps.

We report this species from AB (**NPR**) for the first time here based on BOLD records: Canada, Alberta, Waterton Lakes National Park, Foothills Parkland Region, 1338m el. 49.083, −113.876, aspen forest, Malaise trap, 10-17.VII.2012, Emma Sylvester (CBG) 1 specimen.

Additional records. See Annex 1.

228. *Philhygra varula* (Casey) (Fig. 19.303a–h)

Metaxya varula Casey, 1911. Seevers 1978, Lohse et al. 1990 [cited as *L. stylifera*], Gusarov [unpublished synonymy (online) of *P. stylifera* Lohse with *P. varula*], Klimaszewski et al. 2011 [synonymy accepted], 2018a, Webster et al. 2012.

Diagnosis Body narrowly subparallel, length 3.5–4.0 mm; head, abdomen and antennae dark brown, pronotum and elytra light brown, sometimes entire body almost black (Fig. 19.303a); integument moderately glossy, forebody with fine, dense punctation, and meshed microsculpture; head slightly narrower than pronotum, pronotum and elytra subequal in width, and pronotum as wide as head and narrower than elytra and abdomen; head round, eyes as long as genae in dorsal view; antennae slender, antennomeres IV–VI elongate, and VII–X subquadrate to slightly elongate; pronotum slightly transverse, widest at middle, appearing subquadrate; elytra at suture slightly longer than pronotum; abdomen subparallel, arcuate laterally. MALE. Tergite VIII moderately elongate, with posterior margin truncate (Fig. 19.303d); sternite VIII elongate, parabolic apically (Fig. 19.303e); median lobe of aedeagus in lateral view with tubus narrowly elongate and slightly arcuate, constricted subapically and rounded at apex (Fig. 19.303b); in dorsal view bulbus large, roughly oval, tubus narrowly elongate and spatulate apically in dorsal view (Fig. 19.303c); internal sac structures as illustrated (Fig. 19.303b,c). FEMALE. Tergite VIII irregularly truncate apically (Fig. 19.303f); sternite VIII obtusely triangularly produced apically (Fig. 19.303g); pygidium as illustrated (Fig. 19.303h); spermatheca not illustrated, minute with short sac-shaped capsule without apical invagination, stem short and narrow.

The spermatheca illustrated for this species in Lohse et al. 1990, does not belong to this species.

DISTRIBUTION **Origin**: Nearctic. **CANADA**: LB, MB, NB, NF, QC. **USA**: AK [**NSR**].

COLLECTION AND HABITAT DATA

Habitat: this species is often found in estuaries; in QC, reported from under stones in an estuary above tidal zone and under stones and along a stream in sand, gravel, and organic mats at other localities; in NF, reported from rotting mushrooms and lakeshore debris. NB specimens were reported from coastal habitats; under litter resting on a large log (drift wood) in a freshwater marsh adjacent to a salt marsh, in grass litter in a seepage area with *Carex* and *Spartina patens* (salt grass) along upper margin of a salt marsh, and in decaying sea wrack under alders on the upper margin of a sea beach. **Collecting period**: V–VIII. **Collecting method**: sifting grass litter in freshwater salt marshes, sea wrack on sea beaches, litter among stones, and organic mats.

We report this species from AK (**NSR**) for the first time here, based on the following: USA: Alaska, Cold Bay, 55.22819, −162.73185, ±10km 31.VIII.1956, R. H. Washburn (UAM) 1 specimen.

Genus *Schistoglossa* Kraatz, 1856

(Figs. 19.304–19.307) (Klimaszewski et al. 2009, 2018a, 2020)

Diagnosis Body subparallel, length 1.8–3.5 mm; colour brown; head slightly narrower than pronotum or nearly as broad; eyes small, about one quarter to one third length of genae in dorsal view, infraorbital carina short, feeble or virtually obsolete; antennae approximately reaching posterior margin of elytra, antennomeres VII–X slightly transverse or slightly elongate, ligula completely divided into two separate lobes (Fig. 19.304e,f), mandibles appearing bifid apically but in fact left mandible is bifid and the right mandible bears one subapical tooth and two apical teeth, apical teeth often overlap and appear as one in dorsal view creating an illusion of a bifid mandible, the teeth of left mandible are in the horizontal plane and the two apical teeth of the right mandible are in the vertical plane and can be locked together (Fig. 19.304a–d); maxillary palpi with four palpomeres, terminal palpomere needle-shaped (Fig. 19.304g); pronotum transverse to nearly subquadrate, broadest in middle or in apical one-third, hind angles rounded, hypomera clearly visible in lateral view, integument smooth and moderately glossy, microsculpture weak, punctation fine, pubescence on disc directed entirely or partially posteriad along midline, lateroposteriad elsewhere; mesocoxae contiguous; male tergite VIII with two large lateral teeth and apical crenulation in *Schistoglossa s. str.*, tergite VIII truncate apically and without lateral teeth and apical crenulation in the subgenus *Boreomorpha*; median lobe of aedeagus bearing large bulbus and triangular tubus in dorsal view, crista apicalis small, internal sac with reversed V-shaped structure in dorsal view in *Schistoglossa s. str.*, with semicircular apical structures in the subgenus *Boreomorpha*; spermatheca with capsule tubular (*Schistoglossa s. str.*), or approximately spherical (*Boreomorpha*).

COMMENTS Cold-loving species occurring in arctic and subarctic habitats, and in the temperate regions they are usually confined to bogs, fens, and marshes.

Key to Species of *Schistoglossa*
(Modified from Klimaszewski et al. 2020)

1. Elytra at suture slightly shorter than pronotum along midline, antennomeres VII–X distinctly transverse (Fig. 19.307a); male tergite VIII emarginate apically and finely crenulate (Fig. 19.307d); apex of median lobe of aedeagus curved ventrad in lateral view (Fig. 19.307b); spermatheca S-shaped (Fig. 19.307h) ....
.......***Schistoglossa hampshirensis* Klimaszewski**

– Elytra at suture at least as long as pronotum along midline; antennomeres VII–X subquadrate or elongate; other characters not as above **2**

2. Antennomers VII–X elongate, eyes strongly protruding, elytra broadening posteriorly (Fig. 19.305a); body length 2.8–3.0 mm; median lobe of aedeagus and spermatheca as illustrated (Fig. 19.305b–d, i) ....
...........***Schistoglossa blatchleyi* (Bernhauer and Scheerpeltz)**

– Antennomeres VII–X subquadrate to slightly transverse, eyes not protruding and streamlined with contour of head (Fig. 19.306a); median lobe of aedeagus and spermatheca not as above (Fig. 19.306b,c,h)...
.... ***Schistoglossa campbelli* Klimaszewski**

(Subgenus *Boreomorpha* Klimaszewski and Webster, in Klimaszewski et al. 2009)

(Fig. 19.305) (Klimaszewski et al. 2009, 2018a, 2020)

Diagnosis Antennomeres V–X slightly elongate; eyes moderate-sized, about half as long as genae in dorsal view; infraorbital carina weak, present only basally or obsolete; male tergite VIII broadly arcuate apically, apical margin entire and without lateral teeth and crenulation; median lobe of aedeagus with bulbus and tubus evenly coalescent in dorsal view, internal sac without V-shaped structure typical for *Schistoglossa s. str.* but with two crescent-shaped subapical structures clearly visible in lateral view; spermatheca club-shaped and with approximately spherical capsule.

229. *Schistoglossa (Boreomorpha) blatchleyi* (Bernhauer and Scheerpeltz) (Fig. 19.305a–i)

Atheta blatchleyi Bernhauer and Scheerpeltz, 1926, replacement name for preoccupied *Atheta* (s. str.) *caviceps* Blatchley, 1910; in *Atheta*:

Gusarov 2003a, Gouix and Klimaszewski 2007. As *Boreophilia chillcotti* Lohse, 1990, in Lohse et al. 1990. In *Schistoglossa*: Klimaszewski et al. 2009, 2016d, 2019 [synonymy confirmed].

Description Body length 2.7–2.8 mm, width 0.4 mm, approximately uniformly dark brown with slightly darker head and abdomen and paler legs and mouthparts (Fig. 19.305a); integument with weak microsculpture, sculpticells irregularly hexagonal, pubescence short, except slightly longer on head and abdomen, yellowish-brown in artificial light, moderately dense; head slightly elongate, round, carinate basally, tempora as long as approximately twice the maximal diameter of eye as seen from above; antennae moderately robust with all articles slightly to strongly elongate; pronotum transverse, broadly rounded laterally and posteriorly, broadest in apical half, pubescence directed anteriorly on apical third of the midline and laterally elsewhere; elytra moderately elongate, longer than pronotum, transverse, hind margin nearly straight, pubescence directed obliquely posteriad; abdomen parallelsided, four basal tergites deeply impressed basally. MALE. Tergite VIII transverse, with apical margin broadly arcuate, entire except occasionally with small medial emargination (Fig. 19.305e); sternite VIII elongate and rounded apically, antecostal suture moderately sinuate (Fig. 19.305f); median lobe of aedeagus broadly roughly triangular in dorsal view (Fig. 19.305d); in lateral view venter of tubus sinuate and apex short and blunt (Fig. 19.305b,c), apical sclerites of internal sac semilunar-shaped, and basal sclerites narrowly elongate (Fig. 19.305b,c). FEMALE. Tergite and sternite VIII as illustrated (Fig. 19.305g,h); spermatheca club-shaped with capsule spherical and narrowed basally, bearing fine internal ribbings and with broad apical invagination, stem moderately narrow and straight (Fig. 19.305i).

DISTRIBUTION **Origin**: Nearctic. **CANADA**: MB, NB, NT, ON, QC, SK, YT. **USA**: AK, IN.

COLLECTION AND HABITAT DATA
Habitat: riparian habitats, along margin of *Carex* marshes and fens, in *Sphagnum* hummocks on margin of tamarack (*Larix laricina*) bogs, red maple swamps in moist leaves near vernal pools and leaves at base of trees, wet alder swamps in leaf litter and grass on hummocks, in moss and litter at the base of trees in calcareous eastern white cedar (*Thuja occidentalis*) fens with shrubby cinquefoil (*Pentaphylloides floribunda* A. Love), in flood debris along a river margin. **Collecting period**: III, IV, VI–VIII. **Collecting method**: treading *Sphagnum* and in *Carex* marsh.

COMMENTS This species was listed from NF by Newton (2019) but we are not aware of published voucher data for this record.

(Subgenus *Schistoglossa* Kraatz, 1856)

(Figs. 19.306–19.307) (Klimaszewski et al. 2009, 2018a, 2020)

Diagnosis Antennomeres V–X slightly transverse; eyes small, about one-third as long as genae in dorsal view, infraorbital carinae feeble, present only basally or obsolete; male tergite VIII truncate or emarginate apically, apical margin weakly to coarsely crenulate and bearing two large lateral teeth; median lobe of aedeagus with bulbus and tubus distinctly separated in dorsal view (not evenly coalescent), internal sac with V-shaped structure; spermatheca L- or S-shaped and with tubular capsule.

230. *Schistoglossa (Schistoglossa) campbelli* Klimaszewski (Fig. 19.306a–h)

Schistoglossa (Schistoglossa) campbelli Klimaszewski, in Klimaszewski et al. 2009, 2015a, 2020.

Diagnosis Body length 2.0–2.4 mm, width 0.4 mm, approximately uniformly dark brown with slightly paler tip of abdomen and legs (Fig. 19.306a); integument with weak microsculpture, sculpticells irregularly hexagonal, pubescence short, except slightly longer on head and abdomen, yellowish-brown in artificial light, moderately dense; head slightly elongate, rounded posteriorly, weakly carinate basally, tempora as long as two and a half times the maximal diameter of eye as seen from above; antennae slim, antennomeres V–X slightly transverse; pronotum moderately transverse, broadly rounded laterally and posteriorly, broadest in basal half, pubescence directed anteriorly on apical fourth of the midline and posteriorly and obliquely laterally elsewhere; elytra moderately transverse, and slightly broadened posteriorly, hind margin sinuate, pubescence directed straight posteriad; abdomen parallel-sided, three basal tergites deeply impressed basally. MALE. Tergite VIII with apical margin broadly emarginate, feebly crenulate between two triangular lateral processes (Fig. 19.306d); sternite VIII elongate and parabolic apically, antecostal suture straight (Fig. 19.306e); venter of tubus of median lobe of aedeagus arcuate with apex curved ventrad in lateral view (Fig. 19.306b), in dorsal view bulbus oval in outline and tubus subparallel basally and triangular apically, apical sclerites of internal sac V-shaped (Fig. 19.306c). FEMALE. Tergite VIII with apical margin broadly, shallowly emarginate (Fig. 19.306f); sternite VIII parabolic apically and vaguely obtusely angulate medially (Fig. 19.306g); spermatheca sharply bent at base of neck, capsule spherical apically and tubular basally, with small apical invagination, stem moderately narrow, broadly curved (Fig. 19.306h).

DISTRIBUTION Origin: Nearctic. **CANADA**: AB, BC. **USA**: AK [**NSR**].

COLLECTION AND HABITAT DATA

Habitat: edge of marsh with *Carex* and *Sphagnum*. In AK, found in subalpine forest, *Abies lasiocarpa, Tsuga mertensiana*; forest edge, *Populus* litter; meadow, *Heracleum maximum, Sanguisorba canadensis, Viola langsdorfii*; under rocks at edge of snowfield; subalpine meadow, *Athyrium, Carex macrocephala, Vaccinium ovalifolium*; subalpine *Veratrum, Calamagrostis*; and riparian, *Caltha leptosepala, Hypnum* sp.

Collecting period: VI–VIII. **Collecting method**: treading *Sphagnum* and in *Carex* marsh, ground search.

We report this species from AK (**NSR**) for the first time here, based on the following: USA, Alaska, [Kenai Peninsula] slope above Emerald Lake, 59.6309, –151.049, ±50m 30.VI.2011, M. Bowser (UAM) 2 specimens; Baranof Is. pl.19, 454m el., 56.41512, –134.83815, ±7m meadow, *Heracleum maximum, Sanguisorba canadensis, Viola langsdorfii*, ground search, 29.VII.2013, S. Ridling (UAM) 1 specimen; Dall Isl., 712m el. 54.99826, –133.02086, ±34m edge of snowfield, under rocks 9:20–9:59, hand collected, 16.VII.2011, D.S. Sikes (UAM) 1 specimen; Dall Isl., pl.01, 646m el., 54.99768, –133.01714, ±5m subalpine meadow, *Athyrium, Carex macro.*, hand collected, 14.VII.2011, D.S. Sikes (UAM) 1 specimen; Dall Isl., pl.11, 691m el., 54.99403, –133.0181, ±4m subalpine meadow, *Carex macro., Vacc. ovalifolium*, hand, 14.VII.2011, D.S. Sikes (UAM) 1 specimen; Dall Isl., pl.13, 628m el., 54.99403, –133.01569, ±5m riparian, *Caltha leptosepala, Hypnum* sp., hand collected, 14.VII.2011, D.S. Sikes (UAM) 1 specimen; Dall Isl., pl.19, 544m el., 54.9967, –133.00807, ±5m subalpine forest, *Abies lasiocarpa, Tsuga mert.*, hand, 14.VII.2011, D.S. Sikes (UAM) 1 specimen; Fairbanks: Creamer's Field, 141m el., 64.86711, –147.72794, ±54m forest edge, *Populus* litter, Winkler, 31.VII.2010, D.S. Sikes (UAM) 1 specimen; Prince of Wales Isl. nr Black Lk, 760-981m el., 55.59008, –132.89311, ±503m misc., hand collected, 6-10.VII.2011, C. Bickford (UAM) 1 specimen; Prince of Wales Isl. nr Black Lk pl.20, 812m el., 55.5903, –132.88896, ±2m meadow, *Neph. crista-g., Anem. narc.*, hand, 7.VII.2011, C. Bickford (UAM) 1 specimen, (UAM) 1 specimen; S. Chilkat Range pl.19, 652m el., 58.42509, –135.14229, ±30m subalpine *Veratrum, Calamagrostis*, hand collected, 9. VII.2010, D.S. Sikes (UAM) 1 specimen.

231. *Schistoglossa (Schistoglossa) hampshirensis* Klimaszewski (Fig. 19.307a–h)

Schistoglossa (Schistoglossa) hampshirensis Klimaszewski, in Klimaszewski et al. 2009. Klimaszewski et al. 2015a, 2018a, 2020.

Diagnosis Body length 1.8–2.0 mm, width 0.3 mm; colour virtually uniformly dark brown with tarsi, antennal bases and mouthparts paler (Fig. 19.307a); integument with fine microsculpture, sculpticells irregularly hexagonal, pubescence short, except slightly longer on head and abdomen, yellowish-brown in artificial light, moderately dense; head roughly round, slightly elongate, with feeble infraorbital carina, eyes about one third length of genae in dorsal view; antennae slim, antennomeres V–X moderately transverse; pronotum slightly transverse, broadly rounded laterally and posteriorly, broadest in basal half, pubescence on disc directed anteriad on apical third of midline and obliquely laterad elsewhere; elytra short, at suture about as long as pronotum along midline, distinctly transverse, subparallel, hind margin approximately straight, pubescence on disc directed obliquely or straight posteriad; abdomen parallel-sided, three basal tergites deeply impressed basally. MALE. Tergite VIII with apical margin broadly emarginate, irregularly crenulate between two sublateral triangular processes (Fig. 19.307d); sternite VIII with apical margin broadly V-shaped, broadly rounded apically (Fig. 19.307e); in lateral view median lobe of aedeagus moderately broad, venter of tubus arcuate, apex slightly curved ventrad (Fig. 19.307b), in dorsal view apical sclerites of internal sac V-shaped (Fig. 19.307c), bulbus oval, tubus short, subparallel and triangular apically (Fig. 19.307c). FEMALE. Tergite VIII sinuate apically with a small median emargination (Fig. 19.307f); sternite VIII with apical margin broadly triangular, rounded medially (Fig. 19.307g); spermatheca with tubular capsule, stem narrow, sinuately tapering to a small posterior coil (Fig. 19.307h).

DISTRIBUTION Origin: Nearctic. **CANADA**: AB, BC, QC. **USA**: AK [**NSR**], NH.

COLLECTION AND HABITAT DATA
Habitat: in marshes or near wetlands; in *Carex* marshes in sphagnum hummocks, in sphagnum hummocks with *Carex* and grasses in open wet black spruce bog, in *Carex* hummocks in *Carex* marshes, in grass and sphagnum at the base of alders in a marsh with grasses and alders, and along stream margins among cobblestones under alders or in litter; one swept from *Carex* in small *Carex* marsh. In BC, one female was captured in soil in fire-killed forest with grass. In AK, found in black and white spruce; *Salix*, moss leaf litter; and open tussock and standing water, willow and bog birch. **Collecting period**: IV–VII. **Collecting method**: treading sedge hummocks into water, sifting litter and sphagnum, Berlese funnels, sweeping foliage in *Carex* marsh, hanging Malaise.

We report this species from AK (**NSR**) for the first time here, based on the following: USA, Alaska, Fairbanks, UAF, 166m el., 64.86673, –147.84297, ±10m black & white spruce, hanging Malaise, 5.1m, 31.V-13.VI.2016, J.C. Hagelin (UAM) 1 specimen; Serpentine Hot Springs, 118m el., 65.85183, –164.69884, ±104m *Salix*, moss leaf litter, Berlese, 10.VII. 2016, D.S. Sikes (UAM) 6 specimens; White Mountain Recreation Area, Nome Creek Site 5, 589m el., 65.34077, –146.9275, ±10m open tussock and standing water, willow and bog birch, hanging Malaise, 4.95m, 12.VI-3.VII.2015, J.C. Hagelin (UAM) 1 specimen.

COMMENTS The NH specimens were collected at elevations of 5200–5700 feet and the only QC specimen was found on Mount Albert at 3,700 feet, suggesting this may be an arctic alpine and or subarctic species (Klimaszewski et al. 2018a). The record of this species from NB was based on a female and was considered to be misidentified by Klimaszewski et al. (2018a).

Genus *Strigota* Casey, 1910

(Figs. 19.308–19.309) (Gusarov 2003a, Klimaszewski et al. 2011, 2018a, 2020)

Diagnosis Body elongate, subfusoid (somewhat *Oxypoda*-like); integument of forebody with punctation fine and dense, pubescence dense, pilose, appearing silky particularly on abdomen; antennae thick, antennomeres V–X more or less transverse; pronotum 1.3 times as wide as long; pronotal hypomeron not visible in lateral view; pronotum and elytra subequal in length; isthmus almost as long as process of mesoventrite; in both sexes antecostal suture of abdominal tergite VIII joins basal margin (always separated from basal margin and extended laterad in other athetines); apical margin of male sternite VIII with row of microsetae on dorsal side.

Key to Species of *Strigota*
(Modified from Klimaszewski et al. 2018a)

1. Elytra at suture shorter than pronotal length at midline, body uniformly dark brown to black (Fig. 19.309a); apical margin of male tergite VIII with sharply pointed triangular projection in about medial one-fourth (Fig. 19.309d); tubus of median lobe of aedeagus strongly bent ventrad at basal one-third in lateral view (Fig. 19.309b); spermathecal capsule globose and sharply delimited from neck, stem with broad posterior loop (Fig. 19.309h) ....***Strigota obscurata* Klimaszewski and Brunke**

– Elytra at suture about as long as pronotal length at midline, body bicoloured in most, dark brown to black with elytra paler, usually reddish-brown (Fig. 19.308a); apical margin of male tergite VIII broadly triangularly produced, rounded at apex (Fig. 19.308d); tubus of median lobe of aedeagus slightly bent ventrad in basal one-fourth in lateral view (Fig. 19.308b); spermathecal capsule slightly globose and not sharply delimited from neck, stem with tight posterior loop (Fig. 19.308h)....... ***Strigota ambigua* (Erichson)**

232. *Strigota ambigua* (Erichson) (Fig. 19.308a–h)

Homalota ambigua Erichson, 1839. Bernhauer 1907 [in *Atheta* (*Dilacra*)], Casey 1910, 1911 [as nine synonyms], Notman 1922 [as synonym *Anaduosternum brevipennis*], in *Strigota*: Gusarov 2003a, Majka et al. 2008, Klimaszewski et al. 2011, 2012, Brunke et al. 2012, Klimaszewski et al. 2016d, 2018a, 2020, Webster et al. 2016.

Diagnosis Body elongate-oval, moderately convex, length 2.4–3.0 mm; colour uniformly light brown or dark brown to almost black, or forebody, legs/tarsi, antennae and elytra light brown and abdomen dark brown (Fig. 19.308a); punctation on forebody fine and dense, pubescence fine and silky in appearance; head narrower than pronotum, ratio of maximum width of head to maximum width of pronotum 0.7; antennae robust, antennomeres I–IV elongate, V–X slightly to distinctly transverse; pronotum transverse, sides evenly arcuate, ratio of maximum width to length 1.3, about as wide as elytra; elytra at suture slightly shorter than pronotum; abdomen gradually narrowed posteriad. MALE. Tergite VIII moderately elongate, sinuate basally, apical margin broadly triangularly produced, rounded at apex (not as sharp as in *S. obscurata*), antecostal suture joining basal margin anterad of spiracles (Fig. 19.308d); sternite VIII elongate, broadly rounded apically (Fig. 19.308e); median lobe of aedeagus with ventral margin of tubus straight (Fig. 19.308b), bent closer to base, and not as markedly as that of S. *obscurata*, in dorsal view bulbus narrowly oval and tubus narrowly subparallel baso-medially, triangular before rounded apex (Fig. 19.308c). FEMALE. Tergite VIII slightly triangularly produced, broadly rounded at very apex (Fig. 19.308f); apical margin of sternite VIII broadly arcuate, with dense fringe of coarse microsetae (Fig. 19.308g); spermatheca with club-shaped capsule with a small invagination, neck long and tubular, stem sinuate and coiled posteriorly (Fig. 19.308h). The spermatheca of this species is nearly identical to that of *S. obscurata* except the capsule is less sharply deflexed and more tubular, and the posterior coil is tighter in *S. ambigua*.

DISTRIBUTION **Origin**: Nearctic. **CANADA**: BC, LB, NB, NF, NS, ON, PE, QC, SK, YT. **USA**: CA, CO, CT, IA, KS, MA, MO, NC, NJ, NM, NY, NV, TX.

COLLECTION AND HABITAT DATA
Habitat: open habitats with well-drained soil including sand dunes, lakeshores, beaches, *Dryas* covered limestone barrens, soybean fields, old fields, open gaps in spruce forest with sandy soil mostly covered with *Cladina* lichens, weedy disturbed field, and riverbanks. Adults under stones, cobblestones on moist sand, in organic debris in groundhog burrows, organic debris, flood refuse, and in beach-drift debris. **Collecting period**: IV–IX. **Collecting method**: pitfall traps, Malaise traps, sifting various kinds of litter.

COMMENTS This species was listed from WI by Newton (2019) but we are not aware of published voucher data for this record.

233. *Strigota obscurata* Klimaszewski and Brunke (Fig. 19.309a–h)

Strigota obscurata Klimaszewski and Brunke, in Brunke et al. 2012. Klimaszewski et al. 2016d, 2018a, 2020, Webster et al. 2016.

Diagnosis Body narrowly elongate, length 2.2–2.5 mm; colour dark brown to black, with legs and/or tarsi brown, central disc of elytra sometimes with traces of reddish tinge (Fig. 19.309a); integument moderately glossy, with dense, meshed microsculpture, pubescence short, dense, appearing somewhat silky; head convex, rounded posteriorly, eyes at least slightly shorter than genae in dorsal view, pubescence on disc directed towards midline; antennae stout, antennomeres I–III distinctly elongate, IV–V subquadrate and VI–X moderately transverse; pronotum slightly transverse, widest in basal third, pubescence on disc directed posteriad at midline, obliquely posteriad elsewhere; elytra transverse, at suture shorter than pronotum at midline, pubescence on disc directed straight posteriad; abdomen subparallel with first three visible tergites deeply impressed basally; hind tarsus with basal article as long as two following articles combined. MALE. Tergite VIII with basal margin bisinuate, apical margin subtruncate, with moderate, sharply-pointed triangular projection medially (Fig. 19.309d); sternite VIII elongate with antecostal suture widely separated from and parallel to basal margin, apical margin broadly truncate, rounded laterally (Fig. 19.309e); in lateral view median lobe of aedeagus with bulbus moderate-sized, tubus bent obliquely ventrad before middle of ventral margin, internal sac with several short, inconspicuous sclerites (Fig. 19.309b); in dorsal view median lobe of aedeagus with tubus constricted basally, almost evenly tapering to near blunt apex (Fig. 19.309c). FEMALE. Tergite VIII with basal margin bisinuate, apical margin obtusely produced, with short, obtuse triangular projection medially (Fig. 19.309f); sternite VIII with apical margin broadly arcuate medially, rounded laterally (Fig. 19.309g); spermatheca with spherical capsule bearing a small invagination, stem elbowed near middle, broadly coiled posteriorly (Fig. 19.309h). The spermatheca of this species is nearly identical to that of *S. ambigua* except for the capsule, which is more spherical and sharply deflexed, and the stem is more broadly coiled.

The distinctive shape of tergite VIII of *Strigota obscurata* will easily differentiate it from the remaining congeners.

DISTRIBUTION Origin: Nearctic. **CANADA**: AB [**BOLD**], BC, NB, NF [**BOLD**], ON, QC, SK. **USA**: AK [**NSR**].

COLLECTION AND HABITAT DATA
Habitat: open habitats; in flood debris on a river margin, on soil at the base of grass in a residential lawn, in Lindgren funnel trap in jack pine forest, soybean fields in pitfalls (most common rove beetle in this habitat in southern ON), oak savanna/alvar habitats under stones and in pitfall traps, hedgerow in pitfall traps, and in Malaise pans in a field, in sub-boreal spruce forest in BC. In AK, found in black spruce forest. **Collecting period**: V–X. **Collecting method**:

Malaise traps, pitfall traps, sifting litter, Lindgren funnel trap.

We report this species for the first time from AB (**NPR**), and NF (**NPR**) based on BOLD records, and from AK (**NSR**) based on UAM specimens: Canada, Alberta, Waterton Lakes National Park, Red Rock Parkway, 1335m el., 49.081, –113.879, moraine grassland, pitfall, 21-27.VI.2012, BIOBus 2012 team (CBG) 2 specimens. Newfoundland, Gros Morne National Park, Berry Hill Campground, 104m el., 49.626, –57.922, peat bog, flight-intercept trap, 10-20.VII.2013, BIObus 2013 team (CBG) 1 specimen. USA, Alaska, Dalton Hwy, 403m el., 67.48637, –149.88414, ±24m muskeg, black spruce, dirt road, 1.VI.2008, D.S. Sikes (UAM) 1 specimen.

Genus *Tarphiota* Casey, 1893

(Figs. 19.310–19.311) (Mäklin, 1852, Casey 1893, 1910, Fenyes 1920, Ahn 1996, Klimaszewski et al. 2020)

Diagnosis Body narrowly elongate, dark brown to black; integument of forebody strongly granulose due to fine-meshed, raised reticulation; head without infraorbital carinae; eyes large; pronotum 1.2–1.3 times as wide as long, microsetae evenly distributed, macrosetae absent; elytra longer than pronotum, slightly sinuate apico-laterally; hind wings present; antennomeres VII–X slightly to strongly transverse; mandibles each with blunt apex; ligula narrow, Y-shaped; mesocoxae separated, mesocoxal cavities unmargined, mesoventral process pointed; anterior and middle tibiae each with a row of spines; species inhabit mid to upper littoral zone of fine-grained sandy beaches.

Key to Species of *Tarphiota*
(From Klimaszewski et al. 2020)

1. Body length 2.9–3.1 mm; maximum width of elytra one-fourth wider than maximum width of pronotum, antennomeres VII–X subquadrate (Fig. 19.310a); male and female tergite VIII without noticeable apical emargination (Fig. 19.310c, e); spermathecal stem narrow (Fig. 19.310g) .... ***Tarphiota fucicola* (Mäklin)**

– Body length 2.0–2.6 mm; maximum width of elytra one-sixth wider than maximum width of pronotum, antennomeres VII–X distinctly transverse (Fig. 19.311a); male and female tergite VIII with apical margin broadly shallowly emarginate (Fig. 19.311d,f) .... ***Tarphiota geniculata* (Mäklin)**

234. *Tarphiota fucicola* (Mäklin) (Fig. 19.310a–g)

Tachyusa fucicola Mäklin, 1852. Casey 1893 [as *T. pallidipes*], Casey 1910 [as *T. debilicollis*], Fenyes 1920, Hatch 1957, Lohse and Smetana 1985 [lectotype], Ahn 1996, Ashe in Navarrete-Heredia et al. 2002, Klimaszewski et al. 2006, 2020.

Diagnosis Body length 2.9–3.1 mm; dark brown to black with brown legs and antennae (Fig. 19.310a); integument of forebody strongly granulose; head about 1.2 times as wide as long; antennae slender, antennomeres VII–X subquadrate; pronotum about 1.3 times as wide as long, sides rounded; elytra 1.3 times as long as pronotum; abdominal tergites with evenly distributed pubescence, several macrosetae present. MALE. Tergite VIII truncate apically (Fig. 19.310c); sternite VIII rounded apically (Fig. 19.310d); in lateral view tubus of median lobe of aedeagus gently arched ventrad, tapering to narrow and pointed apex (Fig. 19.310b).

FEMALE. Tergite VIII broadly arcuate apically (Fig. 19.310e); sternite VIII unevenly rounded apically (Fig. 19.310f); spermatheca with medium-sized globular capsule and short elbowed neck, stem very narrow and long (Fig. 19.310g).

DISTRIBUTION Origin: Nearctic. **CANADA**: BC. **USA**: AK, CA, OR, WA. **Mexico** (Baja California, Sonora).

COLLECTION AND HABITAT DATA
Habitat: under driftwood, under rotting seaweed on sandy beaches. **Collecting period**: III, V–VIII, XI. **Collecting method**: sifting seaweed.
Additional records. See Annex 1.

235. *Tarphiota geniculata* (Mäklin) (Fig. 19.311a–i)
Homalota geniculata Mäklin in Mannerheim, 1852. Casey 1910 [as *T. hirsutula, T. iota, T. insolita, T. litorina, T. seditiosa*], Fenyes 1920, Hatch 1957, Lohse and Smetana 1985 [lectotype], Ahn 1996, Ashe in Navarrete-Heredia et al. 2002, Klimaszewski et al. 2006, 2020.

Diagnosis Body length 2.0–2.6 mm; dark brown to black with legs and antennae brown (Fig. 19.311a); integument of forebody strongly granulose; head about 1.2 times as wide as long; antennae slender, antennomeres VII–X distinctly transverse; pronotum about 1.2 times as wide as long, sides rounded; elytra 1.2 times as long as pronotum; abdominal tergites with evenly distributed pubescence, several macrosetae present. MALE. Tergite VIII with apical margin truncate except broadly, shallowly emarginate medially (Fig. 19.311d); sternite VIII with apical margin triangularly produced, rounded at apex (Fig. 19.311e); in lateral view tubus of median lobe of aedeagus with ventral margin curved ventrad in basal half, then gently sinuate to narrow and rounded apex (Fig. 19.311b), in dorsal view bulbus and tubus broad (Fig. 19.311c). FEMALE. Tergite VIII with apical margin slightly sinuate, broadly, shallowly emarginate medially (Fig. 19.311f); sternite VIII rounded apically (Fig. 19.311g); spermatheca with small globular capsule bearing distinct apical invagination, neck arcuate, stem moderately broad and moderately long and sinuate (Fig. 19.311h,i).

DISTRIBUTION Origin: Nearctic. **CANADA**: BC. **USA**: AK, CA, OR, WA. **Mexico** (Baja California).

COLLECTION AND HABITAT DATA
Habitat: sandy beach, under back beach rocks and drift, under driftwood, under rotting seaweed on sandy beaches, running on water swept shore. **Collecting period**: IV–VIII, X. **Collecting method**: sifting seaweed, picking by forceps, colored pan trap.
Additional records. See Annex 1.

Genus *Trichiusa* Casey, 1893

(Fig. 19.312a–h)
(Casey, 1893, 1906, Brunke et al. 2012, Schülke and Smetana 2015, Webster et al. 2016, Klimaszewski et al. 2018a, 2020)

Diagnosis Body compact, broad, on average 3.0 mm long, often with long, dense and shaggy pubescence, on pronotum directed anteriad along midline of disc and often erected; body form distinctive with head and pronotum distinctly narrower than elytra, elytra at least 1.2 times as wide as pronotum; mesocoxae widely separated; abdomen robust, oval in outline, broadest at segment V and at least as broad as elytra at widest point; median lobe of aedeagus in dorsal view with bulbus of moderate-size and tubus elongate, and apically triangular, in lateral view tubus long and arcuate ventrally (Fig. 19.312b); spermatheca with tubular or spherical capsule bearing moderate to large apical invagination, stem thin, sinuate and looped or twisted posteriad.

236. *Trichiusa pilosa* Casey (Fig. 19.312a–h)
Trichiusa pilosa Casey, 1893. Casey 1906 and auct. before 2015 [as synonyms *T. atra, T. parviceps, T. postica*], Klimaszewski et al. 2015a, 2018a, Webster et al. 2016.

Diagnosis Body broad, oval, length 1.5–1.8 mm; colour dark brown with reddish tinge, base of abdomen slightly paler, or uniformly black, appendages usually lighter than rest of body (Fig. 19.312a); sparsely punctate and pubescent; setae straight and erect, particularly on pronotum; integument highly glossy; head slightly narrower than pronotum; pronotum moderately transverse, rounded laterally and basally, distinctly narrower than elytra; elytra broad and at suture about as long as pronotum; abdomen arcuate laterally and broadest at middle of its length. MALE. Tergite VIII strongly transverse, truncate apically (Fig. 19.312d); sternite VIII arcuately obliquely tapering to narrowly truncate apex (Fig. 19.312e); in lateral view tubus of median lobe of aedeagus slightly sinuate and moderately narrow apically with apex rounded (Fig. 19.312b), in dorsal view bulbus narrowly oval and tubus narrowly triangularly shaped, apex rounded (Fig. 19.312c); internal sac structures weakly defined (Fig. 19.312b, c). FEMALE. Tergite VIII broad and truncate apically (Fig. 19.312f); sternite VIII very broadly rounded apically (Fig. 19.312g); spermatheca with broad and spherical capsule with minute, narrow apical invagination, neck elbowed, stem narrow, sinuate with tight spiral posteriorly (Fig. 19.312h).

DISTRIBUTION **Origin**: Nearctic. **CANADA**: AB, BC, NB, NS, ON. **USA**: ID, IN, KS, OH, RI.

COLLECTION AND HABITAT DATA
Habitat: NB specimens from among cobblestones on lakeshore and from a Lindgren funnel trap in canopy of *Populus tremuloides* in a hardwood forest; one individual from bison dung in BC and another from a window trap in a 10-year-old regenerating white spruce stand in AB. **Collecting period**: VI–IX. **Collecting method**: sifting leaf litter, Lindgren funnel traps, window trap.

COMMENTS This species was listed from NY by Newton (2019) but we are not aware of published voucher data for this record.

Athetini with Subtribe Affiliation Uncertain

Genus *Paragoniusa* Maruyama and Klimaszewski, 2004

(Fig. 19.313a–g) (Maruyama and Klimaszewski 2004)

Diagnosis Body broad, sub-limuloid; head distinctly transverse, 1.50–1.59 times as wide as long, temples considerably swollen laterad; antennomeres III–X transverse, in dry specimens gaps between antennomeres inconspicuous; ligula bifid apically, labial palpi with setae α, β, γ and δ, mentum with anterior margin sinuate and anteromedial margin produced; pronotum transverse, 1.8–2.0 times as wide as long, setae on disc directed posteriad medially and posterolaterad elsewhere, pronotal macrosetae absent, hypomera invisible in lateral view; metasternal process very short, only slightly produced; mesolateral macroseta of middle tibia absent; hind tarsomere I shorter than II, one empodial seta present, tarsal formula 4-5-5; abdominal sternites III–VII sparsely covered with macrosetae, which are poorly differentiated from ordinary setae, tergite VIII with short and inconspicuous macrosetae; median lobe of aedeagus in dorsal view with bulbus enlarged and tubus small and triangular, crista apicalis of bulbus large, fin-shaped; spermatheca pipe-shaped, stem short, almost straight, slightly swollen at base. For mouthparts illustrations see Maruyama and Klimaszewski (2004).

237. *Paragoniusa myrmicae* Maruyama and Klimaszewski (Fig. 19.313a–g)
Paragoniusa myrmicae Maruyama and Klimaszewski, 2004. Maruyama and Klimaszewski, 2006, Klimaszewski et al. 2011, 2020, Webster et al. 2016.

Diagnosis Body extremely wide, oval, flattened, length 2.7–3.0 mm; head and abdomen dark brown and pronotum, elytra, legs and antennae light rusty-brown (Fig. 19.313a); integument moderately glossy, forebody with fine and dense punctation, short pubescence; head transverse, about 1/3 narrower than pronotum; antennae robust, antennomeres IV–VI subquadrate, VII–X slightly transverse; pronotum widely oval, transverse, widest at basal third, two times as wide as long, explanate laterally, wider than elytra; elytra about as wide as pronotum, at suture as long as pronotum, posterior margin sinuate; abdomen subparallel, almost as wide as elytra. MALE. Tergite VIII with posterior margin broadly shallowly emarginate medially, obtuse laterally (Fig. 19.313d); sternite VIII with posterior margin truncate; median lobe of aedeagus with tubus narrow, almost straight, with basal tooth in lateral view (Fig. 19.313b); crista apicalis of bulbus triangularly shaped and large (Fig. 19.313b), in ventral view bulbus large, oval, and tubus small, broadly triangular in dorsal view (Fig. 19.313c). FEMALE: tergite VIII arcuate apically (Fig. 19.313e); sternite VIII with apical margin arcuate on each side of shallow medial emargination (Fig. 19.313f); spermatheca J-shaped, capsule roughly spherical, curled at base to face posteriad, stem short and straight (Fig. 19.313g).

Paragoniusa myrmicae is somewhat similar to members of *Goniusa*, but *P. myrmicae* has a larger body and the macrosetae on the abdominal sternites are sparser and inconspicuous. See also diagnosis of *Paragoniusa*.

DISTRIBUTION **Origin**: Nearctic. **CANADA**: AB, BC, LB, NB. **USA**: not recorded.

COLLECTION AND HABITAT DATA
Habitat: probably myrmecophilous; three specimens from nest of *Myrmica alaskensis* Wheeler in a rotten log (BC). Other specimens captured in various traps; one from a pitfall trap in a burned forest (LB), from window traps (flight intercept traps) in unspecified habitats (AB); and from a Lindgren funnel trap in and old eastern white cedar and spruce forest with *Populus* (NB). **Collecting period**: V–VII. **Collecting method**: sifting material in ant nest in log, flight interception trap, Lindgren funnel trap, pitfall traps.

COMMENTS This species was listed for QC by Newton (2019) but we are not aware of published voucher data for this record.

Fig. 19.177 (**a–h**). *Acrotona sphagnorum* Klimaszewski and Webster: (**a**) habitus in dorsal view; (**b**) median lobe of aedeagus in lateral view; (**c**) median lobe of aedeagus in dorsal view; (**d**) male tergite VIII; (**e**) male sternite VIII; (**f**) female tergite VIII; (**g**) female sternite VIII; (**h**) spermatheca. Scale bar for habitus = 1 mm; remaining scale bars = 0.2 mm

Fig. 19.178 (**a–g**). *Acrotona horwoodae* Klimaszewski and Godin: (**a**) habitus in dorsal view; (**b**) median lobe of aedeagus in lateral view; (**c**) male tergite VIII; (**d**) male sternite VIII; (**e**) female tergite VIII; (**f**) female sternite VIII; (**g**) spermatheca. Scale bar for habitus = 1 mm; remaining scale bars = 0.2 mm

Fig. 19.179 (**a–h**). *Adota maritima* (Mannerheim): (**a**) habitus in dorsal view; (**b**) median lobe of aedeagus in lateral view; (**c**) median lobe of aedeagus in dorsal view; (**d**) male tergite VIII; (**e**) male sternite VIII; (**f**) female tergite VIII; (**g**) female sternite VIII; (**h**) spermatheca. Scale bar for habitus = 1 mm; remaining scale bars = 0.2 mm

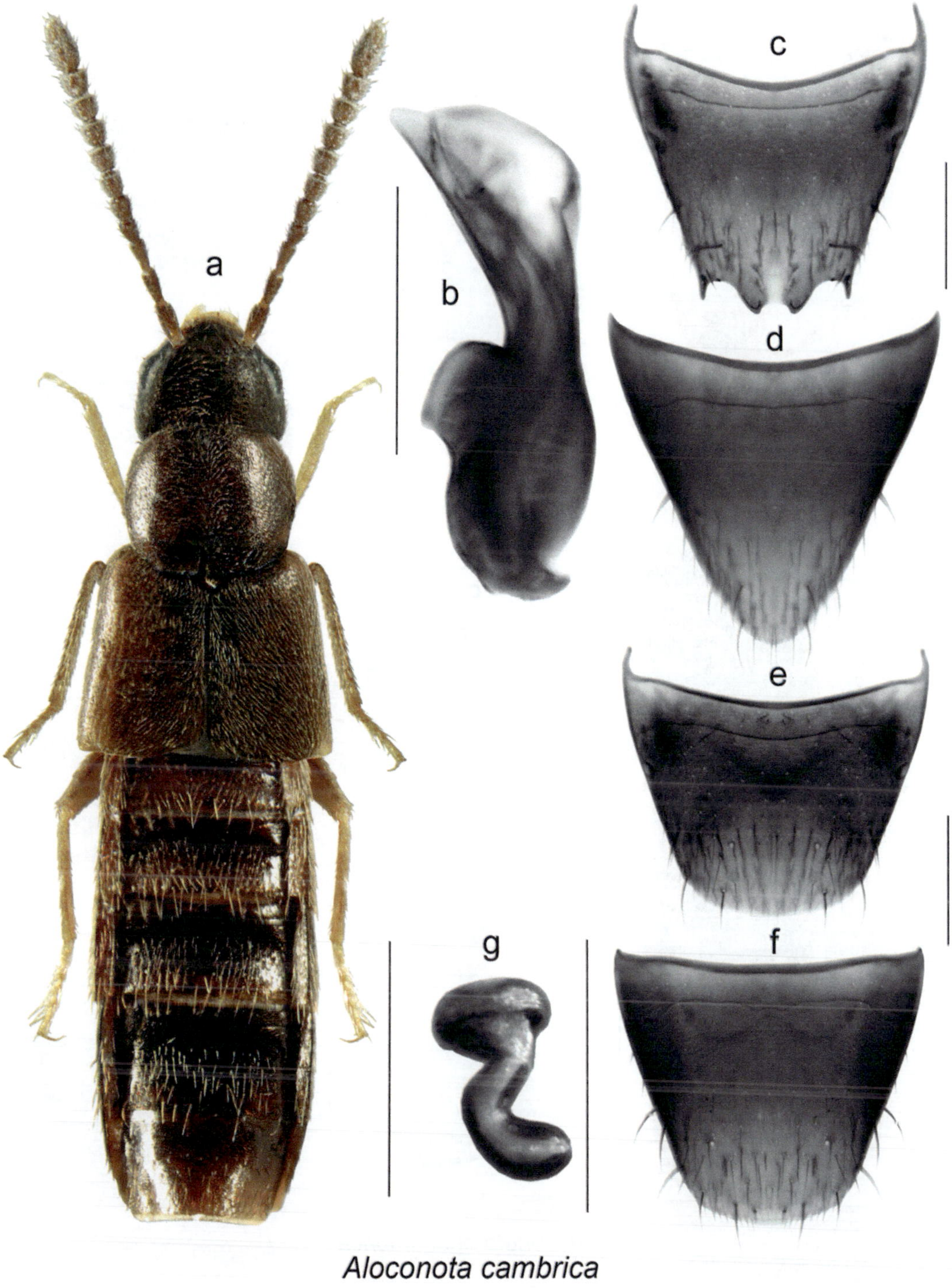

Fig. 19.180 (**a–g**). *Aloconota cambrica* (Wollaston): (**a**) habitus in dorsal view; (**b**) median lobe of aedeagus in lateral view; (**c**) male tergite VIII; (**d**) male sternite VIII; (**e**) female tergite VIII; (**f**) female sternite VIII; (**g**) spermatheca. Scale bar for habitus = 1 mm; remaining scale bars = 0.2 mm

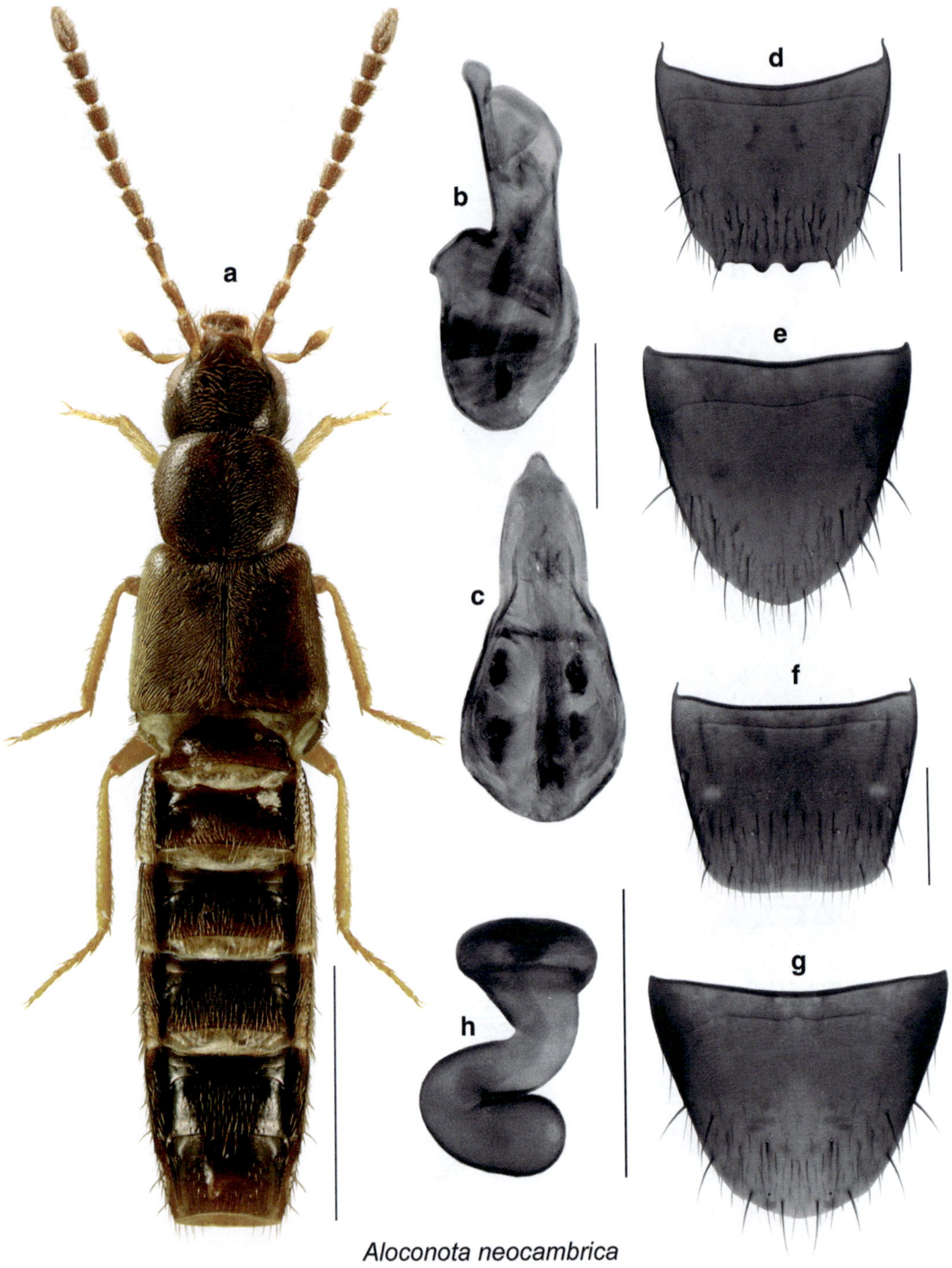

Fig. 19.181 (**a–h**). *Aloconota neocambrica* Klimaszewski and Langor: (**a**) habitus in dorsal view; (**b**) median lobe of aedeagus in lateral view; (**c**) median lobe of aedeagus in dorsal view; (**d**) male tergite VIII; (**e**) male sternite VIII; (**f**) female tergite VIII; (**g**) female sternite VIII; (**h**) spermatheca. Scale bar for habitus = 1 mm; remaining scale bars = 0.2 mm

Fig. 19.182 (**a–d**). *Amischa praelonga* (Casey): (**a**) habitus in dorsal view; (**b**) female tergite VIII; (**c**) female sternite VIII; (**d**) spermatheca. Scale bar for habitus = 1 mm; remaining scale bars = 0.2 mm

Fig. 19.183 (**a–e**). *Amischa tersa* Casey: (**a**) habitus in dorsal view; (**b**) median lobe of aedeagus in lateral view; (**c**) median lobe of aedeagus in dorsal view; (**d**) male tergite VIII; (**e**) male sternite VIII. Scale bar for habitus = 1 mm; remaining scale bars = 0.2 mm

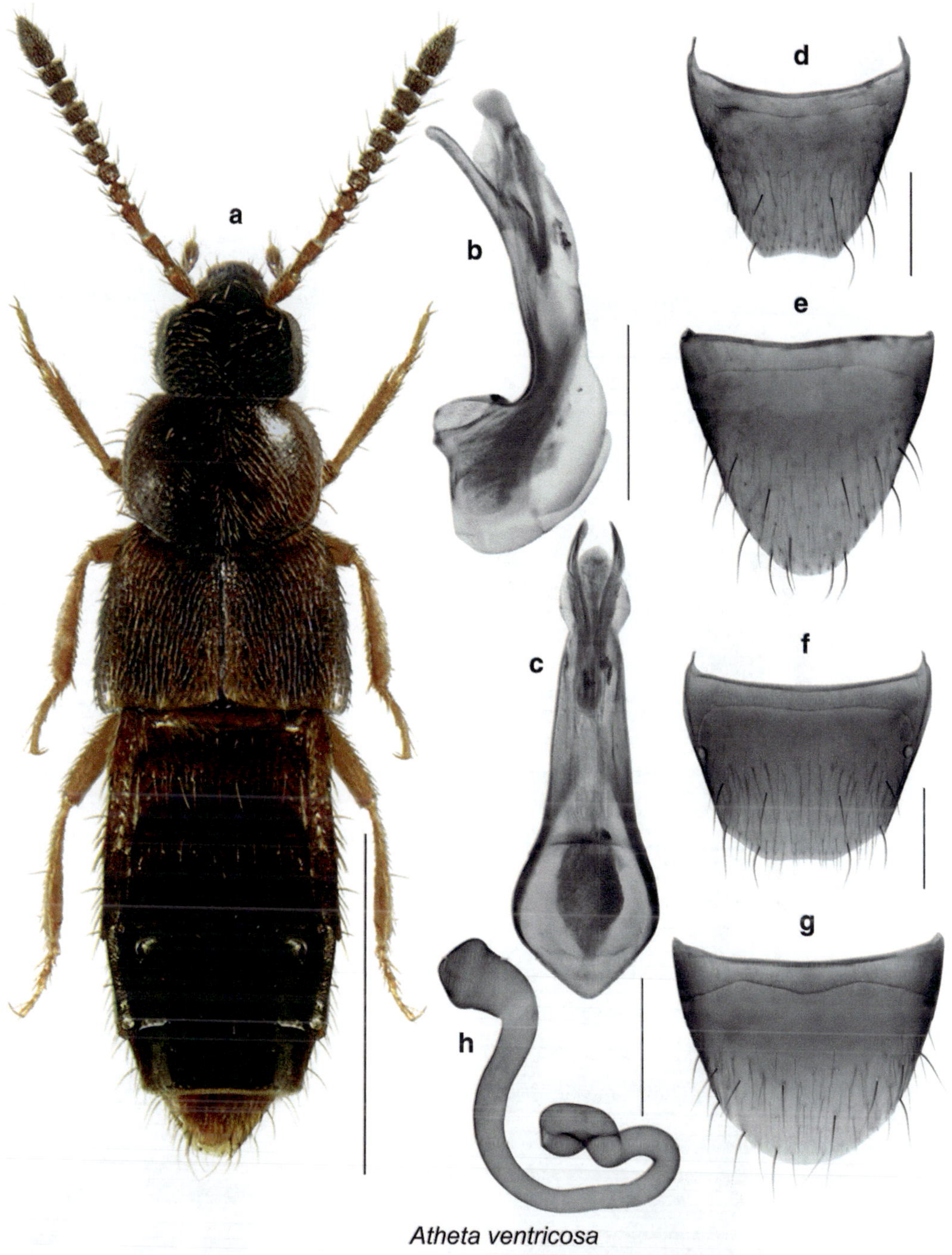

Fig. 19.184 (**a–h**). *Atheta (Alaobia) ventricosa* Bernhauer: (**a**) habitus in dorsal view; (**b**) median lobe of aedeagus in lateral view; (**c**) median lobe of aedeagus in dorsal view; (**d**) male tergite VIII; (**e**) male sternite VIII; (**f**) female tergite VIII; (**g**) female sternite VIII; (**h**) spermatheca. Scale bar for habitus = 1 mm; remaining scale bars = 0.2 mm

Fig. 19.185 (**a–d**). *Atheta (Atheta) borealides* Newton: (**a**) habitus in dorsal view; (**b**) female tergite VIII; (**c**) female sternite VIII; (**d**) spermatheca. Scale bar for habitus = 1 mm; remaining scale bars = 0.2 mm

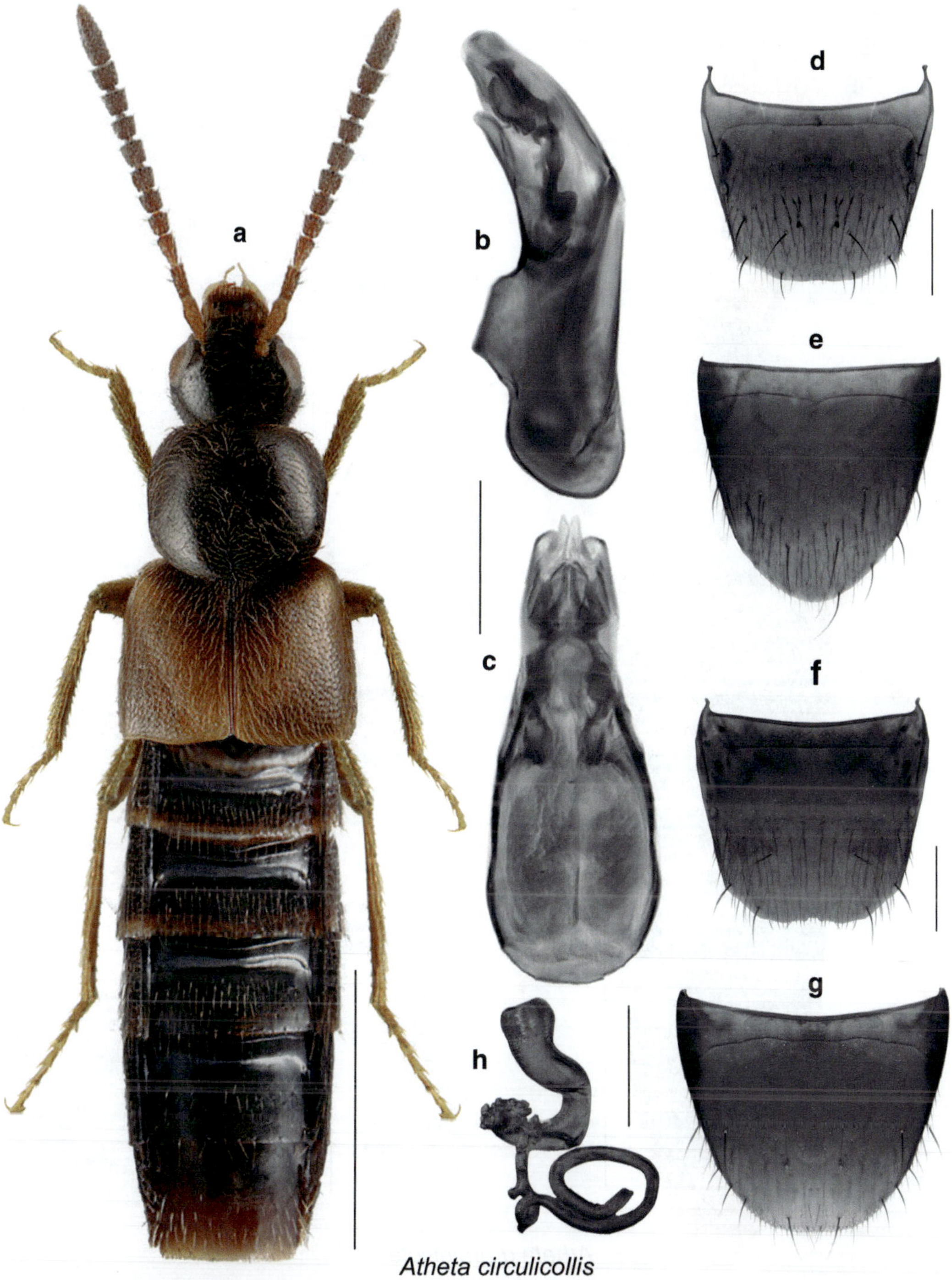

Fig. 19.186 (**a–h**). *Atheta (Atheta) circulicollis* Lohse: (**a**) habitus in dorsal view; (**b**) median lobe of aedeagus in lateral view; (**c**) median lobe of aedeagus in dorsal view; (**d**) male tergite VIII; (**e**) male sternite VIII; (**f**) female tergite VIII; (**g**) female sternite VIII; (**h**) spermatheca. Scale bar for habitus = 1 mm; remaining scale bars = 0.2 mm

Fig. 19.187 (**a–h**). *Atheta (Atheta) graminicola* (Gravenhorst): (**a**) habitus in dorsal view; (**b**) median lobe of aedeagus in lateral view; (**c**) median lobe of aedeagus in dorsal view; (**d**) male tergite VIII; (**e**) male sternite VIII; (**f**) female tergite VIII; (**g**) female sternite VIII; (**h**) spermatheca. Scale bar for habitus = 1 mm; remaining scale bars = 0.2 mm

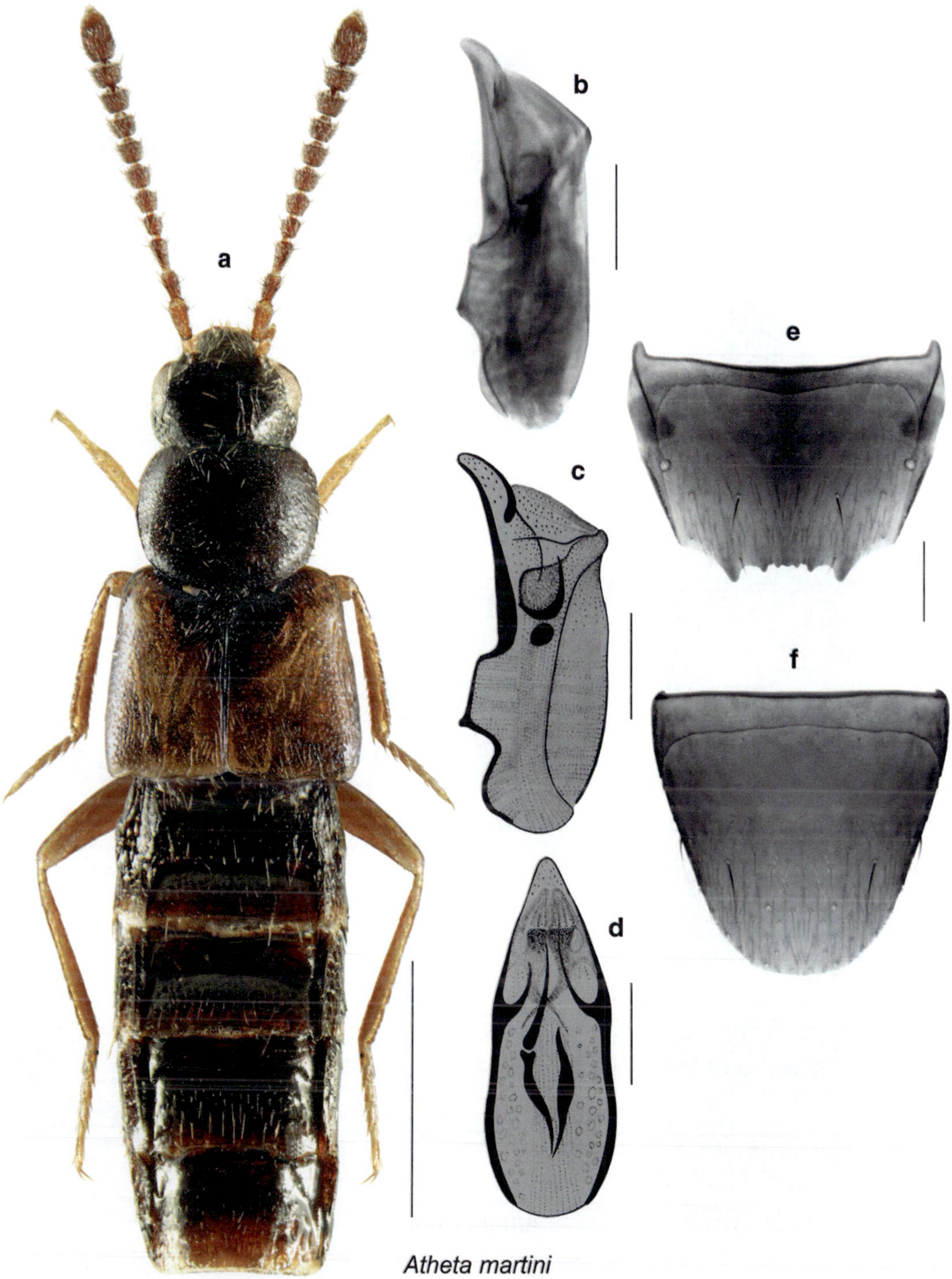

Fig. 19.188 (**a–f**). *Atheta (Atheta) martini* Lohse: (**a**) habitus in dorsal view; (**b, c**) median lobe of aedeagus in lateral view; (**d**) median lobe of aedeagus in dorsal view; (**e**) male tergite VIII; (**f**) male sternite VIII. Scale bar for habitus = 1 mm; remaining scale bars = 0.2 mm

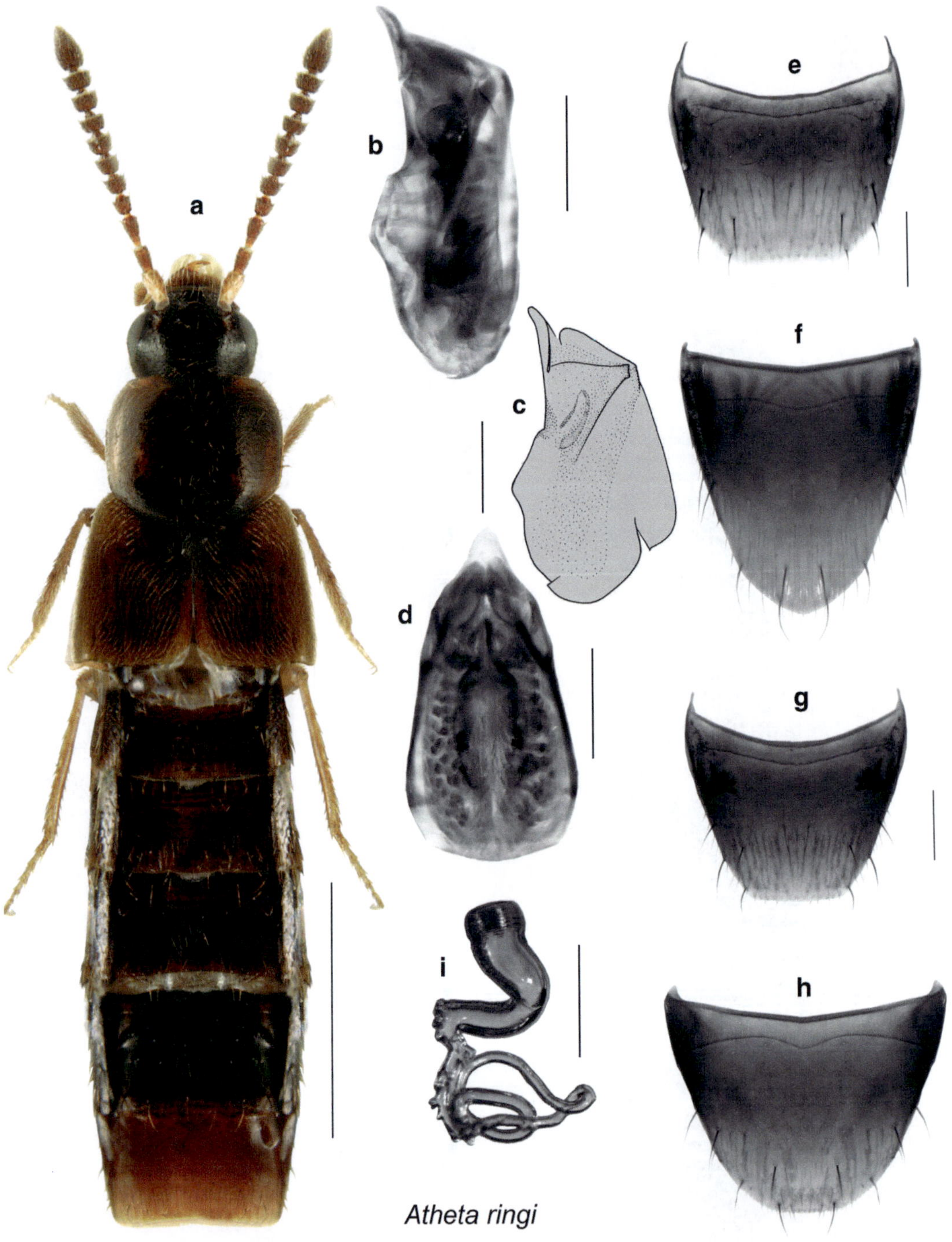

Fig. 19.189 (**a–i**). *Atheta (Atheta) ringi* Klimaszewski: (**a**) habitus in dorsal view; (**b, c**) median lobe of aedeagus in lateral view; (**d**) median lobe of aedeagus in dorsal view; (**e**) male tergite VIII; (**f**) male sternite VIII; (**g**) female tergite VIII; (**h**) female sternite VIII; (**i**) spermatheca. Scale bar for habitus = 1 mm; remaining scale bars = 0.2 mm

Fig. 19.190 (**a–i**). *Atheta (Bessobia) cryptica* (Lohse): (**a**) habitus in dorsal view; (**b, c**) median lobe of aedeagus in lateral view; (**d**) median lobe of aedeagus in dorsal view; (**e**) male tergite VIII; (**f**) male sternite VIII; (**g**) female tergite VIII; (**h**) female sternite VIII; (**i**) spermatheca. Scale bar for habitus = 1 mm; remaining scale bars = 0.2 mm

Fig. 19.191 (**a–g**). *Atheta (Datomicra) chadburnae* Klimaszewski and Godin, sp.n.: (**a**) habitus in dorsal view; (**b**) median lobe of aedeagus in lateral view; (**c**) male tergite VIII; (**d**) male sternite VIII; (**e**) female tergite VIII; (**f**) female sternite VIII; (**g**) spermatheca. Scale bar for habitus = 1 mm; remaining scale bars = 0.2 mm

Fig. 19.192 (**a–h**). *Atheta (Datomicra) dadopora* C.G. Thomson: (**a**) habitus in dorsal view; (**b**) median lobe of aedeagus in lateral view; (**c**) median lobe of aedeagus in dorsal view; (**d**) male tergite VIII; (**e**) male sternite VIII; (**f**) female tergite VIII; (**g**) female sternite VIII; (**h**) spermatheca. Scale bar for habitus = 1 mm; remaining scale bars = 0.2 mm

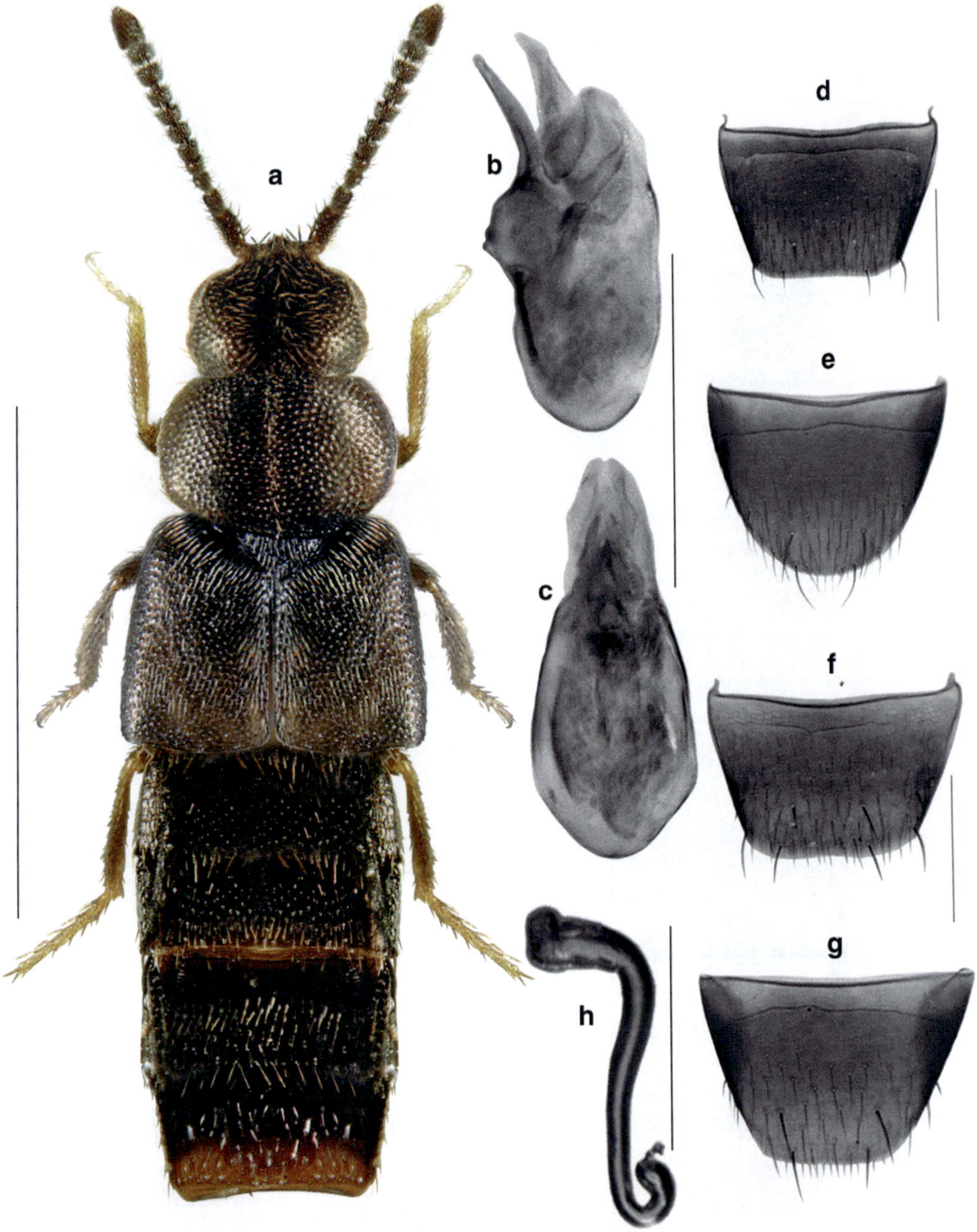

Fig. 19.193 (**a**–**h**). *Atheta (Dimetrota) whitehorsensis* Klimaszewski and Godin: (**a**) habitus in dorsal view; (**b**) median lobe of aedeagus in lateral view; (**c**) median lobe of aedeagus in dorsal view; (**d**) male tergite VIII; (**e**) male sternite VIII; (**f**) female tergite VIII; (**g**) female sternite VIII; (**h**) spermatheca. Scale bar for habitus = 1 mm; remaining scale bars = 0.2 mm

Atheta cadeti

Fig. 19.194 (**a–h**). *Atheta* (*Dimetrota*) *cadeti* Klimaszewski and Godin: (**a**) habitus in dorsal view; (**b**) median lobe of aedeagus in lateral view; (**c**) median lobe of aedeagus in dorsal view; (**d**) male tergite VIII; (**e**) male sternite VIII; (**f**) female tergite VIII; (**g**) female sternite VIII; (**h**) spermatheca. Scale bar for habitus = 1 mm; remaining scale bars = 0.2 mm

Fig. 19.195 (**a–h**). *Atheta (Dimetrota) allocera* Eppelsheim (former *D. campbelli* Lohse): (**a**) habitus in dorsal view; (**b**) median lobe of aedeagus in lateral view; (**c**) median lobe of aedeagus in dorsal view; (**d**) male tergite VIII; (**e**) male sternite VIII; (**f**) female tergite VIII; (**g**) female sternite VIII; (**h**) spermatheca. Scale bar for habitus = 1 mm; remaining scale bars = 0.2 mm

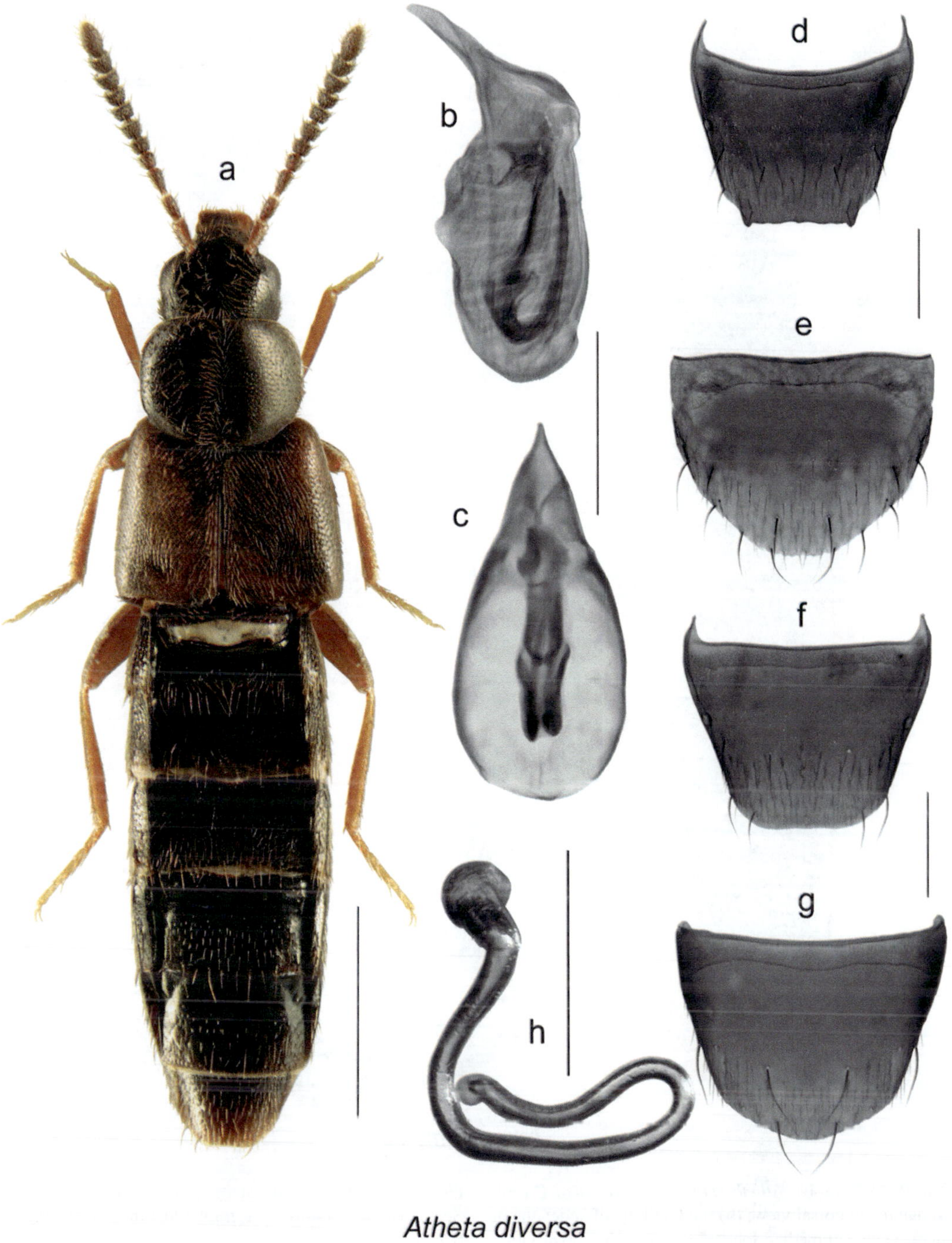

Fig. 19.196 (**a–h**). *Atheta (Dimetrota) diversa* (Sharp) (former *D. regissalmonis* Lohse): (**a**) habitus in dorsal view; (**b**) median lobe of aedeagus in lateral view; (**c**) median lobe of aedeagus in dorsal view; (**d**) male tergite VIII; (**e**) male sternite VIII; (**f**) female tergite VIII; (**g**) female sternite VIII; (**h**) spermatheca. Scale bar for habitus = 1 mm; remaining scale bars = 0.2 mm

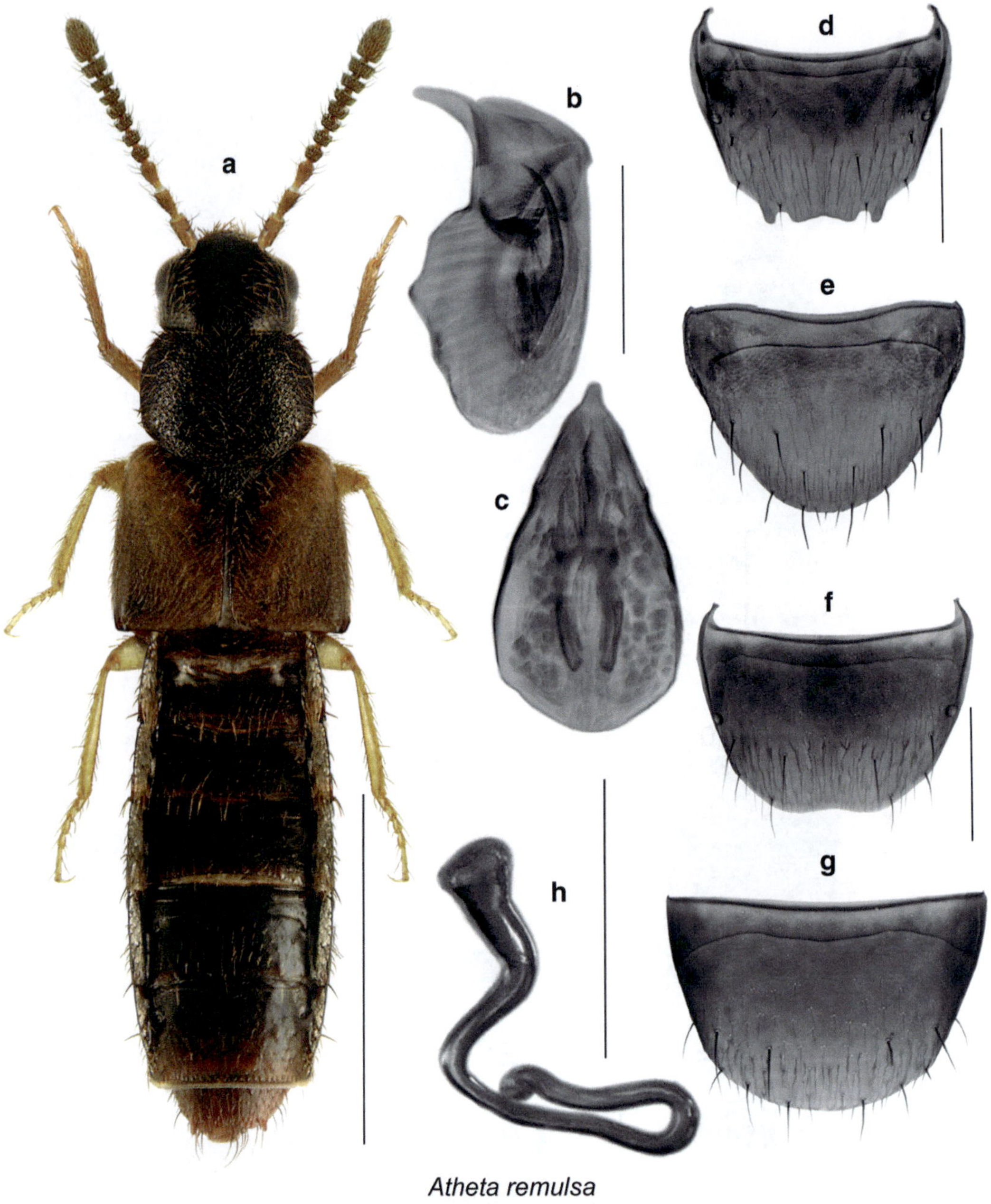

Fig. 19.197 (**a–h**). *Atheta (Dimetrota) remulsa* Casey: (**a**) habitus in dorsal view; (**b**) median lobe of aedeagus in lateral view; (**c**) median lobe of aedeagus in dorsal view; (**d**) male tergite VIII; (**e**) male sternite VIII; (**f**) female tergite VIII; (**g**) female sternite VIII; (**h**) spermatheca. Scale bar for habitus = 1 mm; remaining scale bars = 0.2 mm

Fig. 19.198 (**a–h**). *Atheta (Dimetrota) smetanai* (Lohse) (based on types): (**a**) habitus in dorsal view; (**b**) median lobe of aedeagus in lateral view; (**c**) median lobe of aedeagus in dorsal view; (**d**) male tergite VIII; (**e**) male sternite VIII; (**f**) female tergite VIII; (**g**) female sternite VIII; (**h**) spermatheca. Scale bar for habitus = 1 mm; remaining scale bars = 0.2 mm

Fig. 19.199 (**a–h**). *Atheta (Dimetrota) stercoris* Fenyes (former *A. columbica* Bernhauer): (**a**) habitus in dorsal view; (**b**) median lobe of aedeagus in lateral view; (**c**) median lobe of aedeagus in dorsal view; (**d**) male tergite VIII; (**e**) male sternite VIII; (**f**) female tergite VIII; (**g**) female sternite VIII; (**h**) spermatheca. Scale bar for habitus = 1 mm; remaining scale bars = 0.2 mm

Fig. 19.200 (**a–h**). *Atheta (Dimetrota) pseudometlakatlana* Klimaszewski and Godin: (**a**) habitus in dorsal view; (**b**) median lobe of aedeagus in lateral view; (**c**) median lobe of aedeagus in dorsal view; (**d**) male tergite VIII; (**e**) male sternite VIII; (**f**) female tergite VIII; (**g**) female sternite VIII; (**h**) spermatheca. Scale bar for habitus = 1 mm; remaining scale bars = 0.2 mm

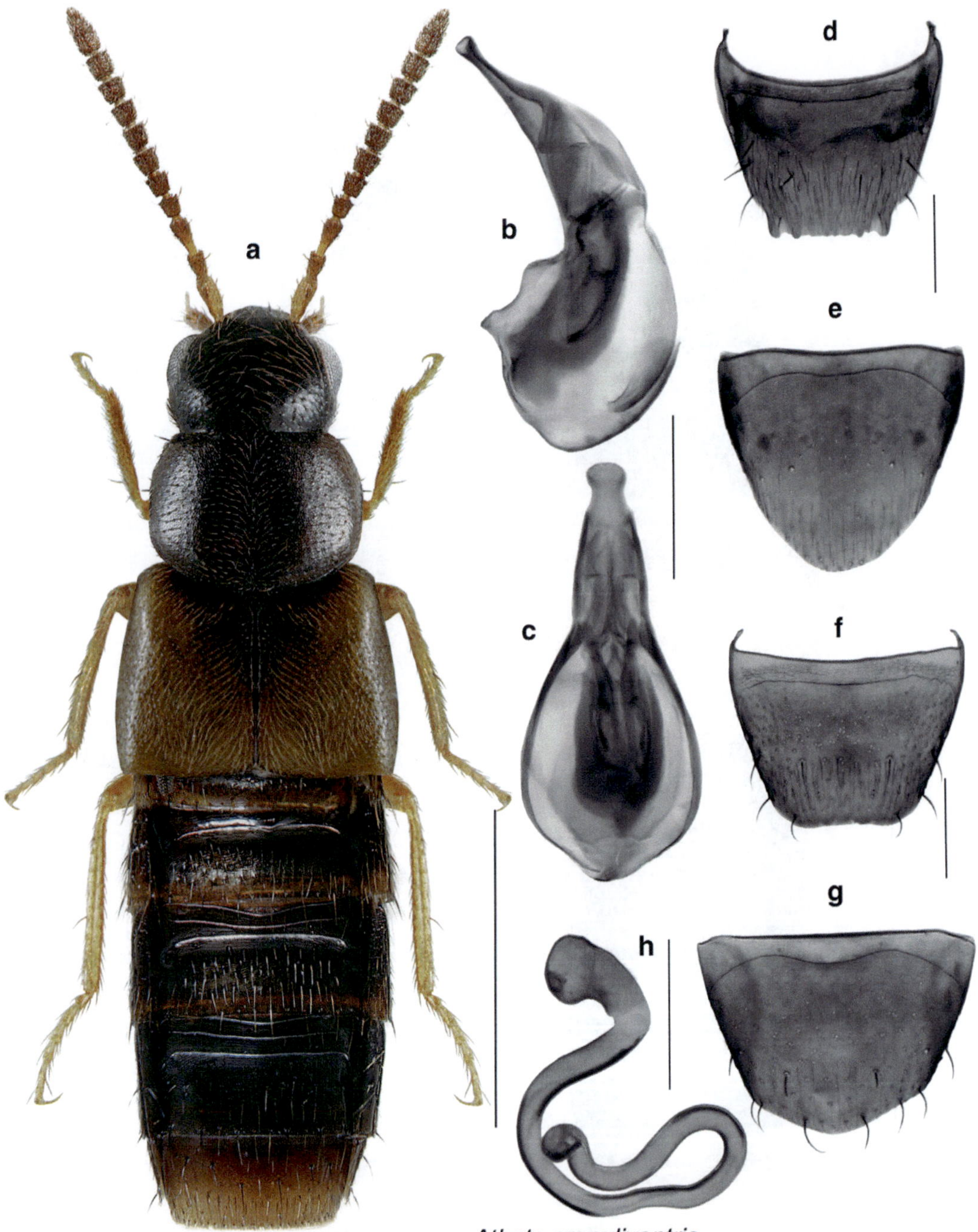

Fig. 19.201 (**a–h**). *Atheta (Dimetrota) crenuliventris* Bernhauer: (**a**) habitus in dorsal view; (**b**) median lobe of aedeagus in lateral view; (**c**) median lobe of aedeagus in dorsal view; (**d**) male tergite VIII; (**e**) male sternite VIII; (**f**) female tergite VIII; (**g**) female sternite VIII; (**h**) spermatheca. Scale bar for habitus = 1 mm; remaining scale bars = 0.2 mm

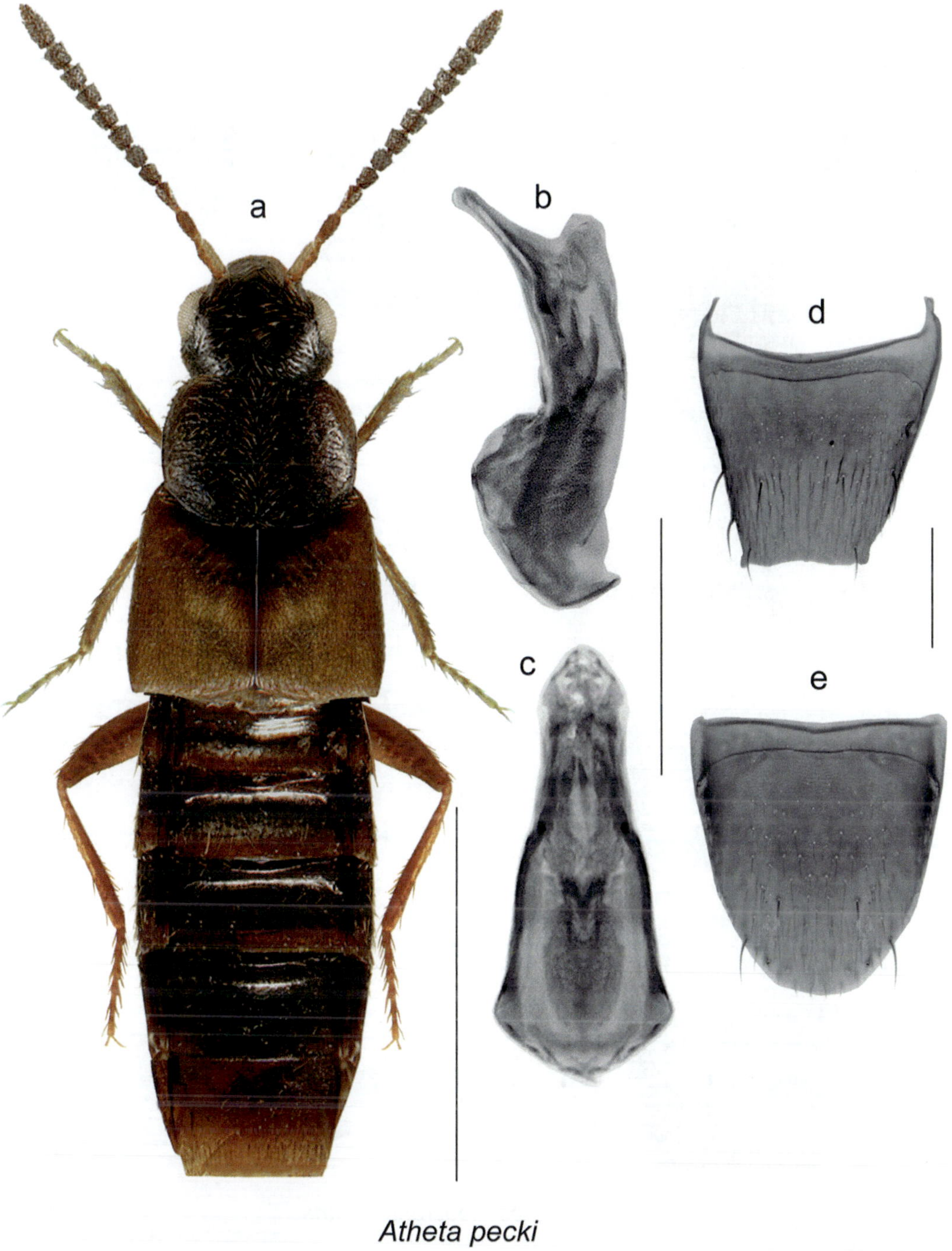

Fig. 19.202 (**a–e**). *Atheta (Dimetrota) pecki* Klimaszewski and Langor: (**a**) habitus in dorsal view; (**b**) median lobe of aedeagus in lateral view; (**c**) median lobe of aedeagus in dorsal view; (**d**) male tergite VIII; (**e**) male sternite VIII. Scale bar for habitus = 1 mm; remaining scale bars = 0.2 mm

Fig. 19.203 (**a–h**). *Atheta (Dimetrota) pseudocrenuliventris* Klimaszewski: (**a**) habitus in dorsal view; (**b**) median lobe of aedeagus in lateral view; (**c**) median lobe of aedeagus in dorsal view; (**d**) male tergite VIII; (**e**) male sternite VIII; (**f**) female tergite VIII; (**g**) female sternite VIII; (**h**) spermatheca. Scale bar for habitus = 1 mm; remaining scale bars = 0.2 mm

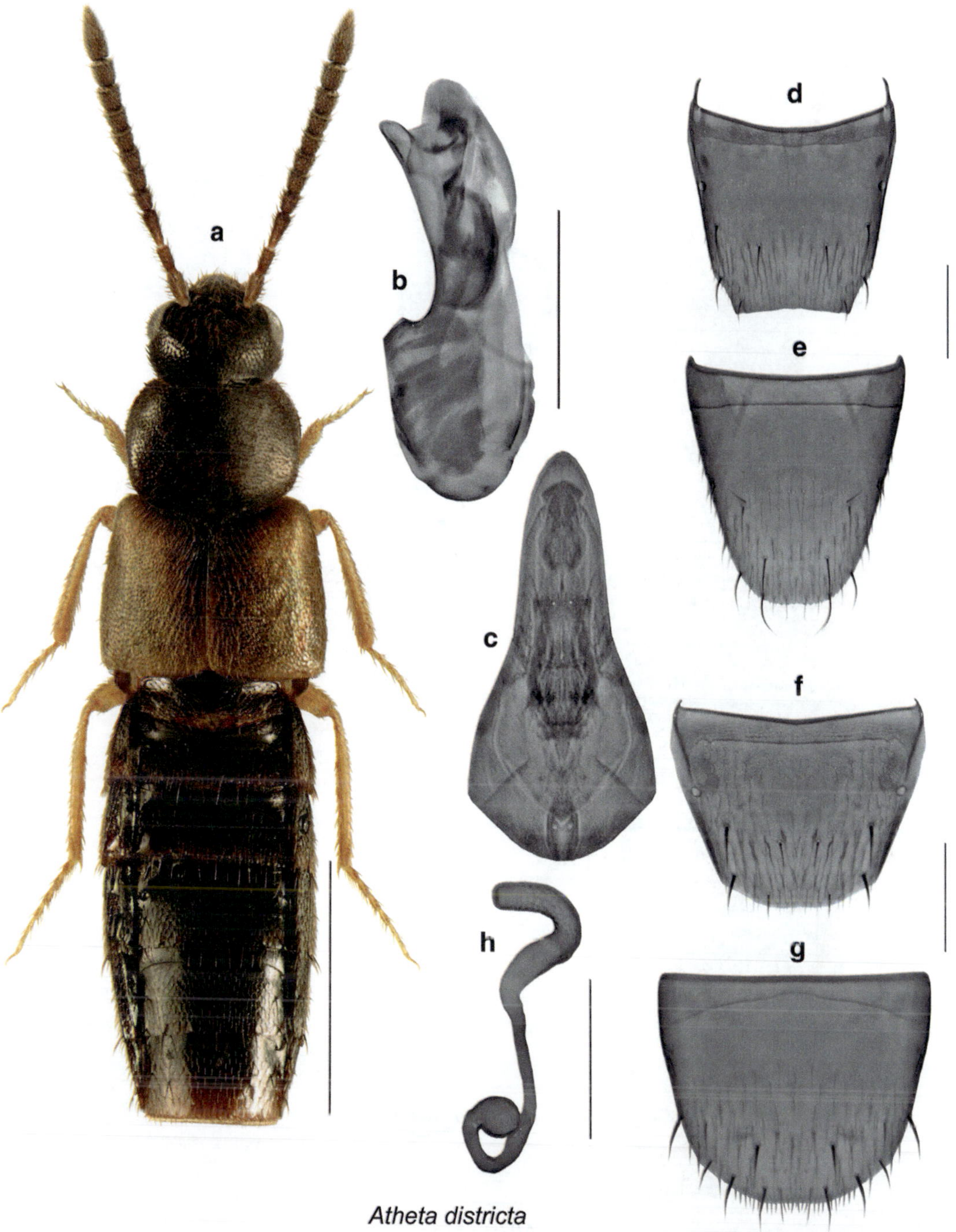

Fig. 19.204 (**a–h**). *Atheta (Dimetrota) districta* Casey: (**a**) habitus in dorsal view; (**b**) median lobe of aedeagus in lateral view; (**c**) median lobe of aedeagus in dorsal view; (**d**) male tergite VIII; (**e**) male sternite VIII; (**f**) female tergite VIII; (**g**) female sternite VIII; (**h**) spermatheca. Scale bar for habitus = 1 mm; remaining scale bars = 0.2 mm

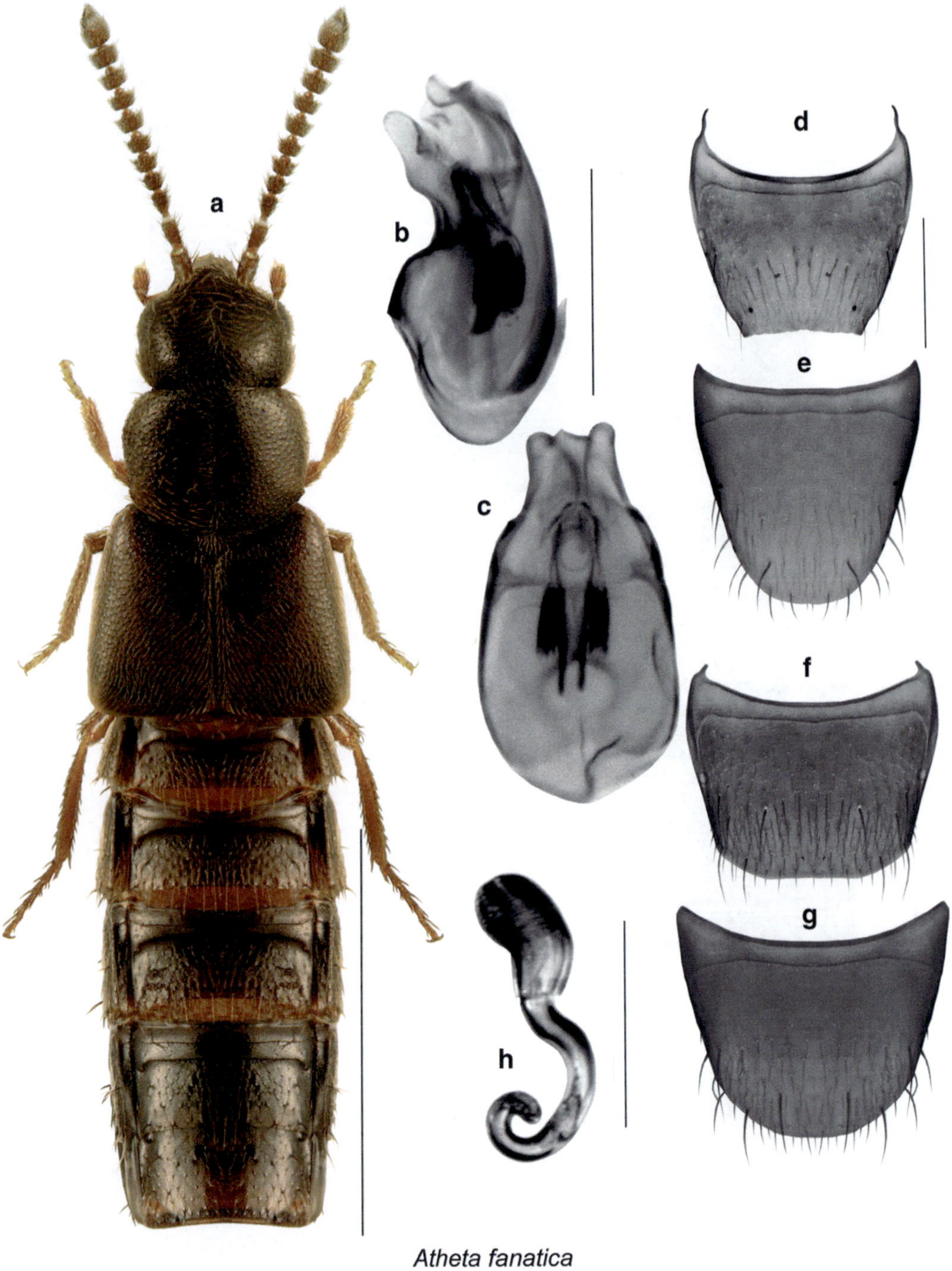

Fig. 19.205 (**a–h**). *Atheta (Dimetrota) fanatica* Casey: (**a**) habitus in dorsal view; (**b**) median lobe of aedeagus in lateral view; (**c**) median lobe of aedeagus in dorsal view; (**d**) male tergite VIII; (**e**) male sternite VIII; (**f**) female tergite VIII; (**g**) female sternite VIII; (**h**) spermatheca. Scale bar for habitus = 1 mm; remaining scale bars = 0.2 mm

Fig. 19.206 (**a–h**). *Atheta (Dimetrota) munsteri* Bernhauer: (**a**) habitus in dorsal view; (**b**) median lobe of aedeagus in lateral view; (**c**) median lobe of aedeagus in dorsal view; (**d**) male tergite VIII; (**e**) male sternite VIII; (**f**) female tergite VIII; (**g**) female sternite VIII; (**h**) spermatheca. Scale bar for habitus = 1 mm; remaining scale bars = 0.2 mm

Fig. 19.207 (**a–h**). *Atheta (Dimetrota) hampshirensis* Bernhauer: (**a**) habitus in dorsal view; (**b**) median lobe of aedeagus in lateral view; (**c**) median lobe of aedeagus in dorsal view; (**d**) male tergite VIII; (**e**) male sternite VIII; (**f**) female tergite VIII; (**g**) female sternite VIII; (**h**) spermatheca. Scale bar for habitus = 1 mm; remaining scale bars = 0.2 mm

Fig. 19.208 (**a–h**). *Atheta (Dimetrota) altaica* Bernhauer: (**a**) habitus in dorsal view; (**b**) median lobe of aedeagus in lateral view; (**c**) median lobe of aedeagus in dorsal view; (**d**) male tergite VIII; (**e**) male sternite VIII; (**f**) female tergite VIII; (**g**) female sternite VIII; (**h**) spermatheca. Scale bar for habitus = 1 mm; remaining scale bars = 0.2 mm

Fig. 19.209 (**a–h**). *Atheta (Dimetrota) atramentaria* (Gyllenhal): (**a**) habitus in dorsal view; (**b**) median lobe of aedeagus in lateral view; (**c**) median lobe of aedeagus in dorsal view; (**d**) male tergite VIII; (**e**) male sternite VIII; (**f**) female tergite VIII; (**g**) female sternite VIII; (**h**) spermatheca. Scale bar for habitus = 1 mm; remaining scale bars = 0.2 mm

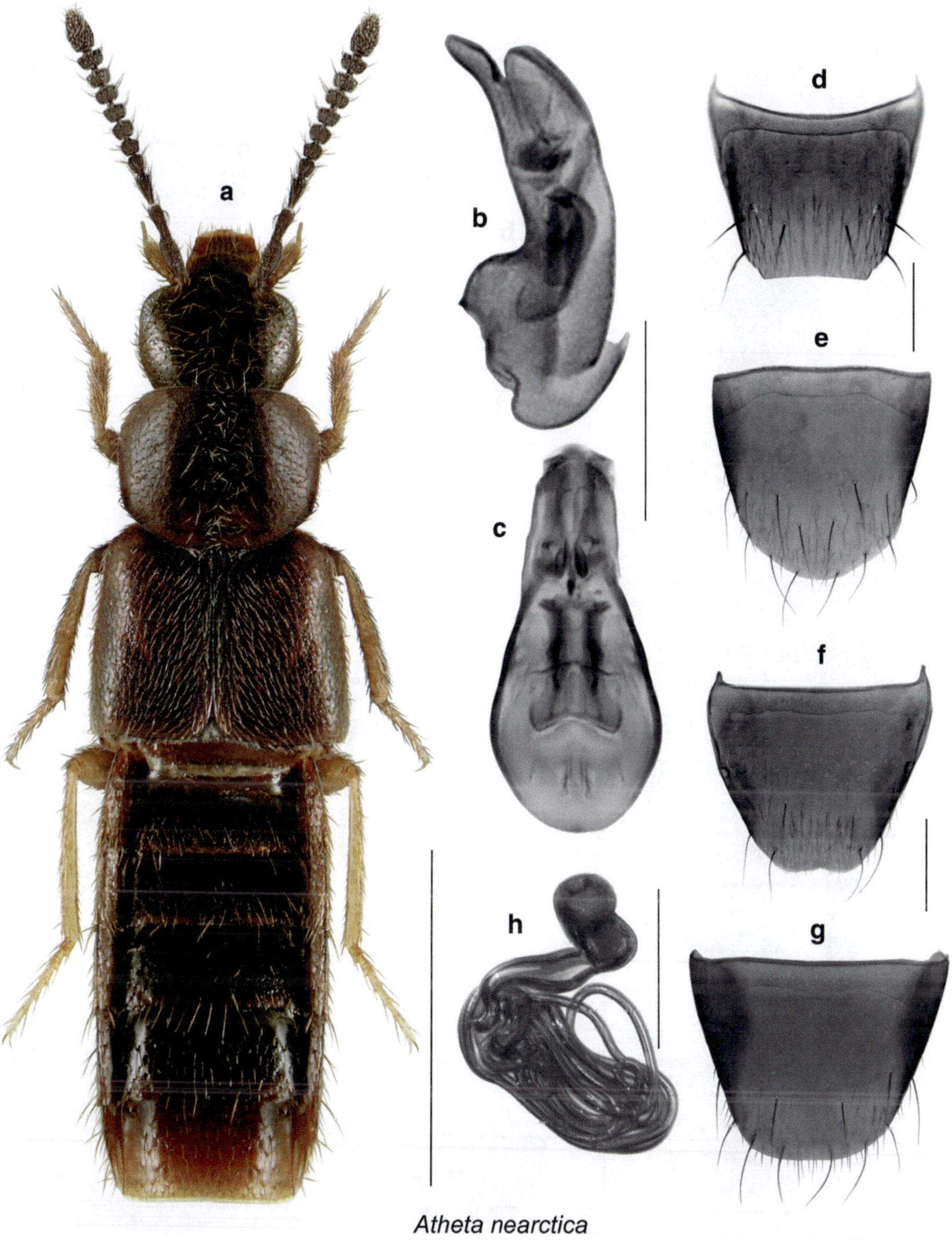

Fig. 19.210 (**a–h**). *Atheta (Dimetrota) nearctica* (Lohse): (**a**) habitus in dorsal view; (**b**) median lobe of aedeagus in lateral view; (**c**) median lobe of aedeagus in dorsal view; (**d**) male tergite VIII; (**e**) male sternite VIII; (**f**) female tergite VIII; (**g**) female sternite VIII; (**h**) spermatheca. Scale bar for habitus = 1 mm; remaining scale bars = 0.2 mm

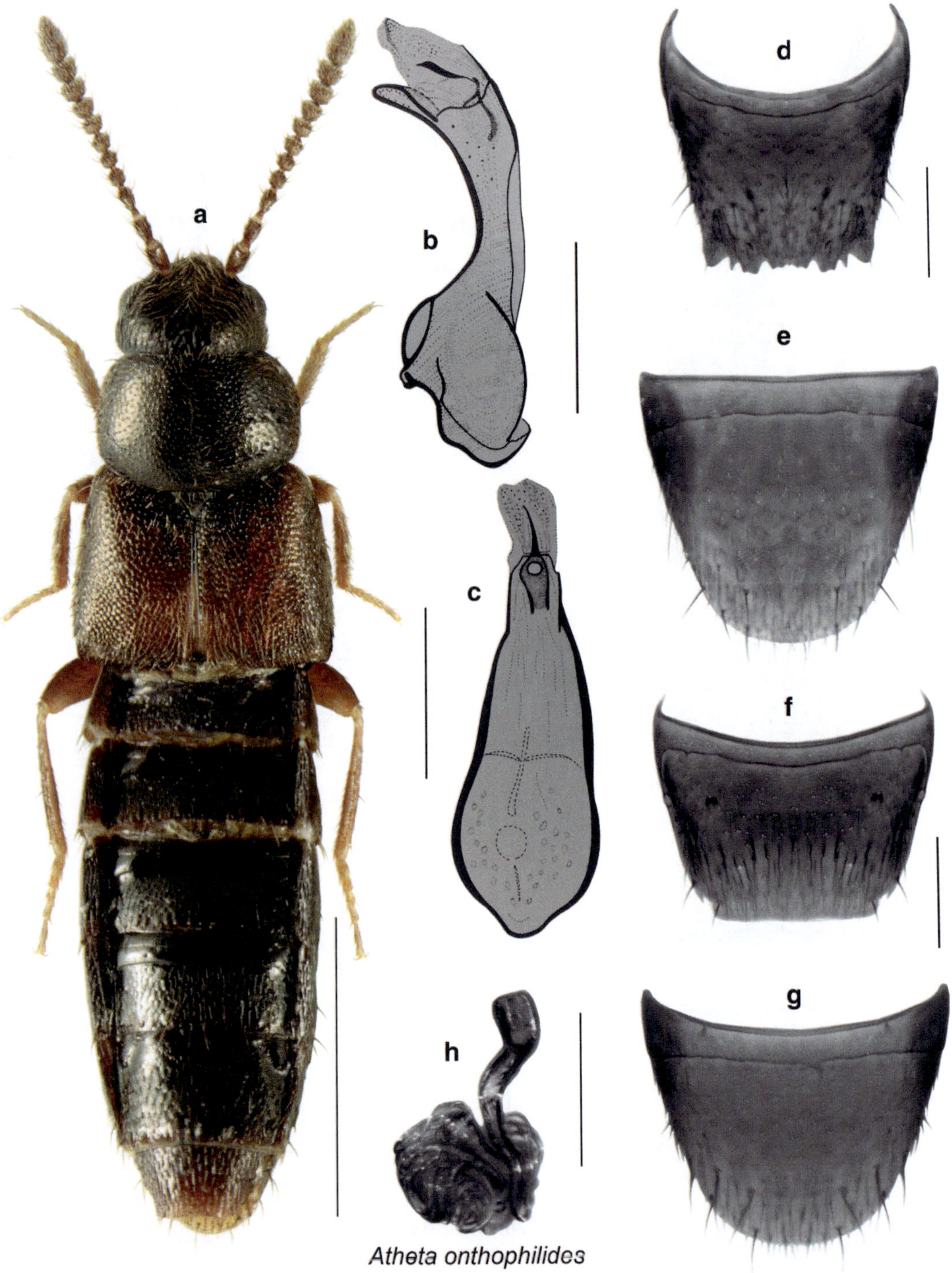

Fig. 19.211 (**a–h**). *Atheta (Dimetrota) onthophilides* (Newton): (**a**) habitus in dorsal view; (**b**) median lobe of aedeagus in lateral view; (**c**) median lobe of aedeagus in dorsal view; (**d**) male tergite VIII; (**e**) male sternite VIII; (**f**) female tergite VIII; (**g**) female sternite VIII; (**h**) spermatheca. Scale bar for habitus = 1 mm; remaining scale bars = 0.2 mm

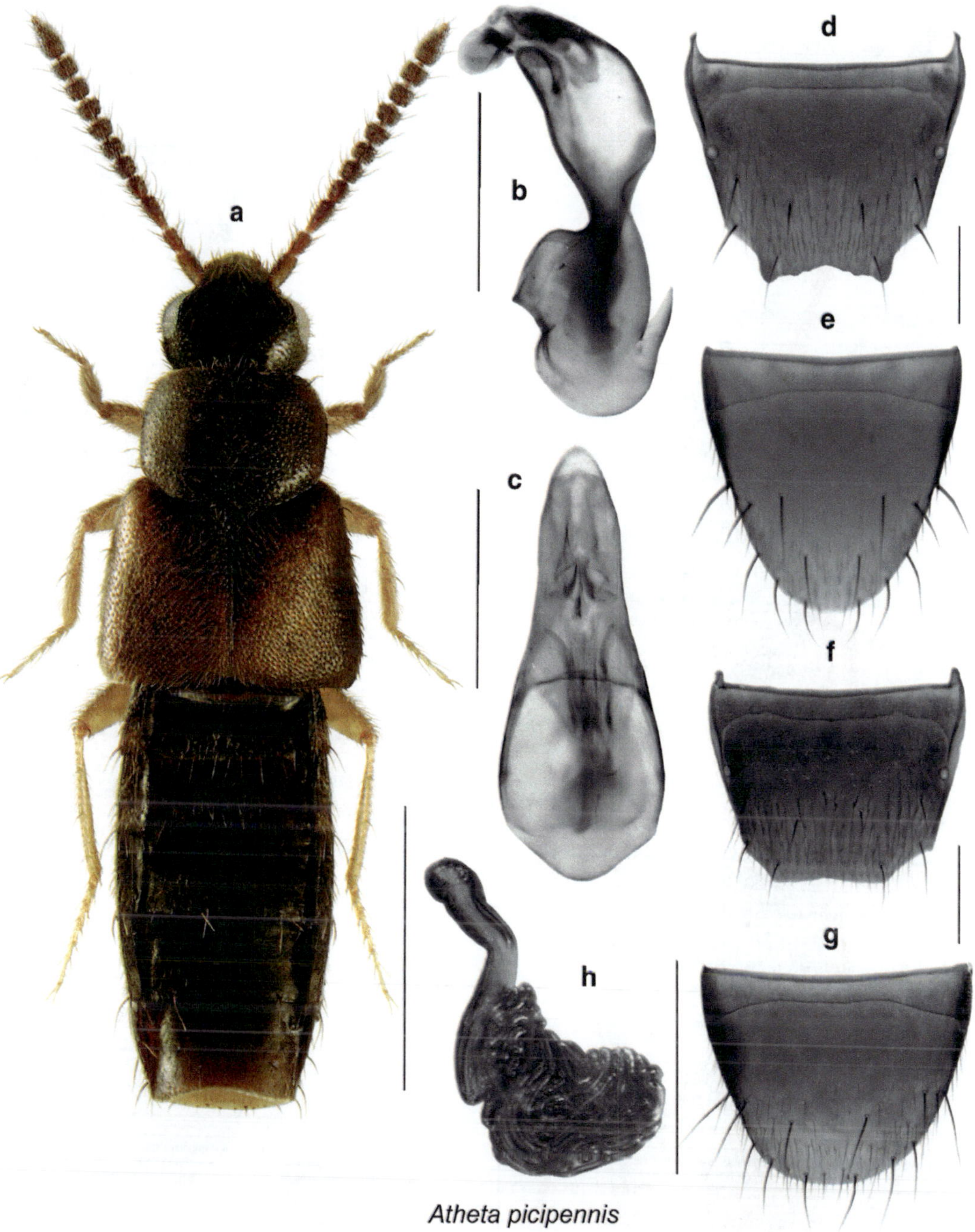

Fig. 19.212 (**a–h**). *Atheta (Dimetrota) picipennis* (Mannerheim), (based on specimens from BC): (**a**) habitus in dorsal view; (**b**) median lobe of aedeagus in lateral view; (**c**) median lobe of aedeagus in dorsal view; (**d**) male tergite VIII; (**e**) male sternite VIII; (**f**) female tergite VIII; (**g**) female sternite VIII; (**h**) spermatheca. Scale bar for habitus = 1 mm; remaining scale bars = 0.2 mm

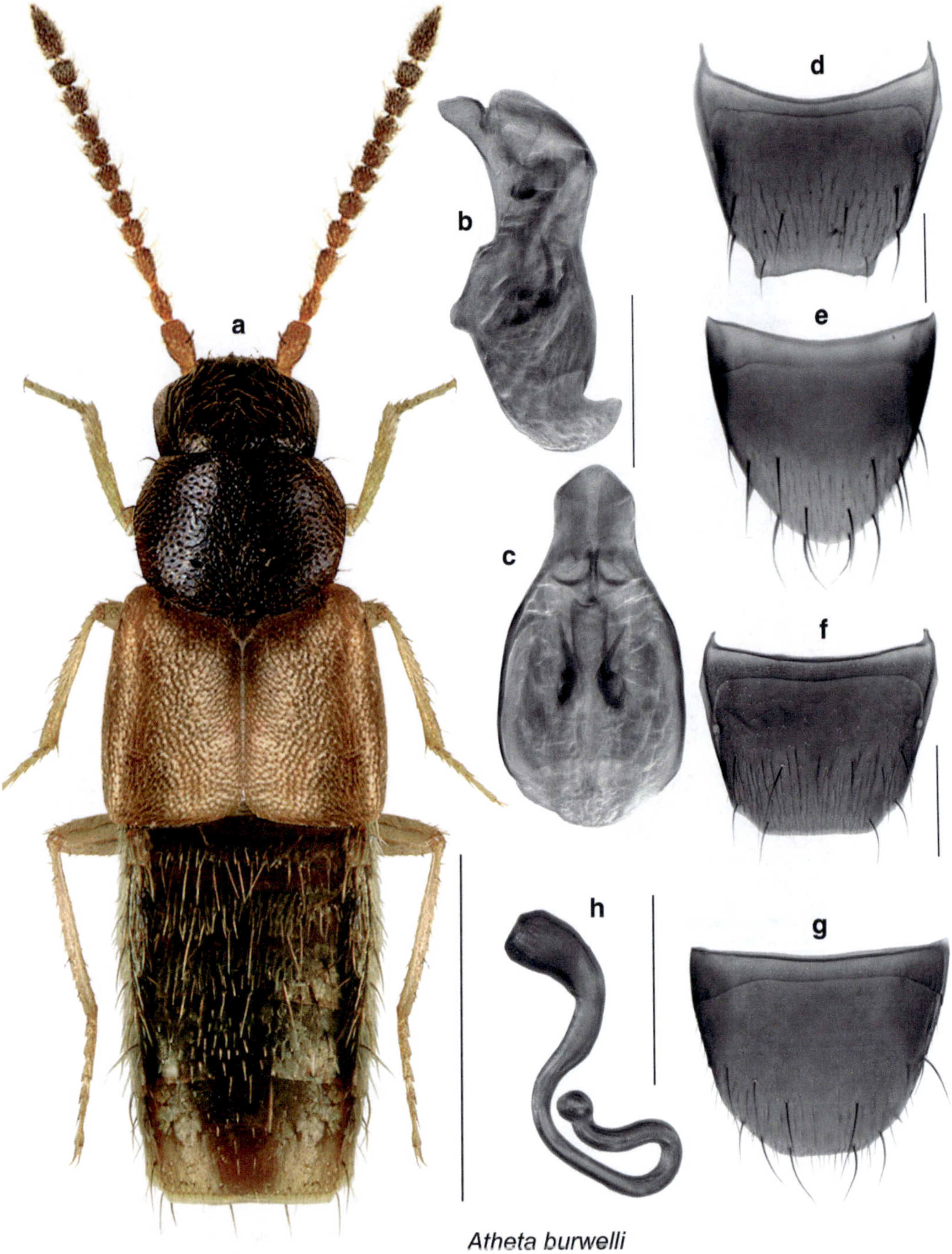

Fig. 19.213 (**a–h**). *Atheta (Dimetrota) burwelli* (Lohse): (**a**) habitus in dorsal view; (**b**) median lobe of aedeagus in lateral view; (**c**) median lobe of aedeagus in dorsal view; (**d**) male tergite VIII; (**e**) male sternite VIII; (**f**) female tergite VIII; (**g**) female sternite VIII; (**h**) spermatheca. Scale bar for habitus = 1 mm; remaining scale bars = 0.2 mm

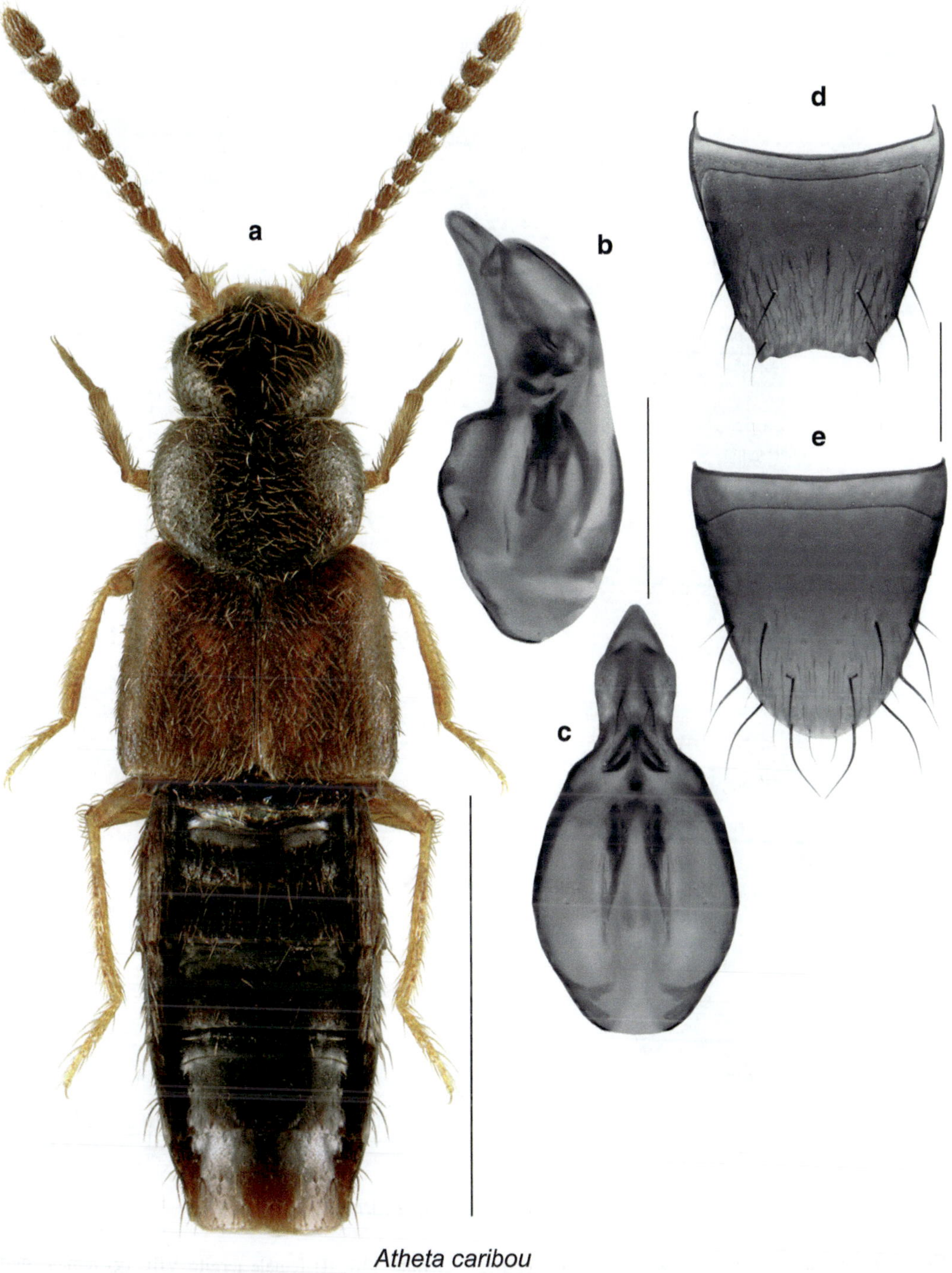

Fig. 19.214 (**a–e**). *Atheta (Dimetrota) caribou* (Lohse): (**a**) habitus in dorsal view; (**b**) median lobe of aedeagus in lateral view; (**c**) median lobe of aedeagus in dorsal view; (**d**) male tergite VIII; (**e**) male sternite VIII. Scale bar for habitus = 1 mm; remaining scale bars = 0.2 mm

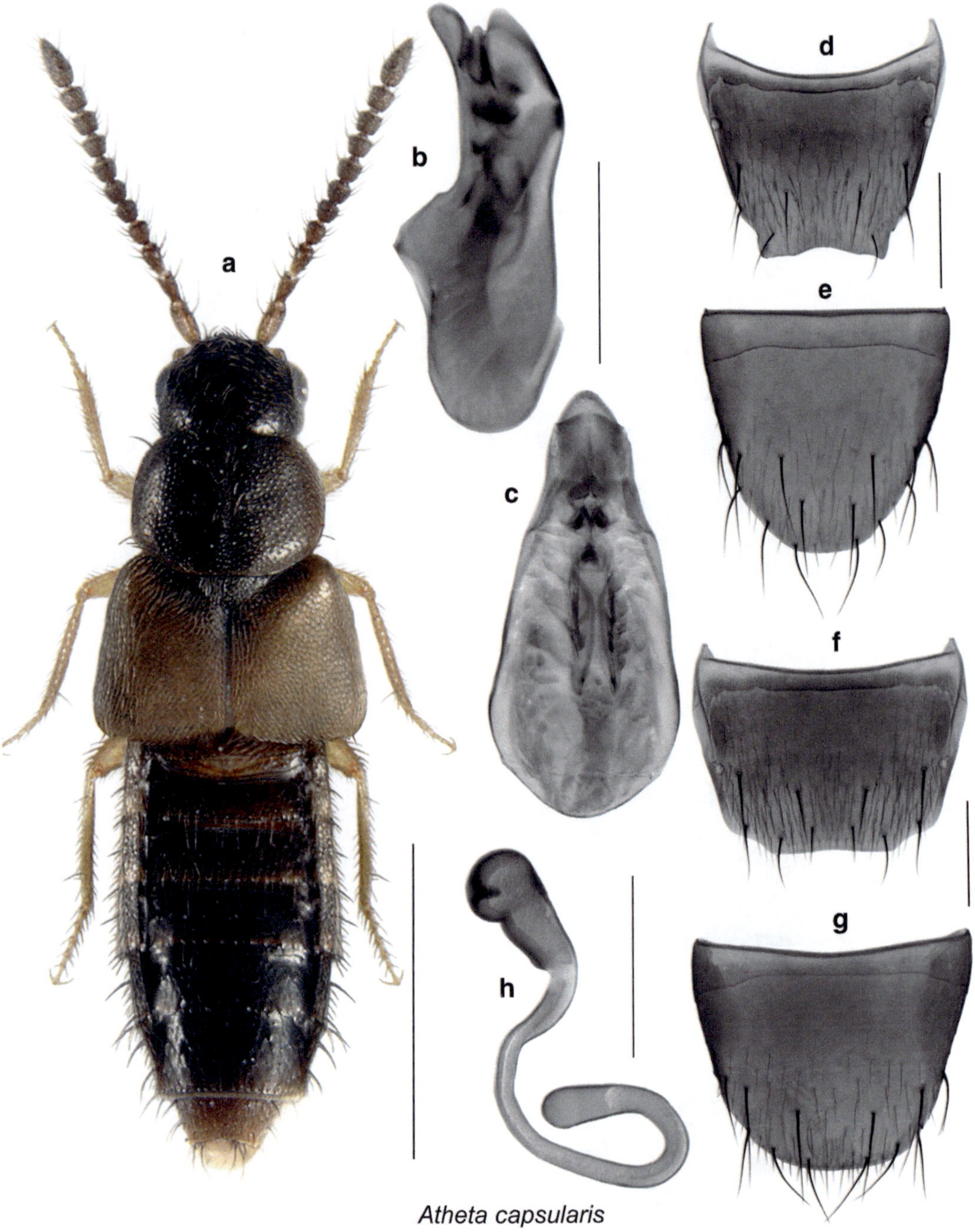

Fig. 19.215 (**a–h**). *Atheta (Dimetrota) capsularis* Klimaszewski: (**a**) habitus in dorsal view; (**b**) median lobe of aedeagus in lateral view; (**c**) median lobe of aedeagus in dorsal view; (**d**) male tergite VIII; (**e**) male sternite VIII; (**f**) female tergite VIII; (**g**) female sternite VIII; (**h**) spermatheca. Scale bar for habitus = 1 mm; remaining scale bars = 0.2 mm

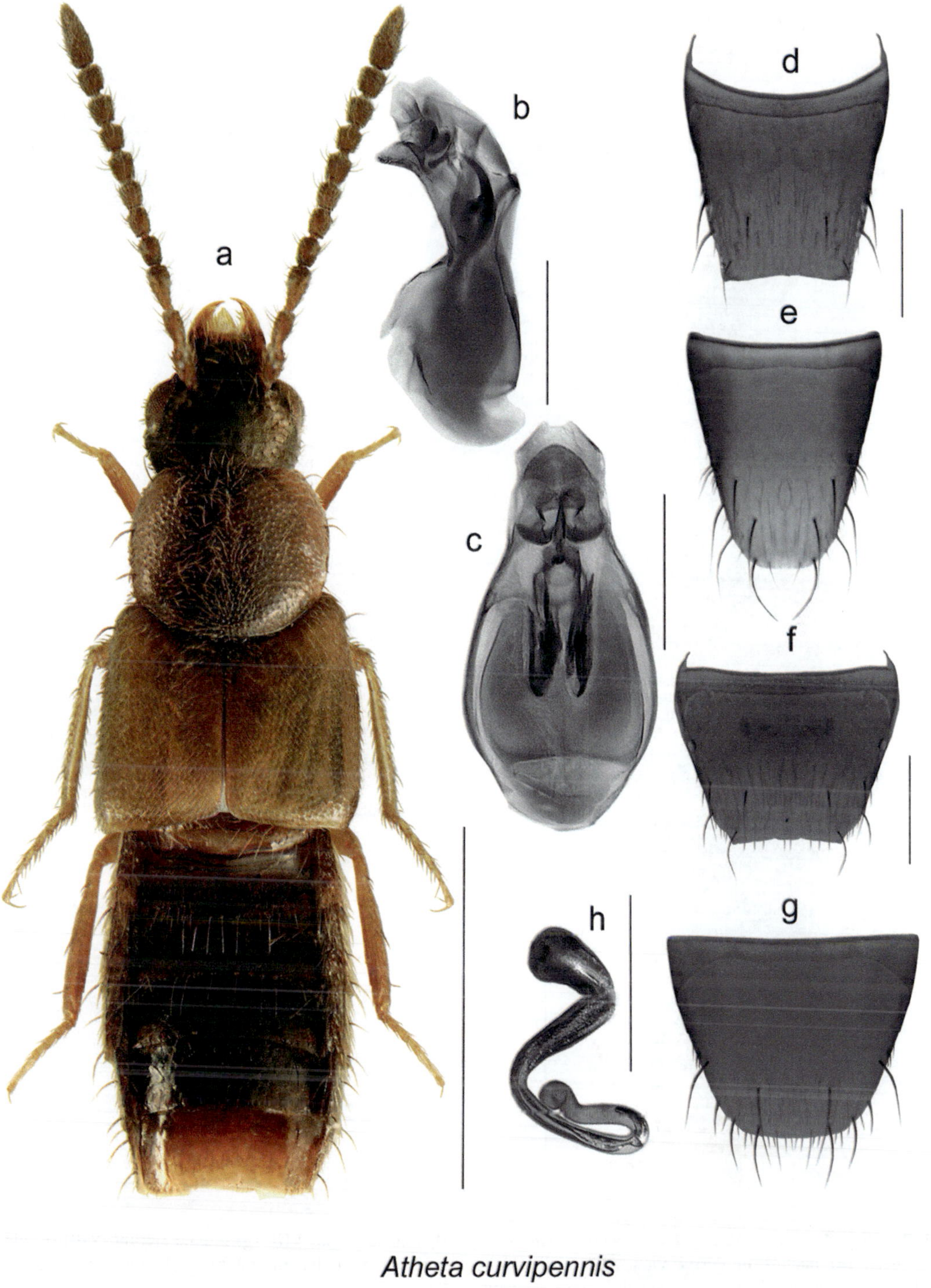

Fig. 19.216 (**a–h**). *Atheta (Dimetrota) curvipennis* Klimaszewski and Langor: (**a**) habitus in dorsal view; (**b**) median lobe of aedeagus in lateral view; (**c**) median lobe of aedeagus in dorsal view; (**d**) male tergite VIII; (**e**) male sternite VIII; (**f**) female tergite VIII; (**g**) female sternite VIII; (**h**) spermatheca. Scale bar for habitus = 1 mm; remaining scale bars = 0.2 mm

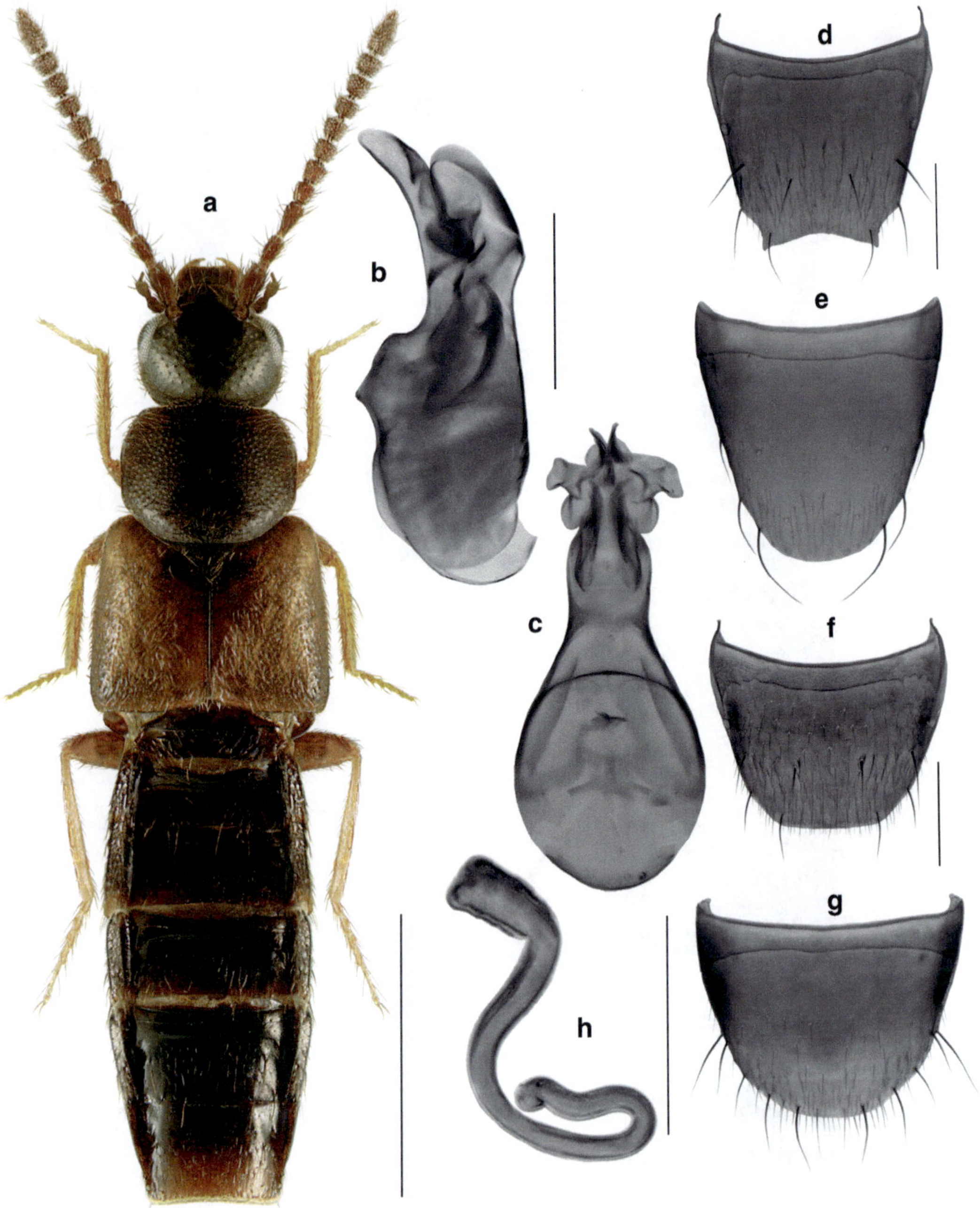

Fig. 19.217 (**a–h**). *Atheta (Dimetrota) prudhoensis* (Lohse): (**a**) habitus in dorsal view; (**b**) median lobe of aedeagus in lateral view; (**c**) median lobe of aedeagus in dorsal view; (**d**) male tergite VIII; (**e**) male sternite VIII; (**f**) female tergite VIII; (**g**) female sternite VIII; (**h**) spermatheca. Scale bar for habitus = 1 mm; remaining scale bars = 0.2 mm

Fig. 19.218 (**a–h**). *Atheta (Dimetrota) picipennoides* H.K. Hanssen (based on misidentified paralectotypes of *Atheta aeneipennis* (Thomson)): (**a**) habitus in dorsal view; (**b**) median lobe of aedeagus in lateral view; (**c**) median lobe of aedeagus in dorsal view; (**d**) male tergite VIII; (**e**) male sternite VIII; (**f**) female tergite VIII; (**g**) female sternite VIII; (**h**) spermatheca. Scale bar for habitus = 1 mm; remaining scale bars = 0.2 mm

Fig. 19.219 (**a–h**). *Atheta (Dimetrota) aeneipennis* (Thomson) [=*A. parapicipennis* Brundin] (based on type series): (**a**) habitus in dorsal view; (**b**) median lobe of aedeagus in lateral view; (**c**) median lobe of aedeagus in dorsal view; (**d**) male tergite VIII; (**e**) male sternite VIII; (**f**) female tergite VIII; (**g**) female sternite VIII; (**h**) spermatheca. Scale bar for habitus = 1 mm; remaining scale bars = 0.2 mm

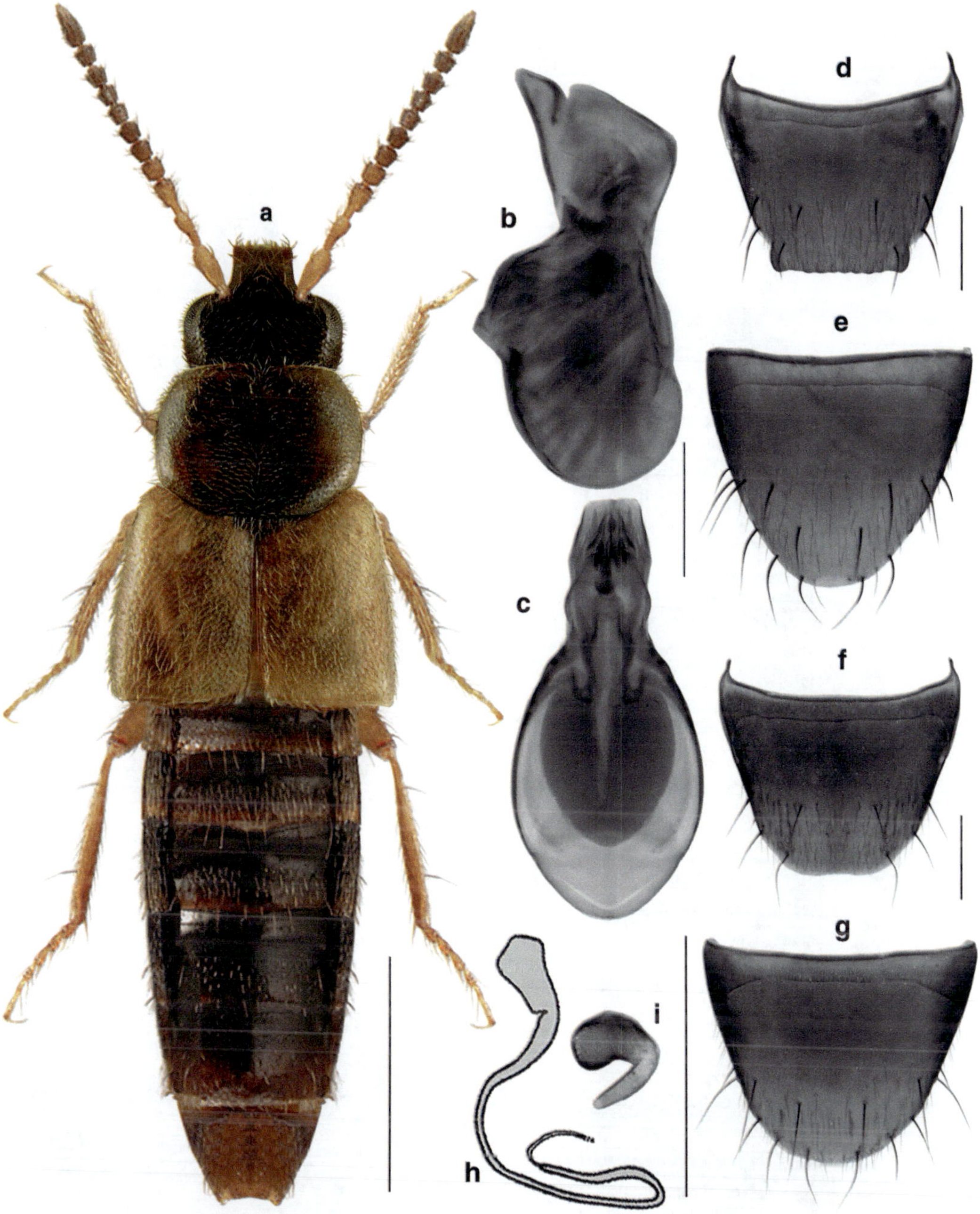

Fig. 19.220 (**a–i**). *Atheta (Dimetrota) strigosula* Casey: (**a**) habitus in dorsal view; (**b**) median lobe of aedeagus in lateral view; (**c**) median lobe of aedeagus in dorsal view; (**d**) male tergite VIII; (**e**) male sternite VIII; (**f**) female tergite VIII; (**g**) female sternite VIII; (**h,i**) spermatheca. Scale bar for habitus = 1 mm; remaining scale bars = 0.2 mm

Fig. 19.221 (**a–d**). *Atheta (Dimetrota) takhini* Klimaszewski and Godin, **sp. n.**: (**a**) habitus in dorsal view; (**b**) median lobe of aedeagus in lateral view; (**c**) male tergite VIII; (**d**) male sternite VIII. Scale bar for habitus = 1 mm; remaining scale bars = 0.2 mm

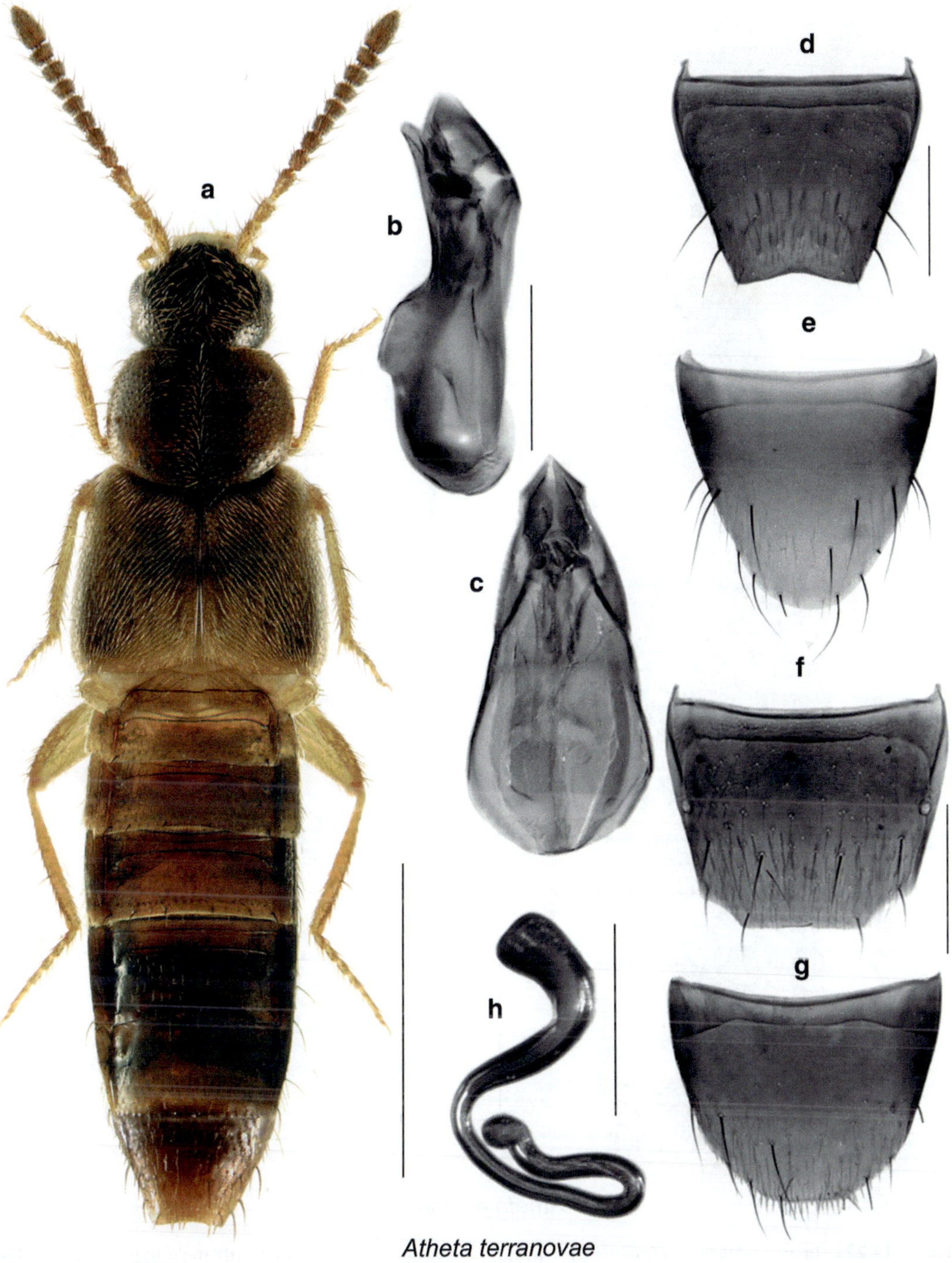

Fig. 19.222 (**a–h**). *Atheta (Dimetrota) terranovae* Klimaszewski and Langor: (**a**) habitus in dorsal view; (**b**) median lobe of aedeagus in lateral view; (**c**) median lobe of aedeagus in dorsal view; (**d**) male tergite VIII; (**e**) male sternite VIII; (**f**) female tergite VIII; (**g**) female sternite VIII; (**h**) spermatheca. Scale bar for habitus = 1 mm; remaining scale bars = 0.2 mm

Fig. 19.223 (**a–e**). *Atheta (Dimetrota) avalon* Klimaszewski and Langor: (**a**) habitus in dorsal view; (**b**) median lobe of aedeagus in lateral view; (**c**) median lobe of aedeagus in dorsal view; (**d**) male tergite VIII; (**e**) male sternite VIII. Scale bar for habitus = 1 mm; remaining scale bars = 0.2 mm

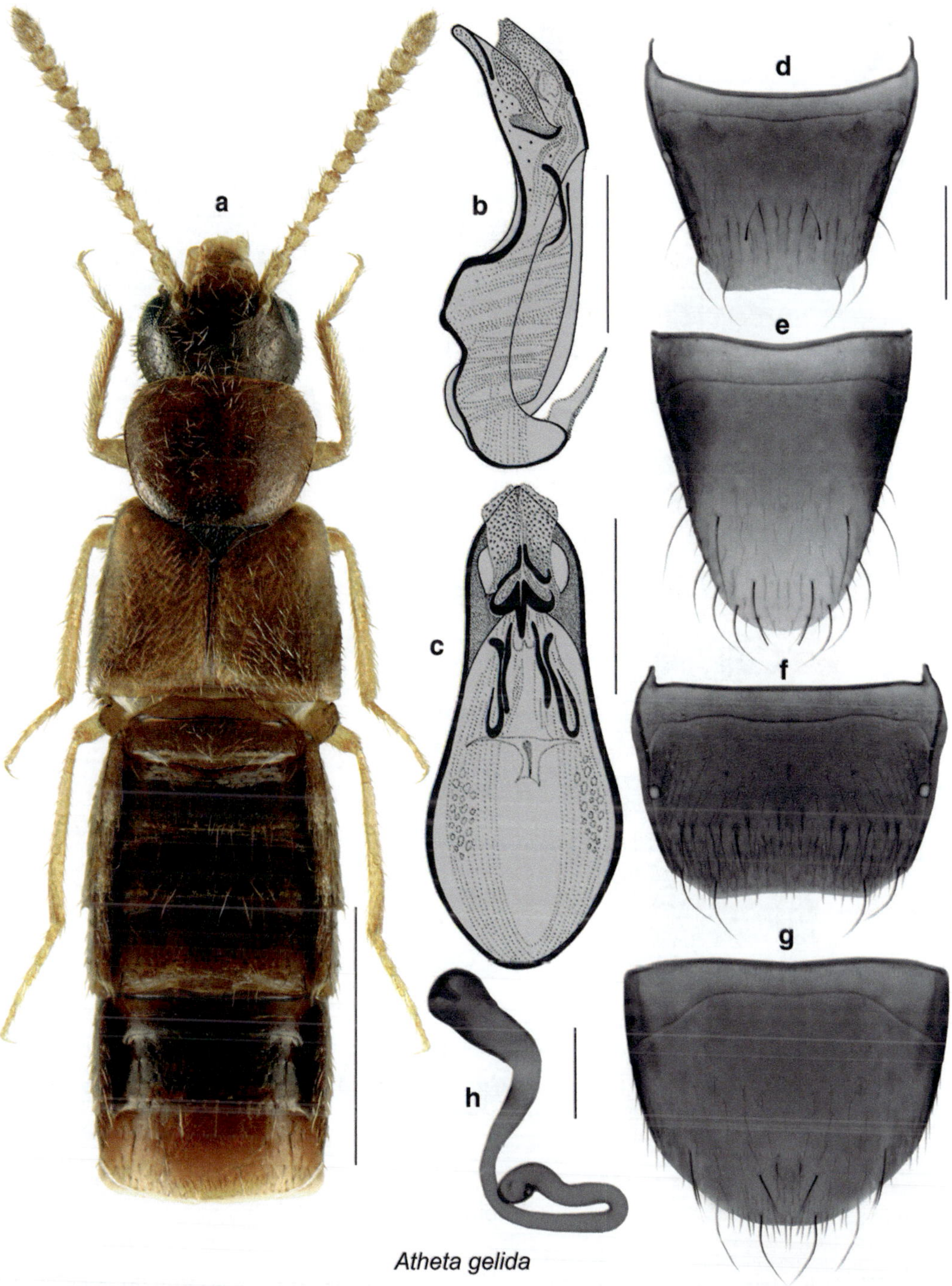

Fig. 19.224 (**a–h**). *Atheta (Dimetrota) gelida* (J. Sahlberg): (**a**) habitus in dorsal view; (**b**) median lobe of aedeagus in lateral view; (**c**) median lobe of aedeagus in dorsal view; (**d**) male tergite VIII; (**e**) male sternite VIII; (**f**) female tergite VIII; (**g**) female sternite VIII; (**h**) spermatheca. Scale bar for habitus = 1 mm; remaining scale bars = 0.2 mm

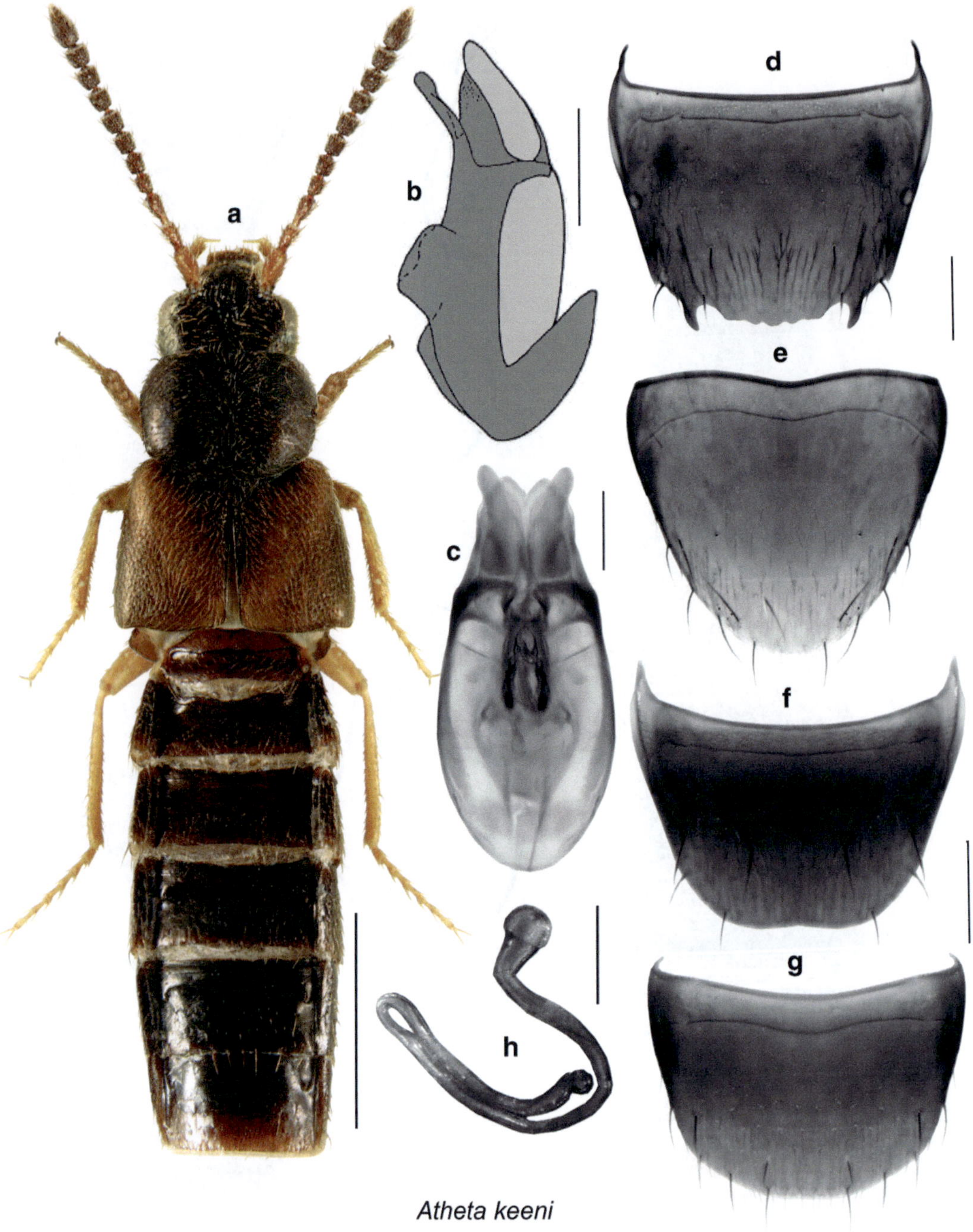

Fig. 19.225 (**a–h**). *Atheta (Lamiota) keeni* Casey: (**a**) habitus in dorsal view; (**b**) median lobe of aedeagus in lateral view; (**c**) median lobe of aedeagus in dorsal view; (**d**) male tergite VIII; (**e**) male sternite VIII; (**f**) female tergite VIII; (**g**) female sternite VIII; (**h**) spermatheca. (**b**) modified from Gusarov (2003a). Scale bar for habitus = 1 mm; remaining scale bars = 0.2 mm

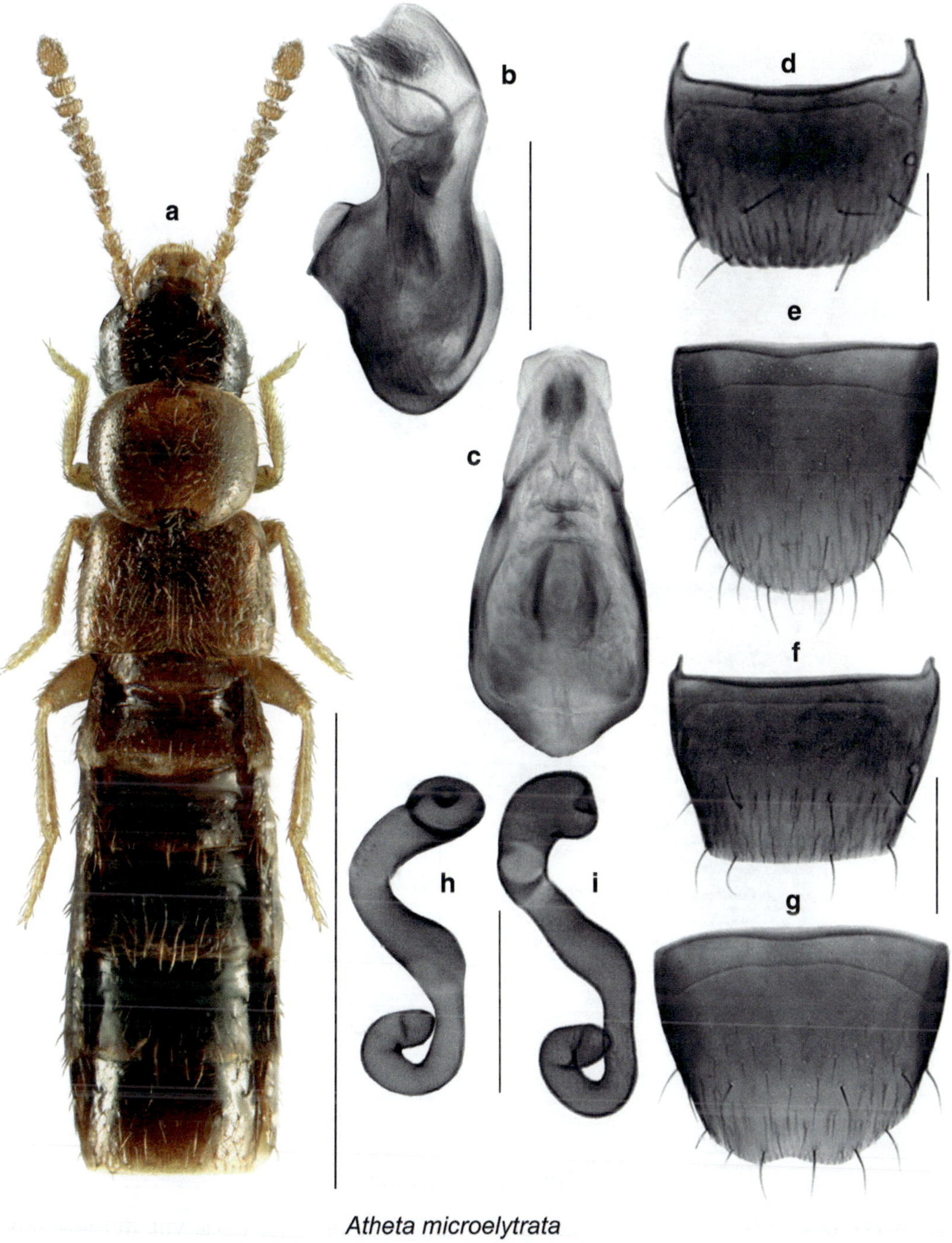

Fig. 19.226 (**a–i**). *Atheta (Microdota) microelytrata* Klimaszewski and Godin: (**a**) habitus in dorsal view; (**b**) median lobe of aedeagus in lateral view; (**c**) median lobe of aedeagus in dorsal view; (**d**) male tergite VIII; (**e**) male sternite VIII; (**f**) female tergite VIII; (**g**) female sternite VIII; (**h, i**) spermatheca. Scale bar for habitus = 1 mm; remaining scale bars = 0.2 mm

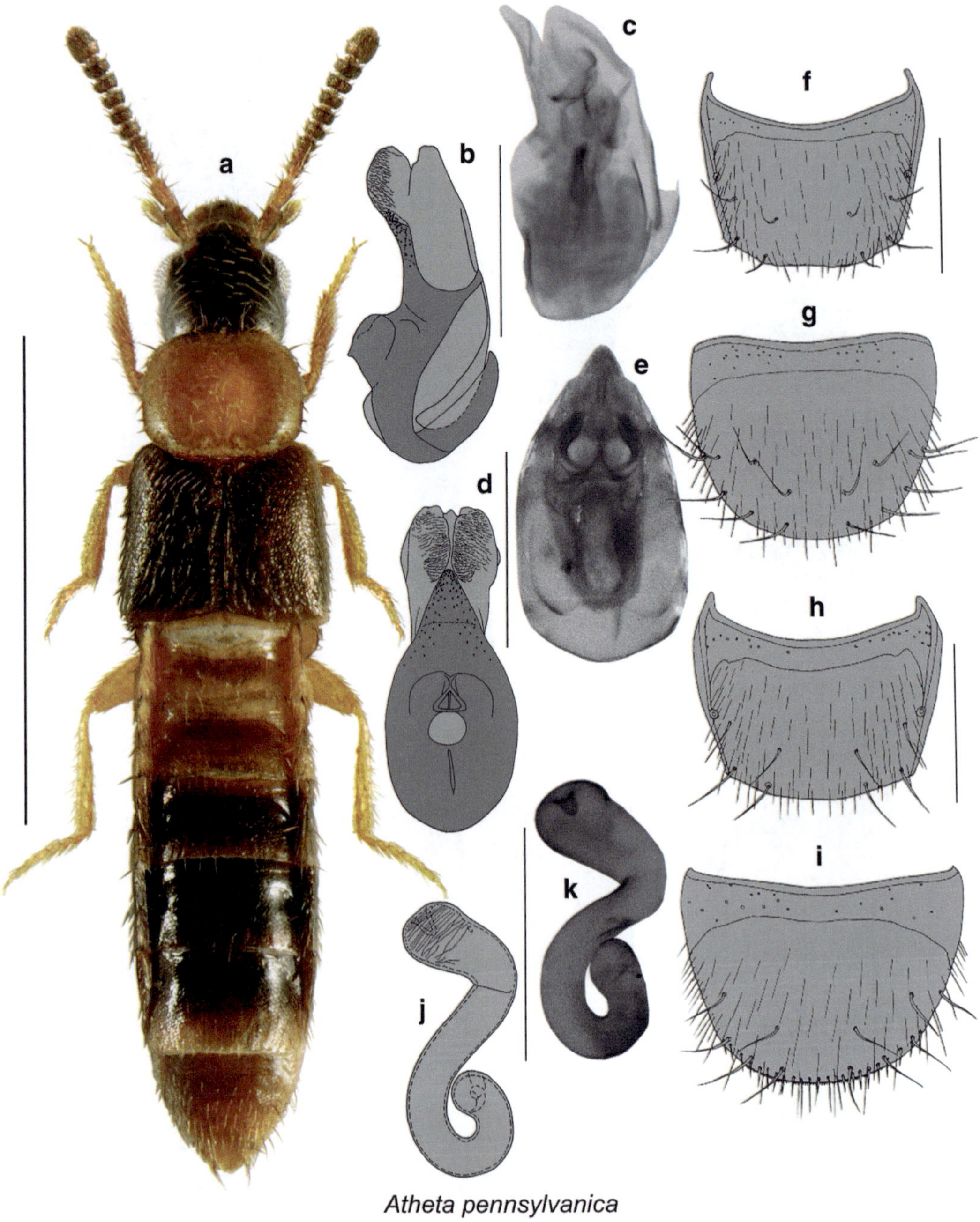

Fig. 19.227 (**a–k**). *Atheta (Microdota) pennsylvanica* Bernhauer: (**a**) habitus in dorsal view; (**b, c**) median lobe of aedeagus in lateral view; (**d, e**) median lobe of aedeagus in ventral and dorsal view; (**f**) male tergite VIII; (**g**) male sternite VIII; (**h**) female tergite VIII; (**i**) female sternite VIII; (**j, k**) spermatheca. Scale bar for habitus = 1 mm; remaining scale bars = 0.2 mm

Fig. 19.228 (**a–h**). *Atheta (Microdota) platonoffi* Brundin: (**a**) habitus in dorsal view; (**b**) median lobe of aedeagus in lateral view; (**c**) median lobe of aedeagus in dorsal view; (**d**) male tergite VIII; (**e**) male sternite VIII; (**f**) female tergite VIII; (**g**) female sternite VIII; (**h**) spermatheca. Scale bar for habitus = 1 mm; remaining scale bars = 0.2 mm

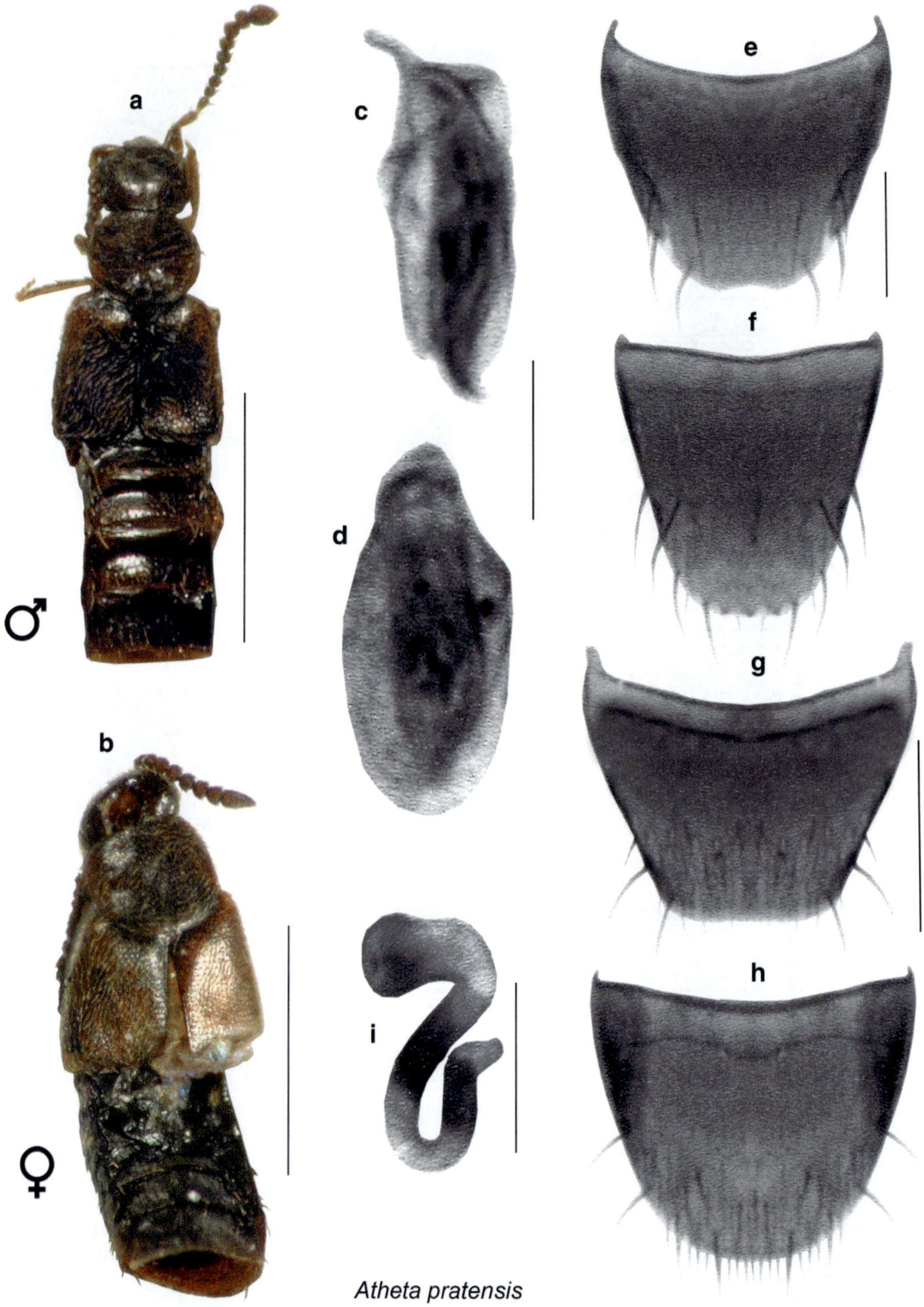

Fig. 19.229 (**a–i**). *Atheta (Microdota) pratensis* (Mäklin) (based on type specimens): (**a,b**) habitus in dorsal view; (**c**) median lobe of aedeagus in lateral view; (**d**) median lobe of aedeagus in dorsal view; (**e**) male tergite VIII; (**f**) male sternite VIII; (**g**) female tergite VIII; (**h**) female sternite VIII; (**i**) spermatheca. Scale bar for habitus = 1 mm; remaining scale bars = 0.2 mm

Fig. 19.230 (**a–c**). *Atheta (Microdota) pratensis* (Mäklin) (based on specimen from LB, possible misidentification for *A. subtilis*): (**a**) habitus in dorsal view; (**b**) median lobe of aedeagus in lateral view; (**c**) median lobe of aedeagus in dorsal view. Scale bar for habitus = 1 mm; remaining scale bars = 0.2 mm

Fig. 19.231 (**a–h**). *Atheta (Microdota) pseudosubtilis* Klimaszewski and Langor: (**a**) habitus in dorsal view; (**b**) median lobe of aedeagus in lateral view; (**c**) median lobe of aedeagus in dorsal view; (**d**) male tergite VIII; (**e**) male sternite VIII; (**f**) female tergite VIII; (**g**) female sternite VIII; (**h**) spermatheca. Scale bar for habitus = 1 mm; remaining scale bars = 0.2 mm

Fig. 19.232 (**a–h**). *Atheta (Microdota) ripariides* Newton: (**a**) habitus in dorsal view; (**b**) median lobe of aedeagus in lateral view; (**c**) median lobe of aedeagus in dorsal view; (**d**) male tergite VIII; (**e**) male sternite VIII; (**f**) female tergite VIII; (**g**) female sternite VIII; (**h**) spermatheca. Scale bar for habitus = 1 mm; remaining scale bars = 0.2 mm

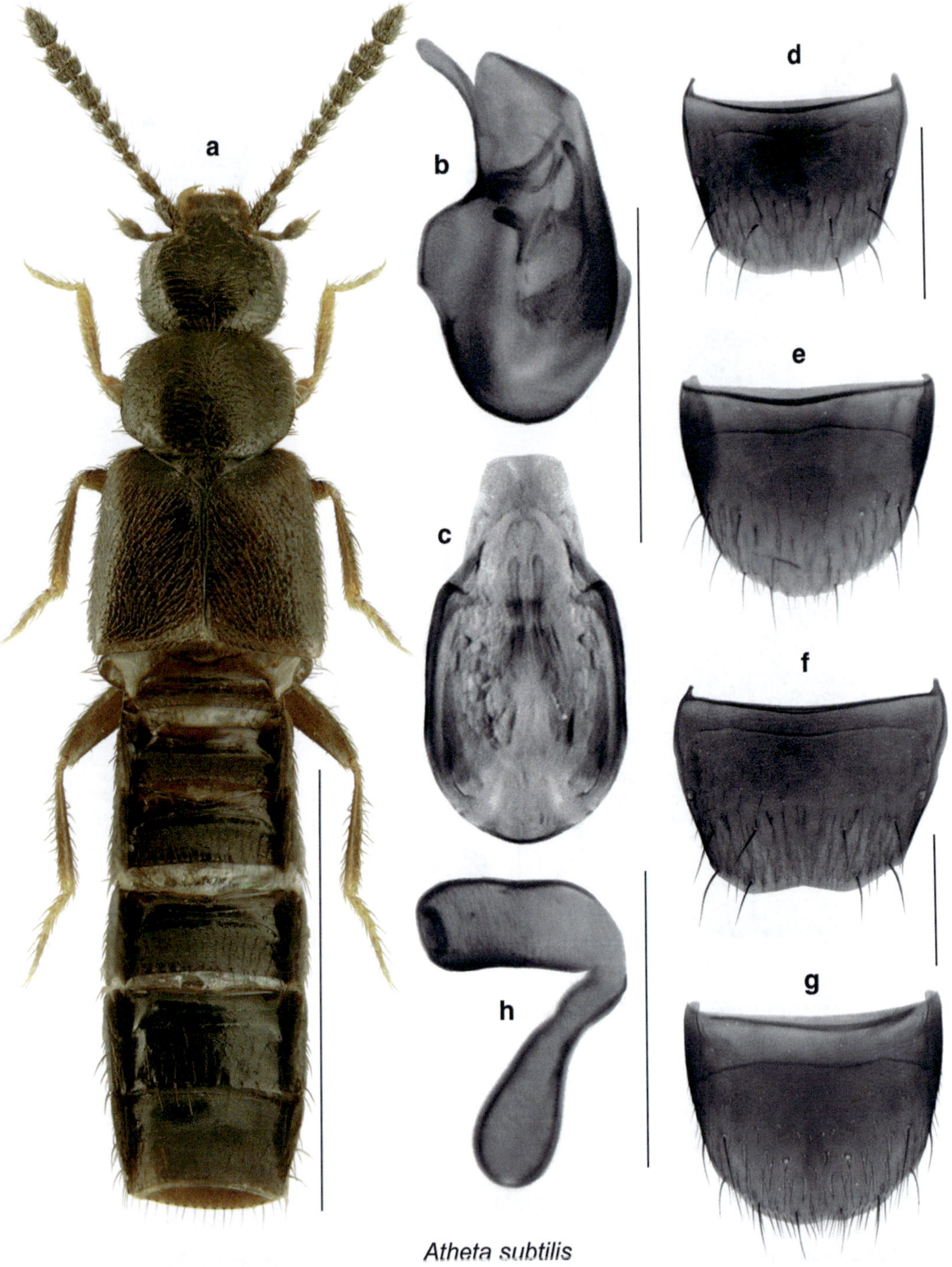

Fig. 19.233 (**a–h**). *Atheta (Microdota) subtilis* (Scriba): (**a**) habitus in dorsal view; (**b**) median lobe of aedeagus in lateral view; (**c**) median lobe of aedeagus in dorsal view; (**d**) male tergite VIII; (**e**) male sternite VIII; (**f**) female tergite VIII; (**g**) female sternite VIII; (**h**) spermatheca. Scale bar for habitus = 1 mm; remaining scale bars = 0.2 mm

Fig. 19.234 (**a–e**). *Atheta (Microdota) surgens* (Casey): (**a**) habitus in dorsal view; (**b**) median lobe of aedeagus in lateral view; (**c**) median lobe of aedeagus in dorsal view; (**d**) male tergite VIII; (**e**) male sternite VIII. Scale bar for habitus = 1 mm; remaining scale bars = 0.2 mm

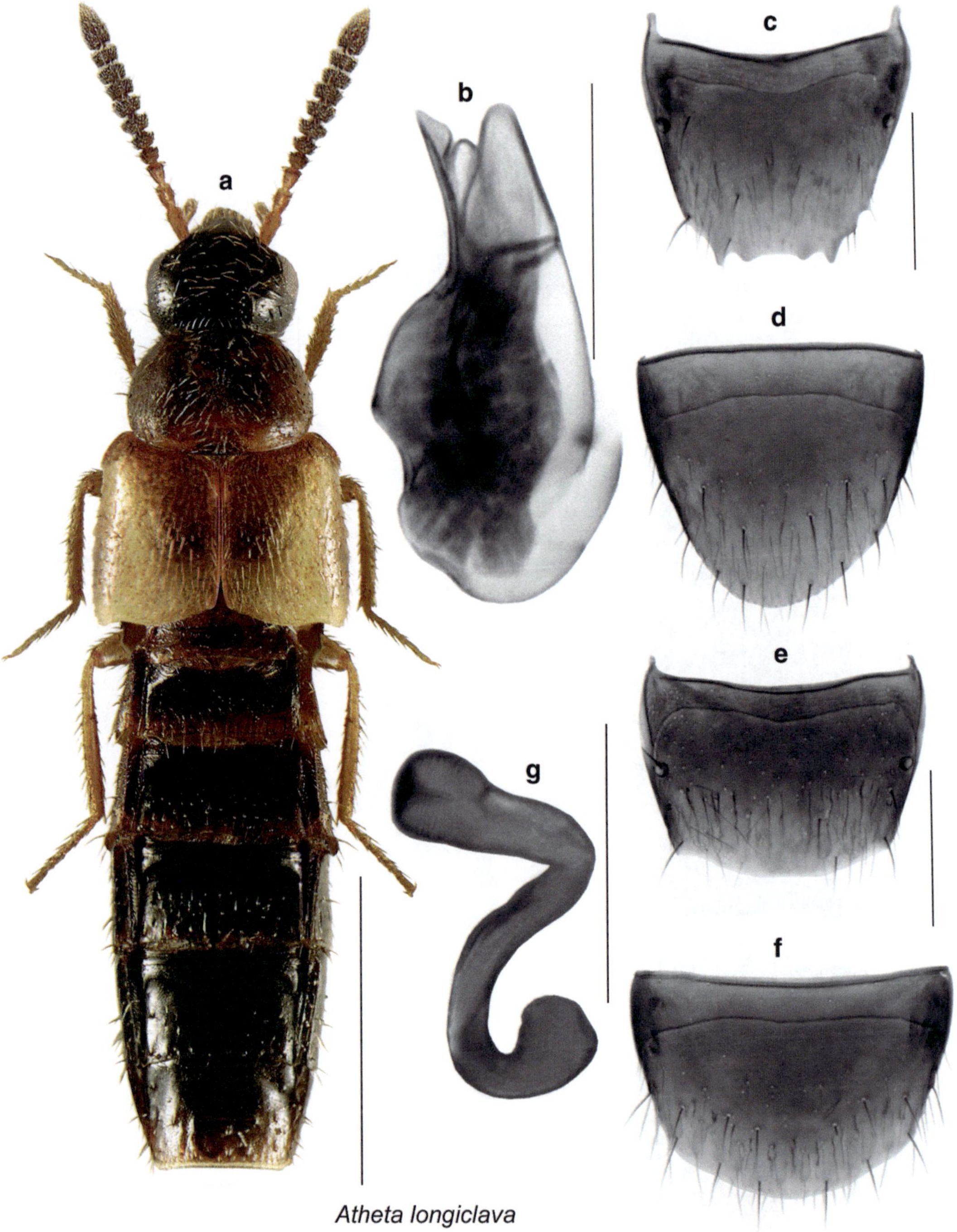

Fig. 19.235 (**a–g**). *Atheta (Pseudota) longiclava* Casey: (**a**) habitus in dorsal view; (**b**) median lobe of aedeagus in lateral view; (**c**) male tergite VIII; (**d**) male sternite VIII; (**e**) female tergite VIII; (**f**) female sternite VIII; (**g**) spermatheca. Scale bar for habitus = 1 mm; remaining scale bars = 0.2 mm

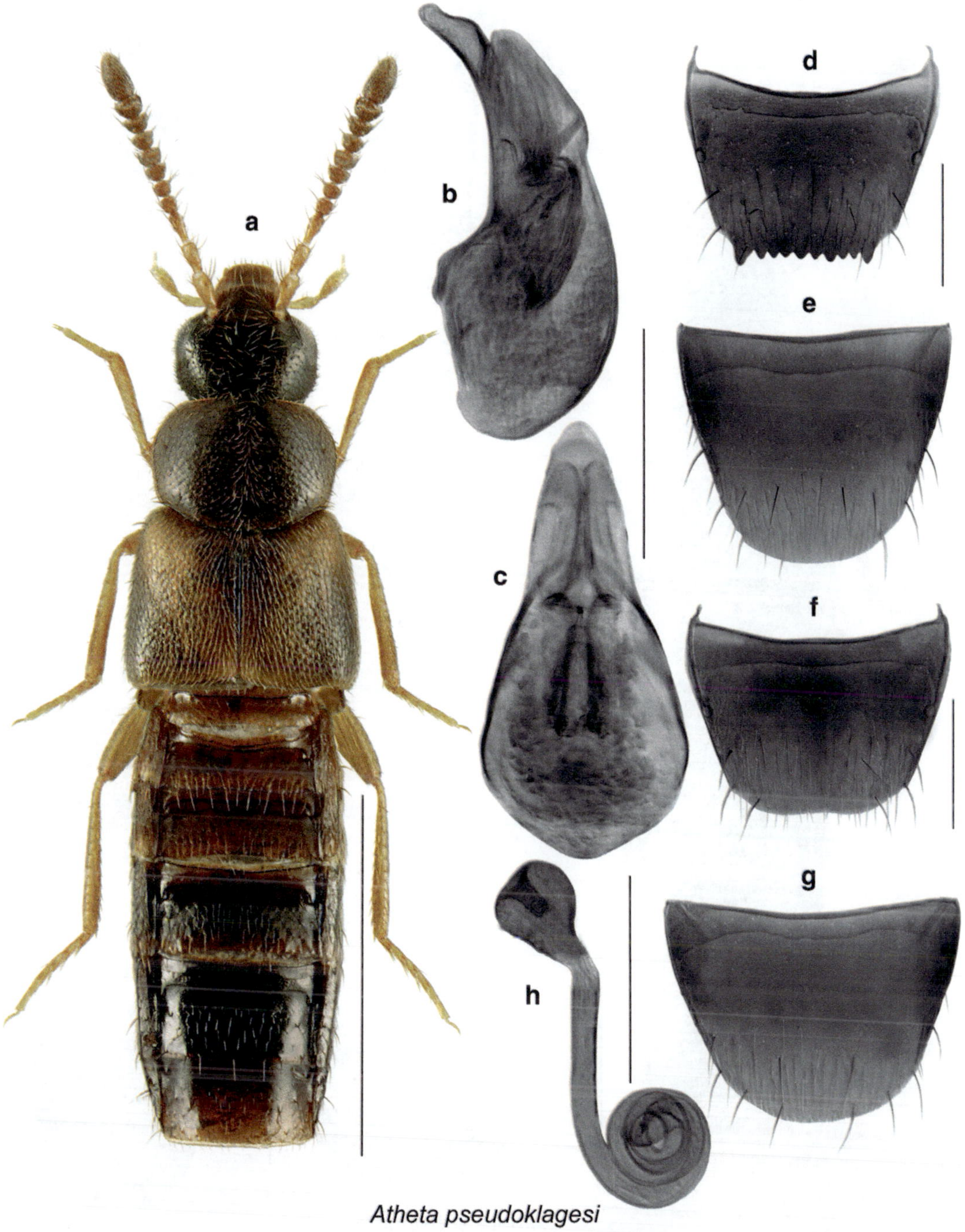

Fig. 19.236 (**a–h**). *Atheta (Pseudota) pseudoklagesi* Klimaszewski and Webster: (**a**) habitus in dorsal view; (**b**) median lobe of aedeagus in lateral view; (**c**) median lobe of aedeagus in dorsal view; (**d**) male tergite VIII; (**e**) male sternite VIII; (**f**) female tergite VIII; (**g**) female sternite VIII; (**h**) spermatheca. Scale bar for habitus = 1 mm; remaining scale bars = 0.2 mm

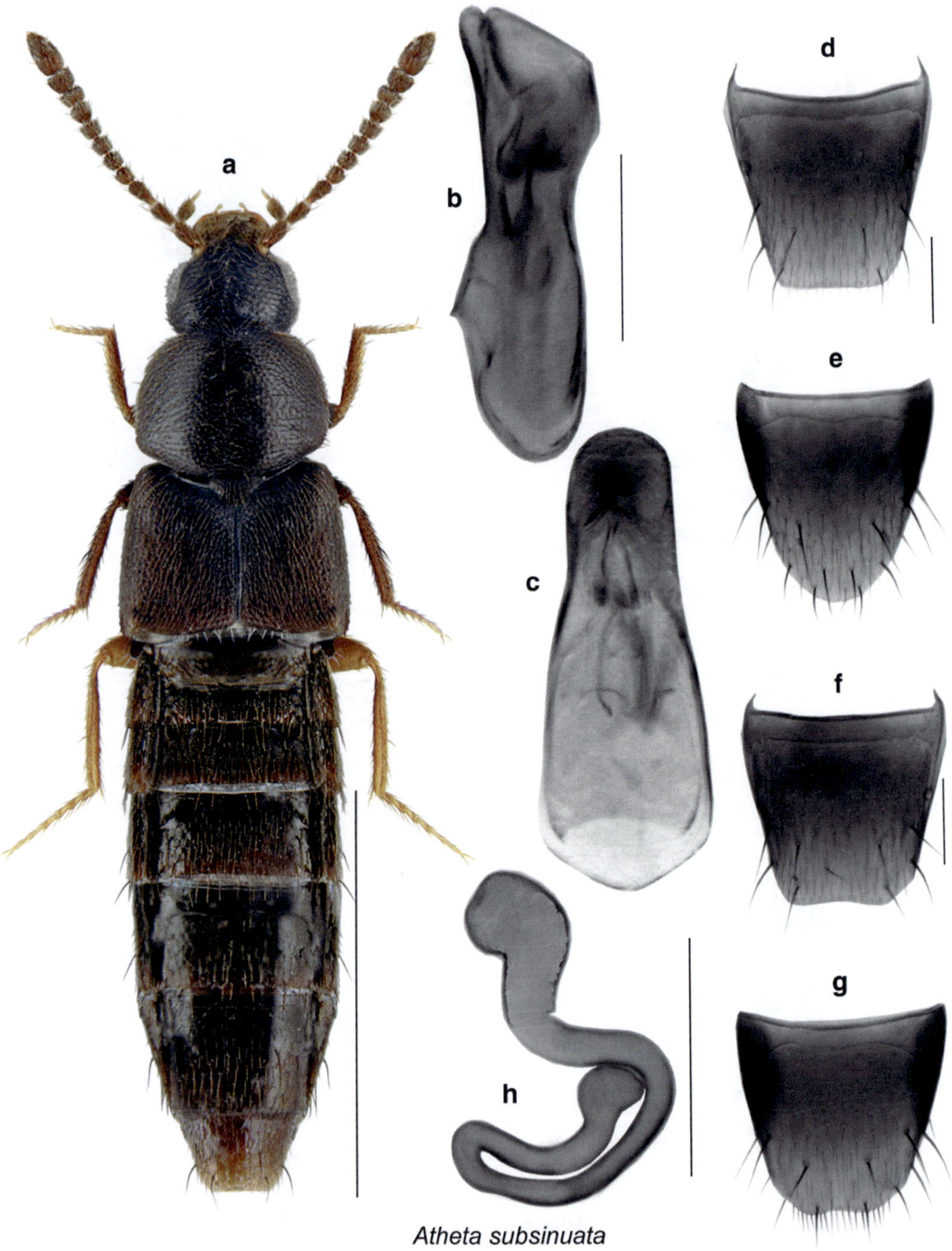

Fig. 19.237 (**a–h**). *Atheta (Rhagocneme) subsinuata* (Erichson): (**a**) habitus in dorsal view; (**b**) median lobe of aedeagus in lateral view; (**c**) median lobe of aedeagus in dorsal view; (**d**) male tergite VIII; (**e**) male sternite VIII; (**f**) female tergite VIII; (**g**) female sternite VIII; (**h**) spermatheca. Scale bar for habitus = 1 mm; remaining scale bars = 0.2 mm

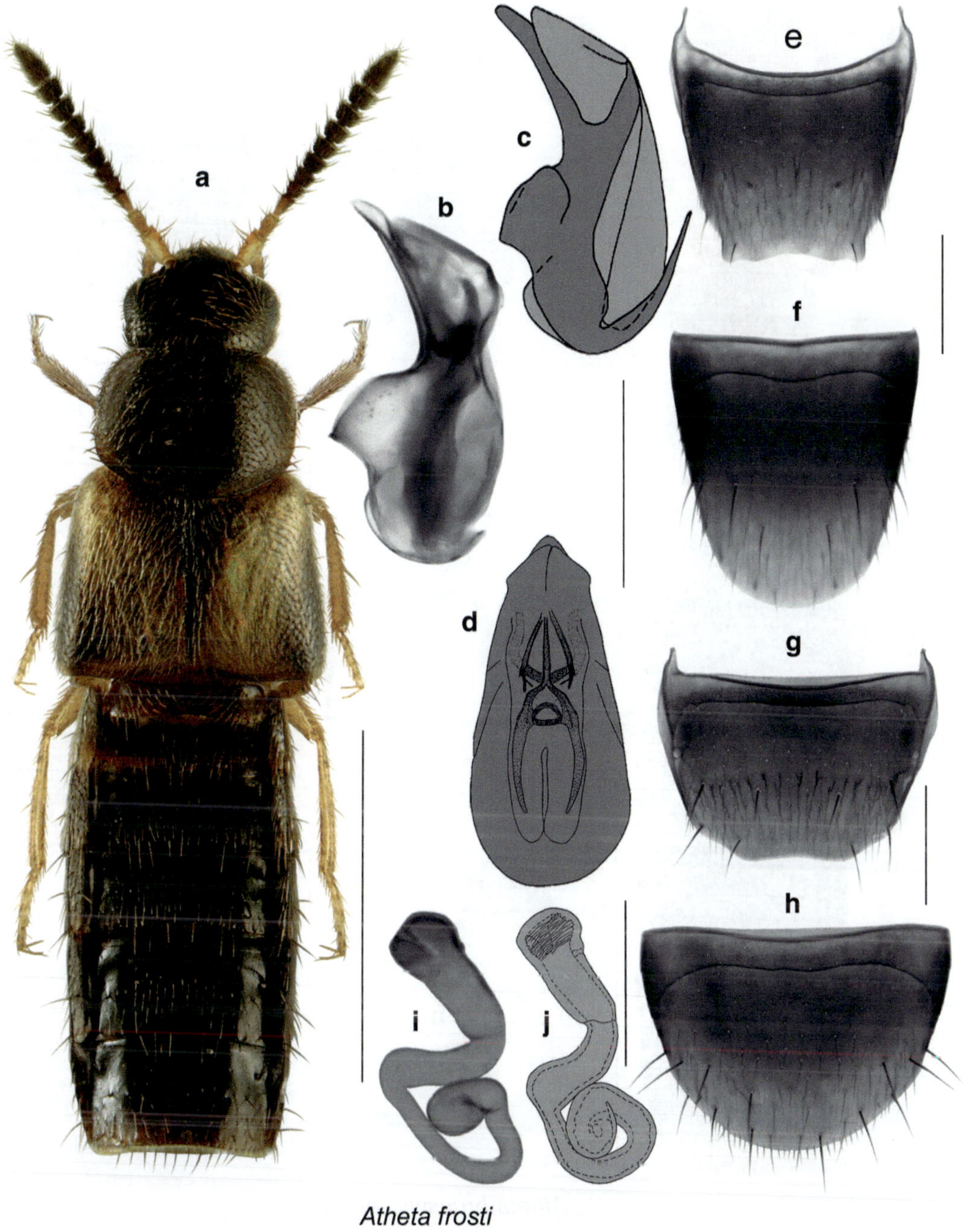

Fig. 19.238 (**a–j**). *Atheta (Teropla) frosti* Bernhauer: (**a**) habitus in dorsal view; (**b, c**) median lobe of aedeagus in lateral view; (**d**) median lobe of aedeagus in dorsal view; (**e**) male tergite VIII; (**f**) male sternite VIII; (**g**) female tergite VIII; (**h**) female sternite VIII; (**i, j**) spermatheca. Scale bar for habitus = 1 mm; remaining scale bars = 0.2 mm

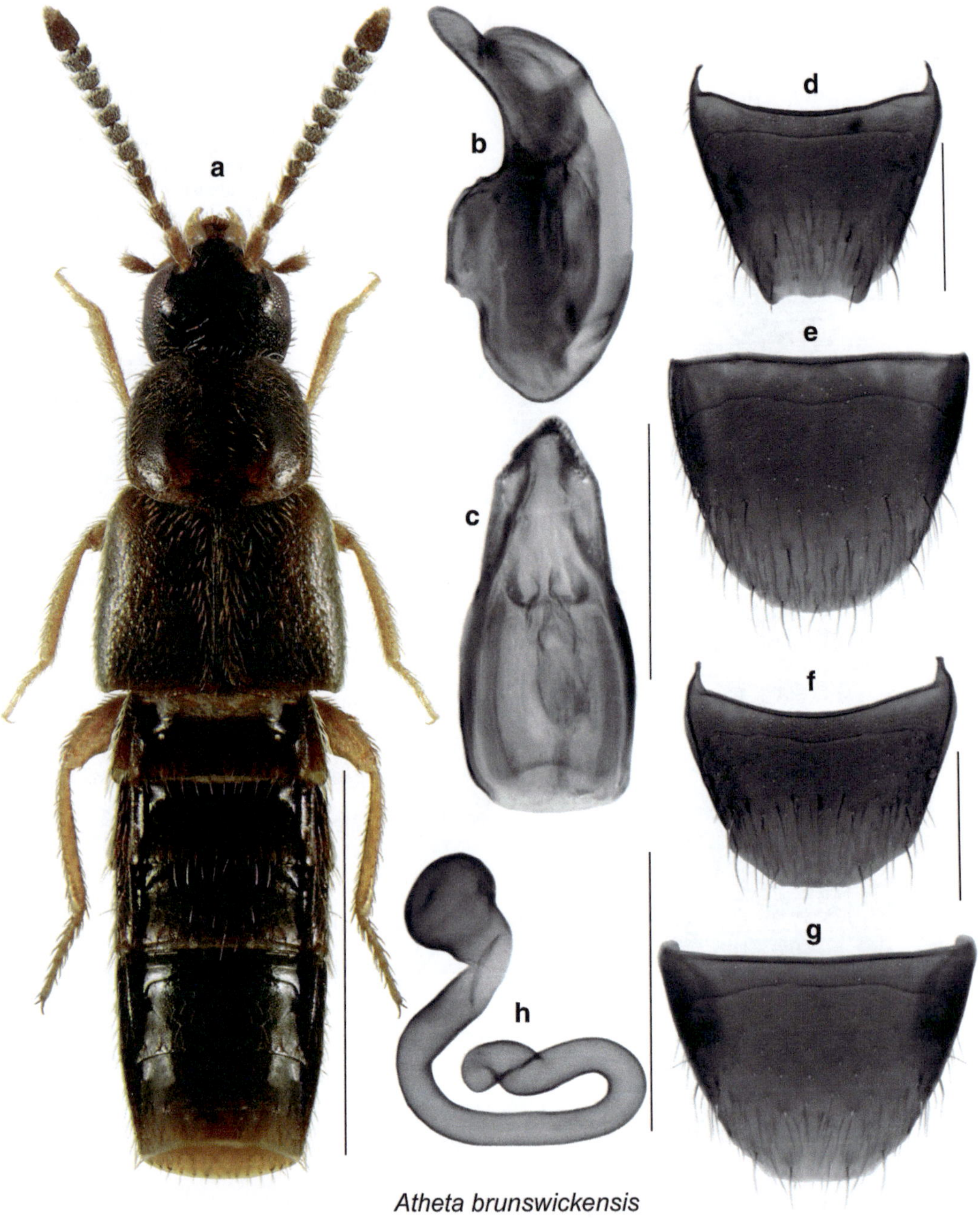

Fig. 19.239 (**a–h**). *Atheta brunswickensis* Klimaszewski: (**a**) habitus in dorsal view; (**b**) median lobe of aedeagus in lateral view; (**c**) median lobe of aedeagus in dorsal view; (**d**) male tergite VIII; (**e**) male sternite VIII; (**f**) female tergite VIII; (**g**) female sternite VIII; (**h**) spermatheca. Scale bar for habitus = 1 mm; remaining scale bars = 0.2 mm

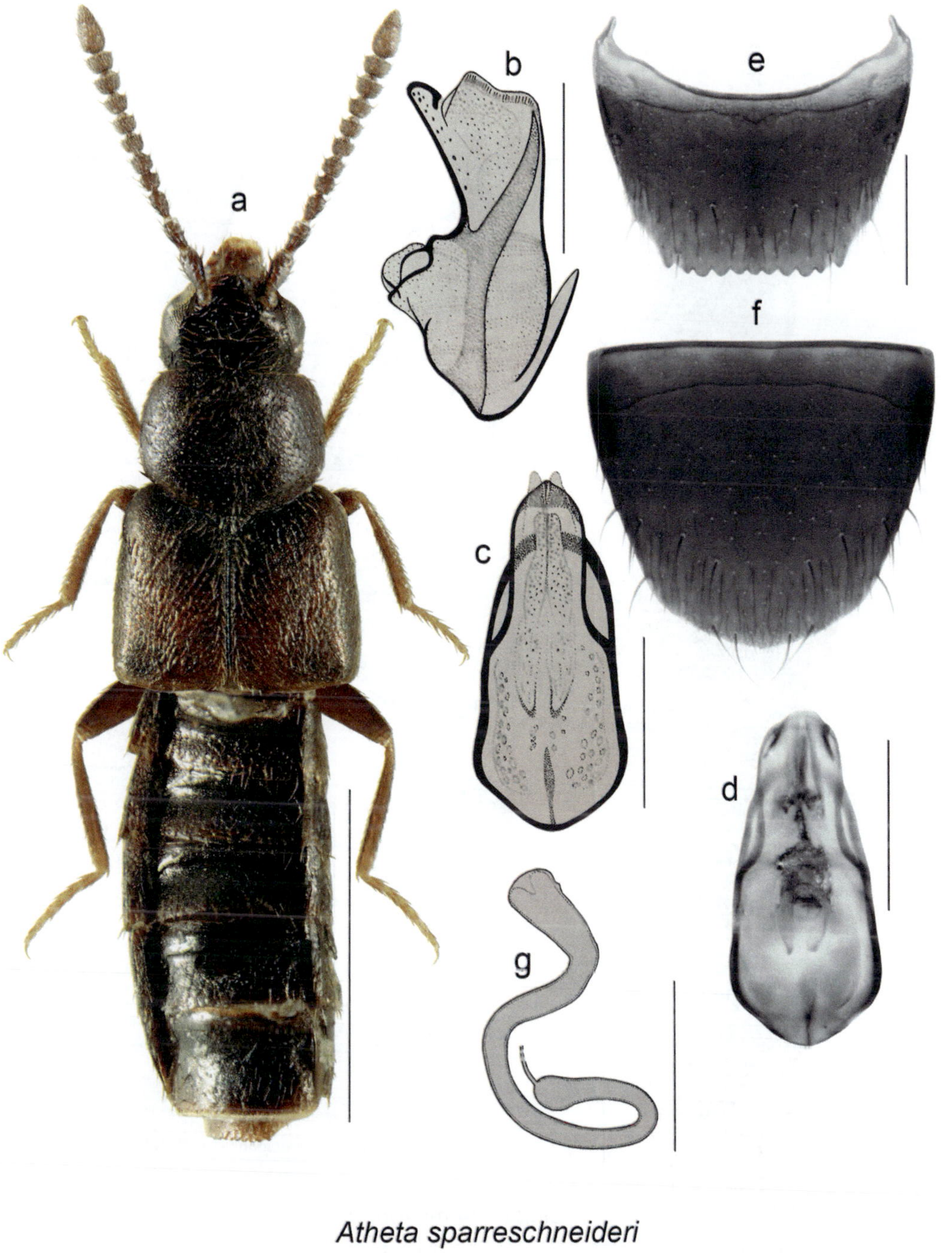

Fig. 19.240 (**a–g**). *Atheta sparreschneideri* Munster: (**a**) habitus in dorsal view; (**b**) median lobe of aedeagus in lateral view; (**c, d**) median lobe of aedeagus in dorsal view; (**e**) male tergite VIII; (**f**) male sternite VIII; (**g**) spermatheca. Scale bar for habitus = 1 mm; remaining scale bars = 0.2 mm

Fig. 19.241 (**a–g**). *Bellatheta microptera* (Lohse): (**a**) habitus in dorsal view; (**b**) median lobe of aedeagus in lateral view; (**c**) male tergite VIII; (**d**) male sternite VIII; (**e**) female tergite VIII; (**f**) female sternite VIII; (**g**) spermatheca. Scale bar for habitus = 1 mm; remaining scale bars = 0.2 mm

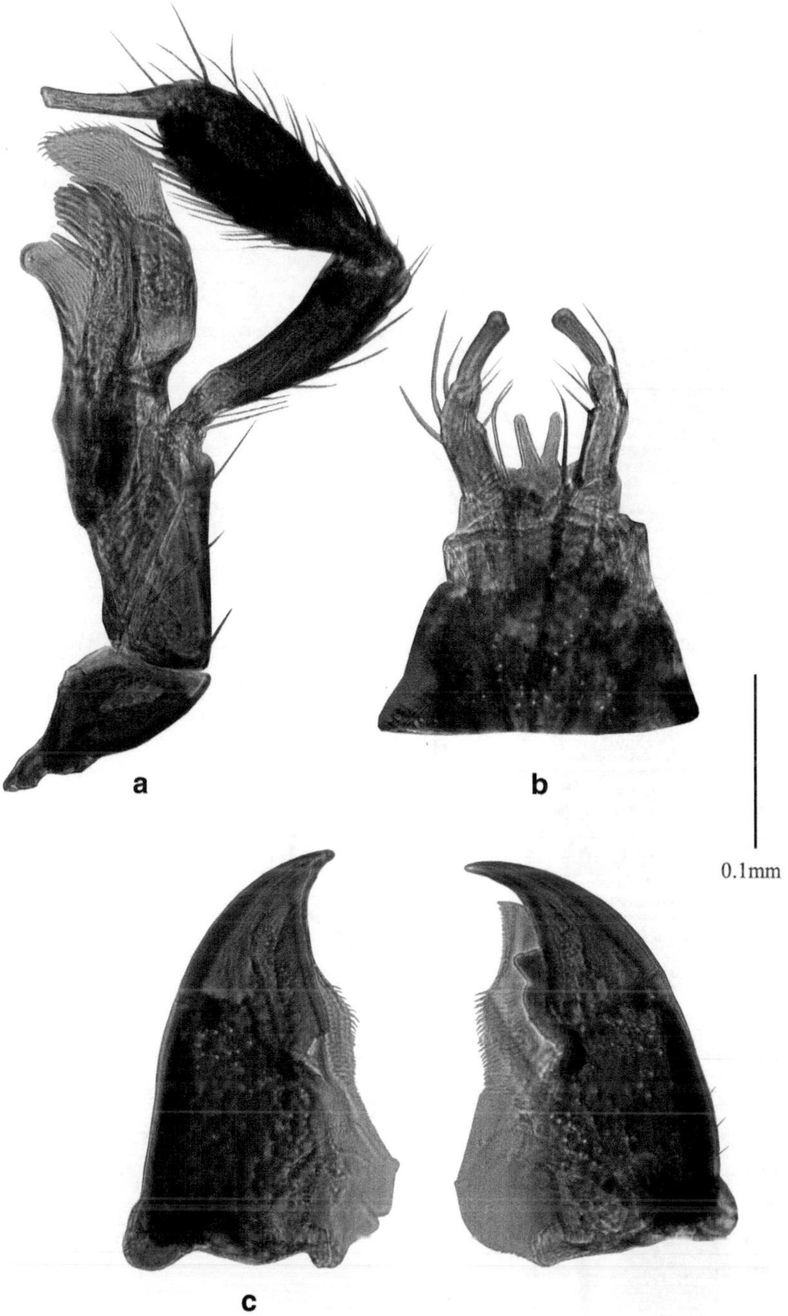

Fig. 19.242 (**a–d**). Mouthparts of *Boreophilia* (based on *B. islandica* (Kraatz)): (**a**) maxillary palpus; (**b**) labial palpi and V-shaped ligula; (**c**) mandibles. Scale bar = 0.1 mm

Fig. 19.243 (**a–j**). *Boreophilia eremita* (Rye): (**a**) habitus in dorsal view; (**b, c**) median lobe of aedeagus in lateral view; (**d, e**) median lobe of aedeagus in dorsal view; (**f**) male tergite VIII; (**g**) male sternite VIII; (**h**) female tergite VIII; (**i**) female sternite VIII; (**j**) spermatheca. (**b, d**) after Lohse et al. (1990). Scale bar for habitus = 1 mm; remaining scale bars = 0.2 mm

Fig. 19.244 (**a–h**). *Boreophilia fusca* (CR Sahlberg): (**a**) habitus in dorsal view; (**b**) median lobe of aedeagus in lateral view; (**c**) median lobe of aedeagus in dorsal view; (**d**) male tergite VIII; (**e**) male sternite VIII; (**f**) female tergite VIII; (**g**) female sternite VIII; (**h**) spermatheca. Scale bar for habitus = 1 mm; remaining scale bars = 0.2 mm

Fig. 19.245 (**a–k**). *Boreophilia hyperborea* (Brundin): (**a**) habitus in dorsal view; (**b, c**) median lobe of aedeagus in lateral view; (**d, e**) median lobe of aedeagus in dorsal view; (**f**) male tergite VIII; (**g**) male sternite VIII; (**h**) female tergite VIII; (**i**) female sternite VIII; (**j, k**) spermatheca. (**b, d, j**) after Lohse et al. (1990). Scale bar for habitus = 1 mm; remaining scale bars = 0.2 mm

Fig. 19.246 (**a–h**). *Boreophilia islandica* (Kraatz): (**a**) habitus in dorsal view; (**b**) median lobe of aedeagus in lateral view; (**c**) median lobe of aedeagus in dorsal view; (**d**) male tergite VIII; (**e**) male sternite VIII; (**f**) female tergite VIII; (**g**) female sternite VIII; (**h**) spermatheca. (**c**) after Lohse et al. (1990). Scale bar for habitus = 1 mm; remaining scale bars = 0.2 mm

Fig. 19.247 (**a–j**). *Boreophilia nearctica* Lohse: (**a**) habitus in dorsal view; (**b, c**) median lobe of aedeagus in lateral view; (**d**) median lobe of aedeagus in dorsal view; (**e**) male tergite VIII; (**f**) male sternite VIII; (**g**) female tergite VIII; (**h**) female sternite VIII; (**i, j**) spermatheca. (**b, j**) after Lohse et al. (1990). Scale bar for habitus = 1 mm; remaining scale bars = 0.2 mm

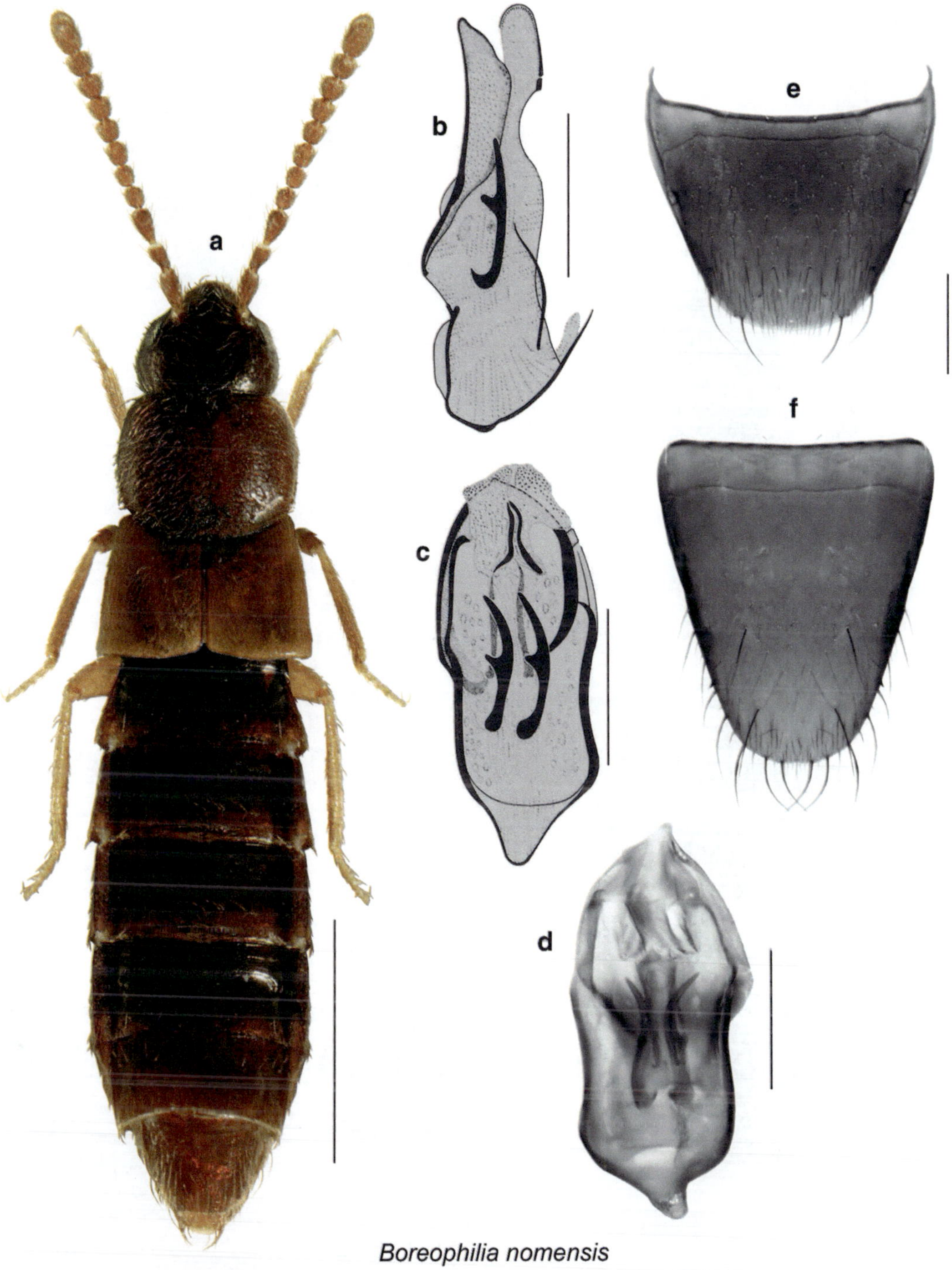

Fig. 19.248 (**a–f**). *Boreophilia nomensis* (Casey): (**a**) habitus in dorsal view; (**b**) median lobe of aedeagus in lateral view; (**c, d**) median lobe of aedeagus in dorsal view; (**e**) male tergite VIII; (**f**) male sternite VIII. (**b, c**) after Lohse et al. (1990). Scale bar for habitus = 1 mm; remaining scale bars = 0.2 mm

Fig. 19.249 (**a–h**). *Boreophilia beringi* Klimaszewski and Brunke: (**a**) habitus in dorsal view; (**b**) median lobe of aedeagus in lateral view; (**c**) median lobe of aedeagus in dorsal view; (**d**) male tergite VIII; (**e**) male sternite VIII; (**f**) female tergite VIII; (**g**) female sternite VIII; (**h**) spermatheca. Scale bar for habitus = 1 mm; remaining scale bars = 0.2 mm

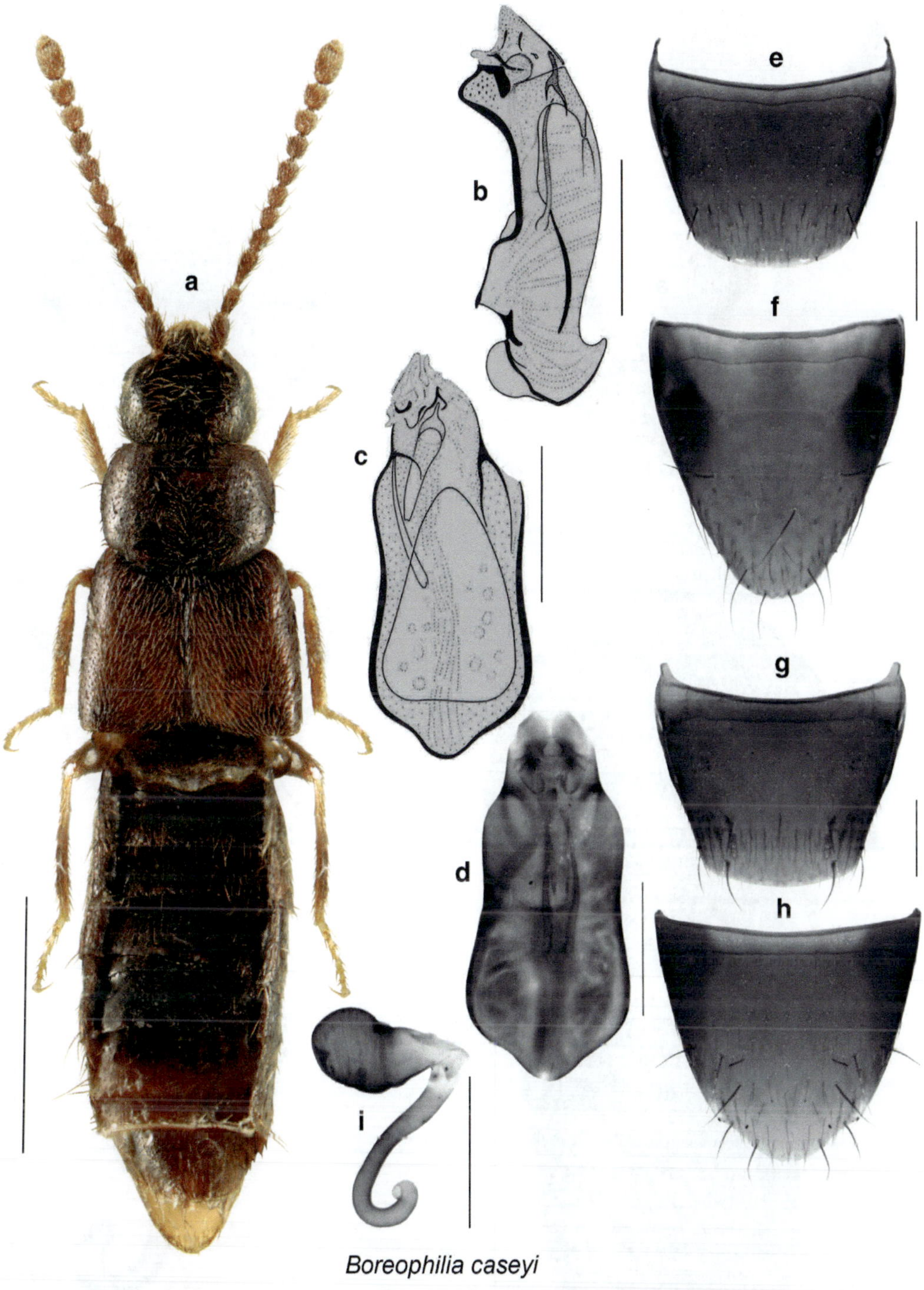

Fig. 19.250 (**a–i**). *Boreophilia caseyi* Lohse: (**a**) habitus in dorsal view; (**b**) median lobe of aedeagus in lateral view; (**c, d**) median lobe of aedeagus in dorsal view; (**e**) male tergite VIII; (**f**) male sternite VIII; (**g**) female tergite VIII; (**h**) female sternite VIII; (**i**) spermatheca. (**b, c**) after Lohse et al. (1990). Scale bar for habitus = 1 mm; remaining scale bars = 0.2 mm

Fig. 19.251 (**a–d**). *Boreophilia davidgei* Klimaszewski and Godin: (**a**) habitus in dorsal view; (**b**) female tergite VIII; (**c**) female sternite VIII; (**d**) spermatheca. Scale bar for habitus = 1 mm; remaining scale bars = 0.2 mm

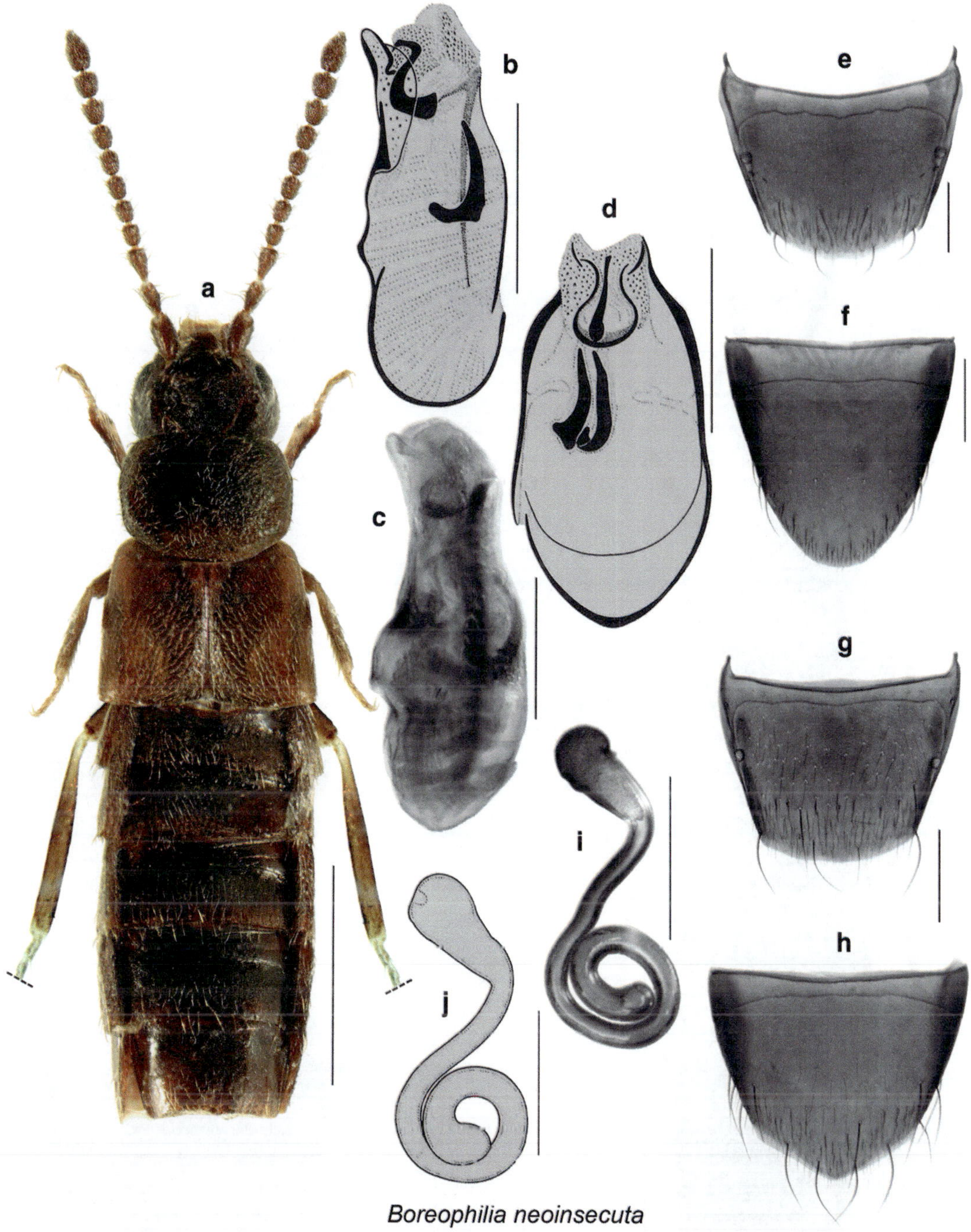

Fig. 19.252 (a–j). *Boreophilia neoinsecuta* Klimaszewski: (**a**) habitus in dorsal view; (**b, c**) median lobe of aedeagus in lateral view; (**d**) median lobe of aedeagus in dorsal view; (**e**) male tergite VIII; (**f**) male sternite VIII; (**g**) female tergite VIII; (**h**) female sternite VIII; (**i, j**) spermatheca. (**b, d, j**) after Lohse et al. (1990). Scale bar for habitus = 1 mm; remaining scale bars = 0.2 mm

Fig. 19.253 (**a–h**). *Boreophilia subplana* (JR Sahlberg): (**a**) habitus in dorsal view; (**b**) median lobe of aedeagus in lateral view; (**c**) median lobe of aedeagus in dorsal view; (**d**) male tergite VIII; (**e**) male sternite VIII; (**f**) female tergite VIII; (**g**) female sternite VIII; (**h**) spermatheca. Scale bar for habitus = 1 mm; remaining scale bars = 0.2 mm

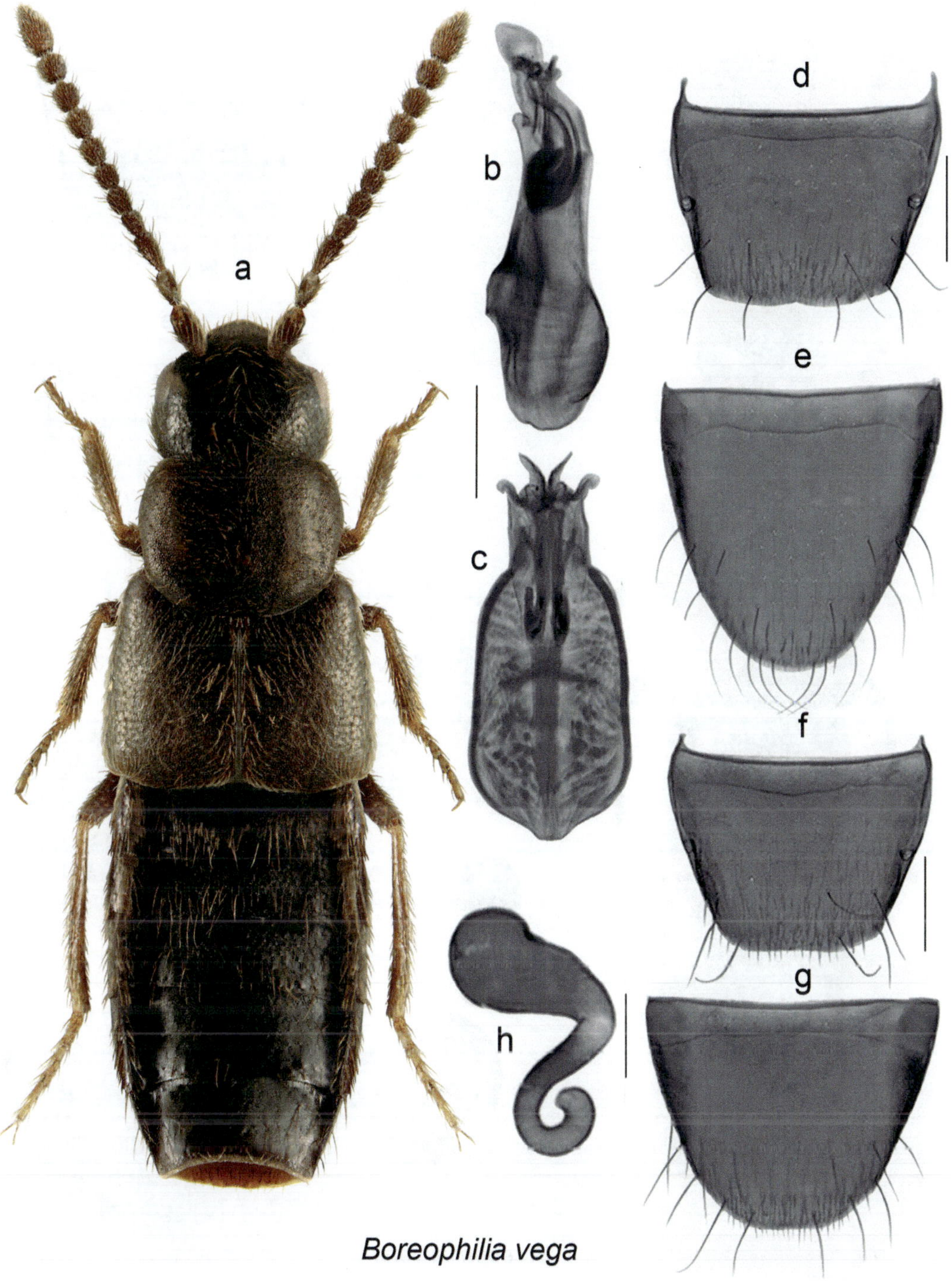

Fig. 19.254 (**a–h**). *Boreophilia vega* (Fenyes): (**a**) habitus in dorsal view; (**b**) median lobe of aedeagus in lateral view; (**c**) median lobe of aedeagus in dorsal view; (**d**) male tergite VIII; (**e**) male sternite VIII; (**f**) female tergite VIII; (**g**) female sternite VIII; (**h**) spermatheca. Scale bar for habitus = 1 mm; remaining scale bars = 0.2 mm

Fig. 19.255 (**a–h**). *Boreophilia venti* (Lohse): (**a**) habitus in dorsal view; (**b**) median lobe of aedeagus in lateral view; (**c**) median lobe of aedeagus in dorsal view; (**d**) male tergite VIII; (**e**) male sternite VIII; (**f**) female tergite VIII; (**g**) female sternite VIII; (**h**) spermatheca. Scale bar for habitus = 1 mm; remaining scale bars = 0.2 mm

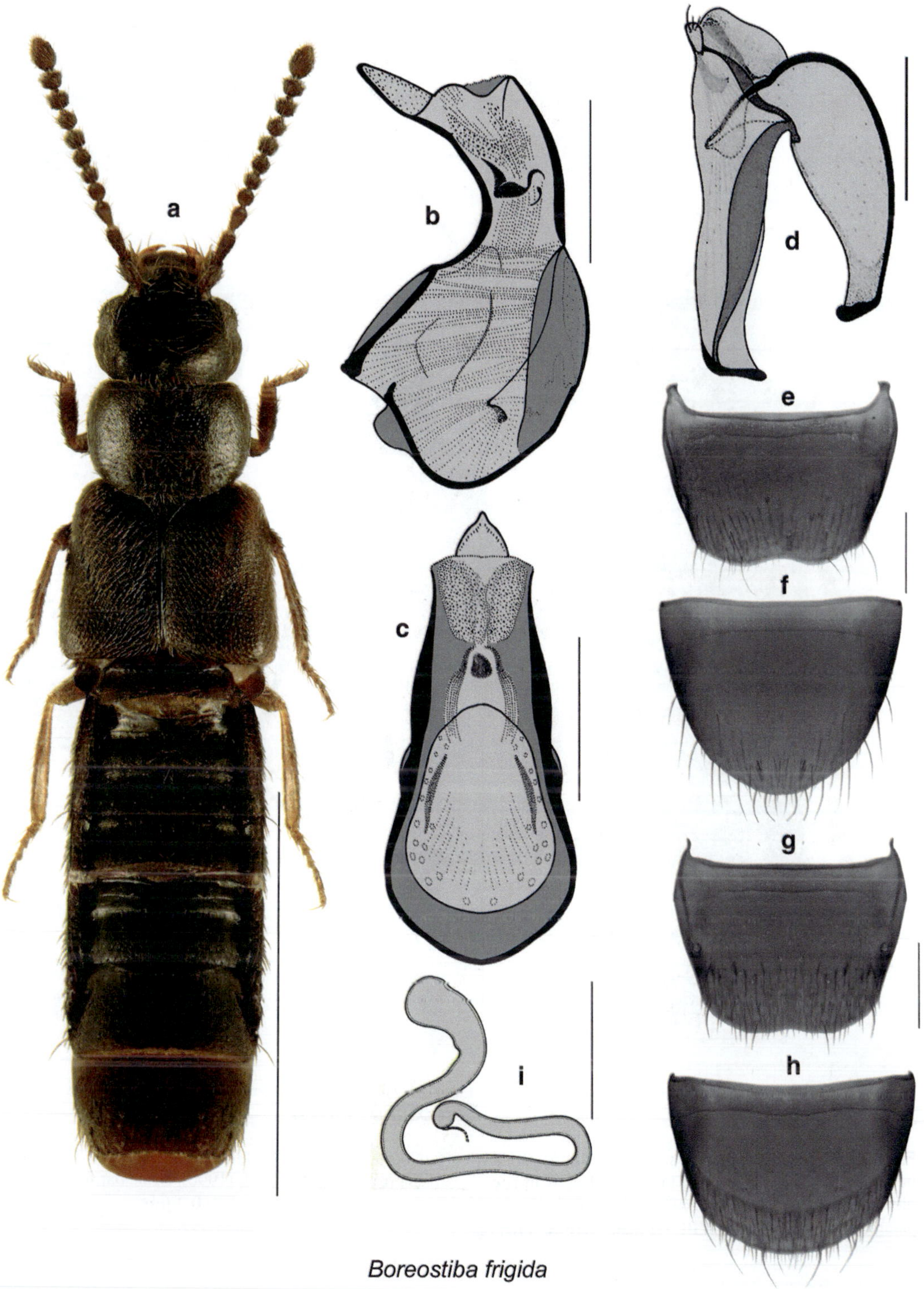

Fig. 19.256 (a–i). *Boreostiba frigida* (JR Sahlberg): (**a**) habitus in dorsal view; (**b**) median lobe of aedeagus in lateral view; (**c**) median lobe of aedeagus in dorsal view; (**d**) paramere; (**e**) male tergite VIII; (**f**) male sternite VIII; (**g**) female tergite VIII; (**h**) female sternite VIII; (**i**) spermatheca. (**b, c, d, i**) after Lohse et al. (1990). Scale bar for habitus = 1 mm; remaining scale bars = 0.2 mm

Fig. 19.257 (**a–i**). *Boreostiba sibirica* (Mäklin): (**a**) habitus in dorsal view; (**b**) median lobe of aedeagus in lateral view; (**c**) median lobe of aedeagus in dorsal view; (**d**) paramere; (**e**) male tergite VIII; (**f**) male sternite VIII; (**g**) female tergite VIII; (**h**) female sternite VIII; (**i**) spermatheca. (**c**) after Lohse et al. (1990). Scale bar for habitus = 1 mm; remaining scale bars = 0.2 mm

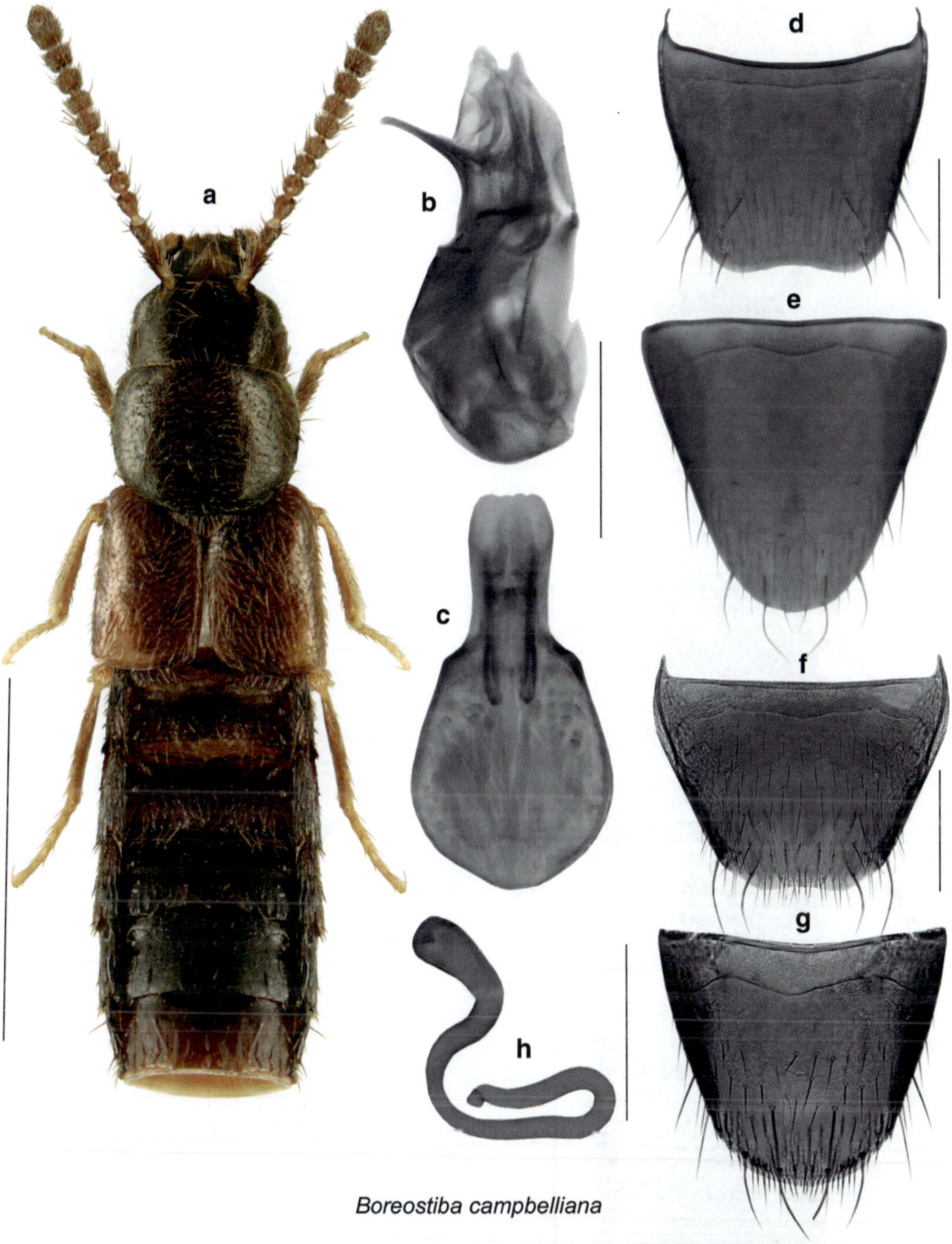

Fig. 19.258 (**a–h**). *Boreostiba campbelliana* Lohse: (**a**) habitus in dorsal view; (**b**) median lobe of aedeagus in lateral view; (**c**) median lobe of aedeagus in dorsal view; (**d**) male tergite VIII; (**e**) male sternite VIII; (**f**) female tergite VIII; (**g**) female sternite VIII; (**h**) spermatheca. Scale bar for habitus = 1 mm; remaining scale bars = 0.2 mm

Fig. 19.259 (**a–d**). *Boreostiba lagunae* Lohse: (**a**) habitus in dorsal view; (**b**) female tergite VIII; (**c**) female sternite VIII; (**d**) spermatheca. Scale bar for habitus = 1 mm; remaining scale bars = 0.2 mm

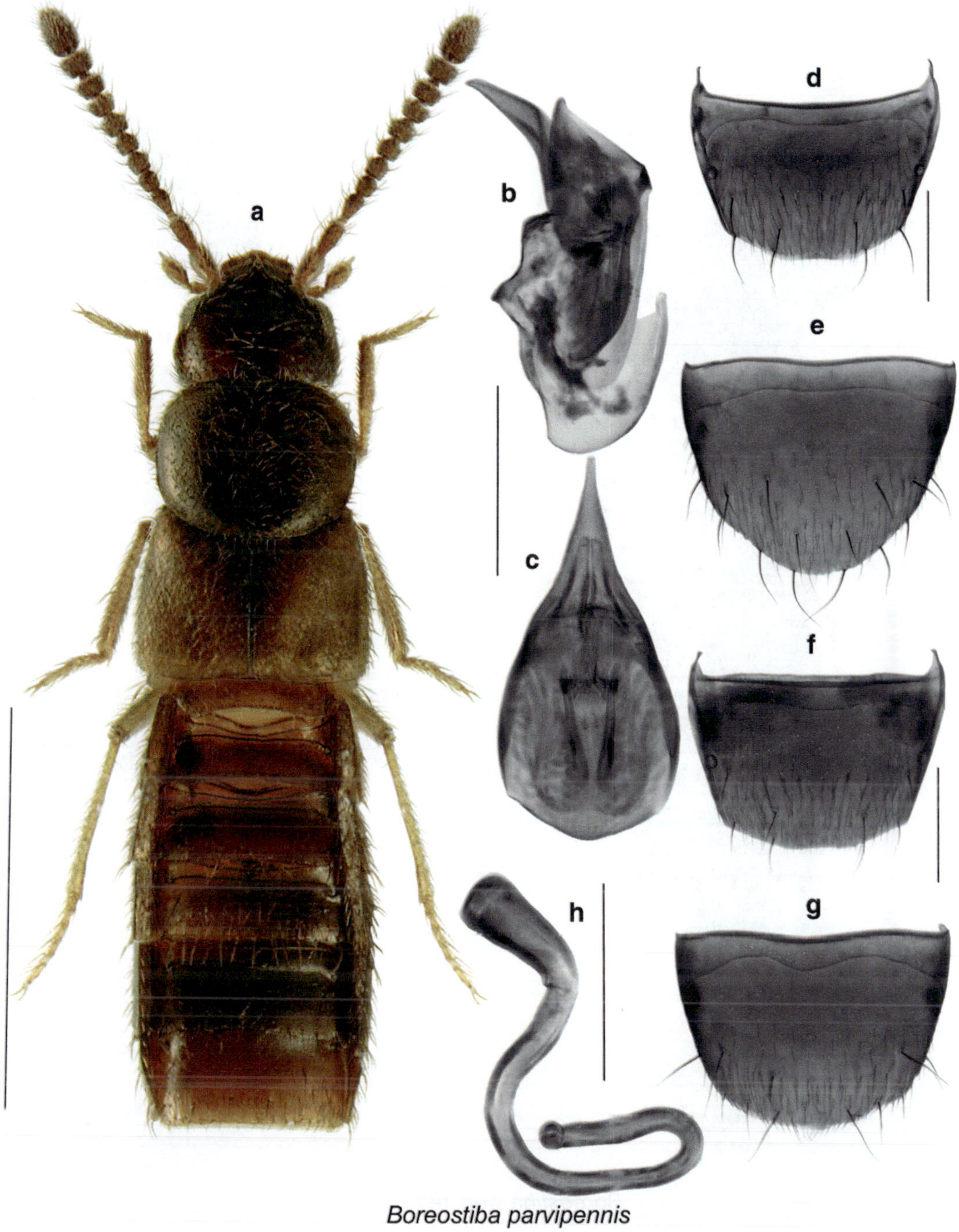

Fig. 19.260 (**a–h**). *Boreostiba parvipennis* Bernhauer: (**a**) habitus in dorsal view; (**b**) median lobe of aedeagus in lateral view; (**c**) median lobe of aedeagus in dorsal view; (**d**) male tergite VIII; (**e**) male sternite VIII; (**f**) female tergite VIII; (**g**) female sternite VIII; (**h**) spermatheca. Scale bar for habitus = 1 mm; remaining scale bars = 0.2 mm

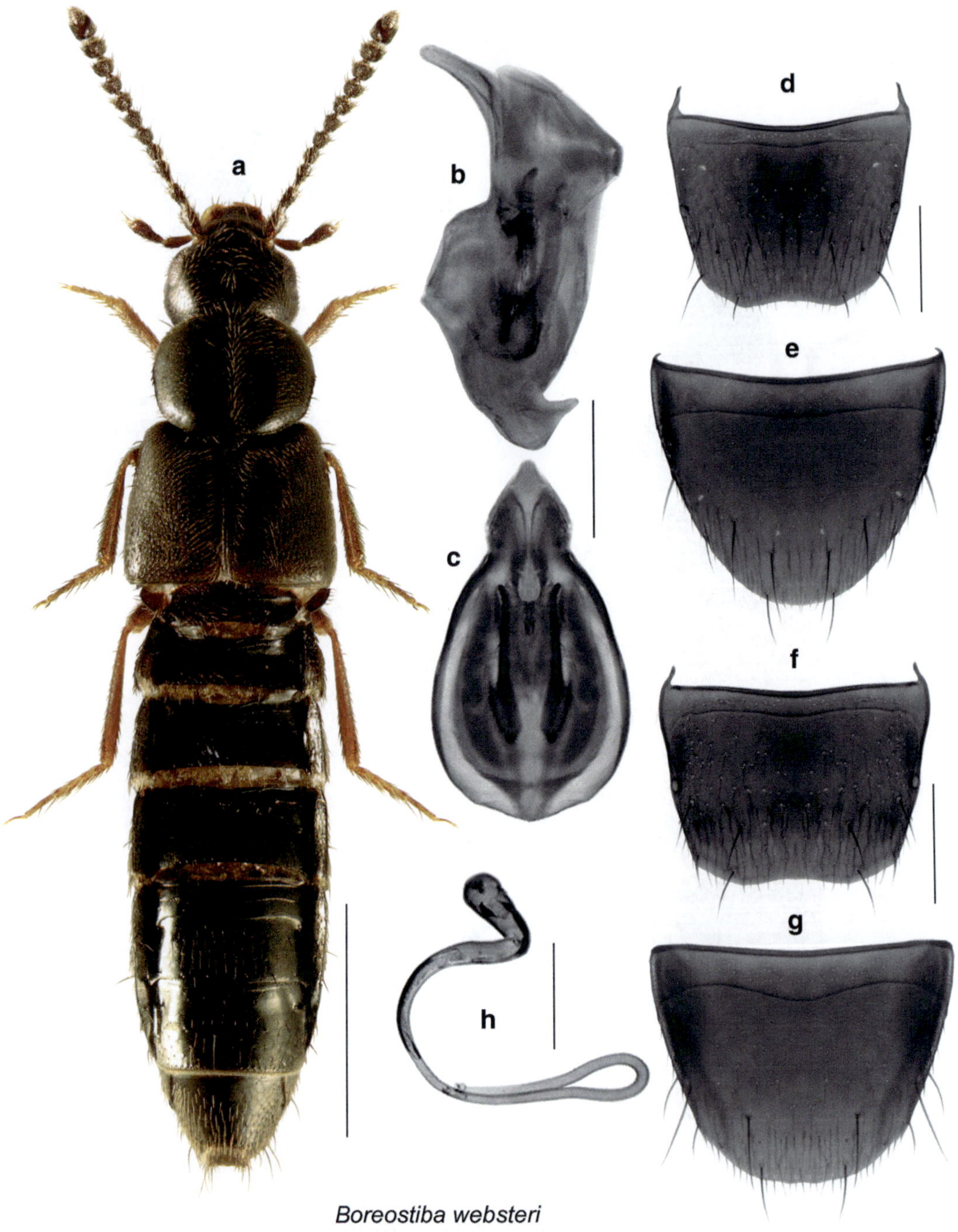

Fig. 19.261 (**a–h**). *Boreostiba websteri* Klimaszewski and Langor: (**a**) habitus in dorsal view; (**b**) median lobe of aedeagus in lateral view; (**c**) median lobe of aedeagus in dorsal view; (**d**) male tergite VIII; (**e**) male sternite VIII; (**f**) female tergite VIII; (**g**) female sternite VIII; (**h**) spermatheca. Scale bar for habitus = 1 mm; remaining scale bars = 0.2 mm

Fig. 19.262 (**a–g**). *Dinaraea angustula* (Gyllenhal): (**a**) habitus in dorsal view; (**b**) median lobe of aedeagus in lateral view; (**c**) male tergite VIII; (**d**) male sternite VIII; (**e**) female tergite VIII; (**f**) female sternite VIII; (**g**) spermatheca. Scale bar for habitus = 1 mm; remaining scale bars = 0.2 mm

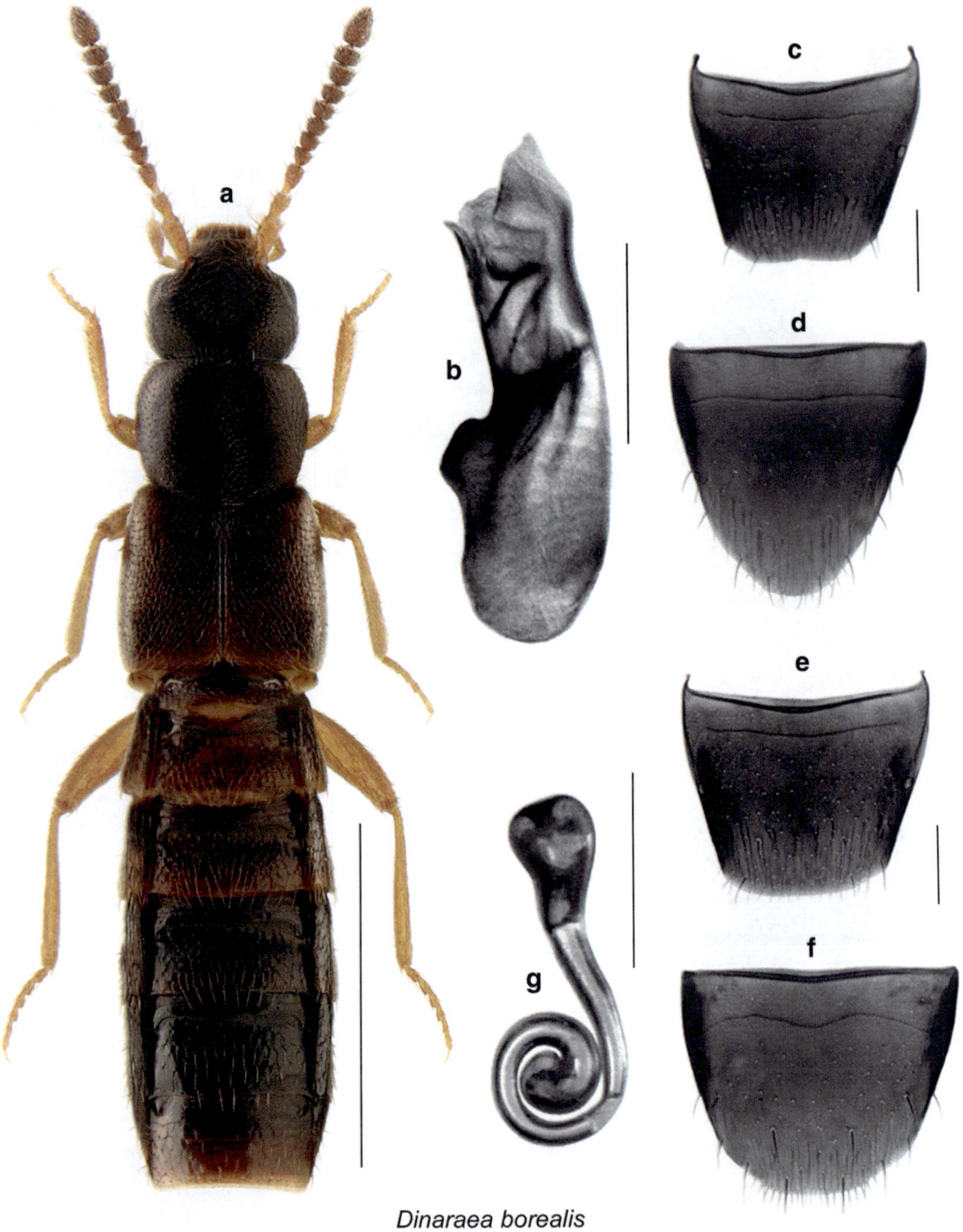

Fig. 19.263 (a–g). *Dinaraea borealis* Lohse: (**a**) habitus in dorsal view; (**b**) median lobe of aedeagus in lateral view; (**c**) male tergite VIII; (**d**) male sternite VIII; (**e**) female tergite VIII; (**f**) female sternite VIII; (**g**) spermatheca. Scale bar for habitus = 1 mm; remaining scale bars = 0.2 mm

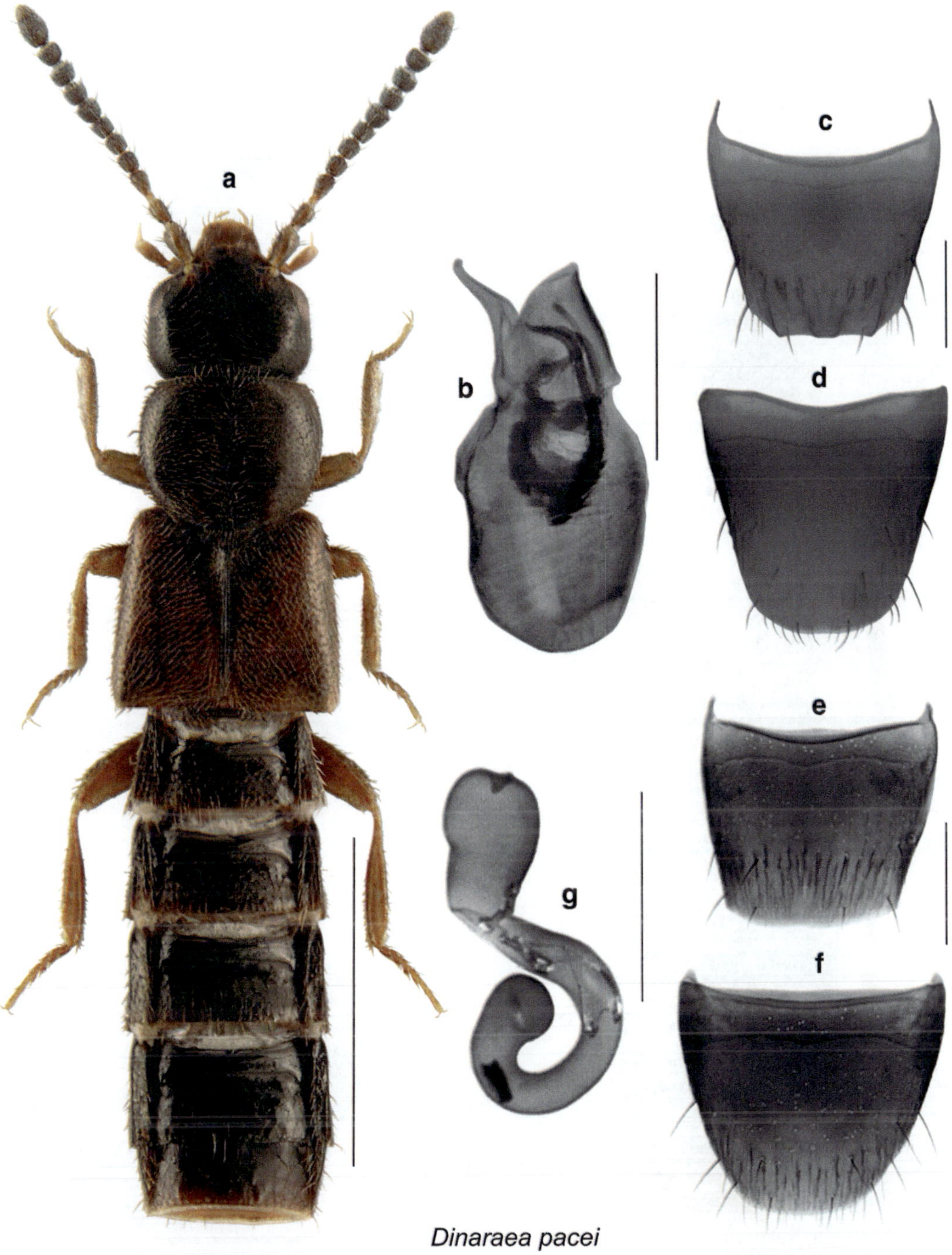

Fig. 19.264 (**a–g**). *Dinaraea pacei* Klimaszewski and Langor: (**a**) habitus in dorsal view; (**b**) median lobe of aedeagus in lateral view; (**c**) male tergite VIII; (**d**) male sternite VIII; (**e**) female tergite VIII; (**f**) female sternite VIII; (**g**) spermatheca. Scale bar for habitus = 1 mm; remaining scale bars = 0.2 mm

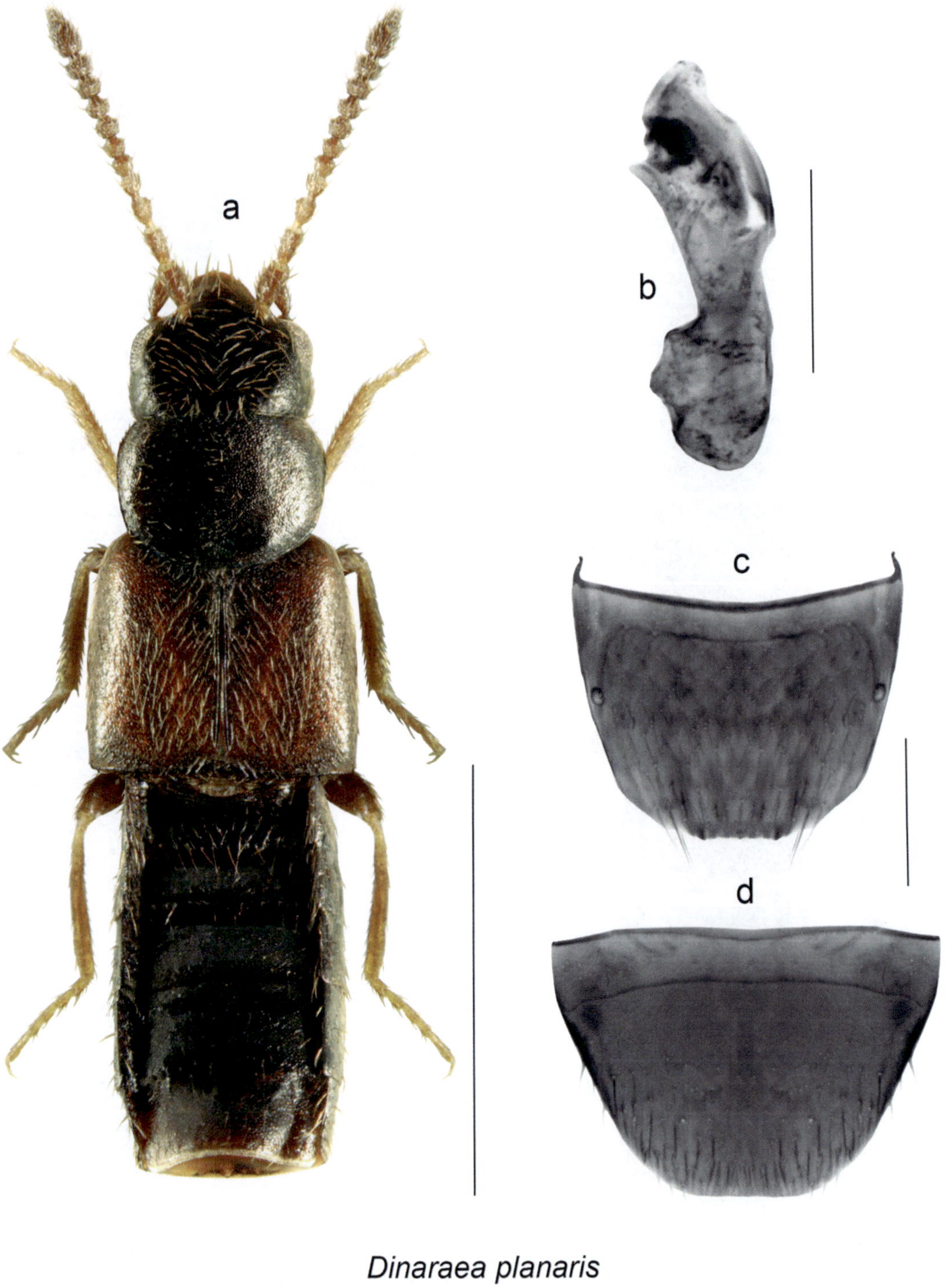

Fig. 19.265 (**a–d**). *Dinaraea planaris* (Mäklin), (based on neotype): (**a**) habitus in dorsal view; (**b**) median lobe of aedeagus in lateral view; (**c**) male tergite VIII; (**d**) male sternite VIII. Scale bar for habitus = 1 mm; remaining scale bars = 0.2 mm

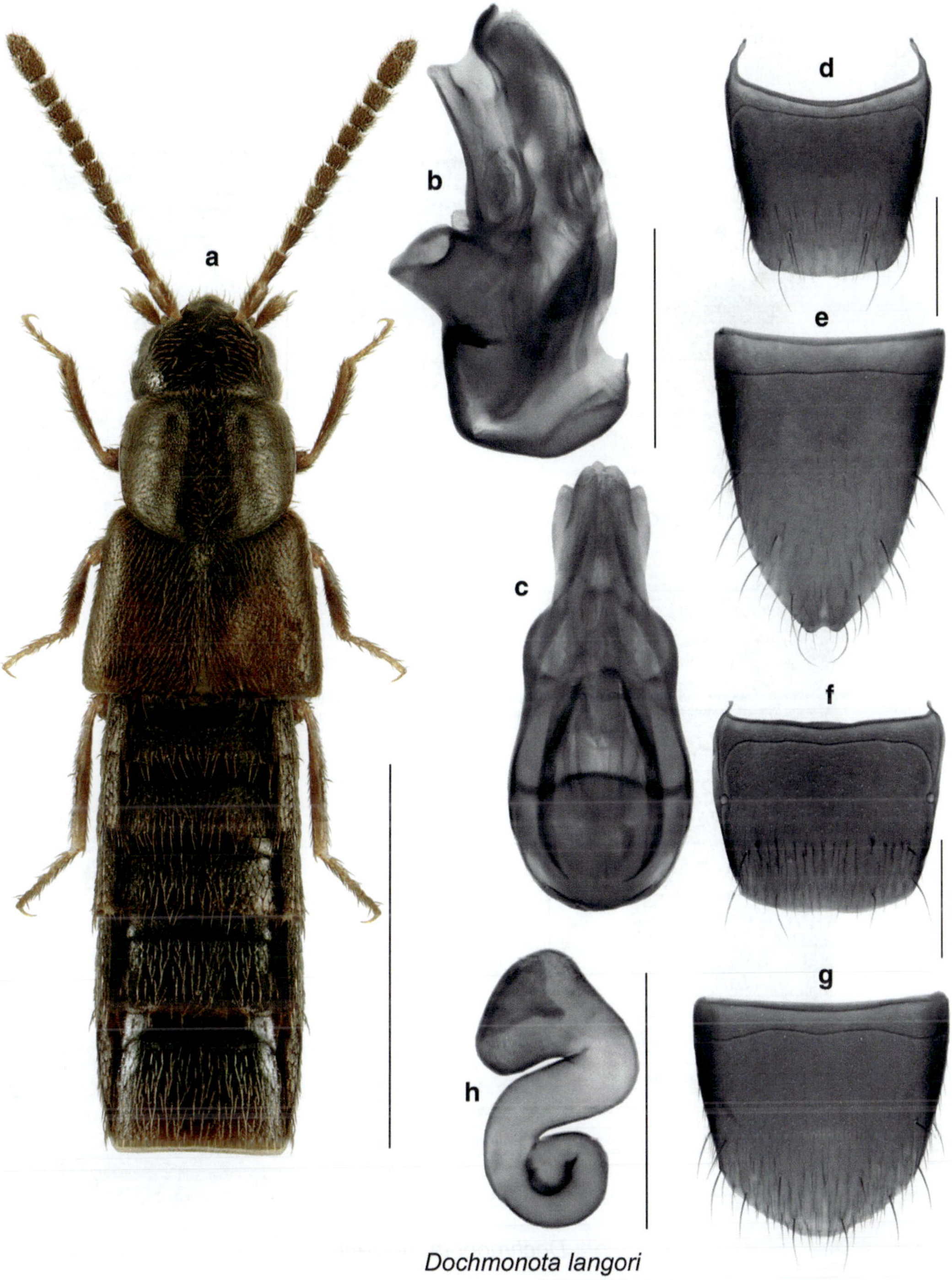

Fig. 19.266 (**a–h**). *Dochmonota langori* Klimaszewski and Larson: (**a**) habitus in dorsal view; (**b**) median lobe of aedeagus in lateral view; (**c**) median lobe of aedeagus in dorsal view; (**d**) male tergite VIII; (**e**) male sternite VIII; (**f**) female tergite VIII; (**g**) female sternite VIII; (**h**) spermatheca. Scale bar for habitus = 1 mm; remaining scale bars = 0.2 mm

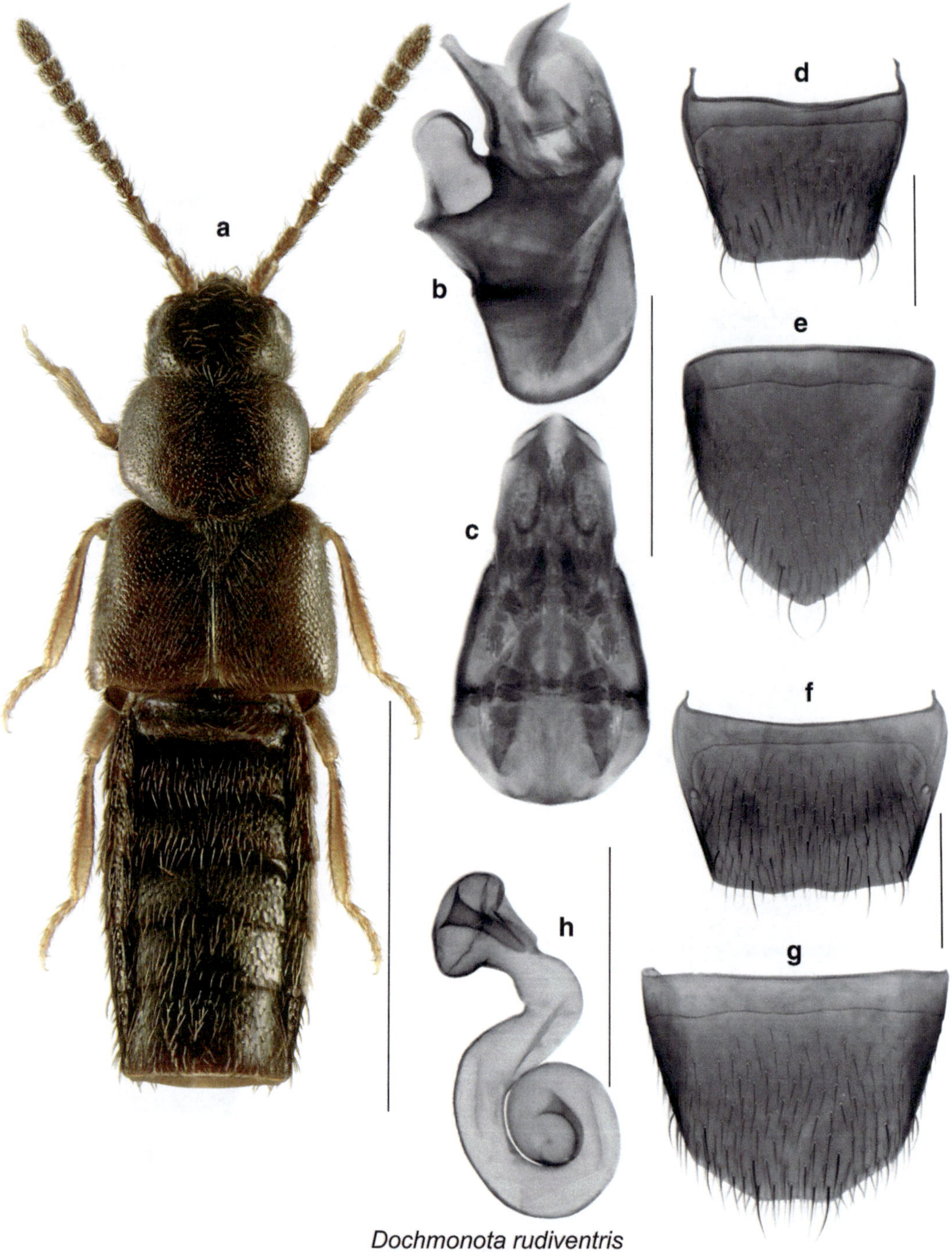

Fig. 19.267 (**a–h**). *Dochmonota rudiventris* (Eppelsheim): (**a**) habitus in dorsal view; (**b**) median lobe of aedeagus in lateral view; (**c**) median lobe of aedeagus in dorsal view; (**d**) male tergite VIII; (**e**) male sternite VIII; (**f**) female tergite VIII; (**g**) female sternite VIII; (**h**) spermatheca. Scale bar for habitus = 1 mm; remaining scale bars = 0.2 mm

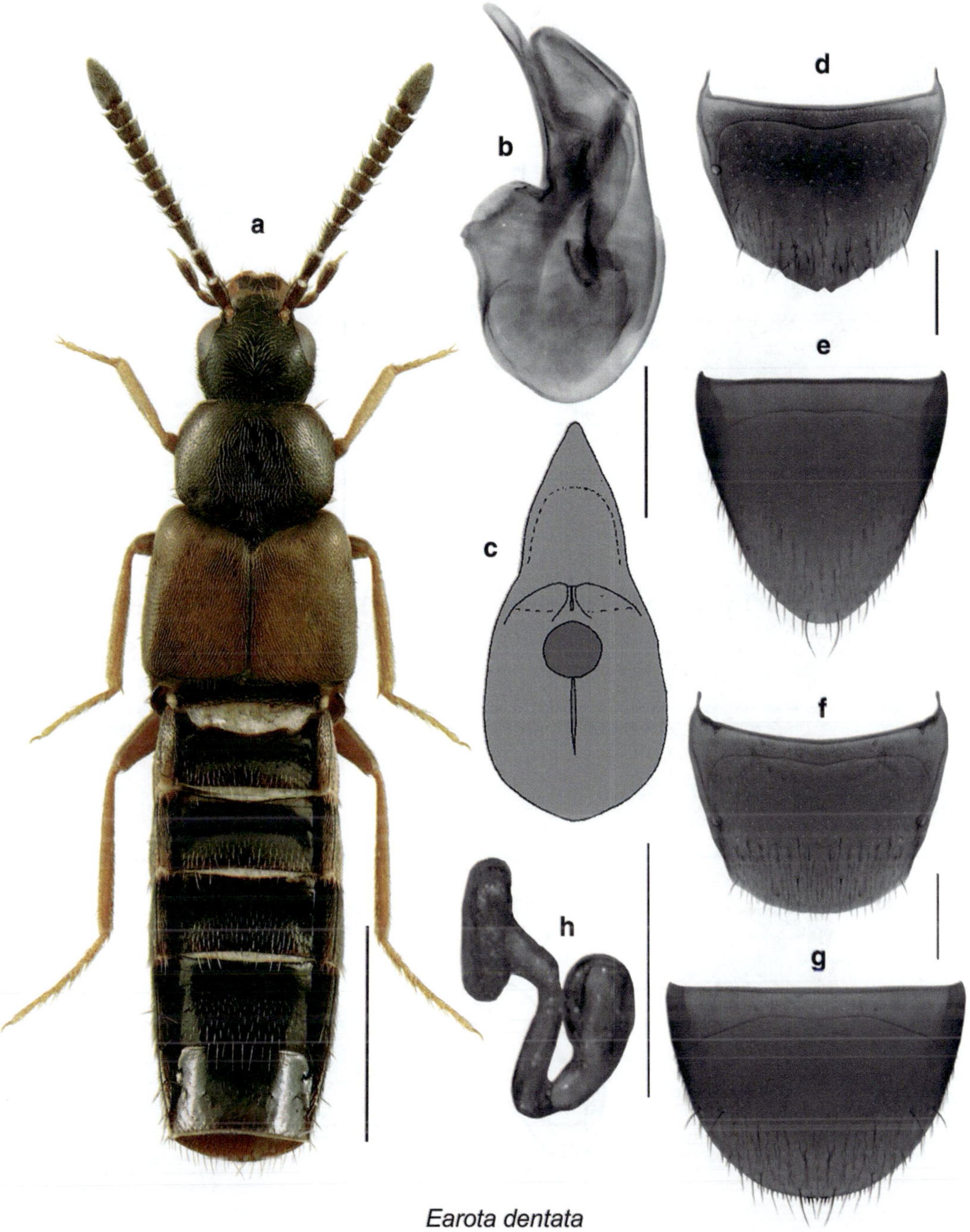

Fig. 19.268 (**a–h**). *Earota dentata* (Bernhauer): (**a**) habitus in dorsal view; (**b**) median lobe of aedeagus in lateral view; (**c**) median lobe of aedeagus in ventral view; (**d**) male tergite VIII; (**e**) male sternite VIII; (**f**) female tergite VIII; (**g**) female sternite VIII; (**h**) spermatheca. Scale bar for habitus = 1 mm; remaining scale bars = 0.2 mm

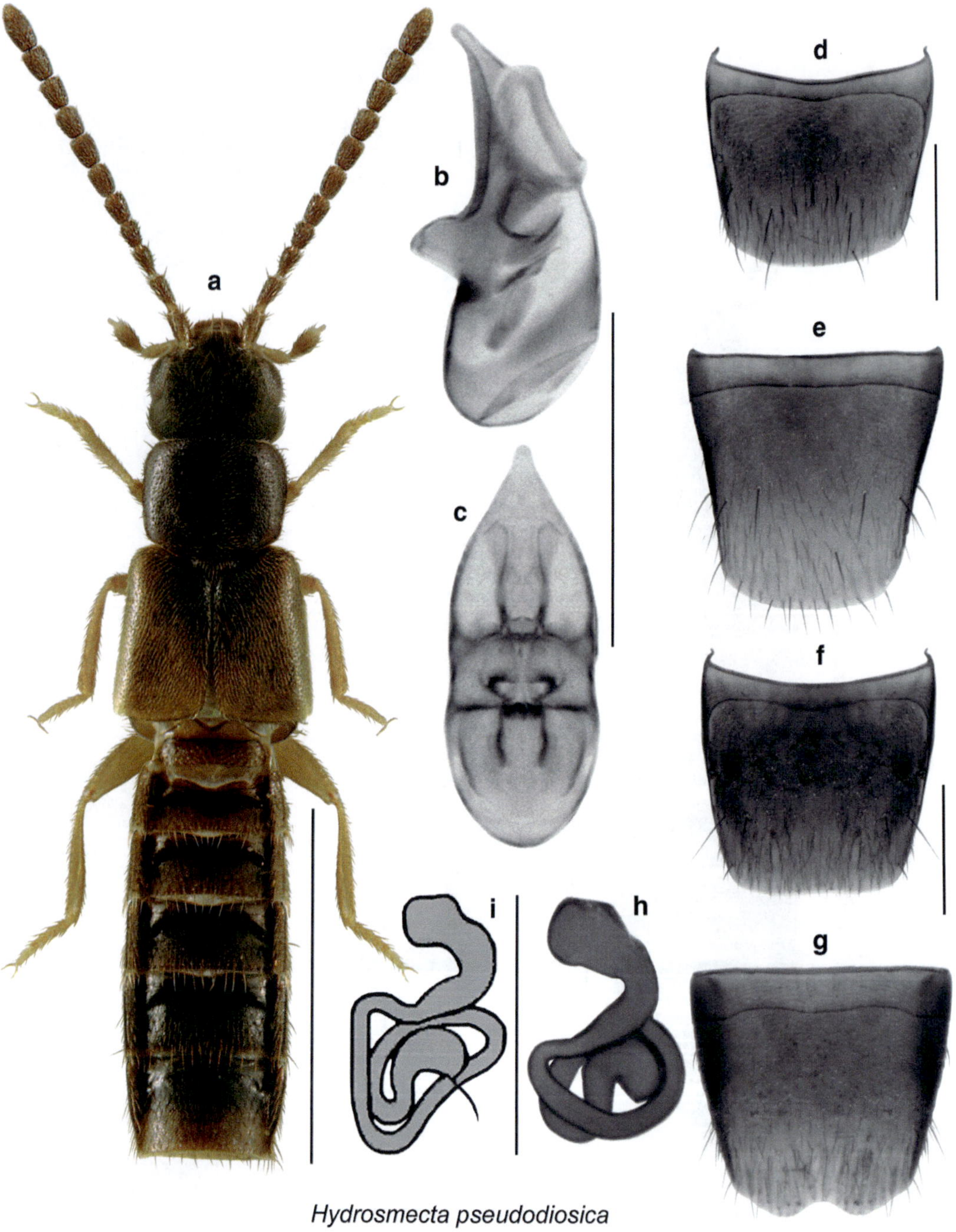

Fig. 19.269 (**a–i**). *Hydrosmecta pseudodiosica* Lohse: (**a**) habitus in dorsal view; (**b**) median lobe of aedeagus in lateral view; (**c**) median lobe of aedeagus in dorsal view; (**d**) male tergite VIII; (**e**) male sternite VIII; (**f**) female tergite VIII; (**g**) female sternite VIII; (**h, i**) spermatheca. (**i**) after Lohse et al. (1990). Scale bar for habitus = 1 mm; remaining scale bars = 0.2 mm

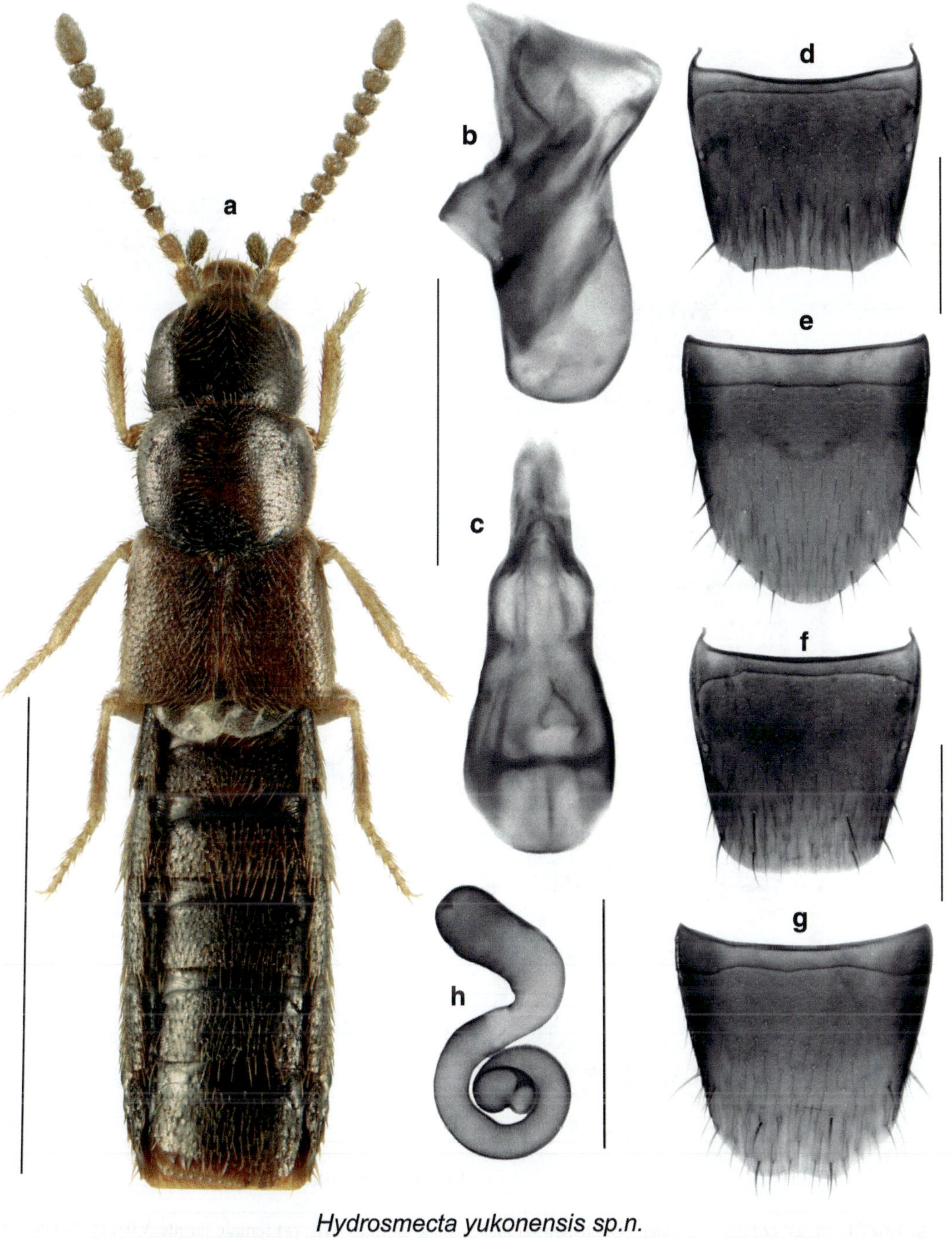

Fig. 19.270 (**a–h**). *Hydrosmecta yukonensis* Klimaszewski and Godin, sp.n.: (**a**) habitus in dorsal view; (**b**) median lobe of aedeagus in lateral view; (**c**) median lobe of aedeagus in dorsal view; (**d**) male tergite VIII; (**e**) male sternite VIII; (**f**) female tergite VIII; (**g**) female sternite VIII; (**h**) spermatheca. Scale bar for habitus = 1 mm; remaining scale bars = 0.2 mm

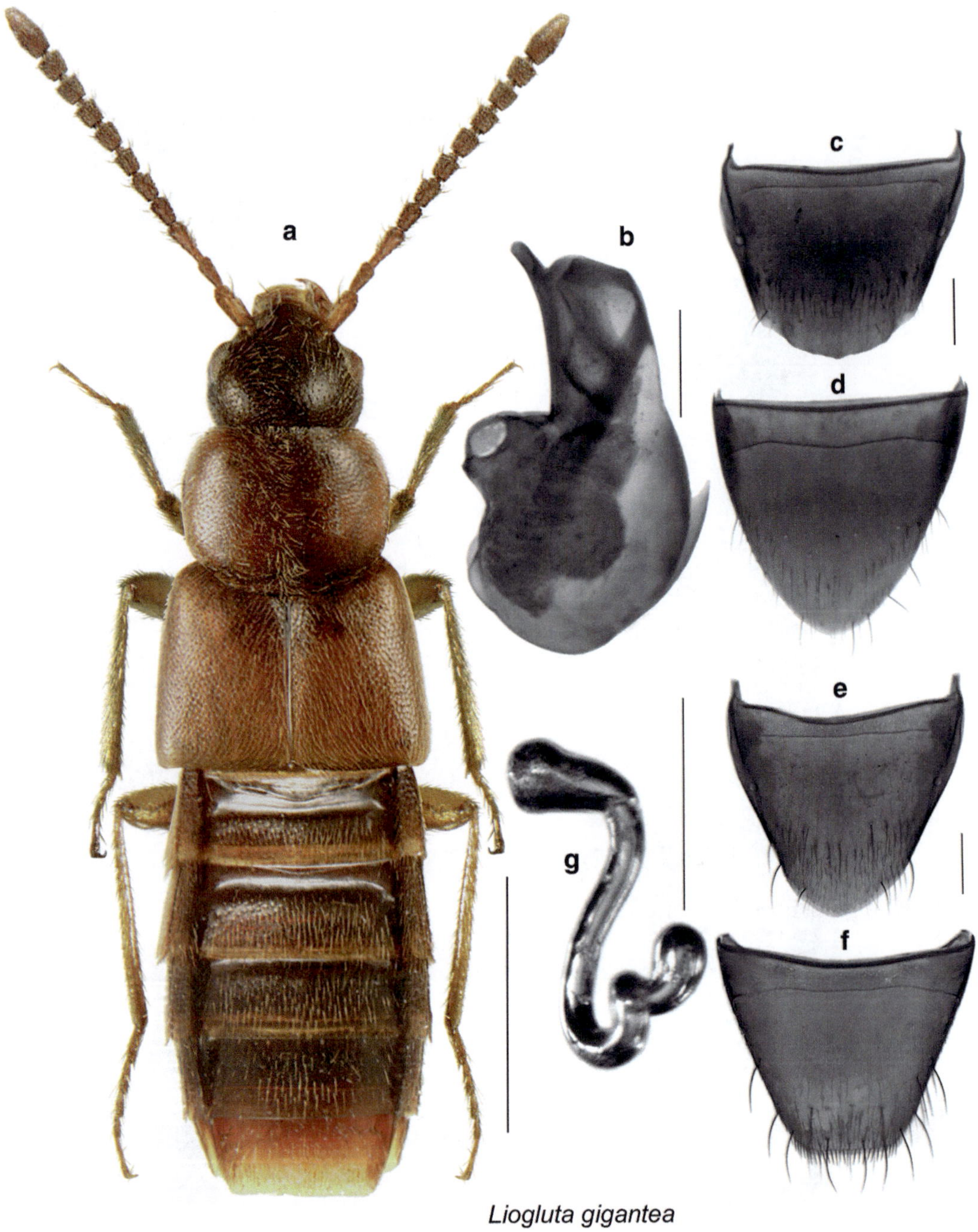

Fig. 19.271 (**a–g**). *Liogluta gigantea* Klimaszewski and Langor: (**a**) habitus in dorsal view; (**b**) median lobe of aedeagus in lateral view; (**c**) male tergite VIII; (**d**) male sternite VIII; (**e**) female tergite VIII; (**f**) female sternite VIII; (**g**) spermatheca. Scale bar for habitus = 1 mm; remaining scale bars = 0.2 mm

Fig. 19.272 (**a–h**). *Liogluta granulosa* Lohse: (**a**) habitus in dorsal view; (**b**) elytron; (**c**) median lobe of aedeagus in lateral view; (**d**) male tergite VIII; (**e**) male sternite VIII; (**f**) female tergite VIII; (**g**) female sternite VIII; (**h**) spermatheca. Scale bar for habitus = 1 mm; remaining scale bars = 0.2 mm

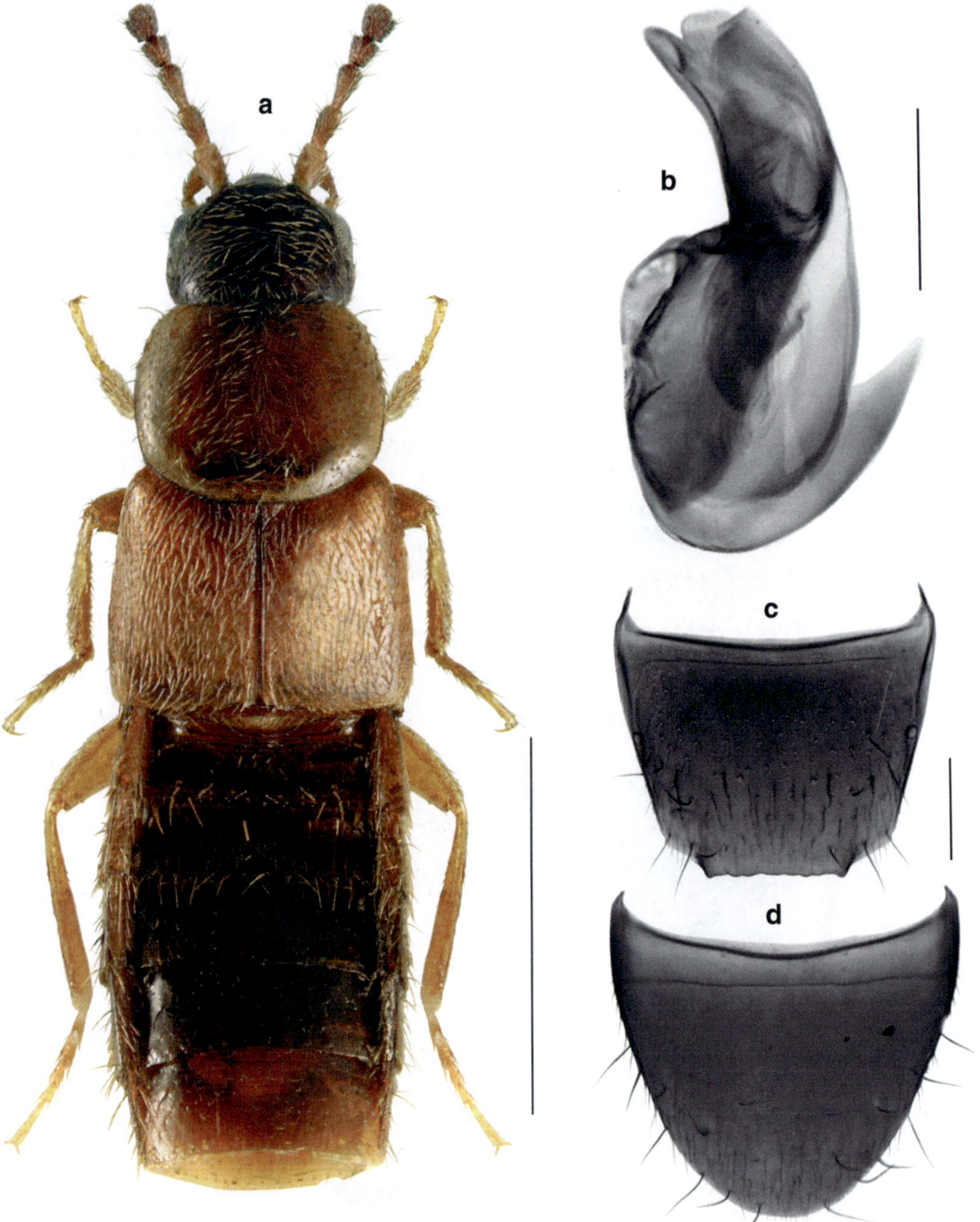

Liogluta atriventis

Fig. 19.273 (**a–d**). *Liogluta atriventris* (Casey) (based on lectotype): (**a**) habitus in dorsal view; (**b**) median lobe of aedeagus in lateral view; (**c**) male tergite VIII; (**d**) male sternite VIII. Scale bar for habitus = 1 mm; remaining scale bars = 0.2 mm

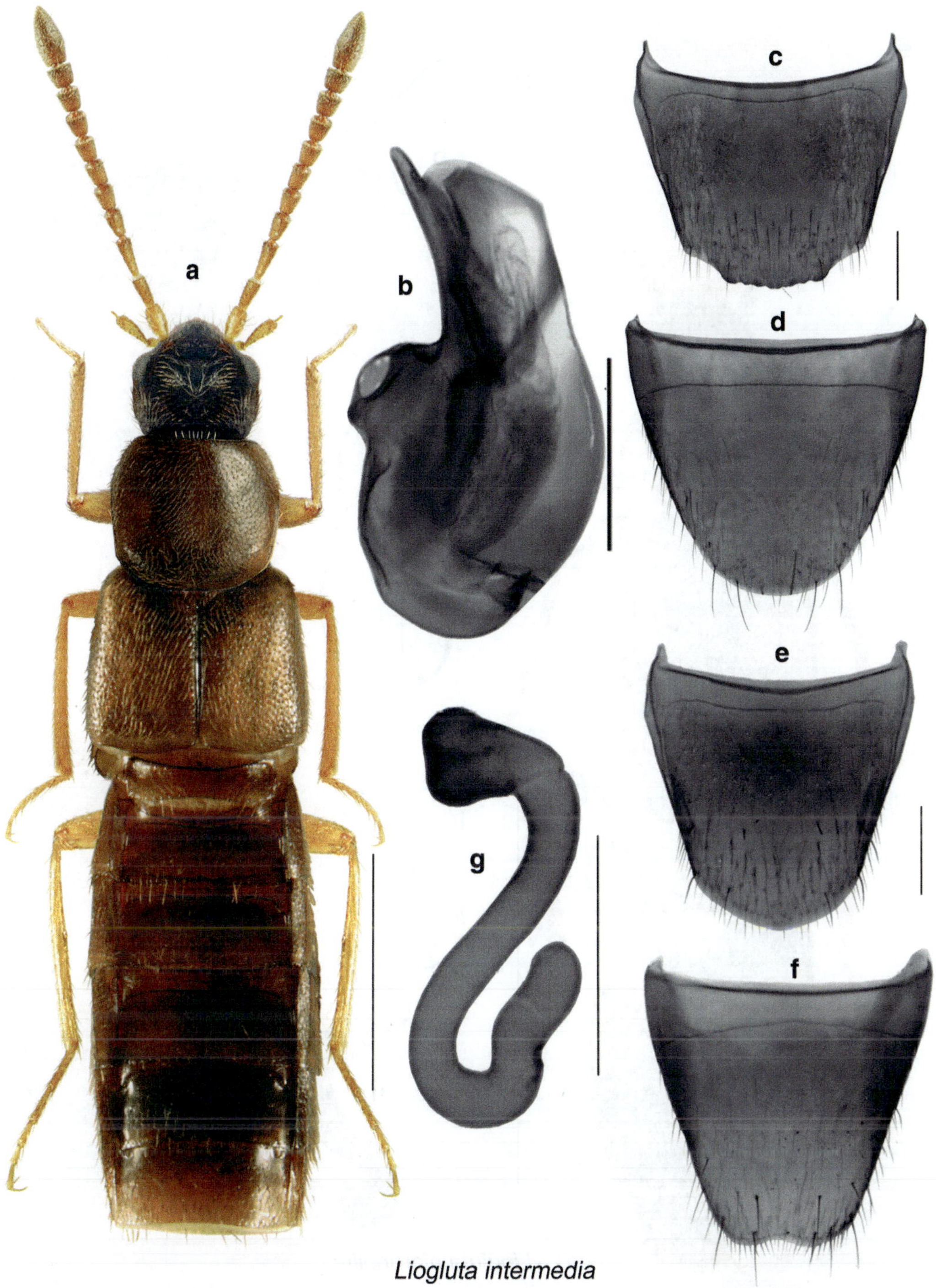

Fig. 19.274 (**a–g**). *Liogluta intermedia* Klimaszewski and Langor: (**a**) habitus in dorsal view; (**b**) median lobe of aedeagus in lateral view; (**c**) male tergite VIII; (**d**) male sternite VIII; (**e**) female tergite VIII; (**f**) female sternite VIII; (**g**) spermatheca. Scale bar for habitus = 1 mm; remaining scale bars = 0.2 mm

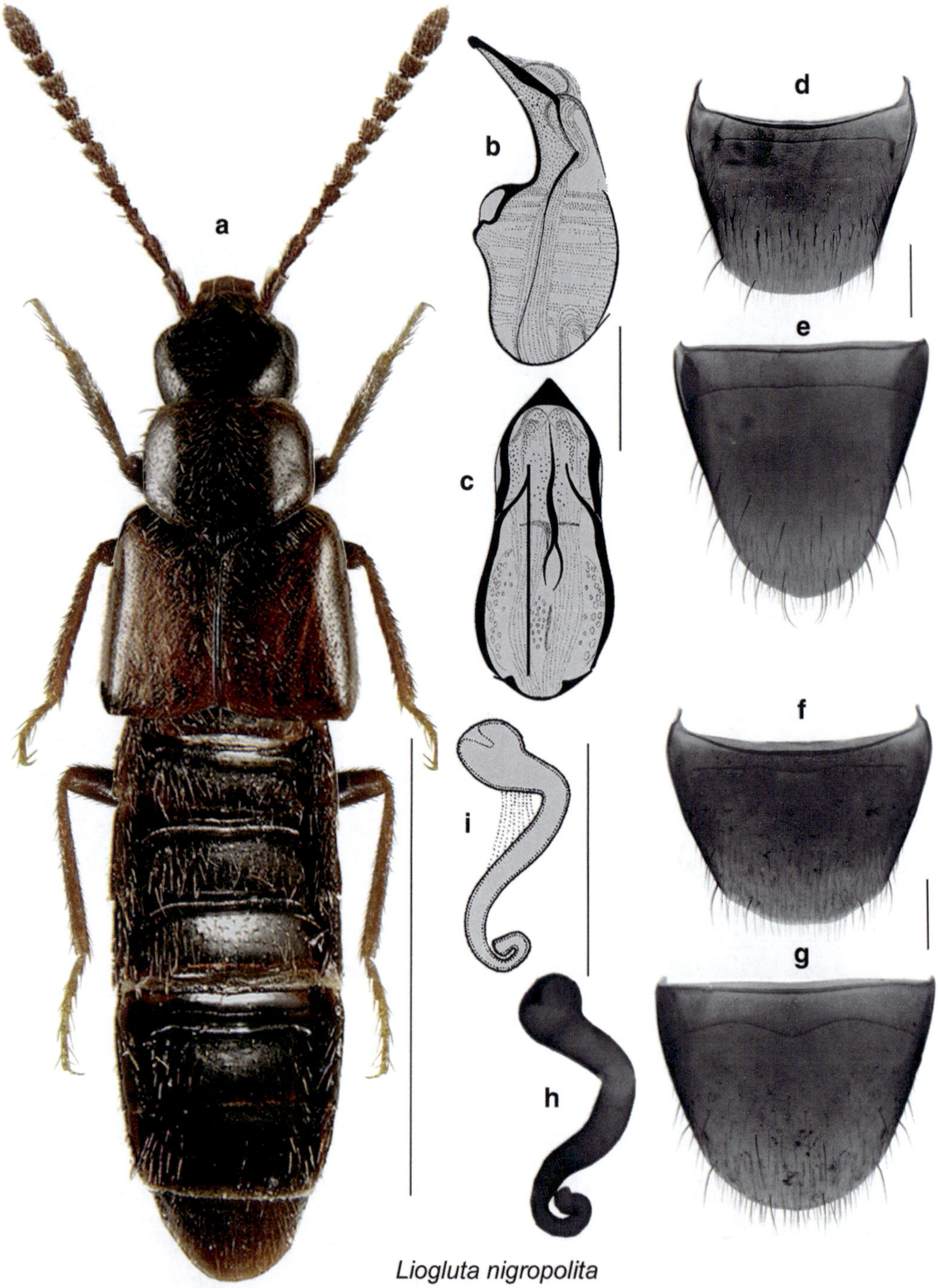

Fig. 19.275 (**a–i**). *Liogluta nigropolita* (Bernhauer): (**a**) habitus in dorsal view; (**b**) median lobe of aedeagus in lateral view; (**c**) median lobe of aedeagus in dorsal view; (**d**) male tergite VIII; (**e**) male sternite VIII; (**f**) female tergite VIII; (**g**) female sternite VIII; (**h, i**) spermatheca. (**b, c, i**) after Lohse et al. (1990). Scale bar for habitus = 1 mm; remaining scale bars = 0.2 mm

Fig. 19.276 (**a–f**). *Liogluta nitens* (Mäklin): (**a**) habitus in dorsal view; (**b**) median lobe of aedeagus in lateral view; (**c**) male tergite VIII; (**d**) male sternite VIII; (**e**) female tergite VIII; (**f**) female sternite VIII. Scale bar for habitus = 1 mm; remaining scale bars = 0.2 mm

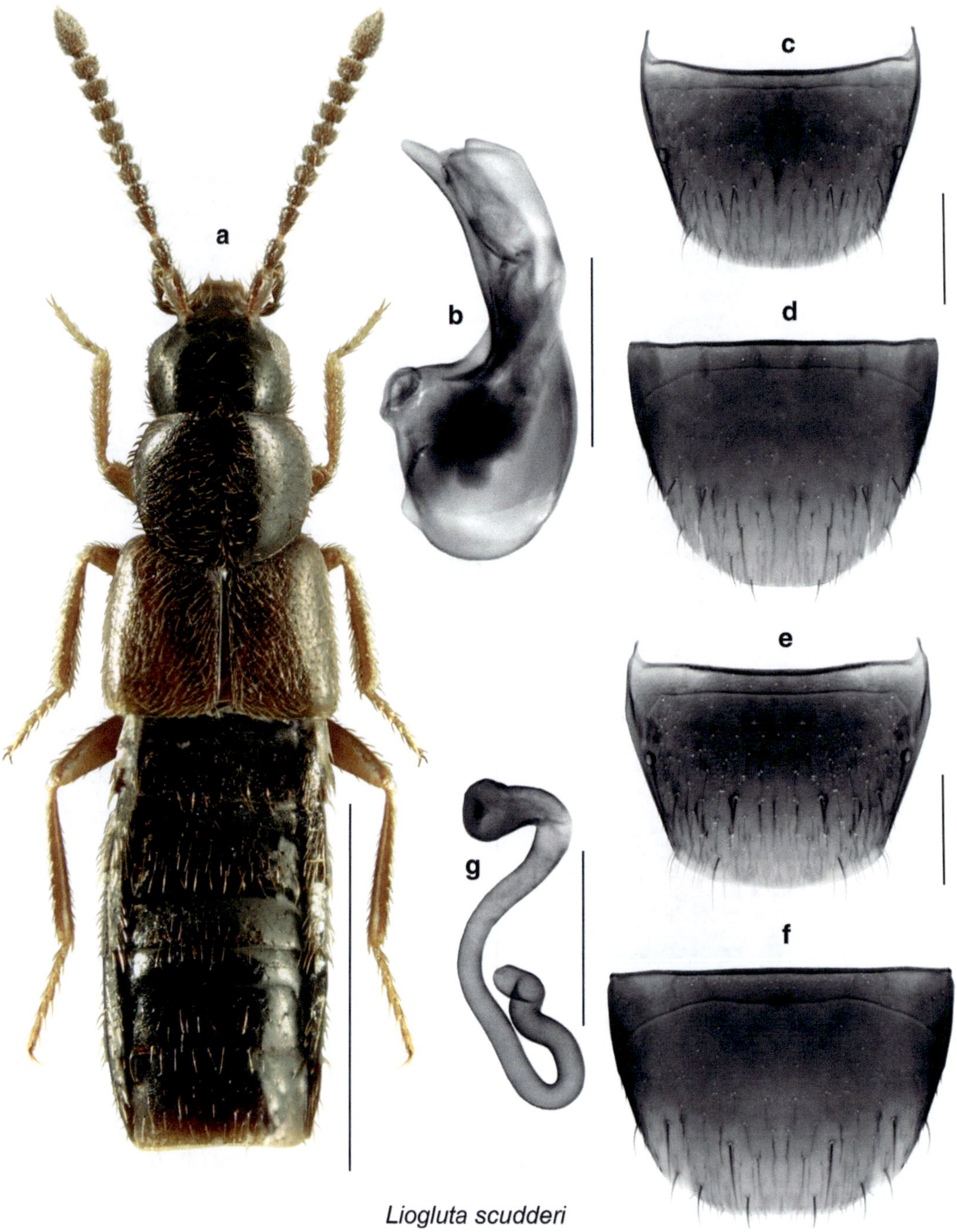

Fig. 19.277 (**a–g**). *Liogluta scudderi* Klimaszewski: (**a**) habitus in dorsal view; (**b**) median lobe of aedeagus in lateral view; (**c**) male tergite VIII; (**d**) male sternite VIII; (**e**) female tergite VIII; (**f**) female sternite VIII; (**g**) spermatheca. Scale bar for habitus = 1 mm; remaining scale bars = 0.2 mm

Fig. 19.278 (**a–g**). *Liogluta terminalis* (Casey): (**a**) habitus in dorsal view; (**b**) median lobe of aedeagus in lateral view; (**c**) male tergite VIII; (**d**) male sternite VIII; (**e**) female tergite VIII; (**f**) female sternite VIII; (**g**) spermatheca. Scale bar for habitus = 1 mm; remaining scale bars = 0.2 mm

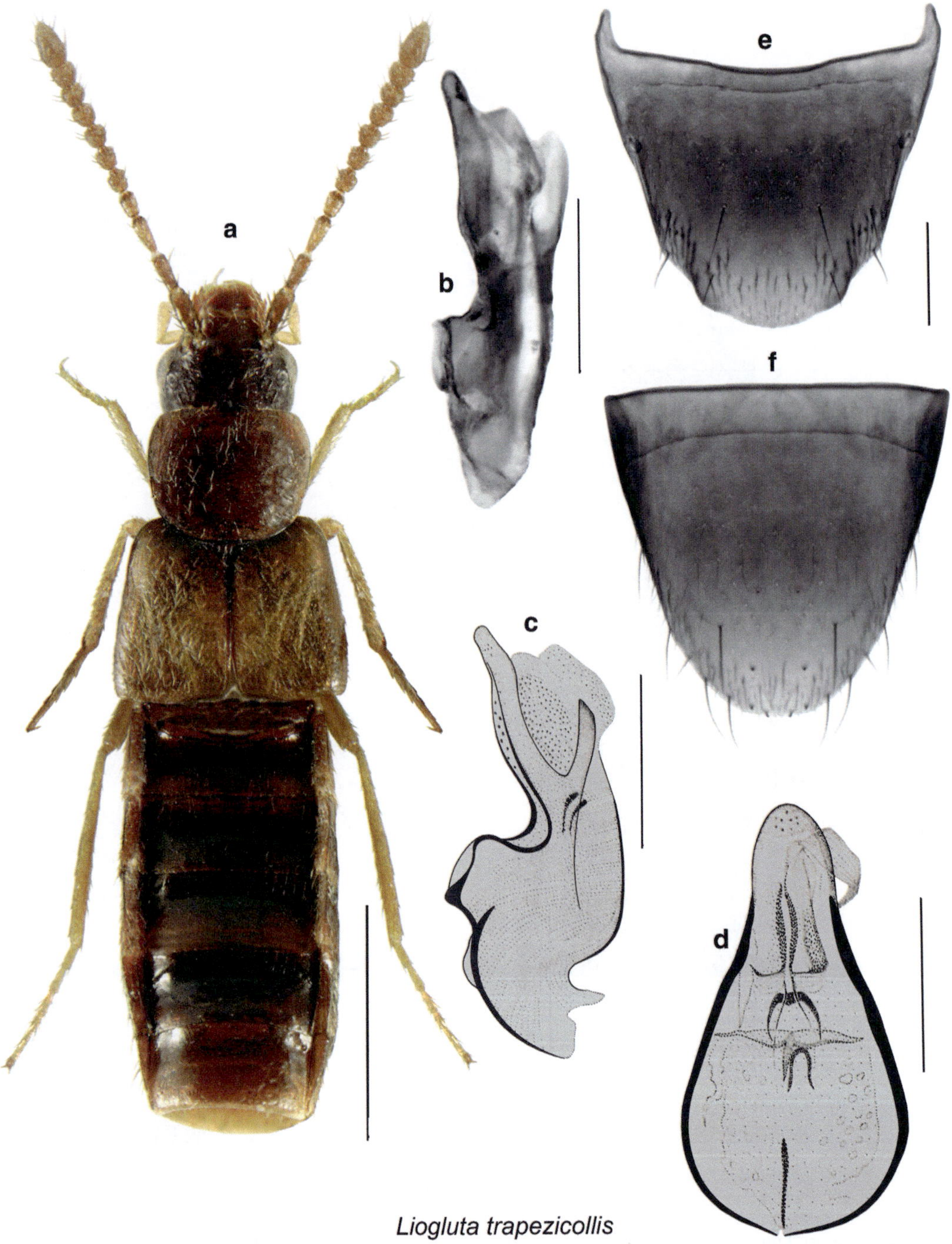

Fig. 19.279 (**a–f**). *Liogluta trapezicollis* Lohse: (**a**) habitus in dorsal view; (**b, c**) median lobe of aedeagus in lateral view; (**d**) median lobe of aedeagus in dorsal view; (**e**) male tergite VIII; (**f**) male sternite VIII. Scale bar for habitus = 1 mm; remaining scale bars = 0.2 mm

Fig. 19.280 (**a–h**). *Liogluta vasta* (Mäklin): (**a**) habitus in dorsal view; (**b,c**) median lobe of aedeagus in lateral view; (**d**) male tergite VIII; (**e**) male sternite VIII; (**f**) female tergite VIII; (**g**) female sternite VIII; (**h**) spermatheca. Scale bar for habitus = 1 mm; remaining scale bars = 0.2 mm

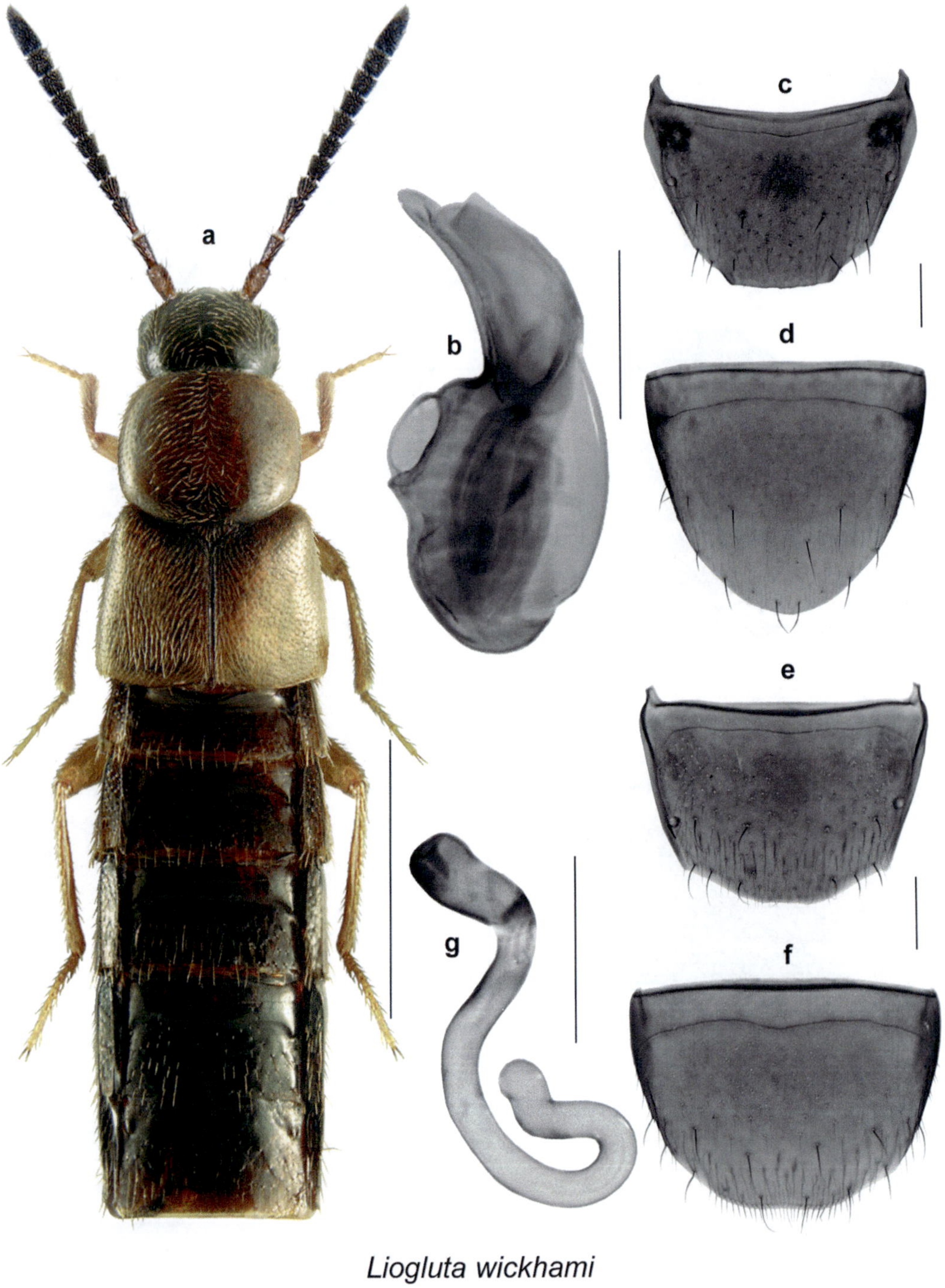

Fig. 19.281 (**a–g**). *Liogluta wickhami* (Casey): (**a**) habitus in dorsal view; (**b**) median lobe of aedeagus in lateral view; (**c**) male tergite VIII; (**d**) male sternite VIII; (**e**) female tergite VIII; (**f**) female sternite VIII; (**g**) spermatheca. Scale bar for habitus = 1 mm; remaining scale bars = 0.2 mm

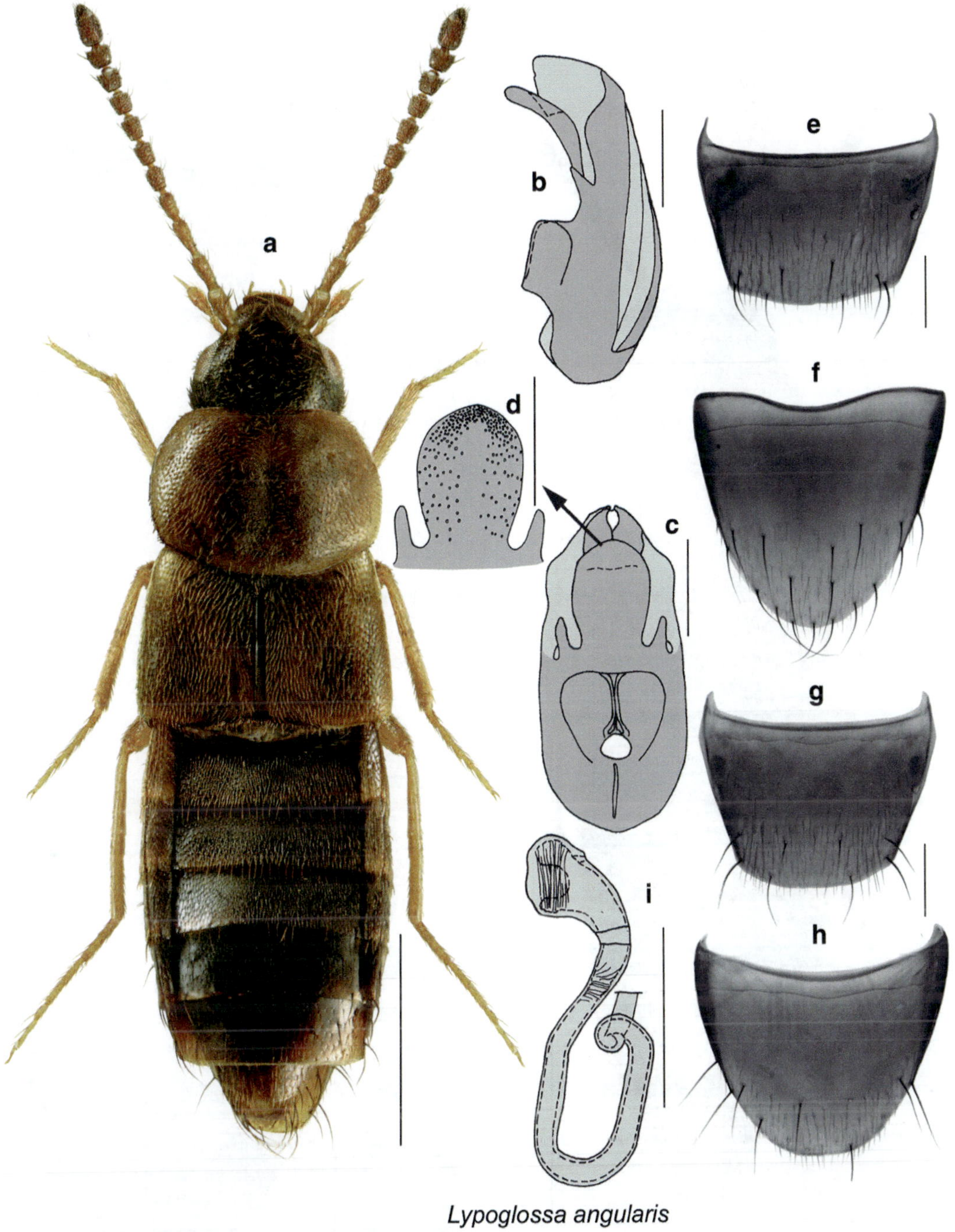

Fig. 19.282 (**a–i**). *Lypoglossa angularis* (Mäklin): (**a**) habitus in dorsal view; (**b**) median lobe of aedeagus in lateral view; (**c**) median lobe of aedeagus in ventral view; (**d**) tubus in ventral view (enlarged); (**e**) male tergite VIII; (**f**) male sternite VIII; (**g**) female tergite VIII; (**h**) female sternite VIII; (**i**) spermatheca. (**b, c, d, i**) after Gusarov (2004). Scale bar for habitus = 1 mm; remaining scale bars = 0.2 mm

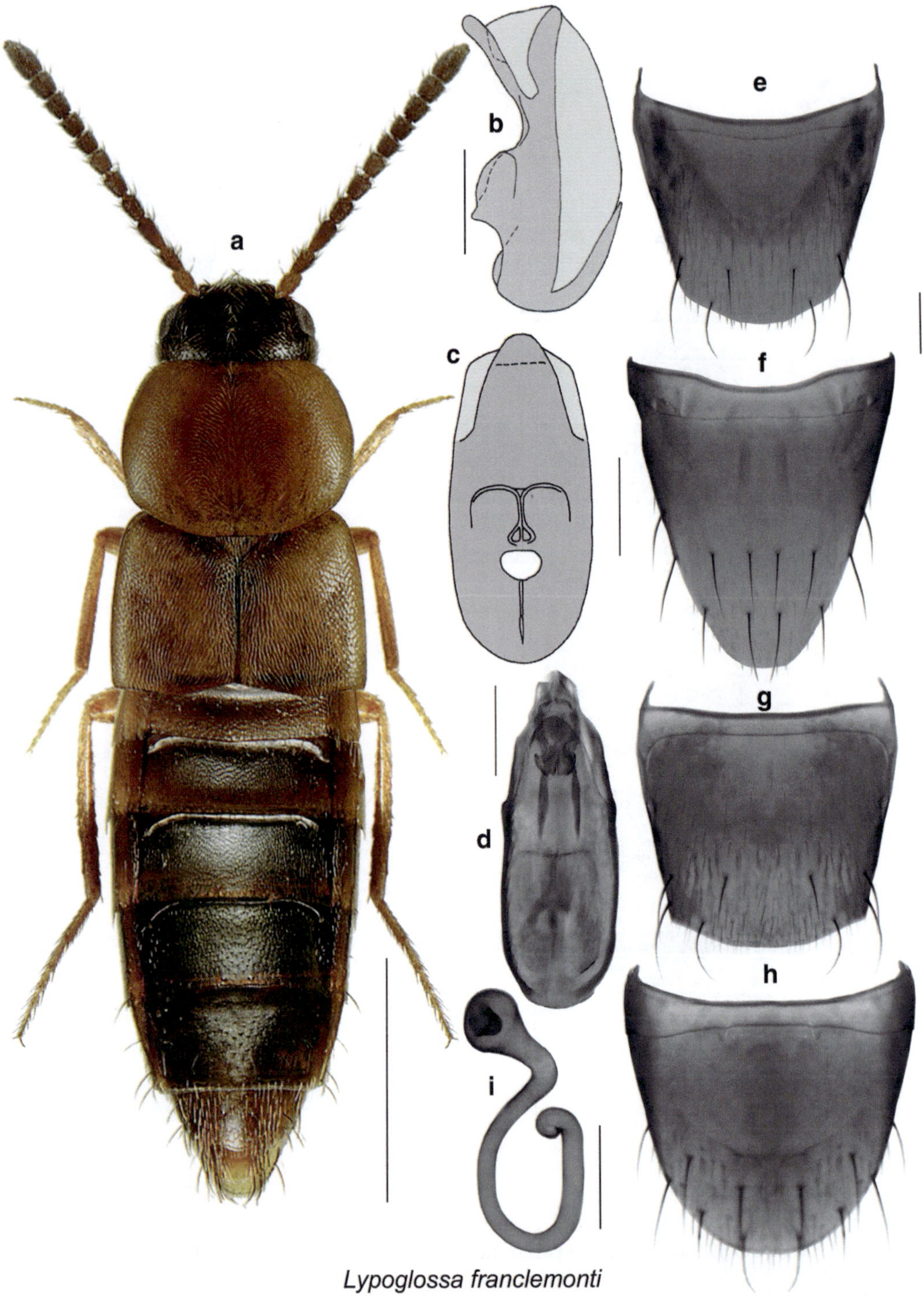

Fig. 19.283 (**a–i**). *Lypoglossa franclemonti* Hoebeke: (**a**) habitus in dorsal view; (**b**) median lobe of aedeagus in lateral view; (**c**) median lobe of aedeagus in ventral view; (**d**) median lobe aedeagus in dorsal view; (**e**) male tergite VIII; (**f**) male sternite VIII; (**g**) female tergite VIII; (**h**) female sternite VIII; (**i**) spermatheca. (**b, c**) after Gusarov (2004). Scale bar for habitus = 1 mm; remaining scale bars = 0.2 mm

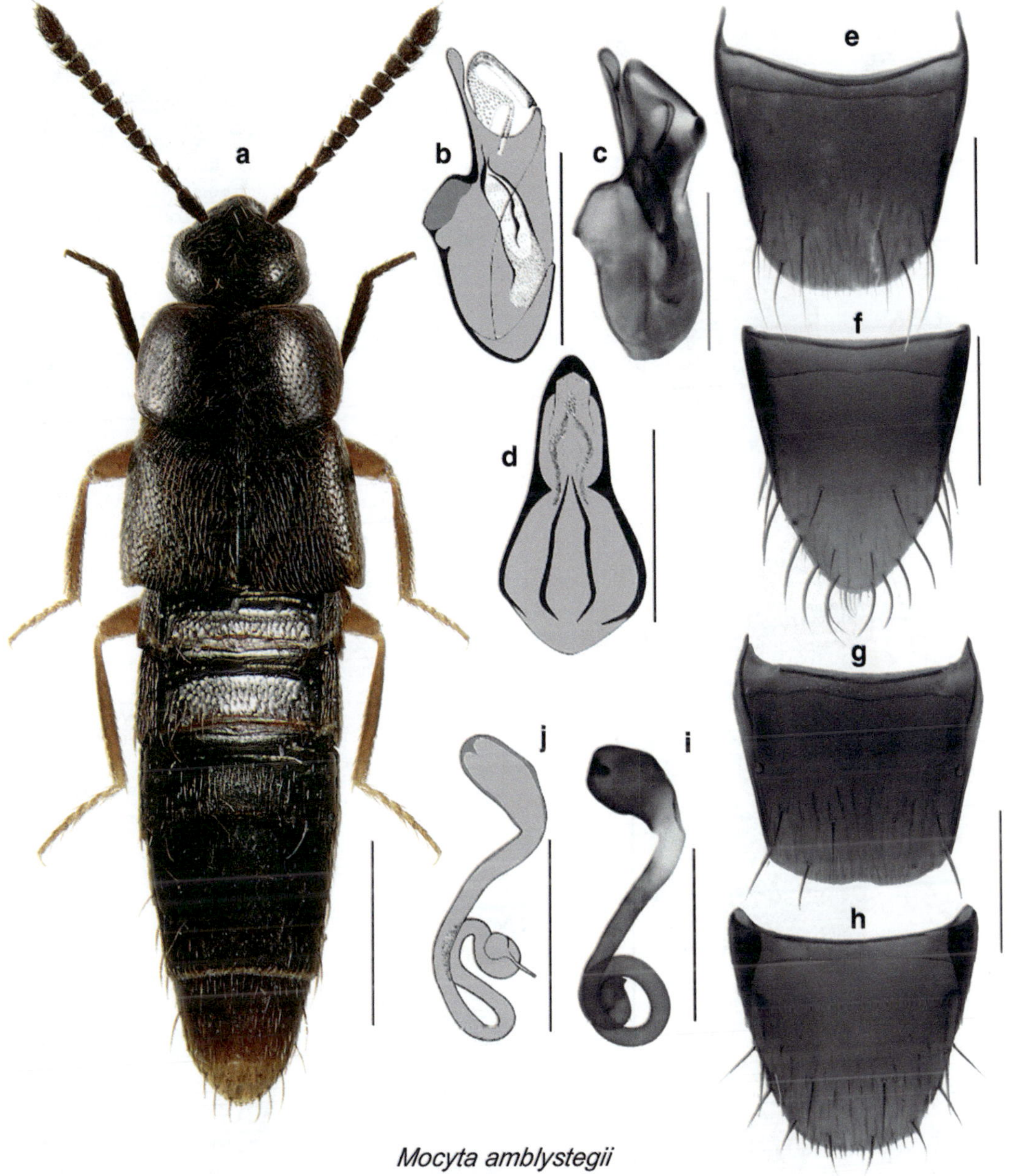

Fig. 19.284 (**a–j**). *Mocyta amblystegii* (Brundin): (**a**) habitus in dorsal view; (**b, c**) median lobe of aedeagus in lateral view; (**d**) median lobe of aedeagus in dorsal view; (**e**) male tergite VIII; (**f**) male sternite VIII; (**g**) female tergite VIII; (**h**) female sternite VIII; (**i, j**) spermatheca. (**b, d, j**) after Lohse et al. (1990). Scale bar for habitus = 1 mm; remaining scale bars = 0.2 mm

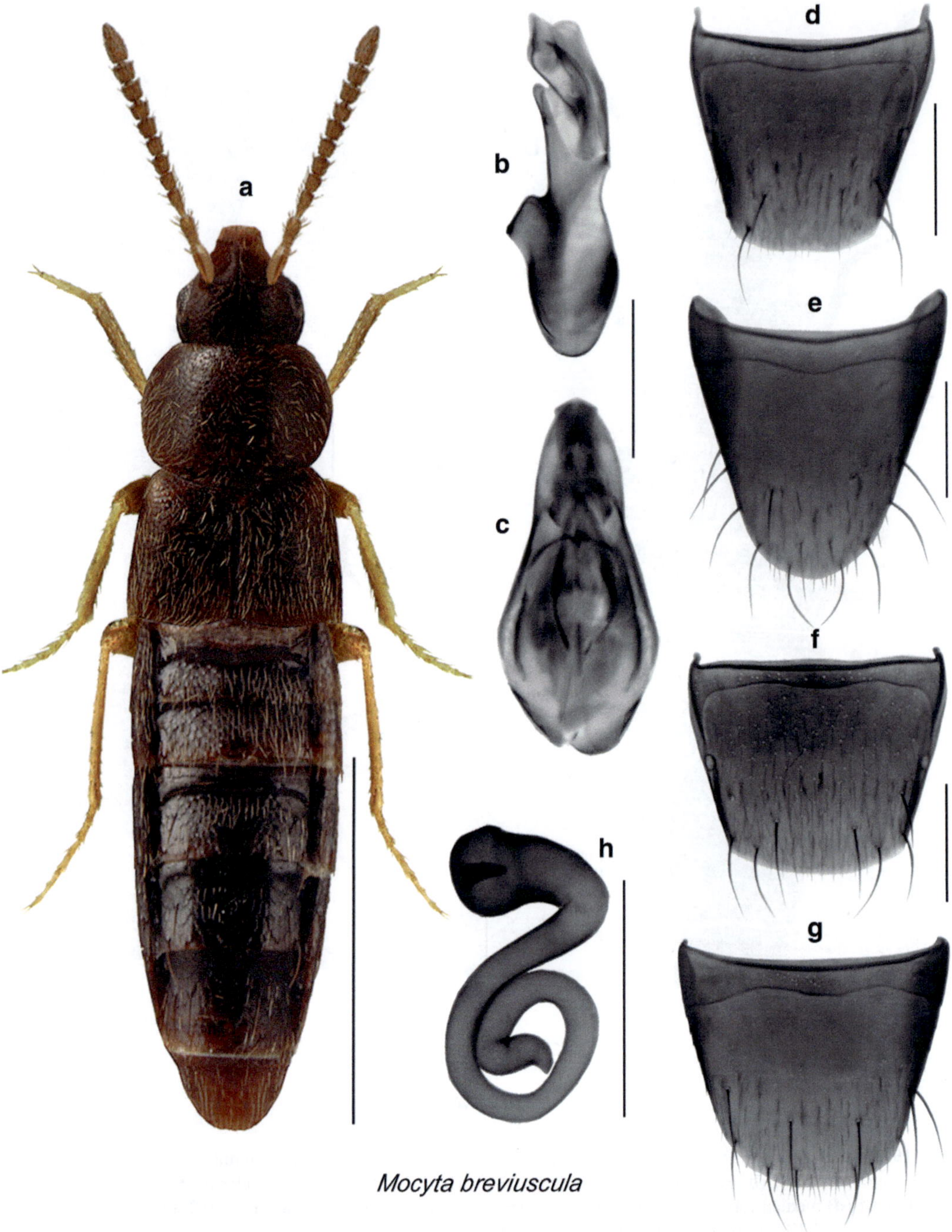

Fig. 19.285 (**a–h**). *Mocyta breviuscula* (Mäklin): (**a**) habitus in dorsal view; (**b**) median lobe of aedeagus in lateral view; (**c**) median lobe of aedeagus in dorsal view; (**d**) male tergite VIII; (**e**) male sternite VIII; (**f**) female tergite VIII; (**g**) female sternite VIII; (**h**) spermatheca. Scale bar for habitus = 1 mm; remaining scale bars = 0.2 mm

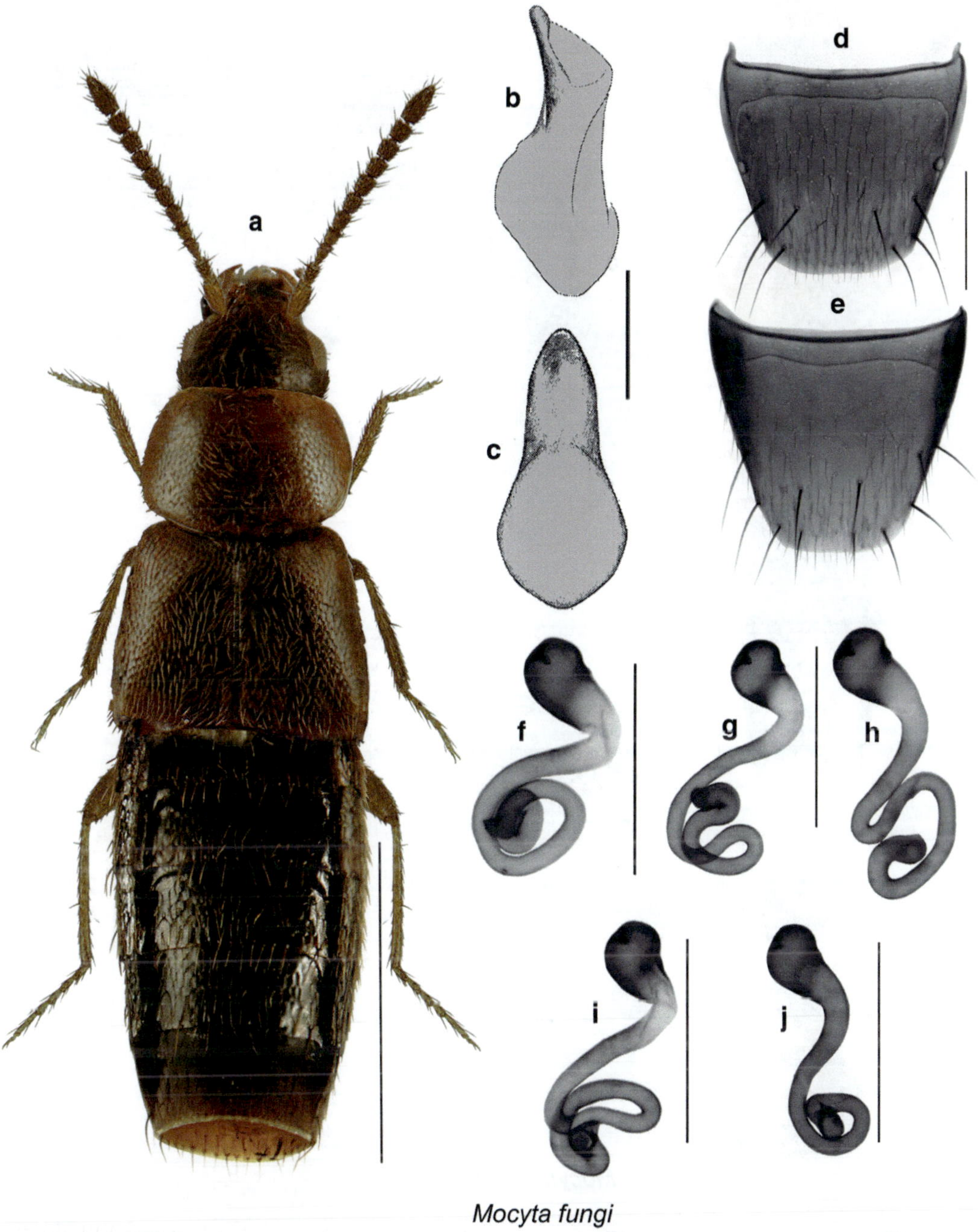

Fig. 19.286 (**a–j**). *Mocyta fungi* (Gravenhorst): (**a**) habitus in dorsal view; (**b**) median lobe of aedeagus in lateral view (based on European specimen); (**c**) median lobe of aedeagus in dorsal view (based on European specimen); (**d**) female tergite VIII; (**e**) female sternite VIII; (**f–j**) spermatheca. Scale bar for habitus = 1 mm; remaining scale bars = 0.2 mm

Fig. 19.287 (**a–h**). *Mocyta sphagnorum* Klimaszewski and Webster: (**a, b**) habitus of male (left) and female (right) in dorsal view; (**c**) median lobe of aedeagus in lateral view; (**d**) male tergite VIII; (**e**) male sternite VIII; (**f**) female tergite VIII; (**g**) female sternite VIII; (**h**) spermatheca. Scale bar for habitus = 1 mm; remaining scale bars = 0.2 mm

Fig. 19.288 (**a–g**). *Philhygra angusticauda* (Bernhauer): (**a**) habitus in dorsal view; (**b**) median lobe of aedeagus in lateral view; (**c**) male tergite VIII; (**d**) male sternite VIII; (**e**) female tergite VIII; (**f**) female sternite VIII; (**g**) pygidium. Scale bar for habitus = 1 mm; remaining scale bars = 0.2 mm

Fig. 19.289 (**a–h**). *Philhygra botanicarum* (Muona): (**a**) habitus in dorsal view; (**b**) median lobe of aedeagus in lateral view; (**c**) median lobe of aedeagus in dorsal view; (**d**) male tergite VIII; (**e**) male sternite VIII; (**f**) female tergite VIII; (**g**) female sternite VIII; (**h**) pygidium. Scale bar for habitus = 1 mm; remaining scale bars = 0.2 mm

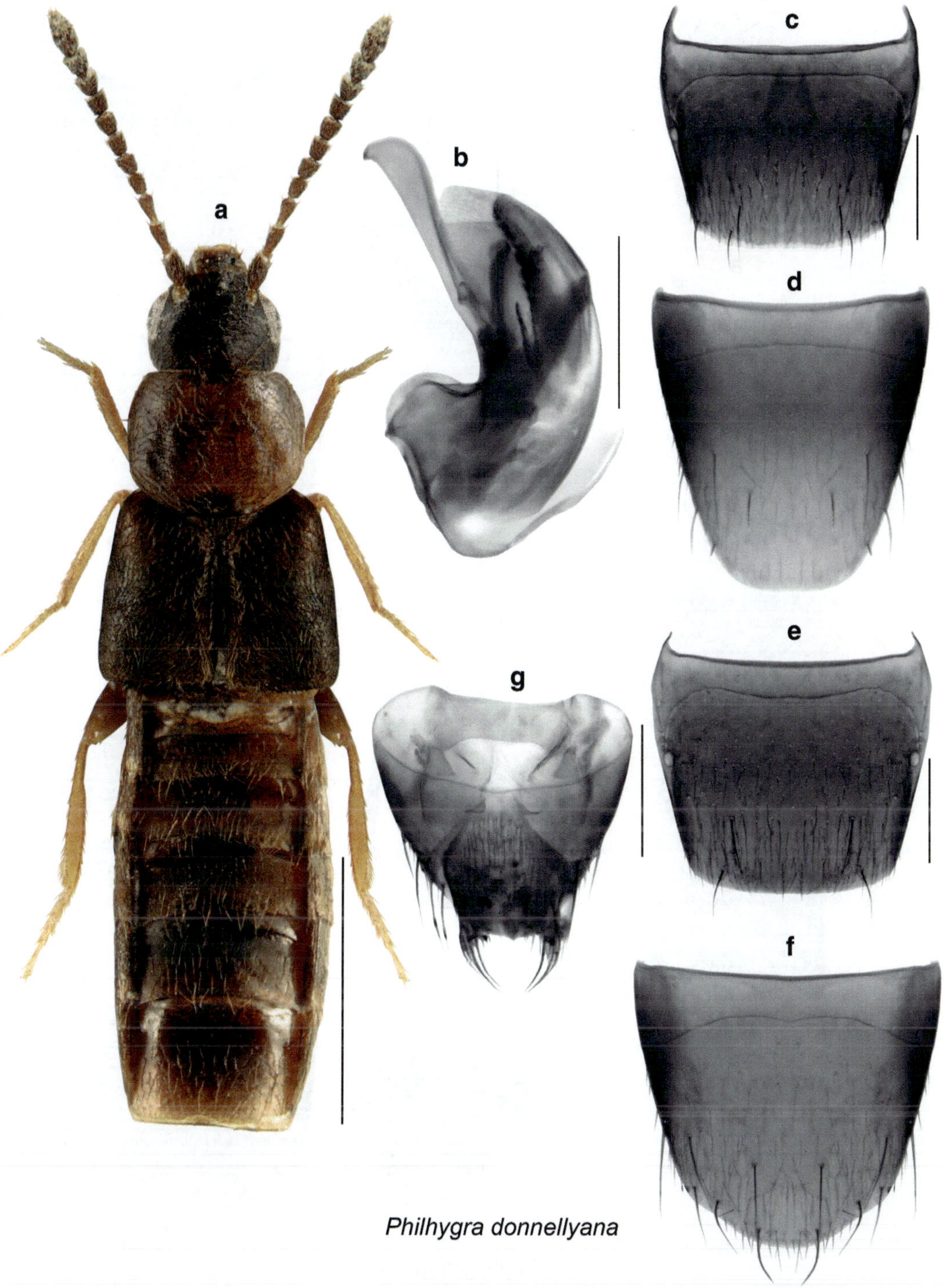

Fig. 19.290 (**a–g**). *Philhygra donnellyana* Klimaszewski, sp.n.: (**a**) habitus in dorsal view; (**b**) median lobe of aedeagus in lateral view; (**c**) male tergite VIII; (**d**) male sternite VIII; (**e**) female tergite VIII; (**f**) female sternite VIII; (**g**) pygidium. Scale bar for habitus = 1 mm; remaining scale bars = 0.2 mm

Fig. 19.291 (**a–h**). *Philhygra falcifera* Lohse: (**a**) habitus in dorsal view; (**b**) median lobe of aedeagus in lateral view; (**c**) median lobe of aedeagus in dorsal view; (**d**) male tergite VIII; (**e**) male sternite VIII; (**f**) female tergite VIII; (**g**) female sternite VIII; (**h**) pygidium. Scale bar for habitus = 1 mm; remaining scale bars = 0.2 mm

Fig. 19.292 (**a–h**). *Philhygra jarmilae* Klimaszewski and Langor: (**a**) habitus in dorsal view; (**b**) median lobe of aedeagus in lateral view; (**c**) median lobe of aedeagus in dorsal view; (**d**) male tergite VIII; (**e**) male sternite VIII; (**f**) female tergite VIII; (**g**) female sternite VIII; (**h**) pygidium. Scale bar for habitus = 1 mm; remaining scale bars = 0.2 mm

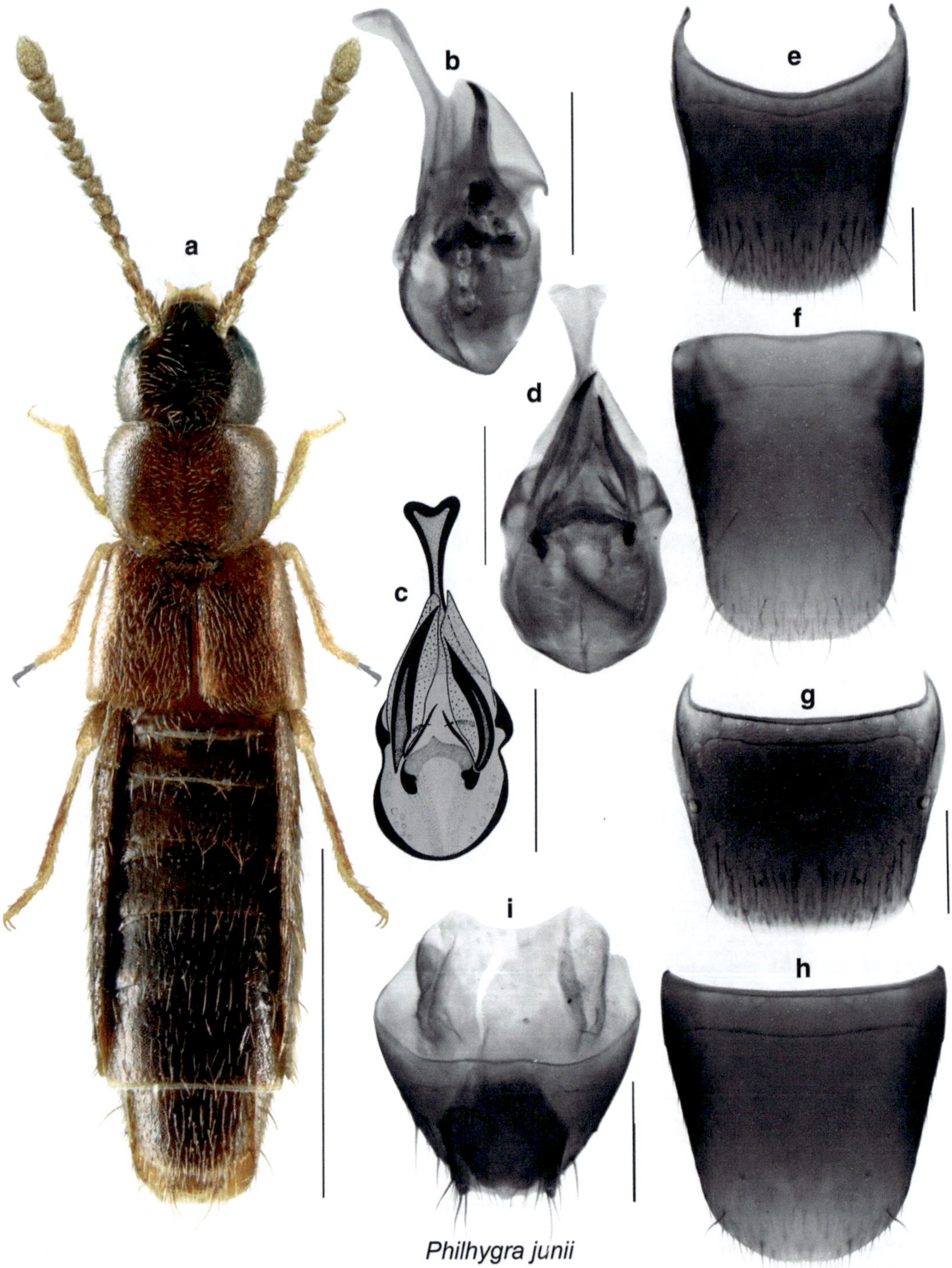

Fig. 19.293 (**a–i**). *Philhygra junii* Lohse: (**a**) habitus in dorsal view; (**b**) median lobe of aedeagus in lateral view; (**c, d**) median lobe of aedeagus in dorsal view; (**e**) male tergite VIII; (**f**) male sternite VIII; (**g**) female tergite VIII; (**h**) female sternite VIII; (**i**) pygidium. (**d**) after Lohse et al. (1990). Scale bar for habitus = 1 mm; remaining scale bars = 0.2 mm

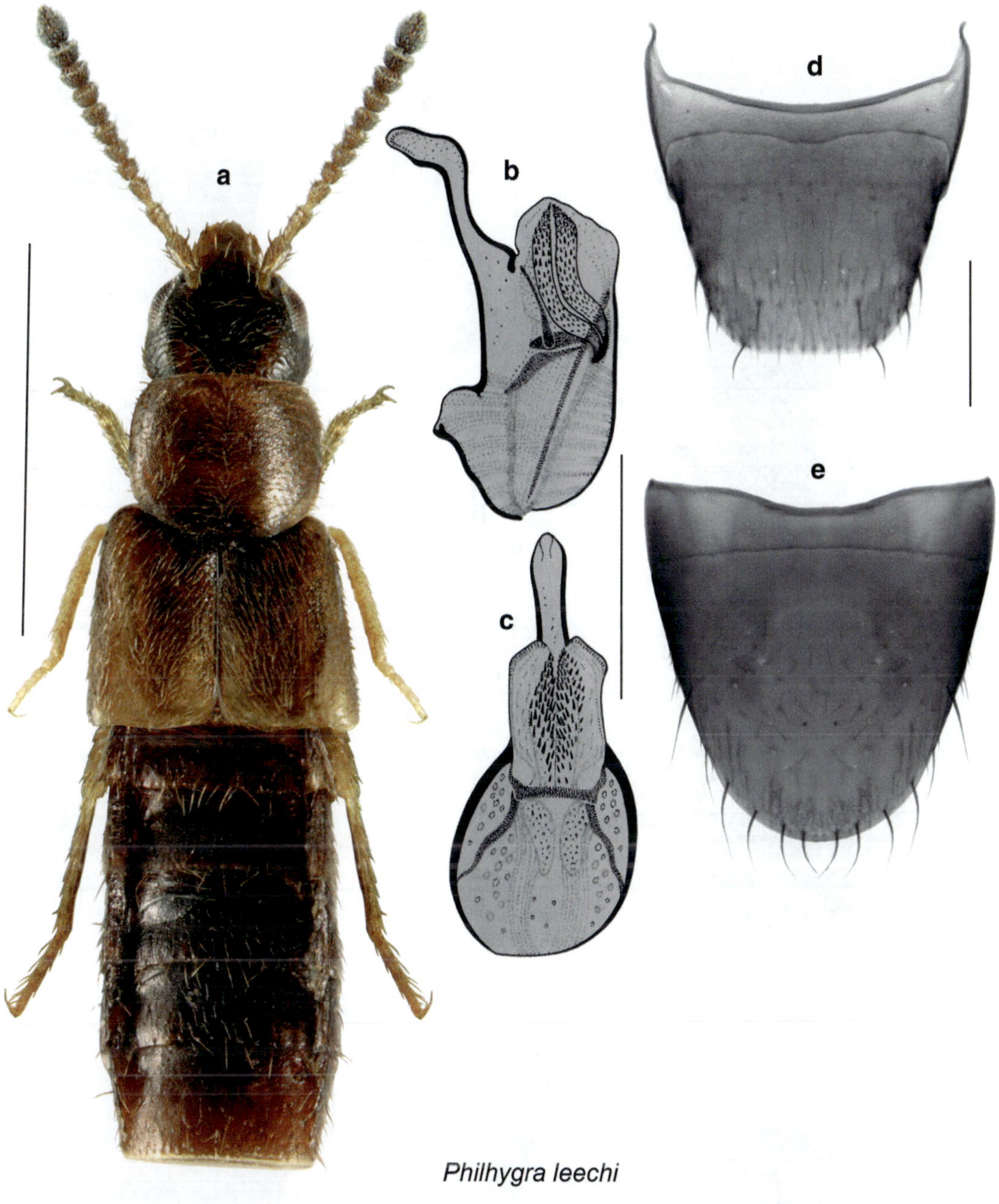

Fig. 19.294 (**a–e**). *Philhygra leechi* Lohse: (**a**) habitus in dorsal view; (**b**) median lobe of aedeagus in lateral view; (**c**) median lobe of aedeagus in dorsal view; (**d**) male tergite VIII; (**e**) male sternite VIII. (**b, c**) after Lohse et al. (1990). Scale bar for habitus = 1 mm; remaining scale bars = 0.2 mm

Fig. 19.295 (**a–h**). *Philhygra malleoides* Lohse: (**a**) habitus in dorsal view; (**b**) median lobe of aedeagus in lateral view; (**c**) median lobe of aedeagus in dorsal view; (**d**) male tergite VIII; (**e**) male sternite VIII; (**f**) female tergite VIII; (**g**) female sternite VIII; (**h**) pygidium. Scale bar for habitus = 1 mm; remaining scale bars = 0.2 mm

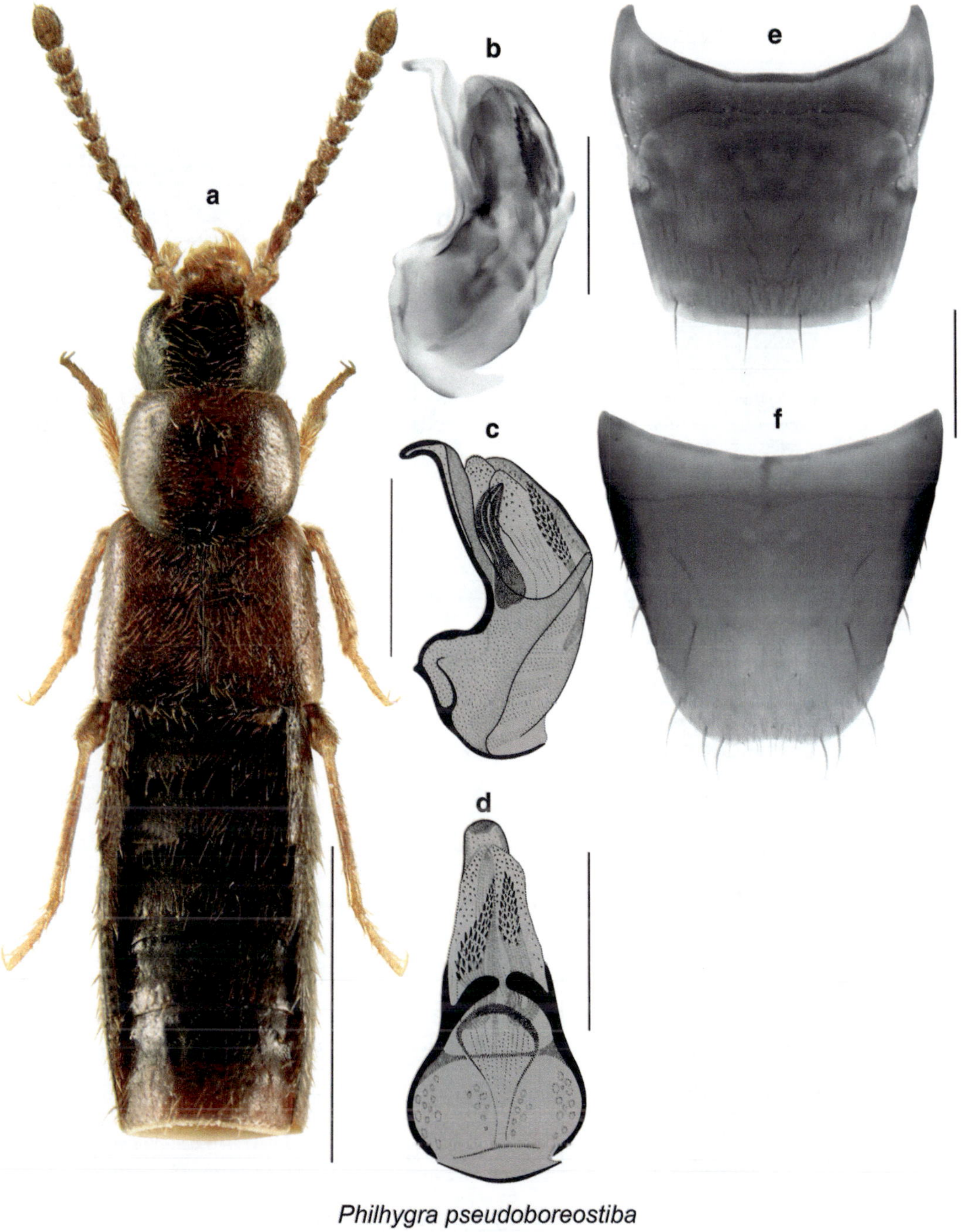

Fig. 19.296 (**a–f**). *Philhygra pseudoboreostiba* Lohse: (**a**) habitus in dorsal view; (**b,c**) median lobe of aedeagus in lateral view; (**d**) median lobe of aedeagus in dorsal view; (**e**) male tergite VIII; (**f**) male sternite VIII. (**c,d**) after Lohse et al. (1990). Scale bar for habitus = 1 mm; remaining scale bars = 0.2 mm

Fig. 19.297 (**a–g**). *Philhygra pseudolarsoni* Klimaszewski and Godin: (**a**) habitus in dorsal view; (**b**) median lobe of aedeagus in lateral view; (**c**) male tergite VIII; (**d**) male sternite VIII; (**e**) female tergite VIII; (**f**) female sternite VIII; (**g**) pygidium. Scale bar for habitus = 1 mm; remaining scale bars = 0.2 mm

Fig. 19.298 (**a–h**). *Philhygra pseudopolaris* Klimaszewski and Langor: (**a**) habitus in dorsal view; (**b**) median lobe of aedeagus in lateral view; (**c**) median lobe of aedeagus in dorsal view; (**d**) male tergite VIII; (**e**) male sternite VIII; (**f**) female tergite VIII; (**g**) female sternite VIII; (**h**) pygidium. Scale bar for habitus = 1 mm; remaining scale bars = 0.2 mm

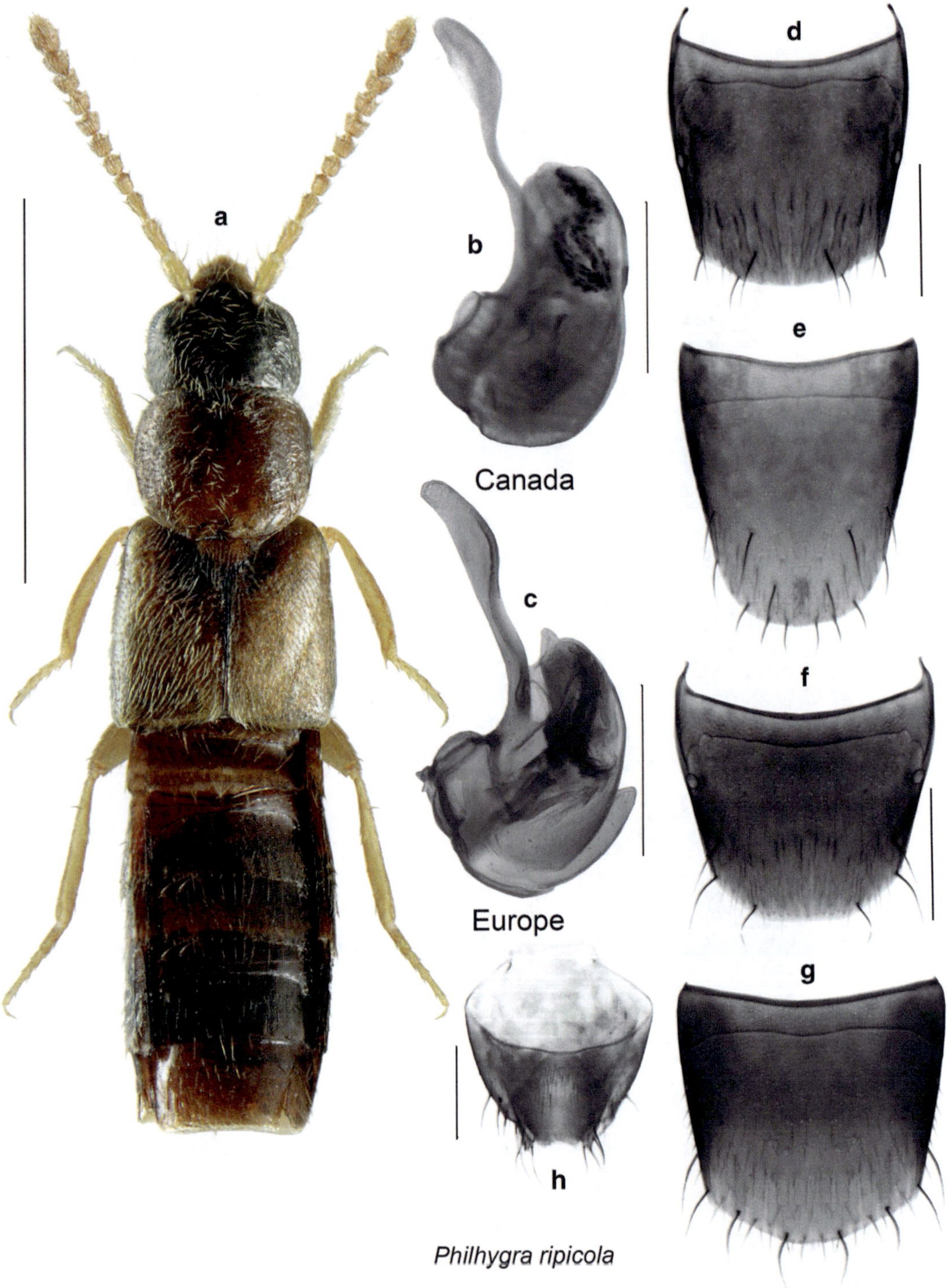

Fig. 19.299 (**a–h**). *Philhygra ripicola* (Hanssen), (former Nearctic *P. ripicoloides* Lohse): (**a**) habitus in dorsal view; (**b, c**) median lobe of aedeagus in lateral view (NT, Canada (**b**); Europe (**c**)); (**d**) male tergite VIII; (**e**) male sternite VIII; (**f**) female tergite VIII; (**g**) female sternite VIII; (**h**) pygidium. Scale bar for habitus = 1 mm; remaining scale bars = 0.2 mm

Fig. 19.300 (**a–h**). *Philhygra rostrifera* Lohse: (**a**) habitus in dorsal view; (**b**) median lobe of aedeagus in lateral view; (**c**) median lobe of aedeagus in dorsal view; (**d**) male tergite VIII; (**e**) male sternite VIII; (**f**) female tergite VIII; (**g**) female sternite VIII; (**h**) pygidium. Scale bar for habitus = 1 mm; remaining scale bars = 0.2 mm

Fig. 19.301 (**a–h**). *Philhygra sinuipennis* Klimaszewski and Langor: (**a**) habitus in dorsal view; (**b**) median lobe of aedeagus in lateral view; (**c**) median lobe of aedeagus in dorsal view; (**d**) male tergite VIII; (**e**) male sternite VIII; (**f**) female tergite VIII; (**g**) female sternite VIII; (**h**) pygidium. Scale bar for habitus = 1 mm; remaining scale bars = 0.2 mm

Fig. 19.302 (**a–g**). *Philhygra terrestris* Klimaszewski and Godin: (**a**) habitus in dorsal view; (**b**) median lobe of aedeagus in lateral view; (**c**) male tergite VIII; (**d**) male sternite VIII; (**e**) female tergite VIII; (**f**) female sternite VIII; (**g**) pygidium. Scale bar for habitus = 1 mm; remaining scale bars = 0.2 mm

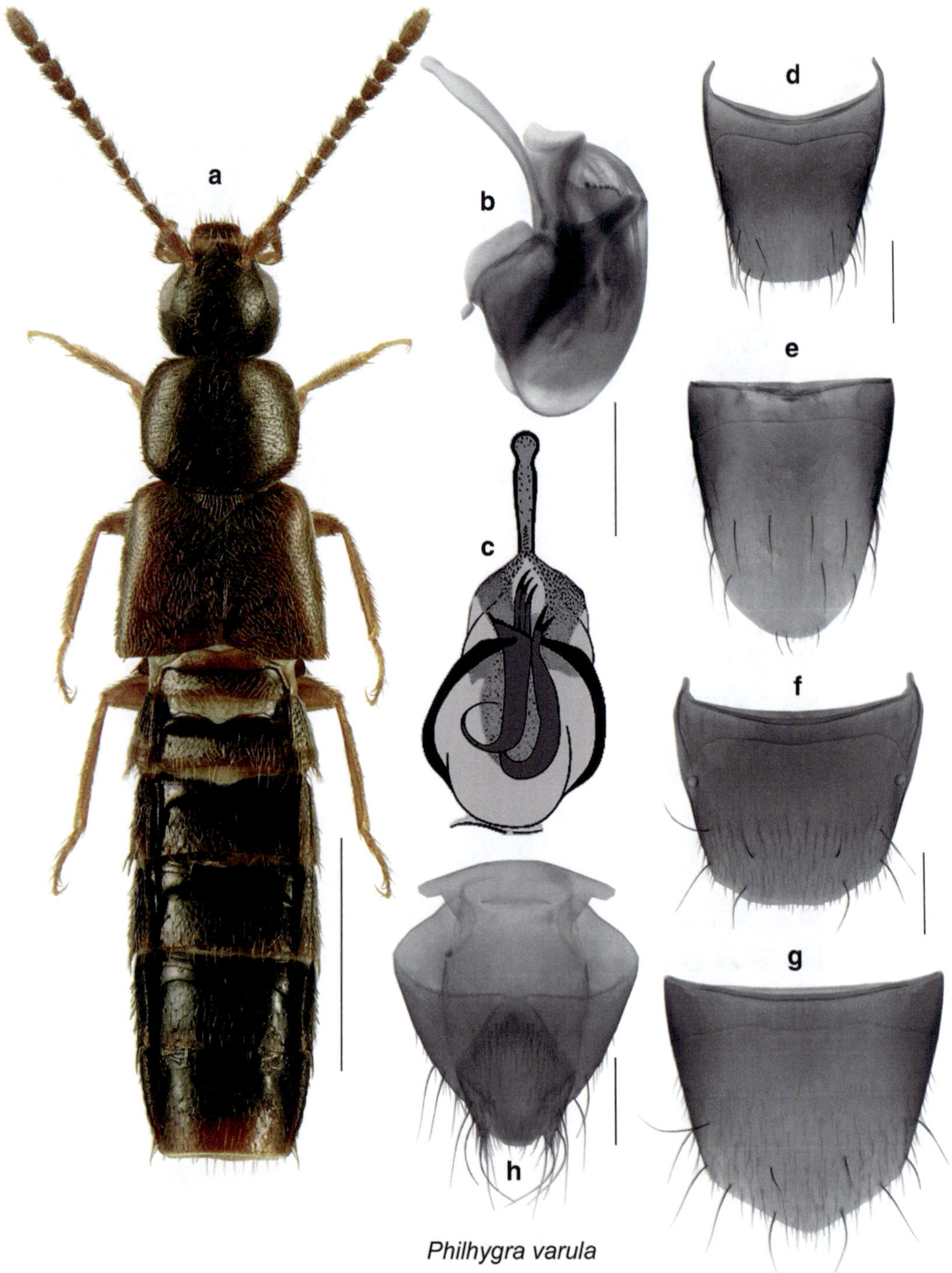

Fig. 19.303 (**a–h**). *Philhygra varula* (Casey): (**a**) habitus in dorsal view; (**b**) median lobe of aedeagus in lateral view; (**c**) median lobe of aedeagus in dorsal view; (**d**) male tergite VIII; (**e**) male sternite VIII; (**f**) female tergite VIII; (**g**) female sternite VIII; (**h**) pygidium. Scale bar for habitus = 1 mm; remaining scale bars = 0.2 mm

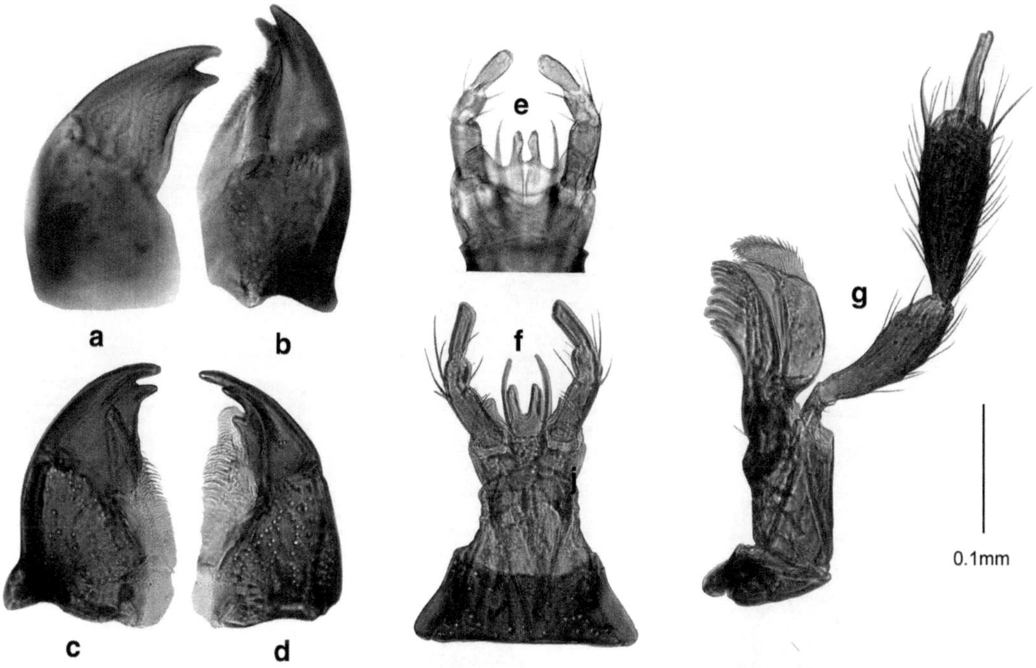

Fig. 19.304 (**a–g**). Mouthparts of *Schistoglossa* species: (**a–d**) mandibles; (**e, f**) labial palps and glossae; (**g**) maxillary palpus. Scale bars = 0.1 mm

Fig. 19.305 (**a–i**). *Schistoglossa blatchleyi* (Bernhauer and Scheerpeltz): (**a**) habitus in dorsal view; (**b, c**) median lobe of aedeagus in lateral view; (**d**) median lobe of aedeagus in dorsal view; (**e**) male tergite VIII; (**f**) male sternite VIII; (**g**) female tergite VIII; (**h**) female sternite VIII; (**i**) spermatheca. Scale bar for habitus = 1 mm; remaining scale bars = 0.2 mm

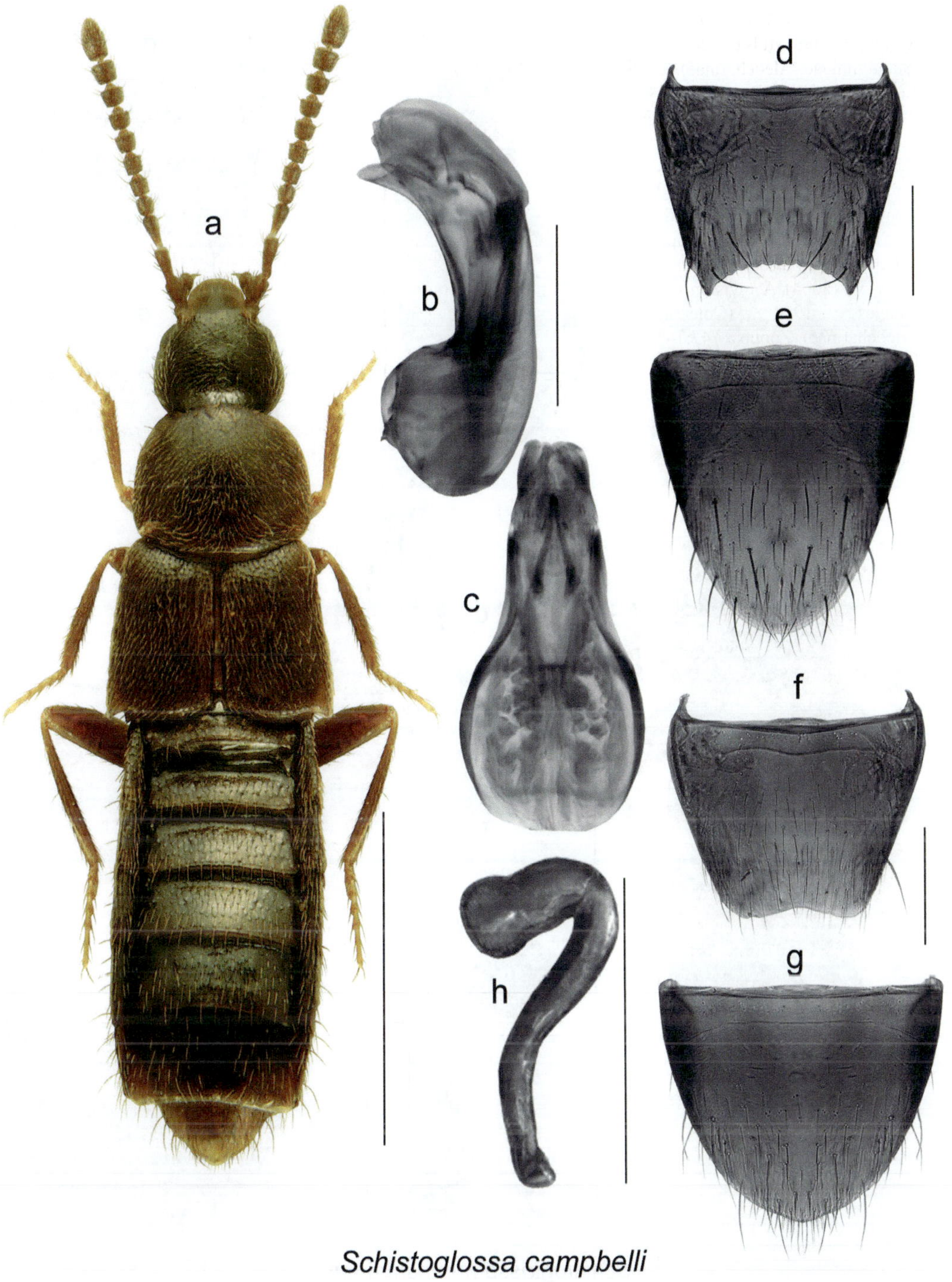

Fig. 19.306 (**a–h**). *Schistoglossa campbelli* Klimaszewski: (**a**) habitus in dorsal view; (**b**) median lobe of aedeagus in lateral view; (**c**) median lobe of aedeagus in dorsal view; (**d**) male tergite VIII; (**e**) male sternite VIII; (**f**) female tergite VIII; (**g**) female sternite VIII; (**h**) spermatheca. Scale bar for habitus = 1 mm; remaining scale bars = 0.2 mm

Fig. 19.307 (**a–h**). *Schistoglossa hampshirensis* Klimaszewski: (**a**) habitus in dorsal view; (**b**) median lobe of aedeagus in lateral view; (**c**) median lobe of aedeagus in dorsal view; (**d**) male tergite VIII; (**e**) male sternite VIII; (**f**) female tergite VIII; (**g**) female sternite VIII; (**h**) spermatheca. Scale bar for habitus = 1 mm; remaining scale bars = 0.2 mm

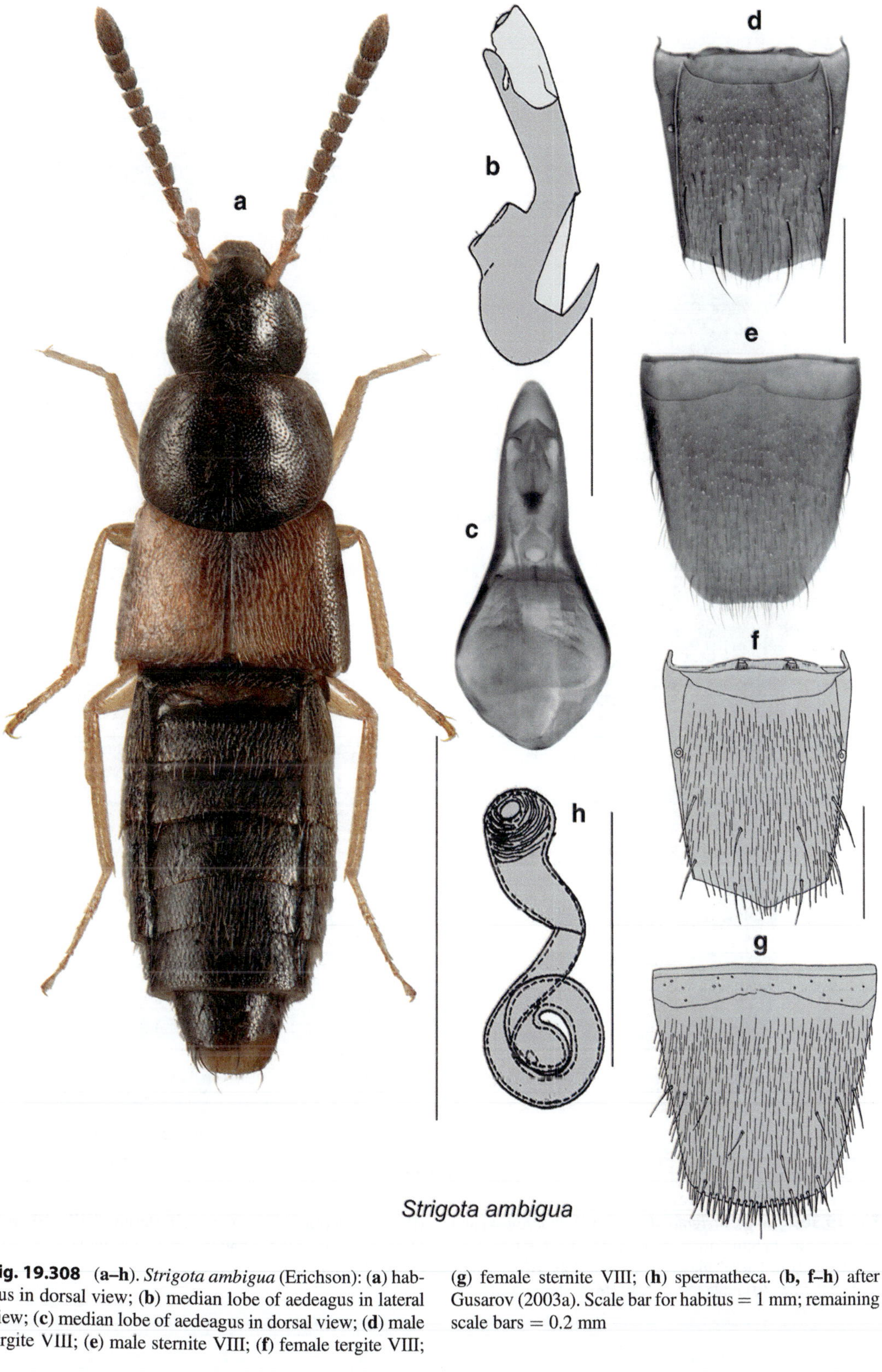

Fig. 19.308 (**a–h**). *Strigota ambigua* (Erichson): (**a**) habitus in dorsal view; (**b**) median lobe of aedeagus in lateral view; (**c**) median lobe of aedeagus in dorsal view; (**d**) male tergite VIII; (**e**) male sternite VIII; (**f**) female tergite VIII; (**g**) female sternite VIII; (**h**) spermatheca. (**b, f–h**) after Gusarov (2003a). Scale bar for habitus = 1 mm; remaining scale bars = 0.2 mm

Fig. 19.309 (**a–h**). *Strigota obscurata* Klimaszewski and Brunke: (**a**) habitus in dorsal view; (**b**) median lobe of aedeagus in lateral view; (**c**) median lobe of aedeagus in dorsal view; (**d**) male tergite VIII; (**e**) male sternite VIII; (**f**) female tergite VIII; (**g**) female sternite VIII; (**h**) spermatheca. Scale bar for habitus = 1 mm; remaining scale bars = 0.2 mm

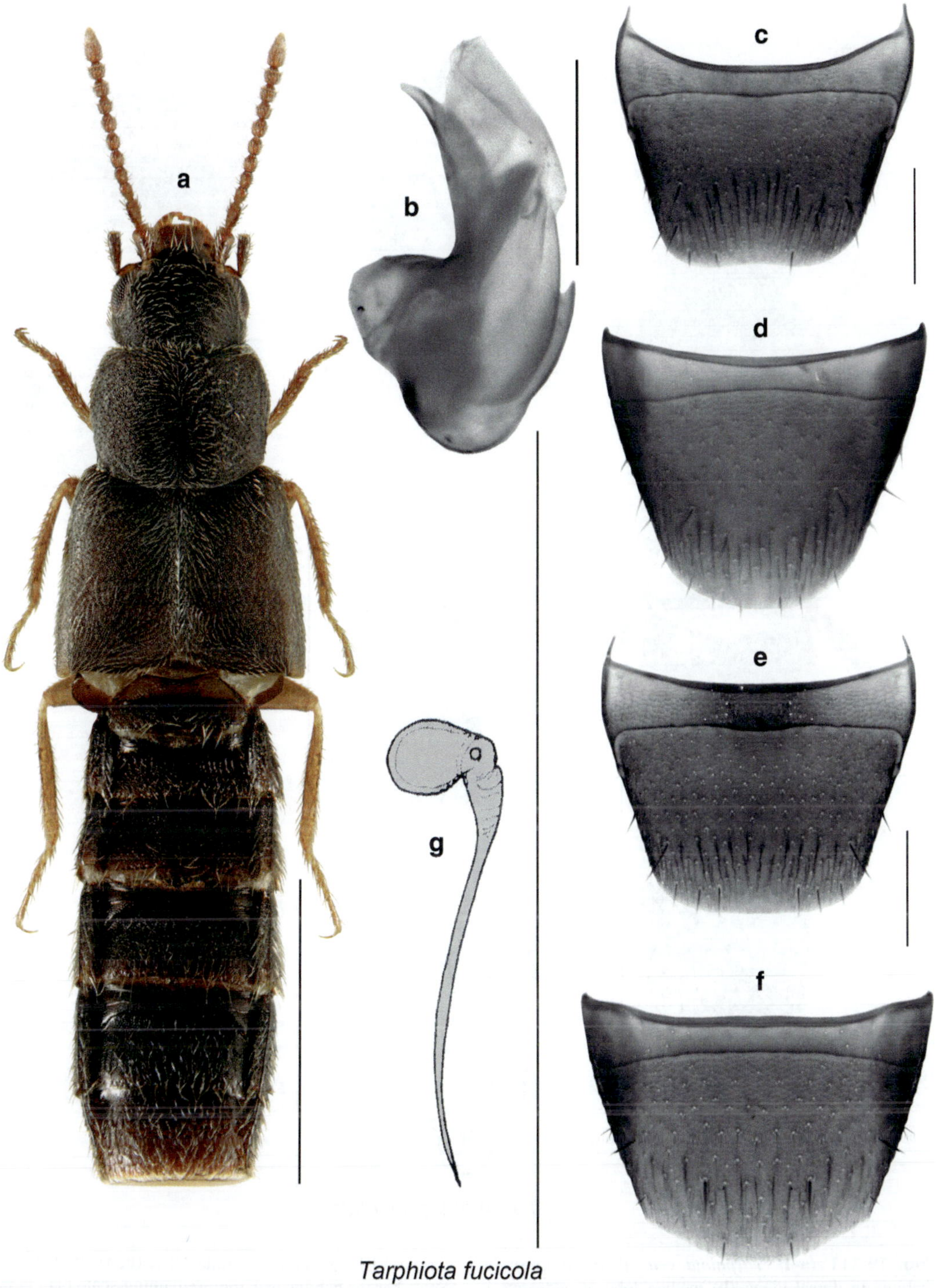

Fig. 19.310 (**a–g**). *Tarphiota fucicola* (Mäklin): (**a**) habitus in dorsal view; (**b**) median lobe of aedeagus in lateral view; (**c**) male tergite VIII; (**d**) male sternite VIII; (**e**) female tergite VIII; (**f**) female sternite VIII; (**g**) spermatheca. Scale bar for habitus = 1 mm; remaining scale bars = 0.2 mm

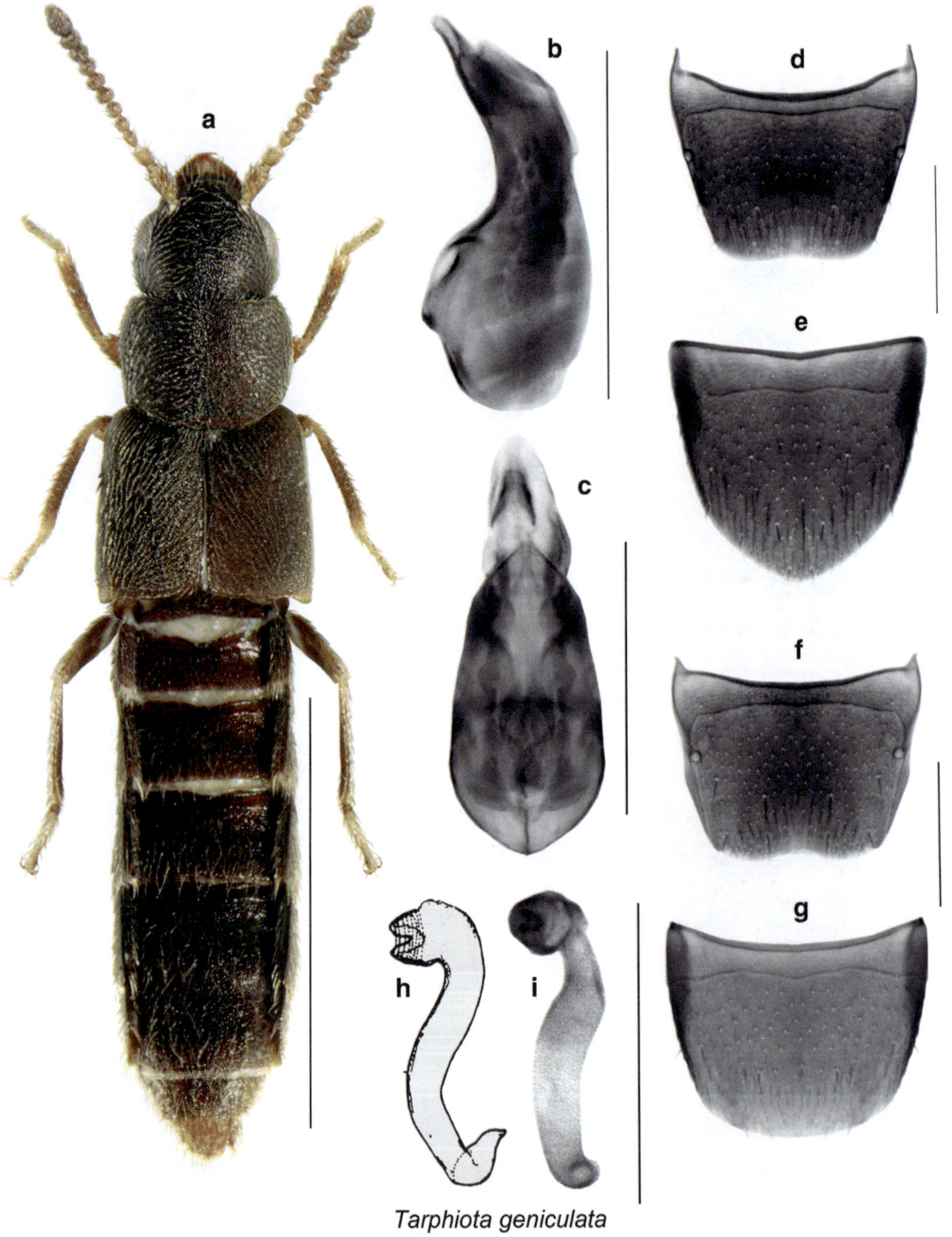

Fig. 19.311 (**a–i**). *Tarphiota geniculata* (Mäklin); (**a**) habitus in dorsal view; (**b**) median lobe of aedeagus in lateral view; (**c**) median lobe of aedeagus in dorsal view; (**d**) male tergite VIII; (**e**) male sternite VIII; (**f**) female tergite VIII; (**g**) female sternite VIII; (**h, i**) spermatheca. Scale bar for habitus = 1 mm; remaining scale bars = 0.2 mm

Fig. 19.312 (**a**–**h**). *Trichiusa pilosa* Casey: (**a**) habitus in dorsal view; (**b**) median lobe of aedeagus in lateral view; (**c**) median lobe of aedeagus in dorsal view; (**d**) male tergite VIII; (**e**) male sternite VIII; (**f**) female tergite VIII; (**g**) female sternite VIII; (**h**) spermatheca. Scale bar for habitus = 1 mm; remaining scale bars = 0.2 mm

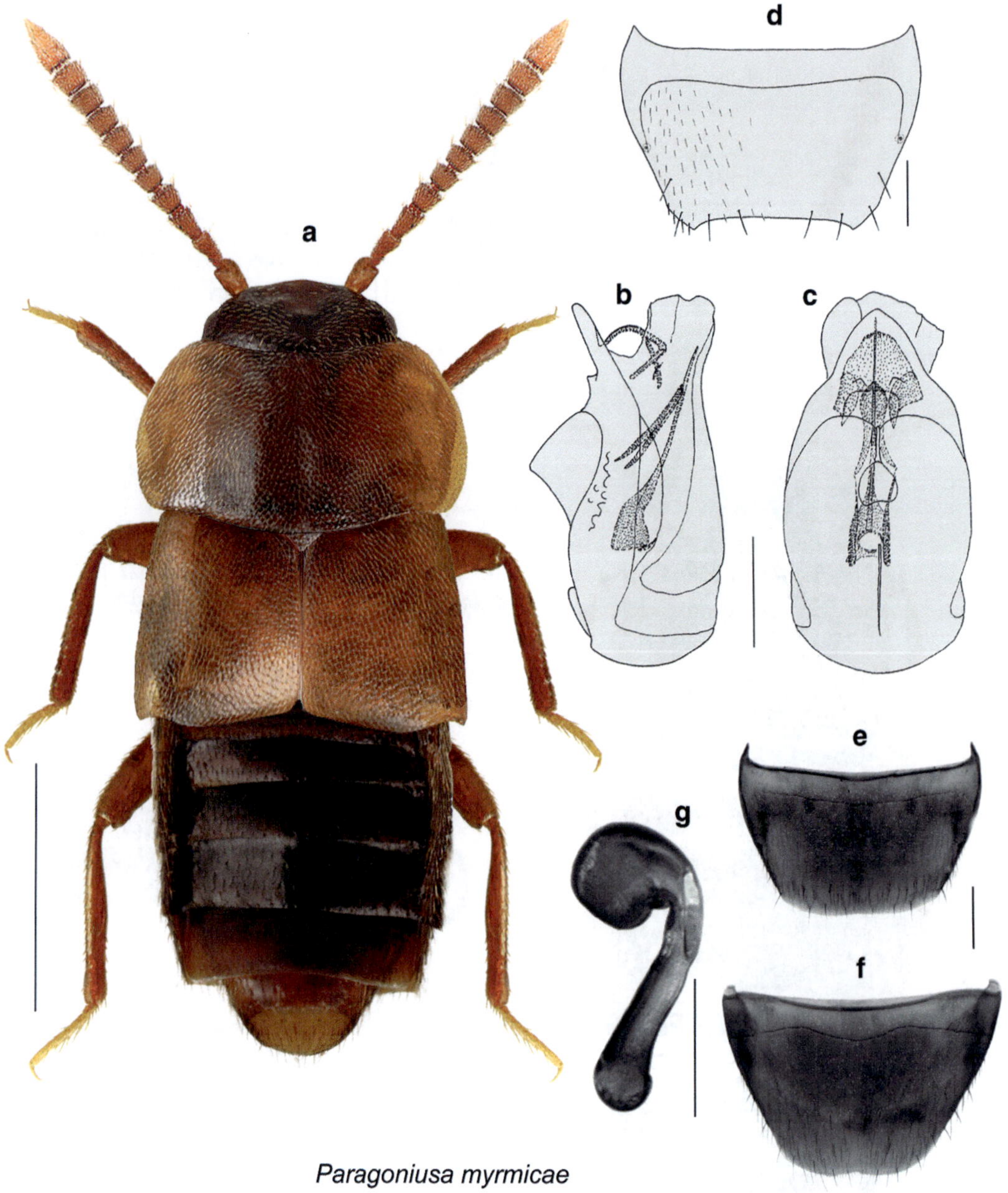

Fig. 19.313 (**a–g**). *Paragoniusa myrmicae* Maruyama and Klimaszewski: (**a**) habitus in dorsal view; (**b**) median lobe of aedeagus in lateral view; (**c**) median lobe of aedeagus in dorsal view; (**d**) male tergite VIII; (**e**) female tergite VIII; (**f**) female sternite VIII; (**g**) spermatheca. (**b, c, d**) after Maruyama and Klimaszewski (2006). Scale bar for habitus = 1 mm; remaining scale bars = 0.2 mm

References

Ahn K-J (1996) Revision of the intertidal aleocharine genus *Tarphiota* (Coleoptera: Staphylinidae). Entomol News 107:177–185

Ashe JS (2001) Key to the tribes and genera of Nearctic Aleocharinae. 8. In: Arnett RH Jr, Thomas CM (eds) American Beetles. Archostemata, Myxophaga, Adephaga, Polyphaga: Staphyliniformia. CRC Press, Washington, DC, pp 299–374. 443 pp.

Ashe JS, Gusarov VI (2004) Review of the genus Sableta Casey, 1910 (Coleoptera: Staphylinidae: Aleocharinae: Athetini). Zootaxa 745:1–15

Benick G (1973) Die von Prof. Dr. Z. Kaszab in der Mongolei gesammelten Arten der Subfamilie Athetae (Col. Staphyl.). Nouvelle Revue d'Entomologie 3 (4):211–217

Benick G, Lohse GA (1974) Tribus 14 (Callicerini). In: Freude H, Harde KW, Lohse A (eds) Die Käfer Mitteleuropas. Band 5: Staphylinidae II (Hypocyphtinae und Aleocharinae) Pselaphidae. Goecke & Evers, Krefeld, pp 72–221. 381 pp

Bernhauer M (1901) Neue Staphyliniden aus Centralasien. Verhandlungen der Kaiserlich-Königlichen Zoologisch-Botanischen Gesellschaft in Wien 51 (2):106–115

Bernhauer M (1902) Beitrag zur Staphylinidenfauna des palaearctischen Gebietes. Münchener Koleopterologische Zeitschrift 1[1902–1903](1): 54–62

Bernhauer M (1906) Neue Aleocharinen aus Nordamerika. (II. Teil). Deutsche Entomologische Zeitschrift 1906 (2):337–348

Bernhauer M (1907) Neue Aleocharini aus Nordamerika. (Col.) (3. Stück.). Deutsche Entomologische Zeitschrift 1907(4):381–405

Bernhauer M (1909) Neue Aleocharini aus Nordamerika. (Col.) (4. Stück.). Deutsche Entomologische Zeitschrift 1909(4):515–528

Bernhauer M, Scheerpeltz O (1926) Staphylinidae VI (Pars 82). In: Junk W, Schenkling S (eds) Coleopterorum catalogus. Volumen V. Staphylinidae. W. Junk, Berlin, pp 499–988. 988 pp

Bishop DJ, Majka CG, Bondrup-Nielsen S, Peck SB (2009) Deadwood and saproxylic beetle diversity in naturally disturbed and managed spruce forests in Nova Scotia. In: Majka CG, Klimaszewski J (Eds) Biodiversity, biosystematics, and ecology of Canadian Coleoptera II. ZooKeys [Special Issue] 22:309–340

Blatchley WS (1910) An illustrated descriptive catalog of the Coleoptera or beetles (exclusive of the Rhynchophora) known to occur in Indiana. The Nature Publishing Co., Indianapolis, IN, p 1386. https://doi.org/10.5962/bhl.title.56580

Brundin L (1940) Zwei neue nordische Atheta-Arten aus der islandica-Gruppe (Col., Staphylinidae). Vorlaufige Mitteilung. Entomologisk Tidskrift 61:131–133

Brundin L (1944) Monographie der palaearktischen Arten der *Atheta* – Untergattung Hygroecia (Coleoptera, Staphylinidae). Ann Naturhist Mus Wien 53 [1942]:129–301. pl. 13–40

Brundin L (1954) Die palaearktischen Arten der *Atheta*-Untergattung *Dimetrota* Muls. et Rey (Col., Staphylinidae). Eine systematische Studie. Arkiv för Zoologi (Andra Serie) 5 [1953–54](7):369–434

Brundin L (1948) *Microdota*-Studien. (Col. Staphylinidae). Entomologisk Tidskrift 69(1-2):8–66

Brundin L (1952) *Acrotona*-Studien (Gattung *Atheta*). Entomologisk Tidskrift 73:93–145

Brundin L (1953) Die palaearktischen Arten der *Atheta*-Untrgattung *Dimetrota* Mulsant and Rey. Ark Zool 5:369–434

Brunke A, Klimaszewski J, Dorval J-A, Bourdon C, Paiero SM, Marshall SA (2012) New species and distributional records of Aleocharinae (Coleoptera, Staphylinidae) from Ontario, Canada, with a checklist of recorded species. In: Klimaszewski J, Anderson R (Eds) Biosystematics and ecology of Canadian Staphylinidae (Coleoptera) II. ZooKeys [Special Issue] 186:119–206

Brunke AJ, Pentinsaari M, Klimaszewski J (2021) Integrative taxonomy of Nearctic and Palaearctic Aleocharinae: new species, synonymies, and records (Coleoptera, Staphylinidae). ZooKeys 1041:27–99

Cameron M (1920) New species of Staphylinidae from Singapore, part III. Trans Entomol Soc Lond 1920:212–284

Cameron M (1939) Coleoptera, Staphylinidae. Vol. IV. Parts I, and II. In: Sewell RBS (ed) Fauna of British India, including Ceylon and Burma. Taylor and Francis, London. xviii, 691 p., pls. 1–3[in 2 vols.]

Casey TL (1885) New genera and species of Californian Coleoptera. Bull Calif Acad Sci 1(4):283–336

Casey TL (1893) Coleopterological notices. V. Ann N Y Acad Sci 7(1894):281–606. [see reference in Chapter 5 for explanation of date]

Casey TL (1906) Observations on the staphylinid groups Aleocharinae and Xantholinini, chiefly of America. Trans Acad Sci St Louis 16(6):125–434

Casey TL (1910) New species of the staphylinid tribe Myrmedoniini. In: Memoirs on the Coleoptera I. New Era Printing Co., Lancaster, PA, pp 1–183. 205 pp

Casey TL (1911) New American species of Aleocharinae and Myllaeninae. In: Memoirs on the Coleoptera. II. The New Era Printing Co., Lancaster, PA, pp 1–245. 259 pp

Clark ME (1949) An annotated list of the Coleoptera taken at or near Terrace, British Columbia. Part 2. Proc Entomol Soc B C 45([1948]):21–24

Dollin PE, Majka CG, Duinker PN (2008) Saproxylic beetle (Coleoptera) communities and forest management practices in coniferous stands in southwestern Nova Scotia, Canada. *In*: Majka CG, Klimaszewski J (Eds.) Biodiversity, biosystematics, and ecology of Canadian Coleoptera II. ZooKeys [Special Issue] 2:291–336

Elven H, Bachmann L, Gusarov VI (2010) Phylogeny of the tribe Athetini (Coleoptera: Staphylinidae) inferred from mitochondrial and nuclear sequence data. Mol Phylogenet Evol 57(1):84–100

Elven H, Bachmann L, Gusarov VI (2012) Molecular phylogeny of the Athetini-Lomechusini-Ecitocharini clade of aleocharine rove beetles (Insecta). Zool Scr 41(6):617–636

Eppelsheim E (1886) Neue Staphylinen vom Amur. Deutsche Entomologische Zeitschrift 30(1):33–46

Eppelsheim E (1893) Beitrag zur Staphylinen-Fauna des sűdwestlichen Baikal-Gebietes. Deutsche Entomologische Zeitschrift 37:17–67

Erichson WF (1839) Genera et species staphylinorum insectorum coleopterorum familiae. Erst Band. FH Morin, Berlin. viii, 400 pp

Ernst CM, Buddle CM (2013) Seasonal patterns in the structure of epigeic beetle (Coleoptera) assemblages in two subarctic habitats in Nunavut, Canada. Can Entomol 145(2):171–183

Fauvel A (1889) Liste des coléoptères communs à l'Europe et a l'Amérique du Nord. D'après le catalogue de M. J. Hamilton. Avec remarques et additions. Revue d'Entomologie 8(3):92–174

Fenyes A (1918) Coleoptera. Family Staphylinidae. Subfamily Aleocharinae. In: Wytsman P (ed) Genera Insectorum, vol XXVII. Fascicule 173A. M. Nijhoff, Den Haag, pp 1–110. 453 pp. + 7 pls

Fenyes A (1920) Coleoptera. Fam. Staphylinidae. Subfam. Aleocharinae. In: Wytsman P (ed) Genera Insectorum, vol XXVII. Fascicule 173B. L. Desmet-Verteneuil, Bruxelles, pp 111–414. 453 pp. + 7 pls

Gouix N, Klimaszewski J (2007) Catalogue of aleocharine rove beetles of Canada and Alaska (Coleoptera, Staphylinidae, Aleocharinae). Pensoft Publishers, Sofia-Moscow. 165 pp

Gravenhorst JLC (1806) Monographia Coleopterorum Micropterorum. Henricus Dieterich, Göttingen. xvi + 236 + [12] pp

Gusarov VI (2002) A revision of Nearctic species of the genus *Earota* Mulsant & Rey, 1874 (Coleoptera: Staphylinidae: Aleocharinae). Zootaxa 92:1–16

Gusarov VI (2003a) Revision of some types of North American aleocharines (Coleoptera: Staphylinidae: Aleocharinae), with synonymic notes. Zootaxa 353:1–134

Gusarov VI (2003b) A revision of Nearctic species of the genera *Adota* Casey, 1910 and *Psammostiba* Yosii & Sawada, 1976 (Coleoptera: Staphylinidae: Aleocharinae). Zootaxa 185:1–35

Gusarov VI (2004) A revision of the genus *Lypoglossa* Fenyes, 1918 (Coleoptera: Staphylinidae: Aleocharinae). Zootaxa 747:1–36

Gyllenhal L (1810) Insecta Svecica. Classis I: Coleoptera sive Eleuterata. Tomi I pars II. F.J. Leverentz, Scaris. xix + [1] + 660 pp

Hanssen HK (1932) Zwei neue Atheten (aus den nördlichen Norwegen). Norsk Entomologisk Tidsskrift 3(1932–1934):72–78

Hatch MS (1957) The Beetles of the Pacific Northwest. Part II: Staphyliniformia. Univ Washing Public Biol 16. IX + 384 pp

Hoebeke ER (1992) Taxonomy and distribution of the athetine genus *Lypoglossa* Fenyes (Coleoptera: Staphylinidae: Aleocharinae) in North America, with description of a new species. J N Y Entomol Soc 100 (2):381–398

Klimaszewski J, Peck SB (1986) A review of the cavernicolous Staphylinidae (Coleoptera) of eastern North America: Part I. Aleocharinae Quaestiones Entomologicae 22(2):51–113

Klimaszewski J, Winchester NN (2002) Aleocharine rove beetles (Coleoptera Staphylinidae) of the ancient Sitka spruce forest on Vancouver Island, British Columbia, Canada. Mémoires de la Société royale belge d'Entomologie 40:3–126

Klimaszewski J, Sweeney J, Price J, Pelletier G (2005) Rove beetles (Coleoptera: Staphylinidae) in red spruce stands, eastern Canada: diversity, abundance, and descriptions of new species. Can Entomol 137(1):1–48

Klimaszewski J, Majka CG, Langor D (2006) Review of the North American *Tarphiota* Casey, with a description of a new seashore-inhabiting *Atheta* species exhibiting convergent characteristics (Coleoptera: Staphylinidae: Aleocharinae). Entomol Sci 9:67–78

Klimaszewski J, Assing V, Majka CG, Pelletier G, Webster RP, Langor D (2007a) Records of adventive aleocharine beetles (Coleoptera: Staphylinidae: Aleocharinae) found in Canada. Can Entomol 139 (1):54–79

Klimaszewski J, Langor D, Savard K, Pelletier G, Chandler DS, Sweeney J (2007b) Rove beetles (Coleoptera: Staphylinidae) in yellow birch-dominated stands of southeastern Quebec, Canada: diversity, abundance, and description of a new species. Can Entomol 139 (6):793–833

Klimaszewski J, Godin B, Pelletier G, Savard K (2008a) Six new species and records of aleocharine beetles from the Yukon and Alaska (Coleoptera: Staphylinidae: Aleocharinae). Can Entomol 140 (3):265–291

Klimaszewski J, Langor DW, Work TT, Hammond JHE, Savard K (2008b) Smaller and more numerous harvesting gaps emulate natural forest disturbances: a biodiversity test case using rove beetles (Coleoptera, Staphylinidae). Divers Distrib 14(6):969–982

Klimaszewski J, Webster RP, Savard K (2009) First record of the genus *Schistoglossa* Kraatz from Canada with descriptions of seven new species (Coleoptera, Staphylinidae, Aleocharinae). In: Majka CG and Klimaszewski J (Eds.) Biodiversity, biosystematics, and ecology of Canadian Coleoptera II. ZooKeys [Special Issue] 22:45–79

Klimaszewski J, Langor D, Pelletier G, Bourdon C, Perdereau L (2011) Aleocharine beetles (Coleoptera, Staphylinidae) of the province of Newfoundland and Labrador, Canada. Pensoft Series Faunistica No. 98. Pensoft, Sophia-Moscow. 313 pp

Klimaszewski J, Godin B, Bourdon C (2012) Further contributions to the aleocharine fauna of the Yukon Territory, Canada (Coleoptera, Staphylinidae). In: Klimaszewski J, Anderson R (Eds.): Biosystematics and ecology of Canadian Staphylinidae (Coleoptera) II. Zookeys [Special Issue] 186:207–237

Klimaszewski J, Brunke A, Assing V, Langor DW, Newton AF, Bourdon C, Pelletier G, Webster RP, Herman L, Perdereau L, Davies A, Smetana A, Chandler DS, Majka C, Scudder GGE (2013a) Synopsis of adventive species of Coleoptera (Insecta) recorded from Canada. Part 2 : Staphylinidae. Pensoft Series Faunistica No. 104. Pensoft, Sofia-Moscow. 360 pp

Klimaszewski J, Webster RP, Langor DW, Bourdon C, Jacobs J (2013b) Review of Canadian species of the genus *Dinaraea* Thomson, with descriptions of six new species (Coleoptera, Staphylinidae, Aleocharinae, Athetini). ZooKeys 327:65–101

Klimaszewski J, Godin B, Langor D, Bourdon C, Lee S-I, Horwood D (2015a) New distribution records for Canadian Aleocharinae (Coleoptera, Staphylinidae), and new synonymies for *Trichiusa*. ZooKeys 498:51–91

Klimaszewski J, Webster RP, Bourdon C, Pelletier G, Godin B, Langor DW (2015b) Review of Canadian species of the genus *Mocyta* Mulsant & Rey (Coleoptera, Staphylinidae, Aleocharinae), with the description of a new species and a new synonymy. ZooKeys 487:111–139

Klimaszewski J, Webster RP, Sikes D, Bourdon C, Labrecque M (2015c) A review of Canadian and Alaskan species of the genera *Clusiota* Casey and *Atheta* Thomson, subgenus *Microdota* Mulsant & Rey (Coleoptera, Staphylinidae, Aleocharinae). ZooKeys 524:103–136

Klimaszewski J, Webster RP, Langor D, Sikes D, Bourdon C, Godin B, Ernst C (2016a) A review of Canadian and Alaskan species of the genus Liogluta Thomson, and descriptions of three new species (Coleoptera, Staphylinidae, Aleocharinae). In: Webster RP, Bouchard P, Klimaszewski J (Eds) The Coleoptera of New Brunswick and Canada: providing baseline biodiversity and natural history data. ZooKeys [Special Issue] 573:217–256

Klimaszewski J, Langor DW, Hammond HEJ, Bourdon C (2016b) A new species of *Anomognathus* and new Canadian and provincial records of aleocharine rove beetles from Alberta, Canada (Coleoptera, Staphylinidae, Aleocharinae). ZooKeys 581:141–164

Klimaszewski J, Langor DW, Bourdon C, Gilbert A, Labrecque M (2016c) Two new species and new provincial records of aleocharine rove beetles from Newfoundland and Labrador, Canada (Coleoptera, Staphylinidae, Aleocharinae). ZooKeys 593:49–89

Klimaszewski J, Larson DJ, Labrecque M, Bourdon C (2016d) Twelve new species and fifty-three new provincial distribution records of Aleocharinae rove beetles of Saskatchewan, Canada (Coleoptera, Staphylinidae). ZooKeys 610:45–112

Klimaszewski J, Webster RP, Davies A (2017) Genus *Hydrosmecta* C.G. Thomson: a review of species occurring in eastern Canada (Coleoptera, Staphylinidae, Aleocharinae). Insecta Mundi 0593:1–17

Klimaszewski J, Webster RP, Langor DW, Brunke A, Davies A, Bourdon C, Labrecque M, Newton AF, Dorval J-A, Frank JH (2018a) Aleocharine rove beetles of eastern Canada (Coleoptera, Staphylinidae, Aleocharinae): a glimpse of megadiversity. Springer, Cham. 902 pp

Klimaszewski J, Godin B, Davies A, Bourdon C, Horwood D (2018b) Forty new records of aleocharine beetles, and two new species in the genera *Acrotona* Thomson and *Atheta* Thomson, for the province of Manitoba, Canada (Coleoptera: Staphylinidae: Aleocharinae). Insecta Mundi 0641:1–33

Klimaszewski J, Sikes DS, Brunke A, Bourdon C (2019) Species review of the genus *Boreophilia* Benick from North America (Coleoptera: Staphylinidae: Aleocharinae: Athetini): systematics, habitat and distribution. ZooKeys 848:57–102

Klimaszewski J, Hoebeke ER, Godin B, Davies A, Perry KI, Bourdon C, Winchester N (2020) Aleocharine rove beetles of British Columbia: a hotspot of Canadian biodiversity (Coleoptera, Staphylinidae). Springer, Cham. 631 pp

Kraatz G (1856) Staphylinii. Naturgeschichte der Insecten Deutschlands. Erste Abtheilung Coleoptera. Zweiter Band. Erste, zweite Lieferung. Bogen 1–24. Nicolai, Berlin, viii + 376 pp

Kraatz G (1857) Die Coleopteren Island's [pp. 282–289] *In*: Staudinger, O.: Reise nach Island zu entomologischen Zwecken unternommen. Entomologische Zeitung (Stettin) 18(7–9):209–289

Lee S-G, Ahn K-J (2015) A taxonomic review of Korean species of the *Atheta* Thomson subgenus *Microdota* Mulsant & Rey, with descriptions of two new species (Coleoptera, Staphylinidae, Aleocharinae). ZooKeys 502:61–97

Lee S-G, Ahn K-J (2017) A taxonomic review of the Korean species of the genus *Datomicra* Mulsant and Rey of the genus *Atheta* Thomson, with description of a new species (Coleoptera, Staphylinidae, Aleocharinae). Zootaxa 4262:508–522

Lee S-G, Yoo I-S, Ahn K-J (2013) Korean species of the coastal genus Adota and a key to the known species (Coleoptera: Staphylinidae: Aleochariane). Fla Entomol 96(3):1062–1072

Lohse GA, Smetana A (1985) Revision of the types of species of Oxypodini and Athetini (*sensu* Seevers) described by Mannerheim and Mäklin from North America (Coleoptera: Staphylinidae). Coleopt Bull 39 (3):281–300

Lohse GA, Klimaszewski J, Smetana A (1990) Revision of arctic Aleocharinae of North America (Coleoptera: Staphylinidae). Coleopt Bull 44(2):121–202

Majka CG, Klimaszewski J (2008a) Introduced Staphylinidae (Coleoptera) in the Maritime Provinces of Canada. Can Entomol 140(1):48–72

Majka CG, Klimaszewski J (2008b) New records of Canadian Aleocharinae (Coleoptera: Staphylinidae). *In*: Majka CG, Klimaszewski J (Eds.) Biodiversity, biosystematics, and ecology of Canadian Coleoptera. ZooKeys [Special Issue] 2:85–114

Majka CG, Klimaszewski J (2008c) Adventive Staphylinidae (Coleoptera) of the Maritime Provinces of Canada: further contributions. *In*: Majka CG, Klimaszewski J (Eds.) Biodiversity, biosystematics, and ecology of Canadian Coleoptera. ZooKeys [Special Issue] 2:151–174

Majka CG, Klimaszewski J (2010) Contributions to the knowledge of the Aleocharinae (Coleoptera, Staphylinidae) in the Maritime Provinces of Canada. ZooKeys 46:15–39

Majka CG, Ogden J (2006) *Brachygluta abdominalis* (Aubé) (Coleoptera: Staphylinidae) newly recorded in Canada with notes on other beach-drift beetles. Proc Entomol Soc Wash 108(4):761–764

Majka CG, Sikes D (2009) Thomas L. Casey and Rhode Island's precinctive beetles: Taxonomic lessons and the utility of distributional checklists. ZooKeys 22:267–283

Mäklin FW (1852) [New species and notes]. *In* Mannerheim, C.G.: Zweiter Nachtrag zur Kaefer-Fauna der Nord-Amerikanischen Laender des Russischen Reiches. Bulletin de la Société Impériale des Naturalistes de Moscou 25(2):283–372

Mäklin FW (1853) [New species and notes]. In: Mannerheim, C.G.: Dritter Nachtrag zur Kaefer-Fauna der Nord-Amerikanischen Laender des Russischen Reiches. Bulletin de la Société Impériale des Naturalistes de Moscou 26(3):95–273

Mäklin FW (1880) Ytterligare diagnoser öfver några nya sibirska Coleopter-arter. Öfversigt af Finska Vetenskaps-Societetens Förhandlingar 22:79–86

Majka CG, Klimaszewski J, Lauff RF (2008) The coastal rove beetles (Coleoptera, Staphylinidae) of Atlantic Canada: a survey and new records. ZooKeys 2:115–150

Mannerheim CG (1843) Beitrag zur Kaefer-fauna der Aleutischen Inseln, der Insel Sitkha und Neu-Californiens. Bulletin de la Société Impériale des Naturalistes de Moscou 16(2):175–314

Maruyama M, Klimaszewski J (2004) A new genus and species of the myrmecophilous Athetini, *Paragoniusa myrmicae* (Coleoptera: Staphylinidae: Aleocharinae) from Canada. Entomological Review of Japan 59 (2):241–248

Maruyama M, Klimaszewski J (2006) Notes on myrmecophilous aleocharines (Insecta, Coleoptera, Staphylinidae) from Canada, with a description of a new species of *Myrmoecia*. Bull Nat Sci Mus, Tokyo (A) 32 (3):125–131

McLean JA, Klimaszewski J, Li A, Savard K (2009) Survey of rove beetles (Coleoptera, Staphylinidae) from Stanley Park, Vancouver, British Columbia, Canada, with new records and description of a new species. Part 1. *In*: Majka CG, Klimaszewski J (eds.) Biodiversity, Biosystematics, and Ecology of Canadian Coleoptera II. ZooKeys [Special Issue] 22:5–17

Moore I, Legner EF (1975) A catalogue of the Staphylinidae of America north of Mexico (Coleoptera). University of California, Division of Agricultural Sciences, Special Publication No. 3015:1-514

Mulsant E, Rey C (1873a) Description de divers coléoptères brévipennes nouveaux ou peu connus. Opuscules Entomologiques 15:147–189

Mulsant E, Rey C (1873b) Histoire naturelle des Coléoptères de France. Brévipennes. Aléochariens (Suite) [Tableau... des Aléochariens + Septième Branche, Myrmédoniaires, part 1]. Deyrolle, Paris. [2] + 695 pp., pls. 1–5

Munster T (1923) Bidrag til kjendskapen om slekten *Atheta* Thoms. (Col., Staph.) Norsk. Entomologisk Tidskrift 1[1920–1923](4):206–208

Muona J (1979) Staphylinidae. In: Silfverberg H (ed) Enumeratio Coleopterorum Fennoscandiae et Daniae. Helsingfors Entomologiska Bytesförening, Helsinki, pp. 14–28

Muona J (1984) Review of Palearctic Aleocharinae also occurring in North America (Coleoptera: Staphylinidae). Entomologica Scandinavica 15 (2):227–231

Muona J (1990) The Fennoscandian and Danish species of the genus *Amischa* Thomson (Coleoptera, Staphylinidae). Entomologisk Tidskrift 11(1-2):17–24

Navarrete-Heredia JL, Newton AL, Thayer MK, Ashe JS, Chandler DS (2002) Guía ilustrada para los géneros de Staphylinidae (Coleoptera) de México. Illustrated guide to the genera of Staphylinidae (Coleoptera) of Mexico. Universidad de Guadalajara & CONABIO, México. xii + 401 pp

Newton AF (2017) Nomenclatural and taxonomic changes in Staphyliniformia (Coleoptera). Insecta Mundi 0595:1–52

Pace R (1982) *Oncosomechusa besucheti* n. gen., n. sp., dell India et *Emmelostiba besuchetti* n. gen., n. sp., del Libano (Coleoptera, Staphylinidae). Rev Suisse Zool 111(3):457–523

Palm T (1970) Svensk Insektfauna. 9. Skalbaggar. Coleoptera. Hafte 6. Kortvingar: Fam. Staphylinidae underfam. Aleocharinae (Atheta). (Nr. 52). Entomologiske Föreningen i Stockholm, Stockholm. 296 pp., 21 pls

Roubal J (1928) Zwei neue Coleopteren aus der Tschechoslovakei. Wiener Entomologische Zeitung 4(1):27–30

Rye EC (1866) Descriptions of new species, &c., of Brachelytra. Entomologist's Monthly Magazine 3 (11):121–125

Sahlberg CR (1831) [Pp. 361-408]. In: Insecta Fennica, dissertationibus academicis, A. 1817-1834 editis. Pars I: A. J.C. Frenckel, Helsingfors, VIII + 519 pp

Sahlberg JR (1880) Bidrag till nordvestra Sibiriens insektfauna. Coleoptera. Insamlade under expeditionerna till obi och Jenessej 1876 och 1877. I. Cicindelidae, Carabidae, Dytiscidae, Hydrophilidae, Gyrinidae, Dryopidae, Georyssidae, Limnichidae, Heteroceridae, Staphylinidae och Micropeplidae Kongliga Svenska Vetenskaps-Akademiens Handlingar (Ny Följd) 17(4):3–115 + 1 pl

Sahlberg JR (1887) Bidrag till Tschuktsch-halföns Insektfauna. Coleoptera och Hemiptera, insamlade under Vega-Expeditionen vid halföns norra och östra kust, 1878-1879. In: Nordenskiöld AE (ed) Vega-expeditionens vetenskapliga iakttagelser bearbetade af deltagare i resan och andra forskare. Fjerde Bandet. F. & G. Beijers, Stockholm, pp 1–42. 582 pp

Sawada K (1970) Aleocharinae (Staphylinidae, Coleoptera) of the IBP-Station in the Shiga Hights, Central Japan (I). Bull Natl Sci Mus (Tokyo) 13:21–64

Sazhnev AS (2018) Materials to the rove beetles fauna (Coleoptera: Staphylinidae) of the Commander Islands (Kamchatka Region, Russia). Cacasian Entomol Bull 14(1):19–23

Schülke M, Smetana A (2015) Family Staphylinidae Latreille, 1802. In: Löbl I, Löbl D (eds) Catalogue of Palaearctic Coleoptera: Hydrophiloidea – Staphylinoidea, vol 2. Revised and updated edition. Brill, Leiden & Boston, pp 304–1134

Scriba W (1866) *Homalota subtilis* Scriba n. sp. Berliner. Entomologische Zeitschrift 10(1-3):128

Seevers CH (1978) A generic and tribal revision of the North American Aleocharinae (Coleoptera: Staphylinidae) [with additions and annotations by Lee H. Herman]. Fieldiana Zoology 71. vi + 289 pp

Sharp D (1869) A revision of the British species of *Homalota*. Trans Entomol Soc Lond XIV:91–272

Smetana A (2004) Family Staphylinidae Latreille, 1802 New nomenclatorial and taxonomic acts, and comments [pp. 29-35], subfamilies Omaliinae-Dasycerinae [pp. 237-272], subfamilies Phloeocharinae-Apateticinae [pp. 329-495], subfamilies Piestinae-Staphylininae [pp. 504-698]. In: Löbl I, Smetana A (eds) Catalogue of Palaearctic Coleoptera, Volume 2. Hydrophiloidea – Histeroidea – Staphylinoidea. Apollo Books, Stenstrup, p 942

Stefani FOP, Klimaszewski J, Morency M-J, Bourdon C, Labrie P, Blais M, Venier L, Séguin A (2016) Fungal Community composition in the gut of rove beetles (Coleoptera: Staphylinidae) from the Canadian boreal forest reveals possible endosymbiotic interactions for dietary needs. Fungal Ecol 23:164–171

Strand A, Vik A (1964) Die Genitalorgane der nordischen Arten der Gattung *Atheta* Thoms. (Col., Staphylinidae). Norsk Entomologisk Tidsskrift 12 (5–8):327–335. pls. 1–21

Thomson CG (1858) Försök till uppställning af Sveriges Staphyliner. Öfversigt af Kongliga Vetenskaps-Akademiens Förhandlingar 15(1):27–40

Thomson CG (1859) Skandinaviens Coleoptera, synoptiskt bearbetade. I. Tom. Berlingska Boktryckeriet, Lund. [3] + 290 pp

Thomson CG (1867) Skandinaviens Coleoptera, synoptiskt bearbetade. Supplementum. Tom IX. Lundbergska Boktryckeriet, Lund. 407 pp

Webster RP, Klimaszewski J, Pelletier G, Savard K (2009) New Staphylinidae (Coleoptera) records with new collection data from New Brunswick, Canada. I. Aleocharinae. *In*: Majka CG, Klimaszewski J (Eds.) Biodiversity, biosystematics, and ecology of Canadian Coleoptera II. ZooKeys [Special Issue] 22:171–248

Webster RP, Klimaszewski J, Sweeney JD, DeMerchant I (2012) New Staphylinidae (Coleoptera) records with new collection data from New Brunswick, and an addition to the fauna of Quebec, Canada: Aleocharinae. In: Klimaszewski J, Anderson R (Eds) Biosystematics and ecology of Canadian Staphylinidae (Coleoptera) II. ZooKeys [Special Issue] 186:83–118

Webster RP, Klimaszewski J, Bourdon C, Sweeney JD, Hughes CC, Labrecque M (2016) Further contribution to the Aleocharinae (Coleoptera, Staphilinidae) fauna of New Brunswick and Canada including description of 27 new species. *In*: Webster RP, Bouchard P, Klimaszewski J (Eds) The Coleoptera of New Brunswick and Canada: providing baseline biodiversity and natural history data. ZooKeys [Special Issue] 573:85–216

Webster RP, deTonnancour P, Sweeney JD, Webster VL, Kostanowicz CA, Hughes C, Anderson RS, Klymko J, Chatal C, Vigneault R (2020) New Coleoptera records from eastern Canada, with additions to the fauna of Manitoba, British Columbia, and Yukon Territory. ZooKeys 946:53–112

Wollaston TV (1855) Descriptions of two coleopterous insects new to the British Fauna. The Zoologist 13 (*Appendix):ccv–ccvii

Work TT, Klimaszewski J, Thiffault E, Bourdon C, Paré D, Bousquet Y, Venier L, Titus B (2013) Initial responses of rove and ground beetles (Coleoptera, Staphylinidae, Carabidae) to removal of logging residues following clearcut harvesting in the boreal forest of Quebec, Canada. ZooKeys 258:31–52

Voinkov AA (2020) Rove beetles (Coleoptera: Staphylinidae) of the Orongo River valley, Western Transbaikalia. Trudy Russkogo Entomologicheskogo Obshchestva 91:77–86

Tribe Lomechusini Fleming, 1821 20

Tribe Lomechusini Fleming, 1821

(Fig. 20.314a–h) (Ashe 2001, Maruyama 2006, Elven et al. 2012, Klimaszewski et al. 2018, 2020)

Diagnosis Epipharynx with sensillum α fully developed; galeae and laciniae moderately to greatly elongate, length of galea equal to or greater than distance to base of galea from cardo; mesocoxae broadly separated by broad processes of meso- and metaventrite, process of mesoventrite broad and short; tarsal formula 4-5-5; median lobe of aedeagus with athetine bridge, compressor plate covered by extension from base of bulbus.

COMMENTS Many taxa of this tribe in other regions are associated with ants.

Subtribe Myrmedoniina C.G. Thomson, 1867 (Maruyama 2006, Klimaszewski et al. 2018, 2020)

Diagnosis This group differs from other subtribes of Lomechusini in having the galeae elongate, the ligula bifid, and the first three visible abdominal segments without trichomes (Maruyama 2006); otherwise, the group is not well defined.

Genus *Zyras* Stephens, 1835

(Fig. 20.314a–h) (Hlaváč and Jászay 2009, Klimaszewski et al. 2005, 2018)

Diagnosis Body robust, 4.0–6.0 mm, integument highly glossy, often with bright contrasting colours, moderately convex, punctation usually large and coarse; pubescence fine and slightly or not adhering to the integument, without silky appearance; head moderately convex; antennomere I as long as II and III combined, II smaller than III, IV slightly transverse, V–X moderately to distinctly transverse and more so toward the apex, terminal antennomere about as long as two preceding ones combined; pronotum broadest in apical one-third, usually about one-fourth broader than long and about one-fourth narrower than elytra, impressed basally, usually with well-defined medial antebasal fovea, punctation moderately large to large, pubescence on disc directed obliquely posteriad along arcuate lines on each side; second and third visible abdominal tergites without tuberosities; median lobe of aedeagus with enlarged bulbus and narrowly conical tubus; spermatheca very small with capsule narrowly elongate, without apparent invagination, stem short, seminal canal with multiple coils forming a bundle.

J. Klimaszewski et al., *A Faunal Review of Aleocharine Beetles in the Rapidly Changing Arctic and Subarctic Regions of North America (Coleoptera, Staphylinidae)*,
https://doi.org/10.1007/978-3-030-68191-3_20

COMMENTS The genus *Zyras* is treated here in a narrow sense following Klimaszewski et al. (2005). Seevers (1978) predicted the generic subdivision of Nearctic *Zyras* (*sensu lato*) would include transfer of many species to the genus *Pella*. Seevers (1978) considered Nearctic *Zyras* in a broad sense, with the exception of the genera *Apalonia* and *Myrmoecia*, and provided a key to twelve Nearctic species. Kistner (1971) treated at least some of the subgenera recognized by Lohse (1974) as distinct genera (e.g., *Pella*).

238. *Zyras obliquus* (Casey) (Fig. 20.314a–h)

Myrmedonia obliqua Casey, 1893. Seevers 1978 [as synonym of *Z. haworthi* (Stephens, 1832)], Klimaszewski and Peck 1986, Klimaszewski and Winchester 2002 [as *Z. pseudohaworthi*], Klimaszewski et al. 2005, 2009, Webster et al. 2009, Majka and Klimaszewski 2010, Klimaszewski et al. 2016, 2018, 2020.

Diagnosis Body narrowly elongate, length 4.0–6.5 mm; head pronotum and abdomen, except for its reddish-brown margins, dark brown to black, elytra bicoloured reddish-brown with base and lateral parts of disc dark brown to black, appendages reddish-brown (Fig. 20.314a); integument highly glossy, without distinct meshed microsculpture; pubescence moderately long and sparse; antennomere I about as long as the two following antennomeres combined, II smaller than III, IV quadrate, V–X moderately transverse and slightly more so toward the apex, and the terminal one about as long as the two preceding antennomeres combined; pronotum approximately trapezoidal in shape, with small basal impression, distinctly narrower than elytra; elytra about twice as broad as length of suture; hind tarsi with basal article slightly longer than two following articles combined. MALE. Apical margin of tergite VIII with one small rounded tooth on each side of midline (Fig. 20.314d); sternite VIII elongate, with apical margin rounded laterally, broadly truncate medially (Fig. 20.314e); median lobe of aedeagus in lateral view with tubus almost straight ventrally, apex narrow and curved slightly ventrad (Fig. 20.314b); in dorsal view bulbus enlarged and tubus conical (Fig. 20.314c), internal sac with two narrowly elongate subparallel structures (Fig. 20.314b,c). FEMALE. Tergite VIII unevenly truncate apically (Fig. 20.314f); sternite VIII broadly rounded apically (Fig. 20.314g); capsule of spermatheca small, narrow, curved, with subapical constriction, stem short, seminal canal very long with multiple coils forming a bundle (Fig. 20.314h).

DISTRIBUTION Origin: Nearctic. **CANADA**: AB, BC, LB, MB, NB, NF, NS, ON, QC, SK. **USA**: MI, MO, NH, NY, OR.

COLLECTION AND HABITAT DATA Habitat: adults from various traps in 3-5-year-old post-harvest Douglas-fir forest, transition zone of Sitka spruce forest, mixed *Populus*/*Salix*/*Alnus* forest, and *Pinus banksiana* forest, under red oak in red oak forest. Microhabitat unknown. **Collecting period**: VI–VIII. **Collecting method**: pitfall traps, window traps, and pan-traps.

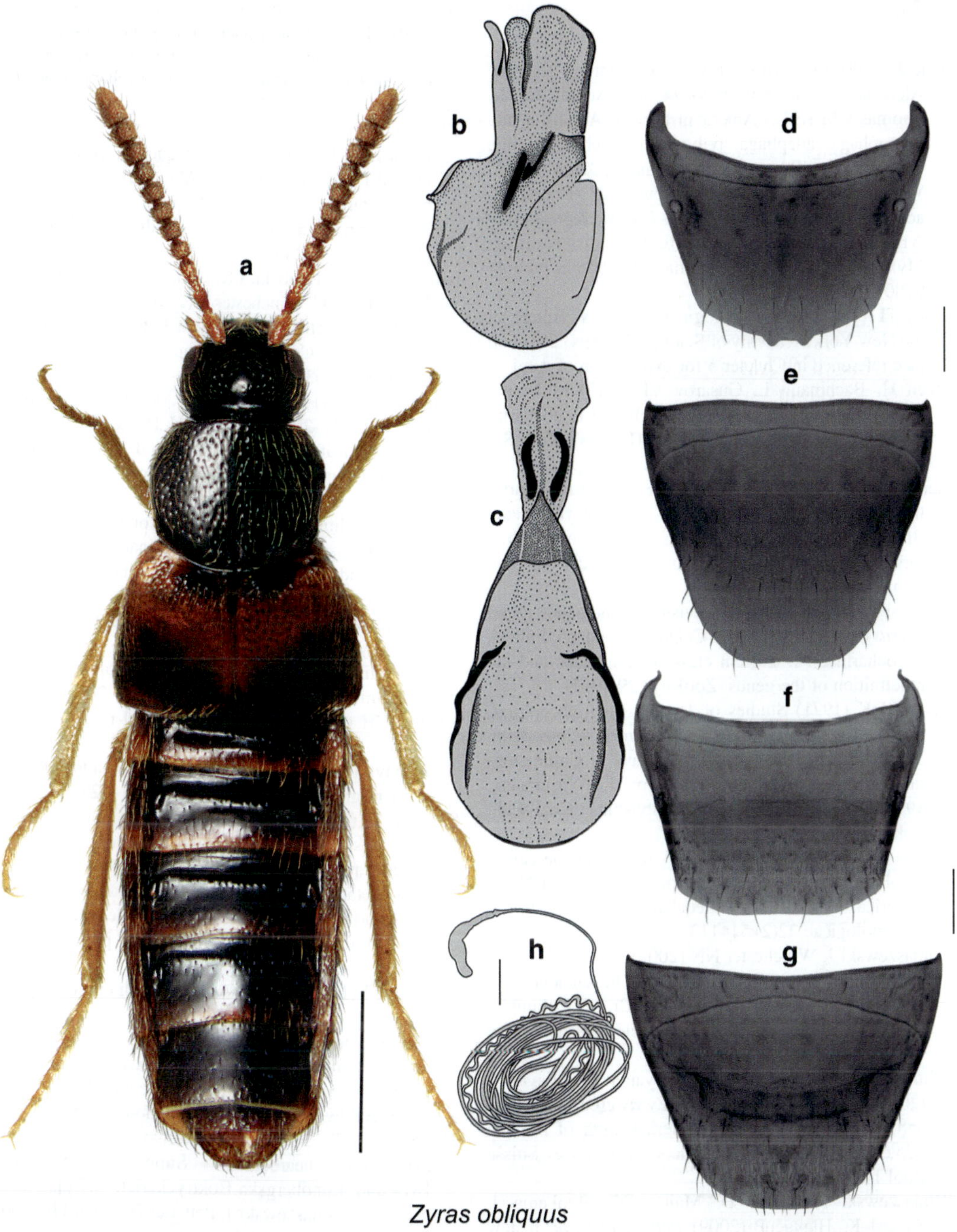

Fig. 20.314 (**a–h**). *Zyras obliquus* (Casey): (**a**) habitus in dorsal view; (**b**) median lobe of aedeagus in lateral view; (**c**) median lobe of aedeagus in dorsal view; (**d**) male tergite VIII; (**e**) male sternite VIII; (**f**) female tergite VIII; (**g**) female sternite VIII; (**h**) spermatheca. Scale bar for habitus = 1 mm; remaining scale bars = 0.2 mm

References

Ashe JS (2001) Key to the tribes and genera of Nearctic Aleocharinae. 8. Pp. 299–374. In: Arnett RH Jr, Thomas CM (Eds.) American beetles. Archostemata, myxophaga, adephaga, polyphaga: staphyliniformia. CRC Press, Boca Raton, London, New York, Washington, D.C. 443 pp. [publ. 28 Dec 2000 acccording to CRC Press: https://www.crcpress.com/American-Beetles-Volume-I-Archostemata-Myxophaga-Adephaga-Polyphaga/Jr-Thomas/p/book/9780849319259]

Casey TL (1893) Coleopterological notices. V. Annals of the New York Academy of Sciences 7 [1894]:281–606 [see reference in Chapter 5 for explanation of date]

Elven H, Bachmann L, Gusarov VI (2012) Molecular phylogeny of the Athetini-Lomechusini-Ecitocharini clade of aleocharine rove beetles (Insecta). Zool Scr 41(6):617–636

Fleming J (1821) Insecta. In: Supplement to the fourth, fifth and sixth editions of the Encyclopaedia Britannica, with preliminary dissertations on the history of sciences, vol 5. A. Constable and Company, Edinburgh, pp 41–56. 584 pp

Hlaváč P, Jászay T (2009) A revision of the genus *Zyras* (*Zyras*) Stephens, 1835 (Coleoptera, Staphylinidae, Aleocharinae). I. Current classification status and the redefinition of the genus. ZooKeys 29:49–71

Kistner DK (1971) Studies of Japanese myrmecophiles. Part I. The genera *Pella* and *Falagria* (Coleoptera, Staphylinidae). Entomological Essays to Commemorate the Retirement of Professor K. Yasumatsu, 1971:141–165. Hokuryukan Publishing Co., Ltd., Tokyo

Klimaszewski J, Peck SB (1986) A review of the cavernicolous Staphylinidae (Coleoptera) of eastern North America: Part I. Aleocharinae. Quaestiones Entomologicae 22(2):51–113

Klimaszewski J, Winchester NN (2002) Aleocharine rove beetles (Coleoptera Staphylinidae) of the ancient Sitka spruce forest on Vancouver Island, British Columbia, Canada. Mémoires de la Société royale belge d'Entomologie 40:3–126

Klimaszewski J, Pelletier G, Maruyama M, Hlaváč P (2005) Canadian species of the *Zyras* group of genera and review of types from America north of Mexico (Coleoptera, Staphylinidae, Aleocharinae). Rev Suisse Zool 112(3):703–733

Klimaszewski J, Lynch D, Majka CG, Renkema J, Savard K, Hlaváč P (2009) *Pella glooscapi*, a new rove beetle, and new records of aleocharines from Nova Scotia, Canada (Coleoptera, Staphylinidae). In: Majka CG, Klimaszewski J (Eds.) Biodiversity, biosystematics, and ecology of Canadian Coleoptera II. ZooKeys [Special Issue] 22:35–44

Klimaszewski J, Larson DJ, Labrecque M, Bourdon C (2016) Twelve new species and fifty-three new provincial distribution records of Aleocharinae rove beetles of Saskatchewan, Canada (Coleoptera, Staphylinidae). ZooKeys 610:45–112

Klimaszewski J, Webster RP, Langor DW, Brunke A, Davies A, Bourdon C, Labrecque M, Newton AF, Dorval J-A, Frank JH (2018) Aleocharine rove beetles of eastern Canada (Coleoptera, Staphylinidae, Aleocharinae): a glimpse of megadiversity. Springer, Cham, xvi + 902 pp

Klimaszewski J, Hoebeke ER, Godin B, Davies A, Perry KI, Bourdon C, Winchester N (2020) Aleocharine rove beetles of British Columbia: a hotspot of Canadian biodiversity (Coleoptera, Staphylinidae). Springer, Cham, xvi + 631 pp

Lohse GA (1974) Hypocyphtinae, Aleocharinae [in part, pp. 7–72, 221–292]. In: Freude H, Harde KW, Lohse A (eds) Die Käfer Mitteleuropas. Band 5: Staphylinidae II (Hypocyphtinae und Aleocharinae), Pselaphidae. Goecke & Evers, Krefeld. 381 pp

Majka CG, Klimaszewski J (2010) Contributions to the knowledge of the Aleocharinae (Coleoptera, Staphylinidae) in the Maritime Provinces of Canada. ZooKeys 46:15–19

Maruyama M (2006) Revision of the Palearctic species of the myrmecophilous genus *Pella* (Coleoptera, Staphylinidae, Aleocharinae). National Science Museum Monographs (Tokyo) 32. iii + 207 pp

Seevers CH (1978) A generic and tribal revision of the North American Aleocharinae (Coleoptera: Staphylinidae) [with additions and annotations by Lee H. Herman]. Fieldiana Zoology 71. vi + 289 pp

Stephens JF (1832) Illustrations of British entomology, or a synopsis of indigenous insects, containing their generic and specific distinctions, with an account of their metamorphoses, times of appearance, localities, food, and economy, as far as practicable. Mandibulata, vol 5. Baldwin & Cradock, London, pp 1–240. pls. 24–26

Stephens JF (1835) Illustrations of British entomology, or a synopsis of indigenous insects, containing their generic and specific distinctions, with an account of their metamorphoses, times of appearance, localities, food, and economy, as far as practicable. Mandibulata, vol 5. Baldwin and Cradock, London, pp 369–477

Thomson CG (1867) Skandinaviens Coleoptera, synoptiskt bearbetade. Supplementum. Tom IX. Lund: Lundbergska Boktryckeriet, 407 pp

Webster RP, Klimaszewski J, Pelletier G, Savard K (2009) New Staphylinidae (Coleoptera) records with new collection data from New Brunswick, Canada. I. Aleocharinae. In: Majka CG, Klimaszewski J (Eds.) Biodiversity, biosystematics, and ecology of Canadian Coleoptera II. ZooKeys [Special Issue] 22:171–248

Correction to: A Faunal Review of Aleocharine Beetles in the Rapidly Changing Arctic and Subarctic Regions of North America (Coleoptera, Staphylinidae)

Correction to:
J. Klimaszewski et al., *A Faunal Review of Aleocharine Beetles in the Rapidly Changing Arctic and Subarctic Regions of North America (Coleoptera, Staphylinidae)*, https://doi.org/10.1007/978-3-030-68191-3

The section heading for Annex 3 was missing and it has been updated in backmatter with this erratum.

The updated online version of the book can be found at
https://doi.org/10.1007/978-3-030-68191-3

J. Klimaszewski et al., *A Faunal Review of Aleocharine Beetles in the Rapidly Changing Arctic and Subarctic Regions of North America (Coleoptera, Staphylinidae)*,
https://doi.org/10.1007/978-3-030-68191-3_21

Annex 1

Additional distribution records for the Arctic and Subarctic species of Aleocharinae in Canada and Alaska. Tribes are listed in taxonomic order and species are listed following the order in the book. The item numbers are followed by the species numbers in parentheses as they are found in the main body of text.

Subfamily Aleocharinae Fleming

Tribe Gymnusini Heer

Genus *Gymnusa* Gravenhorst

1 (3). *Gymnusa atra* Casey *
Additional records. USA, Alaska, Alaska Hwy. Mi 1249, Deadman Lake, 6.VII.1968, A. Smetana (CNC) 3 males, 2 females.

2 (6). *Gymnusa grandiceps* Casey
Additional records. Canada: Ontario, Moosonee, 51.28288, –80.63926, Rep. 2, wet, pitfall trap, 16-19.VI.2010, NBP Field Party M2WP711 (LFC) 1 female.

3 (7). *Gymnusa campbelli* Klimaszewski
Additional records. Canada: Quebec, Schefferville, 54.85350, –66.67043, Rep. 3, wet., pitfall trap, 30.VI-3.VII.2010, NBP Field Party S3WP312 (LFC) 1 male; same data except yellow pan, 3-6.VII.2010, S3WY321 (LFC) 1 female; Rep. 3, wet, yellow pan, 6-9. VII.2010, S3WY831 (LFC) 1 female. USA, Alaska, Alaska Hwy. Mi 1249, Deadman Lake, 6.VII.1968, A. Smetana (CNC) 1 male, 1 female

4 (9). *Gymnusa pseudovariegata* Klimaszewski
Additional records. Canada, Yukon, Riedl.? Yukon Refugium Project, 27.VI.1977, J.V. Matthews (CNC) 2 females. USA: Alaska, Selawik NWR 40m el. 66.4549, –158.13963, ±14m lakeshore grasses, treading, 24.VI.2010, D.S. Sikes (UAM) 5 specimens.

5 (10). *Gymnusa smetanai* * Klimaszewski
Additional records. Canada, Yukon, British Mts., Teal Lake, 400 m, 68.900, –140.3833, J.M. Campbell (CNC) 1 female; Northwest Territories, Tuktoyaktuk, 3-9.VII,1971, D.R. Oliver (CNC) 1 female.

USA: Alaska, Fairbanks, 166m el., 64.86673, –147.84297, ±10m, black and white spruce, hanging Malaise, 5.1m, 13-27.VI.2016, J.C. Hagelin (UAM) 1 specimen.

Tribe Aleocharini Fleming

Genus *Aleochara* Gravenhorst

6 (15). *Aleochara (Calochara) villosa* Mannerheim ‡
Additional records. USA: Alaska, Kenai Mts., 7 mi S Moose Pass, 450 feet, 16.VI.1978, A. Smetana and E. Becker (CNC) 1 male, 2 females. Fairbanks, 930 Fitz Ct., 280 m, 64.901, –147.529, 1.V.2011, D. Sikes (UAM:

J. Klimaszewski et al., *A Faunal Review of Aleocharine Beetles in the Rapidly Changing Arctic and Subarctic Regions of North America (Coleoptera, Staphylinidae)*,
https://doi.org/10.1007/978-3-030-68191-3

Ento:221913), 1 unknown sex (UAM); same except 26.IV.2009, (UAM:Ento:96906), 1 unknown sex; Sterling, Corey St, 90m el., 60.53953, –150.83232, ±150m *P. glauca,* disturbed, pheromone trap, 30.VI.2004, D. M. Collet (UAM) 1 specimen.

7 (19). ***Aleochara (Coprochara) sulcicollis*** **Mannerheim**

Additional records. USA: Alaska, Adak, 1m el., 51.90244, –176.61842, ±7m under drift wood, sandy beach, 13.VIII.2009, D.S. Sikes (UAM) 11 specimens; Little Sitkin, 13m el., 51.98173, −178.48885, ±35m under beach wrack, driftwood, 9.VI.2008, D.S. Sikes (UAM) 2 specimens; Prince of Wales Island, Staney Ck. 1B, 77m el., 55.872, –133.06523, ±26m clear-cut, Lindgren 1, 28.V-11.VI.2012, J. Stockbridge, S. Meierotto, I. MacDougall (UAM) 1 specimen; Wosnesenski, 1m el., 55.21733, –161.36768, ±24m back beach, colored pan trap, 11.VI.2009, D.S. Sikes (UAM) 4 specimens.

8 (20). ***Aleochara (Coprochara) verna*** **Say ***

Additional records. Canada: Yukon, Dempster Hwy. Mi 75, 3400 feet, 23.VII.1978, A. Smetana and J.M. Campbell (CNC) 1 male. USA: Alaska, Delta Jctn, CRP site37, 330-350m el., 64.03138, –145.12108, ±150m pitfall trap C, 8-15.VI.2007, A. M. Hagerty, S. Y. Emmert (UAM) 1 specimen; Denali NPP., Igloo Canyon, 1123m el., 63.59198, –149.6042, ±30m rocky alpine, 4 pitfalls w/ rainroofs, pooled, J.J. Rykken; 20.VII-3.VIII.2016, D.S. Sikes, A. Haberski, K. Chen (UAM) 1 specimen; Matanuska, 1-1000m el., 61.53929, –149.23588, ±10km, 22.VII.1974 (UAM) 1 specimen; same data except, 1.VII.1974, (UAM) 1 specimen; Palmer, USDA Research Station, 5-90m el., 61.56603, –149.25049, ±1km Diaz., *Calamagrostis*, 12.VIII.1977, Richard H. Washburn (UAM) 1 specimen; Palmer, USDA Research Station, 5-90m el., 61.56603, −149.25049, ±1km rotary trap, 23. VIII.1963, (UAM) 1 specimen; Toolik Field Station, 724m el., 68.6286, –149.59772, ±36m arctic tundra, 3 pitfalls, 2.VI-30.VII.2008, D.S. Sikes (UAM) 1 specimen.

9 (22). ***Aleochara (Xenochara) castaneipennis*** **Mannerheim**

Additional records. Canada: Yukon, Dempster Hwy. Mi 75, 23.VII.1978, 3400 feet, A. Smetana and J.M. Campbell (CNC) 3 males; Rampart House, 31.VI.1951, J.E.H. Martin (CNC) 1 male. USA: Alaska, King Salmon, Naknek R., 11.VIII.1952, J.B. Hartley (CNC) 1 male, 2 females; Denali NPP., Mt. Healy, 1249m el., 63.75362, –148.96379, ±30m rocky alpine, 4 yellow pitfalls w/o rainroofs, pooled, J.J. Rykken; D.S. Sikes; A. Haberski; K. Chen, 19.VI-3VIII.2016, (UAM) 1 specimen; Denali NPP., Mt. Healy, 927m el., 63.74016, –148.95346, ±30m shrub, alder, dwarf birch, grass, clubmoss, 4 pitfalls w/ rainroofs, pooled, J.J. Rykken; D.S. Sikes; R. Stewart-Rogers, 3-16.VIII.2016, (UAM) 3 specimens; Fairbanks, 930 Fitz Ct, 280m el., 64.9013, –147.52861, ±30m *Betula* & *P. tremuloides*, flight intercept trap, D.S. Sikes, 16.VIII-22.IX.2007, (UAM) 83 specimens; same data except, 29.V-4.VII.2008, (UAM) 15 specimens; Mi 179 Dalton Hwy, 365m el., 67.29803, –150.15699, ±7m 3/4 Alder, white spruce, *Betula* forest, D.S. Sikes, 5-31.VII.2008, (UAM) 32 specimens; near Kasilof, 60.3375, −151.2744, ±50m in mushrooms, D. Langor, 21.VII.2005, (UAM) 1 specimen; Sitka, Mosquito Cove Trail, 1m el., 57.1356, –135.37059, ±15m old growth Hemlock, coastal rain forest, D.S. Sikes, 19.IX.2008, (UAM) 3 specimens.

Tribe Oxypodini C.G. Thomson

Genus *Calodera* Mannerheim

10 (27). ***Calodera parviceps*** **(Casey)**

Additional records. Canada, Ontario, Moosonee, 51.280, –80.642, Rep. 1 wet yellow pan, 19-23.VI.2010, NBP field party M1WY321 (LFC) 1 female.

Genus *Devia* Blackwelder

11 (28). ***Devia prospera*** **(Erichson) ***
Additional records. Canada, Yukon, Klo-Kut, Old Crow area, VI.1975, trap 2, R.E. Morlan and J.V. Matthews (CNC) 1 male; Stirling bend Stewart R., 6.VI.1975, J.V. Matthews and R.E. Morlan (CNC) 1 male. USA: Alaska, Chena Pump Rd, 136m el., 64.8375, −147.88043, ±30m forest edge *Pop. trem., Pop. bals.*, pitfall, 17-24. VII.2007, B. Mortensen (UAM) 2 specimens; Fairbanks, 930 Fitz Ct, 280m el., 64.9013, −147.52861, ±30m *Betula pap., Pop. trem.*, flight intercept trap, 20.V-3.VI.2007, D.S. Sikes (UAM) 1 specimen; Fairbanks, Creamer's Field, 136-156m el., 64.86719, −147.73433, ±80m *Betula* litter, Berlese, 13.VIII.2007, C. W. O'Brien, L. B. O'Brien (UAM) 3 specimens; Fairbanks, Creamer's Field, 152m el., 64.86808, −147.73764, ±8m wet boreal, pitfall, 15-18.VII.2008, D.S. Sikes, C. Eshleman (UAM) 1 specimen; Fairbanks, Creamer's Field, 152m el., 64.86864, −147.7381, ±6m barrier trap pair, 15-18.VII.2008, D.S. Sikes (UAM) 1 specimen; Fairbanks, Creamer's Field, 154m el., 64.86607, −147.73817, ±8m pitfall, 19-21.VII.2008, D.S. Sikes, R. Spiers (UAM) 6 specimens; Fairbanks, Creamer's Field, 156m el., 64.86852, −147.73779, ±15m mesic forest, pitfall, 19-21.VII.2008, D.S. Sikes, C. Reilly (UAM) 2 specimens; Fairbanks, Creamer's Field, 158m el., 64.86823, −147.73758, ±12m mesic *Betula*, pitfall, 26.VI-18.VII.2008, D.S. Sikes, R. Davidson (UAM) 2 specimens; Fairbanks, Ridgepointe Dr., 253m el., 64.81772, −147.97333, ±61m sparse forest, *Pop. trem.*, pitfall, 17-24.VII.2007, B. Mortensen (UAM) 3 specimens; Fairbanks, UAF 166m el. 64.86673, −147.84297, ±10m black spruce, some white spruce, pollinator vane trap, 0.5m, 27.V-10.VI.2015, J.C. Hagelin (UAM) 1 specimen; Fairbanks: Creamer's Field, 132-152m el., 64.86676, −147.7391, ±20m field edge of *Salix / Betula / Picea* forest & vernal ponds / stream, 28. IV.2007, D.S. Sikes, (UAM) 1 specimen; Fairbanks: Creamer's Field, 132m el., 64.86763, −147.73824, ±500m, birch leaf litter among roots, Berlese, 17.X.2006, M. E. Sikes, M. Ross (UAM) 3 specimens; Fairbanks: Creamer's Field, 141m el., 64.86711, −147.72794, ±54m forest edge, *Populus* litter, Winkler, 31.VI.2010, D.S. Sikes (UAM) 23 specimens; Glenn Hwy, Long Lake, 471m el., 61.80319, −148.23671, ±104m spruce, birch, alder, cottonwood, Winkler, 28.VI.2011, S. Ridling; M. G. Wyatt (UAM) 1 specimen.

Genus *Gnathusa* Fenyes

12 (32). ***Gnathusa tenuicornis*** **Fenyes**
Additional records. USA: Alaska, Donnelly Campground, Richardson Hgwy., 2-10. VIII.1974, J. Matthews and A. Sigowim (CNC) 1 male; 12 Mile Summit, Steese Hwy., 887-907m el., 65.38789, −145.97179, ±50m arctic tundra, Berlese, 23.V.2007, B. Fleshman (UAM) 1 specimen; Prince of Wales Island, Staney Ck. 1B, 77m el., 55.872, −133.06523, ±26m clear-cut, pitfall 4, 11-25.VI.2012, J. Stockbridge, S. Meierotto, I. MacDougall (UAM) 1 specimen.

Genus *Neothetalia* Klimaszewski

13 (35). ***Neothetalia canadiana*** **Klimaszewski**
Additional records. USA: Alaska, 35.5 Mi Dalton Hwy, 271m el., 65.74425, −149.34564, ±4m *Betula, Salix*, white spruce, flight intercept trap, 31.V.-5.VI.2008, D.S. Sikes (UAM) 1 specimen; Denali NP., Mt. Healy, 927m el., 63.74016, −148.95346, ±30m shrub, alder, dwarf birch, grass, clubmoss, 4 yellow pitfalls w/o rainroofs, pooled, 3-16.VIII.2016, J.J. Rykken; D.S. Sikes; R. Stewart-Rogers (UAM) 1 specimen; Fairbanks, 930 Fitz Ct, 280m el., 64.9013, −147.52861, ±30m small stand *Betula* & *Pop. trem.*, flight intercept trap, 29.IV-29.V.2008, D.S. Sikes (UAM) 5 specimens; Prince of Wales Island, Hatchery Ck.2, 134m el., 55.89356, −132.9437, ±34m unthinned 2nd growth, pitfall 1, 9-24.VII.2012, J. Stockbridge, S. Meierotto, I. MacDougall (UAM) 6 specimens; S. Chilkat Range pl.10, 718m el., 58.42621, −135.1575, ±28m subalpine, *Salix arctica, Vacc. caesp.*, pitfall, 8-9.VII.2010, D.S. Sikes (UAM) 1 specimen.

14 (36). ***Neothetalia columbiana*** **(Klimaszewski)**
Additional records. USA: Alaska, Chowiet Isl., 122-152m el., 56.03237, −156.7157, ±50m alpine, under rock, 11.VII.2008, D. M. Collet, G. Humphries (UAM) 1 specimen; Dall Isl., 543m el., 54.99675, −133.01027, ±7m forest, primate dung baited, pitfall, 13-16.VII.2011, D.S. Sikes (UAM) 1 specimen; Juneau, Douglas Isl., 14m el., 58.32316, −134.60909, ±505m old growth hemlock rotting logs, dung, hand, 26. VI.2010, D.S. Sikes (UAM) 1 specimen; Prince of Wales Island, Coffman Cv, 67m el. 55.9795, −132.86256, ±101m thinned 2nd growth, pitfall 2, 13-27.V.2013, J. A. Slowik, A. Hutton (UAM) 2 specimens; same data except, pitfall 4, 13-27.V.2013, J. A. Slowik, A. Hutton (UAM) 1 specimen; pitfall 1, 18-31.V.2011, J. Stockbridge, B. Wong, (UAM) 1 specimen; Lindgren 1, 28.V-11.VI.2012, J. Stockbridge, S. Meierotto, I. MacDougall (UAM) 1 specimen; pitfall 2, 9-24.VII.2012, J. Stockbridge, S. Meierotto, I. MacDougall (UAM) 1 specimen; Prince of Wales Island, Coffman Cv, 78m el., 55.98053, −132.8607, ±15m unthinned 2nd growth, pitfall 1, 10-30.VI.2010, J. Stockbridge, C. Bickford (UAM) 1 specimen; same data except, pitfall 2, 10-30.VI.2010, J. Stockbridge, C. Bickford (UAM) 1 specimen; pitfall 1, 11-25. VI.2012, J. Stockbridge, S. Meierotto, I. MacDougall (UAM) 9 specimens; pitfall 3, 11-25.VI.2012, J. Stockbridge, S. Meierotto, I. MacDougall (UAM) 1 specimen; pitfall 4, 11-25.VI.2012, J. Stockbridge, S. Meierotto, I. MacDougall (UAM) 1 specimen; pitfall 4, 12-27.VII.2011, J. Stockbridge, B. Wong (UAM) 1 specimen; pitfall 3, 12-27.V.2011, J. Stockbridge, B. Wong (UAM) 1 specimen; pitfall 1, 13-27.V.2013, J. A. Slowik, A. Hutton (UAM) 2 specimens; pitfall 2, 13-27.V.2013, J. A. Slowik, A. Hutton (UAM) 2 specimens; pitfall 1, 14-24.VII.2010, J. Stockbridge, C. Bickford (UAM) 2 specimens; pitfall 1-4, 14-28.VI.2011, J. Stockbridge, B. Wong (UAM) 13 specimens; pitfalls 1-4, 18-31.V.2011, J. Stockbridge, B. Wong (UAM) 16 specimens; pitfall 3, 24.VII-7.VIII.2012, J. Stockbridge, S. Meierotto, I. MacDougall (UAM) 5 specimens; pitfalls 1, 3, 25.VI-9.VII.2012 (UAM) 10 specimens; pitfalls 2, 4, 27.VII-7.VIII.2011, J. Stockbridge, B. Wong (UAM) 3 specimens; pitfalls 1, 3, 28.VI-12.VII.2011, J. Stockbridge, B. Wong (UAM) 3 specimens; 30.VI-14. VII.2010 (UAM) 1 specimen; pitfall 2, 30.VI-14.VII.2010, J. Stockbridge, C. Bickford (UAM) 2 specimens; pitfalls 2, 4, 31.V-14. VI.2011, J. Stockbridge, B. Wong (UAM) 3 specimens; pitfalls 1, 3, 4, 9-24.VII.2012, J. Stockbridge, S. Meierotto, I. MacDougall (UAM) 11 specimens; Prince of Wales Island, Hatchery Ck.1, 49m el., 55.92654, −132.95645, ±10m unthinned 2nd growth, Lindgren 1, 13-27. V.2013, J. A. Slowik, A. Hutton (UAM) 1 specimen; same data except, pitfalls 1, 3, 13-27. V.2013, J. A. Slowik, A. Hutton (UAM) 4 specimens; pitfall 4, 15-28.V.2012, J. Stockbridge, S. Meierotto, I. MacDougall (UAM) 1 specimen; pitfall 2, J. 18.V-4.VI.2010, Stockbridge, C. Bickford (UAM) 1 specimen; pitfall 4, 28.V-11.VI.2012, J. Stockbridge, S. Meierotto, I. MacDougall (UAM) 2 specimens; pitfall 3, 30.VII-11.VIII.2010, J. Stockbridge, C. Bickford (UAM) 1 specimen; 4-27.VII.2010 (UAM) 1 specimen; Prince of Wales Island, Hatchery Ck.1, 57m el., 55.92524, −132.95054, ±10m thinned 2nd growth, pitfall 1, 13-27. V.2013, J. A. Slowik, A. Hutton (UAM) 3 specimens; same data except, pitfall 4, 13-27. V.2013, J. A. Slowik, A. Hutton (UAM) 1 specimen; Prince of Wales Island, Hatchery Ck.2, 134m el., 55.89356, −132.9437, ±34m unthinned 2nd growth, Lindgren 2, 10-31.VII.2010, J. Stockbridge, C. Bickford (UAM) 1 specimen; same data except, pitfalls 3, 4, 12-27.VII.2011, J. Stockbridge, B. Wong (UAM) 6 specimens; pitfall 3, 27.VI-12.VII.2011, J. Stockbridge, B. Wong, (UAM) 1 specimen; pitfalls 3, 4, 14-28.V.2012, J. Stockbridge, S. Meierotto, I. MacDougall (UAM) 3 specimens; pitfall 1, 3, 31.V-13.VII.2011, J. Stockbridge, B. Wong, (UAM) 1 specimen; pitfall 1, 12-27.V.2013,

J. A. Slowik, A. Hutton, (UAM) 6 specimens; pitfalls 3, 4, 14-28.V.2012, J. Stockbridge, S. Meierotto, I. MacDougall, (UAM) 2 specimens; pitfalls 3, 4, 18.V-2.VI.2010, J. Stockbridge, C. Bickford, (UAM) 2 specimens; pitfall 3, 25, VI-9.VII.2012, J. Stockbridge, S. Meierotto, I. MacDougall (UAM) 1 specimen; pitfall 3, 27. VII-7.VIII.2011, J. Stockbridge, B. Wong (UAM) 1 specimen; Prince of Wales Island, Hatchery Ck.4, 78m el., 55.88602, –132.8607, ±11m old growth, pitfall 3, 11-25.VI.2012, J. Stockbridge, S. Meierotto, I. MacDougall (UAM) 1 specimen; same data except, pitfall 4, 13-27.V.2013, J. A. Slowik, A. Hutton (UAM) 3 specimens; pitfall 3, 24.VII-6. VIII.2012, J. Stockbridge, S. Meierotto, I. MacDougall (UAM) 5 specimens; pitfall 4, 27.VII-7.VIII.2011, J. Stockbridge, B. Wong (UAM) 1 specimen; pitfall 3, 28.V-11.VI.2012, J. Stoockbridge, S. Meierotto, I. MacDougall (UAM) 1 specimen; 9-24.VII.2012, (UAM) 3 specimens; Prince of Wales Island, Hatchery Ck.4, 82m el. 55.88285, –132.89795, ±27m thinned 2nd growth, pitfall 3, 27.VII-7. VIII.2011, J. Stockbridge, B. Wong (UAM) 1 specimen; pitfall 3, 11-25.VI.2012, J. Stockbridge, S. Meierotto, I. MacDougall (UAM) 1 specimen; pitfall 4, 11-25.VI.2012, J. Stockbridge, S. Meierotto, I. MacDougall (UAM) 1 specimen; pitfall 3, 13-27.V.2013, J. A. Slowik, A. Hutton (UAM) 1 specimen; pitfall 2, 14-30.V.2010, J. Stockbridge, C. Bickford (UAM) 2 specimens; pitfall 1, 15-27.VI.2010, J. Stockbridge, C. Bickford (UAM) 1 specimen; pitfall 3, 15-28.V.2012, J. Stockbridge, S. Meierotto, I. MacDougall (UAM) 1 specimen; Lindgren 2, 25.VI-9.VII.2012, J. Stockbridge, S. Meierotto, I. MacDougall (UAM) 1 specimen; pitfall 1, 27.IV-15.V.2010, J. Stockbridge (UAM) 1 specimen; pitfall 2, 27.VI-14.VII.2010, J. Stockbridge, C. Bickford (UAM) 1 specimen; pitfalls 2, 4, 28.VI-12.VII.2011, J. Stockbridge, B. Wong (UAM) 6 specimens; pitfall 4, 11-25. VI.2012, J. Stockbridge, S. Meierotto, I. MacDougall (UAM) 2 specimens; pitfall 2, 3, 13-27.V.2013, J. A. Slowik, A. Hutton (UAM) 14 specimens; pitfall 4, 15-28.V.2012, J. Stockbridge, S. Meierotto, I. MacDougall (UAM) 5 specimens; pitfall 1, 17-31.V.2011, J. Stockbridge, B. Wong (UAM) 1 specimen; pitfall 3, 24.VII-6.VIII.2012, J. Stockbridge, S. Meierotto, I. MacDougall (UAM) 5 specimens; 25.VII-9.VII.2012, (UAM) 1 specimen; pitfall 2, 27.IV-15.V.2010, J. Stockbridge (UAM) 1 specimen; pitfall 2, 30.V-14.V.2010, J. Stockbridge, C. Bickford (UAM) 1 specimen; pitfall 3, 9-24VII.2012, J. Stockbridge, S. Meierotto, I. MacDougall (UAM) 8 specimens; pitfalls 1, 3, 28.VI-12.VII.2011, J. Stockbridge, B. Wong (UAM) 2 specimens; Prince of Wales Island, Luck Lk. 1 Rd., 101m el., 55.97805, –132.75456, ±10m old growth, pitfall 1, 11-26. VII.2012, J. Stockbridge, S. Meierotto, I. MacDougall (UAM) 3 specimens; pitfall 3, 11-26.VII.2012, J. Stockbridge, S. Meierotto, I. MacDougall (UAM) 6 specimens, 13-27. VI.2012 (UAM) 1 specimen; 16-31.V.2012 (UAM) 9 specimens; pitfall 1, 24.V.-5.VI.2011, J. Stockbridge, B. Wong; pitfall 1, 26.VII-9. VIII.2012, J. Stockbridge, S. Meierotto, I. MacDougall (UAM) 2 specimens, 27.VI-11. VII.2012 (UAM) 1 specimen; pitfalls 3, 4, 27. VI-11.VII.2012, J. Stockbridge, S. Meierotto, I. MacDougall (UAM) 5 specimens pitfall 1, 5-21.VI.2011, J. Stockbridge, B. Wong (UAM) 1 specimen; Prince of Wales Island, Luck Lk. 2 Rd., 105m el., 55.96855, –132.75615, ±10m old growth, pitfalls 1, 2, 11-26.VII.2012, J. Stockbridge, S. Meierotto, I. MacDougall (UAM) 3 specimens; same data except, pitfall 3, 13-27.VI.2012, J. Stockbridge, S. Meierotto, I. MacDougall (UAM) 6 specimens; pitfall 2, 27.VI-11.VII.2012, J. Stockbridge, S. Meierotto, I. MacDougall (UAM) 1 specimen; pitfall 3, 31.V-13.VII.2012, , J. Stockbridge, S. Meierotto, I. MacDougall (UAM) 1 specimen; Lindgren 1, 4-29.VI.2010, J. Stockbridge, C. Bickford (UAM) 1 specimen; Prince of Wales Island, Luck Lk. 3 Rd., 117m el., 55.95347, –132.7708, ±5m old growth, Lindgren 2, pitfall 4, 27.VI-11.VII.2012, , J. Stockbridge, S. Meierotto, I. MacDougall, (UAM) 3 specimens; pitfall 2, 13-27.VI.2012, J. Stockbridge, S. Meierotto, I. MacDougall (UAM) 1 specimen;

pitfall 4, 13-27.VI.2012, J. Stockbridge, S. Meierotto, I. MacDougall (UAM) 5 specimens; pitfall 2, 31.V-13.VI.2012, J. Stockbridge, S. Meierotto, I. MacDougall, (UAM) 1 specimen; pitfall 4, 31.V-13.VI.2012, J. Stockbridge, S. Meierotto, I. MacDougall (UAM) 4 specimens; pitfall 4, 6-19.VII.2011, J. Stockbridge, B. Wong (UAM) 1 specimen; Prince of Wales Island, Luck Point 1A, 174m el., 55.98452, –132.78786, ±26m clear-cut, pitfall 4, 27.VI-11.VII.2012, J. Stockbridge, S. Meierotto, I. MacDougall, (UAM) 1 specimen; pitfall 1, 31.V-14.VII.2012, J. Stockbridge, S. Meierotto, I. MacDougall, (UAM) 1 specimen; Prince of Wales Island, Luck Point 1B, 169m el., 55.98497, –132.787, ±25m clear-cut, pitfall 2, 11-26.VII.2012, J. Stockbridge, S. Meierotto, I. MacDougall (UAM) 1 specimen, 14-27.VI.2012 (UAM) 2 specimens; pitfall 2, 18.V-2.VI.2010, J. Stockbridge, C. Bickford (UAM) 1 specimen; pitfall 2, 31.V-14.VI.2012, J. Stockbridge, S. Meierotto, I. MacDougall (UAM) 1 specimen; pitfall 3, 31.V-14.VII.2012, J. Stockbridge, S. Meierotto, I. MacDougall (UAM) 2 specimens; Prince of Wales Island, Luck Point, 205m el. 55.98261, –132.77986, ±6m thinned 2nd growth, pitfalls 1, 3, 14-26.V.2013, J. A. Slowik, A. Hutton (UAM) 4 specimens; same data except, pitfall 4, 14-27.VI.2012, J. Stockbridge, S. Meierotto, I. MacDougall (UAM) 1 specimen; pitfall 3, 30.VII-11.VIII.2010, J. Stockbridge, C. Bickford (UAM) 1 specimen; pitfall 1, 31.V-14.VII.2012, J. Stockbridge, S. Meierotto, I. MacDougall (UAM) 1 specimen; Prince of Wales Island, Luck Point 2A, 245m el., 55.97953, -132.77156, ±24m clear-cut, pitfall 2, 14-26.V.2013, J. A. Slowik, A. Hutton (UAM) 1 specimen; Prince of Wales Island, Luck Point 2A, 245m el., 55.97953, –132.77156, ±24m clear-cut, pitfall 1, 27.VI-11.VII.2012, J. Stockbridge, S. Meierotto, I. MacDougall (UAM) 3 specimens; same data except, pitfall 4, 31.V-14.VI.2012, J. Stockbridge, S. Meierotto, I. MacDougall (UAM) 1 specimen; Prince of Wales Island, Luck Point 2B, 264m el., 55.97939, –132.77216, ±25m clear-cut, Lindgren 2, 14-27.VI.2012, J. Stockbridge, S. Meierotto, I. MacDougall (UAM) 1 specimen;same data except, Lindgren 1, 31.V-14.VI.2012, J. Stockbridge, S. Meierotto, I. MacDougall (UAM) 3 specimens; pitfall 4, 14-27.VI.2012, J. Stockbridge, S. Meierotto, I. MacDougall (UAM) 1 specimen; pitfall 1, 18.V-2.VI.2010, J. Stockbridge, C. Bickford (UAM) 1 specimen; pitfall 3, 27.VI-11.VII.2012, J. Stockbridge, S. Meierotto, I. MacDougall (UAM) 1 specimen; pitfall 4, 27.VI-11.VII.2012, J. Stockbridge, S. Meierotto, I. MacDougall (UAM) 1 specimen, 31.V-14.VI.2012, (UAM) 3 specimens; pitfall 3, 5-18.VII.2011, J. Stockbridge, B. Wong (UAM) 1 specimen; Prince of Wales Island, Luck Pt., 129m el., 55.98256, –132.77943, ±12m unthinned 2nd growth, pitfall 1, 11-26.VII.2012, J. Stockbridge, S. Meierotto, I. MacDougall (UAM) 2 specimens; same data except, pitfall 2, 11-26.VII.2012, J. Stockbridge, S. Meierotto, I. MacDougall (UAM) 1 specimen; pitfall 3, 4, 11-26.VII.2012, J. Stockbridge, S. Meierotto, I. MacDougall (UAM) 3 specimens; pitfall 4, 14-26.V.2013, J. A. Slowik, A. Hutton (UAM) 2 specimens; pitfall 1, 14-27.VI.2012, J. Stockbridge, S. Meierotto, I. MacDougall (UAM) 2 specimens; pitfalls 12, 4, 18.V-2.VI.2010, J. Stockbridge, C. Bickford, (UAM) 9 specimens pitfall 1, 2-27.VI.2010, J. Stockbridge, C. Bickford (UAM) 3 specimens; pitfall 1, 23.V-5.VI.2011, J. Stockbridge, B. Wong (UAM) 2 specimens; pitfall 1, 26.VII-9.VIII.2012, J. Stockbridge, S. Meierotto, I. MacDougall (UAM) 2 specimens; pitfalls 1, 3, 27.VI-11.VII.2012, J. Stockbridge, S. Meierotto, I. MacDougall (UAM) 1 specimen; pitfalls 1, 2, 27.VI-9.VII.2010, J. Stockbridge, C. Bickford (UAM) 2 specimens; pitfall 2, 31.V-14.VI.2012, J. Stockbridge, S. Meierotto, I. MacDougall (UAM) 1 specimen; pitfall 4, 31.V-14.VI.2012, J. Stockbridge, S. Meierotto, I. MacDougall (UAM) 1 specimen; pitfall 1, 9.VII-1.VIII.2010, J. Stockbridge, C. Bickford (UAM) 1 specimen; pitfall 4, 9.VII-1.VIII.2010, J. Stockbridge, C. Bickford (UAM) 1 specimen; Prince of Wales Island, Staney Ck, 63m el., 55.79723, –133.13467, ±51m unthinned 2nd growth, pitfall

1, 11-25.VI.2012, J. Stockbridge, S. Meierotto, I. MacDougall (UAM) 1 specimen; same data except, pitfall 1, 12-26.V.2013, J. A. Slowik, A. Hutton (UAM) 2 specimens; pitfall 3, 12-26. V.2013, J. A. Slowik, A. Hutton (UAM) 3 specimens; pitfall 1, 13-27.VI.2011, J. Stockbridge, B. Wong (UAM) 1 specimen; pitfall 1, 14-28.V.2012, J. Stockbridge, S. Meierotto, I. MacDougall (UAM) 5 specimens; pitfall 4, 24.VII-6.VIII.2012, J. Stockbridge, S. Meierotto, I. MacDougall (UAM) 1 specimen; pitfall 3, 25.VI-9.VII.2012, J. Stockbridge, S. Meierotto, I. MacDougall (UAM) 1 specimen; pitfall 2, 27.VI-12.VII.2011, J. Stockbridge, B. Wong (UAM) 1 specimen; pitfall 4, 27.VI-12.VII.2011, J. Stockbridge, B. Wong (UAM) 1 specimen; pitfall 1, 28.V-11.VII.2012, J. Stockbridge, S. Meierotto, I. MacDougall (UAM) 1 specimen; Prince of Wales Island, Staney Ck. 1A, 89m el., 55.87134, −133.06755, ±24m clear-cut, pitfall 1, 14-30.V.2010, J. Stockbridge, C. Bickford (UAM) 1 specimen; Prince of Wales Island, Staney Ck. 1B, 77m el., 55.872, −133.06523, ±26m clear-cut, pitfall 1, 11-25VI.2012, J. Stockbridge, S. Meierotto, I. MacDougall (UAM) 1 specimen; same data except, 14-28.V.2012, (UAM) 1 specimen; pitfall 4, 27.VI-12.VII.2011, J. Stockbridge, B. Wong (UAM) 1 specimen; pitfall 1, 9-24.VII.2012, J. Stockbridge, S. Meierotto, I. MacDougall (UAM) 1 specimen; Prince of Wales Island, Staney Ck., 41-45m el., 55.79901, −133.11782, ±20m old growth, pitfall 4, 11-25.VI.2012, J. Stockbridge, S. Meierotto, I. MacDougall (UAM) 1 specimen; same data except, pitfall 1, 12-26.V.2013, J. A. Slowik, A. Hutton (UAM) 1 specimen; pitfall 3, 13-27.VI.2011, J. Stockbridge, B. Wong (UAM) 1 specimen; pitfall 1, 14-28.V.2012, J. Stockbridge, S. Meierotto, I. MacDougall (UAM) 1 specimen; pitfall 4, 24.VII-11.VIII.2010, J. Stockbridge, C. Bickford (UAM) 1 specimen; pitfall 1, 24. VII-6.VIII.2012, , J. Stockbridge, S. Meierotto, I. MacDougall (UAM) 1 specimen; pitfall 3, 27. VI-12.VII.2011, J. Stockbridge, B. Wong (UAM) 1 specimen; Prince of Wales Island, Staney Ck., 50m el., 55.79726, −133.1363, ±20m thinned 2nd growth, Lindgren 1, 11-25.VI.2012, J. Stockbridge, S. Meierotto, I. MacDougall (UAM) 1 specimen; same data except, pitfall 3, 11-25.VI.2012, J. Stockbridge, S. Meierotto, I. MacDougall (UAM) 1 specimen; pitfall 4, 13-27.VI.2011, J. Stockbridge, B. Wong (UAM) 1 specimen; Lindgren 2, 14-28.V.2012, J. Stockbridge, S. Meierotto, I. MacDougall, (UAM) 1 specimen; pitfall 1, 14-28.V.2012, J. Stockbridge, S. Meierotto, I. MacDougall (UAM) 2 specimens; pitfall 4, 27.VI-12. VII.2011, J. Stockbridge, B. Wong (UAM) 1 specimen; Lindgren 2, 28.V-11.VI.2012, J. Stockbridge, S. Meierotto, I. MacDougall (UAM) 1 specimen; S. Chilkat Range pl.10, 718m el. 58.42621, −135.1575, ±28m subalpine, *Salix arctica, Vacc. caesp.*, pitfall, 8-9.VII.2010, D.S. Sikes (UAM) 1 specimen; Tangik Isl., 1m el., 54.14645, −165.48705, ±504m gull nests hilltop, 13.VII.2009, Anaïs Renaud (UAM) 1 specimen.

15 (37). *Neothetalia pallitarsis* (Kirby)

Additional records. USA: Alaska, Denali NP., Igloo Canyon, 970m el., 63.59455, −149.60834, ±30m willow, dwarf birch, shrub, litter, Winkler, 17.VIII.2016, J.J. Rykken, D.S. Sikes, R. Stewart-Rogers (UAM) 1 specimen; same data except, 4 yellow pitfalls w/o rainroofs, pooled, 20.VII-3.VIII.2016, J.J. Rykken, D.S. Sikes, A. Haberski, K. Chen (UAM) 1 specimen; 4 pitfalls w/ rainroofs, pooled, 20.VII-4. VIII.2016, J.J. Rykken, D.S. Sikes, A. Haberski, K. Chen (UAM) 3 specimens; 4 pitfalls w/ rainroofs, pooled, 3-17.VIII.2016, J.J. Rykken, D.S. Sikes, R. Stewart-Rogers (UAM) 2 specimens; litter, Berlese, 4.VIII.2016, J.J. Rykken, D.S. Sikes, A. Haberski, K. Chen (UAM) 1 specimen; Denali NPP., Mt. Healy, 1183m el., 63.75095, −148.96071, ±30m alpine tundra, 4 pitfalls w/ rainroofs, pooled, 19.VII-3. VIII.2016, J.J. Rykken, D.S. Sikes, A. Haberski, K. Chen (UAM) 1 specimen; same data except, 4 pitfalls w/ rainroofs, pooled, 3-16.VIII.2016, J.J. Rykken, D.S. Sikes, R. Stewart-Rogers (UAM) 1 specimen; 4 yellow pitfalls w/o rainroofs, pooled, 19.VII-3.VIII.2016,

J.J. Rykken, D.S. Sikes, A. Haberski, K. Chen (UAM) 2 specimens; Denali NPP., Mt. Healy, 927m el., 63.74016, –148.95346, ±30m shrub, alder, dwarf birch, grass, clubmoss, 10 minute ground search, 16.VIII.2016, D.S. Sikes (UAM) 2 specimens; same data except, litter, Winkler, 16.VIII.2016, J.J. Rykken. D.S. Sikes. R. Stewart-Rogers (UAM) 10 specimens; 4 pitfalls w/ rainroofs, pooled, 19.VII-3.VIII.2016, J.J. Rykken, D.S. Sikes, A. Haberski, K. Chen (UAM) 3 specimens; 4 yellow pitfalls w/o rainroofs, pooled, 19.VII-3.VIII.2016, J.J. Rykken, D.S. Sikes, A. Haberski, K. Chen (UAM) 5 specimens; 4 pitfalls w/ rainroofs, pooled, 2-19.VII2016, J.J. Rykken, D.S. Sikes, A. Haberski, Z. Snelson (UAM) 14 specimens; Denali NPP., Mt. Healy, 927m el., 63.74016, –148.95346, ±30m shrub, alder, dwarf birch, grass, clubmoss, 4 yellow pitfalls w/o rainroofs, pooled, 2.VII-19.VIII.2016, J.J. Rykken, D.S. Sikes, A. Haberski, Z. Snelson (UAM) 1 specimen; same data except, 4 pitfalls w/ rainroofs, pooled, 3-16.VIII.2016, J.J. Rykken, D.S. Sikes, R. Stewart-Rogers (UAM) 13 specimens; litter, Berlese, 3.VIII.2016, J.J. Rykken, D.S. Sikes, A. Haberski, K. Chen (UAM) 2 specimens; Haines, Flower Mtn, 1058m el., 59.37706, -136.33562, ±13m under rocks above snow, hand collected, 18.VII.2010, D.S. Sikes (UAM) 1 specimen; Haines, Flower Mtn., 1010m el., 59.38636, –136.31877, ±1km misc. alpine habitats, hand collected, 15-19.VII.2010, D.S. Sikes (UAM) 1 specimen; Haines, Flower Mtn., 1116m el., 59.37788, –136.32814, ±1km on snow fields, 15-19.VII.2010, D.S. Sikes (UAM) 32 specimens; Heintzleman Ridge, pl.17(2B), 642m el., 58.42809, –134.43529, ±29m subalpine rocks, *Alnus*, hand collected, 30.VI.2010, D.S. Sikes (UAM) 1 specimen; Prince of Wales Island, Staney Ck. 1A, 89m el., 55.87134, –133.06755, ±24m clear-cut, pitfall 2, 9-24.VII.2012, J. Stockbridge, S. Meierotto, I. MacDougall (UAM) 1 specimen; S. Chilkat Range, 712m el., 58.42632, –135.1563, ±104m alpine snowfield, hand collected, 8.VII.2010, D.S. Sikes (UAM) 1 specimen; S. Chilkat Range pl.18, 668m el., 58.42455, –135.14445, ±30m subalpine *Tsuga, Luetkea*, hand collected, 9.VII.2010, D.S. Sikes (UAM) 1 specimen; same data except, pitfall, 8-9.VII.2010, D.S. Sikes (UAM) 3 specimens; S. Chilkat Range pl.20, 659m el., 58.4248, –135.14096, ±29m subalpine *Tsuga, Vaccinium*, pitfall, 8-9.VII.2010, D.S. Sikes (UAM) 5 specimens.

Genus *Ocyusa* Kraatz

16 (38). ***Ocyusa canadensis*** **Lohse**
Additional records. Canada: New Brunswick, Kouchibouguac Nat. Pk., 19.VI.1978, D.B. Lyons, Code 70927 (CNC) 1 female. BC, Rt. 97 near Smart River, 59.9326, –131.7556, 6.VI.2019, R.P. Webster, small pond with emergent vegetation, treading (RWC), 1 male; Yukon, Dempster Hwy, mi. 236, Eagle R., 1400', 21, VII.1978, J.M.Campbell and A. Smetana (CNC) 1 male. USA, Alaska, Alaska Hwy. Mi 1249, Deadman Lake, 6.VII.1968, A. Smetana (CNC) 2 males, 2 females, 1 sex undetermined.

Genus *Oxypoda* Mannerheim

17 (49). ***Oxypoda irrasa*** **Mäklin**
Additional records. Canada, Yukon, Takhini Salt flats, 60.855, –135.717, 663m, alkaline clay, pitfall trap, *Salicornia* habitat, coll. B. Godin and D. Horwood (LFC) 1 male; Yukon, Whitehorse, Granger, Whitehorse, Granger, 60.7087, –135.097, 657m, garden, hand picked, coll. B. Godin (BGC, LFC) 1 male, 1 female. USA, Alaska, 35.5 Mi Dalton Hwy, 271m el., 65.74425, –149.34564, ±4m *Betula, Salix*, white spruce, flight intercept trap, 31.V-5.VI.2008, D.S. Sikes (UAM) 1 specimen; Fairbanks, 930 Fitz CT, 280-293m el., 64.90099, –147.52808, ±15m yard; cut spruce, hand, 1-2.VI.2013, D.S. Sikes (UAM) 2 specimens; Fairbanks, 930 Fitz Ct, 280m el., 64.9013, –147.52861, ±30m *Betula* and *P. tremuloides*, flight intercept trap, 16.VIII-22.IX.2007, D.S. Sikes (UAM) 2 specimens; same data except, 26.VIII-10.IX.2006, (UAM) 1 speciemen; 29.IV-29.V.2008, (UAM) 2 specimens; Prudhoe Bay

Rd., 4 mi SW Chandalar Cmp., 68.0666, −149.6666, 7.VII.1978, 3000', Smetana and Campbell (CNC) 1 female.

18 (50). *Oxypoda hiemalis* Casey

Additional records. Canada: Ontario, Moosonee, 51.24466, –80.67767, Rep. 2 mesic, pitfall trap, 23-26.VI.2010, NBP field party M3MP731 (LFC) 1 female; Northwest Terr., Norman Wells, 65.26694, –126.72827, Rep. 2 wet, pitfall trap, 7-11.VI.2011, NBP field party N2MP811, N2MP812 (LFC) 2 females; same label except: 11-14.VI.2011, N2MP821 (LFC) 1male; 65.30494, –126.70873, Rep. 3 mesic, yellow pan, 14-17.VI.2011, NBP field party N3MY131 (LFC) 1 female; Yukon, Takhini Salt flats, 60.855, –135.717, 1-15.V.2014, 2-16. VI.2014, pitfall trap, alkaline clay, near willow, coll. B. Godin and D. Horwood (LFC) 1 male, 1 female.

USA, Alaska, Kenai, Kenai Pen., 24.VI.1951, W.J. Brown (CNC) 1 male, 1 sex undetermined.

19 (51). *Oxypoda lacustris* Casey

Additional records. Canada, Northwest Territories, Yellowknife, 62.51923, –113.38494, Rep. 2 wet, yellow pan, 15-18.VII.2011, NBP field party Y2WY432, Y2WY433 (LFC) 1 male, 1 female; same label except: 62.52110, –113.38174, 7-11.VII.2011, Y1WP215 (LFC) 1 female; 62.51591 -113.39146, Re. 3 wet, 11-15.VII.2011, Rep. 3 wet, yellow pan, Y3WY121 (LFC) 1 female; Norman Wells, 65.28901, –126.84077, Rep. 3 wet, pitfall trap, 11-14.VI.2011, NBP field party N3WP822 (LFC) 1 male; same label except: yellow pan, N3WY424 (LFC) 1 female; 65.29112, –126.62262, N1WP411 (LFC) 1 female; 65.26694, –126.72827, Rep. 2 mesic, yellow pan, 11-14.VI.2011, NBP field party N2MY621 (LFC) 1 female.Yukon, Stewart River S of McQuesten above Stirling Bend, 63.5047, −137.3283, 7.VII.1975, J.V. Matthews (CNC) 1 male. Quebec, Schefferville, 54.85350, –66.67043, Rep. 3 wet, pitfall, 6-9. VII.2010, NBP field party S3WP531 (LFC) 1 female. Ontario, Moosonee, 51.24690, -80.68102, Rep. 3 mesic, pitfall trap, 17-20. VI.2010, NBP field party M3MP912 (LFC) 1 female; same label except: 23-26.VI.2010, M3MP631 (LFC) 1 female; 51.24466, -80.67767, Rep. 2 mesic, pitfall trap, 20-23. VI.2010, NBP field party M2MP121, M2MP122, M2MP131 (LFC) 2 males, 1 female. USA, Alaska, Becharof NWR, 8m el., 57.85395, –156.51857, ±20m lakeshore sand, detritus, hand collected, 13.VI.2007, S. Clawson (UAM) 2 specimens; Chena Pump Rd, 136m el., 64.8375, –147.88043, ±30m forest edge *Pop. trem., P. bals.*, pitfall, 17-24.VII.2007, B. Mortensen (UAM) 1 specimen; Chena Pump Rd, 136m el., 64.8375, –147.88043, ±30m forest edge *Pop. trem., Pop. bals.*, pitfall, 17-24. VII.2007, B. Mortensen (UAM) 4 specimens; Chichagof Is. pl.04, 728m el., 57.98643, -135.20468, ±4m rich meadow, *Geranium erianthum, Artemisia arctica, Sanguisorba canadensis*, pitfall, 14-15.VII.2013, S. Ridling (UAM) 1 specimen; Eielson AFB, Horseshoe Lake area (site B), 188m el., 64.79488, -147.009, ±10m black spruce, grass, aspen, bottom catch, hanging Malaise 4m, 26.VI-30. VII.2013, S. Busby (UAM) 1 specimen; Fairbanks, 930 Fitz Ct, 280m el., 64.9013, -147.52861, ±30m *Betula pap., Pop. trem.*, flight intercept trap, 20.V-3.VI.2007, D.S. Sikes (UAM) 1 specimen; Fairbanks, Creamer's Field, 152m el., 64.86808, –147.73764, ±8m wet boreal, pitfall, 15-18.VII.2008, D.S. Sikes, C. Eshleman (UAM) 5 specimens; Fairbanks, Creamer's Field, 158m el., 64.86823, -147.73758, ±12m mesic *Betula*, pitfall, 26.VI.-18.VII.2008, D.S. Sikes, R. Davidson (UAM) 2 specimens; Fairbanks, Creamer's Field, Reindeer Lake, 146m el., 64.88118, –147.77643, ±10m b. spruce bog, hanging Malaise, 5.55m, 29.VI-13.VII.2016, J.C. Hagelin (UAM) 2 specimens; Fairbanks/North Pole, Richardson Highway, 157m el., 64.7798, –147.55915, ±10m mixed black & white spruce forest with open/ boggy areas & dry zones, bottom catch, hanging Malaise 4.7m, 19.VI-9.VII.2013, J. Hagelin, A. H. Scurr (UAM) 1 specimen; Glacier View Trail, 116m el., 60.1866, –149.6308, ±50m dry stream, pitfall trap (w/roof), 26-29.VII.2016, J.J. Rykken (UAM) 1 specimen; Kotzebue, 2m el., 66.8923, –162.59494, ±36m stagnant pond, dip net, 6.VI.2016, D.S. Sikes, K.M. Daly (UAM) 1 specimen; Naknek, 2-5m el., 58.73973, –157.0636, ±20m creekside/ocean beach confluence, under boards and drift wood, hand collected, 10.VI.2007, D.S. Sikes (UAM) 2 specimens; nr. Livengood, Elliot Highway, Mile 84.5, 252m el., 65.39532, –148.9097, ±10m birch, b. spruce on n. slope, hanging Malaise, 4.16m, 14-28.VII.2016, J.C. Hagelin (UAM) 1 specimen; Tetlin NWR, Fish Camp lake, 508m el., 63.13509, –142.30251, ±10m open black spruce; blowdown area, hanging Malaise, 7.4m, 14-28.VI.2016, J.C. Hagelin (UAM) 1 specimen; Tetlin NWR, Fish Camp Lake, 510m el., 63.13932, –142.29695, ±10m stunted black spruce; open; willow understory, hanging Malaise, 5.46m, 14-28.VI.2016, J.C. Hagelin (UAM) 1 specimen; White Mountain Recreation Area, Nome Creek Site 5, 589m el., 65.34077, –146.9275, ±10m open tussock and standing water, willow and bog birch, hanging Malaise, 4.95m, 12.VI-3.VII.2015, J.C. Hagelin (UAM) 2 specimens; Wosnesenski, 10m el., 55.21697, –161.36644, ±105m under rocks, dung, wood, meadows w/lupine, 11.VII.2009, D.S. Sikes (UAM) 6 specimens; Mi 1249 Alaska Hwy., Deadman Lake, 6-7.VII.1968, J.M. Campbell and A. Smetana (CNC) 1 female; Mi 1252 Alaska Hwy., 7.VII.1968, J.M. Campbell and A. Smetana (CNC) 1 male; Mi 1259 Alaska Hwy., 7.VII.1968, J.M. Campbell and A. Smetana (CNC) 1 male; Hope, Kenai Pen., 12.VI.1951, W.J. Brown (CNC) 1 female.

20 (52). *Oxypoda demissa* Casey
Additional records. Canada, Yukon, Takhini Salt flats, 60.855, –135.717, 14.V.2014, 16.VI.2014, pitfall trap, alkaline clay, near willow, coll. B. Godin and D. Horwood (LFC) 2 females.

21 (54). *Oxypoda lucidula* Casey
Additional records. Canada, Northwest Territories, Yellowknife, 62.50293, –113.40479, Rep. 2 mesic, yellow pan, 11-15.VII.2011, NBP field party Y2MY622 (LFC) 1 female. USA, Alaska, Anaktuvuk Pass, 647m el., 68.14049, –151.74091, ±250m *Salix* leaf litter, Berlese, 19.V.2016, D.S. Sikes; K.M. Daly (UAM) 1 specimen; Alaska Hwy. Mi. 1249, Deadman Lake, 6.VII.1968, A. Smetana (CNC) 1 male, 1 female, 1 sex undetermined.

22 (55). *Oxypoda robusticornis* Bernhauer
Additional records. USA, New Hampshire, Mt. Washington, Alpine Garden, 5300 feet, 12.IX.1987, dwarf *Salix*, *Vaccinium*, leatherleaf litter, J.M. Campbell and A. Davies (CNC) 1 male.

23 (58). *Oxypoda frigida* Bernhauer
Additional records. USA: Alaska, [Arctic Nat. Wild Ref. (ANWR)] Coleen R., –15 mi NW. Porcupine R., 326-346m el., 67.77621, –142.39573, ±981m gravel bars, mature forest, low bluff with sage, sweep, hand, J.J. Kruse, 25-28.VII.2001, J.J. Kruse (UAM) 1 specimen; Chena Pump Rd, 136m el., 64.8375, –147.88043, ±30m forest edge *Populus tremuloides, P. balsamea,* pitfall, 17-24.VII.2007, B. Mortensen (UAM) 1 specimen; Chichagof Is. pl.03, 702m el., 57.98708, –135.20357, ±3m meadow, *Nephrophyllidium crista-galli, Geranium erianthum, Artemisia arctica*, pitfall, 14-15.VII.2013, S. Ridling (UAM) 1 specimen; Chichagof Is. pl.19, 991m el., 57.98173, –135.21889, ±4m meadow heath, *Phyllodoce glandulifera, Senecio triangularis, Luetkea pectinata*, pitfall, 14-15.VII.2013, S. Ridling (UAM) 4 specimens, 16-17.VII.2013, (UAM) 3 specimens; Denali NP., Mt. Healy, 595m el., 63.73478, –148.92911, ±30m forest, mostly spruce, some aspen, litter, Winkler, 16.VIII.2016, J.J. Rykken, D.S. Sikes, R. Stewart-Rogers (UAM) 3 specimens; same data except, litter, Berlese, 19.VII.2016, J.J. Rykken, D.S. Sikes, A. Haberski, Z. Snelson (UAM) 3 specimens; 4 yellow pitfalls w/o rainroofs, pooled, 2-19.VII.2016, J.J. Rykken, D.S. Sikes, A. Haberski, Z. Snelson (UAM) 2 specimens; litter, Berlese, 3.VIII.2016, J.J. Rykken, D.S. Sikes, A. Haberski, K. Chen (UAM) 3 specimens; Denali NPP., Mt. Healy, 927m el.,

63.74016, –148.95346, ±30m shrub, alder, dwarf birch, grass, clubmoss, litter, Winkler, J.J. Rykken; 16.VIII.2016, D.S. Sikes, R. Stewart-Rogers (UAM) 5 specimens; same data except, litter, Berlese, 19.VII.2016, J.J. Rykken, D.S. Sikes, A. Haberski, Z. Snelson (UAM) 2 specimens; 4 pitfalls w/ rainroofs, pooled, 2-19.VII.2016, J.J. Rykken, D.S. Sikes, A. Haberski, Z. Snelson (UAM) 1 specimen; Denali NPP., nr Teklanika R., 833m el., 63.64261, –149.57181, ±30m spruce litter, Berlese, 4.VIII.2016, J.J. Rykken, D.S. Sikes, A. Haberski, K. Chen (UAM) 2 specimens; same data except, litter, Berlese, 20.VII.2016, J.J. Rykken, D.S. Sikes, A. Haberski, Z. Snelson (UAM) 4 specimens; Denali NPP., Primrose Ridge, 883m el., 63.72686, –149.40852, ±28m shrub, dwarf birch, litter, Berlese, 2.VIII.2016, J.J. Rykken, D.S. Sikes, A. Haberski, K. Chen (UAM) 6 specimens; Denali NPP., Toklat, 960m el., 63.52373, –150.04829, ±30m spruce, litter, Winkler, 3-18.VIII.2016, J.J. Rykken, D.S. Sikes, A. Haberski, R. Stewart-Rogers (UAM) 2 specimens; Denali NPP., Wright Mtn., 772m el., 63.71967, –149.48056, ±30m forest, mostly spruce, some aspen, litter, Berlese, 19.VII.2016, J.J. Rykken, D.S. Sikes, A. Haberski, Z. Snelson (UAM) 1 specimen; litter, Berlese, 2.VIII.2016, J.J. Rykken, D.S. Sikes, A. Haberski, K. Chen (UAM) 2 specimens; Fairbanks, 930 Fitz Ct, 280-293m el., 64.90127, –147.52797, ±30m *Populus, Betula, Alnus* leaf litter, Berlese, 10. X.2013, D.S. Sikes, (UAM) 1 specimen; Fairbanks, Murphy Dome Rd., 635m el., 64.9428, –148.29016, ±5m unglaciated; mixed *Betula, Picea, Alnus* forest, Berlese, 12.IX.2009, G. Henry (UAM) 6 specimens; Fairbanks, Murphy Dome Rd., 638m el., 64.94277, –148.29033, ±5m unglaciated; *Betula, Picea, Alnus* forest, Berlese, 12.IX.2009, G. Henry (UAM) 14 specimens; Fairbanks, Murphy Dome Rd. unglaciated, 630m el., 64.9428, –148.29016, ±5m mixed *Betula, Picea, Alnus* forest, Berlese, 12.IX.2009, G. Henry (UAM) 2 specimens; Fairbanks, Murphy Dome Rd. unglaciated, 633m el., 64.94277, –148.29033, ±5m mixed *Betula, Picea, Alnus* forest, Berlese, 12.IX.2009, G. Henry (UAM) 3 specimens; Glenn Hwy, Long Lake, 471m el., 61.80319, –148.23671, ±104m spruce, birch, alder, cottonwood, Winkler, 28.VI.2011, S. Ridling, M. G. Wyatt (UAM) 1 specimen; Haines, Flower Mtn. pl.04, 917m el., 59.39156, –136.31377, ±29m alpine *Salix, Valariana* shrubs, pitfall, 17-18.VII.2010, D.S. Sikes (UAM) 1 specimen; Haines, Flower Mtn. pl.05, 939m el., 59.39038, –136.31345, ±29m alpine *Salix, Valariana*, shrubs, pitfall, 17-18.VII.2010, D.S. Sikes (UAM) 1 specimen; Prince of Wales Island, Hatchery Ck.1, 50m el., 55.92444, –132.93938, ±4m old growth, Berlese, 10.VII.2010, J. Stockbridge, C. Bickford (UAM) 1 specimen; same data except, Berlese, 27. V.2013, J. A. Slowik, A. Hutton (UAM) 2 specimens; Prince of Wales Island, Hatchery Ck.2, 134m el., 55.89356, –132.9437, ±34m unthinned 2nd growth, pitfall 4, 12-27.V.2013, J. A. Slowik, A. Hutton (UAM) 5 specimens; same data except, pitfall 3, 14-28.V.2012, J. Stockbridge, S. Meierotto, I. MacDougall (UAM) 2 specimens; pitfall 4, 14-28.V.2012, J. Stockbridge, S. Meierotto, I. MacDougall (UAM) 1 specimen; pitfall 2, 14-28.V.2012, J. Stockbridge, S. Meierotto, I. MacDougall (UAM) 1 specimen; pitfall 3, 25.VI-9.VII.2012, J. Stockbridge, S. Meierotto, I. MacDougall (UAM) 1 specimen; pitfall 4, 27.VII-7. VIII.2011, J. Stockbridge, B. Wong (UAM) 1 specimen; pitfall 3, 28.V-11.VI.2012, J. Stockbridge, S. Meierotto, I. MacDougall (UAM) 1 specimen; pitfall 4, 30.VII-11. VIII.2010, J. Stockbridge, C. Bickford (UAM) 1 specimen; pitfall 4, 9-24.VII.2012, J. Stockbridge, S. Meierotto, I. MacDougall (UAM) 1 specimen; Prince of Wales Island, Luck Lk. 1 Rd., 101m el., 55.97805, –132.75456, ±10m old growth, Berlese, 22. V.2012, J. Stockbridge, S. Meierotto, I. MacDougall (UAM) 2 specimens; same data except, Berlese, 5.VI.2011, J. Stockbridge, B. Wong (UAM) 1 specimen; Prince of Wales Island, Staney Ck. 1A, 89m el., 55.87134, –133.06755, ±24m clear-cut, pitfall 4, 16-31. V.2011, J. Stockbridge, B. Wong (UAM)

1 specimen; S. Chilkat Range, 664m el., 58.42464, −135.14903, ±104m alpine meadow litter, 2 mini-Winklers, 8.VII.2010, D.S. Sikes (UAM) 1 specimen; S. Chilkat Range pl.18, 668m el., 58.42455, −135.14445, ±30m subalpine *Tsuga, Luetkea*, hand collected, 9.VII.2010, D.S. Sikes (UAM) 1 specimen; Wosnesenski, 10m el., 55.21697, −161.36644, ±105m under rocks, dung, wood, meadows w/lupine, 11.VII.2009, D.S. Sikes (UAM) 1 specimen.

24 (60). *Oxypoda subpolaris* Casey

Additional records. USA, Alaska, Selawik NWR, Kugarak Riv., 10m el., 66.56171 −158.99939, ±54m *Betula, Alnus* litter, 2 Winklers, 23.VI.2010, D.S. Sikes (UAM) 1 specimen.

25 (61). *Oxypoda sylvia* Casey

Additional records. USA: Alaska, Denali NPP., Mt. Healy, 1183m el., 63.75095, −148.96071, ±30m alpine tundra, 4 pitfalls w/ rainroofs, pooled, 19.VII-3.VIII.2016, J.J. Rykken, D.S. Sikes, A. Haberski, K. Chen (UAM) 1 specimen; Denali NPP., Mt. Healy, 595m el., 63.73478, −148.92911, ±30m forest, mostly spruce, some aspen, 10 minute ground search, 16.VIII.2016, D.S. Sikes (UAM) 2 specimens; same data except, 4 pitfalls w/ rainroofs, pooled, 19.VII-3.VIII.2016, J.J. Rykken, D.S. Sikes, A. Haberski, K. Chen (UAM) 8 specimens; 2-19.VII.2016, J.J. Rykken, D.S. Sikes, A. Haberski, Z. Snelson (UAM) 2 specimens; 3-16.VIII.2016, J.J. Rykken, D.S. Sikes, R. Stewart-Rogers (UAM) 28 specimens; Denali NPP., Mt. Healy, 927m el., 63.74016, −148.95346, ±30m shrub, alder, dwarf birch, grass, clubmoss, 4 pitfalls w/ rainroofs, pooled, 2-19.VII.2016, J.J. Rykken, D.S. Sikes, A. Haberski, Z. Snelson (UAM) 2 specimens; same data except, 2.VII-19.VIII.2016, (UAM) 1 specimen; 3-16.VIII.2016, J.J. Rykken; D.S. Sikes; R. Stewart-Rogers (UAM) 19 specimens; Denali NPP., nr Teklanika R., 833m el., 63.64261, −149.57181, ±30m spruce, 4 pitfalls w/ rainroofs, pooled, 20.VII-4.VIII.2016, J.J. Rykken, D.S. Sikes, A. Haberski, K. Chen (UAM) 1 specimen; same data except, 3-17.VIII.2016, J.J. Rykken, D.S. Sikes, R. Stewart-Rogers (UAM) 1 specimen; Denali NPP., Primrose Ridge, 883m el., 63.72686, −149.40852, ±28m shrub, dwarf birch, 4 pitfalls w/ rainroofs, pooled, 19.VII-2.VIII.2016, J.J. Rykken, D.S. Sikes, A. Haberski, K. Chen (UAM) 5 specimens; same data except, 2-17.VIII.2016, J.J. Rykken, D.S. Sikes, R. Stewart-Rogers (UAM) 4 specimens; 2-20.VII.2016, J.J. Rykken, D.S. Sikes, A. Haberski, Z. Snelson (UAM) 3 specimens; Denali NPP., Toklat, 960m el., 63.52373, −150.04829, ±30m spruce, 4 yellow pitfalls w/o rainroofs, pooled, 4-18.VIII.2016, J.J. Rykken, D.S. Sikes, R. Stewart-Rogers (UAM) 1 specimen; Denali NPP., Wright Mtn., 772m el., 63.71967, −149.48056, ±30m forest, mostly spruce, some aspen, 4 pitfalls w/ rainroofs, 4 yellow pitfalls w/o rainroofs, 2-20.VII.2016, J.J. Rykken, D.S. Sikes, A. Haberski, Z. Snelson (UAM) 10 specimens; Kenai Fjords National Park, Exit Creek, 123m el., 60.1845, −149.6294, ±50m creek edge, pitfall trap (w/roof), 26-29.VII.2016, J.J. Rykken (UAM) 1 specimen; Kenai Fjords National Park, Glacier View Trail, 132m el., 60.1842, −149.6378, ±50m stream, pitfall trap (yellow), 26-29.VII.2016, J.J. Rykken (UAM) 2 specimens; Fairbanks, 930 Fitz Ct, 280m el., 64.9013, −147.52861, ±30m *Betula* & *P. tremulo.*, flight intercept trap, 16.VIII-22.IX.2007, D.S. Sikes (UAM) 13 specimens; Fairbanks, Creamer's Field, 152m el., 64.86864, −147.7381, ±6m barrier trap pair, 15-18.VII.2008, D.S. Sikes (UAM) 5 specimens; Fairbanks, Ester Dome, 719m el., 64.8768, −148.06427, ±61m summit, *Pop. trem.*, pitfall, 17-24.VII.2007, B. Mortensen (UAM) 1 specimen; Mi 179 Dalton Hwy, 365m el., 67.29803, −150.15699, ±7m 3/4 Alder, white spruce, *Betula* forest, 5-31.VII.2008, D.S. Sikes (UAM) 23 specimens; Prince of Wales Island, Coffman Cv, 78m el., 55.98053, −132.8607, ±15m unthinned 2nd growth, pitfalls 2, 3, 4, 24.VII-11.VIII.2010, J. Stockbridge, C. Bickford (UAM) 6 specimens; Prince of Wales Island, Luck Lk. 3 Rd., 117m el., 55.95347, −132.7708, ±5m old growth, pitfall 3, 30.VII-11.VIII.2010, J. Stockbridge, C. Bickford (UAM) 1 specimen; Prince of Wales Island, Luck Pt., 129m el.,

55.98256, −132.77943, ±12m unthinned 2nd growth, pitfall 2, 11-26.VII.2012, J. Stockbridge, S. Meierotto, I. MacDougall (UAM) 1 specimen; Sterling, Corey St, 90m el., 60.53953, −150.83232, ±150m disturbed *P. glauca* forest, uv light, 6.IX.2004, D.M. Collet (UAM) 1 specimen; same data except, pheromone trap, 26.VII.2004, D.M. Collet (UAM) 1 specimen.

Genus *Parocalea* Bernhauer

26 (62). *Parocalea nearctica* Lohse
Additional records. USA, Alaska, Anaktuvuk Pass, 647m el., 68.14049 −151.74091, ±250m under rocks, forceps, aspirator, 19.V.2016, K. Hildebrandt; D.S. Sikes; K.M. Daly (UAM) 1 specimen.

27 (63). *Parocalea pseudobaicalica* Lohse
Additional records. Canada, Yukon, Mile 158, Dempster Hwy, 3.VII.1979, Lot 1, B.F. & J.l. Carr (LFC) 2 males; Carcross, Wheaton River, 60.2042, −135.2817, 934m, 19.VII.2010, sand bank, hand collecting, coll. B. Godin (BGC) 1 male; same data except: 60.2085, −135.2894, 931m, soil sifting, Willow litter, coll. B. Godin (LFC) 1 female; 6 km E Old Crow, Klo-kut Archeol. Site, 67.5666, −139.6833, 15.VI.1975, pitfall trap, MRA-5, J.V. Matthews (CNC) 1 male.

Tribe Tachyusini C.G. Thomson

Genus *Brachyusa* Mulsant and Rey

28 (68). *Brachyusa helenae* (Casey)
Additional records. Canada, Northwest Territories, Reindeer Sta., Caribou Hills, 2. VII.1972, A. Smetana (CNC) 1 male, 4 females.

Genus *Gnypeta* C.G. Thomson

29 (73). *Gnypeta ashei* Klimaszewski
Additional records. Canada, Nunavut, Hazen Camp, 81.82, −71.3538, 22.VI.1966, Corbet, Nettleship, marsh (CNC) 1 male, 2 females. USA, Alaska, Prudhoe Bay Rd., 10 mi N Galbraith Lk., 68.5833, −149.50, 4.VII.1978, 2600 feet, A. Smetana and J.M. Campbell (CNC) 2 males, 1 female.

30 (76). *Gnypeta brincki* Palm *
Additional records. Canada, Northwest Territories, East Channel, MacKenzie Delta, 12. VI.1973, FWI – Pipeline Project EC2 0120673 PT (CNC) 1 male.

USA, Alaska, Selawik NWR, Kugarak R., 8m el., 66.56137, −158.99834, ±304m collections made in an around USFWS bunkhouse, *Spiraea*, dwarf birch, *Equisetum*, 23-24.VI.2010, D.S. Sikes (UAM) 1 specimen; Prudhoe Bay Rd., 10 mi N Galbraith Lk., 68.5833, −149.50, 4.VII.1978, 2600 feet, A. Smetana and J.M. Campbell (CNC) 3 females.

31 (77). *Gnypeta dentata* Klimaszewski
Additional records. Northwest Territories, Reindeer Sta., Caribou Hills, 2.VII.1972, A. Smetana (CNC) 1 female.

32 (79). *Gnypeta sellmani* Brundin *
Additional records. Canada, Labrador, L'Anse au Loup, 9.VIII.1972, J.M. Campbell (CNC) 1 female; Northwest Territories, Reindeer Sta., Caribou Hills, 2.VII.1972, A. Smetana (CNC) 3 males, 2 females; Yukon, Dempster Hwy mi 259, Richardson Mts., 1900 feet, 21.VII.1978, A. Smetana and J.M. Campbell (CNC) 1 male. USA, Alaska, Prudhoe Bay Rd., 10 mi N Galbraith Lk., 68.5833, −149.50, 2600 feet, 4. VII.1978, A. Smetana and J.M. Campbell (CNC) 1 male, 4 females.

Tribe Myllaenini Ganglbauer

Genus *Myllaena* Erichson

33 (86). *Myllaena insomnis* Casey
Additional records. USA: Alaska, Norutak Lake, 229m el., 66.79609, −154.27638, ±19m, on sandy beach and under beach rocks, 23. VI.2010, D.S. Sikes (UAM) 1 specimen; Alaska Hwy. Mi. 1249, Deadman Lake, 6.VII.1968,

A. Smetana (CNC) 1 female, 4 sex undetermined; Prudhoe Bay Rd., 2 mi S Grayling Lk., 66.9166, −150.4166, 9.VII.1978, 1300 feet, J.M. Campbell and A. Smetana (CNC) 3 males, 1 female, 1 sex undetermined; Prudhoe Bay Rd., 0.5 mi N Yukon R., 9.VII.1978, 400 feet, J.M. Campbell and A. Smetana (CNC) 1 female; Prudhoe Bay Rd., 8 mi N South Fork Koyukuk River, 67.2166, −150.1166, 1000 feet, 8.VII.1978, A. Smetana and J.M. Campbell (CNC) 1 sex undetermined; Prudhoe Bay Rd., Bonanza Cr., 900 feet, 66.6666, −150.6666, 2.VII.1978, A. Smetana and J.M. Campbell (CNC) 1 sex undetermined.

Tribe Liparocephalini Fenyes

Genus *Amblopusa* Casey

34 (88). *Amblopusa alaskana* Ahn and Ashe
Additional records. Alaska, Aguttu, 2m el., 52.41932, –173.70123, +/– under driftwood, forceps, 5.VIII.2010, D. Sikes, UAM100336108 (UAM) 1 male.

Genus *Paramblopusa* Ahn and Ashe

35 (92). *Paramblopusa eoa* Ahn and Maruyama *
Additional records. USA: Alaska, Adak Isl., Finger Bay, 1-2m el., 51.83237, –176.62752, ±9m splash zone rocks with *Verrucaria*, in crevices, forceps, 25.VIII.2012, D.S. Sikes (UAM) 5 specimens; Agattu, 2m el., 52.41932 173.70123, ±11m under drift wood, 5.VIII.2010, D.S. Sikes (UAM) 2 specimens; Agattu, 2m el., 52.41998, –173.70213, ±55m splash zone cliff cracks, 5.VIII.2010, D.S. Sikes (UAM) 1 specimen; Amatignak Isl., 2-8m el., 51.25868, –179.07648, ±254m under drift wood, rocks, 10. VI.2008, D.S. Sikes (UAM) 4 specimens; Attu, 4m el., 52.82239, –173.18022, ±5m mid beach/ splash zone boulder with cracks, 4.VIII.2010, D.S. Sikes (UAM) 2 specimens; Kasatochi, 12m el., 52.17177, –175.52917, ±11m under rocks, hand collected, 11.VI.2008, D.S. Sikes (UAM) 1 specimen; Kiska, 10m el., 52.1329, –177.59616, ±35m rock outcrop cracks midbeach, 6.VIII.2010, D.S. Sikes (UAM) 2 specimens; Kiska, 3m el., 52.05022, –177.50598, ±13m splash zone boulder and cliff base cracks, 6. VI.2010, D.S. Sikes (UAM) 1 specimen; St. George Isl., 3m el., 56.57969, –169.68292, ±24m under back beach driftwood, forceps, 9. VIII.2012, D.S. Sikes (UAM) 2 specimens; Wrangell Is., Pats Landing, 56.3448, –132.3496, ±50m under rock, mid-intertidal, 3.III.2019, S. Wise-Eagle (UAM) 1 specimen.

Genus *Liparocephalus* Mäklin

36 (93). *Liparocephalus brevipennis* Mäklin
Additional records. Alaska, Aleutian Is., Little Sitkin, el. 13 m, 51.98173, –178.48886, +/– 35m under beach wrack, driftwood, 9.VI.2008, D. Sikes, (UAM) 2 males; 2 females; 2 sex undetermined; Kasatochi Is., 3m el., 52.16254, –175.52328, +/– 14m rock outcrop, splash zone w/ marine algae, 30.VIII.2017, D. Sikes et al., (UAM) 2 females; Adak, Kuluk Beach, 1-2m el., 51.87835, –176.62871, +/– 23m intertidal rock crevices, forceps, 27.VIII.2017, D. Sikes et al., (UAM) 1 female.

Tribe Homalotini Heer

Genus *Gyrophaena* Mannerheim

37 (96). *Gyrophaena keeni* Casey
Additional records. Canada, Yukon Territory, Upper Liard, Albert Creek, 60.0522, –128.928, 699 m, mushrooms, 17.VIII.2007 (BGC), 2 specimens; Dawson City, 11.VII.1968, Campbell and Smetana (CNC) 5 sex undetermined.

38 (97). *Gyrophaena nana* (Paykull) *
Additional records. Canada: Alberta, Calgary, 13.VIII.1957, Lot 1, B.F. &. J.L. Carr (CNC) 1 male. Ontario, Moosonee, 51.277, –80.647, Rep. 3 wet, yellow pan, 16-19.VI.2010, NBP field party M3WY112 (LFC) 1 female. Yukon Territory, 16 mi. N Minto, 26.VI.1979, Lot 3, in

toadstools, B.F. &. J.L. Carr (CNC) 5 males, 3 females; Rampart House, 19.VIII.1951, J.E.H. Martin (CNC) 2 males, 14 sex undetermined; Dawson City, 16.VII.1968, Campbell and Smetana (CNC) 3 sex undetermined. USA: Alaska, Fairbanks, 930 Fitz Ct, 280m el., 64.9013, –147.52861, ±30m *Betula* & *P. tremulo.*, flight intercept trap, 16.VIII-22.IX.2007, D.S. Sikes, (UAM) 1 specimen; 29.V-4.VII.2008, (UAM) 2 specimens; Taylor Hwy 81, 9 mi N Tetlin Jnct., 16.VII.1978, Campbell and Smetana (CNC) 1 male, 13 sex undetermined.

Genus *Silusa* Erichson

39 (100). ***Silusa californica*** **(Bernhauer)**

Additional records. USA: Alaska, Denali NPP., Mt. Healy, 927m el., 63.74016, –148.95346, ±30m shrub, alder, dwarf birch, grass, clubmoss, 4 yellow pitfalls w/o rainroofs, pooled, 2.VI-19.VIII.2016 (UAM) 2 specimens; Denali NPP., Mt. Healy, 927m el., 63.74016, –148.95346, ±30m shrub, alder, dwarf birch, grass, clubmoss, 4 yellow pitfalls w/o rainroofs, pooled, 3-16.VIII.2016 (UAM) 2 specimens; Fairbanks, 930 Fitz Ct, 280m el., 64.9013, –147.52861, ±30m *Betula* & *P. tremuloides*, flight intercept trap, 16.VIII-22.IX.2007 (UAM) 2 specimens; Fairbanks, Parks Hwy mi. 340, 420m el., 64.76683, –148.27065, ±4m aspen litter, hillside, Berlese, 26.VII.2014 (UAM) 1 specimen; Prince of Wales Island, Coffman Cv, 67m el., 55.9795, –132.86256, ±101m thinned 2nd growth, pitfalls 1, 2, 10-30.VI.2010, (UAM) 2 specimens; same data except, pitfall 3, 11-25.VI.2012 (UAM) 1 specimen; Lindgren 1, 13-27.V.2013 (UAM) 1 specimen; pitfall 2, 24.VII-11.VIII.2010 (UAM) 1 specimen; pitfalls 1, 2, 3, 24.VII-7.VIII.2012 (UAM) 3 specimens; 25.VI-9.VII.2012 (UAM) 1 specimen; pitfall 4, 27.VII-7.VIII.2011 (UAM) 1 specimen; pitfalls 1, 4, 9-24.VII.2012 (UAM) 2 specimens; Prince of Wales Island, Coffman Cv, 78m el., 55.98053, –132.8607, ±15m unthinned 2nd growth, pitfall 2, 10-30.VI.2010 (UAM) 1 specimen; same data except, pitfalls 1, 3, 4, 11-25.VI.2012 (UAM) 5 specimens; pitfalls 2, 4, 12-27.VII.2011 (UAM) 3 specimens; pitfalls 2, 4, 13-26.V.2013 (UAM) 2 specimens; pitfalls 1, 2, 3, 4, 14-28.VI.2011 (UAM) 5 specimens; pitfall 3, 18-31.V.2011 (UAM) 2 specimens; pitfalls 2, 3, 4, 24.VII-11.VIII.2010 (UAM) 14 specimens; pitfalls 1, 2, 3, 24.VII-7.VIII.2012 (UAM) 26 specimens; pitfalls 1, 3, 25.VI-9.VII.2012 (UAM) 5 specimens; pitfalls 1, 3, 4, 27.VII-7.VIII.2011 (UAM) 4 specimens; pitfall 3, 28.VI-12.VII.2011 (UAM) 1 specimen; pitfall 4, 28.V-11.VI.2012 (UAM) 1 specimen; pitfall 2, 30.VI-14.VII.2010 (UAM) 2 specimens; pitfall 4, 31.V.-14.VI.2011 (UAM) 2 specimens; pitfall 1, 31.V.-14.VI.2011 (UAM) 1 specimen; pitfalls 1, 2, 3, 4, 9-24.VII.2012 (UAM) 16 specimens; Prince of Wales Island, Hatchery Ck.1, 49m el., 55.92654, –132.95645, ±10m unthinned 2nd growth, pitfalls 1, 2, 3, 4, 10-30.VII.2010 (UAM) 18 specimens; same data except, pitfall 1, 11-25.VI.2012 (UAM) 1 specimen; pitfalls 1, 2, 3, 13-27.V.2013 (UAM) 3 specimens; pitfall 4, 15-28.V.2012 (UAM) 1 specimen; pitfalls 1, 2, 3, 4, 18.V-4.VI.2010 (UAM) 12 specimens; pitfall 1, 19.VII-2.VIII.2011 (UAM) 1 specimen; pitfall 3, 2-9.VIII.2011 (UAM) 1 specimen; pitfall 1, 21.VI-6.VII.2011 (UAM) 1 specimen; pitfall 3, 24.VII-6.VIII.2012 (UAM) 1 specimen; pitfalls 1, 3, 24.VII-7.VIII.2012 (UAM) 5 specimens; pitfall 4, 24.V-5.VI.2011 (UAM) 1 specimen; pitfalls 1, 2, 3, 4, 27.VI-10.VII.2010 (UAM) 4 specimens; pitfalls 3, 4, 28.V-11.VI.2012, (UAM) 2 specimens; pitfalls 1, 2, 3, 4, 30.VII-11.VIII.2010, (UAM) 11 specimens; pitfalls 3, 4, 4-27.VI.2010 (UAM) 3 specimens; 15-21.VI.2011 (UAM) 1 specimen; pitfalls 1, 3, 9-24.VII.2012 (UAM) 4 specimens; Prince of Wales Island, Hatchery Ck.1, 50m el. 55.92444, –132.93938, ±4m old growth, pitfalls 1, 2, 3, 4, 10-30.VII.2010 (UAM) 11 specimens; same data except, pitfalls 1, 3, 4, 11-25.VI.2012, (UAM) 3 specimens; pitfalls 3, 4, 13-27.V.2013 (UAM) 2 specimens; pitfall 3, 18.V-3.VI.2010 (UAM) 2 specimens; pitfalls 2, 3, 2-9.VIII.2011 (UAM) 2 specimens; pitfalls 1, 3, 21.VI-6.VII.2011 (UAM) 3 specimen; pitfall 1, 24.VII-6.VIII.2012, (UAM) 1 specimen; pitfalls 1, 2, 4, 24.VII-7.VIII.2012 (UAM) 15 specimens;

pitfalls 3, 4, 24.V-5.VI.2011 (UAM) 2 specimens; pitfalls 1, 2, 25.VI-9.VII.2012 (UAM) 2 specimens; pitfall 3, 27.VI-10.VII.2010 (UAM) 1 specimen; pitfalls 1, 2, 3, 4, 28.V-11. VI.2012 (UAM) 5 specimens; pitfalls 1, 2, 3, 4, 30.VII-11.VIII.2010 (UAM) 10 specimens; pitfall 2, 4-27.VI.2010 (UAM) 1 specimen; pitfall 1, 5-21.VI.2011 (UAM) 1 specimen; pitfalls 3, 4, 6-19.VII.2011 (UAM) 5 specimens; pitfalls 1, 2, 3, 4, 9-24.VII.2012 (UAM) 10 specimens; Prince of Wales Island, Hatchery Ck.1, 57m el., 55.92524, −132.95054, ±10m thinned 2nd growth, pitfalls 2, 4, 11-25.VI.2012 (UAM) 2 specimens; pitfall 4, 13-27.V.2013 (UAM) 1 specimen; pitfalls 2, 3, 19.VII-2.VIII.2011 (UAM) 2 specimens; pitfall 1, 28.V-11.VI.2012 (UAM) 1 specimen; pitfalls 2, 3, 30.VII-11. VIII.2010 (UAM) 2 specimens; pitfalls 1, 2, 4, 9-24.VII.2012 (UAM) 3 specimens; Prince of Wales Island, Hatchery Ck.2, 134m el., 55.89356, −132.9437, ±34m unthinned 2nd growth, pitfall 4, 14-28.V.2012, (UAM) 2 specimens; same data except, pitfall 2, 25.VI-9.VII.2012 (UAM) 1 specimen; pitfall 4, 10-31. VII.2010 (UAM) 3 specimens; pitfalls 3, 4, 11-25.VI.2012 (UAM) 3 specimens; pitfalls 3, 4, 12-27.VII.2011 (UAM) 3 specimens; pitfalls 1, 2, 3, 14-28.V.2012 (UAM) 3 specimens; pitfalls 1, 3, 4, 17-31.V.2011 (UAM) 4 specimens; pitfall 4, 18.V-2.VI.2010 (UAM) 1 specimen; pitfalls 3, 24.VII-6.VIII.2012 (UAM) 2 specimens; 25.VI-9.VII.2012 (UAM) 1 specimen; pitfall 3, 27.VII-7.VIII.2011 (UAM) 3 specimens; pitfall 2, 28.V-11.VI.2012 (UAM) 1 specimen; pitfall 3, 29.VI-10.VII.2010 (UAM) 1 specimen; pitfall 1, 30.VII-11.VIII.2010 (UAM) 1 specimen; pitfalls 3, 4, 30.VII-11. VIII.2010 (UAM) 2 specimens; pitfalls 3, 4, 9-24.VII.2012 (UAM) 3 specimens; Prince of Wales Island, Hatchery Ck.4 78m el., 55.88602, -132.8607 ±11m old growth, pitfalls 2, 3, 11-25. VI.2012 (UAM) 2 specimens; pitfall 3, 12-27. VII.2011 (UAM) 1 specimen; pitfalls 1, 2, 3, 4, 14-28.VI.2011 (UAM) 16 specimens; pitfalls 2, 3, 4, 24.VII-6.VIII.2012 (UAM) 10 specimens; pitfall 2, 25.VI-9.VII.2012 (UAM) 2 specimens; pitfalls 2, 4, 27.VII-7.VIII.2011 (UAM) 2 specimens; pitfalls 1, 3, 28.VI-12.VII.2011 (UAM) 5 specimens; pitfalls 1, 2, 3, 4, 9-24. VII.2012 (UAM) 8 specimens; Prince of Wales Island, Hatchery Ck.4, 82m el., 55.88285, − 132.89795, ±27m unthinned 2nd growth, pitfall 1, 11-25.VI.2012 (UAM) 1 specimen; same data except, pitfall 3, 12-27.VII.2011 (UAM) 2 specimens; pitfalls 2, 3, 24.VII-11.VIII.2010 (UAM) 3 specimens; pitfall 1, 24.VII-6. VIII.2012 (UAM) 1 specimen; pitfalls 2, 3, 25. VI-9.VII.2012 (UAM) 2 specimens; Lindgren 2, pitfall 3, 27.VII-7.VIII.2011 (UAM) 4 specimens; pitfalls 1, 2, 3, 4, 28.VI-12. VII.2011 (UAM) 7 specimens; pitfall 1, 9-24. VII.2012 (UAM) 1 specimen; Prince of Wales Island, Hatchery Ck.4, 82m el., 55.88433, − 132.89734, ±26m thinned 2nd growth, pitfall 4, 13-27.V.2013 (UAM) 1 specimen; same data except, pitfalls 1, 3, 24.VII-6.VIII.2012 (UAM) 2 specimens; pitfall 4, 28.V.-11.VI.2012 (UAM) 1 specimen; Prince of Wales Island, Luck Lk. 1 Rd., 101m el., 55.97805, −132.75456, ±10m old growth, pitfall 4, 11-26.VII.2012 (UAM) 1 specimen; same data except, pitfalls 1, 2, 3, 4, 11-26.VII.2012 (UAM) 5 specimens; pitfall 2, 13-27.VI.2012 (UAM) 1 specimen; pitfalls 1, 2, 4, 2-9.VIII.2011, (UAM) 3 specimens; 21.VI-6.VII.2011 (UAM) 1 specimen; pitfall 3, 24.V-5.VI.2011 (UAM) 2 specimens; pitfalls 1, 4, 26.VII-9.VIII.2012 (UAM) 3 specimens; pitfall 2, 30.VII-11. VIII.2010 (UAM) 1 specimen; pitfalls 1, 4, 4-29.VI.2010 (UAM) 2 specimens; pitfall 1, 6-19.VII.2011 (UAM) 1 specimen; pitfalls 1, 2, 8-30.VII.2010 (UAM) 3 specimens; Prince of Wales Island, Luck Lk. 2 Rd., 105m el., 55.96855, −132.75615, ±10m old growth, pitfalls 1, 2, 3, 4, 11-26.VII.2012 (UAM) 7 specimens; same data except, pitfalls 1, 3, 13-27.VI.2012 (UAM) 2 specimens; pitfall 2, 2-9.VIII.2011 (UAM) 1 specimen; pitfalls 1, 3, 4, 26.VII-9.VIII.2012 (UAM) 7 specimens; pitfall 2, 27.VI-11.VII.2012 (UAM) 2 specimens; pitfalls 3, 4, 30.VII-11.VIII.2010 (UAM) 14 specimens; pitfall 1, 31.V-13.VI.2012 (UAM) 1 specimen; pitfall 4, 4-29.VI.2010 (UAM) 1 specimen; pitfalls 1, 3, 8-30.VII.2010

(UAM) 5 specimens; Prince of Wales Island, Luck Lk. 3 Rd., 117m el., 55.95347, −132.7708, ±5m old growth, 21.VI.2011 (UAM) 1 specimen; same data except, pitfalls 1, 2, 3, 4, 11-26. VII.2012 (UAM) 9 specimens; pitfalls 1, 2, 3, 4, 13-27.VI.2012 (UAM) 4 specimens; pitfall 2, 18. V-4.VI.2010 (UAM) 1 specimen; pitfall 1, 21.VI-6.VII.2011 (UAM) 1 specimen; pitfalls 2, 3, 4, 24.V-5.VI.2011 (UAM) 3 specimens; pitfalls 1, 2, 3, 4, 26.VII-9.VIII.2012 (UAM) 13 specimens; pitfalls 3, 4, 27.VI-11.VII.2012 (UAM) 2 specimens; pitfall 3, 29.VI-8.VII.2010 (UAM) 1 specimen; pitfalls 1, 2, 4, 30.VII-11. VIII.2010 (UAM) 14 specimens; pitfall 3, 31.V-13.VI.2012 (UAM) 1 specimen; pitfall 2, 5-21. VI.2011 (UAM) 1 specimen; pitfall 1, 8-30. VII.2010 (UAM) 1 specimen; pitfalls 1, 2, 3, 4, 8-30.VII.2010 (UAM) 41 specimens; Prince of Wales Island, Luck Point 1A, 174m el., 55.98452, −132.78786, ±26m clear-cut, pitfall 4, 11-26.VI.2012 (UAM) 1 specimen; Prince of Wales Island, Luck Point 1B, 169m el., 55.98497, −132.787, ±25m clear-cut, pitfall 2, 31.V-14.VI.2012 (UAM) 1 specimen; Prince of Wales Island, Luck Point, 205m el., 55.98261, −132.77986, ±6m thinned 2nd growth, pitfall 3, 14-26.V.2013 (UAM) 1 specimen; same data except, pitfalls 1, 2, 14-27.VI.2012 (UAM) 2 specimens; pitfall 2, 20.VIJUN-5.VII.2011 (UAM) 1 specimen; same data except, pitfall 2, 9.VII-1.VIII.2010 (UAM) 1 specimen; same data except, pitfall 4, 9.VII-1.VIII.2010 (UAM) 1 specimen; Prince of Wales Island, Luck Point 2°, 245m el., 55.97953, −132.77156, ±24m clear-cut, pitfall 1, 11-26.VI.2012 (UAM) 3 specimens; same data except, pitfall 1, 14-26. V.2013 (UAM) 1 specimen; pitfalls 1, 3, 4, 14-27.VI.2012 (UAM) 3 specimens; pitfall 3, 18.VII-2.VIII.2011 (UAM) 1 specimen; pitfall 4, 23.V-5.VII.2011 (UAM) 1 specimen; pitfalls 1, 3, 26.VII-9.VIII.2012 (UAM) 3 specimens; Prince of Wales Island, Luck Pt., 129m el., 55.98256, −132.77943, ±12m unthinned 2nd growth, pitfall 4, 1-11.VIII.2010 (UAM) 13 specimens; same data except, pitfalls 1, 2, 3, 4, 11-26.VII.2012 (UAM) 1 specimens; pitfalls 1, 3, 14-27VI.2012 (UAM) 2 specimens; pitfalls 1, 18.VII-2.VIII.2011 (UAM) 5 specimens; pitfall 2, 2-27.VI.2010 (UAM) 1 specimen; pitfall 3, 21, VI-5.VII.2011 (UAM) 1 specimen, 23.V-5. VI2011 (UAM) 2 specimens; pitfalls 1, 2, 3, 26. VI-9.VIII.2012 (UAM) 15 specimens; pitfalls 1, 4, 27.VI-11.VII2012 (UAM) 1 specimen; pitfalls 2, 4, 27.VI-9.VII2010 (UAM) 2 specimens; pitfall 3, 5-18.VII.2011 (UAM) 1 specimen; pitfalls 2, 3, 4, 9.VII-1.VIII.2010 (UAM) 9 specimens; Prince of Wales Island, Staney Ck, 63m el., 55.79723, −133.13467, ±51m unthinned 2nd growth, pitfall 4, 10-27. VI.2010 (UAM) 1 specimen; same data except, pitfalls 1, 3, 12-26.V.2013 (UAM) 2 specimens; pitfall 2, 12-27.VII.2011 (UAM) 1 specimen; pitfall 3, 13-27.VI.2011 (UAM) 1 specimen; pitfall 2, 14-24.VII.2010 (UAM) 1 specimen; pitfalls 2, 4, 17-31.V.2011 (UAM) 2 specimens; pitfall 3, 24.VII-6.VIII.2012 (UAM) 1 specimen; pitfall 3, 27.VI-12.VII.2011 (UAM) 1 specimen; 31.V.-13.VI.2011 (UAM) 1 specimen; pitfalls 1, 2, 3, 9-24.VII.2012 (UAM) 3 specimens; Prince of Wales Island, Staney Ck. 1A, 89m el., 55.87134, −133.06755, ±24m clear-cut, Lindgren 1, 27.VII-7.VIII.2011 (UAM) 1 specimen; pitfall 2, 13-27.VI.2011 (UAM) 1 specimen; pitfalls 1, 2, 4, 14-30.V.2010 (UAM) 3 specimens; pitfalls 1, 3, 4, 15-25.VII.2010 (UAM) 4 specimens; pitfalls 1, 3, 4, 24.VII-11. VIII.2010 (UAM) 19 specimens; pitfalls 1, 2, 3, 4, 27.VII-7.VIII.2011 (UAM) 5 specimens; Prince of Wales Island, Staney Ck. 1B, 77m el., 55.872, −133.06523, ±26m clear-cut, Lindgren 2, 13-27.VI.2011 (UAM) 1 specimen; same data except, pitfall 2, 11-25.VI.2012 (UAM) 1 specimen; pitfalls 1, 2, 4, 12-27.VII.2011 (UAM) 3 specimens; pitfall 2, 13-27.VI.2011 (UAM) 1 specimen; pitfalls 2, 4, 14-30.V.2010 (UAM) 3 specimens; pitfalls 1, 2, 3, 4, 15-25.VII.2010 (UAM) 16 specimens; pitfalls 1, 2, 4, 16-31. V.2011 (UAM) 3 specimens; pitfall 1, 24.VII-11.VIII.2010 (UAM) 27 specimens; pitfalls 2, 3, 27.VII-7.VIII.2011 (UAM) 4 specimens; pitfalls 1, 2, 4, 27.VI-12.VII.2011 (UAM) 4 specimens; pitfalls 1, 4, 27.VI-15.VII.2010 (UAM) 9 specimens; pitfalls 1, 2, 31.V-13.VII.2011 (UAM) 2 specimens; pitfall 1, 9-24.VII.2012

(UAM) 1 specimen; Prince of Wales Island, Staney Ck., 41-45m el., 55.79901, −133.11782, ±20m old growth, pitfalls 1, 2, 3, 4, 11-25. VI.2012 (UAM) 4 specimens; same data except, pitfalls 2, 3, 4, 12-26.V.2013 (UAM) 3 specimens; pitfall 1, 12-27.VII.2011 (UAM) 1 specimen; pitfalls 1, 2, 3, 4, 12-27.VII.2011 (UAM) 14 specimens; pitfalls 1, 2, 3, 4, 13-27.VI.2011 (UAM) 8 specimens; pitfalls 1, 2, 3, 4, 14-25. VII.2010 (UAM) 15 specimens; pitfalls 1, 2, 3, 14-28.V.2012 (UAM) 3 specimens; pitfalls 3, 4, 14-30.V.2010 (UAM) 4 specimens; pitfalls 1, 2, 3, 4, 16-31.V.2011 (UAM) 5 specimens; pitfalls 1, 2, 3, 4, 24.VII-11.VIII.2010 (UAM) 51 specimens; pitfalls 1, 2, 3, 4, 24.VII-6. VIII.2012 (UAM) 19 specimens; pitfalls 1, 2, 4, 25.VI-9.VII.2012 (UAM) 4 specimens; pitfalls 1, 2, 3, 4, 27.VII-7.VIII.2011 (UAM) 7 specimens; pitfalls 1, 2, 3, 4, 27.VI-12. VII.2011 (UAM) 26 specimens; pitfalls 2, 3, 27. VI-14.VII.2010 (UAM) 5 specimens; pitfalls 3, 4, 28.V-11.VI.2012 (UAM) 2 specimens; pitfalls 2, 3, 30.V-10.VI.2010 (UAM) 2 specimens; pitfalls 1, 2, 3, 4, 31.V-13.VI.2011 (UAM) 4 specimens; pitfalls 1, 2, 3, 4, 9-24.VII.2012 (UAM) 39 specimens; Prince of Wales Island, Staney Ck., 50m el., 55.79726, −133.1363, ±20m thinned 2nd growth, pitfall 2, 9-24. VII.2012, (UAM) 1 specimen; Prince of Wales Island, Staney Ck., 72m el., 55.87126, −133.06697, ±5m clearcut/ ecotone, pitfall 3, 11-22.V.2010, (UAM) 1 specimen.

COMMENTS The following collector names were removed from label data to save space: C. Bickford, A. Haberski, A. Hutton, I. MacDougall, S. Meierotto, D.S. Sikes, J. A. Slowik, Z. Snelson, R. Stewart-Rogers, J. Stockbridge, J.J. Rykken, and B. Wong.

Tribe Athetini Casey

Genus *Adota* Casey

40 (108). ***Adota maritima*** **(Mannerheim) ***
Additional records. USA, Alaska: Adak, 1m el., 51.90244, −176.61842, ±7m under drift wood, sandy beach, 13.VIII.2009, D. S. Sikes (UAM) 8 specimens; Adak Isl., 1-3m el., 51.89521, −176.62297, ±501m beach, under 7 pieces of driftwood, forceps, 25.VIII.2012, D. S. Sikes (UAM) 1 specimen; Adak Isl., 3m el., 51.87449, −176.62831 ±353m sandy beach, 5 pieces of driftwood, hand, 18.VIII.2011, D. S. Sikes (UAM) 2 specimens; Adak Isl., Kuluk Bay, 2-5m el., 51.87507, −176.62839, ±324m sandy beach, under 8 pieces of driftwood 1m long, forceps, 24.VIII. 2012, D. S. Sikes (UAM) 1 specimen; Attu Island, Site 4, 3 m el., 52.83247, −173.18347, ±20m, 1-30.V.2005, Kevin S. Winker (UAM) 2 specimens; Great Sitkin, 1m el., 52.00438, −176.06184, ±6m beach drift wood, 11.VIII.2009, D. S. Sikes (UAM) 9 specimens; Wosnesenski, 1m el., 55.21733, −161.36768, ±24m back beach, colored pan trap, 11.VII.2009, D. S. Sikes (UAM) 2 specimens; Wosnesenski, 1m el., 55.21733, −161.36768, ±24m under back beach rocks and drift, 11.VII. 2009, D. S. Sikes (UAM) 2 specimens.

Genus *Atheta* C.G. Thomson

41 (113). ***Atheta (Alaobia) ventricosa*** **Bernhauer**
Additional records. USA: Alaska, 35.5 Mi Dalton Hwy, 271m el., 65.74425, −149.34564, ±4m *Betula, Salix*, white spruce, flight intercept trap, 31.V-5.VI.2008 (UAM) 10 specimens; Becharof NWR Island Arm, 3m el., 57.75391, −155.98835, ±20m sweep, 26.VI.2007 (UAM) 1 specimen; Fairbanks, 930 Fitz Ct, 280m el., 64.9013, −147.52861, ±30m *Betula* & *P. tremuloides*, flight intercept trap, 16.VIII-22.IX.2007 (UAM) 2 specimens;same data except, 20.V-3.VI.2007 (UAM) 1 specimen, 29.IV-29.V.2008 (UAM) 5 specimens; Fairbanks: Creamer's Field, 140m el., 64.86883, −147.73065, ±104m *Betula* forest, decaying mushrooms, 31.VI.2010 (UAM) 2 specimens; Fairbanks, Creamer's Field, 141m el., 64.86711, −147.72794, ±54m forest edge, *Populus* litter, Winkler, 31.VII.2010 (UAM) 1 specimen; Prince of Wales Island, Coffman Cv, 67m el., 55.9795, −132.86256, ±101m thinned 2nd growth, pitfall 4, 24.VII-11. VIII.2010 (UAM) 2 specimens; same data except. pitfall 2, 27.IV-15.V.2010, (UAM) 1 specimen; Lindgren 2, 28.VI-12.VII.2011 (UAM)

1 specimen; pitfalls 1, 2, 30.VI-14.VII.2010 (UAM) 2 specimens; pitfall 4, 9-24.VII.2012 (UAM) 1 specimen; Prince of Wales Island, Coffman Cv, 78m el., 55.98053, –132.8607, ±15m unthinned 2nd growth, pitfall 1, 10-30. VI.2010 (UAM) 1 specimen; pitfall 4, 13-27. V.2013 (UAM) 1 specimen; pitfalls 2, 3, 28.VI-12.VII.2011 (UAM) 6 specimens; pitfall 2, 9-24. VII.2012 (UAM) 1 specimen; Prince of Wales Island, Hatchery Ck.1, 49m el., 55.92654, –132.95645, ±10m unthinned 2nd growth, pitfall 1, 10-30.VII.2010 (UAM) 2 specimens; pitfall 3, 11-25.VI.2012 (UAM) 1 specimen; pitfalls 2, 3, 13-27.V.2013 (UAM) 2 specimens; pitfall 3, 14-28.V.2012 (UAM) 1 specimen; pitfall 1, 18. V-4.VI.2010 (UAM) 1 specimen; pitfall 2, 24. VII-6.VIII.2012 (UAM) 1 specimen; pitfall 3, 25.VI-9.VII.2012 (UAM) 3 specimens; pitfall 3, 27.VI-10.VII.2010 (UAM) 2 specimens; pitfall 2, 30.VII-11.VIII.2010 (UAM) 1 specimen; Prince of Wales Island, Hatchery Ck.1, 50m el., 55.92444, –132.93938, ±4m old growth, Lindgren 2, 18.V-3.VI.2010 (UAM) 1 specimen; same data except, pitfalls 1, 2, 10-30.VII.2010 (UAM) 5 specimens; pitfalls 3, 4, 11-25VI.2012 (UAM) 3 specimens; pitfalls 2, 3, 4, 13-27. V.2013 (UAM) 6 specimens; pitfall 1, 18.V-3. VI.2010 (UAM) 1 specimen; pitfall 3, 21.VI-6. VII.2011 (UAM) 1 specimen; pitfalls 1, 2, 4, 24. VII-7.VIII.2012 (UAM) 7 specimens; pitfall 1, 25.VI-9.VII.2012 (UAM) 3 specimens; pitfalls 2, 3, 4, 27VI-10VII.2010 (UAM) 5 specimens; pitfalls 1, 2, 4, 28.V-11.VI.2012 (UAM) 3 specimens; pitfall 4, 30.VII-11.VIII.2010 (UAM) 1 specimen; pitfalls 1, 3, 4-27.VI.2010 (UAM) 2 specimens; pitfall 3, 5-21.VI.2011 (UAM) 1 specimen; pitfall 1, 6-19.VII.2011 (UAM) 1 specimen; pitfall 1, 9-24.VII.2012 (UAM) 1 specimen; Prince of Wales Island, Hatchery Ck.1, 57m el., 55.92524, –132.95054, ±10m thinned 2nd growth, pitfall 4, 15-28. V.2012 (UAM) 1 specimen; Prince of Wales Island, Hatchery Ck.2, 134m el., 55.89356, -132.9437, ±34m unthinned 2nd growth, pitfall 3, 12-27.V.2013 (UAM) 1 specimen; same data except, Lindgren 2, 24.VII-6.VIII.2012 (UAM) 1 specimen; Berlese, 24.V.2012, (UAM) 1 specimen; pitfall 2, 10-31,VII.2010 (UAM) 2 specimens; pitfall 3, 10-31.VIII.2010 (UAM) 2 specimens; pitfall 1, 11-25.VI.2012 (UAM) 3 specimens; pitfall 1, 12-27.V.2013 (UAM) 3 specimens; pitfalls 2, 3, 4, 12-27.V.2013 (UAM) 17 specimens; pitfalls 1, 2, 13-27. VI.2011 (UAM) 3 specimens; pitfalls 1, 2, 14-28.V.2012 (UAM) 3 specimens; pitfalls 2, 3, 4, 17-31.V.2011 (UAM) 3 specimens; pitfall 2, 18.V-2.VI.2010 (UAM) 1 specimen; pitfalls 1, 2, 3, 4, 2-29.VI.2010 (UAM) 11 specimens; pitfalls 1, 2, 3, 4, 25.VI-9.VII.2012 (UAM) 6 specimens; pitfalls 1, 3, 27.VI-12.VII2011 (UAM) 2 specimens; pitfalls 2, 3, 29.VI-10. VII.2010 (UAM) 3 specimens; pitfall 3, 30.VII-11.VIII.2010 (UAM) 1 specimen; pitfalls 1, 4, 31. V.-13.VI.2011 (UAM) 4 specimens; pitfalls 1, 3, 9-24.VII.2012 (UAM) 4 specimens; Prince of Wales Island, Hatchery Ck.4, 78m el., 55.88602, –132.8607, ±11m old growth, pitfall 2, 11-25.VI.2012 (UAM) 1 specimen; same data except, pitfall 4, 14-28.VI.2011 (UAM) 1 specimen; pitfall 3, 9-24.VII.2012 (UAM) 1 specimen; Prince of Wales Island, Hatchery Ck.4, 82m el., 55.88285, –132.89795, ±27m unthinned 2nd growth, pitfalls 1, 2, 11-25.VI.2012 (UAM) 8 specimens; same data except, pitfall 4, 12-27. VII.2011 (UAM) 1 specimen; pitfalls 2, 3, 4, 13-27.V.2013 (UAM) 7 specimens; pitfalls 1, 2, 3, 14-24.VII.2010 (UAM) 10 specimens; pitfalls 2, 3, 14-28.VI.2011 (UAM) 5 specimens; pitfall 1, 14-30.V.2010 (UAM) 2 specimens; pitfalls 1, 2, 3, 4, 15-27.VI.2010 (UAM) 7 specimens; pitfalls 1, 2, 3, 4, 15-28.V.2012 (UAM) 10 specimens; pitfalls 2, 3, 17-31. V.2011 (UAM) 4 specimens; Berlese, 20. VI.2012 (UAM) 1 specimen; pitfalls 1, 2, 3, 4, 24.VII-11.VIII.2010 (UAM) 9 specimens; pitfalls 1, 2, 4, 25.VI-9.VII.2012 (UAM) 16 specimens; pitfalls 1, 2, 3, 4, 27.IV-15.V.2010 (UAM) 14 specimens; Berlese, 2.VI.2011 (UAM) 1 specimen; pitfalls 1, 2, 30.V-15.VI.2010 (UAM) 6 specimens; pitfalls 2, 4, 31.V-14.VI.2011 (UAM) 2 specimens; pitfalls 1, 2, 3, 4, 9-24. VII.2012 (UAM) 13 specimens; Prince of Wales Island, Hatchery Ck.4, 82m el., 55.88433, –132.89734, ±26m thinned 2nd growth, pitfall

4, 11-25.VI.2012 (UAM) 1 specimen; pitfalls 2, 3, 13-27.V.2013 (UAM) 2 specimens; pitfalls 2, 3, 14-27.VI.2010 (UAM) 4 specimens; pitfall 1, 28.V-11.VI.2012 (UAM) 2 specimens; pitfall 1, 30.V-14.VI.2010 (UAM) 1 specimen; Lindgren 2, pitfall 1, 9-24.VII.2012 (UAM) 2 specimens; Prince of Wales Island, Luck Lk. 1 Rd., 101m el., 55.97805, –132.75456, ±10m old growth, pitfalls 2, 3, 4, 11-26. VII.2012 (UAM) 3 specimens; same data except, pitfalls 2, 3, 13-27.VI.2012 (UAM) 3 specimens; 16-31.V.2012 (UAM) 1 specimen; pitfalls 1, 2, 18.V-4.VI.2010 (UAM) 5 specimens; pitfall 1, 2, 21.VI-6.VII.2011 (UAM) 3 specimens; pitfall 4, 26.VII-9.VIII.2012 (UAM) 1 specimen; pitfalls 1, 2, 3, 4, 27.VI-11.VII.2012, (UAM) 4 specimens; pitfall 2, 29.VI-8.VII.2010 (UAM) 1 specimen; pitfall 3, 4-29.VI.2010 (UAM) 1 specimen; pitfall 1, 6-19.VII.2011 (UAM) 1 specimen; pitfall 1, 8-30.VII.2010 (UAM) 2 specimens; pitfall 2, 13-27.VI.2012 (UAM) 1 specimen; pitfall 1, 18.V-4.VI.2010 (UAM) 1 specimen; pitfall 3, 21.VI-6.VII.2011 (UAM) 1 specimen; pitfall 4, J. 5-21.VI.2011 (UAM) 1 specimen; Prince of Wales Island, Luck Lk. 3 Rd., 117m el., 55.95347, -132.7708, ±5m old growth, Berlese, 5.VI.2011 (UAM) 1 specimen; same data except, Lindgren 2, 13-27. VI.2012 (UAM) 2 specimens; Lindgren 2, 30. VII-11.VIII.2010 (UAM) 1 specimen; pitfalls 1, 2, 3, 4, 11-26.VII.2012, (UAM) 15 specimens; pitfalls 1, 2, 3, 13-27.VI.2012 (UAM) 7 specimens; pitfalls 2, 3, 16-31.V.2012 (UAM) 3 specimens; pitfalls 2, 4, 18.V-4.VI.2010 (UAM) 2 specimens; pitfall 2, 19.VII-2. VIII.2011 (UAM) 2 specimens; pitfall 1, 2-9. VIII.2011 (UAM) 1 specimen; pitfalls 2, 4, 21. VI-6.VII.2011 (UAM) 2 specimens; pitfalls 2, 3, 24.V-5.VI.2011 (UAM) 2 specimens; pitfalls 1, 2, 3, 26.VI-9.VIII.2012 (UAM) 6 specimens; pitfalls 2, 3, 4, 27.VI-11.VII.2012 (UAM) 19 specimens; pitfalls 1, 3, 29.VI-8.VII.2010 (UAM) 2 specimens; pitfall 1, 30.VII-11. VIII.2010 (UAM) 3 specimens; pitfalls 2, 3, 30. VII-11.VIII.2010 (UAM) 4 specimens; pitfall 1, 31.V-13.VI.2012 (UAM) 1 specimen; pitfall 4, 4-29.VI.2010 (UAM) 2 specimens; pitfalls 1, 2, 5-21.VI.2011 (UAM) 4 specimens; pitfalls 2, 4, 6-19.VII.2011 (UAM) 4 specimens; pitfalls 1, 2, 3, 4, 8-30.VII.2010 (UAM) 9 specimens; Prince of Wales Island, Luck Point 1A, 174m el., 55.98452, –132.78786, ±26m clear-cut, pitfall 3, 31.V-14.VI.2012 (UAM) 1 specimen; Prince of Wales Island, Luck Point 1B, 169m el., 55.98497, –132.787, ±25m clear-cut, pitfalls 1, 2, 3, 14-26.V.2013 (UAM) 6 specimens; Prince of Wales Island, Luck Point, 205m el., 55.98261, –132.77986, ±6m thinned 2nd growth, pitfall 3, 26.VII-9.VIII.2012 (UAM) 1 specimen; Prince of Wales Island, Luck Point 2A, 245m el., 55.97953, –132.77156, ±24m clear-cut, pitfalls 2, 3, 1-11.VIII.2010 (UAM) 3 specimens; same data except, pitfall 3, 11-26.VII.2012 (UAM) 1 specimen; 14-27.VI.2012 (UAM) 1 specimen; pitfall 1, 27.VI-9.VII.2010 (UAM) 3 specimens; pitfall 1, 31.V-14.VI,2012 (UAM) 1 specimen; pitfall 2, 9.VII-1.VIII.2010 (UAM) 1 specimen; Prince of Wales Island, Luck Point 2B, 264m el., 55.97939, –132.77216, ±25m clear-cut, pitfalls 1, 2, 2-27.VI.2010 (UAM) 2 specimens; pitfalls 1, 4, 31.V-14.VI.2012 (UAM) 2 specimens; Prince of Wales Island, Luck Pt., 129m el., 55.98256, –132.77943, ±12m unthinned 2nd growth, pitfalls 1, 4, 1-11.VIII.2010 (UAM) 2 specimens; same data except, pitfall 4, 27.VI-9.VII.2010 (UAM) 1 specimen; Prince of Wales Island, Staney Ck, 63m el., 55.79723, –133.13467, ±51m unthinned 2nd growth, pitfall 3, 11-25.VI.2012 (UAM) 1 specimen; same data except, pitfalls 1, 3, 14-28.V.2012 (UAM) 2 specimens; Prince of Wales Island, Staney Ck. 1A, 89m el., 55.87134, –133.06755, ±24m clear-cut, pitfall 4, 11-25.VI.2012 (UAM) 2 specimens; same data except, pitfalls 3, 4, 12-26.V.2013 (UAM) 4 specimens; pitfalls 3, 4, 14-28.V.2012 (UAM) 2 specimens; pitfall 3, 27. VI-12.VII.2011 (UAM) 1 specimen; pitfalls 3, 4, 28.V-11.VI.2012 (UAM) 2 specimens; pitfall 3, 31.V-13.VI.2011 (UAM) 1 specimen; Prince of Wales Island, Staney Ck. 1B, 77m el., 55.872, -133.06523, ±26m clear-cut, pitfalls 1, 2, 3, 11-25.VI.2012 (UAM) 4 specimens; pitfalls 1, 3, 4, 12-26.V.2013 (UAM) 6 specimen; pitfalls 1, 3, 12-27.VII.2011 (UAM) 2 specimens; pitfalls

1, 2, 3, 14-28.V.2012 (UAM) 5 specimens; pitfalls 2, 3, 15-25.VII.2010 (UAM) 2 specimens; pitfall 2, 16-31.V.2011 (UAM) 2 specimens; pitfall 2, 24.VII-6.VIII.2012 (UAM) 1 specimen; pitfalls 1, 2, 28.V-11.VI.2012 (UAM) 4 specimens; pitfall 1, 31.V-13.VI.2011 (UAM) 1 specimen; Prince of Wales Island, Staney Ck., 41-45m el., 55.79901, −133.11782, ±20m old growth, pitfalls 1, 4, 10-27.VI.2010 (UAM) 3 specimens; same data except, pitfalls 2, 4, 11-25.VI.2012 (UAM) 2 specimens; pitfalls 1, 2, 12-26.V2013 (UAM) 2 specimens; pitfall 1, 13-27.VI.2011 (UAM) 3 specimens; pitfall 2, 14-28.V.2012 (UAM) 1 specimen; pitfalls 2, 3, 14-30.V.2010 (UAM) 3 specimens; pitfalls 1, 3, 16-31.V.2011 (UAM) 2 specimens; pitfalls 3, 4, 24.VII-11.VIII.2010 (UAM) 2 specimens; pitfalls 1, 3, 27.VI-14.VII.2010 (UAM) 3 specimens; pitfall 4, 30.V-10.VI.2010 (UAM) 1 specimen; Prince of Wales Island, Staney Ck., 50m el., 55.79726, −133.1363, ±20m thinned 2nd growth, pitfall 3, 12-26.V.2013 (UAM) 1 specimen; same data except, Lindgren 2, pitfalls 4, 14-28.V.2012 (UAM) 2 specimens; Sitka, Mosquito Cove Trail, 1m el., 57.1356, −135.37059, ±15m coastal rain forest, 16. IX.2008 (UAM) 1 specimen; Sterling, Corey St, 90m el., 60.53953, −150.83232, ±150m *P. glauca* forest, Lindgren, 26.VIII.2004 (UAM) 1 specimen; Wrangell, near rifle range, 56.47307, −132.34599, ±50m old growth conifers, hand collected, 21.IV-7.V.2009 (UAM) 1 specimen. New Hampshire: Coos Co., Dixville, on snow, 1.IX.2011, T. Murray (University of Guelph Insect Collection) 1 specimen.

COMMENTS The following collector names were removed from label data to save space: C. Bickford, A. Haberski, A. Hutton, I. MacDougall, S. Meierotto, D.S. Sikes, J. A. Slowik, Z. Snelson, R. Stewart-Rogers, J. Stockbridge, J.J. Rykken, S. Wise-Eagle and B. Wong.

42 (116). ***Atheta (Atheta) graminicola*** **(Gravenhorst) ***

Additional records. Canada: Northwest Territories, Yellowknife, 62.51591, −113.39146, Rep. 3, wet, yellow pan, 7-11.VII.2011, NBP Field Party Y3WY712 (LFC) 1 female; Norman Wells, 65.25226, −126.66128, Rep. 2, wet, yellow pan, 14-17.VI.2011, NBP Field Party N2WY332 (LFC) 1 male; Reindeer Sta., Caribou Hills, 30.VI.1972, A. Smetana (CNC) 10 sex undetermined; Aklavik, 9.VI.1956, R.E. Leech (CNC) 5 sex undetermined. Quebec, Koroc River, 3.VIII.1986, in leaf litter beside stream, A.&S. Morgan (CNC) 1 sex undetermined. Yukon, Tombstone Mts., 64.60560, −138.36143, Rep. 1, mesic, yellow pan, NBP Field Party T1MY922 (LFC) 1 male; Richardson Mts., Erebia Cr. on Fish Cr., 67.9666, −136.4833, 29.VI.1987, wet willow litter on Fish Creek flood plain, V. Behan, VB54-87 (CNC) 6 females; Dempster Hwy., Mi 65, 3300 feet, 13.VII.1968, J.M. Campbell and A. Smetana (CNC) 4 sex undetermined; Dempster Hwy., Mi 71, 3300 feet, 12.VII.1968, J.M. Campbell and A. Smetana (CNC) 8 sex undetermined; Dempster Hwy., Mi 74, 3300 feet, 12.VII.1968, J.M. Campbell and A. Smetana (CNC) 1 sex undetermined; Dempster Hwy., Mi 147, 1900 feet, 22.VII.1978, J.M. Campbell and A. Smetana (CNC) 3 sex undetermined; Stewart River S of McQuesten above Stirling Bend, 63.5047, −137.3283, 7.VI.1975, J.V. Matthews (CNC) 1 male; British Mts., Swan Pond, 340 m, 4 km NE Wolf Cr., 69.133, −140.233, 23. VI.1984, 84-25A, treading sedges around edge of pond, J.M. Campbell (CNC) 2 sex undetermined. USA, Alaska, Deering, 27.VII-2. VIII.1968, trap # 23, J. Matthews (CNC) 1 male; Tangle Lakes, 14-21/14-26.1969, J. Matthews (CNC) 2 sex undetermined; Seward Peninsula, 20-23.VII.1968, J. Matthews (CNC) 1 male, 2 females; Unalakleet, 23.VI.1961, B.S. Hemming (CNC) 4 sex undetermined; Agattu Isl., 15m el., 52.47308, −173.59065, ±30m beach, *Heracl. lanatum*, 5 pitfalls, 19. VII.2008, L. Kenney, R. Kaler (UAM) 1 specimen, same data except, 5-19.VI.2008 (UAM) 1 specimen; Aiktak Is., 20m el., 54.18393, −164.84167, ±5m streamside, Leymus, 2 pitfalls, 26.VI-14.VII.2008, B.A. Drummond, M E. McClintock (UAM) 1 specimen; Alaska Peninsula / Becharof NWR, plot 1487 (Island Arm

East), 9m el., 57.76107, –155.98752, ±10m S facing slope in CTS-*Salix*, & fern & cow parsnip, pitfall trap, 29.VI-5.VII.2007, S. Savage (UAM) 2 specimens; Alaska Peninsula / Becharof NWR, plot 1540 (Severson), 17m el., 57.81477, –156.11205, ±7m under CTS-*Salix* near the edge with crowberry, pitfall trap, 27.VI-5.VII.2007, S. Savage (UAM) 17 specimens; same data except, 5.VII-21.VIII.2007 (UAM) 12 specimens; Alaska Peninsula / Becharof NWR, plot 1692 (Becharof Outlet), 20m el., 58.04911, –156.84003, ±4m In EDS-crowberry, ~3-5 m of closed tall shrub to E & W, pitfall trap, 5.VII-15. VIII.2007, S. Savage (UAM) 1 specimen; Anaktuvuk Pass, 647m el., 68.14049 –151.74091, ±250m *Salix* leaf litter, Berlese, 19. V.2016, D.S. Sikes, K.M. Daly (UAM) 16 specimens; Anaktuvuk Pass, 665m el., 68.14059, –151.74266, ±200m *Salix*, tundra hill-side, cobble, pollinator cup/pitfall, 20.V.2016, D.S. Sikes, K.M. Daly (UAM) 2 specimens; Anaktuvuk Pass, Eleanor lake, 649m el., 68.14742, –151.71854, ±100m lake shore and pond, sweep and dipnet, 20.V.2016, D.S. Sikes, K.M. Daly (UAM) 1 specimen; Becharof NWR, 8m el., 57.85395, –156.51857, ±20m lakeshore sand, detritus, hand collected, 13.VI.2007, S. Clawson (UAM) 2 specimens; Fairbanks, 930 Fitz Ct, 280m el., 64.9013, –147.52861, ±30m small stand *Betula* & *Pop. trem.*, flight intercept trap, 29.IV-29.V.2008, D.S. Sikes (UAM) 1 specimen; Fairbanks, Bonanza Creek, 237m el., 64.71922, –148.38416, ±10m wood cutting area, w. spruce, birch, grass, hanging Malaise, 6.15m, 6-13.V.2016, J.C. Hagelin, (UAM) 1 specimen; Fairbanks, Creamer's Field, 143m el,. 64.86714, –147.7281, ±305m forest edge, meadow, grass, forbs, pitfall, 9-10. VII.2011, D.S. Sikes (UAM) 2 specimens; Fairbanks, Creamer's Field, 154m el., 64.866, –147.73788, ±7m pitfall, 15-18.VII.2008, D.S. Sikes, A. Parent (UAM) 1 specimen; Fairbanks, Creamer's Field, Reindeer Lake, 146m el., 64.88118, –147.77643, ±10m b. spruce bog, hanging Malaise, 5.55m, 13-27. VII.2016, J.C. Hagelin (UAM) 2 specimens, 29. VI-13.VII.2016 (UAM) 1 specimen; same data except, 9-31.V.2016 (UAM) 1 specimen; Fairbanks, Goldstream Creek, 213m el., 64.93589, –147.75597, ±10m b. & w. spruce, wild rose, ~20m from creek, hanging Malaise, 3.61m, 14-28.VI.2016, J.C. Hagelin (UAM) 1 specimen; Fairbanks: Creamer's Field, 132m el., 64.86805, –147.73764, ±20m mesic *Betula, Picea*, pitfall, 15-18.VII.2008, D.S. Sikes, C. Eshleman (UAM) 1 specimen; Kanuti NWR Trap 1, 215m el., 66.43871, –151.29975, ±3m *Eriophorum* tussock shrubby wetland, Malaise, 21-23.VI.2013, J. Jorgenson, M.D. Garroutte, R.J. Liebermann (UAM) 1 specimen; Kotzebue, 1m el., 66.89151, –162.5933, ±100m brackish shoreline, under rocks, wrack, 6.VII.2016, D.S. Sikes, K.M. Daly (UAM) 1 specimen; Little Sitkin, 17m el., 51.98182, –178.49165, ±25m back beach, *Leymus* grass, pitfall #1, 9.VI-30. VII.2008, D.S. Sikes (UAM) 1 specimen; Peninsula NWR 1484 Ugashik River, 29m el., 57.702, –156.52213, ±8m pitfall trap, 29.VIII.2007, S. E. Hawks (UAM) 1 specimen; Peninsula NWR, 16m el., 57.5946, –156.64705, ±10m sandy lakeshore, alder detritus, hand collected, 20.VI.2007, S. Clawson (UAM) 1 specimen; Peninsula NWR Dog Salmon River, 7m el., 57.39756, –157.25113, ±7m muddy river bank, hand collected, 27. VI.2007, S. E. Hawks (UAM) 2 specimens; Richardson Hwy. Bathing Beauty Pd., 155m el., 64.71391, –147.19339, ±26m shoreline veg. under log, 18.VI.2008, D.S. Sikes, D. R. Maddison (UAM) 1 specimen; Sanak Isl, 1-22m el., 54.37127, –162.61283, ±180m maritime tundra, pitfall trap, 24.VI-4.VII.2006, J. Cornell (UAM) 1 specimen; Alaska Hwy. mi 1249, Deadman Lake, 6.VII.1968, A. Smetana (CNC) 2 males, 1 female, 14 sex undetermined; Deering, 24. VIII.1967, J. Matthews (CNC) 1 female.

43 (121). *Atheta (Datomicra) dadopora* C.G. Thomson *

Additional records. USA: Alaska, 35.5 Mi Dalton Hwy, 271m el., 65.74425, –149.34564, ±4m *Betula, Salix*, white spruce, flight intercept trap, 31.V-5.VI.2008, D.S. Sikes (UAM) 14 specimens; Alaska Peninsula / Becharof NWR, plot 1428 (Ugashik Lake East), 8m el.,

57.59805, −156.63251, ±9m OTS-*Salix*, crowberry hummock, pitfall trap, 19.VI-8.VIII.2007, S. Savage (UAM) 5 specimens; Fairbanks, 930 Fitz Ct, 280m el., 64.9013, −147.52861, ±30m *Betula* & *P. tremulo*., flight intercept trap, 16.VIII-22.IX.2007, D.S. Sikes (UAM) 19 specimens; same data except, 20.V-3. VI.2007 (UAM) 37 specimens; 29.V-4.VII.2008 (UAM) 16 specimens; 29.IV.-29.V.2008 (UAM) 7 specimens; Mi 179 Dalton Hwy, 365m el., 67.29803, −150.15699, ±7m 3/4 Alder, white spruce, *Betula* forest, 5-31.VII.2008, D.S. Sikes (UAM) 1 specimen; Sterling, Gann site, 40-70m el., 60.51418, −150.79506, ±25m *S. barclayi* meadow, sweep, 19.VI.1993, D.M. Collet (UAM) 1 specimen.

44 (124). *Atheta (Dimetrota) allocera* Eppelsheim *

Additional records. Canada: Yukon, Tombstone Mts., 64.58353, −138.26843, Rep. 3, mesic, yellow pan, 24-27.VI.2011, NBP Field Party T3MY125 (LFC) 1 female. USA, Alaska, 35.5 Mi Dalton Hwy, 271m el., 65.74425, −149.34564, ±4m, *Betula, Salix*, white spruce, flight intercept trap, 31.V-5.VI.2008, D.S. Sikes (UAM) 8 specimens; Aiktak Is., 10-20m el., 54.18549, −164.8432, ±4km maritime tundra, moist valleys, 5 pitfalls, 29.VII-17.VIII.2009, A. L. Larned (UAM) 1 specimen; Mi 179 Dalton Hwy, 365m el., 67.29803, −150.15699, ±7m 3/4 Alder, white spruce, *Betula* forest, 5-31.VII.2008, D.S. Sikes (UAM) 1 specimen; Toolik Field Station, 724m el., 68.6286, −149.59772, ±36m under rocks, etc., 1-3.VI.2008, D.S. Sikes (UAM) 9 specimens.

45 (126). *Atheta (Dimetrota) remulsa* Casey

Additional records. USA: Alaska, near Kasilof, 60.3375, −151.2744, ±50m in mushrooms, 21. VII.2005, D. Langor (UAM) 18 specimens.

46 (135). *Atheta (Dimetrota) munsteri* Bernhauer *

Additional records. Canada: Yukon, Tombstone Mtns., 64.60629, −138.36737, 24-27.VI.2011, rep 1, wet, yellow trap, NBP Field Party T1WY922 (LFC) 1 male. USA, Alaska, Denali NPP., Igloo Canyon, 1047m el., 63.59104, −149.60716, ±30m alpine tundra, 4 yellow pitfalls w/o rainroofs, pooled, 20.VII-3.VIII.2016, J.J. Rykken, D.S. Sikes, A. Haberski, K. Chen (UAM) 1 specimen.

47 (136). *Atheta (Dimetrota) hampshirensis* Bernhauer

Additional records. USA: Alaska, Fairbanks: Creamer's Field, 140m el., 64.86883, −147.73065, ±104m *Betula* forest, decaying mushrooms, 31.VII.2010, D.S. Sikes (UAM) 1 specimen; Prince of Wales Island, (POW) Staney Creek, 50m el., 55.79726, −133.1363, ±20m thinned 2nd growth, forceps, 16-17. V.2013, D.S. Sikes, J. Stockbridge, J. Slowik (UAM) 3 specimens; Prince of Wales Island, Coffman Cv, 67m el., 55.9795, −132.86256, ±101m thinned 2nd growth, Lindgren 1, 28.V-10.VI.2010, J. Stockbridge, C. Bickford (UAM) 1 specimen; Prince of Wales Island, Coffman Cv, 78m el., 55.98053, −132.8607, ±15m unthinned 2nd growth, pitfall 1, 24.VII-11.VIII.2010, J. Stockbridge, C. Bickford (UAM) 1 specimen; Prince of Wales Island, Hatchery Ck.1, 50m el., 55.92444, −132.93938, ±4m old growth, Lindgren 1, 11-25.VI.2012, J. Stockbridge, S. Meierotto, I. MacDougall (UAM) 2 specimens; Prince of Wales Island, Hatchery Ck.1, 57m el., 55.92524, −132.95054, ±10m thinned 2nd growth, pitfall 3, 24.VII-7.VIII.2012, J. Stockbridge, S. Meierotto, I. MacDougall (UAM) 1 specimen; Prince of Wales Island, Luck Lk. 1 Rd., 101m el., 55.97805, −132.75456, ±10m old growth, Lindgren 2, 16-31.V.2012, J. Stockbridge, S. Meierotto, I. MacDougall (UAM) 1 specimen; Prince of Wales Island, Luck Point, 205m el., 55.98261, −132.77986, ±6m thinned 2nd growth, Lindgren 1, 2-9.VIII.2011, J. Stockbridge, B. Wong (UAM) 1 specimen; Prince of Wales Island, Luck Point 2A, 245m el., 55.97953, −132.77156, ±24m clear-cut, Lindgren 1, 27.VI-11.VII.2012, J. Stockbridge, S. Meierotto, I. MacDougall (UAM) 1 specimen; same data except, pitfall 1, 1-11.VIII.2010, J. Stockbridge,

C. Bickford (UAM) 1 specimen; pitfall 4, 18.V-2.VI.2010, J. Stockbridge, C. Bickford (UAM) 1 specimen; Prince of Wales Island, Luck Pt., 129m el., 55.98256, –132.77943, ±12m unthinned 2nd growth, pitfall 2, 26.VII-9.VIII.2012, J. Stockbridge, S. Meierotto, I. MacDougall (UAM) 1 specimen; Prince of Wales Island, Staney Ck, 63m el., 55.79723, –133.13467, ±51m unthinned 2nd growth, pitfall 3, 14-24.VII.2010, J. Stockbridge, C. Bickford (UAM) 1 specimen; same data except, pitfall 1, 30.V-10.VI.2010, J. Stockbridge, C. Bickford (UAM) 1 specimen; Prince of Wales Island, Staney Ck., 41-45m el., 55.79901, –133.11782, ±20m old growth, pitfall 4, 30.V-10.VI.2010, J. Stockbridge, C. Bickford (UAM) 1 specimen; Sitka, Mosquito Cove Trail, 1m el., 57.1356, –135.37059, ±15m old growth Hemlock, coastal rain forest, 19.IX.2008, D.S. Sikes (UAM) 2 specimens; Univ. Alaska Fairbanks, 185m el. 64.8577 –147.8557, ±61m forest edge *Pop. trem.*, pitfall, 17-24.VII.2007, B. Mortensen (UAM) 1 specimen; Kenai Mts., 2 mi S Moose Pass, 500 feet, 30.V.1978, A. Smetana and E.C. Becker (CNC) 1 female; Denali St. Pk., Byers Lk. Cpgd., 26.VI.1978, A. Smetana and E.C. Becker (CNC) 1 male, 1 female.

48 (137). *Atheta (Dimetrota) altaica* Bernhauer *

Additional records. Canada: Yukon, British Mts., Firth R. 250 m, 69.2166, –140.066, 18.VI.1984, sifting under old caribou dung, J.M. Campbell (CNC) 3 males, 3 females. USA, Alaska, Baranof Is. pl.08, 758m el., 56.41191, –134.85774, ±4m wetland, *Eriophorum angustifolium, Nephrophyllidium crista-galli, Carex nigricans*, pitfall, 28-29.VII.2013, S. Ridling (UAM) 7 specimens; Baranof Is. wpt.172, 752m el., 56.41158, –134.85837, ±5m subalpine wet tundra, fresh deer scat, forceps, 27.VII.2013, S. Ridling (UAM) 1 specimen; Denali NPP., Mt. Healy, 595m el., 63.73478, –148.92911, ±30m forest, mostly spruce, some aspen, 4 yellow pitfalls w/o rainroofs, pooled, 2-19.VII.2016, J.J. Rykken, D.S. Sikes, A. Haberski, Z. Snelson (UAM) 3 specimens; 4 yellow pitfalls w/o rainroofs, pooled, 2.VII-19.VIII.2016, J.J. Rykken, D.S. Sikes, A. Haberski, Z. Snelson (UAM) 3 specimens; Denali NPP., Primrose Ridge, 883m el., 63.72686, –149.40852, ±28m shrub, dwarf birch, 4 pitfalls w/ rainroofs, pooled, 2-20.VII.2016, J.J. Rykken, D.S. Sikes, A. Haberski, Z. Snelson (UAM) 1 specimen; Kenai Pen., Silvertip, 60.7397, –149.35428, ±10m *Betula* sp., pruned branch tips, 11.VIII.2014, R. Progar, Susan Bresney (UAM) 1 specimen; Prince of Wales Island, Hatchery Ck.4, 78m el., 55.88602, –132.8607, ±11m old growth, pitfall 3, 24.VII-6.VIII.2012, J. Stockbridge, S. Meierotto, I. MacDougall (UAM) 3 specimens; Prince of Wales Island, Luck Pt., 129m el., 55.98256, –132.77943, ±12m unthinned 2nd growth, pitfall 1, 11-26.VII.2012, J. Stockbridge, S. Meierotto, I. MacDougall (UAM) 1 specimen; S. Baranof Is. E. of Snipe Bay, 688m el., 56.40948, –134.86179, ±17m subalpine tundra, near edge of krummholz patch, primate dung, baited pitfall, 29-30.VII.2013, S. Ridling (UAM) 2 specimens; Tetlin NWR, 822m el., 63.36124, –143.47868, ±10m black spruce, hanging Malaise, 2.9m, 24.VII-10.VIII.2015, J.C. Hagelin (UAM) 1 specimen; White Mountain Recreation Area, Nome Creek Site 5, 589m el., 65.34077, –146.9275, ±10m open tussuck and standing water, some willow and bog birch, 10-24.VII.2015, J.C. Hagelin (UAM) 1 specimen.

49 (139). *Atheta (Dimetrota) nearctica* (Lohse)

Additional records. Canada: Yukon, Tombstone Mts., 64.60629, –138.35637, Rep. 1, wet, yellow pan, 24-27.VI.2011, NBP Field Party T1WY621 (LFC) 1 female. USA: Alaska, Elliot Highway Mile 86 135m el., 65.37669, –148.94533, ±10m black spruce near stream, hanging Malaise, 4.8m, 7-16.VI.2016, J.C. Hagelin (UAM) 1 specimen.

50 (141). *Atheta (Dimetrota) picipennis* (Mannerheim)

Additional records. USA: Alaska, 35.5 Mi Dalton Hwy, 271m el., 65.74425, –149.34564, ±4m *Betula, Salix*, white spruce, flight intercept trap, 31.V-5.VI.2008, D.S. Sikes (UAM) 1 specimen;

Dall Isl., 543m el., 54.99675, −133.01027, ±7m forest, primate dung baited, pitfall, 13-16. VII.2011, D.S. Sikes (UAM) 16 specimens; Etolin Isl., 760m el., 56.14311 −132.33711, ±5m alpine heath, sparse *Tsuga*, primate dung, pitfall, 19-22.VII.2011, D.S. Sikes (UAM) 6 specimens; Fairbanks, 930 Fitz Ct, 280m el., 64.9013 −147.52861, ±30m *Betula pap., Pop. trem.*, flight intercept trap, 29.V-4.VII.2008, D.S. Sikes (UAM) 3 specimens; Fairbanks, Creamer's Field, 152m el., 64.86864, −147.7381, ±6m barrier trap pair, 15-18. VII.2008, D.S. Sikes (UAM) 1 specimen; Mi 179 Dalton Hwy, 365m el., 67.29803, −150.15699, ±7m 3/4 Alder, white spruce, *Betula* forest, 5-31.VII.2008, D.S. Sikes (UAM) 1 specimen; Prince of Wales Isl. nr Black Lk pl.02, 780m el., 55.58988, −132.89548, ±2m wet meadow, *Calt. lept., Athy. filix-fem.*, hand, C. Bickford, 8.VII.2011, (UAM) 1 specimen; Prince of Wales Island, Staney Creek, 50m el., 55.79726, −133.1363, ±20m thinned 2nd growth, forceps, 16-17.V.2013, D.S. Sikes, J. Stockbridge, J. Slowik (UAM) 1 specimen; Prince of Wales Island, Coffman Cv, 67m el., 55.9795, −132.86256, ±101m thinned 2nd growth, Lindgren 1, 14-26.VII.2010, J. Stockbridge, C. Bickford (UAM) 1 specimen; same data except, Lindgren 2, 28.V-10.VI.2010, J. Stockbridge, C. Bickford (UAM) 1 specimen; Prince of Wales Island, Hatchery Ck.1, 57m el. 55.92524, −132.95054, ±10m thinned 2nd growth, Berlese, 5.VI.2011, J. Stockbridge, B. Wong (UAM) 1 specimen; Prince of Wales Island, Hatchery Ck.2, 134m el., 55.89356, −132.9437, ±13m thinned 2nd growth, Lindgren 2, 2-29.VI.2010, J. Stockbridge, C. Bickford (UAM) 2 specimens; same data except, Lindgren 2, 18.V-2.VI.2010, J. Stockbridge, C. Bickford (UAM) 1 specimen; Prince of Wales Island, Hatchery Ck.4, 78m el., 55.88602, −132.8607, ±11m old growth, Lindgren 1, 14-30.V.2010, J. Stockbridge, C. Bickford (UAM) 1 specimen; Prince of Wales Island, Luck Lk. 1 Rd., 101m el., 55.97805, −132.75456, ±10m old growth, pitfall 2, 18.V-4.VI.2010, J. Stockbridge, C. Bickford (UAM) 1 specimen; Prince of Wales Island, Luck Lk. 3 Rd., 117m el., 55.95347, −132.7708, ±5m old growth, pitfall 4, 18.V-4.VI.2010, J. Stockbridge, C. Bickford (UAM) 1 specimen; Prince of Wales Island, Luck Point 1A, 174m el., 55.98452, −132.78786, ±26m clear-cut, Lindgren 1, 2-27.VI.2010, J. Stockbridge, C. Bickford (UAM) 1 specimen; Prince of Wales Island, Luck Point 1B, 169m el., 55.98497, −132.787, ±25m clear-cut, Lindgren 2, 9.VII-1.VIII.2010, J. Stockbridge, C. Bickford (UAM) 1 specimen; same data except, pitfall 3, 18.V-2.VI.2010, J. Stockbridge, C. Bickford (UAM) 1 specimen; Prince of Wales Island, Luck Point 2A, 245m el., 55.97953, −132.77156, ±24m clear-cut, Lindgren 2, 31.V-14.VI.2012, J. Stockbridge, S. Meierotto, I. MacDougall (UAM) 1 specimen; same data except, Lindgren 1, 9.VII-1.VIII.2010, J. Stockbridge, C. Bickford (UAM) 1 specimen; pitfall 1, 1-11.VIII.2010, J. Stockbridge, C. Bickford (UAM) 3 specimens; pitfall 2, 18. V-2.VI.2010, J. Stockbridge, C. Bickford (UAM) 1 specimen; pitfall 2, 27.VI-11.VII.2012, J. Stockbridge, S. Meierotto, I. MacDougall (UAM) 2 specimens; Prince of Wales Island, Luck Point 2B, 264m el., 55.97939, −132.77216, ±25m clear-cut, Lindgren 2, 14-27. VI.2012, J. Stockbridge, S. Meierotto, I. MacDougall (UAM) 1 specimen; same data except, Lindgren 1, 31.V-14.VI.2012, J. Stockbridge, S. Meierotto, I. MacDougall (UAM) 1 specimen; Prince of Wales Island, Luck Pt., 129m el., 55.98256, −132.77943, ±12m unthinned 2nd growth, pitfall 1, 9.VII-1. VIII.2010, J. Stockbridge, C. Bickford (UAM) 1 specimen; pitfall 4, 9.VII-1.VIII.2010, J. Stockbridge, C. Bickford (UAM) 1 specimen; Prince of Wales Island, Staney Ck, 63m el., 55.79723, −133.13467, ±51m unthinned 2nd growth, pitfall 3, 27.VII-7.VIII.2011, J. Stockbridge, B. Wong (UAM) 1 specimen; Prince of Wales Island, Staney Ck. 1B, 77m el., 55.872, −133.06523, ±26m clear-cut, Lindgren 1, 16-31.V.2011, J. Stockbridge, B. Wong (UAM) 1 specimen; Prince of Wales Island, Staney Ck., 41-45m el., 55.79901, −133.11782, ±20m old growth, pitfall 4, 30.V-10.VI.2010,

J. Stockbridge, C. Bickford (UAM) 1 specimen; S. Chilkat Range, 638m el., 58.42384, –135.14883, ±6m subalpine, primate dung, pitfall, 9-10.VII.2010, D.S. Sikes (UAM) 8 specimens; S. Chilkat Range, 658m el., 58.42489, –135.14287, ±6m brown bear scat, alpine, hand collected, 9.VII.2010, D.S. Sikes (UAM) 6 specimens; S. Chilkat Range, 673m el., 58.42423, –135.14789, ±35m subalpine meadow, 24 yellow bowls, 8-9.VII.2010, D.S. Sikes (UAM) 1 specimen; S. Chilkat Range, 780m el., 58.43079, –135.16963, ±24m alpine heath, 24 yellow bowls, 9-10.VII.2010, D.S. Sikes (UAM) 1 specimen; S. Chilkat Range pl.02, 773m el., 58.43019, –135.16829, ±29m alpine *Carex, Cassiope*, pitfall, 8-9.VII.2010, D.S. Sikes (UAM) 1 specimen; S. Chilkat Range pl.13, 687m el., 58.4252, –135.15205, ±29m subalpine *Vaccinium, Cornus*, pitfall, 8-9.VII.2010, D.S. Sikes (UAM) 1 specimen; S. Chilkat Range pl.15, 680m el., 58.42497, –135.1496, ±28m subalpine *Gymnocarpium*, pitfall, 8-9.VII.2010, D.S. Sikes (UAM) 1 specimen; S. Chilkat Range pl.20, 659m el., 58.4248, –135.14096, ±29m subalpine *Tsuga, Vaccinium*, pitfall, 8-9.VII.2010, D.S. Sikes (UAM) 2 specimens; Sitka, Mosquito Cove Trail, 1m el., 57.1356, –135.37059, ±15m old growth Hemlock, coastal rain forest, 19.IX.2008, D.S. Sikes (UAM) 2 specimens; Tangle Lakes, 7.VII.1969, moose dung, J. Matthews (CNC) 7 males, 9 females, 11 sex undetermined; Tangle Lakes, 23.VI-2.VII.1969, J. Matthews (CNC) 1 sex undtermined; Tangle Lakes, 18.VI-VII.1969, J. Matthews (CNC) 1 sex undtermined.

51 (146). *Atheta (Dimetrota) prudhoensis* (Lohse) *

Additional records. USA: Alaska, 35.5 Mi Dalton Hwy, 271m el., 65.74425, –149.34564, ±4m *Betula, Salix*, white spruce, flight intercept trap, 31.V-5.VI.2008, D.S. Sikes (UAM) 2 specimens; Aiktak Is., 10 20m el., 54.18549, –164.8432, ±50m *Leymus, Heracleum*, maritime tundra, 5 pitfalls, 17-31.VIII.2009, A.L. Larned (UAM) 2 specimens; Alaska Peninsula / Becharof NWR, plot 1692 (Becharof Outlet), 20m el., 58.04903, –156.84016, ±5m bottom and at the edge of closed tall shrub-alder, near eds-crowberry, pitfall trap, 5.VII-15.VIII.2007, S. Savage (UAM) 7 specimens; Anaktuvuk Pass, 647m el., 68.14049, –151.74091, ±250m *Salix* leaf litter, Berlese, 19.V.2016, D.S. Sikes, K.M. Daly (UAM) 1 specimen; Anchorage, Triangle Lake, 98m el., 61.27451, –149.80249, ±4m open, black spruce bog, bottom catch, hanging Malaise 6.9m, 20.VI-15.VII.2013, L.H. DeCicco (UAM) 1 specimen; Baranof Is. pl.14, 610m el., 56.41285, –134.84598, ±9m shrubland, *Nephrophyllidium crista-galli, Elliottia pyrolifora, Luetkea pectinata*, pitfall, 27-28.VII.2013, S. Ridling (UAM) 1 specimen; Baranof Is. wpt.172, 752m el., 56.41158, –134.85837, ±5m subalpine wet tundra, fresh deer scat, forceps, 27.VII.2013, S. Ridling (UAM) 1 specimen; Beaver Creek, White Mountains, 418m el., 65.37804, –147.75587, ±7m creek edge, 4.VIII.2007, D.S. Sikes (UAM) 1 specimen; Bering Land Br. N. P., Serpentine Hot Spr., 118-300m el., 65.85183, –164.69884, ±5km tundra, shrubs, 8-10.VII.2016, D.S. Sikes, K.M. Daly (UAM) 1 specimen; Bering Land Bridge N. P., 135m el., 65.84873, –164.69435, ±30m beaver pond shore, dipnet, treading, sweep, 8.VII.2016, D.S. Sikes, K.M. Daly (UAM) 3 specimens; Bering Land Bridge N. P., Serpentine Hot Springs, 134-160m el., 65.8512, –164.70132, ±124m shrub, tundra, colored bowls transect, 8-10.VII.2016, D.S. Sikes, K.M. Daly (UAM) 1 specimen; Buldir Isl., 21m el., 52.37083, –175.88111, ±5m mossy area nr petrel burrows, 5 pitfalls, 26.VI-10.VII.2008, P. Leveque (UAM) 1 specimen; Chichagof Is., 741m el., 57.98567, –135.20573, ±19m subalpine tundra, meadow area, yellow pan trap (1), 14-18.VII.2013, S. Ridling (UAM) 1 specimen; Chichagof Is. pl.03, 702m el., 57.98708, –135.20357, ±3m meadow, *Nephrophyllidium crista-galli, Geranium erianthum, Artemisia arctica*, pitfall, 15-16.VII.2013, S. Ridling (UAM) 1 specimen; Chichagof Is. pl.04, 728m el., 57.98643, –135.20468, ±4m rich meadow, *Geranium erianthum, Artemisia arctica, Sanguisorba canadensis*, pitfall, 17-18.VII.2013, S. Ridling

(UAM) 1 specimen; Chichagof Is. pl.06, 725m el., 57.98495, –135.20358, ±4m krummholz, *Tsuga mertensiana, Elliottia pyroliflora, Vaccinium ovalifolium*, pitfall, 16-17.VII.2013, S. Ridling (UAM) 3 specimens, 17-18.VII.2013 (UAM) 1 specimen; Chichagof Is. pl.12, 851m el., 57.98392, –135.21161, ±4m meadow, *Leptarrhena pyrolifolia, Veratrum viride, Heracleum maximum*, pitfall, 15-16.VII.2013, S. Ridling (UAM) 2 specimens; Chichagof Is. pl.16, 963m el., 57.98159, –135.21707, ±4m rocky pavement, *Petasites frigidus, Parnassia fimbriata, Pedicularis oederi*, pitfall, 15-16. VII.2013, S. Ridling (UAM) 1 specimen, 16-17. VII.2013 (UAM) 1 specimen; Chichagof Is. pl.17, 979m el., 57.9809, -135.21817, ±3m wet meadow, *Senecio triangularis, Ranunculus occidentalis, Valeriana sitchensis*, pitfall, 14-15. VII.2013, S. Ridling (UAM) 1 specimen; Chichagof Is. pl.18, 1015m el., 57.98103, –135.21989, ±4m heath tundra, *Empetrum nigrum, Salix reticulata, Vaccinium caespitosum*, pitfall, 14-15.VII.2013, S. Ridling (UAM) 1 specimen, 16-17.VII.2013 (UAM) 1 specimen; Chichagof Is. pl.19, 991m el., 57.98173, –135.21889, ±4m meadow heath, *Phyllodoce glandulifera, Senecio triangularis, Luetkea pectinata*, pitfall, 14-15.VII.2013, S. Ridling (UAM) 1 specimen; same data except, 16-17. VII.2013 (UAM) 2 specimens;same data except, 17-18.VII.2013 (UAM) 3 specimens; Denali NPP, Primrose Ridge, 1276m el., 63.74095, –149.35963, ±28m rocky alpine, 4 pitfalls w/ rainroofs, pooled, 2-18.VII.2016, J.J. Rykken, D.S. Sikes, A. Haberski, Z. Snelson (UAM) 2 specimens; Denali NPP., Igloo Canyon, 1047m el., 63.59104, –149.60716, ±30m alpine tundra, 4 pitfalls w/ rainroofs, pooled, 2-20. VII.2016, J.J. Rykken, D.S. Sikes, A. Haberski (UAM) 1 specimen; Denali NPP., Igloo Canyon, 970m el., 63.59455, –149.60834, ±30m willow, dwarf birch, shrub, 4 yellow pitfalls w/o rainroofs, pooled, 20.VII-3.VIII.2016, J.J. Rykken, D.S. Sikes, A. Haberski, K. Chen (UAM) 3 specimens; same data except, 4 pitfalls w/ rainroofs, pooled, 3-17.VIII.2016, J.J. Rykken, D.S. Sikes, R. Stewart-Rogers (UAM) 1 specimen; 4 yellow pitfalls w/o rainroofs, pooled, 3-17.VIII.2016, J.J. Rykken, D.S. Sikes, A. Haberski (UAM) 1 specimen; Denali NPP., Mt. Healy, 1183m el., 63.75095, –148.96071, ±30m alpine tundra, 4 yellow pitfalls w/o rainroofs, pooled, 19.VII-3.VIII.2016, J.J. Rykken, D.S. Sikes, A. Haberski, K. Chen (UAM) 1 specimen; Denali NPP., Mt. Healy, 595m el., 63.73478, –148.92911, ±30m forest, mostly spruce, some aspen, 4 pitfalls w/ rainroofs, pooled, 19.VII-3.VIII.2016, J.J. Rykken, D.S. Sikes, A. Haberski, K. Chen (UAM) 1 specimen;same data except, 2-19.VII.2016, J.J. Rykken, D.S. Sikes, A. Haberski, Z. Snelson (UAM) 2 specimens; Denali NPP., Mt. Healy, 927m el., 63.74016, –148.95346, ±30m shrub, alder, dwarf birch, grass, clubmoss, 4 yellow pitfalls w/o rainroofs, pooled, 2.VII-19. VIII.2016, J.J. Rykken, D.S. Sikes, A. Haberski, Z. Snelson (UAM) 2 specimens; same date except, 4 pitfalls w/ rainroofs, pooled, 3-16. VIII.2016, J.J. Rykken, D.S. Sikes, R. Stewart-Rogers (UAM) 1 specimen; Denali NPP., nr Toklat, 1047m el., 63.50232, –150.06764, ±29m shrub, willow, dwarf birch, 4 yellow pitfalls w/o rainroofs, pooled, 4-18.VIII.2016, J.J. Rykken, D.S. Sikes, A. Haberski (UAM) 1 specimen; Denali NPP., nr Toklat, 1179m el., 63.49723, –150.10019, ±28m alpine tundra, 4 pitfalls w/ rainroofs, pooled, 2-21.VII.2016, J.J. Rykken, D.S. Sikes, A. Haberski, Z. Snelson (UAM) 1 specimen; Denali NPP., Primrose Ridge, 1252m el., 63.74163, –149.35381, ±28m alpine tundra, 4 pitfalls w/ rainroofs, pooled, 2-20.VII.2016, J.J. Rykken, D.S. Sikes, A. Haberski, Z. Snelson (UAM) 2 specimens; same data except, 4 yellow pitfalls w/o rainroofs, pooled, 2-20.VII.2016, J.J. Rykken, D.S. Sikes, A. Haberski, Z. Snelson (UAM) 5 specimens; Denali NPP., Primrose Ridge, 883m el., 63.72686, –149.40852, ±28m shrub, dwarf birch, 4 pitfalls w/ rainroofs, pooled, 2-17. VIII.2016, J.J. Rykken, D.S. Sikes, R. Stewart-Rogers (UAM) 1 specimen, 2-20.VII.2016, J.J. Rykken, D.S. Sikes, A. Haberski, Z. Snelson (UAM) 3 specimens; Denali NPP., Primrose Ridge, 883m el., 63.72686, –149.40852, ±28m shrub, dwarf birch, 4 yellow pitfalls w/o rainroofs, pooled, 19.VII-2.VIII.2016, J.J. Rykken, D.S. Sikes, A. Haberski, K. Chen (UAM) 1 specimen; same data except, 2-20. VII.2016, J.J. Rykken, D.S. Sikes, A. Haberski, Z. Snelson (UAM) 2 specimens; Denali NPP., Wright Mtn., 772m el., 63.71967, –149.48056, ±30m forest, mostly spruce, some aspen, 4 yellow pitfalls w/o rainroofs, pooled, 19.VII-2. VIII.2016, J.J. Rykken, D.S. Sikes, A. Haberski, K. Chen (UAM) 1 specimen; same data except, 2-20.VII.2016, J.J. Rykken, D.S. Sikes, A. Haberski, Z. Snelson (UAM) 4 specimens; Etolin Is., 754m el., 56.14261, –132.33534, ±205m heath, stream, krummholz, snowfield, 24 yellow bowls, 21-22.VII.2011, D.S. Sikes (UAM) 1 specimen; Etolin Isl., 760m el., 56.14311, –132.33711, ±5m alpine heath, sparse *Tsuga*, primate dung, pitfall, 19-22.VII.2011, D.S. Sikes (UAM) 2 specimens; Fairbanks, 930 Fitz Ct, 280-293m el., 64.90127, – 147.52797, ±30m *Populus, Betula, Alnus* leaf litter, Berlese, 10.X,2013, D.S. Sikes (UAM) 1 specimen; Fairbanks, 930 Fitz Ct ,280m el., 64.9013, –147.52861, ±30m *Betula* & *P. tremulo*., flight intercept trap, 16.VIII-22. IX,2007, D.S. Sikes (UAM) 19 specimens, 20. V-3.VII.2007 (UAM) 1 specimen; same data except, 29.V-4.VII.2008 (UAM) 8 specimens; pitfall, 13.VIII-22.IX.2007, D.S. Sikes (UAM) 1 specimen; inky cap mushroom, 27.VIII.2009, D.S. Sikes (UAM) 2 specimens; Fairbanks, Creamer's Field, 140m el., 64.86747, – 147.7292, ±55m field/forest edge, sweep, 22. VIII.2010, D.S. Sikes (UAM) 1 specimen; Fairbanks, Creamer's Field, 152m el., 64.86864, –147.7381, ±6m barrier trap pair, 15-18.VII.2008, D.S. Sikes (UAM) 4 specimens; Fairbanks, Creamer's Field, 156m el., 64.86852, –147.73779, ±15m mesic forest, pitfall, D.S. Sikes, C. Reilly, 19-21.VII.2008 (UAM) 1 specimen; Fairbanks, Yak Estates, 230-237m el., 64.84651, –147.8962, ±155m forest, pitfall, 15-17.VIII.1977, M.W. Oswood, H.C. Oswood (UAM) 2 specimens; Fairbanks, Yak Estates, 230-237m el., 64.84651, – 147.8962, ±155m pitfall, 18-20.VIII.1977, M.W. Oswood, H.C. Oswood (UAM) 6 specimens; Fairbanks: Creamer's Field, 140m el., 64.86883, –147.73065, ±104m *Betula* forest, decaying mushrooms, 31.VII.2010, D.S. Sikes (UAM) 1 specimen; Kenai Fjords National Park, Glacier View Trail, 117m el., 60.1871, –149.631, ±50m downed logs, pitfall trap (yellow), 27-29. VII.2016, J.J. Rykken (UAM) 1 specimen; Haines, Flower Mtn., 1000m el., 59.38739, – 136.32097, ±5m alpine, primate dung, pitfall, 15-18.VII.2010, D.S. Sikes (UAM) 3 specimens; Hawthorne Peak pl.04, 796m el., 58.23924, – 134.2419, ±6m meadow, *Lupinus nootkatensis, Nephrophyllidium crista-galli, Valeriana sitchensis*, pitfall, 22-23.VII.2013, S. Ridling (UAM) 1 specimen; Hawthorne Peak pl.10, 821m el., 58.24498, –134.24661, ±5m herbaceous heath, *Luetkea pectinata, Cassiope mertensiana, Lupinus nootkatensis*, pitfall, 19-20.VII.2013, S. Ridling (UAM) 1 specimen; Horner Hot Springs Creek, camp, 60m el., 64.91582, –154.83472, ±5m Malaise, 15-16. VII.2010, Jozef A. Slowik (UAM) 1 specimen; Koyukuk NWR, nr. Natlaratlen River, 69m el., 65.14783, –156.56415, ±3m *Chamaedaphne*, sedge, non-sphagnum moss innundated meadow, Malaise, 14-16.VIII.2014, R.J. Liebermann (UAM) 2 specimens; Mahoney Mt. pl.12, 836m el., 55.4099, –131.57826, ±5m krummholz, *Tsuga mertensiana, Vaccinium ovalifolium, Chamaecyparis nootkatensis*, pitfall, 7-8. VII.2013, S. Ridling (UAM) 1 specimen; Mi 179 Dalton Hwy, 365m el., 67.29803, – 150.15699, ±7m 3/4 Alder, white spruce, *Betula* forest, 5-31.VII.2008, D.S. Sikes (UAM) 5 specimens; near Kasilof, 60.3375, –151.2744, ±50m in mushrooms, 21.VII.2005, D. Langor (UAM) 1 specimen; Prince of Wales Island, Luck Point 2A, 245m el., 55.97953, – 132.77156, ±24m clear cut, Lindgren 2, 26.VII-9.VIII.2012, J. Stockbridge, S. Meierotto, I. MacDougall (UAM) 1 specimen; S. Chilkat Range, 638m el., 58.42384, –135.14883, ±6m subalpine, primate dung, pitfall, 9-10.VII.2010, D.S. Sikes (UAM) 1 specimen; S. Chilkat Range, 673m el., 58.42423, –135.14789, ±35m subalpine meadow, 24 yellow bowls, 8-9.

VII.2010, D.S. Sikes (UAM) 1 specimen; Sitka, Indian R., 24m el., 57.06068, –135.30496, ±50m old growth Hemlock, Devil's Club, ferns, sweep, 18.IX.2008, D.S. Sikes (UAM) 1 specimen; Sitka, Mosquito Cove Trail, 1m el., 57.1356, –135.37059, ±15m coastal rain forest, 16.IX.2008, D.S. Sikes (UAM) 3 specimens; Sitka, Mosquito Cove Trail, 1m el., 57.1356, –135.37059, ±15m old growth Hemlock, coastal rain forest, 19. IX.2008, D.S. Sikes (UAM) 6 specimens; Sterling, Corey St, 90m el., 60.53953, –150.83232, ±150m, 6.VIII.2004, D.M. Collet (UAM) 1 specimen; Talkeetnas, 3.9km N Glenn Hwy, Hicks Crk, 929m el., 61.84389, –147.84134, ±7m *Picea, Populus, Betula, Salix*, pitfall, 13-16. VIII.2008, K. W. May (UAM) 1 specimen; Tetlin NWR, 822m el., 63.36124, –143.47868, ±10m black spruce, hanging Malaise, 2.9m, 24.VII-10. VIII.2015, J.C. Hagelin (UAM) 1 specimen; Tangle Lakes, 7.VII.1969, moose dung, J. Matthews (CNC) 2 males.

52 (147). ***Atheta (Dimetrota) strigosula*** **Casey**
Additional records. USA: Alaska, 12.7km NNW Talkeetna, East-West Express Park trailhead, 62.41323, –150.25362, ±10m *Salix*, pruned tree branch, 12.VIII.2013, B. Shryock, R. Progar, P. Sharma (UAM) 1 specimen; Denali NPP, Primrose Ridge, 1276m el., 63.74095, –149.35963, ±28m rocky alpine, 4 pitfalls w/ rainroofs, pooled, 2-18.VII.2016, J.J. Rykken, D.S. Sikes, A. Haberski, Z. Snelson (UAM) 1 specimen; Fairbanks, 930 Fitz Ct, 280m el., 64.9013, –147.52861, ±30m *Betula* & *P. tremuloides*; flight intercept trap, D.S. Sikes; 16.VIII-22.IX.2007, D.S. Sikes (UAM) 100 specimens; same data except, 26.VIII-10. IX.2006, (UAM) 2 specimens; 29.V-4.VII.2008, (UAM) 36 specimens; 29.IV-29.V. 2008 (UAM) 2 specimens; inky cap mushroom, 27.VIII.2009, D.S. Sikes (UAM) 1 specimen; Fairbanks, Creamer's Field, 156m el., 64.86852, –147.73779, ±15m mesic forest, pitfall, 19-21. VII.2008, D.S. Sikes, C. Reilly (UAM) 1 specimen; Fairbanks: Creamer's Field, 132m el., 64.86805, –147.73764, ±20m mesic *Betula, Picea*, pitfall, 15-18.VII.2008, D.S. Sikes, C. Eshleman (UAM) 1 specimen; Horner Hot Springs Creek, camp, 60m el., 64.91582, –154.83472, ±5m Malaise, J.A. Slowik, 15-16. VII.2010 (UAM) 1 specimen; near Kasilof, 60.3375, –151.2744, ±50m in mushrooms, 21. VII.2005, D. Langor (UAM) 21 specimens; Prince of Wales Island, Hatchery Ck.1, 57m el., 55.92524, –132.95054, ±10m thinned 2nd growth, pitfall 4, 10-30.VII.2010, J. Stockbridge, C. Bickford (UAM) 1 specimen; Prince of Wales Island, Hatchery Ck.4, 82m el., 55.88285, –132.89795, ±27m unthinned 2nd growth, pitfall 2, 9-24.VII.2012, J. Stockbridge, S. Meierotto, I. MacDougall (UAM) 1 specimen; Sterling, Corey St, 90m el., 60.53953, –150.83232, ±150m 6.VIII.2004, D.M. Collet (UAM) 1 specimen; same data except, disturbed *P. glauca* forest, uv light, 6.IX.2004, D.M. Collet, (UAM) 1 specimen; 13.VIII.2004 (UAM) 1 specimen; 31.VIII.2004 (UAM) 2 specimens; 3. VIII.2004 (UAM) 3 specimens.

53 (149). ***Atheta (Dimetrota) terranovae*** **Klimaszewski and Langor**
Additional records. Canada: British Columbia, Glacier National Park, Hemlock Grove Boardwalk, 1094m el., 51.243, –117.649, old-growth hemlock and cedar forest, pitfalls, pan traps and flight-intercept trap, 04-10.VIII.2014, BIOBus 2014 team (CBG) 13 specimens; Glacier National Park, Meeting of the Waters Trail, 1271m el., 51.26, –117.49, spruce forest, pitfall and pan traps, 04-10.VIII.2014, BIOBus 2014 team (CBG) 6 specimens; Kinaskan Lake Provincial Park, Kinaskan Lake Trail, 833m el., 57.532, –130.202, spruce and aspen forest, pitfall, 26.VII.-01.VIII.2014, BIOBus 2014 team (CBG) 1 specimen. Quebec, Abitibi, lac Duparquet, 14-21. VIII.1994, Malaise, bac, Échantillon 1218, 1847, Forêt mixte, P. Paquin (LFC) 1 male; Abitibi, lac Duparquet, 24-31.VII.1994, Malaise, bac, Ėchantillon 950, 1847, Forêt mixte, P. Paquin (LFC) 1 male; Terr. Nouevau-Québec, Chemin Selbaie, Piège-fosse, Ėchantillon P-2149, F 1, Pessière noir, P. Paquin (LFC) 2 females; Terr. Nouevau-Québec, Chemin Casa-berardi, 31.VIII-7.IX.1997, Malaise, pan, Ėchantillon

5396, 300, Pessière noir, P. Paquin (LFC) 2 males. USA: Alaska, [Arctic Nat. Wild Ref. (ANWR)] Coleen R., -15 mi NW. Porcupine R., 326-346m el., 67.77621, −142.39573, ±981m gravel bars, mature forest, low bluff with sage, sweep, hand, 25-28.VII.2001, J.J. Kruse (UAM) 1 specimen; 25 mi S Delta Jct., 637m el., 63.67343, −145.88283, ±4m old glaciated; mixed *Betula, Picea, Alnus* forest, Berlese, 7. IX.2009, G. Henry (UAM) 1 specimen; 25 mi. S Delta Junction, 552m el., 63.67342, −145.88287, ±4m old glaciated; *Populus tremuloides, P. balsamea, Picea*, Berlese, 7.IX.2009, G. Henry, (UAM) 2 specimens; 35.5 Mi Dalton Hwy, 271m el., 65.74425, −149.34564, ±4m *Betula, Salix*, white spruce, flight intercept trap, 31.V-5.VI.2008, D.S. Sikes (UAM) 2 specimens; Alaska Peninsula / Becharof NWR, plot 1692 (Becharof Outlet), 20m el., 58.04903, −156.84016, ±5m bottom and at the edge of closed tall shrub-alder, near eds-crowberry, pitfall trap, 5.VII-15.VIII.2007, S. Savage (UAM) 1 specimen; Denali NPP, nr Toklat, 1263m el., 63.50154, −150.10298, ±28m rocky alpine, 4 pitfalls w/ rainroofs, pooled, 4-18.VIII.2016, J.J. Rykken, D.S. Sikes, R. Stewart-Rogers (UAM) 1 specimen; Denali NPP., Igloo Canyon, 970m el., 63.59455, −149.60834, ±30m willow, dwarf birch, shrub, 4 yellow pitfalls w/o rainroofs, pooled, J.J. Rykken; 20.VII-3. VIII.2016, D.S. Sikes, A. Haberski, K. Chen (UAM) 2 specimens; same data except, 4 pitfalls w/ rainroofs, pooled, 3-17.VIII.2016, J.J. Rykken, D.S. Sikes, R. Stewart-Rogers (UAM) 1 specimen; 4 yellow pitfalls w/o rainroofs, pooled, 3-17.VIII.2016, J.J. Rykken, D.S. Sikes, A. Haberski (UAM) 5 specimens; Denali NPP., Mt. Healy, 1183m el., 63.75095, −148.96071, ±30m alpine tundra, 4 pitfalls w/ rainroofs, pooled,19.VII-3.VIII.2016, J.J. Rykken, D.S. Sikes, A. Haberski; K. Chen (UAM) 2 specimens; same data except, 4 yellow pitfalls w/o rainroofs, pooled, 2-19.VII.2016, J.J. Rykken, D.S. Sikes, A. Haberski, Z. Snelson (UAM) 2 specimens; 3-16.VIII.2016, J.J. Rykken; D.S. Sikes; R. Stewart-Rogers (UAM) 2 specimens; Denali NPP., Mt. Healy, 1249m el., 63.75362, −148.96379, ±30m rocky alpine, 4 pitfalls w/ rainroofs, pooled, 19.VII-3. VIII.2016, J.J. Rykken, D.S. Sikes, A. Haberski, K. Chen (UAM) 4 specimens; Denali NPP., Mt. Healy, 595m el., 63.73478, −148.92911, ±30m forest, mostly spruce, some aspen, 10 minute ground search, 16.VIII.2016, D.S. Sikes (UAM) 2 specimens; Denali NPP., Mt. Healy, 595m el., 63.73478, −148.92911, ±30m forest, mostly spruce, some aspen, litter, Winkler, 16. VIII.2016, J.J. Rykken, D.S. Sikes, R. Stewart-Rogers (UAM) 2 specimens; same data except, 4 pitfalls w/ rainroofs, pooled, 19.VII-3. VIII.2016, , J.J. Rykken, D.S. Sikes, A. Haberski, K. Chen (UAM) 1 specimen; 4 yellow pitfalls w/o rainroofs, pooled, 3-16. VIII.2016, J.J. Rykken, D.S. Sikes, R. Stewart-Rogers (UAM) 8 specimens; 4 pitfalls w/ rainroofs, pooled, 3-16.VIII.2016, J.J. Rykken, D.S. Sikes, A. Haberski, J. Cheney (UAM) 2 specimens; Denali NPP., Mt. Healy, 927m el., 63.74016, −148.95346, ±30m shrub, alder, dwarf birch, grass, clubmoss, litter, Winkler, 16. VIII.2016, J.J. Rykken, D.S. Sikes, R. Stewart-Rogers (UAM) 1 specimen; same data except, 4 pitfalls w/ rainroofs, pooled, 3-16.VIII.2016, J.J. Rykken, D.S. Sikes, R. Stewart-Rogers (UAM) 1 specimen; 4 yellow pitfalls w/o rainroofs, pooled, 3-16.VIII.2016, J.J. Rykken, D.S. Sikes, R. Stewart-Rogers (UAM) 2 specimens; Denali NPP., nr Teklanika R., 833m el., 63.64261, −149.57181, ±30m spruce, 4 yellow pitfalls w/o rainroofs, pooled, 3-17. VIII.2016, J.J. Rykken, D.S. Sikes, R. Stewart-Rogers (UAM) 15 specimens; Denali NPP., nr Toklat, 1047m el., 63.50232, −150.06764, ±29m shrub, willow, dwarf birch, 4 pitfalls w/ rainroofs, pooled, 2-21.VII.2016, J.J. Rykken, D.S. Sikes, A. Haberski, Z. Snelson (UAM) 1 specimen; same data except, 4 yellow pitfalls w/o rainroofs, pooled, 4-18.VIII.2016, J.J. Rykken, D.S. Sikes, A. Haberski (UAM) 1 specimen; Denali NPP., nr Toklat, 1179m el., 63.49723, −150.10019, ±28m alpine tundra, 4 yellow pitfalls w/o rainroofs, pooled, 20.VII-4. VIII.2016, J.J. Rykken, D.S. Sikes, A. Haberski, K. Chen (UAM) 1 specimen; Denali NPP.,

Primrose Ridge, 883m el., 63.72686, –149.40852, ±28m shrub, dwarf birch, 4 pitfalls w/ rainroofs, pooled, 19.VII-2.VIII.2016, J.J. Rykken, D.S. Sikes, A. Haberski, K. Chen (UAM) 7 specimens; same data except. 4 pitfalls w/ rainroofs, pooled, 2-17.VIII.2016, J.J. Rykken, D.S. Sikes, R. Stewart-Rogers (UAM) 5 specimens; 4 pitfalls w/ rainroofs, pooled, 2-20.VII.2016, J.J. Rykken, D.S. Sikes, A. Haberski, Z. Snelson (UAM) 1 specimen; 4 yellow pitfalls w/o rainroofs, pooled, 19.VII-2.VIII.2016, J.J. Rykken, D.S. Sikes, A. Haberski, K. Chen (UAM) 1 specimen; 4 yellow pitfalls w/o rainroofs, pooled, 2-20.VII.2016, J.J. Rykken, D.S. Sikes, A. Haberski, Z. Snelson (UAM) 5 specimens; Denali NPP., Toklat, 960m el., 63.52373, –150.04829, ±30m spruce, 4 yellow pitfalls w/o rainroofs, pooled, 4-18.VIII.2016, J.J. Rykken, D.S. Sikes, R. Stewart-Rogers (UAM) 2 specimens; Denali NPP., Wright Mtn., 772m el., 63.71967, –149.48056, ±30m forest, mostly spruce, some aspen, 4 pitfalls w/ rainroofs, pooled, 19.VII-2.VIII.2016, J.J. Rykken, D.S. Sikes, A. Haberski, K. Chen (UAM) 3 specimens; same data except, 4 yellow pitfalls w/o rainroofs, pooled, 2-17.VIII.2016, J.J. Rykken, D.S. Sikes, R. Stewart-Rogers (UAM) 8 specimens; 4 pitfalls w/ rainroofs, pooled, 2-20.VII.2016, J.J. Rykken, D.S. Sikes, A. Haberski Z. Snelson (UAM) 1 specimen; 4 yellow pitfalls w/o rainroofs, pooled, 2-20.VII.2016, J.J. Rykken, D.S. Sikes, A. Haberski, Z. Snelson (UAM) 4 specimens; Fairbanks, 930 Fitz Ct, 280m el., 64.90114, –147.52774, ±5m woodpile, 30.VIII.2007, D.S. Sikes (UAM) 1 specimen; Fairbanks, 930 Fitz Ct, 280m el., 64.9013, –147.52861, ±30m *Betula* & *P. tremulo.*, flight intercept trap, 16.VIII-22.IX.2007, D.S. Sikes (UAM) 17 specimens; same data except, 29.V-4.VII.2008 (UAM) 1 specimen; Fairbanks, Ester Dome, 719m el,. 64.8768, –148.06427, ±61m summit, *Pop. trem.*, pitfall, 17-24.VII.2007, B. Mortensen (UAM) 2 specimens; Fairbanks; Yak Estates, 230-237m el., 64.84651, –147.8962, ±155m pitfall, 18-20.VIII.1977, M. W. Oswood, H. C. Oswood (UAM) 7 specimens; Fairbanks: Creamer's Field, 140m el,. 64.86883, –147.73065, ±104m *Betula* forest, decaying mushrooms, 31.VII.2010, D.S. Sikes (UAM) 8 specimens; Mi 179 Dalton Hwy, 365m el., 67.29803, –150.15699, ±7m 3/4 Alder, white spruce, *Betula* forest, 5-31.VII.2008, D.S. Sikes (UAM) 11 specimens; mile 217.2 Richardson Highway, Castner Glacier, 781m el., 63.40134, –145.71895, ±5m low *Salix, Alnus, P. balsam*, Berlese, 7.IX.2009, G. Henry (UAM) 1 specimen; Sitka, Mosquito Cove Trail, 1m el., 57.1356, –135.37059, ±15m old growth Hemlock, coastal rain forest, 19.IX.2008, D.S. Sikes (UAM) 1 specimen; Sterling, Corey St, 90m el., 60.53953, –150.83232, ±150m disturbed *P. glauca* forest, uv light, 6.IX.2004, D.M. Collet (UAM) 1 specimen; same data except,13.VIII.2004, (UAM) 1 specimen; 31.VIII.2004, (UAM) 1 specimen; Tetlin NWR, 822m el., 63.36124, –143.47868, ±10m black spruce, hanging Malaise, 2.9m, 24.VI-10.VIII.2015, J.C. Hagelin (UAM) 1 specimen.

54 (151). *Atheta (Dimetrota) gelida* (J. Sahlberg) *

Additional records. Canada: Yukon, British Mts., Alder Mt., 480m, 69.200, −140.05, 23.VI.1984, sifting *Alnus* litter, J.M. Campbell (CNC) 3 males, 7 females. USA, Alaska, Prudhoe Bay Rd., 6 mi S Galbraith Lk., 68.400, −149.416, 5.VII.1978, 2800 feet, J.M. Campbell and A. Smetana (CNC) 1 male; Prudhoe Bay Rd., 10 mi N Galbraith Lk., 68.583, −149.500, 4.VII.1978, 2600 feet, J.M. Campbell and A. Smetana (CNC) 1 sex undetermined; Prudhoe Bay Rd., 15 mi S Galbraith Lk., 68.266, −149.433, 6.VII.1978, 3000 feet, A. Smetana and J.M. Campbell (CNC) 2 males, 2 females, 6 sex undetermined; Prudhoe Bay Rd., 4 mi SW Chandalar Cmp., 68.066, −149.666, 7.VII.1978, 4000 feet, gopher burrow, A. Smetana and J.M. Campbell (CNC) 16 males, 36 females, 26 sex undetermined, (LFC) 5 males, 2 females; Prudhoe Bay Rd., 9 mi N Atigun Pass, 68.233, −149.416, 6.VII.1978, 3100 feet, J.M. Campbell and A. Smetana (CNC) 1 male, 1 sex undetermined; Mt. McKinley N. P., Polychrome Pass, 3500 feet, 21.VI.1978, J.M. Campbell and E. Becker (CNC) 1 male, 1 female.

55 (153). ***Atheta (Microdota) microelytrata*** **Klimaszewski and Godin**
Additional records. Canada, Yukon, Takhini, 60.8769, −135.3596, 715m, 19.IX.2009, hotsprings, alder/willow litter, sifting, coll. B. Godin and D. Horwood (LFC) 2males.

56 (154). ***Atheta (Microdota) pennsylvanica*** **Bernhauer**
Additional records. Canada: Newfoundland, Goose Bay, 53.32035, −60.29268, 18-21. VI.2010, rep 3 mesic, yellow pan, NBP Field Party G3MY121 (LFC) 1 female. Quebec, Schefferville, 54.75970, −66.71120, 6-9. VII.2010, rep 1, wet, yellow pan, NBP Field Party S1WY936 (LFC) 1 female.

57 (155). ***Atheta (Microdota) platonoffi*** **Brundin ***
Additional records. USA: Alaska, Bonanza Crk LTER, 225m el., 64.70963, −148.32647, ±61m forest edge, *Populus tremuloides*, grass, pitfall, 17-24.VII.2007, B. Mortensen (UAM) 1 specimen; Fairbanks, 930 Fitz Ct, 280m el., 64.9013, −147.52861, ±30m *Betula* & *P. tremuloides*, flight intercept trap, 16.VIII.-22. IX.2007, D.S. Sikes, (UAM) 4 specimens; Fairbanks, Creamer's Field, 158m el., 64.86823, −147.73758, ±12m mesic *Betula*, pitfall, 26.VI-18.VI.2008, D.S. Sikes, R. Davidson (UAM) 1 specimen; Fairbanks, Ester Dome, 719m el., 64.8768, −148.06427, ±61m summit, *Populus tremuloides*, pitfall, 17-24.VII.2007, B. Mortensen (UAM) 1 specimen.

58 (158). ***Atheta (Microdota) ripariides*** **Newton**
Additional records. Canada: Saskatchewan, Prince Albert, 53.9386, −106.057, 556m, 14. IX.2014, Mushrooms, sifting, col. B. Godin and D. Horwood (BGC) 1male.

59 (162). ***Atheta (Pseudota) pseudoklagesi*** **Klimaszewski and Webster**
Additional records. Canada: Manitoba, Riding Mountain National Park, Arrowhead Trail, 663m el., 50.676, −99.898, mixed wood, medium stage aspen stand, Malaise trap, 19-26.VI.2012, Parks Canada staff (CBG) 1 specimen. Saskatchewan, Meadow Lake, 54.4826, −109.6584, 513m, 10. IX.2014, Mushrooms, sifting, coll. B. Godin and D. Horwood (LFC) 1 male; same data except: 54.4012, −108.8014, 483m (LFC) 1 male; Prince Albert, 53.9386, −106.057, 556m, 14.IX.2014, Mushrooms, sifting, coll. B. Godin and D. Horwood (BGC, LFC) 2 males, 1 female.

Genus *Bellatheta* Roubal

60 (167). ***Bellatheta microptera*** **(Lohse)**
Additional records. Canada, Yukon, Dempster Hwy. km 220 N, 26.VI. 1987, Tamarack bog, ex *Alnus crispa* and *Picea mariana* litter, V, Behan, VB36-87 (CNC) 1 male. USA: Alaska, Denali NPP., Mt. Healy, 927m el., 63.74016, −148.95346, ±30m shrub, alder, dwarf birch, grass, clubmoss, litter, Winkler, 16.VIII.2016, J.J. Rykken, D.S. Sikes, R. Stewart-Rogers (UAM) 3 specimens; Toolik Field Station, 721m el., 68.62722, −149.5986, ±4m *Betula nana* leaf litter, Berlese, 4.VI.2008, D.S. Sikes (UAM) 1 specimen; Toolik Field Station, 777m el., 68.62685, −149.59798, ±14m *Salix* leaf litter, Berlese, 3.VI.2008, D.S. Sikes (UAM) 3 specimens; Prudhoe Bay rd., 15 mi S Galbraith Lk., 68.2666, −149.4333, 5.VII.1978, 2800–3000 feet, A. Smetana and J.M. Campbell (CNC) 3 males, 4 females; Prudhoe Bay Rd., Bonanza Cr., 900 feet, 66.6666, −150.6666, 2. VII.1972, A. Smetana and J.M. Campbell (CNC) 1 sex undetermined; Prudhoe Bay Rd., Fish Cr., 900 feet, 66.5333, −150.8333, 2.VII.1972, J.M. Campbell and A. Smetana (CNC) 1 sex undetermined.

Genus *Boreophilia* Benick

61 (169). ***Boreophilia fusca*** **(CR Sahlberg) ***
Additional records. Canada: Manitoba, Churchill, 4 km SE Churchill, Dene Village, 5m el., 58.734, −94.112, sweep net, 14.VII.2008, P. Kevan, T. Woodcock (CBG) 1 specimen. Northwest Territories, Aklavik, 9.VI.1956, R.E. Leech (CNC) 2 males, 1 female. USA, Alaska, Unalakleet, 4.VII.1961, R. Madge (CNC) 1 male; Naknek, 3.VII.1952, on tundra, W.R. Mason (CNC) 1 male, 1 female; Alaska Hwy. Mi 1249, Deadman Lake, 6.VII.1968, A. Smetana (CNC) 1 female.

62 (170). *Boreophilia hyperborea* (Brundin) *
Additional records. USA: Canada, Manitoba, Churchill, 12 km S Churchill, Goose Creek Marina, 2m el., 58.663, –94.166, dip net, 09. VIII.2010, J. Witt (CBG) 1 specimen. Alaska, Anaktuvuk Pass, 665m el., 68.14059, –151.74266, ±200m *Salix*, tundra hillside, cobble, pollinator cup/pitfall, 20.V.2016, D.S. Sikes; K.M. Daly (UAM) 1 specimen; Anaktuvuk Pass, 697m el., 68.14957, –151.7746, ±100m *Salix, Betula, Alnus*, grasses, sweeping, 21. V.2016, D.S. Sikes; K.M. Daly (UAM) 1 specimen; Anaktuvuk Pass, Eleanor Lake, 649m el., 68.14742, –151.71854, ±100m lake shore and pond, sweep and dipnet, 20.V.2016, D.S. Sikes, K.M. Daly (UAM) 2 specimens; Toolik Field Station, 724m el., 68.6286, –149.59772, ±36m under rocks, etc., 1–3.VI.2008, D.S. Sikes (UAM) 2 specimens.

63 (171). *Boreophilia islandica* (Kraatz) *
Additional records. Canada: Northwest Territories, Yellowknife, 62.51923, –113.38494, 7-11.VII.2011, rep. 2 wet, pitfall trap, NBP field party Y2WP111 (LFC) 1 male; Norman Wells, 65.29112, –126.62262, 14-17.VI.2011, NBP field party N1WY633 (LFC) 1 female; Yukon Territory, Tombstone Mtns., 64.57942, –138.28212, 27.VI-1.VII.2011, rep. 3, wet, pitfall trap, NBP field party, T3W P937 (LFC) 1 female. USA, Alaska, Agattu Isl., 15m el., 52.47308, –173.59065, ±30m beach, *Heracl. lanatum*, 5 pitfalls, 5-19.VI.2008, L. Kenney, R. Kaler (UAM) 1 specimen; Aiktak Is., 10-20m el., 54.18356, –164.82793, ±1km maritime tundra, *Leymus, Heracleum*, 5 pitfalls, 24.VI-11. VII.2009, S. Sapora (UAM) 2 specimens; Aiktak Is., 10-20m el., 54.18356, –164.82793, ±1km maritime tundra, *Leymus, Heracleum*, 5 pitfalls, 9-24.VI.2009, A. L. Larned (UAM) 1 specimen; Aiktak Is,. 10-20m el., 54.18549, –164.8432, ±50m *Leymus, Heracleum, Angelica*, 5 pitfalls, 11-29.VII.2009, A. L. Larned (UAM) 3 specimens; Aiktak Is., 10-20m el., 54.18549, –164.8432, ±50m *Leymus, Heracleum*, maritime tundra, 5 pitfalls, 17-31.VIII.2009, A. L. Larned (UAM) 1 specimen; Aiktak Is., 20m el., 54.18393, –164.84167, ±5m streamside, *Leymus*, 2 pitfalls, 26.VI-14.VII.2008, B. A. Drummond, M. E. McClintock (UAM) 2 specimens; Anaktuvuk Pass, 647m el., 68.14049, –151.74091, ±250m *Salix*, sweeping, 19.V.2016, D.S. Sikes; K.M. Daly (UAM) 1 specimen; Becharof NWR Island Arm, 3m el., 57.75391, –155.98835, ±20m sweep, 26.VI.2007, G. Milet (UAM) 1 specimen; Kasatochi, 42m el., 52.16869, –175.52475, ±34m under rocks, talus, cliff base, 11.VI.2008, D.S. Sikes (UAM) 2 specimens; Mi 179 Dalton Hwy, 365m el., 67.29803, –150.15699, ±7m 3/4 Alder, white spruce, *Betula* forest, 5-31.VI.2008, D.S. Sikes (UAM) 1 specimen; Nogahabara Dunes [Koyukuk NWR], 156m el., 65.6578, –157.47615, ±23m 1400–1500 h, look down, 12. VI.2010, J. A. Slowik (UAM) 1 specimen; S. Chilkat Range pl.19, 652m el., 58.42509, –135.14229, ±30m subalpine *Veratrum, Calamagrostis*, hand collected, 9.VII.2010, D.S. Sikes (UAM) 1 specimen; White Mountains Recreation Area, Nome Creek Site 4, 180m el., 65.33469, –146.83997, ±10m b. & w. spruce, willow, hanging Malaise, 4.6m, 10-17.VI.2016, J.C. Hagelin (UAM) 1 specimen; Tangle Lakes, 14-21.1969, J. Matthews (CNC) 1 male; Deering, 24.VIII.1961, J. Matthews (CNC) 1 female.

64 (172). *Boreophilia nearctica* Lohse
Additional records. USA: Alaska, Fairbanks, Bonanza Creek, 237m el., 64.71922, –148.38416, ±10m wood cutting area, w. spruce, birch, grass, hanging Malaise, 6.15m, 6-13. V.2016, J.C. Hagelin (UAM) 1 specimen; Galena, Yukon Riv., W of town, 54m el., 64.74198, –156.98022, ±3m in trees, Malaise, 9-11.VII.2010, J.A. Slowik; Naknek, 2-5m el., 58.73973, –157.0636, ±20m creekside/ocean beach confluence, under boards and drift wood, hand collected, 10.VI.2007, D.S. Sikes (UAM) 1 specimen; Selawik NWR, 8m el., 66.56137, –158.99834, ±304m *Spirea*, dwarf birch, 23-24. VI.2010, D.S. Sikes (UAM) 1 specimen.

65 (173). *Boreophilia nomensis* (Casey)
Additional records. USA: Alaska, Denali NPP., Mt. Healy, 1183m el., 63.75095, –148.96071, ±30m alpine tundra, 4 pitfalls w/ rainroofs,

pooled, 2-19.VII.2016, J.J. Rykken, D.S. Sikes, A. Haberski, Z. Snelson (UAM) 1 specimen; Thompson Pass, 796m el., 61.13731, -145.74487, ±105m under rocks nr. snow field, hand, 28.VII.2010, D.S. Sikes, A.B. Sikes (UAM) 1 specimen.

66 (174). ***Boreophilia beringi*** **Klimaszewski and Brunke**
Additional records. USA: Alaska, Bering Land Bridge N. P., 413m el., 65.83713, −164.58995, ±30m snowfield, tundra; under rocks, on moss, 9. VII.2016, D.S. Sikes; K.M. Daly; S. Tocktoo; T. Stenek (UAM) 7 specimens.

67 (175). ***Boreophilia caseyi*** **Lohse**
Additional records. USA, Alaska, Deering, A/25, 27.VII-2.VIII.1968, J. Matthews (CNC) 2 males, 1 female; Deering, Seward Peninsula, 20-23.VII.1968, J. Matthews (CNC) 2 males.

68 (177). ***Boreophilia neoinsecuta*** **Klimaszewski**
Additional records. USA: Alaska, Anaktuvuk Pass, 640-680m el., 68.14049, −151.74091, ±2km in and around village, hand, forceps, 20-22.V.2016, D.S. Sikes, K.M. Daly (UAM) 1 specimen; Anaktuvuk Pass, 647m el., 68.14049, −151.74091, ±250m under rocks, forceps, aspirator, 19.V.2016, K. Hildebrandt, D.S. Sikes, K.M. Daly (UAM) 3 specimens.

69 (178). ***Boreophilia subplana*** **(J.R. Sahlberg) ***
Additional records. USA: Alaska, Anaktuvuk Pass, 647m el., 68.14049, −151.74091, ±250m *Salix* leaf litter, Berlese, 19.V.2016, D.S. Sikes, K.M. Daly (UAM) 7 specimens; Anaktuvuk Pass, 647m el., 68.14049, −151.74091, ±250m *Salix*, sweeping, 19.V.2016, D.S. Sikes, K.M. Daly (UAM) 1 specimen; Anaktuvuk Pass, 665m el., 68.14059, −151.74266, ±200m *Salix*, tundra hillside, cobble, pollinator cup/pitfall, 20.V.2016, D.S. Sikes, K.M. Daly (UAM) 5 specimens; Anaktuvuk Pass, Eleanor lake, 649m el., 68.14742, −151.71854, ±100m lake shore and pond, sweep and dipnet, 20.V.2016, D.S. Sikes, K.M. Daly (UAM) 2 specimens; Kotzebue, 1m el., 66.89151, −162.5933, ±100m brackish shoreline; under rocks, wrack, 6.VII.2016, D.S. Sikes, K.M. Daly (UAM) 1 specimen; Tetlin NWR, 822m el., 63.36124, −143.47868, ±10m black spruce, hanging Malaise, 2.9m, 13-24. VII.2015, J.C. Hagelin (UAM) 1 specimen; Tetlin NWR, Alaska Highway Site 2, 523m el., 63.08456, −141.96761, ±10m black spruce, hanging Malaise, 4m, 13-24.VII.2015, J.C. Hagelin (UAM) 1 specimen; Kotzebue, 14. IX.1967, J. Matthews (CNC) 1 female.

70 (180). Boreophilia venti (Lohse)
Additional records. USA: Alaska, Anaktuvuk Pass, 665m el., 68.14059, −151.74266, ±200m *Salix*, tundra hillside, cobble, pollinator cup/pitfall, 20.V.2016, D.S. Sikes, K.M. Daly (UAM) 4 specimens; Kotzebue, 14.IX.1967, J. Matthews (CNC) 1 female (tentative); Deering, 20-23.VII.1968, J. Matthews (CNC) 1 female (tentative).

Genus *Boreostiba* Lohse

71 (181). ***Boreostiba frigida*** **(J.R. Sahlberg) ***
Additional records. USA: Alaska, Bering Land Bridge N. P., 413m el., 65.83713, −164.58995, ±30m snowfield, tundra; under rocks, on moss, 9. VII.2016, D.S. Sikes, K.M. Daly, S. Tocktoo, T. Stenek (UAM) 19 specimens; Hall Isl., 2-20m el., 60.67879, −173.0688, ±205m splash zone rocks, driftwood, creekside rocks, forceps, 6.VIII.2012, D.S. Sikes, C.E. Bickford (UAM) 1 specimen.

72 (182). ***Boreostiba sibirica*** **(Mäklin) ***
Additional records. Canada: Northwest Territories, Reindeer Sta., Caribou Hills, 2. VII.1972. A. Smetana (CNC) 1 male, 3 females. Yukon, 8 mi NW Keno Mtn., 6000 feet, 19. VII.1968, A. Smetana (CNC, LFC) 14 males, 19 females, 10 sex undetrmined; Blow R., 68.56, −137.06, 7.VII.1987, ex rich *Alnus*, *Ribes*, *Salix* litter on bluffs above river, V. Behan VB138-87 (CNC) 1 male, 1 sex undetermined; Richardson Mts., Dall Ridge Valley to Fish Cr., 67.9833, −136.5333, 2.VII.1987, ex *Ptasites*, *Salix* litter, moss, *Saxifraga* by stream, V. Behan, VB88-87 (CNC) 1 female, 1 sex undetermined; Richardson Mts. near Dall Ridge,

67.9833, −136.5333, 1.VII.1987, V. Behan, ex *Geum glaciale*, *Primula tschuktschorum* by stream from snow melt (CNC) 1 female, 2 sex undetermined.

73 (183). ***Boreostiba campbelliana*** **Lohse ***
Additional records. Canada, Yukon, Richardson Mts., Erebia Cr. on Fish Cr., 67.9666, −136.4833, 29.VI.1987, wet willow litter on Fish Cr. flood plain, V. Behan, VB54-87 (CNC) 1 male.

74 (185). ***Boreostiba parvipennis*** **(Bernhauer)**
Additional records. Canada: Northwest Territories, 32 km W Yellowknife, 62.531, –114.9581, 168 m, 3.VI.2009, 168m, birch/spruce soil sifting, coll. B. Godin (BGC, LFC) 1 male, 2 females; Yukon, Whitehorse, Hotsprings, 60.8769, –135.3596, 715m, 15.VII.2012, willow litter, sifting, coll. B. Godin (LFC) 1 male; Whitehorse, Paddy's pond, 60.7066, –135.0917, 649m, willow litter sifting, B. Godin (BGC, LFC) 3 males; Whitehorse, Grey Mountain, 60.7153, –135.0100, 746m, mushrooms, B. Godin (LFC) 1 male; Whitehorse, McIntyre Mt., 60.6305, –135.1742, 1336m, willow, sifting, B. Godin (LFC) 2 females. USA, Alaska, Fairbanks, 599 Arvita Ct, 137m el., 64.88919, –147.69572, ±100m house, hand collected, 10.IV.2007, K. L. Jeannet (UAM) 1 specimen; mile 217.2 Richardson Hwy, Castner Glacier, 786m el., 63.40134, –145.71895, ±5m Low *Salix, Alnus, P. balsam*, Berlese, 7.IX.2009, G. Henry (UAM) 1 specimen; Denali NPP., Mt. Healy, 595m el., 63.73478, −148.92911, ±30m forest, mostly spruce, some aspen, litter, Berlese, 3.VIII.2016, J.J. Rykken, D.S. Sikes, A. Haberski, K. Chen (UAM) 1 specimen; Denali NPP., Mt. Healy, 927m el., 63.74016, –148.95346, ±30m shrub, alder, dwarf birch, grass, clubmoss, litter, Winkler, 16.VIII.2016, J.J. Rykken, D.S. Sikes, R. Stewart-Rogers (UAM) 3 specimens; same data except, 4 pitfalls w/ rainroofs, pooled, 19.VII-3.VIII.2016, J.J. Rykken, D.S. Sikes, A. Haberski, K. Chen (UAM) 1 specimen.

Genus *Earota* Mulsant and Rey

75 (193). ***Earota dentata*** **(Bernhauer)**
Additional records. USA: Alaska, Fairbanks, 930 Fitz Ct, 280m el., 64.9013, –147.52861, ±30m *Betula* & *Populus*, flight intercept trap, 26.VIII-10.IX.2006, D.S. Sikes (UAM) 1 specimen.

Genus *Liogluta* C.G. Thomson

76 (197). ***Liogluta granulosa*** **Lohse**
Additional records. Canada: Yukon, Chadburn Lake, 60.6543, –134.951, 667m, 28.VI.2014, hand picked, coll. B. Godin and D. Horwood (BGC, LFC) 2 males, 2 females; Richardson Mts., VII.1987, V. Behan (CNC) 1 male. USA, Alaska, Denali NPP., Mt. Healy, 595m el., 63.73478, −148.92911, ±30m forest, mostly spruce, some aspen, 4 pitfalls w/ rainroofs, pooled, 2-19.VII.2016, J.J. Rykken, D.S. Sikes, A. Haberski, Z. Snelson (UAM) 1 specimen; Denali NPP., Mt. Healy, 927m el., 63.74016, –148.95346, ±30m shrub, alder, dwarf birch, grass, clubmoss, 4 pitfalls w/ rainroofs, pooled, 19.VII-3.VIII.2016, J.J. Rykken, D.S. Sikes, A. Haberski, K. Chen (UAM) 3 specimens; same data except, 4 yellow pitfalls w/o rainroofs, pooled, 19.VII-3.VIII.2016, J.J. Rykken, D.S. Sikes, A. Haberski, K. Chen (UAM) 10 specimens; 4 pitfalls w/ rainroofs, pooled, 2-19.VII.2016, J.J. Rykken, D.S. Sikes, A. Haberski, Z. Snelson (UAM) 26 specimens; 4 yellow pitfalls w/o rainroofs, pooled, 2.VII-19.VIII.2016, J.J. Rykken, D.S. Sikes, A. Haberski, Z. Snelson (UAM) 8 specimens; 4 pitfalls w/ rainroofs, pooled, 3-16.VIII.2016, J.J. Rykken, D.S. Sikes, R. Stewart-Rogers (UAM) 7 specimens; 4 yellow pitfalls w/o rainroofs, pooled, 3-16.VIII.2016, J.J. Rykken, D.S. Sikes, R. Stewart-Rogers (UAM) 5 specimens; litter, Berlese, 3.VIII.2016, J.J. Rykken, D.S. Sikes, A. Haberski, K. Chen (UAM) 1 specimen; Naknek, 2-5m el., 58.73973, –157.0636, ±20m creekside/ocean beach confluence, under boards and drift wood, hand collected, 10.VI.2007,

D.S. Sikes (UAM) 1 specimen; Prince of Wales Island, Staney Creek Cabin, 21m el., 55.82005, –133.15384, ±1km salt water marsh, old growth forest edge, grasses, *Heracleum* flowers, sweep and hand, 23.VI.2012, S. L. Meierotto, Ian MacDougall (UAM) 1 specimen.

Genus *Lypoglossa* Fenyes

77 (207). ***Lypoglossa angularis*** **(Mäklin)**
Additional records. USA: Alaska, 4.8km N of Glenn Hwy, Boulder Crk Tr., 853m el., 61.849, –148.22233, ±6m *Betula, Salix, Picea*, 22-24. VI.2009 (UAM) 1 specimen; Alaska Peninsula / Becharof NWR, plot 1540 (Severson), 17m el., 57.81477, –156.11205, ±7m under CTS-*Salix* near the edge with crowberry, pitfall trap, 27.VI-5.VII.2007 (UAM) 1 specimen; same data except, 5.VII-21.VIII.2007, (UAM) 1 specimen; Alaska Peninsula / Becharof NWR, plot 1692 (Becharof Outlet), 20m el., 58.04903, –156.84016, ±5m bottom edge of closed tall shrub-alder, near eds-crowberry, pitfall trap, 15.VI-5.VII.2007 (UAM) 3 specimens; Alaska Peninsula / Becharof NWR, plot 1692 (Becharof Outlet), 20m el., 58.04911, –156.84003, ±4m in EDS-crowberry, ~3–5 m of closed tall shrub to E & W, pitfall trap, 5.VII-15.VIII.2007 (UAM) 4 specimens; Chena Pump Rd, 136m el., 64.8375, –147.88043, ±30m forest edge *Pop. trem., Pop. bals.*, pitfall, B. Mortensen, 17-24.VII.2007, (UAM) 1 specimen; Chichagof Is., 724m el., 57.9845, –135.20499, ±16m herbaceous tundra, *Vaccinium caespitosum, Phyllodoce glanduliflora, Geranium erianthum*, in soil, soil sort, 18.VII.2013 (UAM) 4 specimens; Chichagof Is. pl.01, 642m el., 57.98874, –135.20212, ±4m shrubby krummholz, *Elliottia pyroliflora, Tsuga mertensiana, Nephrophyllidium crista-galli*, pitfall, 15–16.VII.2013 (UAM) 1 specimen; Chichagof Is. pl.02, 670m el., 57.98787, –135.20279, ±4m heath, *Cassiope mertensiana, Luetkea pectinata, Harrimanella stelleriana*, pitfall, 15-16.VII.2013 (UAM) 2 specimens, same data except, 16-17.VII.2013 (UAM) 7 specimens; Chichagof Is. pl.03, 702m el., 57.98708, –135.20357, ±3m meadow, *Nephrophyllidium crista-galli, Geranium erianthum, Artemisia arctica*, pitfall, 14-15.VII.2013 (UAM) 3 specimens, same data except, 15-16.VII.2013 (UAM) 9 specimens, 16-17.VII.2013 (UAM) 3 specimens, 17-18.VII.2013 (UAM) 3 specimens; Chichagof Is. pl.04, 728m el., 57.98643, –135.20468, ±4m rich meadow, *Geranium erianthum, Artemisia arctica, Sanguisorba canadensis*, pitfall, 14-15.VII.2013 (UAM) 4 specimens, same data except. 15-16.VII.2013 (UAM) 1 specimen, 16-17.VII.2013 (UAM) 2 specimens; same data except, 17-18.VII.2013 (UAM) 1 specimen; Chichagof Is. pl.05, 736m el., 57.9856, –135.20456, ±3m heath tundra, *Vaccinium caespitosum, Cassiope mertensiana, Harrimanella stelleriana*, pitfall, 14-15.VII.2013 (UAM) 4 specimens; same data except, 15-16. VII.2013 (UAM) 2 specimens, 16-17.VII.2013 (UAM) 2 specimens, 17-18.VII.2013 (UAM) 1 specimen; Chichagof Is. pl.06, 725m el., 57.98495, –135.20358, ±4m krummholz, *Tsuga mertensiana, Elliottia pyroliflora, Vaccinium ovalifolium*, pitfall, 14-15.VII.2013 (UAM) 2 specimens; same data except, 16-17.VII.2013 (UAM) 3 specimens, 17-18.VII.2013 (UAM) 1 specimen; Chichagof Is. pl.07, 738m el., 57.98589, –135.20624, ±6m krummholz, *Tsuga mertensiana, Rubus spectabilis, Elliottia pyroliflora*, pitfall, 14-15.VII.2013 (UAM) 4 specimens;same data except, 15-16.VII.2013 (UAM) 5 specimens, 16-17.VII.2013 (UAM) 2 specimens, 17-18.VII.2013 (UAM) 1 specimen; Chichagof Is. pl.08, 724m el., 57.9845, –135.20499, ±6m herbaceous tundra, *Vaccinium caespitosum, Phyllodoce glandulifera, Geranium erianthum*, pitfall, 14-15.VII2013 (UAM) 6 specimens; same data except, 15-16.VII.2013 (UAM) 6 specimens, 16-17.VII.2013 (UAM) 2 specimens; Chichagof Is. pl.09, 767m el., 57.9841, –135.20654, ±5m herbaceous heath, *Vaccinium caespitosum, Vaccinium uliginosum, Empetrum nigrum*, pitfall, 14-15.VII.2013 (UAM) 1 specimen; same data except, 15-16. VII.2013 (UAM) 1 specimen; Dall Isl., pl.01, 646m el., 54.99768, –133.01714, ±5m subalpine meadow, *Athyrium, Carex macro.*, pitfall, 13-14. VII.2011 (UAM) 1 specimen; Dall Isl., pl.17, 548m el., 54.99555, –133.01039, ±5m flood meadow, *Athyrium, Caltha leptosepala*, pitfall,

13-14.VII.2011 (UAM) 1 specimen; Denali NPP., Mt. Healy, 595m el., 63.73478, −148.92911, ±30m forest, mostly spruce, some aspen, 4 pitfalls w/ rainroofs, pooled, 2-19. VII.2016 (UAM) 1 specimen; same data except, 4 yellow pitfalls w/o rainroofs, pooled, 3-16. VIII.2016 (UAM) 1 specimen; litter, Berlese, 3. VIII.2016 (UAM) 3 specimens; Denali NPP., Mt. Healy, 927m el., 63.74016, −148.95346, ±30m shrub, alder, dwarf birch, grass, clubmoss, 4 yellow pitfalls w/o rainroofs, pooled, 19.VII-3. VIII.2016 (UAM) 1 specimen; same data except, 4 pitfalls w/ rainroofs, pooled, 2-19.VII.2016 (UAM) 2 specimens; Denali NPP., nr Teklanika R., 833m el., 63.64261, −149.57181, ±30m spruce, litter, Berlese, 20.VII.2016 (UAM) 1 specimen, Etolin Isl., pl.08, 875m el., 56.13867, −132.33047, ±4m krummholtz, *Tsuga mertensiana, Nephrophy.*, pitfall, 19-20. VII.2011 (UAM) 1 specimen; same data except, 21-22.VII.2011, (UAM) 3 specimens; Fairbanks, 930 Fitz Ct, 280-293m el,. 64.90127, −147.52797, ±30m *Populus, Betula, Alnus* leaf litter, Berlese, 10.X.2013 (UAM) 1 specimen; Fairbanks, Creamer's Field, 136-156m el., 64.86719, −147.73433, ±80m *Betula* litter, Berlese, 13.VIII.2007 (UAM) 1 specimen; Fairbanks, Creamer's Field, 143m el., 64.86714, −147.7281, ±305m forest edge, medow, grass, forbs, pitfall, 9–10.VII.2011 (UAM) 1 specimen; Fairbanks: Creamer's Field, 132m el., 64.86805, −147.73764, ±20m mesic *Betula, Picea*, pitfall, 15-18.VII.2008 (UAM) 2 specimens; Glenn Hwy, Long Lake, 471m el., 61.80319, −148.23671, ±104m spruce, birch, alder, cottonwood, Winkler, 28.VI.2011 (UAM) 2 specimens; Hawthorne Peak pl.04, 796m el., 58.23924, −134.2419, ±6m meadow, *Lupinus nootkatensis, Nephrophyllidium crista-galli, Valeriana sitchensis*, pitfall, 20-21.VI.2013 (UAM) 1 specimen; same data except, 22-23. VII.2013 (UAM) 3 specimens; Hawthorne Peak pl.05, 775m el., 58.24007, −134.24252, ±5m meadow, *Nephrophyllidium crista-galli, Carex macrochaeta, Lupinus nootkatensis*, pitfall, 19-20.VII.2013, 22-23.VII.2013 (UAM) 1 specimen; Hawthorne Peak pl.06, 849m el., 58.24242, −134.24146, ±5m low tundra, *Vaccinium uliginosum, Luetkea pectinata, Phyllodoce glandulifera*, pitfall, 19-20.VII.2013 (UAM) 3 specimens; same data except: 20-21. VII.2013 (UAM) 2 specimens, 21-22.VII.2013 (UAM) 1 specimen, 22-23.VII.2013 (UAM) 2 specimens, ground search, 22.VII.2013 (UAM) 1 specimen; Hawthorne Peak pl.10, 821m el., 58.24498, −134.24661, ±5m herbaceous heath, *Luetkea pectinata, Cassiope mertensiana, Lupinus nootkatensis*, pitfall, 19-20.VII.2013 (UAM) 1 specimen; same data except, 20-21.VII.2013 (UAM) 1 specimen; 22-23.VII.2013 (UAM) 3 specimens; Hawthorne Peak pl.11, 688m el., 58.24415, −134.25174, ±9m heath meadow, *Vaccinium uliginosum, Phyllodoce glandulifera, Gymnocarpium disjunctum*, pitfall, 19-20.VII.2013 (UAM) 2 specimens; same data except, 20-21.VII.2013 (UAM) 1 specimen; 21-22.VII.2013 (UAM) 2 specimens; 22-23.VII.2013 (UAM) 1 specimen; Hawthorne Peak pl.12, 722m el., 58.24507, −134.25221, ±5m meadow, *Carex macrochaeta, Geranium erianthum, Salix stolonifera*, pitfall, 19-20.VII.2013 (UAM) 1 specimen; same data except, 20-21.VII.2013 (UAM) 1 specimen; Hawthorne Peak pl.14, 811m el., 58.24704, −134.25217, ±6m wet meadow, *Carex nigricans, Petasites frigidus, Ranunculus occidentalis*, pitfall, 19-20.VII.2013 (UAM) 12 specimens; same data except, 20-21.VII.2013 (UAM) 3 specimens; 21-22.VII.2013 (UAM) 10 specimens; 22-23. VII.2013 (UAM) 2 specimens; Hawthorne Peak pl.20, 953m el., 58.25053, −134.25192, ±5m low heath tundra, *Cassiope mertensiana, Phyllodoce glandulifera, Carex nigricans*, pitfall, 21-22. VII.2013 (UAM) 1 specimen; same data except, 22-23.VII.2013 (UAM) 1 specimen; Heintzleman Ridge, pl.18(3B), 656m el., 58.42829, −134.43436, ±32m subalp rocks *Cassio. Salix*, pitfall, 30.VI-2.VII.2010 (UAM) 3 specimens; Juneau, Mt. Roberts, 648m el., 58.29557, −134.37923, ±50m edge of snowfield, hand, 28. V.2017 (UAM) 1 specimen; Mahoney Mt. area, 785m el., 55.40887, −131.57666, ±6m krummholz near stream, wolf scat, baited pitfall, 8-10. VII.2013 (UAM) 1 specimen; Mahoney Mt. pl.04, 817m el., 55.41855, −131.56485, ±5m elf krummholz, *Tsuga mertensiana,*

Elliottia pyroliflora, Vaccinium ovalifolium, pitfall, 7-8.VII.2013 (UAM) 1 specimen; Mahoney Mt. pl.12, 836m el., 55.4099, –131.57826, ±5m krummholz, *Tsuga mertensiana, Vaccinium ovalifolium, Chamaecyparis nootkatensis*, pitfall, 7-8.VII.2013 (UAM) 6 specimens; same data except, 9-10.VII.2013 (UAM) 2 specimens; Mahoney Mt. pl.13, 789m el., 55.40966, –131.57649, ±4m rocky minimalistic meadow, *Vaccinium uliginosum, Nephrophyllidium crista-galli, Cassiope mertensiana*, pitfall, 8-9.VII.2013 (UAM) 1 specimen; Mahoney Mt. pl.14, 789m el., 55.40879, –131.57684, ±4m krummholz shrub, *Elliottia pyroliflora, Cassiope mertensiana, Phyllodoce glandulifera*, pitfall, 7-8.VII.2013 (UAM) 2 specimens; same data except, 9-10.VII.2013 (UAM) 2 specimens; Mahoney Mt. pl.20, 784m el., 55.40266, –131.57668, ±5m tundra, *Luetkea pectinata, Harrimanella stelleriana, Vaccinium uliginosum*, pitfall, 7-8.VII.2013 (UAM) 1 specimen; same data except, 9-10.VII.2013 (UAM) 1 specimen; Naknek, 2-5m el., 58.73973, –157.0636, ±20m creekside/ocean beach confluence, under boards and drift wood, hand collected, 10.VI.2007 (UAM) 1 specimen; North Pole, 138m el., 64.826 –147.4576, ±50m stand of *Alnus* between house and slough, pitfall, 6-13.V.2007 (UAM) 1 specimen; Park's Hwy., Bonanza Creek area, 310-330m el., 64.76269, –148.29686, ±50m boreal forest, Berlese, 25.V.2007 (UAM) 2 specimens; Peninsula NWR 1484 Ugashik River, 29m el., 57.702, –156.52213, ±8m pitfall trap, 29.VIII.2007 (UAM) 1 specimen; Portage Valley, Williwaw Campground, 60.7844, –148.8753, ±50m 23.VII.2011 (UAM) 1 specimen; Prince of Wales Isl. nr Black Lk pl.02, 780m el., 55.58988, –132.89548, ±2m nr bear dung, *Caltha, Athyrium*, pitfall, 7-8.VII.2011 (UAM) 2 specimens; same data except: 9-10. VII.2011, (UAM) 1 specimen; Prince of Wales Isl. nr Black Lk pl.03, 820m el., 55.58907, –132.895, ±2m *Nephro. crista-galli, Caltha leptos.*, pitfall, 7-8.VII.2011 (UAM) 1 specimen; same data except, 6-7.VII.2011 (UAM) 2 specimens; Prince of Wales Isl. nr Black Lk pl.04, 866m el., 55.58842, –132.89409, ±2m *Vacc. uliginosum, Emp. nigrum*, pitfall, 7-8. VII.2011 (UAM) 1 specimen; Prince of Wales Isl. nr Black Lk pl.05, 898m el., 55.58765, –132.8931, ±2m wet meadow w/hummocks, pitfall, 6-7.VII.2011 (UAM) 2 specimens; Prince of Wales Isl. nr Black Lk pl.09, 925m el., 55.58724, –132.89107, ±3m heath, *Harr. stelleriana, Vacc. caes.*, pitfall, 8-9.VII.2011 (UAM) 1 specimen; same data except, 9-10. VII.2011 (UAM) 1 specimen; Prince of Wales Isl. nr Black Lk pl.14, 881m el., 55.58898, –132.88927, ±2m wet meadow, *Luet. pect., Calt. lept.*, pitfall, 7-8.VII.2011 (UAM) 2 specimens; Prince of Wales Isl. nr Black Lk pl.15, 918m el., 55.58818, –132.88881, ±2m heath, *Harr. stelleriana, Luet. pect.*, pitfall, 7-8.VII.2011 (UAM) 1 specimen; Prince of Wales Isl. nr Black Lk pl.16, 953m el., 55.58732, –132.88821, ±2m heath, *Harr. stell., Luet. pectinata*, pitfall, 8-9.VII.2011 (UAM) 2 specimens; same data except, 9-10.VII.2011 (UAM) 2 specimens; Prince of Wales Isl. nr Black Lk pl.18, 894m el., 55.58872, –132.88701, ±2m heath, *Har. ste., Vac. cae., Lup. noo.*, pitfall, 6-7.VII.2011 (UAM) 2 specimens; Prince of Wales Isl. nr Black Lk pl.19, 845m el., 55.58964, –132.88783, ±2m *Nephro. crista-galli, Luet. pectinata*, pitfall, 7-8. VII.2011 (UAM) 1 specimen; S. Chilkat Range, 638m el., 58.42384, –135.14883, ±6m subalpine, primate dung, pitfall, 9-10.VII.2010 (UAM) 4 specimens; S. Chilkat Range pl.09, 743m el., 58.427, –135.15852, ±30m subalpine heath, *Empetrum, Vaccinium*, pitfall11-12.VII.2010 (UAM) 2 specimens; same data except, 9-10. VII.2010 (UAM) 1 specimen; S. Chilkat Range pl.10, 718m el., 58.42621, –135.1575, ±28m subalpine, *Salix arctica, Vacc. caesp.*, pitfall, 10-11.VII.2010, (UAM) 1 specimen; same data except, 11-12.VII.2010 (UAM) 1 specimen; 8-9. VII.2010 (UAM) 4 specimens; 9-10.VII.2010 (UAM) 2 specimens; S. Chilkat Range pl.13, 687m el., 58.4252, –135.15205, ±29m subalpine *Vaccinium, Cornus*, pitfall, 8-9.VII.2010 (UAM) 1 specimen; S. Chilkat Range pl.15, 680m el., 58.42497, –135.1496, ±28m subalpine *Gymnocarpium*, pitfall, 10-11.VII.2010 (UAM) 1 specimen; same data except, 8-9.VII.2010 (UAM) 12 specimens; 9-10.VII.2010 (UAM) 10 specimens; S. Chilkat Range pl.17, 670m el., 58.42412, –135.146, ±30m subalpine *Rubus*,

Gymnocarpium, pitfall, 11-12.VII.2010 (UAM) 1 specimen; same data except, 8-9.VII.2010 (UAM) 5 specimens; 9-10.VII.2010 (UAM) 1 specimen; hand collected, 9.VII.2010 (UAM) 1 specimen; S. Chilkat Range pl.18, 668m el., 58.42455, −135.14445, ±30m subalpine *Tsuga, Luetkea,* hand collected, 9.VII.2010 (UAM) 1 specimen; same data except, pitfall, 8-9. VII.2010 (UAM) 1 specimen; S. Chilkat Range pl.19, 652m el., 58.42509, −135.14229, ±30m subalpine *Veratrum, Calamagrostis*, pitfall, 8-9. VII.2010 (UAM) 5 specimens, S. Chilkat Range pl.20, 659m el., 58.4248, −135.14096, ±29m subalpine *Tsuga, Vaccinium*, pitfall, 8-9. VII.2010 (UAM) 6 specimens; Serpentine Hot Springs, 118m el., 65.85183, −164.69884, ±104m *Salix*, moss leaf litter, Berlese, 10. VII.2016 (UAM) 5 specimens; White Mountains Recreation Area, Nome Creek Site 4, 180m el., 65.33469, −146.83997, ±10m b. & w. spruce, willow, hanging Malaise, 4.6m, 10-17.VI.2016 (UAM) 1 specimen.

COMMENTS. The following collector's names were removed from label data to save space: C. Bickford, M. Bowser, K. Chen, C. Eshleman, B. Fleshman, A. Haberski, S. E. Hawks, K. W. May, C. W. O'Brien, L. B. O'Brien, S. Savage, D.S. Sikes, Z. Snelson, R. Stewart-Rogers, S. Ridling, and J.J. Rykken.

Genus *Mocyta* Mulsant and Rey

78 (209). ***Mocyta amblystegii*** **(Brundin) ***
Additional records. Canada, Yukon, Dempster Hwy. between N Fork Pass and Chapman Lake, above tree line, 8.VI.1975, J.V. Matthews (CNC) 1 female.

USA: Alaska, Anaktuvuk Pass, 647m el., 68.14049, −151.74091, ±250m *Salix* leaf litter, Berlese, 19.V.2016, D.S. Sikes; K.M. Daly (UAM) 14 specimens; Anaktuvuk Pass, 647m el., 68.14049, −151.74091, ±250m under rocks, forceps, aspirator, 19.V.2016, K. Hildebrandt; D.S. Sikes; K.M. Daly (UAM) 2 specimens; Fairbanks, 930 Fitz Ct, 280m el., 64.9013, −147.52861, +30m birch bark, 25.IV.2011, D.S. Sikes (UAM) 1 specimen; Fairbanks, Creamer's Field, Reindeer Lake, 146m el., 64.88118, −147.77643, ±10m b. spruce bog, hanging Malaise, 5.55m, 9-31.V.2016, J.C. Hagelin (UAM) 1 specimen; Nome Creek site 3, 605m el., 65.33084, −146.80476, ±10m b. & w. spruce, hanging Malaise, 7.92m, 10-17. VI.2016, J.C. Hagelin (UAM) 1 specimen; White Mountain Recreation Area, Nome Creek Site 5, 589m el., 65.34077, −146.9275, ±10m open tussock and standing water, willow and bog birch, hanging Malaise, 4.95m, 12.VI-3.VII.2015, J.C. Hagelin (UAM) 1 specimen; Prudhoe Bay Rd. 9 mi N Atigun Pass, 68.2333, −149.4166, 6. VII.1978, 3100 feet, A. Smetana and J.M. Campbell (CNC) 14 males, 9 females; Prudhoe Bay Rd. 15 mi S Galbraith Lk., 68.2666, −149.4333, 5.VII.1978, 3000 feet, A. Smetana and J.M. Campbell (CNC) 9 males, 19 females; Kotzebue, 14.IX.1967, J. Matthews (CNC) 2 females.

79 (210). ***Mocyta breviuscula*** **(Mäklin)**
Additional records. USA: Alaska, Becharof NWR Upper Kejulik, 82m el., 57.97709, −155.50765, ±50m sweep, 13.VI.2007, G. Milet (UAM) 1 specimen; Chena Hot Springs Rd, 200m el., 64.89227, −147.50857, ±30m forest edge *Pop. trem. Salix* spp., pitfall, 17-24. VII.2007, B. Mortensen (UAM) 4 specimens; Chena Pump Rd, 136m el., 64.8375, −147.88043, ±30m forest edge *Pop. trem., Pop. bals.*, pitfall, 17-24.VII.2007, B. Mortensen (UAM) 5 specimens; Fairbanks, 166m el., 64.88281, −147.78452, ±104m pitfall slotcup, 2-9.VI.2009, J.A. Slowik (UAM) 1 specimen; Fairbanks, 240-340m el., 64.8932, −147.80155, ±30m forest edge *Pop. trem., Rubus id.*, pitfall, 17-24.VII.2007, B. Mortensen (UAM) 6 specimens; Fairbanks, 383m el., 64.9181, −147.71997, ±61m S-facing hill, forest edge *Pop. trem.* grass, pitfall, 17-24.VII.2007, B. Mortensen (UAM) 2 specimens; Fairbanks, 930 Fitz Ct, 280-293m el., 64.90127, −147.52797, ±30m *Populus, Betula, Alnus* leaf litter, Berlese, 10. IX.2013, D.S. Sikes (UAM) 5 specimens; Fairbanks, 930 Fitz Ct, 280m el., 64.9013, −

147.52861, ±30m *Betula pap., Pop. trem.*, flight intercept trap, D.S. Sikes, 15-20.VI.2007 (UAM) 1 specimen; same data except, 20.V.-3.VI.2007, (UAM) 5 specimens; Fairbanks, 930 Fitz Ct, 280m el., 64.9013, −147.52861, ±30m birch bark, 25.IV.2011, D.S. Sikes (UAM) 35 specimens; Fairbanks, Creamer's Field, 152m el., 64.86808, −147.73764, ±8m wet boreal, pitfall, 15-18.VII.2008, D.S. Sikes, C. Eshleman (UAM) 1 specimen; Fairbanks, Creamer's Field, 158m el., 64.86823, −147.73758, ±12m mesic *Betula*, pitfall, 26.VI-18.VII.2008, D.S. Sikes, R. Davidson (UAM) 3 specimens; Fairbanks, Ridgepointe Dr,. 253m el., 64.81772, −147.97333, ±61m sparse forest, *Pop. trem.*, pitfall, 17-24.VII.2007 B. Mortensen (UAM) 5 specimens; Fairbanks, Ridgepointe Dr., 305m el., 64.828, −147.964, ±125m forest edge, *Pop. trem.*, pitfall, 17-24.VII.2007, B. Mortensen (UAM) 2 specimens; Fairbanks: Creamer's Field, 132-152m el., 64.86504, −147.73863, ±200m fireweed, grass, Berlese, 17.IX.2006, M. E. Sikes, M. Ross (UAM) 1 specimen; Fairbanks: Creamer's Field, 132m el., 64.86764, −147.73744, ±100m alder & birch litter near dry vernal pool, Berlese, 17.IX.2006, M. E. Sikes, M. Ross (UAM) 2 specimens; Fairbanks: Creamer's Field, 140m el., 64.86883, −147.73065, ±104m *Betula* forest, decaying mushrooms, 31.VII.2010, D.S. Sikes (UAM) 1 specimen; Fairbanks, Murphy Dome, 810m el., 64.96017, −148.38188, ±3m ant nest under rock, alpine tundra, 13JUN 2009, D.S. Sikes (UAM) 1 specimen; Glenn Hwy, Long Lake, 471m el., 61.80319, −148.23671, ±104m spruce, birch, alder, cottonwood, Winkler, 28.VI.2011, S. Ridling, M. G. Wyatt (UAM) 6 specimens; Haines, Flower Mtn. pl.05, 939m el., 59.39038, −136.31345, ±29m alpine *Salix, Valariana*, shrubs, pitfall, 17-18.VII.2010, D.S. Sikes (UAM) 1 specimen; King Salmon, 15m el., 58.68207, −156.66959, ±7m alder & willow brush, pitfall trap, 7 9.VI.2007, S. Clawson (UAM) 1 specimen; Kotzebue, 1m el., 66.89151, −162.5933, ±100m brackish shoreline; under rocks, wrack, 6.VII.2016, D.S. Sikes, K.M. Daly (UAM) 1 specimen; Naknek, 2-5m el., 58.73973, −157.0636, ±20m creekside/ocean beach confluence, under boards and drift wood, hand collected, 10.VI.2007, D.S. Sikes (UAM) 1 specimen; Nome Creek site 3, 605m el., 65.33084, −146.80476, ±10m black & whie spruce, hanging Malaise, 7.93m, 1-15.VII.2016, J.C. Hagelin (UAM) 1 specimen; North Pole, 2026 Kendall Ave., 135-155m el., 64.82639, −147.44139, ±30m a few feet behind house, on piece of wood on gravel, hand collected, 7. IX.2006, B. Fleshman (UAM) 1 specimen; nr. Livengood, Elliot Highway, Mile 84.5, 252m el., 65.39532, −148.9097, ±10m birch, b. spruce on n. slope, hanging Malaise, 4.16m, 16-30. VI.2016, J.C. Hagelin (UAM) 1 specimen; Parks Hwy (mile 231) Denali Education Center, 569m el., 63.6553, −148.82404, ±210m nest under rock, spruce, aspen, willow forest, hand, 27. VI.2011, D.S. Sikes, J. Tisland (UAM) 1 specimen; Serpentine Hot Springs, 118m el., 65.85183, −164.69884, ±104m *Salix*, moss leaf litter, Berlese, 10.VII.2016, D.S. Sikes (UAM) 12 specimens; Sterling, Gann site, 40-70m el., 60.51418, −150.79506, ±25m *S. barclayi* meadow, sweep, 19.VI.1993, D.M. Collet (UAM) 1 specimen; Sterling, Gann site, 40-70m el., 60.51418, −150.79506, ±25m *S. barclayi* meadow, sweep, 19.VI.1993, D.M. Collet (UAM) 1 specimen; Univ. Alaska Fairbanks, 174m el., 64.85727, −147.85843, ±61m forest edge *Pop. trem.*, pitfall, 17-24.VII.2007, B. Mortensen (UAM) 3 specimens; Wosnesenski, 10m el., 55.21697, −161.36644, ±105m under rocks, dung, wood, meadows w/lupine, 11. VII.2009, D.S. Sikes (UAM) 3 specimens; Kenai Mts., 15 mi N Seward, 400 feet, 29. V.1978, A. Smetana and E.C. Becker (CNC) 1 female, 1 sex undetermined.

80 (211). ***Mocyta fungi*** **(Gravenhorst) †**

Additional records. Canada: Labrador, Red Bay, 8.VIII.1972, J.M. Campbell (CNC) 1 female. USA, Alaska, Fairbanks, 240-340m el., 64.8932, 147.80155, ±30m forest edge *Pop. trem., Rubus ida.*, pitfall, 17-24.VII.2007, B. Mortensen (UAM) 1 specimen; Naknek, 2-5m el., 58.73973, −157.0636, ±20m creekside/ocean beach confluence, under boards and drift wood, hand collected, 10.VI.2007,

D.S. Sikes (UAM) 1 specimen; Prince of Wales Island, Luck Point 1B, 169m el., 55.98497, –132.787, ±25m clear cut, pitfall 3, 14-27. VI.2012, J. Stockbridge, S. Meierotto, I. MacDougall (UAM) 1 specimen; Selawik NWR, 8m el., 66.56137, –158.99834, ±304m *Spirea*, dwarf birch, 23-24.VI.2010, D.S. Sikes (UAM) 1 specimen; Toolik Field Station, 724m el., 68.6286, –149.59772, ±36m under rocks, etc., 1-3.VI.2008, D.S. Sikes (UAM) 1 specimen; Toolik Field Station, 777m el., 68.62685, –149.59798, ±14m *Salix* leaf litter, Berlese, 3. VI.2008, D.S. Sikes (UAM) 6 specimens.

Genus *Philhygra* Mulsant and Rey

81 (213). *Philhygra angusticauda* (Bernhauer) *

Additional records. USA: Alaska, Fairbanks, 383m el., 64.9181, –147.71997, ±61m S-facing hill, forest edge Pop. trem., pitfall, 17-24. VII.2007, B. Mortensen (UAM) 1 specimen; Lowell Pt. [S. of Seward], 1-30m el., 60.06508, –149.44261, ±500m rocks on beach above tide line, 15.VIII.2010, S. Ridling (UAM) 1 specimen; Prince of Wales Island, Staney Ck. 1B, 77m el., 55.872, –133.06523, ±26m clear-cut, Lindgren 1, 13-27.VI.2011, J. Stockbridge, B. Wong (UAM) 1 specimen; same data except, 16-31.V.2011 (UAM) 1 specimen; Prince of Wales Island, Staney Ck. 1B, 77m el., 55.872, –133.06523, ±26m clear-cut, Lindgren 2, 24.VII-11.VIII.2010, J. Stockbridge, C. Bickford (UAM) 1 specimen.

82 (219). *Philhygra leechi* Lohse

Additional records. Canada: Northwest Territories, Reindeer Sta., Caribou Hills, 2. VII.1972, A. Smetana (CNC) 3 males.

83 (220). *Philhygra malleoides* Lohse *

Additional records. USA: Alaska, Anaktuvuk Pass, 647m el., 68.14049, –151.74091, ±250m *Salix* leaf litter, Berlese, 19.V.2016, D.S. Sikes, K.M. Daly (UAM) 7 specimens; Anaktuvuk Pass, 647m el., 68.14049, –151.74091, ±250m under rocks, forceps, aspirator, 19.V.2016, K. Hildebrandt, D.S. Sikes, K.M. Daly (UAM) 1 specimen; Dalton Hwy, 403m el., 67.48637, –149.88414, ±24m muskeg, black spruce, dirt road, 1.VI.2008, D.S. Sikes (UAM) 1 specimen; Fairbanks, Cripple Creek, 241m el., 64.8066, –148.0518, ±10m b. spruce bog, hanging Malaise, 2.3m, 27.VI-11.VII.2016, J.C. Hagelin (UAM) 1 specimen.

84 (223). *Philhygra pseudopolaris* Klimaszewski and Langor

Additional records. Canada: British Columbia, Darkwoods Conservation Area, Grassy Lake, 1750m el., 49.281, –117.043, open marsh with willows, grass and reeds, sweep net, 19. VIII.2014, BIOBus 2014 team (CBG) 4 specimens; same data but 23.VIII.2014 (CBG) 4 specimens; same data but pitfall, 19-25. VIII.2014 (CBG) 5 specimens. Northwest Territories, Reindeer Sta., Caribou Hills, 2. VII.1972, A. Smetana (CNC) 3 males, 8 females, 4 sex undetermined. USA: Alaska, Anaktuvuk Pass, 647m el., 68.14049, –151.74091, ±250m *Salix* leaf litter, Berlese, 19.V.2016, D.S. Sikes; K.M. Daly (UAM) 1 specimen; Anaktuvuk Pass, 647m el., 68.14049, –151.74091, ±250m under rocks, forceps, aspirator19.V.2016, K. Hildebrandt, D.S. Sikes, K.M. Daly (UAM) 1 specimen; Selawik NWR, 40m el., 66.4549, –158.13963, ±14m lakeshore grasses, treading, 24.VI.2010, D.S. Sikes (UAM) 1 specimen; Alaska Hwy. Mi 1249, Deadman Lake, 6. VII.1968, A. Smetana (CNC) 5 females.

85 (224). *Philhygra ripicola (Hanssen)* *

Additional records. Canada: Northwest Territories, Reindeer Sta., Caribou Hills, 2. VII.1972, A. Smetana (CNC) 1 male, 5 females. Saskatchewan, Prince Albert, 53.9804, –106.2800, 4.VI.2013, sand beach, sifting, coll. B. Godin and D. Horwood (LFC) 1 male.

86 (225). *Philhygra rostrifera* Lohse

Additional records. Canada: Northwest Territories, 32 km W Yellowknife, 62.531, –114.9581, 3.VI.2009, 168 m, Birch/spruce soil sifting, coll. B. Godin (LFC) 1 male; Saskatchewan, Cypress Hills, 49.5978, –109.9231, 1134m, 2.IX.2012, wet willow stand, coll. B. Godin and D. Horwood (LFC) 1 female; Lugcreek, 55.1776 106.6885, 6.VI.2013, Spruce/alder litter, sifting, coll. coll. B. Godin and D. Horwood (LFC) 1 female; Prince Albert, 53.9870, –106.2802, 7.VI.2013, ferns and scat, sifting, coll. B. Godin and D. Horwood (LFC) 1 female; La Ronge, 55.1180, –105.2457, 366m, ditch litter, deciduous, sifting, coll. B. Godin and D. Horwood (LFC) 1 female. USA, Alaska, Fairbanks: Creamer's Field, 132m el., 64.86805, –147.73764, ±20m mesic *Betula, Picea*, pitfall, 15-18.VII.2008, D.S. Sikes, C. Eshleman (UAM) 1 specimen.

87 (226). ***Philhygra sinuipennis*** **Klimaszewski and Langor**

Additional records. Canada: Saskatchewan, Prince Albert, 53.9804, –106.2800, 532m, 4. VI.2013, ferns and scat, sifting, coll. B. Godin and D. Horwood (LFC) 1 male.

88 (227). ***Philhygra terrestris*** **Klimaszewski and Godin**

Additional records. Canada: Saskatchewan, Cypress Hills, 49.5978, –109.9231, 1134m, 2. IX.2021, wet willow stand, coll. B. Godin and D. Horwood (LFC) 2 males; La Ronge, 55.1180, –105.2457, 366m, ditch litter, deciduous, sifting, coll. B. Godin and D. Horwood (LFC) 1 female; Meadow Lake, 54.4188, –108.9440, 7.VI.2013, Birch/alder sifting, coll. B. Godin and D. Horwood (BGC) 2 females.

Genus *Tarphiota* Casey

89 (234). ***Tarphiota fucicola*** **(Mäklin)**

Additional records. USA: Alaska: Adak, 2m el., 51.87317, –176.6283, ±55m sandy beach, under driftwood, forceps, 25.VIII.2013, S. Peek, D. Sikes (UAM) 23 specimens; Adak Isl., 1-3m el., 51.89521, –176.62297, ±501m beach, under 7 pieces of driftwood, forceps, 25.VIII.2012, D. S. Sikes (UAM) 12 specimens; Adak Isl., 3m el., 51.87449, –176.62831, ±353m sandy beach, 5 pieces of driftwood, hand, 18.VIII.2011, D. S. Sikes (UAM) 4 specimens; Adak Isl., Kuluk Bay, 2-5m el., 51.87507, –176.62839, ±324m sandy beach, under 8 pieces of driftwood >1m long, forceps, 24.VIII.2012, D. S. Sikes (UAM) 65 specimens; Adak, town beach [Kuluk beach], 3m el., 51.87317, –176.6283 ±55m sandy marine beach, under 2 pieces drift wood, forceps, 24. VIII.2014, D. S. Sikes (UAM) 1 specimens; Great Sitkin, 1m el., 52.00438, –176.06184, ±6m beach drift wood, 11.VIII.2009, D. S. Sikes (UAM) 72 specimens; Kasatochi, E., 5-10m el., 52.16451, –175.49316 ±305m sandy beach, under 21 pieces driftwood, forceps, A. R. DeGange; 23.VIII.2014, D. S. Sikes (UAM) 3 specimens; Kiska, 1m el., 52.05026, –177.50909 ±55m under driftwood, 6.VIII.2010, D. S. Sikes (UAM) 67 specimens; Kiska Harbor, 1-10m el., 51.98248, –177.54501 ±504m under debris, 6.VIII.2010, D. S. Sikes (UAM) 3 specimens.

90 (235). ***Tarphiota geniculata*** **(Mäklin)**

Additional records. USA: Alaska: Adak, 2m el., 51.87317, –176.6283, ±55m sandy beach, under driftwood, forceps, 25.VIII.2013, S. Peek, D. Sikes (UAM) 1 specimen; Adak Isl., 1-3m el., 51.89521, –176.62297, ±501m beach, under 7 pieces of driftwood, forceps, 25.VIII.2012, D. S. Sikes (UAM) 5 specimens; Adak Isl., 3m el., 51.87449, –176.62831, ±353m sandy beach, 5 pieces of driftwood, hand, 18.VIII.2011, D. S. Sikes (UAM) 1 specimen; Adak Isl., Kuluk Bay, 2-5m el., 51.87507, –176.62839, ±324m sandy beach, under 8 pieces of driftwood >1m long, forceps, 24.VIII.2012, D. S. Sikes (UAM) 4 specimens; Adak, town beach [Kuluk beach], 3m el., 51.87317, –176.6283, ±55m sandy marine beach, under 2 pieces drift wood, forceps, 24.VIII.2014, D. S. Sikes (UAM) 1 specimen; Great Sitkin, 1m el., 52.00438, –176.06184, ±6m beach drift wood, 11.VIII.2009, D. S. Sikes (UAM) 1 specimen; Kiska, 1m el., 52.05026, –177.50909, ±55m under driftwood, 6.VIII.2010, D. S. Sikes (UAM) 1 specimen; Wosnesenski, 1m el., 55.21733, –161.36768, ±24m back beach, colored pan trap, 11. VII.2009, D. S. Sikes (UAM) 82 specimens; Wosnesenski, 1m el., 55.21733, –161.36768, ±24m under back beach rocks & drift, 11. VII.2009, D. S. Sikes (UAM) 3 specimens.

Annex 2

Table A.1 List of aleocharine species recorded from Arctic and Subarctic North America with their distribution records in Canada and the United States. * Holarctic species, †adventive species, ‡ species of uncertain distribution status—adventive in Canada or Holarctic, species without marks are native; NPR new provincial record, NTR new territorial record, NSR new state record; records marked [BOLD] are DNA barcoded specimens from The Barcode of Life Data System (http://boldsystems.org/) identified based on the barcode sequences and represent new provincial records. Tribes, genera and species are listed in alphabetical order

Taxonomic name	Canadian distribution	United States distribution
Tribe Aleocharini Fleming		
Aleochara assiniboin Klimaszewski	BC, MB, NT [**NTR**], ON, SK, YT	
Aleochara bimaculata Gravenhorst	AB, BC, LB, MB, NB, NF, NS, NT, ON, QC, SK	AL, AZ, CA, CO, CT, DC, FL, GA, IA, ID, IL, IN, KS, KY, LA, MA, MD, ME, MI, MN, MO, MT, NC, ND, NE, NH, NJ, NM, NV, NY, OK, OR, PA, RI, SC, SD, TN, TX, UT, VA, VT, WA, WI, WV, WY
Aleochara castaneipennis Mannerheim	AB, BC, LB, NB, NF, NS, NT, ON, QC, YT	AK, AR, AZ, CO, WA, WY
Aleochara fumata Gravenhorst †	AB, BC, LB, MB, NB, NF, NS, ON, PE, QC, YT	AL, AR, AZ, CA, GA, ID, MA, MD, ME, MN, MO, NC, NH, NJ, NM, NY, OK, OR, PA, RI, TN, VA, VT, WA, WI, WV, WY. Brazil, Argentina
Aleochara gracilicornis Bernhauer	BC, MB, NB, NF, NS, NT, ON, QC, SK	AR, AZ, CO, FL, IL, IN, KS, LA, MA, MD, ME, MI, MN, MO, MT, NH, NJ, NM, NY, OH, PA, RI, SD, UT, VA, WI
Aleochara morion Gravenhorst †	AB, BC, NB, NF, NS, ON, QC, SK	CA, CT, NV
Aleochara sekanai Klimaszewski	AB, LB, MB, NB, NT, ON, SK, YT	AK
Aleochara shelleyae Klimaszewski and Langor	NF, ON	
Aleochara suffusa (Casey)	AB, BC, MB, QC, SK	AK, AZ, CO, ID, NM, WA, WY
Aleochara sulcicollis Mannerheim	BC	AK, CA, OR, WA. Mexico, Chile
Aleochara tahoensis Casey	AB, BC, MB, NB, NF, NS, NT, ON, SK, YT	AR, CA, CO, ID, MT, NH, NM, NV, OR, SD, WA, WI
Aleochara verna Say *	AB, BC, LB, MB, NB, NF, NS, ON, PE, QC, SK, YT	AK, AZ, CA, CO, DC, FL, GA, IA, ID, IL, IN, KS, MA, ME, MI, MN, MO, MT, NH, NC, NJ, NM, NY, NV, OK, OR, PA, SC, SD, TN, TX, UT, VA, VT, WA, WI, WY
Aleochara villosa Mannerheim ‡	AB, BC, NB, ON, QC, SK	AK, CA, OK, OR, WA
Tribe Athetini Casey		
Acrotona horwoodae Klimaszewski and Godin	YT	
Acrotona sphagnorum Klimaszewski and Webster	LB [**NPR**], NB, ON [**NPR**], QC [**NPR**]	
Adota maritima (Mannerheim) *	BC	AK, CA
Aloconota cambrica (Wollaston) †	BC	AK [**NSR**], CA

(continued)

J. Klimaszewski et al., *A Faunal Review of Aleocharine Beetles in the Rapidly Changing Arctic and Subarctic Regions of North America (Coleoptera, Staphylinidae)*,
https://doi.org/10.1007/978-3-030-68191-3

Table A.1 (continued)

Taxonomic name	Canadian distribution	United States distribution
Aloconota neocambrica Klimaszewski and Langor	LB, NB, NF	
Amischa praelonga (Casey)	YT	WY
Amischa tersa Casey	YT [tentative]	CO
Atheta allocera Eppelsheim * (= *A. campbelli* (Lohse))	AB [**NPR**], LB, NB, NF, ON, QC [**NPR**], YT	AK
Atheta altaica Bernhauer *	AB [**NPR**], BC [**BOLD**], LB, NF, NT, QC[**NPR**], YT	AK
Atheta atramentaria (Gyllenhal) ‡	LB, NF, QC[**NPR**]	
Atheta avalon Klimaszewski and Langor	LB, NF	
Atheta borealides Newton [replacement for *A. borealis* Klimaszewski and Langor]	AB, NF, NU, ON	
Atheta brunswickensis Klimaszewski	BC, MB, NB, NS, NT [**NTR**], ON, QC, SK [**NPR**], YT	AK [**NSR**]
Atheta burwelli (Lohse)	AB [**BOLD**], BC [**BOLD**], LB, NB, NF, NT [**BOLD**], NU, ON, YT	AK [**NSR**]
Atheta cadeti Klimaszewski and Godin	YT	
Atheta capsularis Klimaszewski	AB [**BOLD**], BC, LB, MB, NB, NF, NT [**BOLD**], ON, QC, SK [**NPR**], YT	AK [**NSR**]
Atheta caribou (Lohse)	LB, YT	
Atheta chadburnae Klimaszewski and Godin, **sp. n.**	YT	
Atheta circulicollis Lohse	LB, NB, NF, NT [**NTR**], ON, QC	AK [**NSR**]
Atheta crenuliventris Bernhauer	LB, MB, NB, NF, NS [**BOLD**], ON, QC, SK	ME
Atheta cryptica (Lohse)	AB [**NPR**], BC, NF, QC, YT	AK [**NSR**]
Atheta curvipennis Klimaszewski and Langor	LB, NF	AK [**NSR**]
Atheta dadopora C.G. Thomson *	AB, BC, LB, MB, NB, NF, NS, NT [**BOLD**], ON, PE, QC, SK, YT	AK, NY, PA, RI
Atheta districta Casey	AB, BC, LB, MB, NB, NF, NS, ON, QC, SK	
Atheta diversa (Sharp) * (= *A. regissalmonis* (Lohse))	LB, NF, QC	AK
Atheta fanatica Casey	AB, BC, LB, NB, NS, NT [**BOLD**], ON [**BOLD**], QC, SK, YT	AK [**NSR**], NV
Atheta frosti Bernhauer	BC, LB, MB, NB, NS, ON, QC, SK	AK [**NSR**], MA, NC, NH, NY, PA, RI, VT, WV
Atheta gelida J.R. Sahlberg *	MB, NT, NU, QC, YT	AK
Atheta graminicola (Gravenhorst) *	AB, BC, LB, MB, NB, NF, NT, ON, QC, SK, YT	AK, NH [**NSR**], OR
Atheta hampshirensis Bernhauer	AB, BC, LB, MB, NB, NF, NS, ON, QC	AK, CA, NC, NH, NY, PA, RI, WA
Atheta keeni Casey	AB [**NPR**], BC	AK, OR

(continued)

Table A.1 (continued)

Taxonomic name	Canadian distribution	United States distribution
Atheta longiclava (Casey)	LB [**NPR**], MB [**NPR**], ON, QC[**NPR**]	RI
Atheta martini Lohse	YT	
Atheta microelytrata Klimaszewski and Godin	YT, NT [**NTR**],	AK [**NSR**]
Atheta munsteri Bernhauer *	BC[?], MB, NT, NU [**BOLD**], YT	AK
Atheta nearctica (Lohse)	LB, MB [**NPR**], NT, QC [**NPR**], YT	AK
Atheta onthophilides (Newton 2017) [=*Acrotona onthophila* Lohse], **new generic combination**	YT	AK
Atheta pecki Klimaszewski and Langor	LB	
Atheta pennsylvanica Bernhauer	LB, MB, NB, NF, NS, ON, QC	IN, MA, MN, NY, PA, RI, VA, VT
Atheta picipennis (Mannerheim)	BC	AK, CA, WA
Atheta platonoffi Brundin *	AB, BC, LB, MB [**NPR**], NB, NF, NS, ON, SK, YT	AK
Atheta pratensis (Mäklin)	YT	AK, WA
Atheta prudhoensis (Lohse) *	AB [**BOLD**], BC, LB, NB, NF, NS, ON, QC, SK, YT	AK, NY, VT
Atheta pseudocrenuliventris Klimaszewski	AB[**NPR**], BC [**NPR**], LB, NB, NF, NS, NT [**NTR**], ON, YT	AK [**NSR**]
Atheta pseudoklagesi Klimaszewski and Webster	AB, BC, MB, NB, NF, QC, SK, YT [only reconfirmed records are listed]	AK [**NSR**]
Atheta pseudometlakatlana Klimaszewski and Godin	AB [**BOLD**], BC, SK, YT	AK [**NSR**]
Atheta pseudosubtilis Klimaszewski and Langor	AB, LB, NB, NF	
Atheta remulsa Casey	AB, BC, LB, MB, NB, NF, NS, ON, QC, SK, YT	AK, NY
Atheta ringi Klimaszewski	BC	AK
Atheta ripariides Newton (=*riparia* Klimaszewski and Godin)	BC, MB, SK, YT	
Atheta smetanai (Lohse)	LB, MB, NT, QC, YT	AK
Atheta sparreschneideri Munster *	YT	AK
Atheta stercoris Fenyes	BC	AK [**NSR**]
Atheta strigosula Casey	AB [**BOLD**], BC, LB, MB, NB, NF, ON, QC, SK, YT	AK, NY
Atheta subsinuata (Erichson) †	SK, YT	
Atheta subtilis (Scriba) †	LB, NB, NF, QC	
Atheta surgens (Casey)	BC	AK
Atheta takhini Klimaszewski and Godin, **sp. n.**	YT	
Atheta terranovae Klimaszewski and Langor	AB [**BOLD**], BC, LB, MB, NB, NF, ON, QC, SK, YT	AK
Atheta ventricosa Bernhauer	AB, BC, LB, NB, NF, NS, ON, QC, SK, YT	AK, CA[?], DC, IN, MA[?], NC, NH, NJ, NY, PA, VT, WA[?].
Atheta whitehorsensis Klimaszewski and Godin	NB, YT	

(continued)

Table A.1 (continued)

Taxonomic name	Canadian distribution	United States distribution
Bellatheta microptera (Lohse) [=*Emmelostiba, Pseudosipalia*] **new combination**	YT	AK
Boreophilia beringi Klimaszewski and Brunke	YT [**NTR**]	AK
Boreophilia caseyi Lohse	MB, NU, YT	AK
Boreophilia davidgei Klimaszewski and Godin	AB, YT	
Boreophilia eremita (Rye) *	LB, MB, NB, QC [**NPR**]	AK
Boreophilia fusca (C.R. Sahlberg) *	MB, NT	AK
Boreophilia hyperborea (Brundin) *	MB, NT, NU	AK
Boreophilia islandica (Kraatz) *	AB, NF, NT, NU, YT	AK
Boreophilia nearctica Lohse	AB, LB, MB, NF, QC, YT	AK
Boreophilia neoinsecuta Klimaszewski	MB, NT [**NTR**], YT	AK
Boreophilia nomensis (Casey)	BC, YT	AK
Boreophilia subplana (J.Sahlberg) *	NT, NU	AK, NH
Boreophilia vega (Fenyes) * [=*hershelensis* Klim. and Godin]	NT [**NTR**], NU [**BOLD**], YT	
Boreophilia venti (Lohse)	YT	AK
Boreostiba campbelliana Lohse *	YT, NU [**NTR**]	AK, NH [**NSR**]
Boreostiba frigida (J.R. Sahlberg) *	LB, QC, YT	AK [**NSR**]
Boreostiba lagunae Lohse	YT	
Boreostiba parvipennis (Bernhauer)	AB, BC, LB, MB [**BOLD**], NF, NT, QC, YT	AK, NH
Boreostiba sibirica (Mäklin) *	LB, NT, NU, YT	AK, NH [not verified]
Boreostiba websteri Klimaszewski and Langor	LB, NB	AK [**NSR**]
Dinaraea angustula (Gyllenhal) †	AB, BC, LB, NB, NF, NS, ON, PE, QC, SK, YT	CA, NY, PA
Dinaraea borealis Lohse	NB, NT [BOLD], ON, QC, SK [**BOLD**], YT [**BOLD**]	AK [**NSR**]
Dinaraea pacei Klimaszewski and Langor	BC, LB, NB, NF, QC, SK, YT	AK
Dinarea planaris (Mäklin) [**neotype designation**]	YT	AK
Dochmonota langori Klimaszewski and Larson	MB[**NPR**], SK	
Dochmonota rudiventris (Eppelsheim) ‡	BC, NB, NF, NT ON [**BOLD**], QC, YT	ID, MA
Earota dentata (Bernhauer)	AB, BC, MB, NB, NF, NS, ON, QC, SK, YT	AK, AL, AZ, CA, CO, DC, IA, ID, IL, IN, KS, MA, ME, MD, MT, NC, NH, NJ, NM, NV, NY, OH, OR, PA, VA, WA
Hydrosmecta pseudodiosica Lohse	NB, ON, YT	AK [**NSR**]
Hydrosmecta yukonensis Klimaszewski and Godin, **sp.n.**	YT	
Liogluta atriventris (Casey)	BC	AK [**NSR**]
Liogluta gigantea Klimaszewski and Langor	LB, ON, QC	
Liogluta granulosa Lohse	YT	AK
Liogluta intermedia Klimaszewski and Langor	LB, NF, NS, ON, QC	NH

(continued)

Table A.1 (continued)

Taxonomic name	Canadian distribution	United States distribution
Liogluta nigropolita (Bernhauer)	LB, NF, NT, NU, QC, YT	AK [**NSR**], NH
Liogluta nitens (Mäklin)	AB, BC, YT (YT vouchers not located)	AK, OR, WA
Liogluta scudderi Klimaszewski	BC	
Liogluta terminalis (Casey) (=*aloconotoides* Lohse)	AB, BC, LB, NB, NF, NS, ON, QC, SK [**BOLD**], YT	MT, NH
Liogluta trapezicollis Lohse	YT	
Liogluta vasta (Mäklin)	BC, YT	AK
Liogluta wickhami (Casey)	BC	
Lypoglossa angularis (Mäklin)	AB, BC, LB, NB, NF, NS, QC, SK [**NPR**], YT	AK, AZ, CA, CO, ME, MT, NH, NM, OR, WA, WY
Lypoglossa franclemonti Hoebeke	AB, BC, MB, NB, NF, NS, NT, ON, QC, SK, YT	ME, NH, NY, VT
Mocyta amblystegii (Brundin) *	MB, NT, NU, YT	AK
Mocyta breviuscula (Mäklin)	AB, BC, LB, MB, NB, NF, NS, ON, QC, SK, YT	AK, CA, NV, OR
Mocyta fungi (Gravenhorst) †	AB, BC, LB, MB, NB, NF, NS, NU, ON, PE, QC, SK, YT	AK, CA, MA, ME, MN, NY, OR, RI, WA
Mocyta sphagnorum Klimaszewski and Webster ‡	NB, NF, ON, QC, SK	AK [**NSR**]
Paragoniusa myrmicae Maruyama and Klimaszewski	AB, BC, LB, NB	
Philhygra angusticauda (Bernhauer)	AB, BC, NB	AK, NH
Philhygra botanicarum (Muona) *	AB [**BOLD**], BC, LB, NB, NF, NS, ON, SK, YT	AK [**NSR**]
Philhygra donnellyana Klimaszewski, **sp. n.**		AK
Philhygra falcifera Lohse	MB, SK	
Philhygra jarmilae Klimaszewski and Langor	AB [**BOLD**], MB, NB, NF, NT [**BOLD**], ON, SK, YT	AK [**NSR**]
Philhygra junii Lohse	NU [**NTR**], YT	
Philhygra leechi Lohse	MB, NT, YT	
Philhygra malleoides Lohse *	MB, NF, NT, NU [**BOLD**], QC, YT	AK
Philhygra pseudoboreostiba Lohse	YT	
Philhygra pseudolarsoni Klimaszewski and Godin	AB [**NPR**], MB [**BOLD**], NB, YT	
Philhygra pseudopolaris Klimaszewski and Langor	AB [**BOLD**], BC, MB, NF, NT, QC, YT	AK, CA
Philhygra ripicola (Hanssen) (=*P. ripicoloides* Lohse) *. **New Synonymy**	NF, NT, SK, YT	AK [**NSR**]
Philhygra rostrifera Lohse	LB, MB [**NPR**], NU [**NTR**], NT, SK, YT [**BOLD**]	AK
Philhygra sinuipennis Klimaszewski and Langor	AB [**NPR**], NB, NF, SK, YT	
Philhygra terrestris Klimaszewski and Godin	AB [**BOLD**], BC, NB, SK, YT	AK
Philhygra varula (Casey)	LB, MB, NB, NF, QC	AK [**NSR**]
Schistoglossa blatchleyi (Bernhauer and Scheerpeltz)	MB, NB, NT, ON, QC, SK, YT	AK, IN

(continued)

Table A.1 (continued)

Taxonomic name	Canadian distribution	United States distribution
Schistoglossa campbelli Klimaszewski	AB, BC	AK [**NSR**]
Schistoglossa hampshirensis Klimaszewski	AB, BC, QC	AK [**NSR**], NH
Strigota ambigua (Erichson)	BC, LB, NB, NF, NS, ON, PE, QC, SK, YT	CA, CO, CT, IA, KS, MA, MO, NC, NJ, NM, NY, NV, TX
Strigota obscurata Klimaszewski and Brunke	AB [**BOLD**], BC, NB, NF [**BOLD**], ON, QC, SK	AK [**NSR**]
Tarphiota fucicola (Mäklin)	BC	AK, CA, OR, WA. Mexico (Baja California, Sonora)
Tarphiota geniculata (Mäklin)	BC	AK, CA, OR, WA. Mexico (Baja California)
Trichiusa pilosa Casey	AB, BC, NB, NS, ON	ID, IN, KS, OH, RI
Tribe Boreocyphini Klimaszewski and Langor		
Boreocypha websteri Klimaszewski and Langor	LB, NB, PE	AK [**NSR**]
Tribe Gymnusini Heer		
Deinopsis canadensis Klimaszewski	LB, NB, NF, ON	
Deinopsis harringtoni Casey	LB, MB, NB, NF, NS, ON, QC	AR, CA, MA, ME, MI, NH, WI
Gymnusa atra Casey *	AB, BC, LB, MB, NB, NF, NS, NT, ON, QC, YT	AK, IL, IA, MA, ME, MI, MN, NH, NY. Eastern Russia, northeastern China
Gymnusa brevicollis (Paykull)†	LB, NF, QC	MA, ME. France (St. Pierre et Miquelon). Northern Europe, western Siberia (Russia)
Gymnusa campbelli Klimaszewski	MB, NB, NF, NT, ON, QC, SK, YT	AK, MA
Gymnusa grandiceps Casey	BC, LB, MB, NB, NF, NS, ON, QC	IL, MA, MD, ME, MI, NH, NY, RI
Gymnusa konopackii Klimaszewski	MB, NT, NU, YT	AK
Gymnusa lindrothi Klimaszewski and Langor	LB, NF, NB	
Gymnusa pseudovariegata Klimaszewski	AB, BC, LB, MB, NB, NF, NS, NT, NU [**BOLD**], ON, QC, YT	AK, CA, MA, OR, WA
Gymnusa smetanai Klimaszewski *	MB, NF, NT, ON, YT	AK
Tribe Homalotini Heer		
Gyrophaena insolens Casey	AB [**BOLD**], LB, MB, NB, NF, ON, SK	AK [**NSR**], MI
Gyrophaena keeni Casey	AB, BC, LB, MB, NB, NF, NS, ON, QC, SK [**NPR**], YT	AK [**NSR**], FL, MA, ME, MT, NH, NY, OR, TN, WA, WY
Gyrophaena nana (Paykull) *	AB, BC, MB, NB, NF, NT [**NTR**] , ON, QC [**NPR**], SK [**NPR**], YT	AK, CA, MA, ME, MI, MT, WI, WY
Gyrophaena rugipennis Mulsant et Rey * (=*Gyrophaena neonana* Seevers, **New Synonymy**)	NB, NF, ON, SK [**NPR**], YT [tentative]	IN, NC, PA, WI
Leptusa gatineauensis Klimaszewski and Pelletier	AB, BC, NB, NF, NS, ON, QC, SK	AK [**NSR**]
Silusa californica (Bernhauer)	AB, BC, LB, MB, NB, NF, NS, ON, PE, QC, SK, YT	AK, CA, MN
Silusa densa Fenyes	AB, LB, NB, NF, ON [**NPR**], QC [**NPR**]	CA

(continued)

Table A.1 (continued)

Taxonomic name	Canadian distribution	United States distribution
Tribe Hypocyphtini Laporte		
Cypha inexpectata Klimaszewski and Godin	AB [**BOLD**], NT [**NTR**], MB [**BOLD**], ON, QC [**NPR**], SK, YT	AK [based on female record], CA [**BOLD**]
Tribe Liparocephalini Fenyes		
Amblopusa alaskana Ahn and Ashe		AK
Amblopusa brevipes Casey	BC	AK, CA
Diaulota densissima Casey	BC	AK, CA, OR, WA
Liparocephalus brevipennis Mäklin	BC	AK
Liparocephalus humblei Klimaszeski, **sp. n.**		AK
Paramblopusa borealis (Casey) *	BC	AK, CA, OR, WA. Japan
Paramblopusa eoa Ahn and Maruyama *		AK, Russia (Kuril Is.)
Tribe Lomechusini Fleming		
Zyras obliquus (Casey)	AB, BC, LB, MB, NB, NF, NS, ON, QC, SK	MI, MO, NH, NY, OR
Tribe Myllaenini Ganglbauer		
Myllaena arcana Casey	AB, LB, MB, NB, NF, NS, ON, QC, SK	AL, FL, IA, IL, MA, ME, NH, NJ, TN, WV. Mexico
Myllaena audax Casey	BC, LB, NB, NF, NT, ON, QC	AK, CA, GA, IL, LA, MA, NH, NJ, NY, OR, RI, UT, WA
Myllaena insomnis Casey	AB, BC, LB, MB, NB, NF, NS, NT, ON, QC, SK, YT	AK, ID, MA, MI, MN, MT, NY, WI
Myllaena ludificans Casey	NB, QC, ON	AR, MA, OK, RI, WI
Tribe Oxypodini C.G. Thomson		
Alfocalea montana Klimaszewski	AB, BC	
Calodera caseyi Assing [male undescribed]	NB	AK [**NSR**], PA
Calodera parviceps (Casey)	BC, NB, NS, NT [**NTR**], ON, YT	AK [**NSR**], RI
Devia prospera (Erichson) *	AB, BC, LB, MB, NB, NT, ON, QC, SK, YT	AK, CA, CO, MA, MI, MN, MT, NH, NM, OR, SD, UT, WA, WI, WY
Gnathusa alfacaribou Klimazewski and Langor	LB	
Gnathusa caribou Lohse * [new status]	NT, NU [**NTR**], YT	AK, Finland [new record]
Gnathusa eva Fenyes	AB, BC, SK, YT	CA
Gnathusa tenuicornis Fenyes	AB, BC, MB [**BOLD**], NB [?], SK [**BOLD**], YT	AK, CA, OR
Hylota ochracea Casey	NB, NS, NT, ON, QC, SK	NY, VT
Meotica pallens (Redtenbacher) †	BC, NB, NS, ON, YT	NJ, RI
Mniusa yukonensis (Klimaszewski and Godin)	BC, NB, NS, ON [**NPR**], QC, YT	AK [**NSR**]
Neothetalia canadiana Klimaszewski	AB [**BOLD**], BC, LB, NB, QC, YT	AK
Neothetalia columbiana (Klimaszewski)	AB [**BOLD**], BC	AK
Neothetalia pallitarsis (Kirby)	AB, BC	AK

(continued)

Table A.1 (continued)

Taxonomic name	Canadian distribution	United States distribution
Ocyusa canadensis Lohse	BC, NB, NF, NT [**NTR**], ON, SK, YT	AK
Oxypoda amica Casey	MB, NB, NS, NT, ON, QC	IA, IN, KS, KY, LA, MO, NC, NJ, NY, TN
Oxypoda brachyptera (Stephens) †	NB, NF, NS, NU [**NTR**], ON, QC	
Oxypoda canadensis Klimaszewski	AB, BC, LB, MB, NF, NT, ON, QC, SK [**NPR**], YT	AK, NH
Oxypoda convergens Casey	AB, BC, LB, MB, NB, NF, NS, ON, QC, SK [**BOLD**]	IA, MO, NY
Oxypoda demissa Casey	AB [**BOLD**], LB, MB [**BOLD**], NB, NF, NS, ON, QC, SK, YT	AK [**NSR**]
Oxypoda frigida Bernhauer	AB, BC, LB, NB, NF, NS, NT, ON, QC, SK [**BOLD**], YT	AK, NH
Oxypoda gatosensis Bernhauer	BC	CA
Oxypoda hiemalis Casey	AB, LB, MB, NB, NF, NS, NT, ON, QC, YT	AK, NH
Oxypoda inimica Casey	NB, NF, NT, ON, QC	MA
Oxypoda irrasa Mäklin	AB, BC, NT [**BOLD**], ON [**BOLD**], SK, YT	AK, OR
Oxypoda lacustris Casey	AB, BC, LB, MB, NB, NF, NS [**BOLD**], NT, ON, QC, SK, YT	AK, WA
Oxypoda leechi Lohse	YT	
Oxypoda lucidula Casey	AB, LB, MB, NB, NF, NT, NU [**BOLD**], ON, QC, YT	AK, IA, MO, NH, NY
Oxypoda manitobae Casey	BC, MB, SK	CO
Oxypoda nelsoni Lohse	YT [**NTR**]	AK
Oxypoda operta Sjöberg *	AB, LB, NS, ON, QC, YT	NH
Oxypoda orbicollis Casey	AB, BC, LB, NB, NS, ON, QC, SK, YT	WI
Oxypoda pseudoconvergens Klimaszewski and Godin	AB [**BOLD**], YT	
Oxypoda robusticornis Bernhauer [=*madgei* Lohse]		AK, NH
Oxypoda subpolaris Casey [=*hemingi* Lohse]		AK
Oxypoda sylvia Casey [=*grandipennis* (Casey)]	AB, BC, LB, MB, NB, NF, NS, ON, QC, SK, YT	AK, ME, NH
Oxypoda volkeri Klimaszewski	NT	
Oxypoda yukonensis Klimaszewski and Godin	YT	AK [**NSR**]
Parocalea nearctica Lohse	NT, YT	AK
Parocalea pseudobaicalica Lohse	BC, NT, NU [**BOLD**], QC, YT	AK, NH
Phloeopora arctica Lohse	AB, BC, NT, ON, YT	
Phloeopora canadensis Klimaszewski and Langor	BC, NF, NB, NS, YT [**BOLD**]	AK [**NSR**]
Phloeopora gilbertae Klimaszewski and Webster	NB, NT, QC [**NPR**]	AK [**NSR**]

(continued)

Table A.1 (continued)

Taxonomic name	Canadian distribution	United States distribution
Parocyusa fuliginosa (Casey)	LB, NB, ON, SK	MA, NC, PA
Tribe Placusini Mulsant and Rey		
Placusa pseudosuecica Klimaszewski	AB, BC, ON, QC, SK	AK [**NSR**]
Placusa tachyporoides (Waltl) ‡ [new distributional status]	AB, BC, LB [**NPR**], NB, NS, ON, QC, SK	AK [**NSR**], CA, MA, MN
Placusa tacomae Casey	AB, BC, NB, NF, NS, NT, ON, QC, SK, YT	AK [**NSR**], AZ, CO, MA, ME, MN, NY, VA, WA, WI
Placusa vaga Casey	AB, BC, NB, NS, NT, ON, QC, SK, YT	AK [**NSR**], CA
Tribe Tachyusini C.G. Thomson		
Brachyusa helenae (Casey)	BC, LB, NB, NF, NT, ON, SK, YT	AK, MT
Dasygnypeta velata (Erichson) * (=*Gnypeta minuta* Klimaszewski and Webster)	LB, MB, NB, NT, SK	AK
Gnypeta ashei Klimaszewski	MB, NT, NU, QC, YT	AK
Gnypeta atrolucens Casey	LB, NB, NF, QC	NH, NY, VT
Gnypeta brevicornis Casey	BC	
Gnypeta brincki Palm *	NT, QC, YT	AK
Gnypeta caerulea (C.R. Sahlberg) *	AB, BC, LB, MB, NB, NF, NS, NT, ON, PE, QC, SK, YT	AK
Gnypeta carbonaria (Mannerheim) *	AB, MB, NB, NF, NT, ON, QC, SK	AK
Gnypeta dentata Klimaszewski	AB, MB [**NPR**], NT, SK,	
Gnypeta groenlandica Lohse	MB, NU, YT	AK, Greenland
Gnypeta lohsei Klimaszewski	AB, BC	AK, WA
Gnypeta sellmani Brundin *	LB, MB, NF, NT QC, SK, YT	AK
Paradilacra densissima (Bernhauer)	AB, BC, NB, ON, SK	CA, MT, NV, ND, OR, UT
Tachyusa americanoides Paśnik	AB, BC, MB, NB, NF, NS, NT, ON, QC	IL, MA, MT, NH, NY, PA, VT

Annex 3

Corrections to Klimaszewski et al. 2020

The genus *Neoisoglossa* (pp. viii, xiv, 12, 63, 65, 143, 174, 599, 617) was incorrectly attributed to Casey, it was actually proposed by Klimaszewski and Pelletier (2004: 466), apparently as an unnecessary replacement name for *Isoglossa* Casey 1893. Blackwelder (1952) cited *Isoglossa* Casey, 1893 as a junior homonym of *Isoglossa* Newman, 1833 [and as a synonym of *Ocalea*], but this name could not be found, so *Isoglossa* Casey stands as the valid name with *Neoisoglossa* as a synonym. This synonymy, along with two other synonyms, was correctly presented in the catalog of Gouix and Klimaszewski (2007: 35).

The name *Gnypeta baranowskii* was spelled incorrectly as "*baranowski*" in the species heading (p. 241), correctly elsewhere (pp. 238, 253); we here act as First Revisers and select *Gnypeta baranowskii* as the correct original spelling [ICZN Article 24.2.3]. The species was subsequently transferred to the genus *Dasygnypeta* Lohse by Brunke et al. (2021: 47) without citing both spellings.

The names *Tetralaucopora americana* (Casey) and *T. fuliginosa* (Casey) were cited in the list on p. 619. The synonymy of the genus *Tetralaucopora* Bernhauer, 1928 with *Parocyusa* Bernhauer, 1902 was re-established by Assing (2021: 105), who placed *T. fuliginosa* in the genus *Parocyusa*. He also tentatively placed *T. americana* as a synonym of *Parocyusa rubicunda* (Erichson, 1837), this synonymy was confirmed by Brunke et al. (2021: 40).

- pp. viii, 68, 395, 397, 594, 604: change *Atheta lucifera* to *Atheta luctifera*
- pp. viii, 307: change "Nomen nudum" for *Stictalia notata* to "Nomen dubium" [there was a type, we just cannot find it or be sure of the application of the name as a result]
- pp. viii, 599, 617: change *Neoisoglossa agnita* to *Isoglossa agnita*
- pp. viii, 67, 367 400: change *Datomicra wrangleri* to *Datomicra wrangeli*, remove from list of New synonyms on p. viii
- pp. viii, 68, 391, 392, 595, 606: change *Atheta relicata* to *Atheta relicta* and (*Lamiota*) *relicata* to (*Lamiota*) *relicta*
- p. viii: after *Atheta subrugosa* change Kiessenwetter to Kiesenwetter
- p. 2: change "Bousquet's checklist" to "Campbell & Davies (1991)"
- p. 3: add reference [was deleted]:
  Campbell JM, Davies A (1991) Family Staphylinidae rove beetles [pp. 86-124] In: Bousquet Y (editor): Checklist of beetles of Canada and Alaska. Publication 1861/E, Research Branch, Agriculture Canada, Ottawa. vi + 430 pp
- p. 6: change "Wytman's" to "Wytsman's"
- p. 6: (R. Wenzel, "in Smetana and Herman 2001" change to "in Forward to above [1978]")
- p. 8: Ashe, J.S. (1984a): add: "[pp. 227–255]" In:

J. Klimaszewski et al., *A Faunal Review of Aleocharine Beetles in the Rapidly Changing Arctic and Subarctic Regions of North America (Coleoptera, Staphylinidae)*,
https://doi.org/10.1007/978-3-030-68191-3

p. 8: Ashe, J.S. (1984b): change "Quest Entomol" to "Quaest Entomol"

p. 8: add reference [was deleted]:
Seevers CH, Herman LH (1978) A generic and tribal revision of the North American Aleocharinae (Coleoptera: Staphylinidae). Fieldiana: Zoology 71.: vi + 289 pp

p. 15: to reference Klimaszewski J, Pelletier G, Germain C, Hebert C (2001) add authors: "Humble LM, Winchester NN"

p. 16: reference Mäklin FG: change initials to FW and remove "(ed)" in both; to (1853) add fascicle "26(3)" after "Moscou"

p. 16: to reference Mannerheim CG (1843) add fascicle "16(2)" after "Moscou"

p. 16: to reference Seevers CH (1978) after title add: (with additions and annotations by Lee H. Herman)

p. 17: change Yamamoto and Maruyama (2017) to "(2018)"

p. 46: after the citation "Klimaszewski 2000" add ", present book" [because the references cited do not bring the total number of BC species to 227]

p. 58: change *Atheta onthophaga* (Lohse) to *Atheta onthophilides* Newton

p. 65: change *N. agnita* to *I. agnita*

p. 67: change *Datomicra wrangleri* to *Datomicra wrangeli* and remove "New synonym,"

pp. 73, 294: reference LeConte JL (1880) change volume 12 to 8(3-4)

pp. 74, 570: reference Wollaston TV (1855), journal is "Zoologist 13(Appendix): ccv-ccvii"

p. 91: change 54.46.22° to "54°46.22'", 127.09.13° to "127°09.13'"

p. 105: change *Aleochara verna* Say, 1836 to "1833"

p. 149: change *Gnathusa eva* 1909 to "1910"

p. 161: *Oxypoda famula*: Distribution [United States]: AK based on Leng Catalog (1920: 126), repeated by Moore and Legner (1975: 464), neither listed the type locality: British Columbia.

p. 178: *Neothetalia nimia*: Distribution [United States]: AK listed without records, list p. 618 only has BC

p. 246: *Tachyusa obsoleta*: Distribution [United States]:
should be: AZ, AR, CO, DC, IL, IN, KS, KY, MA, MN, MO, MT, NC, ND, NE, NH, NJ, NM, NY, OH, PA, SD, TX, VT, WY [records for CA, NV, OK, UT, WI could not be found]

p. 301: reference Hoebeke ER (1985): change volume number to 93

p. 367: change *Datomicra wrangler* and *wrangleri* to *Datomicra wrangeli* [4 times] and remove "New synonymy" [it was synonymized with *celata* by Majka, Klimaszewski & Lauf (2006: 37)]

p. 368: change *Datomicra decolourata* to *Datomicra decolorata*

p. 376: change Hatch 1957 "[as *Stethusa*]" to "[as *Ischnopoda* (*Stethusa*)]"

p. 379: this was the first published synonymy of *Atheta vicaria* with *A. fanatica*; the first published synonymy of *Atheta omissa* with *A. fanatica* was by Klimaszewski, Langor & al. (2011: 152)

p. 382: under the Distribution for *Atheta hampshirensis* add: USA: WA

p. 382: under *A. picipennis*: add reference "Sawada 1982" [only reference to synonymy of *cursor*]

p. 384: after "Hatch, 1957" add: "[as *Ischnopoda* (*Atheta*)]"

p. 393: under *A. relicta*: change to: "known only from the male lectotype (although evidently from the original description Casey had male and female specimens)"

p. 396: under *Atheta holmbergi* add reference: "Klimaszewski et al. 2015c [as *Atheta* (*Dimetrota*)]"

p. 400: under *Atheta nescia* after "As *Atheta* (*Pancota*)" add: "Fenyes 1920, "

p. 401: under *Atheta concessa* after Hatch, 1957 add: [as *Ischnopoda* (*Liogluta*)]

p. 402: under *Atheta psuedoklagesi*: add reference Klimaszewski et al. 2018c [Manitoba records]: Klimaszewski J, Godin B, Davies AE, Bourdon C, Horwood D (2018c) Forty new records of aleocharine beetles, and two new species in the genera *Acrotona* Thomson and *Atheta* Thomson, for the province of Manitoba, Canada (Coleoptera: Staphylinidae: Aleocharinae). Insecta Mundi 0641: 1-33.

p. 406: in Comments under *Atheta formalis* change *nomen nudum* to *nomen dubium*

p. 406: under *Atheta irrupta* after Hatch, 1957 add: [as *Ischnopoda* (*Pancota*)]

p. 413: under Distribution for *Dalotia coriaria* change "before 1870 (Crotch 1870)" back to "before 1873 (Crotch 1874)" [there is no mention of this species or of any distributions in 1870, and see p. 566 below]

p. 415: Distribution under *Atheta inexpectata* remove: ", LB, NB, NF, QC, SK, YT. USA: AK" [known only from the holotype]

p. 418: under *Atheta dentata* add reference: Bernhauer (1907) [USA records]

p. 440: under *Philhygra angusticauda* replace "[cited as *Atheta* (*Brundinia*)]" with "[cited as *Ischnopoda* (*Brundinia*)]"

p. 443: under *Philhygra laevicollis* after Hatch, 1957 add: [as *Ischnopoda* (*Brundinia*)]

p. 445: remove "Gusarov, in" from *Philhygra terrivaga* Klimaszewski and Winchester, 2002, add "(attributed to "Gusarov in press")" [specimens from Washington state were also cited in 2002 (p. 45) based on Gusarov's prospective publication]

p. 445: wetland S of Hwy 16°54′22″N should read: S of Hwy 16, 54°46.22′N

p. 448: under *Psammostiba kenai*: "Paratypes: labelled as the holotype" should read: "Paratypes: same label data as the holoytpe"

p. 566: change reference Crotch GR (1870) to: Crotch GR (1874) Checklist of the Coleoptera of America, north of Mexico. Naturalists' Agency, Salem, MA, 136 pp.

p. 567: under Klimaszewski J, Langor DW [et al.] 2008b change Hammond JHE to Hammond HEJ

p. 568: modify reference LeConte JL (1863) to: New species of North American Coleoptera. Prepared for the Smithsonian Institution. Part I. Smithson Misc Collections 6(167):1–86

p. 569: reference Moore I, Legner EF (1976), after "Marine insects" add "North-Holland Publishing Co."

p. 572: after *Myrmecopora... californica* Scheerpeltz change 1964 to 1965

p. 573: under *Bryobiota bicolor* change "with *Ashella giulianii* (Moore) as the generic type" to "with *Salinamexus giulianii* Moore as the generic type" [ICZN (1999) Recommendation 67B]

p. 576: under *Myrmecocephalus arizonicus* add references [which provided records]: Fenyes 1912; Hoebeke 1985; Klimaszewski, Larson et al. 2016 [SK records]
Klimaszewski J, Larson DJ, Labrecque M, Bourdon C (2016) Twelve new species and fifty-three new provincial distribution records of Aleocharinae rove beetles of Saskatchewan, Canada (Coleoptera, Staphylinidae). ZooKeys 610:45–112.
under Habitat add: pine clearcut, fire guard, recently dead white spruce, sifting moss and pine litter. "Elsewhere, specimens were collected from under bark of logs, from leaf litter, flood debris and wet moss, from soil along a stream, from fungus (Fomitopsis pinicola, Fomes robineae), and from a squirrel midden" (Hoebeke 1985).

p. 582: modify reference Scheerpeltz O (1965) to: Eine neue Art der Gattung *Myrmecopora* Saulcy von der pazifischen Küste Nordamerikas (Col., Staph.). Deut Entomol Z (Neue Folge) 12(1–2): 49–51

p. 584: under *Xenodusa reflexa* change "Wasmann 1894" to "Wasmann 1893", add reference below (p. 592)

p. 592: add reference [cited p. 584]:
Wasmann E (1893) Neue Myrmekophilen. Erstes Stück. Deut Entomol Z 1893(1):97-112, pl 5

p. 598: under *Calodera bennetti* : remove "[NSR]" [this is a new species]

p. 600 (list): add DC, MO, NY; remove WA [error for DC]

p. 601: under *Aleochara bimaculata* [United States] add: CT, DC

p. 602: under *Aleochara verna* [United States] add: OK

p. 604: under *Atheta hampshirensis* [United States] add: WA [as in Table A.1]

p. 605: under *Atheta newfoundlandica* move TN to United States distribution

p. 606: under *Atheta recondita* [Untited States] add AZ [this species placed tentatively in

Acrotona by Klimaszewski, Godin et al. (2015: 55)]

p. 607: under *Boreophilia caseyi* remove NT [the "NT" paratypes were actually from Nunavut]

p. 608: under *Earota dentata* [United States] add AZ

p. 609: under *Lypoglossa angularis* add: NS [as *angularis obtusa* in Majka & Klimaszewski (2010: 21)]; [United States] WY [as *angularis fenyesi* in Gusarov (2004[214]: 214)]

p. 611: change *Schistoglossa hamshiresis* to *hampshirensis*

p. 611: change *Strophogastra pencillata* to *Strophogastra penicillata*

p. 614: under *Gyrophaena lobata* [United States] change KY to KS

p. 615: change *Oligota chrysopyga* to *Holobus chrysopygus*

p. 616: under *Xenodusa cava* [United States] add MD

p. 617: under *Crataraea suturalis* remove YT

p. 619: under *Tetralaucopora americana* [United States] remove PA

p. 620: change *Brachyusa saskatchewani* to *saskatchewanae*

p. 620: change *Gnypeta groenlandicae* to *groenlandica*

p. 620: change *Gnypeta punctatula* to *punctulata*, add [United States]: CA, CO, UT

p. 621: under *Tachyusa faceta* [United States] add CA, CO, DC, IL, IN, KS, MT, NE, NM, NV, OK, SD, UT, WI, WY [Pasnik, 2006: 139]

Index

J. Klimaszewski et al., *A Faunal Review of Aleocharine Beetles in the Rapidly Changing Arctic and Subarctic Regions of North America (Coleoptera, Staphylinidae)*,
https://doi.org/10.1007/978-3-030-68191-3

GPSR Compliance

The European Union's (EU) General Product Safety Regulation (GPSR) is a set of rules that requires consumer products to be safe and our obligations to ensure this.

If you have any concerns about our products, you can contact us on ProductSafety@springernature.com

In case Publisher is established outside the EU, the EU authorized representative is:

Springer Nature Customer Service Center GmbH
Europaplatz 3
69115 Heidelberg, Germany

Batch number: 10755735

Printed by Printforce, the Netherlands